P9-BZS-872

L Brooks
CRL. 201

Adaptive Filter Theory

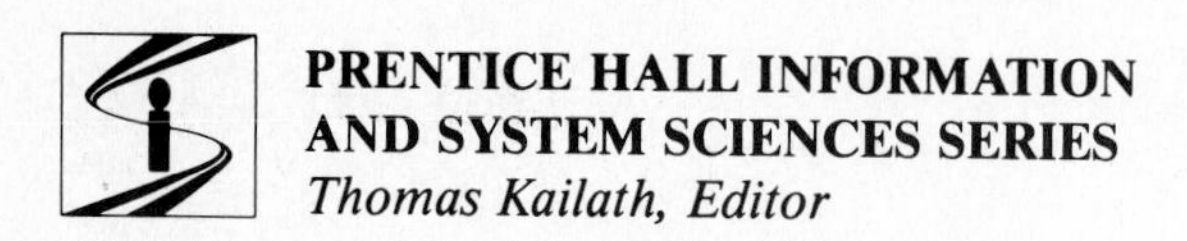

ANDERSON & MOORE	*Optimal Control: Linear Quadratic Methods*
ANDERSON & MOORE	*Optimal Filtering*
ASTROM & WITTENMARK	*Computer-Controlled Systems: Theory and Design, 2/E*
BOYD	*Linear Controller Design, 1/E 1991*
DICKINSON	*Systems: Analysis, Design and Computation*
GARDNER	*Statistical Spectral Analysis: A Nonprobabilistic Theory*
GOODWIN & SIN	*Adaptive Filtering, Prediciton, and Control*
GRAY & DAVISSON	*Random Processes: A Mathematical Approach for Engineers*
HAYKIN	*Adaptive Filter Theory, 2/E*
JAIN	*Fundamentals of Digital Image Processing*
JOHNSON	*Lectures on Adaptive Parameter Estimation*
KAILATH	*Linear Systems*
KUMAR & VARAIYA	*Stochastic Systems: Estimation, Indentification, and Adaptive Control*
KUNG	*VLSI Array Processors*
KUNG, WHITEHOUSE, & KAILATH, EDS.	*VLSI and Modern Signal Processing*
KWAKERNAAK & SIVAN	*Modern Signals and Systems*
LANDAU	*System Identification and Control Design Using PIM+ Software*
LIUNG	*System Identification: Theory for the User*
MACOVSKI	*Medical Imaging Systems*
MIDDLETON & GOODWIN	*Digital Control and Estimation: A Unified Approach*
NARENDRA & ANNASWAMY	*Stable Adaptive Systems*
SASTRY & BODSON	*Adaptive Control: Stability, Convergence, and Robustness*
SOLIMAN & SRINATH	*Continuous and Discrete Signals and Systems*
SPILKER	*Digital Communications by Satellite*
WILLIAMS	*Designing Digital Filters*

Adaptive Filter Theory

Second Edition

SIMON HAYKIN
McMaster University

PRENTICE HALL
Englewood Cliffs, NJ 07632

Library of Congress Cataloging-in-Publication Data

Haykin, Simon S.
Adaptive filter theory / by Simon Haykin. -- 2nd ed.
p. cm.
Includes bibliographical references and index.
ISBN 0-13-013236-5
1. Adaptive filters. I. Title.
TK7872.F5H368 1991
621.381'5324--dc20 90-20418
CIP

Editorial/production supervision and
interior design: Jennifer Wenzel
Acquisition editor: Peter Janzow
Cover design: Wanda Lubelska Design
Manufacturing buyers: Linda Behrens/Patrice Fraccio
Logo design on series page: A. M. Bruckstein

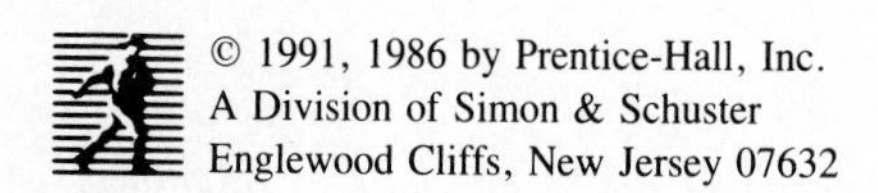

© 1991, 1986 by Prentice-Hall, Inc.
A Division of Simon & Schuster
Englewood Cliffs, New Jersey 07632

All rights reserved. No part of this book may be
reproduced, in any form or by any means,
without permission in writing from the publisher.

Printed in the United States of America

10 9 8 7 6 5 4 3 2 1

ISBN 0-13-013236-5

Prentice-Hall International (UK) Limited, *London*
Prentice-Hall of Australia Pty. Limited, *Sydney*
Prentice-Hall Canada Inc., *Toronto*
Prentice-Hall Hispanoamericana, S.A., *Mexico*
Prentice-Hall of India Private Limited, *New Delhi*
Prentice-Hall of Japan, Inc., *Tokyo*
Simon & Schuster Asia Pte. Ltd., *Singapore*
Editora Prentice-Hall do Brasil, Ltda., *Rio de Janeiro*

To the many researchers
whose contributions have made it possible
to write this book,
and the many readers, reviewers, and students
who have helped me
improve it.

Contents

18 QR DECOMPOSITION-BASED LEAST SQUARES LATTICE 642

Part IV Limitations, Extensions, and Discussions 679

19 FINITE-PRECISION AND OTHER PRACTICAL EFFECTS 680

20 BLIND DECONVOLUTION 722

Preface

The subject of adaptive filters has matured to the point where it now constitutes an important part of statistical signal processing. Whenever there is a requirement to process signals that result from operation in an environment of unknown statistics, the use of an adaptive filter offers an attractive solution to the problem as it usually provides a significant improvement in performance over the use of a fixed filter designed by conventional methods. Furthermore, the use of adaptive filters provides new signal processing capabilities that would not be possible otherwise. We thus find that adaptive filters are successfully applied in such diverse fields as communications, control, radar, sonar, seismology, and biomedical engineering.

The aim of the book is three-fold: (1) to develop the mathematical theory of various realizations of *linear adaptive filters with finite-duration impulse response,* (2) to illustrate this theory with examples and computer experiments, and (3) to introduce the ideas of nonlinearity and high-order statistics in the context of nonlinear adaptive filtering. There is no unique solution to the adaptive filtering problem. Rather, we have a "kit of tools" represented by a variety of recursive algorithms, each of which offers desirable features of its own.

Historical notes are added in the book in the hope that they provide a source of motivation for the interested reader to plough through the rich history of the subject. Each chapter of the book, except for Chapters 1 and 21, ends with problems that are designed to extend the theory and to challenge the readers to undertake computer experiments of their own.

In Chapter 1, we discuss, in general terms, the operation of adaptive filters and their practical applications. We end this chapter with some historical notes on the subject and related issues.

The next 20 chapters are organized in four parts. Specifically, Chapters 2 through 4 constitute the first part of the book devoted to *discrete-time wide-sense stationary stochastic processes*, and their statistical characterization. Chapters 5 through 7 constitute the second part of the book devoted to *linear optimum filtering*, namely, Wiener filtering, linear prediction, and Kalman filtering. Chapters 8 through 19 constitute the third and larger part of the book devoted to *linear finite-duration impulse response (FIR) adaptive filtering*. The fourth part of the book consists of Chapters 19 through 21, dealing with limitations, extensions and discussions. In Chapter 19 we study finite-precision effects. Chapter 20 is devoted to blind deconvolution based on the use of nonlinear estimation and higher-order statistics. Chapter 21 is the concluding chapter of the book, presenting (1) a summary of linear adaptive FIR filtering, and (2) unexplored issues relevant to adaptive filtering.

The third part of the book is itself organized in four parts:

1. Chapters 8 and 9 present a detailed treatment of the ubiquitous *least-mean-square (LMS) algorithm* and its relation to the method of steepest descent.
2. Chapters 10 through 12 present a detailed treatment of the classical method of least squares, and its numerical computation using the powerful method of *singular value decomposition*.
3. Chapters 13 and 14 present detailed treatments of the standard *recursive least-squares* (RLS) algorithm, and its implementation using the powerful method of *QR-decomposition,* respectively.
4. Chapters 15 through 18 are devoted to *fast* recursive least-squares algorithms. In particular, Chapter 14 develops the background theory of fast algorithms. Chapter 15 develops the *fast transversal filter (FTF) algorithm*. Chapter 16 develops the recursive *least-squares lattice (LSL) algorithms*. Chapter 17 develops *QR decomposition-based least-squares lattice (QRD-LSL) algorithms*.

Appendices are included on seven topics: complex variables, differentiation with respect to a complex-valued vector, the method of Lagrange multipliers, the method of maximum likelihood estimation, conditional mean estimation, the maximum entropy method, and a fast algorithm for computing the minimum variance distortionless response spectrum. These appendices provide supporting material for the book.

To help the reader, a Glossary for the book is included. It consists of a list of definitions, notations and conventions, a list of abbreviations, and a list of principal symbols used in the book.

Another item of interest is the extensive Bibliography and References included at the end of the book, before the Index. All publications referred to in the text are compiled in this list. Each reference is identified in the text by the name(s) of the author(s) and the year of publication. The list also includes many other references that have been added for completeness.

The book is written at a level suitable for use in graduate courses on adaptive signal processing. In this context, it is noteworthy that the organization of the material covered in the book offers a great deal of flexibility in the selection of a suitable list of topics for such a graduate course.

The following material is available from the publisher as aids to the teacher of the course:

1. Solutions manual that presents detailed solutions to all of the problems in the book.
2. Master transparencies for all the figures and summaries of algorithms used in the book.

It is hoped that the book will also be useful to engineers in industry working on problems relating to the theory and application of adaptive filters.

Simon Haykin

Acknowledgments

I am deeply indebted to Dr. Kevin Buckley, University of Minnesota, for critically reviewing the entire manuscript of this book, for making so many valuable suggestions to improve the book, and for making available to me many enlightening questions with complete solutions. In this review, he was helped by Dr. Robert Stewart, University of Strathclyde, Scotland. I am truly grateful to both Dr. Buckley and Dr. Stewart for their kind help.

I am also most grateful to Dr. W. F. McGee, University of Ottawa, Ontario, and Dr. C. L. Nikias, Northeastern University, Boston, Massachusetts, for their complete reviews of an early version of the manuscript. They both made numerous corrections, and suggestions for improving the book. Helpful inputs from three other anonymous reviews are also appreciated.

I am indebted to Dr. Ian Proudler and Dr. Terry Shepherd, both of the RSRE, Great Malvern, England, Dr. D. T. M. Slock of Philips, Brussels, Belgium, Dr. M. H. Verhaegen of Technische Universiteit Delft, The Netherlands, Dr. J. P. Reilly and Dr. S. Qiao, both of McMaster University, Dr. David Thomson of Bell Telephone Laboratories, Murray Hill, N.J., Dr. Frank Luk, Dr. C. R. Johnson, Jr., and Dr. Adam Bojanczyk of Cornell University, Dr. Neil Bershad of the University of California at Irvine, and Dr. S. Bellini of Politecnico di Milano, Italy, for their critical reviews of selected chapters of the book, and for many helpful inputs.

I am most grateful to Dr. P. A. Regalia of the Institute National des Télécommunications, France, Dr. J. Demmel of the Courant Institute, New York, and Dr. Q. T. Zhang of McMaster University, for reviewing selected sections of the book, and for their helpful comments. Inputs from Dr. Gene Golub of Stanford University, and J. P. Ardouin, Defense Research Establishment Valcartier, Quebec are also appreciated.

I am grateful to the Institute of Electrical and Electronic Engineers (New York), Automatica (London), and RSRE (Great Malvern, England) for giving me permission to reproduce certain figures. I also wish to thank Dr. J. Zeidler of the University of California at San Diego for giving me permission to reproduce Figure 1.24, from one of his papers, and for reviewing the related material.

I am particularly indebted to my graduate students Jim Orlando, Tasos Drosopoulos, and Henry Leung for performing various computer experiments included in the book.

I am grateful to my editors Tim Bozik and Pete Janzow for their guidance and encouragement, and to Jennifer Wenzel and other staff members of Prentice Hall for their help in the production of the book.

Last, but by no means least, I am most thankful to my secretary, Lola Brooks, for typing so many different versions of the manuscript of the book, and always doing it in a highly committed and cheerful way.

Simon Haykin

Adaptive Filter Theory

1 Introduction

1.1 THE LINEAR FILTERING PROBLEM

The term *filter* is often used to describe a device in the form of a piece of physical hardware or computer software that is applied to a set of noisy data in order to extract information about a prescribed quantity of interest. The noise may arise from a variety of sources. For example, the data may have been derived by means of noisy sensors or may represent a useful signal component that has been corrupted by transmission through a communication channel. In any event, we may use a filter to perform three basic information-processing operations:

1. *Filtering,* which means the extraction of information about a quantity of interest at time t by using data measured up to and including time t.
2. *Smoothing,* which differs from filtering in that information about the quantity of interest need not be available at time t, and data measured later than time t can be used in obtaining this information. This means that in the case of smoothing there is a delay in producing the result of interest. Since in the smoothing process we are able to use data obtained not only up to time t, but also data obtained after time t, we would expect it to be more accurate in some sense than the filtering process.
3. *Prediction,* which is the forecasting side of information processing. The aim here is to derive information about what the quantity of interest will be like at some time $t + \tau$ in the future, for some $\tau > 0$, by using data measured up to and including time t.

We say that the filter is *linear* if the filtered, smoothed, or predicted quantity of interest at the output of the device is a *linear function of the observations applied to the filter input*.

In the statistical approach to the solution of the *linear filtering problem* as classified above, we assume the availability of certain statistical parameters (i.e., *mean and correlation functions*) of the useful signal and unwanted additive noise, and the requirement is to design a linear filter with the noisy data as input so as to minimize the effects of noise at the filter output according to some statistical criterion. A useful approach to this filter-optimization problem is to minimize the mean-square value of the *error signal* that is defined as the difference between some desired response and the actual filter output. For stationary inputs, the resulting solution is *commonly* known as the *Wiener filter*, which is said to be *optimum in the mean-square sense*.

The Wiener filter is inadequate for dealing with situations in which *nonstationarity* of the signal and/or noise is intrinsic to the problem. In such situations, the optimum filter has to assume a *time-varying* form. A highly successful solution to this more difficult problem is found in the *Kalman filter*, a powerful device with a wide variety of engineering applications.

Linear filter theory, encompassing both Wiener and Kalman filters, has been developed fully in the literature for *continuous-time* as well as *discrete-time* signals. However, for technical reasons influenced by the wide availability of digital computers and the ever-increasing use of digital signal-processing devices, we find in practice that the discrete-time representation is often the preferred method. Accordingly, in subsequent chapters, we will only consider the discrete-time version of Wiener and Kalman filters; Wiener filters are considered in Chapters 5 and 6, and Kalman filters in Chapter 7. In this method of representation, the input and output signals, as well as the characteristics of the filters themselves, are all defined at discrete instants of time. In any case, a continuous-time signal may always be represented by a *sequence of samples* that are derived by observing the signal at uniformly spaced instants of time. No loss of information is incurred during this conversion process provided, of course, we satisfy the well-known *sampling theorem,* according to which the sampling rate has to be greater than twice the highest frequency component of the continuous-time signal. We may thus represent a continuous-time signal $u(t)$ by the sequence $\{u(n)\}$, $n = 0, \pm 1, = \pm 2, \ldots$, where for convenience we have normalized the sampling period to unity.

1.2 ADAPTIVE FILTERS

The design of a Wiener filter requires *a priori* information about the statistics of the data to be processed. The filter is optimum only when the statistical characteristics of the input data match the a priori information on which the design of the filter is based. When this information is not known completely, however, it may not be possible to design the Wiener filter or else the design may no longer be optimum. A straightforward approach that we may use in such situations is the "estimate and plug" procedure. This is a two-stage process whereby the filter first "estimates" the statistical parameters of the relevant signals and then "plugs" the results so obtained into a *nonrecursive* formula for computing the filter parameters. For *real-time* operation, this procedure has the disadvantage of requiring excessively elaborate and costly hardware. A more efficient method is to use an *adaptive filter*. By such a device we mean one that is *self-designing* in that the adaptive filter relies for its operation on a *recursive algorithm,* which makes it possible for the filter to perform satisfactorily in an environment where complete knowledge of the relevant signal characteristics is not available. The algorithm starts from some predetermined set of *initial conditions,* representing complete ignorance about the environment. Yet, in a stationary environment, we find that after successive iterations of the algorithm it *converges* to the optimum Wiener solution in some statistical sense. In a nonstationary environment, the algorithm offers a *tracking* capability, whereby it can track time variations in the statistics of the input data, provided that the variations are sufficiently slow.

As a direct consequence of the application of a recursive algorithm, whereby the parameters of an adaptive filter are updated from one iteration to the next, the parameters become *data dependent*. This, therefore, means that an adaptive filter is a *nonlinear device in the sense that it does not obey the principle of superposition*.

In another context, an adaptive filter is often referred to as linear in the sense that the estimate of a quantity of interest is obtained adaptively (at the output of the filter) as a *linear combination of the available set of observations applied to the filter input*.

A wide variety of recursive algorithms have been developed in the literature for the operation of adaptive filters. In the final analysis, the choice of one algorithm over another is determined by various factors:

1. *Rate of convergence*. This is defined as the number of iterations required for the algorithm, in response to stationary inputs, to converge "close enough" to the optimum Wiener solution in the mean-square sense. A fast rate of convergence allows the algorithm to adapt rapidly to a stationary environment of unknown statistics.
2. *Misadjustment*. For an algorithm of interest, this parameter provides a quantitative measure of the amount by which the final value of the mean-squared error, averaged over an ensemble of adaptive filters, deviates from the minimum mean-squared error that is produced by the Wiener filter.
3. *Tracking*. When an adaptive filtering algorithm operates in a nonstationary environment, the algorithm is required to *track* statistical variations in the environment. The tracking performance of the algorithm, however, is influenced by two contradictory features: (a) rate of convergence, and (b) steady-state fluctuation due to algorithm noise.
4. *Robustness*. In one context, robustness refers to the ability of the algorithm to operate satisfactorily with ill-conditioned input data. We say that a data sequence is *ill conditioned* when the *condition number* of the underlying correlation matrix is large. The term robustness is also used in the context of numerical behavior; see point 7.
5. *Computational requirements*. Here the issues of concern include (a) the number of operations (i.e., multiplications, divisions, and additions/subtractions) required to make one complete iteration of the algorithm, (b) the size of memory locations required to store the data and the program, and (c) the investment required to program the algorithm on a computer.

6. *Structure*. This refers to the structure of information flow in the algorithm, determining the manner in which it is implemented in hardware form. For example, an algorithm whose structure exhibits high modularity, parallelism, or concurrency is well suited for implementation using very large scale integration (VLSI).[1]
7. *Numerical properties*. When an algorithm is implemented numerically, inaccuracies are produced due to *quantization errors*. The quantization errors are due to analog-to-digital conversion of the input data and digital representation of internal calculations. Ordinarily, it is the latter source of quantization errors that poses a serious design problem. In particular, there are two basic issues of concern: numerical stability and numerical accuracy. *Numerical stability* is an inherent characteristic of an adaptive filtering algorithm. *Numerical accuracy,* on the other hand, is determined by the number of *bits* (i.e., *bi*nary digi*ts*) used in the numerical representation of data samples and filter coefficients. An adaptive filtering algorithm is said to be numerically robust when it is insensitive to variations in the wordlength used in its digital implementation.

1.3 FILTER STRUCTURES

The operation of an adaptive filtering algorithm involves two basic processes: (1) a *filtering* process designed to produce an output in response to a sequence of input data, and (2) an *adaptive* process, the purpose of which is to provide a mechanism for the *adaptive control* of an *adjustable* set of parameters used in the filtering process. These two processes work interactively with each other. Naturally, the choice of a structure for the filtering process has a profound effect on the operation of the algorithm as a whole.

There are three types of filter structures that distinguish themselves in the context of an adaptive filter with *finite memory* or, equivalently, *finite(-duration) impulse response*. The three filter structures are as follows:

1. *Transversal filter*. The *transversal filter,*[2] also referred to as a *tapped-delay line filter,* consists of three basic elements, as depicted in Fig. 1.1: (a) *unit-delay element,* (b) *multiplier,* and (c) *adder*. The number of delay elements used in the filter determines the duration of its impulse response. The number of delay elements, shown as $M - 1$ in Fig. 1.1, is commonly referred to as the order of the filter. In

[1] VLSI technology favors the implementation of algorithms that possess high modularity, parallelism, or concurrency. We say that a structure is *modular* when it consists of similar stages connected in cascade. By *parallelism* we mean a large number of operations being performed side by side. By *concurrency* we mean a large number of *similar* computations being performed at the same time.

[2] The transversal filter was first described by Kallmann as a continuous-time device whose output is formed as a linear combination of voltages taken from uniformly spaced taps in a nondispersive delay line (Kallmann, 1940). In recent years, the transversal filter has been implemented using digital circuitry, charge-coupled devices, or surface-acoustic wave devices. Owing to its versatility and ease of implementation, the transversal filter has emerged as an essential signal-processing structure in a wide variety of applications.

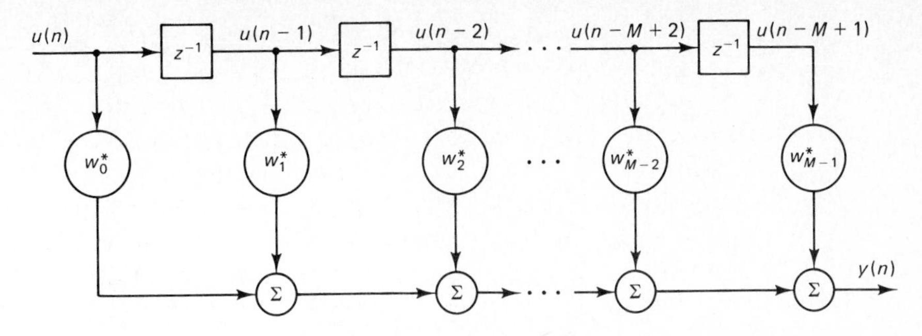

Figure 1.1 Transversal filter.

this figure, the delay elements are each identified by the *unit-delay operator* z^{-1}. In particular, when z^{-1} operates on the input $u(n)$, the resulting output is $u(n-1)$. The role of each multiplier in the filter is to multiply the *tap input* (to which it is connected) by a filter coefficient referred to as a *tap weight*. Thus a multiplier connected to the kth tap input $u(n-k)$ produces the *inner product* $w_k^* u(n-k)$, where w_k is the respective tap weight and $k = 0, 1, \ldots, M-1$. The asterisk denotes *complex conjugation,* which assumes that the tap inputs and therefore the tap weights are all *complex valued*. The combined role of the adders in the filter is to sum the individual multiplier outputs (i.e., inner products) and produce an overall filter output. For the transversal filter described in Fig. 1.1, the filter output is given by

$$y(n) = \sum_{k=0}^{M-1} w_k^* u(n-k) \tag{1.1}$$

Equation (1.1) is called a finite *convolution sum* in the sense that it *convolves* the finite impulse response of the filter $\{w_n^*\}$ with the filter input $\{u(n)\}$ to produce the filter output $y(n)$.

2. *Lattice predictor*. A *lattice predictor*[3] is *modular* in structure in that it consists of a number of individual stages, each of which has the appearance of a lattice, hence the name "lattice" as a structural descriptor. Figure 1.2 depicts a lattice predictor consisting of $M-1$ stages; the number $M-1$ is referred to as the *predictor order*. The mth stage of the lattice predictor in Fig. 1.2 is described by the pair of input–output relations (assuming the use of complex-valued, wide-sense stationary input data):

$$f_m(n) = f_{m-1}(n) + \Gamma_m^* b_{m-1}(n-1) \tag{1.2}$$

$$b_m(n) = b_{m-1}(n-1) + \Gamma_m f_{m-1}(n) \tag{1.3}$$

where $m = 1, 2, \ldots, M-1$, and $M-1$ is the *final* predictor order. The variable $f_m(n)$ is the mth *forward prediction error,* and $b_m(n)$ is the mth *backward prediction error*. The coefficient Γ_m is called the mth *reflection coefficient*. The forward predic-

[3] The development of the lattice predictor is credited to Itakura and Saito (1972). According to Markel and Gray (1976), the work of Itakura and Saito in Japan on the linear prediction of speech had in actual fact been presented in 1969.

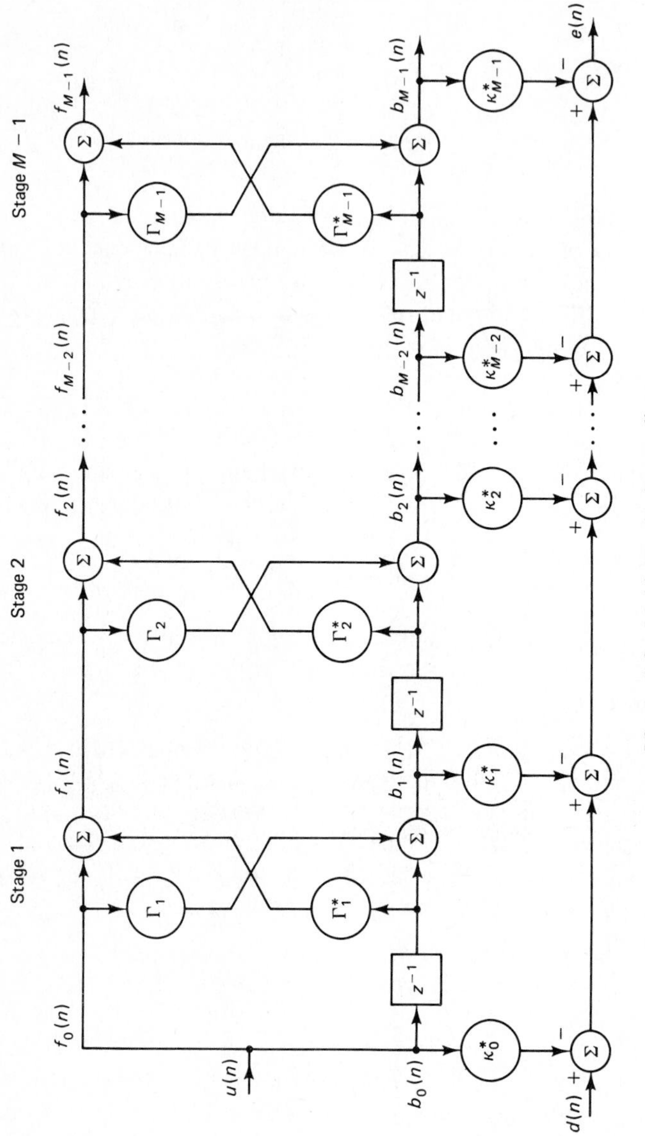

Figure 1.2 Multistage lattice predictor.

tion error $f_m(n)$ is defined as the difference between the input $u(n)$ and its *one-step predicted* value; the latter is based on the set of *m past inputs* $u(n-1), \ldots, u(n-m)$. Correspondingly, the backward prediction error $b_m(n)$ is defined as the difference between the input $u(n-m)$ and its "backward" prediction based on the set of m "future" inputs $u(n), \ldots, u(n-m+1)$. Considering the conditions at the input of stage 1 in Fig. 1.2, we have

$$f_0(n) = b_0(n) = u(n) \tag{1.4}$$

where $u(n)$ is the lattice predictor input at time n. Thus, starting with the *initial conditions* of Eq. (1.4) and given the set of reflection coefficients $\Gamma_1, \Gamma_2, \ldots, \Gamma_{M-1}$, we may determine the pair of outputs $f_{M-1}(n)$ and $b_{M-1}(n)$ by moving through the lattice predictor, stage by stage.

For a *correlated* input sequence $u(n), u(n-1), \ldots, u(n-M+1)$ drawn from a stationary process, the backward prediction errors $b_0, b_1(n), \ldots, b_{M-1}(n)$ form a sequence of *uncorrelated* random variables. Moreover, there is a one-to-one correspondence between these two sequences of random variables in the sense that if we are given one of them, we may uniquely determine the other, and vice versa. Accordingly, a linear combination of the backward prediction errors $b_0(n), b_1(n), \ldots, b_{M-1}(n)$ may be used to provide an *estimate* of some desired response $d(n)$, as depicted in the lower half of Fig. 1.2. The arithmetic difference between $d(n)$ and the estimate so produced represents the estimation error $e(n)$. The process described herein is referred to as a *joint-process estimation*. Naturally, we may use the original input sequence $u(n), u(n-1), \ldots, u(n-M+1)$ to produce an estimate of the desired response $d(n)$ directly. The indirect method depicted in Fig. 1.2, however, has the advantage of simplifying the computation of the tap weights $\kappa_0, \kappa_1, \ldots, \kappa_{M-1}$ by exploiting the uncorrelated nature of the corresponding backward prediction errors used in the estimation.

3. *Systolic array*. A *systolic array*[4] represents a *parallel computing* network ideally suited for *mapping* a number of important linear algebra computations, such as *matrix multiplication, triangularization,* and *back substitution*. Two basic types of processing elements may be distinguished in a systolic array: *boundary cells* and *internal cells*. Their functions are depicted in Fig. 1.3(a) and (b), respectively. In each case, the parameter r represents a value *stored* within the cell. The function of the boundary cell is to produce an output equal to the input u divided by the number r stored in the cell. The function of the internal cell is twofold: (a) to multiply the input z (coming in from the top) by the number r stored in the cell, subtract the product rz from the second input (coming in from the left), and thereby produce the difference $u - rz$ as an output from the right side of the cell, and (b) to transmit the first input z downward without alteration.

[4] The systolic array was pioneered by Kung and Leiserson (1979). Its development provided a practical means for overcoming the *von Neumann bottleneck*, which is a major drawback of serial computing. In particular, the use of systolic arrays has made it possible to achieve the high throughput required for many advanced signal processing algorithms to operate in *real time*. Indeed, it may be justifiably said that systolic arrays have provided the light at the end of the tunnel in the context of parallel computing for advanced signal processing (McWhirter, 1989). For an introductory treatment of systolic arrays, see Kung (1985) and McWhirter (1989).

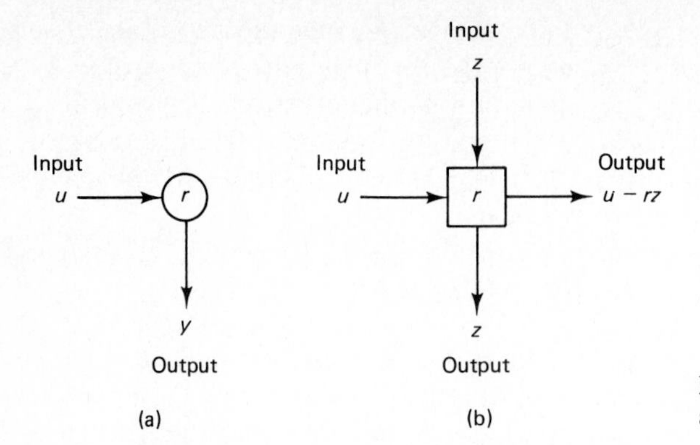

Figure 1.3 Two basic cells of a systolic array: (a) boundary cell; (b) internal cell.

Consider, for example, the 4-by-3 systolic array shown in Fig. 1.4. Let the elements of the 3-by-1 vector $\mathbf{u} = [u_1, u_2, u_3]^T$ denote the inputs applied to this systolic array from its left side. Let the elements of the 4-by-1 vector $\mathbf{z} = [z_1, z_2, z_3, z_4]^T$ denote the inputs applied to the systolic array of Fig. 1.4 from its top side. The output vector $\mathbf{v} = [v_1, v_2, v_3]^T$ produced on the right side of the systolic array is computed as

$$\mathbf{v} = \mathbf{u} - \mathbf{R}^T\mathbf{z} \tag{1.5}$$

where the elements of the *transposed* matrix $\mathbf{R}^T$ are the respective internal cell contents of the rectangular array. All the internal cells in Fig. 1.4 operate on the input data and produce output data uniformly in one single time unit. The input data (i.e.,

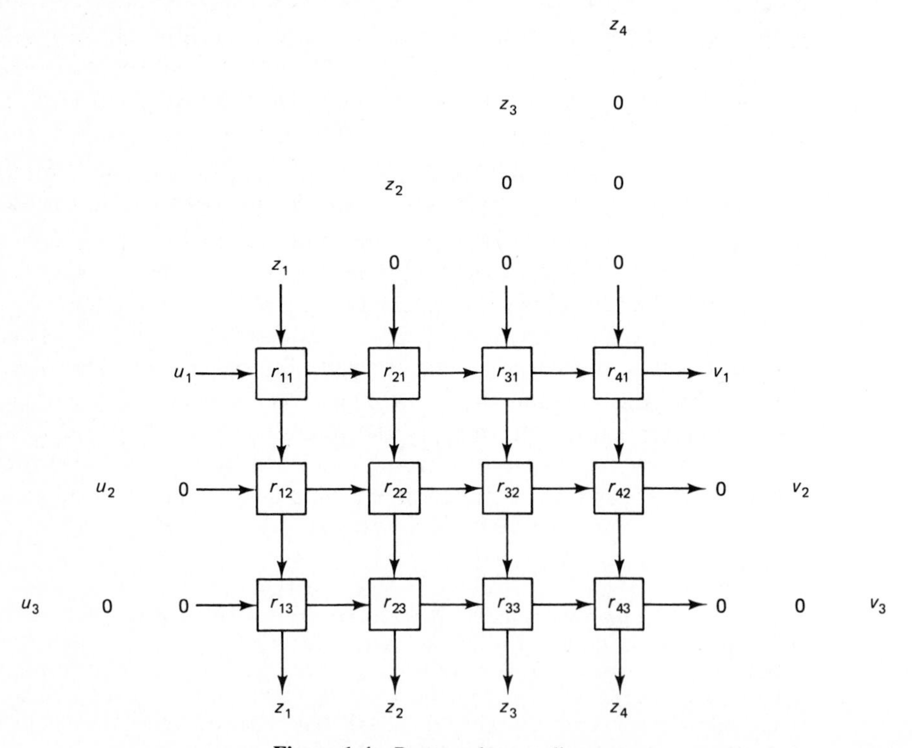

Figure 1.4 Rectangular systolic array.

the elements of the vector **u** entering the array from the left side and the vector **z** entering it from the top side) are prearranged in an orderly *skewed* manner. This is done so as to perform the computations in a *pipelined* fashion. In particular, the individual inputs from the vectors **u** and **z** are *delayed* by a certain number of time units before arriving at the array. This explains the reason for the introduction of some extra "zeros" in Fig. 1.4.

For a second example, consider the 3-by-3 triangular array shown in Fig. 1.5. This second systolic array involves a combination of boundary and internal cells. In this case, the triangular array computes an output vector **y** related to the input vector **u** as follows:

$$\mathbf{y} = \mathbf{R}^{-T}\mathbf{z} \tag{1.6}$$

where the $\mathbf{R}^{-T}$ is the *inverse* of the transposed matrix $\mathbf{R}^T$. The elements of $\mathbf{R}^T$ are the respective cell contents of the triangular array. The zeros added to the inputs of the array in Fig.1.5 are intended to provide the delays necessary for pipelining the computation described in Eq. (1.6).

A systolic array architecture, as described herein, offers the desirable features of *modularity, local interconnections,* and highly *pipelined* and *synchronized* parallel processing; the synchronization is achieved by means of a global *clock*.

We note that the transversal filter of Fig. 1.1, the joint-process estimator of Fig. 1.2 based on a lattice predictor, and the triangular systolic array of Fig. 1.5 have a common property: All three of them are characterized by an impulse response of finite duration. In other words, they are examples of a *finite-duration impulse response (FIR) filter,* whose structures contain *feedforward* paths only. On the other hand, the filter structure shown in Fig. 1.6 is an example of an *infinite-duration impulse response (IIR) filter*. The feature that distinguishes an IIR filter from an FIR filter is the inclusion of *feedback* paths. Indeed, it is the presence of feedback that makes the duration of the impulse response of an IIR filter infinitely long. Furthermore, the presence of feedback introduces a new problem, namely, that of *stability*. In particular, it is possible for an IIR filter to become unstable (i.e., break

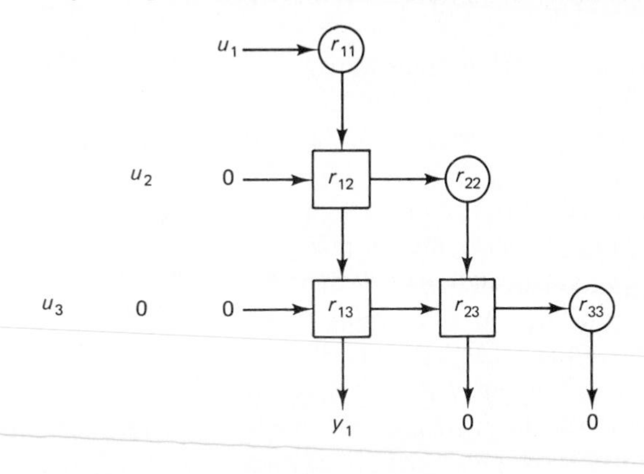

Figure 1.5
Triangular systolic array.

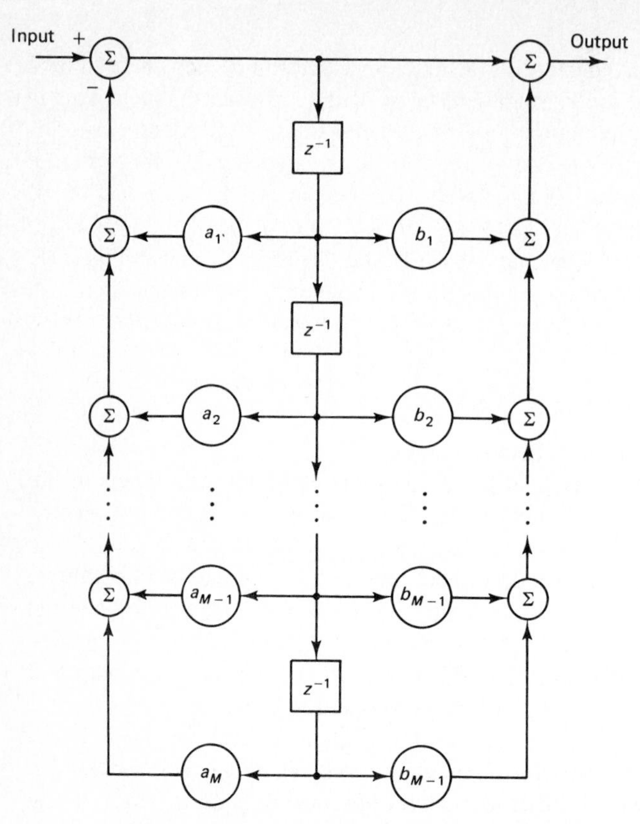

Figure 1.6 IIR filter.

into oscillation), unless special precaution is taken in the choice of feedback coefficients. By contrast, an FIR filter is inherently *stable*. This explains the reason for the popular use of FIR filters, in one form or another, as the structural basis for the design of adaptive filters.

Real and Complex Forms of Adaptive Filtering

In the development of adaptive filtering algorithms, it is customary to assume that the input data are in *baseband* form. The term "baseband" is used to designate the band of frequencies representing the original (message) signal as received by the source of information.

In such applications as communications, radar, and sonar, the information-bearing signal component of the receiver input typically consists of a message signal *modulated* onto a carrier wave. The bandwidth of the message signal is usually small compared to the carrier frequency, which means that the modulated signal is a *narrow-band signal*. To obtain the baseband representation of a narrow-band signal, the signal is translated down in frequency in such a way that the effect of the carrier wave is completely removed, yet the information content of the message signal (in both amplitude and phase) is fully preserved. In general, the baseband signal so obtained is *complex*. In other words, a sample $u(n)$ of the signal may be written as (Haykin, 1988)

$$u(n) = u_I(n) + ju_Q(n) \tag{1.7}$$

where $u_I(n)$ is the *in-phase* (real) *component*, and $u_Q(n)$ is the *quadrature* (imaginary) *component*. Equivalently, we may express $u(n)$ as

$$u(n) = |u(n)|e^{j\phi(n)} \tag{1.8}$$

where $|u(n)|$ is the *magnitude* and $\phi(n)$ is the *phase angle*.

Accordingly, the theory of adaptive filters developed in subsequent chapters of the book assumes the use of complex signals. An adaptive filtering algorithm so developed is said to be in *complex form*.

If the signals to be processed are *real,* we naturally use the *real form* of the adaptive filtering algorithm of interest. Given the complex form of an adaptive filtering algorithm, it is straightforward to deduce the corresponding real form of the algorithm. Specifically, we do two things:

1. The operation of *complex conjugation,* wherever in the algorithm, is simply removed.
2. The operation of *Hermitian transposition* (i.e., conjugate transposition) of a matrix, wherever in the algorithm, is replaced by ordinary transposition.

1.4 APPROACHES TO THE DEVELOPMENT OF ADAPTIVE FILTER THEORY

There is no unique solution to the adaptive filtering problem. Rather, we have a "kit of tools" represented by a variety of recursive algorithms, each of which offers desirable features of its own. The challenge facing the user of adaptive filtering is, first, to understand the capabilities and limitations of various adaptive filtering algorithms, and second, to use this understanding in the selection of the appropriate algorithm for the application at hand.

Basically, we may identify three distinct methods for deriving recursive algorithms for the operation of adaptive filters, as discussed next.

Approach Based on Wiener Filter Theory

Here we use a tapped-delay line or transversal filter as the structural basis for implementing the adaptive filter. For the case of stationary inputs, the *mean-squared error* (i.e., the mean-square value of the difference between the desired response and the transversal filter output) is precisely a second-order function of the tap weights in the transversal filter. The dependence of the mean-squared error on the unknown tap weights may be viewed to be in the form of a *multidimensional paraboloid* (i.e., punch bowl) with a uniquely defined bottom or *minimum point*. We refer to this paraboloid as the *error performance surface*. The tap weights corresponding to the minimum point of the surface define the optimum Wiener solution.

To develop a recursive algorithm for updating the tap weights of the adaptive transversal filter, we proceed in two stages. We first modify the system of *Wiener-Hopf equations* (i.e., the matrix equation defining the optimum Wiener solution) through the use of the *method of steepest descent,* a well-known technique in optimization theory. This modification requires the use of a *gradient vector*, the value of which depends on two parameters: the *correlation matrix* of the tap inputs in the transversal filter and the *cross-correlation vector* between the desired response and the same tap inputs. Next, we use instantaneous values for these correlations so as to derive an *estimate* for the gradient vector. The resulting algorithm is widely known as the *least-mean-square (LMS) algorithm*. This algorithm is simple and yet capable of achieving satisfactory performance under the right conditions. Its major limitations are a relatively slow rate of convergence and a sensitivity to variations in the condition number of the correlation matrix of the tap inputs. Nevertheless, the LMS algorithm is highly popular and widely used in a variety of applications.

In a nonstationary environment, the orientation of the error-performance surface varies continuously with time. In this case, the LMS algorithm has the added task of continually *tracking* the bottom of the error-performance surface. Indeed, tracking will occur provided that the input data vary slowly compared to the *learning rate* of the LMS algorithm.

In Chapter 9, we derive the LMS algorithm and study its behavior in stationary and nonstationary environments. In that chapter, we also derive a related algorithm called the *gradient lattice algorithm* (GAL). The LMS algorithm uses a transversal filter structure, whereas the GAL algorithm uses a *lattice* structure. The latter structure is modular in form, with each stage having the appearance of a lattice, hence the name. The LMS and GAL algorithms belong to a broadly defined class of *stochastic gradient algorithms*.

Approach Based on Kalman Filter Theory

Depending on whether the requirement is to operate in a stationary or nonstationary environment, we may exploit the Kalman filter as the basis for deriving an adaptive filtering algorithm appropriate to the situation.

The Kalman filtering problem for a linear dynamic system is formulated in terms of two basic equations: the *process equation* that describes the dynamics of the system in terms of the *state vector*, and the *measurement equation* that describes *measurement errors* incurred in the system. The solution to the problem is expressed as a set of time-update recursions that are expressed in matrix form. To apply these recursions to solve the adaptive filtering problem, however, the theory requires that we postulate a *model* of the optimum operating conditions, which serves as the frame of reference for the Kalman filter to *track*. With a transversal filter used to provide the structural basis for the adaptive filter, we may identify the tap-weight vector in the filter as the state vector. Thus, for a stationary environment, we use a *fixed-state model* in which the tap weight or state vector of the model assumes a constant value. For a nonstationary environment, we use a *noisy-state model* in which, for example, the tap weight or state vector of the model executes a *random walk* around some mean value. Thus, by adopting one or the other of these idealized state models and also by making some other identifications, we may use the recursive solution to the Kalman filtering problem to derive different recursive algorithms for updating the tap-weight vector of the adaptive transversal filter. These algorithms are powerful in that usually they can provide a much faster rate of convergence than that attainable by the LMS algorithm. They are *robust* in that their rate of convergence is essentially insensitive to the eigenvalue-spread problem. Moreover, we may structure the Kalman algorithm to deal with a stationary or nonstationary environment. The basic limitation of these algorithms, however, is their computational complexity, which is a direct consequence of the matrix formulation of the solution to the Kalman filtering problem. The development of Kalman filter theory is discussed in Chapter 7.

Method of Least Squares

The two procedures described above for deriving adaptive filtering algorithms follow from the Wiener filter and Kalman filter, respectively. The theory of both filters is based on statistical concepts. The approach based on the classical *method of least squares* differs from these two in that it involves the use of *time averages*. According to the method of least squares, we minimize an index of performance that consists of the *sum of weighted error squares,* where the *error* or *residual* is defined as the difference between some desired response and the actual filter output. The method of least squares may be formulated with *block estimation* or *recursive estimation* in mind. In block estimation the input data stream is arranged in the form of blocks of equal length (duration), and the filtering of input data proceeds on a block-by-block basis. In recursive estimation, on the other hand, the estimates of interest (e.g., tap weights of a transversal filter) are *updated* on a sample-by-sample basis. Ordinarily, a recursive estimator requires less storage than a block estimator. Issues relating to the block estimation approach are discussed in Chapters 10 through 12. As for the recursive estimation approach we may identify three basically different classes of adaptive filtering algorithms that originate from the method of least squares:

1. *Recursive least-squares algorithm.* As with adaptive filtering algorithms derived from the Wiener and Kalman filters, the *recursive least-squares (RLS) algorithm* also assumes the use of a transversal filter as the structural basis of the adaptive filter. The derivation of the algorithm, presented in Chapter 13, relies on a basic result in linear algebra known as the *matrix-inversion lemma.* The RLS algorithm is a special case of the Kalman algorithm for adaptive transversal filters. As such, the RLS algorithm enjoys the same virtues and suffers from the same limitation (computational complexity) as the Kalman algorithm. The intimate relationship between the RLS algorithm and Kalman filtering is also discussed in Chapter 13.
2. *QR-decomposition-based recursive least-squares (QRD-RLS) algorithm.* This class of adaptive filtering algorithms is based on the *QR-decomposition* of the data matrix. Two well-known techniques for performing this decomposition are the *Householder transformation* and the *Givens rotation.* These are data-adaptive transformations that involve *orthogonal triangularization* of the incoming data matrix. Indeed, they are both highly popular in modern numerical analysis. At this point in the discussion, the important point to note is that a recursive least-squares algorithm based on the Householder transformation or Givens rotation is numerically stable and robust. The theory of Householder transforms is considered in Chapter 12. An attraction of Givens rotations is

that their use is amenable to recursive computation, with the rotation parameters being updated on a sample-by-sample basis. The latter form of computation leads naturally to the use of *systolic arrays* for implementing recursive least-squares estimation, which is demonstrated in Chapter 14.

3. *Fast algorithms*. The standard RLS algorithm (using a transversal structure) and its counterpart based on the QR-decomposition (using a systolic array) have a computational complexity that increases as the square of M, where M is the number of adjustable weights (i.e., the number of degrees of freedom) in the algorithm. Such algorithms are often referred to as *$O(M^2)$ algorithms*, where $O(\cdot)$ denotes "order of." By contrast, the LMS algorithm is an $O(M)$ algorithm in that its computational complexity increases linearly with M. When M is large, the computational complexity of the $O(M^2)$ algorithms may become objectionable from a hardware implementation point of view. There is therefore a strong motivation to modify the formulation of recursive linear least-squares algorithms in such a way that the computational complexity assumes an $O(M)$ form. This objective is indeed achievable, first, by virtue of the inherent *redundancy* in the *Toeplitz structure* of the input data matrix, and second, by exploiting this redundancy through the use of *linear least-squares prediction in both the forward and backward directions*. The resulting algorithms are known as *fast algorithms* that combine the desirable characteristics of recursive linear least-squares estimation with an $O(M)$ computational complexity. The background theory of fast algorithms is covered in Chapter 15. Three types of fast algorithms may be identified, depending on the filter structure or technique employed:
   (a) *Fast transversal filters (FTF) algorithm*. This algorithm involves a parallel combination of four transversal filters, each one of which has an assigned task to perform. The FTF algorithm is derived in Chapter 16.
   (b) *Recursive least-squares lattice (LSL) algorithms*. This second type of fast algorithm relies on the use of a lattice structure to perform the forward and backward forms of linear least-squares prediction. There are four distinct forms of the recursive LSL algorithm; details are given in Chapter 17.
   (c) *QR-decomposition-based least-squares lattice (QRD-LSL) algorithm*. In this fast algorithm, the forward and backward forms of linear least-squares prediction are exploited for the purpose of reducing the computational complexity in performing the recursive QR-decomposition of the input data matrix. This algorithm is derived in Chapter 18.

The removal of redundancy from a recursive linear least-squares algorithm may leave the algorithm vunerable to finite-precision effects. In particular, this remark applies to the FTF algorithm and the two "conventional" forms of the recursive LSL algorithm. These algorithms are known to suffer from a numerical accuracy/stability problem. The other two forms of the recursive LSL algorithm maintain their robustness to finite-precision effects by employing error feedback in the computation of their coefficients. The QR-decomposition-based least-squares lattice (QRD-LSL) algorithm is numerically stable because of the good numerical properties inherently associated with the QR-decomposition method.

1.5 STATE-SPACE MODEL

The *state-space model* provides a description of the internal and external characteristics of a *linear, finite-dimensional system*. The *state* of a system represents the minimum amount of information on the past behavior of the system. It is therefore sufficient to predict the future response of the system. The state-space model of a linear, finite-dimensional stochastic system is described by the pair of equations

$$\mathbf{x}(n+1) = \mathbf{\Phi}(n+1, n)\mathbf{x}(n) + \mathbf{v}_1(n) \tag{1.9}$$

$$\mathbf{y}(n) = \mathbf{C}(n)\mathbf{x}(n) + \mathbf{v}_2(n) \tag{1.10}$$

where n denotes time, $\mathbf{x}(n)$ is the *state vector*, $\mathbf{y}(n)$ is the *observation vector*, and $\mathbf{v}_1(n)$ and $\mathbf{v}_2(n)$ are statistically independent *noise vectors*. The matrix $\mathbf{\Phi}(n+1, n)$ is the *state transition matrix*, relating the state of the system at time $n+1$ to that at time n. The matrix $\mathbf{C}(n)$ is the *measurement matrix*. Equations (1.9) and (1.10) are called the *process equation* and the *measurement equation* of the system, respectively.

Consider, for example, a stochastic system described by the difference equation

$$\begin{aligned} u(n) + a_1 u(n-1) + \cdots + a_M u(n-M) \\ = v(n) + b_1 v(n-1) + \cdots + b_M v(n-M) \end{aligned} \tag{1.11}$$

where the a's and the b's are constant coefficients. Equation (1.11) is represented by the signal-flow graph shown in Fig. 1.6, where the input equals $v(n)$ and the output equals $u(n)$. We assume that $\{v(n)\}$ is a white-noise process. In such a case, $\{u(n)\}$ is said to be an *autoregressive-moving average (ARMA) process*. Equation (1.11) may be recast in the following state-space form:

$$\mathbf{x}(n+1) = \mathbf{A}\mathbf{x}(n) + \mathbf{b}v(n) \tag{1.12}$$

$$u(n) = \mathbf{C}\mathbf{x}(n) \tag{1.13}$$

where $\mathbf{x}(n)$ is an $(M+1)$-by-1 state vector, and

$$\mathbf{A} = \begin{bmatrix} 1 & 1 & 0 & \cdots & 0 \\ -a_1 & 0 & 1 & \cdots & 0 \\ -a_2 & 0 & 0 & \cdots & 0 \\ \vdots & \vdots & \vdots & \ddots & \vdots \\ -a_{M-1} & 0 & 0 & \cdots & 1 \\ -a_M & 0 & 0 & \cdots & 0 \end{bmatrix} \tag{1.14}$$

$$\mathbf{b} = \begin{bmatrix} 1 \\ b_1 \\ b_2 \\ \vdots \\ b_{M-1} \\ b_M \end{bmatrix} \tag{1.15}$$

$$\mathbf{C} = [1, 0, 0, \ldots, 0, 0] \tag{1.16}$$

By successive substitution, it is a straightforward matter to show that the combination of Eqs. (1.12) and (1.13) is equivalent to the difference equation (1.11).

The state-space model is fundamental to the study of Kalman filters. This study is taken up in Chapter 7.

1.6 COST FUNCTIONS

The concept of a cost function is basic to the *optimization* of filters. A discrete-time filter is characterized by a *coefficient vector* that may refer to a tap-weight vector as in a transversal filter, or a vector of reflection coefficients as in a multistage lattice filter. A *cost function* defines a transformation from a *vector space* spanned by the elements of the coefficient vector into the space of a real scalar.[5] Naturally, there is no unique definition for a cost function. Nevertheless, we may identify some commonly used cost functions, depending on the approach used and the application of interest.

A highly popular cost function is the *mean-square error criterion,* defined as the mean-square value of an *estimation error*. The estimation error, denoted by $e(n)$, is itself defined by the difference between some desired response $d(n)$ and the actual filter output $y(n)$, as shown by

$$e(n) = d(n) - y(n) \tag{1.17}$$

For example, in the case of a transversal filter characterized by the tap-weight vector $\mathbf{w}$, the filter output is given by the *inner product* of tap-weight vector $\mathbf{w}$ and the tap-input vector $\mathbf{u}(n)$. Assuming that $\mathbf{u}(n)$ and $\mathbf{w}$ are *complex valued,* we write

$$y(n) = \mathbf{w}^H\mathbf{u}(n) \tag{1.18}$$

where the superscript H indicates *Hermitian transposition* (i.e., conjugate transposition). The use of Hermitian transposition is preferred over ordinary transposition as it simplifies the appearance of the equations that define the characterization of optimum filters. We may thus define a cost function for the optimum design of the transversal filter as

$$\begin{aligned} J(\mathbf{w}) &= E[|e(n)|^2] \\ &= E[|d(n) - \mathbf{w}^H\mathbf{u}(n)|^2] \end{aligned} \tag{1.19}$$

where E is the *expectation operator*. The filter is optimized by finding the tap-weight vector that *minimizes* the cost function J. We refer to such an approach as the *minimum mean-square error criterion,* which forms the basis of *Wiener filters*.

The cost function defined in Eq. (1.19) is *probabilistic* in nature in the sense that it involves *ensemble averaging* represented by the expectation operator E. Alternatively, we may define a cost function by considering the idea of *time averaging,* as described next.

[5] In the general definition of a function, we speak of a transformation from a vector space into the space of real (or complex) scalars (Luenberger, 1969; Dorny, 1975). A cost function provides a quantitative measure for assessing the quality of performance; hence the restriction of it to a real scalar.

Continuing with the example of a transversal filter, suppose we are given a *set of observations* represented by the tap-input vector $\mathbf{u}(i)$, where the time index $i = 1, 2, \ldots, n$. We assume that the tap-weight vector $\mathbf{w}$ of the transversal filter is held *constant* for the *entire* observation interval. At time i, we define the estimation error as

$$e(i) = d(i) - \mathbf{w}^H \mathbf{u}(i) \tag{1.20}$$

where $d(i)$ is the corresponding value of the desired response. In this case, we may define the cost function as

$$\begin{aligned} \mathscr{E}(\mathbf{w}, n) &= \sum_{i=1}^{n} |e(i)|^2 \\ &= \sum_{i=1}^{n} |d(i) - \mathbf{w}^H \mathbf{u}(i)|^2 \end{aligned} \tag{1.21}$$

Note that whereas the ensemble-averaged cost function $J(\mathbf{w})$ in Eq. (1.19) depends only on the tap-weight vector $\mathbf{w}$, the time-averaged cost function $\mathscr{E}(\mathbf{w}, n)$ depends on $\mathbf{w}$ and the observation interval n. Accordingly, the minimization of $\mathscr{E}(\mathbf{w}, n)$ with respect to $\mathbf{w}$ yields a solution for the tap-weight vector that varies with the observation interval. This second approach forms the basis of the *method of least squares*.

The cost functions $J(\mathbf{w})$ and $\mathscr{E}(\mathbf{w}, n)$ are both *convex* with a unique minimum point. Accordingly, their use yields a unique solution for the tap-weight vector of the transversal filter. A qualification in the context of the method of least squares, however, is in order. In particular, in Eq. (1.21) it is assumed that the number of observations n is larger than the number of tap weights constituting the vector $\mathbf{w}$; that is, we have an *overdetermined* system with more equations than unknowns.

A limitation of second-order statistics (e.g., the mean-square-error criterion) is that they are *phase blind*. We may overcome this limitation by the use of a *nonlinear* cost function. By so doing, the filter is enabled to extract information (particularly phase information) from the input signal in a more efficient manner. For this to be possible, however, the input signal must have *non-Gaussian* statistics. The use of such an approach provides the basis for an important class of *nonlinear* adaptive filtering algorithms that can perform *blind deconvolution,* blind in the sense that the algorithms do *not* require a desired response. The issue of blind deconvolution (and its use in blind equalization) are covered in Chapter 20.

1.7 APPLICATIONS

The desirable features of an adaptive filter, namely, the ability to operate satisfactorily in an unknown environment and also track time variations of input statistics, make the adaptive filter a powerful device for signal-processing and control applications.

Indeed, adaptive filtering has been successfully applied in such diverse fields as communications, radar, sonar, seismology, and biomedical engineering. Although these applications are indeed quite different in nature, nevertheless, they have one basic common feature: An input vector and a desired response are used to compute an estimation error, which is in turn used to control the values of a set of adjustable filter coefficients. The adjustable coefficients may take the form of tap weights, reflection coefficients, or rotation parameters, depending on the filter structure employed. However, the essential difference between the various applications of adaptive filtering arises in the manner in which the desired response is extracted. In this context, we may distinguish four basic classes of adaptive filtering applications, as depicted in Fig. 1.7. For convenience of presentation, the following notations are used in this figure:

$$u = \text{input applied to the adaptive filter}$$

$$y = \text{output of the adaptive filter}$$

$$d = \text{desired response}$$

$$e = d - y = \text{estimation error.}$$

The functions of the four basic classes of adaptive filtering applications depicted herein are as follows:

I. *Identification* [Fig. 1.7(a)]. The notion of a *mathematical model* is fundamental to sciences and engineering. In the class of applications dealing with identification, an adaptive filter is used to provide a linear model that represents the best fit (in some sense) to an *unknown plant*. The plant and the adaptive filter are driven by the same input. The plant output supplies the desired response for the adaptive filter. If the plant is dynamic in nature, the model will be time varying.

II. *Inverse modeling* [Fig. 1.7(b)]. In this second class of applications, the function of the adaptive filter is to provide an *inverse model* that represents the best fit (in some sense) to an *unknown noisy plant*. Ideally, the inverse model has a transfer function equal to the *reciprocal (inverse)* of the plant's transfer function. A delayed version of the plant (system) input constitutes the desired response for the adaptive filter. In some applications, the plant input is used without delay as the desired response.

III. *Prediction* [Fig. 1.7(c)]. Here the function of the adaptive filter is to provide the best *prediction* (in some sense) of the present value of a random signal. The present value of the signal thus serves the purpose of a desired response for the adaptive filter. Past values of the signal supply the input applied to the adaptive filter. Depending on the application of interest, the adaptive filter output or the estimation (prediction) error may serve as the system output. In the first case, the system operates as a *predictor*; in the latter case, it operates as a *prediction-error filter*.

IV. *Interference cancelling* [Fig. 1.7(d)]. In this final class of applications, the adaptive filter is used to cancel *unknown interference* contained (alongside an information-bearing signal component) in a *primary signal,* with the cancellation being optimized in some sense. The primary signal serves as the desired response for the adaptive filter. A *reference (auxiliary) signal* is employed as the input to the adaptive filter. The reference signal is derived from a sensor or set

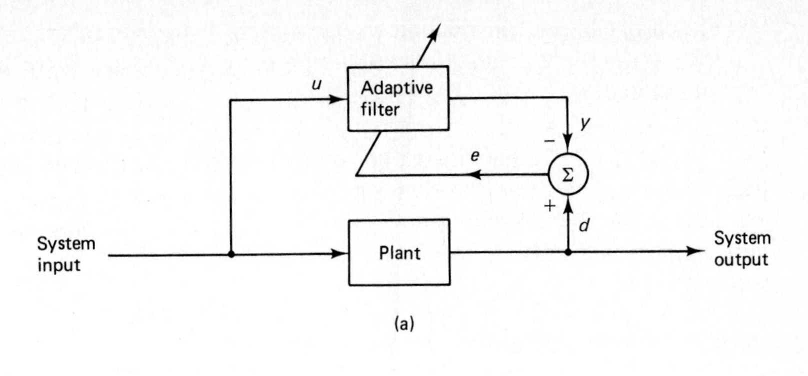

(a)

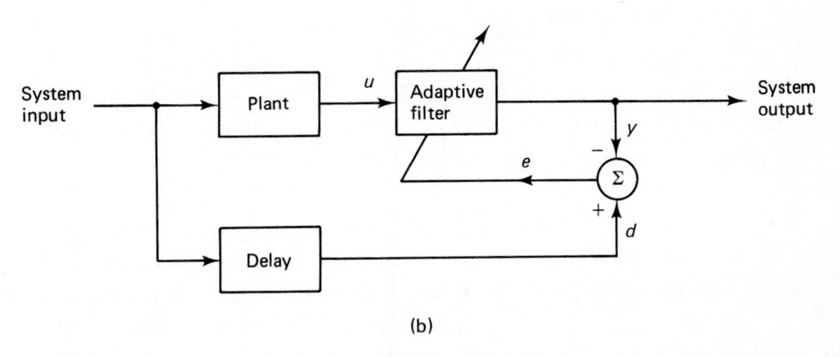

(b)

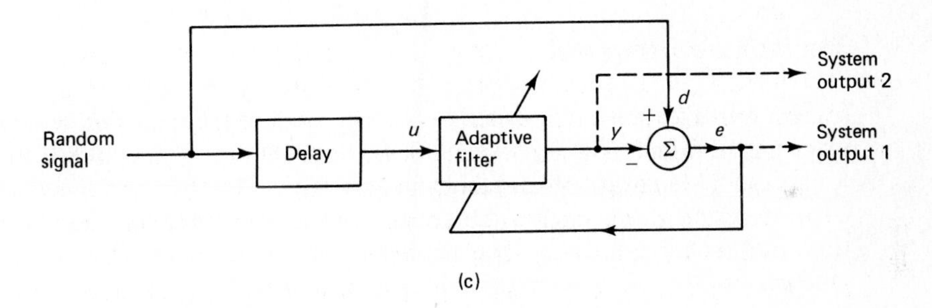

(c)

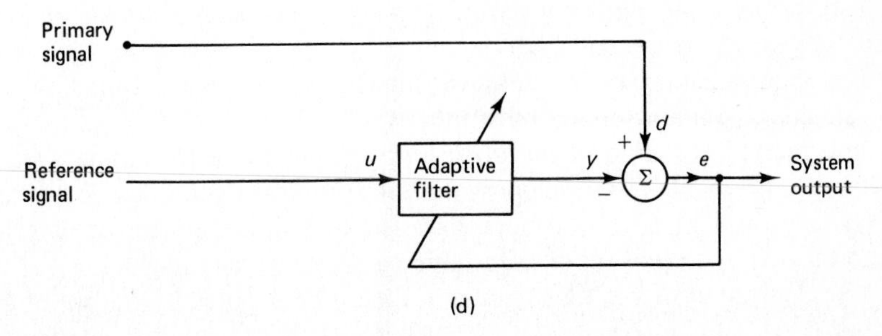

(d)

Figure 1.7 Four basic classes of adaptive filtering applications: (a) class I: identification; (b) class II: inverse modeling; (c) class III: prediction; (d) class IV: interference cancelling.

of sensors located in relation to the sensor(s) supplying the primary signal in such a way that the information-bearing signal component is weak or essentially undetectable.

In Table 1.1 we have listed some applications that are illustrative of the four basic classes of adaptive filtering applications. These applications, totaling twelve, are drawn from the fields of control systems, seismology, electrocardiography, communications, and radar.[6] They are described individually in the remainder of this section.

TABLE 1.1 APPLICATIONS OF ADAPTIVE FILTERING

Class of adaptive filtering	Application
I. Identification	System identification
	Layered earth modeling
II. Inverse modeling	Predictive deconvolution
	Adaptive equalization
III. Prediction	Linear predictive coding
	Adaptive differential pulse-code modulation
	Autoregressive spectrum analysis
	Signal detection
IV. Interference cancelling	Adaptive noise cancelling
	Echo cancellation
	Radar polarimetry
	Adaptive beamforming

System Identification

System identification is the experimental approach to the modeling of a process or a plant (Åström and Wittenmark, 1990; Söderström and Stoica, 1988; Ljung, 1987; Ljung and Söderström, 1983; Goodwin and Payne, (1977). It involves the following steps: experimental planning, the selection of a model structure, parameter estimation, and model validation. The procedure of system identification, as pursued in practice, is iterative in nature in that we may have to go back and forth between these steps until a satisfactory model is built. Here we discuss briefly the idea of adaptive filtering algorithms for estimating the parameters of an unknown plant modeled as a transversal filter.

Suppose we have an unknown dynamic plant that is linear and time varying. The plant is characterized by a *real-valued* set of discrete-time measurements that describe the variation of the plant output in response to a known stationary input. The requirement is to develop an *on-line transversal filter model* for this plant, as illustrated in Fig. 1.8. The model consists of a finite number of unit-delay elements and a corresponding set of adjustable parameters (tap weights).

Let the available input signal at time n be denoted by the set of samples: $u(n)$, $u(n-1), \ldots, u(n-M+1)$, where M is the number of parameters in the

[6] For additional applications of adaptive filtering, see Widrow and Stearns (1985), Cowan and Grant (1985), and the special issues on adaptive filters listed in the References and Bibliography at the end of the book.

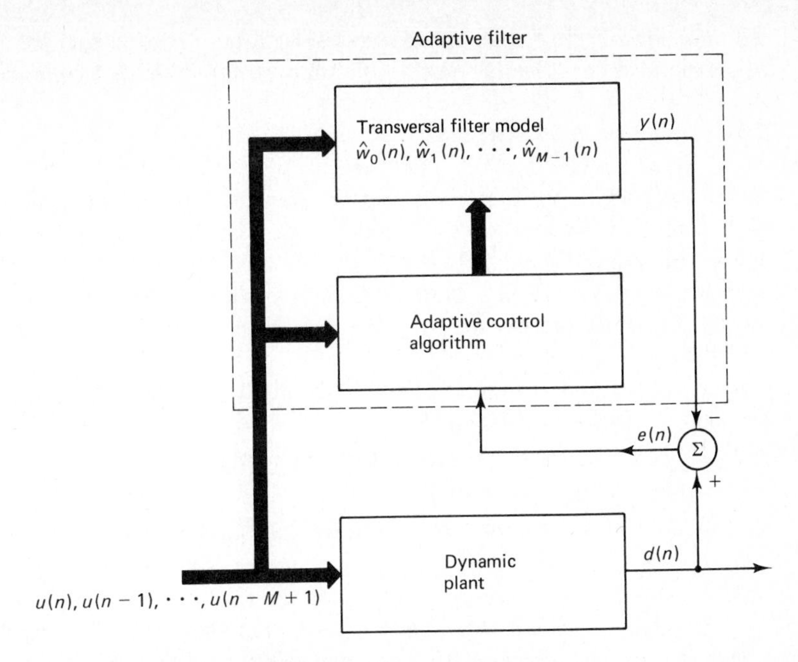

Figure 1.8 System identification.

model. This input signal is applied simultaneously to the plant and the model. Let their respective outputs be denoted by $d(n)$ and $y(n)$. The plant output $d(n)$ serves the purpose of a desired response for the adaptive filtering algorithm employed to adjust the model parameters. The model output is given by

$$y(n) = \sum_{k=0}^{M-1} \hat{w}_k(n)u(n-k) \tag{1.22}$$

where $\hat{w}_0(n)$, $\hat{w}_1(n)$, . . . , and $\hat{w}_{M-1}(n)$ are the estimated model parameters. The model output $y(n)$ is compared with the plant output $d(n)$. The difference between them, $d(n) - y(n)$, defines the *modeling (estimation) error*. Let this error be denoted by $e(n)$.

Typically, at time n, the modeling error $e(n)$ is nonzero, implying that the model deviates from the plant. In an attempt to account for this deviation, the error $e(n)$ is applied to an *adaptive control algorithm*. The samples of the input signal, $u(n)$, $u(n-1)$, . . . , $u(n-M+1)$, are also applied to the algorithm. The combination of the transversal filter and the adaptive control algorithm constitutes the adaptive filtering algorithm. The algorithm is designed to control the adjustments made in the values of the model parameters. As a result, the model parameters assume a new set of values for use on the next iteration. Thus, at time $n + 1$, a new model output is computed, and with it a new value for the modeling error. The operation described is then repeated. This process is continued for a sufficiently large number of iterations (starting from time $n = 0$), until the deviation of the model from the plant, measured by the magnitude of the modeling error $e(n)$, becomes sufficiently small in some statistical sense.

When the plant is time varying, the plant output is *nonstationary*, and so is the desired response presented to the adaptive filtering algorithm. In such a situation, the adaptive filtering algorithm has the task of not only keeping the modeling error small but also continually tracking the time variations in the dynamics of the plant.

Layered Earth Modeling

In *exploration seismology,* we usually think of a layered model of the earth (Robinson and Durrani, 1986; Mendel, 1986; Justice, 1985; Robinson and Treitel, 1980). In order to collect (record) seismic data for the purpose of characterizing such a model and thereby unraveling the complexities of the earth's surface, it is customary to use the *method of reflection seismology* that involves the following:

1. A *source of seismic energy* (e.g., dynamite, air gun) that is typically activated on the surface of the earth.
2. *Propagation* of the seismic signal away from the source and deep into the earth's crust.
3. *Reflection* of seismic waves from the interfaces between the earth's geological layers.
4. *Picking up* and *recording* the seismic returns (i.e., reflections of seismic waves from the interfaces) that carry *information* about the subsurface structure. On land, *geophones* (consisting of small sensors implanted into the earth) are used to pick up the seismic returns.

The *method of reflection seismology,* combined with a lot of signal processing, is capable of supplying a two- or three-dimensional "picture" of the earth's subsurface, down to about 20,000 to 30,000 ft and with high enough accuracy and resolution. This picture is then examined by an "interpreter" to see if it is likely that the part of the earth's subsurface (under exploration) contains hydrocarbon (petroleum) reservoirs. Accordingly, a decision is made whether or not to drill a well, which (in the final analysis) is the only way of knowing if petroleum is actually present.

A seismic wave is similar in nature to an acoustic wave, except that the earth permits the propagation of shear waves as well as compressional waves. (In an acoustic medium, only compressional waves are supported.) The earth tends to act like an *elastic medium* for the propagation of seismic waves. The property of elasticity means that a fluid or solid body resists changes in size and shape due to the applications of an external force, and that the body is restored to its original size and shape upon removal of the force. It is this property that permits the propagation of seismic waves through the earth.

An important issue in exploration seismology is the interpretation of seismic returns from the different geological layers of the earth. This interpretation is fundamental to the *identification* of crusted regions such as depth rocks, sand layers, or sedimentary layers. The sedimentary layers are of particular interest because they may contain hydrocarbon reservoirs. The idea of a layered earth model plays a key role here.

The *layered earth model* is based on the physical fact that seismic-wave motion in each layer is characterized by two components propagating in opposite directions (Robinson and Durrani, 1986). This phenomenon is illustrated in Fig. 1.9. To understand the interaction between downgoing and upgoing waves, we have reproduced a

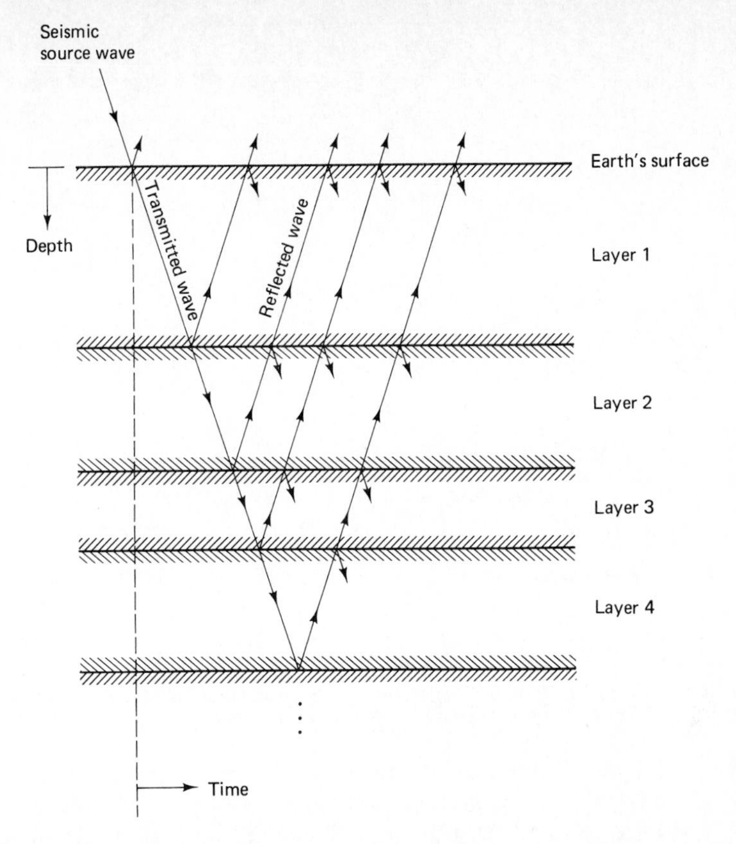

Figure 1.9 Upgoing and downgoing waves in different layers of the layered earth model. *Note:* The layers are unevenly spaced to add a sense of realism.

portion of this diagram in Fig 1.10(a), which pertains to the kth interface. The picture shown in Fig. 1.10(a) is decomposed into two parts, as depicted in Fig. 1.10(b) and (c). We thus observe the following:

1. In layer k, there is an *upgoing* wave that consists of the superposition of the reflection of a downgoing wave incident on the kth interface (i.e., boundary) and the transmission of an upgoing (incident) wave from layer $k + 1$.
2. In layer $k + 1$, there is a *downgoing* wave that consists of the superposition of the transmission of a downgoing (incident) wave from layer k and the reflection of an upgoing wave incident on the kth interface.

Lattice Model. Let c_k denote the upward *reflection coefficient* of the kth interface [see Fig. 1.10(b)]. Let $d_k(n)$ and $u_k(n)$ denote the downgoing and upgoing waves, respectively, at the *top* of layer k, and let $d_k'(n)$ and $u_k'(n)$ denote the downgoing and upgoing waves respectively at the *bottom* of layer k, as depicted in Fig. 1.10(a). The index n denotes time. Ideally, the waves propagate through the medium without distortion, or absorption. Accordingly, we have from Fig. 1.10(a),

$$d_k'(n) = d_k(n - \tfrac{1}{2}) \tag{1.23}$$

and

$$u_k'(n) = u_k(n + \tfrac{1}{2}) \tag{1.24}$$

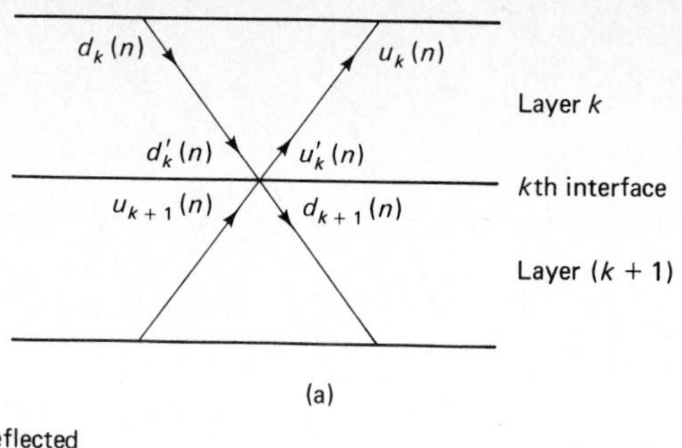

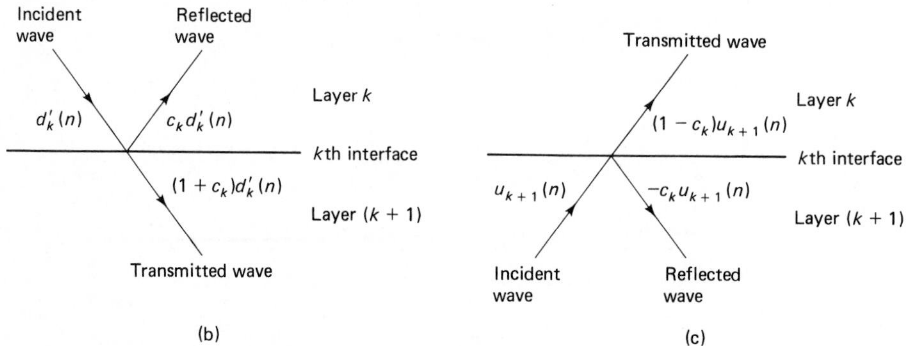

Figure 1.10 (a) Propagation of seismic waves through a pair of adjacent layers; (b) effects of downgoing incident wave; (c) effects of upgoing incident wave.

where the travel time from the top of a layer to its bottom (or vice versa) is assumed to be one-half of a time unit. The superposition of the pictures depicted in Fig. 1.10(b) and (c), and comparison with that of Fig. 1.10(a) yields the following interactions between the downgoing and upgoing waves:

$$d_{k+1}(n) = -c_k u_{k+1}(n) + (1 + c_k)d'_k(n) \tag{1.25}$$

and

$$u'_k(n) = c_k d'_k(n) + (1 - c_k)u_{k+1}(n) \tag{1.26}$$

The upward *transmission coefficient*[7] of the kth interface is defined by [see Fig. 1.10(c)]

$$\tau'_k = 1 - c_k \tag{1.27}$$

Thus, using this definition in Eq. (1.26), and also using this equation to eliminate $u_{k+1}(n)$ from Eq. (1.25), we obtain

$$u'_k(n) = c_k d'_k(n) + \tau'_k u_{k+1}(n) \tag{1.28}$$

and

$$d_{k+1}(n) = \frac{1}{\tau'_k} d'_k(n) - \frac{c_k}{\tau'_k} u'_k(n) \tag{1.29}$$

Using this pair of equations, we may construct a *lattice model* for layer k, as shown in Fig. 1.11(a) (Robinson and Durrani, 1987). Moreover, we may extend this idea to

[7] The prime in the upward transmission coefficient τ'_k is used to distinguish it from the *downward* transmission coefficient [see Fig. 1.11(a)], given by

$$\tau_k = 1 + c_k$$

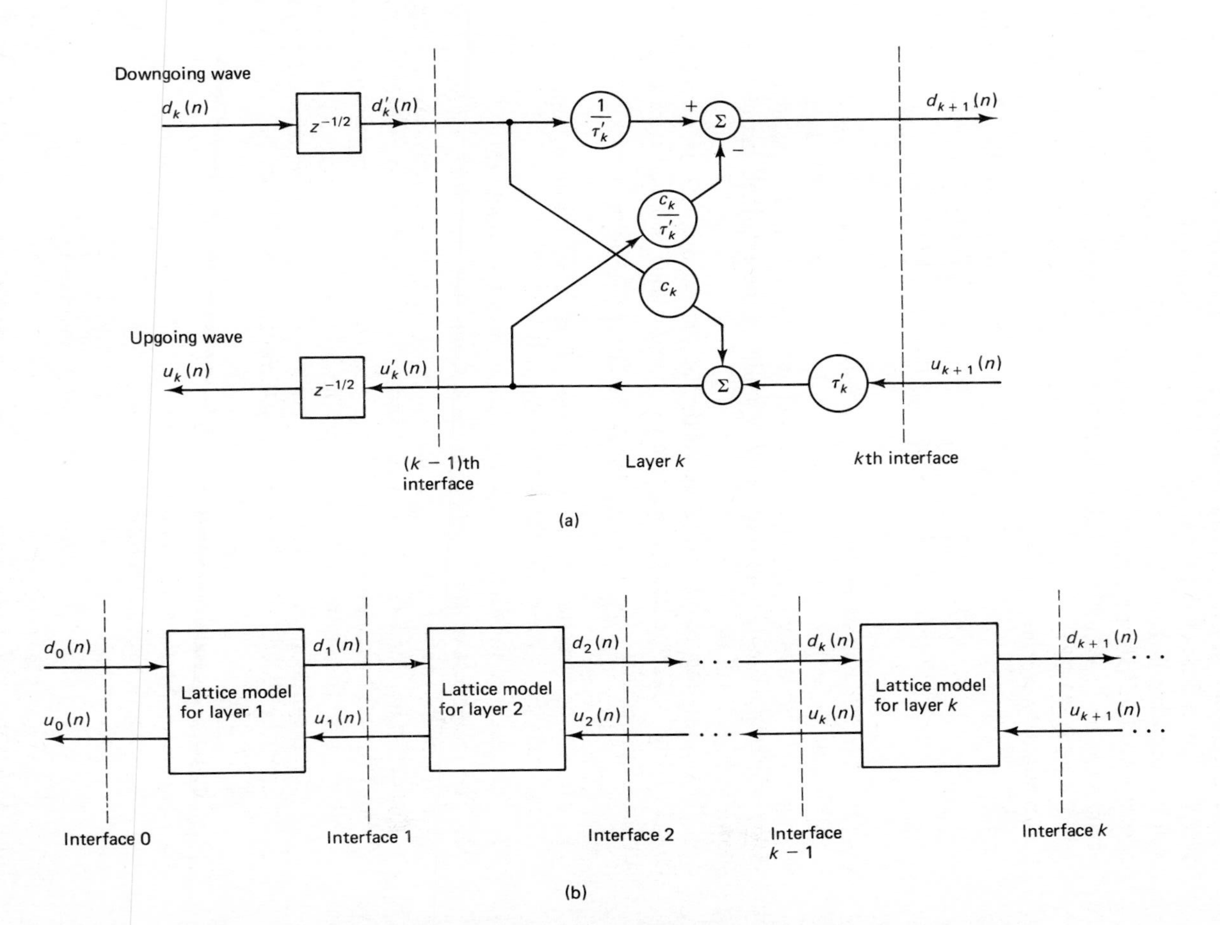

Figure 1.11 (a) Lattice model for layer k; (b) multistage lattice model of the layered earth model, with each lattice configuration having the form given in part (a).

develop a *multistage lattice model,* shown in block diagram form in Fig. 1.11(b), which depicts the propagation of waves through several layers of the medium. The lattice model for each layer has the details given in Fig. 1.11(a). The combined use of these two figures provides a great deal of physical insight into the interaction of downgoing and upgoing waves as they propagate from one layer to the next.

Examination of Eq. (1.28) reveals that the evaluation of $u_k'(n)$ at the bottom of layer k requires knowledge of $u_{k+1}(n)$ at the top of layer $k + 1$. But $u_{k+1}(n)$ is not available until the layer $k + 1$ has been dealt with. The lattice model of Fig. 1.11 is therefore of limited practical use. To overcome this limitation, we may use the z-transform to modify this model. Specifically, applying the z-transform to Eqs. (1.23), (1.24), (1.28), and (1.29) and manipulating them into matrix form, we get the so-called *scattering equation:*

$$\begin{bmatrix} D_{k+1}(z) \\ U_{k+1}(z) \end{bmatrix} = \frac{z^{1/2}}{\tau_k'} \begin{bmatrix} z^{-1} & -c_k \\ -c_k z^{-1} & 1 \end{bmatrix} \begin{bmatrix} D_k(z) \\ U_k(z) \end{bmatrix} \tag{1.30}$$

where z^{-1} is the unit-delay operator. The 2-by-2 matrix on the right side of Eq. (1.30) is called the *scattering matrix.* Thus, based on Eq. (1.30), we may construct the *modified lattice model* for layer k, shown in Fig. 1.12(a) (Robinson and Durrani, 1986). Correspondingly, the multistage version of the modified lattice model is as shown in Fig. 1.12(b).

The following points are noteworthy in the context of the modified lattice model of Fig. 1.12 for the propagation of compressional seismic waves in the subsurface of the earth:

1. The lattice structure of the model has *physical* significance, since it follows naturally from the notion of a layered earth.
2. The structure for each layer (state) of the model is *symmetric.*
3. The reciprocal of the transmission coefficient for each layer merely plays the role of a *scaling factor* insofar as input–output relations are concerned. Specifically, for layer k, we may remove $1/\tau_k'$ from the top path of the model in Fig. 1.12(a) simply by absorbing it in $D_{k+1}(z)$. Similarly, we may remove $1/\tau_k'$ from the bottom path by absorbing it in $U_{k+1}(z)$. Moreover, the values of the transmission coefficients $\tau_1', \tau_2', \ldots, \tau_k', \ldots$ are determined from the respective values of the refection coefficients $c_1, c_2, \ldots, c_k, \ldots$ by using Eq. (1.27).
4. The overall model for layers 1, 2, . . . , k, . . . is *uniquely determined by the sequence of reflection coefficients* $c_1, c_2, \ldots, c_k, \ldots$.

A case of special interest arises when

$$u_{k+1}(n) = 0, \qquad k\text{: deepest layer} \tag{1.31}$$

This case corresponds to the case when the *final* interface [i.e., the $(k + 1)$th interface] acts as a *perfect absorber*. In other words, there is no outgoing wave from the deepest layer, so Eq. (1.31) follows. This equation thus represents the *boundary*

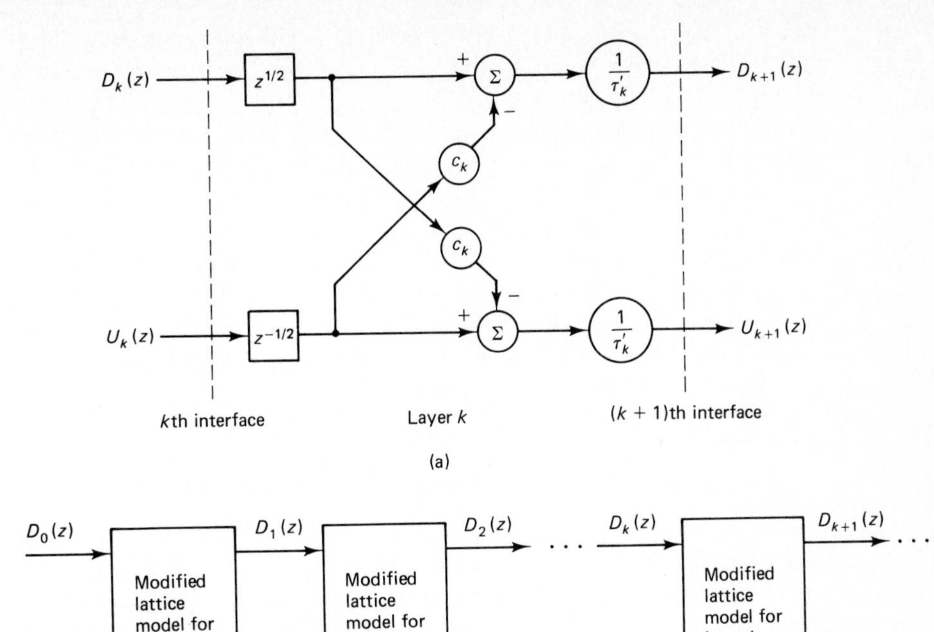

Figure 1.12 (a) Modified lattice model for layer k; (b) multistage version of modified lattice model, with each stage having the representation shown in part (a).

condition on the lattice model of Fig. 1.11. The corresponding boundary condition for the modified lattice model of Fig. 1.12 is

$$U_{k+1}(z) = 0, \qquad k\text{: deepest layer} \tag{1.32}$$

Given this boundary condition and the sequence of reflection coefficients c_1, c_2, . . . , c_k, . . . , we may then use the modified lattice model of Fig. 1.12(b) (in a stage-by-stage fashion) to determine $U_0(z)$, the z-transform of the output (outgoing) seismic wave $u_0(n)$ at the earth's surface, in terms of $D_0(z)$, the z-transform of the input (downgoing) seismic wave $d_0(n)$.

Tapped-Delay-Line (Transversal Model). Figure 1.13 depicts a *tapped-delay-line model* for a layered earth. It provides a local parameterization of the propagation (scattering) phenomenon in the earth's subsurface. According to this model, the input (downgoing) seismic $d_0(n)$ and the output (upgoing) seismic wave $u_0(n)$ are, in general, linearly related by the *infinite convolution sum*

$$u_0(n) = \sum_{k=0}^{\infty} w_k d_0(n-k) \tag{1.33}$$

where the infinite sequence of tap weights, $\{w_n\}$, represents the *spatial mapping* of the medium's weighting or the *impulse response* of the medium. Equation (1.33)

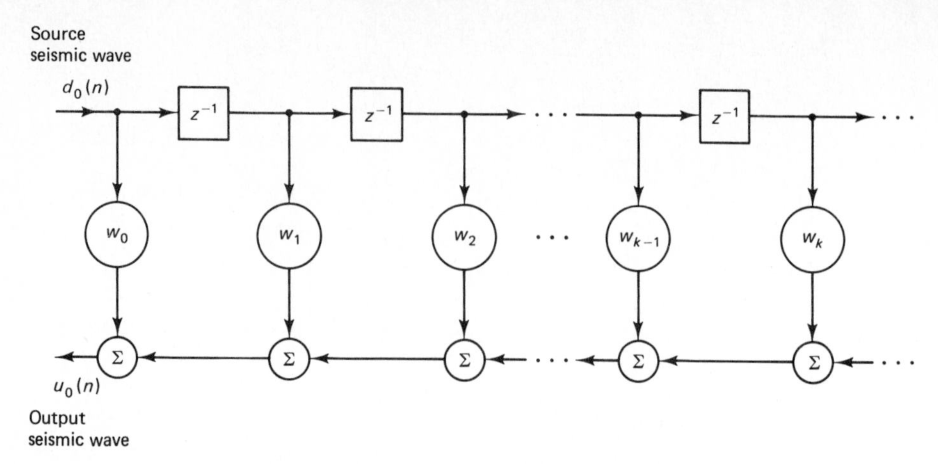

Figure 1.13 Tapped-delay-line model of layered earth.

states that the output $u_0(n)$ is an infinite series of time-delayed and scaled replicas of the input $d_0(n)$.

There is a *one-to-one correspondence* between the impulse response $\{w_n\}$ that characterizes the tapped-delay-line model of Fig. 1.13 and the sequence of reflection coefficients $\{c_n\}$ that characterizes the lattice model of Fig. 1.11, as shown by

$$\{w_n\} \rightleftharpoons \{c_n\} \tag{1.34}$$

In other words, given the c_n we may uniquely determine the w_n, and vice versa.

In reflection seismology, the model of Fig. 1.13 is referred to as the *convolutional model,* in view of the convolution of the impulse response of the medium with the input. This model is the starting point of seismic deconvolution (described in the next application).

Parameter Estimation.[8] The seismic wave $d_0(n)$ generated by the source of energy acts as a "probing" wave that is transmitted into the earth. Correspondingly, the seismic wave $u_0(n)$ is the output evoked by the propagation of $d_0(n)$ in the earth's subsurface. A *recorded* trace of the output $u_0(n)$ for varying time n is called a *seismogram*. Thus, given digital recordings of the probing wave $d_0(n)$ and the resulting seismogram $u_0(n)$, we may apply an adaptive filtering algorithm to estimate the impulse response $\{w_n\}$ of the layered earth. This computation is performed *off-line,* with the probing wave $d_0(n)$ used as input to the adaptive filtering algorithm and the seismogram $u_0(n)$ serving the role of desired response for the algorithm.

[8] For a survey of different parameter estimation procedures applicable to reflection seismology, see Mendel (1986). This paper also discusses other related issues, namely, *representation* (i.e., how something should be modeled), *measurement* (which physical parameters should be measured and how they should be measured), and *validation* (i.e., demonstrating confidence in the model). For a deterministic approach applicable to reflection seismology, see Bruckstein and Kailath (1987). The approach taken here is based on an *inverse scattering* framework for determining the parameters of a layered wave propagation medium from measurements taken at the boundary.

Predictive Deconvolution

Convolution is fundamental to the analysis of linear time-invariant systems. Specifically, the output of a linear time-invariant system is the convolution of the input with the impulse response of the system. Convolution is *commutative*. We may therefore also say that the output of the system is the convolution of the impulse response of the system with the input. Moreover, convolution is a linear operation; it therefore holds regardless of the type of signal used as the system input.

Consider the convolutional model for reflection seismology depicted in Fig. 1.13. We may express the input–output relation of this model simply as

$$u_0(n) = w_n * d_0(n) \tag{1.35}$$

where $d_0(n)$ is the input, w_n is the impulse response, and $u_0(n)$ is the output. The symbol $*$ is shorthand for convolution. The important point to note is that given the values of w_n and $d_0(n)$ for varying n, we may determine the corresponding values of $u_0(n)$.

Deconvolution is a linear operation that *removes* the effect of some previous convolution performed on a given data record (time series). Suppose that we are given the input $d_0(n)$ and the output $u_0(n)$. We may then use deconvolution to determine the impulse response w_n. In symbolic form we may thus write

$$w_n = u_0(n) * d_0^{-1}(n) \tag{1.36}$$

where $d_0^{-1}(n)$ denotes the *inverse* of $d_0(n)$. Note, however, that $d_0^{-1}(n)$ is *not* the reciprocal of $d_0(n)$; rather, the use of the superscript -1 is merely a flag indicating "inverse."

In *seismic deconvolution,* we are given the seismogram $u_0(n)$ and the requirement is to unravel it so as to obtain an estimate of the impulse response w_n of a layered earth model. The problem, however, is complicated by the fact that in the general case of reflection seismology we do not have an estimate of the input seismic wave (also referred to as the seismic wavelet) $d_0(n)$. To overcome this practical uncertainty, we may use an elegant statistical procedure known as *predictive deconvolution* (Robinson, 1954; Robinson and Durrani, 1986). The term "predictive" arises from the fact that the procedure relies on the use of linear prediction. The derivation of predictive deconvolution rests on two simplifying hypotheses for seismic wave propagation with normal incidence:

1. The input wave $d_0(n)$, generated by the source of seismic energy, is the *impulse response of an all-pole feedback system,* and is thus minimum phase.
2. The impulse response w_n of the layered earth model has the properties of a *white-noise process*.

Condition 1 is referred to as the *feedback hypothesis,* and condition 2 is referred to as the *random hypothesis*. Geophysical experience over three decades has shown that it is indeed possible to satisfy these two hypotheses (Robinson, 1984). As a result, predictive deconvolution is used routinely on all seismic records in every exploration program.

The implication of the feedback hypothesis is that we may express the present value $d_0(n)$ of the input wave as a *linear combination of the past values*, as shown by

$$d_0(n) = -\sum_{k=1}^{M} a_k d_0(n-k) \tag{1.37}$$

where the a_k are the *feedback coefficients*, and M is the *order* of the all-pole feedback system. The order M may be fixed in advance; alternatively, it may be determined by a mean-square-error criterion.

According to the random hypothesis, the impulse response w_n has the properties of a white-noise process. We therefore expect the estimate $\hat{w}_n$ produced by the deconvolution filter in Fig. 1.14 to have similar properties. In other words, the deconvolution filter acts as a *whitening filter*. Furthermore, the deconvolution filter is an *all-zero* filter with a transfer function equal to the reciprocal of the transfer function of the all-pole feedback system used to model $d_0(n)$. This means that if we express the transfer function of the feedback system [i.e., the z-transform of $d_0(n)$] as

$$D_0(z) = \frac{1}{1 + a_1 z^{-1} + a_2 z^{-2} + \cdots + a_M z^{-M}} \tag{1.38}$$

where $a_1, a_2, \ldots, a_M$ are the *feedback coefficients*, and ignore the additive noise $v(n)$ in the model of Fig. 1.14, then the transfer function of the deconvolution filter is

$$\begin{aligned} A(z) &= \frac{1}{D_0(z)} \\ &= 1 + a_1 z^{-1} + a_2 z^{-2} + \cdots + a_M z^{-M} \end{aligned} \tag{1.39}$$

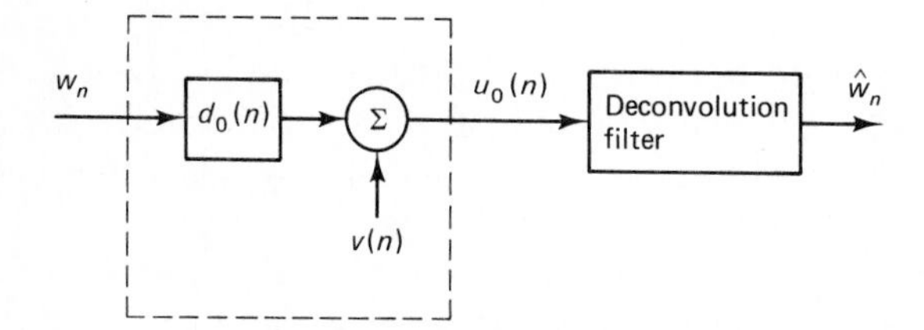

Figure 1.14 Block diagram illustrating seismic deconvolution.

To evaluate $A(z)$, we may use a block processing approach based on the *augmented matrix form of the Wiener–Hopf equations for linear prediction*; for details, see Chapter 6. This relation consists of a system of $(M + 1)$ simultaneous equations that involve the following:

1. A set of $(M + 1)$ known quantities represented by the *estimates* $\hat{r}(0)$, $\hat{r}(1)$, $\ldots$, $\hat{r}(M)$ of the autocorrelation function of the seismogram $u_0(n)$ for varying lags $0, 1, \ldots, M$, respectively. To get these values, we may use the formula for a *biased estimate* of the autocorrelation function:

$$\hat{r}(l) = \frac{1}{N} \sum_{n=l+1}^{N} d_0(n) d_0(n-l), \qquad l = 0, 1, \ldots, M \tag{1.40}$$

where N is the *record length* of the seismogram. Typically, N is very large compared to M.

2. A set of $(M + 1)$ unknowns, made up of the feedback coefficients a_1, a_2, $\ldots$, a_M and the variance σ^2 of the white-noise process assumed to model w_n.

Given the seismogram $u_0(n)$, we may therefore uniquely determine the feedback coefficients $a_1, a_2, \ldots, a_M$ and the variance σ^2 by solving this system of equations.

From Eq. (1.39), we see that the impulse response of the deconvolution filter consists of the sequence $a_1, a_2, \ldots, a_M$. Accordingly, the convolution of this impulse response with $u_0(n)$ yields the desired estimate $\hat{w}_n$, as shown by (see Fig. 1.14)

$$\hat{w}_n = \sum_{k=0}^{M} a_k u_0(n - k) \tag{1.41}$$

where $a_0 = 1$. Equation (1.41) is a description of the deconvolution process. Note, however, the wave $d_0(n)$ generated by the source of seismic energy does not enter this description directly as in the idealized representation of Eq. (1.37). Rather, the physical nature of $d_0(n)$ influences the deconvolution process by modeling $d_0(n)$ as the impulse response of an all-pole feedback system.

An alternative procedure for constructing the deconvolution filter is to use an adaptive filtering algorithm, as illustrated in Fig. 1.15. In this application, the present value $u_0(n)$ of the seismic output serves the purpose of a desired response for the algorithm, and the past values $u_0(n - 1), u_0(n - 2), \ldots, u_0(n - M)$ are used as elements of the input vector. The prediction error controls the adaptation of the M tap weights of the transversal filter component of the algorithm. When the algorithm has converged, the tap weights of the transversal filter provide estimates of the feedback coefficients $a_1, a_2, \ldots, a_M$.

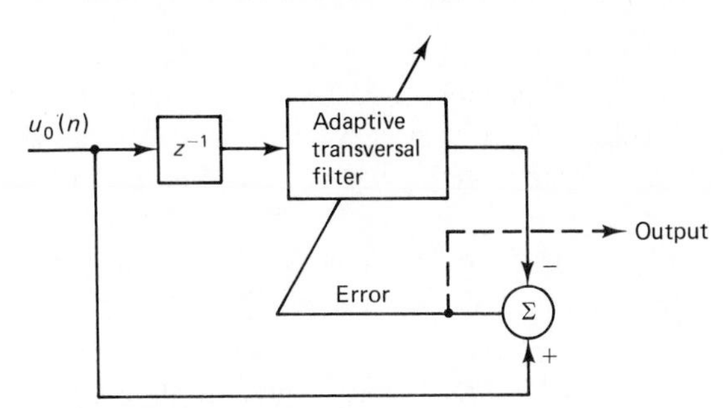

Figure 1.15 Adaptive filtering scheme for computing the impulse response of the deconvolution filter.

Adaptive Equalization

During the past three decades, a considerable effort has been devoted to the study of data-transmission systems that utilize the available channel bandwidth efficiently. The objective here is to design the system so as to accommodate the highest possible rate of data transmission, subject to a specified reliability that is usually measured in terms of the error rate or average probability of symbol error. The transmission of digital data through a linear communication channel is limited by two factors:

1. *Intersymbol interference* (ISI). This is caused by dispersion in the transmit filter, the transmission medium, and the receive filter.

2. *Additive thermal noise*. This is generated by the receiver at its front end.

For bandwidth-limited channels (e.g., voice-grade telephone channels), we usually find that intersymbol interference is the chief determining factor in the design of high-data-rate transmission systems.

Figure 1.16 shows the equivalent baseband model of a binary *pulse-amplitude modulation* (PAM) *system*. The signal applied to the input of the transmitter part of the system consists of a *binary data sequence* $\{b_k\}$, in which the binary symbol b_k consists of 1 or 0. This sequence is applied to a pulse generator, the output of which is filtered first in the transmitter, then by the medium, and finally in the receiver. Let $u(k)$ denote the sampled output of the receive filter in Fig. 1.16; the sampling is performed in synchronism with the pulse generator in the transmitter. This output is compared to a *threshold* by means of a *decision device*. If the threshold is exceeded, the receiver makes a decision in favor of symbol 1. Otherwise, it decides in favor of symbol 0.

Let a scaling factor a_k be defined by

$$a_k = \begin{cases} +1, & \text{if the input bit } b_k \text{ consists of symbol 1} \\ -1, & \text{if the input bit } b_k \text{ consists of symbol 0} \end{cases} \tag{1.42}$$

Then, in the absence of noise, we may express $u(k)$ as

$$\begin{aligned} u(k) &= \sum_n a_n p(k-n) \\ &= a_k p(0) + \sum_{\substack{n \\ n \neq k}} a_n p(k-n) \end{aligned} \tag{1.43}$$

where $p(n)$ is the sampled version of the impulse response of the cascade connection of the transmit filter, the transmission medium, and the receive filter. The first term on the right side of Eq. (1.43) defines the desired symbol, whereas the remaining series represents the intersymbol interference caused by the *channel* (i.e., the combination) of the transmit filter, the medium, and the receive filter). This intersymbol interference, if left unchecked, can result in erroneous decisions when the sampled signal at the channel output is compared with some preassigned threshold by means of a decision device.

To overcome the intersymbol interference problem, control of the time-sampled function $p(n)$ is required. In principle, if the characteristics of the transmission medium are known precisely, then it is virtually always possible to design a pair of transmit and receive filters that will make the effect of intersymbol interference

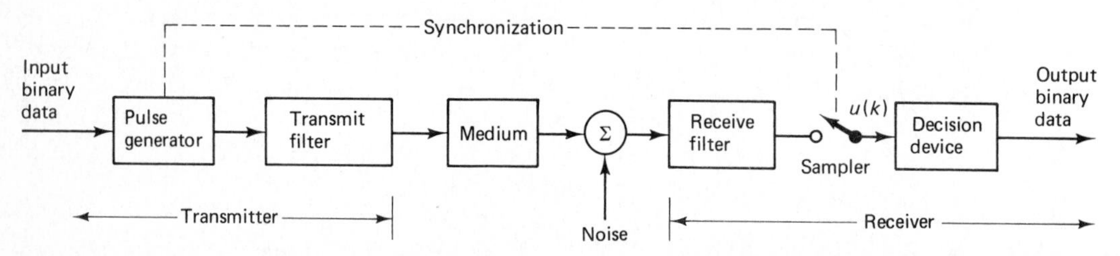

Figure 1.16 Block diagram of a baseband data transmission system (without equalization).

(at sampling times) arbitrarily small. This is achieved by proper sharing of the overall response of the channel in accordance with Nyquist's classic work on telegraph transmission theory. The overall frequency response consists of a *flat portion* and a *roll-off portion* that has a cosine form (Haykin, 1988). Correspondingly, the overall impulse response attains its maximum value at time $n = 0$ and is zero at all other sampling instants; the intersymbol interference is therefore zero. In practice we find that the channel is *time varying,* due to variations in the transmission medium, which makes the received signal *nonstationary*. Accordingly, the use of a fixed pair of transmit and receive filters, designed on the basis of average channel characteristics, may not adequately reduce intersymbol interference. This suggests the need for an *adaptive equalizer* that provides precise control over the time response of the channel (Lucky, 1965, 1966; Lucky et al., 1968; Proakis, 1975; Qureshi, 1985).

Among the basic philosophies for equalization of data-transmission systems are preequalization at the transmitter and postequalization at the receiver. Since the former technique requires the use of a feedback path, we will only consider equalization at the receiver, where the adaptive equalizer is placed after the receive filter–sampler combination as in Fig. 1.16. In theory, the effect of intersymbol interference may be made arbitrarily small by making the number of adjustable coefficients (tap weights) in the adaptive equalizer infinitely large.

An adaptive filtering algorithm requires knowledge of the "desired" response so as to form the error signal needed for the adaptive process to function. In theory, the transmitted sequence (originating at the transmitter output) is the "desired" response for adaptive equalization. In practice, however, with the adaptive equalizer located in the receiver, the equalizer is physically separated from the origin of its ideal desired response. There are two methods in which a *replica (facsimile)* of the desired response may be generated locally in the receiver:

1. *Training method*. In the first method, a replica of the desired response is *stored* in the receiver. Naturally, the generator of this stored reference has to be electronically *synchronized* with the known transmitted sequence. A widely used *test (probing) signal* consists of a *pseudonoise (PN) sequence* (also known as a *maximal-length sequence*) with a broad and even power spectrum. The PN sequence has noiselike properties. Yet it has a deterministic waveform that repeats periodically. For the generation of a PN sequence, we may use a *linear feedback shift register* that consists of a number of consecutive two-state memory stages (flip-flops) regulated by a single timing clock (Golomb, 1964). A *feedback* term, consisting of the modulo-2 sum of the outputs of various memory stages, is applied to the first memory stage of the shift register and thereby prevents it from emptying.
2. *Decision-directed method*. Under normal operating conditions, a good facsimile of the transmitted sequence is being produced at the output of the *decision device* in the receiver. Accordingly, if this output were the correct transmitted sequence, it may be used as the "desired" response for the purpose of adaptive equalization. Such a method of learning is said to be *decision directed,* because the receiver attempts to learn by employing its own decisions (Lucky et al., 1968). If the average probability of symbol error is small (less than 10 percent, say), the decisions made by the receiver are correct enough for the *estimates of*

the error signal (used in the adaptive process) to be *accurate most of the time*. This means that, in general, the adaptive equalizer is able to improve the tap-weight settings by virtue of the correlation procedure built into its feedback control loop. The improved tap-weight settings will, in turn, result in a lower average probability of symbol error and therefore more accurate estimates of the error signal for adaptation, and so it goes on. However, it is also possible for the reverse effect to occur, in which case the tap-weight settings of the equalizer lose acquisition of the channel.

With a known training sequence, as in the first method, the adaptive filtering algorithm used to adjust the equalizer coefficients corresponds mathematically to searching for the unique minimum of a quadratic error-performance surface. The *unimodal* nature of this surface assures convergence of the algorithm. In the decision-direct method, on the other hand, the use of estimated and unreliable data modifies the error performance into a *multimodal* one, in which case complex behavior may result (Mazo, 1980). Specifically, the error performance surface now exhibits two types of local minima:

1. *Desired local minima,* whose positions correspond to coefficient (tap-weight) settings that yield the same performance as that obtained with a known training sequence
2. *Undesired (extraneous) local minima,* whose positions correspond to coefficient settings that yield inferior equalizer performance

A poor choice of the initial coefficient settings may cause the adaptive equalizer to converge to an undesirable local minimum and stay there.

The most significant point to note from this discussion is that, in general, a *linear* adaptive equalizer must be trained before it is switched to the decision-directed mode of operation if we are to be sure of delivering high performance.

Figure 1.17 shows the details (in block diagram form) of an adaptive equalizer built around a transversal filter that is highly popular. In this figure we have an odd number of tap weights, a practice widely followed in the literature on adaptive equalization. The total number of tap weights is $M = 2L + 1$. To *initialize* the equalizer, all the tap weights may be set equal to zero, except for the one at the center tap, which is set equal to unity. This ensures that initially the adaptive equalizer has unity gain. For the training period, the switch in Fig. 1.17 is placed in position 1. When the training process is completed, the switch is moved to position 2, and the decision-directed mode of operation begins. It is noteworthy that the decision-directed method of supplying the desired response is well suited for *tracking* relatively slow variations in channel characteristics during the course of data transmission.

A final comment pertaining to performance evaluation is in order. A popular experimental technique for assessing the performance of a data transmission system involves the use of an *eye pattern*. This pattern is obtained by applying (1) the received wave to the vertical deflection plates of an oscilloscope, and (2) a sawtooth wave at the transmitted symbol rate to the horizontal deflection plates. The resulting display is called an *eye pattern* because of its resemblance to the human eye for bi-

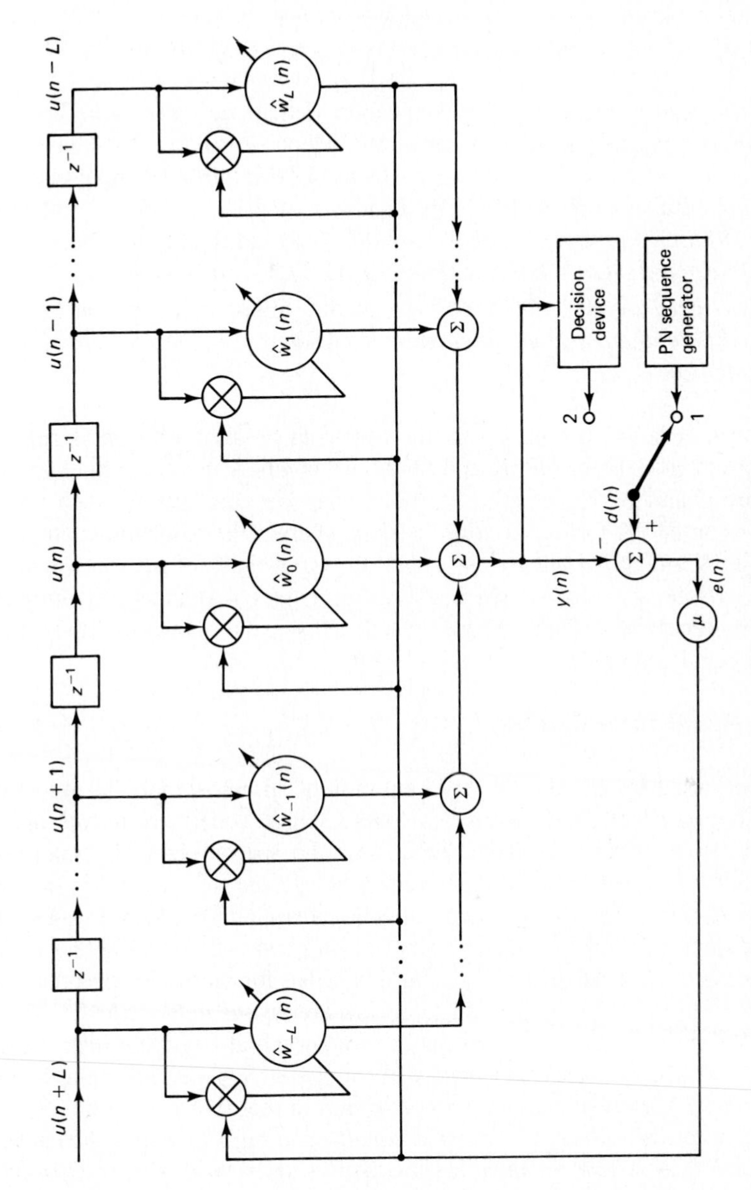

Figure 1.17 Block diagram of an adaptive equalizer using a transversal filter (assuming real-valued data).

nary data. Thus, in a system using adaptive equalization, the equalizer attempts to correct for intersymbol interference in the system and thereby open the eye pattern as far as possible.

Thus far we have only discussed adaptive equalizers for baseband PAM systems. However, voice-band data transmission systems employ modulation–demodulation schemes that are commonly known as *modems*. Depending on the speed of operation, we may categorize modems as follows (Qureshi, 1985):

1. *Low-speed* (2400 to 4800 bits/s) modems that use *phase-shift keying* (PSK); PSK is a digital modulation scheme in which the phase of a sinusoidal carrier wave is shifted by $2\pi k/M$ radians in accordance with the input data, where M is the number of phase levels used and $k = 0, 1, \ldots, M - 1$. Specific values of M used in practice are $M = 2$ and 4, representing *binary phase-shift keying* (BPSK) and *quadriphase-shift keying* (QPSK), respectively.
2. *High-speed* (4800 to 9600 or even 16,800 bits/s) modems that use combined *amplitude and phase modulation* or, equivalently, *quadrature amplitude modulation* (QAM).

The important point to note is that the baseband model for BPSK is real, whereas the baseband models for QPSK and QAM are complex involving both in-phase and quadrature channels. Hence, the baseband adaptive equalizer for data transmission systems using BPSK (or its variation) is *real,* whereas the baseband adaptive equalizers for QPSK and QAM are *complex* (i.e., the tap weights of the transversal filter are complex). Note also that a real equalizer processes real inputs to produce a real equalized output, whereas a complex equalizer processes complex inputs to produce complex equalized outputs.

Linear Predictive Coding

The coders used for the digital representation of speech signals fall into two broad classes: *source coders* and *waveform coders*. Source coders are *model dependent* in that they use a priori knowledge about how the speech signal is generated at the source. Source coders for speech are generally referred to as *vocoders* (a contraction for voice coders). They can operate at low coding rates; however, they provide a synthetic quality, with the speech signal having lost substantial naturalness. Waveform coders, on the other hand, essentially strive for facsimile reproduction of the speech waveform. In principle, these coders are *signal independent*. They may be designed to provide telephone-toll quality for speech at relatively high coding rates. In this subsection we describe a special form of source coder known as a linear predictive coder. Waveform coders are considered in the next subsection.

In the context of speech, *linear predictive coding* (LPC) strives to produce digitized voice data at low bit rates (as low as 2.4 kb/s), with two important motivations in mind. First, the use of linear predictive coding permits the transmission of digitized voice over a *narrow-band channel* (having a bandwidth of approximately 3 kHz). Second, the realization of a low-bit rate makes the *encryption* of voice signals easier and more reliable than would be the case otherwise; encryption is an essential requirement for *secure communications* (as in a military environment). Note that a bit rate of 2.4 kb/s is less than 5 percent of the 64 kb/s used typically for the standard pulse code modulation (PCM); see the next subsection.

Linear predictive coding achieves a low bit rate for the digital representation of speech by exploiting the special properties of a classical model of the speech production process, which is described next.

Figure 1.18 shows a simplified block diagram of the classical model for the speech production process. It assumes that the sound-generating mechanism (i.e., the source of excitation) is linearly separable from the intelligence-modulating vocal-tract filter. The precise form of the excitation depends on whether the speech sound is voiced or unvoiced:

1. A *voiced* speech sound (such as[9] /i/ in eve) is generated from quasi-periodic vocal-cord sound. In the model of Fig. 1.18 the impulse-train generator produces a sequence of impulses (i.e., very short pulses), which are spaced by a fundamental period equal to the *pitch period*. This signal, in turn, excites a linear filter whose impulse response equals the vocal-cord sound pulse.
2. An *unvoiced* speech sound (such as /f/ in fish) is generated from random sound produced by turbulent airflow. In this case the excitation consists simply of a *white* (i.e., broad spectrum) noise source. The probability distribution of the noise samples does not appear to be critical.

The frequency response of the vocal-tract filter for unvoiced speech or that of the vocal tract multiplied by the spectrum of the vocal-cord sound pulses determines the short-time spectral envelope of the speech signal.

At first sight, it may appear that the speech production model falls under class I of adaptive filtering application (i.e., identification). In reality, however, this is not so. As may be seen from Fig. 1.18, there is *no* access to the input signal of the vocal tract.

The method of *linear predictive coding* (LPC) is an example of source coding. This method is important because it provides not only a powerful technique for the

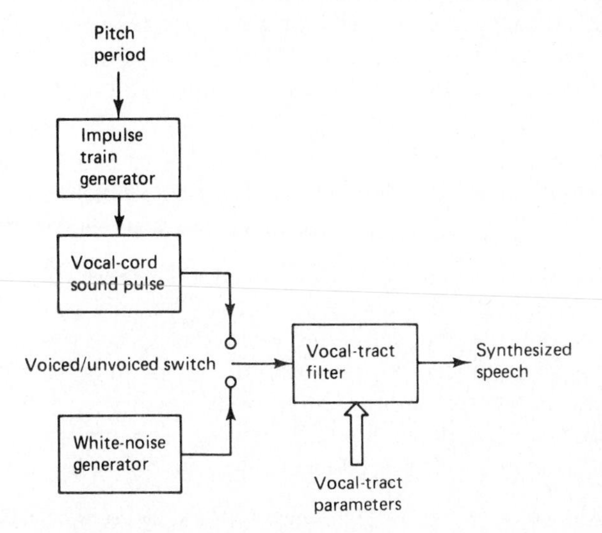

Figure 1.18 Block diagram of simplified model for the speech production process.

[9] The symbol / / is used to denote the *phenome*, a basic linguistic unit.

digital transmission of speech at low bit rates but also accurate estimates of basic speech parameters.

The development of LPC relies on the model of Fig. 1.18 for the speech-production process. The frequency response of the vocal tract for unvoiced speech or that of the vocal tract multiplied by the spectrum of the vocal sound pulse for voiced speech is described by the *transfer function*

$$H(z) = \frac{G}{1 + \sum_{k=1}^{M} a_k z^{-k}} \tag{1.44}$$

where G is a gain parameter and z^{-1} is the unit-delay operator. The form of excitation applied to this filter is changed by switching between voiced and unvoiced sounds. Thus, the filter with transfer function $H(z)$ is excited by a sequence of impulses to generate voiced sounds or a white-noise sequence to generate unvoiced sounds. In this application, the input data are real valued; hence the filter coefficients, a_k, are likewise real valued.

In linear predictive coding, as the name implies, linear prediction is used to estimate the speech parameters. Given a set of past samples of a speech signal, $u(n-1), u(n-2), \ldots, u(n-M)$, a linear prediction of $u(n)$, the present sample value of the signal, is defined by

$$\hat{u}(n) = \sum_{k=1}^{M} \hat{w}_k u(n-k) \tag{1.45}$$

The predictor coefficients, $\hat{w}_1, \hat{w}_2, \ldots, \hat{w}_M$, are optimized by minimizing the mean-square value of the prediction error, $e(n)$, defined as the difference between $u(n)$ and $\hat{u}(n)$. The use of the minimum-mean-squared-error criterion for optimizing the predictor may be justified for two basic reasons:

1. If the speech signal satisfies the model described by Eq. (1.44) and if the mean-square value of the error signal $e(n)$ is minimized, then we find that $e(n)$ equals the excitation $u(n)$ multiplied by the gain parameter G in the model of Fig. 1.18 and $a_k = -\hat{w}_k$, $k = 1, 2, \ldots, M$.[10] Thus, the estimation error $e(n)$ consists of quasi-periodic pulses in the case of voiced sounds or a white-noise sequence in the case of unvoiced sounds. In either case, the estimation error $e(n)$ would be small most of the time.
2. The use of the minimum-mean-squared-error criterion leads to tractable mathematics.

Figure 1.19 shows the block diagram of an LPC vocoder. It consists of a transmitter and a receiver. The transmitter first applies a *window* (typically 10 to 30 ms long) to the input speech signal, thereby identifying a block of speech samples for processing. This window is short enough for the vocal-tract shape to be nearly stationary, so the parameters of the speech-production model in Fig. 1.19 may be treated as essentially constant for the duration of the window. The transmitter then analyzes the input speech signal in an adaptive manner, block by block, by perform-

[10] The relationship between the set of predictor coefficients, $\{\hat{w}_k\}$, and the set of all-pole filter coefficients, $\{a_k\}$, is derived in Chapter 6.

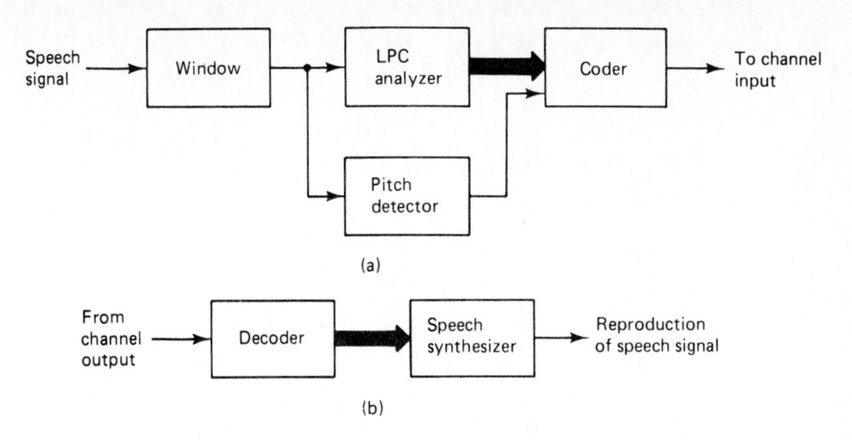

Figure 1.19 Block diagram of LPC vocoder: (a) transmitter, (b) receiver.

ing a linear prediction and pitch detection. Finally, it codes the parameters made up of (1) the set of predictor coefficients, (2) the pitch period, (3) the gain parameter, and (4) the voiced–unvoiced parameter, for transmission over the channel. The receiver performs the inverse operations, by first decoding the incoming parameters. In particular, it computes the values of the predictor coefficients, the pitch period, and the gain parameter, and determines whether the segment of interest represents voiced or unvoiced sound. Finally, the receiver uses these parameters to synthesize the speech signal by utilizing the model of Fig. 1.18.

Adaptive Differential Pulse-Code Modulation

In *pulse-code modulation,* which is the standard technique for waveform coding, three basic operations are performed on the speech signal. The three operations are *sampling* (time discretization), *quantization* (amplitude discretization), and *coding* (digital representation of discrete amplitudes). The operations of sampling and quantization are designed to preserve the shape of the speech signal. As for coding, it is merely a method of translating the discrete sequence of sample values into a more appropriate form of signal representation.

The rationale for sampling follows from a basic property of all speech signals: They are bandlimited. This means that a speech signal can be sampled in time at a finite rate in accordance with the sampling theorem. For example, commercial telephone networks designed to transmit speech signals occupy a bandwidth from 200 to 3200 Hz. To satisfy the sampling theorem, a conservative sampling rate of 8 kHz is commonly used in practice.

Quantization is justified on the following grounds. Although a speech signal has a continuous range of amplitudes (and therefore its samples also have a continuous amplitude range), nevertheless, it is not necessary to transmit the exact amplitudes of the samples. Basically, the human ear (as ultimate receiver) can only detect finite amplitude differences.

In PCM, as used in telephony, the speech signal (after low-pass filtering) is sampled at the rate of 8 kHz, nonlinearly (e.g., logarithmically) quantized, and then coded into 8-bit words, as in Fig. 1.20(a). The result is a good signal-to-quantiza-

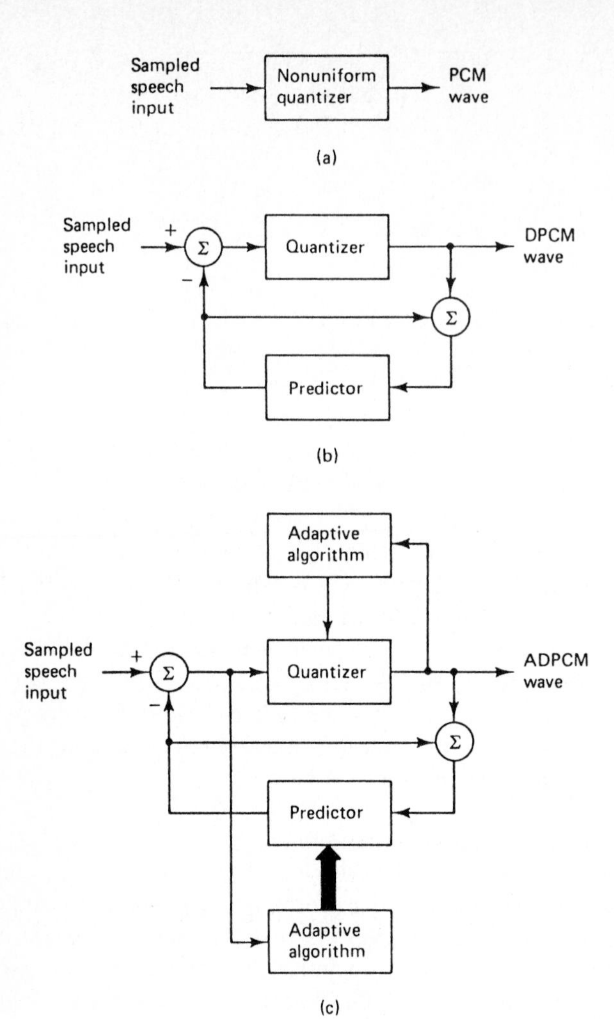

Figure 1.20 Waveform coders: (a) PCM, (b) DPCM, (c) ADPCM.

tion-noise ratio over a wide dynamic range of input signal levels. This method requires a bit rate of 64 kb/s.

Differential pulse-code modulation (DPCM), another example of waveform coding, involves the use of a predictor as in Fig. 1.20(b). The predictor is designed to exploit the correlation that exists between adjacent samples of the speech signal, in order to realize a reduction in the number of bits required for the transmission of each sample of the speech signal and yet maintain a prescribed quality of performance. This is achieved by quantizing and then coding the prediction error that results from the subtraction of the predictor output from the input signal. If the prediction is optimized, the variance of the prediction error will be significantly smaller than that of the input signal, so a quantizer with a given number of levels can be adjusted to produce a quantizing error with a smaller variance than would be possible if the input signal were quantized directly as in a standard PCM system. Equivalently, for a quantizing error of prescribed variance, DPCM requires a smaller number of quantizing levels (and therefore a smaller bit rate) than PCM. Differential pulse-code modulation uses a fixed quantizer and a fixed predictor. A further reduction in the transmission rate can be achieved by using an adaptive quantizer together with an adaptive predictor of sufficiently high order, as in Fig. 1.20(c). This type of waveform coding is called *adaptive differential pulse-code modulation* (ADPCM), where A denotes adaptation of both quantizer and predictor algorithms. An adaptive predictor is used in order to account for the nonstationary nature of speech signals. ADPCM can digitize speech with toll quality (8-bit PCM quality) at 32 kb/s. It can realize this level of quality with a 4-bit quantizer.[11]

Adaptive Autoregressive Spectrum Analysis

The *power spectrum* provides a quantitative measure of the second-order statistics of a discrete-time stochastic process as a function of frequency. In *parametric spectrum analysis,* we evaluate the power spectrum of the process by assuming a *model* for the process. In particular, the process is modeled as the output of a linear filter that is excited by a *white-noise process,* as in Fig. 1.21. By definition, a white-noise process has a constant power spectrum. A model that is of practical utility is the *autoregressive (AR) model,* in which the transfer function of the filter is assumed to consist of poles only. Let this transfer function be denoted by

$$\begin{aligned} H(e^{j\omega}) &= \frac{1}{1 + a_1 e^{-j\omega} + \cdots + a_M e^{-jM\omega}} \\ &= \frac{1}{1 + \sum_{k=1}^{M} a_k e^{-jk\omega}} \end{aligned} \tag{1.46}$$

where the a_k are called the *autoregressive (AR) parameters,* and M is the *model order*. Let σ_v^2 denote the constant power spectrum of the white-noise process $\{v(n)\}$ applied to the filter input. Accordingly, the power spectrum of the filter output $\{u(n)\}$ equals

$$S_{\text{AR}}(\omega) = \sigma_v^2 |\text{H}(e^{j\omega})|^2 \tag{1.47}$$

We refer to $S_{\text{AR}}(\omega)$ as the *autoregressive (AR) power spectrum*. Equation (1.46) assumes that the AR process $\{u(n)\}$ is real, in which case the AR parameters themselves assume real values.

When the AR model is time varying, the model parameters become time dependent, as shown by $a_1(n), a_2(n), \ldots, a_M(n)$. In this case, we express the power spectrum of the time-varying AR process as

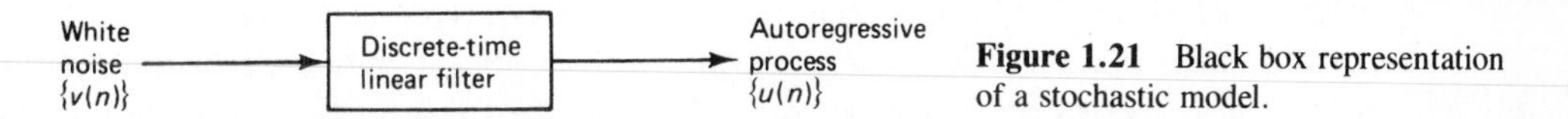

Figure 1.21 Black box representation of a stochastic model.

[11] The International Telephone and Telegraph Consultative Committee (CCITT) has adopted the 32-kb/s ADPCM as an international standard. The adaptive predictor used herein has a transfer function consisting of two poles and six zeros. A two-pole configuration was chosen, because it permits control of decoder stability in the presence of transmission errors. Six zeros were combined with the two poles in order to improve performance. The eight coefficients of the predictor are adapted by using a simplified version of the LMS (stochastic gradient) algorithm; for details, see Benevenuto et al. (1986) and Nishitani et al. (1987).

$$S_{AR}(\omega, n) = \frac{\sigma_v^2}{\left| 1 + \sum_{k=1}^{M} a_k(n) e^{-jk\omega} \right|^2} \tag{1.48}$$

We may determine the AR parameters of the time-varying model by applying the process $\{u(n)\}$ to an *adaptive prediction-error filter*, as indicated in Fig. 1.22. The filter consists of a transversal filter with adjustable tap weights. In the adaptive scheme of Fig. 1.22 the prediction error produced at the output of the filter is used to control the adjustments applied to the tap weights of the filter.

The *adaptive AR model* provides a practical means for measuring the *instantaneous frequency* of a frequency-modulated process. In particular, we may do this by measuring the frequency at which the AR power spectrum $S_{AR}(\omega, n)$ attains its peak value for varying time n.

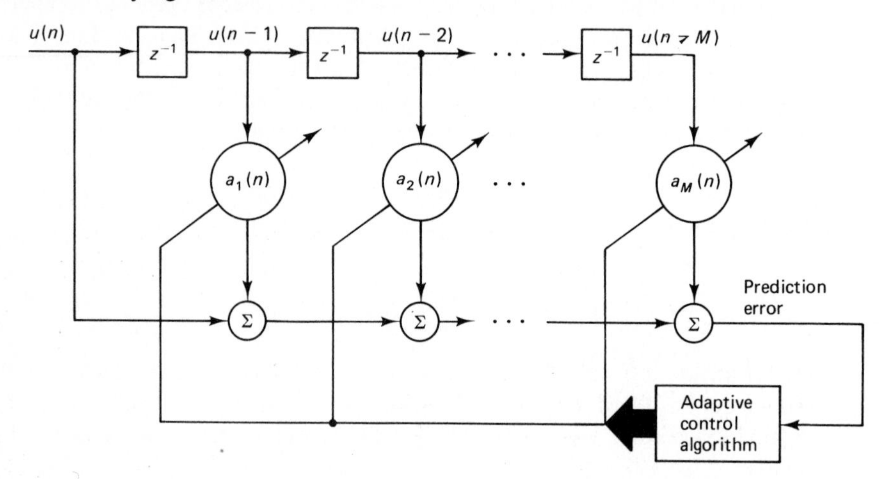

Figure 1.22 Adaptive prediction-error filter for real-valued data.

Signal Detection

The *detection problem,* that is, the problem of detecting an information-bearing signal in noise, may be viewed as one of *hypothesis testing* with deep roots in *statistical decision theory* [Van Trees (1968)]. In the statistical formulation of hypothesis testing, there are two criteria of most interest: the *Bayes criterion* and the *Neyman–Pearson criterion.* In the Bayes test, we minimize the *average cost* or *risk* of the experiment of interest, which incorporates two sets of parameters: (1) *a priori probabilities* that represent the observer's information about the source of information before the experiment is conducted, and (2) a set of *costs* assigned to the various possible courses of action. As such, the Bayes criterion is directly applicable to digital communications. In the Neyman–Pearson test, on the other hand, we maximize the *probability of detection* subject to the constraint that the *probability of false alarm* does *not* exceed some preassigned value. Accordingly, the Neyman–Pearson criterion is directly applicable to radar or sonar. An idea of fundamental importance that emerges in hypothesis testing is that, for a Bayes criterion or Neyman–Pearson criterion, the optimum test consists of two distinct operations: (1) processing the observed data to compute a test statistic called the *likelihood ratio,* and (2) computing the likelihood ratio with a *threshold* to make a *decision* in favor of one of the two hypotheses. The choice of one criterion or the other merely affects the value assigned to the threshold. Let H_1 denote the hypothesis that the observed data consist of noise alone and H_2 denote the hypothesis that the data consist of signal plus noise. The likelihood ratio is defined as the ratio of two maximum likelihood functions, the numerator assuming that hypothesis H_2 is true and the denominator assuming that hypothesis H_1 is true. If the likelihood ratio exceeds the threshold, the decision is made in favor of hypothesis H_2; otherwise, the decision is made in favor of hypothesis H_1.

In simple binary hypothesis testing, it is assumed that the signal is known, and the noise is both white and Gaussian. In this case, the likelihood ratio test yields a *matched filter* (matched in the sense that its impulse response equals the time-reversed version of the known signal). When the additive noise is a *colored Gaussian noise* of known mean and correlation matrix, the likelihood ratio test yields a filter that consists of two sections: a *whitening filter* that transforms the colored noise component at the input into a white Gaussian noise process, and a *matched filter* that is matched to the new version of the known signal as modified by the whitening filter.

However, in some important operational environments such as *communications, radar,* and *active sonar,* there may be inadequate information on the signal and noise statistics to design a fixed optimum detector. For example, in a sonar environment it may be difficult to develop a precise *model* for the received sonar signal, one that would account for the following factors completely:

1. Loss in the signal strength of a *target echo* from an object of interest (e.g., enemy vessel), due to oceanic propagation effects and reflection loss at the target.
2. Statistical variations in the additive *reverberation* component, produced by reflections of the transmitted signal from scatterers such as the ocean surface, ocean floor, biologies, and inhomogeneities within the ocean volume.
3. Potential sources of *noise* such as biological, shipping, oil drilling, seismic, and oceanographic.

In situations of this kind, the use of adaptivity offers an attractive approach to solve the target (signal) detection problem. Typically, the design of an *adaptive detector* proceeds by exploiting some knowledge of general characteristics of the signal and noise, and designing the detector in such a way that its internal structure is adjustable in response to changes in the received signal. In general, the incorporation of this adjustment makes the performance analysis of an adaptive detector much more difficult to undertake than that of a fixed detector.

Fixed and Adaptive Detectors. Figure 1.23(a) shows the block diagram of a conventional detector based on the *discrete Fourier transform* (DFT) for the detection of narrow-band signals in white Gaussian noise [Williams and Ricker (1972)]. The DFT may be viewed as a bank of nonoverlapping narrow-band filters whose passbands span the frequency range of interest. In the detector of Fig. 1.23(a) the magnitude of each complex output of the DFT is squared to form a *sufficient*

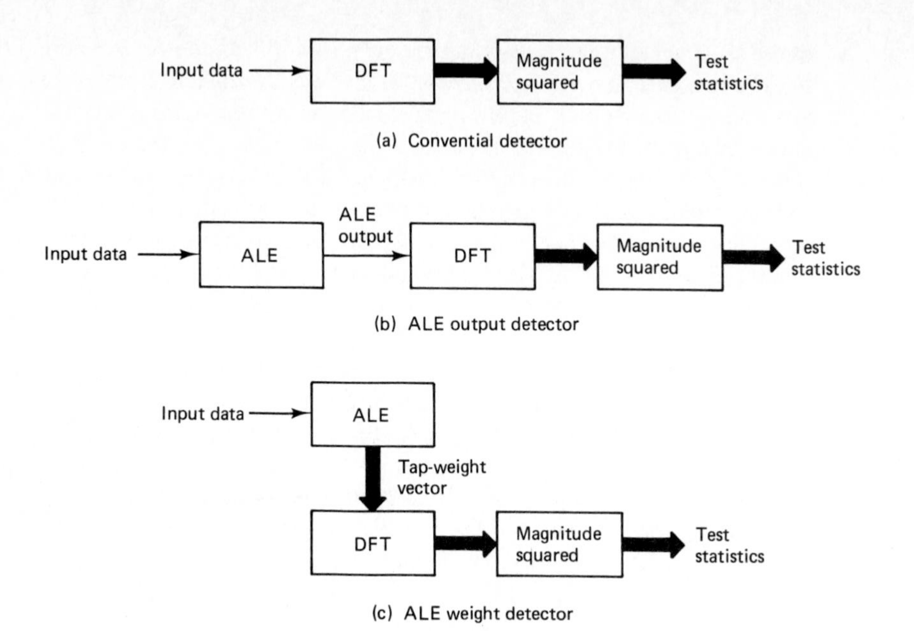

Figure 1.23 Fixed and adaptive detection schemes: (a) conventional detector. (b) ALE output detector. (c) ALE weight detector.

statistic. This statistic is optimum (in the Neyman-Pearson sense) for detecting a sinusoid of known frequency (centered in the pertinent passband of the DFT) but unknown phase, and in the presence of white Gaussian noise. The detector output is compared to a threshold. If the threshold is exceeded, the detector decides in favor of the narrow-band signal; otherwise, the detector declares the signal to be absent.

The performance of this conventional noncoherent detector may be improved by using an *adaptive line enhancer* (ALE) as a *prefilter* (preprocessor) to the detector [Widrow et al. (1975)]. The ALE is a special form of adaptive noise canceller that is designed to suppress the wide-band noise component of the input, while passing the narrow-band signal component with little attenuation. Figure 1.24 depicts the block diagram of an ALE. It consists of the interconnection of a delay element

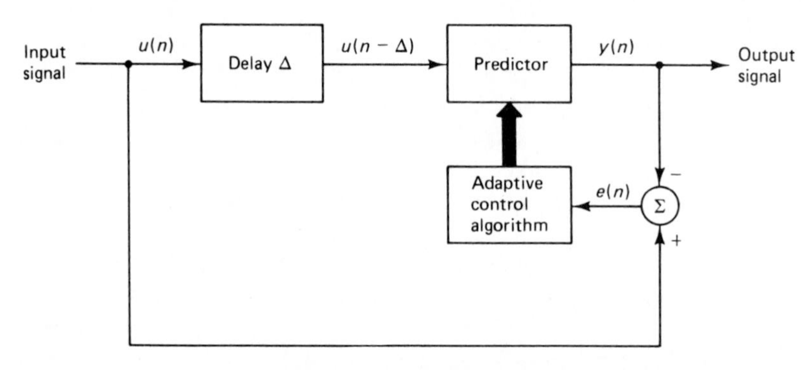

Figure 1.24 Adaptive line enhancer.

and a linear predictor. The predictor output $y(n)$ is subtracted from the input signal $u(n)$ to produce the estimation error $e(n)$. This estimation error is, in turn, used to adaptively control the tap weights of the predictor. The predictor input equals $u(n - \Delta)$, where the delay Δ is equal to or greater than the sampling period. The main function of Δ is to remove the correlation between the noise component in the original input signal $u(n)$ and the delayed predictor input $u(n - \Delta)$. It is for this reason that the delay Δ is called the *decorrelation parameter* of the ALE.

Two types of ALE detection structures have been proposed in the literature (Zeidler (1990)]:

1. *ALE output detector*. In this adaptive detector shown in Fig. 1.23(b), the output of an ALE of length L is applied to a DFT of length K. The magnitude of the resulting DFT output is squared to produce the sufficient statistic for the detector.
2. *ALE weight detector*. In this second adaptive detector, shown in Fig. 1.23(c), the tap-weight vector of an ALE of length L is applied to a DFT of length L. The magnitude of the DFT output is squared as before to produce the sufficient statistic.

In both cases, the ALE processes N input data points, with the ALE length L small compared to N.

Detection Performance in Stationary Noise. Using extensive computer simulation, Zeidler (1990) has studied the performance of these two ALE detectors and compared them to the conventional detector. The parameters used in the study are described as follows (using Zeidler's terminology):

Total processing time interval, T
Sampling frequency, f_s
Signal bandwidth, α
Signal-to-noise ratio, SNR
ALE filter length, L
DFT filter length, K
Adaptive time constant, τ_A
Decorrelation parameter, Δ
Postdetection integration time constant, T_s

For the study, the parameters T and f_s were fixed, and the parameters α, SNR, L, τ_A, Δ, and T_s were varied. In all cases, the values $T = 128$ seconds and $f_s = 1024$ Hz were used, the input noise bandwidth was 512 Hz, and the center frequency of the narrow-band signal was bin-centered with respect to the DFT output. The ALE resolution β_1 and the DFT resolution β_2 are given by

$$\beta_1 = \frac{f_s}{L} \tag{1.49}$$

and

$$\beta_2 = \frac{f_s}{K} \tag{1.50}$$

The detector input consisted of the sampled version of a zero-mean white Gaussian noise process added to a narrow-band signal. The narrow-band signal was itself generated by passing a zero-mean white Gaussian noise process through a single-pole low-pass filter; the filtered output was then translated to the desired frequency.

The performances of the ALE output detector, the ALE weight detector, and the conventional detector were computed for a fixed probability of detection P_D equal to 0.5 and a fixed probability of false alarm P_{FA} equal to 10^{-4}. The results of this computation are shown plotted as a function of the input signal bandwidth in Fig. 1.25(a). From this figure, we can make the following observations [Zeidler (1990)]:

1. The performance of all three detectors, measured in terms of SNR, is degraded as the signal bandwidth α is increased.
2. The performance of the adaptive and conventional detectors remains nearly the same if $\beta_1 = \alpha$.
3. The performance of the ALE weight detector and the ALE output detector is essentially the same for sinusoidal inputs, but the performance of the ALE weight detector degrades more rapidly with increasing signal bandwidth, particularly when $\beta_1 = \alpha$.
4. A relative loss of 1.8 dB in performance between the ALE output detector and the conventional detector occurs as the signal bandwidth α increases from 0 to 4 HZ for $\beta_1 = 1$ Hz. The equivalent relative loss in performance between the ALE weight detector and the conventional detector is approximately 3.7 dB as α increases from 0 to 4 Hz for $\beta_1 = \beta_2 = 1$ Hz. The slight improvement in both the ALE weight detector and the ALE output detector for $\alpha = 0$ is attributed to implementation noise in the DFT processor.

The results of Fig. 1.25(a) were obtained for a stationary Gaussian noise background. In an overall sense, these results indicate that in such a noise background the conventional (fixed) DFT detector provides an almost optimum detection performance. The real benefit of the ALE is realized in a nonstationary noise background, as shown next.

Detection Performance in Nonstationary Noise. The incorporation of the ALE in the detector produces a radical change in the probability density function of the detector output in a noise-only input situation. This is true for an unintegrated or integrated DFT output. In either case, both the mean and the mean-to-variance ratio of the detector output due to a noise-only input are significantly reduced for the ALE/DFT detector, as compared with the conventional DFT detector. In a stationary noise background, this effect is of no serious concern, because fairly accurate thresholds can be established for both detectors from a knowledge of the probability density function. In a nonstationary noise background, on the other hand, the re-

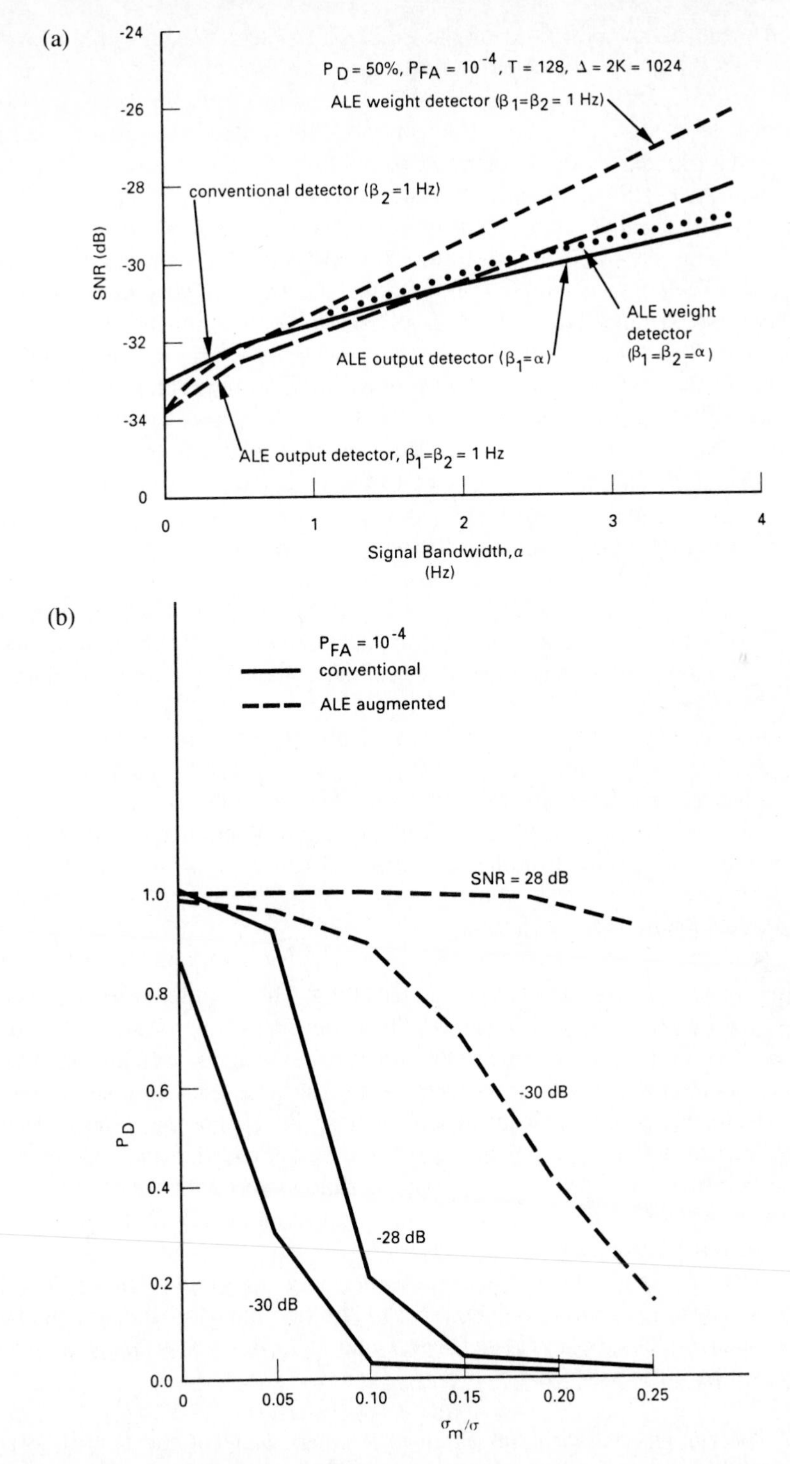

Figure 1.25 (a) Comparative detection performance for ALE output detector, ALE weight detector and conventional detector: $P_D = 0.5$, $P_{FA} = 10^{-4}$. (b) Comparison of relative degradation in probability of detection P_D is a function of noise nonstationarity σ_m/σ, for conventional and adaptive detectors. (J. Zeidler, "Performance Analysis of LMS Adaptive Prediction Filters," © 1990, IEEE. Reprinted with permission.)

duced mean-to-variance ratio of the ALE/DFT detector output may prove to be of practical value if the use of an adaptive threshold is ineffective or impractical.

Figure 1.25(b), for a nonstationary situation, presents a comparison of the degradation in the probability of detection P_D (for a fixed signal-to-noise ratio) plotted versus the degree of nonstationarity of the additive noise for the conventional DFT and ALE/DFT detectors with incoherent detection [Zeidler (1990)]. The *degree of nonstationarity* is denoted by σ_m/σ, where σ_m and σ are standard deviations that are themselves defined as follows. The variance (i.e., square of the standard deviation) σ^2 refers to a sample $v(n)$ of a wide-sense stationary Gaussian noise process, and the variance σ_m^2 refers to a sample $m(k)$ of an independent process of the same type. The process $\{m(k)\}$ is chosen in such a way that it changes value every T samples of the first process $\{v(n)\}$. Moreover, the noise process $\{m(k)\}$ is amplitude-modulated by the first noise process $\{v(n)\}$. The resulting process, denoted by $\{v'(n)\}$ is block-stationary and non-Gaussian. However, with respect to each integration interval of length T, the variance of the modulated noise sample $v'(n)$ is a random variable equal to $|1 + m(k)|^2\sigma^2$, conditional on $m(k)$. Hence, the process $\{v'(n)\}$ is nonstationary insofar as the detector is concerned. Returning to the results presented in Fig. 1.25(b), we see that the performance of the conventional (fixed) detector degrades more rapidly, and at higher SNR values, than the adaptive detector. This is to be expected since the higher mean-to-variance ratio of the detector output in a noise-only input situation exhibits a greater sensitivity to changes in the input noise power.

The practical value of an ALE as a preprocessor to a conventional matched filter has been demonstrated by Nielson and Thomas (1988) as a means of improving the performance of the detector in the presence of Arctic ocean noise. This type of noise is known to have highly non-Gaussian and nonstationary characteristics; hence the benefit to be gained from the use of an ALE.

Adaptive Noise Cancelling

As the name implies, adaptive noise cancelling relies on the use of *noise cancelling* by subtracting noise from a received signal, an operation controlled in an *adaptive* manner for the purpose of improved signal-to-noise ratio. Ordinarily, it is inadvisable to subtract noise from a received signal, because such an operation could produce disastrous results by causing an increase in the average power of the output noise. However, when proper provisions are made, and filtering and subtraction are controlled by an adaptive process, it is possible to achieve a superior system performance compared to direct filtering of the received signal (Widrow et al., 1975b; Widrow and Stearns, 1985).

Basically, an adaptive noise canceller is a *dual-input, closed-loop adaptive control system* as illustrated in Fig. 1.26. The two inputs of the system are derived from a pair of sensors: a *primary sensor* and a *reference (auxiliary) sensor*. Specifically, we have the following:

1. The primary sensor receives an *information-bearing signal* $s(n)$ corrupted by *additive noise* $v_0(n)$, as shown by

$$d(n) = s(n) + v_0(n) \tag{1.51}$$

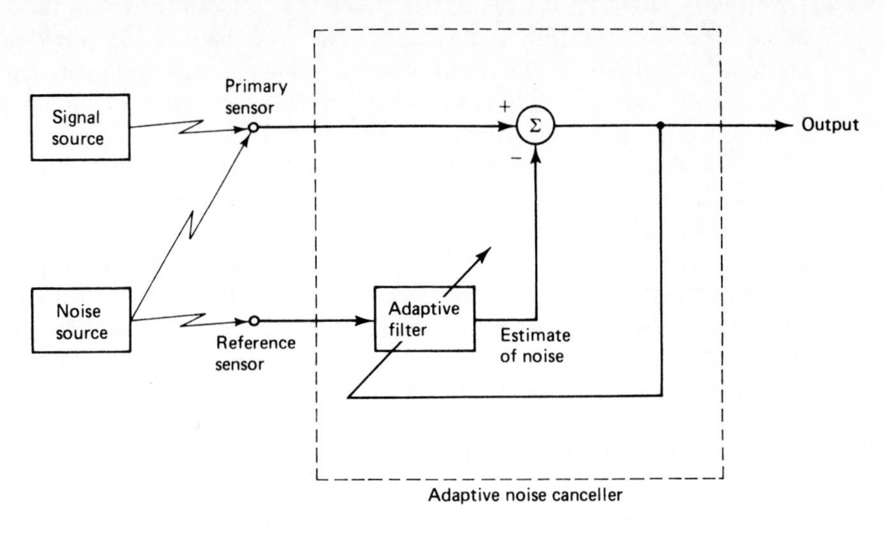

Figure 1.26 Adaptive noise cancellation.

The signal $s(n)$ and the noise $v_0(n)$ are uncorrelated with each other; that is,

$$E[s(n)v_0(n-k)] = 0, \qquad \text{for all } k \tag{1.52}$$

where $s(n)$ and $v_0(n)$ are assumed to be real valued.

2. The reference sensor receives a noise $v_1(n)$ that is *uncorrelated* with the signal $s(n)$ but *correlated* with the noise $v_0(n)$ in the primary sensor output in an *unknown* way; that is,

$$E[s(n)v_1(n-k)] = 0, \qquad \text{for all } k \tag{1.53}$$

and

$$E[v_0(n)v_1(n-k)] = p(k) \tag{1.54}$$

where, as before, the signals are real valued and $p(k)$ is an unknown cross-correlation for lag k.

The reference signal $v_1(n)$ is processed by an adaptive filter to produce the output signal:

$$y(n) = \sum_{k=0}^{M-1} \hat{w}_k(n)v_1(n-k) \tag{1.55}$$

where the $\hat{w}_k(n)$ are the adjustable (real) tap weights of the adaptive filter. The filter output $y(n)$ is subtracted from the primary signal $d(n)$, serving as the "desired" response for the adaptive filter. The error signal is defined by

$$e(n) = d(n) - y(n) \tag{1.56}$$

Thus, substituting Eq. (1.51) in (1.56), we get

$$e(n) = s(n) + v_0(n) - y(n) \tag{1.57}$$

The error signal is in turn used to adjust the tap weights of the adaptive filter, and the control loop around the operations of filtering and subtraction is thereby closed. Note that the information-bearing signal $s(n)$ is indeed part of the error signal $e(n)$, as indicated in Eq. (1.57).

The error signal $e(n)$ constitutes the overall *system output*. From Eq. (1.57) we see that the noise component in the system output is $v_0(n) - y(n)$. Now, the adaptive filter attempts to minimize the mean-square value (i.e., average power) of the error signal $e(n)$. The information-bearing signal $s(n)$ is essentially unaffected by the adaptive noise canceller. Hence, minimizing the mean-square value of the error signal $e(n)$ is equivalent to minimizing the mean-square value of the output noise $v_0(n) - y(n)$. With the signal $s(n)$ remaining essentially constant, it follows that *the minimization of the mean-square value of the error signal is indeed the same as the maximization of the output signal-to-noise ratio of the system.*

The signal-processing operation described herein has two limiting cases that are noteworthy:

1. The adaptive filtering operation is *perfect* in the sense that

$$y(n) = v_0(n)$$

   In this case, the system output is *noise free* and the noise cancellation is perfect. Correspondingly, the output signal-to-noise ratio is infinitely large.
2. The reference signal $v_1(n)$ is *completely uncorrelated* with both the signal and noise components of the primary signal $d(n)$; that is,

$$E[d(n)v_1(n - k)] = 0, \qquad \text{for all } k$$

   In this case, the adaptive filter "switches itself off," resulting in a zero value for the output $y(n)$. Hence, the adaptive noise canceller has *no* effect on the primary signal $d(n)$, and the output signal-to-noise ratio remains unaltered.

The effective use of adaptive noise cancelling therefore requires that we place the reference sensor in the noise field of the primary sensor with two specific objectives in mind. First, the information-bearing signal component of the primary sensor output is *undetectable* in the reference sensor output. Second, the reference sensor output is *highly correlated* with the noise component of the primary sensor output. Moreover, the adaptation of the adjustable filter coefficients must be near optimum.

In the remainder of this subsection, we describe three useful applications of the adaptive noise cancelling operation:

1. *Cancelling 60-Hz interference in electrocardiography*. In *electrocardiography* (ECG), commonly used to monitor heart patients, an *electrical discharge* radiates energy through a human *tissue* and the resulting output is received by an *electrode*. The electrode is usually positioned in such a way that the received energy is maximized. Typically, however, the electrical discharge involves very low potentials. Correspondingly, the received energy is very small. Hence extra care has to be exercised in minimizing signal degradation due to external *interference*. By far, the strongest form of interference is that of a 60-Hz periodic waveform picked up by the

receiving electrode (acting like an antenna) from nearby electrical equipment (Hutha and Webster, 1973). Needless to say, this interference has undesirable effects in the interpretation of electrocardiograms. Widrow et al. (1975b) have demonstrated the use of adaptive noise cancelling (based on the LMS algorithm) as a method for reducing this form of interference. Specifically, the primary signal is taken from the ECG preamplifier, and the reference signal is taken from a wall outlet with proper attenuation. Figure 1.27 shows a block diagram of the adaptive noise canceller used by Widrow et al. (1975b). The adaptive filter has two adjustable weights, $\hat{w}_0(n)$ and $\hat{w}_1(n)$. One weight, $\hat{w}_0(n)$, is fed directly from the reference point. The second weight, $\hat{w}_1(n)$, is fed from a 90°-phase-shifted version of the reference input. The sum of the two weighted versions of the reference signal is then subtracted from the ECG output to produce an error signal. This error signal together with the weighted inputs are applied to the LMS algorithm, which in turn controls the adjustments applied to the two weights. In this application, the adaptive noise canceller acts as a variable "notch filter." The frequency of the sinusoidal interference in the ECG output is presumably the same as that of the sinusoidal reference signal. However, the amplitude and phase of the sinusoidal interference in the ECG output are unknown. The two weights $\hat{w}_0(n)$ and $\hat{w}_1(n)$ provide the two *degrees of freedom* required to control the amplitude and phase of the sinusoidal reference signal so as to cancel the 60-Hz interference contained in the ECG output.

2. *Reduction of acoustic noise in speech.* At a noisy site (e.g., the cockpit of a military aircraft), voice communication is affected by the presence of *acoustic*

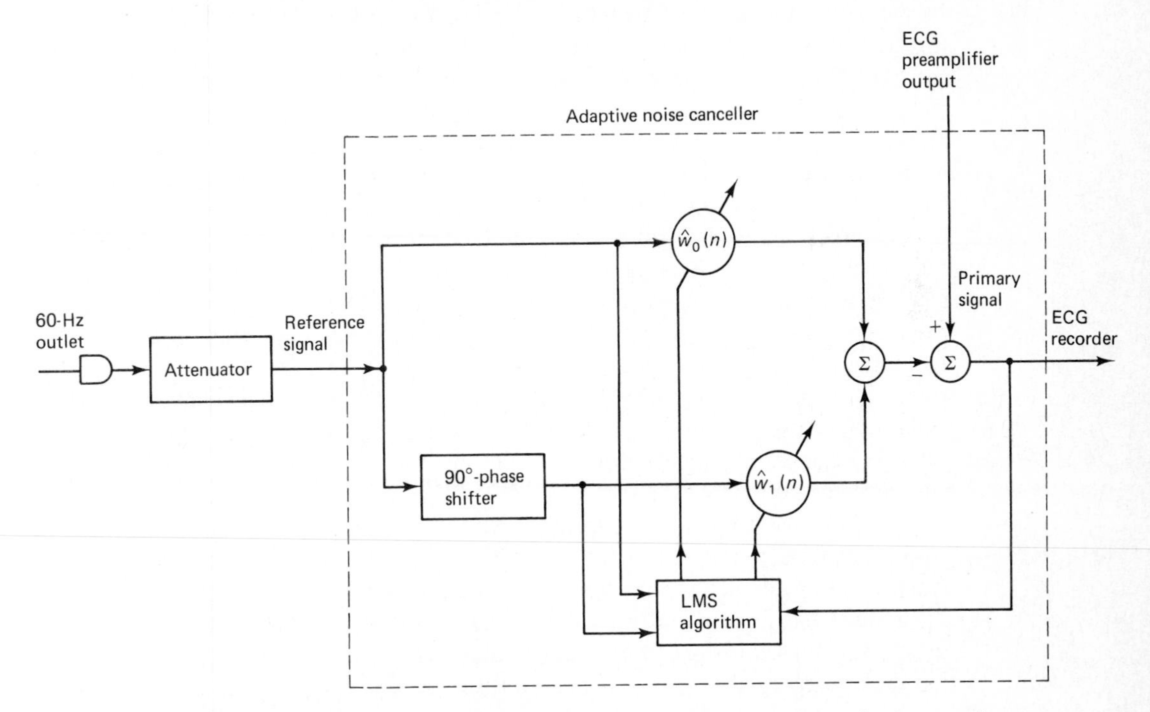

Figure 1.27 Adaptive noise canceller for suppressing 60-Hz interference in electrocardiography. [After Widrow et al. (1975b)].

noise. This effect is particularly serious when linear predictive coding (LPC) is used for the digital representation of voice signals at low bit rates; LPC was discussed earlier. To be specific, high-frequency acoustic noise severely affects the estimated LPC spectrum in both the low- and high-frequency regions. Consequently, the intelligibility of digitized speech using LPC often falls below the minimum acceptable level. Kang and Fransen (1987) describe the use of an adaptive noise canceller, based on the LMS algorithm, for reducing acoustic noise in speech. The noise-corrupted speech is used as the primary signal. To provide the reference signal (noise only), a reference microphone is placed in a location where there is sufficient isolation from the source of speech (i.e., the known location of the speaker's mouth). In the experiments described by Kang and Fransen, a reduction of 10 to 15 dB in the acoustic noise floor is achieved, without degrading voice quality. Such a level of noise reduction is significant in improving voice quality, which may be unacceptable otherwise.

3. *Adaptive speech enhancement*. Consider the situation depicted in Fig. 1.28. The requirement is to listen to the voice of the desired speaker in the presence of background noise, which may be satisfied through the use of adaptive noise cancelling. Specifically, *reference microphones* are added at locations far enough away from the desired speaker such that their outputs contain *only* noise. As indicated in Fig. 1.28, a weighted sum of the auxiliary microphone outputs is subtracted from the output of the desired speech-containing microphone, and an adaptive filtering algorithm (e.g., the LMS algorithm) is used to adjust the weights so as to minimize the average output power. A useful application of the idea described herein is in the adaptive noise cancellation for hearing aids[12] (Chazan et al., 1988). The so-called "cocktail party effect" severely limits the usefulness of hearing aids. The cocktail party phenomenon refers to the ability of a person with normal hearing to focus on a conversation taking place at a distant location in a crowded room. This ability is lacking in a person who wears hearing aids, because of extreme sensitivity to the presence of *background noise*. This sensitivity is attributed to two factors: (a) the loss of directional cues, and (b) the limited channel capacity of the ear caused by the reduction in both dynamic range and frequency response. Chazan et al. (1988) describe an adaptive noise cancelling technique aimed at overcoming this problem. The technique involves the use of an *array of microphones* that exploit the difference in spatial characteristics between the desired signal and the noise in a crowded room. The approach taken by Chazan et al. is based on the fact that each microphone output may be viewed as the sum of the signals produced by the individual speakers engaged in conversations in the room. Each signal contribution in a particular microphone output is essentially the result of a speaker's speech signal having passed through the *room filter*. In other words, each speaker (including the desired speaker) produces a signal at the microphone output that is the sum of the direct transmission of his/her speech signal and its reflections from the walls of the room. The requirement is to reconstruct the desired speaker signal, including its room reverberations, while cancelling out the source of noise. In general, the transformation undergone by the speech signal from the desired speaker is not known. Also,

[12] This idea is similar to that of adaptive spatial filtering in the context of antennas, which is considered later in this section.

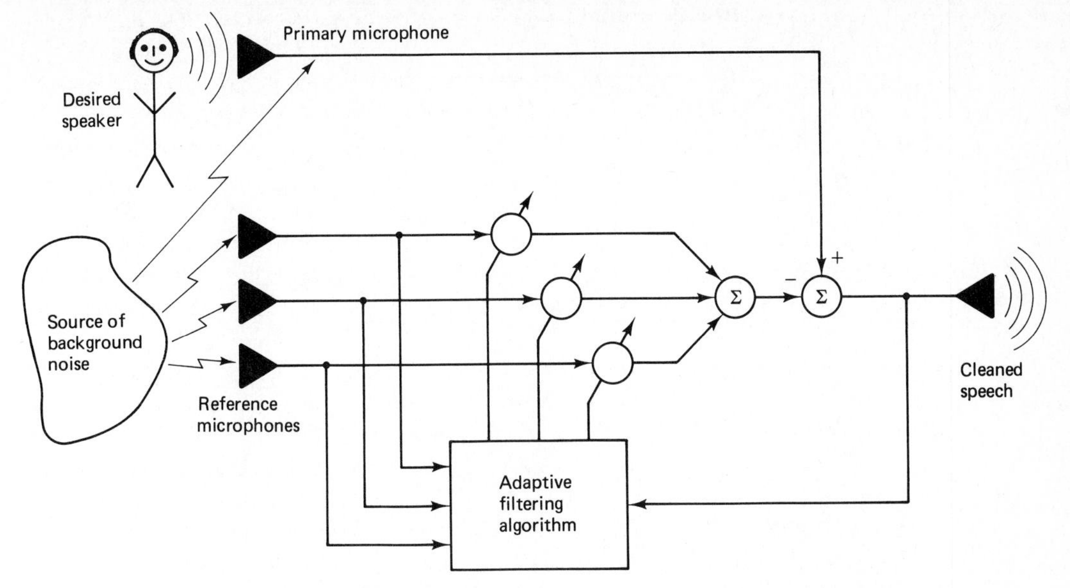

Figure 1.28 Block diagram of an adaptive noise canceller for speech.

the characteristics of the background noise are variable. We thus have a signal-processing problem for which adaptive noise cancelling offers a feasible solution.

Echo Cancellation

Almost all conversations are conducted in the presence of *echoes*. An echo may be nonnoticeable or distinct, depending on the time delay involved. If the delay between the speech and the echo is short, the echo is not noticeable but perceived as a form of spectral distortion or reverberation. If, on the other hand, the delay exceeds a few tens of milliseconds, the echo is distinctly noticeable. Distinct echoes are annoying.

Echoes may also be experienced on a telephone circuit (Sondhi and Berkley, 1980). When a speech signal encounters an *impedance mismatch* at any point on a telephone circuit, a portion of that signal is reflected (returned) as an echo. An echo represents an *impairment* that can be annoying subjectively as the more obvious impairments of low volume and noise.

To see how echoes occur, consider a long-distance telephone circuit depicted in Fig. 1.29. Every telephone set in a given geographical area is connected to a central office by a *two-wire line* called the *customer loop*; the two-wire line serves the need for communications in either direction. However, for circuits longer than about 35 miles, a separate path is necessary for each direction of transmission. Accordingly, there has to be provision for connecting the two-wire circuit to the four-wire circuit. This connection is accomplished by means of a *hybrid transformer*, commonly referred to as a *hybrid*. Basically, a hybrid is a bridge circuit with three ports (terminal pairs), as depicted in Fig. 1.30. If the bridge is *not* perfectly balanced, the

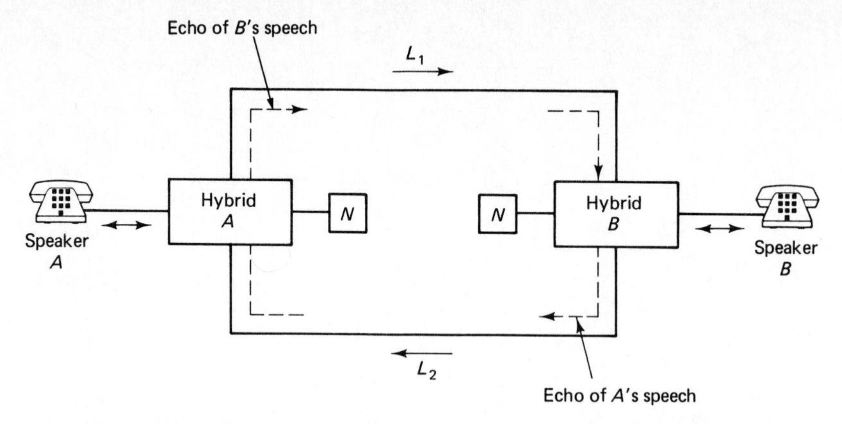

Figure 1.29 Long-distance telephone circuit; the boxes marked N are balancing impedances.

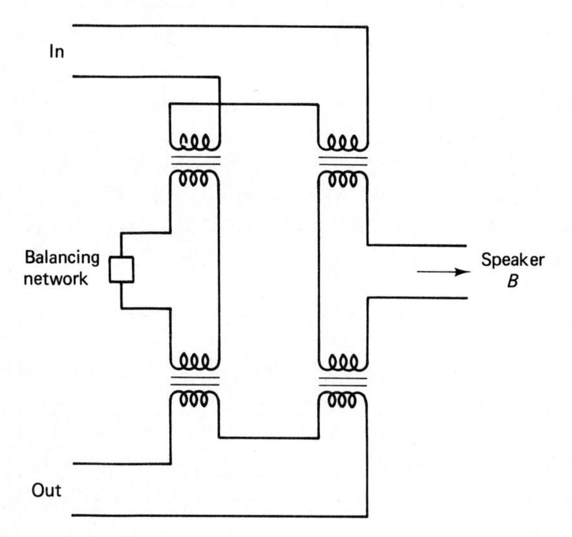

Figure 1.30 Hybrid circuit.

"in" port of the hybrid becomes coupled to the "out" port, thereby giving rise to an echo.

Echoes are noticeable when a long-distance call is made on a telephone circuit, particularly one that includes a *geostationary satellite*. Due to the high altitude of such a satellite, there is a one-way travel time of about 270 ms between a ground station and the satellite. Thus, the *round-trip delay* in a satellite link (including telephone circuits) can be as long as 600 ms. Generally speaking, the longer the echo delay, the more it must be attenuated before it becomes noticeable.

The question to be answered is: How do we exercise echo control? It appears that the idea with the greatest potential for echo control is that of *adaptive echo cancellation* (Sondhi and Prasti, 1966; Sondhi, 1967; Sondhi and Berkley, 1980; Messerschmitt, 1984; Murano et al., 1990). The basic principle of echo cancellation is to *synthesize a replica of the echo and subtract it from the returned signal*. This principle is illustrated in Fig. 1.31 for only one direction of transmission (from

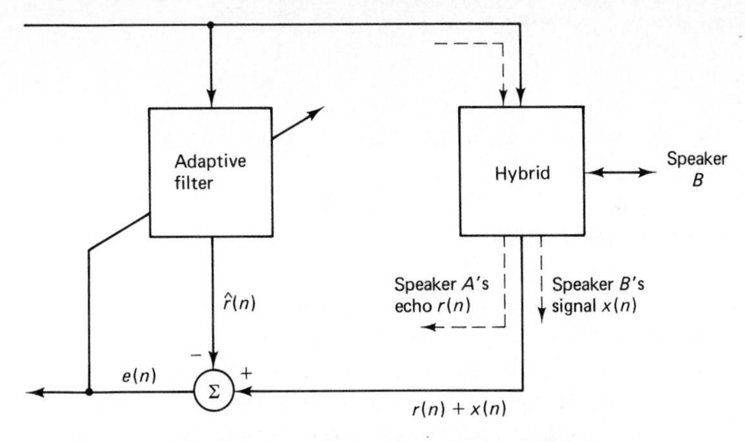

Figure 1.31 Signal definitions for echo cancellation.

speaker A on the *far* left of the hybrid to speaker B on the right). The adaptive canceller is placed in the four-wire path *near* the origin of the echo. The synthetic echo, denoted by $\hat{r}(n)$, is generated by passing the speech signal from speaker A (i.e., the "reference" signal for the adaptive canceller) through an adaptive filter that ideally matches the transfer function of the echo path. The reference signal, passing through the hybrid, results in the echo signal $r(n)$. This echo, together with a near-end talker signal $x(n)$ (i.e., the speech signal from speaker B) constitutes the "desired" response for the adaptive canceller. The synthetic echo $\hat{r}(n)$ is subtracted from the desired response $r(n) + x(n)$ to yield the canceller error signal

$$e(n) = r(n) - \hat{r}(n) + x(n) \tag{1.58}$$

Note that the error signal $e(n)$ also contains the near-end talker signal $x(n)$. In any event, the error signal $e(n)$ is used to control the adjustments made in the coefficiencies (tap weights) of the adaptive filter.

In practice, the echo path is highly variable, depending on the distance to the hybrid, the characteristics of the two-wire circuit, and so on. These variations are taken care of by the adaptive control loop built into the canceller. The control loop continuously adapts the filter coefficiencies to take care of fluctuations in the echo path.

For the adaptive echo cancellation circuit to operate satisfactorily, the impulse response of the adaptive filter should have a length greater than the longest echo path that needs to be accommodated. Let T_s be the sampling period of the digitized speech signal, M be the number of adjustable coefficients (tap weights) in the adaptive filter, and τ be the longest echo delay to be accommodated. We must then choose

$$MT_s > \tau \tag{1.59}$$

As mentioned previously (when discussing adaptive differential pulse-code modulation), the sampling rate for speech signals on the telephone network is conservatively chosen as 8 kHz, that is,

$$T_s = 125 \ \mu\text{s}$$

Suppose, for example, that the echo delay $\tau = 30$ ms. Then we must choose

$$M > 240 \text{ taps}$$

Thus, the use of an echo canceller with $M = 256$ taps, say, is satisfactory for this situation.

Radar Polarimetry

The *polarization* of an electromagnetic wave is defined as the direction of vibration of the electric field component of the wave (Skolnik, 1982). Various forms of polarization are possible to transmit and receive: *linear* horizontal or vertical polarization, left-hand or right-hand *circular* polarization, or *elliptical* polarization that is a combination of linear and circular polarizations. The application we wish to describe in this subsection relates to a novel radar technique for *precise navigation along a confined waterway* (Haykin, 1986).

For its operation, the system used for this application relies on two basic components:

1. *Noncoherent radar with dual-polarized antenna*. The radar antenna transmits a (linear) horizontally polarized signal, and receives both horizontal and vertical polarizations.
2. *Polarimetric retro-reflector*. This is an electromagnetic device that rotates the polarization of the incident electromagnetic signal by 90° (Macikunas, et al., 1988). For example, a horizontally polarized electromagnetic signal incident on the reflector is retransmitted as a vertically polarized signal. Moreover, this rotation is achieved with only a small power loss and over a fairly wide range of angles.

The system operates as follows. A series of polarimetric retro-reflectors are located at strategic points along the shores of the waterway. The dual-polarized radar is located on a ship that navigates along the waterway. Thus, by transmitting a horizontally polarized electromagnetic signal and responding to a vertically polarized echo produced by a polarimetric retro-reflector on the shore, the radar is capable of measuring its distance from the shoreline fairly accurately. Moreover, the combination of a horizontally polarized transmission and a vertically polarized reception has the effect of reducing the effect of ground *clutter* (i.e., reflections produced by natural and manufactured objects located along the shores).

Typically, however, the system suffers from a cross-polarized signal (leakage) problem that arises due to performance limitations in the microwave parts of the radar system, as well as depolarization of the returned signal by the environment. Depolarization (appearing in the cross-polarized channel) due to ground backscatter has been reported to be as high as -3 dB relative to a like-polarized channel (Giuli, 1986); returns of 0 dB have been observed from manufactured objects. This leakage problem may be overcome by using a dual-input, *adaptive cross-polar canceller* (Ukrainec and Haykin, 1989).

Specifically, the output of the *cross-polarized* (horizontal transmit-vertical receive) channel provides the "primary" signal. This output contains the "desired" echo from the polarimetric retro-reflector as well as clutter (undesired echoes) from the surrounding environment. The output of the *like-polarized* (horizontal transmit-horizontal receive) channel provides the "reference" signal. The clutter in the like-polarized channel is correlated with that in the cross-polarized channel, and for all practical purposes there is *no* trace of the echo from the polarimetric retro-reflector in the like-polarized channel. Furthermore, the polarization leakage phenomenon responsible for the appearance of clutter on the cross-polarized channel may be modelled as a *zero-order system with varying gain*. Accordingly, for the implementation of the adaptive polar-interference canceller, we may use a single-weight adaptive filter placed in the like-polarized channel, as shown in Fig. 1.32. In this figure, we have assumed the use of baseband representations for the narrow-band signals at the two channel outputs.

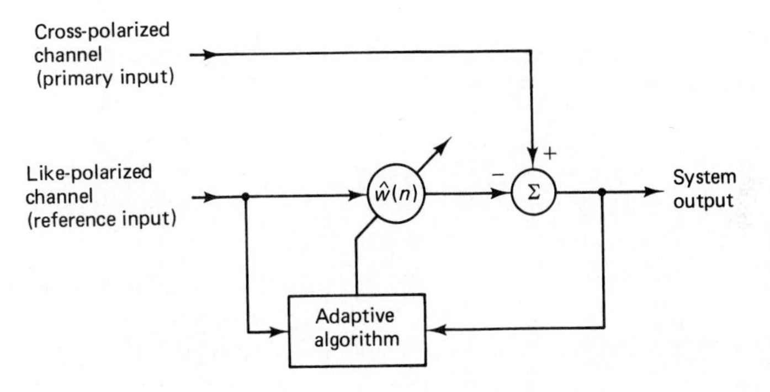

Figure 1.32 Adaptive cross-polar canceller.

The three images presented in Fig. 1.33 illustrate the effectiveness of this simple cross-polar canceller. These images were obtained from an experimental radar system operating in a highly cluttered industrial environment; they are displayed in *B-scan* (range versus azimuth) format. The images in Fig. 1.33(a) and (b) are displays for the like-polarized and cross-polarized channel outputs, operating normally. The image in Fig. 1.33(c) is for the same channel output, incorporating an adaptive cross-polar canceller; the recursive least-squares (RLS) algorithm was used for the adaptation of the weight in the canceller. A comparison of the images in Fig. 1.33(b) and (c) clearly reveals the improvement in the quality of the image in Fig. 1.33(c) evidenced by the more pronounced appearance of the echo from the polarimetric retro-reflector, which is attained through the use of a single-weight *adaptive cross-polar canceller*.

Adaptive Beamforming

For our last application, we describe a *spatial* form of adaptive signal processing that finds practical use in radar, sonar, communications, geophysical exploration, astrophysical exploration, and biomedical signal processing.

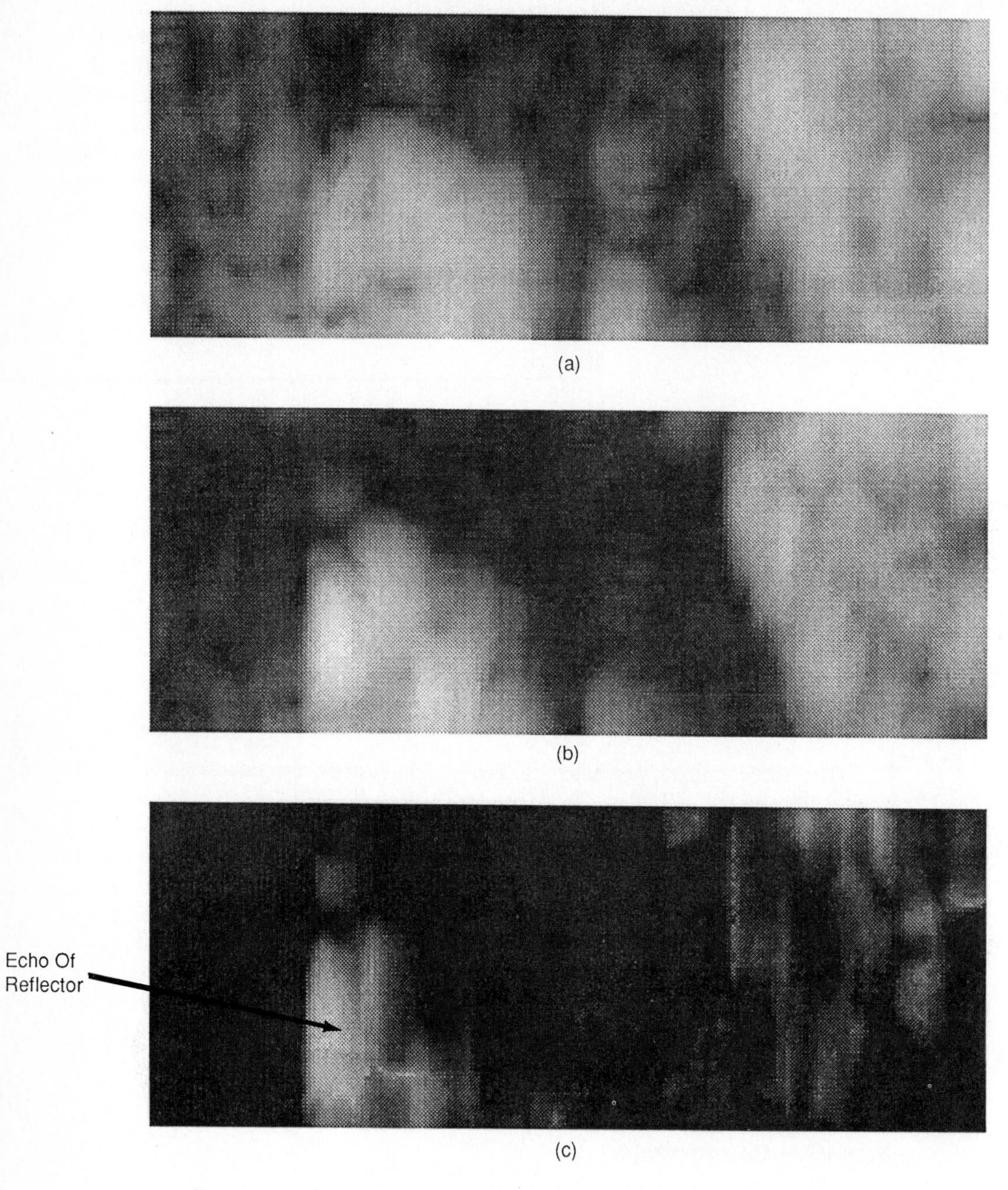

Figure 1.33 Polarimetric radar returns from an industrial site: (a) horizontal polar return; (b) vertical polar return: (c) cross-polar cancelled return. For each photograph, the horizontal axis represents range and the vertical axis represents azimuth.

In the particular type of spatial filtering of interest to us in this book, a number of independent *sensors* are placed at different points in space to "listen" to the received signal. In effect, the sensors provide a means of *sampling* the received signal *in space*. The set of sensor outputs collected at a particular instant of time constitutes a *snapshot*. Thus, a snapshot of data in spatial filtering (for the case when the sensors lie on a straight line) plays a role analogous to that of a set of consecutive tap inputs that exist in a transversal filter at a particular instant of time.[13]

In radar, the sensors consist of antenna elements (e.g., dipoles, horns, slotted waveguides) that respond to incident electromagnetic waves. In sonar, the sensors consist of hydrophones designed to respond to acoustic waves. In any event, spatial filtering, known as *beamforming*, is used in these systems to distinguish between the spatial properties of signal and noise. The device used to do the beamforming is called a *beamformer*. The term "beamformer" is derived from the fact that the early forms of antennas (spatial filters) were designed to form *pencil beams*, so as to receive a signal radiating from a specific direction and attenuate signals radiating from other directions of no interest (Van Veen and Buckley, 1988). Note that beamforming applies to the radiation (transmission) or reception of energy.

In a primitive type of spatial filtering, known as the *delay-and-sum beamformer*, the various sensor outputs are delayed (by appropriate amounts to align signal components coming from the direction of a target) and then summed, as in Fig. 1.34. Thus, for a single target, the average power at the output of the delay-and-sum beamformer is maximized when it is steered toward the target. A major limitation of the delay-and-sum beamformer, however, is that it has no provisions for dealing with sources of *interference*.

In order to enable a beamformer to respond to an unknown interference environment, it has to be made *adaptive* in such a way that it places *nulls* in the direc-

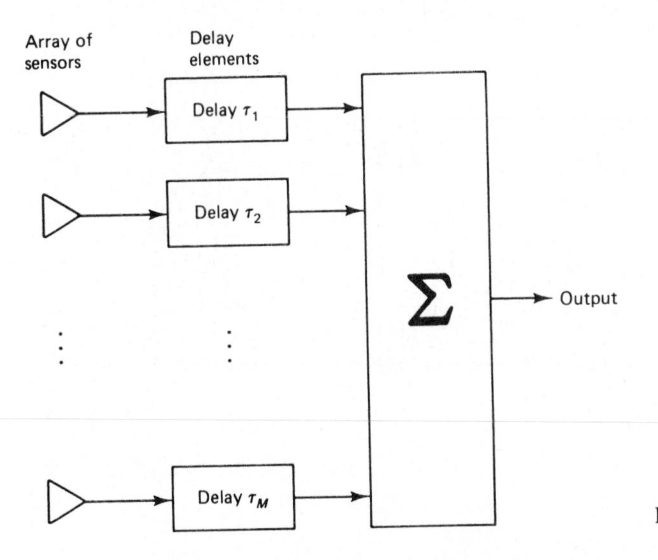

Figure 1.34 Delay-and-sum beamformer.

[13] For a discussion of the analogies between time- and space-domain forms of signal processing, see Bracewell (1978) and Van Veen and Buckley (1988).

tion(s) of the source(s) of interference automatically and in real time. By so doing, the output signal-to-noise ratio of the system is increased, and the *directional response* of the system is thereby improved. Below, we consider two examples of *adaptive beamformers* that are well suited for use with narrow-band signals in radar and sonar systems.

Adaptive Beamformer with Minimum-Variance Distortionless Response. Consider an adaptive beamformer that uses a linear array of M identical sensors, as in Fig. 1.35. The individual sensor outputs, assumed to be in *baseband* form, are weighted and then summed. The beamformer has to satisfy two requirements: (1) a *steering* capability whereby the target signal is always protected, and (2) the effects of sources of interference are minimized. One method of providing for these two requirements is to minimize the variance (i.e., average power) of the

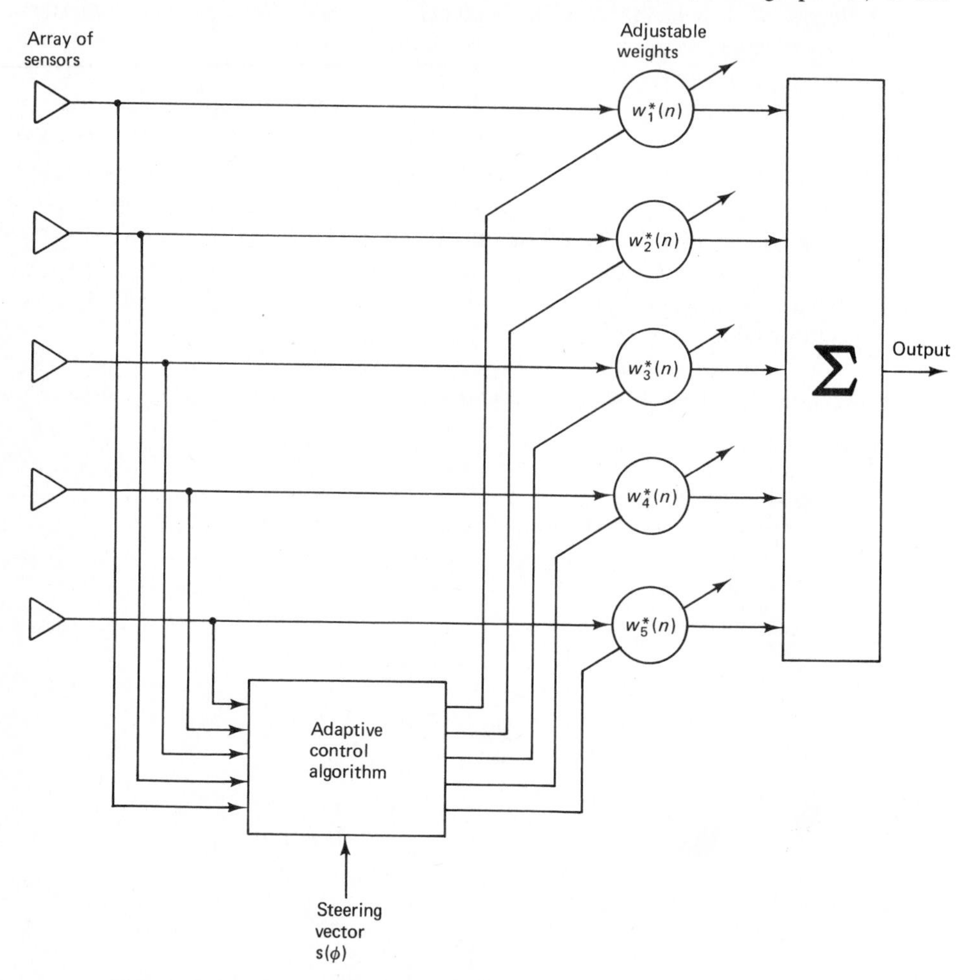

Figure 1.35 Adaptive MVDR beamformer for an array of five sensors. The sensor outputs (in baseband form) are complex valued; hence, the weights are complex valued.

beamformer output, subject to the *constraint* that, during the process of adaptation, the weights satisfy the condition:

$$\mathbf{w}^H(n)\mathbf{s}(\phi) = 1, \qquad \text{for all } n, \text{ and } \phi = \phi_t \tag{1.60}$$

where $\mathbf{w}(n)$ is the M-by-1 weight vector and $\mathbf{s}(\phi)$ is an M-by-1 *steering vector*. The superscript H denotes Hermitian transposition (i.e., transposition combined with complex conjugation). In this application, the baseband data are complex valued; hence, the need for complex conjugation. The value of *electrical angle* $\phi = \phi_t$ is determined by the direction of the target. The angle ϕ is itself measured with sensor 1 (at the top end of the array) treated as the point of reference.

The dependence of vector $\mathbf{s}(\phi)$ on the angle ϕ is defined by

$$\mathbf{s}^T(\phi) = [1, e^{-j\phi}, \ldots, e^{-j(M-1)\phi}]$$

The angle ϕ is itself related to incidence angle θ of a plane wave, measured with respect to the normal to the linear array, as follows[14]

$$\phi = \frac{2\pi d}{\lambda}\sin\theta \tag{1.61}$$

where d is the spacing between adjacent sensors of the array, and λ is the wavelength (see Fig. 1.36). The incidence angle θ lies inside the range $-\pi/2$, to $\pi/2$. The permissible values that the angle ϕ may assume lie inside the range $-\pi$ to π. This means that we must choose the spacing $d < \lambda/2$, so that there is a one-to-one correspondence between the values of θ and ϕ without ambiguity. The condition $d < \lambda/2$ may be viewed as the spatial analog of the sampling theorem.

The imposition of the *signal-protection constraint* in Eq. (1.60) ensures that, for a prescribed look direction, the response of the array is maintained constant (i.e., equal to 1), no matter what values are assigned to the weights. An algorithm that minimizes the variance of the beamformer output, subject to this constraint, is therefore referred to as the *minimum-variance distortionless response (MVDR) beamforming algorithm* (Capon, 1969; Owsley, 1985). The imposition of the constraint described in Eq. (1.60) reduces the number of "degrees of freedom" available to the MVDR algorithm to $M - 2$, where M is the number of sensors in the array. This means that the number of independent nulls produced by the MVDR algorithm (i.e., the number of independent interferences that can be cancelled) is $M - 2$.

The MVDR beamforming is a special case of *linearly constrained minimum variance (LCMV) beamforming*. In the latter case, we minimize the variance of the beamformer output, subject to the constraint

$$\mathbf{w}^H(n)\mathbf{s}(\phi) = g, \qquad \text{for all } n \text{ and } \phi = \phi_t \tag{1.62}$$

where g is a complex constant. The LCMV beamformer linearly constrains the weights, such that any signal coming from electrical angle ϕ_t is passed to the output with response (gain) g. Comparing the constraint of Eq. (1.60) with that of Eq.

[14] When a plane wave impinges on a linear array as in Fig. 1.35 there is a spatial delay of $d \sin\theta$ between the signals received at any pair of adjacent sensors. With a wavelength of λ, this spatial delay is translated into an electrical angular difference defined by $\phi = 2\pi(d \sin\theta/\lambda)$.

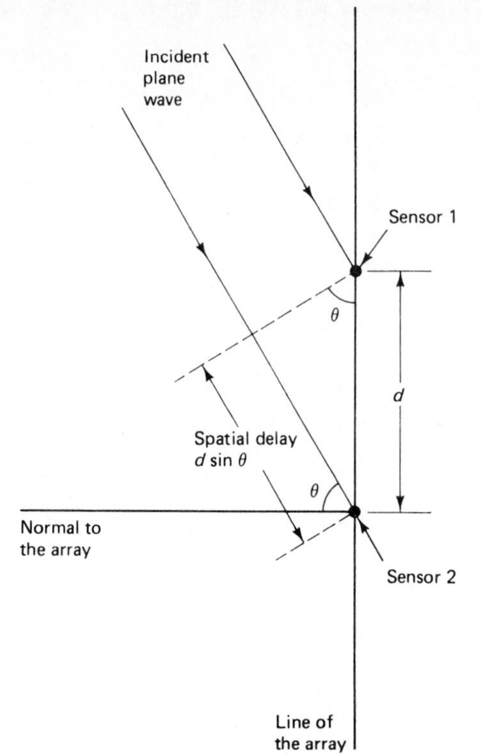

Figure 1.36 Spatial delay incurred when a plane wave impinges on a linear array.

(1.62), we see that the MVDR beamformer is indeed a special case of the LCMV beamformer for $g = 1$.

Adaptation in Beam Space. The MVDR beamformer performs adaptation directly in the *data space*. The adaptation process for interference cancellation may also be performed in *beam space*. To do so, the input data (received by the array of sensors) are transformed into the beam space by means of an *orthogonal multiple-beamforming network,* as illustrated in the block diagram of Fig. 1.37. The resulting output is processed by a *multiple sidelobe canceller* so as to cancel interference(s) from unknown directions.

The beamforming network is designed to generate a set of *orthogonal* beams. The multiple outputs of the beamforming network are referred to as *beam ports*. Assume that the sensor outputs are equally weighted and have a *uniform* phase. Under this condition, the response of the array produced by an incident plane wave arriving at the array along direction θ, measured with respect to the normal to the array, is given by

$$A(\phi, \alpha) = \sum_{n=-N}^{N} e^{jn\phi} e^{-jn\alpha} \tag{1.63}$$

where $M = (2N + 1)$ is the total number of sensors in the array, with the sensor at the mid–point of the array treated as the point of reference. The electrical angle ϕ is related to θ by Eq. (1.61), and α is a constant called the *uniform phase factor*. The

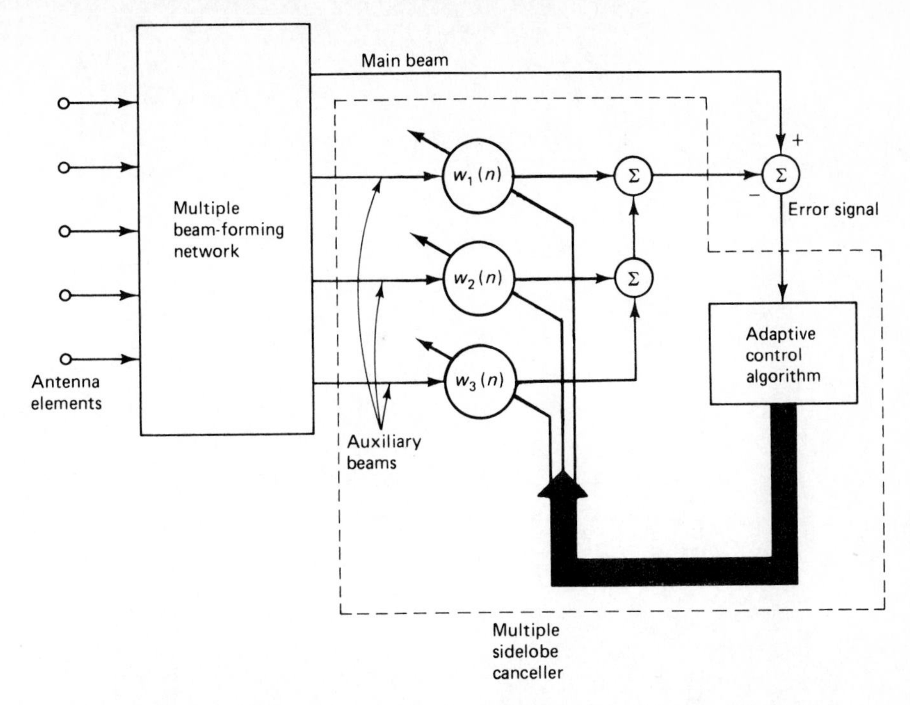

Figure 1.37 Block diagram of adaptive combiner with fixed beams; owing to the symmetric nature of the multiple beamforming network, final values of the weights are real valued.

quantity $A(\phi, \alpha)$ is called the *array pattern*. For $d = \lambda/2$, we find from Eq. (1.61) that

$$\phi = \pi \sin\theta$$

Summing the geometric series in Eq. (1.63), we may express the array pattern as

$$A(\phi, \alpha) = \frac{\sin[\frac{1}{2}(2N + 1)(\phi - \alpha)]}{\sin[\frac{1}{2}(\phi - \alpha)]} \tag{1.64}$$

By assigning different values to α, the main beam of the antenna is thus scanned across the range $-\pi \leq \phi \leq \pi$. To generate an orthogonal set of beams, equal to $2N$ in number, we assign the following discrete values to the uniform phase factor

$$\alpha = \frac{\pi}{2N + 1}k, \qquad k = \pm 1, \pm 3, \ldots, \pm 2N - 1 \tag{1.65}$$

Figure 1.38 illustrates the variations of the magnitude of the array pattern $A(\phi, \alpha)$ with ϕ for the case of $2N + 1 = 5$ elements and $\alpha = \pm\pi/5, \pm 3\pi/5$. Note that owing to the symmetric nature of the beamformer, the final values of the weights are real valued.

The orthogonal beams generated by the beamforming network represent $2N$ independent *look directions,* one per beam. Depending on the target direction of interest, a particular beam in the set is identified as the *main beam* and the remainder are viewed as *auxiliary beams*. We note from Fig. 1.38 that each of the auxiliary

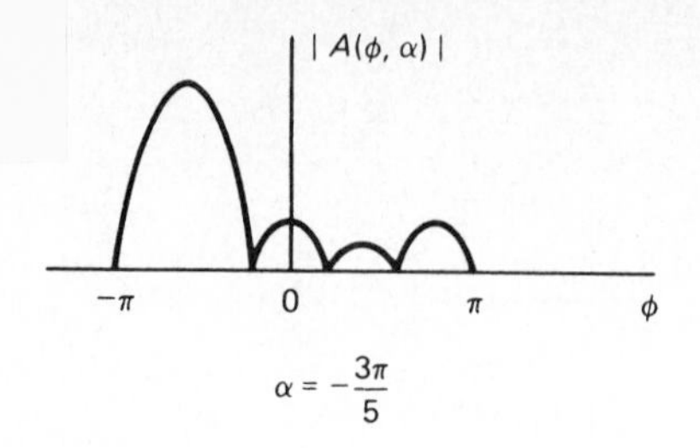

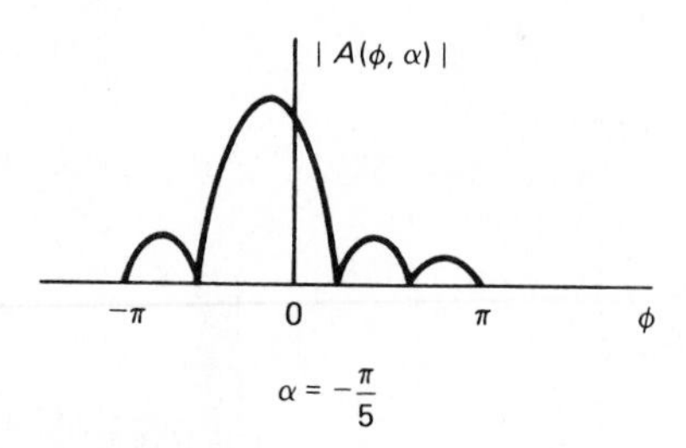

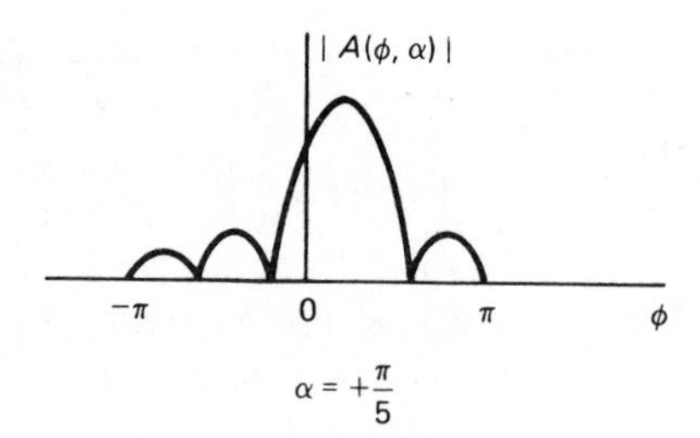

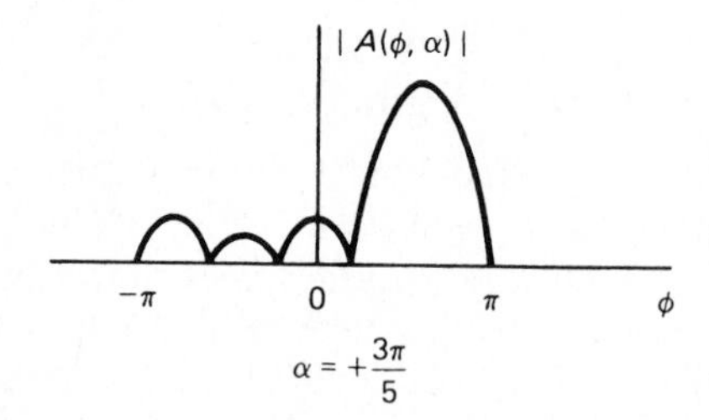

Figure 1.38 Variations of the magnitude of the array pattern $A(\phi, \alpha)$ with ϕ and α.

beams has a *null in the look direction of the main beam*. The auxiliary beams are adaptively weighted by the multiple sidelobe canceller so as to form a cancellation beam that is subtracted from the main beam. The resulting estimation error is fed back to the multiple sidelobe canceller so as to control the corrections applied to its adjustable weights.

Since all the auxiliary beams have nulls in the look direction of the main beam, and the main beam is excluded from the multiple sidelobe canceller, the overall output of the adaptive beamformer is constrained to have a constant response in the look direction of the main beam (i.e., along the direction of the target). Moreover, with $(2N - 1)$ degrees of freedom (i.e., the number of available auxiliary beams) the system is capable of placing up to $(2N - 1)$ nulls along the (unknown) directions of independent interferences.

Note that with an array of $(2N + 1)$ sensors, we may produce a beamforming network with $(2N + 1)$ orthogonal beam ports by assigning the uniform phase factor the following set of values:

$$\alpha = \frac{k\pi}{2N + 1}, \qquad k = 0, \pm 2, \ldots, \pm 2N \tag{1.66}$$

In this case, a small fraction of the main lobe of the beam port at either end lies in the nonvisible region. Nevertheless, with one of the beam ports providing the main beam and the remaining $2N$ ports providing the auxiliary beams, the adaptive beamformer is now capable of producing up to $2N$ independent nulls.

1.8 SOME HISTORICAL NOTES

To understand a science it is necessary to know its history.
Auguste Comte

We complete this introductory chapter by presenting a brief historical review of developments in three areas that are closely related insofar as the subject matter of this book is concerned. The areas are linear estimation theory, adaptive filtering algorithms, and signal-processing applications of adaptive filters. It should, however, be emphasized that our interest in adaptive filters as presented in this book is primarily restricted to those with two characteristics: an impulse response that has a *finite* duration, and an estimate of some desired response that is a *linear* combination of input data samples.[15]

Linear Estimation Theory[16]

The earliest stimulus for the development of estimation theory was apparently provided by astonomical studies in which the motion of planets and comets was studied using telescopic measurement data. The beginnings of a "theory" of estimation in which attempts are made to minimize various functions of errors can be attributed to Galileo Galilei in 1632. However, the origin of linear estimation theory is credited to Gauss who, at the age of 18 in 1795, invented the *method of least squares* to study the motion of heavenly bodies (Gauss, 1809). Nevertheless, in the early nineteenth century, there was considerable controversy regarding the actual inventor of the method of least squares. The controversy arose because Gauss did not publish his discovery in 1795. Rather, it was first published by Legendre in 1805, who independently invented the method (Legendre, 1810).

[15] In Chapter 21 we present a summary of pertinent topics not covered in the book. In Chapter 20, we study blind deconvolution that is an example of nonlinear adaptive filtering.

[16] The notes presented on linear estimation are influenced by the following review papers: Sorenson (1970), Kailath (1974), and Makhoul (1975).

The first studies of minimum mean-square estimation in stochastic processes were made by Kolmogorov, Krein, and Wiener during the late 1930s and early 1940s (Kolmogorov, 1939; Krein, 1945; Wiener, 1949). The works of Kolmogorov and Krein were independent of Wiener's, and while there was some overlap in the results, their aims were rather different. There were many conceptual differences (as one would expect after 140 years) between Gauss's problem and the problem treated by Kolmogorov, Krein, and Wiener.

Kolmogorov, inspired by some early work of Wold on discrete-time stationary processes (Wold, 1938), developed a comprehensive treatment of the linear prediction problem for discrete-time stochastic processes. Krein noted the relationship of Kolmogorov's results to some early work by Szegö on orthogonal polynomials (Szegö, 1939; Grenander and Szegö, 1958) and extended the results to continuous time by clever use of a bilinear transformation.

Wiener, independently, formulated the continuous-time linear prediction problem and derived an explicit formula for the optimum predictor. Wiener also considered the "filtering" problem of estimating a process corrupted by an additive "noise" process. The explicit formula for the optimum estimate required the solution of an integral equation known as the *Wiener–Hopf equation* (Wiener and Hopf, 1931).

In 1947, Levinson formulated the Wiener filtering problem in discrete time. In the case of discrete-time signals, the Wiener–Hopf equation takes on a matrix form described by

$$\mathbf{R}\mathbf{w}_o = \mathbf{p} \tag{1.67}$$

where $\mathbf{w}_o$ is the tap-weight vector of the optimum Wiener filter structured in the form of a transversal filter, $\mathbf{R}$ is the correlation matrix of the tap inputs, and $\mathbf{p}$ is the cross-correlation vector between the top inputs and the desired response. For stationary inputs, the correlation matrix $\mathbf{R}$ assumes a special structure known as *Toeplitz,* so named after the mathematician O. Toeplitz. By exploiting the properties of a Toeplitz matrix, Levinson derived an elegant recursive procedure for solving the matrix form of the Wiener-Hopf equation (Levinson, 1947). In 1960, Durbin rediscovered Levinson's recursive procedure as a scheme for recursive fitting of autoregressive models to scalar time-series data (Durbin, 1960). The problem considered by Durbin is a special case of Eq. (1.67) in that the column vector $\mathbf{p}$ comprises the same elements found in the correlation matrix $\mathbf{R}$. In 1963, Whittle showed there is a close relationship between the Levinson–Durbin recursion and that for Szegö's orthogonal polynomials and also derived a multivariate generalization of the Levinson–Durbin recursion (Whittle, 1963).

Wiener and Kolmogorov assumed an infinite amount of data and assumed the stochastic processes to be stationary. During the 1950s, some generalizations of the Wiener–Kolmogorov filter theory were made by various authors to cover the estimation of stationary processes given only for a finite observation interval and to cover the estimation of nonstationary processes. However, there were dissatisfactions with the most significant of the results of this period because they were rather complicated, difficult to update with increases in the observations interval, and difficult to modify for the vector case. These last two difficulties became particularly evident in the late 1950s in the problem of determining satellite orbits. In this application, there were generally vector observations of some combinations of position and velocity, and also there were large amounts of data sequentially accumulated with each pass of the satellite over a tracking station. Swerling was one of the first to tackle this problem by presenting some useful recursive algorithms (Swerling, 1958). For different reasons, Kalman independently developed a somewhat more restricted algorithm than Swerling's, but it was an algorithm that seemed particularly matched to the dynamical estimation problems that were brought by the advent of the space age (Kalman, 1960). After Kalman had published his paper and it had attained considerable fame, Swerling wrote a letter claiming priority for the Kalman filter equations (Swerling, 1963). However, history shows that Swerling's plea has fallen on deaf ears. It is ironic that orbit determination problems provided the stimulus for both Gauss's method of least squares and the Kalman filter and that there were squabbles concerning their inventors. Kalman's original formulation of the linear filtering problem was derived for discrete-time processes. The continuous-time filter was derived by Kalman in his subsequent collaboration with Bucy; this latter solution is sometimes referred to as the *Kalman–Bucy filter* (Kalman and Bucy, 1961).

In a series of stimulating papers, Kailath reformulated the solution to the linear filtering problem by using the *innovations* approach (Kailath, 1968, 1970; Kailath and Frost, 1968; Kailath and Geesey, 1973). In this approach, a stochastic process $\{u(n)\}$ is represented as the output of a causal and causally invertible filter driven by a white-noise process $\{v(n)\}$. The requirement that the filter be causally invertible ensures that the white-noise process $\{v(n)\}$, termed the *innovations process,* is probabilistically equivalent to the original process $\{u(n)\}$. This probabilistic equivalence is the reason for calling the white-noise process $\{v(n)\}$ the "innovations process." Innovation denotes "newness," and this quality is represented by the whiteness of the process $\{v(n)\}$ where any redundant information in the form of correlation in the process $\{u(n)\}$ has been removed. Hence only new information is retained in the innovations process $\{v(n)\}$. According to Kailath, the name "innovations process" was apparently first used by Wiener and Masani in the mid-1950s (Kailath, 1974).

Adaptive Filtering Algorithms

The earliest work on adaptive filters may be traced back to the late 1950s, during which time a number of researchers were working independently on different applications of adaptive filters. From this early work, the *least-mean-square (LMS) algorithm* emerged as a simple algorithm for the operation of adaptive transversal filters. The LMS algorithm was devised by Widrow and Hoff in 1959 in their study of a pattern recognition scheme known as the adaptive linear threshold logic element (Widrow and Hoff, 1960; Widrow, 1970). The LMS algorithm is a stochastic gradient algorithm in that it iterates each tap weight in the transversal filter in the direction of the gradient of the squared amplitude of an error signal with respect to that tap weight. As such, the LMS algorithm is closely related to the concept of *stochastic approximation* developed by Robbins and Monroe in statistics for solving certain sequential parameter estimation problems (Robbins and Monroe, 1951). The primary difference between them is that the LMS algorithm uses a fixed step-size parameter to control the correction applied to the tap weight from one iteration to the next, whereas in stochastic approximation methods the step-size parameter is made inversely proportional to time n or to a power of n. Another algorithm, closely related to the LMS algorithm, is the *gradient adaptive lattice* (GAL) algorithm (Griffiths, 1977, 1978); the difference between them is structural in that the GAL algorithm is lattice-based, whereas the LMS algorithm uses a transversal filter.

Another major contribution to the development of adaptive filtering algorithms was made by Godard in 1974. He used Kalman filter theory to propose a new class of adaptive filtering algorithms for obtaining rapid convergence of the tap weights of a transversal filter to their optimum settings (Godard, 1974). Although, prior to this date, several investigators had applied Kalman filter theory to solve the adaptive filtering problem, Godard's approach is widely accepted as the most successful. This algorithm is referred to in the literature as the *Kalman algorithm* or *Godard algorithm*; we will use the former terminology.

The Kalman algorithm is closely related to the recursive least-squares (RLS) algorithm that follows from the method of least squares. The RLS algorithm has been derived independently by several investigators. However, the original reference on the RLS algorithm appears to be Plackett (1950).

As mentioned in Section 1.4, the Kalman or RLS algorithm usually provides a much faster rate of convergence than the LMS algorithm at the expense of increased computational complexity. The desire to reduce computational complexity to a level comparable to that of the simple LMS algorithm prompted the search for computationally efficient RLS algorithms. Various forms of such algorithms have been introduced in the literature. In particular, mention should be made of three classes of computationally efficient RLS algorithms: One involves the use of transversal filters (Falconer and Ljung, 1978; Carayannis et al., 1983; Cioffi and Kailath, 1984); another involves the use of lattice predictors (Morf et al., 1977; Lee et al., 1982; Ling and Proakis, 1984; Ling et al., 1986); the third one is mentioned in the next paragraph. The contributions made by Ling and Proakis (1984) involve the use of *error feedback* that has the important effect of stabilizing the lattice predictor in the presence of quantization errors. In one form or another, the development of these fast algorithms can be traced back to results that were derived by Morf in 1974 for solving the deterministic counterpart of the stochastic problem solved by Levinson for stationary inputs (Morf, 1974).

In 1981, Gentleman and Kung introduced yet another highly efficient algorithm for solving the linear least-squares problem by using an iterative open-loop two-stage process, a procedure well suited for implementation using systolic arrays (Gentleman and Kung, 1981). This algorithm, called the *QR-decomposition-based-recursive least-squares algorithm,* involves orthogonal triangularization of the input data matrix through direct application of a special form of QR-decomponsition known as *Givens rotations* (Givens, 1958). In 1983, McWhirter described a modification of this algorithm by avoiding computation of the least-squares weight vector (McWhirter, 1983). For applications that do not require explicit computation of the least-squares weight vector, the modification made by McWhirter offers an adaptive filtering algorithm that is attractive. The McWhirter structure consists solely of a triangular systolic array. Using the *modified Gram–Schmidt orthogonalization algorithm* (Björck, 1967), a similar triangular structure for recursively solving the linear least-squares problem has been derived by Ling et al. (1986) and Manolakis et al. (1987). The conventional forms of QR-decomposition-based or modified

Gram–Schmidt orthogonalized-based RLS algorithms are computationally intensive. Cioffi (1988, 1990), and Proudler et al. (1989) developed fast realizations of the QR-decomposition-based RLS algorithm. A similar fast realization, using the modified Gram–Schmidt orthogonalization, has been derived by Ling (1989).

Adaptive Signal-Processing Applications

Adaptive Equalization. Until the early 1960s, the equalization of telephone channels to combat the degrading effects of intersymbol interference on data transmission was performed by using either fixed equalizers (resulting in a performance loss) or equalizers whose parameters were adjusted manually (a rather cumbersome procedure). In 1965, Lucky made a major breakthrough in the equalization problem by proposing a *zero-forcing algorithm* for automatically adjusting the tap weights of a transversal equalizer (Lucky, 1965). A distinguishing feature of the work by Lucky was the use of a *mini-max* type of performance criterion. In particular, he used a performance index called *peak distortion,* which is directly related to the maximum value of intersymbol interference that can occur. The tap weights in the equalizer are adjusted to minimize the peak distortion. This has the effect of *forcing* the intersymbol interference due to those adjacent pulses that are contained in the transversal equalizer to become *zero*; hence, the name of the algorithm. A sufficient, but not necessary, condition for optimality of the zero-forcing algorithm is that the *initial distortion* (the distortion that exists at the equalizer input) be less than unity. In a subsequent paper published in 1966, Lucky extended the use of the zero-forcing algorithm to the tracking mode of operation. In 1965, DiToro independently used adaptive equalization for combatting the effect of intersymbol interference on data transmitted over high-frequency links.

The pioneering work by Lucky inspired many other significant contributions to different aspects of the adaptive equalization problem in one way or another. Gersho, and Proakis and Miller, independently reformulated the adaptive equalization problem using a mean-square-error criterion (Gersho, 1969; Proakis and Miller, 1969). In 1972, Ungerboeck presented a detailed mathematical analysis of the convergence properties of an adaptive transversal equalizer using the LMS algorithm. In 1974, Godard used Kalman filter theory to derive a powerful algorithm for adjusting the tap weights of a transversal equalizer. In 1978, Falconer and Ljung presented a modification of this algorithm that simplified its computational complexity to a level comparable to that of the simple LMS algorithm. Satorius and Alexander in 1979 and Satorius and Pack in 1981 demonstrated the usefulness of lattice-based algorithms for adaptive equalization of dispersive channels.

This brief historical review pertains to the use of adaptive equalizers for *linear synchronous receivers*; by "synchronous" we mean that the equalizer in the receiver has its taps spaced at the reciprocal of the symbol rate. Even though our interest in adaptive equalizers is restricted to this class of receivers, nevertheless, such a historical review would be incomplete without some mention of fractionally spaced equalizers and decision-feedback equalizers.

In a *fractionally spaced equalizer* (FSE), the equalizer taps are spaced closer than the reciprocal of the symbol rate. An FSE has the capability of compensating for delay distortion much more effectively than a conventional synchronous equalizer. Another advantage of the FSE is the fact that data transmission may begin with an arbitrary sampling phase. However, mathematical analysis of the FSE is much more complicated than for a conventional synchronous equalizer. It appears that early work on the FSE was initiated by Brady (1970). Other contributions to the subject include subsequent work by Ungerboeck (1976) and Gitlin and Weinstein (1981).

A *decision-feedback equalizer* consists of a feedforward section and a feedback section connected as shown in Fig. 1.39. The feedforward section itself consists of a transversal filter whose taps are spaced at the reciprocal of the symbol rate. The data sequence to be equalized is applied to the input of this section. The feedback section consists of another transversal filter whose taps are also spaced at the reciprocal of the symbol rate. The input applied to the feedback section is made up of decisions on previously detected symbols. The function of the feedback section is to subtract out that portion of intersymbol interference produced by previously detected symbols from the estimates of future symbols. This cancellation is an old idea known as the *bootstrap technique*. A decision-feedback equalizer yields good performance in the presence of severe intersymbol interference as experienced in fading radio channels, for example. The first report on decision-feedback equalization was published by Austin (1967), and the optimization of the decision-feedback receiver for minimum mean-squared error was first accomplished by Monsen (1971).

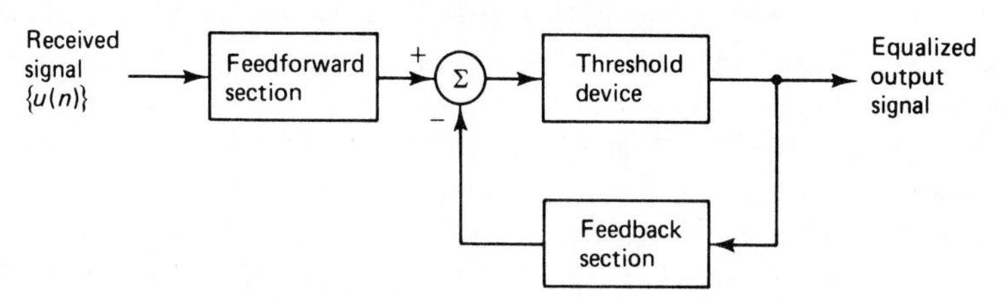

Figure 1.39 Block diagram of decision feedback equalizer.

Coding of Speech. In 1966, Saito and Itakura used a *maximum likelihood*[17] approach for the application of prediction to speech. A standard assumption in the application of the maximum likelihood principle is that the input process is Gaussian. Under this condition, the exact application of the maximum likelihood principle yields a set of nonlinear equations for the parameters of the predictor. To overcome this difficulty, Itakura and Saito utilized approximations based on the assumption that the number of available data points greatly exceeds the prediction order. The use of this assumption makes the result obtained from the maximum likelihood principle assume an approximate form that is the same as the *autocorrelation method*[18] of linear prediction. The application of the maximum likelihood principle is justified on the assumption that speech is a stationary Gaussian process, which seems reasonable in the case of unvoiced sounds.

In 1970, Atal presented the first use of the term "linear prediction" for speech analysis. Details of this new approach, linear predictive coding (LPC), to speech analysis and synthesis were published by Atal and Hanauer in 1971, in which the

[17] For a discussion of the maximum likelihood principle, see Appendix D.

[18] The autocorrelation method of linear prediction is considered in Chapter 10.

speech wave form is represented directly in terms of time-varying parameters related to the transfer function of the vocal tract and the characteristics of the excitation. The predictor coefficients are determined by minimizing the mean-squared error, with the error defined as the difference between the actual and predicted values of the speech samples. In the work by Atal and Hanauer, the speech wave was sampled at 10 kHz and then analyzed by predicting the present speech sample as a linear combination of the 12 previous samples. Thus 15 parameters [the 12 parameters of the predictor, the pitch period, a binary parameter indicating whether the speech is voiced or unvoiced, and the root-mean-square (rms) value of the speech samples] were used to describe the speech analyzer. For the speech synthesizer, an all-pole filter was used, with a sequence of quasi-periodic pulses or a white-noise source providing the excitation.

Another significant contribution to the linear prediction of speech was made in 1972 by Itakura and Saito; they used partial correlation techniques to develop a new structure, the lattice, for formulating the linear prediction problem[19]. The parameters that characterize the lattice predictor are called *reflection coefficients* or *partial correlation (PARCOR) coefficients*. Although by that time the essence of the lattice structure had been considered by several other investigators, the invention of the lattice predicitor is credited to Saito and Itakura. In 1973, Wakita showed that the filtering actions of the lattice predictor model and an acoustic tube model of speech are identical, with the reflection coefficients in the acoustic tube model as common factors. This discovery made possible the extraction of the reflection coefficients by the use of a lattice predictor.

Early designs of a lattice predictor were based on a *block processing* approach (Burg, 1967). In 1981, Makhoul and Cossell used an *adaptive* approach for designing the lattice predictor for applications in speech analysis and synthesis. They showed that the convergence of the adaptive lattice predictor is fast enough for its performance to equal that of the optimal (but more expensive) adaptive autocorrelation method.

This historical review on speech coding relates to LPC vocoders. We next present a historical review of the adaptive predictive coding of speech, starting with ordinary pulse-code modulation (PCM).

PCM was invented in 1937 by Reeves (1975). This was followed by the invention of differential pulse-code modulation (DPCM) by Cutler (1952). The early use of DPCM for the predictive coding of speech signals was limited to linear predictors with *fixed* parameters (McDonald, 1966). However, due to the nonstationary nature of speech signals, a fixed predictor cannot predict the signal values efficiently at all times. In order to respond to the nonstationary characteristics of speech signals, the predictor has to be adaptive (Atal and Schroeder, 1967). In 1970, Atal and Schroeder described a sophisticated scheme for adaptive predictive coding of speech. The scheme recognizes that there are two main causes of redundancy in speech (Schroeder, 1966): (1) quasi-periodicity during voiced segments, and (2) lack of flatness of the short-time spectral envelope. Thus, the predictor is designed to remove signal redundancy in two stages. The first stage of the predictor removes the quasi-periodic nature of the signal. The second stage removes formant information

[19] According to Markel and Gray (1976), the work of Itakura and Saito in Japan on the PARCOR formulation of linear prediction had been presented in 1969.

from the spectral envelope. The scheme achieves dramatic reductions in bit rate at the expense of a significant increase in circuit complexity. Atal and Schroeder (1970) report that the scheme can transmit speech at 10 kb/s, which is several times less than the bit rate required for logarithmic-PCM encoding with comparable speech quality.

PCM and ADPCM are both time-domain coders in that the speech signal is processed in the time domain as a single full-band signal. In *adaptive sub-band coding* (ASBC), on the other hand, the number of bits used to encode each sub-band of the speech signal is varied dynamically and shared with other sub-bands. The result is that the encoding accuracy is always placed where it is needed in the frequency-domain characterization of the speech signal (Jayant, 1986; Jayant and Noll, 1984). Indeed, sub-bands of the speech signal with little or no energy may not be encoded at all. Subjective measurements show that the best 16 kb/s adaptive sub-band coders approach the higher bit rate (i.e., 64 kb/s) PCM coders in quality (Jayant, 1986). However, the complexity of the adaptive sub-band coder is typically 100 times that of the PCM coder for about the same reproduction quality.

Spectrum Analysis. At the turn of the twentieth century, Schuster introduced the *periodogram* for analyzing the power spectrum[20] of a time series (Schuster, 1898). The periodogram is defined as the squared amplitude of the discrete Fourier transform of the time series. The periodogram was originally used by Schuster to detect and estimate the amplitude of a sine wave of known frequency that is buried in noise. Until the work of Yule in 1927, the periodogram was the only numerical method available for spectrum analysis. However, the periodogram suffers from the limitation that when it is applied to empirical time series observed in nature the results obtained are very erratic. This led Yule to introduce a new approach based on the concept of a *finite parameter model* for a stationary stochastic process in his investigation of the periodicities in time series with special reference to Wolfer's sunspot number (Yule, 1927). Yule, in effect, created a stochastic feedback model in which the present sample value of the time series is assumed to consist of a linear combination of past sample values plus an error term. This model is called an *autoregressive model* in that a sample of the time series regresses on its own past values, and the method of spectrum analysis based on such a model is accordingly called autoregressive spectrum analysis. The name "autoregressive" was coined by Wold in his 1938 thesis (Wold, 1938).

Interest in the autoregressive method was reinitiated by Burg (1967, 1975). Burg introduced the term *maximum-entropy method* to describe an algorithmic approach for estimating the power spectrum directly from the available time series. The idea behind the maximum-entropy method is to extrapolate the autocorrelation function of the time series in such a way that the *entropy* of the corresponding probability density function is maximized at each step of the extrapolation. In 1971, Van den Bos showed that the maximum-entropy method is equivalent to least-squares fitting of an autoregressive model to the known autocorrelation sequence.

[20] For a fascinating historical account of the concept of power spectrum, its origin and its estimation, see Robinson (1982).

Another related problem that has attracted a great deal of attention in the literature is the estimation of the frequencies of superimposed sine waves that are corrupted by additive white noise.[21] The origin of the problem can be traced back to 1795, the year Gaspard Riche, Baron de Prony, published his work on the fitting of superimposed exponentials to (noiseless) data (Prony, 1795). A useful solution to the problem for noisy data was developed independently by Ulrych and Clayton and Nuttal, using the method of forward-backward linear prediction (formulated in terms of least squares) to evaluate the autoregressive parameters (Ulrych and Clayton, 1976; Nuttall, 1976). To reduce the effects of noise, we may modify the method of least squares by manipulating the eigenvalue decomposition of the pertinent correlation matrix. The minimum-norm method (Kumaresan and Tufts, 1983) and the MUSIC algorithm (Schmidt, 1979, 1981) exploit such an approach in different ways.

Another important contribution made to the literature on spectrum analysis is that due to Thomson (1982). His *method of multiple windows,* based on the prolate spheroidal wave functions, represents a nonparametric method for spectrum estimation.

Adaptive Noise Cancellation Schemes. The initial work on adaptive echo cancellers started around 1965. It appears that Kelly of Bell Telephone Laboratories was the first to propose the use of an adaptive filter for echo cancellation, with the speech signal itself utilized in performing the adaptation; Kelly's contribution is recognized in the paper by Sondhi (1967). This invention and its refinement are described in the patents by Kelly and Logan (1970) and Sondhi (1970).

The adaptive line enhancer was originated by Widrow and his co-workers at Stanford University. An early version of this device was built in 1965 to cancel 60-Hz interference at the output of an electrocardiographic amplifier and recorder. This work is described in the paper by Widrow et al. (1975b). The adaptive line enhancer and its application as an adaptive detector are patented by McCool et al. (1980).

The adaptive echo canceller and the adaptive line enhancer, although intended for different applications, may be viewed as examples of the *adaptive noise canceller* discussed by Widrow et al. (1975). This scheme operates on the outputs of two sensors: a *primary sensor* that supplies a desired signal of interest buried in noise, and a *reference sensor* that supplies noise alone, as illustrated in Fig. 1.26. It is assumed that (1) the signal and noise at the output of the primary sensor are uncorrelated, and (2) the noise at the output of the reference sensor is correlated with the noise component of the primary sensor output.

The adaptive noise canceller consists of an adaptive filter that operates on the reference sensor output to produce an *estimate* of the noise, which is subtracted from the primary sensor output. The overall output of the canceller is used to control the adjustments applied to the tap weights in the adaptive filter. The adaptive canceller

[21] For a detailed historical account of the superimposed signals problem, see Wax (1985). In this dissertation, an optimal solution to the joint estimation of the (unknown) number of signals and their parameters is derived. The signals of interest represent sine waves of unknown frequencies in harmonic analysis or plane waves of unknown directions in its spatial counterpart.

tends to minimize the mean-square value of the overall output, thereby causing the output to be the best estimate of the desired signal in the minimum-mean-square sense.

Adaptive Beamforming. The development of adaptive beamforming technology may be traced back to the invention of the *intermediate frequency (IF) sidelobe canceller* by Howells in the late 1950s. In a paper published in the 1976 Special Issue of the IEEE Transactions on Antennas and Propagation, Howells describes his personal observations on early work on adaptive antennas at the General Electric and Syracuse University Research Corporation (Howells, 1976). According to this historic report, Howells had developed by mid-1957 a sidelobe canceller capable of automatically nulling out the effect of one jammer. The sidelobe canceller uses a *primary* (high-gain) antenna and a *reference omni-directional* (low-gain) antenna to form a two-element array with one degree of freedom that makes it possible to steer a deep null anywhere in the sidelobe region of the combined antenna pattern. In particular, a null is placed in the direction of the jammer, with only a minor perturbation of the main lobe. Subsequently, Howells patented the sidelobe canceller (Howells, 1965).

The second major contribution to adaptive array antennas was made by Applebaum in 1966. In a classic report, he derived the *control law* governing the operation of an adaptive array antenna, with a control loop for each element of the array (Applebaum, 1966). The algorithm derived by Applebaum was based on maximizing the signal-to-noise ratio (SNR) at the array antenna output for any type of noise environment. Applebaum's theory included the sidelobe canceller as a special case. His 1966 classic report was reprinted in the 1976 Special Issue of IEEE Transactions on Antennas and Propagation.

Another algorithm for the weight adjustment in adaptive array antennas was advanced independently in 1967 by Widrow and his co-workers at Stanford University. They based their theory on the simple and yet effective LMS algorithm (Widrow et al., 1967). The LMS algorithm itself had been invented by Widrow and Hoff in 1960. The 1967 paper by Widrow et al. was not only the first publication in the open literature on adaptive array antenna systems, but also it is considered to be another classic of that era.

It is noteworthy that the maximum SNR algorithm (used by Applebaum) and the LMS algorithm (used by Widrow and his co-workers) for adaptive array antennas are rather similar. Both algorithms derive the control law for adaptive adjustment of the weights in the array antenna by sensing the correlation between element signals. Indeed, they both converge toward the optimum Wiener solution for stationary inputs (Gabriel, 1976).

A different method for solving the adaptive beamforming problem was proposed by Capon (1969). Capon realized that the poor performance of the delay-and-sum beamformer is due to the fact that its response along a direction of interest depends not only on the power of the incoming target signal but also undesirable contributions received from other sources of interference. To overcome this limitation of the delay-and-sum beamformer, Capon proposed a new beamformer in which the weight vector $\mathbf{w}(n)$ is chosen so as to *minimize the variance* (i.e., average power) of the beamformer output, subject to the constraint $\mathbf{w}^H(n)\mathbf{s}(\phi) = 1$ for all n, where $\mathbf{s}(\phi)$ is a prescribed *steering vector*. This constrained minimization yields an adaptive beamformer with *minimum-variance distortionless response* (MVDR).

McWhirter has proposed the use of systolic arrays for the design of adaptive beamformers (McWhirter, 1983; Ward et al., 1986). Systolic array implementations of the adaptive MVDR beamformer have also been proposed by Schreiber (1986) and McWhirter and Shepherd (1989). These structures are based on modifications of original ideas due to Kung and Gentleman for the use of systolic arrays to solve the linear least-squares problem (Kung and Leiserson, 1978; Gentleman and Kung, 1981).

Part I

Discrete-Time Wide-Sense Stationary Stochastic Processes

Part I consists of Chapters 2 through 4. In this part of the book we study the characteristics of discrete-time stochastic processes that are wide-sense stationary, and thereby lay a foundation for the rest of the book. In particular, Chapter 2 covers time-domain characteristics of such processes, and the development of stochastic models for their mathematical description. Chapter 3 covers frequency-domain characteristics with particular emphasis on the power spectrum of a discrete-time wide-sense stationary process. The material covered on power spectrum is very pertinent, considering that spectrum estimation is used throughout the book, in examples and computer problems, to illustrate adaptive and optimum filtering algorithms. In Chapter 4 we study the eigenvalue problem, which is central to a detailed mathematical description of discrete-time wide-sense stationary processes.

2

Stationary Processes and Models

The term *stochastic process* or *random process* is used to describe the time evolution of a statistical phenomenon according to probabilistic laws. The time evolution of the phenomenon means that the stochastic process is a function of time, defined on some observation interval. The statistical nature of the phenomenon means that, before conducting an experiment, it is not possible to define exactly the way it evolves in time. Examples of a stochastic process include speech signals, television signals, radar signals, digital computer data, the output of a communication channel, seismological data, and noise.

The form of a stochastic process that is of interest to us is one that is defined at *discrete and uniformly spaced instants of time* (Box and Jenkins, 1976; Priestley, 1981). Such a restriction may arise naturally in practice, as in the case of radar signals or digital computer data. Alternatively, the stochastic process may be defined originally for a continuous range of real values of time; however, before processing, it is *sampled uniformly* in time, with the sampling rate chosen to be greater than twice the highest frequency component of the process (Haykin, 1988).

A stochastic process is *not* just a single function of time; rather, it represents, in theory, an infinite number of *different* realizations of the process. One particular realization of a discrete-time stochastic process is called a *discrete-time series* or simply *time series*. For convenience of notation, we will *normalize time with respect to the sampling period*. For example, the sequence $u(n), u(n-1), \ldots, u(n-M)$ represents a time series that consists of the *present* observation $u(n)$ made at time n and M past observations of the process made at times $n-1, \ldots, n-M$.

We say that a stochastic process is *strictly stationary* if its statistical properties are *invariant* to a shift of time. Specifically, for a discrete-time stochastic process represented by the time series $u(n), u(n-1), \ldots, u(n-M)$ to be strictly stationary, the *joint probability density function* of these observations made at times n, $n-1, \ldots, n-M$ must remain the same no matter what values we assign to n for fixed M.

2.1 PARTIAL CHARACTERIZATION OF A DISCRETE-TIME STOCHASTIC PROCESS

In practice, we usually find that it is not possible to determine (by means of suitable measurements) the joint probability density function for an arbitrary set of observations made on a stochastic process. Accordingly, we must content ourselves with a *partial* characterization of the process by specifying its first and second moments.

Consider a discrete-time stochastic process represented by the time series $u(n), u(n-1), \ldots, u(n-M)$, which may be complex valued. We define the *mean-value function* of the process as

$$\mu(n) = E[u(n)] \tag{2.1}$$

where E denotes the *statistical expectation operator*. We define the *autocorrelation function* of the process as

$$r(n, n-k) = E[u(n)u^*(n-k)], \qquad k = 0, \pm 1, \pm 2, \ldots, \tag{2.2}$$

where the asterisk denotes *complex conjugation*. We define the *autocovariance function* of the process as

$$c(n, n-k) = E[(u(n) - \mu(n))(u(n-k) - \mu(n-k))^*], \qquad k = 0, \pm 1, \pm 2, \ldots, \tag{2.3}$$

From Eqs. (2.1) to (2.3), we see that the mean-value, autocorrelation, and autocovariance functions of the process are related by

$$c(n, n-k) = r(n, n-k) - \mu(n)\mu^*(n-k) \tag{2.4}$$

For a partial characterization of the process, we therefore need to specify (1) the mean-value function $\mu(n)$, and (2) the autocorrelation function $r(n, n-k)$ or the autocovariance function $c(n, n-k)$ for various values of n and k that are of interest.

This form of partial characterization offers two important advantages:

1. It lends itself to practical measurements.
2. It is well suited to *linear* operations on stochastic processes.

For a discrete-time stochastic process that is strictly stationary, all three quantities defined in Eqs. (2.1) to (2.3) assume simpler forms. In particular, we find that the mean-value function of the process is a constant μ (say), so we may write

$$\mu(n) = \mu, \qquad \text{for all } n \tag{2.5}$$

We also find that both the autocorrelation and autocovariance functions depend only on the *difference* between the observation times n and $n - k$, that is, k, as shown by

$$r(n, n-k) = r(k) \tag{2.6}$$

and

$$c(n, n-k) = c(k) \tag{2.7}$$

Note that when $k = 0$, corresponding to a time difference or *lag* of zero, $r(0)$ equals the *mean-square value* of $u(n)$:

$$r(0) = E[|u(n)|^2] \tag{2.8}$$

and $c(0)$ equals the *variance* of $u(n)$:

$$c(0) = \sigma_u^2 \tag{2.9}$$

The conditions of Eqs. (2.5) to (2.7) are *not* sufficient to guarantee that the discrete-time stochastic process is strictly stationary. However, a discrete-time stochastic process that is not strictly stationary, but for which these conditions hold, is said to be *wide-sense stationary*, or *stationary to the second order*. A strictly stationary process $\{u(n)\}$ is stationary in the wide sence if and only if (Doob, 1953)

$$E[|u(n)|^2] < \infty, \qquad \text{for all } n$$

This condition is ordinarily satisfied by stochastic processes encountered in the physical sciences and engineering. It is also of interest to note that if a *Gaussian* process is wide-sense stationary, the process is strictly stationary. When the mean μ is zero, the autocorrelation and autocovariance functions of a weakly stationary process assume the same value.

2.2 MEAN ERGODIC THEOREM

The *expectations* or *ensemble averages* of a stochastic process are averages "across the process." Clearly, we may also define *long-term sample averages* or *time averages* that are averages "along the process." Indeed, time averages may be used to build a *stochastic model* of a physical process by *estimating* unknown parameters of the model. For such an approach to be rigorous, however, we have to show that time averages converge to corresponding ensemble averages of the process in some statistical sense. A popular criterion for convergence is that of mean square error, as described next.

To be specific, consider a discrete-time stochastic process $\{u(n)\}$ that is wide-sense stationary. Let a constant μ denote the mean of the process, and $c(k)$ denote its autocovariance function for lag k. For an estimate of the mean μ, we may use the time average

$$\hat{\mu}(N) = \frac{1}{N}\sum_{n=0}^{N-1} u(n) \tag{2.10}$$

where N is the total number of samples used in the estimation. Note that the estimate $\hat{\mu}(N)$ is a random variable with a mean and variance of its own. In particular, we readily find from Eq. (2.10) that the mean (expectation) of $\hat{\mu}(N)$ is

$$E[\hat{\mu}(n)] = \mu, \qquad \text{for all } N \tag{2.11}$$

It is in the sense of Eq. (2.11) that we say the time average $\hat{\mu}(N)$ is an *unbiased* estimator of the ensemble average (mean) of the process.

Moreover, we say that the process $\{u(n)\}$ is *mean ergodic in the mean square error sense* if the mean-square value of the error between the ensemble average μ and the time average $\hat{\mu}(N)$ approaches zero as the number of samples N approaches infinity; that is,

$$\lim_{N\to\infty} [(\mu - \hat{\mu}(N))^2] = 0$$

Using the time average formula of Eq. (2.10), we may write

$$\begin{aligned}
E\left[\left(\mu - \hat{\mu}(N)\right)^2\right] &= E\left[\left(\mu - \frac{1}{N}\sum_{n=0}^{N-1} u(n)\right)^2\right] \\
&= \frac{1}{N^2} E\left[\left(\sum_{n=0}^{N-1} (u(n) - \mu)\right)^2\right] \\
&= \frac{1}{N^2} E\left[\sum_{n=0}^{N-1}\sum_{k=0}^{N-1} (u(n) - \mu)(u(k) - \mu)\right] \qquad (2.12) \\
&= \frac{1}{N^2} \sum_{n=0}^{N-1}\sum_{k=0}^{N-1} E[(u(n) - \mu)(u(k) - \mu)] \\
&= \frac{1}{N^2} \sum_{n=0}^{N-1}\sum_{k=0}^{N-1} c(n - k)
\end{aligned}$$

Let $l = n - k$. We may then simplify the double summation in Eq. (2.12) as follows

$$E[(\mu - \hat{\mu}(N))^2] = \frac{1}{N} \sum_{l=N+1}^{N-1} \left(1 - \frac{|l|}{N}\right) c(l)$$

Accordingly, we may state that the necessary and sufficient condition for the process $\{u(n)\}$ to be mean ergodic in the mean square error sense is that

$$\lim_{N\to\infty} \frac{1}{N} \sum_{l=N+1}^{N-1} \left(1 - \frac{|l|}{N}\right) c(l) = 0 \qquad (2.13)$$

In other words, if the process $\{u(n)\}$ is asymptotically uncorrelated in the sense of Eq. (2.13), then the time average $\hat{\mu}(N)$ of the process converges to the ensemble average μ in the mean square error sense. This is the statement of a particular form of the *mean ergodic theorem* (Gray and Davisson, 1986).

The use of the mean ergodic theorem may be extended to other time averages of the process. Consider, for example, the following time average used to estimate the autocorrelation function of a wide-sense stationary process:

$$\hat{r}(k, N) = \frac{1}{N} \sum_{n=0}^{N-1} u(n)u(n - k), \qquad 0 \le k \le N - 1 \qquad (2.14)$$

The process $\{u(n)\}$ is said to be *correlation ergodic* in the mean square error sense if the mean-square value of the difference between the true value $r(k)$ and the estimate $\hat{r}(k, n)$ approaches zero as the number of samples N approaches infinity. Let $\{z(n, k\}$ denote a new discrete-time stochastic process related to the original process $\{u(n)\}$ as follows:

$$z(n, k) = u(n)u(n - k) \qquad (2.15)$$

Hence, by substituting $z(n, k)$ for $u(n)$, we may use the mean ergodic theorem to establish the conditions for $z(n, k)$ to be mean ergodic or, equivalently, for $u(n, k)$ to be correlation ergodic.

2.3 CORRELATION MATRIX

Let the M-by-1 *observation vector* $\mathbf{u}(n)$ represent the elements of the time series $u(n), u(n-1), \ldots, u(n-M+1)$. To show the composition of the vector $\mathbf{u}(n)$ explicitly, we write

$$\mathbf{u}^T(n) = [u(n), u(n-1), \ldots, u(n-M+1)] \tag{2.16}$$

where the superscript T denotes *transposition*. We define the *correlation matrix* of a stationary discrete-time stochastic process represented by this time series as *the expectation of the outer product of the observation vector* $\mathbf{u}(n)$ *with itself*. Let $\mathbf{R}$ denote the M-by-M correlation matrix defined in this way. We thus write

$$\mathbf{R} = E\,[\mathbf{u}(n)\,\mathbf{u}^H(n)] \tag{2.17}$$

where the superscript H denotes *Hermitian transposition* (i.e., the operation of transposition combined with complex conjugation). By substituting Eq. (2.16) in (2.17) and using the condition of wide-sense stationarity, we may express the correlation matrix $\mathbf{R}$ in the expanded form:

$$\mathbf{R} = \begin{bmatrix} r(0) & r(1) & \cdots & r(M-1) \\ r(-1) & r(0) & \cdots & r(M-2) \\ \vdots & \vdots & \ddots & \vdots \\ r(-M+1) & r(-M+2) & \cdots & r(0) \end{bmatrix} \tag{2.18}$$

The element $r(0)$ on the main diagonal is always real-valued. For complex-valued data, the remaining elements of $\mathbf{R}$ assume complex values.

Properties of the Correlation Matrix

The correlation matrix $\mathbf{R}$ plays a key role in the statistical analysis and design of discrete-time filters. It is therefore important that we understand its various properties and their implications. In particular, using the definition of Eq. (2.17), we find that the correlation matrix of a stationary discrete-time stochastic process has the following properties:

Property 1. *The correlation matrix of a stationary discrete-time stochastic process is Hermitian.*

We say that a *complex-valued* matrix is *Hermitian* if it is equal to its *conjugate transpose*. We may thus express the Hermitian property of the correlation matrix $\mathbf{R}$ by writing

$$\mathbf{R}^H = \mathbf{R} \tag{2.19}$$

This property follows directly from the definition of Eq. (2.17).

Another way of stating the Hermitian property of the correlation matrix $\mathbf{R}$ is to write

$$r(-k) = r^*(k) \tag{2.20}$$

where $r(k)$ is the autocorrelation function of the stochastic process for a lag of k. Accordingly, for a wide-sense stationary process we only need M values of the autocorrelation function $r(k)$ for $k = 0, 1, \ldots, M - 1$ in order to completely define the correlation matrix $\mathbf{R}$. We may thus rewrite Eq. (2.18) as follows:

$$R = \begin{bmatrix} r(0) & r(1) & \cdots & r(M-1) \\ r^*(1) & r(0) & \cdots & r(M-2) \\ \vdots & \vdots & \ddots & \vdots \\ r^*(M-1) & r^*(M-2) & \cdots & r(0) \end{bmatrix} \tag{2.21}$$

From here on, we will use this representation for the expanded matrix form of the correlation matrix of a wide-sense stationary discrete-time stochastic process. Note that for the special case of *real-valued data,* the autocorrelation function $r(k)$ is real for all k, and the correlation matrix $\mathbf{R}$ is *symmetric*.

Property 2. *The correlation matrix of a stationary discrete-time stochastic process is Toeplitz.*

We say that a square matrix is *Toeplitz* if all the elements on its main diagonal are equal, and if the elements on any other diagonal parallel to the main diagonal are also equal. From the expanded form of the correlation matrix $\mathbf{R}$ given in Eq. (2.21), we see that all the elements on the main diagonal are equal to $r(0)$, all the elements on the first diagonal above the main diagonal are equal to $r(1)$, all the elements along the first diagonal below the main diagonal are equal to $r^*(1)$, and so on for the other diagonals. We conclude therefore that the correlation matrix $\mathbf{R}$ is Toeplitz.

It is important to recognize, however, that the Toeplitz property of the correlation matrix $\mathbf{R}$ is a direct consequence of the assumption that the discrete-time stochastic process represented by the observation vector $\mathbf{u}(n)$ is wide-sense stationary. Indeed, we may state that if the discrete-time stochastic process is wide-sense stationary, then its correlation matrix $\mathbf{R}$ must be Toeplitz, and, conversely, if the correlation matrix $\mathbf{R}$ is Toeplitz, then the discrete-time stochastic process must be wide-sense stationary.

Property 3. *The correlation matrix of a discrete-time stochastic process is always nonnegative definite and almost always positive definite.*

Let $\mathbf{x}$ be an arbitrary (nonzero) M-by-1 complex-valued vector. Define the scalar random variable y as the *inner product* of $\mathbf{x}$ and the observation vector $\mathbf{u}(n)$, as shown by

$$y = \mathbf{x}^H \mathbf{u}(n)$$

Taking the Hermitian transpose of both sides and recognizing that y is a scalar, we get

$$y^* = \mathbf{u}^H(n)\mathbf{x}$$

where the asterisk denotes *complex conjugation*. The mean-square value of the random variable y equals

$$\begin{aligned} E[|y|^2] &= E[yy^*] \\ &= E[\mathbf{x}^H\mathbf{u}(n)\mathbf{u}^H(n)\mathbf{x}] \\ &= \mathbf{x}^H E[\mathbf{u}(n)\mathbf{u}^H(n)]\mathbf{x} \\ &= \mathbf{x}^H\mathbf{R}\mathbf{x} \end{aligned}$$

where $\mathbf{R}$ is the correlation matrix as defined in Eq. (2.17). The expression $\mathbf{x}^H\mathbf{R}\mathbf{x}$ is called a *Hermitian form*. Since

$$E[|y|^2] \geq 0$$

it follows that

$$\mathbf{x}^H\mathbf{R}\mathbf{x} \geq 0 \tag{2.22}$$

A Hermitian form that satisfies this condition is said to be *nonnegative definite* or *positive semidefinite*. Accordingly, we may state that the correlation matrix of a wide-sense stationary process is always nonnegative definite.

If the Hermitian form $\mathbf{x}^H\mathbf{R}\mathbf{x}$ satisfies the condition

$$\mathbf{x}^H\mathbf{R}\mathbf{x} > 0$$

for every nonzero $\mathbf{x}$, we say that the correlation matrix $\mathbf{R}$ is *positive definite*. This condition is satisfied for a wide-sense stationary process unless there are linear dependencies between the random variables that constitute the M elements of the observation vector $\mathbf{u}(n)$. Such a situation arises essentially only when the process $\{u(n)\}$ consists of the sum of K sinusoids with $K \leq M$; see Section 2.4 for more details. In practice, we find that this idealized situation is so rare in occurrence that the correlation matrix $\mathbf{R}$ is almost always positive definite.

The positive definiteness of a correlation matrix implies that its determinant and all principal minors are greater than zero. For example, for $M = 2$, we must have

$$\begin{vmatrix} r(0) & r(1) \\ r^*(1) & r(0) \end{vmatrix} > 0$$

Similarly, for $M = 3$, we must have

$$\begin{vmatrix} r(0) & r(1) \\ r^*(1) & r(0) \end{vmatrix} > 0$$

$$\begin{vmatrix} r(0) & r(2) \\ r^*(2) & r(0) \end{vmatrix} > 0$$

$$\begin{vmatrix} r(0) & r(1) & r(2) \\ r^*(1) & r(0) & r(1) \\ r^*(2) & r^*(1) & r(0) \end{vmatrix} > 0$$

and so on for higher values of M. These conditions, in turn, imply that the correlation matrix is nonsingular. We say that a matrix is *nonsingular* if its inverse exists;

otherwise, it is *singular*. Accordingly, we may state that a correlation matrix is almost always nonsingular.

Property 4. *When the elements that constitute the observation vector of a stationary discrete-time stochastic process are rearranged backward, the effect is equivalent to the transposition of the correlation matrix of the process.*

Let $\mathbf{u}^B(n)$ denote the M-by-1 vector obtained by rearranging the elements that constitute the observation vector $\mathbf{u}(n)$ *backward*. We illustrate this operation by writing

$$\mathbf{u}^{BT}(n) = [u(n - M + 1), u(n - M + 2), \ldots, (u(n)] \tag{2.23}$$

where the superscript B denotes the backward rearrangement of a vector. The correlation matrix of the vector $\mathbf{u}^{\mathbf{B}}(n)$ equals, by definition,

$$E[\mathbf{u}^B(n)\mathbf{u}^{BH}(n)] = \begin{bmatrix} r(0) & r^*(1) & \cdots & r^*(M-1) \\ r(1) & r(0) & \cdots & r^*(M-2) \\ \vdots & \vdots & \ddots & \vdots \\ r(M-1) & r(M-2) & \cdots & r(0) \end{bmatrix} \tag{2.24}$$

Hence, comparing the expanded correlation matrix of Eq. (2.24) with that of Eq. (2.21), we see that

$$E\ [\mathbf{u}^B(n)\ \mathbf{u}^{BH}(n)] = \mathbf{R}^T \tag{2.25}$$

which is the desired result.

Property 5. *The correlation matrices $\mathbf{R}_M$ and $\mathbf{R}_{M+1}$ of a stationary discrete-time stochastic process, pertaining to M and $M + 1$ observations of the process, respectively, are related by*

$$\mathbf{R}_{M+1} = \left[\begin{array}{c|c} r(0) & \mathbf{r}^H \\ \hline \mathbf{r} & \mathbf{R}_M \end{array}\right] \tag{2.26}$$

or equivalently,

$$\mathbf{R}_{M+1} = \left[\begin{array}{c|c} \mathbf{R}_M & \mathbf{r}^{B*} \\ \hline \mathbf{r}^{BT} & r(0) \end{array}\right] \tag{2.27}$$

where $r(0)$ is the autocorrelation of the process for a lag of zero, and

$$\mathbf{r}^H = [r(1), r(2), \ldots, r(M)] \tag{2.28}$$

and

$$\mathbf{r}^{BT} = [r(-M), r(-M + 1), \ldots, r(-1)] \tag{2.29}$$

Note that in describing Property 5 we have added a subscript (M or $M + 1$) to the symbol for the correlation matrix in order to display dependence on the number of observations used to define this matrix. We will follow such a practice (in the context of the correlation matrix and other vector quantities) *only* when the issue at hand involves dependence on the number of observations or dimension of the matrix.

To prove the relation of Eq. (2.26), we express the correlation matrix $\mathbf{R}_{M+1}$ in its expanded form, partitioned as follows:

$$\mathbf{R}_{M+1} = \left[\begin{array}{c|cccc} r(0) & r(1) & r(2) & \cdots & r(M) \\ \hline r^*(1) & r(0) & r(1) & \cdots & r(M-1) \\ r^*(2) & r^*(1) & r(0) & \cdots & r(M-2) \\ \vdots & \vdots & \vdots & \ddots & \vdots \\ r^*(M) & r^*(M-1) & r^*(M-2) & \cdots & r(0) \end{array}\right] \tag{2.30}$$

Using Eqs. (2.18), (2.20), (2.28) in (2.30), we get the result given in Eq. (2.26). Note that according to this relation the observation vector $\mathbf{u}_{M+1}(n)$ is *partitioned* in the form

$$\mathbf{u}_{M+1}(n) = \left[\begin{array}{c} u(n) \\ \hline u(n-1) \\ u(n-2) \\ \vdots \\ u(n-M) \end{array}\right] = \left[\begin{array}{c} u(n) \\ \hline \mathbf{u}_M(n-1) \end{array}\right] \tag{2.31}$$

where the subscript $M + 1$ is intended to denote the fact that the vector $\mathbf{u}_{M+1}(n)$ has $M + 1$ elements, and likewise for $\mathbf{u}_M(n)$.

To prove the relation of Eq. (2.27), we express the correlation matrix $\mathbf{R}_{M+1}$ in its expanded form, partitioned as follows:

$$\mathbf{R}_{M+1} = \left[\begin{array}{cccc|c} r(0) & r(1) & \cdots & r(M-1) & r(M) \\ r^*(1) & r(0) & \cdots & r(M-2) & r(M-1) \\ \vdots & \vdots & \ddots & \vdots & \vdots \\ r^*(M-1) & r^*(M-2) & \cdots & r(0) & r(1) \\ \hline r^*(M) & r^*(M-1) & \cdots & r^*(1) & r(0) \end{array}\right] \tag{2.32}$$

Here again, using Eqs. (2.18), (2.20), and (2.29) in (2.32), we get the result given in Eq. (2.27). Note that according to this second relation the observation vector $\mathbf{u}_{M+1}(n)$ is partitioned in the alternative form

$$
\mathbf{u}_{M+1}(n) = \begin{bmatrix} u(n) \\ u(n-1) \\ \vdots \\ u(n-M+1) \\ \hdashline u(n-M) \end{bmatrix}
$$

$$
= \begin{bmatrix} \mathbf{u}_M(n) \\ \hdashline u(n-M) \end{bmatrix} \tag{2.33}
$$

2.4 CORRELATION MATRIX OF SINE WAVE PLUS NOISE

A time series of special interest is one that consists of a *complex sinusoid corrupted by additive noise*. Such a time series is representative of several important signal processing applications. In the *temporal context*, for example, this time series represents the composite signal at the input of a receiver, with the complex sinusoid representing a *target signal* and the noise representing *thermal noise* generated at the front end of the receiver. In the *spatial context*, it represents the received signal in a linear array of sensors, with the complex sinusoid representing a *plane wave* produced by a remote source (emitter) and the noise representing *sensor noise*.

Let α denote the amplitude of the complex sinusoid, and ω denote its angular frequency. Let $v(n)$ denote a sample of the noise, assumed to have zero mean. We may then write a corresponding sample of the time series that consists of the complex sinusoid plus noise as follows:

$$
u(n) = \alpha \exp(j\omega n) + v(n), \qquad n = 0, 1, \ldots, N-1 \tag{2.34}
$$

The sources of the complex sinusoid and the noise are independent of each other. Since the noise component $v(n)$ has zero mean, by assumption, we see from Eq. (2.34) that the mean of $u(n)$ is equal to $\alpha \exp(j\omega n)$.

To calculate the autocorrelation function of the process $\{u(n)\}$, we clearly need to know the autocorrelation function of the noise process $\{v(n)\}$. To proceed, then, we assume a special form of noise characterized by the autocorrelation function:

$$
E[v(n)v^*(n-k)] = \begin{cases} \sigma_v^2, & k = 0 \\ 0, & k \neq 0 \end{cases} \tag{2.35}
$$

Such a form of noise is commonly referred to as *white noise;* more will be said on it in Chapter 3. Since the sources responsible for the generation of the complex sinusoid and the noise are independent and, therefore, uncorrelated, it follows that the autocorrelation function of the process $\{u(n)\}$ equals the sum of the autocorrelation functions of its two individual components. Accordingly, using Eqs. (2.34) and (2.35), we find that the autocorrelation function of the process $\{u(n)\}$ for a lag k is given by

$$r(k) = E[u(n)u^*(n-k)]$$
$$= \begin{cases} |\alpha|^2 + \sigma_v^2, & k = 0 \\ |\alpha|^2 \exp(j\omega k), & k \neq 0 \end{cases} \tag{2.36}$$

where $|\alpha|$ is the magnitude of the complex amplitude α. Note that for a lag $k \neq 0$, the autocorrelation function $r(k)$ varies with k in the same sinusoidal fashion as the sample $u(n)$ varies with n, except for a change in amplitude. Given the series of samples $u(n), u(n-1), \ldots, u(n-M+1)$, we may thus express the correlation matrix of $\{u(n)\}$ as

$$\mathbf{R} = |\alpha|^2 \begin{bmatrix} 1 + \frac{1}{\rho} & \exp(j\omega) & \cdots & \exp(j\omega(M-1)) \\ \exp(-j\omega) & 1 + \frac{1}{\rho} & \cdots & \exp(j\omega(M-2)) \\ \vdots & \vdots & \ddots & \vdots \\ \exp(-j\omega(M-1)) & \exp(-j\omega(M-2)) & \cdots & 1 + \frac{1}{\rho} \end{bmatrix} \tag{2.37}$$

where ρ is the *signal-to-noise ratio,* defined by

$$\rho = \frac{|\alpha|^2}{\sigma_v^2} \tag{2.38}$$

The correlation matrix $\mathbf{R}$ of Eq. (2.37) has all the properties described in Section 2.3; the reader is invited to verify them.

Equation (2.36) provides the mathematical basis of a two-step practical procedure for estimating the parameters of a complex sinusoid in the presence of additive noise:

1. We measure the mean-square value $r(0)$ of the process $\{u(n)\}$. Hence, given the noise variance σ_v^2, we may determine the magnitude $|\alpha|$.
2. We measure the autocorrelation function $r(k)$ of the process $\{u(n)\}$ for a lag $k \neq 0$. Hence, given $|\alpha|^2$ from step 1, we may determine the angular frequency ω.

Note that this estimation procedure is *invariant to the phase of* α, which is a direct consequence of the definition of the autocorrelation function $r(k)$.

Example 1

Consider the idealized case of a noiseless sinusoid of angular frequency ω. For the purpose of illustration, we assume that the time series of interest consists of three uniformly spaced samples drawn from this sinusoid. Hence, setting the signal-to-noise ratio $\rho = \infty$ and the number of samples $M = 3$, we find from Eq. (2.37) that the correlation matrix of the time series so obtained has the following value

$$\mathbf{R} = |\alpha|^2 \begin{bmatrix} 1 & \exp(j\omega) & \exp(j2\omega) \\ \exp(-j\omega) & 1 & \exp(j\omega) \\ \exp(-j2\omega) & \exp(-j\omega) & 1 \end{bmatrix}$$

From this expression we readily see that the determinant of **R** and all principal minors are identically zero. Hence, this correlation matrix is singular.

We may generalize the result of this example by stating that when a process $\{u(n)\}$ consists of M samples drawn from the sum of K sinusoid with $K < M$, and there is *no* additive noise, then the correlation matrix of that process is singular.

2.5 STOCHASTIC MODELS

The term *model* is used for any hypothesis that may be applied to explain or describe the hidden laws that are supposed to govern or constrain the generation of some data of interest. The representation of a stochastic process by a model dates back to an idea that was originated by Yule (1927). The idea is that a time series, $\{u(n)\}$, consisting of highly correlated observations may be generated by applying a series of statistically independent "shocks" to a linear filter, $\{v(n)\}$, as in Fig. 2.1. The shocks are random variables drawn from a fixed distribution that is usually assumed to be *Gaussian* with zero mean and constant variance. Such a series of random variables constitutes a purely random process that is commonly referred to as *white Gaussian noise*. Specifically, we may describe the input $\{v(n)\}$ in Figure 2.1 in statistical terms as follows:

$$E[v(n)] = 0, \qquad \text{for all } n \tag{2.39}$$

and

$$E[v(n)v^*(k)] = \begin{cases} \sigma_v^2, & k = n \\ 0, & \text{otherwise} \end{cases} \tag{2.40}$$

where σ_v^2 is the noise variance. Equation (2.39) follows from the zero-mean assumption, and Eq. (2.40) follows from the white assumption. The implication of the Gaussian assumption is discussed in Section 2.11.

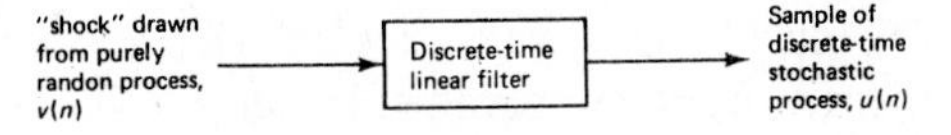

Figure 2.1 Stochastic model.

In general, the time-domain description of the input–output relation for the stochastic model of Fig. 2.1 may be written as follows:

$$\begin{pmatrix}\text{present value}\\ \text{of model output}\end{pmatrix} + \begin{pmatrix}\text{linear combination}\\ \text{of past values}\\ \text{of model output}\end{pmatrix} = \begin{pmatrix}\text{linear combination of}\\ \text{present and past values}\\ \text{of model input}\end{pmatrix} \tag{2.41}$$

The structure of the linear filter in Fig. 2.1 is determined by the manner in which the two linear combinations indicated in Eq. (2.41) are formulated. We may thus identify three popular types of stochastic models:

1. Autoregressive models, in which only the present value of the model input is used.
2. Moving average models, in which no past values of the model output are used.

3. Mixed autoregressive-moving average models, in which the description of Eq. (2.41) applies in its entire form. Hence, this class of stochastic models includes autoregressive and moving average models as special cases.

These models are described next, in order.

Autoregressive Models

We say that the time series $u(n)$, $u(n-1)$, . . . , $u(n-M)$ represents the realization of an *autoregressive process (AR) of order M* if it satisfies the difference equation

$$u(n) + a_1^* u(n-1) + a_M^* u(n-M) = v(n) \tag{2.42}$$

where a_1, a_2, . . . , a_M are constants called the AR *parameters,* and $\{v(n)\}$ is a white-noise process. The term $a_k^*\, u(n-k)$ is an *inner product* of a_k and $u(n-k)$, where $k = 1, \ldots, M$.

To explain the reason for the name "autoregressive," we rewrite Eq. (2.42) in the form

$$u(n) = w_1^* u(n-1) + w_2^* u(n-2) + \cdots + w_M^* u(n-M) + v(n) \tag{2.43}$$

where $w_k = -a_k$. We thus see that the present value of the process, that is, $u(n)$, equals a *finite linear combination of past values* of the process, $u(n-1)$, . . . , $u(n-M)$, plus an *error term* $v(n)$. We now see the reason for the name "autoregressive." Specifically, a linear model

$$y = \sum_{k=1}^{M} w_k^* x_k + v$$

relating a *dependent* variable y to a set of *independent* variables x_1, x_2, . . . , x_M plus an error term v is often referred to as a *regression model,* and y is said to be "regressed" on x_1, x_2, . . . , x_M. In Eq. (2.43), the variable $u(n)$ *is regressed* on previous values of *itself*; hence, the name "autoregressive."

The left side of Eq. (2.42) represents the *convolution* of the input sequence $\{u(n)\}$ and the sequence of parameters $\{a_n^*\}$. To highlight this point, we rewrite Eq. (2.42) in the form of a convolution sum:

$$\sum_{k=0}^{M} a_k^* u(n-k) = v(n) \tag{2.44}$$

where $a_0 = 1$. By taking the *z-transform*[1] of both sides of Eq. (2.44), we transform the convolution sum on the left side of the equation into a multiplication of the z-transforms of the two sequences $\{u(n)\}$ and $\{a_n^*\}$. Let $H_A(z)$ denote the z-transform of the sequence $\{a_n^*\}$:

$$H_A(z) = \sum_{n=0}^{M} a_n^* z^{-n} \tag{2.45}$$

[1] For a discussion of the z-transform and its properties, see Oppenheim and Schafer (1975), Roberts and Mullis (1987), and Proakis and Manolakis (1988).

Let $U(z)$ denote the z-transform of the input sequence $\{u(n)\}$:

$$U(z) = \sum_{n=0}^{\infty} u(n)\, z^{-n} \tag{2.46}$$

where z is a *complex variable*. We may thus transform the difference equation (2.42) into the equivalent form

$$H_A(z)U(z) = V(z) \tag{2.47}$$

where

$$V(z) = \sum_{n=0}^{\infty} v(n) z^{-n} \tag{2.48}$$

The z transform of Eq. (2.47) offers two interpretations, depending on whether the AR process $\{u(n)\}$ is viewed as the input or output of interest:

1. Given the AR process $\{u(n)\}$, we may use the filter shown in Fig. 2.2(a) to produce the white noise process $\{v(n)\}$ as output. The parameters of this filter bear a one-to-one correspondence with those of the AR process $\{u(n)\}$. Accordingly, this filter represents a *process analyzer* with discrete transfer function $H_A(z) = V(z)/U(z)$. The impulse response of the AR process analyzer, that is, the inverse z-transform of $H_A(z)$ has *finite duration*.
2. With the white noise $\{v(n)\}$ acting as input, we may use the filter shown in Fig. 2.2(b) to produce the AR process $\{u(n)\}$ as output. Accordingly, this second filter represents a *process generator*, whose transfer function equals

$$\begin{aligned} H_G(z) &= \frac{U(z)}{V(z)} \\ &= \frac{1}{H_A(z)} \\ &= \frac{1}{\sum_{n=0}^{M} a_n^* z^{-n}} \end{aligned} \tag{2.49}$$

The impulse response of the AR process generator, that is, the inverse z-transform of $H_G(z)$ has *infinite duration*.

The filters shown in Fig. 2.2 consist of an interconnection of functional blocks that are used to represent three basic operations: (1) *addition*, (2) *multiplication* by a scalar, and (3) *storage*. The storage is represented by blocks labeled z^{-1}, the *unit-sample delay*. By applying z^{-1} to $u(n)$, we get $u(n-1)$; similarly, by applying z^{-1} to $u(n-1)$, we get $u(n-2)$, and so on. Thus, by using a cascade of M unit-sample delays, we provide storage for the past values $u(n-1), u(n-2), \ldots, u(n-M)$.

The AR process analyzer of Fig. 2.2(a) is an *all-zero filter*. It is so called because its transfer function $H_A(z)$ is completely defined by specifying the locations of its *zeros*. This filter is inherently stable.

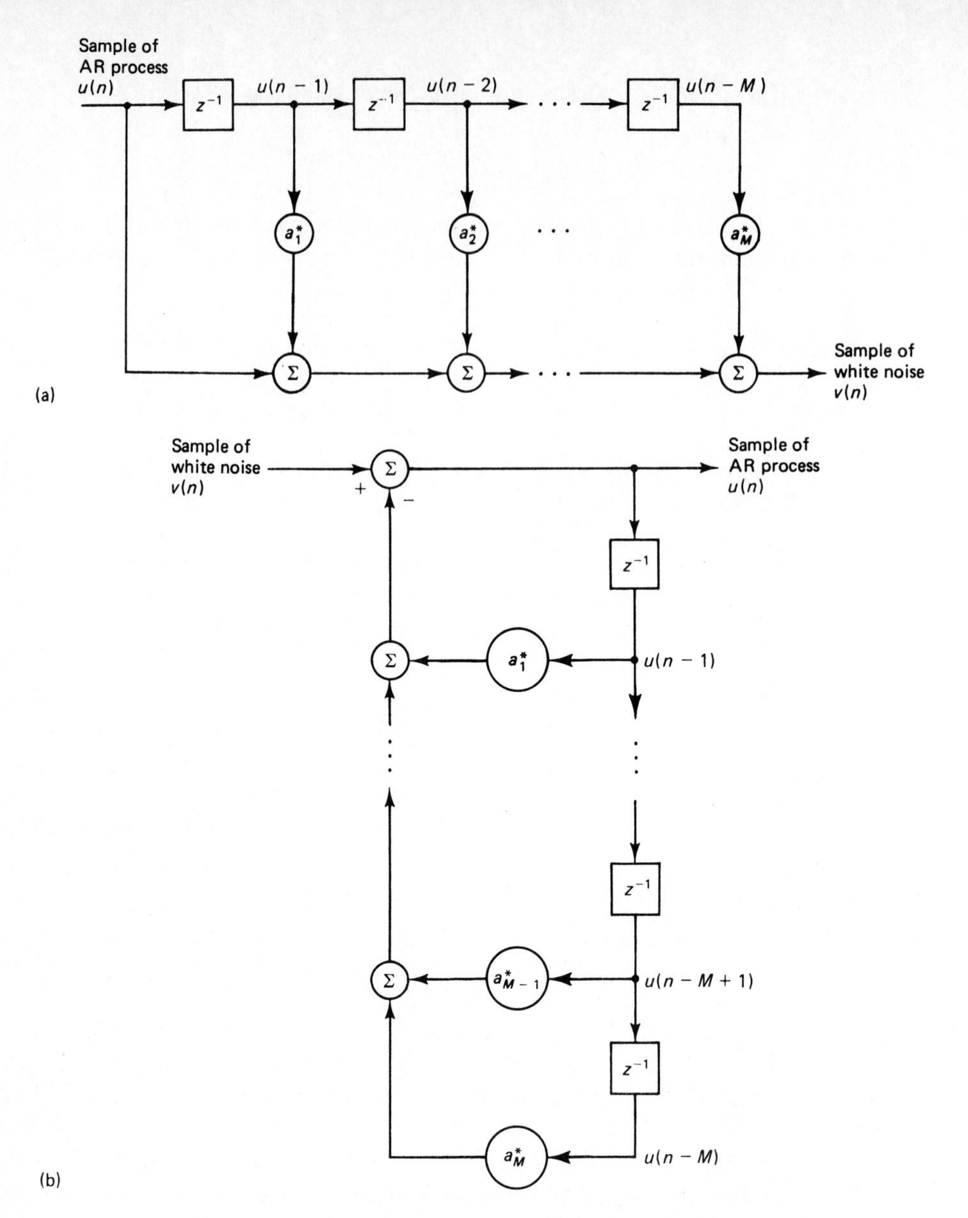

Figure 2.2 (a) AR process analyzer; (b) AR process generator.

The AR process generator of Fig. 2.2(b) is an *all-pole filter*. It is so called because its transfer function $H_G(z)$ is completely defined by specifying the locations of its *poles,* as shown by

$$H_G(z) = \frac{1}{(1 - p_1 z^{-1})(1 - p_2 z^{-1}) \cdots (1 - p_M z^{-1})} \tag{2.50}$$

The parameters $p_1, p_2, \ldots, p_M$ are *poles* of $H_G(z)$; they are defined by the roots of the *characteristic equation*

$$1 + a_1^* z^{-1} + a_2^* z^{-2} + \cdots + a_M^* z^{-M} = 0 \tag{2.51}$$

We say that *a filter is stable if and only if its output is bounded for every bounded input*. In the case of a *causal* discrete-time linear filter, causal in the sense that its impulse response is zero for negative time, this definition of stability leads to the following frequently used criterion (Tretter, 1976):

> A causal filter with discrete transfer function $H(z)$ is stable if and only if all the poles of $H(z)$ lie inside the unit circle in the z-plane.

Accordingly, for the all-pole AR process generator of Fig. 2.2(b) to be stable, the roots of the characteristic equation (2.51) must all lie inside the unit circle in the z-plane. This is also a necessary and sufficient condition for wide-sense stationarity of the AR process produced by the model of Fig. 2.2(b). We will have more to say on the issue of stationarity in Section 2.7.

Moving Average Models

In a *moving average (MA) model*, the discrete-time linear filter of Fig. 2.1 consists of an *all-zero filter* driven by white noise. The resulting process $\{u(n)\}$, produced at the filter output, is described by the difference equation:

$$u(n) = v(n) + b_1^* v(n-1) + \cdots + b_K^* v(n-K) \tag{2.52}$$

where $b_1, \ldots, b_K$ are constants called the *MA parameters,* and $\{v(n)\}$ is a white-noise process of zero mean and variance σ_v^2. Except for $v(n)$, each term on the right side of Eq. (2.52) represents an inner product. The *order* of the MA process equals K. The term moving average is a rather quaint one; nevertheless, its use is firmly established in the literature. Its usage arose in the following way: If we are given a complete realization of $\{v(n)\}$, we may compute $u(n)$ by constructing a *weighted average* of $v(n), v(n-1), \ldots, v(n-K)$.

From Eq. (2.52), we readily obtain the MA model (i.e., process generator) depicted in Fig. 2.3. Specifically, we start with a white-noise process $\{v(n)\}$ at the model input and generate an MA process $\{u(n)\}$ of order K at the model output. To proceed in the reverse manner, that is, to produce the white noise process $\{v(n)\}$, given the MA process $\{u(n)\}$, we require the use of an *all-pole filter*. In other words, the filters used in the generation and analysis of an MA process are the *opposite* of those used in the case of an AR process.

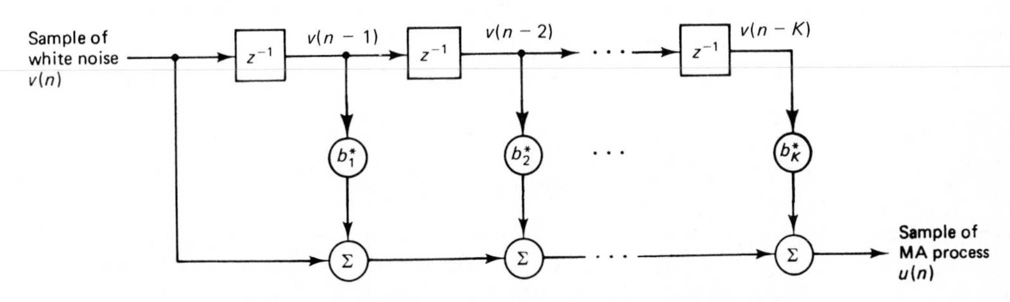

Figure 2.3 Moving average model (process generator).

Autoregressive-Moving Average Models

To generate a mixed *autoregressive-moving average (ARMA) process* $\{u(n)\}$, we use a discrete-time linear filter in Fig. 2.1 with a transfer function that contains *both poles and zeros*. Accordingly, given a white-noise process $\{v(n)\}$ as the filter input, the ARMA process $\{u(n)\}$ produced at the filter output is described by the difference equation

$$u(n) + a_1^* u(n-1) + \cdots + a_M^* u(n-M)$$
$$= v(n) + b_1^* v(n-1) + \cdots + b_K^* v(n-K) \qquad (2.53)$$

where $a_1, \ldots, a_M$ and $b_1, \ldots, b_K$ are called the *ARMA parameters*. Except for $u(n)$ on the left side and $v(n)$ on the right side of Eq. (2.53), all the other terms represent inner products. The *order* of the ARMA process equals (M, K).

From Eq. (2.53), we readily deduce the ARMA model (i.e., process generator) depicted in Fig. 2.4. Comparing this figure with the corresponding ones of Figs. 2.2(b) and 2.3, we clearly see that AR and MA models are indeed special cases of an ARMA model.

The transfer function of the ARMA process generator in Fig. 2.4 has both poles and zeros. Similarly, the ARMA analyzer used to generate a white-noise process $\{v(n)\}$, given an ARMA process $\{u(n)\}$, is characterized by a transfer function containing both poles and zeros.

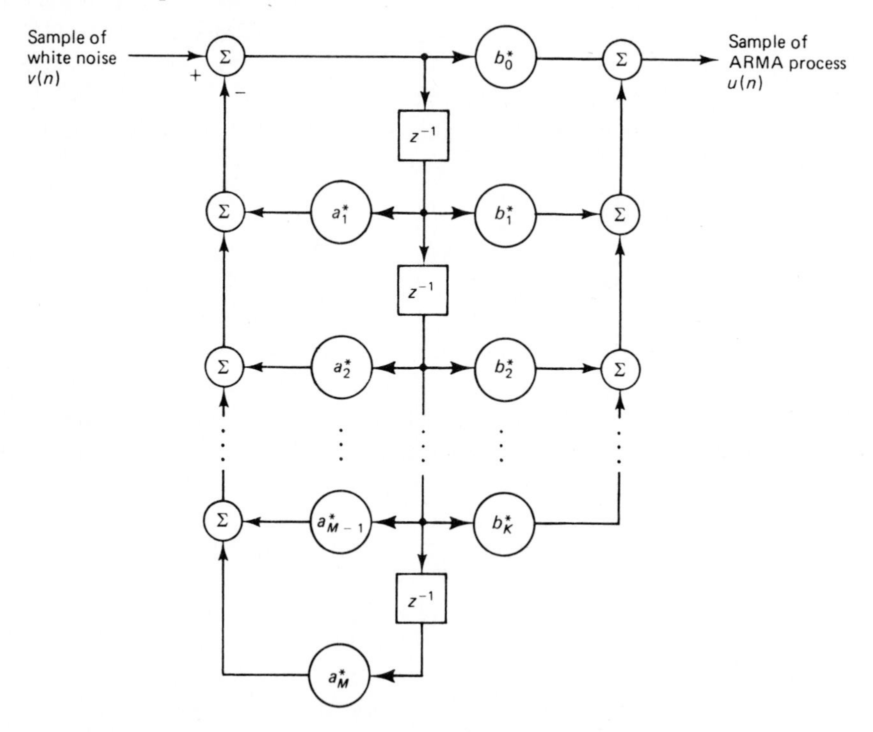

Figure 2.4 ARMA model (process generator) of order (M, K), assuming that $M > K$.

Discussion

From a computational viewpoint, the AR model has an advantage over the MA and ARMA models. Specifically, the computation of the AR coefficients in the model of Fig. 2.2(a) involves a system of *linear equations* known as the Yule–Walker equations, details of which are given in Section 2.8. On the other hand, the computation of the MA coefficients in the model of Fig. 2.3 and the computation of the ARMA coefficients in the model of Fig. 2.4 are much more complicated. Both of these computations require solving systems of *nonlinear equations*. It is for this reason that, in practice, we find that the use of AR models is more popular than MA and ARMA models. The wide application of AR models may also be justified by virtue of a fundamental theorem of time series, which is discussed next.

2.6 WOLD DECOMPOSITION

Wold (1938) proved a fundamental theorem which states that any stationary discrete-time stochastic process may be decomposed into the sum of a *general linear process* and a *predicatable process,* with these two processes being uncorrelated with each other. More precisely, Wold proved the following result:

Any stationary discrete-time stochastic process $\{x(n)\}$ may be expressed in the form

$$x(n) = u(n) + s(n) \tag{2.54}$$

where

1. $\{u(n)\}$ and $\{s(n)\}$ are uncorrelated processes,
2. $\{u(n)\}$ is a general linear process represented by

$$u(n) = \sum_{k=0}^{\infty} b_k^* v(n-k) \tag{2.55}$$

with $b_0 = 1$, and

$$\sum_{k=0}^{\infty} |b_k|^2 < \infty,$$

and where $\{v(n)\}$ is a white-noise process uncorrelated with $s(n)$; that is,

$$E[v(n)s^*(k)] = 0, \qquad \text{for all } n, k$$

3. $\{s(n)\}$ is a predictable process; that is, the process can be predicted from its own past with zero prediction variance.

This result is known as *Wold's decomposition theorem*. A proof of this theorem is given in Priestley (1981).

According to Eq. (2.55), the general linear process $\{u(n)\}$ may be generated by feeding an *all-zero filter* with the white-noise process $\{v(n)\}$ as in Fig. 2.5(a). The zeros of the transfer function of this filter equal the roots of the equation:

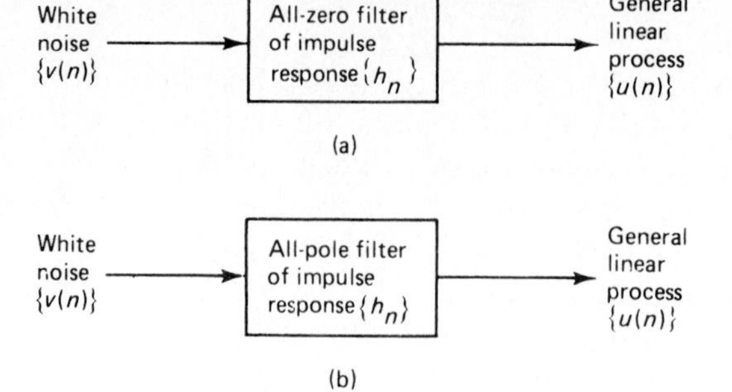

Figure 2.5 (a) Model, based on all-zero filter, for generating linear process $\{u(n)\}$; (b) model, based on all-pole filter, for generating general linear process $\{u(n)\}$. Both filters have the same impulse response.

$$\sum_{n=0}^{\infty} b_n^* z^{-n} = 0$$

A solution of particular interest is an all-zero filter that is *minimum phase*. We say that this filter is minimum phase if the zeros of its transfer function lie inside the unit circle (Oppenheim and Schafer, 1975). In such a case we may replace the all-zero filter with an *equivalent* all-pole filter that has the same impulse response $\{h_n\} = \{b_n^*\}$, as in Fig. 2.5(b). This means that except for a predictable component, a stationary discrete-time stochastic process may also be represented as an AR process of sufficiently high order.

2.7 ASYMPTOTIC STATIONARITY OF AN AUTOREGRESSIVE PROCESS

Equation (2.42) represents a *linear, constant coefficient, difference equation of order M,* in which $v(n)$ plays the role of *input* or *driving function* and $u(n)$ that of *output* or *solution*. By using the *classical method*[2] for solving such an equation, we may formally express the solution $u(n)$ as the sum of a *complementary function,* $u_c(n)$, and a *particular solution,* $u_p(n)$, as follows:

$$u(n) = u_c(n) + u_p(n) \tag{2.56}$$

The evaluation of the solution $u(n)$ may thus proceed in two stages:

1. The complementary function $u_c(n)$ is the solution of the *homogeneous equation*:

$$u(n) + a_1^* u(n-1) + a_2^* u(n-2) + \cdots + a_M^* u(n-M) = 0$$

In general, the complementary function $u_c(n)$ will therefore be of the form

$$u_c(n) = B_1 p_1^n + B_2 p_2^n + \cdots + B_M p_M^n \tag{2.57}$$

where $B_1, B_2, \ldots, B_M$ are arbitrary constants, and $p_1, p_2, \ldots, p_M$ are roots of the characteristic equation (2.51).

[2] We may also use the z-transform method to solve the difference equation (2.42). However, for the discussion here we find it more informative to use the classical method.

2. The particular solution $u_p(n)$ is defined by

$$u_p(n) = H_G(D)[v(n)] \tag{2.58}$$

where D is the *unit-delay operator*, and the operator $H_G(D)$ is obtained by substituting D for z^{-1} in the discrete-transfer function of Eq. (2.49). The unit-delay operator D has the property

$$D^k[u(n)] = u(n-k), \qquad k = 0, 1, 2 \ldots \tag{2.59}$$

The constants $B_1, B_2, \ldots, B_M$ are determined by the choice of *initial conditions* that equal M in number. It is customary to set

$$\begin{aligned} u(0) &= 0 \\ u(-1) &= 0 \\ &\vdots \\ u(-M+1) &= 0 \end{aligned} \tag{2.60}$$

This is equivalent to setting the output of the model in Fig. 2.2(b) as well as the succeeding $(M-1)$ tap inputs equal to zero at time $n = 0$. Thus, by substituting these initial conditions into Eqs. (2.56)–(2.58), we obtain a set of M simultaneous equations that can be solved for the constants $B_1, B_2, \ldots, B_M$.

The result of imposing the initial conditions of Eq. (2.60) on the solution $u(n)$ is to make the discrete-time stochastic process represented by this solution nonstationary. On reflection, it is clear that this must be so, since we have given a "special status" to the time point $n = 0$, and the property of *invariance under a shift of time origin* cannot hold, even for second-order moments. If, however, the solution $u(n)$ is able to "forget" its initial conditions, the resulting process is asymptotically stationary in the sense that it settles down to a stationary behavior as n approaches infinity (Priestley, 1981). This requirement may be achieved by choosing the parameters of the AR model in Fig. 2.2(b) such that the complementary function $u_c(n)$ decays to zero as n approaches infinity. From Eq. (2.57) we see that, for arbitrary constants in the equation, this requirement can be met if and only if

$$|p_k| < 1, \qquad \text{for all } k$$

Hence, *for asymptotic stationarity of the discrete-time stochastic process represented by the solution $u(n)$, we require that all the poles of the filter in the AR model lie inside the unit circle in the z-plane.*

Correlation Function of an Asymptotically Stationary AR Process

Assuming that the condition for asymptotic stationarity is satisfied, we may derive an important recursive relation for the autocorrelation function of the resulting AR process $\{u(n)\}$ as follows. We first multiply both sides of Eq. (2.42) by $u^*(n-1)$ and then apply the expectation operator, thereby obtaining

$$E\left[\sum_{k=0}^{M} a_k^* u(n-k)u^*(n-l)\right] = E[v(n)u^*(n-l)] \tag{2.61}$$

Next, we simplify the left side of Eq. (2.61) by interchanging the expectation and summation, and recognizing that the expectation $E[u(n-k)u^*(n-l)]$ equals the autocorrelation function of the AR process for a lag of $l-k$. We simplify the right side by observing that the expectation $E[v(n)u^*(n-l)]$ is zero for $l>0$, since $u(n-l)$ only involves samples of the white-noise process at the filter input in Fig. 2.2(b) up to time $n-l$, which are uncorrelated with the white-noise sample $v(n)$. Accordingly, we simplify Eq. (2.61) as follows:

$$\sum_{k=0}^{M} a_k^* r(l-k) = 0, \qquad l > 0 \tag{2.62}$$

where $a_0 = 1$. We thus see that the autocorrelation function of the AR process satisfies the difference equation

$$r(l) = w_1^* r(l-1) + w_2^* r(l-2) + \cdots + w_M^* r(l-M), \qquad l > 0 \tag{2.63}$$

where $w_k = -a_k$, $k = 1, 2, \cdots, M$. Note that Eq. (2.63) is analogous to the difference equation satisfied by the AR process $\{u(n)\}$ itself.

We may express the general solution of Eq. (2.63) as follows:

$$r(m) = \sum_{k=1}^{M} C_k p_k^m \tag{2.64}$$

where $C_1, C_2, \ldots, C_M$ are constants, and $p_1, p_2, \ldots, p_M$ are roots of the characteristic equation (2.51). Note that when the AR model of Fig. 2.2(b) satisfies the condition for asymptotic stationarity, $|p_k| < 1$ for all k, in which case the autocorrelation function $r(m)$ approaches zero as the lag m approaches infinity.

The exact form of the contribution made by a pole p_k in Eq. (2.64) depends on whether the pole is real or complex. When p_k is real, the corresponding contribution decays geometrically to zero as the lag m increases. We refer to such a contribution as a *damped exponential*. On the other hand, complex poles occur in conjugate pairs, and the contribution of a complex-conjugate pair of poles is in the form of a *damped sine wave*. We thus find that, in general, the autocorrelation function of an asymptotically stationary AR process consists of a mixture of damped exponentials and damped sine waves.

2.8 CHARACTERIZATION OF AN AUTOREGRESSIVE MODEL

In order to uniquely define the AR model of order M, depicted in Fig. 2.2(b), we need to specify two sets of model parameters:

1. The AR coefficients $a_1, a_2, \ldots, a_M$
2. The variance σ_v^2 of the white noise $\{v(n)\}$ used as excitation

We now address these two issues in turn.

Yule–Walker Equations

Writing Eq. (2.63) for $l = 1, 2, \ldots, M$, we get a set of M simultaneous equations with the values $r(0), r(1), \ldots, r(M)$ of the autocorrelation function of the AR process as the known quantities and the AR parameters $a_1, a_2, \ldots, a_M$ as the unknowns. This set of equations may be expressed in the expanded matrix form

$$\begin{bmatrix} r(0) & r(1) & \cdots & r(M-1) \\ r^*(1) & r(0) & \cdots & r(M-2) \\ \vdots & \vdots & \ddots & \vdots \\ r^*(M-1) & r^*(M-2) & \cdots & r(0) \end{bmatrix} \begin{bmatrix} w_1 \\ w_2 \\ \vdots \\ w_M \end{bmatrix} = \begin{bmatrix} r^*(1) \\ r^*(2) \\ \vdots \\ r^*(M) \end{bmatrix} \tag{2.65}$$

where we have $w_k = -a_k$. The set of equations (2.65) is called the *Yule–Walker equations* (Yule, 1927; Walker, 1931).

We may express the Yule–Walker equations in the compact matrix form

$$\mathbf{R}\mathbf{w} = \mathbf{r} \tag{2.66}$$

and its solution as (assuming that the correlation matrix $\mathbf{R}$ is nonsingular)

$$\mathbf{w} = \mathbf{R}^{-1}\mathbf{r} \tag{2.67}$$

where $\mathbf{R}^{-1}$ is the inverse of matrix $\mathbf{R}$, and the vector $\mathbf{w}$ is defined by

$$\mathbf{w}^T = [w_1, w_2, \ldots, w_M]$$

The correlation matrix $\mathbf{R}$ is defined by Eq. (2.21), and vector $\mathbf{r}$ is defined by Eq. (2.28). From these two equations, we see that we may uniquely determine both the matrix $\mathbf{R}$ and the vector $\mathbf{r}$, given the autocorrelation sequence $r(0), r(1), \ldots, r(M)$. Hence, using Eq. (2.67) we may compute the coefficient vector $\mathbf{w}$ and, therefore, the AR coefficients $a_k = -w_k$, $k = 1, 2, \ldots, a_M$. In other words, there is a unique relationship between the coefficients $a_1, a_2, \ldots, a_M$ of the AR model and the *normalized* correlation coefficients $\rho_1, \rho_2, \ldots, \rho_M$ of the AR process $\{u(n)\}$, as shown by

$$\{a_1, a_2, \ldots, a_M\} \rightleftharpoons \{\rho_1, \rho_2, \ldots, \rho_M\} \tag{2.68}$$

where the *correlation coefficient* ρ_k is defined by

$$\rho_k = \frac{r(k)}{r(0)} \tag{2.69}$$

Variance of the White Noise

For $l = 0$, we find that the expectation on the right side of Eq. (2.61) assumes the special form

$$\begin{aligned} E[v(n)u^*(n)] &= E[v(n)v^*(n)] \\ &= \sigma_v^2 \end{aligned} \tag{2.70}$$

where σ_v^2 is the variance of the zero-mean white noise $\{v(n)\}$. Accordingly, setting $l = 0$ in Eq. (2.61) and complex-conjugating both sides, we get the following formula for the variance of the white-noise process:

$$\sigma_v^2 = \sum_{k=0}^{M} a_k r(k) \tag{2.71}$$

where $a_0 = 1$. Hence, given the autocorrelations $r(0), r(1), \ldots, r(M)$, we may determine the white-noise variance σ_v^2.

2.9 COMPUTER EXPERIMENT: AUTOREGRESSIVE PROCESS OF ORDER 2

To illustrate the theory developed above for the modeling of an AR process, we consider the example of a second-order AR process that is real valued.[3] Figure 2.6 shows the block diagram of the model used to generate this process. Its time-domain description is governed by the second-order difference equation

$$u(n) + a_1 u(n-1) + a_2 u(n-2) = v(n) \tag{2.72}$$

where $v(n)$ is drawn from a white-noise process of zero mean and variance σ_v^2. Figure 2.7(a) shows one realization of this white-noise process. The variance σ_v^2 is chosen to make the variance of $u(n)$ equal unity.

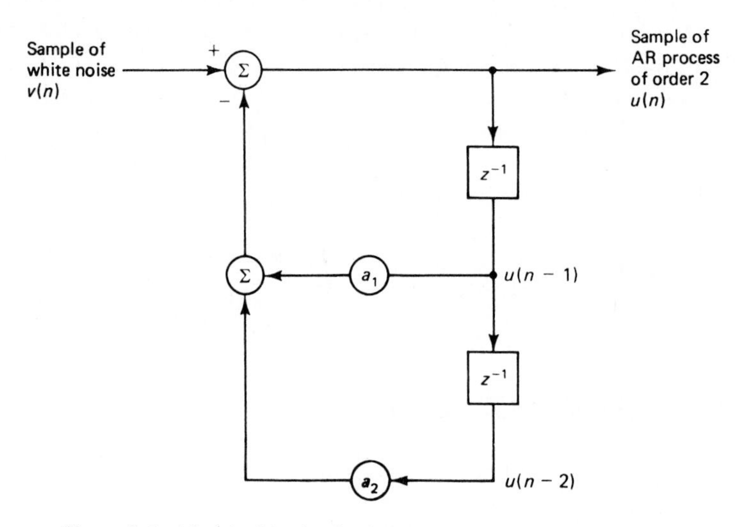

Figure 2.6 Model of (real-valued) autoregressive process of order 2.

[3] In this example, we follow the approach described by Box and Jenkins (1976).

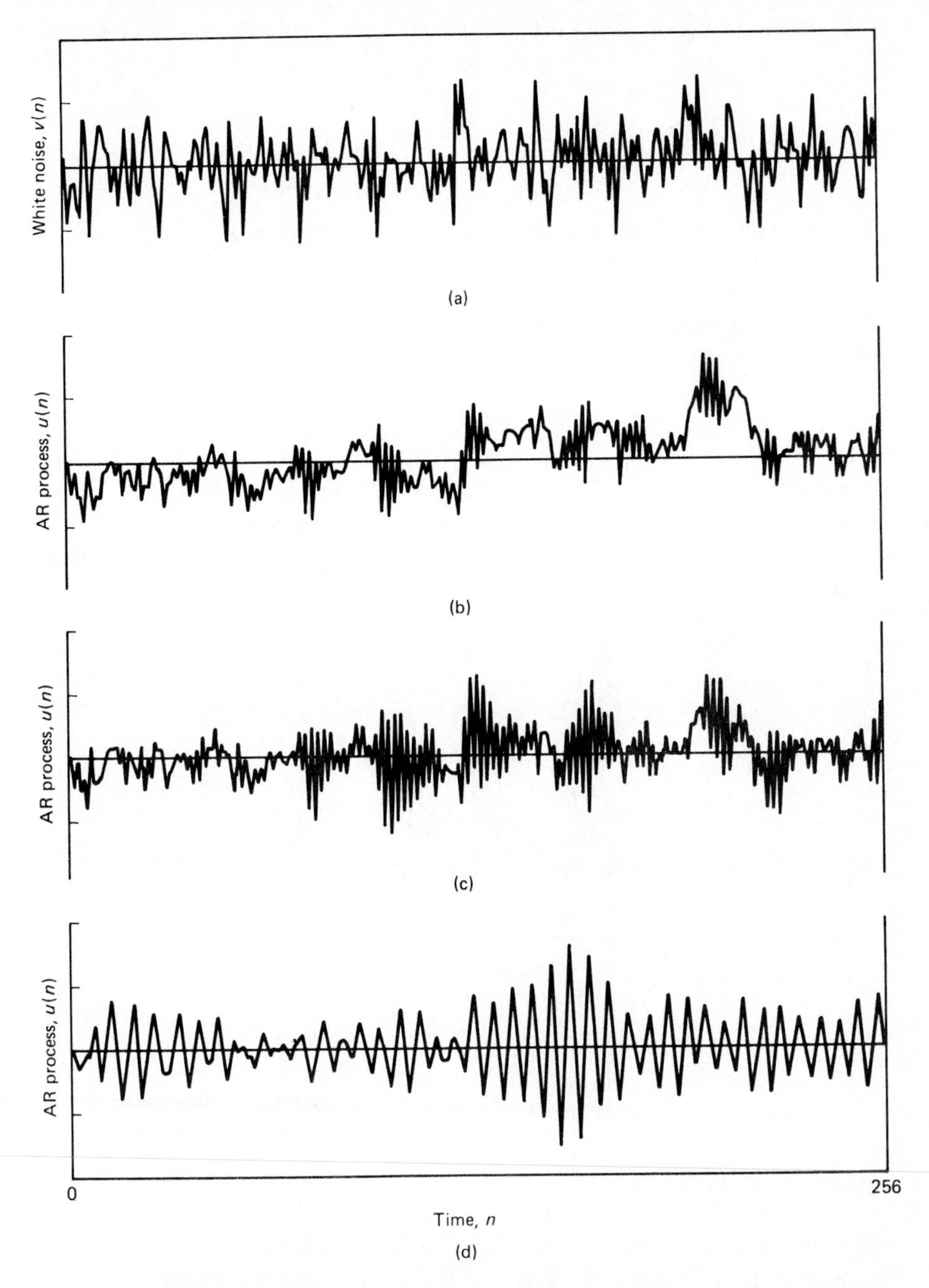

Figure 2.7 (a) One realization of white-noise input; (b), (c), (d) corresponding outputs of AR model of order 2 for parameters of Eqs. (2.79), (2.80), and (2.81), respectively.

Conditions for Asymptotic Stationarity

The second-order AR process $\{u(n)\}$ is described by the characteristic equation

$$1 + a_1 z^{-1} + a_2 z^{-2} = 0 \tag{2.73}$$

Let p_1 and p_2 denote the two roots of this equation:

$$p_1, p_2 = \tfrac{1}{2}\left(-a_1 \pm \sqrt{a_1^2 - 4a_2}\right) \tag{2.74}$$

To ensure the asymptotic stationarity of the AR process $\{u(n)\}$, we require that these two roots lie inside the unit circle in the z-plane. That is, both p_1 and p_2 must have a magnitude less than 1. This, in turn, requires that the AR parameters a_1 and a_2 lie in the triangular region defined by

$$\begin{aligned} -1 &\le a_2 + a_1 \\ -1 &\le a_2 - a_1 \\ -1 &\le a_2 \le 1 \end{aligned} \tag{2.75}$$

as shown in Fig. 2.8.

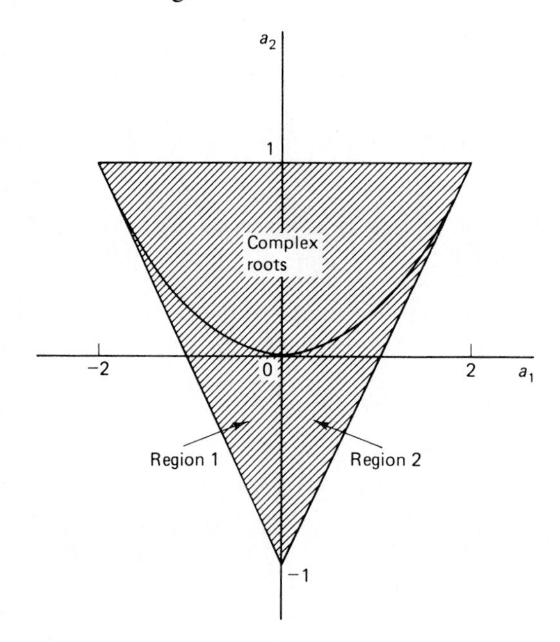

Figure 2.8 Permissible region for the AR parameters a_1 and a_2.

Autocorrelation Function

The autocorrelation function $r(m)$ of an asymptotically stationary AR process for lag m satisfies the difference equation (2.62). Hence, using this equation, we obtain the following second-order difference equation for the autocorrelation function of a second-order AR process:

$$r(m) + a_1 r(m-1) + a_2 r(m-2) = 0, \qquad m > 0 \tag{2.76}$$

For the initial values, we have (as will be explained later)

$$r(0) = \sigma_u^2$$

$$r(1) = \frac{-a_1}{1 + a_2}\sigma_u^2 \tag{2.77}$$

Thus, solving Eq. (2.76) for $r(m)$, we get (for $m > 0$)

$$r(m) = \sigma_u^2\left[\frac{p_1(p_2^2 - 1)}{(p_2 - p_1)(p_1p_2 + 1)}p_1^m - \frac{p_2(p_1^2 - 1)}{(p_2 - p_1)(p_1p_2 + 1)}p_2^m\right] \tag{2.78}$$

where p_1 and p_2 are defined by Eq. (2.74).

There are two specific cases to be considered, depending on whether the roots p_1 and p_2 are real or complex valued, as described next.

Case 1: *Real Roots.* This case occurs when

$$a_1^2 - 4a_2 > 0$$

which corresponds to regions 1 and 2 below the parabolic boundary in Fig. 2.8. In region 1, the autocorrelation function remains positive as it damps out, corresponding to a positive dominant root. This situation is illustrated in Fig. 2.9(a) for the AR parameters

$$\begin{aligned} a_1 &= -0.10 \\ a_2 &= -0.8 \end{aligned} \tag{2.79}$$

In Fig. 2.7(b), we show the time variation of the output of the model in Fig. 2.6 [with a_1 and a_2 assigned the values given in Eq. (2.79)]. This output is produced by the white-noise input shown in Fig. 2.7(a).

In region 2 of Fig. 2.8, the autocorrelation function alternates in sign as it damps out, corresponding to a negative dominant root. This situation is illustrated in Fig. 2.9(b) for the AR parameters

$$\begin{aligned} a_1 &= 0.1 \\ a_2 &= -0.8 \end{aligned} \tag{2.80}$$

In Fig. 2.7(c) we show the time variation of the output of the model in Fig. 2.6 [with a_1 and a_2 assigned the values given in Eq. (2.80)]. This output is also produced by the white-noise input shown in Fig. 2.7(a).

Case 2: *Complex-Conjugate Roots.* This occurs when

$$a_1^2 - 4a_2 < 0$$

which corresponds to the shaded region shown in Fig. 2.8 above the parabolic boundary. In this case, the autocorrelation function displays a pseudoperiodic behavior, as illustrated in Fig. 2.9(c) for the AR parameters

$$\begin{aligned} a_1 &= -0.975 \\ a_2 &= 0.95 \end{aligned} \tag{2.81}$$

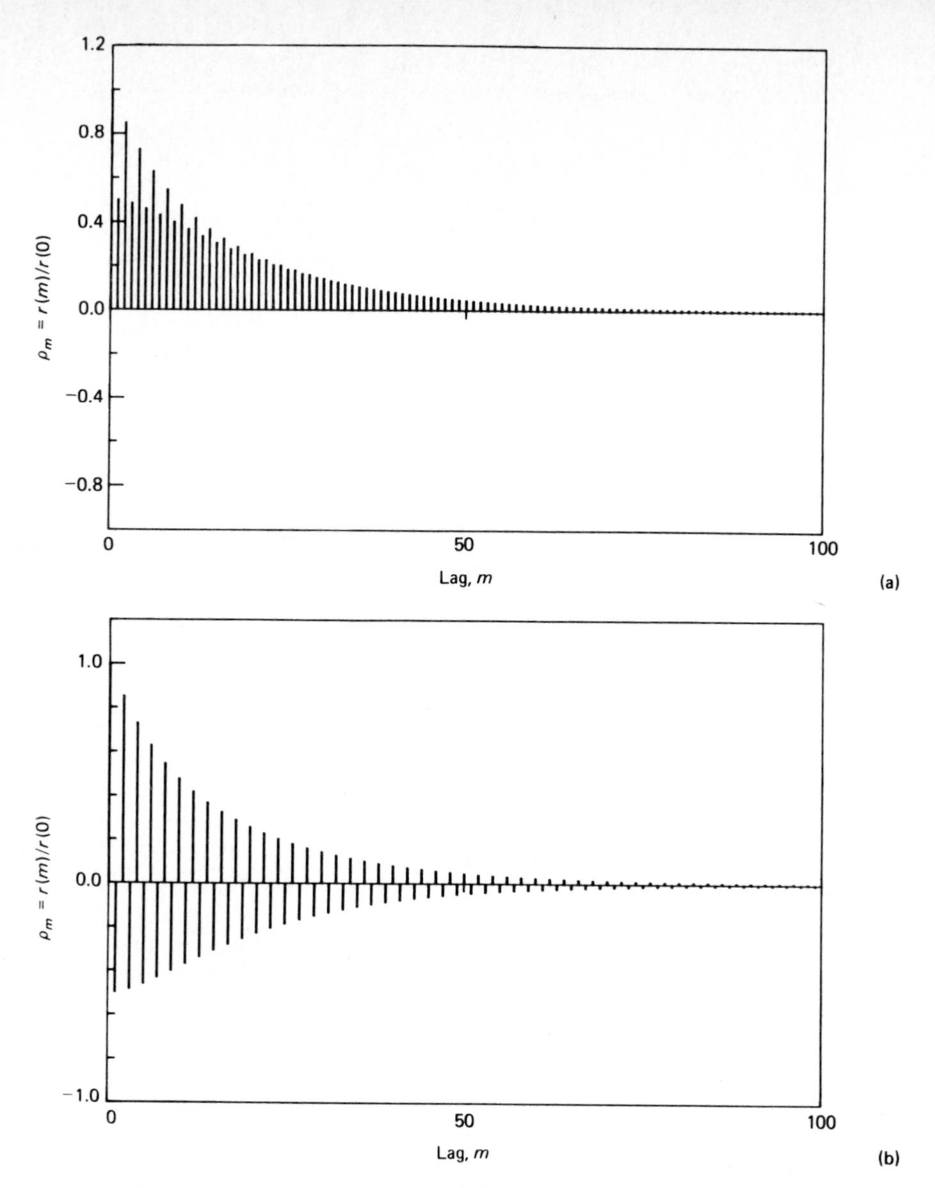

Figure 2.9 Plots of normalized autocorrelation function of real-valued AR(2) process with eigenvalue spread $\chi(\mathbf{R}) = 3$: (a) $r(1) > 0$; (b) $r(1) < 0$; (c) conjugate roots.

In Fig. 2.7(d) we show the time variation of the output of the model in Fig. 2.6 [with a_1 and a_2 assigned the values given in Eq. (2.81)], which is produced by the white-noise input shown in Fig. 2.7(a).

Yule–Walker Equations

Substituting the value $M = 2$ for the AR model order in Eq. (2.65), we get the following Yule–Walker equations for the second-order AR process:

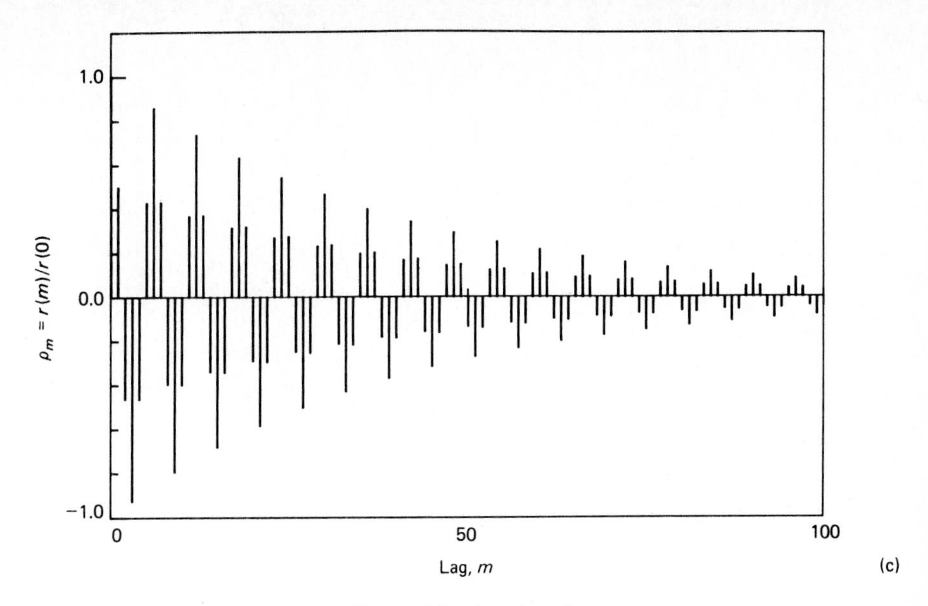

Figure 2.9 *(continued)*

$$\begin{bmatrix} r(0) & r(1) \\ r(1) & r(0) \end{bmatrix} \begin{bmatrix} w_1 \\ w_2 \end{bmatrix} = \begin{bmatrix} r(1) \\ r(2) \end{bmatrix} \tag{2.82}$$

where we have used the fact that $r(-1) = r(1)$ for a real-valued process. Solving Eq. (2.82) for w_1 and w_2, we get

$$\begin{aligned} w_1 &= -a_1 = \frac{r(1)[r(0) - r(2)]}{r^2(0) - r^2(1)} \\ w_2 &= -a_2 = \frac{r(0)r(2) - r^2(1)}{r^2(0) - r^2(1)} \end{aligned} \tag{2.83}$$

We may also use Eq. (2.82) to express $r(1)$ and $r(2)$ in terms of the AR parameters a_1 and a_2 as follows:

$$\begin{aligned} r(1) &= \frac{-a_1}{1 + a_2}\sigma_u^2 \\ r(2) &= \left(-a_2 + \frac{a_1^2}{1 + a_2}\right)\sigma_u^2 \end{aligned} \tag{2.84}$$

where $\sigma_u^2 = r(0)$. This solution explains the initial values for $r(0)$ and $r(1)$ that were quoted in Eq. (2.77).

The conditions for asymptotic stationarity of the second-order AR process are given in terms of the AR parameters a_1 and a_2 in Eq. (2.75). Using the expressions for $r(1)$ and $r(2)$ in terms of a_1 and a_2, given in Eq. (2.84), we may reformulate the conditions for asymptotic stationarity as follows:

$$-1 < \rho_1 < 1$$
$$-1 < \rho_2 < 1 \tag{2.85}$$
$$\rho_1^2 < \frac{1}{2}(1 + \rho_2)$$

where ρ_1 and ρ_2 are the normalized *correlation coefficients* defined by

$$\left.\begin{aligned} \rho_1 &= \frac{r(1)}{r(0)} \\ \text{and} \qquad & \\ \rho_2 &= \frac{r(2)}{r(0)} \end{aligned}\right\} \tag{2.86}$$

Figure 2.10 shows the admissible region for ρ_1 and ρ_2.

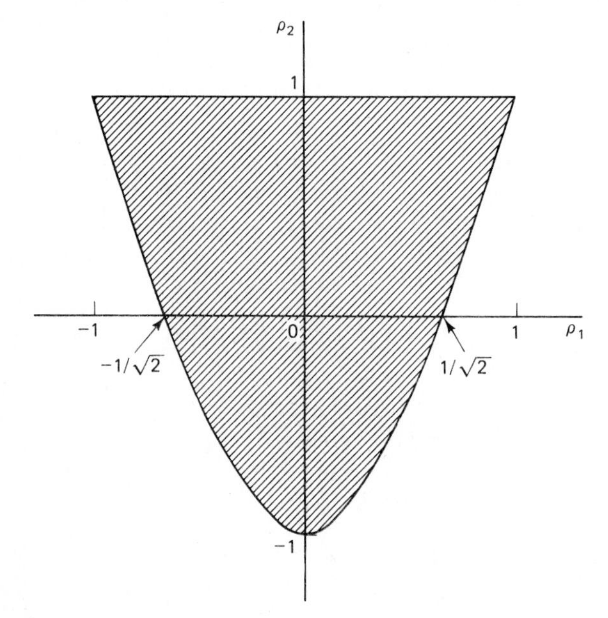

Figure 2.10 Permissible region for parameters of second-order AR process in terms of normalized correlation coefficients ρ_1 and ρ_2.

Variance of the White-Noise Process

Putting $M = 2$ in Eq. (2.71), we may express the variance of the white-noise process $\{v(n)\}$ as

$$\sigma_v^2 = r(0) + a_1 r(1) + a_2 r(2) \tag{2.87}$$

Next, substituting Eq. (2.84) in (2.87), and solving for $\sigma_u^2 = r(0)$, we get

$$\sigma_u^2 = \left(\frac{1 + a_2}{1 - a_2}\right) \frac{\sigma_v^2}{[(1 + a_2)^2 - a_1^2]} \tag{2.88}$$

For the three sets of AR parameters considered previously, we thus find that the variance of the white noise $\{v(n)\}$ has the values given in Table 2.1, assuming that $\sigma_u^2 = 1$.

TABLE 2.1 AR PARAMETER VARIANCE

a_1	a_2	σ_v^2
−0.10	−0.8	0.27
0.1	−0.8	0.27
−0.975	0.95	0.0731

2.10 SELECTING THE ORDER OF THE MODEL

The representation of a stochastic process by a linear model may be used for synthesis or analysis. In *synthesis,* we generate a desired time series by assigning a prescribed set of values to the parameters of the model and feeding it with white noise of zero mean and prescribed variance. In *analysis,* on the other hand, we *estimate* the parameters of the model by processing a given time series of finite length. Insofar as the estimation is statistical, we need an appropriate measure of the fit between the model and the observed data. This implies that unless we have some prior information, the estimation procedure should include a criterion for selecting the *model order* (i.e., the number of independently adjusted parameters in the model). In the case of an AR process defined by Eq. (2.42), the model order equals M. In the case of an MA process defined by Eq. (2.52), the model order equals K. In the case of an ARMA process defined by Eq. (2.53), the model order equals (M, K). Various criteria for model-order selection are described in the literature (Priestley, 1981; Kay, 1988). In this section we describe two important criteria for selecting the order of the model, one of which was pioneered by Akaike (1973, 1974) and the other by Rissanen (1978) and Schwartz (1978). Akaike's criterion, being the older of the two, is more widely known. Both criteria, however, result from the use of information-theoretic arguments.

An Information-Theoretic Criterion

Let $u_i = u(i)$, $i = 1, 2, \ldots, N$, denote the data obtained by N independent observations of a stationary discrete-time stochastic process, and $g(u_i)$ denote the probability density function of u_i. Let $f_U(u_i|\hat{\boldsymbol{\theta}}_m)$ denote the conditional probability density function of u_i, given $\hat{\boldsymbol{\theta}}_m$, where $\hat{\boldsymbol{\theta}}_m$ is the *estimated* vector of parameters that model the process. Let m be the order of the model so that we may write

$$\hat{\boldsymbol{\theta}}_m = \begin{bmatrix} \hat{\theta}_{1m} \\ \hat{\theta}_{2m} \\ \vdots \\ \hat{\theta}_{mm} \end{bmatrix} \tag{2.89}$$

We thus have several models that compete with each other to represent the process of interest. *An information-theoretic criterion* (AIC) proposed by Akaike selects the model for which the quantity

$$\text{AIC}(m) = -2L(\hat{\boldsymbol{\theta}}_m) + 2m \tag{2.90}$$

is a minimum. The function $L(\hat{\boldsymbol{\theta}}_m)$ is defined by

$$L(\hat{\boldsymbol{\theta}}_m) = \max \sum_{i=1}^{N} \ln f_U(u_i|\hat{\boldsymbol{\theta}}_m) \tag{2.91}$$

where ln denotes the natural logarithm. The criterion of Eq. (2.91) is derived by minimizing the *Kullback–Leibler mean information,*[4] which is used to provide a measure of the separation or distance between the "unknown" true probability density function $g(u)$ and the conditional probability density function $f_U(u|\hat{\boldsymbol{\theta}}_m)$ given by the model in the light of the observed data.

The function $L(\hat{\boldsymbol{\theta}}_m)$, constituting the first term on the right side of Eq. (2.90), except for a scalar, is recognized as the *log-likelihood* of the *maximum-likelihood estimates*[5] of the parameters in the model. The second term, $2m$, represents a *model complexity penalty* that makes AIC(m) an estimate of the Kullback-Leibler mean information.

The first term of Eq. (2.90) tends to decrease rapidly with model order m. On the other hand, the second term increases linearly with m. The result is that if we plot AIC (m) versus model order m the graph will, in general, show a definite minimum value, and the *optimum order* of the model is determined by that value of m at which AIC(m) attains its minimum value. The minimum value of AIC is called MAIC (minimum AIC).

Minimum Description Length Criterion

Rissanen (1978, 1989) and Schwartz (1978) question the theoretical validity of the structure-dependent term, $2m$, in Akaike's information-theoretic criterion of Eq. (2.90). They use entirely different approaches to solve the statistical model identification problem.

Rissanen uses information-theoretic ideas, starting with the notion that the number of digits (i.e., *length*) required to *encode* a set of observations $u(1)$, $u(2)$, . . . , $u(N)$ depends on the model that is assumed to have generated the observed data. Accordingly, Rissanen selects the model that minimizes the code length of the observed data.[6]

Schwartz, on the other hand, uses a *Bayesian* approach. In particular, he studies the asymptotic behavior of Bayes estimators under a special class of *priors*. These

[4] In Akaike (1973, 1974, 1977) and Ulrych and Ooe (1983), the criterion of Eq. (2.90) is derived from the principle of minimizing the expectation $E[I(g;f(.|\hat{\boldsymbol{\theta}}_m)]$, where

$$I(g;f(.|\hat{\boldsymbol{\theta}}_m)) = \int_{-\infty}^{\infty} g(u) \ln g(u)\, du - \int_{-\infty}^{\infty} g(u) \ln f_U(u|\hat{\boldsymbol{\theta}}_m)\, du$$

We refer to $I(g;f(.|\hat{\boldsymbol{\theta}}))$ as the *Kullback–Leibler mean information* for discrimination between $g(u)$ and $f_U(u|\hat{\boldsymbol{\theta}}_m)$ (Kullback and Leibler, 1951). The idea is to minimize the information added to the time series by modeling it as an AR, MA, or ARMA process of finite order, since any information added is virtually false information in a real-world situation. Since $g(u)$ is fixed and unknown, the problem reduces to one of maximizing the second term that makes up $I(g;f(.|\hat{\boldsymbol{\theta}}_m))$.

[5] For a discussion of the method of maximum likelihood, see Appendix D.

[6] The *minimum description length* of individual recursively definable objects has been studied by Kolmogorov (1968) and others.

priors put positive probability on the subspaces that correspond to the competing models. The decision is made by selecting the model that yields the *maximum a posteriori probability*.

It turns out that, in the large-sample limit, the two approaches taken by Rissanen and Schwartz yield essentially the same criterion, defined by (Wax and Kailath, 1985):

$$\text{MDL}(m) = -L(\hat{\boldsymbol{\theta}}_m) + \tfrac{1}{2} m \ln N \tag{2.92}$$

where m is the number of independently adjusted parameters in the model, and N is the number of observations. As with Akaike's information-theoretic criterion, the function $L(\hat{\boldsymbol{\theta}}_m)$ is the log-likelihood of the maximum-likelihood estimates of the parameters in the model. In Eq. (2.92), we have chosen to refer to the criterion as MDL (for *minimum description length*) in recognition of the fact that Rissanen's derivation is more general; Schwartz's derivation is restricted to the case that the observations are independent and come from an exponential distribution.

Discussion

Basically, the minimum description length criterion of Eq. (2.92) differs from Akaike's information-theoretic criterion in Eq. (2.90) only in that the number of independently adjusted parameters in the model is multiplied by $\frac{1}{2} \ln N$, where N is the number of observations. Accordingly, we find that in general the minimum description length criterion leans more than Akaike's criterion toward lower-order models.

It has been shown by Rissanen (1978) and Wax and Kailath (1985) that the minimum description length criterion is a *consistent* model-order estimator, whereas Akaike's information-theoretic criterion is not. In this context we say that the criterion is consistent if it converges to the true model order as the sample size (i.e., the number of observations N) increases. For small sample sizes, however, the consistency of these two criteria is not as well defined; nevertheless, the results of computer simulation experiments appear to show that Akaike's information criterion may yield more accurate model-order estimates than the minimum description length criterion.

2.11 COMPLEX GAUSSIAN PROCESSES

Gaussian stochastic processes, or simply *Gaussian processes*, are frequently encountered in both theoretical and applied analysis. In this section we present a summary of some important properties of Gaussian processes that are *complex valued*.[7]

Let $\{u(n)\}$ denote a complex Gaussian process consisting of N samples. For the first- and second-order statistics of this process, we assume the following:

1. A *mean* of zero as shown by

$$\mu = E[u(n)] = 0, \qquad \text{for } n = 1, 2, \ldots, N \tag{2.93}$$

[7] For a detailed treatment of complex Gaussian processes, see the book by Miller (1974). Properties of complex Gaussian processes are also discussed in Kelly et al. (1960), Reed (1962), and McGee (1971).

2. An *autocorrelation function* denoted by

$$r(k) = E[u(n)u^*(n-k)], \qquad k = 0, 1, \ldots, N-1 \tag{2.94}$$

Equation (2.94) implies wide-sense stationarity of the process. Knowledge of the mean μ and the autocorrelation function $r(k)$ for varying values of lag k is indeed sufficient for the complete characterization of the complex Gaussian process $\{u(n)\}$. In particular, it may be shown that the *joint probability density function* of N samples of the process is as follows (Kelly et al., 1960):

$$\mathbf{f}_{\mathbf{U}}(\mathbf{u}) = \frac{1}{(2\pi)^N \det \mathbf{\Lambda}} \exp\left(-\frac{1}{2}\mathbf{u}^H\mathbf{\Lambda}^{-1}\mathbf{u}\right) \tag{2.95}$$

where $\mathbf{u}$ is the N-by-1 data vector, that is,

$$\mathbf{u}^{\mathrm{T}} = [u(1), u(2), \ldots, u(N)] \tag{2.96}$$

and $\mathbf{\Lambda}$ is the N-by-N Hermitian-symmetric *moment matrix* of the process, defined in terms of the correlation matrix $\mathbf{R} = \{r(k)\}$ as

$$\begin{aligned}\mathbf{\Lambda} &= \tfrac{1}{2}E[\mathbf{u}\mathbf{u}^H] \\ &= \tfrac{1}{2}\mathbf{R}\end{aligned} \tag{2.97}$$

Note that the joint probability density function $f_{\mathrm{U}}(\mathbf{u})$ is $2N$-dimensional, where the factor 2 accounts for the fact that each of the N samples of the process has a real and an imaginary part. Note also that the probability density function of a single sample $u(n)$ of the process, which is a special case of Eq. (2.95), is given by

$$f_U(u) = \frac{1}{\pi\sigma^2} \exp\left(-\frac{|u|^2}{\sigma^2}\right) \tag{2.98}$$

where $|u|$ is the magnitude of the sample $u(n)$ and σ^2 is its variance.

Based on the representation described herein, we may now summarize some important properties of a *zero-mean complex Gaussian process* $\{u(n)\}$ *that is wide-sense stationary* as follows:

1. The process $\{u(n)\}$ is *stationary in the strict sense*.
2. The process $\{u(n)\}$ is *circularly complex* in the sense that any two different samples $u(n)$ and $u(k)$ of the process satisfy the condition:

$$E[u(n)u(k)] = 0, \qquad \text{for } n \neq k \tag{2.99}$$

It is for this reason that the process $\{u(n)\}$ is often referred to as a *circularly complex Guassian process*.

3. Suppose that $u_n = u(n)$, for $n = 1, 2, \ldots, N$, are samples picked from a zero-mean, complex Gaussian process $\{u(n)\}$. We may thus state Property 3 in two parts (Reed, 1962):

(a) If $k \neq l$, then

$$E[u_{s_1}^* u_{s_2}^* \cdots u_{s_k}^* u_{t_1} u_{t_2} \cdots u_{t_l}] = 0 \tag{2.100}$$

where s_i and t_j are integers selected from the available set $\{1, 2, \ldots, N\}$.

(b) If the total number of samples N is an even integer, and $k = l = N/2$, then

$$E[u_{s_1}^* u_{s_2}^* \cdots u_{s_l}^* u_{t_1} u_{t_2} \cdots u_{t_l}] = E[u_{s_{\pi(1)}}^* u_{t_1}] E[u_{s_{\pi(2)}}^* u_{t_2}] \cdots E[u_{s_{\pi(l)}}^* u_{t_l}] \tag{2.101}$$

where π is a permutation of the set of integers $\{1, 2, \ldots, l\}$, and $\pi(j)$ is the jth element of that permutation. For the set of integers $\{1, 2, \ldots, l\}$ we have a total of $l!$ possible permutations. This means that the right side of Eq. (2.101) consists of the product of $l!$ expectation product terms. Equation (2.101) is called the *Gaussian moment factoring theorem*.

Example 2

Consider the case of $N = 4$, for which the complex Gaussian process $\{u(n)\}$ consists of the four samples u_1, u_2, u_3, and u_4. Hence, the use of the Gaussian moment factoring theorem given in Eq. (2.101) yields the following useful identity:

$$E[u_1^* u_2^* u_3 u_4] = E[u_1^* u_3]\, E[u_2^* u_4] + E[u_2^* u_3]\, E[u_1^* u_4] \tag{2.102}$$

For other useful identities derived from this theorem, see Problem 12.

PROBLEMS

1. The sequences $\{y(n)\}$ and $\{u(n)\}$ are related by the difference equation

$$y(n) = u(n + a) - u(n - a)$$

where a is a constant. Evaluate the autocorrelation function of $\{y(n)\}$ in terms of that of $\{u(n)\}$.

2. Consider a correlation matrix $\mathbf{R}$ for which the inverse matrix $\mathbf{R}^{-1}$ exists. Show that $\mathbf{R}^{-1}$ is Hermitian.

3. **(a)** Equation (2.26) relates the $(M + 1)$-by-$(M + 1)$ correlation matrix $\mathbf{R}_{M+1}$, pertaining to the observation vector $\mathbf{u}_{M+1}(n)$ taken from a stationary stochastic process, to the M-by-M correlation matrix $\mathbf{R}_M$ of the observation vector $\mathbf{u}_M(n)$ taken from the same process. Evaluate the inverse of the correlation matrix $\mathbf{R}_{M+1}$ in terms of the inverse of the correlation matrix $\mathbf{R}_M$.

(b) Repeat your evaluation using Eq. (2.27).

4. A first-order autoregressive (AR) process $\{u(n)\}$ that is real-valued satisfies the real-valued difference equation

$$u(n) + a_1\, u(n - 1) = v(n)$$

where a_1 is a constant, and $\{v(n)\}$ is a white-noise process of variance σ_v^2. Such a process is also referred to as a *first-order Markov process*.

(a) Show that if $\{v(n)\}$ has a nonzero mean the AR process $\{u(n)\}$ is nonstationary.

(b) For the case when $\{v(n)\}$ has zero mean, and the constant a_1 satisfies the condition $|a_1| < 1$, show that the variance of $\{u(n)\}$ equals

$$\text{Var}[u(n)] = \frac{\sigma_v^2}{1 - a_1^2}$$

(c) For the conditions specified in part (b), find the autocorrelation function of the AR process $\{u(n)\}$. Sketch this autocorrelation function for the two cases $0 < a_1 < 1$ and $-1 < a_1 < 0$.

5. Consider an autoregressive process $\{u(n)\}$ of order 2, described by the difference equation

$$u(n) = u(n-1) - 0.5u(n-2) + v(n)$$

where $\{v(n)\}$ is a white-noise process of zero mean and variance 0.5.

(a) Write the Yule–Walker equations for the process.

(b) Solve these two equations for the autocorrelation function values $r(1)$ and $r(2)$.

(c) Find the variance of $\{u(n)\}$.

6. Consider a wide-sense stationary process that is modeled as an AR process $\{u(n)\}$ of order M. The set of parameters made up of the average power P_0 and the AR coefficients $a_1, a_2, \ldots, a_M$ bear a one-to-one correspondence with the autocorrelation sequence $r(0), r(1), r(2), \ldots, r(M)$, as shown by

$$\{r(0), r(1), r(2), \ldots, r(M)\} \rightleftharpoons \{P_0, a_1, a_2, \ldots, a_M\}$$

Justify the validity of this statement.

7. Evaluate the transfer functions of the following two stochastic models:

(a) The MA model of Fig. 2.3.

(b) The ARMA model of Fig. 2.4.

(c) Specify the conditions for which the transfer function of the ARMA model of Fig. 2.4 reduces (1) to that of an AR model, and (2) to that of an MA model.

8. Consider an MA process $\{x(n)\}$ of order 2 described by the difference equation

$$x(n) = v(n) + 0.75v(n-1) + 0.25v(n-2)$$

where $\{v(n)\}$ is a zero-mean white-noise process of unit variance. The requirement is to approximate this process by an AR process $\{u(n)\}$ of order M. Do this approximation for the following orders:

(a) $M = 2$

(b) $M = 5$

(c) $M = 10$

Comment on your results.

9. A time series $\{u(n)\}$ from a wide-sense stationary stochastic process of zero mean and correlation matrix $\mathbf{R}$ is applied to an FIR filter of impulse response $\{w_n\}$. This impulse response defines the coefficient vector $\mathbf{w}$.

(a) Show that the average power of the filter output is equal to $\mathbf{w}^H\mathbf{R}\mathbf{w}$.

(b) How is the result in part (a) modified if the stochastic process at the filter input is a white noise of variance σ^2?

10. A discrete-time stochastic process $\{x(n)\}$ that is real valued consists of an AR process $\{u(n)\}$ and additive white noise process $\{v_2(n)\}$. The AR component is described by the difference equation

$$u(n) + \sum_{k=1}^{M} a_k u(n-k) = v_1(n)$$

where $\{a_k\}$ are the set of AR parameters and $\{v_1(n)\}$ is a white noise process that is independent of $\{v_2(n)\}$. Show that $\{x(n)\}$ is an ARMA process described by

$$x(n) = -\sum_{k=1}^{M} a_k x(n-k) + \sum_{k=1}^{M} b_k e(n-k) + e(n)$$

where $\{e(n)\}$ is a white noise process. How are the MA parameters $\{b_k\}$ defined? How is the variance of $e(n)$ defined?

11. A general linear complex-valued process $\{u(n)\}$ is described by

$$u(n) = \sum_{k=0}^{\infty} b_k^* v(n-k)$$

where $\{v(n)\}$ is a white noise process, and b_k is a complex coefficient. Justify the following statements:

(a) If the process $\{v(n)\}$ is Gaussian, then the original process $\{u(n)\}$ is also Gaussian.

(b) Conversely, a Gaussian process $\{u(n)\}$ implies that the process $\{v(n)\}$ is Gaussian.

12. Consider a complex Gaussian process $\{u(n)\}$. Let $u(n) = u_n$. Using the Gaussian moment factoring theorem, demonstrate the following identities:

(a) $E[(u_1^* u_2)^k] = k!\,(E[u_1^* u_2])^k$

(b) $E[|u|^{2k}] = k!\,(E[|u|^2])^k$

Computer-Oriented Problems

13. Consider the second-order autoregressive (AR) process $\{u(n)\}$ described by

$$u(n) + a_1 u(n-1) + a_2 u(n-2) = v(n)$$

where a_1 and a_2 are real-valued constant coefficients, and $\{v(n)\}$ is a white-noise process of zero mean. The AR process $\{u(n)\}$ is normalized to have unit variance. In this problem the AR coefficients are varied in such a way that the characteristic equation of the process has complex conjugate roots. In particular, we wish to investigate the use of three sets of AR coefficients:

$$(1)\ a_1 = -0.195, \qquad a_2 = 0.95$$

$$(2)\ a_1 = -1.5955, \qquad a_2 = 0.95$$

$$(3)\ a_1 = -1.9114, \qquad a_2 = 0.95$$

For each set of parameters, perform the following computations:

(a) The variance of the noise process $\{v(n)\}$.

(b) Using a random-noise generator for $\{v(n)\}$, plot the corresponding time variation of the AR process $\{u(n)\}$.

(c) Plot the autocorrelation function of the AR process $\{u(n)\}$ for varying lags.

14. Repeat Problem 13 for the case when the coefficients are assigned each of the following three sets of values, in turn:

$$(1)\ a_1 = -0.020, \qquad a_2 = -0.8$$

$$(2)\ a_1 = -0.1636, \qquad a_2 = -0.8$$

$$(3)\ a_1 = -0.1960, \qquad a_2 = -0.8$$

In this case, the characteristic equation of the AR process has unequal roots.

3

Spectrum Analysis

The autocorrelation function represents the *time-domain description* of the second-order statistics of a stochastic process. The *frequency-domain description* of the second-order statistics of such a process is represented by the *power spectral density,* which is also commonly referred to as the *power spectrum* or simply *spectrum*. Indeed, the power spectral density of a stochastic process is firmly established as the most useful description of the time series commonly encountered in engineering and physical sciences.

This chapter is devoted to the definition of the power spectral density of a wide-sense stationary discrete-time stochastic process, the properties of power spectral density, and methods for its estimation. We begin the discussion by establishing a mathematical definition of the power spectral density of a stationary process in terms of the Fourier transform of a time series that represents a single realization of the process.

3.1 BASIC DEFINITION OF POWER SPECTRAL DENSITY

Consider a time series $\{u(n)\}$ that represents a *single realization* of a wide-sense stationary discrete-time stochastic process. For convenience of presentation, we will assume that the process has zero mean. Let the elements of the time series $\{u(n)\}$, assumed to be infinite length, be denoted by $u(0)$, $u(1)$, $u(2)$, ... Initially, we focus our attention on a *windowed* portion of this time series, defined by

$$u_N(n) = \begin{cases} u(n), & n = 0, 1, \ldots, N-1 \\ 0, & n > N \end{cases} \tag{3.1}$$

where N is the *total length (duration of the window)*. Clearly, we have

$$u(n) = \lim_{N \to \infty} u_N(n) \tag{3.2}$$

By definition, the *discrete-time Fourier transform* of the windowed time series $\{u(n)\}$ is given by

$$U_N(\omega) = \sum_{n=0}^{N-1} u_N(n) e^{-j\omega n} \tag{3.3}$$

where ω is the *angular frequency*, lying in the interval $[-\pi, \pi]$. In general, $U_N(\omega)$ is complex valued; specifically, its *complex conjugate* is given by

$$U_N^*(\omega) = \sum_{k=0}^{N-1} u_N^*(k) e^{j\omega k} \tag{3.4}$$

where the asterisk denotes complex conjugation. In Eq. (3.4) we have used the variable k to denote discrete time for reasons that will become apparent immediately. In particular, we may multiply Eq. (3.3) by (3.4) to express the squared magnitude of $U_N(n)$ as follows:

$$|U_N(\omega)|^2 = \sum_{n=0}^{N-1} \sum_{k=0}^{N-1} u_N(n) u_N^*(k) e^{-j\omega(n-k)} \tag{3.5}$$

Taking the statistical expectation of both sides of Eq. (3.5), and interchanging the order of expectation and double summation, we get

$$E[|U_N(\omega)|^2] = \sum_{n=0}^{N-1} \sum_{k=0}^{N-1} E[u_N(n) u_N^*(k)] e^{-j\omega(n-k)} \tag{3.6}$$

We now recognize that for the wide-sense stationary discrete-time stochastic process under discussion,

$$r_N(n-k) = E[u_N(n) u_N^*(k)] \tag{3.7}$$

where $r_N(n-k)$ is the autocorrelation function of the windowed process for lag $n-k$. Accordingly, we may rewrite Eq. (3.6) as

$$E[|U_N(\omega)|^2] = \sum_{n=0}^{N-1} \sum_{k=0}^{N-1} r_N(n-k) e^{-j\omega(n-k)} \tag{3.8}$$

Let

$$l = n - k$$

We may then rewrite Eq. (3.8) as follows:

$$\frac{1}{N} E[|U_N(\omega)|^2] = \sum_{l=-N+1}^{N-1} \left(1 - \frac{|l|}{N}\right) r_N(l) e^{-j\omega l} \tag{3.9}$$

Equation (3.9) may be interpreted as the discrete-time Fourier transform of the product of two time functions: the autocorrelation function $r_N(l)$ for lag l, and a triangular window $w_B(l)$ known as the *Bartlett window*. The latter function is defined by

$$w_B(l) = \begin{cases} 1 - \dfrac{|l|}{N}, & |l| \leq N - 1 \\ 0, & |l| \geq N \end{cases} \tag{3.10}$$

As N approaches infinity, the Bartlett window $w_B(l)$ approaches unity for all l. Correspondingly, we may write

$$\lim_{N\to\infty} \frac{1}{N} E[|U_N(\omega)|^2] = \sum_{l=-\infty}^{\infty} r(l)e^{-j\omega l} \tag{3.11}$$

where $r(l)$ is the autocorrelation function of the original time series $\{u(n)\}$, assumed to have infinite length. The quantity $U_N(\omega)$ is the discrete-time Fourier transform of a rectangular *windowed* portion of this time series that has length N.

Equation (3.11) leads us to define the quantity

$$S(\omega) = \lim_{N\to\infty} \frac{1}{N} E[|U_N(\omega)|^2] \tag{3.12}$$

where the quantity $|U_N(\omega)|^2/N$ is called the *periodogram* of the windowed time series $\{u_N(n)\}$. Note that the order of expectation and limiting operations indicated in Eq. (3.12) cannot be changed.

When the limit in Eq. (3.12) exists, the quantity $S(\omega)$ has the following interpretation (Priestley, 1981):

$S(\omega)\, d\omega$ = average of the contribution to the total power from components of a wide-sense stationary stochastic process with angular frequencies located between ω and $\omega + d\omega$; the average is taken over all possible realizations of the process (3.13)

Accordingly, the quantity $S(\omega)$ is termed the *power spectral density* of the process. Thus, equipped with the definition of power spectral density given in Eq. (3.12), we may now rewrite Eq. (3.11) as

$$S(\omega) = \sum_{l=-\infty}^{\infty} r(l)e^{-j\omega l} \tag{3.14}$$

In summary, Eq. (3.12) gives a basic definition of the power spectral density of a wide-sense stationary stochastic process, and Eq. (3.14) defines the mathematical relationship between the autocorrelation function and the power spectral density of such a process.

3.2 PROPERTIES OF POWER SPECTRAL DENSITY

Property 1. *The autocorrelation function and power spectral density of a wide-sense stationary stochastic process form a Fourier transform pair.*

Consider a wide-sense stationary stochastic process represented by the time series $\{u(n)\}$, assumed to be of infinite length. Let $r(l)$ denote the autocorrelation function of such a process for lag l, and let $S(\omega)$ denote its power spectral density. According to Property 1, these two quantities are related by the pair of relations:

$$S(\omega) = \sum_{l=-\infty}^{\infty} r(l)e^{-j\omega l}, \qquad -\pi \leq \omega \leq \pi \tag{3.15}$$

and

$$r(l) = \frac{1}{2\pi}\int_{-\pi}^{\pi} S(\omega)e^{j\omega l}\, d\omega, \qquad l = 0, \pm 1, \pm 2, \ldots \tag{3.16}$$

Equation (3.15) states that *the power spectral density is the discrete-time Fourier transform of the autocorrelation function*. On the other hand, Eq. (3.16) states that *the autocorrelation function is the inverse discrete-time Fourier transform of the power spectral density*. This fundamental pair of equations constitutes the *Einstein–Wiener–Khintchine relations*.

In a way, we already have a proof of this property. Specifically, Eq. (3.15) is merely a restatement of Eq. (3.14), previously established in Section 3.1. Equation (3.16) follows directly from this result by invoking the formula for the inverse discrete-time Fourier transform.

Property 2. *The power spectral density of a stationary discrete-time stochastic process is periodic;* that is,

$$S(\omega + 2k\pi) = S(\omega) \tag{3.17}$$

This shows that the power spectral density $S(\omega)$ is a periodic function of the angular frequency ω, with the period equal to 2π.

Property 3. *The power spectral density of a stationary discrete-time stochastic process is real.*

To prove this property, we rewrite Eq. (3.15) as

$$S(\omega) = r(0) + \sum_{k=1}^{\infty} r(k)e^{-j\omega k} + \sum_{k=-\infty}^{-1} r(k)e^{-j\omega k}$$

Replacing k with $-k$ in the third term on the right side of this equation, and recognizing that $r(-k) = r^*(k)$, we get

$$\begin{aligned} S(\omega) &= r(0) + \sum_{k=1}^{\infty} [r(k)e^{-j\omega k} + r^*(k)e^{j\omega k}] \\ &= r(0) + 2\sum_{k=1}^{\infty} \text{Re}[r(k)e^{-j\omega k}] \end{aligned} \tag{3.18}$$

where Re denotes the *real part operator*. Equation (3.18) shows that the power spectral density $S(\omega)$ is a real-valued function of ω.

Property 4. *The power spectral density of a real-valued stationary discrete-time stochastic process is even (i.e., symmetric); if the process is complex-valued, its power spectral density is not even.*

For a real-valued stochastic process, we find that $S(-\omega) = S(\omega)$, indicating that $S(\omega)$ is an even function of ω; that is, it is symmetric above the origin. If, however, the process is complex-valued, then $r(-k) = r^*(k)$, in which case we find that $S(-\omega) \neq S(\omega)$, and $S(\omega)$ is *not* an even function of ω.

Property 5. *The mean-square value of a stationary discrete-time stochastic process equals, except for the scaling factor* $1/2\pi$, *the area under the power spectral density curve for* $-\pi \leq \omega \leq \pi$.

This property follows directly from Eq. (3.16), evaluated for $l = 0$. For this condition, we may thus write

$$r(0) = \frac{1}{2\pi} \int_{-\pi}^{\pi} S(\omega)\, d\omega \tag{3.19}$$

Since $r(0)$ equals the mean-square value of the process, we see that Eq. (3.19) is a mathematical description of Property 5.

Property 6. *The power spectral density of a stationary discrete-time stochastic process is nonnegative.*

That is,

$$S(\omega) \geq 0, \qquad \text{for all } \omega \tag{3.20}$$

This property follows directly from the basic formula of Eq. (3.12), reproduced here for convenience:

$$S(\omega) = \lim_{N\to\infty} \frac{1}{N} E[|U_N(\omega)|^2] \tag{3.21}$$

We first note that $|U_N(\omega)|^2$, representing the squared magnitude of the discrete-time Fourier transform of a windowed portion of the time series $\{u(n)\}$, is nonnegative for all ω. The expectation $E[|U_N(\omega)|^2]$ is also nonnegative for all ω. Thus, using the basic definition of $S(\omega)$ in terms of $U_N(\omega)$, the property described by Eq. (3.20) follows immediately.

3.3 TRANSMISSION OF A STATIONARY PROCESS THROUGH A LINEAR FILTER

Consider a discrete-time filter that is both *linear* and *time invariant*. Let the filter be characterized by the *discrete transfer function* $H(z)$, defined as *the ratio of the z-transform of the filter output to the z-transform of the filter input*. Suppose that we feed the filter with a stationary discrete-time stochastic process of power spectral density $S(\omega)$, as in Fig. 3.1. Let $S_o(\omega)$ denote the power spectral density of the filter output. We may then write

$$S_o(\omega) = |H(e^{j\omega})|^2 S(\omega) \tag{3.22}$$

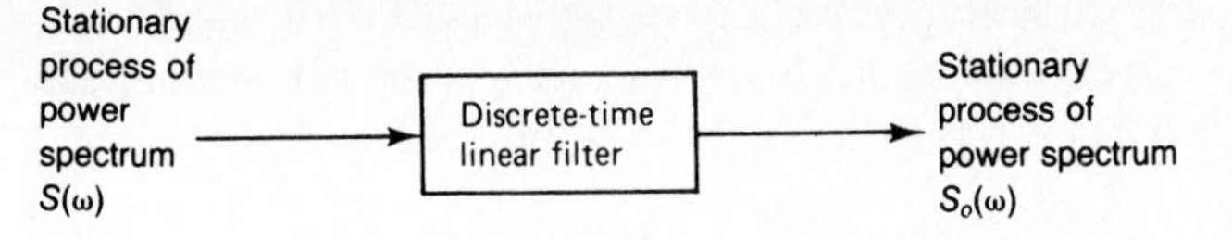

Figure 3.1 Transmission of stationary process through a discrete-time linear filter.

where $H(e^{j\omega})$ is the *frequency response* of the filter. The frequency response $H(e^{j\omega})$ equals the discrete transfer function $H(z)$ evaluated on the unit circle in the z-plane. The important feature of this result is that the value of the output spectral density at angular frequency ω depends purely on the squared *amplitude response* of the filter and the input power spectral density at the same angular frequency ω.

Equation (3.22) is a fundamental relation in stochastic process theory. To prove it, we may proceed as follows. Let the time series $\{y(n)\}$ denote the filter output in Fig. 3.1, produced in response to the time series $\{u(n)\}$ applied to the filter input. Assuming that $\{u(n)\}$ represents a single realization of a wide-sense-stationary discrete-time stochastic process, we find that $\{y(n)\}$ also represents a single realization of a wide-sense stationary discrete-time stochastic process modified by the filtering operation (see Problem 2). From the definition of the frequency response $H(e^{j\omega})$, we have

$$Y(\omega) = H(e^{j\omega})U(\omega) \tag{3.23}$$

where $U(\omega)$ and $Y(\omega)$ are the discrete-time Fourier transforms of the filter input $\{u(n)\}$ and the filter output $\{y(n)\}$, respectively. Similarly, for windowed portions of these two time series, we may write

$$Y_N(\omega) = H(e^{j\omega})U_N(\omega) \tag{3.24}$$

where the subscript N is the total length of the window. Writing the basic formula of Eq. (3.21) in terms of the pertinent quantities that refer to the filter output, we have

$$S_o(\omega) = \lim_{N\to\infty} \frac{1}{N} E[|Y_N(\omega)|^2] \tag{3.25}$$

Substituting Eq. (3.24) in (3.25), and recognizing that the squared frequency response $|H(e^{j\omega})|^2$ is a constant insofar as the operation of expectation is concerned, we get

$$\begin{aligned} S_o(\omega) &= |H(e^{j\omega})|^2 \lim_{N\to\infty} \frac{1}{N} E[|U_N(\omega)|^2] \\ &= |H(e^{j\omega})|^2 S(\omega) \end{aligned}$$

This completes the proof of Eq. (3.22).

Power Spectrum Analyzer

Suppose that the discrete-time linear filter in Fig. 3.1 is designed to have a bandpass characteristic. That is, the amplitude response of the filter is defined by

$$|H(e^{j\omega})| = \begin{cases} 1, & |\omega - \omega_c| \le \Delta\omega \\ 0, & \text{remainder of the interval } -\pi \le \omega \le \pi \end{cases} \tag{3.26}$$

This amplitude response is depicted in Fig. 3.2. We assume that the *angular bandwidth* of the filter, $2\Delta\omega$, is small enough for the spectrum inside this bandwidth to be essentially constant. Then using Eq. (3.22) we may write

$$S_o(\omega) = \begin{cases} S(\omega_c) & |\omega - \omega_c| \le \Delta\omega \\ 0, & \text{remainder of the interval } -\pi \le \omega \le \pi \end{cases}$$

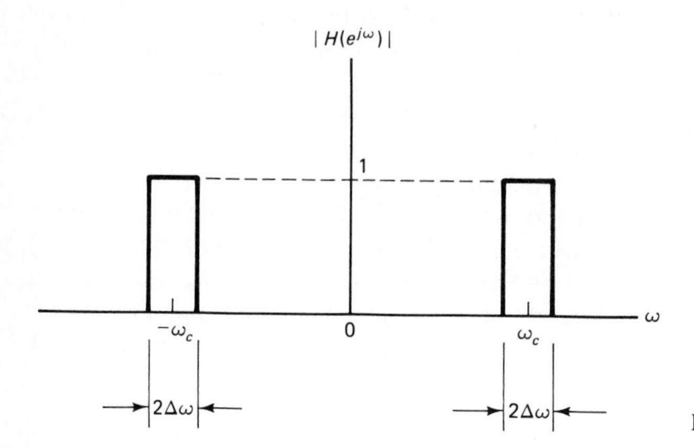

Figure 3.2 Ideal bandpass characteristic.

Next, using Property 5 of the power spectral density, described by Eq. (3.19), we may express the mean-square value of the filter output

$$\begin{aligned} P_o &= \frac{1}{2\pi}\int_{-\pi}^{\pi} S_o(\omega)d\omega \\ &= \frac{2}{\pi} S(\omega_c)\,\Delta\omega \end{aligned} \tag{3.27}$$

Equivalently, we may write

$$S(\omega_c) = \frac{\pi P_o}{2\Delta\omega} \tag{3.28}$$

Equation (3.28) states that the value of the power spectral density of the filter input $\{u(n)\}$, measured at the center frequency ω_c of the filter, is equal to the mean-square value P_o of the filter output, scaled by a constant factor. We may thus use Eq. (3.28) as the mathematical basis for building a *power spectrum analyzer,* as depicted in Fig. 3.3. Ideally, the discrete-time bandpass filter employed here should satisfy two requirements: *fixed bandwidth* and *adjustable center frequency*. Clearly, in a practical filter design, we can only approximate these two ideal requirements. Note also that the reading of the *power meter* at the output end of Fig. 3.3 has the same physical significance as the mean-square value P_o, except for a scale factor.

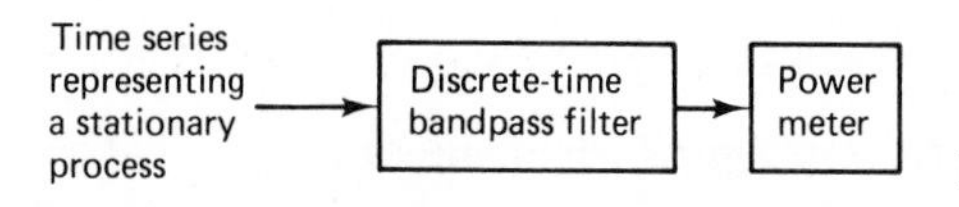

Figure 3.3 Power spectrum analyzer.

Example 1: White Noise

A stochastic process of zero mean is said to be white if its power spectral density $S(\omega)$ is constant for all frequencies, as shown by

$$S(\omega) = \sigma^2, \qquad \text{for } -\pi \le \omega \le \pi$$

where σ^2 is the variance of a sample taken from the process. Suppose that this process is passed through a bandpass discrete-time filter, characterized as in Fig. 3.2. Hence, from Eq. (3.27), we find that the mean-square value of the filter output is

$$P_o = \frac{2\sigma^2 \Delta\omega}{\pi}$$

The autocorrelation function of the white noise at the filter input is

$$r(\tau) = \sigma^2 \delta_{\tau,0}$$

where $\delta_{\tau,0}$, is the Kronecker delta:

$$\delta_{\tau,0} = \begin{cases} 1, & \tau = 0 \\ 0, & \text{otherwise} \end{cases}$$

3.4 CRAMÉR SPECTRAL REPRESENTATION FOR A STATIONARY PROCESS

Equation (3.12) provides one way of defining the power spectral density of a wide-sense stationary process. Another way of defining the power spectral density is to use the *Cramér spectral representation for a stationary process*. According to this representation, a sample $u(n)$ of a discrete-time stochastic process is written as an inverse Fourier transform (Thomson, 1982, 1988):

$$u(n) = \frac{1}{2\pi} \int_{-\pi}^{\pi} e^{j\omega n} dZ(\omega) \tag{3.29}$$

If the process represented by the time series $\{u(n)\}$ is wide-sense stationary with no periodic components, then the *increment process* $\{dZ(\omega)\}$ has the following three properties:

1. The mean of the increment process $\{dZ(\omega)\}$ is zero; that is,

$$E[dZ(\omega)] = 0, \qquad \text{for all } \omega \tag{3.30}$$

2. The energy of the increment process $\{dZ(\omega)\}$ at different frequencies is uncorrelated; that is,

$$E(dZ(\omega)dZ^*(\nu)] = 0, \qquad \text{for } \nu \neq \omega \tag{3.31}$$

3. The expected value of $|dZ(\omega)|^2$ defines the spectrum $S(\omega)\, d\omega$; that is,

$$E[|dZ(\omega)|^2] = S(\omega)\, d\omega \tag{3.32}$$

In other words, for a wide-sense stationary discrete-time stochastic process represented by the time series $\{u(n)\}$, the increment process $\{dZ(\omega)\}$ defined by Eq. (3.29) is a *zero-mean orthogonal process.*

Equation (3.32), in conjunction with Eq. (3.30), provides another basic definition for the power spectral density $S(\omega)$. We justify the validity of this statement by proceeding as follows. Let $U(\omega)$ denote the *discrete-time Fourier transform* of the time series $\{u(n)\}$, as shown by

$$U(\omega) = \sum_{n=0}^{\infty} u(n)^{-j\omega n}, \qquad -\pi \leq \omega \leq \pi \tag{3.33}$$

Then, by definition, the time series $\{u(n)\}$ is recovered from $U(\omega)$ by applying the *inverse discrete-time Fourier transform:*

$$u(n) = \frac{1}{2\pi}\int_{-\pi}^{\pi} U(\omega)e^{j\omega n}d\omega, \qquad n = 0, 1, \ldots \tag{3.34}$$

Given $U(\omega)$, we define the *increment variable*

$$dZ(\omega) = U(\omega)\, d\omega \tag{3.35}$$

We may then rewrite Eq. (3.34) as

$$u(n) = \frac{1}{2\pi}\int_{-\pi}^{\pi} e^{j\omega n}\, dZ(\omega) \tag{3.36}$$

where it is understood that $n = 0, 1, 2, \ldots$. Complex-conjugating both sides of Eq. (3.36) and using ν in place of ω, we get

$$u^*(n) = \frac{1}{2\pi}\int_{-\pi}^{\pi} e^{-j\nu n}\, dZ^*(\nu) \tag{3.37}$$

Hence, multiplying Eq. (3.36) by (3.37), we may express the squared magnitude of $u(n)$ as

$$|u(n)|^2 = \frac{1}{(2\pi)^2}\int_{-\pi}^{\pi}\int_{-\pi}^{\pi} e^{jn(\omega-\nu)}\, dZ(\omega)\, dZ^*(\nu) \tag{3.38}$$

Next, taking the statistical expectation of Eq. (3.38), and interchanging the order of expectation and double integration, we get

$$E[|u(n)|^2] = \frac{1}{(2\pi)^2}\int_{-\pi}^{\pi}\int_{-\pi}^{\pi} e^{jn(\omega-\nu)}\, E[dZ(\omega)\, dZ^*(\nu)] \tag{3.39}$$

If we now use the two basic properties of the increment process $\{dZ(\omega)\}$ described by Eqs. (3.31) and (3.32), we may simplify Eq. (3.39) into the form

$$E[|u(n)|^2] = \frac{1}{2\pi}\int_{-\pi}^{\pi} S(\omega)\, d\omega \tag{3.40}$$

The expectation $E[|u(n)|^2]$ on the left side of Eq. (3.40) is recognized as the mean-square value of the complex sample $u(n)$. The right side of this equation equals the total area under the curve of the power spectral density $S(\omega)$, scaled by the factor

$1/2\pi$. Accordingly, Eq. (3.40) is merely a restatement of Property 5 of the power spectral density $S(\omega)$, described by Eq. (3.19).

3.5 THE FUNDAMENTAL EQUATION

Consider a time series $u(0), u(1), \ldots, u(N-1)$, consisting of N observations (samples) of a wide-sense stationary stochastic process. The discrete-time Fourier transform of this time series is given by

$$U_N(\omega) = \sum_{n=0}^{N-1} u(n)e^{-j\omega n} \tag{3.41}$$

According to the Cramér spectral representation of the process, the observation $u(n)$ is given by Eq. (3.29). Hence, using the dummy variable ν in place of ω in Eq. (3.29), and then substituting the result in Eq. (3.41), we get

$$U_N(\omega) = \frac{1}{2\pi}\int_{-\pi}^{\pi}\sum_{n=0}^{N-1}(e^{-j(\omega-\nu)n})\,dZ(\nu) \tag{3.42}$$

where we have interchanged the order of summation and integration. Define

$$K_N(\omega) = \sum_{n=0}^{N-1} e^{-j\omega n} \tag{3.43}$$

which is known as the *Dirichlet kernel*. The kernel $K_N(\omega)$ represents a geometric series with a first term of unity, a common ratio of $e^{-j\omega}$, and a total number of terms equal to N. Summing this series, we may redefine the kernel $K_N(\omega)$ as follows:

$$\begin{aligned} K_N(\omega) &= \frac{1 - e^{-j\omega N}}{1 - e^{-j\omega}} \\ &= \frac{\sin(N\omega/2)}{\sin(\omega/2)}\exp\left[-\frac{1}{2}j\omega(N-1)\right] \end{aligned} \tag{3.44}$$

Note that $K_N(0) = N$. Returning to Eq. (3.42), we may use the definition of the Dirichlet kernel $K_N(\omega)$ given in Eq. (3.43) to rewrite $U_N(\omega)$ as follows:

$$U_N(\omega) = \frac{1}{2\pi}\int_{-\pi}^{\pi} K_N(\omega - \nu)\,dZ(\nu) \tag{3.45}$$

The integral equation (3.45) is a *linear* relation, referred to as the *fundamental equation* of power spectrum analysis.

An integral equation is one that involves an *unknown* function under the integral sign. In the context of power spectrum analysis as described by Eq. (3.45), the increment variable $dz(\omega)$ is the unknown function, and $U_N(\omega)$ is known. Accordingly, Eq. (3.45) may be viewed as an example of a *Fredholm integral equation of the first kind* (Whittaker and Watson, 1965; Morse and Feshbach, 1953).

Note that $U_N(\omega)$ may be inverse Fourier transformed to recover the original data. It follows therefore that $U_N(\omega)$ is a *sufficient statistic* of the data. This property makes the use of Eq. (3.45) for spectrum analysis all the more important.

3.6 POWER SPECTRUM ESTIMATION

In this final section of the chapter, we discuss the practical issue of how to *estimate* the power spectral density of a wide-sense stationary process. Unfortunately, this important issue is complicated by the fact that there is a bewildering array of power spectrum estimation procedures, with each procedure purported to have or to show some optimum property. The situation is made worse by the fact that unless care is taken in the selection of the right method, we may end up with misleading conclusions.

Two philosophically different families of power spectrum estimation methods may be identified in the literature: *parametric methods* and *nonparametric methods*. The basic ideas behind these methods are discussed in the sequel.

Parametric Methods. In parametric methods of spectrum estimation we begin by postulating a *stochastic model* for the situation at hand. Depending on the specific form of stochastic model adopted, we may identify three different parametric approaches for spectrum estimation:

1. *Model identification procedures*. In this case of parametric methods a rational function or a polynomial in $e^{-j\omega}$ is assumed for the transfer function of the model, and a white noise source is used to drive the model, as depicted in Fig. 3.4. The power spectrum of the resulting model output provides the desired spectrum estimate. Depending on the application of interest, we may adopt one of the following models [Kay and Marple (1981); Marple (1987); Kay (1988)]:
   (i) *Autoregressive (AR) model* with an all-pole transfer function.
   (ii) *Moving average (MA) model* with an all-zero transfer function.
   (iii) *Autoregressive-moving average (ARMA) model* with a pole-zero transfer function.

   The resulting power spectra measured at the outputs of these models are referred to as AR, MA, and ARMA spectra, respectively. With reference to the input-output relation of Eq. (3.22), let the power spectrum $S(\omega)$ of the model input be put equal to the white noise variance σ^2. We then find that the power spectrum $S_o(\omega)$ of the model output is equal to the squared amplitude response $|H(e^{j\omega})|^2$ of the model, multiplied by σ^2. The problem thus becomes one of estimating the model parameters [i.e., parametrizing the transfer function $H(e^{j\omega})$] such that the process produced at the model output provides an acceptable representation (in some statistical sense) of the stochastic process under study. Such an approach to power spectrum estimation may indeed be viewed as a problem in *model (system) identification*.

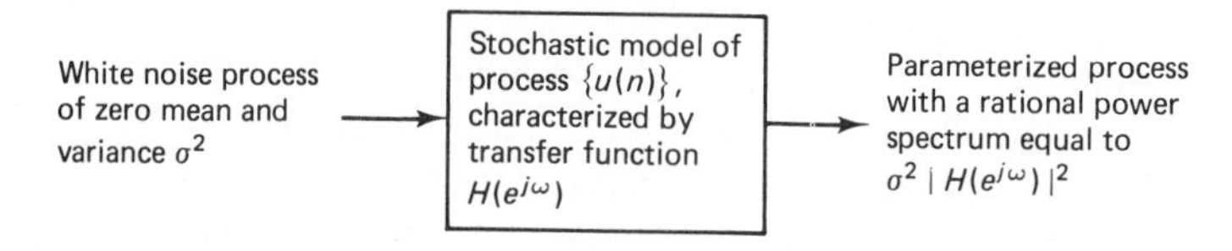

Figure 3.4 Rationale of model identification procedure for power spectrum estimation.

Among the model-dependent spectra defined herein, the AR spectrum is by far the most popular. The reason for this popularity is twofold: (1) the *linear* form of the system of simultaneous equations involving the unknown AR model parameters, and (2) the availability of efficient algorithms for computing the solution. Moreover, it may be shown that the AR spectrum is indeed equivalent to the value resulting from the application of the so-called *maximum entropy method* (MEM). The theory of the AR spectrum is discussed in detail in Chapter 6, and its relation to the MEM spectrum is discussed in Appendix F. The computation of the AR spectrum using the method of least squares is presented in Chapter 10.

2. *Minimum variance distortionless response method.* To describe this second parametric approach for power spectrum estimation, consider the situation depicted in Fig. 3.5. The process $\{u(n)\}$ is applied to a transversal filter (i.e., discrete-time filter with an all-zero transfer function). In the *minimum variance distortionless response (MVDR) method* the filter coefficients are chosen so as to minimize the variance (which is the same as average power for a zero-mean process) of the filter output, subject to the constraint that the frequency response of the filter is equal to unity at some angular frequency ω_0. Under this constraint, the process $\{u(n)\}$ is passed through the filter with *no distortion* at the angular frequency ω_0. Moreover, signals at angular frequencies other than ω_0 tend to be attenuated.

   The theory of the MVDR spectrum, is developed in Chapter 4, and a fast algorithm for its computation is described in Appendix G. The computation of the MVDR spectrum using the method of least squares is described in Chapter 10.

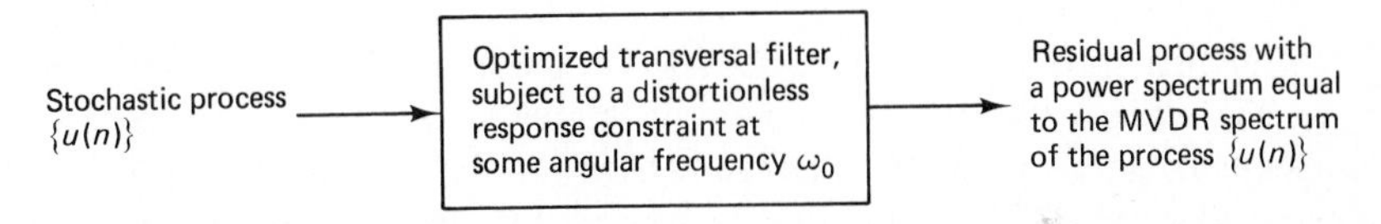

Figure 3.5 Rationale of MVDR procedure for power spectrum estimation.

3. *Eigendecomposition-based methods.* In this final class of parametric spectrum estimation methods, the eigendecomposition of the ensemble-averaged correlation matrix $\mathbf{R}$ of the processs $\{u(n)\}$ is used to define two disjoint subspaces: *signal plus noise subspace* and *noise subspace.* This form of partitioning in then exploited to derive two popular algorithms:
   (i) Multiple signal classification (MUSIC) algorithm.
   (ii) Minimum norm algorithm.

   The eigenvalue analysis of the ensemble-averaged correlation matrix is covered in Chapter 4. The derivations of the MUSIC and minimum norm algorithms and their computations are presented in Chapter 12.

   Naturally, for a given process $\{u(n)\}$, the AR, MVDR, MUSIC, and minimum norm spectra have distinctively different forms. Yet, these spectra are intimately related to each other. The relationships between these spectra are discussed in Chapter 12.

Nonparametric Methods. In nonparametric methods of power spectrum estimation, on the other hand, no assumptions are made with respect to the stochastic process under study. The starting point in the discussion is the fundamental equation (3.45). Depending on the way in which this equation is interpreted, we may distinguish two different nonparametric approaches, as discussed next.

Periodogram-Based Methods

Traditionally, the fundamental equation (3.45) is treated as a *convolution* of two frequency functions. One frequency function, $U(\omega)$, represents the discrete-time Fourier transform of an *infinitely long* time series, $\{u(n)\}$; this function arises from the definition of the increment variable $dZ(\omega)$ as the product of $U(\omega)$ and the frequency increment $d\omega$ {see Eq. (3.35)}. The other frequency function is the kernel $K_N(\omega)$, defined by Eq. (3.44). This approach leads us to consider Eq. (3.12) as the basic definition of the power spectral density $S(\omega)$, and therefore the *periodogram* $|U_N(\omega)|^2/N$ as the starting point for the data analysis. However, the periodogram suffers from a *serious limitation in the sense that it is not a sufficient statistic for the data.* This implies that the phase information ignored in the use of the periodogram is essential. Consequently, the statistical insufficiency of the periodogram is inherited by any estimate that is based on or equivalent to the periodogram.

Multiple-Window Method

A more constructive nonparametric approach is to treat the fundamental equation (3.45) as a *Fredholm integral equation of the first kind* for the increment variable $dZ(\omega)$; the goal here is to obtain an *approximate solution* for the equation with statistical properties that are close to those of $dZ(\omega)$ in some sense (Thomson, 1982). The key to the attainment of this important goal is the use of a set of special sequences known as the Slepian sequences. The *Slepian sequences*[1], also known as the *discrete prolate spheroidal sequences,* are fundamental to the study of *time- and frequency-limited systems*. Briefly, they have the following properties, making them absolutely unique:

Property 1. *A Slepian sequence is its own Fourier transform.* The discrete-time Fourier transform of a Slepian sequence is known as a *Slepian function,* or a *discrete prolate spheroidal wave function.* Property 1 states that a Slepian sequence (defined as a function of time) and the corresponding Slepian function (defined as a function of frequency) have a similar mathematical description.

Property 2. *The Slepian sequences are doubly orthogonal.* Specifically, they are *orthogonal* over the open interval $(-\infty, \infty)$ and *orthonormal* over the closed interval $[0, N-1]$, where N is the total data length. In addition to the orthogonality, orthonormality implies *normalization* of each Slepian sequence to have unit energy.

[1] Detailed information on Slepian sequences is given in Slepian (1978). A method for computing them, for large data length, is given in the Appendix of the paper by Thomson (1982). For additional information, see the references listed in Thomson's paper. Mullis and Scharf (1991) also present an informative discussion of the role of Slepian sequences in spectrum analysis.

Property 3. *The Slepian functions are doubly orthogonal.* Specifically, they are orthogonal over the *main lobe* denoted by $[-\Omega, \Omega]$ with Ω measued in radians, and orthonormal over $[-\pi, \pi]$. Moreover, the Slepian functions represent *spectral windows that are the most concentrated in frequency*. Accordingly, they are ideally suited as *coefficients in the expansion of a frequency function*.

Property 4. *The Slepian functions constitute a complete set*. To explain the meaning of this property, suppose that a frequency function is expanded with the Slepian functions as the expansion coefficients. Then, Property 4 implies that if we are willing to use more Slepian functions in the expansion, the representation error decreases. In general, we want to be able to decrease the energy contained in the representation error to any desired value by letting the number of expansion coefficients become large enough (Van Trees, 1968). Property 4, in effect, says that the Slepian functions do satisfy the requirement, if so desired.

In a remarkable series of papers, the most recent of which being that of Slepian (1978), it has been shown that the Slepian functions, or discrete prolate spheroidal wave functions, are indeed the *eigenfunctions* of the Dirichlet kernel $K_N(\omega)$ given in Eq. (3.44). Let $\Phi_i(\omega)$ denote the ith Slepian function. We may then write

$$\lambda_i \Phi_i(\omega) = \int_{-\Omega}^{\Omega} K_N(\omega - \nu)\Phi_i(\nu)\, d\nu, \qquad -\pi \leq \omega \leq \pi \tag{3.46}$$

The functions $\Phi_i(\omega)$ are called the *eigenfunctions,* and the numbers λ_i are called the *eigenvalues* (Van Trees, 1968). Note that with $K_N(\omega)$ being dimensionless, λ_i is measured in *radians*. Moreover, the kernel $K_N(\omega - \nu)$ may be expanded in the series

$$K_N(\omega - \nu) = \sum_{i=0}^{\infty} \lambda_i \Phi_i(\omega)\Phi_i^*(\nu), \qquad -\pi \leq \omega, \nu \leq \pi \tag{3.47}$$

where the convergence of the series is uniform for $-\pi \leq \omega, \nu \leq \pi$. The series expansion described in Eq. (3.47) is called *Mercer's theorem*.

Returning to the issue of solving the fundamental equation (3.45), we may now contemplate an "approximate" solution for some local interval about ω, say $(\omega - \Omega, \omega + \Omega)$ using the Slepian functions as a *basis*. We shall refer to this interval as the *inner* or *local domain,* and the remainder of the principal frequency domain $(-\pi, \pi)$ as the *outer domain*. Because the Slepian functions constitute a complete set (Property 4), we may assume that the *observable* portion of dZ has the approximate expansion (Thomson, 1982, 1989)

$$dZ(\omega - \nu) \simeq \sum_{i=0}^{\infty} X_i(\omega)\Phi_i^*(\nu)\, d\nu \tag{3.48}$$

for $|\nu| < \Omega$. With the assumed wide-sense stationarity of the stochastic process represented by the time series $\{u_N(n)\}$, we find that the part of dZ in the inner domain is uncorrelated with the part of dZ in the outer domain; see Eq. (3.31). For the purpose of our discussion here, it is helpful to think of the "inner" energy as the "signal" of interest, and the rest of the energy as "noise." Using the fundamental equation (3.45) and the orthogonality properties of the Slepian functions, we obtain the *eigencoefficients*

$$X_i(\omega) = \sum_{n=0}^{N-1} \phi_i(n)u(n)e^{-j\omega n}, \qquad i = 0, 1, 2, \ldots \tag{3.49}$$

where the $u(n)$ are the input data of length N, and the $\phi_i(n)$ are the Slepian sequences (i.e., the inverse discrete-time Fourier transforms of the Slepian functions). According to Eq. (3.49), the *eigencoefficients* $X_i(\omega)$ *in the expansion of Eq. (3.48) are obtained by first windowing the input data with the discrete-time Fourier transform of the product*. This procedure is identical to that used in obtaining *standard* windowed spectrum estimates, with one major difference. Conventionally, only the first term (i.e., $i = 0$) in Eq. (3.49) is used. On the other hand, in *multiple-window power spectrum* estimates a multiplicity of windows represented by Slepian sequences are used; hence the name of the method.

Discussion

In general, a discrete-time stochastic process $\{u(n)\}$ has a *mixed spectrum* in that its power spectrum contains two components: a deterministic component and a continuous component. The *deterministic component* represents the *first moment* of the increment process $\{dZ(\omega)\}$; it is explicitly given by

$$E[dZ(\omega)] = \sum_k a_k \delta(\omega - \omega_k)\, d\omega \tag{3.50}$$

where $\delta(\omega)$ is the *Dirac delta function* defined in the frequency domain. The ω_k are the angular frequencies of *periodic* or *line components* contained in the process $\{u(n)\}$, and the a_k are their amplitudes. The continuous component, on the other hand, represents the *second central moment* of the increment process $\{dZ(\omega)\}$, as shown by

$$E[|\,dZ(\omega) - E[dZ(\omega)]\,|^2] = S(\omega)\, d\omega \tag{3.51}$$

It is important that the distinction between the first and second moments is carefully noted.

Spectra computed using the parametric methods (i.e., AR, MVDR, MUSIC, and minimum norm spectra) tend to have sharper peaks and higher resolution than those obtained from the nonparametric (classical) methods. The application of these parametric methods is therefore well suited for estimating the deterministic component and, in particular, for locating the frequencies of periodic components in additive white noise when the signal-to-noise ratio is high. Another well-proven technique for estimating the deterministic component is to use the classical method of maximum likelihood, which is discussed in Appendix D. Of course, if the physical laws governing the generation of a process match a stochastic model (e.g., AR model) in an exact manner or approximately in some statistical sense, then the parametric method corresponding to that model may be used to estimate the power spectrum of the process. If, however, the stochastic process of interest has a purely continuous power spectrum, and the underlying physical mechanism responsible for the generation of the process is unknown, then the recommended procedure is the nonparametric method of multiple windows.

PROBLEMS

1. Consider the definition given in Eq. (3.12) for the power spectral density. Is it permissible to interchange the operation of taking the limit and that of the expectation in this equation? Justify your answer.
2. A wide-sense stationary discrete-time stochastic process is applied to a linear time-invariant discrete-time filter. Show that the stochastic process produced at the filter output is also wide-sense stationary.
3. The mean-square value of the filter output given in Eq. (3.27) assumes that the bandwidth of the filter is small compared to its center frequency. Is this assumption necessary for the corresponding result obtained in Example 1 for a white-noise process? Justify your answer.
4. A white-noise process with a variance of 0.1 V squared is applied to a low-pass discrete-time filter, whose bandwidth is 1 Hz. The process is real.
   (a) Calculate the variance of the filter output.
   (b) Assuming that the input is Gaussian, determine the probability density function of the filter output.
5. Equation (3.22) defines the input–output relation of a linear discrete-time filter in terms of frequency-domain concepts. Derive the time-domain counterpart of this relation by defining the autocorrelation function of the filter output in terms of the autocorrelation function of the filter input.
6. Justify the fact that the expectation of $|dZ(\omega)|^2$ has the physical significance of energy.
7. For convenience of presentation, the Dirichlet kernel $K_N(\omega)$ is sometimes defined as

$$K_N(\omega) = \sum_{n=0}^{N-1} \exp\left[-j\omega\left(n - \frac{N-1}{2}\right)\right]$$

   Would this change in notation affect the spectrum estimate derived from the fundamental equation (3.45)? Justify your answer.
8. In Section 3.6 we stated that the treatment of the fundamental equation (3.45) as a convolution integral is equivalent to the use of the basic definition given in Eq. (3.12). Demonstrate the validity of this statement.
9. Let $S_{\text{PER}}(\omega)$ denote the periodogram of a time series $\{u(n)\}$. Show that the expectation of $S_{\text{PER}}(\omega)$ is defined by

$$E[S_{\text{PER}}(\omega)] = \frac{1}{2\pi}\int_{-\pi}^{\pi} W_B(\nu - \omega)S(\nu)\,d\nu$$

   where $S(\omega)$ is the true power spectrum, and $W_B(\omega)$ is the discrete-time Fourier transform of the pertinent Bartlett window $w_B(l)$. How is $W_B(\omega)$ related to the Dirichlet kernel $K_N(\omega)$?

   It is said that the periodogram $S_{\text{PER}}(\omega)$ is a biased estimate of the true power spectrum $S(\omega)$ for a data record of finite length N, and that the estimate becomes unbiased as N approaches infinity. Use the integral formula for $E[S_{\text{PER}}(\omega)]$ to justify these two statements.

Computer-Oriented Problem

10. Continue with the computer-oriented Problems 13 and 14 of Chapter 2. Specifically, compute and plot the power spectra of the AR processes described in these two problems.

4

Eigenanalysis

In Section 3.6 we briefly considered the use of eigenvalues and eigenfunctions for the representation of the Dirichlet kernel $K_N(\omega)$ that is a continuous function of the angular frequency ω. In this chapter we take up the *eigenanalysis* of Hermitian matrices in some detail; the Hermitian matrix of interest to us is represented by the correlation matrix of a wide-sense stationary discrete-time stochastic process. Our motivation here is the fact that eigenanalysis is basic to the study of digital signal processing.

We begin the discussion by outlining the eigenvalue problem in the context of the correlation matrix. We then study the properties of eigenvalues and eigenvectors of the correlation matrix; and a related optimum filtering problem. We finish the discussion by briefly describing canned routines for eigenvalue computations and some related issues.

4.1 THE EIGENVALUE PROBLEM

Let the Hermitian matrix $\mathbf{R}$ denote the M-by-M correlation matrix of a wide-sense stationary discrete-time stochastic process represented by the M-by-1 observation vector $\mathbf{u}(n)$. In general, this matrix may contain complex elements. We wish to find an M-by-1 vector $\mathbf{q}$ that satisfies the condition

$$\mathbf{R}\mathbf{q} = \lambda\mathbf{q} \tag{4.1}$$

for some constant λ. This condition states that the vector $\mathbf{q}$ is linearly transformed to the vector $\lambda\mathbf{q}$ by the Hermitian matrix $\mathbf{R}$. Since λ is a constant, the vector $\mathbf{q}$ therefore has special significance in that it is left *invariant in direction* (in the M-dimensional space) by a linear transformation. For a typical M-by-M matrix $\mathbf{R}$, there will be M such vectors. To show this, we first rewrite Eq. (4.1) in the form

$$(\mathbf{R} - \lambda\mathbf{I})\mathbf{q} = \mathbf{0} \tag{4.2}$$

where $\mathbf{I}$ is the M-by-M identity matrix, and $\mathbf{0}$ is the M-by-1 null vector. The matrix $\mathbf{R} - \lambda\mathbf{I}$ has to be singular. Hence Eq. (4.2) has a nonzero solution in the vector $\mathbf{q}$ if and only if the determinant of the matrix $(\mathbf{R} - \lambda\mathbf{I})$ equals zero; that is,

$$\det(\mathbf{R} - \lambda\mathbf{I}) = 0 \tag{4.3}$$

This determinant, when expanded, is clearly a polynomial in λ of degree M. We thus find that, in general, Eq. (4.3) has M distinct roots. Correspondingly, Eq. (4.3) has M solutions in the vector $\mathbf{q}$.

Equation (4.3) is called the *characteristic equation* of the matrix $\mathbf{R}$. Let $\lambda_1, \lambda_2, \ldots, \lambda_M$ denote the M roots of this equation. These roots are called the *eigenvalues* of the matrix $\mathbf{R}$. Although the M-by-M matrix $\mathbf{R}$ has M eigenvalues, they need not be distinct. When the characteristic equation (4.3) has multiple roots, the matrix $\mathbf{R}$ is said to have *degenerate* eigenvalues. Note that, in general, the use of root finding in the characteristic equation (4.3) is a poor method for computing the eigenvalues of the matrix $\mathbf{R}$; the issue of eigenvalue computations is considered later in Section 4.4.

Let λ_i denote an eigenvalue of the matrix $\mathbf{R}$. Also, let $\mathbf{q}_i$ be a nonzero vector such that

$$\mathbf{R}\mathbf{q}_i = \lambda_i\mathbf{q}_i \tag{4.4}$$

The vector $\mathbf{q}_i$ is called the *eigenvector* associated with λ_i. An eigenvector can correspond to only one eigenvalue. However, an eigenvalue may have many eigenvectors. For example, if $\mathbf{q}_i$ is an eigenvector associated with eigenvalue λ_i, then so is $a\mathbf{q}_i$ for any scalar $a \neq 0$.

Example 1: White Noise

Consider the M-by-M correlation matrix of a white-noise process that is described by the diagnoal matrix

$$\mathbf{R} = \text{diag}(\sigma^2, \sigma^2, \ldots, \sigma^2)$$

where σ^2 is the variance of a sample of the process. *This correlation matrix* $\mathbf{R}$ *has a single degenerate eigenvalue equal to the variance* σ^2 *with multiplicity* M. *Any* M*-by-1 random vector qualifies as the associated eigenvector, confirming the random nature of white noise.*

Example 2: Complex Sinusoid

Consider next the M-by-M correlation matrix of a time series whose elements are samples of a complex sinusoid with random phase and unit power. This correlation matrix may be written as

$$\mathbf{R} = \begin{bmatrix} 1 & e^{j\omega} & \cdots & e^{j(M-1)\omega} \\ e^{-j\omega} & 1 & \cdots & e^{j(M-2)\omega} \\ \vdots & \vdots & \ddots & \vdots \\ e^{-j(M-1)\omega} & e^{-j(M-2)\omega} & \cdots & 1 \end{bmatrix}$$

where ω is the angular frequency of the complex sinusoid. The M-by-1 vector

$$\mathbf{q} = [1, e^{-j\omega}, \ldots, e^{-j(M-1)\omega}]^T$$

is an eigenvector of the correlation matrix $\mathbf{R}$, and the corresponding eigenvalue is M (i.e., the dimension of the matrix $\mathbf{R}$). In other words, a complex sinusoidal vector represents an eigenvector of its own correlation matrix, except for the trivial operation of complex conjugation.

Note that the correlation matrix $\mathbf{R}$ has *rank* 1, which means that any column of $\mathbf{R}$ may be expressed as a linear combination of the remaining columns (i.e., the matrix $\mathbf{R}$ has only one independent column).

4.2 PROPERTIES OF EIGENVALUES AND EIGENVECTORS

In this section we discuss the various properties of the eigenvalues and eigenvectors of the correlation matrix $\mathbf{R}$ of a stationary discrete-time stochastic process. Some of the properties derived here are direct consequences of the Hermitian property and the nonnegative definiteness of the correlation matrix $\mathbf{R}$, which were established in Section 2.3.

Property 1. *If $\lambda_1, \lambda_2, \ldots, \lambda_M$ denote the eigenvalues of the correlation matrix $\mathbf{R}$, then the eigenvalues of the matrix $\mathbf{R}^k$ equal $\lambda_1^k, \lambda_2^k, \ldots, \lambda_M^k$ for any integer $k > 0$.*

Repeated premultiplication of both sides of Eq. (4.1) by the matrix $\mathbf{R}$ yields

$$\mathbf{R}^k\mathbf{q} = \lambda^k\mathbf{q} \tag{4.5}$$

This shows that (1) if λ is an eigenvalue of $\mathbf{R}$, then λ^k is an eigenvalue of $\mathbf{R}^k$, which is the desired result, and (2) every eigenvector of $\mathbf{R}$ is also an eigenvector of $\mathbf{R}^k$.

Property 2. *Let $\mathbf{q}_1, \mathbf{q}_2, \ldots, \mathbf{q}_M$ be the eigenvectors corresponding to the distinct eigenvalues $\lambda_1, \lambda_2, \ldots, \lambda_M$ of the M-by-M correlation matrix $\mathbf{R}$, respectively. Then the eigenvectors $\mathbf{q}_1, \mathbf{q}_2, \ldots, \mathbf{q}_M$ are linearly independent.*

We say that the eigenvectors $\mathbf{q}_1, \mathbf{q}_2, \ldots, \mathbf{q}_M$ are *linearly dependent* if there are scalars $v_1, v_2, \ldots, v_M$, not all zero, such that

$$\sum_{i=1}^{M} v_i\mathbf{q}_i = \mathbf{0} \tag{4.6}$$

If no such scalars exist, we say that the eigenvectors are *linearly independent*.

We will prove the validity of Property 2 by contradiction. Suppose that Eq. (4.6) holds for certain not all zero scalars v_i. Repeated multiplication of Eq. (4.6) by matrix $\mathbf{R}$ and the use of Eq. (4.1) yield the following set of M equations:

$$\sum_{i=1}^{M} v_i\lambda_i^{k-1}\mathbf{q}_i = \mathbf{0}, \qquad k = 1, 2, \ldots, M \tag{4.7}$$

This set of equations may be written in the form of a single matrix equation:

$$[v_1\mathbf{q}_1, v_2\mathbf{q}_2, \ldots, v_M\mathbf{q}_M]\mathbf{S} = \mathbf{0} \tag{4.8}$$

where

$$\mathbf{S} = \begin{bmatrix} 1 & \lambda_1 & \lambda_1^2 & \cdots & \lambda_1^{M-1} \\ 1 & \lambda_2 & \lambda_2^2 & \cdots & \lambda_2^{M-1} \\ \vdots & \vdots & \vdots & \ddots & \vdots \\ 1 & \lambda_M & \lambda_M^2 & \cdots & \lambda_M^{M-1} \end{bmatrix} \tag{4.9}$$

The matrix $\mathbf{S}$ is called a *Vandermonde matrix* (Strang, 1980). When the λ_i are distinct, the Vandermonde matrix $\mathbf{S}$ is nonsingular. Therefore, we may postmultiply Eq. (4.8) by the inverse matrix $\mathbf{S}^{-1}$, obtaining

$$[v_1\mathbf{q}_1, v_2\mathbf{q}_2, \ldots, v_M\mathbf{q}_M] = \mathbf{0}$$

Hence, each column $v_i\mathbf{q}_i = \mathbf{0}$. Since the eigenvectors $\mathbf{q}_i$ are not zero, this condition can be satisfied if and only if the v_i are all zero. This contradicts the assumption that the scalars v_i are not all zero. In other words, the eigenvectors are linearly independent.

We may put this property to an important use by having the linearly independent eigenvectors $\mathbf{q}_1, \mathbf{q}_2, \ldots, \mathbf{q}_M$ serve as a *basis* for the representation of an arbitrary vector $\mathbf{w}$ with the same dimension as the eigenvectors themselves. In particular, we may express the arbitrary vector $\mathbf{w}$ as a linear combination of the eigenvectors $\mathbf{q}_1, \mathbf{q}_2, \ldots, \mathbf{q}_M$ as follows:

$$\mathbf{w} = \sum_{i=1}^{M} v_i\mathbf{q}_i \tag{4.10}$$

where $v_1, v_2, \ldots, v_M$ are constants. Suppose now we apply a linear transformation to the vector $\mathbf{w}$ by premultiplying it by the matrix $\mathbf{R}$, obtaining

$$\mathbf{Rw} = \sum_{i=1}^{M} v_i\mathbf{R}\mathbf{q}_i \tag{4.11}$$

By definition, we have $\mathbf{R}\mathbf{q}_i = \lambda_i\mathbf{q}_i$. Therefore, we may express the result of this linear transformation in the equivalent form

$$\mathbf{Rw} = \sum_{i=1}^{M} v_i\lambda_i\mathbf{q}_i \tag{4.12}$$

We thus see that when a linear transformation is applied to an arbitrary vector the eigenvectors remain independent of each other, and the effect of the transformation is simply to multiply each eigenvector by its respective eigenvalue.

Property 3. *Let $\lambda_1, \lambda_2, \ldots, \lambda_M$ be the eigenvalues of the M-by-M correlation matrix* $\mathbf{R}$. *Then all these eigenvalues are real and nonnegative.*

To prove this property, we first use Eq. (4.1) to express the condition on the ith eigenvalue λ_i as

$$\mathbf{R}\mathbf{q}_i = \lambda_i\mathbf{q}_i, \qquad i = 1, 2, \ldots, M \tag{4.13}$$

Premultiplying both sides of this equation by $\mathbf{q}_i^H$, the Hermitian transpose of eigenvector $\mathbf{q}_i$, we get

$$\mathbf{q}_i^H \mathbf{R} \mathbf{q}_i = \lambda_i \mathbf{q}_i^H \mathbf{q}_i, \qquad i = 1, 2, \ldots, M \tag{4.14}$$

The inner product $\mathbf{q}_i^H \mathbf{q}_i$ is a positive scalar, representing the squared Euclidean length of the eigenvector $\mathbf{q}_i$ that is, $\mathbf{q}_i^H \mathbf{q}_i > 0$. We may therefore divide both sides of Eq. (4.14) by $\mathbf{q}_i^H \mathbf{q}_i$ and so express the ith eigenvalue λ_i as the ratio

$$\lambda_i = \frac{\mathbf{q}_i^H \mathbf{R} \mathbf{q}_i}{\mathbf{q}_i^H \mathbf{q}_i}, \qquad i = 1, 2, \ldots, M \tag{4.15}$$

Since the correlation matrix $\mathbf{R}$ is always nonnegative definite, the Hermitian form $\mathbf{q}_i^H \mathbf{R} \mathbf{q}_i$ in the numerator of this ratio is always real and nonegative; that is, $\mathbf{q}_i^H \mathbf{R} \mathbf{q}_i \geq 0$. Therefore, it follows from Eq. (4.15) that $\lambda_i \geq 0$ for all i. That is, all the eigenvalues of the correlation matrix $\mathbf{R}$ are always real and nonnegative.

When, however, the correlation matrix $\mathbf{R}$ is positive definite, which is almost always the case, we have $\mathbf{q}_i^H \mathbf{R} \mathbf{q}_i > 0$ and, correspondingly, $\lambda_i > 0$ for all i. That is, the eigenvalues of the correlation matrix $\mathbf{R}$ are almost always real and positive.

The ratio of the Hermitian form $\mathbf{q}_i^H \mathbf{R} \mathbf{q}_i$ to the inner product $\mathbf{q}_i^H \mathbf{q}_i$ on the right side of Eq. (4.15) is called the *Rayleigh quotient* of the vector $\mathbf{q}_i$. We may thus state that an eigenvalue of the correlation matrix equals the Rayleigh quotient of the corresponding eigenvector.

Property 4. *Let $\mathbf{q}_1, \mathbf{q}_2, \ldots, \mathbf{q}_M$ be the eigenvectors corresponding to the distinct eigenvalues $\lambda_1, \lambda_2, \ldots, \lambda_M$ of the M-by-M correlation matrix $\mathbf{R}$, respectively. Then the eigenvectors $\mathbf{q}_1, \mathbf{q}_2, \ldots, \mathbf{q}_M$ are orthogonal to each other.*

Let $\mathbf{q}_i$ and $\mathbf{q}_j$ denote any two eigenvectors of the correlation matrix $\mathbf{R}$. We say that these two eigenvectors are *orthogonal* to each other if

$$\mathbf{q}_i^H \mathbf{q}_j = 0, \qquad i \neq j \tag{4.16}$$

Using Eq. (4.1), we may express the conditions on the eigenvectors $\mathbf{q}_i$ and $\mathbf{q}_j$ as follows, respectively,

$$\mathbf{R}\mathbf{q}_i = \lambda_i \mathbf{q}_i \tag{4.17}$$

and

$$\mathbf{R}\mathbf{q}_j = \lambda_j \mathbf{q}_j \tag{4.18}$$

Premultiplying both sides of Eq. (4.17) by the Hermitian-transposed vector $\mathbf{q}_j^H$, we get

$$\mathbf{q}_j^H \mathbf{R} \mathbf{q}_i = \lambda_i \mathbf{q}_j^H \mathbf{q}_i \tag{4.19}$$

Since the correlation matrix $\mathbf{R}$ is Hermitian, we have $\mathbf{R}^H = \mathbf{R}$. Also, from Property 3 we know that the eigenvalue λ_j is real for all j. Hence, taking the Hermitian transpose of both sides of Eq. (4.18) and using these two properties, we get

$$\mathbf{q}_j^H \mathbf{R} = \lambda_j \mathbf{q}_j^H \tag{4.20}$$

Postmultiplying both sides of Eq. (4.20) by the vector $\mathbf{q}_i$,

$$\mathbf{q}_j^H \mathbf{R} \mathbf{q}_i = \lambda_j \mathbf{q}_j^H \mathbf{q}_i \tag{4.21}$$

Subtracting Eq. (4.21) from (4.19);

$$(\lambda_i - \lambda_j)\mathbf{q}_j^H\mathbf{q}_i = 0 \tag{4.22}$$

Since the eigenvalues of the correlation matrix $\mathbf{R}$ are assumed to be distinct, we have $\lambda_i \neq \lambda_j$. Accordingly, the condition of Eq. (4.22) holds if and only if

$$\mathbf{q}_j^H\mathbf{q}_i = 0, \qquad i \neq j \tag{4.23}$$

which is the desired result. That is, the eigenvectors $\mathbf{q}_i$ and $\mathbf{q}_j$ are *orthogonal* to each other for $i \neq j$.

Property 5: Unitary Similarity Transformation. *Let $\mathbf{q}_1, \mathbf{q}_2, \ldots, \mathbf{q}_M$ be the eigenvectors corresponding to the distinct eigenvalues $\lambda_1, \lambda_2, \ldots, \lambda_M$ of the M-by-M correlation matrix $\mathbf{R}$, respectively. Define the M-by-M matrix*

$$\mathbf{Q} = [\mathbf{q}_1, \mathbf{q}_2, \ldots, \mathbf{q}_M]$$

where

$$\mathbf{q}_i^H\mathbf{q}_j = \begin{cases} 1, & i = j \\ 0, & i \neq j \end{cases}$$

Define the M-by-M diagonal matrix

$$\mathbf{\Lambda} = \text{diag}(\lambda_1, \lambda_2, \ldots, \lambda_M)$$

Then the original matrix $\mathbf{R}$ may be diagonalized as follows:

$$\mathbf{Q}^H\mathbf{R}\mathbf{Q} = \mathbf{\Lambda}$$

The condition that $\mathbf{q}_i^H\mathbf{q}_i = 1$ for $i = 1, 2, \ldots, M$, requires that each eigenvector be *normalized* to have a *length of* 1. The *squared length* or *squared norm* of a vector $\mathbf{q}_i$ is defined as the inner product $\mathbf{q}_i^H\mathbf{q}_i$. The orthogonality condition that $\mathbf{q}_i^H\mathbf{q}_j = 0$, for $i \neq j$, follows from Property 4. When both of these conditions are simultaneously satisfied, that is,

$$\mathbf{q}_i^H\mathbf{q}_j = \begin{cases} 1, & i = j \\ 0, & i \neq j \end{cases} \tag{4.24}$$

we say the eigenvectors $\mathbf{q}_1, \mathbf{q}_2, \ldots, \mathbf{q}_M$ form an *orthonormal* set. By definition, the eigenvectors $\mathbf{q}_1, \mathbf{q}_2, \ldots, \mathbf{q}_M$ satisfy the equations [see Eq. (4.1)]

$$\mathbf{R}\mathbf{q}_i = \lambda_i\mathbf{q}_i, \qquad i = 1, 2, \ldots, M \tag{4.25}$$

The M-by-M matrix $\mathbf{Q}$ has as its columns the orthonormal set of eigenvectors $\mathbf{q}_1, \mathbf{q}_2, \ldots, \mathbf{q}_M$; that is,

$$\mathbf{Q} = [\mathbf{q}_1, \mathbf{q}_2, \ldots, \mathbf{q}_M] \tag{4.26}$$

The M-by-M diagonal matrix $\mathbf{\Lambda}$ has the eigenvalues $\lambda_1, \lambda_2, \ldots, \lambda_M$ for the elements of its main diagonal:

$$\mathbf{\Lambda} = \text{diag}(\lambda_1, \lambda_2, \ldots, \lambda_M) \tag{4.27}$$

Accordingly, we may rewrite the set of M equations (4.25) as a single matrix equation:

$$\mathbf{R}\mathbf{Q} = \mathbf{Q}\mathbf{\Lambda} \tag{4.28}$$

Owing to the orthonormal nature of the eigenvectors, as defined in Eq. (4.24), we find that

$$\mathbf{Q}^H\mathbf{Q} = \mathbf{I}$$

Equivalently, we may write

$$\mathbf{Q}^{-1} = \mathbf{Q}^H \tag{4.29}$$

That is, the matrix $\mathbf{Q}$ is nonsingular with an inverse $\mathbf{Q}^{-1}$ equal to the Hermitian transpose of $\mathbf{Q}$. A matrix that has this property is called a *unitary matrix*.

Thus, premultiplying both sides of Eq. (4.28) by the Hermitian-transposed matrix $\mathbf{Q}^H$ and using the property of Eq. (4.29), we get the desired result:

$$\mathbf{Q}^H\mathbf{R}\mathbf{Q} = \mathbf{\Lambda} \tag{4.30}$$

This transformation is called the *unitary similarity transformation*.

We have thus proved an important result. The correlation matrix $\mathbf{R}$ may be *diagonalized* by a unitary similarity transformation. Furthermore, the matrix $\mathbf{Q}$ that is used to diagonalize $\mathbf{R}$ has as its columns an orthonormal set of eigenvectors for $\mathbf{R}$. The resulting diagonal matrix $\mathbf{\Lambda}$ has as its diagonal elements the eigenvalues of $\mathbf{R}$.

By postmultiplying both sides of Eq. (4.28) by the inverse matrix $\mathbf{Q}^{-1}$ and then using the property of Eq. (4.29), we may also write

$$\begin{aligned}\mathbf{R} &= \mathbf{Q}\mathbf{\Lambda}\mathbf{Q}^H \\ &= \sum_{i=1}^{M} \lambda_i \mathbf{q}_i \mathbf{q}_i^H \end{aligned} \tag{4.31}$$

where M is the dimension of matrix $\mathbf{R}$. The term $\mathbf{q}_i\mathbf{q}_i^H$ is the outer product of the ith eigenvector with itself. Equation (4.31) states that the correlation matrix of a wide-sense stationary process equals a linear combination of all such outer products, with each outer product being weighted by the respective eigenvalue. This equation is known as *Mercer's theorem*. It is also referred to as the *spectral theorem*.

Property 6. *Let $\lambda_1, \lambda_2, \ldots, \lambda_M$ be the eigenvalues of the M-by-M correlation matrix* $\mathbf{R}$. *Then the sum of these eigenvalues equals the trace of matrix* $\mathbf{R}$.

The *trace* of a square matrix is defined as the sum of the diagonal elements of the matrix. Taking the trace of both sides of Eq. (4.30), we may write

$$\text{tr}[\mathbf{Q}^H\mathbf{R}\mathbf{Q}] = \text{tr}[\mathbf{\Lambda}] \tag{4.32}$$

The diagonal matrix $\mathbf{\Lambda}$ has as its diagonal elements the eigenvalues of $\mathbf{R}$. Hence, we have

$$\text{tr}[\mathbf{\Lambda}] = \sum_{i=1}^{M} \lambda_i \tag{4.33}$$

Using a rule in matrix algebra, we may write[1]

[1] This result follows from the following rule in matrix algebra. Let $\mathbf{A}$ be an M-by-N matrix and $\mathbf{B}$ be an N-by-M matrix. The trace of the matrix product $\mathbf{AB}$ equals the trace of $\mathbf{BA}$.

$$\mathrm{tr}[\mathbf{Q}^H\mathbf{R}\mathbf{Q}] = \mathrm{tr}[\mathbf{R}\mathbf{Q}\mathbf{Q}^H]$$

However, $\mathbf{Q}\mathbf{Q}^H$ equals the identity matrix $\mathbf{I}$ [this follows from Eq. (4.29)]. Hence we have

$$\mathrm{tr}[\mathbf{Q}^H\mathbf{R}\mathbf{Q}] = \mathrm{tr}[\mathbf{R}]$$

Accordingly, we may rewrite Eq. (4.32) as

$$\mathrm{tr}[\mathbf{R}] = \sum_{i=1}^{M} \lambda_i \tag{4.34}$$

We have thus shown that the trace of the correlation matrix $\mathbf{R}$ equals the sum of its eigenvalues. Although in proving this result we used a property that requires the matrix $\mathbf{R}$ to be Hermitian with distinct eigenvalues, nevertheless, the result applies to any square matrix.

Property 7. *The correlation matrix* $\mathbf{R}$ *is ill conditioned if the ratio of the largest eigenvalue to the smallest eigenvalue of* $\mathbf{R}$ *is large.*

To appreciate the impact of Property 7, it is important that we recognize the fact that the development of an algorithm for the effective solution of a signal processing problem and the understanding of associated *perturbation theory* go hand-in-hand [Van Loan (1989)]. We may illustrate the synergism between these two fields by considering the following linear system of equations:

$$\mathbf{A}\mathbf{w} = \mathbf{d}$$

where the matrix $\mathbf{A}$ and the vector $\mathbf{d}$ are data-dependent quantities, and $\mathbf{w}$ is a coefficient vector characterizing a linear FIR filter of interest. An elementary formulation of perturbation theory tells us that if the matrix $\mathbf{A}$ and vector $\mathbf{d}$ are perturbed by small amounts $\delta\mathbf{A}$ and $\delta\mathbf{d}$, respectively, and if $\|\delta\mathbf{A}\|/\|\mathbf{A}\|$ and $\|\delta\mathbf{d}\|/\|\mathbf{d}\|$ are both on the order of some ϵ with $\epsilon << 1$, we then have [Golub and Van Loan (1989)]

$$\frac{\|\delta\mathbf{w}\|}{\|\mathbf{w}\|} \leq \epsilon\chi(\mathbf{A})$$

where $\delta\mathbf{w}$ is the change produced in $\mathbf{w}$, and $\chi(\mathbf{A})$ is the *condition number* of matrix $\mathbf{A}$ with respect to inversion. The condition number is so-called because it describes the ill condition or bad behavior of matrix $\mathbf{A}$ quantitatively. In particular, it is defined as follows [Wilkinson (1963); Strang (1980); Golub and Van Loan (1989)].

$$\chi(\mathbf{A}) = \|\mathbf{A}\|\,\|\mathbf{A}^{-1}\| \tag{4.35}$$

where $\|\mathbf{A}\|$ is a *norm* of matrix $\mathbf{A}$, and $\|\mathbf{A}^{-1}\|$ is the corresponding norm of the inverse matrix $\mathbf{A}^{-1}$. The norm of a matrix is a number assigned to the matrix that is in some sense a measure of the magnitude of the matrix. We find it natural to require that the norm of a matrix satisfy the following conditions:

1. $\|\mathbf{A}\| \geq 0$, $\|\mathbf{A}\| = 0$ if and only if $\mathbf{A} = \mathbf{0}$.
2. $\|c\,\mathbf{A}\| = |c|\,\|\mathbf{A}\|$, where c is any real number and $|c|$ is its magnitude.

3. $\|\mathbf{A} + \mathbf{B}\| \leq \|\mathbf{A}\| + \|\mathbf{B}\|$.
4. $\|\mathbf{AB}\| \leq \|\mathbf{A}\| \, \|\mathbf{B}\|$.

Condition 3 is the *triangle inequality*, and condition 4 is the *mutual consistency*. There are several ways of defining the norm $\|\mathbf{A}\|$, which satisfy the preceding conditions (Ralston, 1965). For our present discussion, however, we find it convenient to use the *spectral norm* defined as the square root of the largest eigenvalue of the matrix product $\mathbf{A}^H\mathbf{A}$, where $\mathbf{A}^H$ is the Hermitian transpose of $\mathbf{A}$; that is,

$$\|\mathbf{A}\| = (\text{largest eigenvalue of } \mathbf{A}^H\mathbf{A})^{1/2} \tag{4.36}$$

Since for any matrix $\mathbf{A}$ the product $\mathbf{A}^H\mathbf{A}$ is always Hermitian and nonnegative definite, it follows that the eigenvalues of $\mathbf{A}^H\mathbf{A}$ are all real and nonnegative, as required. Moreover, from Eq. (4.15) we note that an eigenvalue of $\mathbf{A}^H\mathbf{A}$ equals the Rayleigh coefficient of the corresponding eigenvector. Squaring both sides of Eq. (4.36) and using this property, we may therefore write[2]

$$\begin{aligned} \|\mathbf{A}\|^2 &= \max \frac{\mathbf{x}^H\mathbf{A}^H\mathbf{A}\mathbf{x}}{\mathbf{x}^H\mathbf{x}} \\ &= \max \frac{\|\mathbf{A}\mathbf{x}\|^2}{\|\mathbf{x}\|^2} \end{aligned}$$

where $\|\mathbf{x}\|^2$ is the inner product of vector $\mathbf{x}$ with itself, and likewise for $\|\mathbf{A}\mathbf{x}\|^2$. We refer to $\|\mathbf{x}\|$ as the *Euclidean norm* or *length* of vector $\mathbf{x}$. We may thus express the norm of matrix $\mathbf{A}$ in the equivalent form

$$\|\mathbf{A}\| = \max \frac{\|\mathbf{A}\mathbf{x}\|}{\|\mathbf{x}\|} \tag{4.37}$$

According to this relation, the norm of $\mathbf{A}$ measures the largest amount by which any vector (eigenvector or not) is amplified by matrix multiplication, and the vector that is amplified the most is the eigenvector that corresponds to the largest eigenvalue of $\mathbf{A}^H\mathbf{A}$ (Strang, 1980).

Consider now the application of the definition in Eq. (4.36) to the correlation matrix $\mathbf{R}$. Since $\mathbf{R}$ is Hermitian, we have $\mathbf{R}^H = \mathbf{R}$. Hence, from Property 1 we deduce that if $\lambda_{\max}$ is the largest eigenvalue of $\mathbf{R}$, the largest eigenvalue of $\mathbf{R}^H\mathbf{R}$ equals $\lambda_{\max}^2$. Accordingly, the spectral norm of the correlation matrix $\mathbf{R}$ is

$$\|\mathbf{R}\|_S = \lambda_{\max} \tag{4.38}$$

Similarly, we may show that the spectral norm of $\mathbf{R}^{-1}$, the inverse of the correlation matrix, is

$$\|\mathbf{R}^{-1}\|_S = \frac{1}{\lambda_{\min}} \tag{4.39}$$

[2]Note that the vector $\mathbf{x}$ is one of the eigenvectors. Hence, at this stage, we can only say that $\|\mathbf{A}\|^2$ is the *maximum Rayleigh quotient* of the eigenvectors. However, this may be extended to any vector after the minimax theorem is proved; see Property 9.

where $\lambda_{\min}$ is the smallest eigenvalue of $\mathbf{R}$. Thus, by adopting the spectral norm as the basis of the condition number, we have shown that the condition number of the correlation matrix $\mathbf{R}$ equals

$$\chi(\mathbf{R}) = \frac{\lambda_{\max}}{\lambda_{\min}} \tag{4.40}$$

This ratio is commonly referred to as the *eigenvalue spread* or the *eigenvalue ratio* of the correlation matrix. Note that we always have $\chi(\mathbf{R}) \geq 1$.

Suppose that the correlation matrix $\mathbf{R}$ is *normalized* so that the magnitude of the largest element, $r(0)$, equals 1. Then, if the condition number or eigenvalue spread of the correlation matrix $\mathbf{R}$ is large, we find that the inverse matrix $\mathbf{R}^{-1}$ contains some very large elements. This behavior may cause trouble in solving a system of equations involving $\mathbf{R}^{-1}$. In such a case, we say that the correlation matrix $\mathbf{R}$ is *ill conditioned*; hence, the justification of Property 7.

Property 8. *The eigenvalues of the correlation matrix of a discrete-time stochastic process are bounded by the minimum and maximum values of the power spectral density of the process.*

Let λ_i and $\mathbf{q}_i$, $i = 1, 2, \ldots, M$, denote the eigenvalues of the M-by-M correlation matrix $\mathbf{R}$ of a discrete-time stochastic process $\{u(n)\}$ and their associated eigenvectors, respectively. From Eq. (4.15), we have

$$\lambda_i = \frac{\mathbf{q}_i^H \mathbf{R} \mathbf{q}_i}{\mathbf{q}_i^H \mathbf{q}_i}, \qquad i = 1, 2, \ldots, M \tag{4.41}$$

The Hermitian form in the numerator may be expressed in its expanded form as follows

$$\mathbf{q}_i^H \mathbf{R} \mathbf{q}_i = \sum_{k=1}^{M} \sum_{l=1}^{M} q_{ik}^* r(l-k) q_{il} \tag{4.42}$$

where q_{ik}^* is the kth element of the row vector $\mathbf{q}_i^H$, $r(l-k)$ is the k,lth element of the matrix $\mathbf{R}$, and q_{il} is the lth element of the column vector $\mathbf{q}_i$. Using the Einstein-Wiener-Khintchine relation of Eq. (3.16), we may write

$$r(l-k) = \frac{1}{2\pi} \int_{-\pi}^{\pi} S(\omega) e^{j\omega(l-k)}\, d\omega \tag{4.43}$$

where $S(\omega)$ is the power spectral density of the process $\{u(n)\}$. Hence, we may rewrite Eq. (4.42) as

$$\begin{aligned} \mathbf{q}_i^H \mathbf{R} \mathbf{q}_i &= \frac{1}{2\pi} \sum_{k=1}^{M} \sum_{l=1}^{M} q_{ik}^* q_{il} \int_{-\pi}^{\pi} S(\omega) e^{j\omega(l-k)}\, d\omega \\ &= \frac{1}{2\pi} \int_{-\pi}^{\pi} d\omega\, S(\omega) \sum_{k=1}^{M} q_{ik}^* e^{-j\omega k} \sum_{l=1}^{M} q_{il} e^{j\omega l} \end{aligned} \tag{4.44}$$

Let the discrete-time Fourier transform of $q_{i1}^*, q_{i2}^*, \ldots, q_{iM}^*$ be denoted by

$$Q_i'(e^{j\omega}) = \sum_{k=1}^{M} q_{ik}^* e^{-j\omega k} \tag{4.45}$$

Therefore, using Eq. (4.45) in (4.44), we get

$$\mathbf{q}_i^H \mathbf{R} \mathbf{q}_i = \frac{1}{2\pi} \int_{-\pi}^{\pi} |Q_i'(e^{j\omega})|^2 S(\omega) d\omega \tag{4.46}$$

Similarly, we may show that

$$\mathbf{q}_i^H \mathbf{q}_i = \frac{1}{2\pi} \int_{-\pi}^{\pi} |Q_i'(e^{j\omega})|^2 d\omega \tag{4.47}$$

Accordingly, we may use Eq. (4.15) to redefine the eigenvalue λ_i of the correlation matrix $\mathbf{R}$ in terms of the associated power spectral density as

$$\lambda_i = \frac{\int_{-\pi}^{\pi} |Q_i'(e^{j\omega})|^2 S(\omega) d\omega}{\int_{-\pi}^{\pi} |Q_i'(e^{j\omega})|^2 d\omega} \tag{4.48}$$

Let $S_{\min}$ and $S_{\max}$ denote the absolute minimum and maximum values of the power spectral density $S(\omega)$, respectively. Then it follows that

$$\int_{-\pi}^{\pi} |Q_i'(e^{j\omega})|^2 S(\omega) d\omega \geq S_{\min} \int_{-\pi}^{\pi} |Q_i'(e^{j\omega})|^2 d\omega \tag{4.49}$$

and

$$\int_{-\pi}^{\pi} |Q_i'(e^{j\omega})|^2 S(\omega) d\omega \leq S_{\max} \int_{-\pi}^{\pi} |Q_i'(e^{j\omega})|^2 d\omega \tag{4.50}$$

Hence, we deduce that the eigenvalues λ_i are bounded by the maximum and minimum values of the associated power spectral density as follows:

$$S_{\min} \leq \lambda_i \leq S_{\max}, \qquad i = 1, 2, \ldots, M \tag{4.51}$$

Correspondingly, the eigenvalue spread $\chi(\mathbf{R})$ is bounded as

$$\chi(\mathbf{R}) = \frac{\lambda_{\max}}{\lambda_{\min}} \leq \frac{S_{\max}}{S_{\min}} \tag{4.52}$$

It is of interest to note that as the dimension M of the correlation matrix approaches infinity, the maximum eigenvalue $\lambda_{\max}$ approaches $S_{\max}$, and the minimum eigenvalue $\lambda_{\min}$ approaches $S_{\min}$. Accordingly, the eigenvalue spread $\chi(\mathbf{R})$ of the correlation matrix $\mathbf{R}$ approaches the ratio $S_{\max}/S_{\min}$ as the dimension M of the matrix $\mathbf{R}$ approaches infinity.

Property 9. Minimax Theorem. *Let the M-by-M correlation matrix* $\mathbf{R}$ *have eigenvalues* $\lambda_1, \lambda_2, \ldots, \lambda_M$ *that are ordered as follows:*

$$\lambda_1 \leq \lambda_2 \leq \cdots \leq \lambda_M$$

The minimax theorem states that

$$\lambda_k = \min_{\dim(\mathcal{S})=k} \max_{\substack{\mathbf{x}\in\mathcal{S} \\ \mathbf{x}\neq\mathbf{0}}} \frac{\mathbf{x}^H\mathbf{R}\mathbf{x}}{\mathbf{x}^H\mathbf{x}} \qquad k = 1, 2, \ldots M \tag{4.53}$$

where $\mathcal{S}$ is a subspace of the vector space of all M-by-1 complex vectors, dim ($\mathcal{S}$) denotes the dimension of subspace $\mathcal{S}$, and $\mathbf{x} \in \mathcal{S}$ *signifies that the vector* $\mathbf{x}$ *(assumed to be nonzero) varies over the subspace $\mathcal{S}$.*

Let $\mathbb{C}^M$ denote a complex vector space of dimension M. For the purpose of our present discussion, we define the *complex (linear) vector space* $\mathbb{C}^M$ as the set of all complex vectors that can be expressed as a linear combination of M *basis vectors*. Specifically, we may write

$$\mathbb{C}^M = \{\mathbf{y}\} \tag{4.54}$$

where $\mathbf{y}$ is any complex vector defined by

$$\mathbf{y} = \sum_{i=1}^{M} a_i \mathbf{q}_i \tag{4.55}$$

The $\mathbf{q}_i$ are the basis vectors, and the a_i are scalars. For the basis vectors we may use any *orthonormal set* of vectors $\mathbf{q}_1, \mathbf{q}_2, \ldots, \mathbf{q}_M$ that satisfy two requirements:

$$\mathbf{q}_i^H \mathbf{q}_j = \begin{cases} 1, & i = j \\ 0, & i \neq j \end{cases} \tag{4.56}$$

In other words, each basis vector is *normalized* to have a *Euclidean length* or *norm* of unity, and it is *orthogonal* to every other basis vector in the set. The *dimension M* of the complex vector space $\mathbb{C}^M$ is the minimum number of basis vectors required to span the entire space.

The basis functions define the "coordinates" of a complex vector space. Any complex vector of compatible dimension may then be represented simply as a "point" in that space. Indeed, the idea of a complex vector space is a natural generalization of Euclidean geometry. Central to this idea is that of a *subspace*. We say that $\mathcal{S}$ is a subspace of the complex vector space $\mathbb{C}^M$ if it involves a *subset* of the M basis vectors that define $\mathbb{C}^M$. In other words, a subspace of dimension k is defined as the set of complex vectors that can be written as a linear combination of the basis vectors $\mathbf{q}_1, \mathbf{q}_2, \ldots, \mathbf{q}_k$, as shown by

$$\mathbf{x} = \sum_{i=1}^{k} a_i \mathbf{q}_i \tag{4.57}$$

Obviously, we have $k \leq M$. Note, however, that the dimension of the vector $\mathbf{x}$ is M.

These ideas are illustrated in the three-dimensional (real) vector space depicted in Fig. 4.1. The $\mathbf{q}_1,\mathbf{q}_2$-plane, shown shaded, represents a subspace $\mathcal{S}$ of dimension 2. A vector $\mathbf{x}$ lying in the subspace $\mathcal{S}$ is represented by a point at the tip of the arrow labeled $\mathbf{x}$.

Returning to the issue at hand, namely, a proof of the *minimax theorem* described in Eq. (4.53), we may proceed as follows. We first use the spectral theorem of Eq. (4.31) to decompose the M-by-M correlation matrix $\mathbf{R}$ as

$$\mathbf{R} = \sum_{i=1}^{M} \lambda_i \mathbf{q}_i \mathbf{q}_i^H$$

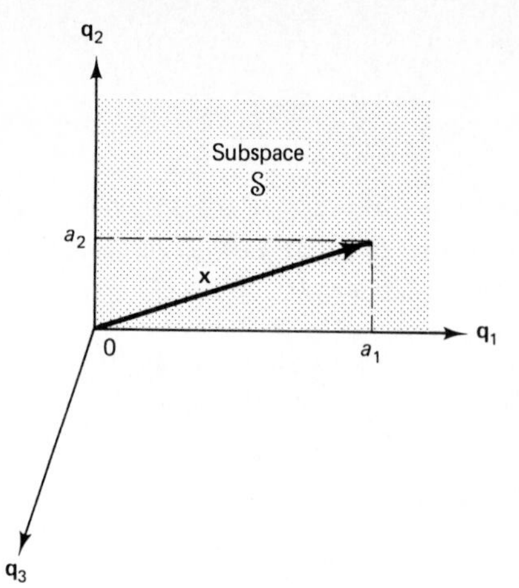

Figure 4.1 Three-dimensional (real) vector space. The 3-by-1 vector **x** lying in subspace $\mathcal{S}$ equals $[a_1, a_2, 0]^T$.

where the λ_i are the eigenvalues of the correlation matrix $\mathbf{R}$ and the $\mathbf{q}_i$ are the associated eigenvectors. In view of the orthonormality conditions of Eq. (4.24) satisfied by the eigenvectors $\mathbf{q}_1, \mathbf{q}_2, \ldots, \mathbf{q}_M$, we may adopt them as the M basis vectors of the complex vector space $\mathbb{C}^M$. Let an M-by-1 vector $\mathbf{x}$ be constrained to lie in a subspace $\mathcal{S}$ of dimension k, as defined in Eq. (4.57). Then, using Eq. (4.31), we may express the Rayleigh quotient of the vector $\mathbf{x}$ as

$$\frac{\mathbf{x}^H \mathbf{R} \mathbf{x}}{\mathbf{x}^H \mathbf{x}} = \frac{\sum_{i=1}^{k} a_i^2 \lambda_i}{\sum_{i=1}^{k} a_i^2} \tag{4.58}$$

Equation (4.58) states that the Rayleigh quotient of a vector $\mathbf{x}$ lying in the subspace $\mathcal{S}$ of dimension k (i.e., the subspace spanned by the eigenvectors $\mathbf{q}_1, \mathbf{q}_2, \ldots, \mathbf{q}_k$) is a *weighted mean* of the eigenvalues $\lambda_1, \lambda_2, \ldots, \lambda_k$. Since, by assumption, we have $\lambda_1 \leq \lambda_2 \leq \cdots \leq \lambda_k$, it follows that for any subspace $\mathcal{S}$ of dimension k,

$$\max_{\substack{\mathbf{x} \in \mathcal{S} \\ \mathbf{x} \neq \mathbf{0}}} \frac{\mathbf{x}^H \mathbf{R} \mathbf{x}}{\mathbf{x}^H \mathbf{x}} \leq \lambda_k$$

This result implies that

$$\min_{\dim(\mathcal{S})=k} \max_{\substack{\mathbf{x} \in \mathcal{S} \\ \mathbf{x} \neq \mathbf{0}}} \frac{\mathbf{x}^H \mathbf{R} \mathbf{x}}{\mathbf{x}^H \mathbf{x}} \leq \lambda_k \tag{4.59}$$

We next prove that for any subspace $\mathcal{S}$ of dimension k, spanned by the eigenvectors $\mathbf{q}_{i_1}, \mathbf{q}_{i_2}, \ldots, \mathbf{q}_{i_k}$ where $\{i_1, i_2, \ldots, i_k\}$ is a subset of $\{1, 2, \ldots, M\}$, there exists at least one nonzero vector $\mathbf{x}$ common to $\mathcal{S}$ and the subspace $\mathcal{S}'$ spanned by the eigenvectors $\mathbf{q}_k, \mathbf{q}_{k+1}, \ldots, \mathbf{q}_M$. To do so, we consider the system of M homogeneous equations:

$$\sum_{j=1}^{k} a_j \mathbf{q}_{i_j} = \sum_{i=k}^{M} b_i \mathbf{q}_i \tag{4.60}$$

where the $(M + 1)$ unknowns are made up as follows:

1. A total of k scalars, namely $a_1, a_2, \ldots, a_k$, on the left side.
2. A total of $M - k + 1$ scalars, namely $b_k, b_{k+1}, \ldots, b_M$ on the right side.

Hence the system of equations (4.60) will always have a nontrivial solution. Moreover, we know from Property 2 that the eigenvectors $\mathbf{q}_{i_1}, \mathbf{q}_{i_2}, \ldots, \mathbf{q}_{i_k}$ are *linearly independent*, as are the eigenvectors $\mathbf{q}_k, \mathbf{q}_{k+1}, \ldots, \mathbf{q}_M$. It follows therefore that there is at least one *nonzero vector* $\mathbf{x} = \sum_{j=1}^{k} a_j \mathbf{q}_{i_j}$ that is *common* to the space of $\mathbf{q}_{i_1}, \mathbf{q}_{i_2}, \ldots, \mathbf{q}_{i_k}$ and the space of $\mathbf{q}_k, \mathbf{q}_{k+1}, \ldots, \mathbf{q}_M$. Thus, using Eqs. (4.60), (4.57), and (4.31), we may also express the Rayleigh quotient of the vector $\mathbf{x}$ as a weighted mean of the eigenvalues $\lambda_k, \lambda_{k+1}, \ldots, \lambda_M$, as shown by

$$\frac{\mathbf{x}^H \mathbf{R} \mathbf{x}}{\mathbf{x}^H \mathbf{x}} = \frac{\sum_{i=k}^{M} b_i^2 \lambda_i}{\sum_{i=k}^{M} b_i^2} \tag{4.61}$$

Since, by assumption, we have $\lambda_k \leq \lambda_{k+1} \leq \cdots \leq \lambda_M$, and since $\mathbf{x}$ is also a vector in the subspace $\mathcal{S}$, we may write

$$\max_{\substack{\mathbf{x} \in \mathcal{S} \\ \mathbf{x} \neq \mathbf{0}}} \frac{\mathbf{x}^H \mathbf{R} \mathbf{x}}{\mathbf{x}^H \mathbf{x}} \geq \lambda_k$$

Therefore,

$$\min_{\dim(\mathcal{S})=k} \max_{\substack{\mathbf{x} \in \mathcal{S} \\ \mathbf{x} \neq \mathbf{0}}} \frac{\mathbf{x}^H \mathbf{R} \mathbf{x}}{\mathbf{x}^H \mathbf{x}} \geq \lambda_k \tag{4.62}$$

because $\mathcal{S}$ is an arbitrary subspace of dimension k.

All that remains for us to do is to combine the results of (4.59) and (4.62), and the minimax theorem of Eq. (4.53) describing Property 9 follows immediately.

From Property 9, we may make two important observations:

1. The minimax theorem as stated in Eq. (4.53) does not require any special knowledge of the eigenstructure (i.e., eigenvalues and eigenvectors) of the correlation matrix $\mathbf{R}$. Indeed, it may be adopted as the basis for defining the eigenvalues λ_k for $k = 1, 2, \ldots, M$.

2. The minimax theorem points to a unique two-fold feature of the eigenstructure of the correlation matrix: (a) the eigenvectors represent the particular basis for an M-dimensional space that is most efficient in the energy sense, and (b) the eigenvalues are the energies of the M-by-1 input (observation) vector $\mathbf{u}(n)$. This issue is pursued in greater depth under Property 10.

Another noteworthy point is that Eq. (4.53) may also be formulated in the following alternative but equivalent form:

$$\lambda_k = \max_{\dim(\mathcal{S}')=M-k+1} \min_{\substack{\mathbf{x} \in \mathcal{S}' \\ \mathbf{x} \neq \mathbf{0}}} \frac{\mathbf{x}^H \mathbf{R} \mathbf{x}}{\mathbf{x}^H \mathbf{x}} \tag{4.63}$$

Equation (4.63) is referred to as the *maximin theorem*.

From Eqs. (4.53) and (4.63) we may readily deduce the following two special cases:

1. For $k = M$, the subspace $\mathcal{S}$ occupies the complex vector space $\mathbb{C}^M$ entirely. Under this condition, Eq. (4.53) reduces to

$$\lambda_M = \max_{\substack{\mathbf{x} \in \mathbb{C}^M \\ \mathbf{x} \neq \mathbf{0}}} \frac{\mathbf{x}^H \mathbf{R} \mathbf{x}}{\mathbf{x}^H \mathbf{x}} \tag{4.64}$$

where λ_M is the *largest eigenvalue* of the correlation matrix $\mathbf{R}$.

2. For $k = 1$, the subspace $\mathcal{S}'$ occupies the complex vector space $\mathbb{C}^M$ entirely. Under this condition, Eq. (4.63) reduces to

$$\lambda_1 = \min_{\substack{\mathbf{x} \in \mathbb{C}^M \\ \mathbf{x} \neq \mathbf{0}}} \frac{\mathbf{x}^H \mathbf{R} \mathbf{x}}{\mathbf{x}^H \mathbf{x}} \tag{4.65}$$

where λ_1 is the *smallest eigenvalue* of the correlation matrix $\mathbf{R}$.

Property 10. Karhunen-Lóeve expansion *Let* $\mathbf{u}(n)$ *denote an M-by-1 random input (observation) vector of zero mean and correlation matrix* $\mathbf{R}$. *Let* $\mathbf{q}_1, \mathbf{q}_2, \ldots, \mathbf{q}_M$ *be eigenvectors of the matrix* $\mathbf{R}$. *The vector* $\mathbf{u}(n)$ *may be expanded as a linear combination of these eigenvectors as follows:*

$$\mathbf{u}(n) = \sum_{i=1}^{M} c_i(n)\mathbf{q}_i \tag{4.66}$$

The coefficients of the expansion are zero-mean, uncorrelated random variables defined by the inner product

$$c_i(n) = \mathbf{q}_i^H \mathbf{u}(n), \qquad i = 1, 2, \ldots, M \tag{4.67}$$

The representation of the random vector $\mathbf{u}(n)$ described by Eqs. (4.66) and (4.66) is the discrete-time version of the *Karhunen-Lóeve expansion*. In particular, Eq. (4.67) is the "analysis" part of the expansion in that it defines the $c_i(n)$ in terms of the input vector $\mathbf{u}(n)$. On the other hand, Eq. (4.66) is the "synthesis" part of the expansion in that it reconstructs the original input vector $\mathbf{u}(n)$ from the $c_i(n)$. Given the expansion of Eq. (4.66), the definition of $c_i(n)$ in Eq. (4.67) follows directly from the fact that the eigenvectors $\mathbf{q}_1, \mathbf{q}_2, \ldots, \mathbf{q}_M$ form an orthonormal set, assuming they are all normalized to have unit length. Conversely, this same property may be used to derive Eq. (4.66), given (4.67).

The coefficients of the expansion are random variables characterized as follows:

$$E[c_i(n)] = 0, \qquad i = 1, 2, \ldots, M \tag{4.68}$$

and

$$E[c_i(n)c_j^*(n)] = \begin{cases} \lambda_i & i = j \\ 0, & i \neq j \end{cases} \tag{4.69}$$

Equation (4.68) states that all the coefficients of the expansion have zero mean; this follows directly from (Eq. 4.67) and the fact the random vector $\mathbf{u}(n)$ is itself assumed to have zero mean. Equation (4.69) states that the coefficients of the expansion are uncorrelated, and that each one of them has a mean square value equal to the respective eigenvalue. This second equation is readily obtained by using the expansion of Eq. (4.66) in the definition of the correlation matrix $\mathbf{R}$ as the expectation of the outer product $\mathbf{u}(n)\mathbf{u}^H(n)$, and then invoking the unitary similarity transformation (i.e., Property 5).

For a physical interpretation of the Karhunen-Loéve expansion, we may view the eigenvectors $\mathbf{q}_1, \mathbf{q}_2, \ldots, \mathbf{q}_M$ as the coordinates of an M-dimensional space, and thus represent the random vector $\mathbf{u}(n)$ by the set of its projections $c_1(n), c_2(n), \ldots, c_M(n)$ onto these axes, respectively. Moreover, we deduce from Eq. (4.66) that

$$\sum_{i=1}^{M} |c_i(n)|^2 = \|\mathbf{u}(n)\|^2 \tag{4.70}$$

where $\|\mathbf{u}(n)\|$ is the Euclidean norm of $\mathbf{u}(n)$. That is to say, the coefficient $c_i(n)$ has an energy equal to that of the observation vector $\mathbf{u}(n)$ measured along the ith coordinate. Naturally, this energy is a random variable whose mean value equals the ith eigenvalue, as shown by

$$E[|c_i(n)|^2] = \lambda_i, \qquad i = 1, 2, \ldots, M \tag{4.71}$$

This result follows directly from Eqs. (4.67) and (4.69).

Note. Properties 1 through 10 are of a general nature in that they apply to the eigenvalues and eigenvectors of the correlation matrices of wide-sense stationary discrete-time stochastic processes, be they real or complex valued. When the process of interest is real valued, however, the correlation matrix becomes *doubly symmetric,* also known as *parsymmetric* or *symmetric centrosymmetric*. By a "doubly symmetric matrix" we mean a *square matrix that is symmetric about both the main diagonal and the secondary diagonal*. Clearly, all symmetric Toeplitz matrices (e.g., the correlation matrix of a real-valued stationary process) are doubly symmetric, but the converse is not necessarily true. Symmetric Toeplitz matrices by virtue of their specific structure, possess additional properties concerning their eigenvectors (Makhoul, 1981; Reddi, 1984). For a discussion of properties of the eigenvectors of doubly symmetric matrices and those of symmetric Toeplitz matrices, and related issues, see Problems 18 through 20.

4.3 EIGENFILTERS

A fundamental issue in communication theory is that of determining an optimum finite (length) impulse response (FIR) filter, with the optimization criterion being that of maximizing the output signal-to-noise ratio. In this section we show that this filter optimization is linked to an eigenvalue problem.

Consider a linear FIR filter whose impulse response is denoted by the sequence $\{w_n\}$. The sequence $\{x(n)\}$ applied to the filter input consists of a useful *signal* component $\{u(n)\}$ plus an additive *noise* component $\{v(n)\}$. The signal $\{u(n)\}$ is drawn from a wide-sense stationary stochastic process of zero mean and correlation matrix $\mathbf{R}$. The noise component $\{v(n)\}$ is white with a constant power spectral density of variance σ^2. The filter output is denoted by $y(n)$. The situation described herein is depicted in Fig. 4.2.

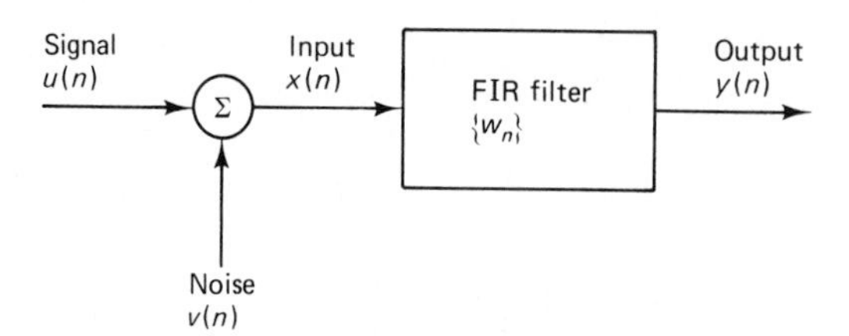

Figure 4.2 Linear filtering.

Since the filter is linear, the principle of superposition applies. We may therefore consider the effects of signal and noise separately. Let P_o denote the average power of the signal component of the filter output $y(n)$. We may therefore show that (see Problem 9 of Chapter 2):

$$P_o = \mathbf{w}^H \mathbf{R} \mathbf{w} \tag{4.72}$$

where the elements of the vector $\mathbf{w}$ are the filter coefficients, and $\mathbf{R}$ is the correlation matrix of the signal component $\{u(n)\}$ in the filter input.

Consider next the effect of noise acting alone. Let N_o denote the average power of the noise component in the filter output $y(n)$. This is a special case of Eq. (4.72), as shown by

$$N_o = \sigma^2 \mathbf{w}^H \mathbf{w} \tag{4.73}$$

where σ^2 is the variance of the white noise in the filter input.

Let $(\text{SNR})_o$ denote the *output signal-to-noise ratio*. Dividing Eq. (4.72) by (4.73), we may thus write

$$\begin{aligned} (SNR)_o &= \frac{P_o}{N_o} \\ &= \frac{\mathbf{w}^H \mathbf{R} \mathbf{w}}{\sigma^2 \mathbf{w}^H \mathbf{w}} \end{aligned} \tag{4.74}$$

The optimum problem may now be stated as follows:

> Determine the coefficient vector w of an FIR filter so as to maximize the output signal-to-noise ratio $(\text{SNR})_o$ subject to the constraint
>
> $$\mathbf{w}^H \mathbf{w} = 1$$

Equation (4.74) shows that except for the scaling factor $1/\sigma^2$, the output signal-to-noise ratio $(\text{SNR})_o$ is equal to the Rayleigh quotient of the coefficient vector $\mathbf{w}$ of the FIR filter. We see therefore that the optimum filtering problem, as stated herein, may be viewed as an eigenvalue problem. Indeed, the solution to the problem follows directly from the minimax theorem. Specifically, using the special form of the minimax theorem given in Eq. (4.64), we may state the following:

1. The maximum value of the output signal-to-noise ratio is given by

$$(\text{SNR})_{o,\max} = \frac{\lambda_{\max}}{\sigma^2} \tag{4.75}$$

   where $\lambda_{\max}$ is the largest eigenvalue of the correlation matrix $\mathbf{R}$. Note that $\lambda_{\max}$ and σ^2 have the same units but different physical interpretations.
2. The coefficient vector of the optimum FIR filter that yields the maximum output signal-to-noise ratio of Eq. (4.75) is defined by

$$\mathbf{w}_o = \mathbf{q}_{\max} \tag{4.76}$$

   where $\mathbf{q}_{\max}$ is the eigenvector associated with the largest eigenvalue $\lambda_{\max}$ of the correlation matrix $\mathbf{R}$. The correlation matrix $\mathbf{R}$ belongs to the signal component $\{u(n)\}$ of the filter input.

An FIR filter whose impulse response has coefficients equal to the elements of an eigenvector is called an *eigenfilter* (Makhoul, 1981). Accordingly, we may state that *the maximum eigenfilter* (*i.e., the eigenfilter associated with the largest eigenvalue of the correlation matrix of the signal component at the filter input*) *is the optimum filter*. It is important to note that the optimum filter described in this way is uniquely characterized by an eigendecomposition of the correlation matrix of the signal component of the filter input. The power spectrum of the white noise at the filter input merely affects the maximum value of the output signal-to-noise ratio. In particular, we may proceed as follows:

1. An eigendecomposition of the correlation matrix $\mathbf{R}$ is performed.
2. Only the largest eigenvalue $\lambda_{\max}$, and the associated eigenvector $\mathbf{q}_{\max}$ is retained.
3. The eigenvector $\mathbf{q}_{\max}$ defines the impulse response of the optimum filter. The eigenvalue $\lambda_{\max}$, divided by the noise variance σ^2, defines the maximum value of the output signal-to-noise ratio.

Such a procedure represents an example of *principal component analysis,* with the eigenvector $\mathbf{q}_{\max}$ associated with the largest eigenvalue $\lambda_{\max}$ being the only principal component of interest in the specification of the optimum filter.

The optimum filter so characterized may be viewed as the "stochastic" counterpart of a matched filter. The optimum filter described herein maximizes the output signal-to-noise ratio for a *random signal* (i.e., a sample function of a discrete-time wide-sense stationary stochastic process) in additive white noise. A *matched filter,* on the other hand, maximizes the output signal-to-noise ratio for a *known signal* in additive white noise (Haykin, 1988).

4.4 EIGENVALUE COMPUTATIONS

The computation of the eigenvalues of a square matrix can, in general, be a complicated issue. Special and cooperative efforts by a group of experts (during 1958 to 1970) have resulted in the development of several *canned routines* that are widely available for matrix eigenvalue computations (Parlett, 1985). Special mention should be made of the following program libraries:

1. EISPACK
2. LINPACK
3. IMSL
4. MATLAB

The canned *eigen-routines* in these libraries are well documented and well tested.

The origin of almost all canned eigen-routines may be traced back to routines published in Volume II, *Linear Algebra,* of the *Handbook for Automatic Computation* co-edited by Wilkinson and Reinsch (1971). This reference is the Bible of eigenvalue computations.

Another useful source of routines, written in the *C programming language,* is the book by Press et al. (1988); a companion book by the same authors, with routines written in FORTRAN and Pascal, is also available. The eigen-routines written in C can only handle real matrices. It is, however, a straightforward matter to extend the use of these eigen-routines to deal with Hermitian matrices, as shown next.

Let $\mathbf{A}$ denote an M-by-M Hermitian matrix, written in terms of its real and imaginary parts as follows:

$$\mathbf{A} = \mathbf{A}_r + j\mathbf{A}_i \tag{4.77}$$

Correspondingly, let an associated M-by-1 eigenvector $\mathbf{q}$ be written as

$$\mathbf{q} = \mathbf{q}_r + j\mathbf{q}_i \tag{4.78}$$

The M-by-M complex eigenvalue problem

$$(\mathbf{A}_r + j\mathbf{A}_i)(\mathbf{q}_r + j\mathbf{q}_i) = \lambda(\mathbf{q}_r + j\mathbf{q}_i) \tag{4.79}$$

may then be reformulated as the $2M$-by-$2M$ real eigenvalue problem:

$$\begin{bmatrix} \mathbf{A}_r & -\mathbf{A}_i \\ \mathbf{A}_i & \mathbf{A}_r \end{bmatrix} \begin{bmatrix} \mathbf{q}_r \\ \mathbf{q}_i \end{bmatrix} = \lambda \begin{bmatrix} \mathbf{q}_r \\ \mathbf{q}_i \end{bmatrix} \tag{4.80}$$

where the eigenvalue λ is a real number. The Hermitian property

$$\mathbf{A}^{\mathrm{H}} = \mathbf{A}$$

is equivalent to $\mathbf{A}_r^T = \mathbf{A}_r$ and $\mathbf{A}_i^T = -\mathbf{A}_i$. Accordingly, the $2M$-by-$2M$ matrix in Eq. (4.80) is not only real but also symmetric. Note, however, that for a given eigenvalue λ, the vector

$$\begin{bmatrix} -\mathbf{q}_i \\ \mathbf{q}_r \end{bmatrix}$$

is also an eigenvector. This means that if $\lambda_1, \lambda_2, \ldots, \lambda_M$ are the eigenvalues of the M-by-M Hermitian matrix $\mathbf{A}$, then the eigenvalues of the $2M$-by-$2M$ symmetric matrix of Eq. (4.80) are $\lambda_1, \lambda_1, \lambda_2, \lambda_2, \ldots, \lambda_M, \lambda_M$. We may therefore make two observations:

1. Each eigenvalue of the matrix in Eq. (4.80) has a multiplicity of 2.
2. The associated eigenvectors consist of pairs, each of the form $\mathbf{q}_r + j\mathbf{q}_i$ and $j(\mathbf{q}_r + j\mathbf{q}_i)$, differing merely by a rotation through 90°.

Thus, to solve the M-by-M complex eigenvalue problem of Eq. (4.79) with the aid of real eigen-routines, we choose one eigenvalue and eigenvector from each pair associated with the augmented $2M$-by-$2M$ real eigenvalue problem of Eq. (4.80).

Strategies for Matrix Eigenvalue Computations

There are two different "strategies" behind practically all modern eigen-routines: *diagonalization,* and *triangularization*. Since not all matrices can be diagonalized through a sequence of unitary similarity transformations, the diagonalization strategy applies only to Hermitian matrices such as a correlation matrix. On the other hand, the triangularization strategy is general in that it applies to any square matrix. These two strategies are described in the sequel.

Diagonalization. The idea behind this strategy is to nudge a Hermitian matrix $\mathbf{A}$ toward a diagonal form by the repeated application of unitary similarity transformations:

$$\begin{aligned} \mathbf{A} &\rightarrow \mathbf{Q}_1^H \mathbf{A} \mathbf{Q}_1 \\ &\rightarrow \mathbf{Q}_2^H \mathbf{Q}_1^H \mathbf{A} \mathbf{Q}_1 \mathbf{Q}_2 \\ &\rightarrow \mathbf{Q}_3^H \mathbf{Q}_2^H \mathbf{Q}_1^H \mathbf{A} \mathbf{Q}_1 \mathbf{Q}_2 \mathbf{Q}_3 \end{aligned} \tag{4.81}$$

and so on. This sequence of unitary similarity transformations is in theory, infinitely long. In practice, however, it is continued until we are close to a diagonal matrix. The elements of the diagonal matrix so obtained define the eigenvalues of the original Hermitian matrix $\mathbf{A}$. The associated eigenvectors are the column vectors of the accumulated sequence of transformations, as shown by

$$\mathbf{Q} = \mathbf{Q}_1 \mathbf{Q}_2 \mathbf{Q}_3 \cdots \tag{4.82}$$

One method for implementing the diagonalization strategy of Eq. (4.81) is to use *Jacobi rotations*. This method is discussed in Chapter 11.

Triangularization. The idea behind this second strategy is to reduce a Hermitian matrix $\mathbf{A}$ to a triangular form by a sequence of unitary similarity transformations. The resulting iterative procedure is called the *QL algorithm*.[2] Suppose that

[2] The QL algorithm uses a lower triangular matrix. There is a companion algorithm, called the QR algorithm, which uses an upper triangular matrix. The QR algorithm is not to be confused with the QR-decomposition.

we are given an M-by-M Hermitian matrix $\mathbf{A}_n$, where the subscript n refers to a particular step in the iterative procedure. Let the matrix $\mathbf{A}_n$ be *factored* in the form

$$\mathbf{A}_n = \mathbf{Q}_n\mathbf{L}_n \tag{4.83}$$

where $\mathbf{Q}_n$ is a *unitary matrix* and $\mathbf{L}_n$ is a *lower triangular matrix* (i.e., the elements of the matrix $\mathbf{L}_n$ located above the main diagonal are all zero). At step $n + 1$ in the iterative procedure, we use the known matrices $\mathbf{Q}_n$ and $\mathbf{L}_n$ to compute a new M-by-M matrix

$$\mathbf{A}_{n+1} = \mathbf{L}_n\mathbf{Q}_n \tag{4.84}$$

Note that the factorization in Eq.(4.84) is written in the opposite order to that in Eq. (4.83). Since $\mathbf{Q}_n$ is a unitary matrix, we have $\mathbf{Q}_n^{-1} = \mathbf{Q}_n^H$, so we may rewrite Eq. (4.83) as

$$\begin{aligned}\mathbf{L}_n &= \mathbf{Q}_n^{-1}\mathbf{A}_n \\ &= \mathbf{Q}_n^H\mathbf{A}_n\end{aligned} \tag{4.85}$$

Therefore, substituting Eq. (4.85) into (4.84), we get

$$\mathbf{A}_{n+1} = \mathbf{Q}_n^H\mathbf{A}_n\mathbf{Q}_n \tag{4.86}$$

Equation (4.86) shows that the Hermitian matrix $\mathbf{A}_{n+1}$ at iteration $n + 1$ is indeed unitarily related to the Hermitian matrix $\mathbf{A}_n$ at iteration n.

The QL algorithm thus consists of a sequence of unitary similarity transformations, summarized by writing

$$\begin{aligned}\mathbf{A}_n &= \mathbf{Q}_n\mathbf{L}_n \\ \mathbf{A}_{n+1} &= \mathbf{L}_n\mathbf{Q}_n\end{aligned}$$

where $n = 0, 1, 2, \ldots$. The algorithm is *initialized* by setting

$$\mathbf{A}_0 = \mathbf{A}$$

where $\mathbf{A}$ is the given M-by-M Hermitian matrix.

For general matrix $\mathbf{A}$, the following theorem is the basis of the QL algorithm[3]

> If matrix $\mathbf{A}$ has eigenvalues of different absolute values, then the matrix $\mathbf{A}_n$ approaches a lower triangular form as the number of iterations n approaches infinity.

The eigenvalues of the original matrix $\mathbf{A}$ appear on the main diagonal of the lower triangular matrix resulting from the QL algorithm in increasing order of absolute value.

To implement the factorization in Eq. (4.84), we may use Jacobi rotations. Here again, however, we defer our discussion of the method to Chapter 11. In that chapter we discuss computations for the singular value decomposition of a general matrix, which includes eigendecomposition as a special case.

[3] For a proof of this theorem, see Stoer and Bulirsch (1980). See also Stewart (1973), Golub and Van Loan (1989), and Press et al. (1988) for an improved version of the QL algorithm.

PROBLEMS

1. The correlation matrix $\mathbf{R}$ of a wide-sense stationary process $\{u(n)\}$ has the following values for its two eigenvalues:

$$\lambda_1 = 0.5$$
$$\lambda_2 = 1.5$$

(a) Find the trace of matrix $\mathbf{R}$.

(b) Write an expression for the decomposition of matrix $\mathbf{R}$ in terms of its two eigenvalues and associated eigenvectors. Comment on the uniqueness of this decomposition.

2. Show that the eigenvalues of a triangular matrix equal the diagonal elements of the matrix.

3. Consider the $2M$-by-$2M$ real eigenvalue problem described in Eq. (4.80). Show that if $\mathbf{q}_r + j\mathbf{q}_i$ is an eigenvector of the matrix described herein, so is $\mathbf{q}_i - j\mathbf{q}_r$, with both eigenvectors being associated with the same eigenvalue.

4. Let $\lambda_1, \lambda_2, \ldots, \lambda_M$ denote the eigenvalues of the correlation matrix of an observation vector $\mathbf{u}(n)$ taken from a stationary process of zero mean and variance σ_u^2. Show that

$$\sum_{i=1}^{M} \lambda_i = M\sigma_u^2$$

5. An M-by-M correlation matrix $\mathbf{R}$ is represented in terms of its eigenvalues $\lambda_1, \lambda_2, \ldots, \lambda_M$ and their associated eigenvectors $\mathbf{q}_1, \mathbf{q}_2, \ldots, \mathbf{q}_M$ as follows:

$$\mathbf{R} = \sum_{i=1}^{M} \lambda_i \mathbf{q}_i \mathbf{q}_i^H$$

(a) Show that the corresponding representation for the *square root* of matrix $\mathbf{R}$ is

$$\mathbf{R}^{1/2} = \sum_{i=1}^{M} \lambda_i^{1/2} \mathbf{q}_i \mathbf{q}_i^H$$

(b) By definition, we have $\mathbf{R} = \mathbf{R}^{1/2}\,\mathbf{R}^{1/2}$. Using this result, describe a procedure for computing the square root of a square matrix.

6. Consider a stationary process $\{u(n)\}$ whose M-by-M correlation matrix equals $\mathbf{R}$. Show that the determinant of the correlation matrix $\mathbf{R}$ equals

$$\det(\mathbf{R}) = \prod_{i=1}^{M} \lambda_i$$

where $\lambda_1, \lambda_2, \ldots, \lambda_M$ are the eigenvalues of $\mathbf{R}$.

7. Consider an M-by-1 observation vector $\mathbf{u}(n)$ that consists of two superimposed uncorrelated sinusoidal processes plus additive white noise. The vector $\mathbf{u}(n)$ is written as

$$\begin{aligned}\mathbf{u}(n) &= [u(n), u(n-1), \ldots, u(n-M+1)]^T \\ &= \alpha_1 \mathbf{s}_1(n) + \,+\alpha_2 \mathbf{s}_2(n) + \mathbf{v}(n)\end{aligned}$$

where

$$\mathbf{s}_i(n) = [1, e^{-j\omega_i}, \ldots, e^{-j\omega_i(M-1)}]^T \qquad i = 1, 2$$

Assume that

$$E[|\alpha_i|^2] = \sigma_i^2, \qquad i = 1, 2$$

and

$$E[\mathbf{v}(n)\mathbf{v}^H(n)] = \sigma_v^2\mathbf{I}$$

You may also assume that

$$\sigma_1^2 > \sigma_2^2$$

$$\omega_1 = \frac{2\pi}{M}$$

and

$$\omega_2 = \frac{4\pi}{m}$$

(a) Determine the eigenvalues of the correlation matrix

$$\mathbf{R} = E[\mathbf{u}(n)\mathbf{u}^H(n)]$$

(b) What are the eigenvectors associated with the largest two eigenvalues of the matrix $\mathbf{R}$? How is the result modified if $\sigma_1^2 = \sigma_2^2$?

8. **(a)** Show that the product of two unitary matrices is also a unitary matrix.
(b) Show that the inverse of a unitary matrix is also a unitary matrix.

9. Let $\mathbf{A}$ be an M-by-M matrix. The *Schur decomposition theorem* states there exists a unitary matrix $\mathbf{Z}$ such that

$$\mathbf{Z}^H\mathbf{A}\mathbf{Z} = \mathbf{T}$$

where $\mathbf{T}$ is an upper triangular matrix. The theorem also states that:

(i) The diagonal of matrix $\mathbf{T}$ is made up of the eigenvalues of the matrix $\mathbf{A}$.

(ii) If $\mathbf{Z} = [\mathbf{z}_1, \mathbf{z}_2, \ldots, \mathbf{z}_M]$, then span $(\mathbf{z}_1, \mathbf{z}_2, \ldots, \mathbf{z}_k)$ is an invariant subspace associated with the eigenvalues $t_{11}, t_{22}, \ldots, t_{kk}$ where $k \le M$.

(a) Apply the Schur decomposition to the correlation matrix $\mathbf{R}$ of a wide-stationary stochastic process. Hence, show that in this case the matrix $\mathbf{T}$ is a diagonal matrix.

(b) What is the implication of the statement under (ii) in the context of the correlation matrix $\mathbf{R}$?

10. Consider the factorization

$$\mathbf{A}_n - k_n\mathbf{I} = \mathbf{Q}_n\mathbf{L}_n$$

where $\mathbf{A}_n$ is an M-by-M Hermitian matrix, $\mathbf{I}$ is the M-by-M identity matrix, $\mathbf{Q}_n$ is an M-by-M unitary matrix, $\mathbf{L}_n$ is an M-by-M lower triangular matrix, and k_n is a scalar. Define the matrix

$$\mathbf{A}_{n+1} = \mathbf{L}_n\mathbf{Q}_n + k_n\mathbf{I}$$

Hence, show that

$$\mathbf{A}_{n+1} = \mathbf{Q}_n^H\mathbf{A}_n\mathbf{Q}_n$$

11. In this problem we consider a *Fourier analyzer* for a single channel. The *Fourier basis* is described by

$$\mathbf{v}_i = \frac{1}{\sqrt{M}}[1, e^{j2\pi i/M}, e^{j4\pi i/M}, \ldots, e^{j(M-1)2\pi i/M}]^T$$

where $i = 0, 1, \ldots, M - 1$. Let an arbitrary M-by-1 vector $\mathbf{u}(n)$ be expanded in terms of this orthonormal set as follows

$$\mathbf{u}(n) = \sum_{i=0}^{M-1} c_i(n)\mathbf{v}_i$$

(a) Evaluate the *Fourier coefficients* $c_0(n), c_1(n), \ldots, c_{M-1}(n)$ in terms of the vector $\mathbf{u}(n)$.

(b) Are the Fourier coefficients correlated? Justify your answer.

(c) What does the expectation of $|c_i(n)|^2$ approximate?

12. Show that the condition number of matrix $\mathbf{A}$ is unchanged when this matrix is multiplied by a unitary matrix of compatible dimensions.

13. Consider an L-by-M matrix $\mathbf{A}$. Show that the M-by-M matrix $\mathbf{A}^H\mathbf{A}$ and the L-by-L matrix $\mathbf{A}\mathbf{A}^H$ have the same nonzero eigenvalues.

14. A stochastic process $\{v(n)\}$ with a wide-band power spectrum is applied to a discrete-time linear filter whose amplitude response $|H(e^{j\omega})|$ is nonuniform. The maximum and minimum values of this response are denoted by $H_{\max}$ and $H_{\min}$, respectively. Let $\chi(\mathbf{R})$ denote the eigenvalue spread of the correlation matrix $\mathbf{R}$ of the stochastic process $\{u(n)\}$ produced at the output of the filter. Show that

$$\chi(\mathbf{R}) \simeq \left(\frac{H_{\max}}{H_{\min}}\right)^2$$

15. *Szegö's theorem* states that, if $g(\cdot)$ is a continuous function, then

$$\lim_{M\to\infty} \frac{g(\lambda_1) + g(\lambda_2) + \cdots + g(\lambda_M)}{M} = \frac{1}{2\pi}\int_{-\pi}^{\pi} g[S(\omega)]\, d\omega$$

where $S(\omega)$ is the power spectral density of a stationary discrete-time stochastic process $\{u(n)\}$, and $\lambda_1, \lambda_2, \ldots, \lambda_M$ are the eigenvalues of the associated correlation matrix $\mathbf{R}$. It is assumed that the process $\{u(n)\}$ is limited to the interval $-\pi \le \omega \le \pi$. Using this theorem, show that

$$\lim_{M\to\infty} [\det(\mathbf{R})]^{1/M} = \exp\left(\frac{1}{2\pi}\int_{-\pi}^{\pi} \ln\,[S(\omega)]\, d\omega\right)$$

16. Consider a linear system of equations described by

$$\mathbf{R}\mathbf{w}_o = \mathbf{p}$$

where $\mathbf{R}$ is an M-by-M matrix, and $\mathbf{w}_o$ and $\mathbf{p}$ are M-by-1 vectors. The vector $\mathbf{w}_o$ represents the set of unknown parameters. Due to a combination of factors (e.g., measurement inaccuracies, computational errors), the matrix $\mathbf{R}$ is perturbed by a small amount $\delta\mathbf{R}$, producing a corresponding change $\delta\mathbf{w}$ in the vector of unknowns.

(a) Show that

$$\frac{\|\delta\mathbf{w}\|}{\|\mathbf{w}_o\|} \le \chi(\mathbf{R})\frac{\|\delta\mathbf{R}\|}{\|\mathbf{R}\|}$$

where $\chi(\mathbf{R})$ is the condition number of $\mathbf{R}$, and $\|\cdot\|$ denotes the norm of the quantity enclosed within.

(b) Develop the corresponding formula for a small change in the vector $\mathbf{p}$.
Hint: Use the inequality

$$\|\mathbf{A}\mathbf{x}\| \le \|\mathbf{A}\|\,\|\mathbf{x}\|$$

17. Consider the three-dimensional vector space of Fig. 4.1. Let the subspace $\mathcal{S}$ denote the $\mathbf{q}_{i_1}$, $\mathbf{q}_{i_2}$-plane where $\{i_1, i_2\}$ is a subset of $\{1, 2, 3\}$. Let the subspace $\mathcal{S}'$ denote the $\mathbf{q}_2$, $\mathbf{q}_3$-plane.

(a) Specify a vector $\mathbf{x}$ of unit length that is common to the subspaces $\mathcal{S}$ and $\mathcal{S}'$.

(b) What is the Rayleigh coefficient of the vector $\mathbf{x}$ specified in part (a)? Justify your answer in light of the minimax theorem.

18. Consider an M-by-M doubly symmetric matrix $\mathbf{R}$ that is symmetric about both the main diagonal and the secondary diagonal. Let $\mathbf{J}$ denote an M-by-M matrix that consists of 1's along the secondary diagonal and zeros everywhere else. The matrix $\mathbf{J}$ is called a *reverse operator* or *exchange matrix* because $\mathbf{JR}$ reverses the rows of matrix $\mathbf{R}$, $\mathbf{RJ}$ reverses the columns of $\mathbf{R}$, and $\mathbf{JRJ}$ reverses both the rows and columns of $\mathbf{R}$.

(a) Show that for the matrix $\mathbf{R}$ to be doubly symmetric, a necessary and sufficient condition is

$$\mathbf{JRJ} = \mathbf{R}$$

Noting that $\mathbf{J}^{-1} = \mathbf{J}$, show that the inverse of matrix $\mathbf{R}$ is also doubly symmetric, as shown by

$$\mathbf{JR}^{-1}\mathbf{J} = \mathbf{R}^{-1}$$

(b) Assume that the doubly symmetric matrix $\mathbf{R}$ has distinct eigenvalues. Hence show that the matrix $\mathbf{R}$ has $\lfloor (M + 1)/2 \rfloor$ symmetric eigenvectors and $\lfloor M/2 \rfloor$ skew-symmetric eigenvectors, where $\lfloor X \rfloor$ denotes the largest integer less than or equal to X. An eigenvector $\mathbf{q}$ is said to be *symmetric* if

$$\mathbf{Jq} = \mathbf{q}$$

and *skew symmetric* if

$$\mathbf{Jq} = -\mathbf{q}$$

where $\mathbf{J}$ is the reverse operator.

(c) Let $A(z)$ denote the transfer function of an eigenfilter of the doubly symmetric matrix $\mathbf{R}$. Show that if $A(z)$ is associated with a symmetric eigenvector or skew-symmetric eigenvector of $\mathbf{R}$ and if z_i is a zero of $A(z)$, then so is $1/z_i$.

19. Let $\mathbf{R}$ denote an M-by-M nonsingular symmetric Toeplitz matrix. Naturally, the properties described in Problem 18 apply to the matrix $\mathbf{R}$. Note, however, that the inverse matrix $\mathbf{R}^{-1}$ is not Toeplitz, in general. But owing to the special structure of a Toeplitz matrix, the matrix $\mathbf{R}$ has two additional properties:

(a) Let $\lambda_{\max}$ denote the largest eigenvalue of the matrix $\mathbf{R}$, which is assumed to be distinct. Show that the discrete transfer function of the eigenfilter associated with $\lambda_{\max}$ has all of its zeros located on the unit circle in the z-plane.

(b) Let $\lambda_{\min}$ denote the smallest eigenvalue of the matrix $\mathbf{R}$, which is assumed to be distinct. Show that the discrete transfer function of the eigenfilter associated with $\lambda_{\min}$ has all of its zeros located on the unit circle in the z-plane.

20. Consider the *normalized* 3-by-3 correlation matrix

$$\mathbf{R} = \begin{bmatrix} 1 & \rho_1 & \rho_2 \\ \rho_1 & 1 & \rho_1 \\ \rho_2 & \rho_1 & 1 \end{bmatrix}$$

where

$$\rho_i = \frac{r(i)}{r(0)}, \qquad i = 1, 2$$

(a) Using properties (b) and (c) of Problem 18, demonstrate the following results:
(1) The matrix **R** has a single skew-symmetric eigenvector of the form

$$\mathbf{q}_i = \frac{1}{\sqrt{2}}[1, 0, -1]^T$$

that is associated with the eigenvalue

$$\lambda_1 = 1 - \rho_2$$

(2) The matrix **R** has two symmetric eigenvectors of the form

$$\mathbf{q}_i = \frac{1}{\sqrt{1 + c_i^2}}[1, c_i, 1]^T, \qquad i = 2, 3$$

where c_i is related to the corresponding eigenvalue λ_i by

$$c_i = \frac{2\rho_1}{\lambda_i - 1} = \frac{\lambda_i - 1 - \rho_2}{\rho_1}, \qquad i = 2, 3$$

Hence, complete the specification of the eigenvalues and the eigenvectors of the matrix **R**.

(b) Given that the eigenvalues of matrix **R** are distinct and ordered as $\lambda_1 > \lambda_2 > \lambda_3$, and given that the eigenfilters associated with λ_1 and λ_3 have their zeros on the unit circle in accordance with properties (a) and (b) of Problem 19, respectively, find the conditions that the coefficients c_2 and c_3 must satisfy for the following situations to occur:
(1) The eigenfilter associated with eigenvalue λ_2 will *also* have its zeros on the unit circle.
(2) The eigenfilter associated with eigenvalue λ_3 will *not* have its zeros on the unit circle.

Illustrate both of these situations with selected values for the correlation coefficients ρ_2 and ρ_3.

21. To solve the optimum filtering problem described in Section 4.3, we selected an eigenfilter associated with the largest eigenvalue of the correlation matrix of the signal component at the filter input. What would be the result of selecting an eigenfilter associated with the smallest eigenvalue of this correlation matrix? Justify your answer.

Part II

Linear Optimum Filtering

Part II of the book consists of Chapters 5 through 7. It is devoted to a detailed treatment of linear optimum filter theory for discrete-time wide-sense stationary stochastic processes. Adaptive filters are derived from this theory. Chapter 5 covers the classical Wiener filter. Chapter 6 builds on the Wiener filter theory to solve the linear prediction problem. Chapter 7 covers the classical Kalman filter for solving the optimum filtering problem (formulated in terms of a state vector) in a recursive manner.

5

Wiener Filters

This chapter deals with a class of *optimum linear discrete-time filters* known collectively as *Wiener filters*. The theory for a Wiener filter is formulated for the general case of *complex-valued* time series with the filter specified in terms of its impulse response. The reason for using complex-valued time series is that in many practical situations (e.g., communications, radar, sonar) the *baseband* signal of interest appears in complex form; the term *baseband* is used to designate a band of frequencies representing the original signal as delivered by a source of information. The case of real-valued time series may of course be considered as a special case of this theory.

We begin our study of Wiener filters by outlining the optimum linear filtering problem, and setting the stage for the rest of the chapter.

5.1 OPTIMUM LINEAR FILTERING: PROBLEM STATEMENT

Consider the block diagram of Fig. 5.1 built around a *linear discrete-time filter*. The filter *input* consists of a *time series* $u(0)$, $u(1)$, $u(2)$, . . . , and the filter is itself characterized by the *impulse response* w_0, w_1, w_2, ... At some *discrete time n*, the

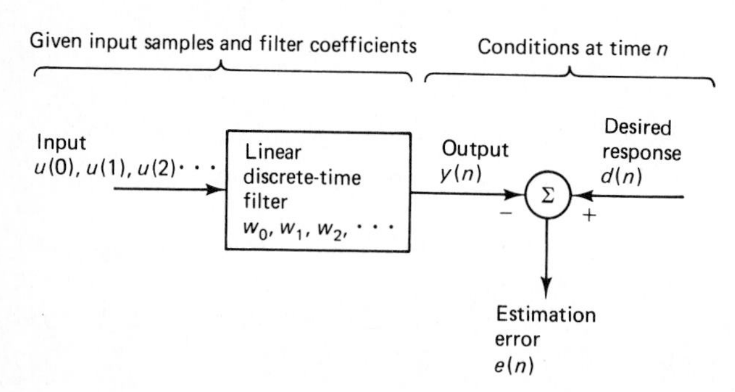

Figure 5.1 Block diagram representation of the statistical filtering problem.

filter produces an *output* denoted by $y(n)$. This output is used to provide an *estimate* of a *desired response* denoted by $d(n)$. With the filter input and the desired response representing single realizations of respective stochastic processes, the estimation is accompanied by an error with statistical characteristics of its own. In particular, the *estimation error*, denoted by $e(n)$, is defined as the difference between the desired response $d(n)$ and the filter output $y(n)$. The requirement is to make the estimation error $e(n)$ "as small as possible" in some statistical sense.

Two restrictions have so far been placed on the filter:

1. The filter is *linear*, which makes the mathematical analysis easy to handle.
2. The filter operates in *discrete time*, which makes it possible for the filter to be implemented using digital hardware.

The final details of the filter specification, however, depend on two other choices that have to be made:

1. Whether the impulse response of the filter has *finite* or *infinite* duration.
2. The type of *statistical criterion* used for the optimization.

The choice of a *finite-duration impulse response* (FIR) or an *infinite-duration impulse response* (IIR) for the filter is dictated by *practical considerations*. The choice of a statistical criterion for optimizing the filter design is influenced by *mathematical tractability*. These two issues are considered in turn.

For the initial development of the Wiener filter theory, we will assume an IIR filter; the theory so developed includes that for FIR filters as a special case. However, for much of the material presented in this chapter, and indeed in the rest of the book, we will confine our attention to the use of FIR filters. We do so for the following reason. An FIR filter is *inherently stable*, because its structure involves the use of *forward paths only*. In other words, the only mechanism for input–output interaction in the filter is via forward paths from the filter input to its output. Indeed, it is this form of signal transmission through the filter that limits its impulse response to a finite duration. On the other hand, an IIR filter involves *both feedforward and feedback*. The presence of feedback means that portions of the filter output and possibly other *internal* variables in the filter are fed back to the input. Consequently, unless it is properly controlled, feedback in the filter can indeed make it *unstable* with the result that the filter *oscillates*; this kind of operation is clearly unacceptable when the requirement is that of filtering for which stability is a "must." By itself, the stability problem in IIR filters is manageable in both theoretical and practical terms. However, when the filter is required to be *adaptive*, bringing with it stability problems of its own, the inclusion of adaptivity combined with feedback that is inherently present in an IIR filter makes a difficult problem that much more difficult to handle. It is for this reason that we find that in the majority of applications requiring the use of adaptivity, the use of an FIR filter is preferred over an IIR filter even though the latter is less demanding in computational requirements.

Turning next to the issue of what criterion to choose for statistical optimization, there are indeed several criteria that suggest themselves. Specifically, we may consider optimizing the filter design by *minimizing a cost function*, or *index of performance*, selected from the following short list of possibilities:

1. Mean-square value of the estimation error
2. Expectation of the absolute value of the estimation error
3. Expectation of third or higher powers of the absolute value of the estimation error

Option 1 has a clear advantage over the other two, because it leads to tractable mathematics. In particular, the choice of *the mean-square error criterion results in a second-order dependence for the cost function on the unknown coefficients in the impulse response of the filter. Moreover, the cost function has a distinct minimum that uniquely defines the optimum statistical design of the filter.*

We may now summarize the essence of the filtering problem by making the following statement:

> Design a linear discrete-time filter whose output $y(n)$ provides an estimate of a desired response, given a set of input samples $u(0)$, $u(1)$, $u(2)$, . . . , such that the mean-square value of the estimation error $e(n)$, defined as the difference between the desired response $d(n)$ and the actual response $y(n)$, is minimized.

We may develop the mathematical solution to this statistical optimization problem by following two entirely different approaches that are complementary. One approach leads to the development of an important theorem commonly known as the principle of orthogonality. The other approach highlights the error performance surface that describes the second-order dependence of the cost function on the filter coefficients. We will proceed by deriving the principle of orthogonality first, because the derivation is relatively simple, and because the principle of orthogonality is highly insightful.

5.2 PRINCIPLE OF ORTHOGONALITY

Consider again the statistical filtering problem described in Fig. 5.1. The filter input is denoted by the time series $u(0)$, $u(1)$, $u(2)$, . . . , and the impulse response of the filter is denoted by w_0, w_1, w_2, . . . , both of which are assumed to have *complex values* and *infinite duration*. The filter output $y(n)$ at discrete time n is defined by the *linear convolution sum:*

$$y(n) = \sum_{k=0}^{\infty} w_k^* u(n-k), \qquad n = 0, 1, 2, \ldots \tag{5.1}$$

where the asterisk denotes *complex conjugation*. Note that the term $w_k^* u(n-k)$ represents an *inner product of the filter coefficient w_k and the filter input $u(n-k)$*. Figure 5.2 illustrates the linear discrete-time form of convolution described in Eq. (5.1).

The purpose of the filter in Fig. 5.1 is to produce an estimate of the desired response $d(n)$. We assume that the filter input and the desired response are single realizations of *jointly wide-sense stationary stochastic processes,* both with zero mean. Accordingly, the estimation of $d(n)$ is accompanied by an error, defined by the difference

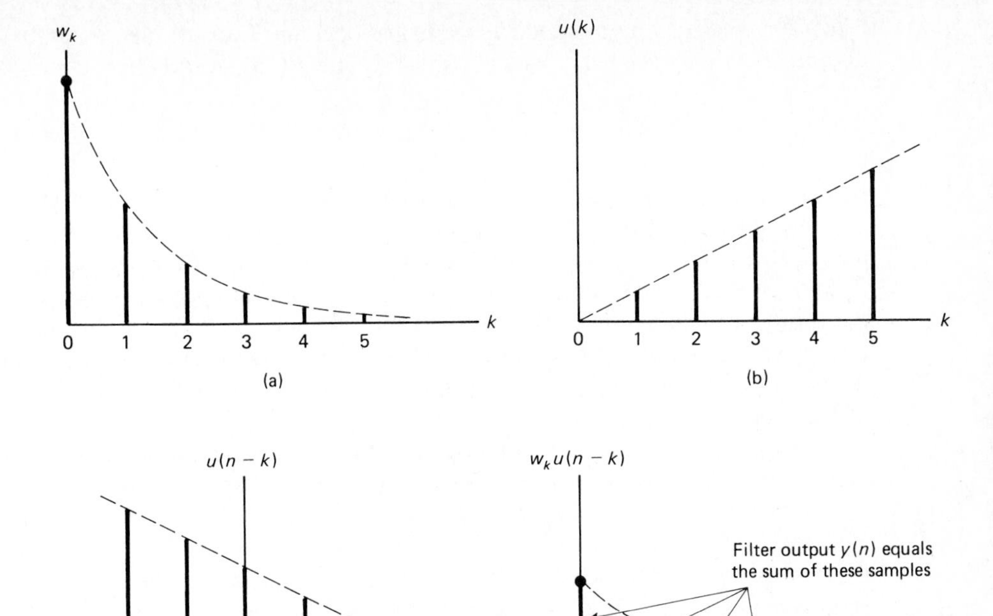

Figure 5.2 Linear convolution: (a) impulse response; (b) filter input; (c) time-reversed and shifted version of filter input; (d) calculation of filter output at time $n = 3$.

$$e(n) = d(n) - y(n) \tag{5.2}$$

The estimation error $e(n)$ is the sample value of a random variable. *To optimize the filter design, we choose to minimize the mean-square value of the estimation error* $e(n)$. We may thus define the cost function as the *mean-squared error*:

$$\begin{aligned} J &= E[e(n)e^*(n)] \\ &= E[|e(n)|^2] \end{aligned} \tag{5.3}$$

where E denotes the *statistical expectation operator*. The problem is therefore to determine the operating conditions for which J attains its minimum value.

For complex input data, the filter coefficients are in general complex, too. Let the kth filter coefficient w_k be denoted in terms of its real and imaginary parts as follows:

$$w_k = a_k + jb_k, \qquad k = 0, 1, 2, \ldots \tag{5.4}$$

Correspondingly, we may define a *gradient operator* ∇, the kth element of which is written in terms of first-order partial derivatives with respect to the real part a_k and the imaginary part b_k, for the kth filter coefficient as

$$\nabla_k = \frac{\partial}{\partial a_k} + j\frac{\partial}{\partial b_k}, \qquad k = 0, 1, 2, \ldots \tag{5.5}$$

Thus, for the situation at hand, applying the operator $\boldsymbol{\nabla}$ to the cost function J, we obtain a multidimensional complex *gradient vector* $\boldsymbol{\nabla}(J)$, the kth element of which is

$$\nabla_k(J) = \frac{\partial J}{\partial a_k} + j\frac{\partial J}{\partial b_k}, \qquad k = 0, 1, 2, \ldots \tag{5.6}$$

Equation (5.6) represents a natural extension of the customary definition of a gradient for a function of real coefficients to the more general case of a function of complex coefficients.[1] Note that for the definition of the complex gradient given in Eq. (5.6) to be valid, it is essential that J be *real*. The gradient operator is always used in the context of finding the *stationary points* of a function of interest. This means that a complex constraint must be converted to a pair of *real* constraints. In Eq. (5.6), the pair of real constraints are obtained by setting both the real and imaginary parts of $\nabla_k(J)$ equal to zero.

For the cost function J to attain its minimum value, all the elements of the gradient vector $\boldsymbol{\nabla}(J)$ must be simultaneously equal to zero, as shown by

$$\nabla_k(J) = 0, \qquad k = 0, 1, 2, \ldots \tag{5.7}$$

Under this set of conditions, the filter is said to be *optimum in the mean-squared-error sense.*[2]

According to Eq. (5.3), the cost function J is a scalar independent of time n. Hence, substituting the first line of Eq. (5.3) in (5.6), we get

$$\nabla_k(J) = E\left[e^*(n)\frac{\partial e(n)}{\partial a_k} + e(n)\frac{\partial e^*(n)}{\partial a_k} + je^*(n)\frac{\partial e(n)}{\partial b_k} + je(n)\frac{\partial e^*(n)}{\partial b_k}\right] \tag{5.8}$$

Using Eqs. (5.2) and (5.4), we get the following partial derivatives:

$$\begin{aligned}
\frac{\partial e(n)}{\partial a_k} &= -u(n-k) \\
\frac{\partial e(n)}{\partial b_k} &= ju(n-k) \\
\frac{\partial e^*(n)}{\partial a_k} &= -u^*(n-k) \\
\frac{\partial e^*(n)}{\partial b_k} &= -ju^*(n-k)
\end{aligned} \tag{5.9}$$

Thus, substituting these partial derivatives in Eq. (5.8) and then cancelling common

[1] The concept of a gradient is commonly discussed in books on optimization (see, e.g., Dorny, 1975). For the complex case, it is discussed in Widrow et al. (1975a) and Monzingo and Miller (1980).

Note that the cost function J, for the general case of complex data, is *not* an analytic function (see Problem 1). Hence the definition of the derivative of the cost function J with respect to a filter coefficient w_k, say, requires particular attention. This issue is discussed in Appendix B. In this appendix we also discuss the relationship between the concepts of a gradient and a derivative for the case of complex coefficients.

[2] Note that in Eq. (5.7), we have presumed optimality at a stationary point. In the linear filtering problem, finding a stationary point assures global optimization of the filter by virtue of the quadratic nature of the error performance surface; see Section 5.5.

terms, we finally get the result

$$\nabla_k(J) = -2E[u(n-k)e^*(n)] \tag{5.10}$$

We are now ready to specify the operating conditions required for minimizing the cost function J. Let e_o *denote the special value of the estimation error that results when the filter operates in its optimum condition.* We then find that the conditions specified in Eq. (5.7) are indeed equivalent to

$$E[u(n-k)e_o^*(n)] = 0, \qquad k = 0, 1, 2, \ldots \tag{5.11}$$

In words, Eq. (5.11) states the following:

> The necessary and sufficient condition for the cost function J to attain its minimum value is that the corresponding value of the estimation error $e_o(n)$ is orthogonal to each input sample that enters into the estimation of the desired response at time n.

Indeed, this statement constitutes the *principle of orthogonality*; it represents one of the most elegant theorems in the subject of optimum filtering.

Corollary to the Principle of Orthogonality

There is a corollary to the principle of orthogonality that we may derive by examining the correlation between the filter output $y(n)$ and the estimation error $e(n)$. Using Eq. (5.1), we may express this correlation as follows:

$$\begin{aligned} E[y(n)e^*(n)] &= E\left[\sum_{k=0}^{\infty} w_k^* u(n-k)e^*(n)\right] \\ &= \sum_{k=0}^{\infty} w_k^* E[u(n-k)e^*(n)] \end{aligned} \tag{5.12}$$

Let y_o *denote the output produced by the filter optimized in the mean-squared-error sense, with* $e_o(n)$ *denoting the corresponding estimation error*. Hence, using the principle of orthogonality described by Eq. (5.11), we get the desired result:

$$E[y_o(n)e_o^*(n)] = 0 \tag{5.13}$$

We may thus state the corollary to the principle of orthogonality as follows:

> When the filter operates in its optimum condition, the estimate of the desired response produced at the filter output, $y_o(n)$, and the corresponding estimation error, $e_o(n)$, are orthogonal to each other.

Let $\hat{d}(n|\mathcal{U}_n)$ *denote the estimate of the desired response that is optimized in the mean-squared-error sense, given the input data that span the space* $\mathcal{U}_n$ *up to and including time* n.[3] We may then write

[3] If a space $\mathcal{U}_n$ consists of all linear combinations of random variables, $u_1, u_2, \ldots, u_n$, then these random variables *span* the space. In other words, every random variable in $\mathcal{U}_n$ can be expressed as some combination of the u's, as shown by

$$u = w_1^* u_1 + \cdots + w_n^* u_n$$

for some coefficients w_n. This assumes that the space $\mathcal{U}_n$ has a finite dimension.

$$\hat{d}(n|\mathcal{U}_n) = y_o(n) \tag{5.14}$$

Note that the estimate $\hat{d}(n|\mathcal{U}_n)$ has zero mean, because the tap inputs are assumed to have zero mean. This condition matches the assumed zero mean of the desired response $d(n)$.

Geometric Interpretation of the Corollary to the Principle of Orthogonality

Equation (5.13) offers an interesting geometric interpretation of the conditions that exist at the output of the optimum filter as illustrated in Fig. 5.3. In this figure, the desired response, the filter output, and the corresponding estimation error are represented by vectors labeled $\mathbf{d}$, $\mathbf{y}_o$, and $\mathbf{e}_o$, respectively; the subscript o in $\mathbf{y}_o$ and $\mathbf{e}_o$ refers to the optimum condition. We see that for the optimum filter the vector representing the estimation error is *normal* (i.e., perpendicular) to the vector representing the filter output. It should, however, be emphasized that the situation depicted in Fig. 5.3 is merely an *analogy,* where random variables and expectations are replaced with vectors and vector inner products, respectively.

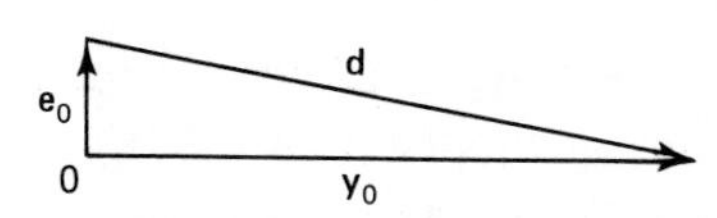

Figure 5.3 Geometric interpretation of the relationship between the desired response, the estimate at the filter output, and the estimation error.

5.3 MINIMUM MEAN-SQUARED ERROR

When the linear discrete-time filter in Fig. 5.1 operates in its optimum condition, Eq. (5.2) takes on the following special form

$$\begin{aligned} e_o(n) &= d(n) - y_o(n) \\ &= d(n) - \hat{d}(n|\mathcal{U}_n) \end{aligned} \tag{5.15}$$

where, in the second line, we have made use of Eq. (5.14). Rearranging the terms in Eq. (5.15), we have

$$d(n) = \hat{d}(n|\mathcal{U}_n) + e_o(n) \tag{5.16}$$

Let $J_{\min}$ denote the minimum mean-squared error, defined by

$$J_{\min} = E[|e_o(n)|^2] \tag{5.17}$$

Hence, evaluating the mean-square values of both sides of Eq. (5.16), and applying to it the corollary to the principle of orthogonality described by Eq. (5.13), we get

$$\sigma_d^2 = \sigma_{\hat{d}}^2 + J_{\min} \tag{5.18}$$

where σ_d^2 is the variance of the desired response, and $\sigma_{\hat{d}}^2$ is the variance of the estimate $\hat{d}(n|\mathcal{U}_n)$; both of them are assumed to be of zero mean. Solving Eq. (5.18) for the minimum mean-squared error, we get

$$J_{\min} = \sigma_d^2 - \sigma_{\hat{d}}^2 \tag{5.19}$$

This relation shows that for the optimum filter the minimum mean-squared error equals the difference between the variance of the desired response and the variance of the estimate that the filter produces at its output.

It is convenient to normalize the expression in Eq. (5.19) in such a way that the minimum value of the mean-squared error always lies between zero and one. We may do this by dividing both sides of Eq. (5.19) by σ_d^2, obtaining

$$\frac{J_{\min}}{\sigma_d^2} = 1 - \frac{\sigma_{\hat{d}}^2}{\sigma_d^2} \tag{5.20}$$

Clearly, this is possible because σ_d^2 is never zero, except in the trivial case of a desired response $d(n)$ that is zero for all n. Let

$$\epsilon = \frac{J_{\min}}{\sigma_d^2} \tag{5.21}$$

The quantity ϵ is called the *normalized mean-squared error*, in terms of which we may rewrite Eq. (5.20) in the form

$$\epsilon = 1 - \frac{\sigma_{\hat{d}}^2}{\sigma_d^2} \tag{5.22}$$

We note that (1) the ratio ϵ can never be negative, and (2) the ratio $\sigma_{\hat{d}}^2/\sigma_d^2$ is always positive. We therefore have

$$0 \le \epsilon \le 1 \tag{5.23}$$

If ϵ is zero, the optimum filter operates perfectly in the sense that there is complete agreement between the estimate $\hat{d}(n|\mathcal{U}_n)$ at the filter output and the desired response $d(n)$. On the other hand, if ϵ is unity, there is no agreement whatsoever between these two quantities; this corresponds to the worst possible situation.

5.4 WIENER–HOPF EQUATIONS

The principle of orthogonality, described in Eq. (5.11), specifies the necessary and sufficient condition for the optimum operation of the filter. We may reformulate the necessary and sufficient condition for optimality by substituting Eqs. (5.1) and (5.2) in (5.11). In particular, we may write

$$E\left[u(n-k)\left(d^*(n) - \sum_{i=0}^{\infty} w_{oi} u^*(n-i)\right)\right] = 0, \qquad k = 0, 1, 2, \ldots$$

where w_{oi} is the ith coefficient in the impulse response of the optimum filter. Expanding this equation and rearranging terms, we get

$$\sum_{i=0}^{\infty} w_{oi} E[u(n-k)u^*(n-i)] = E[u(n-k)d^*(n)], \qquad k = 0, 1, 2, \ldots \tag{5.24}$$

The two expectations in Eq. (5.24) may be interpreted as follows:

1. The expectation $E[u(n-k)u^*(n-i)]$ is equal to the *autocorrelation function of the filter input* for a lag of $i - k$. We may thus express this expectation as

$$r(i-k) = E[u(n-k)u^*(n-i)] \tag{5.25}$$

2. The expectation $E[u(n-k)d^*(n)]$ is equal to the cross-correlation between the filter input $u(n-k)$ and the desired response $d(n)$ for a lag of $-k$. We may thus express this second expectation as

$$p(-k) = E[u(n-k)d^*(n)] \tag{5.26}$$

Accordingly, using the definitions of Eqs. (5.25) and (5.26) in (5.24), we get the following system of equations as the necessary and sufficient condition for the optimality of the filter:

$$\sum_{i=0}^{\infty} w_{oi} r(i-k) = p(-k), \qquad k = 0, 1, 2, \ldots \tag{5.27}$$

The system of equation (5.27) defines the optimum filter coefficients in terms of two correlation functions: the autocorrelation function of the filter input, and the cross-correlation between the filter input and the desired response. These equations are called the *Wiener–Hopf equations.*[4]

Solution of the Wiener–Hopf Equations for Linear Transversal Filters

The solution of the set of Wiener–Hopf equations is greatly simplified for the special case when a *linear transversal filter,* or FIR filter, is used to perform the estimation of desired response $d(n)$ in Fig. 5.1. Consider then the structure shown in Fig. 5.4. The transversal filter involves a combination of three basic operations: *storage, multiplication,* and *addition,* as described next:

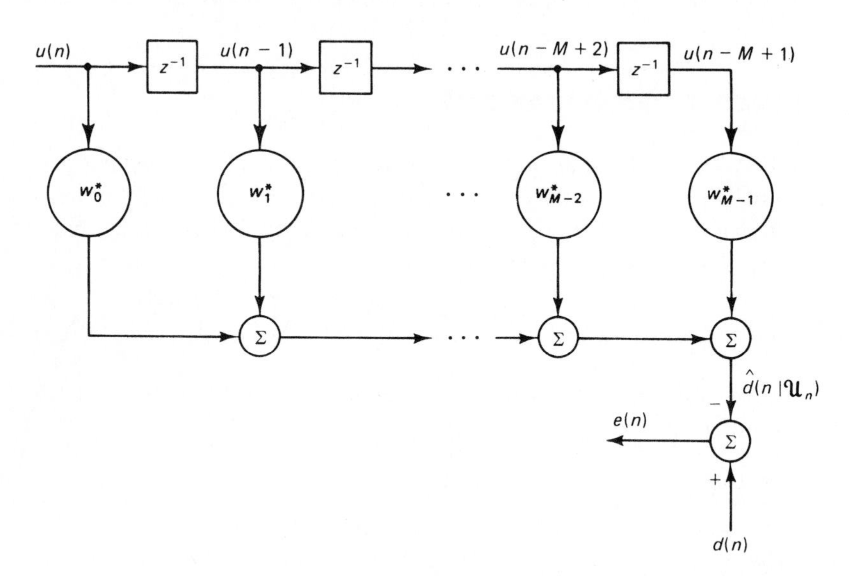

Figure 5.4 Transversal filter.

[4] In order to solve the Wiener–Hopf equations for the optimum filter coefficients, we need to use a special technique known as *spectral factorization.* For a description of this technique and its use in solving the Wiener–Hopf equations (5.27), the interested reader is referred to Haykin (1989).

1. The storage is represented by a cascade of $M - 1$ *one-sample delays,* with the block for each such unit labeled as z^{-1}. We refer to the various points at which the one-sample delays are accessed as *tap points*. The tap inputs are denoted by $u(n), u(n-1), \ldots, u(n-M+1)$. Thus, with $u(n)$ viewed as the *current* value of the filter input, the remaining $M - 1$ tap inputs, $u(n-1), \ldots,$ $u(n-M+1)$, represent *past* values of the input.
2. The *inner products* of tap inputs $u(n), u(n-1), \ldots, u(n-M+1)$ and *tap weights* $w_0, w_1, \ldots, w_{M-1}$, respectively, are formed by using a corresponding set of multipliers. In particular, the multiplication involved in forming the inner product of $u(n)$ and w_0 is represented by a block labeled w_0^*, and so on for the other inner products.
3. The function of the adders is to sum the multiplier outputs to produce an overall output.

The impulse response of the transversal filter in Fig. 5.4 is defined by the finite set of tap weights $w_0, w_1, \ldots, w_{M-1}$. Accordingly, the Wiener–Hopf equations (5.27) reduce to a *system of M simultaneous equations,* as shown by

$$\sum_{i=0}^{M-1} w_{oi} r(i-k) = p(-k), \qquad k = 0, 1, \ldots, M-1 \tag{5.28}$$

where $w_{o0}, w_{o1}, \ldots, w_{o,M-1}$ are the optimum values of the tap weights of the filter.

Matrix Formulation of the Wiener–Hopf Equations

Let $\mathbf{R}$ denote the M-by-M *correlation matrix* of the tap inputs $u(n), u(n-1), \ldots,$ $u(n-M+1)$ in the transversal filter of Fig. 5.4:

$$\mathbf{R} = E[\mathbf{u}(n)\mathbf{u}^H(n)] \tag{5.29}$$

where $\mathbf{u}(n)$ is the M-by-1 tap-input vector:

$$\mathbf{u}(n) = [u(n), u(n-1), \ldots, u(n-M+1)]^T \tag{5.30}$$

In expanded form, we have

$$\mathbf{R} = \begin{bmatrix} r(0) & r(1) & \cdots & r(M-1) \\ r^*(1) & r(0) & \cdots & r(M-2) \\ \vdots & \vdots & \ddots & \vdots \\ r^*(M-1) & r^*(M-2) & \cdots & r(0) \end{bmatrix} \tag{5.31}$$

Correspondingly, let $\mathbf{p}$ denote the M-by-1 *cross-correlation vector* between the tap inputs of the filter and the desired response $d(n)$:

$$\mathbf{p} = E[\mathbf{u}(n)d^*(n)] \tag{5.32}$$

In expanded form, we have

$$\mathbf{p} = [p(0), p(-1), \ldots, p(1-M)]^T \tag{5.33}$$

Note that the lags used in the definition of $\mathbf{p}$ are either zero or else negative. We may thus rewrite the Wiener–Hopf equations (5.28) in the compact matrix form:

$$\mathbf{R}\mathbf{w}_o = \mathbf{p} \tag{5.34}$$

where $\mathbf{w}_o$ denotes the M-by-1 *optimum tap-weight vector* of the transversal filter; that is,

$$\mathbf{w}_o = [\mathbf{w}_{o0}, \mathbf{w}_{o1}, \ldots, \mathbf{w}_{o,M-1}]^T \tag{5.35}$$

To solve the Wiener–Hopf equations (5.34) for $\mathbf{w}_o$, we assume that the correlation matrix $\mathbf{R}$ is nonsingular. We may then premultiply both sides of Eq. (5.34) by $\mathbf{R}^{-1}$, the *inverse* of the correlation matrix, obtaining

$$\mathbf{w}_o = \mathbf{R}^{-1}\mathbf{p} \tag{5.36}$$

The computation of the optimum tap-weight vector $\mathbf{w}_o$ requires knowledge of two quantities: (1) the correlation matrix $\mathbf{R}$ of the tap-input vector $\mathbf{u}(n)$, and (2) the cross-correlation vector $\mathbf{p}$ between the tap-input vector $\mathbf{u}(n)$ and the desired response $d(n)$.

5.5 ERROR-PERFORMANCE SURFACE

The Wiener–Hopf equations (5.34), as derived herein, are traceable to the principle of orthogonality, which itself was derived in Section 5.2. We may also derive the Wiener–Hopf equations by examining the dependence of the cost function J on the tap weights of the transversal filter in Fig. 5.4. First, we write the estimation error $e(n)$ as follows:

$$e(n) = d(n) - \sum_{k=0}^{M-1} w_k^* u(n-k) \tag{5.37}$$

where $d(n)$ is the desired response; $w_0, w_1, \ldots, w_{M-1}$ are the tap weights of the filter; and $u(n), u(n-1), \ldots, u(n-M+1)$ are the tap inputs. Accordingly, we may define the cost function for the transversal filter structure of Fig. 5.4 as

$$\begin{aligned} J &= E[e(n)e^*(n)] \\ &= E[|d(n)|^2] - \sum_{k=0}^{M-1} w_k^* E[u(n-k)d^*(n)] - \sum_{k=0}^{M-1} w_k E[u^*(n-k)d(n)] \\ &\quad + \sum_{k=0}^{M-1}\sum_{i=0}^{M-1} w_k^* w_i E[u(n-k)u^*(n-i)] \end{aligned} \tag{5.38}$$

We may now recognize the four expectations on the right side of the second line in Eq. (5.38), as follows:

1. For the first expectation, we have

$$\sigma_d^2 = E[|d(n)|^2] \tag{5.39}$$

where σ_d^2 is the *variance* of the desired response $d(n)$, assumed to be of zero mean.

2. For the second and third expectations, we have, respectively,

$$p(-k) = E[u(n-k)d^*(n)] \tag{5.40}$$

and

$$p^*(-k) = E[u^*(n-k)d(n)] \tag{5.41}$$

where $p(-k)$ is the cross-correction between the tap input $u(n-k)$ and the desired response $d(n)$.

3. Finally, for the fourth expectation, we have

$$r(i-k) = E[u(n-k)u^*(n-i)] \tag{5.42}$$

where $r(i-k)$ is the autocorrelation function of the tap inputs for lag $i-k$. We may thus rewrite Eq. (5.38) in the form

$$J = \sigma_d^2 - \sum_{k=0}^{M-1} w_k^* p(-k) - \sum_{k=0}^{M-1} w_k p^*(-k) + \sum_{k=0}^{M-1}\sum_{i=0}^{M-1} w_k^* w_i r(i-k) \tag{5.43}$$

Equation (5.43) states that for the case when the tap inputs of the transversal filter and the desired response are jointly stationary, the cost function, or mean-squared error, J is precisely a *second-order function of the tap weights in the filter*. Consequently, we may visualize the dependence of the cost function J on the tap weights $w_0, w_1, \ldots, w_{M-1}$ as a *bowl-shaped* $(M+1)$*-dimensional surface* with M *degrees of freedom represented by the tap weights of the filter*. This surface is characterized by a unique minimum. We refer to the surface so described as the *error-performance surface* of the transversal filter in Fig. 5.4.

At the *bottom* or *minimum point* of the error-performance surface, the cost function J attains its *minimum value* denoted by $J_{\min}$. At this point, the *gradient vector* $\nabla(J)$ is identically zero. In other words,

$$\nabla_k(J) = 0, \qquad k = 0, 1, \ldots, M-1 \tag{5.44}$$

where $\nabla_k(J)$ is the kth element of the gradient vector. As before, we write the kth tap weight w_k as

$$w_k = a_k + jb_k$$

Hence, using Eq. (5.43), we may express $\nabla_k(J)$ as

$$\begin{aligned}\nabla_k(J) &= \frac{\partial J}{\partial a_k} + j\frac{\partial J}{\partial b_k} \\ &= -2p(-k) + 2\sum_{i=0}^{M-1} w_i r(i-k)\end{aligned} \tag{5.45}$$

Applying the necessary and sufficient condition of Eq. (5.44) for optimality to Eq. (5.45), we find that the optimum tap weight $w_{o0}, w_{o1}, \ldots, w_{o,M-1}$ for the transversal filter in Fig. 5.4 are defined by the system of equations:

$$\sum_{i=0}^{M-1} w_{oi} r(i-k) = p(-k), \qquad k = 0, 1, \ldots, M-1$$

This system of equations is identical to the Wiener–Hopf equations (5.28) derived in Section 5.4.

Minimum Mean-Squared Error

Let $\hat{d}(n|\mathcal{U}_n)$ denote the estimate of the desired response $d(n)$, produced at the output of the transversal filter in Fig. 5.4 that is optimized in the mean-squared-error sense, given the tap inputs $u(n), u(n-1), \ldots, u(n-M+1)$ that span the space $\mathcal{U}_n$. Then from Fig. 5.4 we deduce that

$$\begin{aligned}\hat{d}(n|\mathcal{U}_n) &= \sum_{k=0}^{M-1} w_{ok}^* u(n-k) \\ &= \mathbf{w}_o^H \mathbf{u}(n)\end{aligned} \tag{5.46}$$

where $\mathbf{w}_o$ is the tap-weight vector of the optimum filter with elements $w_{o0}, w_{o1}, \ldots, w_{o,M-1}$, and $\mathbf{u}(n)$ is the tap-input vector defined in Eq. (5.30). Note that $\mathbf{w}_o^H\mathbf{u}(n)$ denotes an inner product of the optimum tap-weight vector $\mathbf{w}_o$ and the tap-input vector $\mathbf{u}(n)$. We assume that $\mathbf{u}(n)$ has zero mean, also making the estimate $\hat{d}(n|\mathcal{U}_n)$ have zero mean. Hence we may use Eq. (5.46) to evaluate the variance of $\hat{d}(n|\mathcal{U}_n)$, obtaining

$$\begin{aligned}\sigma_{\hat{d}}^2 &= E[\mathbf{w}_o^H \mathbf{u}(n)\mathbf{u}^H(n)\mathbf{w}_o] \\ &= \mathbf{w}_o^H E[\mathbf{u}(n)\mathbf{u}^H(n)]\mathbf{w}_o \\ &= \mathbf{w}_o^H \mathbf{R}\mathbf{w}_o\end{aligned} \tag{5.47}$$

where $\mathbf{R}$ is the correlation matrix of the tap-weight vector $\mathbf{u}(n)$, as defined in Eq. (5.29). We may eliminate the dependence of the variance $\sigma_{\hat{d}}^2$ on the optimum tap-weight vector $\mathbf{w}_o$ by using Eq. (5.34). In particular, we may rewrite Eq. (5.47) as

$$\begin{aligned}\sigma_{\hat{d}}^2 &= \mathbf{w}_o^H \mathbf{p} \\ &= \mathbf{p}^H \mathbf{w}_o\end{aligned} \tag{5.48}$$

To evaluate the minimum mean-squared error produced by the transversal filter in Fig. 5.4, we may use Eq. (5.48) in (5.19), obtaining

$$\begin{aligned}J_{\min} &= \sigma_d^2 - \mathbf{p}^H \mathbf{w}_o \\ &= \sigma_d^2 - \mathbf{p}^H \mathbf{R}^{-1}\mathbf{p}\end{aligned} \tag{5.49}$$

which is the desired result.

Canonical Form of the Error-Performance Surface

Equation (5.43) defines the expanded form of the mean-squared error J produced by the transversal filter in Fig. 5.4. We may rewrite this equation in matrix form, by using the definitions for the correlation matrix $\mathbf{R}$ and the cross-correlation vector $\mathbf{p}$ given in Eqs. (5.31) and (5.33), respectively, as shown by

$$J = \sigma_d^2 - \mathbf{w}^H\mathbf{p} - \mathbf{p}^H\mathbf{w} + \mathbf{w}^H\mathbf{R}\mathbf{w} \tag{5.50}$$

Subtracting the first line of Eq. (5.49) from (5.50) and then using the optimum solution of Eq. (5.34) to eliminate the cross-correlation vector $\mathbf{p}$ from the result of this subtraction, we find that the mean-squared error may be expressed in the quadratic form (see Problem 5)

$$J = J_{\min} + (\mathbf{w} - \mathbf{w}_o)^H \mathbf{R}(\mathbf{w} - \mathbf{w}_o) \tag{5.51}$$

This equation shows explicitly the unique optimality of the minimizing tap-weight vector $\mathbf{w}_o$.

Although the quadratic form on the right side of Eq. (5.51) is quite informative, nevertheless, it is desirable to change the basis on which it is defined so that the representation of the error-performance surface is considerably simplified. To do this, we recall from Chapter 4 that the correlation matrix $\mathbf{R}$ of the tap-input vector may be expressed in terms of eigenvalues and eigenvectors as follows:

$$\mathbf{R} = \mathbf{Q}\mathbf{\Lambda}\mathbf{Q}^H \tag{5.52}$$

where $\mathbf{\Lambda}$ is a diagonal matrix consisting of the eigenvalues $\lambda_1, \lambda_2, \ldots, \lambda_M$ of the correlation matrix, and the matrix $\mathbf{Q}$ has for its columns the eigenvectors $\mathbf{q}_1$, $\mathbf{q}_2, \ldots, \mathbf{q}_M$ associated with these eigenvalues, respectively. Hence, substituting Eq. (5.52) into (5.51), we get

$$J = J_{\min} + (\mathbf{w} - \mathbf{w}_o)^H \mathbf{Q}\mathbf{\Lambda}\mathbf{Q}^H(\mathbf{w} - \mathbf{w}_o) \tag{5.53}$$

Define a *transformed* version of the difference between the tap-weight vector $\mathbf{w}$ and the optimum solution $\mathbf{w}_o$ as

$$\boldsymbol{\nu} = \mathbf{Q}^H(\mathbf{w} - \mathbf{w}_o) \tag{5.54}$$

Then we may put the quadratic form of Eq. (5.53) into its *canonical form* defined by

$$J = J_{\min} + \boldsymbol{\nu}^H \mathbf{\Lambda} \boldsymbol{\nu} \tag{5.55}$$

This new formulation of the mean-squared error contains no cross-product terms, as shown by

$$\begin{aligned} J &= J_{\min} + \sum_{k=1}^{M} \lambda_k \nu_k \nu_k^* \\ &= J_{\min} + \sum_{k=1}^{M} \lambda_k |\nu_k|^2 \end{aligned} \tag{5.56}$$

where ν_k is the kth component of the vector $\boldsymbol{\nu}$. The feature that makes the canonical form of Eq. (5.56) a rather useful representation of the error-performance surface is the fact that the components of the transformed coefficient vector $\boldsymbol{\nu}$ constitute the *principal axes* of the error-performance surface. The practical significance of this result will become apparent in later chapters.

5.6 NUMERICAL EXAMPLE

To illustrate the filtering theory developed in the preceding, we consider the example depicted in Fig. 5.5. The desired response $\{d(n)\}$ is modeled as an AR process of order 1; that is, it may be produced by applying a white-noise process $\{v_1(n)\}$ of zero mean and variance $\sigma_1^2 = 0.27$ to the input of an all-pole filter of order 1, whose transfer function equals [see Fig. 5.5(a)]

$$H_1(z) = \frac{1}{1 + 0.8458z^{-1}}$$

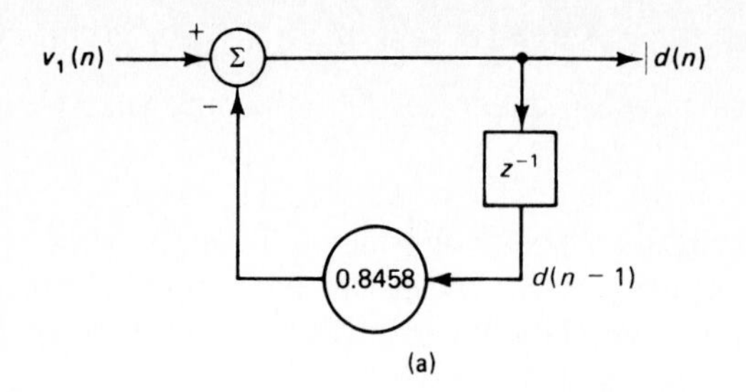

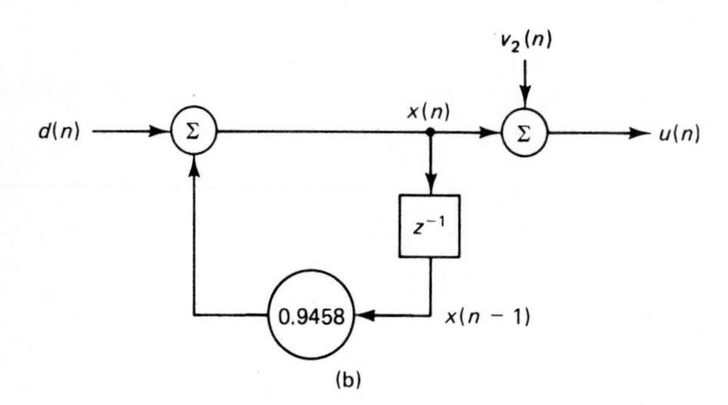

Figure 5.5 (a) Autoregressive model of desired response $\{d(n)\}$; (b) model of noisy communication channel.

The process $\{d(n)\}$ is applied to a communication channel modeled by the all-pole transfer function

$$H_2(z) = \frac{1}{1 - 0.9458z^{-1}}$$

The channel output $\{x(n)\}$ is corrupted by an additive white-noise process $\{v_2(n)\}$ of zero mean and variance $\sigma_2^2 = 0.1$, so a sample of the received signal $\{u(n)\}$ equals [see Fig. 5.5(b)]

$$u(n) = x(n) + v_2(n) \tag{5.57}$$

The white-noise processes $\{v_1(n)\}$ and $\{v_2(n)\}$ are uncorrelated. It is also assumed that $d(n)$ and $u(n)$, and therefore $v_1(n)$ and $v_2(n)$, are all real valued.

The requirement is to specify a Wiener filter consisting of a transversal filter with two taps, which operates on the received signal $\{u(n)\}$ so as to produce an estimate of the desired response that is optimum in the mean-square sense.

Statistical Characterization of the Desired Response $\{d(n)\}$ and the Received Signal $\{u(n)\}$

We begin the analysis by considering the difference equations that characterize the various processes described by the models of Fig. 5.5. First, the generation of the desired response $\{d(n)\}$ is governed by the first-order difference equation

$$d(n) + a_1 d(n-1) = v_1(n) \tag{5.58}$$

where $a = 0.8458$. The variance of the process $\{d(n)\}$ equals (see Problem 4 of Chapter 2)

$$\begin{aligned}\sigma_d^2 &= \frac{\sigma_1^2}{1 - a_1^2} \\ &= \frac{0.27}{1 - (0.8458)^2} \\ &= 0.9486\end{aligned} \tag{5.59}$$

The process $\{d(n)\}$ acts as input to the channel. Hence, from Fig. 5.5(b), we find that the channel output $\{x(n)\}$ is related to the channel input $\{d(n)\}$ by the first-order difference equation

$$x(n) + b_1 x(n - 1) = d(n) \tag{5.60}$$

where $b_1 = -0.9458$. We also observe from the two parts of Fig. 5.5 that the channel output $\{x(n)\}$ may be generated by applying the white-noise process $\{v_1(n)\}$ to a second-order all-pole filter whose transfer function equals

$$\begin{aligned}H(z) &= H_1(z)H_2(z) \\ &= \frac{1}{(1 + 0.8458z^{-1})(1 - 0.9458z^{-1})}\end{aligned} \tag{5.61}$$

Accordingly, $\{x(n)\}$ is a second-order AR process described by the difference equation

$$x(n) + a_1 x(n - 1) + a_2 x(n - 2) = v(n) \tag{5.62}$$

where $a_1 = -0.1$ and $a_2 = -0.8$. Note that both AR processes $\{d(n)\}$ and $\{x(n)\}$ are stationary.

To characterize the Wiener filter, we need to solve the Wiener–Hopf equations (5.34). This set of equations requires knowledge of two quantities: (1) the correlation matrix $\mathbf{R}$ pertaining to the received signal $\{u(n)\}$, and (2) the cross-correlation vector $\mathbf{p}$ between $\{u(n)\}$ and the desired response $\{d(n)\}$. In our example, $\mathbf{R}$ is a 2-by-2 matrix and $\mathbf{p}$ is a 2-by-1 vector, since the transversal filter used to implement the Wiener filter is assumed to have two taps.

The received signal $\{u(n)\}$ consists of the channel output $\{x(n)\}$ plus the additive white noise $\{v_2(n)\}$. Since the processes $\{x(n)\}$ and $\{v_2(n)\}$ are uncorrelated, it follows that the correlation matrix $\mathbf{R}$ equals the correlation matrix of $\{x(n)\}$ plus the correlation matrix of $\{v_2(n)\}$. That is,

$$\mathbf{R} = \mathbf{R}_x + \mathbf{R}_2 \tag{5.63}$$

For the correlation matrix $\mathbf{R}_x$, we write [since the process $\{x(n)\}$ is real valued]

$$\mathbf{R}_x = \begin{bmatrix} r_x(0) & r_x(1) \\ r_x(1) & r_x(0) \end{bmatrix}$$

where $r_x(0)$ and $r_x(1)$ are the autocorrelation functions of the received signal $\{x(n)\}$ for lags of 0 and 1, respectively. From Section 2.9 we have

$$r_x(0) = \sigma_x^2$$

$$= \left(\frac{1 + a_2}{1 - a_2}\right) \frac{\sigma_1^2}{[(1 + a_2)^2 - a_1^2]}$$

$$= \left(\frac{1 - 0.8}{1 + 0.8}\right) \frac{0.27}{[(1 - 0.8)^2 - (0.1)^2]}$$

$$= 1$$

$$r_x(1) = \frac{-a_1}{1 + a_2}$$

$$= \frac{0.1}{1 - 0.8}$$

$$= 0.5$$

Hence,

$$\mathbf{R}_x = \begin{bmatrix} r_x(0) & r_z(1) \\ r_x(1) & r_x(0) \end{bmatrix} = \begin{bmatrix} 1 & 0.5 \\ 0.5 & 1 \end{bmatrix} \tag{5.64}$$

Next we observe that since $\{v_2(n)\}$ is a white-noise process of zero mean and variance $\sigma_2^2 = 0.1$, the 2-by-2 correlation matrix $\mathbf{R}_2$ of this process equals

$$\mathbf{R}_2 = \begin{bmatrix} 0.1 & 0 \\ 0 & 0.1 \end{bmatrix} \tag{5.65}$$

Thus, substituting Eqs. (5.64) and (5.65) in (5.63), we find that the 2-by-2 correlation matrix of the received signal $\{x(n)\}$ equals

$$\mathbf{R} = \begin{bmatrix} 1.1 & 0.5 \\ 0.5 & 1.1 \end{bmatrix} \tag{5.66}$$

For the 2-by-1 cross-correlation vector $\mathbf{p}$, we write

$$\mathbf{p} = \begin{bmatrix} p(0) \\ p(1) \end{bmatrix}$$

where $p(0)$ and $p(1)$ are the cross-correlation functions between $\{d(n)\}$ and $\{u(n)\}$ for lags of 0 and 1, respectively. Since these two processes are real valued, we have $p(k) = p(-k)$, where

$$p(-k) = E[u(n - k)d(n)], \qquad k = 0, 1 \tag{5.67}$$

Substituting Eqs. (5.57) and (5.60) into (5.67), and recognizing that the channel output $\{x(n)\}$ is uncorrelated with the white-noise process $\{v_2(n)\}$, we get

$$p(k) = r_x(k) + b_1 r_x(k - 1), \qquad k = 0, 1$$

Putting $b_1 = -0.9458$ and using the element values for the correlation matrix $\mathbf{R}_x$

given in Eq. (5.64), we obtain

$$\begin{aligned} p(0) &= r_x(0) + b_1 r_x(-1) \\ &= 1 - 0.9458 \times 0.5 \\ &= 0.5272 \\ p(1) &= r_x(1) + b_1 r_x(0) \\ &= 0.5 - 0.9458 \times 1 \\ &= -0.4458 \end{aligned}$$

Hence,

$$\mathbf{p} = \begin{bmatrix} 0.5272 \\ -0.4458 \end{bmatrix} \tag{5.68}$$

Error-Performance Surface

The dependence of the mean-squared error on the 2-by-1 tap-weight vector $\mathbf{w}$ is defined by Eq. (5.50). Hence, substituting Eqs. (5.59), (5.66), and (5.68) into (5.50), we get

$$\begin{aligned} J(w_0, w_1) &= 0.9486 - 2[0.5272, -0.4458] \begin{bmatrix} w_0 \\ w_1 \end{bmatrix} + [w_0, w_1] \begin{bmatrix} 1.1 & 0.5 \\ 0.5 & 1.1 \end{bmatrix} \begin{bmatrix} w_0 \\ w_1 \end{bmatrix} \\ &= 0.9486 - 1.0544 w_0 + 0.8916 w_1 + w_0 w_1 + 1.1(w_0^2 + w_1^2) \end{aligned}$$

Using a three-dimensional computer plot, the mean-squared error $J(w_0, w_1)$ is plotted versus the tap weights w_0 and w_1. The result is shown in Fig. 5.6.

Figure 5.7 shows contour plots of the tap weight w_1 versus w_0 for varying values of the mean-squared error J. We see that the locus of w_1 versus w_0 for a fixed J is in the form of an ellipse. The elliptical locus shrinks in size as the mean-squared error J approaches the minimum value $J_{\min}$. For $J = J_{\min}$, the locus reduces to a point with coordinates w_{o0} and w_{o1}.

Wiener Filter

The 2-by-1 optimum tap-weight vector $\mathbf{w}_o$ of the Wiener filter is defined by Eq. (5.36). In particular, it consists of the inverse matrix $\mathbf{R}^{-1}$ multiplied by the cross-correlation vector $\mathbf{p}$. Inverting the correlation matrix $\mathbf{R}$ of Eq. (5.66), we get

$$\begin{aligned} \mathbf{R}^{-1} &= \begin{bmatrix} r(0) & r(1) \\ r(1) & r(0) \end{bmatrix}^{-1} \\ &= \frac{1}{r^2(0) - r^2(1)} \begin{bmatrix} r(0) & -r(1) \\ -r(1) & r(0) \end{bmatrix} \\ &= \begin{bmatrix} 1.1456 & -0.5208 \\ -0.5208 & 1.1456 \end{bmatrix} \end{aligned} \tag{5.69}$$

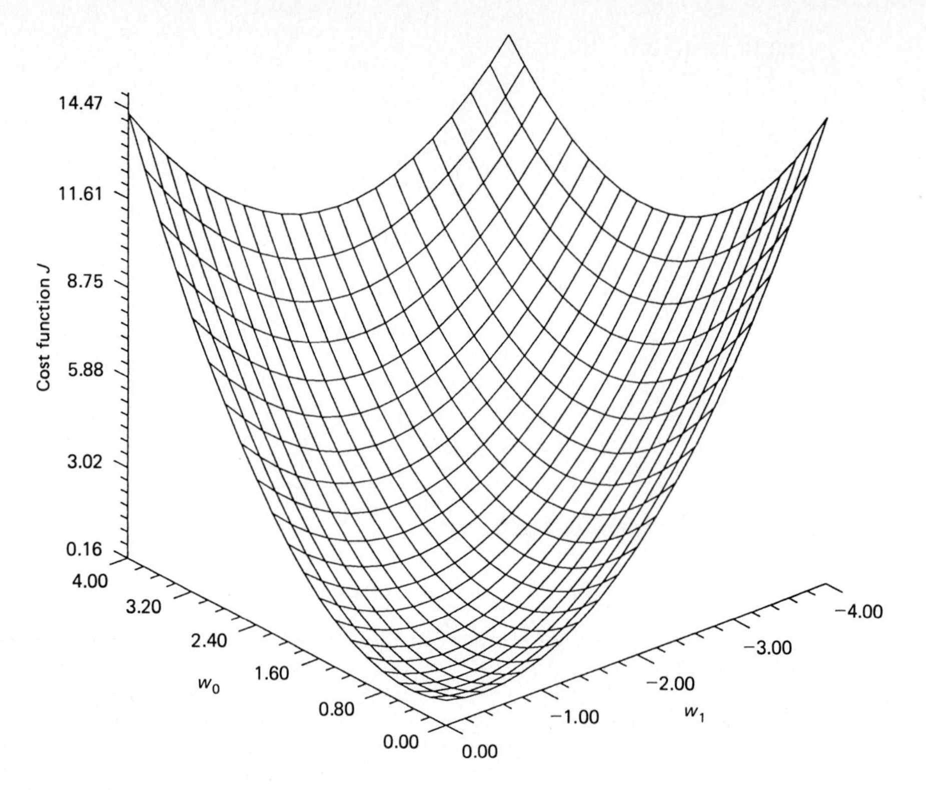

Figure 5.6 Error performance surface of the two-tap transversal filter described in the numerical example.

Hence, substituting Eqs. (5.68) and (5.69) into (5.36), we get the desired result:

$$\mathbf{w}_o = \begin{bmatrix} 1.1456 & -0.5208 \\ -0.5208 & 1.1456 \end{bmatrix} \begin{bmatrix} 0.5272 \\ -0.4458 \end{bmatrix} = \begin{bmatrix} 0.8360 \\ -0.7853 \end{bmatrix} \tag{5.70}$$

Minimum Mean-Squared Error

To evaluate the minimum value of the mean-squared error, $J_{\min}$, which results from the use of the optimum tap-weight vector $\mathbf{w}_o$, we use Eq. (5.49). Hence, substituting Eqs. (5.59), (5.68), and (5.70) into (5.49), we get

$$J_{\min} = 0.9486 - [0.5272, \ -0.4458] \begin{bmatrix} 0.8360 \\ -0.7853 \end{bmatrix} = 0.1579 \tag{5.71}$$

The point represented jointly by the optimum tap-weight vector $\mathbf{w}_o$ of Eq. (5.70) and the minimum mean-squared error of Eq. (5.71) defines the bottom of the error-performance surface.

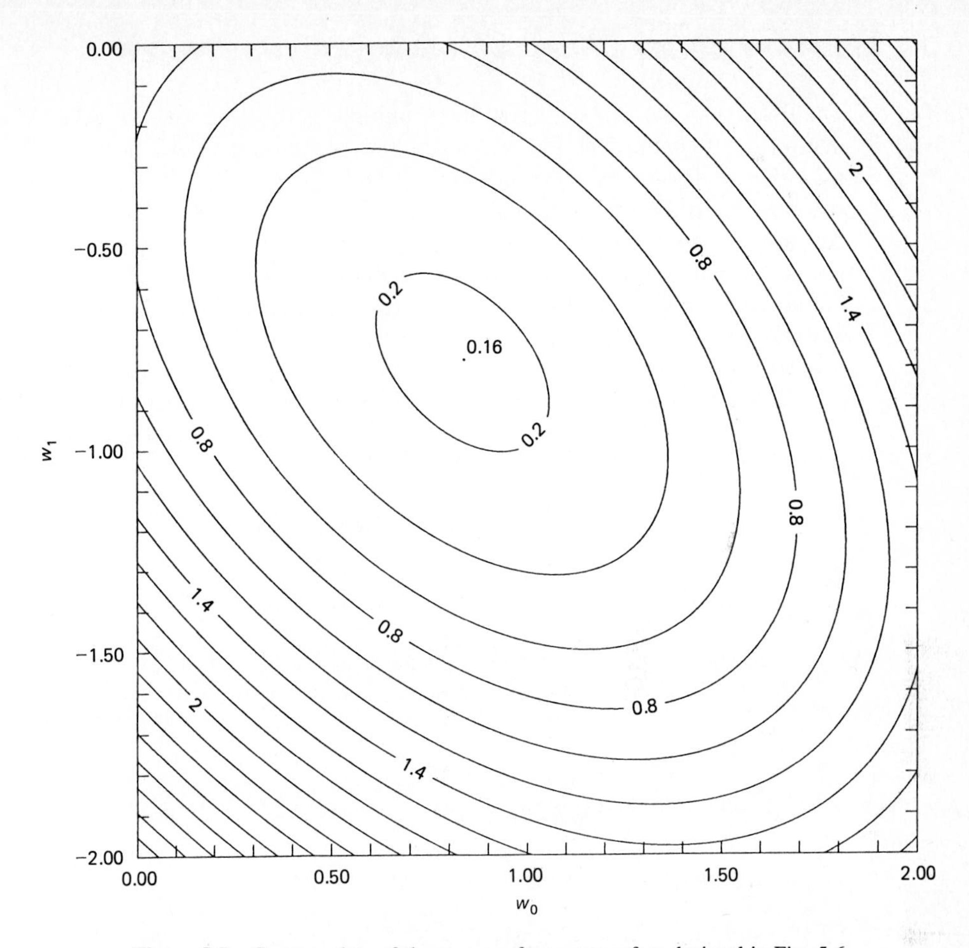

Figure 5.7 Contour plots of the error performance surface depicted in Fig. 5.6.

Canonical Error Performance Surface

The characteristic equation of the 2-by-2 correlation matrix **R** of Eq. (5.66) is

$$(1.1 - \lambda)^2 - (0.5)^2 = 0$$

The two eigenvalues of the correlation matrix **R** are therefore

$$\lambda_1 = 1.6$$

$$\lambda_2 = 0.6$$

The canonical error performance surface is therefore defined by [see Eq. (5.56)]

$$J(\nu_1, \nu_2) = J_{\min} + 1.6\nu_1^2 + 0.6\nu_2^2 \tag{5.71}$$

The locus of ν_2 versus ν_1, as defined in Eq. (5.71), traces an *ellipse* for a fixed value of $J - J_{\min}$. In particular, the ellipse has a minor axis of $[(J - J_{\min})/\lambda_1]^{1/2}$ along the ν_1-coordinate and a major axis of $[(J - J_{\min}/\lambda_2)]^{1/2}$ along the ν_2-coordinate; this assumes that $\lambda_1 > \lambda_2$.

The essense of a Wiener filter is that it minimizes the mean-square value of an estimation error, defined as the difference between a desired response and the actual filter output. In solving this optimization (minimization) problem, there are *no* constraints imposed on the solution. In some filtering applications, however, it may be desirable (or even mandatory) to design a filter that minimizes a mean-square criterion, subject to a specific *constraint*. For example, the requirement may be that of minimizing the average output power of a linear filter while the response of the filter measured at some specific frequency of interest is constrained to remain constant. In this section, we consider one such solution.

Consider a linear transversal filter, as in Fig. 5.8. The filter output, in response to the tap inputs $u(n), u(n-1), \ldots, u(n-M+1)$, assumed to have zero mean, is given by

$$y(n) = \sum_{k=0}^{M-1} w_k^* u(n-k) \tag{5.72}$$

The *constrained optimization problem* we wish to solve may be stated as follows:

> Find the optimum set of filter coefficients $w_{o0}, w_{o1}, \ldots, w_{o,M-1}$ that minimizes the mean-square value of the filter output $y(n)$, subject to the linear constraint
>
> $$\sum_{k=0}^{M-1} w_k^* e^{-j\omega_0 k} = g \tag{5.73}$$
>
> where ω_0 is some angular frequency, and g is a complex-valued gain.

The mean-square value of the filter output is given by

$$E[|y(n)|^2] = \sum_{k=0}^{M-1} \sum_{i=0}^{M-1} w_k^* w_i r(i-k) \tag{5.74}$$

where $r(i-k)$ is the autocorrelation function of the tap inputs for lag $i-k$.

To solve this constrained optimization problem, we use the *method of Lagrange multipliers.*[5] We begin by defining a *real-valued* cost function J that combines the two parts of the constrained optimization problem, given in Eqs. (5.73) and (5.74). Specifically, we write

$$J = \underbrace{\sum_{k=0}^{M-1} \sum_{i=0}^{M-1} w_k^* w_i r(i-k)}_{\text{output power}} + \underbrace{\text{Re}\left[\lambda^*\left(\sum_{k=0}^{M-1} w_k^* e^{-j\omega_0 k} - g\right)\right]}_{\text{linear constraint}} \tag{5.75}$$

where λ is a *complex Lagrange multiplier*. Note that there is no desired response in the definition of the cost function J; rather, it includes a linear constraint that has to be satisfied at the angle frequency ω_0.

We wish to solve for the optimum values of the tap weights that minimize J, defined in Eq. (5.75). To do so, we may determine the gradient vector $\nabla(J)$, and

[5] The method of Lagrange multipliers is described in Appendix C.

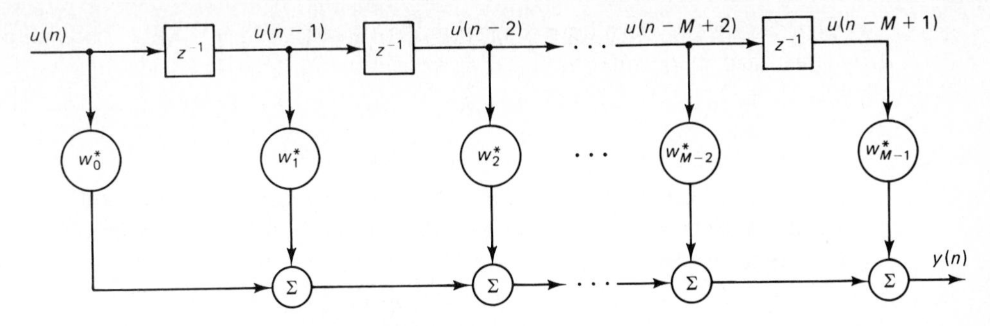

Figure 5.8 Linear transversal filter.

then set it equal to zero. Thus, proceeding in a manner similar to that described in Section 5.2, we find that the kth element of the gradient vector $\nabla(J)$ is

$$\nabla_k(J) = 2 \sum_{i=0}^{M-1} w_i r(i-k) + \lambda^* e^{-j\omega_0 k} \tag{5.76}$$

Let w_{oi} be the ith element of the optimum tap-weight vector $\mathbf{w}_o$. Then the condition for optimality of the filter is described by

$$\sum_{i=0}^{M-1} w_{oi} r(i-k) = -\frac{\lambda^*}{2} e^{-j\omega_0 k}, \qquad k = 0, 1, \ldots, M-1 \tag{5.77}$$

This system of M simultaneous equations defines the optimum values of the tap weights. It has a form somewhat similar to that of the Wiener–Hopf equations (5.28).

At this point in the analysis, we find it convenient to switch to matrix notation. In particular, we may rewrite the system of M simultaneous equations given in (5.77) simply as

$$\mathbf{R}\mathbf{w}_o = -\frac{\lambda^*}{2}\mathbf{s}(\omega_0) \tag{5.78}$$

where $\mathbf{R}$ is the M-by-M correlation matrix, and $\mathbf{w}_o$ is the M-by-1 optimum weight vector of the constrained filter. The M-by-1 *fixed frequency vector* $\mathbf{s}(\omega_0)$ is defined by

$$\mathbf{s}(\omega_0) = [1, e^{-j\omega_0}, \ldots, e^{-j(M-1)\omega_0}]^T \tag{5.79}$$

Solving Eq. (5.78) for $\mathbf{w}_o$, we thus have

$$\mathbf{w}_o = -\frac{\lambda^*}{2}\mathbf{R}^{-1}\mathbf{s}(\omega_0) \tag{5.80}$$

where $\mathbf{R}^{-1}$ is the inverse of the correlation matrix $\mathbf{R}$, assuming that $\mathbf{R}$ is nonsingular.

The solution for the optimum weight vector $\mathbf{w}_o$ given in Eq. (5.80) is not quite complete, as it involves the unknown Lagrange multiplier λ (or its complex conjugate to be precise). To eliminate λ^* from this expression, we first use the linear constraint of Eq. (5.73) to write

$$\mathbf{w}_o^H \mathbf{s}(\omega_0) = g \tag{5.81}$$

Hence, taking the Hermitian transpose of both sides of Eq. (5.80), postmultiplying by $\mathbf{s}(\omega_0)$, and then using the linear constraint of Eq. (5.81), we get

$$\lambda = -\frac{2g}{\mathbf{s}^H(\omega_0)\mathbf{R}^{-1}\mathbf{s}(\omega_0)} \tag{5.82}$$

where we have used the fact that $\mathbf{R}^{-H} = \mathbf{R}^{-1}$. The quadratic form $\mathbf{s}^H(\omega_0)\mathbf{R}^{-1}\mathbf{s}(\omega_0)$ is real valued. Hence, substituting Eq. (5.82) in (5.80), we get the desired formula for the optimum weight vector:

$$\mathbf{w}_o = g^* \frac{\mathbf{R}^{-1}\mathbf{s}(\omega_0)}{\mathbf{s}^H(\omega_0)\mathbf{R}^{-1}\mathbf{s}(\omega_0)} \tag{5.83}$$

Note that by minimizing the output power, subject to the constraint of Eq. (5.73), signals at angular frequencies other than ω_0 tend to be attenuated.

For obvious reasons, a filter characterized by the tap-weight vector $\mathbf{w}_o$ is referred to as a *linearly constrained minimum variance* (LCMV) *filter*. For a zero-mean input and therefore zero-mean output, "minimum variance" and "minimum mean-square value" are indeed synonymous.

Minimum Variance Distortionless Response Spectrum

The complex constant g defines the response of an LCMV filter at the angular frequency ω_0. For the special case of $g = 1$, the optimum solution given in Eq. (5.83) simplfies as

$$\mathbf{w}_o = \frac{\mathbf{R}^{-1}\mathbf{s}(\omega_0)}{\mathbf{s}^H(\omega_0)\mathbf{R}^{-1}\mathbf{s}(\omega_0)} \tag{5.84}$$

Now, the minimum mean-square value (average power) of the optimum filter output may be expressed as the quadratic form

$$J_{\min} = \mathbf{w}_o^H \mathbf{R} \mathbf{w}_o \tag{5.85}$$

Hence, substituting Eq. (5.84) in (5.85) and simplifying, we get the result

$$J_{\min} = \frac{1}{\mathbf{s}^H(\omega_0)\mathbf{R}^{-1}\mathbf{s}(\omega_0)} \tag{5.86}$$

Equation (5.86) defines the minimum mean-square value of the output of a linear filter whose response is constrained to equal unity at the angular frequency ω_0. In other words, the filter is constrained to produce a *distortionless response* at ω_0. Clearly, the choice of ω_0 influences the value of $J_{\min}$. We may therefore generalize this result by introducing a new power spectrum defined as follows:

$$S_{\text{MVDR}}(\omega) = \frac{1}{\mathbf{s}^H(\omega)\mathbf{R}^{-1}\mathbf{s}(\omega)} \tag{5.87}$$

where

$$\mathbf{s}(\omega) = [1, e^{-j\omega}, \ldots, e^{-j\omega(M-1)}] \tag{5.88}$$

The M-by-1 vector $\mathbf{s}(\omega)$ is called a *variable frequency vector* or *frequency scanning vector*. By definition, $S_{\text{MVDR}}(\omega)$ has the dimension of power. Its dependence on the angular frequency ω at the filter input therefore justifies referring to it as a power spectrum estimate. Indeed, it is commonly referred to as the *minimum variance distortionless response* (MVDR) *spectrum*.[6] Note that at any ω, power due to other angular frequencies is minimized. Accordingly, the MVDR spectrum tends to have sharper peaks and higher resolution, compared to nonparametric (classical) methods based on the definition of power spectrum discussed in Chapter 3.

In Appendix G, we present a fast algorithm for computing the MVDR spectrum when the correlation matrix $\mathbf{R}$ is known, which is frequently the case in time-series analysis. Also, it should be noted that the MVDR spectrum is an important member of a family of super-resolution spectrum analysis techniques; other members of this family are developed as we progress through the book. A justification for the terminology "super-resolution" or "high resolution" is deferred to Chapter 12.

PROBLEMS

1. A necessary condition for a function $f(z)$ of the complex variable $z = x + jy$ to be an analytic function is that its real part $u(x, y)$ and imaginary part $v(x, y)$ must satisfy the Cauchy–Riemann equations. For a discussion of complex variables, see Appendix A. Demonstrate that the cost function J defined in Eq. (5.43) is *not* an analytic function by doing the following:
   (a) Show that the inner product $w_k p^*(-k)$ is an analytic function of the complex tap-weight (filter coefficient) w_k.
   (b) Show that the second inner product $w_k^* p(-k)$ is *not* an analytic function.
2. Consider a Wiener filtering problem characterized as follows: The correlation matrix $\mathbf{R}$ of the tap-input vector $\mathbf{u}(n)$ is

$$\mathbf{R} = \begin{bmatrix} 1 & 0.5 \\ 0.5 & 1 \end{bmatrix}$$

   The cross-correlation vector $\mathbf{p}$ between the tap-input vector $\mathbf{u}(n)$ and the desired response $d(n)$ is

$$\mathbf{p} = \begin{bmatrix} 0.5 \\ 0.25 \end{bmatrix}$$

   (a) Evaluate the tap weights of the Wiener filter.
   (b) What is the minimum mean-squared error produced by this Wiener filter?
   (c) Formulate a representation of the Wiener filter in terms of the eigenvalues of matrix $\mathbf{R}$ and associated eigenvectors.
3. The tap-weight vector of a transversal filter is defined by

$$\mathbf{u}(n) = \alpha(n)\mathbf{s}(n) + \mathbf{v}(n)$$

[6] The formula given in Eq. (5.87) is credited to Capon (1969). It is also referred to in the literature as the *maximum-likelihood method* (MLM). In reality, however, this formula has no bearing on the classical principle of maximum likelihood. The use of the terminology MLM for this formula is therefore not recommended.

where

$$\mathbf{s}(\omega) = [1, e^{-j\omega}, \ldots, e^{-j\omega(M-1)}]^T$$

and

$$\mathbf{v}(n) = [v(n), v(n-1), \ldots, v(n-M+1)]^T$$

The complex amplitude of the sinusoidal vector $\mathbf{s}(\omega)$ is a random variable with zero mean and variance $\sigma_\alpha^2 = E[|\alpha(n)|^2]$.

(a) Determine the correlation matrix of the tap-input vector $\mathbf{u}(n)$.

(b) Suppose that the desired response $d(n)$ is uncorrelated with $\mathbf{u}(n)$. What is the value of the tap-weight vector of the corresponding Wiener filter?

(c) Suppose that the variance σ_α^2 is zero, and the desired response is defined by

$$d(n) = v(n-k)$$

where $0 \le k \le M-1$. What is the new value of the tap-weight vector of the Wiener filter?

(d) Determine the tap-weight vector of the Wiener filter for a desired response defined by

$$d(n) = \alpha(n)e^{-j\omega\tau}$$

where τ is a prescribed delay.

4. Show that the Wiener–Hopf equations (5.34), defining the tap-weight vector $\mathbf{w}_o$ of the Wiener filter, and Eq. (5.49), defining the minimum mean-squared error $J_{\min}$, may be combined into a single matrix relation

$$\mathbf{A}\begin{bmatrix} 1 \\ -\mathbf{w}_o \end{bmatrix} = \begin{bmatrix} J_{\min} \\ \mathbf{0} \end{bmatrix}$$

The matrix $\mathbf{A}$ is the correlation matrix of the augmented vector

$$\begin{bmatrix} d(n) \\ \mathbf{u}(n) \end{bmatrix}$$

where $d(n)$ is the desired response and $\mathbf{u}(n)$ is the tap-input vector of the Wiener filter.

5. Equation (5.50) defines the mean-squared error $J(\mathbf{w})$ as a function of the tap-weight vector $\mathbf{w}$. Show that this expression may be reformulated as follows:

$$J(\mathbf{w}) = J_{\min} + (\mathbf{w} - \mathbf{w}_o)^H\mathbf{R}(\mathbf{w} - \mathbf{w}_o)$$

where $J_{\min}$ is the minimum mean-squared error, $\mathbf{w}_o$ is the optimum tap-weight vector, and $\mathbf{R}$ is the correlation matrix of the tap-input vector.

6. The minimum mean-squared error $J_{\min}$ is defined by [see Eq. (5.49)]

$$J_{\min} = \sigma_d^2 - \mathbf{p}^H\mathbf{R}^{-1}\mathbf{p}$$

where σ_d^2 is the variance of the desired response $d(n)$, $\mathbf{R}$ is the correlation matrix of the tap-input vector $\mathbf{u}(n)$, and $\mathbf{p}$ is the cross-correlation vector between $\mathbf{u}(n)$ and $d(n)$. By applying the unitary similarity transformation to the inverse of the correlation matrix, that is, $\mathbf{R}^{-1}$, show that

$$J_{\min} = \sigma_d^2 - \sum_{k=1}^{M}\frac{|\mathbf{q}_k^H\mathbf{p}|^2}{\lambda_k}$$

where λ_k is the kth eigenvalue of the correlation matrix $\mathbf{R}$, and $\mathbf{q}_k$ is the corresponding eigenvector. Note that $\mathbf{q}_k^H\mathbf{p}$ is a scalar.

7. In this problem we explore an application of Wiener filtering to *radar*. The sampled form of the transmitted radar signal is $A_0 e^{j\omega_0 n}$ where ω_0 is the transmitted angular frequency, and A_0 is the transmitted complex amplitude. The received signal is

$$u(n) = A_1 e^{j\omega_1 n} + v(n)$$

where $|A_1| < |A_0|$ and ω_1 differs from ω_0 by virtue of the *Doppler* shift produced by the motion of a target of interest, and $v(n)$ is a sample of white noise.

(a) Show that the correlation matrix of the time series $\{u(n)\}$, made up of M elements, may be written as

$$\mathbf{R} = \sigma_v^2 \mathbf{I} + \sigma_1^2 \mathbf{s}(\omega_1)\mathbf{s}^H(\omega_1)$$

where σ_v^2 is the variance of the zero-mean white noise $v(n)$, and

$$\sigma_1^2 = E[|A_1|^2]$$

and

$$\mathbf{s}(\omega_1) = [1, e^{-j\omega_1}, \ldots, e^{-j\omega_1(M-1)}]^T$$

(b) The time series $\{u(n)\}$ is applied to an M-tap Wiener filter with the cross-correlation vector $\mathbf{p}$ between $\{u(n)\}$ and the desired response $d(n)$ preset to

$$\mathbf{p} = \sigma_0^2 \mathbf{s}(\omega_0)$$

where

$$\sigma_0^2 = E[|A_0|^2]$$

and

$$\mathbf{s}(\omega_0) = [1, e^{-j\omega_0}, \ldots, e^{-j\omega_0(M-1)}]^T$$

Derive an expression for the tap-weight vector of the Wiener filter.

8. An array processor consists of a primary sensor and a reference sensor interconnected with each other. The output of the reference sensor is weighted by w and then subtracted from the output of the primary sensor. Show that the mean-square value of the output of the array processor is minimized when the weight w attains the optimum value

$$w_o = \frac{E[u_1(n)u_2^*(n)]}{E[|u_2(n)|^2]}$$

where $u_1(n)$ and $u_2(n)$ are the primary- and reference-sensor outputs at time n, respectively.

9. A linear array consists of M uniformly spaced sensors. The individual sensor outputs are weighted and then summed, producing the output

$$e(n) = \sum_{k=1}^{M} w_k^* u_k(n)$$

where $u_k(n)$ is the output of sensor k at time n, and w_k is the associated weight. The weights are chosen to minimize the mean-square value of $e(n)$, subject to the constraint

$$\mathbf{w}^H \mathbf{s} = 1$$

where $\mathbf{s}$ is a prescribed steering vector.

By using the method of Lagrange multipliers, show that the optimum value of the vector $\mathbf{w}$ is defined by the spatial version of the MVDR formula

$$\mathbf{w}_o = \frac{\mathbf{R}^{-1}\mathbf{s}}{\mathbf{s}^H\mathbf{R}^{-1}\mathbf{s}}$$

where $\mathbf{R}$ is the spatial correlation matrix of the linear array.

10. Consider a discrete-time stochastic process $\{u(n)\}$ that consists of K (uncorrelated) complex sinusoids plus additive white noise of zero mean and variance σ^2. That is,

$$u(n) = \sum_{k=1}^{K} A_k e^{j\omega_k n} + v(n)$$

where the terms $A_k \exp(j\omega_k n)$ and the $v(n)$ refer to the kth sinusoid and noise, respectively. The process $\{u(n)\}$ is applied to a transversal filter with M taps, producing the output

$$e(n) = \mathbf{w}^H\mathbf{u}(n)$$

Assume that $M > K$. The requirement is to choose the tap-weight vector $\mathbf{w}$ so as to minimize the mean-square value of $e(n)$, subject to the multiple signal-protection constraint

$$\mathbf{S}^H\mathbf{w} = \mathbf{D}^{1/2}\mathbf{1}$$

where $\mathbf{S}$ is the M-by-K signal matrix whose kth column has 1, $\exp(j\omega_k)$, . . . , $\exp[j\omega_k(M - 1)]$ for its elements, $\mathbf{D}$ is the K-by-K diagonal matrix whose nonzero elements equal the average powers of the individual sinusoids, and the K-by-1 vector $\mathbf{1}$ has 1's for all its K elements. Using the method of Lagrange multipliers, show that the value of the optimum weight vector that results from this constrained optimization equals

$$\mathbf{w}_o = \mathbf{R}^{-1}\mathbf{S}(\mathbf{S}^H\mathbf{R}^{-1}\mathbf{S})^{-1}\mathbf{D}^{1/2}\mathbf{1}$$

where $\mathbf{R}$ is the correlation matrix of the M-by-1 tap-input vector $\mathbf{u}(n)$. This formula represents a temporal generalization of the MVDR formula.

11. Consider the problem of detecting a known signal in the presence of additive noise. The noise is assumed to be Gaussian, to be independent of the signal, and to have zero mean and a positive definite correlation matrix $\mathbf{R}_N$. The aim of the problem is to show that under these conditions the three criteria: minimum mean-squared error, maximum signal-to-noise ratio, and the likelihood ratio test yield identical designs for the transversal filter.

Let $\{u(n)\}$, $n = 1, 2, \ldots, M$, denote a set of M complex-valued data samples. Let $\{v(n)\}$, $n = 1, 2, \ldots, M$, denote a set of samples taken from a Gaussian noise process of zero mean. Finally, let $\{s(n)\}$, $n = 1, 2, \ldots, M$, denote samples of the signal. The detection problem is to determine whether the input consists of signal plus noise or noise alone. That is, the two hypotheses to be tested for are

$$\begin{aligned}
&\text{hypothesis } H_2\text{:} \quad u(n) = s(n) + v(n), \qquad n = 1, 2, \ldots, M \\
&\text{hypothesis } H_1\text{:} \quad u(n) = v(n), \qquad\qquad\quad\; n = 1, 2, \ldots, M
\end{aligned}$$

(a) The *Wiener filter* minimizes the mean-squared error. Show that this criterion yields an optimum tap-weight vector for estimating s_k, the kth component of signal vector $\mathbf{s}$, that equals

$$\mathbf{w}_o = \frac{s_k}{1 + \mathbf{s}^H\mathbf{R}_N^{-1}\mathbf{s}}\mathbf{R}_N^{-1}\mathbf{s}$$

Hint: To evaluate the inverse of the correlation matrix of $\{u(n)\}$ under hypothesis H_2, you may use the matrix inversion lemma. Let

$$\mathbf{A} = \mathbf{B}^{-1} + \mathbf{C}\mathbf{D}^{-1}\mathbf{C}^H$$

where **A**, **B** and **D** are positive-definite matrices. Then

$$\mathbf{A}^{-1} = \mathbf{B} - \mathbf{BC}(\mathbf{D} + \mathbf{C}^H\mathbf{BC})^{-1}\mathbf{C}^H\mathbf{B}$$

(b) The *maximum signal-to-noise ratio filter* maximizes the ratio

$$\rho = \frac{\text{average power of filter output due to signal}}{\text{average power of filter output due to noise}}$$

$$= \frac{E[(\mathbf{w}^H\mathbf{s})^2]}{E[(\mathbf{w}^H\mathbf{v})^2]}$$

Show that the tap-weight vector for which the output signal-to-noise ratio ρ is at maximum equals

$$\mathbf{w}_{SN} = \mathbf{R}_N^{-1}\mathbf{s}$$

Hint: Since $\mathbf{R}_N$ is positive definite, you may use $\mathbf{R}_N = \mathbf{R}_N^{1/2}\mathbf{R}_N^{1/2}$.

(c) The *likelihood ratio processor* computes the log-likelihood ratio and compares it to a threshold. If the threshold is exceeded, it decides in favor of hypothesis H_2; otherwise, it decides in favor of hypothesis H_1. The likelihood ratio is defined by

$$\Lambda = \frac{\mathbf{f}_\mathbf{U}(\mathbf{u}\,|\,H_2)}{\mathbf{f}_\mathbf{U}(\mathbf{u}\,|\,H_1)}$$

where $\mathbf{f}_\mathbf{U}(\mathbf{u}\,|\,H_i)$ is the conditional joint probability density function of the observation vector $\mathbf{u}$, given that hypothesis H_i is true, where $i = 1, 2$. Show that the likelihood ratio test is equivalent to the test

$$\mathbf{w}_{ml}^H\mathbf{u} \underset{H_1}{\overset{H_2}{\gtrless}} \eta$$

where η is the threshold and

$$\mathbf{w}_{ml} = \mathbf{R}_N^{-1}\mathbf{s}$$

Hint: Refer to Section 2.11 for the joint probability function of the M-by-1 Gaussian noise vector $\mathbf{v}$ with zero mean and correlation matrix $\mathbf{R}_v$.

12. This problem is a continuation of the numerical example presented in Section 5.6. In particular, consider the canonical error performance surface described in Eq. (5.71). Plot the loci of ν_2 versus ν_1 for

$$J - J_{\min} = 0,\ 1.6,\ 6.4,\ 14.4$$

How do these loci compare with those obtained by plotting w_2 versus w_1 for varying $J - J_{\min}$?

13. In this problem, we explore the extent of the improvement that may result from using a more complex Wiener filter for the environment described in Section 5.6. To be specific, the new formulation of the Wiener filter has three taps.

(a) Find the 3-by-3 correlation matrix of the tap inputs of this filter and the 3-by-1 cross-correlation vector between the desired response and the tap inputs.

(b) Compute the 3-by-1 tap-weight vector of the Wiener filter, and also compute the new value for the minimum mean-squared error.

6

Linear Prediction

One of the most celebrated problems in time-series analysis is that of *predicting* a future value of a stationary discrete-time stochastic process, given a set of past samples of the process. To be specific, consider the time series $u(n), u(n-1), \ldots, u(n-M)$, representing $(M+1)$ samples of such a process up to and including time n. The operation of prediction may, for example, involve using the samples $u(n-1), u(n-2), \ldots, u(n-M)$ to make an estimate of $u(n)$. Let $\mathcal{U}_{n-1}$ denote the M-dimensional space spanned by the samples $u(n-1), u(n-2), \ldots, u(n-M)$, and use $\hat{u}(n|\mathcal{U}_{n-1})$ to denote the *predicted value* of $u(n)$ given this set of samples. In *linear prediction,* we express this predicted value as a linear combination of the samples $u(n-1), u(n-2), \ldots, u(n-M)$. This operation corresponds to one-step prediction into the future, measured with respect to time $n-1$. Accordingly, we refer to this form of prediction as *one-step linear prediction in the forward direction* or simply *forward linear prediction*. In another form of prediction, we use the samples $u(n), u(n-1), \ldots, u(n-M+1)$ to make a prediction of the *past* sample $u(n-M)$. We refer to this second form of prediction as *backward linear prediction.*[1]

In this chapter, we study forward linear prediction (FLP) as well as backward linear prediction (BLP). In particular, we use the Wiener filter theory of Chapter 5 to optimize the design of a forward or backward *predictor* in the mean-square sense for the case of a wide-sense stationary discrete-time stochastic process. As explained in Chapter 2, the correlation matrix of such a process has a Toeplitz structure. We will put this Toeplitz structure to good use in developing algorithms that are computationally efficient.

[1] The term "backward prediction" is somewhat of a misnomer. A more appropriate description for this operation is "hindsight." Correspondingly, the use of "forward" in the associated operation of forward prediction is superfluous. Nevertheless, the terms "forward prediction" and "backward prediction" have become deeply embedded in the literature on linear prediction.

6.1 FORWARD LINEAR PREDICTION

Figure 6.1(a) shows a *forward predictor* that consists of a linear transversal filter with M tap weights $w_{o1}, w_{o2}, \ldots, w_{oM}$ and tap inputs $u(n-1), u(n-2), \ldots, u(n-M)$, respectively. We assume that these tap inputs are drawn from a wide-sense stationary stochastic process of zero mean. We further assume that the tap weights are optimized in the mean-square sense in accordance with the Wiener filter theory. The predicted value $\hat{u}(n|\mathcal{U}_{n-1})$ is defined by

$$\hat{u}(n|\mathcal{U}_{n-1}) = \sum_{k=1}^{M} w_{ok}^* u(n-k) \tag{6.1}$$

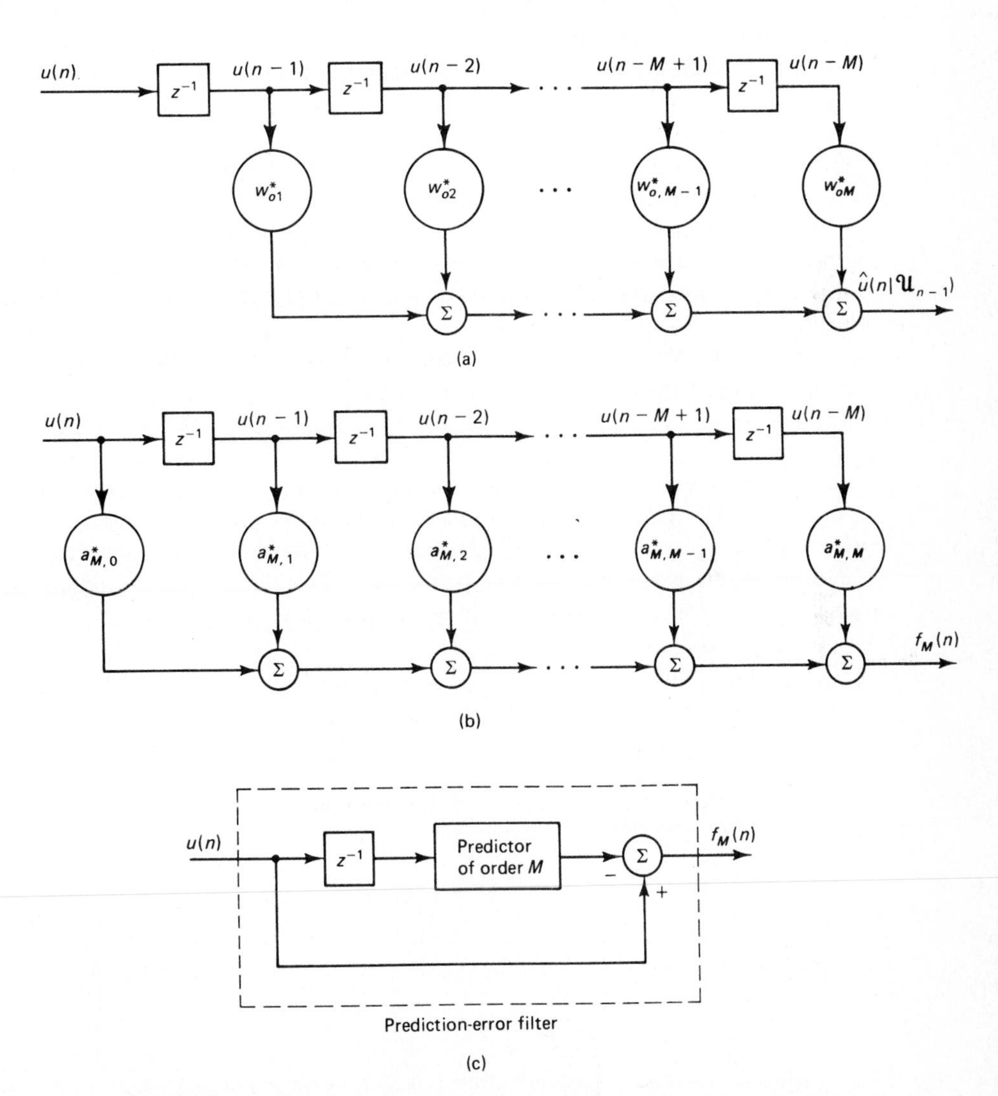

Figure 6.1 (a) One-step predictor; (b) prediction-error filter: (c) relationship between the predictor and the prediction-error filter.

For the stituation described herein, the desired response $d(n)$ equals $u(n)$, representing the actual sample of the input process at time n. We may thus write

$$d(n) = u(n) \tag{6.2}$$

The *forward prediction error* equals the difference between the input sample $u(n)$ and its predicted value $\hat{u}(n|\mathcal{U}_{n-1})$. We denote the forward prediction error by $f_M(n)$ and thus write

$$f_M(n) = u(n) - \hat{u}(n|\mathcal{U}_{n-1}) \tag{6.3}$$

The subscript M in the symbol for the forward prediction error signifies *order* of the predictor, defined as *the number of unit-delay elements needed to store the given set of samples used to make the prediction*. The reason for using the subscript will become apparent later in the chapter.

Let P_M denote the *minimum mean-squared prediction error*:

$$P_M = E[|f_M(n)|^2] \tag{6.4}$$

With the tap inputs assumed to have zero mean, the forward prediction error $f_M(n)$ will likewise have zero mean. Under this condition, P_M will also equal the variance of the forward prediction error. Yet another interpretation for P_M is that it may be viewed as the ensemble-averaged *forward prediction error power,* assuming that $f_M(n)$ is developed across a 1-Ω load. We will use the latter description to refer to P_M.

Let $\mathbf{w}_o$ denote the M-by-1 optimum tap-weight vector of the forward predictor in Fig. 6.1(a). We write it in expanded form as

$$\mathbf{w}_o = [w_{o1}, w_{o2}, \ldots, w_{oM}]^T \tag{6.5}$$

To solve the Wiener–Hopf equations for the vector $\mathbf{w}_o$, we require knowledge of two quantities: (1) the M-by-M correlation matrix of the tap inputs $u(n-1)$, $u(n-2), \ldots, u(n-M)$, and (2) the M-by-1 cross-correlation vector between these tap inputs and the desired response $u(n)$. To evaluate P_M, we require a third quantity, the variance of $u(n)$. We now consider these three quantities, one by one:

1. The tap inputs $u(n-1), u(n-2), \ldots, u(n-M)$ define the M-by-1 tap-input vector, $\mathbf{u}(n-1)$, as shown by

$$\mathbf{u}(n-1) = [u(n-1), u(n-2), \ldots, u(n-M)]^T \tag{6.6}$$

Hence, the correlation matrix of the tap inputs equals

$$\mathbf{R} = E[\mathbf{u}(n-1)\mathbf{u}^H(n-1)]$$
$$= \begin{bmatrix} r(0) & r(1) & \cdots & r(M-1) \\ r^*(1) & r(0) & \cdots & r(M-2) \\ \vdots & \vdots & \ddots & \vdots \\ r^*(M-1) & r^*(M-2) & \cdots & r(0) \end{bmatrix} \tag{6.7}$$

where $r(k)$ is the autocorrelation function of the input process for lag k, where $k = 0, 1, \ldots, M-1$. Note that the symbol used for the correlation matrix of the tap inputs in Fig. 6.1(a) is the same as that for the correlation matrix of the tap inputs in the transversal filter of Fig. 5.4. We are justified to do this since the input process in both cases is assumed to be wide-sense stationary, so the correlation matrix of the process is invariant to a time shift.

2. The cross-correlation vector between the tap inputs $u(n-1), \ldots, u(n-M)$ and the desired response $u(n)$ equals

$$\mathbf{r} = E[\mathbf{u}(n-1)u^*(n)] = \begin{bmatrix} r^*(1) \\ r^*(2) \\ \vdots \\ r^*(M) \end{bmatrix} = \begin{bmatrix} r(-1) \\ r(-2) \\ \vdots \\ r(-M) \end{bmatrix} \tag{6.8}$$

3. The variance of $u(n)$ equals $r(0)$, since $u(n)$ has zero mean.

In Table 6.1, we summarize the various quantities pertaining to the Wiener filter of Fig. 5.4 and the corresponding quantities pertaining to the forward predictor of Fig. 6.1(a). The last column of this table pertains to the backward predictor, on which more will be said later.

TABLE 6.1 SUMMARY OF WIENER FILTER VARIABLES

Quantity	Wiener filter of Fig. 5.4	Forward predictor of Fig. 6.1(a)	Backward predictor of Fig. 6.2(a)
Tap-input vector	$\mathbf{u}(n)$	$\mathbf{u}(n-1)$	$\mathbf{u}(n)$
Desired response	$d(n)$	$u(n)$	$u(n-M)$
Tap-weight vector	$\mathbf{w}_o$	$\mathbf{w}_o$	$\mathbf{g}$
Estimation error	$e(n)$	$f_M(n)$	$b_M(n)$
Correlation matrix of tap inputs	$\mathbf{R}$	$\mathbf{R}$	$\mathbf{R}$
Cross-correlation vector between tap inputs and desired response	$\mathbf{p}$	$\mathbf{r}$	$\mathbf{r}^{B*}$
Minimum mean-squared error	$J_{\min}$	P_M	P_M

Thus, using the correspondences of this table, we may adapt the Wiener–Hopf equations (5.34) to solve the forward linear prediction (FLP) problem for stationary inputs and so write

$$\mathbf{R}\mathbf{w}_o = \mathbf{r} \tag{6.9}$$

Similarly, the use of Eq. (5.49), together with Eq. (6.8), yields the following expression for the forward prediction-error power:

$$P_M = r(0) - \mathbf{r}^H\mathbf{w}_o \tag{6.10}$$

From Eqs. (6.8) and (6.9), we see that the M-by-1 tap-weight vector of the forward predictor and the forward prediction-error power are determined solely by the set of $(M+1)$ autocorrelation function values of the input process for lags $0, 1, \ldots, M$.

Relation between Linear Prediction and Autoregressive Modeling

It is highly informative to compare the Wiener–Hopf equations (6.9) for linear prediction with the Yule–Walker equations (2.66) for an autoregressive (AR) model. We see that these two systems of simultaneous equations are of exactly the same mathematical form. Furthermore, Eq. (6.10) defining the average power (i.e., variance) of the forward prediction error is also of the same mathematical form as Eq. (2.71) defining the variance of the white-noise process used to excite the autoregressive model. For the case of an AR process for which we know the model order M, we may thus state that when a forward predictor is optimized in the mean-square sense, its tap weights take on the same values as the corresponding parameters of the process. This relationship should not be surprising since the equation defining the forward prediction error and the difference equation defining the autoregressive model have the same mathematical form. When the process is not autoregressive, however, the use of a predictor provides an approximation to the process.

Forward Prediction-Error Filter

The forward predictor of Fig. 6.1(a) consists of M unit-delay elements and M tap weights $w_{o1}, w_{o2}, \ldots, w_{oM}$ that are fed with the respective samples $u(n-1)$, $u(n-2), \ldots, u(n-M)$ as inputs. The resultant output is the predicted value of $u(n)$, which is defined by Eq. (6.1). Hence, substituting Eq. (6.1) in (6.3), we may express the forward prediction error as

$$f_M(n) = u(n) - \sum_{k=1}^{M} w_{ok}^* u(n-k) \tag{6.11}$$

Let $a_{M,k}$, $k = 0, 1, \ldots, M$, denote the tap weights of a new transversal filter, which are related to the tap weights of the forward predictor as follows:

$$a_{M,k} = \begin{cases} 1, & k = 0 \\ -w_{ok}, & k = 1, 2, \ldots, M \end{cases} \tag{6.12}$$

Then we may combine the two terms on the right side of Eq. (6.11) into a single summation as follows:

$$f_M(n) = \sum_{k=0}^{M} a_{M,k}^* u(n-k) \tag{6.13}$$

This input–output relation is represented by the transversal filter shown in Fig. 6.1(b). A filter that operates on the set of samples $u(n), u(n-1), \ldots, u(n-M)$ to produce the forward prediction error $f_M(n)$ at its output is called a forward *prediction-error filter* (PEF).

The relationship between the forward prediction-error filter and the forward predictor is illustrated in block diagram form in Fig. 6.1(c). Note that the length of the prediction-error filter exceeds the length of the one-step prediction filter by 1. However, both filters have the same order, M, as they both involve the same number of delay elements for the storage of past data.

Augmented Wiener–Hopf Equations for Forward Prediction

The Wiener–Hopf equations (6.9) define the tap-weight vector of the forward predictor, while Eq. (6.10) defines the resulting forward prediction-error power P_M. We may combine these two equations into a single matrix relation as follows:

$$\begin{bmatrix} r(0) & \mathbf{r}^H \\ \mathbf{r} & \mathbf{R} \end{bmatrix} \begin{bmatrix} 1 \\ -\mathbf{w}_o \end{bmatrix} = \begin{bmatrix} P_M \\ \mathbf{0} \end{bmatrix} \tag{6.14}$$

where $\mathbf{0}$ is the M-by-1 null vector. The M-by-M correlation matrix $\mathbf{R}$ is defined in Eq. (6.7), and the M-by-1 correlation vector $\mathbf{r}$ is defined in Eq. (6.8). The partitioning of the $(M + 1)$-by-$(M + 1)$ correlation matrix on the left side of Eq. (6.14) into the form shown therein was discussed in Section 2.3. Note that this $(M + 1)$-by-$(M + 1)$ matrix equals the correlation matrix of the tap inputs $u(n)$, $u(n - 1), \ldots, u(n - M)$ in the prediction-error filter of Fig. 6.1(b). Moreover, the $(M + 1)$-by-1 coefficient vector on the left side of Eq. (6.14) equals the *forward prediction-error filter vector*:

$$\mathbf{a}_M = \begin{bmatrix} 1 \\ -\mathbf{w}_o \end{bmatrix} \tag{6.15}$$

We may also express the matrix relation of Eq. (6.14) as a system of $(M + 1)$ simultaneous equations as follows:

$$\sum_{l=0}^{M} a_{M,l}\, r(l - i) = \begin{cases} P_M, & i = 0 \\ 0, & i = 1, 2, \ldots, M \end{cases} \tag{6.16}$$

We refer to Eq. (6.14) or (6.16) as the *augmented Wiener–Hopf equations* of a forward prediction-error filter of order M.

Example 1

For the case of a prediction-error filter of order $M = 1$, Eq. (6.14) yields a pair of simultaneous equations described by

$$\begin{bmatrix} r(0) & r(1) \\ r^*(1) & r(0) \end{bmatrix} \begin{bmatrix} a_{1,0} \\ a_{1,1} \end{bmatrix} = \begin{bmatrix} P_1 \\ 0 \end{bmatrix}$$

Solving for $a_{1,0}$ and $a_{1,1}$, we get

$$a_{1,0} = \frac{P_1}{\Delta_r} r(0)$$

$$a_{1,1} = -\frac{P_1}{\Delta_r} r(-1)$$

where Δ_r is the determinant of the correlation matrix; thus

$$\begin{aligned} \Delta_r &= \begin{vmatrix} r(0) & r(1) \\ r^*(1) & r(0) \end{vmatrix} \\ &= r^2(0) - |r(1)|^2 \end{aligned}$$

But $a_{1,0}$ equals 1. Hence

$$P_1 = \frac{\Delta_r}{r(0)}$$

$$a_{1,1} = -\frac{r^*(1)}{r(0)}$$

Consider next the case of a prediction-error filter of order $M = 2$. Equation (6.14) yields a system of three simultaneous equations, as shown by

$$\begin{bmatrix} r(0) & r(1) & r(2) \\ r^*(1) & r(0) & r(1) \\ r^*(2) & r^*(1) & r(0) \end{bmatrix} \begin{bmatrix} a_{2,0} \\ a_{2,1} \\ a_{2,2} \end{bmatrix} = \begin{bmatrix} P_2 \\ 0 \\ 0 \end{bmatrix}$$

Solving for $a_{2,0}$, $a_{2,1}$ and $a_{2,2}$, we get

$$a_{2,0} = \frac{P_2}{\Delta_r}[r^2(0) - |r(1)|^2]$$

$$a_{2,1} = -\frac{P_2}{\Delta_r}[r^*(1)r(0) - r(1)r^*(2)]$$

$$a_{2,2} = \frac{P_2}{\Delta_r}[(r^*(1))^2 - r(0)r^*(2)]$$

where Δ_r is the determinant of the correlation matrix:

$$\Delta_r = \begin{vmatrix} r(0) & r(1) & r(2) \\ r^*(1) & r(0) & r(1) \\ r^*(2) & r^*(1) & r(0) \end{vmatrix}$$

The coefficient $a_{2,0}$ equals 1. Accordingly, we may express the prediction-error power P_2 as

$$P_2 = \frac{\Delta_r}{r^2(0) - |r(1)|^2}$$

and the prediction-error filter coefficients $a_{2,1}$ and $a_{2,2}$ as

$$a_{2,1} = -\frac{r^*(1)(r(0) - r(1)r^*(2))}{r^2(0) - |r(1)|^2}$$

$$a_{2,2} = \frac{(r^*(1))^2 - r(0)r^*(2)}{r^2(0) - |r(1)|^2}$$

6.2 BACKWARD LINEAR PREDICTION

The form of linear prediction considered in Section 6.1 is said to be in the *forward* direction. That is, given the time series $u(n)$, $u(n-1)$, . . . , $u(n-M)$, we use the subset of M samples $u(n-1)$, $u(n-2)$, . . . , $u(n-M)$ to make a prediction of the sample $u(n)$. This operation corresponds to *one-step linear prediction into the future*, measured with respect to time $n - 1$. Naturally, we may also operate on this time series in the *backward* direction. That is, we may use the subset of M samples $u(n)$, $u(n-1)$, . . . , $u(n-M+1)$ to make a prediction of the sample

$u(n-M)$. This second operation corresponds to *backward linear prediction by one step*, measured with respect to time $n - M + 1$.

Let $\mathcal{U}_n$ denote the M-dimensional space spanned by $u(n)$, $u(n-1), \ldots,$ $u(n-M+1)$ that are used in making the backward prediction. Then, using this set of samples as tap inputs, we make a linear prediction of the sample $u(n-M)$, as shown by

$$\hat{u}(n-M|\mathcal{U}_n) = \sum_{k=1}^{M} g_k^* u(n-k+1) \tag{6.17}$$

where $g_1, g_2, \ldots, g_M$ are the tap weights. Figure 6.2(a) shows a representation of the backward predictor as described by Eq. (6.17). We assume that these tap weights are optimized in the mean-square sense in accordance with the Wiener filter theory.

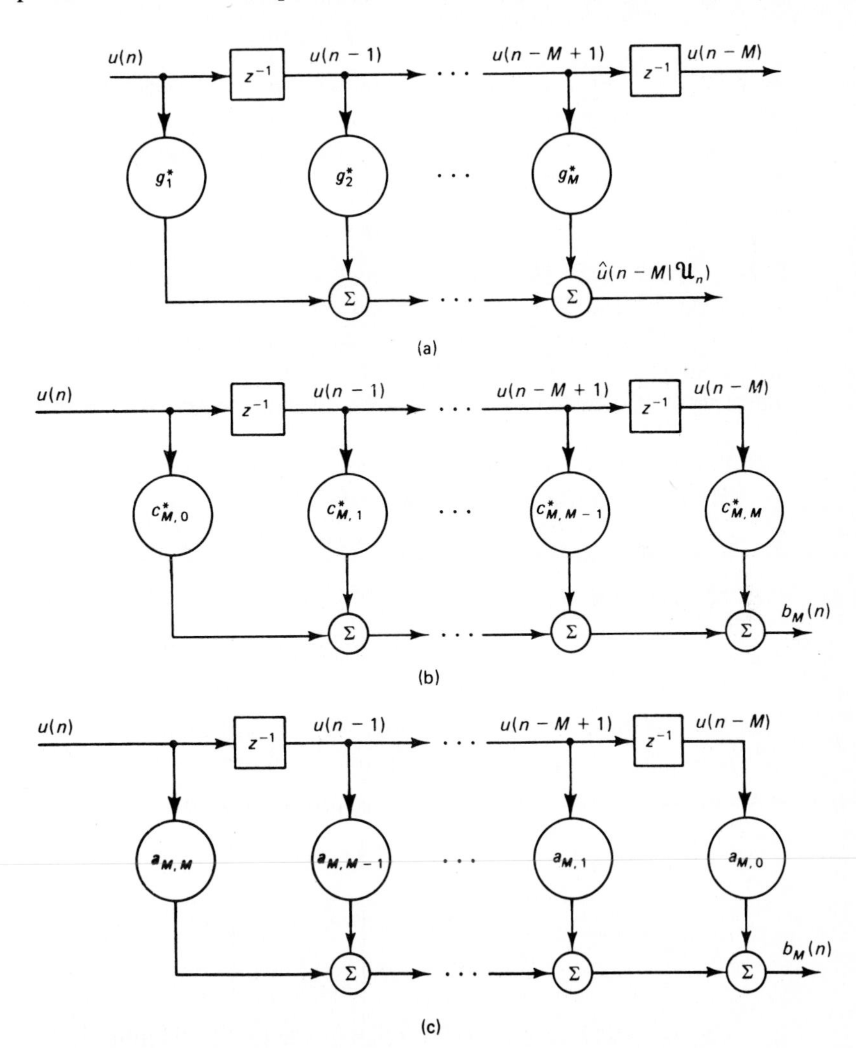

Figure 6.2 (a) Backward one-step predictor; (b) backward prediction-error filter; (c) backward prediction-error filter defined in terms of the tap weights of the corresponding forward prediction-error filter.

In the case of backward prediction, the desired response equals

$$d(n) = u(n - M) \tag{6.18}$$

The *backward prediction error* equals the difference between the actual sample value $u(n - M)$ and its predicted value $\hat{u}(n - M \mid \mathcal{U}_n)$. We denote the backward prediction error by $b_M(n)$ and thus write

$$b_M(n) = u(n - M) - \hat{u}(n - M \mid \mathcal{U}_n) \tag{6.19}$$

Here, again, the subscript M in the symbol for the backward prediction error $b_M(n)$ signifies the number of unit-delay elements needed to store the given set of samples used to make the prediction; that is, M is the order of the predictor.

Let P_M denote the *minimum mean-squared prediction error*

$$P_M = E[\mid b_M(n) \mid^2] \tag{6.20}$$

We may also view P_M as the ensemble-averaged *backward prediction-error power*, assuming that $b_M(n)$ is developed across a 1-Ω load.

Let $\mathbf{g}$ denote the M-by-1 optimum tap-weight vector of the backward predictor in Fig. 6.2(a). We express it in the expanded form

$$\mathbf{g} = [g_1, g_2, \ldots, g_M]^T \tag{6.21}$$

To solve the Wiener–Hopf equations for the vector $\mathbf{g}$, we require knowledge of two quantities: (1) the M-by-M correlation matrix of the tap inputs $u(n)$, $u(n - 1)$, $\ldots$, $u(n - M + 1)$, and (2) the M-by-1 cross-correlation vector between the desired response $u(n - M)$ and these tap inputs. To evaluate P_M, we need a third quantity, the variance of $u(n - M)$. We consider these three quantities in turn:

1. Let $\mathbf{u}(n)$ denote the M-by-1 tap-input vector in the backward predictor of Fig. 6.2(a). We write it in expanded form as

$$\mathbf{u}(n) = [u(n), u(n - 1), \ldots, u(n - M + 1)]^T \tag{6.22}$$

The M-by-M correlation matrix of the tap inputs in Fig. 6.2(a) thus equals

$$\mathbf{R} = E[\mathbf{u}(n)\mathbf{u}^H(n)]$$

The expanded form of the correlation matrix $\mathbf{R}$ is given in Eq. (6.7).

2. The M-by-1 cross-correlation vector between the tap inputs $u(n)$, $u(n - 1), \ldots, u(n - M + 1)$ and the desired response $u(n - M)$ equals

$$\begin{aligned} \mathbf{r}^{B*} &= E[\mathbf{u}(n)u^*(n - M)] \\ &= \begin{bmatrix} r(M) \\ r(M - 1) \\ \vdots \\ r(1) \end{bmatrix} \end{aligned} \tag{6.23}$$

The expanded form of the correlation vector $\mathbf{r}$ is given in Eq. (6.8). As usual, the superscript B denotes backward arrangement and the asterisk denotes complex conjugation.

3. The variance of the desired response $u(n - M)$ equals $r(0)$.

In the last column of Table 6.1, we summarize the various quantities pertaining to the backward predictor of Fig. 6.2(a).

Accordingly, using the correspondences of Table 6.1, we may adapt the Wiener–Hopf equations (5.34) to solve the backward linear prediction (BLP) problem for stationary inputs and so write

$$\mathbf{R}\mathbf{g} = \mathbf{r}^{B*} \tag{6.24}$$

Similarly, the use of Eq. (5.49), together with Eq. (6.24), yields the following expression for the backward prediction-error power:

$$P_M = r(0) - \mathbf{r}^{BT}\mathbf{g} \tag{6.25}$$

Here again we see that the M-by-1 tap-weight vector $\mathbf{g}$ of a backward predictor and the backward prediction-error power P_M are uniquely defined by knowledge of the set of autocorrelation function values of the process for lags 0, 1, . . . , M.

Relations between Backward and Forward Predictors

In comparing the two sets of Wiener–Hopf equations (6.9) and (6.24), pertaining to forward prediction and backward prediction, respectively, we see that the vector on the right side of Eq. (6.24) differs from that of Eq. (6.9) in two respects: (1) its elements are arranged backward, and (2) they are complex conjugated. To correct for the first difference, we reverse the order in which the elements of the vector on the right side of Eq. (6.24) are arranged. This operation has the effect of replacing the left side of Eq. (6.24) by $\mathbf{R}^T\mathbf{g}^B$, where $\mathbf{R}^T$ is the transpose of the correlation matrix $\mathbf{R}$ and $\mathbf{g}^B$ is the backward version of the tap-weight vector $\mathbf{g}$ (see Problem 3). We may thus write

$$\mathbf{R}^T\mathbf{g}^B = \mathbf{r}^* \tag{6.26}$$

To correct for the remaining difference, we complex-conjugate both sides of Eq. (6.26), obtaining

$$\mathbf{R}^H\mathbf{g}^{B*} = \mathbf{r}$$

Since the correlation matrix $\mathbf{R}$ is Hermitian, that is, $\mathbf{R}^H = \mathbf{R}$, we may thus reformulate the Wiener–Hopf equations for backward prediction as

$$\mathbf{R}\mathbf{g}^{B*} = \mathbf{r} \tag{6.27}$$

Now we may compare Eq. (6.27) with (6.9) and thus deduce the following fundamental relationship between the tap-weight vectors of a backward predictor and the corresponding forward predictor:

$$\mathbf{g}^{B*} = \mathbf{w}_o \tag{6.28}$$

Equation (6.28) states that *we may modify a backward predictor into a forward predictor by reversing the sequence in which its tap weights are positioned and also complex-conjugating them.*

Next we wish to show that the ensemble-averaged error powers for backward prediction and forward prediction have exactly the same value. To do this, we first observe that the product $\mathbf{r}^{BT}\mathbf{g}$ equals $\mathbf{r}^T\mathbf{g}^B$, so we may rewrite Eq. (6.25) as

$$P_M = r(0) - \mathbf{r}^T\mathbf{g}^B \tag{6.29}$$

Taking the complex conjugate of both sides of Eq. (6.29), and recognizing that both P_M and $r(0)$ are unaffected by this operation, since they are real-valued scalars, we get

$$P_M = r(0) - \mathbf{r}^H\mathbf{g}^{B*} \tag{6.30}$$

Comparing this result with Eq. (6.10) and using the equivalence of Eq. (6.28), we find that the backward prediction-error power has exactly the same value as the forward prediction-error power. Indeed, it is in anticipation of this equality that we have used the same symbol P_M to denote both the forward prediction-error power and the backward prediction-error power.

Backward Prediction-Error Filter

The backward prediction error $b_M(n)$ equals the difference between the desired response $u(n - M)$ and the linear prediction of it, given the samples $u(n)$, $u(n - 1)$, . . . , $u(n - M + 1)$. This prediction is defined by Eq. (6.17). Therefore, substituting Eq. (6.17) in (6.19), we get

$$b_M(n) = u(n - M) - \sum_{k=1}^{M} g_k^* u(n - k + 1) \tag{6.31}$$

Define the tap weights of the backward prediction-error filter in terms of the corresponding backward predictor as follows:

$$c_{M,k} = \begin{cases} -g_{k+1}, & k = 0, 1, \ldots, M - 1 \\ 1, & k = M \end{cases} \tag{6.32}$$

Hence, we may rewrite Eq. (6.31) as [see Fig. 6.2(b)]

$$b_M(n) = \sum_{k=0}^{M} c_{M,k}^* u(n - k) \tag{6.33}$$

Equation (6.28) defines the tap-weight vector of the backward predictor in terms of that of the forward predictor. We may express the scalar version of this relation as

$$g_{M-k+1}^* = w_{ok}, \qquad k = 1, 2, \ldots, M$$

or, equivalently

$$g_k = w_{o,M-k+1}^*, \qquad k = 1, 2, \ldots, M \tag{6.34}$$

Hence, substituting Eq. (6.34) in (6.32), we get

$$c_{M,k} = \begin{cases} -w_{o,M-k}^*, & k = 0, 1, \ldots, M - 1 \\ 1, & k = M \end{cases} \tag{6.35}$$

Thus, using the relationship between the tap weights of the forward prediction-error filter and those of the forward predictor as given in Eq. (6.12), we may write

$$c_{M,k} = a^*_{M,M-k}, \qquad k = 0, 1, \ldots, M \tag{6.36}$$

Accordingly, we may express the input–output relation of the backward prediction-error filter in the equivalent form

$$b_M(n) = \sum_{k=0}^{M} a_{M,M-k} u(n-k) \tag{6.37}$$

The input–output relation of Eq. (6.37) is depicted in Fig. 6.2(c). Comparison of this representation for a backward prediction-error filter with that of Fig. 6.1(b) for the corresponding forward prediction-error filter reveals that these two forms of a prediction-error filter for *stationary inputs* are uniquely related to each other. In particular, *we may modify a forward prediction-error filter into the corresponding backward prediction-error filter by reversing the sequence in which the tap weights are positioned and complex-conjugating them.* Note that in both figures, the respective tap inputs have the same values.

Augmented Wiener–Hopf Equations for Backward Prediction

The set of Wiener–Hopf equations for backward prediction is defined by Eq. (6.24), and the resultant backward prediction-error power is defined by Eq. (6.25). We may combine these two equations into a single relation as follows:

$$\begin{bmatrix} \mathbf{R} & \mathbf{r}^{B*} \\ \mathbf{r}^{BT} & r(0) \end{bmatrix} \begin{bmatrix} -\mathbf{g} \\ 1 \end{bmatrix} = \begin{bmatrix} \mathbf{0} \\ P_M \end{bmatrix} \tag{6.38}$$

where $\mathbf{0}$ is the M-by-1 null vector. The M-by-M matrix $\mathbf{R}$ is the correlation matrix of the M-by-1 tap-input vector $\mathbf{u}(n)$; it has the expanded form shown in the second line of Eq. (6.7) by virtue of the assumed wide-sense stationarity of the input process. The M-by-1 vector $\mathbf{r}^{B*}$ is the cross-correlation vector between the input vector $\mathbf{u}(n)$ and the desired response $u(n - M)$; here again, the assumed wide-sense stationarity of the input process means that the vector $\mathbf{r}$ has the expanded form shown in the second line of Eq. (6.8). The $(M + 1)$-by-$(M + 1)$ matrix on the left side of Eq. (6.38) equals the correlation matrix of the tap inputs in the backward prediction-error filter of Fig. 6.2(c). The partitioning of this $(M + 1)$-by-$(M + 1)$ into the form shown in Eq. (6.38) was discussed in Section 2.3.

We may also express the matrix relation of Eq. (6.38) as a system of $(M + 1)$ simultaneous equations

$$\sum_{l=0}^{M} a^*_{M,M-l} r(l-i) = \begin{cases} 0, & i = 0, \ldots, M-1 \\ P_M, & i = M \end{cases} \tag{6.39}$$

We refer to Eq. (6.38) or (6.39) as the *augmented Wiener–Hopf equations* of a backward prediction-error filter of order M.

Note that in the matrix form of the augmented Wiener–Hopf equations for backward prediction defined by Eq. (6.38) the correlation matrix of the tap inputs is equivalent to that in the corresponding equation (6.14). This is merely a restatement of the fact that the tap inputs in the backward prediction-error filter of Fig. 6.2(c) are exactly the same as those in the forward prediction-error filter of Fig. 6.1(b).

6.3 LEVINSON–DURBIN ALGORITHM

We now describe a direct method for computing the prediction-error filter coefficients and prediction-error power by solving the augmented Wiener-Hopf equations. The method is recursive in nature and makes particular use of the Toeplitz structure of the correlation matrix of the tap inputs of the filter. It is known as the *Levinson–Durbin algorithm,* so named in recognition of its use first by Levinson (1947) and then its independent reformulation at a later date by Durbin (1960). Basically, the procedure utilizes the solution of the augmented Wiener–Hopf equations for a prediction-error filter of order $m - 1$ to compute the corresponding solution for a prediction-error filter of order m (i.e., one order higher). The order $m = 1, 2, \ldots, M$, where M is the *final* order of the filter. The important virtue of the Levinson–Durbin algorithm is its computational efficiency, in that its use results in a big saving in the number of operations (multiplications or divisions) and storage locations compared to standard methods such as the Gauss elimination method (Makhoul, 1975). To derive the Levinson–Durbin recursive procedure, we will use the matrix formulation of both forward and backward predictions in an elegant way (Burg, 1968, 1975).

Let the $(m + 1)$-by-1 vector $\mathbf{a}_m$ denote the tap-weight vector of a forward prediction-error filter of order m. The $(m + 1)$-by-1 tap-weight vector of the corresponding backward prediction-error filter is obtained by backward rearrangement of the elements of vector $\mathbf{a}_m$ and their complex conjugation. We denote the combined effect of these two operations by $\mathbf{a}_m^{B*}$. Let the m-by-1 vectors $\mathbf{a}_{m-1}$ and $\mathbf{a}_{m-1}^{B*}$ denote the tap-weight vectors of the corresponding forward and backward prediction-error filters of order $m - 1$, respectively. The Levinson–Durbin recursion may be stated in one of two equivalent ways:

1. The tap-weight vector of a *forward* prediction-error filter may be order-updated as follows

$$\mathbf{a}_m = \begin{bmatrix} \mathbf{a}_{m-1} \\ 0 \end{bmatrix} + \Gamma_m \begin{bmatrix} 0 \\ \mathbf{a}_{m-1}^{B*} \end{bmatrix} \tag{6.40}$$

where Γ_m is a constant. The scalar version of this *order update* is

$$a_{m,k} = a_{m-1,k} + \Gamma_m a_{m-1,m-k}^{*}, \qquad k = 0, 1, \ldots, m \tag{6.41}$$

where $a_{m,k}$ is the kth tap weight of a forward prediction-error filter of order m, and likewise for $a_{m-1,k}$. The element $a_{m-1,m-k}^{*}$ is the kth tap weight of a backward prediction-error filter of order $m - 1$. In Eq. (6.41), note that $a_{m-1,0} = 1$ and $a_{m-1,m} = 0$.

2. The tap-weight vector of a *backward* prediction-error filter may be order-updated as follows:

$$\mathbf{a}_m^{B*} = \begin{bmatrix} 0 \\ \mathbf{a}_{m-1}^{B*} \end{bmatrix} + \Gamma_m^* \begin{bmatrix} \mathbf{a}_{m-1} \\ 0 \end{bmatrix} \tag{6.42}$$

The scalar version of this order update is

$$a_{m,m-k}^* = a_{m-1,m-k}^* + \Gamma_m^* a_{m-1,k}, \qquad k = 0, 1, \ldots, m \tag{6.43}$$

where $a_{m,m-k}^*$ is the kth tap weight of the backward prediction-error filter of order m, and the other elements are as defined previously.

The Levinson–Durbin recursion is usually formulated in the context of forward prediction, in vector form as in Eq. (6.40) or scalar form as in Eq. (6.41). The formulation of the recursion in the context of backward prediction, in vector form as in Eq. (6.42) or scalar form as in Eq. (6.43), follows directly from that of Eq. (6.40) or (6.41), respectively, through a combination of backward rearrangement and complex conjugation (see Problem 8).

To establish the condition that the constant Γ_m has to satisfy in order to justify the validity of the Levinson–Durbin algorithm, we proceed in four stages as follows:

1. We premultiply both sides of Eq. (6.40) by $\mathbf{R}_{m+1}$, the $(m + 1)$-by-$(m + 1)$ correlation matrix of the tap inputs $u(n), u(n - 1), \ldots, u(n - m)$ in the forward prediction-error filter of order m. For the left side of Eq. (6.40), we thus get [see Eq. (6.14)]

$$\mathbf{R}_{m+1}\mathbf{a}_m = \begin{bmatrix} P_m \\ \mathbf{0}_m \end{bmatrix} \tag{6.44}$$

where P_m is the forward prediction-error power, and $\mathbf{0}_m$ is the m-by-1 null vector. The subscripts in the matrix $\mathbf{R}_{m+1}$ and the vector $\mathbf{0}_m$ refer to their dimensions, whereas the subscripts in the vector $\mathbf{a}_m$ and the scalar P_m refer to prediction order.

2. For the first term on the right side of Eq. (6.40), we use the following partitioned form of the correlation matrix $\mathbf{R}_{m+1}$ (see Section 2.3).

$$\mathbf{R}_{m+1} = \begin{bmatrix} \mathbf{R}_m & \mathbf{r}_m^{B*} \\ \mathbf{r}_m^{BT} & r(0) \end{bmatrix}$$

where $\mathbf{R}_m$ is the m-by-m correlation matrix of the tap inputs $u(n), u(n - 1), \ldots, u(n - m + 1)$, and $\mathbf{r}_m^{B*}$ is the cross-correlation vector between these tap inputs and desired response $u(n - m)$. We may thus write

$$\begin{aligned} \mathbf{R}_{m+1}\begin{bmatrix} \mathbf{a}_{m-1} \\ 0 \end{bmatrix} &= \begin{bmatrix} \mathbf{R}_m & \mathbf{r}_m^{B*} \\ \mathbf{r}_m^{BT} & r(0) \end{bmatrix}\begin{bmatrix} \mathbf{a}_{m-1} \\ 0 \end{bmatrix} \\ &= \begin{bmatrix} \mathbf{R}_m\mathbf{a}_{m-1} \\ \mathbf{r}_m^{BT}\mathbf{a}_{m-1} \end{bmatrix} \end{aligned} \tag{6.45}$$

The set of augmented Wiener–Hopf equations for the forward prediction-error filter of order $m - 1$ is

$$\mathbf{R}_m\mathbf{a}_{m-1} = \begin{bmatrix} P_{m-1} \\ \mathbf{0}_{m-1} \end{bmatrix} \tag{6.46}$$

where P_{m-1} is the prediction-error power for this filter, and $\mathbf{0}_{m-1}$ is the $(m-1)$-by-1 null vector. Define the scalar

$$\begin{aligned}\Delta_{m-1} &= \mathbf{r}_m^{BT}\mathbf{a}_{m-1} \\ &= \sum_{k=0}^{m-1} a_{m-1,k}r(k-m)\end{aligned} \tag{6.47}$$

Substituting Eqs. (6.46) and (6.67) in (6.45), we may therefore write

$$\mathbf{R}_{m+1}\begin{bmatrix}\mathbf{a}_{m-1} \\ 0\end{bmatrix} = \begin{bmatrix}P_{m-1} \\ \mathbf{0}_{m-1} \\ \Delta_{m-1}\end{bmatrix} \tag{6.48}$$

3. For the second term on the right side of Eq. (6.40), we use the following partitioned form of the correlation matrix $\mathbf{R}_{m+1}$ (see Section 2.3):

$$\mathbf{R}_{m+1} = \begin{bmatrix}r(0) & \mathbf{r}_m^H \\ \mathbf{r}_m & \mathbf{R}_m\end{bmatrix}$$

where $\mathbf{R}_m$ is the m-by-m correlation matrix of the tap inputs $u(n-1)$, $u(n-2)$, . . . , $u(n-m)$, and $\mathbf{r}_m$ is the m-by-1 cross-correlation vector between these tap inputs and desired response $u(n)$. We may thus write

$$\begin{aligned}\mathbf{R}_{m+1}\begin{bmatrix}0 \\ \mathbf{a}_{m-1}^{B*}\end{bmatrix} &= \begin{bmatrix}r(0) & \mathbf{r}_m^H \\ \mathbf{r}_m & \mathbf{R}_m\end{bmatrix}\begin{bmatrix}0 \\ \mathbf{a}_{m-1}^{B*}\end{bmatrix} \\ &= \begin{bmatrix}\mathbf{r}_m^H\mathbf{a}_{m-1}^{B*} \\ \mathbf{R}_m\mathbf{a}_{m-1}^{B*}\end{bmatrix}\end{aligned} \tag{6.49}$$

The scalar $\mathbf{r}_m^H\mathbf{a}_{m-1}^{B*}$ equals

$$\begin{aligned}\mathbf{r}_m^H\mathbf{a}_{m-1}^{B*} &= \sum_{l=1}^{m} r(l)a_{m-1,m-1}^{*} \\ &= \Delta_{m-1}^{*}\end{aligned} \tag{6.50}$$

Also, the set of augmented Wiener–Hopf equations for the backward prediction-error filter of order $m-1$ states that

$$\mathbf{R}_m\mathbf{a}_{m-1}^{B*} = \begin{bmatrix}\mathbf{0}_{m-1} \\ P_{m-1}\end{bmatrix} \tag{6.51}$$

Substituting Eqs. (6.50) and (6.51) in (6.49), we may therefore write

$$\mathbf{R}_{m+1}\begin{bmatrix}0 \\ \mathbf{a}_{m-1}^{B*}\end{bmatrix} = \begin{bmatrix}\Delta_{m-1}^{*} \\ \mathbf{0}_{m-1} \\ P_{m-1}\end{bmatrix} \tag{6.52}$$

4. Summarizing the results obtained in stages 1, 2, and 3 and, in particular, using Eqs. (6.44), (6.48), and (6.52), we may now state that the premultiplication of both sides of Eq. (6.40) by the correlation matrix $\mathbf{R}_{m+1}$ yields

$$\begin{bmatrix} P_m \\ \mathbf{0}_m \end{bmatrix} = \begin{bmatrix} P_{m-1} \\ \mathbf{0}_{m-1} \\ \Delta_{m-1} \end{bmatrix} + \Gamma_m \begin{bmatrix} \Delta^*_{m-1} \\ \mathbf{0}_{m-1} \\ P_{m-1} \end{bmatrix} \tag{6.53}$$

We conclude therefore that, if the order-update recursion of Eq. (6.40) holds, the results described by Eq. (6.53) are direct consequences of this recursion. Conversely, we may state that, if the conditions described by Eq. (6.53) apply, the tap-weight vector of a forward prediction-error filter may be order-updated as in Eq. (6.40).

From Eq. (6.53), we may make two important deductions:

1. By considering the first elements of the vectors on the left and right sides of Eq. (6.53), we have

$$P_m = P_{m-1} + \Gamma_m \Delta^*_{m-1} \tag{6.54}$$

2. By considering the last elements of the vectors on the left and right sides of Eq. (6.53), we have

$$0 = \Delta_{m-1} + \Gamma_m P_{m-1} \tag{6.55}$$

From Eq. (6.55), we see that the constant Γ_m has the value

$$\Gamma_m = -\frac{\Delta_{m-1}}{P_{m-1}} \tag{6.56}$$

where Δ_{m-1} is itself defined by Eq. (6.47). Furthermore, eliminating Δ_{m-1} between Eqs. (6.54) and (6.55), we get the following relation for the order update of the prediction-error power:

$$P_m = P_{m-1}(1 - |\Gamma_m|^2) \tag{6.57}$$

If the order m of the prediction-error filter increases, the corresponding value of the prediction-error power P_m normally decreases or else remains the same. Of course, P_m can never be negative. Hence, we must always have

$$0 \le P_m \le P_{m-1}, \qquad m \ge 1 \tag{6.58}$$

For the elementary case of a prediction-error filter of order zero, we have

$$P_0 = r(0)$$

where $r(0)$ is the autocorrelation function of the input process for lag zero.

Starting with $m = 0$, and increasing the filter order by 1 at a time, we find that through the repeated application of Eq. (6.57) the prediction-error power for a prediction-error filter of *final* order M equals

$$P_M = P_0 \prod_{m=1}^{M} (1 - |\Gamma_m|^2) \tag{6.59}$$

Interpretations of the Parameters Γ_m and Δ_{m-1}

The parameters Γ_m, $1 \le m \le M$, resulting from the application of the Levinson–Durbin recursion to a prediction-error filter of final order M, are called *reflection coefficients*. The use of this term comes from the analogy of Eq. (6.57) with transmission line theory, where (in the latter context) Γ_m may be considered as the reflection coefficient at the boundary between two sections with different characteristic impedances. Note that the condition on the reflection coefficient corresponding to that of Eq. (6.58) is

$$|\Gamma_m| \le 1, \qquad \text{for all } m \tag{6.60}$$

From Eq. (6.41), we see that for a prediction-error filter of order m, the reflection coefficient Γ_m equals the *last* tap-weight $a_{m,m}$ of the filter. That is,

$$\Gamma_m = a_{m,m}$$

As for the parameter Δ_{m-1}, it may be interpreted as a cross-correlation between the forward prediction error $f_{m-1}(n)$ and the delayed backward prediction error $b_{m-1}(n-1)$. Specifically, we may write (see Problem 9)

$$\Delta_{m-1} = E[b_{m-1}(n-1)f^*_{m-1}(n)] \tag{6.61}$$

where $f_{m-1}(n)$ is produced at the output of a forward prediction-error filter of order $m-1$ in response to the tap inputs $u(n)$, $u(n-1)$ . . . , $u(n-m+1)$, and $b_{m-1}(n-1)$ is produced at the output of a backward prediction-error filter of order $m-1$ in response to the tap inputs $u(n-1)$, $u(n-2)$, . . . , $u(n-m)$.

Note that

$$f_0(n) = b_0(n) = u(n)$$

where $u(n)$ is the prediction-error filter input at time n. Accordingly, from Eq. (6.61) we find that the cross-correlation parameter has the *zero-order value*

$$\begin{aligned}\Delta_0 &= E[b_0(n-1)f_0^*(n)]\\ &= E[u(n-1)u^*(n)]\\ &= r^*(1)\end{aligned}$$

where $r(1)$ is the autocorrelation function of the input for a lag of 1.

We may also use Eqs. (6.56) and (6.61) to develop a second interpretation for the parameter Γ_m. In particular, since P_{m-1} may be viewed as the mean-square value of the forward prediction error $f_{m-1}(n)$, we may write

$$\Gamma_m = -\frac{E[b_{m-1}(n-1)f^*_{m-1}(n)]}{E[|f_{m-1}(n)|^2]} \tag{6.62}$$

The right side of Eq. (6.62), except for the minus sign, is referred to as a *partial correlation (PARCOR) coefficient*. This terminology is widely used in the statistical literature (Box and Jenkins, 1976). Hence, the reflection coefficient, as defined here, is the negative of the PARCOR coefficient.

Application of the Levinson–Durbin Algorithm

There are two possible ways of applying the Levinson–Durbin algorithm to compute the prediction-error filter coefficients $a_{M,k}$, $k = 0, 1, \ldots, M$, and the prediction-error power P_M for a final prediction order M:

1. We have explicit knowledge of the autocorrelation function of the input process; in particular, we have $r(0), r(1), \ldots, r(M)$, denoting the values of the autocorrelation function for lags, $0, 1, \ldots, M$, respectively. For example, we may compute *biased* estimates of these parameters by means of the *time-average formula*

$$\hat{r}(k) = \frac{1}{N} \sum_{n=1+k}^{N} u(n)u^*(n-k), \qquad k = 0, 1, \ldots, M \tag{6.63}$$

   where N is the total length of the input time series, with $N >> M$. There are, of course, other estimators that we may use.[2] In any event, given $r(0)$, $r(1)$, $\ldots$, $r(M)$, the computation proceeds by using Eq. (6.47) for Δ_{m-1} and Eq. (6.57) for P_m. The recursion is initiated with $m = 0$, for which we have $P_0 = r(0)$ and $\Delta_0 = r^*(1)$. Note also that $a_{m,0}$ equals 1 for all m, and $a_{m,k}$ is zero for all $k > m$. The computation is terminated when $m = M$. The resulting estimates of the prediction-error filter coefficients and prediction error power obtained by using this procedure are known as the *Yule–Walker estimates*.
2. We have explicit knowledge of the reflection coefficients $\Gamma_1, \Gamma_2, \ldots, \Gamma_M$ and the autocorrelation function $r(0)$ for a lag of zero. Later in the chapter we describe a procedure for estimating these reflection coefficients directly from the given data. In this second application of the Levinson–Durbin recursion, we only need the pair of relations

$$a_{m,k} = a_{m-1,k} + \Gamma_m a^*_{m-1,m-k}, \qquad k = 0, 1, \ldots, m$$

$$P_m = P_{m-1}(1 - |\Gamma_m|^2)$$

   Here, again, the recursion is initiated with $m = 0$ and stopped when the order m reaches the final value M.

Example 2

To illustrate the second method for the application of the Levinson–Durbin recursion, suppose we are given the reflection coefficients Γ_1, Γ_2, Γ_3 and average power P_0. The problem we wish to solve is to use these parameters to determine the corresponding tap weights $a_{3,1}$, $a_{3,2}$, $a_{3,3}$ and the prediction-error power P_3 for a prediction-error filter of order 3. The application of the Levinson–Durbin recursion, described by Eqs. (6.41) and (6.57), yields the following results for $m = 1, 2, 3$:

[2] In practice, the biased estimate of Eq. (6.63) is preferred over an unbiased estimate because it yields a much lower variance for the estimate $\hat{r}(k)$ for values of the lag k close to the data length N; for more details, see Box and Jenkins (1976) and Haykin (1989).

1. Prediction-error filter order $m = 1$:

$$a_{1,0} = 1$$
$$a_{1,1} = \Gamma_1$$
$$P_1 = P_0(1 - |\Gamma_1|^2)$$

2. Prediction-error filter order $m = 2$:

$$a_{2,0} = 1$$
$$a_{2,1} = \Gamma_1 + \Gamma_2\Gamma_1^*$$
$$a_{2,2} = \Gamma_2$$
$$P_2 = P_1(1 - |\Gamma_2|^2)$$

where P_1 is as defined above.

3. Prediction-error filter order $m = 3$:

$$a_{3,0} = 1$$
$$a_{3,1} = a_{2,1} + \Gamma_3\Gamma_2^*$$
$$a_{3,2} = \Gamma_2 + \Gamma_3 a_{2,1}^*$$
$$a_{3,3} = \Gamma_3$$
$$P_3 = P_2(1 - |\Gamma_3|^2)$$

where $a_{2,1}$ and P_2 are as defined above.

The interesting point to observe from this example is that the Levinson–Durbin recursion yields not only the values of the tap weights and prediction-error power for the prediction-error filter of final order M but also the corresponding values of these parameters for the prediction-error filters of intermediate orders $M - 1, \ldots, 1$.

Inverse Levinson–Durbin Algorithm

In the normal application of the Levinson–Durbin recursion, as illustrated in Example 2, we are given the set of reflection coefficients $\Gamma_1, \Gamma_2, \ldots, \Gamma_M$ and the requirement is to compute the corresponding set of tap weights $a_{M,1}, a_{M,2}, \ldots, a_{M,M}$ for a prediction-error filter of final order M. Of course, the remaining coefficient of the filter, $a_{M,0} = 1$. Frequently, however, the need arises to solve the following *inverse* problem: Given the set of tap weights $a_{M,1}, a_{M,2}, \ldots, a_{M,M}$, solve for the corresponding set of reflection coefficients $\Gamma_1, \Gamma_2, \ldots, \Gamma_M$. We may solve this problem by applying the inverse form of Levinson–Durbin recursion, which we refer to simply as the *inverse recursion*.

To derive the inverse recursion, we first combine Eqs. (6.41) and (6.43), representing the scalar versions of the Levinson–Durbin recursion for forward and backward prediction-error filters, respectively, in matrix form as follows:

$$\begin{bmatrix} a_{m,k} \\ a_{m,m-k}^* \end{bmatrix} = \begin{bmatrix} 1 & \Gamma_m \\ \Gamma_m^* & 1 \end{bmatrix} \begin{bmatrix} a_{m-1,k} \\ a_{m-1,m-k}^* \end{bmatrix}, \qquad k = 0, 1, \ldots, m \tag{6.64}$$

where the order $m = 1, 2, \ldots, M$. Then, solving Eq. (6.64) for the tap weight $a_{m-1,k}$, we get

$$a_{m-1,k} = \frac{a_{m,k} - a_{m,m}a^*_{m,m-k}}{1 - |a_{m,m}|^2}, \qquad k = 0, 1, \ldots, m \tag{6.65}$$

where we have used the fact that $\Gamma_m = a_{m,m}$. We may now describe the procedure. Starting with the set of tap weights $\{a_{M,k}\}$ for which the prediction-error filter order equals M, we use the inverse recursion, Eq. (6.65), with decreasing filter order $m = M, M - 1, \ldots, 2$ to compute the tap weights of the corresponding prediction-error filters of order $M - 1, M - 2, \ldots, 1$, respectively. Finally, knowing the tap weights of all the prediction-error filters of interest (whose order ranges all the way from M down to 1), we use the fact that

$$\Gamma_m = a_{m,m}, \qquad m = M, M - 1, \ldots, 1$$

to determine the desired set of reflection coefficients $\Gamma_M, \Gamma_{M-1}, \ldots, \Gamma_1$. Example 3 illustrates the application of the inverse recursion.

Example 3

Suppose we are given the tap weights $a_{3,1}, a_{3,2}, a_{3,3}$ of a prediction-error filter of order 3, and the requirement is to determine the corresponding reflection coefficients $\Gamma_1, \Gamma_2, \Gamma_3$. Application of the inverse recursion, described by Eq. (6.65), for filter order $m = 3, 2$ yields the following sets of tap weights:

1. Prediction-error filter of order 2 [corresponding to $m = 3$ in Eq. (6.65)]:

$$a_{2,1} = \frac{a_{3,1} - a_{3,3}a^*_{3,2}}{1 - |a_{3,3}|^2}$$

$$a_{2,2} = \frac{a_{3,2} - a_{3,3}a^*_{3,1}}{1 - |a_{3,3}|^2}$$

2. Prediction-error filter of order 1 [corresponding to $m = 2$ in Eq. (6.65)]:

$$a_{1,1} = \frac{a_{2,1} - a_{2,2}a^*_{2,1}}{1 - |a_{2,2}|^2}$$

where $a_{2,1}$ and $a_{2,2}$ are as defined above. Thus, the desired reflection coefficients are given by

$$\Gamma_3 = a_{3,3}$$

$$\Gamma_2 = a_{2,2}$$

$$\Gamma_1 = a_{1,1}$$

where $a_{3,3}$ is given, and $a_{2,2}$ and $a_{1,1}$ are computed as shown above.

6.4 RELATIONSHIPS BETWEEN THE AUTOCORRELATION FUNCTION AND THE REFLECTION COEFFICIENTS

It is customary to represent the second-order statistics of a stationary time series in terms of its autocorrelation function or, equivalently, the power spectrum. The autocorrelation function and power spectrum form a discrete-time Fourier transform pair (see Chapter 3). Another way of describing the second-order statistics of a stationary time series is to use the set of numbers $P_0, \Gamma_1, \Gamma_2, \ldots, \Gamma_M$, where $P_0 = r(0)$ is the value of the autocorrelation function of the process for a lag of zero, and $\Gamma_1, \Gamma_2, \ldots, \Gamma_M$ are the reflection coefficients for a prediction-error filter of final order M. This is a consequence of the fact that the set of numbers $P_0, \Gamma_1, \Gamma_2, \ldots, \Gamma_M$ uniquely determines the corresponding set of autocorrelation function values $r(0)$, $r(1), \ldots, r(M)$.

To prove this relationship, we first eliminate Δ_{m-1} between Eqs. (6.47) and (6.55), obtaining

$$\sum_{k=0}^{m-1} a_{m-1,k} r(k-m) = -\Gamma_m P_{m-1} \tag{6.66}$$

Solving Eq. (6.66) for $r(m) = r^*(-m)$ and recognizing that $a_{m-1,0}$ equals 1, we get

$$r(m) = -\Gamma_m^* P_{m-1} - \sum_{k=1}^{m-1} a_{m-1,k}^* r(m-k) \tag{6.67}$$

This is the desired recursive relation. If we are given the set of numbers $r(0), \Gamma_1, \Gamma_2, \ldots, \Gamma_M$, then by using Eq. (6.67), together with the Levinson–Durbin recursive equations (6.41) and (6.57), we may recursively generate the corresponding set of numbers $r(0), r(1), \ldots, r(M)$.

For the case when $|\Gamma_m| \leq 1$, we find from Eq. (6.67) that the permissible region for $r(m)$, the value of the autocorrelation function of the input signal for a lag of m, is the interior (including circumference) of a circle of radius P_{m-1} and center at

$$-\sum_{k=1}^{m-1} a_{m-1,k}^* r(m-k)$$

This is illustrated in Fig. 6.3.

Suppose now that we are given the set of autocorrelation function values $r(1), \ldots, r(M)$. Then we may recursively generate the corresponding set of numbers $\Gamma_1, \Gamma_2, \ldots, \Gamma_M$ by using

$$\Gamma_m = -\frac{1}{P_{m-1}} \sum_{k=0}^{m-1} a_{m-1,k} r(k-m) \tag{6.68}$$

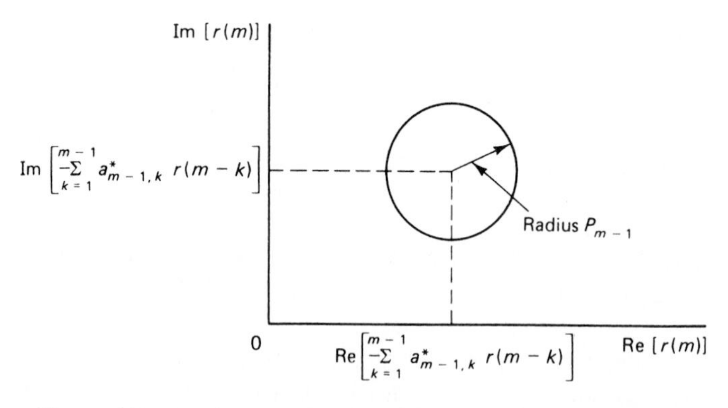

Figure 6.3 Permissible region for $r(m)$ for the case when $\Gamma_m \leq 1$.

which is obtained by solving Eq. (6.66) for Γ_m. In Eq. (6.68), it is assumed that P_{m-1} is nonzero. If P_{m-1} is zero, this would have been the result of $|\Gamma_{m-1}| = 1$, and the sequence of reflection coefficients $\Gamma_1, \Gamma_2, \ldots, \Gamma_{m-1}$ is terminated.

We conclude, therefore, that there is a one-to-one correspondence between the two sets of quantities $\{P_0, \Gamma_1, \Gamma_2, \ldots, \Gamma_M\}$ and $\{r(0), r(1), r(2), \ldots, r(M)\}$, in that if we are given the one we may uniquely determine the other in a recursive manner.

Example 4

Suppose that we are given P_0, and $\Gamma_1, \Gamma_2, \Gamma_3$ and the requirement is to compute $r(0)$, $r(1)$, $r(2)$, and $r(3)$. We start with $m = 1$, for which Eq. (6.67) yields

$$r(1) = -P_0\Gamma_1^*$$

where

$$P_0 = r(0)$$

For $m = 2$, the use of Eq. (6.67) yields

$$r(2) = -P_1\Gamma_2^* - r(1)\Gamma_1^*$$

where

$$P_1 = P_0(1 - |\Gamma_1|^2)$$

Finally, for $m = 3$, the use of Eq. (6.67) yields

$$r(3) = -P_2\Gamma_3^* - [a_{2,1}^* r(2) + \Gamma_2^* r(1)]$$

where

$$P_2 = P_1(1 - |\Gamma_2|^2)$$

$$a_{2,1} = \Gamma_1 + \Gamma_2\Gamma_1^*$$

6.5 TRANSFER FUNCTION OF A FORWARD PREDICTION-ERROR FILTER

Let $H_{f,m}(z)$ denote the *transfer function* of a forward prediction-error filter of order m, and whose impulse response is defined by the sequence of numbers $a_{m,k}^*$, $k = 0, 1, \ldots, m$, as illustrated in Fig. 6.1(b) for $m = M$. By definition, the transfer function of a discrete-time filter equals the *z-transform* of its impulse response (Oppenheim and Schafer, 1975). We may therefore write

$$H_{f,m}(z) = \sum_{k=0}^{m} a_{m,k}^* z^{-k} \tag{6.69}$$

where z^{-1} is the unit-sample delay. Based on the Levinson–Durbin recursion, in particular Eq. (6.41), we may relate the coefficients of this filter of order m to those of a corresponding prediction-error filter of order $m - 1$ (i.e., one order smaller). Therefore, substituting Eq. (6.41) into (6.69), we get

$$H_{f,m}(z) = \sum_{k=0}^{m} a^*_{m-1,k} z^{-k} + \Gamma^*_m \sum_{k=0}^{m} a_{m-1,m-k} z^{-k}$$
$$= \sum_{k=0}^{m-1} a^*_{m-1,k} z^{-k} + \Gamma^*_m z^{-1} \sum_{k=0}^{m-1} a_{m-1,m-1-k} z^{-k} \tag{6.70}$$

where, in the second line, we have used the fact that $a_{m-1,m} = 0$. The sequence of numbers $a^*_{m-1,k}$, $k = 0, 1, \ldots, m-1$, defines the impulse response of a forward prediction-error filter of order $m-1$. Hence, we may write

$$H_{f,m-1}(z) = \sum_{k=0}^{m-1} a^*_{m-1,k} z^{-k} \tag{6.71}$$

The sequence of numbers $a_{m-1,m-1-k}$, $k = 0, 1, \ldots, m-1$, defines the impulse response of a backward prediction-error filter of order $m-1$; this is illustrated in Fig. 6.2(c) for the case of prediction order $m = M$. Hence, the second summation on the right side of Eq. (6.70) represents the transfer function of this backward prediction-error filter. Let $H_{b,m-1}(z)$ denote this transfer function, as shown by

$$H_{b,m-1}(z) = \sum_{k=0}^{m-1} a_{m-1,m-1-k} z^{-k} \tag{6.72}$$

Hence, substituting Eqs. (6.71) and (6.72) in (6.70), we may write

$$H_{f,m}(z) = H_{f,m-1}(z) + \Gamma^*_m z^{-1} H_{b,m-1}(z) \tag{6.73}$$

The order update recursion of Eq. (6.73) shows that given the reflection coefficient Γ_m and the transfer functions of the forward and backward prediction-error filters of order $m-1$, the transfer function of the corresponding forward prediction-error filter of order m is uniquely determined.

Minimum-Phase Property

On the unit circle in the z-plane (i.e., for $|z| = 1$), we find that

$$|H_{f,m-1}(z)| = |H_{b,m-1}(z)|, \qquad |z| = 1$$

This is readily proved by substituting $z = \exp(j\omega)$, $-\pi \le \omega \le \pi$, in Eqs. (6.71) and (6.72). Suppose that the reflection coefficient Γ_m satisfies the requirement $|\Gamma_m| < 1$ for all m. Then we find that on the unit circle in the z-plane the magnitude of the second term in the right side of Eq. (6.73) satisfies the conditions

$$|\Gamma^*_m z^{-1} H_{b,m-1}(z)| < |H_{b,m-1}(z)| = |H_{f,m-1}(z)|, \qquad |z| = 1 \tag{6.74}$$

At this stage in our discussion, it is useful to recall *Rouché's theorem* from the theory of complex variables.[3] Rouché's theorem states:

> If a function $F(z)$ is analytic upon a contour C in the z-plane and within the region enclosed by this contour, and if a second function $G(z)$, in addition to satisfying the same analyticity conditions, also fulfills the condition $|G(z)| < |F(z)|$ on the contour C, then the function $F(z) + G(z)$ has the same number of zeros within the region enclosed by the contour C as does the function $F(z)$.

[3] For a review of complex variable theory, including Rouché's theorem, see Appendix A.

Ordinarily, the enclosed contour C is transversed in the *counterclockwise* direction, and the region enclosed by the contour lies to the *left* of it, as illustrated in Fig. 6.4. We say that a function is analytic upon the contour C and within the region enclosed by it if the function has a continuous derivative everywhere upon the contour C and within the region enclosed by this contour. For this requirement to be satisfied, the function must have no poles upon the contour C or inside the region enclosed by the contour.

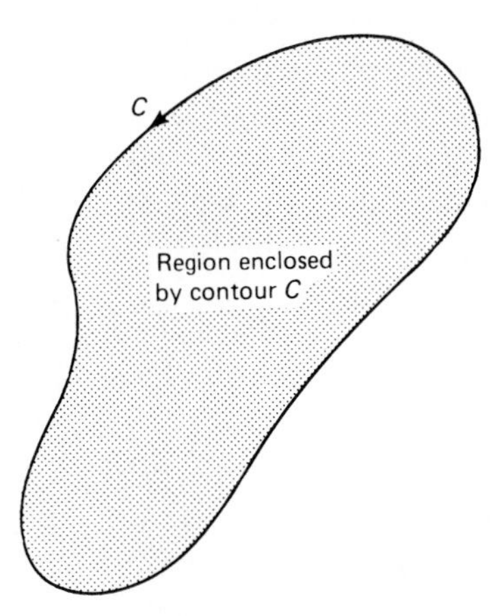

Figure 6.4 Contour C (traversed in the counterclockwise direction) and the region enclosed by it.

Let the contour C be the unit circle in the z-plane, which is traversed in the *clockwise* direction, as in Fig. 6.5. According to the convention described herein, this assumption implies that the region enclosed by the contour C is represented by the entire part of the z-plane that lies *outside* the unit circle.

Let the functions $F(z)$ and $G(z)$ be identified with the two terms in the right side of Eq. (6.73), as shown by

$$F(z) = H_{f,m-1}(z) \tag{6.75}$$

$$G(z) = \Gamma_m^* z^{-1} H_{b,m-1}(z) \tag{6.76}$$

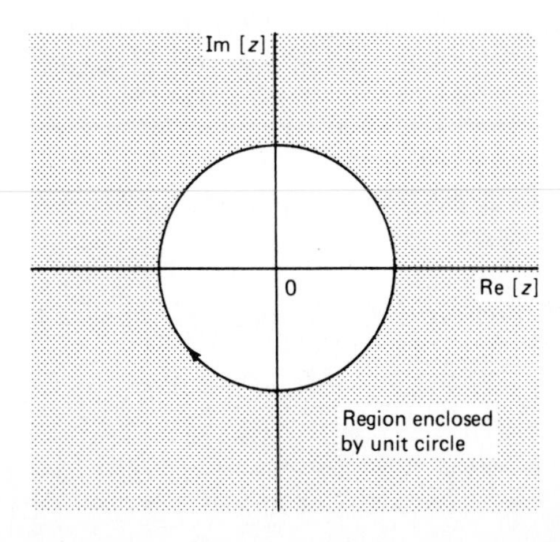

Figure 6.5 Unit circle (traversed in the clockwise direction) used as contour C.

We observe that:

1. The functions $F(z)$ and $G(z)$ have no poles inside the contour C defined in Fig. 6.5. Indeed, their derivatives are continuous throughout the region enclosed by this contour. Therefore, they are both analytic everywhere upon the unit circle and the region outside it.
2. In view of Eq. (6.74), we have $|G(z)| < |F(z)|$ on the unit circle.

Accordingly, the functions $F(z)$ and $G(z)$ defined by Eqs. (6.75) and (6.76), respectively, satisfy all the conditions required by Rouché's theorem with respect to the contour C defined as the unit circle in Fig. 6.5.

Suppose that $H_{f,m-1}(z)$ and therefore $F(z)$ are known to have no zeros outside the unit circle in the z-plane. Then, by applying Rouché's theorem, we find that $F(z) + G(z)$, or, equivalently, $H_{f,m}(z)$ also has no zeros on or outside the unit circle in the z-plane.

In particular, for $m = 0$, the transfer function $H_{f,0}(z)$ is a constant equal to 1; therefore, it has no zeros at all. Using the result just derived, we may state that since $H_{f,0}(z)$ has no zeros outside the unit circle, then $H_{f,1}(z)$ will also have no zeros in this region of the z-plane, provided that $|\Gamma_1| < 1$. Indeed, we can easily prove this result by noting that

$$\begin{aligned} H_{f,1}(z) &= a_{1,0}^* + a_{1,1}^* z^{-1} \\ &= 1 + \Gamma_1^* z^{-1} \end{aligned}$$

Hence, $H_{f,1}(z)$ has a single zero located at $z = -\Gamma_1^*$ and a pole at $z = 0$. With the reflection coefficient Γ_1 constrained by the condition $|\Gamma_1| < 1$, it follows that this zero must lie inside the unit circle. In other words, $H_{f,1}(z)$ has no zeros on or outside the unit circle. If $H_{f,1}(z)$ has no zeros on or outside the unit circle, then $H_{f,2}(z)$ will also have no zeros on or outside the unit circle provided that $|\Gamma_2| < 1$, and so on.

We may thus state that the transfer function $H_{f,m}(z)$ of a forward prediction-error filter of order m has no zeros on or outside the unit circle in the z-plane for all values of m, if and only if the reflection coefficients satisfy the condition $|\Gamma_m| < 1$ for all m. Such a filter is said to be *minimum phase* in the sense that, for a specified amplitude response, it has the minimum phase response possible for all values of z on the unit circle (Oppenheim and Schafer, 1975). Moreover, the amplitude response and phase response of the filter are uniquely related to each other.

6.6 TRANSFER FUNCTION OF A BACKWARD PREDICTION-ERROR FILTER

Consider a backward prediction-error filter of order m whose impulse response is denoted by the sequence of numbers $a_{m,m-k}$, $k = 0, 1, \ldots, m$, as illustrated in Fig. 6.2(c) for $m = M$. Let $H_{b,m}(z)$ denote the transfer function of this filter:

$$H_{b,m}(z) = \sum_{k=0}^{m} a_{m,m-k} z^{-k} \tag{6.77}$$

Equation (6.43) describes the scalar version of the Levinson–Durbin recursion for a backward prediction-error filter of order m. Hence, substituting Eq. (6.43) in (6.77), we get

$$H_{b,m}(z) = z^{-1}H_{b,m-1}(z) + \Gamma_m H_{f,m-1}(z) \tag{6.78}$$

where $H_{b,m-1}$ and $H_{f,m-1}(z)$ are defined by Eqs. (6.72) and (6.71) for order $m - 1$, respectively. The order-update recursion of Eq. (6.78) shows that, given the reflection coefficient Γ_m and the transfer functions of forward and backward prediction-error filters of order $m - 1$, the transfer function of the corresponding backward prediction-error filter of order m is uniquely determined.

Relationship between the Transfer Functions of Backward and Forward Prediction-Error Filters

We also find it useful to develop the relationship between the transfer functions of backward and forward prediction-error filters of the same order, such that if we are given one we may determine the other. To do this, we first evaluate $H_{f,m}^*(z)$, the complex conjugate of the transfer function of a forward prediction-error filter of order m, and so write [see Eq. (6.69)]

$$H_{f,m}^*(z) = \sum_{k=0}^{m} a_{m,k}(z^*)^{-k} \tag{6.79}$$

Replacing z by the reciprocal of its complex conjugate z^*, we may rewrite Eq. (6.79) as

$$H_{f,m}^*\left(\frac{1}{z^*}\right) = \sum_{k=0}^{m} a_{m,k}z^k$$

Next, replacing k by $m - k$, we get

$$H_{f,m}^*\left(\frac{1}{z^*}\right) = z^m \sum_{k=0}^{m} a_{m,m-k}z^{-k} \tag{6.80}$$

The summation on the right of Eq. (6.80) equals $H_{b,m}(z)$, the transfer function of a backward prediction-error filter of order m. We thus get the desired relation:

$$H_{b,m}(z) = z^{-m}H_{f,m}^*\left(\frac{1}{z^*}\right) \tag{6.81}$$

where $H_{f,m}^*(1/z^*)$ is obtained by complex-conjugating $H_{f,m}(z)$, the transfer function of a forward prediction-error filter of order m, and replacing z by the reciprocal of z^*. Equation (6.81) states that multiplication of the new function obtained in this way by z^{-m} yields $H_{b,m}(z)$, the transfer function of the corresponding backward prediction-error filter.

Maximum-Phase Property

Let the transfer function $H_{f,m}(z)$ be expressed in its factored form as follows:

$$H_{f,m}(z) = \prod_{i=1}^{m} (1 - z_i z^{-1}) \tag{6.82}$$

where z_i, $i = 1, 2, \ldots, m$, denote the zeros of the forward prediction-error filter. Hence, substituting Eq. (6.82) into (6.81), we may express the transfer function of the corresponding backward prediction-error filter in the factored form

$$\begin{aligned} H_{b,m}(z) &= z^{-m} \prod_{i=1}^{m} (1 - z_i^* z) \\ &= \prod_{i=1}^{m} (z^{-1} - z_i^*) \end{aligned} \tag{6.83}$$

The zeros of this transfer function are located at $1/z_i^*$, $i = 1, 2, \ldots, m$. That is, the zeros of the backward and forward prediction-error filters are the *inverse* of each other with respect to the unit circle in the z-plane. The geometric nature of this relationship is illustrated for $m = 1$ in Fig. 6.6. The forward prediction-error filter has a zero at $z = -\Gamma_1^*$, as in Fig. 6.6(a), and the backward prediction-error filter has a zero at $z = -1/\Gamma_1$, as in Fig. 6.6(b). In this figure, it is assumed that the reflection coefficient Γ_1 has a complex value.

Consequently, backward prediction-error filters have all their zeros located outside the unit circle in the z-plane, because $|\Gamma_m| < 1$ for all m. Such filters are said to be *maximum phase* (Oppenheim and Schafer, 1975).

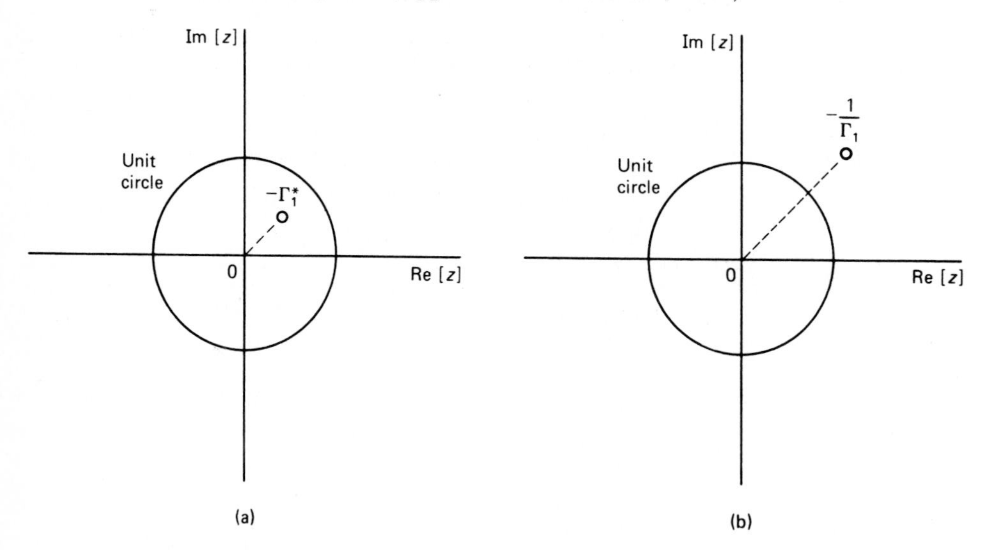

Figure 6.6 (a) Zero of forward prediction-error filter at $z = -\Gamma_1^*$; (b) corresponding zero of backward prediction-error filter at $z = -1/\Gamma_1$.

Relationship between the Backward and Forward Prediction Errors

The transfer function $H_{b,m}(z)$ of a backward prediction-error filter of order m equals (by definition)

$$H_{b,m}(z) = \frac{B_m(z)}{U(z)} \tag{6.84}$$

where $B_m(z)$ is the z-transform of the backward prediction error sequence $\{b_m(n)\}$ at the output of the filter, and $U(z)$ is the z-transform of the signal $\{u(n)\}$ applied to the filter input; see Fig. 6.2(c) for $m = M$. Similarly, the transfer function $H_{f,m}(z)$ of the corresponding forward prediction-error filter equals

$$H_{f,m}(z) = \frac{F_m(z)}{U(z)} \tag{6.85}$$

where $F_m(z)$ is the z-transform of the forward prediction-error sequence $\{f_m(n)\}$ at the output of the filter, and $U(z)$ is, as before, the z-transform of the filter input; see Fig. 6.2(b) for $m = M$. Solving Eqs. (6.84) and (6.85) for $B_m(z)$, we get

$$B_m(z) = \frac{H_{b,m}(z)}{H_{f,m}(z)} F_m(z) \tag{6.86}$$

To evaluate the ratio $H_{b,m}(z)/H_{f,m}(z)$, we divide Eq. (6.83) by (6.82), obtaining the result

$$\frac{H_{b,m}(z)}{H_{f,m}(z)} = \prod_{i=1}^{m} \left(\frac{z^{-1} - z_i^*}{1 - z_i z^{-1}} \right) \tag{6.87}$$

where z_i, $i = 1, 2, \ldots, m$, denote the zeros of the transfer function of the forward prediction-error filter. Owing to the minimum-phase property of this filter, we observe that $|z_i| < 1$ for all i. Hence, the poles and zeros of the transfer function in Eq. (6.87) are the inverse of each other with respect to the unit circle in the z-plane, with the poles located inside the unit circle and the zeros located outside. Figure 6.7 shows the geometry of this relationship for a typical pole-zero factor. A filter with such a transfer function is called an *all-pass filter*. It has the property that its amplitude response equals 1 for all values of z on the unit circle. In other words, it passes all frequencies with no amplitude distortion; hence, the name.

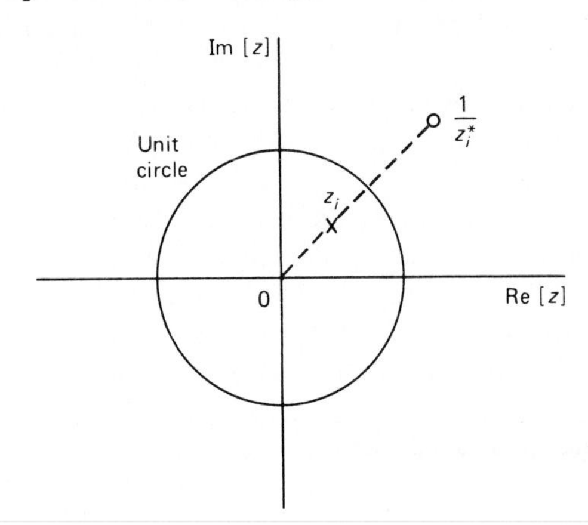

Figure 6.7 Geometry of typical pole-zero factor for all-pass filter.

Accordingly, we may convert a forward prediction-error sequence $\{f_m(n)\}$ into the corresponding backward prediction-error sequence $\{b_m(n)\}$ by passing it through an all-pass filter of order m, as in Fig. 6.8. This filter is stable, since all its poles are located inside the unit circle in the z-plane. Note, however, the opposite conversion from $\{b_m(n)\}$ to $\{f_m(n)\}$ is not practically feasible, because the all-pass filter would then have its poles located outside the unit circle and, therefore, be unstable.

Note also that, since the amplitude response of the all-pass filter in Fig. 6.8 equals 1 for all frequencies, the forward prediction-error sequence $\{f_m(n)\}$ and the

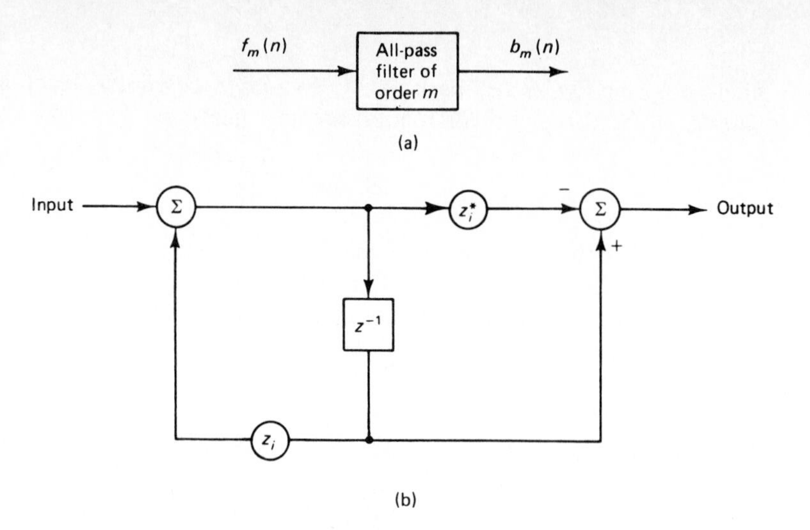

Figure 6.8 (a) Conversion of forward prediction-error sequence $\{f_m(n)\}$ into backward prediction-error sequence $\{b_m(n)\}$: (b) ith section of all-pole filter, $i = 1, 2, \ldots, m$.

backward prediction-error sequence $\{b_m(n)\}$ have exactly the same power spectral density. This is intuitively satisfying since both prediction-error sequences have the same average power for stationary inputs.

6.7 SCHUR–COHN TEST

The test described in Section 6.5 for the minimum-phase condition of a forward prediction-error of order M is relatively simple to apply if we know the associated set of reflection coefficients $\Gamma_1, \Gamma_2, \ldots, \Gamma_M$. For the filter to be minimum phase [i.e., for all the zeros of the transfer function of the filter, $H_{f,m}(z)$, to lie inside the unit circle], we simply require that $|\Gamma_m| < 1$ for all m. Suppose, however, that instead of these reflection coefficients we are given the tap weights of the filter, $a_{M,1}$, $a_{M,2}, \ldots a_{M,M}$. In this case we first apply the inverse recursion [described by Eq. (6.65)] to compute the corresponding set of reflection coefficients $\Gamma_1, \Gamma_2, \ldots, \Gamma_M$. Then, as before, we check whether or not $|\Gamma_m| < 1$ for all m.

The method just described for determining whether or not $H_{f,m}(z)$ has zeros inside the unit circle, given the coefficients $a_{M,1}, a_{M,2}, \ldots, a_{M,M}$, is essentially the same as the *Schur–Cohn test*.[4]

To formulate the Schur–Cohn test, let

$$x(z) = a_{M,M}z^M + a_{M,M-1}z^{M-1} + \cdots + a_{M,0} \tag{6.88}$$

which is a polynomial in z, with $x(0) = a_{M,0} = 1$. Define

$$\begin{aligned} x'(z) &= z^M x^*(1/z^*) \\ &= a^*_{M,M} + a^*_{M,M-1}z + \cdots + a^*_{M,0}z^M \end{aligned} \tag{6.89}$$

[4] The classical Schur–Cohn test is discussed in Marden (1949) and Tretter (1976). The origin of the test can be traced back to Schur (1917) and Cohn (1922), hence the name. The test is also referred to as the Lehmer–Schur method (Ralston, 1965); this is in recognition of the application of Schur's theorem by Lehmer (1961).

which is the *reciprocal polynomial* associated with $x(z)$. The polynomial $x(z)$ is so called since its zeros are the reciprocals of the zeros of $x(z)$. For $z = 0$, we have $x'(0) = a^*_{M,M}$. Next, define the linear combination

$$T[x(z)] = a^*_{M,0}x(z) - a_{M,M}x'(z) \tag{6.90}$$

so that, in particular, the value

$$\begin{aligned} T[x(0)] &= a^*_{M,0}x(0) - a_{M,M}x'(0) \\ &= 1 - |a_{M,M}|^2 \end{aligned} \tag{6.91}$$

is real. Note also that $T[x(z)]$ has no term in z^M. Repeat this operation as far as possible, so that if we define

$$T^i[x(z)] = T\{T^{i-1}[x(z)]\} \tag{6.92}$$

we generate a finite sequence of polynomials in z of *decreasing* order. Let it be assumed that:

1. The coefficient $a_{M,0}$ is nonzero.
2. The polynomial $x(z)$ has no zero on the unit circle.
3. The integer m is the smallest for which

$$T^m[x(z)] = 0, \text{ where } m \le M + 1$$

Then, we may state the Schur–Cohn theorem as follows (Lehmer, 1961):

If for some i such that $1 \le i < m$, we have $T^i[x(0)] < 0$, then $x(z)$ has at least one zero inside the unit circle. If, on the other hand, $T^i[x(0)] > 0$ for $1 \le i < m$, and $T^{m-1}[x(z)]$ is a constant, then no zero of $x(z)$ lies inside the unit circle.

To apply this theorem to determine whether or not the polynomial $x(z)$ of Eq. (6.88), with $a_{M,0} \ne 0$, has a zero inside the unit circle, we proceed as follows (Ralston, 1965):

1. Calculate $T[x(z)]$. Is $T[x(0)]$ negative? If so, there is a zero inside the unit circle; if not, proceed to step 2.
2. Calculate $T^i[x(z)]$, $i = 1, 2, \ldots,$ until $T^i[x(0)] < 0$ for $i < m$, or $T^i[x(0)] > 0$ for $i < m$. If the former occurs, there is a zero inside the unit circle. If the latter occurs, and if $T^{m-1}[x(z)]$ is a constant, then there is no zero inside the unit circle.

Note that when the polynomial $x(z)$ has zeros inside the unit circle, this algorithm does not tell us how many; rather, it only confirms their existence.

The connection between the Schur–Cohn method and the inverse recursion is readily established by observing that (see Problem 10):

1. The polynomial $x(z)$ is related to the transfer function of a backward prediction-error filter of order M as follows:

$$x(z) = z^M H_{b,M}(z) \tag{6.93}$$

Accordingly, if the Schur–Cohn test indicates that $x(z)$ has zero(s) inside the unit circle, we may conclude that the transfer function $H_{b,M}(z)$ is *not* maximum phase.

2. The reciprocal polynomial $x'(z)$ is related to the transfer function of the corresponding forward prediction-error filter of order M as follows:

$$x'(z) = z^M H_{f,M}(z) \tag{6.94}$$

Accordingly, if the Schur–Cohn test indicates that the original polynomial $x(z)$, with which $x'(z)$ is associated, has zero(s) inside the unit circle, we may then conclude that the transfer function $H_{f,M}(z)$ is *not* minimum phase.

3. In general, we have

$$T^i[x(0)] = \prod_{j=0}^{i-1} (1 - |a_{M-j,M-j}|^2), \qquad 1 \leq i \leq M \tag{6.95}$$

and

$$H_{b,M-i}(z) = \frac{z^{i-M} T^i[x(z)]}{T^i[x(0)]} \tag{6.96}$$

where $H_{b,M-i}(z)$ is the transfer function of the backward prediction-error filter of order $M - i$.

6.8 WHITENING PROPERTY OF PREDICTION-ERROR FILTERS

By definition, a *white-noise* process consists of a sequence of uncorrelated random variables. Thus, assuming that such a process, denoted by $\{v(n)\}$, has zero mean and variance σ_v^2, we may write (see Section 2.5)

$$E[v(k)v^*(n)] = \begin{cases} \sigma_v^2, & k = n \\ 0, & k \neq n \end{cases} \tag{6.97}$$

Accordingly, we say that white noise is purely unpredictable in the sense that the value of the process at time n is uncorrelated with all past values of the process up to and including time $n - 1$ (and, indeed, with all future values of the process, too).

We may now state another important property of a prediction-error filter. *In theory,* a prediction-error filter is capable of whitening a stationary discrete-time stochastic process applied to its input, provided that the order of the filter is high enough. Basically, prediction relies on the presence of correlation between adjacent samples of the input process. The implication of this is that, as we increase the order of the prediction-error filter, we successively reduce the correlation between adjacent samples of the input process, until ultimately we reach a point at which the filter has a high enough order to produce an output process that consists of a sequence of uncorrelated samples. The whitening of the original process applied to the filter input will have thereby been accomplished.

6.9 IMPLICATIONS OF THE WHITENING PROPERTY OF A PREDICTION-ERROR FILTER AND THE AUTOREGRESSIVE MODELING OF A STATIONARY STOCHASTIC PROCESS

The whitening property of a prediction-error filter, operating on a stationary discrete-time stochastic process, is intimately related to the *autoregressive modeling* of the process. Indeed, we may view these two operations as *complementary*, as illustrated in Fig. 6.9. Part (a) of the figure depicts a forward prediction-error filter of order M, whereas part (b) depicts the corresponding autoregressive model. We

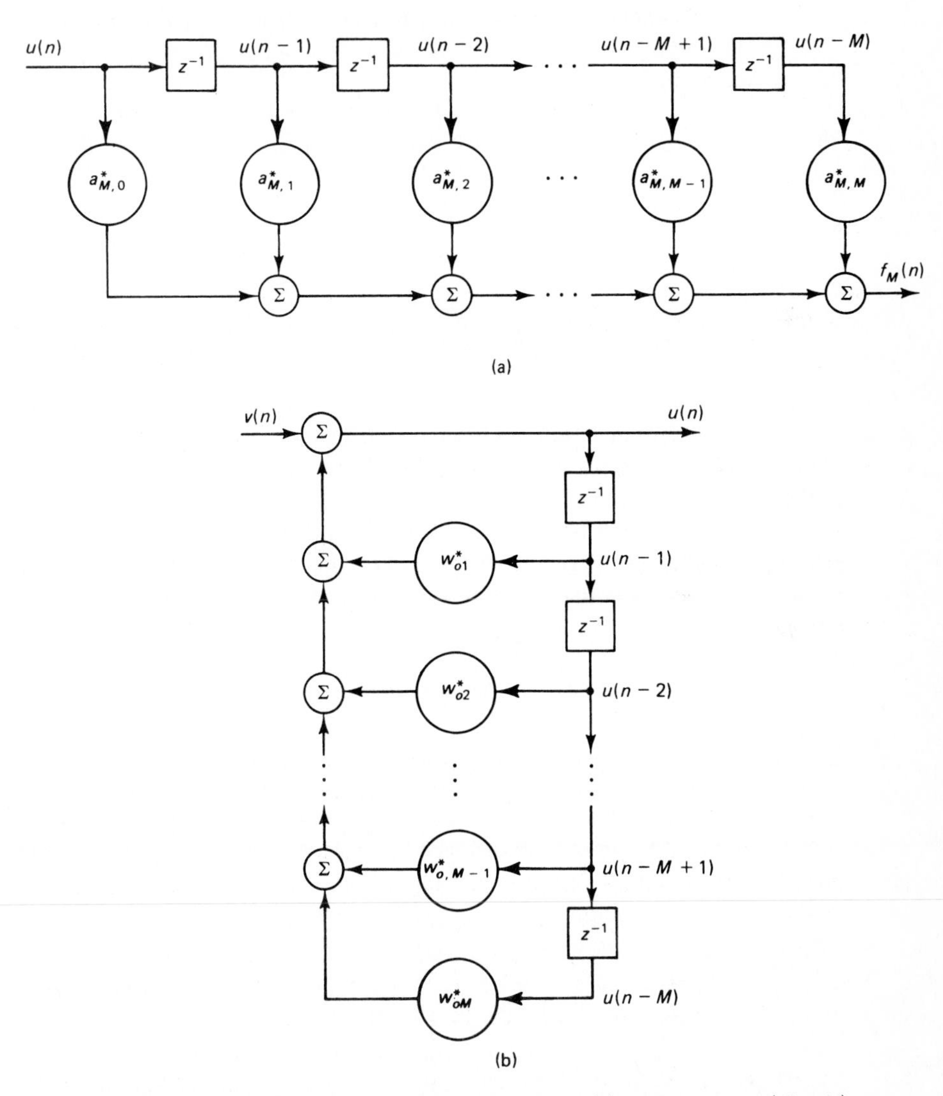

Figure 6.9 (a) Prediction-error (all-zero) filter; (b) autoregressive (all-pole) model with $w_{ok} = -a_{M,k}$, where $k = 1, 2, \ldots, M$; the input $\{v(n)\}$ is white noise.

may make the following observations:

1. We may view the operation of prediction-error filtering applied to a stationary process $\{u(n)\}$ as one of *analysis*. In particular, we may use such an operation to whiten the process $\{u(n)\}$ by choosing the prediction-error filter order M sufficiently large, in which case the prediction error process $\{f_M(n)\}$ at the filter output consists of uncorrelated samples. When this unique condition has been established, the original stochastic process $\{u(n)\}$ is represented by the tap weights of the filter, $\{a_{M,k}\}$, and the prediction error power, P_M.
2. We may view the autoregressive (AR) modeling of the stationary process $\{u(n)\}$ as one of *synthesis*. In particular, we may generate the AR process $\{u(n)\}$ by applying a white-noise process $\{v(n)\}$ of zero mean and variance σ_v^2 to the input of an *inverse* filter whose parameters are set equal to the AR parameters $\{w_{ok}\}$.

The two filter structures of Fig. 6.9 constitute a *matched pair*, with their parameters related as follows:

$$a_{M,k} = -w_{ok}, \qquad k = 1, 2, \ldots, M$$

and

$$P_M = \sigma_v^2$$

The prediction-error filter of Fig. 6.9(a) is an *all-zero filter with an impulse response of finite duration*. On the other hand, the inverse filter in the AR model of Fig. 6.9(b) is an *all-pole filter with an impulse response of infinite duration*. The prediction-error filter is minimum phase, with the zeros of its transfer function located at exactly the same positions (inside the units circle in the z-plane) as the poles of the transfer function of the inverse filter in part (b) of the figure. This assures the stability of the inverse filter or, equivalently, the asymptotic stationarity of the AR process generated at the output of this filter. The principles described herein provide the basis of *linear predictive coding* (LPC) *vocorders* for the transmission and reception of digitized speech; see Section 1.7.

Equivalence of the Autoregressive and Maximum Entropy Power Spectra

Consider the AR model of Fig. 6.9(b). The process $\{v(n)\}$ applied to the model input consists of a white noise of zero mean and variance σ^2. To find the power spectral density of the AR process $\{u(n)\}$ produced at the model output, we multiply the power spectral density of the model input $\{v(n)\}$ by the squared amplitude response of the model (see Chapter 2). Let $S_{\text{AR}}(\omega)$ denote the power spectral density of the AR process $\{u(n)\}$. We may therefore write (Priestley, 1981)

$$S_{\text{AR}}(\omega) = \frac{\sigma^2}{\left|1 - \sum_{k=1}^{M} w_{ok}^* e^{-j\omega k}\right|^2} \tag{6.98}$$

The formula of Eq. (6.98) is called the *autoregressive power spectrum* or simply the *AR spectrum*.

A power spectrum of closely related interest is obtained using the *maximum entropy method (MEM)*. Suppose that we are given $2M + 1$ values of the autocorrelation function of a wide-sense stationary stochastic process $\{u(n)\}$. The essence of the maximum entropy method is to determine the particular power spectrum of the process that corresponds to the most random time series whose autocorrelation function is consistent with the set of $2M + 1$ known values (Burg, 1968, 1975). The result so obtained is referred to as the *maximum entropy power spectrum* or simply the *MEM spectrum;* see Appendix F. Let $S_{\text{MEM}}(\omega)$ denote the MEM spectrum. The determination of $S_{\text{MEM}}(\omega)$ is linked with the characterization of a prediction-error filter of order M, as shown in Appendix F. There it is shown that

$$S_{\text{MEM}}(\omega) = \frac{P_M}{\left| 1 + \sum_{k=1}^{M} a_{M,k}^* e^{-jk\omega} \right|^2} \tag{6.99}$$

where the $a_{M,k}$ are the prediction-error filter coefficients, and P_M is the prediction error power, all of which correspond to a prediction order M.

In view of the one-to-one correspondence that exists between the prediction-error filter of Fig. 6.9(a) and the AR model of Fig. 6.9(b), we have

$$a_{M,k} = -w_{ok}, \qquad k = 1, 2, \ldots, M \tag{6.100}$$

and

$$P_M = \sigma_M^2 \tag{6.101}$$

Accordingly, the formulas given in Eqs. (6.98) and (6.99) are one and the same. In other words, for the case of a wide-sense stationary stochastic process, the AR spectrum (for model order M) and the MEM spectrum (for prediction order M) are indeed *equivalent* (Van den Bos, 1971).

6.10 EIGENVECTOR REPRESENTATIONS OF PREDICTION-ERROR FILTERS

The study of a prediction-error filter would be incomplete without considering its representation in terms of the eigenvalues (and associated eigenvectors) of the correlation matrix of the tap inputs in the filter. To develop such a representation, we first rewrite the augmented Wiener–Hopf equations (6.14), pertaining to a forward prediction-error filter of order M, in the compact matrix form

$$\mathbf{R}_{M+1}\mathbf{a}_M = P_M \mathbf{i}_{M+1} \tag{6.102}$$

where $\mathbf{R}_{M+1}$ is the $(M + 1)$-by-$(M + 1)$ correlation matrix of the tap inputs $u(n)$, $u(n - 1), \ldots, u(n - M)$ in the filter of Fig. 6.1(b), $\mathbf{a}_M$ is the $(M + 1)$-by-1 tap-weight vector of the filter, and the scalar P_M is the prediction error power. The $(M + 1)$-by-1 vector $\mathbf{i}_{M+1}$ is called the *first coordinate vector*; it has unity for its first element and zero for all the others. We illustrate this by writing

$$\mathbf{i}_{M+1} = [1, 0, \ldots, 0]^T \tag{6.103}$$

Solving Eq. (6.102) for $\mathbf{a}_M$, we get

$$\mathbf{a}_M = P_M \mathbf{R}_{M+1}^{-1} \mathbf{i}_{M+1} \tag{6.104}$$

where $\mathbf{R}_{M+1}^{-1}$ is the inverse of the correlation matrix $\mathbf{R}_{M+1}$; it is assumed that the matrix $\mathbf{R}_{M+1}$ is nonsingular, so that its inverse exists. Using the eigenvalue–eigenvector representation of a correlation matrix, which was discussed in Section 4.2, we may express the inverse matrix $\mathbf{R}_{M+1}^{-1}$ as follows:

$$\mathbf{R}_{M+1}^{-1} = \mathbf{Q}\mathbf{\Lambda}^{-1}\mathbf{Q}^H \tag{6.105}$$

where $\mathbf{\Lambda}$ is an $(M + 1)$-by-$(M + 1)$ diagonal matrix consisting of the eigenvalues of the correlation matrix $\mathbf{R}_{M+1}$, and $\mathbf{Q}$ is an $(M + 1)$-by-$(M + 1)$ matrix whose columns are the associated eigenvectors. That is,

$$\mathbf{\Lambda} = \text{diag}[\lambda_0, \lambda_1, \ldots, \lambda_M] \tag{6.106}$$

and

$$\mathbf{Q} = [\mathbf{q}_0, \mathbf{q}_1, \ldots, \mathbf{q}_M] \tag{6.107}$$

where $\lambda_0, \lambda_1, \ldots, \lambda_M$ are the real-valued eigenvalues of the correlation matrix $\mathbf{R}_{M+1}$, and $\mathbf{q}_0, \mathbf{q}_1, \ldots, \mathbf{q}_M$ are the respective eigenvectors. Thus, substituting Eqs. (6.105), (6.106), and (6.107) in (6.104), we get

$$\begin{aligned}
\mathbf{a}_M &= P_M \mathbf{Q}\mathbf{\Lambda}^{-1}\mathbf{Q}^H \mathbf{i}_{M+1} \\
&= P_M[\mathbf{q}_0, \mathbf{q}_1, \ldots, \mathbf{q}_M]\text{diag}[\lambda_0^{-1}, \lambda_1^{-1}, \ldots, \lambda_M^{-1}]\begin{bmatrix} \mathbf{q}_0^H \\ \mathbf{q}_1^H \\ \vdots \\ \mathbf{q}_M^H \end{bmatrix}\begin{bmatrix} 1 \\ 0 \\ \vdots \\ 0 \end{bmatrix} \\
&= P_M \sum_{k=0}^{M} \left(\frac{q_{k0}^*}{\lambda_k}\right)\mathbf{q}_k
\end{aligned} \tag{6.108}$$

where q_{k0} is the first element of the kth eigenvector of the correlation matrix $\mathbf{R}_{M+1}$.

We note that the first element of the forward prediction-error filter vector $\mathbf{a}_M$ equals 1. Therefore, using this fact, we find from Eq. (6.108) that the prediction-error power equals

$$P_M = \frac{1}{\sum_{k=0}^{M} |q_{k0}|^2 \lambda_k^{-1}} \tag{6.109}$$

Thus, we see from Eqs.(6.108) and (6.109) that the tap-weight vector of a forward prediction-error filter of order M and the resultant prediction-error power are uniquely defined by specifying the $(M + 1)$ eigenvalues and the corresponding $(M + 1)$ eigenvectors of the correlation matrix of the tap inputs of the filter.

From Eq. (6.108), we readily find that the tap-weight vector of the corresponding backward prediction-error filter may be represented in the eigenvalue–eigenvector form as

$$\mathbf{a}_M^{B*} = P_M \sum_{k=0}^{M} \left(\frac{q_{k0}}{\lambda_k}\right)\mathbf{q}_k^{B*} \tag{6.110}$$

where $\mathbf{q}_k^{B*}$ is obtained by rearranging the elements of the eigenvector $\mathbf{q}_k$ in reverse order and complex-conjugating them.

6.11 CHOLESKY FACTORIZATION

Consider a stack of backward prediction-error filters of orders 0 to M that are connected in parallel as in Fig. 6.10. The filters are all fed at time n, with the same input, denoted by $u(n)$. Note that for the case of zero prediction order, we simply have a direct connection as shown at the top end of Fig. 6.10. Let $b_0(n)$, $b_1(n)$, $\ldots$, $b_M(n)$ denote the sequence of backward prediction errors produced by these filters. We may express these errors in terms of the respective filter inputs and filter coefficients as follows [see Fig. 6.2(c)]:

$$\begin{aligned} b_0(n) &= u(n) \\ b_1(n) &= a_{1,1}u(n) + a_{1,0}u(n-1) \\ b_2(n) &= a_{2,2}u(n) + a_{2,1}u(n-1) + a_{2,0}u(n-2) \\ &\vdots \\ b_M(n) &= a_{M,M}u(n) + a_{M,M-1}u(n-1) + \cdots + a_{M,0}u(n-M) \end{aligned} \tag{6.111}$$

Combining this system of $(M+1)$ simultaneous linear equations into a compact matrix form, we thus have

$$\mathbf{b}(n) = \mathbf{L}\mathbf{u}(n) \tag{6.112}$$

where $\mathbf{u}(n)$ is the $(M+1)$-by-1 input vector:

$$\mathbf{u}(n) = [u(n), u(n-1), \ldots, u(n-M)]^T \tag{6.113}$$

and $\mathbf{b}(n)$ is the $(M+1)$-by-1 output vector of backward prediction errors:

$$\mathbf{b}(n) = [b_0, b_1(n), \ldots, b_M(n)]^T \tag{6.114}$$

The $(M+1)$-by-$(M+1)$ coefficient matrix on the right side of Eq. (6.112) is defined in terms of the backward prediction-error filter coefficients of orders 0 to M

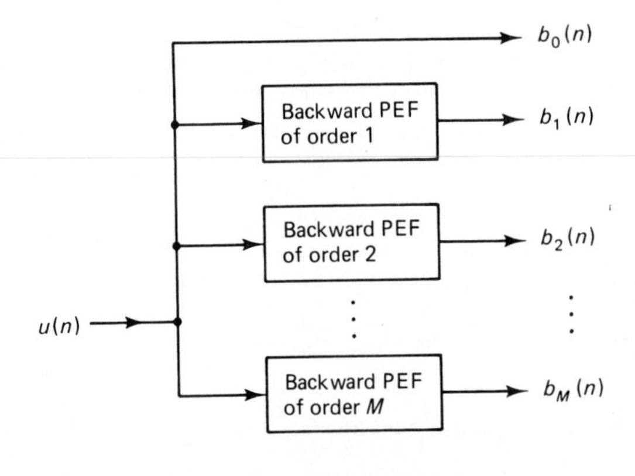

Figure 6.10 Parallel connection of a stack of backward prediction-error filters of orders 0 to M. Note: The direct connection represents a PEF with a single coefficient equal to unity.

as follows:

$$\mathbf{L} = \begin{bmatrix} 1 & 0 & \cdots & 0 \\ a_{1,1} & 1 & \cdots & 0 \\ \vdots & \vdots & \ddots & \vdots \\ a_{M,M} & a_{M,M-1} & \cdots & 1 \end{bmatrix} \tag{6.115}$$

The matrix **L** has three useful properties:

1. The matrix **L** is a *lower triangular matrix* with 1's along its main diagonal; all of its elements above the main diagonal are zero.
2. The determinant of matrix **L** is unity; hence, it is nonsingular (i.e., its inverse exists).
3. The nonzero elements of each row of the matrix **L** except for complex conjugation, equal the weights of a backward prediction-error filter whose order corresponds to the position of that row in the matrix.

The transformation of Eq. (6.112) is known as the *Gram–Schmidt orthogonalization algorithm.*[5] According to this algorithm, there is a *one-to-one correspondence between* the input vector $\mathbf{u}(n)$ and the backward prediction-error vector $\mathbf{b}(n)$. In particular, given $\mathbf{u}(n)$ we may obtain $\mathbf{b}(n)$ by using Eq. (6.112). Conversely, given $\mathbf{b}(n)$, we may obtain the corresponding vector $\mathbf{u}(n)$ by using the inverse of Eq. (6.115), as shown by

$$\mathbf{u}(n) = \mathbf{L}^{-1}\mathbf{b}(n) \tag{6.116}$$

where $\mathbf{L}^{-1}$ is the inverse of the matrix **L**.

Orthogonality of Backward Prediction Errors

The backward prediction errors $b_0(n), b_1(n), \ldots, b_M(n)$, constituting the elements of the vector $\mathbf{b}(n)$, have an important property: *They are all orthogonal to each other, as shown by*

$$E[b_m(n)b_i^*(n)] = \begin{cases} P_m, & i = m \\ 0, & i \neq m \end{cases} \tag{6.117}$$

To prove this property, we may proceed as follows. First of all, without loss of generality, we may assume that $m \geq i$. To prove Eq. (6.117), we express the backward prediction error $b_i(n)$ in terms of the input $\{u(n)\}$ as the linear convolution sum

$$b_i(n) = \sum_{k=0}^{i} a_{i,\,i-k}u(n-k) \tag{6.118}$$

where $a^*_{i,i-k}$, $k = 0, 1, \ldots, i$, are the coefficients of a backward prediction error filter of order i. Hence, we may write

$$E[b_m(n)b_i^*(n)] = E[b_m(n)\sum_{k=0}^{i} a^*_{i,i-k}u^*(n-k)] \tag{6.119}$$

[5] For a full discussion of the Gram–Schmidt algorithm and the various methods for its implementation, see Haykin (1989).

From the principle of orthogonality, we have

$$E[b_m(n)u^*(n-k)] = 0, \qquad 0 \le k \le m-1 \tag{6.120}$$

For the case when $m > i$, and with $0 \le k \le i$, we therefore find that all the expectation terms inside the summation on the right side of Eq. (6.119) are zero. Correspondingly,

$$E[b_m(n)b_i^*(n)] = 0, \qquad m \ne i$$

When $m = i$, Eq. (6.119) reduces to

$$\begin{aligned} E[b_m(n)b_i^*(n)] &= E[b_m(n)\,b_m^*(n)] \\ &= P_m, \qquad m = i \end{aligned}$$

This completes the proof of Eq. (6.117). It is important, however, to note that *this property holds only for wide-sense stationary input data.*

We thus see that the Gram–Schmidt orthogonalization algorithm of Eq. (6.112) transforms the input vector $\mathbf{u}(n)$ consisting of correlated samples into an equivalent vector $\mathbf{b}(n)$ of uncorrelated backward prediction error.[6]

Factorization of the Inverse of Correlation Matrix R

Having equipped ourselves with the property that the backward prediction errors are indeed orthogonal to each other, we may return to the transformation described by the Gram–Schmidt algorithm in Eq. (6.112). Specifically, using this transformation, we may express the correlation matrix of the backward prediction-error vector $\mathbf{b}(n)$ in terms of the correlation matrix of the input vector $\mathbf{u}(n)$ as follows:

$$\begin{aligned} E[\mathbf{b}(n)\mathbf{b}^H(n)] &= E[\mathbf{L}\mathbf{u}(n)\mathbf{u}^H(n)\mathbf{L}^H] \\ &= \mathbf{L}E[\mathbf{u}(n)\mathbf{u}^H(n)]\mathbf{L}^H \end{aligned} \tag{6.121}$$

Let $\mathbf{D}$ denote the correlation matrix of the backward prediction-error vector $\mathbf{b}(n)$, as shown by

$$\mathbf{D} = E[\mathbf{b}(n)\mathbf{b}^H(n)] \tag{6.122}$$

As before, the correlation matrix of the input vector $\mathbf{u}(n)$ is denoted by matrix $\mathbf{R}$. We may therefore rewrite Eq. (6.121) as

$$\mathbf{D} = \mathbf{L}\mathbf{R}\mathbf{L}^H \tag{6.123}$$

[6] Two random variables X and Y are said to be *orthogonal* to each other if

$$E[XY^*] = 0$$

They are said to be *uncorrelated* with each other if

$$E[(X - E[X])(Y - E[Y])^*] = 0$$

If one or both of X and Y have zero mean, then these two conditions become one and the same. For the discussion presented herein, the input data, and therefore the backward predicted errors, are assumed to have zero mean. Under this assumption, we may interchange orthogonality with uncorrelatedness when we refer to the backward prediction error.

We now make two observations:

1. When the correlation matrix $\mathbf{R}$ of the input vector $\mathbf{u}(n)$ is positive definite and therefore its inverse exists, the correlation matrix $\mathbf{D}$ of the backward prediction-error vector $\mathbf{b}(n)$ is also positive definite and likewise its inverse exists.
2. The correlation matrix $\mathbf{D}$ is a diagonal matrix, because $\mathbf{b}(n)$ consists of elements that are all orthogonal to each other. In particular, we may express the correlation matrix $\mathbf{D}$ in the expanded form:

$$\mathbf{D} = \text{diag}(P_0, P_1, \ldots, P_M) \tag{6.124}$$

where P_i is the average power of the ith backward prediction error $b_i(n)$; that is,

$$P_i = E[|b_i(n)|^2], \qquad i = 0, 1, \ldots, M \tag{6.125}$$

The inverse of matrix $\mathbf{D}$ is also a diagonal matrix, as shown by

$$\mathbf{D}^{-1} = \text{diag}(P_0^{-1}, P_1^{-1}, \ldots, P_M^{-1}) \tag{6.126}$$

Accordingly, we may use Eq. (6.123) to express the inverse of the correlation matrix $\mathbf{R}$ as follows:

$$\begin{aligned} \mathbf{R}^{-1} &= \mathbf{L}^H\mathbf{D}^{-1}\mathbf{L} \\ &= (\mathbf{D}^{-1/2}\mathbf{L})^H(\mathbf{D}^{-1/2}\mathbf{L}) \end{aligned} \tag{6.127}$$

which is the desired result.

The inverse matrix $\mathbf{D}^{-1}$, in the first line of Eq. (6.127), is a diagonal matrix defined by Eq. (6.126). The matrix $\mathbf{D}^{-1/2}$, the *square root* of $\mathbf{D}^{-1}$, in the second line of Eq. (6.127) is also a diagonal matrix defined by

$$\mathbf{D}^{-1/2} = \text{diag}(P_0^{-1/2}, P_1^{-1/2}, \ldots, P_M^{-1/2})$$

The transformation of Eq. (6.127) is called the *Cholesky factorization of the inverse matrix* $\mathbf{R}^{-1}$ (Stewart, 1973). Note that the matrix $\mathbf{D}^{-1/2}\mathbf{L}$ is a lower triangular matrix; however, it differs from the lower triangular matrix $\mathbf{L}$ of Eq. (6.115) in that its diagonal elements are different from 1. Note also that the Hermitian-transposed matrix product $(\mathbf{D}^{-1/2}\mathbf{L})^H$ is an upper triangular matrix whose diagonal elements are different from 1. Thus, *according to the Cholesky factorization, the inverse matrix* $\mathbf{R}^{-1}$ *may be factored into the product of an upper triangular matrix and a lower triangular matrix that are the Hermitian transpose of each other*.

6.12 LATTICE PREDICTORS

To implement the Gram–Schmidt algorithm of Eq. (6.112) for transforming an input vector $\mathbf{u}(n)$ consisting of correlated samples into an equivalent vector $\mathbf{b}(n)$ consisting of uncorrelated backward prediction errors, we may use the parallel connection of a direct path and an appropriate number of backward prediction-error filters, as in Fig. 6.10. The vectors $\mathbf{b}(n)$ and $\mathbf{u}(n)$ are said to be "equivalent" in the sense that they contain the same amount of information (see Problem 18). A much more efficient method of implementing the Gram–Schmidt orthogonalization algorithm, however, is to use a ladder structure known as a *lattice predictor*. This device combines several forward and backward prediction-error filtering operations into a single structure. Specifically, a lattice predictor consists of a cascade connection of elementary units (stages), all of which have a structure similar to that of a lattice, hence the name. The number of stages in a lattice predictor equals the prediction order. Thus, for a prediction-error filter of order m, there are m stages in the lattice realization of the filter.

Order-Update Recursions for the Prediction Errors

The input–output relations that characterize a lattice predictor may be derived in various ways, depending on the particular form in which the Levinson–Durbin algorithm is utilized. For the derivation presented here, we start with the matrix formulations of this algorithm given by Eqs. (6.40) and (6.42) that pertain to the forward and backward operations of a prediction-error filter, respectively. For convenience of presentation, we reproduce these two relations here:

$$\mathbf{a}_m = \begin{bmatrix} \mathbf{a}_{m-1} \\ 0 \end{bmatrix} + \Gamma_m \begin{bmatrix} 0 \\ \mathbf{a}_{m-1}^{B*} \end{bmatrix} \tag{6.128}$$

$$\mathbf{a}_m^{B*} = \begin{bmatrix} 0 \\ \mathbf{a}_{m-1}^{B*} \end{bmatrix} + \Gamma_m^* \begin{bmatrix} \mathbf{a}_{m-1} \\ 0 \end{bmatrix} \tag{6.129}$$

The $(m + 1)$-by-1 vector $\mathbf{a}_m$ and the m-by-1 vector $\mathbf{a}_{m-1}$ refer to forward prediction-error filters of order m and $m - 1$, respectively. The $(m + 1)$-by-1 vector $\mathbf{a}_m^{B*}$ and the m-by-1 vector $\mathbf{a}_{m-1}^{B*}$ refer to the corresponding backward prediction-error filters of order m and $m - 1$, respectively. The scalar Γ_m is the associated reflection coefficient.

Consider first the forward prediction-error filter of order m, with its tap inputs denoted by $u(n)$, $u(n - 1)$, . . . , $u(n - m)$. We may partition $\mathbf{u}_{m+1}(n)$, the $(m + 1)$-by-1 tap-input vector of this filter, in the form

$$\mathbf{u}_{m+1}(n) = \begin{bmatrix} \mathbf{u}_m(n) \\ \hline u(n-m) \end{bmatrix} \tag{6.130}$$

or in the equivalent form

$$\mathbf{u}_{m+1}(n) = \begin{bmatrix} u(n) \\ \hline \mathbf{u}_m(n-1) \end{bmatrix} \tag{6.131}$$

Form the inner product of the $(m + 1)$-by-1 vectors $\mathbf{a}_m$ and $\mathbf{u}_{m+1}(n)$. This is done by premultiplying $\mathbf{u}_{m+1}(n)$ by the Hermitian transpose of $\mathbf{a}_m$. Thus, using Eq. (6.128) for $\mathbf{a}_m$, we may treat the terms resulting from this multiplication, as follows:

1. For the left side of Eq. (6.128), premultiplication of $\mathbf{u}_{m+1}(n)$ by $\mathbf{a}_m^H$ yields

$$f_m(n) = \mathbf{a}_m^H \mathbf{u}_{m+1}(n) \tag{6.132}$$

where $f_m(n)$ is the forward prediction error produced at the output of the forward prediction-error filter of order m.

2. For the first term on the right side of Eq. (6.128), we use the partitioned form of $\mathbf{u}_{m+1}(n)$ given in Eq. (6.130). We may therefore write

$$\begin{aligned}
[\mathbf{a}_{m-1}^H \;\vdots\; 0]\mathbf{u}_{m+1}(n) &= [\mathbf{a}_{m-1}^H \;\vdots\; 0]\begin{bmatrix} \mathbf{u}_m(n) \\ \text{--------} \\ u(n-m) \end{bmatrix} \\
&= \mathbf{a}_{m-1}^H \mathbf{u}_m(n) \qquad (6.133) \\
&= f_{m-1}(n)
\end{aligned}$$

where $f_{m-1}(n)$ is the forward prediction error produced at the output of the forward prediction-error filter of order $m-1$.

3. For the second matrix term on the right side of Eq. (6.128), we use the partitioned form of $\mathbf{u}_{m+1}(n)$ given in Eq. (6.131). We may therefore write

$$\begin{aligned}
[0 \;\vdots\; \mathbf{a}_{m-1}^{BT}]\mathbf{u}_{m+1}(n) &= [0 \;\vdots\; \mathbf{a}_{m-1}^{BT}]\begin{bmatrix} u(n) \\ \text{---------} \\ \mathbf{u}_m(n-1) \end{bmatrix} \\
&= \mathbf{a}_{m-1}^{BT} \mathbf{u}_m(n-1) \qquad (6.134) \\
&= b_{m-1}(n-1)
\end{aligned}$$

where $b_{m-1}(n-1)$ is the *delayed* backward prediction error produced at the output of the backward prediction-error filter of order $m-1$.

Combining the results of the multiplication, described by Eqs. (6.132), (6.133), and (6.134), we may thus write

$$f_m(n) = f_{m-1}(n) + \Gamma_m^* b_{m-1}(n-1) \qquad (6.135)$$

Consider next the backward prediction-error filter of order m, with its tap inputs denoted by $u(n), u(n-1), \ldots, u(n-m)$. Here again we may express $\mathbf{u}_{m+1}(n)$, the $(m+1)$-by-1 tap-input vector of this filter, in the partitioned form of Eq. (6.130) or that of Eq. (6.131). In this case, the terms resulting from the formation of the inner product of the vectors $\mathbf{a}_m^{B*}$ and $\mathbf{u}_{m+1}(n)$ are treated as follows:

1. For the left side of Eq. (6.129), premultiplication of $\mathbf{u}_{m+1}(n)$ by the Hermitian transpose of $\mathbf{a}_m^{B*}$ yields

$$b_m(n) = \mathbf{a}_m^{BT} \mathbf{u}_{m+1}(n) \qquad (6.136)$$

where $b_m(n)$ is the backward prediction error produced at the output of the backward prediction-error filter of order m.

2. For the first term on the right side of Eq. (6.129), we use the partitioned form of the tap-input vector $\mathbf{u}_{m+1}(n)$ given in Eq. (6.131). Thus, multiplying the Hermitian transpose of this first term by $\mathbf{u}_{m+1}(n)$, we get

$$
[0 \mid \mathbf{a}_{m-1}^{BT}]\,\mathbf{u}_{m+1}(n) = [0 \mid \mathbf{a}_{m-1}^{BT}]\begin{bmatrix} u(n) \\ \hdashline \mathbf{u}_m(n-1) \end{bmatrix}
$$

$$
= \mathbf{a}_{m-1}^{BT}\mathbf{u}_m(n-1) \tag{6.137}
$$

$$
= b_{m-1}(n-1)
$$

3. For the second matrix term on the right side of Eq. (6.129), we use the partitioned form of the tap-input vector $\mathbf{u}_{m+1}(n)$ given in Eq. (6.130). Thus, multiplying the Hermitian transpose of this second term by $\mathbf{u}_{m+1}(n)$, we get

$$
[\mathbf{a}_{m-1}^{H} \mid 0]\,\mathbf{u}_{m+1}(n) = [\mathbf{a}_{m-1}^{H} \mid 0]\begin{bmatrix} \mathbf{u}_m(n) \\ \hdashline u(n-m) \end{bmatrix}
$$

$$
= \mathbf{a}_{m-1}^{H}\mathbf{u}_m(n) \tag{6.138}
$$

$$
= f_{m-1}(n)
$$

Combining the results of Eqs. (6.136), (6.137), and (6.138), we thus find that the inner product of $\mathbf{a}_m^{B*}$ and $\mathbf{u}_{m+1}(n)$ yields

$$
b_m(n) = b_{m-1}(n-1) + \Gamma_m f_{m-1}(n) \tag{6.139}
$$

Equations (6.135) and (6.139) are the desired pair of *order-update recursions* that characterize stage m of the lattice predictor. They are reproduced here in matrix form as follows:

$$
\begin{bmatrix} f_m(n) \\ b_m(n) \end{bmatrix} = \begin{bmatrix} 1 & \Gamma_m^* \\ \Gamma_m & 1 \end{bmatrix}\begin{bmatrix} f_{m-1}(n) \\ b_{m-1}(n-1) \end{bmatrix}, \qquad m = 1, 2, \ldots, M \tag{6.140}
$$

We may view $b_{m-1}(n-1)$ as the result of applying the unit-delay operator z^{-1} to the backward prediction error $b_{m-1}(n)$; that is,

$$
b_{m-1}(n-1) = z^{-1}[b_{m-1}(n)] \tag{6.141}
$$

Thus, using Eqs. (6.140) and (6.141), we may represent stage m of the lattice predictor by the signal-flow graph shown in Fig. 6.11(a). Except for the branch pertaining to the block labeled z^{-1}, this signal-flow graph has the appearance of a lattice, hence the name "lattice predictor."[7] Note also that the parameterization of stage m of the lattice predictor is uniquely defined by the reflection coefficient Γ_m.

For the elementary case of $m = 0$, we get the *initial conditions*:

$$
f_0(n) = b_0(n) = u(n) \tag{6.142}
$$

[7] The first application of lattice filters in on-line adaptive signal processing was apparently made by Itakura and Saito (1971) in the field of speech analysis. Equivalent lattice-filter models, however, were familiar in geophysical signal processing as "layered earth models" (Robinson, 1967; Burg, 1968). It is also of interest to note that such lattice filters have been well studied in network theory, especially in the cascade synthesis of multiport networks (Dewilde, 1969).

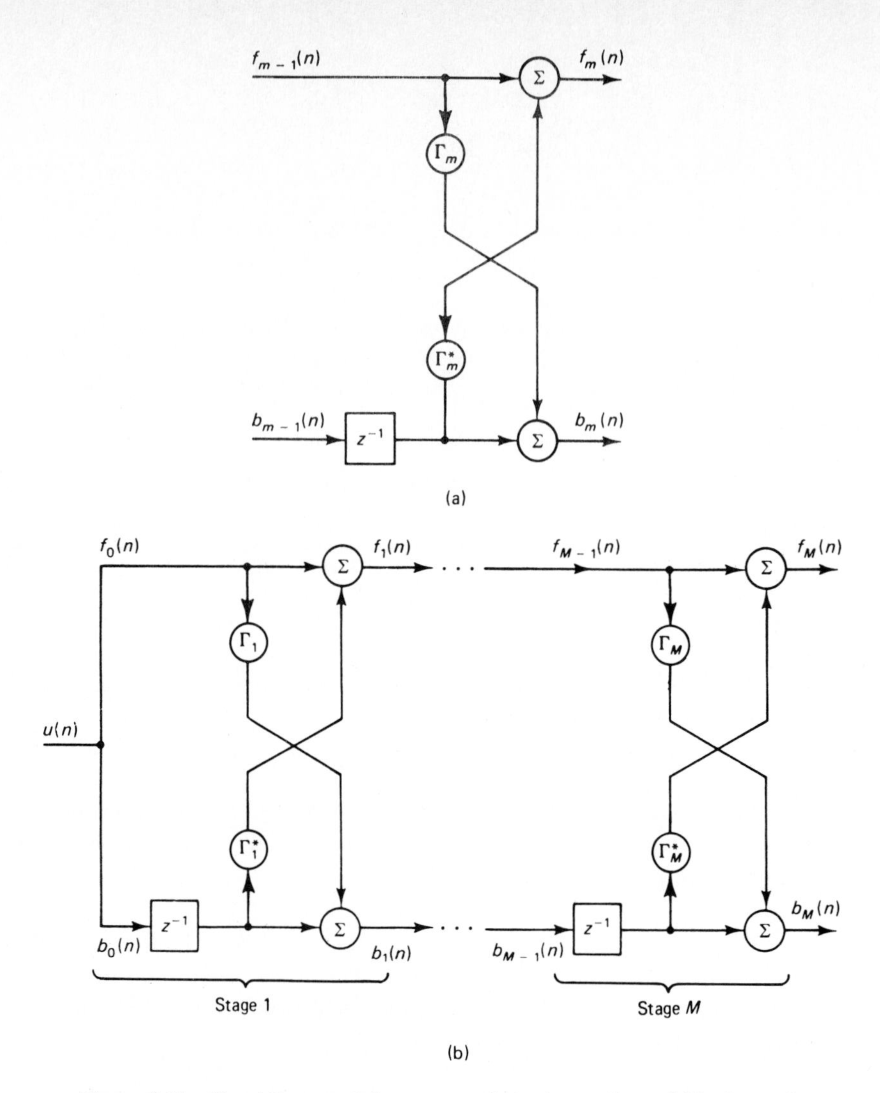

Figure 6.11 Signal-flow graph for stage m of a lattice predictor; (b) lattice equivalent model of prediction-error filter of order M.

where $u(n)$ is the input signal at time n. Therefore, starting with $m = 0$, and progressively increasing the order of the filter by 1, we obtain the *lattice equivalent model* shown in Fig. 6.11(b) for a prediction-error filter of final order M. In this figure we merely require knowledge of the complete set of reflection coefficients Γ_1, $\Gamma_2, \ldots, \Gamma_M$, one for each stage of the filter.

The lattice filter, depicted in Fig. 6.11(b), offers the following attractive features:

1. A lattice filter is a highly efficient structure for generating the sequence of forward prediction errors and the corresponding sequence of backward prediction errors simultaneously.
2. The various stages of a lattice predictor are "decoupled" from each other. This decoupling property was indeed derived in Section 6.12, where it was shown

that the backward prediction errors produced by the various stages of a lattice predictor are "orthogonal" to each other for wide-sense stationary input data.

3. The lattice filter is *modular* in structure; hence, if the requirement calls for increasing the order of the predictor, we simply add one or more stages (as desired) without affecting earlier computations.
4. All the stages of a lattice predictor have a similar structure; hence, it lends itself to the use of *very large scale integration* (VLSI) technology if the use of this technology is considered beneficial to the application of interest.

6.13 INVERSE FILTERING USING THE LATTICE STRUCTURE

The multistage lattice predictor of Fig. 6.11(b) may be viewed as an *analyzer*. That is, it enables us to represent an autoregressive (AR) process $\{u(n)\}$ by a corresponding sequence of reflection coefficients $\{\Gamma_m\}$. By rewiring this multistage lattice predictor in the manner depicted in Fig. 6.12, we may use this new structure as a *synthesizer* or *inverse filter*. That is, given the sequence of reflection coefficients $\{\Gamma_m\}$, we may reproduce the original AR process by applying a white-noise process $\{v(n)\}$ to the input of the structure in Fig. 6.12.

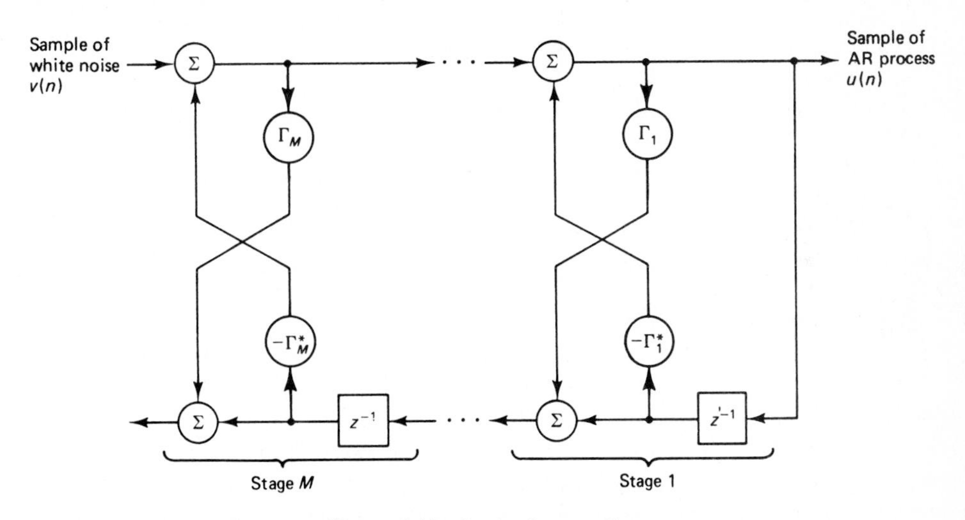

Figure 6.12 Lattice inverse filter.

We illustrate the operation of the lattice inverse filter with an example.

Example 5

Consider the two-stage lattice inverse filter of Fig. 6.13. There are four possible paths that can contribute to the makeup of the sample $u(n)$ at the output, as illustrated in Fig. 6.14. In particular, we have

$$\begin{aligned} u(n) &= v(n) - \Gamma_1^* u(n-1) - \Gamma_1 \Gamma_2^* u(n-1) - \Gamma_2^* u(n-2) \\ &= v(n) - (\Gamma_1^* + \Gamma_1 \Gamma_2^*) u(n-1) - \Gamma_2^* u(n-2) \end{aligned}$$

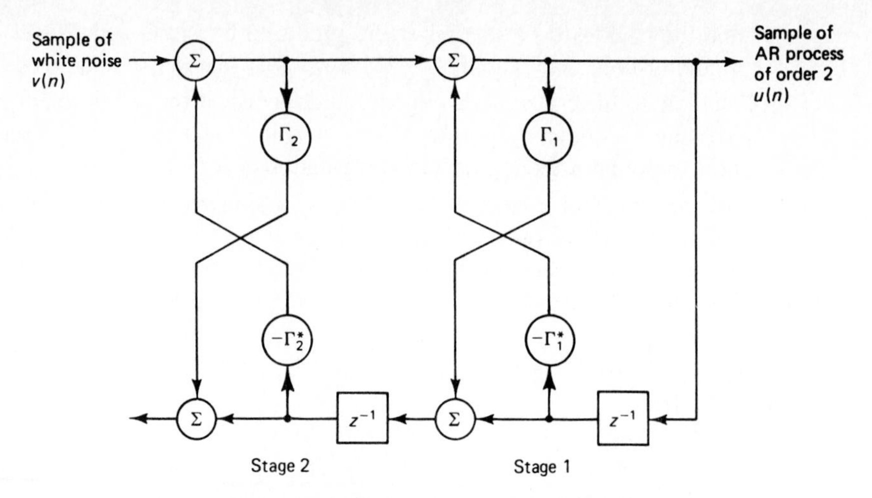

Figure 6.13 Lattice inverse filter of order 2.

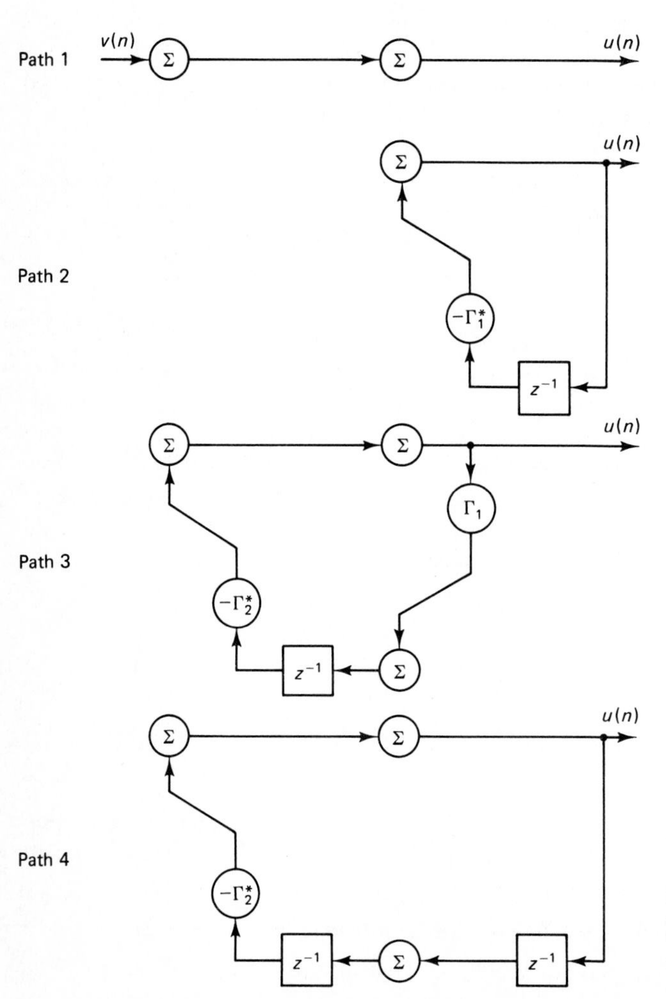

Figure 6.14 The four possible paths that contribute to the makeup of the output $u(n)$ in the lattice inverse filter of Fig. 6.13.

From Example 2, we recall that

$$a_{2,1} = \Gamma_1 + \Gamma_1^* \Gamma_2$$

$$a_{2,2} = \Gamma_2$$

We may therefore express the mechanism governing the generation of process $\{u(n)\}$ as follows:

$$u(n) + a_{2,1}^* u(n - 1) + a_{2,2}^* u(n - 2) = v(n)$$

which is recognized as the difference equation of a second-order AR process.

6.14 JOINT-PROCESS ESTIMATION

In this section, we use the lattice predictor, as a subsystem, to solve a *joint-process estimation problem* that is optimal in the mean-square sense (Griffiths, 1978; Makhoul, 1978). In particular, we consider the minimum mean-square estimation of a process $\{d(n)\}$, termed the desired response, by using a set of *observables* derived from a related process $\{u(n)\}$. We assume that the processes $\{d(n)\}$ and $\{u(n)\}$ are jointly stationary. This estimation problem is similar to that considered in Chapter 5, with one basic difference. In Chapter 5 we used samples of the process $\{u(n)\}$ as the observables directly. Our approach here is different in that for the observables we use the set of backward prediction errors obtained by feeding the input of a multistage lattice predictor with samples of the process $\{u(n)\}$. The fact that the backward prediction errors are orthogonal to each other simplifies the solution to the problem significantly.

The structure of the *joint-process estimator* is shown in Fig. 6.15. This device performs two optimum estimations jointly:

1. The *lattice predictor*, consisting of a cascade of M stages, characterized individually by the reflection coefficients $\Gamma_1, \Gamma_2, \ldots, \Gamma_M$, performs predictions (of varying orders) on the input. In particular, it transforms the sequence of (correlated) input samples $u(n), u(n - 1), \ldots, u(n - M)$ into a corresponding sequence of (uncorrelated) backward prediction errors $b_0(n), b_1(n), \ldots, b_M(n)$.
2. The *multiple regression filter*, characterized by the set of weights $\kappa_0, \kappa_1, \ldots, \kappa_M$, operates on the sequence of backward prediction errors $b_0(n), b_1(n), \ldots, b_M(n)$ as inputs, respectively, to produce an *estimate* of the desired response $d(n)$. The resulting estimate is defined as the sum of the respective inner products of these two sets of quantities, as shown by

$$\hat{d}(n|\mathcal{U}_n) = \sum_{i=0}^{M} \kappa_i^* b_i(n) \tag{6.143}$$

where $\mathcal{U}_n$ is the space spanned by the inputs $u(n), u(n - 1), \ldots, u(n - M)$. We may rewrite Eq. (6.143) in matrix form as follows:

$$\hat{d}(n|\mathcal{U}_n) = \boldsymbol{\kappa}^H \mathbf{b}(n) \tag{6.144}$$

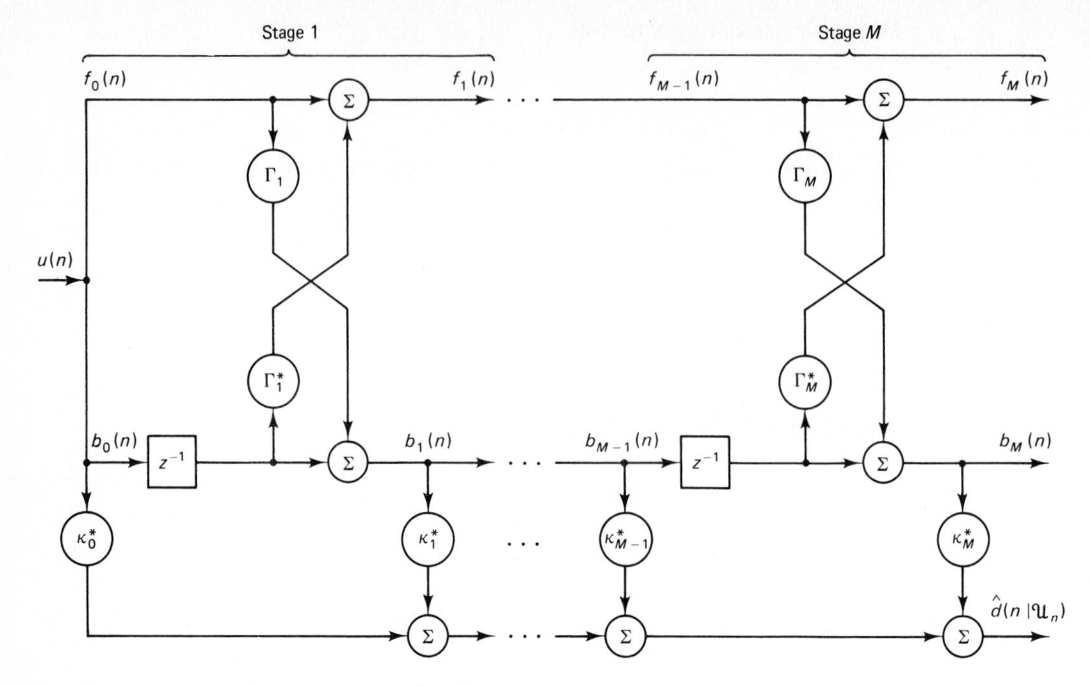

Figure 6.15 Lattice-based structure for joint-process estimation.

where $\boldsymbol{\kappa}$ is an $(M + 1)$-by-1 vector defined by

$$\boldsymbol{\kappa}^T = [\kappa_0, \kappa_1, \ldots, \kappa_M] \tag{6.145}$$

We refer to $\kappa_0, \kappa_1, \ldots, \kappa_M$ as the *regression coefficients* of the estimator, and to $\boldsymbol{\kappa}$ as the *regression vector*.

Let $\mathbf{D}$ denote the $(M + 1)$-by-$(M + 1)$ correlation matrix of $\mathbf{b}(n)$, the $(M + 1)$-by-1 vector of backward prediction errors, defined in Eq. (6.124). Let $\mathbf{s}$ denote the $(M + 1)$-by-1 cross-correlation vector between the backward prediction errors and the desired reponse:

$$\mathbf{s} = E[\mathbf{b}(n)d^*(n)] \tag{6.146}$$

Therefore, applying the Wiener–Hopf equations to our present situation, we find that the optimum tap-weight vector $\boldsymbol{\kappa}_o$ is defined by

$$\mathbf{D}\boldsymbol{\kappa}_o = \mathbf{s} \tag{6.147}$$

Solving for $\boldsymbol{\kappa}_o$, we get

$$\boldsymbol{\kappa}_o = \mathbf{D}^{-1}\mathbf{s} \tag{6.148}$$

where the inverse matrix $\mathbf{D}^{-1}$ is a diagonal matrix, defined in terms of various prediction-error powers as in Eq. (6.126). Note that, unlike the ordinary transversal filter realization of the Wiener filter, the computation of the tap-weight vector $\boldsymbol{\kappa}_o$ in the joint-process estimator of Fig. 6.15 is relatively simple to accomplish.

Relationship between the Regression Vector $\boldsymbol{\kappa}_o$ and the Wiener Solution $\mathbf{w}_o$

From the Cholesky factorization given in Eq. (6.127), we deduce that

$$\mathbf{D}^{-1} = \mathbf{L}\mathbf{R}^{-1}\mathbf{L}^H \tag{6.149}$$

Hence, substituting Eq. (6.149) in (6.148) yields

$$\boldsymbol{\kappa}_o = \mathbf{L}\mathbf{R}^{-1}\mathbf{L}^H\mathbf{s} \tag{6.150}$$

Moreover, from Eq. (6.112) we note that

$$\mathbf{b}(n) = \mathbf{L}\mathbf{u}(n) \tag{6.151}$$

Therefore, substituting Eq. (6.151) in (6.146) yields

$$\begin{aligned} \mathbf{s} &= \mathbf{L}E[\mathbf{u}(n)d^*(n)] \\ &= \mathbf{L}\mathbf{p} \end{aligned} \tag{6.152}$$

where $\mathbf{p}$ is the cross-correlation vector between the tap-input vector $\mathbf{u}(n)$ and the desired response $d(n)$. Thus, using Eq. (6.152) in (6.150), we finally obtain

$$\begin{aligned} \boldsymbol{\kappa}_o &= \mathbf{L}\mathbf{R}^{-1}\mathbf{L}^H\mathbf{L}\mathbf{p} \\ &= \mathbf{L}\mathbf{R}^{-1}\mathbf{p} \\ &= \mathbf{L}\mathbf{w}_o \end{aligned} \tag{6.153}$$

where $\mathbf{L}$ is a lower triangular matrix defined in terms of the backward prediction-error filter coefficients, as in Eq. (6.115). Equation (6.153) is the desired relationship between the regression vector $\boldsymbol{\kappa}_o$ and the Wiener solution $\mathbf{w}_o = \mathbf{R}^{-1}\mathbf{p}$.

6.15 BLOCK ESTIMATION

In this chapter, we have discussed two basic structures for building a *linear predictor* or its natural extension, a *prediction-error filter*; the two structures are a *transversal filter* and a *lattice filter*.[8] The transversal filter is characterized by a set of *tap weights,* whereas the lattice filter is characterized by a corresponding set of *reflection coefficients*. In both cases, the filter coefficients provide the designer with *degrees of freedom,* the number of which equals the *prediction order*. The mathematical link between the tap weights of a transversal predictor and the reflection coefficients of a lattice predictor is provided by the Levinson–Durbin algorithm.

[8] In actual fact, there is a third structure for building a linear predictor that is based on the *Schur algorithm* [Schur (1917)]. Like the Levinson–Durbin algorithm, the Schur algorithm provides a procedure for computing the reflection coefficients from a known autocorrelation sequence. The Schur algorithm, however, lends itself to parallel implementation, with the result that it achieves a throughput rate higher than that obtained using the Levinson–Durbin algorithm. For a discussion of the Schur algorithm, including mathematical details and implementational considerations, see Haykin (1989).

Regardless of the particular structure chosen, we clearly need a procedure for *estimating* the filter coefficients. To carry out this estimation we have two approaches to consider:

1. Block estimation
2. Adaptive estimation

In *block estimation,* the available data are divided into individual *blocks,* each of length N, say. The *block length* N is usually chosen short enough to ensure pseudostationarity of the input data over the length N. Under this assumption, the filter coefficients of interest are then computed on a *block-by-block basis*. Typically, the filter coeffficients vary from one block of data to another.

The block estimation algorithms may be categorized as follows:

1. *Indirect methods*. For each block of data, estimates of the autocorrelation function of the input are computed for different lags. The Levinson–Durbin algorithm is then used to compute the corresponding set of tap weights for a transversal predictor, or the corresponding set of reflection coefficients for a lattice predictor, depending on the application of interest.
2. *Direct methods*. For each block of data, estimates of the reflection coefficients for the different stages of a lattice predictor are computed directly from the data. The reflection coefficients lend themselves to direct computation because of the decoupling property of a multistage lattice predictor for wide-sense stationary inputs.

From a computational viewpoint, direct methods are more efficient than indirect methods if the application of interest requires knowledge of the reflection coefficients. If, on the other hand, the application is that of system identification in terms of the tap weights of a transversal filter, as in autoregressive modeling, for example, the computational advantage of direct methods over indirect methods may well disappear. In this section, we develop a popular algorithm known as the *Burg algorithm,* which may be classified as a direct block estimation method.[9]

Consider a lattice predictor consisting of M stages connected in cascade as in Fig. 6.11(b). For wide-sense stationary input data, the M stages of this lattice predictor are decoupled from each other by virtue of the orthogonality of its backward prediction-error outputs; see Section 6.11. This decoupling property makes it possible for us to accomplish *the global optimization of a multistage lattice predictor as a sequence of local optimizations, one at each stage of the structure*. Moreover, it is a straightforward matter to increase the order of the predictor by simply adding one or more stages, as required, without affecting the earlier design computations. For example, suppose that we have optimized the design of a lattice predictor consisting of M stages. To increase the order of the predictor by 1, we simply add a new stage that is locally optimized, leaving the optimum design of the earlier stages unchanged.

[9] For a more complete discussion of block estimation, including both indirect and direct methods, see Marple (1987), Kay (1988), and Haykin (1989).

Consider stage m of the lattice predictor shown in Fig. 6.11(b), for which the input–output relations are given in matrix form in Eq. (6.140). For convenience of presentation, these relations are reproduced here in the expanded form:

$$f_m(n) = f_{m-1}(n) + \Gamma_m^* b_{m-1}(n-1) \tag{6.154}$$

$$b_m(n) = b_{m-1}(n-1) + \Gamma_m f_{m-1}(n) \tag{6.155}$$

where $m = 1, 2, \ldots, M$, and M is the final order of the predictor. Several criteria may be used to optimize the design of this stage (Makhoul, 1977). However, one particular criterion yields a design with interesting properties that conform to the lattice predictor theory. Specifically, the reflection coefficient Γ_m is chosen so as to minimize the sum of the mean-squared values of the forward and backward prediction errors. Let the cost function J_m denote this sum at the output of stage m of the lattice predictor:

$$J_m = E[|f_m(n)|^2] + E[|b_m(n)|^2] \tag{6.156}$$

Substituting Eqs. (6.154) and (6.155) in (6.156), we get

$$\begin{aligned} J_m = {} & \{E[|f_{m-1}(n)|^2] + E[|b_{m-1}(n-1)|^2]\}[1 + |\Gamma_m|^2] \\ & + 2\Gamma_m E[f_{m-1}(n) b_{m-1}^*(n-1)] \\ & + 2\Gamma_m^* E[b_{m-1}(n-1) f_{m-1}^*(n)] \end{aligned} \tag{6.157}$$

In general, the reflection coefficient Γ_m is complex valued, as shown by

$$\Gamma_m = \alpha_m + j\beta_m \tag{6.158}$$

Therefore, differentiating the cost function J_m with respect to both the real and imaginary parts of Γ_m, we get the complex-valued *gradient*:

$$\begin{aligned} \nabla(J_m) &= \frac{\partial J_m}{\partial \alpha_m} + j\frac{\partial J_m}{\partial \beta_m} \\ &= 2\Gamma_m\{E[|f_{m-1}(n)|^2] + E[|b_{m-1}(n-1)|^2]\} \\ &\quad + 4E[b_{m-1}(n-1) f_{m-1}^*(n)] \end{aligned} \tag{6.159}$$

Putting this gradient to zero, we find that the optimum value of the reflection coefficient, for which the cost function J_m is minimum, equals

$$\Gamma_{m,o} = -\frac{2E[b_{m-1}(n-1) f_{m-1}^*(n)]}{E[|f_{m-1}(n)|^2 + |b_{m-1}(n-1)|^2]}, \qquad m = 1, 2, \ldots, M \tag{6.160}$$

Equation (6.160) for the reflection coefficient is known as the *Burg formula* (Burg, 1968).[10] Its use offers two interesting properties (see Problem 23):

1. The reflection coefficient $\Gamma_{m,o}$ satisfies the condition

$$|\Gamma_{m,o}| \le 1, \qquad \text{for all } m \tag{6.161}$$

In other words, the Burg formula always yields a minimum-phase design for the lattice predictor.

[10] The 1968 paper by Burg is reproduced in the book by Childers (1978).

2. The mean-square values of the forward and backward prediction errors at the output of stage m are related to those at the input as follows, respectively:

$$E[|f_m(n)|^2] = (1 - |\Gamma_{m,o}|^2)E[|f_{m-1}(n)|^2] \tag{6.162}$$

and

$$E[|b_m(n)|^2] = (1 - |\Gamma_{m,o}|^2)E[|b_{m-1}(n-1)|^2] \tag{6.163}$$

The Burg formula, as described in Eq. (6.160), involves the use of ensemble averaging. Assuming that the input $\{u(n)\}$ is *ergodic,* we may substitute time averages for the expectations in the dominator and denominator of this equation. We thus get the *Burg estimate*[11] for the reflection coefficient of stage m in the lattice predictor:

$$\hat{\Gamma}_m = -\frac{2\sum_{n=m+1}^{N} b_{m-1}(n-1)f_{m-1}^*(n)}{\sum_{n=m+1}^{N}[|f_{m-1}(n)|^2 + |b_{m-1}(n-1)|^2]}, \qquad m = 1, 2, \ldots, M \tag{6.164}$$

where N is the length of a block of input data, and $f_0(n) = b_0(n) = u(n)$.

With a lattice predictor of m stages, each of which contains a single unit-delay element, and with the input $u(n)$ zero for $n \le 0$, we find that *all* the samples in the input data contribute to the outputs of stage m in the predictor for the first time at $n = m + 1$; hence, the use of this value for the lower limits of the summation terms in Eq. (6.164). Note also that the estimate $\hat{\Gamma}_m$ for the mth reflection coefficient is dependent on data length N. The choice of N is usually dictated by two conflicting factors. First, it should be large enough to smooth out the effects of noise in computing the time averages in the numerator and denominator of Eq. (6.164). Second, it should be small enough to ensure quasi-statistical stationarity of the input data during the computations, and thereby justify the application of Burg's formula.

The block-estimation approach usually requires a large amount of computation, as well as a large amount of storage. Furthermore, in this approach, we find that for any stage of the predictor the estimate of the reflection coefficient at time $n + 1$ does not depend in a simple way on its previous estimate at time n. This behavior is to be contrasted with the adaptive estimation procedure described in subsequent chapters of the book.

PROBLEMS

1. The augmented Wiener–Hopf equations (6.14) of a forward prediction-error filter were derived by first optimizing the linear prediction filter in the mean-square sense and then combining the two resultants: the Wiener–Hopf equations for the tap-weight vector and

[11] For some practical applications of the Burg estimator given in Eq. (6.164), see Haykin et al. (1982) and Swingler and Walker (1989). The first of these two papers describes a (temporal) procedure based on this estimator for classifying the different forms of radar clutter (e.g., radar returns from different targets) as encountered in an air traffic control environment. The second paper presents a demonstration of a linear-array beamformer based on a spatial interpretation of the Burg estimator for a sonar environment. The studies reported in both of these papers employ real-life data.

the minimum mean-squared prediction error. This problem addresses the issue of deriving Eq. (6.14) directly by proceeding as follows:

(a) Formulate the expression for the mean-square value of the forward prediction error as a function of the tap-weight vector of the forward prediction-error filter.

(b) Minimize this mean-squared prediction error, subject to the constraint that the leading element of the tap-weight vector of the forward prediction-error filter equals 1.

Hint: Use the method of Lagrange multipliers to solve the constrained optimization problem. For details of this method, see Appendix C. This hint also applies to part (b) of Problem 2.

2. The augmented Wiener–Hopf equations (6.38) of a backward prediction-error filter were derived indirectly in Section 6.2. This problem addresses the issue of deriving Eq. (6.38) directly by proceeding as follows:

(a) Formulate the expression for the mean-square value of the backward prediction error in terms of the tap-weight vector of the backward prediction-error filter.

(b) Minimize this mean-squared prediction error, subject to the constraint that the last element of the tap-weight vector of the backward prediction-error filter equals 1.

3. **(a)** Equation (6.24) defines the Wiener–Hopf equations for backward linear prediction. This system of equations is reproduced here for convenience:

$$\mathbf{R}\mathbf{g} = \mathbf{r}^{B*}$$

where $\mathbf{g}$ is the tap-weight vector of the predictor, $\mathbf{R}$ is the correlation matrix of the tap inputs $u(n)$, $u(n-1)$, . . . , $u(n-M+1)$, and $\mathbf{r}^{B*}$ is the cross-correlation vector between these tap inputs and the desired response $u(n-M)$. Show that if the elements of the column vector $\mathbf{r}^{B*}$ are rearranged in reverse order the effect of this reversal is to modify the Wiener–Hopf equations as

$$\mathbf{R}^T\mathbf{g}^B = \mathbf{r}^*$$

(b) Show that the inner products $\mathbf{r}^{BT}\mathbf{g}$ and $\mathbf{r}^T\mathbf{g}^B$ are equal.

4. Consider a wide-sense stationary process $\{u(n)\}$ whose autocorrelation function has the following values for different lags:

$$r(0) = 1$$

$$r(1) = 0.8$$

$$r(2) = 0.6$$

$$r(3) = 0.4$$

(a) Use the Levinson–Durbin recursion to evaluate the reflection coefficients Γ_1, Γ_2, and Γ_3.

(b) Set up a three-stage lattice predictor for this process, using the values for the reflection coefficients found in part (a).

(c) Evaluate the average power of the prediction error produced at the output of each of the three stages in this lattice predictor. Hence, make a plot of prediction-error power versus prediction order. Comment on your results.

5. Consider the filtering structure described in Fig. P6.1, where the delay Δ is an integer greater than one. The requirement is to choose the weight vector $\mathbf{w}$ so as to minimize the mean-square value of the estimation error $e(n)$. Find the optimum value of $\mathbf{w}(n)$.

6. Consider the linear prediction of a stationary autoregressive process $\{u(n)\}$, generated from the first-order difference equation

$$u(n) = 0.9u(n-1) + v(n)$$

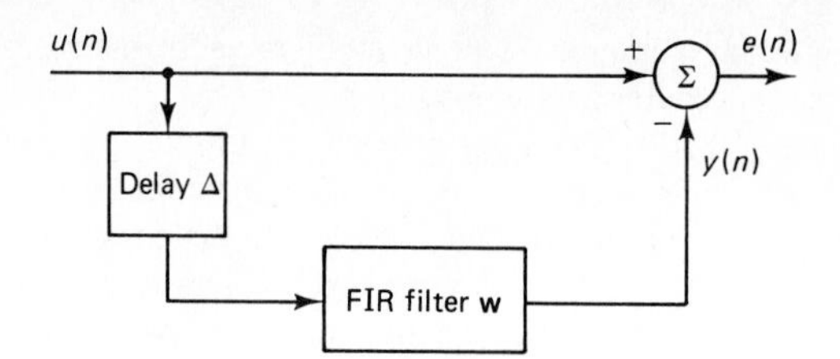

Figure P6.1

where $\{v(n)\}$ is a white noise process of zero mean and unit variance. The prediction order is two.

(a) Determine the tap weights a_1 and a_2 of the forward prediction-error filter.

(b) Determine the reflection coefficients Γ_1 and Γ_2 of the corresponding lattice predictor.

Comment on your results in parts (a) and (b).

7. **(a)** A time series $\{u_1(n)\}$ consists of a single sinusoidal process of complex amplitude α and angular frequency ω in additive white noise of zero mean and variance σ_v^2, as shown by

$$u_1(n) = \alpha e^{j\omega n} + v(n)$$

where

$$E[|\alpha|^2] = \sigma_\alpha^2$$

and

$$E[|v(n)|^2] = \sigma_v^2$$

The time series $\{u(n)\}$ is applied to a linear predictor of order M, optimized in the Wiener sense. Do the following:

(i) Determine the tap weights of the prediction-error filter of order M, and the final value of the prediction error power P_M.

(ii) Determine the reflection coefficients $\Gamma_1, \Gamma_2, \ldots, \Gamma_M$ of the corresponding lattice predictor.

(iii) How are the results in part (i) and (ii) modified when we let the noise variance σ_v^2 approach zero?

(b) Consider next an AR process $\{u_2(n)\}$ described by

$$u_2(n) = -\alpha e^{j\omega} u_2(n-1) + v(n)$$

where, as before, $\{v(n)\}$ is an additive white noise process of zero mean and variance σ_v^2. Assume that $0 < |\alpha| < 1$ but very close to 1. The time series $\{u_2(n)\}$ is also applied to a linear predictor of order M, optimized in the Wiener sense.

(i) Determine the tap weights of the new prediction-error filter of order M.

(ii) Determine the reflection coefficients $\Gamma_1, \Gamma_2, \ldots, \Gamma_M$ of the corresponding lattice predictor.

(c) Use your results in parts (a) and (b) to compare the similarities and differences between the linear prediction of the time series $\{u_1(n)\}$ and $\{u_2(n)\}$.

8. Equation (6.40) defines the Levinson–Durbin recursion for forward linear prediction. By rearranging the elements of the tap-weight vector $\mathbf{a}_m$ backward and then complex-conjugating them, reformulate the Levinson–Durbin recursion for backward linear prediction as in Eq. (6.42).

9. Starting with the definition of Eq. (6.47) for Δ_{m-1}, show that Δ_{m-1} equals the cross-correlation between the delayed backward prediction error $b_{m-1}(n-1)$ and the forward prediction error $f_{m-1}(n)$.

10. Develop in detail the relationship between the Schur–Cohn method and the inverse recursion as outlined by Eqs. (6.93) through (6.96).

11. Consider an autoregressive process $\{u(n)\}$ of order 2, described by the difference equation

$$u(n) = u(n-1) - 0.5u(n-2) + v(n)$$

where $\{v(n)\}$ is a white-noise process of zero mean and variance 0.5.

(a) Find the average power of $\{u(n)\}$.

(b) Find the reflection coefficients Γ_1 and Γ_2.

(c) Find the average prediction-error powers P_1 and P_2.

12. Using the one-to-one correspondence between the two sequences of numbers $\{P_0, \Gamma_1, \Gamma_2\}$ and $\{r(0), r(1), r(2)\}$, compute the autocorrelation function values $r(1)$ and $r(2)$ that correspond to the reflection coefficients Γ_1 and Γ_2 found in Problem 11 for the second-order autoregresssive process $\{u(n)\}$.

13. In Section 6.5, we presented a proof of the minimum-phase property of a prediction-error filter by using Rouché's theorem. In this problem, we explore another proof of this property by contradiction. Consider Fig. P6.2, which shows the prediction-error filter (of order M) represented as the cascade of two functional blocks, one with transfer function $C_i(z)$ and the other with its transfer function equal to the zero factor $(1 - z_i z^{-1})$. Let $S(\omega)$ denote the power spectral density of the process $\{u(n)\}$ applied to the input of the prediction-error filter.

(a) Show that the mean-square value of the forward prediction error $f_M(n)$ equals

$$\epsilon = \int_{-\pi}^{\pi} S(\omega)\,|C_i(e^{j\omega})|^2[1 - 2\rho_i \cos(\omega - \omega_i) + \rho_i^2]d\omega$$

where $z_i = \rho_i e^{j\omega_i}$. Hence, evaluate the derivative $\partial\epsilon/\partial\rho_i$.

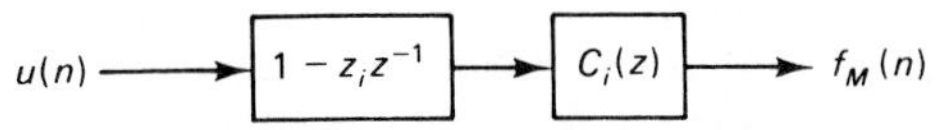

Figure P6.2

(b) Suppose that $\rho_i > 1$ so that the complex zero lies outside the unit circle. Hence, show that under this condition $\partial\epsilon/\partial\rho_i > 0$. Is such a condition possible and at the same time the filter operates at its optimum condition? What conclusion can you draw from your answers?

14. When an autoregressive process of order M is applied to a forward prediction-error filter of order M, the output consists of white noise. Show that when such a process is applied to a backward prediction-error filter of order M, the output consists of an anticausal realization of white noise.

15. Consider a prediction-error filter characterized by a real-valued set of coefficients $a_{m,1}, a_{m,2}, \ldots, a_{m,m}$. Define a polynomial $\phi_m(z)$ as follows:

$$\sqrt{P_m}\,\phi_m(z) = z^m + a_{m,1}z^{m-1} + \cdots + a_{m,m}$$

where P_m is the average prediction-error power of order m, and z^{-1} is the unit-delay operator. [Note the difference between the definition of $\phi_m(z)$ and that of the corresponding transfer function $A_m(z)$ of the filter.] The filter coefficients bear a one-to-one correspondence with the sequence of autocorrelations $r(0), r(1), \ldots, r(m)$. Define

$$S(z) = \sum_{i=-m}^{m} r(i)z^{-i}$$

Show that

$$\frac{1}{2\pi j}\oint_C \phi_n(z)\phi_k(z^{-1})S(z)\,dz = \delta_{mk}$$

where δ_{mk} is the Kronecker delta:

$$\delta_{mk} = \begin{cases} 1, & k = m \\ 0, & k \neq m \end{cases}$$

and the contour C is the unit circle. The polynomial $\phi_m(z)$ is referred to as a *Szegö polynomial*.

16. **(a)** Construct the two-stage lattice predictor for the second-order autoregressive process $\{u(n)\}$ considered in Problem 11.

(b) Given a white-noise process $\{v(n)\}$, construct the two-stage lattice synthesizer for generating the autoregressive process $\{u(n)\}$. Check your answer against the second-order difference equation for the process $\{u(n)\}$ that was considered in Problem 11.

17. In a *normalized* lattice predictor, the forward and backward prediction errors at the various stages of the predictor are all normalized to have *unit variance*. Such an operation makes it possible to utilize the full dynamic range of multipliers used in the hardware implementation of a lattice predictor. For stage m of the normalized lattice predictor, the normalized forward and backward prediction errors are defined as follows, respectively:

$$\bar{f}_m(n) = \frac{f_m(n)}{P_m^{1/2}}$$

and

$$\bar{b}_m(n) = \frac{b_m(n)}{P_m^{1/2}}$$

where P_m is the average power (or variance) of the forward prediction error $f_m(n)$ or that of the backward prediction error $b_m(n)$. Show that the structure of stage m of the normalized lattice predictor is as shown in Fig. P6.3 for real-valued data.

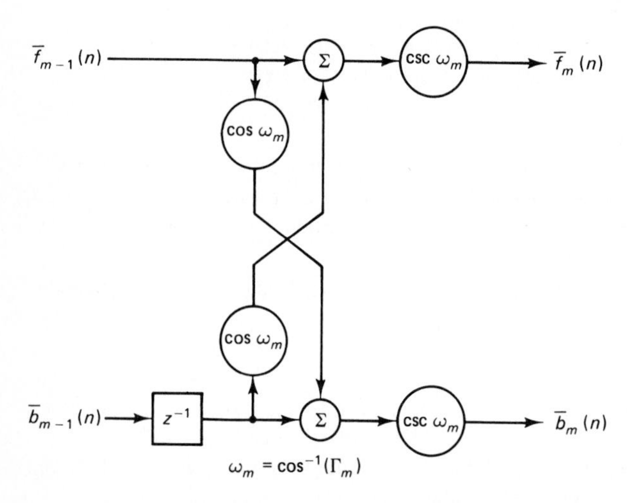

Figure P6.3

18. **(a)** Consider the matrix product **LR** that appears in the decomposition of Eq. (6.123), where the $(M + 1)$-by-$(M + 1)$ lower triangular matrix **L** is defined in Eq. (6.115) and $\mathbf{R}$ is the $(M + 1)$-by-$(M + 1)$ correlation matrix. Let $\mathbf{Y}$ denote this matrix product, and let y_{mk} denote the mkth element of $\mathbf{Y}$. Hence, show that

$$y_{mm} = P_m, \qquad m = 0, 1, \ldots, M$$

where P_m is the prediction-error power for order m.

(b) Show that the mth column of matrix $\mathbf{Y}$ is obtained by passing the autocorrelation sequence $\{r(0), r(1), \ldots, r(m)\}$ through a corresponding sequence of backward prediction-error filters represented by the transfer functions $H_{b,0}(z), H_{b,1}(z), \ldots, H_{b,m}(z)$.

(c) Suppose that we apply the autocorrelation sequence $\{r(0), r(1), \ldots, r(m)\}$ to the input of a lattice predictor of order m. Show that the variables appearing at the various points on the lower line of the predictor at time m equal the elements of the mth column of matrix $\mathbf{Y}$.

(d) For the situation described in part (c), show that the lower output of stage m in the predictor at time m equals P_m, and that the upper output of this same stage at time $m + 1$ equals Δ_m^*. How is the ratio of these two outputs related to the reflection coefficient of stage $m + 1$?

(e) Use the results of part (d) to develop a recursive procedure for computing the sequence of reflection coefficients from the autocorrelation sequence.

19. In Section 6.13, we considered the use of an inverse lattice filter as the generator of an autoregressive process. The lattice inverse filter may also be used to efficiently compute the autocorrelation sequence $r(1), r(2), \ldots, r(m)$ normalized with respect to $r(0)$. The procedure involves initializing the states (i.e., unit-delay elements) of the lattice inverse filter to $1, 0, \ldots, 0$ and then allowing the filter to operate with zero input. In effect, this procedure provides a lattice interpretation of Eq. (6.67) that relates the autocorrelation sequence $\{r(0), r(1), \ldots, r(M)\}$ and the augmented sequence of reflection coefficients $\{P_0, \Gamma_1, \ldots, \Gamma_M\}$. Demonstrate the validity of this procedure for the following values of final order M:

(a) $M = 1$
(b) $M = 2$
(c) $M = 3$

20. Prove the following correlation properties of lattice filters:

(a) $E[f_m(n)u^*(n - k)] = 0, \quad 1 \le k \le m$
$E[b_m(n)u^*(n - k)] = 0, \quad 0 \le k \le m - 1$

(b) $E[f_m(n)u^*(n)] = E[b_m(n)u^*(n - m)] = P_m$

(c) $E[b_m(n)b_i^*(n)] = \begin{cases} P_m, & m = i \\ 0, & m \ne i \end{cases}$

(d) $E[f_m(n)f_i^*(n - l)] = E[f_m(n + l)f_i^*(n)] = 0, \quad 1 \le l \le m - i, \quad m > i$
$E[b_m(n)b_i^*(n - l)] = E[b_m(n + l)b_i^*(n)] = 0, \quad 0 \le l \le m - i - 1, \quad m > i$

(e) $E[f_m(n + m)f_i^*(n + i)] = \begin{cases} P_m, & m = i \\ 0, & m \ne i \end{cases}$
$E[b_m(n + m)b_i^*(n + i)] = P_m, \quad m \ge i$

(f) $E[f_m(n)b_i^*(n)] = \begin{cases} \Gamma_i^* P_m, & m \ge i \\ 0, & m < i \end{cases}$

21. **(a)** The *entropy* of a random input vector $\mathbf{u}(n)$ of joint probability density function $f_{\mathbf{U}}(\mathbf{u})$ is defined by

$$H_u = -\int_{-\infty}^{\infty} f_{\mathbf{U}}(\mathbf{u}) \ln [f_{\mathbf{U}}(\mathbf{u})]\, d\mathbf{u}$$

Given that the backward prediction-error vector $\mathbf{b}(n)$ is related to $\mathbf{u}(n)$ by the Gram–Schmidt algorithm of Eq. (6.112), show that the vectors $\mathbf{b}(n)$ and $\mathbf{u}(n)$ are equivalent in the sense that they have the same entropy, and therefore the same amount of information.

(b) Use the result of part (a) to show that the entropy of the vector $\mathbf{b}$ is given by

$$H_b = \frac{1}{2} \sum_{k=0}^{M} \ln P_k$$

where P_k is the average power of element $b_k(n)$ in the vector $\mathbf{b}(n)$.

22. Consider the problem of optimizing stage m of the lattice predictor. The cost function to be used in the optimization is described by

$$J_m(\Gamma_m) = aE[|f_m(n)|^2] + (1 - a)E[|b_m(n)|^2]$$

where a is a constant that lies between zero and one; $f_m(n)$ and $b_m(n)$ denote the forward and backward prediction errors at the output of stage m, respectively.

(a) Show that the optimum value of the reflection coefficient Γ_m for which J_m is at minimum equals

$$\Gamma_{m,o}(a) = -\frac{E[b_{m-1}(n-1)f^*_{m-1}(n)]}{(1-a)E[|f_{m-1}(n)|^2] + aE[|b_{m-1}(n-1)|^2]}$$

(b) Evaluate $\Gamma_{m,o}(a)$ for each of the following three special conditions:

(1) $a = 1$

(2) $a = 0$

(3) $a = \frac{1}{2}$

Notes: When the parameter $a = 1$, the cost function reduces to

$$J_m(\Gamma_m) = E[|f_m(n)|^2]$$

We refer to this criterion as the *forward method.*

When the parameter $a = 0$, the cost function reduces to

$$J_m(\Gamma_m) = E[|b_m(n)|^2]$$

We refer to this criterion as the *backward method.*

When the parameter $a = \frac{1}{2}$, the formula for $\Gamma_{m,o}(a)$ reduces to the Burg formula.

23. Let $\Gamma_{m,o}(1)$ and $\Gamma_{m,o}(0)$ denote the optimum values of the reflection coefficient Γ_m for stage m of the lattice predictor using the forward method and backward method, respectively, as determined in Problem 22.

(a) Show that the optimum value of $\Gamma_{m,o}$ obtained from the Burg formula equals the harmonic mean of the two values $\Gamma_{m,o}(1)$ and $\Gamma_{m,o}(0)$, as shown by

$$\frac{2}{\Gamma_{m,o}} = \frac{1}{\Gamma_{m,o}(1)} + \frac{1}{\Gamma_{m,o}(0)}$$

(b) Using the result of part (a), show that

$$|\Gamma_{m,o}| \le 1, \qquad \text{for all } m$$

(c) For the case of a lattice predictor using the Burg formula, show that the mean-square values of the forward and backward prediction errors at the output of stage m are related to those at the input as follows, respectively:

$$E[|f_m(n)|^2] = (1 - |\Gamma_{m,o}|^2)E[|f_{m-1}(n)|^2]$$

and

$$E[|b_m(n)|^2] = (1 - |\Gamma_{m,o}|^2)E[|b_{m-1}(n - 1)|^2]$$

24. Consider a linear array that consists of a total of N uniformly spaced antenna elements. A *subaperture* is formed by using a contiguous subset of $M + 1$ elements. Assume that $N > M + 1$ so that it becomes possible to set up several such subapertures.

(a) Show that a prediction-error filter operating on the $M + 1$ element outputs of a subaperture is identical in configuration to a sidelobe canceller with one of the end elements of the subaperture providing the main beam and the remaining M elements of the subaperture treated as auxiliaries.

(b) Set up a lattice-equivalent model for the prediction-error filter that operates on the $M + 1$ element outputs of a subaperture.

(c) Suppose that a total of K snapshots of data are available for processing, with a *snapshot* defined as one simultaneous sampling of the aperture outputs at all elements of the array. Assume that the environment in which the array operates is stationary. Hence, modify the Burg formula for estimating the reflection coefficients of the lattice-equivalent model so that it includes not only spatial averaging but also temporal averaging.

In parts (a) and (b), illustrate your answer for $M = 3$.

7

Kalman Filters

In this chapter we complete our study of optimum linear discrete-time filters by developing the basic ideas of *Kalman filtering*. A distinctive feature of a Kalman filter is that its mathematical formulation is described in terms of *state-space concepts*. Another novel feature of a Kalman filter is that its solution is computed *recursively*. In particular, each updated estimate of the state is computed from the previous estimate and the new input data, so only the previous estimate requires storage. In addition to eliminating the need for storing the entire past observed data, a Kalman filter is computationally more efficient than computing the estimate directly from the entire past observed data at each step of the filtering process. The Kalman filter is thus ideally suited for implementation on a digital computer. Moreover, it has been applied successfully to many practical problems in diverse fields, particularly in aerospace and aeronautical applications.

To pave the way for the development of the Kalman filter, we begin by solving the *recursive minimum mean-squared estimation problem* for the simple case of scalar random variables. For this solution, we use the *innovations approach* that exploits the correlation properties of a special stochastic process known as the *innovations process* (Kailath, 1968, 1970).

7.1 RECURSIVE MINIMUM MEAN-SQUARE ESTIMATION FOR SCALAR RANDOM VARIABLES

Let us assume that, based on a complete set of observed random variables $y(1)$, $y(2), \ldots, y(n-1)$, starting with the first observation at time 1 and extending up to and including time $n-1$, we have found the minimum mean-square estimate $\hat{x}(n-1 \mid \mathcal{Y}_{n-1})$ of a related random variable $x(n-1)$. We are assuming that the observation at (or before) $n = 0$ is zero. The space spanned by the observations $y(1)$,

$\ldots, y(n-1)$ is denoted by $\mathcal{Y}_{n-1}$. Suppose that we now have an additional observation $y(n)$ at time n, and the requirement is to compute an *updated* estimate $\hat{x}(n|\mathcal{Y}_n)$ of the related random variable $x(n)$, where $\mathcal{Y}_n$ denotes the space spanned by $y(1), \ldots, y(n)$. We may do this computation by storing the *past* observations, $y(1), y(2), \ldots, y(n-1)$, and then redoing the whole problem with the available data $y(1), y(2,), \ldots, y(n-1), y(n)$, including the new observation. Computationally, however, it is much more efficient to use a *recursive estimation procedure*. In this procedure we *store* the previous estimate $\hat{x}(n-1|\mathcal{Y}_{n-1})$ and exploit it to compute the updated estimate $\hat{x}(n|\mathcal{Y}_n)$ in the light of the new observation $y(n)$. There are several ways of developing the algorithm to do this recursive estimation. We will use the notion of *innovations* (Kailath, 1968, 1970).

Define the forward prediction error

$$f_{n-1}(n) = y(n) - \hat{y}(n|\mathcal{Y}_{n-1}), \quad n = 1, 2, \ldots \tag{7.1}$$

where $\hat{y}(n|\mathcal{Y}_{n-1})$ is the *one-step prediction* of the observed random variable $y(n)$ at time n, using *all* past observations available up to and including time $n-1$. The past observations used in this estimation are $y(1), y(2), \ldots, y(n-1)$, so the order of the prediction equals $n-1$. We may view $f_{n-1}(n)$ as the output of a forward prediction-error filter of order $n-1$, and with the filter input fed by the time series $y(1), y(2), \ldots, y(n)$. Note that the *prediction order $n-1$ increases linearly with n.* According to the principle of orthogonality, the prediction error $f_{n-1}(n)$ is orthogonal to all past observations $y(1), y(2), \ldots, y(n-1)$ and may therefore be regarded as a *measure* of the new information in the random variable $y(n)$ observed at time n; hence, the name "innovation." The fact is that the observation $y(n)$ does not itself convey completely new information, since the predictable part, $\hat{y}(n|\mathcal{Y}_{n-1})$, is already completely determined by the past observations $y(1), y(2), \ldots, y(n-1)$. Rather, the part of the observation $y(n)$ that is really new is contained in the forward prediction error $f_{n-1}(n)$. We may therefore refer to this prediction error as the *innovation,* and for simplicity of notation write

$$\alpha(n) = f_{n-1}(n), \quad n = 1, 2, \ldots \tag{7.2}$$

The innovation $\alpha(n)$ has several important properties, as follows:

Property 1. *The innovation $\alpha(n)$, associated with the observed random variable $y(n)$, is orthogonal to the past observations $y(1), y(2), \ldots, y(n-1)$, as shown by*

$$E[\alpha(n)y^*(k)] = 0, \quad 1 \le k \le n-1 \tag{7.3}$$

This is simply a restatement of the principle of orthogonality.

Property 2. *The innovations $\alpha(1), \alpha(2), \ldots, \alpha(n)$ are orthogonal to each other, as shown by*

$$E[\alpha(n)\alpha^*(k)] = 0, \quad 1 \le k \le n-1 \tag{7.4}$$

This is a restatement of the fact that [see part (e) of Problem 20, Chapter 6]:

$$E[f_{n-1}(n)f_{k-1}^*(k)] = 0, \quad 1 \le k \le n-1$$

Equation (7.4), in effect, states that the innovation process $\{\alpha(n)\}$, described by Eqs. (7.1) and (7.2), is *white*.

Property 3. *There is a one-to-one correspondence between the observed data $\{y(1), y(2), \ldots, y(n)\}$ and the innovations $\{\alpha(1), \alpha(2), \ldots, \alpha(n)\}$ in that the one sequence may be obtained from the other by means of a causal and causally invertible filter, without any loss of information. We may thus write*

$$\{y(1), y(2), \ldots, y(n)\} \rightleftharpoons \{\alpha(1), \alpha(2), \ldots, \alpha(n)\} \tag{7.5}$$

To prove this property, we use a form of the Gram-Schmidt orthogonalization procedure (described in Chapter 6). The procedure assumes that the observations $y(1)$, $y(2), \ldots, y(n)$ are linearly independent in an algebraic sense. We first put

$$\alpha(1) = y(1) \tag{7.6}$$

where it is assumed that $\hat{y}(1 \mid \mathcal{Y}_0)$ is zero. Next, we put

$$\alpha(2) = y(2) + a_{1,1}y(1) \tag{7.7}$$

The coefficient $a_{1,1}$ is chosen such that the innovations $\alpha(1)$ and $\alpha(2)$ are orthogonal, as shown by

$$E[\alpha(2)\alpha^*(1)] = 0 \tag{7.8}$$

This requirement is satisfied by choosing

$$a_{1,1} = -\frac{E[y(2)y^*(1)]}{E[y(1)y^*(1)]} \tag{7.9}$$

Hence, except for a minus sign, $a_{1,1}$ is a partial correlation coefficient in that it equals the cross-correlation between the observations $y(2)$ and $y(1)$, normalized with respect to the mean-square value of $y(1)$.

Next we put

$$\alpha(3) = y(3) + a_{2,1}y(2) + a_{2,2}y(1) \tag{7.10}$$

where the coefficients $a_{2,1}$ and $a_{2,2}$ are chosen such that $\alpha(3)$ is orthogonal to both $\alpha(1)$ and $\alpha(2)$, and so on. Thus, in general, we may express the transformation of the observed data $y(1), y(2), \ldots, y(n)$ into the innovations $\alpha(1), \alpha(2), \ldots, \alpha(n)$ by writing

$$\begin{bmatrix} \alpha(1) \\ \alpha(2) \\ \vdots \\ \alpha(n) \end{bmatrix} = \begin{bmatrix} 1 & 0 & \cdots & 0 \\ a_{1,1} & 1 & \cdots & 0 \\ \vdots & \vdots & \ddots & \vdots \\ a_{n-1,n-1} & a_{n-1,n-2} & \cdots & 1 \end{bmatrix} \begin{bmatrix} y(1) \\ y(2) \\ \vdots \\ y(n) \end{bmatrix} \tag{7.11}$$

The nonzero elements of row k of the *lower triangular transformation matrix* on the right side of Eq. (7.11) equal $a_{k-1,k-1}, a_{k-1,k-2}, \ldots, 1$, where $k = 1, 2, \ldots, n$. These elements represent the coefficients of a *backward prediction-error filter* of order $k - 1$. Note that $a_{k,0} = 1$ for all k. Accordingly, given the observed data $y(1)$, $y(2), \ldots, y(n)$, we may compute the innovations $\alpha(1), \alpha(2), \ldots, \alpha(n)$. There is no loss of information in the course of this transformation, since we may recover the original observed data $y(1), y(2), \ldots, y(n)$ from the innovations $\alpha(1), \alpha(2), \ldots, \alpha(n)$. This we do by premultiplying both sides of Eq. (7.11) by the inverse of the lower triangular transformation matrix. This matrix is nonsingular since its determinant equals 1 for all n. The transformation is therefore reversible.

Using Eq. (7.5), we may thus write

$$\hat{x}(n|\mathcal{Y}_n) = \text{minimum mean-square estimate of } x(n) \text{ given the observed data } y(1), y(2), \ldots, y(n)$$

or, equivalently,

$$\hat{x}(n|\mathcal{Y}_n) = \text{minimum mean-square estimate of } x(n) \text{ given the innovations } \alpha(1), \alpha(2), \ldots, \alpha(n)$$

Define the estimate $\hat{x}(n|\mathcal{Y}_n)$ as a linear combination of the innovations $\alpha(1), \alpha(2), \ldots, \alpha(n)$:

$$\hat{x}(n|\mathcal{Y}_n) = \sum_{k=1}^{n} b_k \alpha(k) \tag{7.12}$$

where the b_k are to be determined. With the innovations $\alpha(1), \alpha(2), \ldots, \alpha(n)$ orthogonal to each other, and the b_k chosen to minimize the mean-square value of the estimation error $x(n) - \hat{x}(n|\mathcal{Y}_n)$, we find that

$$b_k = \frac{E[x(n)\alpha^*(k)]}{E[\alpha(k)\alpha^*(k)]}, \quad 1 \le k \le n \tag{7.13}$$

We rewrite Eq. (7.12) in the form

$$\hat{x}(n|\mathcal{Y}_n) = \sum_{k=0}^{n-1} b_k \alpha(k) + b_n \alpha(n) \tag{7.14}$$

where

$$b_n = \frac{E[x(n)\alpha^*(n)]}{E[\alpha(n)\alpha^*(n)]} \tag{7.15}$$

However, by definition, the summation term on the right side of Eq. (7.14) equals the previous estimate $\hat{x}(n-1|\mathcal{Y}_{n-1})$. We may thus express the desired recursive estimation algorithm as

$$\hat{x}(n|\mathcal{Y}_n) = \hat{x}(n-1|\mathcal{Y}_{n-1}) + b_n \alpha(n) \tag{7.16}$$

where b_n is defined by Eq. (7.15). Thus, by adding a *correction term* $b_n\alpha(n)$ to the previous estimate $\hat{x}(n-1|\mathcal{Y}_{n-1})$, with the correction proportional to the innovation $\alpha(n)$, we get the updated estimate $\hat{x}(n|\mathcal{Y}_n)$.

The simple formulas of Eqs. (7.12) and (7.16) are the basis of all recursive estimation schemes. Equipped with these simple and yet basic ideas, we are ready to study the more general Kalman filtering problem.

7.2 STATEMENT OF THE KALMAN FILTERING PROBLEM

Let an M-dimensional parameter vector $\mathbf{x}(n)$ denote the *state* of a discrete-time, linear, dynamical system, and let an N-dimensional vector $\mathbf{y}(n)$ denote the *observed data* of the system, both measured at time n. In general, the vectors $\mathbf{x}(n)$ and $\mathbf{y}(n)$ consist of vector random variables. Thus, the system *model* is described by two equations represented in the form of a signal-flow graph, as in Fig. 7.1. They are as follows:

1. A *process equation*

$$\mathbf{x}(n+1) = \mathbf{\Phi}(n+1, n)\mathbf{x}(n) + \mathbf{v}_1(n) \tag{7.17}$$

where $\mathbf{\Phi}(n+1, n)$ is a known M-by-M *state transition matrix* relating the states of the system at times $n+1$ and n. The M-by-1 vector $\mathbf{v}_1(n)$ represents *process noise*. The vector $\mathbf{v}_1(n)$ is modeled as zero-mean, white-noise processes whose correlation matrix is defined by

$$E[\mathbf{v}_1(n)\mathbf{v}_1^H(k)] = \begin{cases} \mathbf{Q}_1(n), & n = k \\ \mathbf{0}, & n \neq k \end{cases} \tag{7.18}$$

2. A *measurement equation*, describing the observation vector as follows:

$$\mathbf{y}(n) = \mathbf{C}(n)\mathbf{x}(n) + \mathbf{v}_2(n) \tag{7.19}$$

where $\mathbf{C}(n)$ is a known N-by-M *measurement matrix*. The N-by-1 vector $\mathbf{v}_2(n)$ is called *measurement noise*. It is modeled as zero-mean, white-noise processes whose correlation matrix is

$$E[\mathbf{v}_2(n)\mathbf{v}_2^H(k)] = \begin{cases} \mathbf{Q}_2(n), & n = k \\ \mathbf{0}, & n \neq k \end{cases} \tag{7.20}$$

The noise vectors $\mathbf{v}_1(n)$ and $\mathbf{v}_2(n)$ are statistically independent, so we may write

$$E[\mathbf{v}_1(n)\mathbf{v}_2^H(k)] = \mathbf{0}, \quad \text{for all } n \text{ and } k \tag{7.21}$$

Note that the state transition matrix $\mathbf{\Phi}(n+1, n)$ and the measurement matrix $\mathbf{C}(n)$ are both assumed *known*. The problem is to use the observed data, consisting of the vectors $\mathbf{y}(1), \mathbf{y}(2), \ldots, \mathbf{y}(n)$, to find for each $n \geq 1$ the minimum mean-square estimates of the components of the state $\mathbf{x}(i)$. The problem is called the

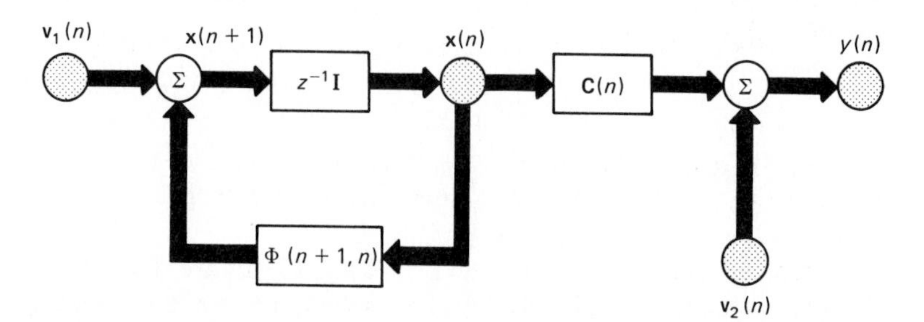

Figure 7.1 Signal-flow-graph representation of discrete-time, linear, dynamical system.

filtering problem if $i = n$, the *prediction* problem if $i > n$, and the *smoothing* problem if $1 \leqslant i < n$. We will only be concerned with the filtering and prediction problems, which are closely related. As remarked earlier in the introduction, we will solve the Kalman filtering problem by using the innovations approach (Kailath, 1968, 1970, 1981; Tretter, 1976).

7.3 THE INNOVATIONS PROCESS

Let the vector $\hat{\mathbf{y}}(n|\mathcal{Y}_n - 1)$ denote the minimum mean-square estimate of the observed data $\mathbf{y}(n)$ at time n, given all the past values of the observed data starting at time $n = 1$ and extending up to and including time $n - 1$. These past values are represented by the vectors $\mathbf{y}(1), \mathbf{y}(2), \ldots, \mathbf{y}(n-1)$, which span the vector space $\mathcal{Y}_{n-1}$. We define the *innovations process* associated with $\mathbf{y}(n)$ as

$$\boldsymbol{\alpha}(n) = \mathbf{y}(n) - \hat{\mathbf{y}}(n|\mathcal{Y}_{n-1}), \qquad n = 1, 2, \ldots \tag{7.22}$$

The M-by-1 vector $\boldsymbol{\alpha}(n)$ represents the new information in the observed data $\mathbf{y}(n)$.

Generalizing the results of Eqs. (7.3), (7.4) and (7.5), we find that the innovations process $\boldsymbol{\alpha}(n)$ has the following properties:

1. The innovations process $\boldsymbol{\alpha}(n)$, associated with the observed data $\mathbf{y}(n)$ at time n, is orthogonal to all past observations $\mathbf{y}(1), \mathbf{y}(2), \ldots, \mathbf{y}(n-1)$ as shown by

$$E[\boldsymbol{\alpha}(n)\mathbf{y}^H(k)] = \mathbf{0}, \quad 1 \le k \le n - 1 \tag{7.23}$$

2. The innovations process consists of a sequence of vector random variables that are orthogonal to each other, as shown by

$$E[\boldsymbol{\alpha}(n)\boldsymbol{\alpha}^H(k)] = \mathbf{0}, \quad 1 \le k \le n - 1 \tag{7.24}$$

3. There is a one-to-one correspondence between the sequence of vector random variables $\{\mathbf{y}(1), \mathbf{y}(2), \ldots, \mathbf{y}(n)\}$ representing the observed data and the sequence of vector random variables $\{\boldsymbol{\alpha}(1), \boldsymbol{\alpha}(2), \ldots, \boldsymbol{\alpha}(n)\}$ representing the innovations process, in that the one sequence may be obtained from the other by means of linear stable operators without loss of information. Thus, we may state that

$$\{\mathbf{y}(1), \mathbf{y}(2), \ldots, \mathbf{y}(n)\} \rightleftharpoons \{\boldsymbol{\alpha}(1), \boldsymbol{\alpha}(2), \ldots, \boldsymbol{\alpha}(n)\} \tag{7.25}$$

To form the sequence of vector random variables defining the innovations process, we may use a Gram-Schmidt orthogonalization procedure similar to that described in Section 7.1, except that the procedure is now formulated in terms of vectors and matrices (see Problem 1).

Correlation Matrix of the Innovations Process

To determine the correlation matrix of the innovations process $\boldsymbol{\alpha}(n)$, we first solve the state equation (7.17) recursively to obtain

$$\mathbf{x}(k) = \mathbf{\Phi}(k, 0)\mathbf{x}(0) + \sum_{i=1}^{k-1} \mathbf{\Phi}(k, i + 1)\mathbf{v}_1(i) \tag{7.26}$$

where we have made use of the following assumptions and properties:

1. The initial value of the state vector $\mathbf{x}(0)$.
2. As previously assumed, the observed data [and therefore the noise vector $\mathbf{v}_1(n)$] are zero for $n \leq 0$.
3. The state transition matrix has the properties

$$\mathbf{\Phi}(k, k-1)\mathbf{\Phi}(k-1, k-2)\ldots\mathbf{\Phi}(i+1, i) = \mathbf{\Phi}(k, i)$$

and

$$\mathbf{\Phi}(k, k) = \mathbf{I}$$

where $\mathbf{I}$ is the identity matrix.

Equation (7.26) shows that $\mathbf{x}(k)$ is a linear combination of $\mathbf{x}(0)$ and $\mathbf{v}_1(1)$, $\mathbf{v}_1(2)$, $\ldots$, $\mathbf{v}_1(k-1)$.

By hypothesis, the measurement noise vector $\mathbf{v}_2(n)$ is uncorrelated with both the initial state vector $\mathbf{x}(0)$ and the process noise vector $\mathbf{v}_1(n)$. Accordingly, premultiplying both sides of Eq. (7.26) by $\mathbf{v}_2^H(n)$, and taking expectations, we deduce that

$$E[\mathbf{x}(k)\mathbf{v}_2^H(n)] = \mathbf{0}, \quad k, n \geq 0 \tag{7.27}$$

Correspondingly, we deduce from the measurement equation (7.19) that

$$E[\mathbf{y}(k)\mathbf{v}_2^H(n)] = \mathbf{0}, \quad 0 \leq k \leq n-1 \tag{7.28}$$

Moreover, we may write

$$E[\mathbf{y}(k)\mathbf{v}_1^H(n)] = \mathbf{0}, \quad 0 \leq k \leq n \tag{7.29}$$

Given the past observations $\mathbf{y}(1), \ldots, \mathbf{y}(n-1)$ that span the space $\mathcal{Y}_{n-1}$, we also find from the measurement equation (7.19) that the minimum mean-square estimate of the present value $\mathbf{y}(n)$ of the observation vector equals

$$\hat{\mathbf{y}}(n|\mathcal{Y}_{n-1}) = \mathbf{C}(n)\,\hat{\mathbf{x}}(n|\mathcal{Y}_{n-1}) + \hat{\mathbf{v}}_2(n|\mathcal{Y}_{n-1})$$

However, the estimate $\hat{\mathbf{v}}_2(n|\mathcal{Y}_{n-1})$ of the measurement noise vector is zero since $\mathbf{v}_2(n)$ is orthogonal to the past observations $\mathbf{y}(1), \ldots, \mathbf{y}(n-1)$; see Eq. (7.28). Hence, we may simply write

$$\hat{\mathbf{y}}(n|\mathcal{Y}_{n-1}) = \mathbf{C}(n)\hat{\mathbf{x}}(n|\mathcal{Y}_{n-1}) \tag{7.30}$$

Therefore, using Eqs. (7.22) and (7.30), we may express the innovations process in the form

$$\boldsymbol{\alpha}(n) = \mathbf{y}(n) - \mathbf{C}(n)\hat{\mathbf{x}}(n|\mathcal{Y}_{n-1}) \tag{7.31}$$

Substituting the measurement equation (7.19) in (7.31), we get

$$\boldsymbol{\alpha}(n) = \mathbf{C}(n)\boldsymbol{\epsilon}(n, n-1) + \mathbf{v}_2(n) \tag{7.32}$$

where $\boldsymbol{\epsilon}(n, n-1)$ is the *predicted state-error vector* at time n, using data up to time

$n - 1$. That is, $\boldsymbol{\epsilon}(n, n-1)$ is the difference between the state vector $\mathbf{x}(n)$ and the one-step prediction vector $\hat{\mathbf{x}}(n|\mathcal{Y}_{n-1})$, as shown by

$$\boldsymbol{\epsilon}(n, n-1) = \mathbf{x}(n) - \hat{\mathbf{x}}(n|\mathcal{Y}_{n-1}) \tag{7.33}$$

Note that the predicted state-error vector is orthogonal to both the process noise vector $\mathbf{v}_1(n)$ and the measurement noise vector $\mathbf{v}_2(n)$; see Problem 2.

The correlation matrix of the innovations process $\boldsymbol{\alpha}(n)$ is defined by

$$\boldsymbol{\Sigma}(n) = E[\boldsymbol{\alpha}(n)\boldsymbol{\alpha}^H(n)] \tag{7.34}$$

Therefore, substituting Eq. (7.32) in (7.34), expanding the pertinent terms, and then using the fact that the vectors $\boldsymbol{\epsilon}(n, n-1)$ and $\mathbf{v}_2(n)$ are orthogonal, we obtain the desired result:

$$\boldsymbol{\Sigma}(n) = \mathbf{C}(n)\mathbf{K}(n, n-1)\mathbf{C}^H(n) + \mathbf{Q}_2(n) \tag{7.35}$$

where $\mathbf{Q}_2(n)$ is the correlation matrix of the noise vector $\mathbf{v}_2(n)$. The M-by-M matrix $\mathbf{K}(n, n-1)$ is called the *predicted state-error correlation matrix*, defined by

$$\mathbf{K}(n, n-1) = E[\boldsymbol{\epsilon}(n, n-1)\boldsymbol{\epsilon}^H(n, n-1)] \tag{7.36}$$

where $\boldsymbol{\epsilon}(n, n-1)$ is the predicted state-error vector. The matrix $\mathbf{K}(n, n-1)$ is used as the statistical description of the error in the predicted estimate $\hat{\mathbf{x}}(n|\mathcal{Y}_{n-1})$.

7.4 ESTIMATION OF THE STATE USING THE INNOVATIONS PROCESS

Consider next the problem of deriving the minimum mean-square estimate of the state $\mathbf{x}(i)$ from the innovations process. From the discussion presented in Section 7.1, we deduce that this estimate may be expressed as a linear combination of the sequence of innovations processes $\boldsymbol{\alpha}(1), \boldsymbol{\alpha}(2), \ldots, \boldsymbol{\alpha}(n)$ [see Eq. (7.12)]:

$$\hat{\mathbf{x}}(i|\mathcal{Y}_n) = \sum_{k=1}^{n} \mathbf{B}_i(k)\boldsymbol{\alpha}(k) \tag{7.37}$$

where $\{\mathbf{B}_i(k)\}$ is a set of M-by-N matrices to be determined. According to the principle of orthogonality, the predicted state-error vector is orthogonal to the innovation process, as shown by

$$\begin{aligned} E[\boldsymbol{\epsilon}(i, n)\boldsymbol{\alpha}^H(m)] &= E\{[\mathbf{x}(i) - \hat{\mathbf{x}}(i|\mathcal{Y}_n)]\boldsymbol{\alpha}^H(m)\} \\ &= \mathbf{0}, \quad m = 1, 2, \ldots, n \end{aligned} \tag{7.38}$$

Substituting Eq. (7.37) in (7.38) and using the orthogonality property of the innovations process, namely Eq. (7.24), we get

$$\begin{aligned} E[\mathbf{x}(i)\boldsymbol{\alpha}^H(m)] &= \mathbf{B}_i(m)E[\boldsymbol{\alpha}(m)\boldsymbol{\alpha}^H(m)] \\ &= \mathbf{B}_i(m)\boldsymbol{\Sigma}(m) \end{aligned} \tag{7.39}$$

Hence, postmultiplying both sides of Eq. (7.39) by $\boldsymbol{\Sigma}^{-1}(m)$, we find that $\mathbf{B}_i(m)$ is given by

$$\mathbf{B}_i(m) = E[\mathbf{x}(i)\boldsymbol{\alpha}^H(m)]\,\boldsymbol{\Sigma}^{-1}(m) \tag{7.40}$$

Finally, substituting Eq. (7.40) in (7.37), we get the desired value for the minimum mean-square estimate $\hat{\mathbf{x}}(i \mid \mathcal{Y}_n)$ as follows:

$$\begin{aligned}\hat{\mathbf{x}}(i \mid \mathcal{Y}_n) &= \sum_{k=1}^{n} E[\mathbf{x}(i)\boldsymbol{\alpha}^H(k)]\, \boldsymbol{\Sigma}^{-1}(k)\, \boldsymbol{\alpha}(k) \\ &= \sum_{k=1}^{n-1} E[\mathbf{x}(i)\boldsymbol{\alpha}^H(k)]\, \boldsymbol{\Sigma}^{-1}(k)\, \boldsymbol{\alpha}(k) \\ &\quad + E[\mathbf{x}(i)\boldsymbol{\alpha}^H(n)]\, \boldsymbol{\Sigma}^{-1}(n)\boldsymbol{\alpha}(n)\end{aligned}$$

For $i = n + 1$, we may therefore write

$$\begin{aligned}\hat{\mathbf{x}}(n+1 \mid \mathcal{Y}_n) &= \sum_{k=1}^{n-1} E[\mathbf{x}(n+1)\boldsymbol{\alpha}^H(k)]\, \boldsymbol{\Sigma}^{-1}(k)\boldsymbol{\alpha}(k) \\ &\quad + E[\mathbf{x}(n+1)\boldsymbol{\alpha}^H(n)]\, \boldsymbol{\Sigma}^{-1}(n)\boldsymbol{\alpha}(n)\end{aligned} \tag{7.41}$$

However, the state $\mathbf{x}(n+1)$ at time $n+1$ is related to the state $\mathbf{x}(n)$ at time n by Eq. (7.17). Therefore, using this relation, we may write for $0 \le k \le n$:

$$\begin{aligned}E\{\mathbf{x}(n+1)\boldsymbol{\alpha}^H(k)] &= E\{[\boldsymbol{\Phi}(n+1, n)\mathbf{x}(n) + \mathbf{v}_1(n)]\boldsymbol{\alpha}^H(k)\} \\ &= \boldsymbol{\Phi}(n+1, n)E[\mathbf{x}(n)\boldsymbol{\alpha}^H(k)]\end{aligned} \tag{7.42}$$

where we have made use of the fact that $\boldsymbol{\alpha}(k)$ depends only on the observed data $\mathbf{y}(1), \ldots, \mathbf{y}(k)$, and therefore from Eq. (7.29) we see that $\mathbf{v}_1(n)$ and $\boldsymbol{\alpha}(k)$ are orthogonal for $0 \le k \le n$. We may thus rewrite the summation term on the right side of Eq. (7.41) as follows:

$$\begin{aligned}\sum_{k=1}^{n-1} E[\mathbf{x}(n+1)\boldsymbol{\alpha}^H(k)]\, \boldsymbol{\Sigma}^{-1}(k)\boldsymbol{\alpha}(k) &= \boldsymbol{\Phi}(n+1, n) \sum_{k=1}^{n-1} E[\mathbf{x}(n)\boldsymbol{\alpha}^H(k)]\, \boldsymbol{\Sigma}^{-1}(k)\boldsymbol{\alpha}(k) \\ &= \boldsymbol{\Phi}(n+1, n)\hat{\mathbf{x}}(n \mid \mathcal{Y}_{n-1})\end{aligned} \tag{7.43}$$

Kalman Gain

Define the M-by-N matrix:

$$\mathbf{G}(n) = E[\mathbf{x}(n+1)\boldsymbol{\alpha}^H(n)]\, \boldsymbol{\Sigma}^{-1}(n) \tag{7.44}$$

Then, using this definition and the result of Eq. (7.43), we may rewrite Eq. (7.41) as follows:

$$\hat{\mathbf{x}}(n+1 \mid \mathcal{Y}_n) = \boldsymbol{\Phi}(n+1, n)\hat{\mathbf{x}}(n \mid \mathcal{Y}_{n-1}) + \mathbf{G}(n)\boldsymbol{\alpha}(n) \tag{7.45}$$

Equation (7.45) is of fundamental significance. It shows that we may compute the minimum mean-square estimate $\hat{\mathbf{x}}(n+1 \mid \mathcal{Y}_n)$ of the state of a linear dynamical system by adding to the previous estimate $\hat{\mathbf{x}}(n|\mathcal{Y}_{n-1})$ that is premultiplied by the state transition matrix $\boldsymbol{\Phi}(n+1, n)$ a correction term equal to $\mathbf{G}(n)\ \boldsymbol{\alpha}(n)$. The correction term equals the innovations process $\boldsymbol{\alpha}(n)$ premultiplied by the matrix $\mathbf{G}(n)$. Accordingly, and in recognition of the pioneering work by Kalman, the matrix $\mathbf{G}(n)$ is called the *Kalman gain*.

There now remains only the problem of expressing the Kalman gain $\mathbf{G}(n)$ in a form convenient for computation. To do this, we first use Eqs. (7.32) and (7.42) to express the expectation of the product of $\mathbf{x}(n+1)$ and $\boldsymbol{\alpha}^{H}(n)$ as follows:

$$\begin{aligned} E[\mathbf{x}(n+1)\boldsymbol{\alpha}^{H}(n)] &= \boldsymbol{\Phi}(n+1, n)E[\mathbf{x}(n)\boldsymbol{\alpha}^{H}(n)] \\ &= \boldsymbol{\Phi}(n+1, n)E[\mathbf{x}(n)(\mathbf{C}(n)\boldsymbol{\epsilon}(n, n-1) + \mathbf{v}_2(n))^{H}] \\ &= \boldsymbol{\Phi}(n+1, n)E[\mathbf{x}(n)\boldsymbol{\epsilon}^{H}(n, n-1)]\mathbf{C}^{H}(n) \end{aligned} \tag{7.46}$$

where we have used the fact that the state $\mathbf{x}(n)$ and noise vector $\mathbf{v}_2(n)$ are uncorrelated [see Eq. (7.27)]. We further note that the predicted state–error vector $\boldsymbol{\epsilon}(n, n-1)$ is orthogonal to the estimate $\hat{\mathbf{x}}(n|\mathcal{Y}_{n-1})$. Therefore, the expectation of the product of $\hat{\mathbf{x}}(n|\mathcal{Y}_{n-1})$ and $\boldsymbol{\epsilon}^{H}(n, n-1)$ is zero, and so we may rewrite Eq. (7.46) by replacing the multiplying factor $\mathbf{x}(n)$ by the predicted state-error vector $\boldsymbol{\epsilon}(n, n-1)$ as follows:

$$E[\mathbf{x}(n+1)\boldsymbol{\alpha}^{H}(n)] = \boldsymbol{\Phi}(n+1, n)E[\boldsymbol{\epsilon}(n, n-1)\boldsymbol{\epsilon}^{H}(n, n-1)]\mathbf{C}^{H}(n) \tag{7.47}$$

From Eq. (7.36), we see that the expectation on the right side of Eq. (7.47) equals the predicted state-error correlation matrix. Hence, we may rewrite Eq. (7.47) as follows:

$$E[\mathbf{x}(n+1)\boldsymbol{\alpha}^{H}(n)] = \boldsymbol{\Phi}(n+1, n)\,\mathbf{K}(n, n-1)\,\mathbf{C}^{H}(n) \tag{7.48}$$

We may now redefine the Kalman gain. In particular, substituting Eq. (7.48) in (7.44), we get

$$\mathbf{G}(n) = \boldsymbol{\Phi}(n+1, n)\mathbf{K}(n, n-1)\mathbf{C}^{H}(n)\,\boldsymbol{\Sigma}^{-1}(n) \tag{7.49}$$

where the correlation matrix $\boldsymbol{\Sigma}(n)$ itself is defined in Eq. (7.35).

The block diagram of Fig. 7.2 shows the signal-flow graph representation of Eq. (7.49) for computing the Kalman gain $\mathbf{G}(n)$. Having computed the Kalman gain $\mathbf{G}(n)$, we may then use Eq. (7.45) to update the one-step prediction, that is, to compute $\hat{\mathbf{x}}(n+1|\mathcal{Y}_n)$ given its old value $\hat{\mathbf{x}}(n|\mathcal{Y}_{n-1})$, as illustrated in Fig. 7.3.

Riccati Equation

As it stands, the formula of Eq. (7.49) is not particularly useful for computing the Kalman gain $\mathbf{G}(n)$, since it requires that the predicted state-error correlation matrix $\mathbf{K}(n, n-1)$ be known. To overcome this difficulty, we derive a formula for the recursive computation of $\mathbf{K}(n, n-1)$.

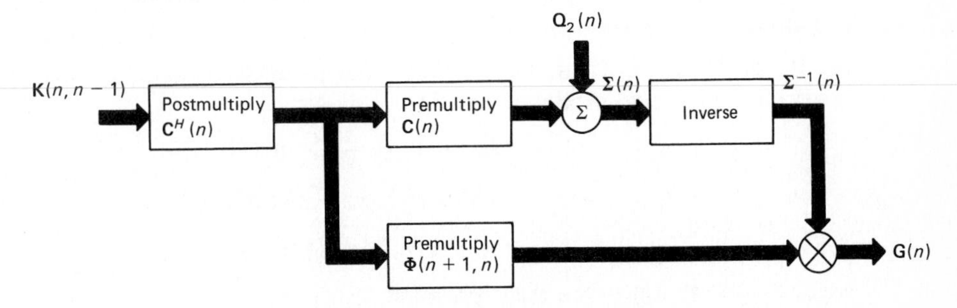

Figure 7.2 Kalman gain computer.

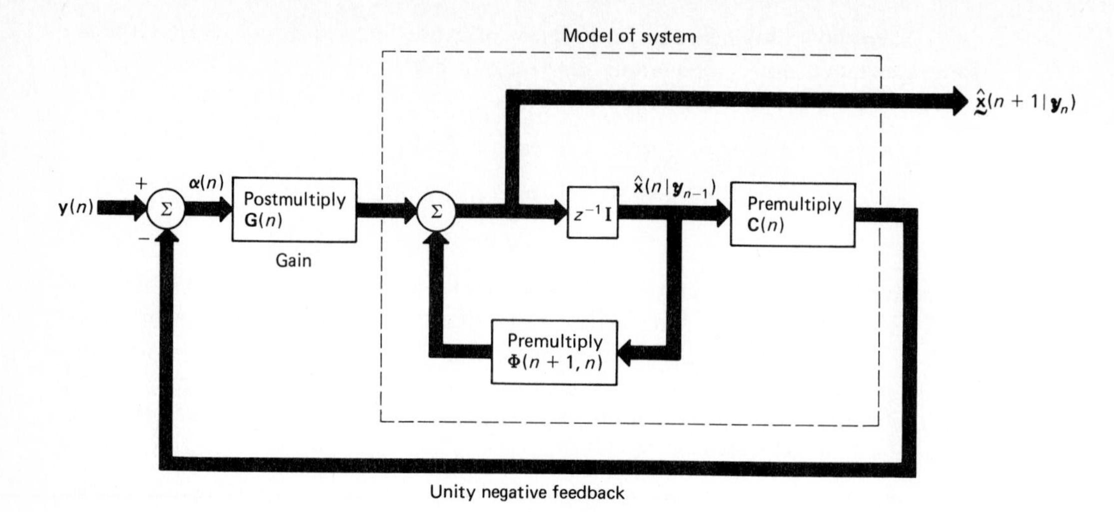

Figure 7.3 One-step predictor.

The predicted state-error vector $\boldsymbol{\epsilon}(n+1, n)$ equals the difference between the state $\mathbf{x}(n+1)$ and the one-step prediction $\hat{\mathbf{x}}(n+1|\mathcal{Y}_n)$[see Eq. (7.33)]:

$$\boldsymbol{\epsilon}(n+1, n) = \mathbf{x}(n+1) - \hat{\mathbf{x}}(n+1|\mathcal{Y}_n) \tag{7.50}$$

Substituting Eqs. (7.17) and (7.45) in (7.50), and using Eq. (7.31) for the innovations process $\boldsymbol{\alpha}(n)$, we get

$$\begin{aligned}\boldsymbol{\epsilon}(n+1, n) = \boldsymbol{\Phi}(n+1, n)&[\mathbf{x}(n) - \hat{\mathbf{x}}(n|\mathcal{Y}_{n-1})] \\ &- \mathbf{G}(n)\,[\mathbf{y}(n) - \mathbf{C}(n)\hat{\mathbf{x}}(n|\mathcal{Y}_{n-1})] + \mathbf{v}_1(n)\end{aligned} \tag{7.51}$$

Next, using the measurement equation (7.19) to eliminate $\mathbf{y}(n)$ in Eq. (7.51), we get the following difference equation for recursive computation of the predicted state-error vector:

$$\begin{aligned}\boldsymbol{\epsilon}(n+1, n) = {}& [\boldsymbol{\Phi}(n+1, n) - \mathbf{G}(n)\,\mathbf{C}(n)]\,\boldsymbol{\epsilon}(n, n-1) \\ &+ \mathbf{v}_1(n) - \mathbf{G}(n)\,\mathbf{v}_2(n)\end{aligned} \tag{7.52}$$

The correlation matrix of the predicted state-error vector $\boldsymbol{\epsilon}(n+1, n)$ equals [see Eq. (7.36)]

$$\mathbf{K}(n+1, n) = E[\boldsymbol{\epsilon}(n+1, n)\,\boldsymbol{\epsilon}^H(n+1, n)] \tag{7.53}$$

Substituting Eq. (7.52) in (7.53), and recognizing that the error vector $\boldsymbol{\epsilon}(n, n-1)$ and the noise vectors $\mathbf{v}_1(n)$ and $\mathbf{v}_2(n)$ are mutually uncorrelated, we may express the predicted state-error correlation matrix as follows:

$$\begin{aligned}\mathbf{K}(n+1, n) = {}& [\boldsymbol{\Phi}(n+1, n) - \mathbf{G}(n)\mathbf{C}(n)]\,\mathbf{K}(n, n-1) \\ &\cdot [\boldsymbol{\Phi}(n+1, n) - \mathbf{G}(n)\mathbf{C}(n)]^H \\ &+ \mathbf{Q}_1(n) + \mathbf{G}(\mathbf{n})\,\mathbf{Q}_2(n)\,\mathbf{G}^H(n)\end{aligned} \tag{7.54}$$

where $\mathbf{Q}_1(n)$ and $\mathbf{Q}_2(n)$ are the correlation matrices of $\mathbf{v}_1(n)$ and $\mathbf{v}_2(n)$, respectively. By expanding the right side of Eq. (7.54), and then using Eqs. (7.49) and (7.35) for the Kalman gain, we get the *Riccati difference equation*[1] for the recursive computation of the predicted state-error correlation matrix:

$$\mathbf{K}(n+1, n) = \mathbf{\Phi}(n+1, n)\, \mathbf{K}(n)\mathbf{\Phi}^H(n+1, n) + \mathbf{Q}_1(n) \tag{7.55}$$

The M-by-M matrix $\mathbf{K}(n)$ is described by the recursion:

$$\mathbf{K}(n) = \mathbf{K}(n, n-1) - \mathbf{\Phi}(n, n+1)\, \mathbf{G}(n)\, \mathbf{C}(n)\, \mathbf{K}(n, n-1) \tag{7.56}$$

Here we have used the fact that

$$\mathbf{\Phi}(n+1, n)\, \mathbf{\Phi}(n, n+1) = \mathbf{I} \tag{7.57}$$

where $\mathbf{I}$ is the identity matrix. This property follows from the definition of the transition matrix. The mathematical significance of the matrix $\mathbf{K}(n)$ in Eq. (7.56) will be explained in Section 7.5.

The block diagram of Fig. 7.4 is a signal-flow graph representation of Eqs. (7.56) and (7.55), in that order. This diagram may be viewed as the representation of the *Riccati equation solver* in that, given $\mathbf{K}(n, n-1)$, it computes the updated value $\mathbf{K}(n+1, n)$.

Equations (7.49), (7.35), (7.31), (7.45), (7.56), and (7.55), in that order, define Kalman's one-step prediction algorithm.

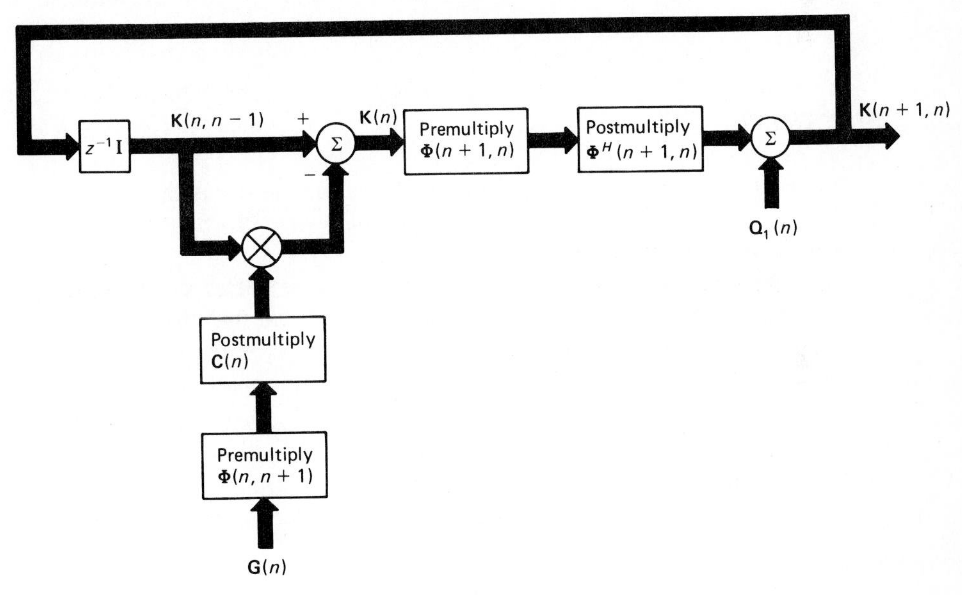

Figure 7.4 Riccati equation solver.

Comments

The process applied to the input of the Kalman filter consists of the observed data $\mathbf{y}(1), \mathbf{y}(2), \ldots, \mathbf{y}(n)$ that span the space $\mathcal{Y}_n$. The resulting filter output equals the predicted state vector $\hat{\mathbf{x}}(n+1 \mid \mathcal{Y}_n)$. Given that the matrices $\mathbf{\Phi}(n+1, n)$, $\mathbf{C}(n)$,

[1] The Riccati difference equation is named in honor of Count Jacopo Francisco Riccati. This equation has become of particular importance in the study of control theory.

$\mathbf{Q}_1(n)$, and $\mathbf{Q}_2(n)$ are all known quantities, we find from Eqs. (7.44), (7.55), and (7.56) that the predicted state-error correlation matrix $\mathbf{K}(n+1, n)$ is actually independent of the input $\mathbf{y}(n)$. Therefore, no one set of measurements helps more than any other to eliminate some uncertainty about the state vecor $\mathbf{x}(n)$ (Anderson and Moore, 1979). The Kalman gain $\mathbf{G}(n)$ is also independent of the input $\mathbf{y}(n)$. Consequently, the predicted state-error correlation matrix $\mathbf{K}(n+1, n)$ and the Kalman gain $\mathbf{G}(n)$ may be computed before the Kalman filter is actually put into operation. With the correlation matrix $\mathbf{K}(n+1, n)$ providing a statistical description of the error in the predicted state vector $\hat{\mathbf{x}}(n+1 \mid \mathcal{Y}(n))$, we may examine this matrix before actually using the Kalman filter to produce a realization of a physical system of interest; in this way, we may determine whether the solution supplied by the Kalman filter is indeed satisfactory.

As already mentioned, the Kalman filter theory assumes knowledge of the matrices $\mathbf{\Phi}(n+1, n)$, $\mathbf{C}(n)$, $\mathbf{Q}_1(n)$ and $\mathbf{Q}_2(n)$. However, the theory may be *generalized* to include a situation where one or more of these matrices may assume values that depend on the input $\mathbf{y}(n)$. In such a situation, we find that although $\hat{\mathbf{x}}(n+1 \mid \mathcal{Y}_n)$ and $\mathbf{K}(n+1, n)$ are still given by Eqs. (7.45) and (7.55), respectively, the Kalman gain $\mathbf{G}(n)$ and the predicted state-error correlation matrix $\mathbf{K}(n+1, n)$ are *not* precomputable (Anderson and Moore, 1979). Rather, they both now depend on the input $\mathbf{y}(n)$. This means that $\mathbf{K}(n+1, n)$ is a *conditional* error correlation matrix, conditional on the input $\mathbf{y}(n)$. An example of this situation is described in Section 7.8.

7.5 FILTERING

The next signal-processing operation we wish to consider is that of filtering. In particular, we wish to compute the *filtered estimate* $\hat{\mathbf{x}}(n \mid \mathcal{Y}_n)$ by using the one-step prediction algorithm described previously.

We first note that the state $\mathbf{x}(n)$ and the noise vector $\mathbf{v}_1(n)$ are independent of each other. Hence, from the state equation (7.17) we find that the minimum mean-square estimate of the state $\mathbf{x}(n+1)$ at time $n+1$, given the observed data up to and including time n [i.e., given $\mathbf{y}(1), \ldots, \mathbf{y}(n)$], equals

$$\hat{\mathbf{x}}(n+1 \mid \mathcal{Y}_n) = \mathbf{\Phi}(n+1, n)\, \hat{\mathbf{x}}(n \mid \mathcal{Y}_n) + \mathbf{v}_1(n \mid \mathcal{Y}_n) \tag{7.58}$$

Since the noise vector $\mathbf{v}_1(n)$ is independent of the observed data $\mathbf{y}(1), \ldots, \mathbf{y}(n)$, it follows that the corresponding minimum mean-square estimate $\hat{\mathbf{v}}_1(n \mid \mathcal{Y}_n)$ is zero. Accordingly, Eq. (7.58) simplifies as

$$\hat{\mathbf{x}}(n+1 \mid \mathcal{Y}_n) = \mathbf{\Phi}(n+1, n)\, \hat{\mathbf{x}}(n \mid \mathcal{Y}_n) \tag{7.59}$$

To find the filtered estimate $\hat{\mathbf{x}}(n \mid \mathcal{Y}_n)$, we premultiply both sides of Eq. (7.59) by the inverse of the transition matrix $\mathbf{\Phi}(n+1, n)$, and thus write

$$\hat{\mathbf{x}}(n \mid \mathcal{Y}_n) = \mathbf{\Phi}^{-1}(n+1, n)\, \hat{\mathbf{x}}(n+1 \mid \mathcal{Y}_n) \tag{7.60}$$

Using the property of the state transition matrix given in Eq. (7.57), we have

$$\mathbf{\Phi}^{-1}(n+1, n) = \mathbf{\Phi}(n, n+1) \tag{7.61}$$

We may therefore rewrite Eq. (7.60) in the form

$$\hat{\mathbf{x}}(n|\mathcal{Y}_n) = \mathbf{\Phi}(n, n+1)\,\hat{\mathbf{x}}(n+1|\mathcal{Y}_n) \tag{7.62}$$

This shows that knowing the solution to the one-step prediction problem, that is, the minimum mean-square estimate $\hat{\mathbf{x}}(n+1|\mathcal{Y}_n)$, we may determine the corresponding filtered estimate $\hat{\mathbf{x}}(n|\mathcal{Y}_n)$ simply by multiplying $\hat{\mathbf{x}}(n+1|\mathcal{Y}_n)$ by the state transition matrix $\mathbf{\Phi}(n, n+1)$.

Thus, we may reformulate Eq. (7.31) to express the innovations process $\boldsymbol{\alpha}(n)$ in terms of the filtered estimate of the state vector at time $n-1$ as follows:

$$\boldsymbol{\alpha}(n) = \mathbf{y}(n) - \mathbf{C}(n)\,\mathbf{\Phi}(n, n-1)\,\hat{\mathbf{x}}(n-1|\mathcal{Y}_{n-1}) \tag{7.63}$$

Similarly, we may reformulate Eq. (7.45) to express the recursion for time-updating the filtered estimate of the state vector as follows:

$$\hat{\mathbf{x}}(n|\mathcal{Y}_n) = \mathbf{\Phi}(n, n-1)\,\hat{\mathbf{x}}(n-1|\mathcal{Y}_{n-1}) + \mathbf{\Phi}(n, n+1)\,\mathbf{G}(n)\,\boldsymbol{\alpha}(n) \tag{7.64}$$

Equations (7.49), (7.63), (7.64), (7.56), and (7.55) collectively and in that order represent another way of describing the Kalman filter, based on the filtered estimate of the state vector.

Filtered State-Error Correlation Matrix

Earlier we introduced the M-by-M matrix $\mathbf{K}(n)$ in the formulation of the Riccati difference equation (7.55). We conclude our present discussion of the Kalman filter theory by showing that this matrix equals the correlation matrix of the error inherent in the filtered estimate $\hat{\mathbf{x}}(n|\mathcal{Y}_n)$.

Define the *filtered state-error vector* $\boldsymbol{\epsilon}(n)$ as the difference between the state $\mathbf{x}(n)$ and the filtered estimate $\hat{\mathbf{x}}(n|\mathcal{Y}_n)$, as shown by

$$\boldsymbol{\epsilon}(n) = \mathbf{x}(n) - \hat{\mathbf{x}}(n|\mathcal{Y}_n) \tag{7.65}$$

Substituting Eqs. (7.45) and (7.62) in (7.65), and recognizing that the product of $\mathbf{\Phi}(n, n+1)$ and $\mathbf{\Phi}(n+1, n)$ equals the identity matrix, we get

$$\begin{aligned}\boldsymbol{\epsilon}(n) &= \mathbf{x}(n) - \hat{\mathbf{x}}(n|\mathcal{Y}_{n-1}) - \mathbf{\Phi}(n, n+1)\,\mathbf{G}(n)\,\boldsymbol{\alpha}(n)\\ &= \boldsymbol{\epsilon}(n, n-1) - \mathbf{\Phi}(n, n+1)\,\mathbf{G}(n)\,\boldsymbol{\alpha}(n)\end{aligned} \tag{7.66}$$

where $\boldsymbol{\epsilon}(n, n, -1)$ is the predicted state-error vector at time n, using data up to time $n-1$, and $\boldsymbol{\alpha}(n)$ is the innovations process.

By definition, the correlation matrix of the filtered state-error vector $\boldsymbol{\epsilon}(n)$ equals the expectation $E[\boldsymbol{\epsilon}(n)\boldsymbol{\epsilon}^H(n)]$. Hence, using Eq. (7.66), we may express this expectation as follows:

$$\begin{aligned}E[\boldsymbol{\epsilon}(n)\boldsymbol{\epsilon}^H(n)] = {}& E[\boldsymbol{\epsilon}(n, n-1)\,\boldsymbol{\epsilon}^H(n, n-1)]\\ &+ \mathbf{\Phi}(n, n+1)\,\mathbf{G}(n)\,E[\boldsymbol{\alpha}(n)\boldsymbol{\alpha}^H(n)]\,\mathbf{G}^H(n)\,\mathbf{\Phi}^H(n, n+1)\\ &- 2E[\boldsymbol{\epsilon}(n, n-1)\boldsymbol{\alpha}^H(n)]\,\mathbf{G}^H(n)\,\mathbf{\Phi}^H(n, n+1)\end{aligned} \tag{7.67}$$

Examining the right side of Eq. (7.67), we find that the three expectations contained in it may be interpreted individually as follows:

1. The first expectation equals the predicted state-error correlation matrix:

$$\mathbf{K}(n, n-1) = E[\boldsymbol{\epsilon}(n, n-1)\, \boldsymbol{\epsilon}^H(n, n-1)]$$

2. The expectation in the second term equals the correlation matrix of the innovations process $\boldsymbol{\alpha}(n)$:

$$\boldsymbol{\Sigma}(n) = E[\boldsymbol{\alpha}(n)\boldsymbol{\alpha}^H(n)]$$

3. The expectation in the third term may be expressed as follows:

$$\begin{aligned} E[\boldsymbol{\epsilon}(n, n-1)\, \boldsymbol{\alpha}^H(n)] &= E[(\mathbf{x}(n) - \hat{\mathbf{x}}(n|\mathcal{Y}_{n-1}))\boldsymbol{\alpha}^H(n)] \\ &= E[\mathbf{x}(n)\, \boldsymbol{\alpha}^H(n)] \end{aligned}$$

where, in the last line, we have used the fact that the estimate $\hat{\mathbf{x}}(n|\mathcal{Y}_{n-1})$ is orthogonal to the innovations process $\boldsymbol{\alpha}(n)$, acting as input. Next, from Eq. (7.42) we see, by putting $k = n$ and premultiplying both sides by the inverse matrix $\boldsymbol{\Phi}^{-1}(n+1, n) = \boldsymbol{\Phi}(n, n+1)$, that

$$\begin{aligned} E[\mathbf{x}(n)\, \boldsymbol{\alpha}^H(n)] &= \boldsymbol{\Phi}(n, n+1)\, E[\mathbf{x}(n+1)\boldsymbol{\alpha}^H(n)] \\ &= \boldsymbol{\Phi}(n, n+1)\, \mathbf{G}(n)\, \boldsymbol{\Sigma}(n) \end{aligned}$$

where, in the last line, we have made use of Eq. (7.44). Hence,

$$E[\boldsymbol{\epsilon}(n, n-1)\, \boldsymbol{\alpha}^H(n)] = \boldsymbol{\Phi}(n, n+1)\, \mathbf{G}(n)\, \boldsymbol{\Sigma}(n)$$

We may now use these results in Eq. (7.67), and so obtain

$$\begin{aligned} E[\boldsymbol{\epsilon}(n)\boldsymbol{\epsilon}^H(n)] = \mathbf{K}(n, n-1) \\ - \boldsymbol{\Phi}(n, n+1)\, \mathbf{G}(n)\, \boldsymbol{\Sigma}(n)\mathbf{G}^H(n)\, \boldsymbol{\Phi}^H(n, n+1) \end{aligned} \tag{7.68}$$

We may further simplify this result by noting that [see Eq. (7.49)]

$$\mathbf{G}(n)\, \boldsymbol{\Sigma}(n) = \boldsymbol{\Phi}(n+1, n)\, \mathbf{K}(n, n-1)\, \mathbf{C}^H(n) \tag{7.69}$$

Accordingly, using Eqs. (7.68) and (7.69), and recognizing that the product of $\boldsymbol{\Phi}(n, n+1)$ and $\boldsymbol{\Phi}(n+1, n)$ equals the identity matrix, we get the desired result for the filtered state-error correlation matrix:

$$E[\boldsymbol{\epsilon}(n)\boldsymbol{\epsilon}^H(n)] = \mathbf{K}(n, n-1) - \mathbf{K}(n, n-1)\mathbf{C}^H(n)\, \mathbf{G}^H(n)\, \boldsymbol{\Phi}^H(n, n+1) \tag{7.70}$$

Equivalently, using the Hermitian property of $E[\boldsymbol{\epsilon}(n)\boldsymbol{\epsilon}^H(n)]$ and that of $\mathbf{K}(n, n-1)$, we may write

$$E[\boldsymbol{\epsilon}(n)\boldsymbol{\epsilon}^H(n)] = \mathbf{K}(n, n-1) - \boldsymbol{\Phi}(n, n+1)\, \mathbf{G}(n)\, \mathbf{C}(n)\, \mathbf{K}(n, n-1) \tag{7.71}$$

Comparing Eq. (7.71) with (7.56), we see that

$$E[\boldsymbol{\epsilon}(n)\, \boldsymbol{\epsilon}^H(n)] = \mathbf{K}(n)$$

This shows that the matrix $\mathbf{K}(n)$ used in the Riccati difference equation (7.55) is in fact the *filtered state-error correlation matrix*. The matrix $\mathbf{K}(n)$ is used as the statistical description of the error in the filtered estimate $\hat{\mathbf{x}}(n|\mathcal{Y}_n)$.

7.6 INITIAL CONDITIONS

To operate the one-step prediction and filtering algorithms described in Sections 7.4 and 7.5, we need to specify the *initial conditions*. We now address this issue.

The initial state of a Kalman filter is not known precisely. Rather, it is usually described by its mean and correlation matrix. In the absence of any observed data at time $n = 0$, we may choose the *initial filtered estimate* as

$$\hat{\mathbf{x}}(0 \mid \mathcal{Y}_0) = E[\hat{\mathbf{x}}(0)] \tag{7.72}$$

and its *correlation matrix* as

$$\mathbf{K}(0) = E[\mathbf{x}((0)\mathbf{x}^H(0)] = \mathbf{P}_0 \tag{7.73}$$

This choice for the initial conditions is not only intuitively satisfying but also has the advantage of yielding a filtered estimate of the state $\hat{\mathbf{x}}(n \mid \mathcal{Y}_n)$ that is *unbiased* (see Problem 10).

Assuming that the state vector $\mathbf{x}(n)$ has *zero mean*, we may simplify Eq. (7.72) by setting

$$\hat{\mathbf{x}}(0 \mid \mathcal{Y}_0) = \mathbf{0}$$

In other words, the initial value of the filtered estimate $\hat{\mathbf{x}}(n \mid \mathcal{Y}_n)$ is set equal to the *null vector*.

Note that, given the initial conditions of Eqs. (7.72) and (7.73), it is a straightforward matter to compute

$$\begin{aligned}\hat{\mathbf{x}}(1 \mid \mathcal{Y}_0) &= \mathbf{\Phi}(1, 0)\, \hat{\mathbf{x}}(0 \mid \mathcal{Y}_0) \\ &= \mathbf{\Phi}(1, 0)\, E[\hat{\mathbf{x}}(0)]\end{aligned}$$

and, using Eq. (7.55),

$$\begin{aligned}\mathbf{K}(1, 0) &= \mathbf{\Phi}(1, 0)\, \mathbf{K}(0)\, \mathbf{\Phi}^H(1, 0) + \mathbf{Q}_1(0) \\ &= \mathbf{\Phi}(1, 0)\, \mathbf{P}_0\, \mathbf{\Phi}^H(1, 0) + \mathbf{Q}_1(0)\end{aligned}$$

Computation of the algorithm may then proceed.

7.7 SUMMARY AND DISCUSSION

The Kalman filter is a *linear, discrete-time, finite-dimensional system,* the implementation of which is well suited for a digital computer. Table 7.1 presents a summary of the variables used to formulate the solution to the Kalman filtering problem. The input of the filter is the vector process $\{\mathbf{y}(n)\}$, represented by the vector space $\mathcal{Y}_n$, and the output is the filtered estimate $\hat{\mathbf{x}}(n \mid \mathcal{Y}_n)$ of the state vector. In Table 7.2, we present a summary of the Kalman filter (including initial conditions) based on the one-step prediction algorithm.

TABLE 7.1 SUMMARY OF KALMAN FILTER VARIABLES

Variable	Definition	Dimension
$\mathbf{x}(n)$	State vector at time n	M-by-1
$\mathbf{y}(n)$	Observation vector at time n	N-by-1
$\mathbf{\Phi}(n+1, n)$	State transition matrix from time n to $n+1$	M-by-M
$\mathbf{C}(n)$	Measurement matrix at time n	N-by-M
$\mathbf{Q}_1(n)$	Correlation matrix of process noise vector $\mathbf{v}_1(n)$	M-by-M
$\mathbf{Q}_2(n)$	Correlation matrix of measurement noise vector $\mathbf{v}_2(n)$	N-by-N
$\hat{\mathbf{x}}(n+1\vert\mathcal{Y}_n)$	Predicted estimate of the state vector at time $n+1$, given the observation vectors $\mathbf{y}(1), \mathbf{y}(2), \ldots, \mathbf{y}(n)$	M-by-1
$\hat{\mathbf{x}}(n\vert\mathcal{Y}_n)$	Filtered estimate of the state vector at time n, given the observation vectors $\mathbf{y}(1), \mathbf{y}(2), \ldots, \mathbf{y}(n)$	M-by-1
$\mathbf{G}(n)$	Kalman gain at time n	M-by-N
$\boldsymbol{\alpha}(n)$	Innovations vector at time n	N-by-1
$\mathbf{\Sigma}(n)$	Correlation matrix of the innovations vector $\boldsymbol{\alpha}(n)$	N-by-N
$\mathbf{K}(n+1, n)$	Correlation matrix of the error in $\hat{\mathbf{x}}(n+1\vert\mathcal{Y}_n)$	M-by-M
$\mathbf{K}(n)$	Correlation matrix of the error in $\hat{\mathbf{x}}(n\vert\mathcal{Y}_n)$	M-by-M

TABLE 7.2 SUMMARY OF THE KALMAN FILTER BASED ON ONE-STEP PREDICTION

Input vector process

Observations $= \{\mathbf{y}(1), \mathbf{y}(2), \ldots, \mathbf{y}(n)\}$

Known parameters

State transition matrix $= \mathbf{\Phi}(n+1, n)$

Measurement matrix $= \mathbf{C}(n)$

Correlation matrix of process noise vector $= \mathbf{Q}_1(n)$

Correlation matrix of measurement noise vector $= \mathbf{Q}_2(n)$

Computation: $n = 1, 2, 3, \ldots$

$$\mathbf{G}(n) = \mathbf{\Phi}(n+1, n)\mathbf{K}(n, n-1)\mathbf{C}^H(n)[\mathbf{C}(n)\mathbf{K}(n, n-1)\mathbf{C}^H(n) + \mathbf{Q}_2(n)]^{-1}$$
$$\boldsymbol{\alpha}(n) = \mathbf{y}(n) - \mathbf{C}(n)\hat{\mathbf{x}}(n|\mathcal{Y}_{n-1})$$
$$\hat{\mathbf{x}}(n+1|\mathcal{Y}_n) = \mathbf{\Phi}(n+1, n)\hat{\mathbf{x}}(n|\mathcal{Y}_{n-1}) + \mathbf{G}(n)\boldsymbol{\alpha}(n)$$
$$\hat{\mathbf{x}}(n|\mathcal{Y}_n) = \mathbf{\Phi}(n, n+1)\hat{\mathbf{x}}(n+1|\mathcal{Y}_n)$$
$$\mathbf{K}(n) = \mathbf{K}(n, n-1) - \mathbf{\Phi}(n, n+1)\mathbf{G}(n)\mathbf{C}(n)\mathbf{K}(n, n-1)$$
$$\mathbf{K}(n+1, n) = \mathbf{\Phi}(n+1, n)\mathbf{K}(n)\mathbf{\Phi}^H(n+1, n) + \mathbf{Q}_1(n)$$

Initial conditions

$$\hat{\mathbf{x}}(0|\mathcal{Y}_0) = E[\mathbf{x}(0)]$$
$$\mathbf{K}(0) = E[\mathbf{x}(0)\mathbf{x}^H(0)] = \mathbf{P}_0$$

The block diagram of Fig. 7.5 is a representation of the Kalman filter, based on three components:

1. Kalman gain computer
2. One-step predictor
3. Riccati equation solver

The details of these three components are shown in Figs. 7.2, 7.3, and 7.4, respectively.

A key property of the Kalman filter is that it leads to minimization of the trace of the filtered state-error correlation matrix $\mathbf{K}(n)$. This means that the Kalman filter is the *linear minimum variance estimator* of the state vector $\mathbf{x}(n)$ (Goodwin and Sin, 1984; Anderson and Moore, 1979).

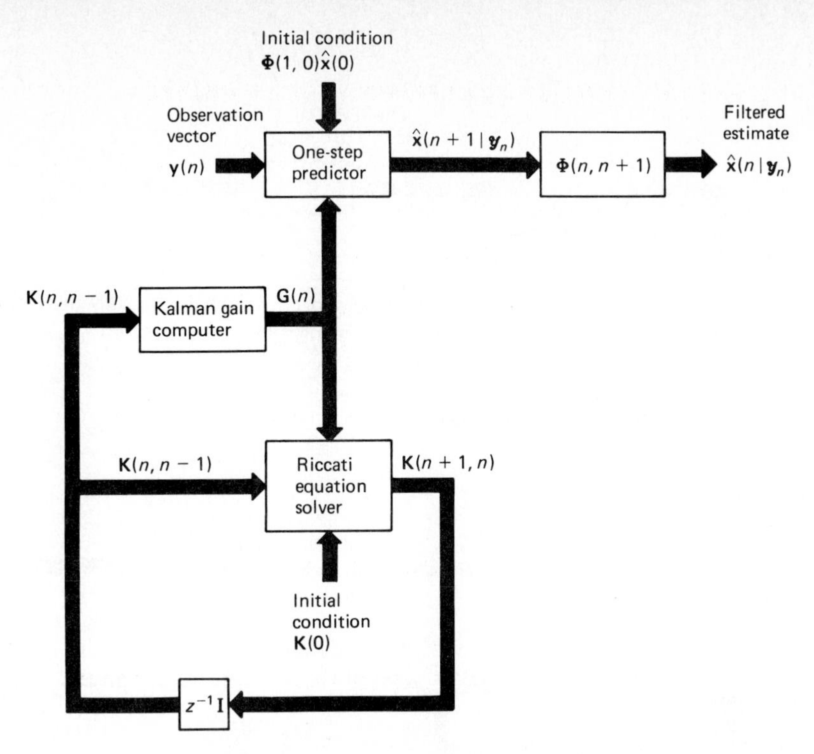

Figure 7.5 Block diagram of Kalman filter.

The input–output relation of the Kalman filter is depicted in Fig. 7.6(a), representing a signal-flow graph interpretation of Eqs. (7.63) and (7.64) and the fact that $\hat{\mathbf{x}}(n-1\,|\,\mathcal{Y}_{n-1})$ is the delayed version of $\hat{\mathbf{x}}(n\,|\,\mathcal{Y}_n)$. In this model, $\mathbf{y}(n)$ is the input and $\hat{\mathbf{x}}(n\,|\,\mathcal{Y}_n)$ is the output.

We may rearrange Eq. (7.63) in the form

$$\mathbf{y}(n) = \boldsymbol{\alpha}(n) + \mathbf{C}(n)\mathbf{\Phi}(n, n-1)\hat{\mathbf{x}}(n-1\,|\,\mathcal{Y}_{n-1}) \tag{7.74}$$

Accordingly, we may represent Eqs. (7.64) and (7.74) by the signal-flow graph shown in Fig. 7.6(b). Here again we have included a branch in the graph to represent the fact that $\hat{\mathbf{x}}(n-1\,|\,\mathcal{Y}_{n-1})$ is the delayed version of $\hat{\mathbf{x}}(n\,|\,\mathcal{Y}_n)$. We may view Fig. 7.6(b) as a model for generating the process $\{\mathbf{y}(n)\}$ by driving it with the innovations process $\{\boldsymbol{\alpha}(n)\}$. The model of Fig. 7.6(b) is known as the *inverse model*. In this model, the innovations process $\{\boldsymbol{\alpha}(n)\}$ acts as the input, and the process $\{\mathbf{y}(n)\}$ *is* the output of interest.

We may develop a third model by viewing the observations process $\{\mathbf{y}(n)\}$ as the input and the innovations process $\{\boldsymbol{\alpha}(n)\}$ as the output. The signal-flow graph representation of this model is the same as that shown in Fig. 7.6(a), except that now $\boldsymbol{\alpha}(n)$ represents the desired output at time n. The resulting model is known as the *whitening filter*.

We conclude therefore that the Kalman filter, the inverse model, and the whitening filter merely represent three different, and yet equivalent, ways of viewing the optimum linear filter.

The Kalman filter has been successfully applied to solve many real-world problems as can be seen in the literature on control systems. Invariably, assumptions are made so as to manipulate the problem of interest into a form amenable to the appli-

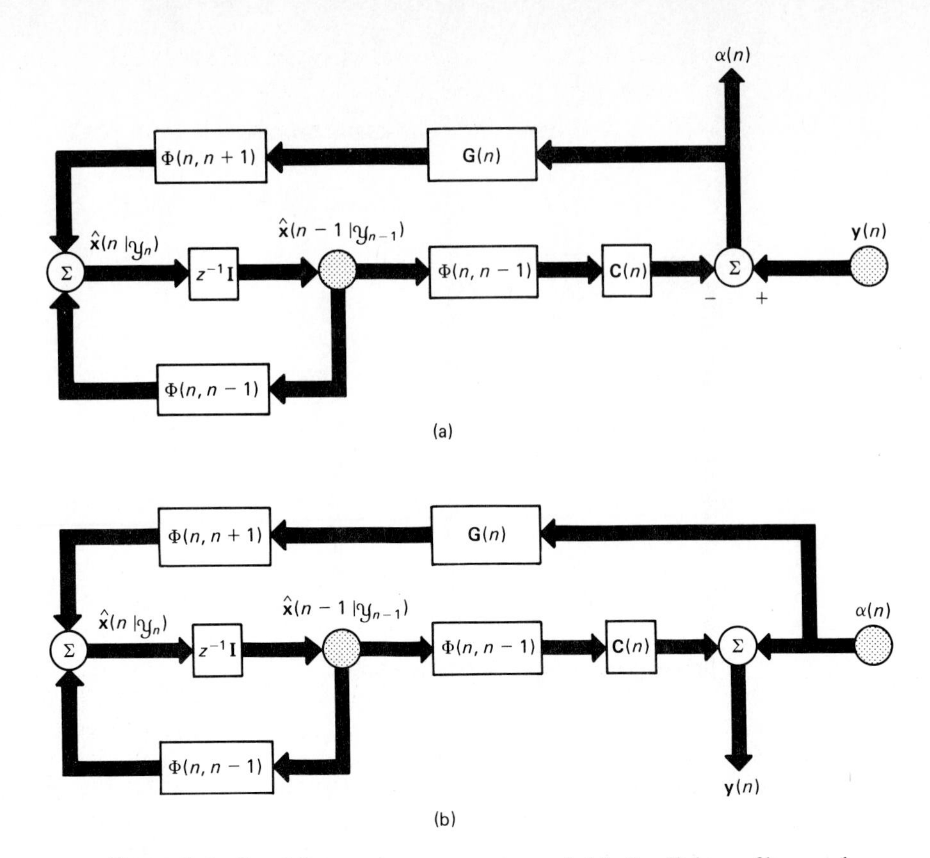

Figure 7.6 Signal-flow-graph representations of (a) the Kalman filter, and (b) the inverse model.

cation of the Kalman filter theory, the aim being to produce a near-optimum and yet workable solution.

Our interest in the Kalman filter theory in this book is that it provides a general framework for the development of various adaptive filtering algorithms with a fast rate of convergence. The application of Kalman filter theory to adaptive filtering was apparently first considered by Lawrence and Kaufman (1971); see Problem 9. This was followed by Godard (1974), who used a differend approach. In particular, Godard formulated the adaptive filtering problem (using a transversal structure) as the estimation of a state vector in Gaussian noise, a classical Kalman filtering problem. Godard's approach is considered in Chapter 13.

7.8 EXAMPLE: DYNAMIC AUTOREGRESSIVE MODEL

To illustrate the Kalman filtering theory described herein, consider an *autoregressive* (AR) *process* $\{u(n)\}$ of order M, which is described by the difference equation

$$\sum_{k=0}^{M} a_k^*(n)u(n-k) = e_o(n) \tag{7.75}$$

where $a_0 = 1$, and $a_1(n), \ldots, a_M(n)$ are time-varying AR parameters, and $\{e_o(n)\}$ is a stationary noise process of zero mean and variance σ^2. The dependence of the AR parameters on time n signifies the dynamic nature of the process. We may rewrite Eq. (7.75) in the form

$$u(n) = \mathbf{u}^T(n-1)\,\mathbf{w}_o^*(n) + e_o(n) \tag{7.76}$$

where the M-by-1 vector $\mathbf{u}(n-1)$ has $u(n-1), \ldots, u(n-M)$ as elements, and the kth element of the M-by-1 vector $\mathbf{w}_o(n)$ equals $w_{o,k}(n) = -a_k(n)$, $k = 1, 2, \ldots, M$. We may view Eq. (7.76) as the measurement equation of the dynamic AR model.

Assuming a *random-walk model* for the process equation of the AR model, we write

$$\mathbf{w}_o^*(n+1) = \mathbf{w}_o^*(n) + \mathbf{v}^*(n) \tag{7.77}$$

where $\{\mathbf{v}(n)\}$ is a stationary noise process of zero mean and correlation matrix $\mathbf{Q}(n) = q\mathbf{I}$. The processes $\{e_o(n)\}$ and $\{\mathbf{v}(n)\}$ are statistically independent. We may view Eq. (7.77) a the process equation of the dynamic AR model.

Comparing Eqs. (7.77) and (7.76) of the dynamic AR model with the process equation (7.17) and the measurement equation (7.19) of the Kalman filter, respectively, we may deduce the identifications listed in Table 7.3. Since, for this example, the state transition matrix $\mathbf{\Phi}(n, n+1)$ equals the identity matrix, the predicted estimate $\hat{\mathbf{x}}(n+1|\mathcal{Y}_n)$ and the filtered estimate $\hat{\mathbf{x}}(n|\mathcal{Y}_n)$ of the state vector assume the same value. In addition, the predicted state-error correlation matrix $\mathbf{K}(n, n-1)$ and the filtered state-error correlation matrix $\mathbf{K}(n)$ assume the same value. Accordingly, through the use of the identifications listed in Table 7.3 and the Kalman filter summarized in Table 7.2, we get the following algorithm for the dynamic AR model described herein:

$$\mathbf{G}(n) = \mathbf{K}(n, n-1)\mathbf{u}(n-1)[\mathbf{u}^H(n-1)\mathbf{K}(n, n-1)\mathbf{u}(n-1) + \sigma^2]^{-1} \tag{7.78}$$

$$\alpha(n) = u(n) - \mathbf{u}^T(n-1)\hat{\mathbf{w}}^*(n-1) \tag{7.79}$$

$$\hat{\mathbf{w}}(n) = \hat{\mathbf{w}}(n-1) + \mathbf{G}(n)\alpha^*(n) \tag{7.80}$$

TABLE 7.3 SUMMARY OF RELATIONS BETWEEN THE KALMAN FILTER AND DYNAMIC AR MODEL

Parameter	State model of the Kalman filter	Dynamic AR model
State vector	$\mathbf{x}(n)$	$\mathbf{w}_o^*(n)$
Transition matrix	$\mathbf{\Phi}(n+1, n)$	$\mathbf{I}$ (identity matrix)
Process noise vector	$\mathbf{v}_1(n)$	$\mathbf{v}(n)$
Covariance matrix of process noise vector	$\mathbf{Q}_1(n)$	$q\mathbf{I}$
Observation vector	$\mathbf{y}(n)$	$u(n)$
Measurement matrix	$\mathbf{C}(n)$	$\mathbf{u}^T(n-1)$
Measurement noise vector	$\mathbf{v}_2(n)$	$e_o(n)$
Covariance matrix of measurement noise vector	$\mathbf{Q}_2(n)$	σ^2

$$\mathbf{K}(n) = \mathbf{K}(n, n-1) - \mathbf{G}(n)\mathbf{u}^H(n-1)\mathbf{K}(n, n-1) \tag{7.81}$$

$$\mathbf{K}(n+1, n) = \mathbf{K}(n) + q\mathbf{I} \tag{7.82}$$

Thus, starting with the initial condition

$$\hat{\mathbf{w}}(0) = \mathbf{0} \tag{7.83}$$

$$\mathbf{K}(0) = c\mathbf{I} \qquad c = \text{small positive constant} \tag{7.84}$$

we may use this algorithm to model a nonstationary process $\{u(n)\}$ as an adaptive AR process, the parameters of which vary randomly about some mean values (Nau and Oliver, 1979). The use of this algorithm assumes prior knowledge of (1) the variance σ^2 of the residual $e_o(n)$ contained in Eq. (7.75) describing the AR equation of dynamic behavior, and (2) the variance q of each element of the noise vector $\mathbf{v}(n)$ that governs the random walk of the AR parameters with time.

7.9 SQUARE-ROOT KALMAN FILTERING ALGORITHMS

The basic form of the Kalman filter described in Section 7.5 suffers from a *numerical instability problem* that originates from the recursive formula used to compute the filtered state-error correlation matrix $\mathbf{K}(n)$ [see Eq. (7.56)]. We see that in this equation and in all the special forms of it the correlation matrix $\mathbf{K}(n)$ is computed as the difference between two nonnegative definite matrices. Hence, unless the numerical accuracy used at every iteration is high enough, the matrix $\mathbf{K}(n)$ resulting from this computation may *not* be nonnegative definite, as required. Another factor that contributes to numerical instability of the Kalman filter is the integrated effect of roundoff errors, especially when the dimension M of the state vector is high. As a result of the numerical degeneration of the classical Kalman filter, the matrix $\mathbf{K}(n)$ may assume an indefinite form, having both positive and negative eigenvalues; clearly, this is unacceptable. Yet another cause of the divergence phenomenon in a Kalman filter is the *inaccurate modeling* of the system under consideration.

To overcome these numerical problems, several modifications of the Kalman filter have been presented in the literature.[2] Two different procedures deserve special mention:

1. *Square root filtering* (SRF). In this procedure, numerically stable unitary transformations are used in each recursion step of the Kalman filtering algorithm (Potter, 1963; Kaminski et al. 1971). In particular, the filtered state error correlation matrix $\mathbf{K}(n)$ is propagated in a square root form in the absence of process noise by using the *Cholesky factorization* (Stewart, 1973):

$$\mathbf{K}(n) = \mathbf{K}^{1/2}(n)\mathbf{K}^{H/2}(n) \tag{7.85}$$

   where $\mathbf{K}^{1/2}(n)$ is reserved for a lower triangular matrix, and $\mathbf{K}^{H/2}$ is its Hermitian transpose. In linear algebra, the Cholesky factor $\mathbf{K}^{1/2}(n)$ is commonly referred to as the *square root* of the matrix $\mathbf{K}(n)$; hence, the terminology "square

[2] For a review of the numerical aspects of different Kalman filter implementations, see Verhaegen and Van Dooren (1986).

root filtering". Since the product of any square matrix and its Hermitian transpose is always positive definite, it follows that the matrix product $\mathbf{K}^{1/2}(n)\mathbf{K}^{H/2}(n)$ can never become indefinite. Indeed even in the presence of roundoff errors, the numerical conditioning of the Cholesky factor $\mathbf{K}^{1/2}(n)$ is generally much better than that of $\mathbf{K}(n)$, as shown later in this section.

2. *UD-factorization*. The SRF implementation of a Kalman filter requires more computation than the conventional Kalman filter. This problem of computational efficiency led to the development of a modified version of the SRF filter known as the *UD-factorization algorithm* (Bierman, 1977). In this second approach, the filtered state error correlation matrix $\mathbf{K}(n)$ is factored into an upper triangular matrix $\mathbf{U}(n)$ with 1's along its main diagonal and a real diagonal matrix $\mathbf{D}(n)$, as shown by

$$\mathbf{K}(n) = \mathbf{U}(n)\mathbf{D}(n)\mathbf{U}^H(n) \tag{7.86}$$

Equivalently, the factorization may be written as

$$\mathbf{K}(n) = (\mathbf{U}(n)\mathbf{D}^{1/2}(n))\,(\mathbf{U}(n)\mathbf{D}^{1/2}(n))^H \tag{7.87}$$

where $\mathbf{D}^{1/2}(n)$ is the *square root* of $\mathbf{D}(n)$. Accordingly, the Kalman filtering algorithm resulting from the use of UD-factorization is also referred to in the literature as a square-root Kalman filter even though the computation of square roots is not actually required. In any event, the nonnegative definiteness of the computed matrix $\mathbf{K}(n)$ is guaranteed by updating the factors $\mathbf{U}(n)$ and $\mathbf{D}(n)$ instead of $\mathbf{K}(n)$ itself.

A detailed derivation of the standard square root Kalman filtering algorithm for real data is given in Kaminski et al. (1971). For details of the square root Kalman filtering algorithm based on UD-factorization, the reader is referred to Bierman (1977) for the case of real data and Hsu (1982) for the more general case of complex data.

Numerical Conditioning Considerations

Although a Kalman filter based on UD-factorization is computationally more efficient than a standard square-root Kalman filter, the latter form of implementation offers a better numerical behavior (Stewart and Chapman, 1990). The *condition number* of a matrix $\mathbf{K}(n)$ is defined by (see Chapter 4):

$$\chi(\mathbf{K}) = \frac{\lambda_{\max}}{\lambda_{\min}} \tag{7.88}$$

where $\lambda^2_{\max}$ is the largest eigenvalue of the matrix product $\mathbf{K}^H(n)\mathbf{K}(n)$, and $\lambda^2_{\min}$ is its smallest eigenvalue. When the computation is performed in m-bit arithmetic, numerical problems arise as the condition number $\chi(\mathbf{K})$ approaches 2^m. To appreciate the numerical advantage of using the SRF approach, we may use Eq. (7.85) to express the condition number $\chi(\mathbf{K})$ as

$$\begin{aligned}\chi(\mathbf{K}) &= \chi(\mathbf{K}^{1/2}\mathbf{K}^{H/2}) \\ &= (\chi(\mathbf{K}^{1/2}))^2\end{aligned} \tag{7.89}$$

Equation (7.89) shows that the condition number of the Cholesky factor $\mathbf{K}^{1/2}(n)$ is the square root of the condition number of the original matrix $\mathbf{K}(n)$. Therefore, the SRF implementation of a Kalman filter will not experience numerical difficulties until $\chi(\mathbf{K}) = 2^{2m}$. In other words, the numerical precision has been effectively *doubled* through the use of standard square root filtering.

Consider next the use of UD-factorization. In this case we find from Eq. (7.87) that the condition number of the matrix $\mathbf{K}(n)$ is given by

$$\begin{aligned} \chi(\mathbf{K}) &= \chi(\mathbf{UDU}^H) \\ &\leq \chi(\mathbf{U})\chi(\mathbf{D})\chi(\mathbf{U}^H) \end{aligned} \tag{7.90}$$

The eigenvalues of an upper triangular matrix are the same as the diagonal elements of the matrix. For the situation at hand, the upper triangular matrix $\mathbf{U}(n)$ has 1's for all of its diagonal elements. Hence, the $\lambda_{\max}$ and $\lambda_{\min}$ of $\mathbf{U}(n)$ are both equal to one. Correspondingly, the condition number of the diagonal matrix $\mathbf{D}(n)$ is

$$\chi(\mathbf{D}) \geq \frac{\lambda_{\max}}{\lambda_{\min}} = \chi(\mathbf{K}) \tag{7.91}$$

where $\lambda_{\max}^2$ and $\lambda_{\min}^2$ are the maximum and minimum eigenvalues of the matrix product $\mathbf{K}^H(n)\mathbf{K}(n)$. Previously we stated that numerical problems may arise when computing with an m-bit arithmetic. From Eq. (7.91) we note that the condition number of the diagonal matrix $\mathbf{D}(n)$ is equal to or greater than that of the original matrix $\mathbf{K}(n)$. It follows therefore that numerical problems can be expected with the UD-factorization approach when $\chi(\mathbf{K}) = 2^m$, whereas the standard SRF approach will not experience numerical problems until $\chi(\mathbf{K}) = 2^{2m}$. In other words, a Kalman filter based on the UD-factorization does *not* possess the numerical advantage of a standard SRF-based Kalman filter. Moreover, a Kalman filter using UD-factorization may suffer from serious overflow/underflow problems (Stewart and Chapman, 1990).

With the ever-increasing improvements in digital technology, the old argument that square roots are expensive and awkward to calculate is no longer as compelling as it used to be. Indeed, with the availability of fast square-root arrays, there are *no* advantages to be gained by using the square-root free UD-factorization approach over the Cholesky factorization approach (Stewart and Chapman, 1990).

PROBLEMS

1. The Gram-Schmidt orthogonalization procedure enables the set of observation vectors $\mathbf{y}(1), \mathbf{y}(2), \ldots, \mathbf{y}(n)$ to be transformed into the set of innovations processes $\boldsymbol{\alpha}(1), \boldsymbol{\alpha}(2), \ldots, \boldsymbol{\alpha}(n)$ without loss of information, and vice versa. Illustrate this procedure for $n = 2$, and comment on the procedure for $n > 2$.
2. The predicted state-error vector is defined by

$$\boldsymbol{\epsilon}(n, n-1) = \mathbf{x}(n) - \hat{\mathbf{x}}(n \mid \mathcal{Y}_{n-1})$$

where $\hat{\mathbf{x}}(n \mid \mathcal{Y}_{n-1})$ is the minimum mean-square estimate of the state $\mathbf{x}(n)$, given the space $\mathcal{Y}_{n-1}$ that is spanned by the observed data $\mathbf{y}(1), \ldots, \mathbf{y}(n-1)$. Let $\mathbf{v}_1(n)$ and

$\mathbf{v}_2(n)$ denote the process noise and measurement noise vectors, respectively. Show that $\boldsymbol{\epsilon}(n, n-1)$ is orthogonal to both $\mathbf{v}_1(n)$ and $\mathbf{v}_2(n)$; that is,

$$E[\boldsymbol{\epsilon}(n, n-1)\mathbf{v}_1^H(n)] = \mathbf{0}$$

and

$$E[\boldsymbol{\epsilon}(n, n-1)\mathbf{v}_2^H(n)] = \mathbf{0}$$

3. Consider a set of scalar observations $\{y(n)\}$ of zero mean, which is transformed into the corresponding set of innovations $\{\alpha(n)\}$ of zero mean and variance $\sigma_\alpha^2(n)$. Let the estimate of the state vector $\mathbf{x}(i)$, given this set of data, be expressed as

$$\hat{\mathbf{x}}(i \mid \mathcal{Y}_n) = \sum_{k=1}^{n} \mathbf{b}_i(k)\alpha(k)$$

where $\mathcal{Y}_n$ is the space spanned by $y(1), \ldots, y(n)$, and $\{\mathbf{b}_i(k)\}$ is a set of vectors to be determined. The requirement is to choose the $\mathbf{b}_i(k)$ so as to minimize the expected value of the squared norm of the estimated state-error vector

$$\boldsymbol{\epsilon}(i \mid \mathcal{Y}_n) = \mathbf{x}(i) - \hat{\mathbf{x}}(i \mid \mathcal{Y}_n)$$

Show that this minimization yields the result

$$\hat{\mathbf{x}}(i \mid \mathcal{Y}_n) = \sum_{k=1}^{n} E[\mathbf{x}(i)\phi^*(k)]\phi(k)$$

where $\phi(k)$ is the normalized innovation

$$\phi(k) = \frac{\alpha(k)}{\sigma_\alpha(k)}$$

This result may be viewed as a special case of Eqs. (7.37) and (7.40).

4. The Kalman gain $\mathbf{G}(n)$ defined in Eq. (7.49) involves the inverse matrix $\boldsymbol{\Sigma}^{-1}(n)$. The matrix $\boldsymbol{\Sigma}(n)$ is itself defined in Eq. (7.35), reproduced here for convenience:

$$\boldsymbol{\Sigma}(n) = \mathbf{C}(n)\mathbf{K}(n, n-1)\mathbf{C}^H(n) + \mathbf{Q}_2(n)$$

The matrix $\mathbf{C}(n)$ is nonnegative definite but not necessarily nonsingular.

(a) Why is $\boldsymbol{\Sigma}(n)$ nonnegative definite?

(b) What prior condition would you impose on the matrix $\mathbf{Q}_2(n)$ to ensure that the inverse matrix $\boldsymbol{\Sigma}^{-1}(n)$ exists?

5. In many cases the predicted state-error correlation matrix $\mathbf{K}(n+1, n)$ converges to the steady-state value $\mathbf{K}$ as time n approaches infinity. Show that the limiting value $\mathbf{K}$ satisfies the *algebraic Riccati equation*

$$\mathbf{K}\mathbf{C}^H(\mathbf{C}\mathbf{K}\mathbf{C}^H + \mathbf{Q}_2)^{-1}\mathbf{C}\mathbf{K} - \mathbf{Q}_1 = \mathbf{0}$$

where it is assumed that the state transition matrix equals the identity matrix, and the matrices $\mathbf{C}$, $\mathbf{Q}_1$ and $\mathbf{Q}_2$ are the limiting values of $\mathbf{C}(n)$, $\mathbf{Q}_1(n)$ and $\mathbf{Q}_2(n)$, respectively.

6. Consider a stochastic process $\{y(n)\}$ represented by an autoregressive-moving average (ARMA) model of order (1, 1):

$$y(n) + ay(n-1) = v(n) + bv(n-1)$$

where a and b are the ARMA parameters and $\{v(n)\}$ is a zero-mean white-noise process of variance σ^2.

(a) Show that a state-space representation of this model is

$$\mathbf{x}(n+1) = \begin{bmatrix} -a & 1 \\ 0 & 0 \end{bmatrix} \mathbf{x}(n) + \begin{bmatrix} 1 \\ b \end{bmatrix} v(n+1)$$

$$y(n) = [1 \quad 0]\, \mathbf{x}(n)$$

where $\mathbf{x}(n)$ is a 2-by-1 state vector.

(b) Assume the applicability of the algebraic Riccati equation described in Problem 5. Hence, show that the solution of this equation is

$$\mathbf{K} = \sigma^2 \begin{bmatrix} 1+c & b \\ b & b^2 \end{bmatrix}$$

where c is a scalar that satisfies the second-order equation

$$c = (b-a)^2 + a^2 c - \frac{(b-a-ac)^2}{1+c}$$

What are the two values of c that satisfy this equation? Determine the corresponding values of the matrix $\mathbf{K}$.

(c) Show that the Kalman gain is

$$\mathbf{G} = \frac{b-a-ac}{1+c} \begin{bmatrix} 1 \\ 0 \end{bmatrix}$$

Determine the values of $\mathbf{G}$ that correspond to the solutions for the scalar c found in part (b).

7. In this problem we examine the nonuniqueness of the state-space representation for a stochastic model by revisiting Problem 6. In particular, for the ARMA model of Problem 6, do the following:

(a) Show that the model may be represented by an alternative set of state-space equations:

$$x(n+1) = -ax(n) + (b-a)v(n)$$

$$y(n) = x(n) + v(n)$$

where the scalar $x(n)$ is the state of the system.

(b) Show that the corresponding solution of the algebraic Riccati equation is a scalar k given by

$$k = a^2 k + (b-a)^2 - \frac{(-ak+b-a)^2}{k+1}$$

Hence, demonstrate that this solution is indeed equivalent to that found for the matrix $\mathbf{K}$ in part (b) of Problem 6.

(c) Show that the corresponding value of the Kalman gain is given by the scalar

$$g = \frac{b-a-ak}{1+k}$$

Compare this result with that found for the Kalman gain $\mathbf{G}$ in part (c) of Problem 6.

8. In this problem we consider the general case of *time-varying real-valued ARMA process* $\{y(n)\}$ described by the difference equation:

$$y(n) + \sum_{k=1}^{M} a_k(n) y(n-k) = \sum_{k=1}^{N} a_{M+k}(n) v(n-k) + v(n)$$

where $a_1(n)$, $a_2(n)$, . . . , $a_M(n)$, $a_{M+1}(n)$, $a_{M+2}(n)$, . . . , $a_{M+N}(n)$ are the ARMA coefficients, the process $\{v(n)\}$ is the input, and process $\{y(n)\}$ is the output. The process $\{v(n)\}$ is a white Gaussian noise process of zero mean and variance σ^2. The ARMA coefficients are subject to random fluctuations, as shown in the model

$$a_k(n+1) = a_k(n) + w_k(n), \qquad k = 1, \ldots, M+N$$

where $\{w_k(n)\}$ is a zero-mean, white Gaussian noise process that is independent of $\{w_j(n)\}$ for $j \neq k$, and also independent of $\{v(n)\}$. The issue of interest is to provide a technique based on the Kalman filter for identifying the coefficients of the ARMA process. To do this, we define an $(M + N)$-dimensional state vector:

$$\mathbf{x}(n) = [a_1(n), \ldots, a_M(n), \ldots, a_{M+N}(n)]^T$$

We also define the measurement matrix (actually, a row vector):

$$\mathbf{C}(n) = [-y(n-1), \ldots, -y(n-M), v(n-1), \ldots, v(n-N)]$$

On this basis, do the following:

(a) Formulate the state-space equations for the ARMA process.

(b) Find an algorithm for computing predicted value of the state vector $\mathbf{x}(n+1)$, given the observation $y(n)$.

(c) How would you initialize the algorithm in part (b)?

9. Consider a communication channel modeled as an FIR filter of known impulse response. The channel output $y(n)$ is defined by

$$y(n) = \mathbf{h}^T\mathbf{x}(n) + w(n)$$

where $\mathbf{h}$ is an M-by-1 vector representing the channel impulse response, $\mathbf{x}(n)$ is an M-by-1 vector representing the present value $u(n)$ of the channel input and $(M - 1)$ previous transmissions, and $\{w(n)\}$ is a white Gaussian noise process of zero mean and variance σ_w^2. At time n, the channel input $\{u(n)\}$ consists of a coded binary sequence of zeros and ones, statistically independent of $\{w(n)\}$. This model suggests that we may view $\mathbf{x}(n)$ as the state vector, in which case the state equation is written as[3]

$$\mathbf{x}(n+1) = \mathbf{A}\mathbf{x}(n) + \mathbf{b}v(n)$$

where $\{v(n)\}$ is a white Gaussian noise process of zero mean and variance σ_v^2 that is independent of $\{w(n)\}$. The matrix $\mathbf{A}$ is an M-by-M matrix whose ijth element is defined by

$$a_{ij} = \begin{cases} 1, & i = j+1 \\ 0, & \text{otherwise} \end{cases}$$

The vector $\mathbf{b}$ is an M-by-1 vector whose ith element is defined by

$$b_i = \begin{cases} 1, & i = 1 \\ 0, & i = 2, \ldots, M \end{cases}$$

We may now state the problem: Given the foregoing channel model and a sequence $\{y(n)\}$ of noisy measurements made at the channel output, use the Kalman filter to construct an equalizer that yields a good estimate of the channel input $u(n)$ at some delayed time $(n + D)$, whose $0 \leq D \leq M - 1$. Show that the equalizer so constructed is an IIR filter whose coefficients are determined by two distinct sets of parameters: (a) the M-by-1 channel impulse response vector, and (b) the Kalman gain, which (in this problem) is an M-by-1 vector.

[3] This problem is adapted from Lawrence and Kaufman (1971).

10. Using the initial conditions described in Eqs. (7.72) and (7.73), show that the resulting filtered estimate $\hat{\mathbf{x}}(n|\mathcal{Y}_n)$ produced by the Kalman filter is unbiased; that is,

$$E[\hat{\mathbf{x}}(n)|\mathcal{Y}_n)] = \mathbf{x}(n)$$

Computer-Oriented Problem

11. *Computer experiment on the adapative autoregressive modeling of a nonstationary process:* Consider a real-valued nonstationary environment described by the model shown in Fig. P7.1. The weights of the unknown system vary according to a *first-order Markov process:*

$$\mathbf{w}_o(n) = a\mathbf{w}_o(\mathrm{n} - 1) + \mathbf{v}(\mathrm{n})$$

where $\mathbf{v}(n)$ is a white-noise vector of zero mean and correlation matrix

$$E[\mathbf{v}(n)\mathbf{v}^T(n)] = \mathbf{I}$$

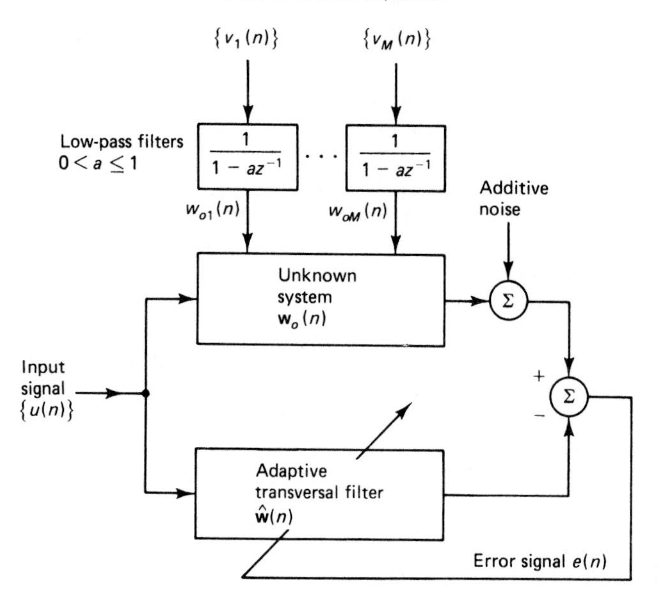

Figure P7.1

The constant a controls the degree of nonstationarity of the environment. In particular, we define

$$\text{nonstationarity constant} = \frac{1}{1 - a}$$

Use a computer to simulate the system of Fig. P7.1 for the following conditions:

(1) The input process $\{u(n)\}$ is white noise with zero mean and unit variance.

(2) The coefficient $a = 0.95$, or nonstationarity constant $= 20$.

(3) The adaptive transversal filter has the same order as the unknown system described by $\mathbf{w}_o(n)$. In particular, assume the order to equal 5.

(4) The additive noise at the output of the unknown system has zero mean and variance of 0.2.

(5) The minimum mean-squared error $J_{\min} = 0.3$.

(6) In the random-walk model for the state equation, assume that $q = 0.1$.

(7) For the initial conditions, set

$$\mathbf{K}(0) = 0.5\mathbf{I}$$

and

$$\hat{\mathbf{w}}(0) = \mathbf{0}$$

(a) Plot a single realization of the first (unknown) tap weight $w_{o,1}(n)$ and the corresponding value $\hat{w}_1(n)$ in the adaptive transversal filter versus time n over the interval $0 \leq n \leq 120$.

(b) By averaging your simulation results over 120 independent trials of the experiment, plot the ensemble-averaged value of the first (unknown) tap weight $w_{o,1}(n)$ and the corresponding value $\hat{w}_1(n)$ in the adaptive transversal filter over the interval $0 \leq n \leq 120$.

Part III

Linear FIR Adaptive Filtering

Part III, by far the largest portion of the book, consists of Chapters 8 through 18. It is devoted to a detailed treatment of linear finite-duration impulse response (FIR) adaptive filters. *This part of the book is itself divided into four parts:*

- *Gradient-based adaptation (Chapters 8 and 9)*
- *Linear least-squares estimation (Chapters 10 through 12)*
- *Recursive least-squares (RLS) estimation (Chapters 13 and 14)*
- *Fast RLS algorithms (Chapters 15 through 18)*

In Chapter 8, we develop the method of steepest descent for computing the tap-weight vector of the Wiener filter in a recursive fashion. In Chapter 9, we use the method of steepest descent to derive the least-mean-square (LMS) algorithm and study its important characteristics. In this chapter we also consider other related stochastic gradient algorithms.

Chapter 10 covers the fundamentals of linear least-squares estimation. In this chapter we illustrate the use of the method of least squares for computing AR and MVDR spectra. Chapter 11 develops the singular value decomposition *(SVD), which provides a powerful tool for solving the linear least-square estimation problem and related issues. Chapter 12 develops the theory of two other popular spectrum estimation techniques, the* multiple signal classification *(MUSIC) and the* minimum norm algorithms, *which follow naturally from the SVD theory.*

In Chapter 13, we derive the standard recursive least squares (RLS) algorithm, *which may be viewed as a special case of the Kalman filter. In Chapter 14, we study the use of the numerically stable* QR-decomposition *technique for solving the RLS problem.*

In Chapter 15, we develop the mathematical background for "fast" solutions of the RLS problem. This is followed by the development of specific forms of fast RLS algorithms, In particular, the fast transversal filters (FTF) *algorithm is developed in Chapter 16, the* recursive least-squares lattice (LSL) algorithm *in Chapter 17, and the lattice implementation of the* QR-decomposition-based recursive least-squares (QRD-RLS) algorithm *in Chapter 18.*

8

Method of Steepest Descent

In this chapter we begin our study of gradient-based adaptation by describing an old optimization technique known as the *method of steepest descent*. This method is basic to the understanding of the various ways in which gradient-based adaptation is implemented in practice. The method of steepest descent is *recursive* in the sense that starting from some initial (arbitrary) value for the tap-weight vector, it improves with the increased number of iterations. The final value so computed for the tap-weight vector converges to the Wiener solution. The important point to note is that the method of steepest descent is descriptive of a *multiparameter closed-loop deterministic control system* that finds the minimum point of the ensemble-averaged error-performance surface without knowledge of the surface itself. Accordingly, it provides some heuristics for writing the recursions that describe the least-mean-square (LMS) algorithm, an issue that is taken up in the next chapter.

8.1 SOME PRELIMINARIES

Consider a transversal filter with tap inputs $u(n), u(n-1), \ldots, u(n-M+1)$ and a corresponding set of *tap weights* $w_0(n), w_1(n), \ldots, w_{M-1}(n)$. The tap inputs represent samples drawn from a wide-sense stationary stochastic process of zero mean and correlation matrix $\mathbf{R}$. In addition to these inputs, the filter is supplied with a *desired response* $d(n)$ that provides a frame of reference for the optimum filtering action. Figure 8.1 depicts the filtering action described herein.

The vector of tap inputs at time n is denoted by $\mathbf{u}(n)$, and the corresponding *estimate* of the desired response at the filter output is denoted by $\hat{d}(n|\mathcal{U}_n)$, where $\mathcal{U}_n$ is the space spanned by the tap inputs $u(n), u(n-1), \ldots, u(n-M+1)$. By

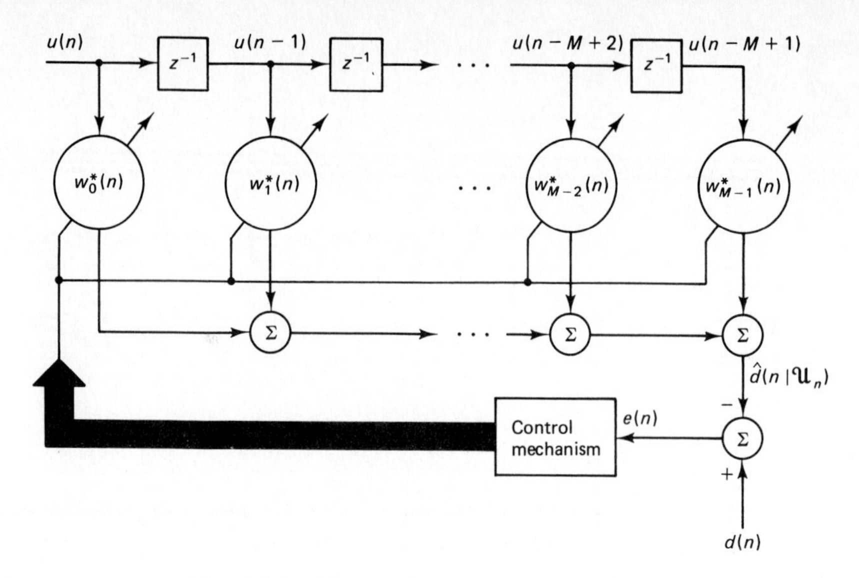

Figure 8.1 Structure of adaptive transversal filter.

comparing this estimate with the actual value of the desired response $d(n)$, we produce an *estimation error* denoted by $e(n)$. We may thus write

$$\begin{aligned} e(n) &= d(n) - \hat{d}(n|\mathcal{U}_n) \\ &= d(n) - \mathbf{w}^H(n)\mathbf{u}(n) \end{aligned} \tag{8.1}$$

where the term $\mathbf{w}^H(n)\mathbf{u}(n)$ is the inner product of the tap-weight vector $\mathbf{w}(n)$ and the tap input vector $\mathbf{u}(n)$. The expanded form of the tap-weight vector is described by

$$\mathbf{w}(n) = [w_0(n), w_1(n), \ldots, w_{M-1}(n)]^T \tag{8.2}$$

and that of the tap-input vector by

$$\mathbf{u}(n) = [u(n), u(n-1), \ldots, u(n-M+1)]^T \tag{8.3}$$

If the tap-input vector $\mathbf{u}(n)$ and the desired $d(n)$ are jointly stationary, then the *mean-squared error* or *cost function* $J(n)$ at time n is a quadratic function of the tap-weight vector, so we may write [see Eq. (5.50)]

$$J(n) = \sigma_d^2 - \mathbf{w}^H(n)\mathbf{p} - \mathbf{p}^H\mathbf{w}(n) + \mathbf{w}^H(n)\mathbf{R}\mathbf{w}(n) \tag{8.4}$$

where σ_d^2 = variance of the desired response $d(n)$

$\mathbf{p}$ = cross-correlation vector between the tap-input vector $\mathbf{u}(n)$ and the desired response $d(n)$

$\mathbf{R}$ = correlation matrix of the tap-input vector $\mathbf{u}(n)$

Equation (8.4) defines the mean-squared error that would result if the tap-weight vector in the transversal filter were *fixed* at the value $\mathbf{w}(n)$. Since $\mathbf{w}(n)$ varies with time n, it is only natural that the mean-squared error varies with time n in a corresponding fashion, hence, the use of $J(n)$ for the mean-squared error in Eq. (8.4). The variation of the mean-squared error $J(n)$ with time n signifies the fact that the estimation error process $\{e(n)\}$ is nonstationary.

We may visualize the dependence of the mean-squared error $J(n)$ on the elements of the tap-weight vector $\mathbf{w}(n)$ as a *bowl-shaped surface* with a unique minimum. We refer to this surface as the *error-performance surface* of the adaptive filter. The adaptive process has the task of continually seeking the *bottom* or *minimum point* of this surface. At the minimum point of the error-performance surface, the tap-weight vector takes on the optimum value $\mathbf{w}_o$, which is defined by the Wiener-Hopf equations (5.34), reproduced here for convenience,

$$\mathbf{R}\mathbf{w}_o = \mathbf{p} \tag{8.5}$$

The minimum mean-squared error equals [see Eq. (5.49)]

$$J_{\min} = \sigma_d^2 - \mathbf{p}^H \mathbf{w}_o \tag{8.6}$$

8.2 STEEPEST-DESCENT ALGORITHM

The requirement that an adaptive transversal filter has to satisfy is to find a solution for its tap-weight vector that satisfies the Wiener-Hopf equations (8.5). One way of doing this would be to solve this system of equations by some analytical means. Although, in general, this procedure is quite straightforward, nevertheless, it presents serious computational difficulties, especially when the filter contains a large number of tap weights and when the input data rate is high.

An alternative procedure is to use the *method of steepest descent*, which is one of the oldest methods of optimization.[1] To find the minimum value of the mean-squared error, $J_{\min}$, by the steepest-descent algorithm, we proceed as follows:

1. We begin with an initial value $\mathbf{w}(0)$ for the tap-weight vector, which is chosen arbitrarily. The value $\mathbf{w}(0)$ provides an initial guess as to where the minimum point of the error-performance surface may be located. Typically, $\mathbf{w}(0)$ is set equal to the null vector.
2. Using this initial or present guess, we compute the *gradient vector*, the real and imaginary parts of which are defined as the derivative of the mean-squared error $J(n)$, evaluated with respect to the real and imaginary parts of the tap-weight vector $\mathbf{w}(n)$ at time n (i.e., the nth iteration).
3. We compute the next guess at the tap-weight vector by making a change in the initial or present guess in a direction opposite to that of the gradient vector.
4. We go back to step 2 and repeat the process.

It is intuitively reasonable that successive corrections to the tap-weight vector in the direction of the negative of the gradient vector (i.e., in the direction of the steepest descent of the error-performance surface) should eventually lead to the minimum mean-squared error $J_{\min}$, at which point the tap-weight vector assumes its optimum value $\mathbf{w}_o$.

[1] The steepest-descent algorithm belongs to a family of *iterative methods of optimization* (Luenberger, 1969); it provides a method of searching a multidimensional performance surface. Another method in this family that may be used for this purpose is *Newton's algorithm*, which is primarily a method for finding the zeros of a function. In the context of adaptive filtering, the use of the method of steepest descent results in an algorithm that is much simpler, but slower, than Newton's method (Widrow and Stearns, 1985).

Let $\boldsymbol{\nabla}(J(n))$ denote the value of the *gradient vector* at time n. Let $\mathbf{w}(n)$ denote the value of the tap-weight vector at time n. According to the method of steepest descent, the updated value of the tap-weight vector at time $n + 1$ is computed by using the simple recursive relation

$$\mathbf{w}(n + 1) = \mathbf{w}(n) + \tfrac{1}{2}\mu[-\boldsymbol{\nabla}(J(n))] \tag{8.7}$$

where μ is a positive real-valued constant. The factor $\frac{1}{2}$ is used merely for convenience.

From Chapter 4 we find that the gradient vector $\nabla(J(n))$ is given by

$$\boldsymbol{\nabla}(J(n)) = \begin{bmatrix} \dfrac{\partial J(n)}{\partial a_0(n)} + j\dfrac{\partial J(n)}{\partial b_0(n)} \\ \dfrac{\partial J(n)}{\partial a_1(n)} + j\dfrac{\partial J(n)}{\partial b_1(n)} \\ \vdots \\ \dfrac{\partial J(n)}{\partial a_{M-1}(n)} + j\dfrac{\partial J(n)}{\partial b_{M-1}(n)} \end{bmatrix}$$

$$= -2\mathbf{p} + 2\mathbf{R}\mathbf{w}(n) \tag{8.8}$$

where, in the first line, $\partial J(n)/\partial a_k(n)$ and $\partial J(n)/\partial b_k(n)$ are the partial derivatives of the cost function $J(n)$ with respect to the real part $a_k(n)$ and the imaginary part $b_k(n)$ of the kth tap weight $w_k(n)$, respectively. For the application of the steepest-descent algorithm, we assume that in Eq. (8.8) the correlation matrix $\mathbf{R}$ and the cross-correlation vector $\mathbf{p}$ are known so that we may compute the gradient vector $\boldsymbol{\nabla}(n)$ for a given value of the tap-weight vector $\mathbf{w}(n)$. Thus, substituting Eq. (8.8) in (8.7), we may compute the updated value of the tap-weight vector $\mathbf{w}(n + 1)$ by using the simple recursive relation

$$\mathbf{w}(n + 1) = \mathbf{w}(n) + \mu[\mathbf{p} - \mathbf{R}\mathbf{w}(n)], \qquad n = 0, 1, 2, \ldots, \tag{8.9}$$

We observe that the parameter μ controls the size of the incremental correction applied to the tap-weight vector as we proceed from one iteration cycle to the next. We therefore refer to μ as the *step-size parameter* or *weighting constant*. Equation (8.9) describes the mathematical formulation of the steepest-descent algorithm.

According to Eq. (8.9), the correction $\delta\mathbf{w}(n)$ applied to the tap-weight vector at time $n + 1$ is equal to $\mu[\mathbf{p} - \mathbf{R}\mathbf{w}(n)]$. This correction may also be expressed as μ times the expectation of the inner product of the tap-input vector $\mathbf{u}(n)$ and the estimation error $e(n)$; see Problem 4. This suggests that we may use a bank of cross-correlators to compute the correction $\delta\,\mathbf{w}(\mathrm{n})$ applied to the tap-weight vector $\mathbf{w}(\mathrm{n})$ as indicated in Fig. 8.2. In this figure, the elements of the correction vector $\delta\,\mathbf{w}(\mathrm{n})$ are denoted by $\delta w_0(n), \delta w_1(n), \ldots, \delta w_{M-1}(n)$.

Another point of interest is that we may view the steepest-descent algorithm of Eq. (8.9) as a *feedback model,* as illustrated by the *signal-flow graph* shown in Fig. 8.3. This model is multidimensional in the sense that the "signals" at the *nodes* of the graph consist of vectors and that the *transmittance* of each branch of the graph is a scalar or a square matrix. For each branch of the graph, the signal vector flowing

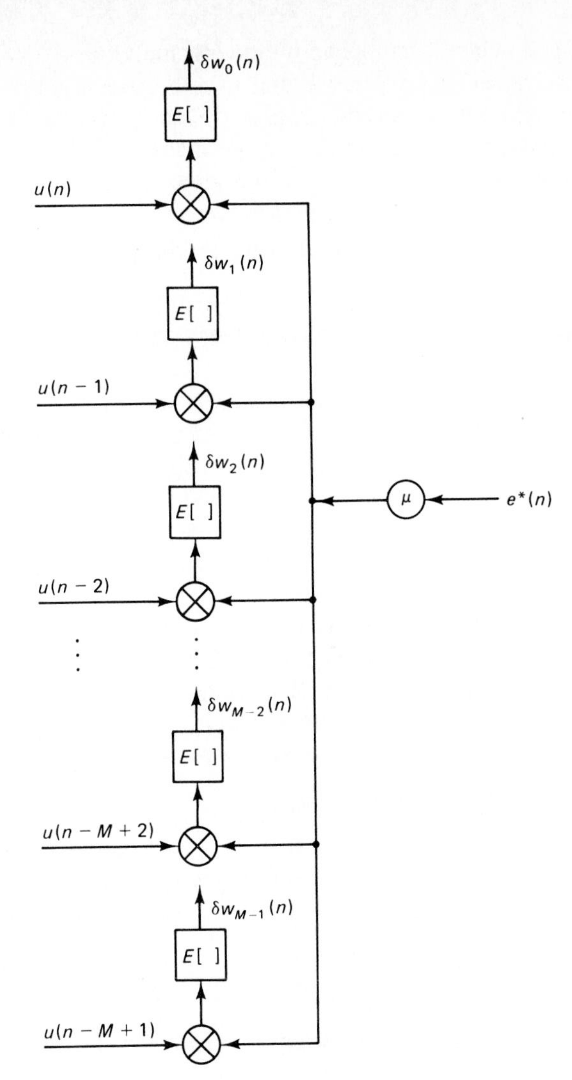

Figure 8.2 Bank of cross-correlators for computing the corrections to the elements of the tap-weight vector at time $n + 1$.

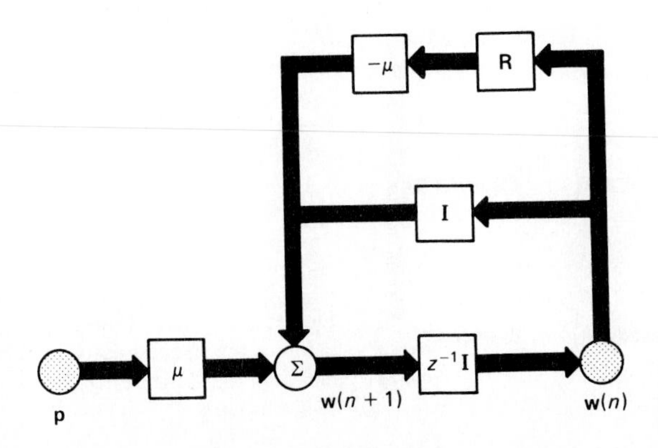

Figure 8.3 Signal-flow-graph representation of the steepest-descent algorithm.

out equals the signal vector flowing in multiplied by the transmittance matrix of the branch. For two branches connected in parallel, the overall transmittance matrix equals the sum of the transmittance matrices of the individual branches. For two branches connected in cascade, the overall transmittance matrix equals the product of the individual transmittance matrices arranged in the same order as the pertinent branches. Finally, the symbol z^{-1} is the unit-delay operator, and $z^{-1}\mathbf{I}$ is the transmittance matrix of a unit-delay branch representing a delay of one iteration cycle.

8.3 STABILITY OF THE STEEPEST-DESCENT ALGORITHM

Since the steepest-descent algorithm involves the presence of *feedback*, as exemplified by the model of Fig. 8.3, the algorithm is subject to the possibility of it becoming *unstable*. From the feedback model of Fig. 8.3, we observe that the *stability performance* of the steepest-descent algorithm in determined by two factors: (1) the step-size parameter μ, and (2) the correlation matrix $\mathbf{R}$ of the tap-input vector $\mathbf{u}(n)$, as these two parameters completely control the transfer function of the *feedback loop*.

To determine *the condition for the stability* of the steepest-descent algorithm, we examine the *natural modes* of the algorithm (Widrow, 1970). In particular, we use the representation of the correlation matrix $\mathbf{R}$ in terms of its eigenvalues and eigenvectors to define a transformed version of the tap-weight vector.

We begin the analysis by defining a *weight-error vector* at time n as

$$\mathbf{c}(n) = \mathbf{w}(n) - \mathbf{w}_o \tag{8.10}$$

where $\mathbf{w}_o$ is the optimum value of the tap-weight vector, as defined by the Wiener-Hopf equations (8.5). Then, eliminating the cross-correlation vector $\mathbf{p}$ between Eqs. (8.5) and (8.9), and rewriting the result in terms of the weight-error vector $\mathbf{c}(n)$, we get

$$\mathbf{c}(n + 1) = (\mathbf{I} - \mu\mathbf{R})\mathbf{c}(n) \tag{8.11}$$

where $\mathbf{I}$ is the identity matrix. Equation (8.11) is represented by the feedback model shown in Fig. 8.4. This diagram further emphasizes the fact that the stability performance of the steepest-descent algorithm is controlled exclusively by μ and $\mathbf{R}$.

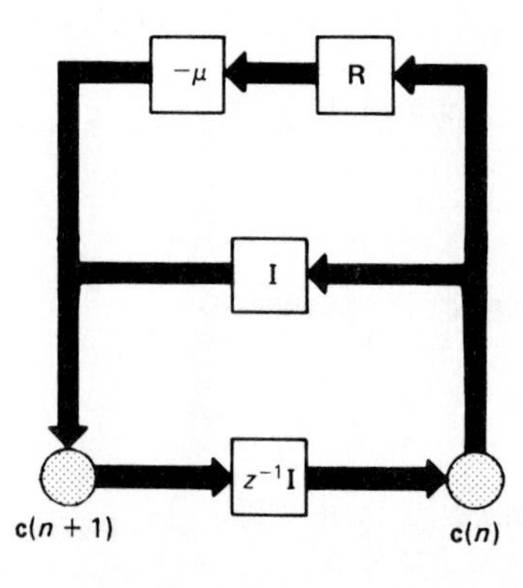

Figure 8.4 Signal-flow-graph representation of the steepest-descent algorithm based on the weight-error vector.

Using the unitary similarity transformation, we may express the correlation matrix $\mathbf{R}$ as follows (see Chapter 4):

$$\mathbf{R} = \mathbf{Q}\mathbf{\Lambda}\mathbf{Q}^H \tag{8.12}$$

The matrix $\mathbf{Q}$ has as its columns an orthogonal set of *eigenvectors* associated with the eigenvalues of the matrix $\mathbf{R}$. The matrix $\mathbf{Q}$ is called the *unitary matrix* of the transformation. The matrix $\mathbf{\Lambda}$ is a *diagonal* matrix and has as its diagonal elements the eigenvalues of the correlation matrix $\mathbf{R}$. These eigenvalues, denoted by $\lambda_1, \lambda_2, \ldots, \lambda_M$, are all positive and real. Each eigenvalue is associated with a corresponding eigenvector or column of matrix $\mathbf{Q}$. Substituting Eq. (8.12) in (8.11), we get

$$\mathbf{c}(n+1) = (\mathbf{I} - \mu\mathbf{Q}\mathbf{\Lambda}\mathbf{Q}^H)\mathbf{c}(n) \tag{8.13}$$

Premultiplying both sides of this equation by $\mathbf{Q}^H$ and using the property of the unitary matrix $\mathbf{Q}$ that $\mathbf{Q}^H$ equals the inverse $\mathbf{Q}^{-1}$ (see Chapter 4), we get

$$\mathbf{Q}^H\mathbf{c}(n+1) = (\mathbf{I} - \mu\mathbf{\Lambda})\mathbf{Q}^H\mathbf{c}(n) \tag{8.14}$$

We now define a new set of coordinates as follows:

$$\begin{aligned}\boldsymbol{\nu}(n) &= \mathbf{Q}^H\mathbf{c}(n)\\ &= \mathbf{Q}^H[\mathbf{w}(n) - \mathbf{w}_o]\end{aligned} \tag{8.15}$$

Accordingly, we may rewrite Eq. (8.13) in the desired form:

$$\boldsymbol{\nu}(n+1) = (\mathbf{I} - \mu\mathbf{\Lambda})\boldsymbol{\nu}(n) \tag{8.16}$$

The initial value of $\boldsymbol{\nu}(n)$ equals

$$\boldsymbol{\nu}(0) = \mathbf{Q}^H[\mathbf{w}(0) - \mathbf{w}_o] \tag{8.17}$$

Assuming that the initial tap-weight vector is zero, Eq. (8.17) reduces to

$$\boldsymbol{\nu}(0) = -\mathbf{Q}^H\mathbf{w}_o \tag{8.18}$$

For the kth *natural mode* of the steepest descent algorithm, we thus have

$$\nu_k(n+1) = (1 - \mu\lambda_k)\nu_k(n), \qquad k = 1, 2, \ldots, M \tag{8.19}$$

where λ_k is the kth eigenvalue of the correlation matrix $\mathbf{R}$. This equation is represented by the scalar-valued feedback model of Fig. 8.5, where z^{-1} is the unit-delay operator. Clearly, the structure of this model is much simpler than that of the original matrix-valued feedback model of Fig. 8.3. These two models represent different and yet equivalent ways of viewing the steepest-descent algorithm.

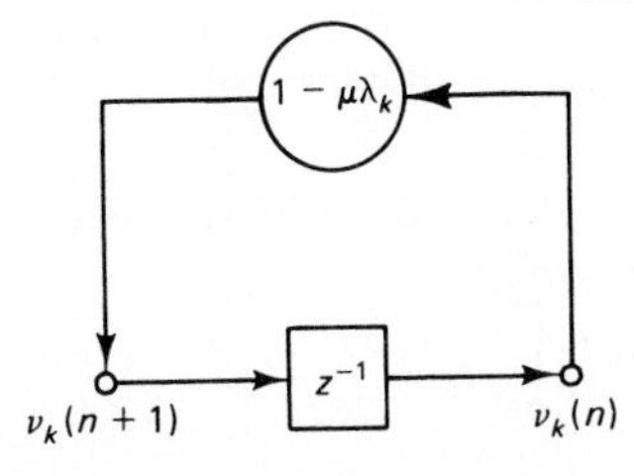

Figure 8.5 Signal-flow-graph representation of the kth natural mode of the steepest-descent algorithm.

Equation (8.19) is a homogeneous *difference equation of the first order*. Assuming that $\nu_k(n)$ has the initial value $\nu_k(0)$, we readily obtain the solution

$$\nu_k(n) = (1 - \mu\lambda_k)^n \nu_k(0), \qquad k = 1, 2, \ldots, M \tag{8.20}$$

Since all eigenvalues of the correlation matrix **R** are positive and real, the response $\nu_k(n)$ will exhibit no oscillations. Furthermore, as illustrated in Fig. 8.6, the numbers generated by Eq. (8.20) represent a *geometric series* with a geometric ratio equal to $1 - \mu\lambda_k$. For *stability* or *convergence* of the steepest-descent algorithm, the magnitude of this geometric ratio must be less than 1 for all k. That is, provided we have

$$-1 < 1 - \mu\lambda_k < 1, \qquad \text{for all } k$$

then as the number of iterations, n, approaches infinity, all the natural modes of the steepest-descent algorithm die out, irrespective of the initial conditions. This is equivalent to saying that the tap-weight vector $\mathbf{w}(n)$ approaches the optimum solution $\mathbf{w}_o$ as n approaches infinity.

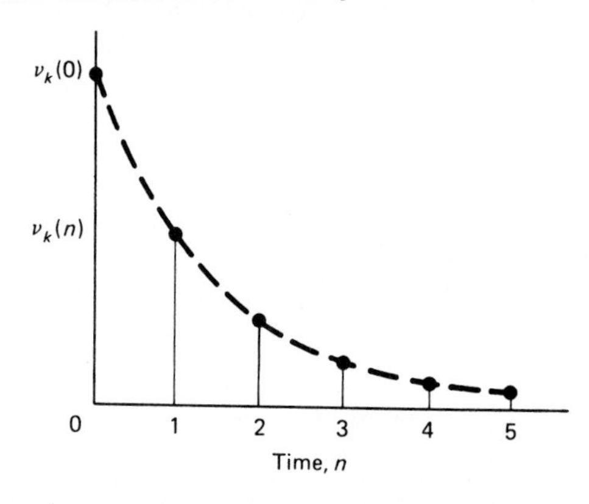

Figure 8.6 Variation of the kth natural mode of the steepest-descent algorithm with time, assuming that the magnitude of $1 - \mu\lambda_k$ is less than 1.

Since the eigenvalues of the correlation matrix **R** are all real and positive, it therefore follows that the necessary and sufficient condition for the convergence or stability of the steepest-descent algorithm is that the step-size parameter μ satisfy the following condition:

$$0 < \mu < \frac{2}{\lambda_{\max}} \tag{8.21}$$

where $\lambda_{\max}$ is the largest eigenvalue of the correlation matrix **R**.

Referring to Fig. 8.6, we see that an exponential envelope of *time constant* τ_k can be fitted to the geometric series by assuming the unit of time to be the duration of one iteration cycle and by choosing the time constant τ_k such that

$$1 - \mu\lambda_k = \exp\left(-\frac{1}{\tau_k}\right)$$

Hence, the kth time constant τ_k can be expressed in terms of the step-size parameter μ and the kth eigenvalue as follows:

$$\tau_k = \frac{-1}{\ln(1 - \mu\lambda_k)} \tag{8.22}$$

The time constant τ_k defines the time required for the amplitude of the kth natural mode $\nu_k(n)$ to decay to $1/e$ of its initial value $\nu_k(0)$, where e is the base of the natural logarithm. For the special case of slow adaptation, for which the step-size parameter μ is small, we may approximate the time constant τ_k as

$$\tau_k \simeq \frac{1}{\mu\lambda_k}, \qquad \mu << 1 \tag{8.23}$$

We may now formulate the transient behavior of the original tap-weight vector $\mathbf{w}(n)$. In particular, premultiplying both sides of Eq. (8.15) by $\mathbf{Q}$, using the fact that $\mathbf{Q}\mathbf{Q}^H = \mathbf{I}$, and solving for $\mathbf{w}(n)$, we get the desired result:

$$\begin{aligned}\mathbf{w}(n) &= \mathbf{w}_o + \mathbf{Q}\boldsymbol{\nu}(n) \\ &= \mathbf{w}_o + [\mathbf{q}_1, \mathbf{q}_2, \ldots, \mathbf{q}_M]\begin{bmatrix}\nu_1(n)\\ \nu_2(n)\\ \vdots \\ \nu_M(n)\end{bmatrix} \\ &= \mathbf{w}_o + \sum_{k=1}^{M} \mathbf{q}_k \nu_k(n)\end{aligned} \tag{8.24}$$

where $\mathbf{q}_1, \mathbf{q}_2, \ldots, \mathbf{q}_M$ are the eigenvectors associated with the eigenvalues $\lambda_1, \lambda_2, \ldots, \lambda_M$ of the correlation matrix $\mathbf{R}$, respectively, and the kth natural mode $\nu_k(n)$ is defined by Eq. (8.20). Thus, substituting Eq. (8.20) in (8.24), we find that the transient behavior of the ith tap weight is described by (Griffiths, 1975)

$$w_i(n) = w_{oi} + \sum_{k=1}^{N} q_{ki}\nu_k(0)(1 - \mu\lambda_k)^n, \qquad i = 1, 2, \ldots, M \tag{8.25}$$

where w_{oi} is the optimum value of the ith tap weight, and q_{ki} is the ith element of the kth eigenvector $\mathbf{q}_k$.

Equation (8.25) shows that each tap weight in the steepest-descent algorithm converges as the weighted sum of exponentials of the form $(1 - \mu\lambda_k)^n$. The time τ_k required for each term to reach $1/e$ of its initial value is given by Eq. (8.22). However, the *overall time constant*, τ_a, defined as the time required for the summation term in Eq. (8.25) to decay to $1/e$ of its initial value, cannot be expressed in a simple closed form similar to Eq. (8.22). Nevertheless, the *slowest rate of convergence* is attained when $q_{ki}\nu_k(0)$ is zero for all k except for that corresponding to the smallest eigenvalue $\lambda_{\min}$ of matrix $\mathbf{R}$, so the upper bound on τ_a is defined by $-1/\ln(1 - \mu\lambda_{\min})$. The *fastest rate of convergence* is attained when all the $q_{ki}\nu_k(0)$ are zero except for that corresponding to the largest eigenvalue $\lambda_{\max}$, and so the lower bound on τ_a is defined by $-1/\ln(1 - \mu\lambda_{\max})$. Accordingly, the overall time constant τ_a for any tap weight of the steepest-descent algorithm is bounded as follows (Griffiths, 1975):

$$\frac{-1}{\ln(1 - \mu\lambda_{\max})} \le \tau_a \le \frac{-1}{\ln(1 - \mu\lambda_{\min})} \tag{8.26}$$

We see therefore that, when the eigenvalues of the correlation matrix **R** are widely spread (i.e., the correlation matrix of the tap inputs is ill conditioned), the settling time of the steepest-descent algorithm is limited by the smallest eigenvalues or the slowest modes.

8.4 TRANSIENT BEHAVIOR OF THE MEAN-SQUARED ERROR

We may develop further insight into the operation of the steepest-descent algorithm by examining the transient behavior of the mean-squared error. At any time n, the value of the mean-squared error $J(n)$ is given by [see Eq. (5.56)]

$$J(n) = J_{\min} + \sum_{k=1}^{M} \lambda_k |\nu_k(n)|^2 \tag{8.27}$$

where $J_{\min}$ is the minimum mean-squared error. The transient behavior of the kth natural mode, $\nu_k(n)$, is defined by Eq. (8.20). Hence substituting Eq. (8.20) into (8.27), we get

$$J(n) = J_{\min} + \sum_{k=1}^{M} \lambda_k(1 - \mu\lambda_k)^{2n} |\nu_k(0)|^2 \tag{8.28}$$

where $\nu_k(0)$ is the initial value of $\nu_k(n)$. When the steepest-descent algorithm is convergent, that is, the step-size parameter μ is chosen within the bounds defined by Eq. (8.21), we see that, irrespective of the initial conditions,

$$\lim_{n\to\infty} J(n) = J_{\min} \tag{8.29}$$

The curve obtained by plotting the mean-squared error $J(n)$ versus the number of iterations, n is called a *learning curve*. From Eq. (8.28), we see that *the learning curve of the steepest-descent algorithm consists of a sum of exponentials, each of which corresponds to a natural mode of the algorithm*. In general, the number of natural modes equals the number of tap weights. In going from the initial value $J(0)$ to the final value $J_{\min}$, the exponential decay for the kth natural mode has a time constant equal to

$$\tau_{k,\text{mse}} \simeq \frac{-1}{2\ln(1 - \mu\lambda_k)} \tag{8.30}$$

For small values of the step-size parameter μ, we may approximate this time constant as

$$\tau_{k\text{mse}}, \simeq \frac{1}{2\mu\lambda_k} \tag{8.31}$$

Equation (8.31) shows that the smaller the step-size paramater μ, the slower will be the rate of decay of each natural mode of the LMS algorithm.

8.5 EXAMPLE: EFFECTS OF EIGENVALUE SPREAD AND STEP-SIZE PARAMETER ON THE STEEPEST-DESCENT ALGORITHM

In this example, we examine the transient behavior of the steepest-descent algorithm applied to a predictor that operates on a real-valued autoregressive (AR) process. Figure 8.7 shows the structure of the predictor, assumed to contain two tap weights that are denoted by $w_1(n)$ and $w_2(n)$; the dependence of these tap weights on time n emphasizes the transient condition of the predictor. The AR process $\{u(n)\}$ is described by the second-order difference equation

$$u(n) + a_1 u(n-1) + a_2 u(n-2) = v(n) \tag{8.32}$$

where the sample $v(n)$ is drawn from a white-noise process of zero mean and variance σ_v^2. The AR parameters a_1 and a_2 are chosen so that the roots of the characteristic equation

$$1 + a_1 z^{-1} + a_2 z^{-2} = 0$$

are complex; that is, $a_1^2 < 4a_2$. The particular values assigned to a_1 and a_2 are determined by the desired eigenvalue spread $\chi(\mathbf{R})$. For specified values of a_1 and a_2, the variance σ_v^2 of the white-noise process is chosen to make the process $\{u(n)\}$ have variance $\sigma_u^2 = 1$.

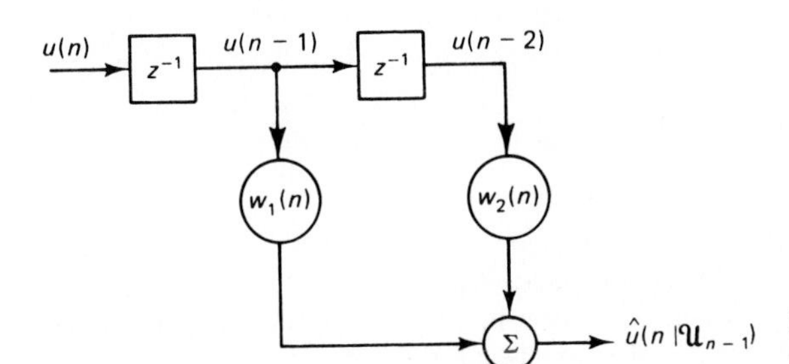

Figure 8.7 Two-tap predictor for real-valued input.

The requirement is to evaluate the transient behavior of the steepest-descent algorithm for the following two conditions:

1. Varying eigenvalue spread $\chi(\mathbf{R})$ and fixed step-size parameter μ
2. Varying step-size parameter μ and fixed eigenvalue spread $\chi(\mathbf{R})$

Characterization of the AR Process

Since the predictor of Fig. 8.7 has two tap weights and the AR process $\{u(n)\}$ is real valued, it follows that the correlation matrix $\mathbf{R}$ of the tap inputs is a 2-by-2 symmetric matrix:

$$\mathbf{R} = \begin{bmatrix} r(0) & r(1) \\ r(1) & r(0) \end{bmatrix}$$

where

$$r(0) = \sigma_u^2$$

$$r(1) = \frac{-a_1}{1 + a_2} \sigma_u^2$$

$$\sigma_u^2 = \left(\frac{1 + a_2}{1 - a_2}\right) \frac{\sigma_v^2}{(1 + a_2)^2 - a_1^2}$$

The two eigenvalues of $\mathbf{R}$ are

$$\lambda_1 = \left(1 - \frac{a_1}{1 + a_2}\right) \sigma_u^2$$

$$\lambda_2 = \left(1 + \frac{a_1}{1 + a_2}\right) \sigma_u^2$$

Hence, the eigenvalue spread equals (assuming that a_1 is negative)

$$\chi(\mathbf{R}) = \frac{\lambda_1}{\lambda_2} = \frac{1 - a_1 + a_2}{1 + a_1 + a_2}$$

The eigenvectors $\mathbf{q}_1$ and $\mathbf{q}_2$ associated with the respective eigenvalues λ_1 and λ_2 are

$$\mathbf{q}_1 = \frac{1}{\sqrt{2}} \begin{bmatrix} 1 \\ 1 \end{bmatrix}$$

$$\mathbf{q}_2 = \frac{1}{\sqrt{2}} \begin{bmatrix} 1 \\ -1 \end{bmatrix}$$

both of which are normalized to unit length.

Experiment 1: Varying Eigenvalue Spread

In this experiment, the step-size parameter μ is fixed at 0.3, and the evaluations are made for the four sets of AR parameters given in Table 8.1.

For a given set of parameters, we use a two-dimensional plot of the transformed tap-weight error $\nu_1(n)$ versus $\nu_2(n)$ to display the transient behavior of the steepest-descent algorithm. In particular, the use of Eq. (8.20) yields

$$\boldsymbol{\nu}(n) = \begin{bmatrix} \nu_1(n) \\ \nu_2(n) \end{bmatrix}$$

$$= \begin{bmatrix} (1 - \mu\lambda_1)^n \nu_1(0) \\ (1 - \mu\lambda_2)^n \nu_2(0) \end{bmatrix}, \qquad n = 1, 2, \ldots \tag{8.33}$$

TABLE 8.1 SUMMARY OF PARAMETER VALUES CHARACTERIZING THE SECOND-ORDER AR MODELING PROBLEM

	AR parameters		Eigenvalues		Eigenvalue spread,	Minimum mean-squared error,
Case	a_1	a_2	λ_1	λ_2	$\chi = \lambda_1/\lambda_2$	$J_{\min} = \sigma_v^2$
1	−0.1950	0.95	1.1	0.9	1.22	0.0965
2	−0.9750	0.95	1.5	0.5	3	0.0731
3	−1.5955	0.95	1.818	0.182	10	0.0322
4	−1.9114	0.95	1.957	0.0198	100	0.0038

To calculate the initial value $\boldsymbol{\nu}(0)$, we use Eq. (8.18), assuming that the initial value $\mathbf{w}(0)$ of the tap-weight vector $\mathbf{w}(n)$ is zero. This equation requires knowledge of the optimum tap-weight vector $\mathbf{w}_o$. Now when the two-tap predictor of Fig. 8.7 is optimized, with the second-order AR process of Eq. (8.32) supplying the tap inputs, we find that the optimum tap-weight vector equals

$$\mathbf{w}_o = \begin{bmatrix} -a_1 \\ -a_2 \end{bmatrix}$$

and the minimum mean-squared error equals

$$J_{\min} = \sigma_v^2$$

Accordingly, the use of Eq. (8.18) yields the initial value:

$$\begin{aligned} \boldsymbol{\nu}(0) &= \begin{bmatrix} \nu_1(0) \\ \nu_2(0) \end{bmatrix} \\ &= \frac{-1}{\sqrt{2}} \begin{bmatrix} 1 & 1 \\ 1 & -1 \end{bmatrix} \begin{bmatrix} -a_1 \\ -a_2 \end{bmatrix} \\ &= \frac{1}{\sqrt{2}} \begin{bmatrix} a_1 + a_2 \\ a_1 - a_2 \end{bmatrix} \end{aligned} \tag{8.34}$$

Thus, for specified parameters, we use Eq. (8.34) to compute the initial value $\boldsymbol{\nu}(0)$, and then use Eq. (8.33) to compute $\boldsymbol{\nu}(1)$, $\boldsymbol{\nu}(2)$, ... By joining the points defined by these values of $\boldsymbol{\nu}(n)$ for varying time n, we obtain a *trajectory* that describes the transient behavior of the steepest-descent algorithm for the particular set of parameters.

It is informative to include in the two-dimentional plot of $\nu_1(n)$ versus $\nu_2(n)$ loci representing Eq. (8.27) for fixed values of n. For our example, Eq. (8.27) yields

$$J(n) - J_{\min} = \lambda_1 \nu_1^2(n) + \lambda_2 \nu_2^2(n) \tag{8.35}$$

When $\lambda_1 = \lambda_2$ and time n is fixed, Eq. (8.35) represents a circle with center at the origin and radius equal to the square root of $([J(n) - J_{\min}]/\lambda$, where λ is the common value of the two eigenvalues. When, on the other hand, $\lambda_1 \neq \lambda_2$, Eq. (8.35) represents (for fixed n) an ellipse with major axis equal to the square root of $[J(n) - J_{\min}]/\lambda_1$ and minor axis equal to the square root of $[J(n) - J_{\min}]/\lambda_2$.

Case 1: Eigenvalue Spread $\chi(R) = 1.22$. For the parameter values given for Case 1 in Table 8.1, the eigenvalue spread $\chi(\mathbf{R})$ equals 1.22; that is, the eigenvalues λ_1 and λ_2 are approximately equal. The use of these parameter values in Eqs. (8.33) and (8.34) yields the trajectory of $[\nu_1(n), \nu_2(n)]$ shown in Fig. 8.8(a), with n as running parameter, and their use in Eq. (8.35) yields the (approximately) circular loci shown for fixed values of $J(n)$, corresponding to $n = 0, 1, 2, 3, 4, 5$.

We may also display the transient behavior of the steepest-descent algorithm by plotting the tap weight $w_1(n)$ versus $w_2(n)$. In particular, for our example the use of Eq. (8.24) yields the tap-weight vector

$$\begin{aligned} \mathbf{w}(n) &= \begin{bmatrix} w_1(n) \\ w_2(n) \end{bmatrix} \\ &= \begin{bmatrix} -a_1 + (\nu_1(n) + \nu_2(n))/\sqrt{2} \\ -a_2 + (\nu_1(n) - \nu_2(n))/\sqrt{2} \end{bmatrix} \end{aligned} \tag{8.36}$$

The corresponding trajectory of $[w_1(n), w_2(n)]$, with n as a running parameter, obtained by using Eq. (8.36), is shown plotted in Fig. 8.9(a). Here again we have in-

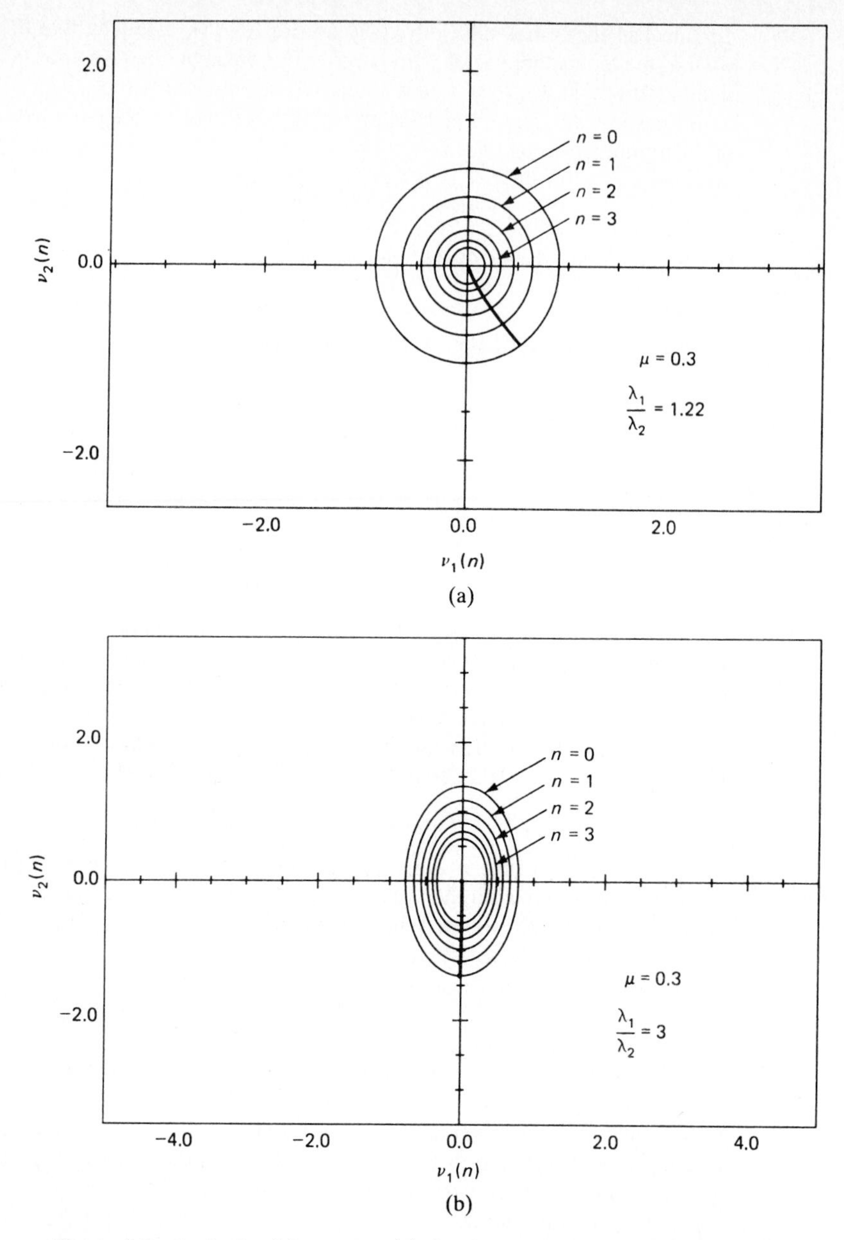

Figure 8.8 Loci of $\nu_1(n)$ versus $\nu_2(n)$ for the steepest-descent algorithm with step-size parameter $\mu = 0.3$ and varying eigenvalue spread: (a) $\chi(\mathbf{R}) = 1.22$; (b) $\chi(\mathbf{R}) = 3$; (c) $\chi(\mathbf{R}) = 10$; (d) $\chi(\mathbf{R}) = 100$.

cluded the loci of $[w_1(n), w_2(n)]$ for fixed values of $J(n)$ corresponding to $n = 0, 1, 2, 3, 4, 5$. Note that these loci, unlike Fig. 8.8(a), are ellipsoidal.

Case 2: Eigenvalue Spread $\chi(R) = 3$. The use of the parameter values for Case 2 in Eqs. (8.33) and (8.34) yields the trajectory of $[\nu_1(n), \nu_2(n)]$ shown in Fig. 8.8(b), with n as running parameter, and their use in Eq. (8.35) yields the ellipsoidal

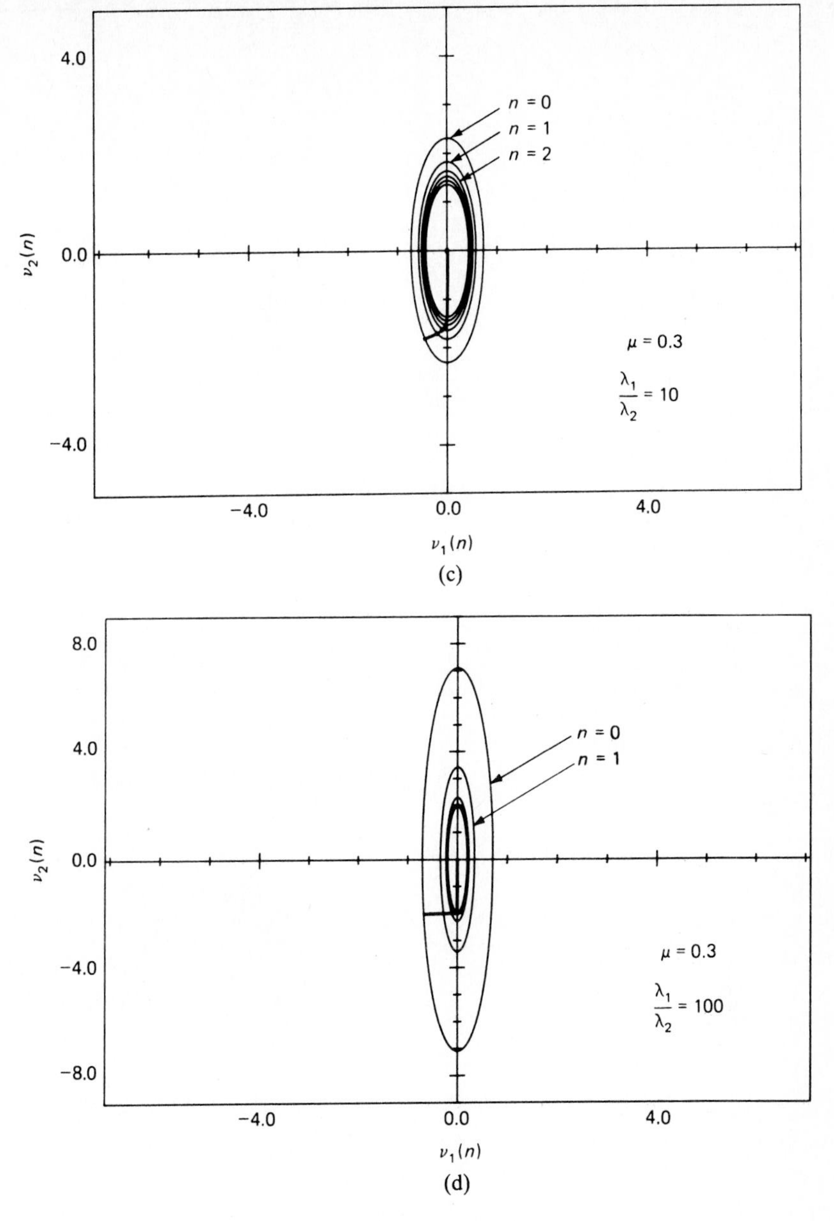

(c)

(d)

Figure 8.8 *(continued)*

loci shown for the fixed values of $J(n)$ for $n = 0, 1, 2, 3, 4, 5$. Note that for this set of parameter values the initial value $\nu_2(0)$ is approximately zero, so the initial value $\boldsymbol{\nu}(0)$ lies practically on the ν_1-axis.

The corresponding trajectory of $[w_1(n), w_2(n)]$, with n as running parameter, is shown in Fig. 8.9(b).

Case 3: Eigenvalue Spread $\chi(\mathbf{R}) = 10$. For this case, the application of Eqs. (8.33) and (8.34) yields the trajectory of $[\nu_1(n), \nu_2(n)]$ shown in Fig. 8.8(c), with n as

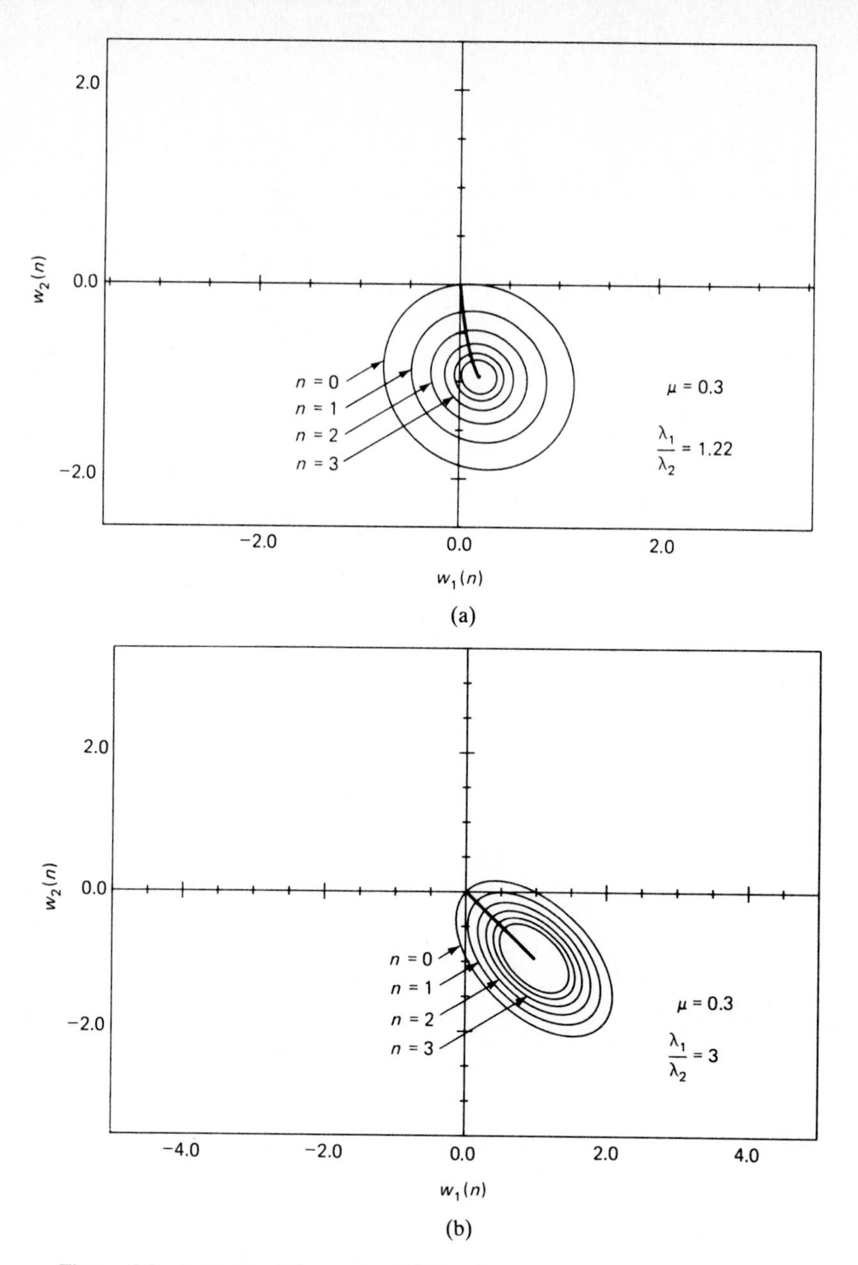

Figure 8.9 Loci of $w_1(n)$ versus $w_2(n)$ for the steepest-descent algorithm with step-size parameter $\mu = 0.3$ and varying eigenvalue spread: (a) $\chi(\mathbf{R}) = 1.22$; (b) $\chi(\mathbf{R}) = 3$; (c) $\chi(\mathbf{R}) = 10$; (d) $\chi(\mathbf{R}) = 100$.

running parameter, and the application of Eq. (8.35) yields the ellipsoidal loci included in this figure for fixed values of $J(n)$ for $n = 0, 1, 2, 3, 4, 5$. The corresponding trajectory of $[w_1(n), w_2(n)]$, with n as running parameter, is shown in Fig. 8.9(c).

Case 4: Eigenvalue Spread $\chi(R) = 100$. For this case, application of the preceding equations yields the results shown in Fig. 8.8(d) for the trajectory of $[\nu_1(n),$

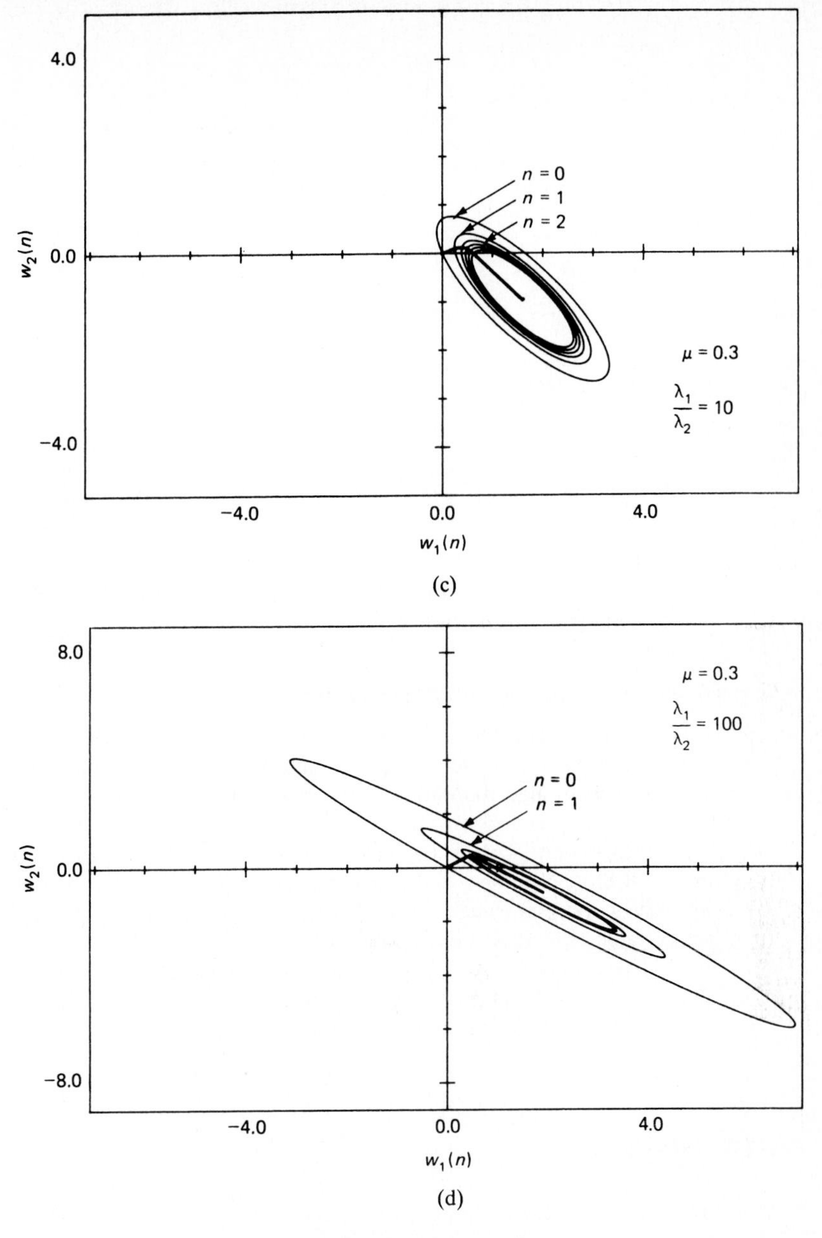

Figure 8.9 *(continued)*

$\nu_2(n)$] and the ellipsoidal loci for fixed values of $J(n)$. The corresponding trajectory of $[w_1(n), w_2(n)]$ is shown in Fig. 8.9(d).

In Fig. 8.10 we have plotted the mean-squared error $J(n)$ versus n for the four eigenvalue spreads 1.22, 3, 10, and 100. We see that as the eigenvalue spread increases (and the input process becomes more correlated), the minimum mean-squared error $J_{\min}$ decreases. This makes intuitive sense: The predictor should do a better job tracking a highly correlated input process than a weakly correlated one.

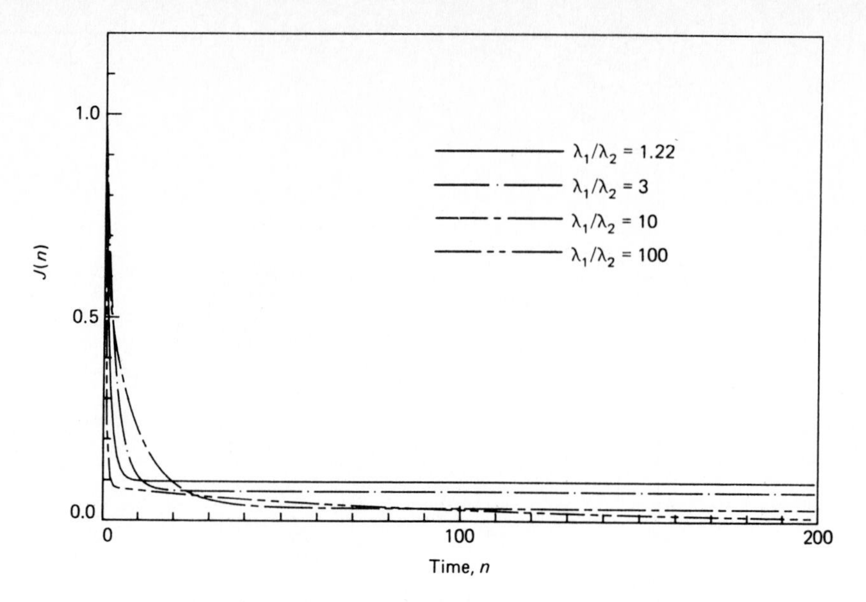

Figure 8.10 Learning curves of steepest-descent algorithm with step-size parameter $\mu = 0.3$ and varying eigenvalue spread.

Experiment 2: Varying Step-Size Parameter

In this experiment the eigenvalue spread is fixed at $\chi(\mathbf{R}) = 10$, and the step-size parameter μ is varied. In particular, we examine the transient behavior of the steepest-descent algorithm for $\mu = 0.3, 1.0$. The corresponding results in terms of the transformed tap-weight errors $\nu_1(n)$ and $\nu_2(n)$ are shown in parts (a) and (b) of Fig. 8.11, respectively. The results included in part (a) of this figure are the same as those in Fig. 8.8(c). Note also that in accordance with Eq. (8.21), the critical value of the step-size parameter equals $\mu_{\max} = 2/\lambda_{\max} = 1.1$, which is slightly in excess of the actual value $\mu = 1$ used in Fig. 8.11(b).

The results for $\mu = 0.3, 1.0$ in terms of the tap weights $w_1(n)$ and $w_2(n)$ are shown in parts (a) and (b) of Fig. 8.12, respectively. Here again, the results included in part (a) of the figure are the same as those in Fig. 8.9(c).

DISCUSSION

Based on the results presented for Experiments 1 and 2, we may make the following observations:

1. The trajectory of $[\nu_1(n), \nu_2(n)]$, with the number of iterations n as running parameter, is normal to the locus of $[\nu_1(n), \nu_2(n)]$ for fixed $J(n)$. Similarly, the trajectory of $[w_1(n), w_2(n)]$ for fixed $J(n)$.
2. When the eigenvalues λ_1 and λ_2 are equal, the trajectory of $[\nu_1(n), \nu_2(n)]$ or that of $[w_1(n), w_2(n)]$, with n as running parameter, is a straight line. This is illustrated in Fig. 8.8(a) or 8.9(a) for which the eigenvalues λ_1 and λ_2 are approximately equal.

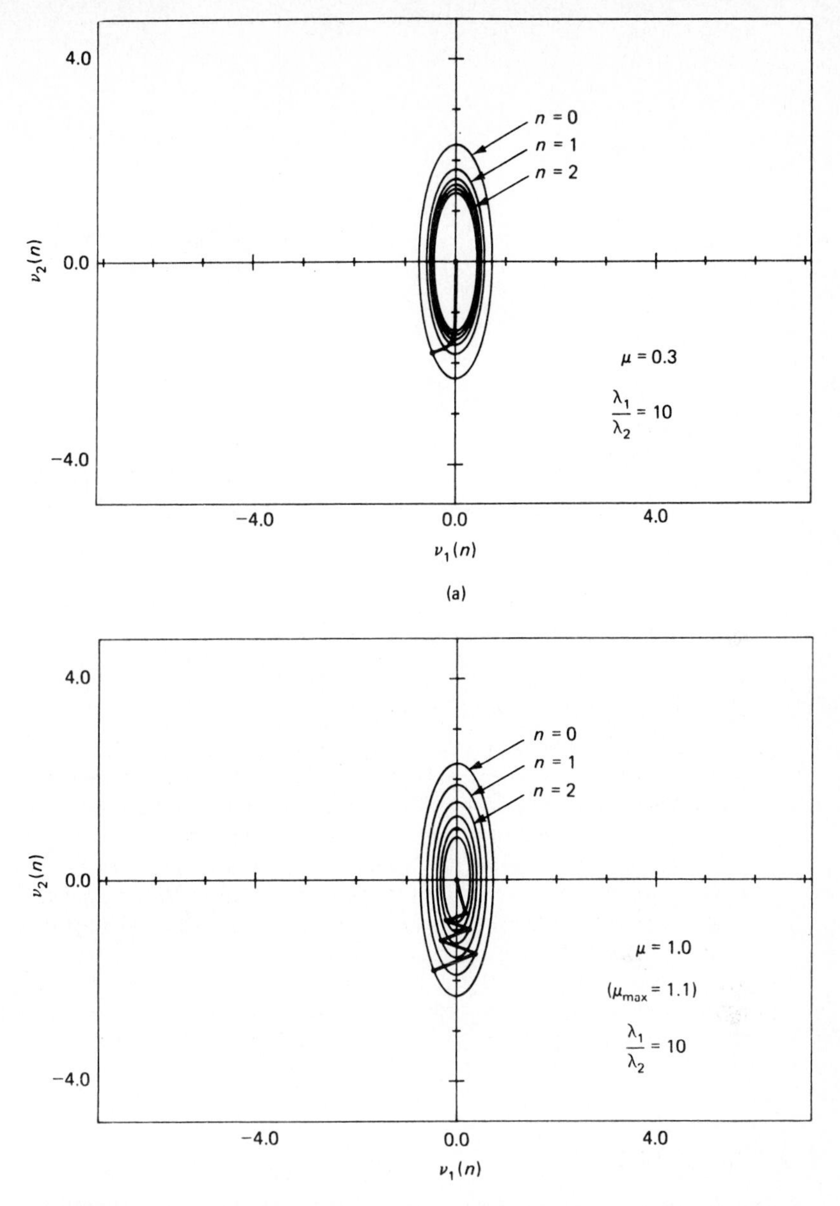

Figure 8.11 Loci of $\nu_1(n)$ versus $\nu_2(n)$ for the steepest-descent algorithm with eigenvalue spread $\chi(\mathbf{R}) = 10$ and varying step-size parameters: (a) overdamped, $\mu = 0.3$; (b) underdamped, $\mu = 1.0$.

3. When the conditions are right for the initial value $\boldsymbol{\nu}(0)$ of the transformed tap-weight error vector $\boldsymbol{\nu}(n)$ to lie on the ν_1-axis or ν_2-axis, the trajectory of $[\nu_1(n), \nu_2(n)]$, with n as running parameter, is a straight line. This is illustrated in Fig. 8.8(b), where $\nu_1(0)$ is approximately zero. Correspondingly, the trajectory of $[w_1(n), w_2(n)]$, with n as running parameter, is also a straight line, as illustrated in Fig. 8.9(b).

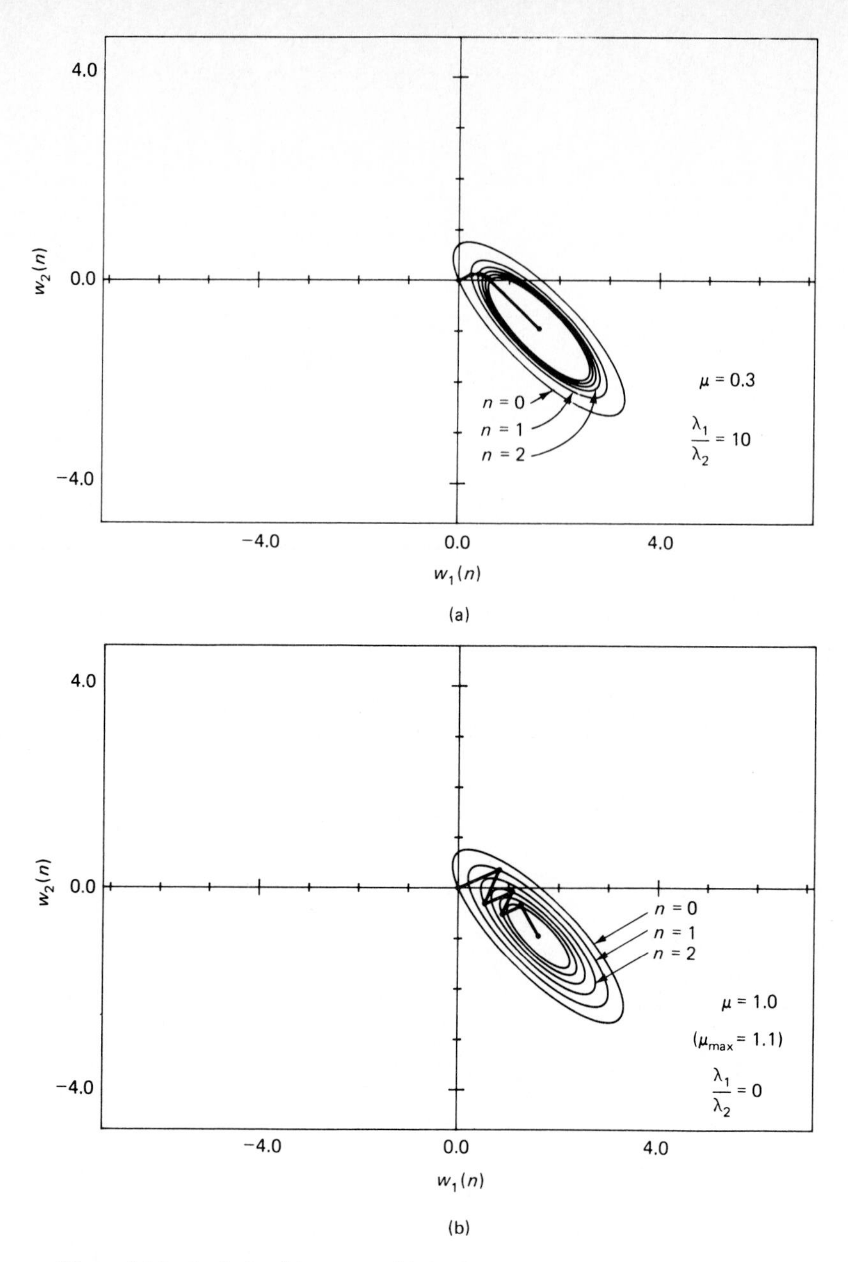

Figure 8.12 Loci of $w_1(n)$ versus $w_2(n)$ for the steepest-descent algorithm with eigenvalue spread $\chi(\mathbf{R}) = 10$ and varying step-size parameters: (a) overdamped, $\mu = 0.3$; underdamped, $\mu = 1.0$.

4. Except for two special cases, (1) equal eigenvalues, and (2) the right choice of initial conditions, the trajectory of $[\nu_1(n), \nu_2(n)]$, with n as running parameter, follows a curved path, as illustrated in Fig. 8.8(c). Correspondingly, the trajectory of $[w_1(n), w_2(n)]$, with n as running parameter, also follows a curved path as illustrated in Fig. 8.9(c). When the eigenvalue spread is very high (i.e., the input data are very highly correlated), two things happen:

(a) The error performance surface assumes the shape of a deep valley.

(b) The trajectories of $[\nu_1(n), \nu_2(n)]$ and $[w_1(n), w_2(n)]$ develop distinct bends. Both of these points are well illustrated in Figs. 8.8(d) and 8.9(d), respectively, for the case of $\chi(\mathbf{R}) = 100$.

5. The steepest-descent algorithm converges fastest when the eigenvalues λ_1 and λ_2 are equal or the starting point of the algorithm is chosen right, for which cases the trajectory formed by joining the points $\nu(0), \nu(1), \nu(2), \ldots$, is a straight line, the shortest possible path.

6. For fixed step-size parameter μ, as the eigenvalue spread $\chi(\mathbf{R})$ increases (i.e, the correlation matrix $\mathbf{R}$ of the tap inputs becomes more ill conditioned), the ellipsoidal loci of $[\nu_1(n), \nu_2(n)]$ for fixed values of $J(n)$, for $n = 0, 1, 2, \ldots$, become increasingly narrower (i.e., the minor axis becomes smaller) and more crowded.

7. When the step-size parameter μ is small, the transient behavior of the steepest-descent algorithm is *overdamped* in that the trajectory formed by joining the points $\boldsymbol{\nu}(0), \boldsymbol{\nu}(1), \boldsymbol{\nu}(2), \ldots$, follows a continuous path. When, on the other hand μ approaches the maximum allowable value, $\mu_{\max} = 2/\lambda_{\max}$, the transient behavior of the steepest-descent algorithm is *underdamped* in that this trajectory exhibits oscillations. These two different forms of transient behavior are illustrated in parts (a) and (b) of Fig. 8.11 in terms of $\nu_1(n)$ and $\nu_2(n)$. The corresponding results in terms of $w_1(n)$ and $w_2(n)$ are presented in parts (a) and (b) of Fig. 8.12.

The conclusion to be drawn from these observations is that the transient behavior of the steepest-descent algorithm is highly sensitive to variations in the step-size parameter μ and the eigenvalue spread of the correlation matrix of the tap inputs.

PROBLEMS

1. Consider a Wiener filtering problem characterized by the following values for the correlation matrix $\mathbf{R}$ of the tap-input vector $\mathbf{u}(n)$ and the cross-correlation vector $\mathbf{p}$ between $\mathbf{u}(n)$ and the desired response $d(n)$:

$$\mathbf{R} = \begin{bmatrix} 1 & 0.5 \\ 0.5 & 1 \end{bmatrix}$$

$$\mathbf{p} = \begin{bmatrix} 0.5 \\ 0.25 \end{bmatrix}$$

(a) Suggest a suitable value for the step-size parameter μ that would ensure convergence of the method of steepest descent, based on the given value for matrix $\mathbf{R}$.

(b) Using the value proposed in part (a), determine the recursions for computing the elements $w_1(n)$ and $w_2(n)$ of the tap-weight vector $\mathbf{w}(n)$. For this computation, you may assume the initial values

$$w_1(0) = w_2(0) = 0$$

(c) Investigate the effect of varying the step-size parameter μ on the trajectory of the tap-weight vector $\mathbf{w}(n)$ as n varies from zero to infinity.

2. Start with the formula for the estimation error:

$$e(n) = d(n) - \mathbf{w}^H(n)\mathbf{u}(n)$$

where $d(n)$ is the desired response, $\mathbf{u}(n)$ is the tap-input vector, and $\mathbf{w}(n)$ is the tap-weight vector in the transversal filter. Hence, show that the gradient of the instantaneous squared error $|e(n)|^2$ equals

$$\hat{\boldsymbol{\nabla}}(n) = -2\mathbf{u}(n)d^*(n) + 2\mathbf{u}(n)\mathbf{u}^H(n)\mathbf{w}(n)$$

3. In this problem we explore another way of deriving the steepest-descent algorithm of Eq. (8.9) used to adjust the tap-weight vector in a transversal filter. The inverse of a positive-definite matrix may be expanded in a series as follows:

$$\mathbf{R}^{-1} = \mu \sum_{k=0}^{\infty} (\mathbf{I} - \mu\mathbf{R})^k$$

where $\mathbf{I}$ is the identity matrix, and μ is a positive constant. To ensure convergence of the series, the constant μ must lie inside the range

$$0 < \mu < \frac{2}{\lambda_{\max}}$$

where $\lambda_{\max}$ is the largest eigenvalue of the matrix $\mathbf{R}$. By using this series expansion for the inverse of the correlation matrix in the Wiener-Hopf equations, develop the recursion

$$\mathbf{w}(n + 1) = \mathbf{w}(n) + \mu[\mathbf{p} - \mathbf{R}\mathbf{w}(n)]$$

where $\mathbf{w}(n)$ is the approximation to the Wiener solution for the tap-weight vector:

$$\mathbf{w}(n) = \mu \sum_{k=0}^{n-1} (\mathbf{I} - \mu\mathbf{R})^k\mathbf{p}$$

4. In the method of steepest descent, show that the correction applied to the tap-weight vector at time $n + 1$ may be expressed as follows:

$$\delta\mathbf{w}(n + 1) = \mu E[\mathbf{u}(n)e^*(n)]$$

where $\mathbf{u}(n)$ is the tap-input vector and $e(n)$ is the estimation error. What happens to this correction at the minimum point of the error performance surface? Discuss your answer in light of the principle of orthogonality.

5. Consider the method of steepest descent involving a single weight $w(n)$. Do the following:
   (a) Determine the mean-squared error $J(n)$ as a function of $w(n)$.
   (b) Find the Wiener solution w_o, and the minimum mean-squared error $J_{\min}$.
   (c) Sketch a plot of $J(n)$ versus $w(n)$.

6. Equation (8.28) defines the transient behavior of the mean-squared error $J(n)$ for varying n that is produced by the steepest-descent algorithm. Let $J(0)$ and $J(\infty)$ denote the initial and final values of $J(n)$. Suppose that we approximate this transient behavior with a single exponential, as follows:

$$J_{\text{approx}}(n) = [J(0) - J(\infty)]e^{-n/\tau} + J(\infty)$$

where τ is termed the *effective initial time constant*. Let τ be chosen such that

$$J_{\text{approx}}(1) = J(1)$$

Hence, show that *the initial rate of convergence* of the steepest-descent algorithm, defined as the inverse of τ, is given by

$$\frac{1}{\tau} = \ln\left[\frac{J(0) - J(\infty)}{J(1) - J(\infty)}\right]$$

Using Eq. (8.28), find the value of $1/\tau$. Assume that the initial value $\mathbf{w}(0)$ is zero, and that the step-size parameter μ is small.

7. Consider an autoregressive (AR) process $\{u(n)\}$ of order 1, described by the difference equation

$$u(n) = -au(n - 1) + v(n)$$

where a is the AR parameter of the process, and $\{v(n)\}$ is a zero-mean white-noise process of variance σ_v^2.

(a) Set up a linear predictor of order 1 to compute the parameter a. To be specific, use the method of steepest descent for the recursive computation of the Wiener solution for the parameter a.

(b) Plot the error-performance curve for this problem, identifying the minimum point of the curve in terms of known parameters.

(c) What is the condition on the step-size parameter μ to ensure stability? Justify your answer.

Computer-Oriented Problems

8. An autoregressive (AR) process $\{u(n)\}$ of order 1 is described by the difference equation

$$u(n) + a_1 u(n - 1) = v(n)$$

where $\{v(n)\}$ is a white-noise process of zero mean and variance σ_v^2. The AR parameter a_1 and variance σ_v^2 are assigned one of the following four sets of values:

$$(1)\quad a_1 = -0.1000, \qquad \sigma_v^2 = 0.9900$$

$$(2)\quad a_1 = -0.5000, \qquad \sigma_v^2 = 0.7500$$

$$(3)\quad a_1 = -0.8182, \qquad \sigma_v^2 = 0.3305$$

$$(4)\quad a_1 = -0.9802, \qquad \sigma_v^2 = 0.0392$$

The AR process $\{u(n)\}$ is applied to a two-tap linear predictor.

(a) Calculate the eigenvalues of the 2-by-2 correlation matrix of the tap inputs of the predictor for each parameter set (1) through (4). Hence, determine the corresponding permissible ranges for the step-size parameter μ in the steepest-descent algorithm.

(b) Using the two elements of the transformed tap-weight error vector [*i.e.*, $\nu_1(n)$ and $\nu_2(n)$] as the variables in the steepest-descent algorithm, construct loci for constant values of mean-squared error for each of the parameter sets (1), (2), and (3). Superimpose on each of the three figures the trajectory that describes the change in the coordinates $\nu_1(n)$ and $\nu_2(n)$ with time n, assuming that the step-size parameter $\mu = 0.3$ and that the initial values of the two tap weights in the predictor both equal zero.

(c) Repeat the computations in part (b), using the tap weights $w_1(n)$ and $w_2(n)$ themselves as the variables in the steepest-descent algorithm.

(d) Plot the learning curve [i.e., the mean-squared error $J(n)$ versus time n] for each of the four parameters sets (1) through (4). In each case, repeat the computations for the following step-size parameters:

$$\mu = 0.02$$

$$\mu = 0.05$$

$$\mu = 0.2$$

(e) Discuss the implications of your results.

9. An autoregressive (AR) process $\{u(n)\}$ of order 2 is described by the difference equation

$$u(n) + a_1 u(n-1) + a_2 u(n-2) = v(n)$$

where $\{v(n)\}$ is a white-noise process of zero mean and variance σ_v^2. The parameters a_1 and a_2 and the variance σ_v^2 are assigned one of the following four sets of values:

$$(1) \quad a_1 = -0.10025, \quad a_2 = 0.0025, \quad \sigma_v^2 = 0.9900$$

$$(2) \quad a_1 = -0.5359, \quad a_2 = 0.0718, \quad \sigma_v^2 = 0.7461$$

$$(3) \quad a_1 = -1.0390, \quad a_2 = 0.2699, \quad \sigma_v^2 = 0.3065$$

$$(4) \quad a_1 = -1.6364, \quad a_2 = 0.6694, \quad \sigma_v^2 = 0.0216$$

The AR process $\{u(n)\}$ is applied to a two-tap linear predictor. Repeat the computations made in parts (a) to (d) of Problem 8, and discuss the implications of your results.

9

Stochastic Gradient-Based Algorithms

In this chapter we develop the theory of a widely used algorithm known as the *least-mean-square* (*LMS*) *algorithm* (Widrow and Hoff, 1960). The LMS algorithm is an important member of the family of *stochastic gradient-based algorithms*. A significant feature of the LMS algorithm is its *simplicity;* it does not require measurements of the pertinent correlation functions, nor does it require matrix inversion. Indeed, it is the simplicity of the LMS algorithm that has made it the *standard* against which other adaptive filtering algorithms are benchmarked.

The material presented in this chapter on the LMS algorithm also includes a detailed analysis of its convergence behavior. Other topics covered in the chapter include a *normalized* version of the LMS algorithm, and the *gradient adaptive lattice* (GAL) *algorithm*. We begin our study of the LMS algorithm by presenting an overview of its structure and operation.

9.1 OVERVIEW OF THE STRUCTURE AND OPERATION OF THE LEAST-MEAN-SQUARE ALGORITHM

The operation of the *least-mean-square* (LMS) *algorithm* is descriptive of a *feedback control system*. Basically, it consists of a combination of two basic processes:

1. An *adaptive process,* which involves the automatic adjustment of a set of tap weights.
2. A *filtering process,* which involves (a) forming the inner product of a set of tap inputs and the corresponding set of tap weights emerging from the adaptive process to produce an estimate of a desired response, and (b) generating an estimation error by comparing this estimate with the actual value of the desired response. The estimation error is in turn used to actuate the adaptive process, thereby closing the feedback loop.

Correspondingly, we may identify two basic components in the structural constitution of the LMS algorithm, as illustrated in the block diagram of Fig. 9.1(a). First, we have a transversal filter, around which the LMS algorithm is built; this component is responsible for performing the filtering process. Second, we have a mechanism for performing the adaptive control process on the tap weights of the transversal filter; hence, the designation "adaptive weight-control mechanism" in Fig. 9.1(a).

Details of the *transversal filter* component are presented in Fig. 9.1(b). The tap inputs $u(n)$, $u(n-1)$, . . ., $u(n-M+1)$ form the elements of the M-by-1 *tap-input vector* $\mathbf{u}(n)$, where $M-1$ is the number of delay elements. The series of tap inputs exhibits the temporal structure of the filter input. Correspondingly, the tap weights $\hat{w}_0(n)$, $\hat{w}_1(n)$, . . . , $\hat{w}_{M-1}(n)$ form the elements of the M-by-1 tap-weight vector $\hat{\mathbf{w}}(n)$. The value computed for the tap-weight vector $\hat{\mathbf{w}}(n)$ using the LMS algorithm represents an estimator whose expected value approaches the Wiener solution $\mathbf{w}_o$ (for a wide-sense stationary environment) as the number of iterations n approaches infinity.

During the filtering process the *desired response* $d(n)$ is supplied for processing, alongside the tap-input vector $\mathbf{u}(n)$. Given this input, the transversal filter produces an output $\hat{d}(n)|\mathcal{U}_n)$ used as an *estimate* of the desired response $d(n)$. Accordingly, we may set up an *estimation error* $e(n)$ as the difference between the desired response and the filter output, as indicated in the output end of Fig. 9.1(b). The estimation error $e(n)$ and the tap-input vector $\mathbf{u}(n)$ are applied to the control mechanism, and the feedback loop around the tap weights is thereby closed.

Figure 9.1(c) presents details of the *adaptive weight-control mechanism*. Specifically, a scaled version of the *inner product* of the estimation error $e(n)$ and the tap input $u(n-k)$ is computed for $k = 0, 1, 2, \ldots, M-2, M-1$. The result so obtained defines the *correction* $\delta\hat{w}_k(n)$ applied to the tap weight $\hat{w}_k(n)$ at time $n+1$. The scaling factor used in this computation is denoted by μ in Fig. 9.1(c). It is called the *adaptation constant* or *step-size parameter;* in the sequel, we use both terminologies interchangeably.

Comparing the control mechanism of Fig. 9.1(c) for the LMS algorithm with that of Fig. 8.2 for the method of steepest descent, we see that the LMS algorithm uses the inner product $u(n-k)e^*(k)$ as an estimator of element k in the gradient vector $\nabla(J(n))$ that characterizes the method of steepest descent. In other words, the expectation operator is missed out from all the paths in Fig. 9.1(c). Accordingly, the recursive computation of each tap weight in the LMS algorithm suffers from a *gradient noise*.

For a stationary environment, we know from Chapter 8 that the method of steepest descent computes a tap-weight vector $\mathbf{w}(n)$ that moves down the ensemble-averaged error-performance surface along a deterministic trajectory, which terminates on the Wiener solution, $\mathbf{w}_o$. The LMS algorithm, on the other hand, behaves differently because of the presence of gradient noise. Rather than terminating on the Wiener solution, the tap-weight vector $\hat{\mathbf{w}}(n)$ [different from $\mathbf{w}(n)$] computed by the LMS algorithm executes a *random motion* around the minimum point of the error-performance surface. This random motion gives rise to two forms of convergence behavior for the LMS algorithm:

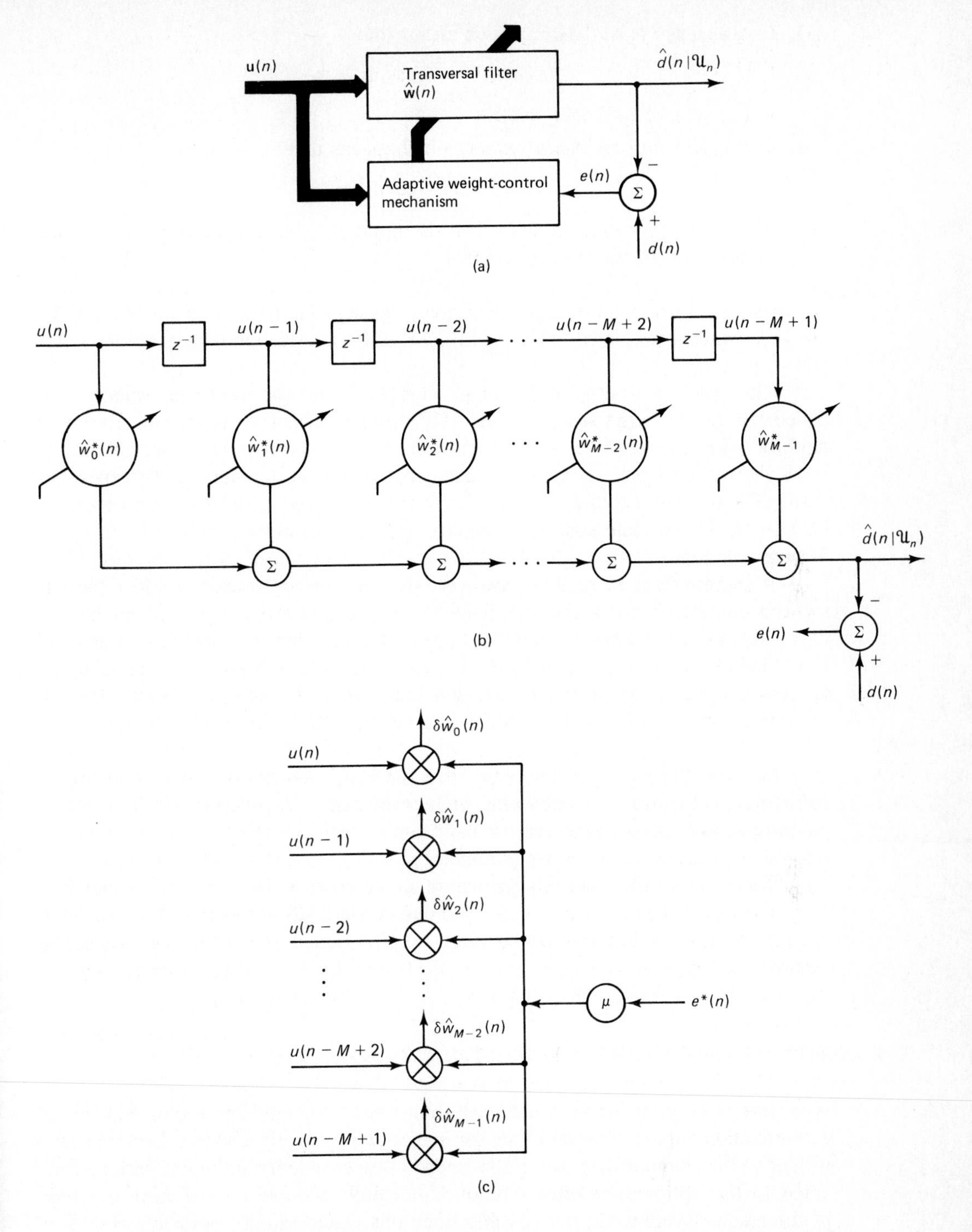

Figure 9.1 (a) Block diagram of adaptive transversal filter. (b) Detailed structure of the transversal filter component. (c) Detailed structure of the adaptive weight-control mechanism.

1. *Convergence in the mean,* which means that

$$E[\hat{\mathbf{w}}(n)] \to \mathbf{w}_o \text{ as } n \to \infty$$

where $\mathbf{w}_o$ is the Wiener solution.

2. *Convergence in the mean square,* which means that

$$J(n) \to J(\infty) \text{ as } n \to \infty$$

where $J(\infty)$ is always greater than the minimum mean-squared error $J_{\min}$ that corresponds to the Wiener solution.

For these two forms of convergence to hold, the step-size parameter μ has to satisfy different conditions related to the eigenvalues of the correlation matrix of the tap inputs.

The difference between the final value $J(\infty)$ and the minimum value $J_{\min}$ is called the *excess mean-squared error* $J_{\text{ex}}(\infty)$. This difference represents the price paid for using the adaptive (stochastic) mechanism to control the tap weights in the LMS algorithm in place of a deterministic approach as in the method of steepest descent. The ratio of $J_{\text{ex}}(\infty)$ to $J_{\min}$ is called the *misadjustment,* which is a measure of how far the steady-state solution computed by the LMS algorithm is away from the Wiener solution.

It is important to realize, however, that the misadjustment is under the designer's control. In particular, the feedback loop acting around the tap weights behaves like a *low-pass filter,* whose "average" time constant is *inversely* proportional to the step-size parameter μ. Hence, by assigning a small value to μ the adaptive process is made to progress slowly, and the effects of gradient noise on the tap weights are largely filtered out. This in turn has the effect of reducing the misadjustment.

We may therefore justifiably say that the LMS algorithm is simple in implementation, yet capable of delivering high performance by adapting to its external environment. To do so, however, we have to pay particular attention to the choice of a suitable value for the step-size parameter μ.

The factors influencing the choice of μ are covered in Sections 9.4 and 9.7. First and foremost, however, we wish to derive the LMS algorithm. This we do in the next section by building on our previous knowledge of the method of steepest descent.

9.2 LEAST-MEAN-SQUARE ADAPTATION ALGORITHM

If it were possible to make exact measurements of the gradient vector $\nabla(J(n))$ at each iteration, and if the step-size parameter μ is suitably chosen, then the tap-weight vector computed by using the steepest-descent algorithm would indeed converge to the optimum Wiener solution. In reality, however, exact measurements of the gradient vector are not possible since this would require prior knowledge of both the correlation matrix $\mathbf{R}$ of the tap inputs and the cross-correlation vector $\mathbf{p}$ between the tap inputs and the desired response [see Eq. (8.8)]. Consequently, the gradient vector must be *estimated* from the available data. In other words, the tap-weight vector is updated in accordance with an algorithm that *adapts to the incoming data*. One such algorithm is the *least-mean-square (LMS) algorithm* (Widrow and Hoff, 1960; Widrow, 1970). A significant feature of the LMS algorithm is its simplicity; it does not require measurements of the pertinent correlation functions, nor does it require matrix inversion.

To develop an estimate of the gradient vector $\nabla(J(n))$, the most obvious strategy is to substitute estimates of the correlation matrix $\mathbf{R}$ and the cross-correlation vector $\mathbf{p}$ in the formula of Eq. (8.8), which is reproduced here for convenience:

$$\nabla(J(n)) = -2\mathbf{p} + 2\mathbf{R}\mathbf{w}(n) \tag{9.1}$$

The *simplest* choice of estimators for $\mathbf{R}$ and $\mathbf{p}$ is to use *instantaneous estimates* that are based on sample values of the tap-input vector and desired response, as defined by, respectively,

$$\hat{\mathbf{R}}(n) = \mathbf{u}(n)\mathbf{u}^H(n) \tag{9.2}$$

and

$$\hat{\mathbf{p}}(n) = \mathbf{u}(n)d^*(n) \tag{9.3}$$

Correspondingly, the instantaneous estimate of the gradient vector is as follows:

$$\hat{\nabla}(J(n)) = -2\mathbf{u}(n)d^*(n) + 2\mathbf{u}(n)\mathbf{u}^H(n)\hat{\mathbf{w}}(n) \tag{9.4}$$

Generally speaking, this estimate is *biased* because the tap-weight estimate vector $\hat{\mathbf{w}}(n)$ is a random vector that depends on the tap-input vector $\mathbf{u}(n)$. Note that the estimate $\hat{\nabla}(J(n))$ may also be viewed as the gradient operator ∇ applied to the instantaneous squared error $|e(n)|^2$.

Substituting the estimate of Eq. (9.4) for the gradient vector $\nabla(J(n))$ in the steepest-descent algorithm as described in Eq. (8.7), we get a new recursive relation for updating the tap-weight vector:

$$\hat{\mathbf{w}}(n+1) = \hat{\mathbf{w}}(n) + \mu\mathbf{u}(n)[d^*(n) - \mathbf{u}^H(n)\hat{\mathbf{w}}(n)] \tag{9.5}$$

Here we have used a hat over the symbol for the tap-weight vector to distinguish it from the value obtained by using the steepest-descent algorithm. Equivalently, we may write the result in the form of three basic relations as follows:

1. *Filter output*:

$$y(n) = \hat{\mathbf{w}}^H(n)\mathbf{u}(n) \tag{9.6}$$

2. *Estimation error*:

$$e(n) = d(n) - y(n) \tag{9.7}$$

3. *Tap-weight adaptation*:

$$\hat{\mathbf{w}}(n+1) = \hat{\mathbf{w}}(n) + \mu\mathbf{u}(n)e^*(n) \tag{9.8}$$

Equations (9.6) and (9.7) define the estimation error $e(n)$, the computation of which is based on the *current estimate* of the tap-weight vector, $\hat{\mathbf{w}}(n)$. Note also that the second term, $\mu\mathbf{u}(n)e^*(n)$, on the right side of Eq. (9.8) represents the *correction* that is applied to the current estimate of the tap-weight vector, $\hat{\mathbf{w}}(n)$. The iterative procedure is started with the initial guess $\hat{\mathbf{w}}(0)$. A convenient choice for this initial guess is the null vector; we may thus set $\hat{\mathbf{w}}(0) = \mathbf{0}$.

The algorithm described by Eqs. (9.6) to (9.8) is the complex form of the adaptive *least-mean-square* (LMS) *algorithm* (Widrow et al., 1975).[1] It is a member of a family of *stochastic gradient algorithms*. In the LMS algorithm the allowed set of directions along which we "step" from one iteration cycle to the next is quite random and cannot therefore be thought of as being gradient directions.

Figure 9.2 shows a signal-flow graph representation of the LMS algorithm in the form of a feedback model. This model bears a close resemblance to the feedback model of Fig. 8.3 describing the steepest-descent algorithm. The signal-flow graph of Fig. 9.2 clearly illustrates the simplicity of the LMS algorithm. In particular, we see from this figure that the LMS algorithm requires only $2M + 1$ *complex multiplications and* $2M$ *complex additions per iteration,* where M is the number of tap weights used in the adaptive transversal filter.

The instantaneous estimates of $\mathbf{R}$ and $\mathbf{p}$ given in Eqs. (9.2) and (9.3), respectively, have relatively large variances. At first sight, it may therefore seem that the LMS algorithm is incapable of good performance since the algorithm uses these instantaneous estimates. However, we must remember that the LMS algorithm is recursive in nature, with the result that the algorithm itself effectively averages these estimates, in some sense, during the course of adaptation.

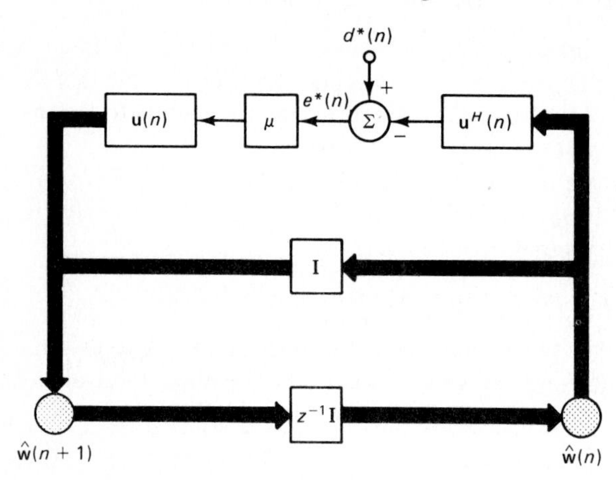

Figure 9.2 Signal-flow graph representation of the LMS algorithm.

9.3 EXAMPLES

Before proceeding further with a convergence analysis of the LMS algorithm, it is instructive to develop an appreciation for the versatility of this important algorithm. We do this by presenting four examples that relate to different applications of the LMS algorithm.

[1] The complex form of the LMS algorithm, as originally proposed by Widrow et al. (1975b), differs slightly from that described in Eqs. (9.6) to (9.8). Widrow and other authors have based their derivation on the definition $\mathbf{R} = E[\mathbf{u}^*(n)\mathbf{u}^T(n)]$ for the correlation matrix. On the other hand, the LMS algorithm described by Eqs. (9.6) to (9.8) is based on the definition $\mathbf{R} = E[\mathbf{u}(n)\mathbf{u}^H(n)]$ for the correlation matrix. The adoption of the latter definition for the correlation matrix of complex-valued data is the natural extension of the definition for real-valued data.

Example 1: Canonical Model of the Complex LMS Algorithm

The LMS algorithm described in Eqs. (9.6) to (9.8) is *complex* in the sense that the input and output data as well as the tap weights are all complex valued. To emphasize the complex nature of the algorithm, we use complex notation to express the data and tap weights as follows:

Tap-input vector:

$$\mathbf{u}(n) = \mathbf{u}_I(n) + j\mathbf{u}_Q(n) \tag{9.9}$$

Desired response:

$$d(n) = d_I(n) + jd_Q(n) \tag{9.10}$$

Tap-weight vector:

$$\hat{\mathbf{w}}(n) = \hat{\mathbf{w}}_I(n) + j\hat{\mathbf{w}}_Q(n) \tag{9.11}$$

Transversal filter output:

$$y(n) = y_I(n) + jy_Q(n) \tag{9.12}$$

Estimation error:

$$e(n) = e_I + je_Q(n) \tag{9.13}$$

The subscripts I and Q denote "in-phase" and "quadrature" components, respectively. Using these definitions in Eqs. (9.6) to (9.8), expanding terms, and then equating real and imaginary parts, we get

$$y_I(n) = \hat{\mathbf{w}}_I^T(n)\mathbf{u}_I(n) + \hat{\mathbf{w}}_Q^T(n)\mathbf{u}_Q(n) \tag{9.14}$$

$$y_Q(n) = \hat{\mathbf{w}}_I^T(n)\mathbf{u}_Q(n) - \hat{\mathbf{w}}_Q^T(n)\mathbf{u}_I(n) \tag{9.15}$$

$$e_I(n) = d_I(n) - y_I(n) \tag{9.16}$$

$$e_Q(n) = d_Q(n) - y_Q(n) \tag{9.17}$$

$$\hat{\mathbf{w}}_I(n+1) = \hat{\mathbf{w}}_I(n) + \mu[e_I(n)\mathbf{u}_I(n) + e_Q(n)\mathbf{u}_Q(n)] \tag{9.18}$$

$$\hat{\mathbf{w}}_Q(n+1) = \hat{\mathbf{w}}_Q(n) + \mu[e_I(n)\mathbf{u}_Q(n) - e_Q(n)\mathbf{u}_I(n)] \tag{9.19}$$

Equations (9.14) to (9.17), defining the error and output signals, are represented by the cross-coupled signal-flow graph shown in Fig. 9.3(a). The update equations (9.18) and (9.19) are likewise represented by the cross-coupled signal-flow graph shown in Fig. 9.3(b). The combination of this pair of signal-flow graphs constitutes the *canonical model* of the complex LMS algorithm. This canonical model clearly illustrates that a complex LMS algorithm is equivalent to a set of four real LMS algorithms with *cross-coupling* between them.

Example 2: Adaptive Deconvolution for Processing of Time-Varying Seismic Data

In Section 1.7 we described the idea of *predictive deconvolution* as a method for processing reflection seismograms. In this example we discuss the use of the LMS algorithm for the adaptive implementation of predictive deconvolution. It is assumed that the seismic data are stationary over the design gate used to generate the deconvolution operation.

Let the set of *real* data points in the seismogram to be processed be denoted by $u(n)$, $n = 1, 2, \ldots, N$, where N is the data length. The real LMS-based adaptive deconvolution method proceeds as follows (Griffiths et al., 1977):

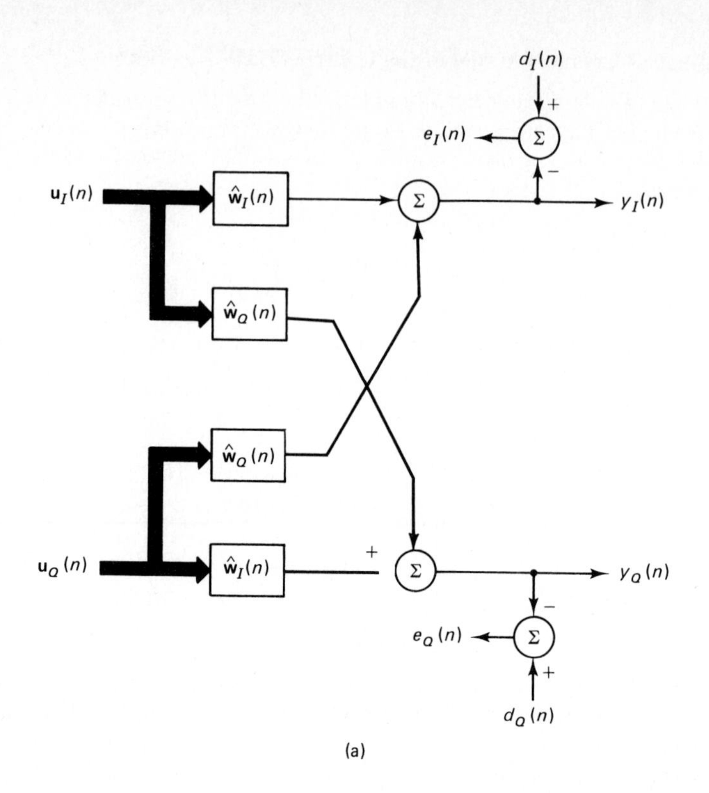

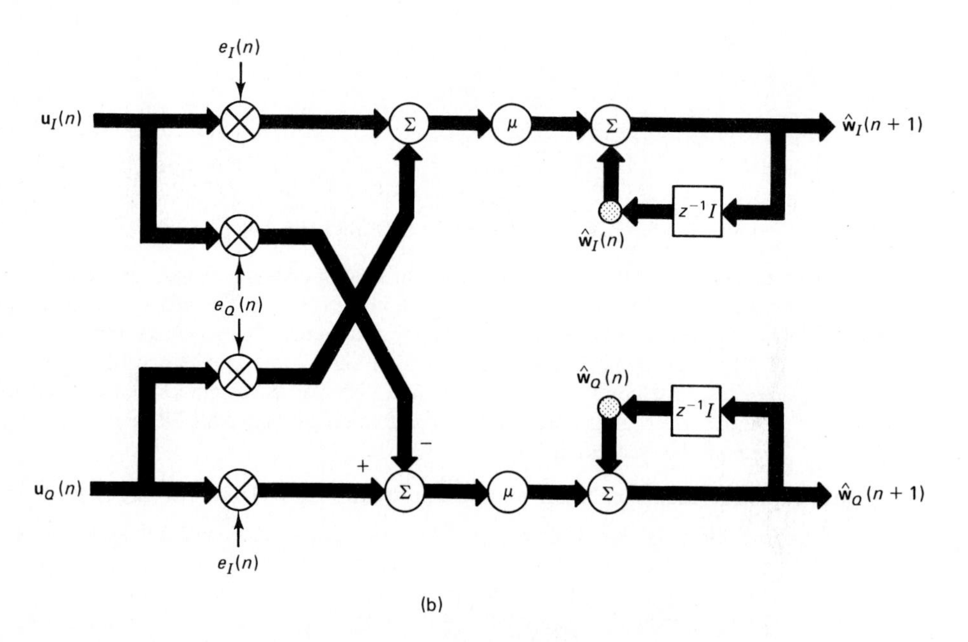

Figure 9.3 Canonical signal-flow graph representation of the complex LMS algorithm. (a) Error and output signals. (b) Tap-weight update equation.

1. An M-dimensional operator $\hat{\mathbf{w}}(n)$ is used to generate a predicted trace from the data, that is,

$$\hat{u}(n + n_0) = \hat{\mathbf{w}}^T(n)\mathbf{u}(n) \tag{9.20}$$

where

$$\hat{\mathbf{w}}(n) = [\hat{w}_0(n), \hat{w}_1(n), \ldots, \hat{w}_{M-1}(n)]^T$$
$$\mathbf{u}(n) = [u(n), u(n-1), \ldots, u(n-M+1)]^T$$

and $n_0 \geq 1$ is the *prediction distance*.

2. The deconvolved trace $y(n)$ defining the difference between the input and predicted traces is evaluated:

$$y(n) = u(n) - \hat{u}(n)$$

3. The operator $\hat{\mathbf{w}}(n)$ is updated:

$$\hat{\mathbf{w}}(n+1) = \hat{\mathbf{w}}(n) + \mu[u(n+n_0) - \hat{u}(n+n_0)]\mathbf{u}(n) \tag{9.21}$$

Equations (9.20) and (9.21) constitute the LMS-based adaptive seismic deconvolution algorithm. The adaptation is begun with an arbitrary initial guess $\hat{\mathbf{w}}(0)$.

Example 3: Instantaneous Frequency Measurement

In this example we study the use of the LMS algorithm as the basis for estimating the frequency content of a *narrow-band signal* characterized by a rapidly varying power spectrum (Griffiths, 1975). In so doing we illustrate the linkage between three basic ideas: an autoregressive (AR) model for describing a stochastic process (studied in Chapter 2), a linear predictor for analyzing the process (studied in Chapter 6), and the LMS algorithm for estimating the AR parameters.

By a "narrow-band signal" we mean a signal whose bandwidth Ω is small compared to the midband frequency ω_c, as illustrated in Fig. 9.4. A *frequency-modulated (FM) signal* is an example of a narrow-band signal, provided that the carrier frequency is high enough. The *instantaneous frequency* (defined as the derivative of phase with respect to time) of an FM signal varies linearly with the modulating signal. Consider then a narrow-band process $\{u(n)\}$ described by a *time varying* AR model of order M, as shown by the difference equation (assuming *real* data):

$$u(n) = -\sum_{k=1}^{M} a_k(n)u(n-k) + v(n) \tag{9.22}$$

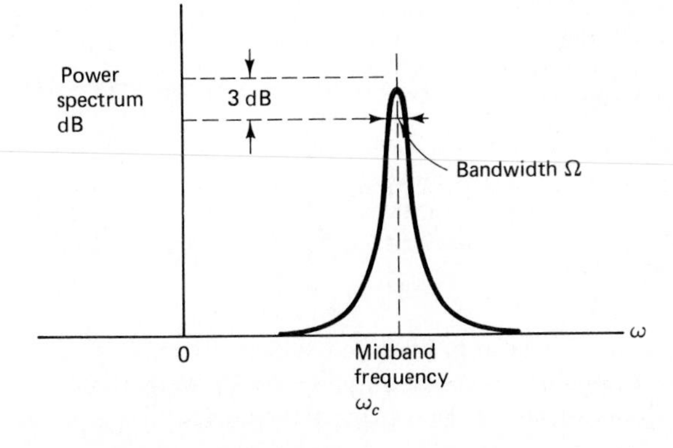

Figure 9.4 Definition of a narrow-band signal in terms of its spectrum.

where the $a_k(n)$ are the time-varying model parameters, and $\{v(n)\}$ is zero-mean white-noise process of time-varying variance $\sigma_v^2(n)$. The *time-varying AR (power) spectrum* of the narrow-band process $\{u(n)\}$ is given by

$$S_{AR}(\omega; n) = \frac{\sigma_v^2(n)}{\left|1 + \sum_{k=1}^{M} a_k(n)e^{-j\omega k}\right|^2}, \qquad -\pi \le \omega \le \pi \tag{9.23}$$

Note that an AR process whose poles are near the unit circle in the z-plane has the characteristics of a narrow-band process, and vice versa.

To estimate the model parameters, we use an adaptive transversal filter employed as a linear predictor of order M. Let the tap weights of the predictor be denoted by $\hat{w}_k(n)$, $k = 1, 2, \ldots, M$. The tap weights are adapted continuously as the input signal $\{u(n)\}$ is received. In particular, we use the LMS algorithm for adapting the tap weights, as shown by

$$\hat{w}_k(n+1) = \hat{w}_k(n) + \mu u(n-k) f_M(n), \qquad k = 1, 2, \ldots, M \tag{9.24}$$

where $f_M(n)$ is the *prediction error:*

$$f_M(n) = u(n) - \sum_{k=1}^{M} \hat{w}_k(n) u(n-k) \tag{9.25}$$

The tap weights of the adaptive predictor are related to the AR model parameters as follows:

$$-\hat{w}_k(n) = \text{estimate of } a_k(n) \text{ at time } n, \qquad \text{for } k = 1, 2, \ldots, M.$$

Moreover, the average power of the prediction error $f_M(n)$ provides an estimate of the noise variance $\sigma_v^2(n)$. Our interest is in locating the frequency of a narrow-band signal. Accordingly, in the sequel, we ignore the estimation of $\sigma_v^2(n)$. Specifically, we only use the tap weights of the adaptive predictor to define the *time-varying frequency function:*

$$F(\omega; n) = \frac{1}{\left|1 - \sum_{k=1}^{M} \hat{w}_k(n)e^{-j\omega k}\right|^2} \tag{9.26}$$

Given the relationship between $\hat{w}_k(n)$ and $a_k(n)$, we see that the essential difference between the frequency function $F(\omega; n)$ in Eq. (9.26) and the AR power spectrum $S_{AR}(\omega; n)$ in Eq. (9.23) lies in their numerator scale factors. The numerator of $F(\omega; n)$ is a constant equal to 1, whereas that of $S_{AR}(\omega; n)$ is a time-varying constant equal to $\sigma_v^2(n)$. The advantages of the frequency function $F(\omega; n)$ over the AR spectrum $S_{AR}(\omega; n)$ are twofold. First, the 0/0 indeterminancy inherent in the narrow-band spectrum of Eq. (9.23) is replaced by a "computationally tractable" limit of 1/0 in Eq. (9.26). Second, the frequency function $F(\omega; n)$ is not affected by amplitude scale changes in the input signal $\{u(n)\}$, with the result that the peak value of $F(\omega; n)$ is related directly to the spectral width of the input signal.

We may use the function $F(\omega; n)$ to measure the instantaneous frequency of a frequency-modulated signal $\{u(n)\}$, provided that the following assumptions are justified (Griffiths, 1975):

1. The adaptive predictor has been in operation sufficiently long, so as to ensure that any transients caused by the initialization of the tap weights have died out.
2. The step-size parameter μ is chosen correctly for the adaptive predictor to track well; that is to say, the prediction error $f_M(n)$ is small for all n.

3. The modulating signal is essentially constant over the sampling range of the adaptive predictor, which extends from time $(n - M)$ to time $(n - 1)$.

Given the validity of these assumptions, we find that the frequency function $F(\omega; n)$ has a peak at the instantaneous frequency of the input signal $\{u(n)\}$. Moreover, the LMS algorithm will track the time variation of the instantaneous frequency.

Example 4: Adaptive Noise Cancelling Applied to a Sinusoidal Interference

The traditional method of suppressing a *sinusoidal interference* corrupting an information-bearing signal is to use a fixed *notch filter* tuned to the frequency of the interference. To design the filter, we naturally need to know the precise frequency of the interference. But what if the notch is required to be very sharp and the interfering sinusoid is known to *drift* slowly? Clearly, then, we have a problem that calls for an *adaptive* solution. One such solution is provided by the use of adaptive noise cancelling.

Figure 9.5 shows the block diagram of a dual-input *adaptive noise canceller*. The primary input supplies an information-bearing signal and a sinusoidal interference that are uncorrelated with each other. The reference input supplies a correlated version of the sinusoidal interference. For the adaptive filter, we may use a transversal filter whose tap weights are adapted by means of the LMS algorithm. The filter uses the reference input to provide (at its output) an estimate of the sinusoidal interfering signal contained in the primary input. Thus, by subtracting the adaptive filter output from the primary input, the effect of the sinusoidal interference is diminished. In particular, an adaptive noise canceller using the LMS algorithm has two important characteristics (Widrow et al., 1976; Glover, 1977):

1. The canceller behaves as an adaptive notch filter whose null point is determined by the angular frequency ω_0 of the sinusoidal interference. Hence, it is tunable, and the tuning frequency moves with ω_0.
2. The notch in the frequency response of the canceller can be made very sharp at precisely the frequency of the sinusoidal interference by choosing a small enough value for the step-size parameter μ.

In the example considered here, the input data are assumed to be real valued:

1. *Primary input:*

$$d(n) = s(n) + A_0 \cos(\omega_0 n + \phi_0) \tag{9.27}$$

where $s(n)$ is an information-bearing signal; A_0 is the amplitude of the sinusoidal interference, ω_0 is the *normalized* angular frequency, and ϕ_0 is the phase.

2. *Reference input:*

$$u(n) = A \cos(\omega_0 n + \phi) \tag{9.28}$$

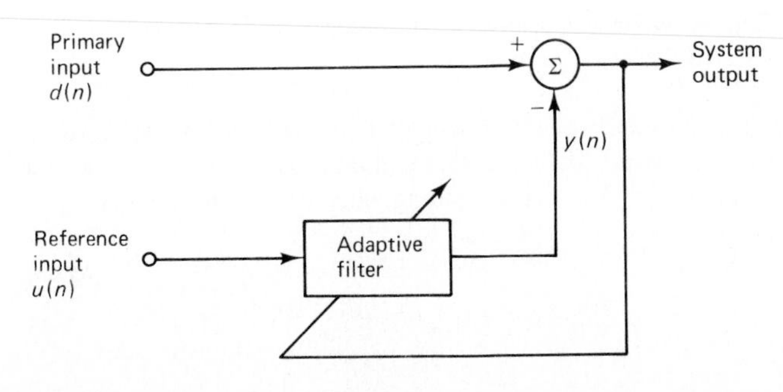

Figure 9.5 Block diagram of adaptive noise canceller.

where the amplitude A and the phase ϕ are different from those in the primary input, but the angular frequency ω_0 is the same.

Using the real, expanded form of the LMS algorithm, the tap-weight update is described by the following equations:

$$y(n) = \sum_{i=0}^{M-1} \hat{w}_i(n)u(n-i) \tag{9.29}$$

$$e(n) = d(n) - y(n) \tag{9.30}$$

$$\hat{w}_i(n+1) = \hat{w}_i(n) + \mu u(n-i)e(n), \qquad i = 0, 1, \ldots, M-1 \tag{9.31}$$

where M is the total number of tap weights in the transversal filter, and the constant μ is the step-size parameter. Note that the sampling period in the input data and all other signals in the LMS algorithm is assumed to be unity for convenience of presentation; this practice is indeed followed throughout the book.

With a sinusoidal excitation as the input of interest, we restructure the block diagram of the adaptive noise canceller as in Fig. 9.6(a). According to this new representation, we may lump the sinusoidal input $u(n)$, the transversal filter, and the weight-update equation of the LMS algorithm into a single (open-loop) system defined by a transfer function $G(z)$, as in the equivalent model of Fig. 9.6(b). The transfer function $G(z)$ is

$$G(z) = \frac{Y(z)}{E(z)}$$

where $Y(z)$ and $E(z)$ are the z-transforms of the reference input $u(n)$ and the estimation error $e(n)$, respectively. Given $E(z)$, our task is to find $Y(z)$, and therefore $G(z)$.

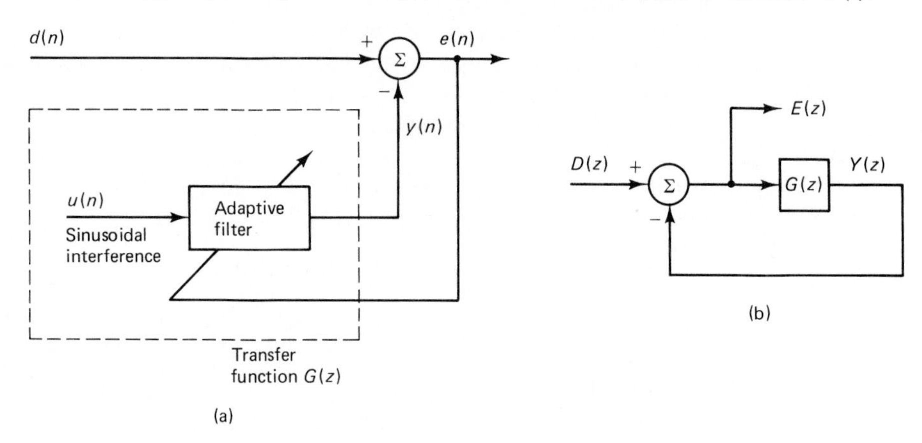

Figure 9.6 (a) New representation of adaptive noise canceller. (b) Equivalent model in the z-domain.

To do so, we use the detailed signal-flow-graph representation of the LMS algorithm depicted in Fig. 9.7 (Glover, 1977). In this diagram, we have singled out the ith tap weight for specific attention. The corresponding value of the tap input is

$$\begin{aligned} u(n-i) &= A\cos[\omega_0(n-i)+\phi] \\ &= \frac{A}{2}[e^{j(\omega_0 n+\phi_i)} + e^{-j(\omega_0 n+\phi_i)}] \end{aligned} \tag{9.32}$$

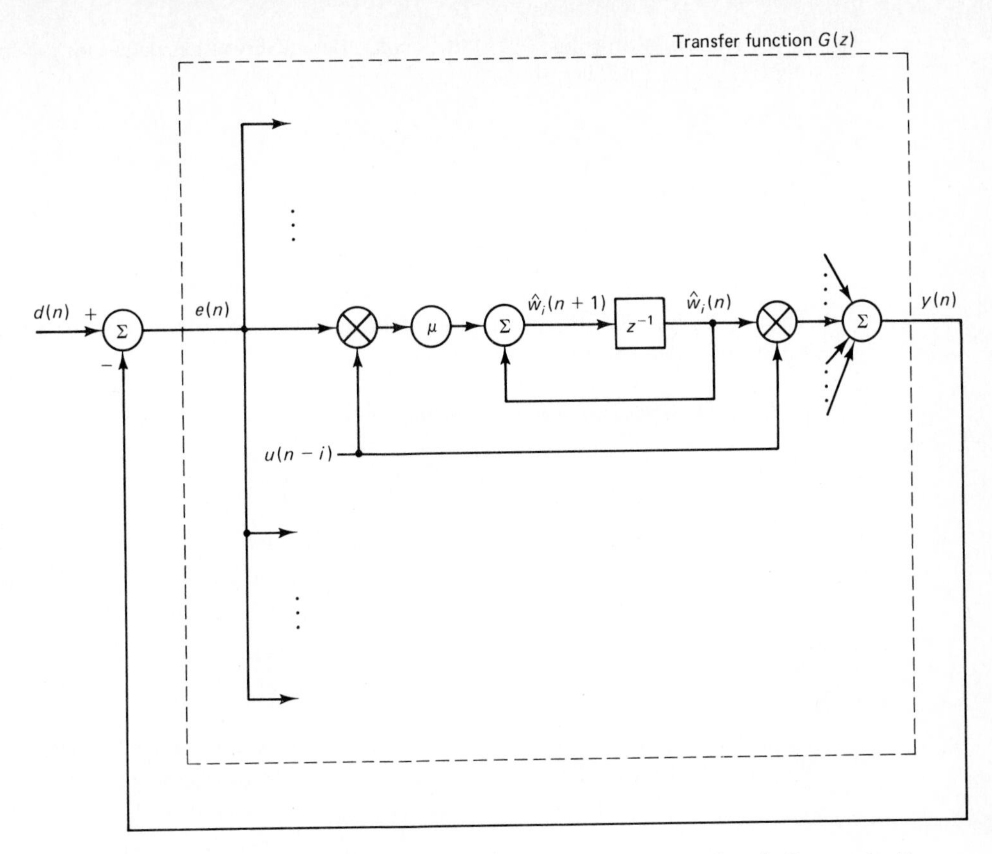

Figure 9.7 Signal-flow graph representation of adaptive noise canceller, singling out the ith tap weight for detailed attention.

where

$$\phi_i = \phi - \omega_0 i$$

In Fig. 9.7, the input $u(n-i)$ is multiplied by the estimation error $e(n)$. Hence, taking the z-transform of the product $u(n-i)e(n)$, we obtain

$$\begin{aligned} Z[u(n-i)e(n)] &= \frac{A}{2}e^{j\phi_i}Z[e(n)e^{j\omega_0 n}] + \frac{A}{2}e^{-j\phi_i}Z[e(n)e^{-j\omega_0 n}] \\ &= \frac{A}{2}e^{j\phi_i}E(ze^{-j\omega_0}) + \frac{A}{2}e^{-j\phi_i}E(ze^{j\omega_0}) \end{aligned} \tag{9.33}$$

where $E(ze^{-j\omega_0})$ is the z-transform $E(z)$ rotated counterclockwise around the unit circle through the angle ω_0. Similarly, $E(ze^{j\omega_0})$ represents a clockwise rotation through ω_0.

Next, taking the z-transform of Eq. (9.31), we get

$$z\hat{W}_i(z) = \hat{W}_i(z) + \mu Z[u(n-i)e(n)] \tag{9.34}$$

where $\hat{W}_i(z)$ is the z-transform of $\hat{w}_i(n)$. Solving Eq. (9.34) for $\hat{W}_i(z)$ and using the z-transform given in Eq. (9.33), we get

$$\hat{W}_i(z) = \frac{\mu A}{2}\frac{1}{z-1}[e^{j\phi_i}E(ze^{-j\omega_0}) + e^{-j\phi_i}E(ze^{j\omega_0})] \tag{9.35}$$

We turn next to Eq. (9.29) that defines the adaptive filter output $y(n)$. Substituting Eq. (9.32) in (9.29), we get

$$y(n) = \frac{A}{2} \sum_{i=0}^{M-1} \hat{w}_i(n)[e^{j(\omega_0 n+\phi_i)} + e^{-j(\omega_0 n+\phi_i)}]$$

Hence, evaluating the z-transform of $y(n)$, we obtain

$$Y(z) = \frac{A}{2} \sum_{i=0}^{M-1} [e^{j\phi_i}\hat{W}_i(ze^{-j\omega_0}) + e^{-j\phi_i}\hat{W}_i(ze^{j\omega_0})] \tag{9.36}$$

Thus, using Eq. (9.35) in (9.36), we obtain an expression for $Y(z)$ that consists of the sum of two components (Glover, 1977):

1. A *time-invariant component* defined by

$$\frac{\mu M A^2}{4}\left(\frac{1}{ze^{-j\omega_0} - 1} + \frac{1}{ze^{j\omega_0} - 1}\right)$$

which is independent of the phase ϕ_i and, therefore, the time index i.

2. A *time-varying component* that is dependent on the phase ϕ_i; hence, the variation with time i. This second component is scaled in amplitude by the factor

$$\beta(\omega_0, M) = \frac{\sin(M\omega_0)}{\sin \omega_0}$$

For a given angular frequency ω_0, we assume that the total number of tap weights M in the transversal filter is large enough to satisfy the following approximation:

$$\frac{\beta(\omega_0, M)}{M} = \frac{\sin(M\omega_0)}{M \sin \omega_0} \approx 0 \tag{9.37}$$

Accordingly, we may justifiably ignore the time-varying component of the z-transform $Y(z)$, and so approximate $Y(z)$ by retaining the time-invariant component only:

$$Y(z) \approx \frac{\mu M A^2}{4} E(z)\left(\frac{1}{ze^{-j\omega_0} - 1} + \frac{1}{ze^{j\omega_0} - 1}\right) \tag{9.38}$$

The open-loop transfer function $G(z)$ is therefore

$$\begin{aligned} G(z) &= \frac{Y(z)}{E(z)} \\ &\approx \frac{\mu M A^2}{4}\left(\frac{1}{ze^{-j\omega_0} - 1} + \frac{1}{ze^{j\omega_0} - 1}\right) \\ &\approx \frac{\mu M A^2}{2}\left(\frac{z \cos \omega_0 - 1}{z^2 - 2z \cos \omega_0 + 1}\right) \end{aligned} \tag{9.39}$$

The transfer function $G(z)$ has two complex-conjugate poles on the unit circle at $z = e^{\pm j\omega_0}$ and a real zero at $z = 1/\cos \omega_0$, as illustrated in Fig. 9.8(a). In other words, the adaptive noise canceller has a null point determined by the angular frequency ω_0 of the sinusoidal interference, as stated previously (see characteristic 1). Indeed, according to Eq. (9.39), we may view $G(z)$ as a pair of *integrators* that have been rotated by $\pm\omega_0$. In actual fact, we see from Fig. 9.7 that it is the input that is first shifted in frequency by an amount $\pm\omega_0$ due to the first multiplication by the reference sinusoid $u(n)$,

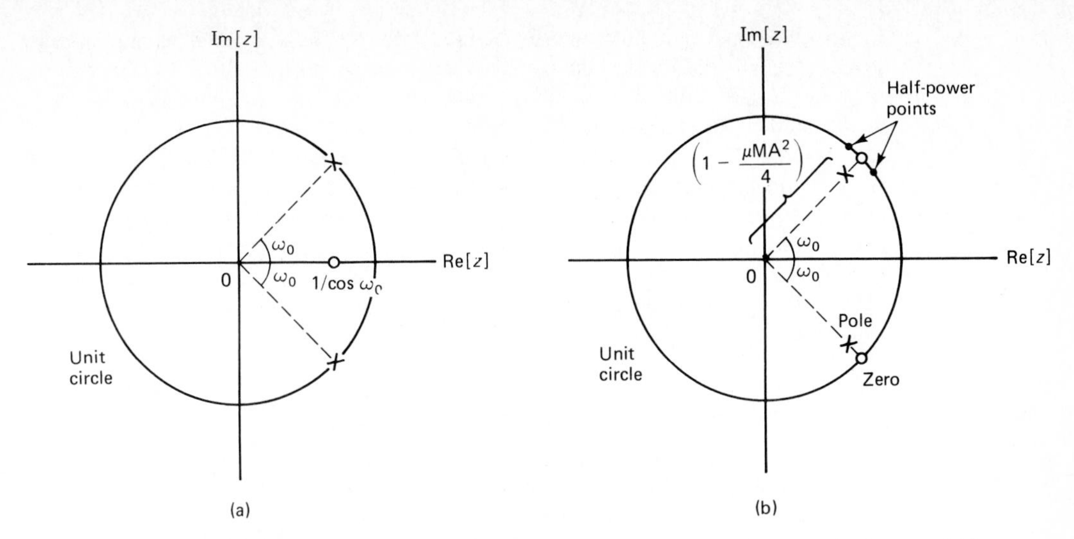

Figure 9.8 Approximate pole-zero patterns of (a) the open-loop transfer function $G(z)$, and (b) the close-loop transfer function $H(z)$.

digitally integrated at dc, and then shifted back again by the second multiplication. This overall operation is similar to a well-known technique in communications for obtaining a resonant filter that involves the combined use of two low-pass filters and heterodyning with sine and cosine at the resonant frequency (Glover, 1977; Wozencraft and Jacobs, 1965).

The model of Fig. 9.6 is recognized as a *closed-loop feedback system* whose transfer function $H(z)$ is related to the open-loop transfer function $G(z)$ as follows:

$$\begin{aligned} H(z) &= \frac{E(z)}{D(z)} \\ &= \frac{1}{1 + G(z)} \end{aligned} \tag{9.40}$$

where $E(z)$ is the z-transform of the system output $e(n)$, and $D(z)$ is the z-transform of the system input $d(n)$. Accordingly, substituting Eq. (9.39) in (9.40), we get the approximate result

$$H(z) \approx \frac{z^2 - 2z\cos\omega_0 + 1}{z^2 - 2(1 - \mu MA^2/4)z\cos\omega_0 + (1 - \mu MA^2/2)} \tag{9.41}$$

Equation (9.41) is the transfer function of a *second-order digital notch filter* with a notch at the normalized angular frequency ω_0. The zeros of $H(z)$ are at the poles of $G(z)$; that is, they are located on the unit circle at $z = e^{\pm j\omega_0}$. For a small value of the step-size parameter μ (i.e., a slow adaptation rate), such that

$$\frac{\mu MA^2}{4} \ll 1$$

we find that the poles of $H(z)$ are approximately located at

$$z \approx \left(1 - \frac{\mu MA^2}{4}\right)e^{\pm j\omega_0} \tag{9.42}$$

In other words, the two poles of $H(z)$ lie inside the unit circle, a radial distance approximately equal to $\mu MA^2/4$ behind the zeros, as indicated in Fig. 9.8(b). The fact that the poles of $H(z)$ lie inside the unit circle means that the adaptive noise canceller is stable, as it should be for practical use in real time.

Figure 9.8(b) also includes the *half-power points* of $H(z)$. Since the zeros of $H(z)$ lie on the unit circle, the adaptive noise canceller has (in theory) a notch of infinite depth (in dB) at $\omega = \omega_0$. The *sharpness* of the notch is determined by the closeness of the poles of $H(z)$ to its zeros. The *3-dB bandwidth, B*, is determined by locating the two half-power points on the unit circle that are $\sqrt{2}$ times as far from the poles as they are from the zeros. Using this geometric approach, we find that the 3-dB bandwidth of the adaptive noise canceller is approximately

$$B \approx \frac{\mu MA^2}{2} \text{ radians} \tag{9.43}$$

The smaller we therefore make μ, the smaller B is, and therefore the sharper the notch is. This confirms characteristic 2 of the adaptive noise canceller that was mentioned previously. Its analysis is thereby completed.

9.4 STABILITY ANALYSIS OF THE LMS ALGORITHM

From the signal-flow graph of Fig. 9.2 we see that the LMS algorithm is an example of a *multivariable nonlinear stochastic feedback system:*

1. The algorithm is of a multivariable nature because, in general, the tap-weight vector $\hat{\mathbf{w}}(n)$ has a dimension M greater than 1.
2. The nonlinearity arises because the (outer) feedback loop depends on the tap-input vector $\mathbf{u}(n)$.
3. Although the initial tap-weight vector $\hat{\mathbf{w}}(0)$ is usually a known constant, the application of the LMS algorithm results in the propagation of randomness of the data processes (consisting of the tap-input vector and the desired response) into the tap-weight vector $\hat{\mathbf{w}}(n)$ as a random vector for $n > 0$.

The combined presence of nonlinearity and randomness makes the stability (convergence) analysis of the LMS algorithm a difficult mathematical task. In any event, the step-size parameter μ, which is part of the (outer) feedback loop in Fig. 9.2, plays a critical role in this analysis. In particular, for the LMS algorithm to be stable (convergent), we have to choose μ so that the following two forms of convergence are satisfied:

1. *Convergence in the mean,* which means that the expectation of the tap-weight vector $\hat{\mathbf{w}}(n)$ approaches the (optimum) Wiener solution $\mathbf{w}_o$ as the number of iterations n approaches infinity
2. *Convergence in the mean square,* which means that the final (steady-state) value $J(\infty)$ of the mean-squared error is finite

Each of these two criteria for convergence imposes a condition of its own on the step-size parameter μ.

To make the convergence analysis of the LMS algorithm mathematically tractable, and hence identify the necessary and sufficient conditions for convergence, we introduce a *fundamental assumption* that consists of four parts:

1. The tap-input vectors $\mathbf{u}(1), \mathbf{u}(2), \ldots \mathbf{u}(\mathrm{n})$ constitute a sequence of statistically independent vectors.
2. At time n, the tap-input vector $\mathbf{u}(n)$ is statistically independent of all previous samples of the desired response, namely, $d(1), d(2), \ldots, d(n-1)$.
3. At time n, the desired response $d(n)$ is dependent on the corresponding tap-input vector $\mathbf{u}(n)$, but statistically independent of all previous samples of the desired response.
4. The tap-input vector $\mathbf{u}(n)$ and the desired response $d(n)$ consist of *mutually Gaussian-distributed* random variables for all n.

The statistical analysis of the LMS algorithm based on the fundamental assumption is called the *independence theory*.

From Eq. (9.5), we observe that the tap-weight vector $\hat{\mathbf{w}}(n+1)$ at time $n+1$ depends *only* on three inputs:

1. The previous sample vectors of the input process, $\mathbf{u}(n), \mathbf{u}(n-1), \ldots, \mathbf{u}(1)$.
2. The previous samples of the desired response, $d(n), d(n-1), \ldots, d(1)$.
3. The initial value of the tap-weight vector, $\hat{\mathbf{w}}(0)$.

Accordingly, in view of points 1 and 2 in the fundamental assumption, we find that the tap-weight vector $\hat{\mathbf{w}}(n+1)$ is independent of both $\mathbf{u}(n+1)$ and $d(n+1)$. This is a very useful observation and one that will be used repeatedly in the sequel. The significance of the other two points in the fundamental assumption will become apparent as we proceed with the analysis.

Clearly, there are many practical problems for which the input process and the desired response do not satisfy the fundamental assumption. Nevertheless, experience with the LMS algorithm has shown that the independence theory retains sufficient information about the structure of the adaptive process for the results of the theory to serve as reliable design guidelines, even for some problems having highly dependent data samples.

Another point that is noteworthy is the fact that, in the fundamental assumption, we have emphasized statistical independence of samples, rather than uncorrelatedness. It is indeed true that independent samples are equivalent to *uncorrelated samples whenever the samples are Gaussian distributed*. In other words, in view of point 4 in the fundamental assumption, points 1 and 2 are indeed equivalent to the following *conditions of uncorrelatedness*, respectively:

$$E[\mathbf{u}(n)\mathbf{u}^H(k)] = \mathbf{0}, \qquad k = 0, 1, \ldots, n-1 \tag{9.44}$$

and

$$E[\mathbf{u}(n)d^*(k)] = \mathbf{0}, \qquad k = 0, \ldots, n-1 \tag{9.45}$$

Nevertheless, the independence theory may be extended to certain other distributions by making corrections for higher-order moments, provided that the samples are independent (Senne, 1968).

In the next two sections, we use the fundamental assumption to derive the first two moments of the estimate of the tap-weight vector: (1) the average value and (2) the correlation matrix. We find it more convenient, however, to work with the weight-error vector rather than the tap-weight vector itself. To avoid confusion with the notation used in the study of the steepest-descent algorithm, we define the weight-error vector for the LMS algorithm as follows:

$$\boldsymbol{\epsilon}(n) = \hat{\mathbf{w}}(n) - \mathbf{w}_o \tag{9.46}$$

where, as before, $\mathbf{w}_o$ denotes the optimum Wiener solution for the tap-weight vector, and $\hat{\mathbf{w}}(n)$ is the estimate produced by the LMS algorithm at time n.

Knowledge of the first two moments of the weight-error vector $\boldsymbol{\epsilon}(n)$, and therefore those of the tap-weight estimate vector $\hat{\mathbf{w}}(n)$, is important in establishing conditions that ensure convergence of the LMS algorithm. In particular, the first moment of $\boldsymbol{\epsilon}(n)$, namely, the expectation $E[\boldsymbol{\epsilon}(n)]$, helps us determine a condition necessary for convergence of the LMS algorithm in the mean. The second moment of $\boldsymbol{\epsilon}(n)$, namely, the expectation $E[\boldsymbol{\epsilon}(n)\boldsymbol{\epsilon}^H(n)]$, helps us determine a condition necessary for convergence of the LMS algorithm in the mean square. This second condition provides a *stronger* convergence indicator than the convergence of the mean; from a practical point of view, it is therefore a more useful indicator of the performance of the LMS algorithm.

9.5 AVERAGE TAP-WEIGHT BEHAVIOR

Subtracting the optimum tap-weight vector $\mathbf{w}_o$ from both sides of Eq. (9.5), and using the definition of Eq. (9.46) to eliminate $\hat{\mathbf{w}}(n)$ from the correction term on the right side of Eq. (9.5), we may rewrite the LMS algorithm in terms of the weight-error vector $\boldsymbol{\epsilon}(n)$ as follows:

$$\boldsymbol{\epsilon}(n+1) = [\mathbf{I} - \mu\mathbf{u}(n)\mathbf{u}^H(n)]\boldsymbol{\epsilon}(n) + \mu\mathbf{u}(n)e_o^*(n) \tag{9.47}$$

where $\mathbf{I}$ is the identity matrix, and $e_o(n)$ is the *estimation error produced in the optimum Wiener solution.* As a consequence of the fundamental assumption, we observe that the tap-weight vector $\hat{\mathbf{w}}(n)$ is independent of the input vector $\mathbf{u}(n)$, which was justified in Section 9.4. Correspondingly, the weight error vector $\boldsymbol{\epsilon}(n)$ is also independent of $\mathbf{u}(n)$. Hence taking the mathematical expectation of both sides of Eq. (9.47), we get

$$E[\boldsymbol{\epsilon}(n+1)] = E[\mathbf{I} - \mu\mathbf{u}(n)\mathbf{u}^H(n))\boldsymbol{\epsilon}(n)] + \mu E[\mathbf{u}(n)e_o^*(n)] \tag{9.48}$$

For the first expectation term on the right side of Eq. (9.48), we use the independence of $\boldsymbol{\epsilon}(n)$ from $\mathbf{u}(n)$ to rewrite this term as

$$\begin{aligned} E[(\mathbf{I} - \mu\mathbf{u}(n)\mathbf{u}^H(n))\boldsymbol{\epsilon}(n)] &= (\mathbf{I} - \mu E[\mathbf{u}(n)\mathbf{u}^H(n)])E[\boldsymbol{\epsilon}(n)] \\ &= (\mathbf{I} - \mu\mathbf{R})E[\boldsymbol{\epsilon}(n)] \end{aligned} \tag{9.49}$$

where $\mathbf{R}$ denotes the correlation matrix of the tap-input vector $\mathbf{u}(n)$. For the second expectation term on the right side of Eq. (9.48), we invoke the principle of orthogonality (see Chapter 5):

$$E[\mathbf{u}(n)e_o^*(n)] = \mathbf{0} \tag{9.50}$$

Hence, this second expectation term is identically zero. Accordingly, we may simplify Eq. (9.48) as follows:

$$E[\boldsymbol{\epsilon}(n+1)] = (\mathbf{I} - \mu\mathbf{R})E[\boldsymbol{\epsilon}(n)] \tag{9.51}$$

Comparing Eqs. (8.11) and (9.15), we observe that they are of exactly the same mathematical form. In particular, the average weight-error vector, $E[\boldsymbol{\epsilon}(n)]$, in Eq. (9.51) pertaining to the LMS algorithm has the same role as the weight-error vector $\mathbf{c}(n)$ in Eq. (8.11) describing the steepest-descent algorithm. From our study of the steepest-descent algorithm in Section 8.3, we recall that $\mathbf{c}(n)$ converges to zero as n approaches infinity, provided that Eq. (8.21) is satisfied. From the analogy between Eqs. (8.11) and (9.51), we therefore deduce that for the LMS algorithm the mean of $\boldsymbol{\epsilon}(n)$ converges to zero as n approaches infinity, provided that the following condition is satisfied (Widrow, 1970; Ungerboeck, 1972):

$$0 < \mu < \frac{2}{\lambda_{\max}} \tag{9.52}$$

where $\lambda_{\max}$ is the largest eigenvalue of the correlation matrix $\mathbf{R}$. In other words, provided that the step-size parameter μ is set within the bounds defined by Eq. (9.52), then the mean of the tap-weight vector $\hat{\mathbf{w}}(n)$ computed by using the LMS algorithm converges to the optimum Weiner solution $\mathbf{w}_o$ as the number of iterations, n, approaches infinity. Under this condition, we say that the LMS algorithm is *convergent in the mean.*

Also, as with the steepest-descent algorithm, we find that when the eigenvalues of the correlation matrix $\mathbf{R}$ are widely spread, the time taken by the average tap-weight vector to converge is primarily limited by the smallest eigenvalues.

9.6 WEIGHT-ERROR CORRELATION MATRIX

We next wish to develop a recursive equation for the time evolution of the correlation matrix of the weight-error vector $\boldsymbol{\epsilon}(n)$. Using $\mathbf{K}(n)$ to denote this correlation matrix at time n, we have, by definition,

$$\mathbf{K}(n) = E[\boldsymbol{\epsilon}(n)\boldsymbol{\epsilon}^H(n)] \tag{9.53}$$

The procedure we will use to derive the recursive equation is simply to write the definition of $\mathbf{K}(n+1)$ by using Eq. (9.53), substitute Eq. (9.47) for $\boldsymbol{\epsilon}(n+1)$, and do the ensemble average using the fundamental assumption.

Taking the Hermitian transpose of both sides of Eq. (9.47), we get

$$\boldsymbol{\epsilon}^H(n+1) = \boldsymbol{\epsilon}^H(n)[\mathbf{I} - \mu\mathbf{u}(n)\mathbf{u}^H(n)] + \mu e_o(n)\mathbf{u}^H(n) \tag{9.54}$$

where, as before, $e_o(n)$ is the estimation error produced in the optimum Wiener solution. To evaluate the correlation matrix $\mathbf{K}(n+1)$, we take the expectation of the outer product $\boldsymbol{\epsilon}(n+1)\boldsymbol{\epsilon}^H(n+1)$. In doing this, various cross terms arise as a result of multiplication, and the computations naturally fall into four steps, as follows:

Step 1. We consider first the term

$$
\begin{aligned}
&E[(\mathbf{I} - \mu\mathbf{u}(n)\mathbf{u}^H(n))\boldsymbol{\epsilon}(n)\boldsymbol{\epsilon}^H(n)(\mathbf{I} - \mu\mathbf{u}(n)\mathbf{u}^H(n))] \\
&\quad = E[\boldsymbol{\epsilon}(n)\boldsymbol{\epsilon}^H(n)] - \mu E[\mathbf{u}(n)\mathbf{u}^H(n)\boldsymbol{\epsilon}(n)\boldsymbol{\epsilon}^H(n)] - \mu E[\boldsymbol{\epsilon}(n)\boldsymbol{\epsilon}^H(n)\mathbf{u}(n)\mathbf{u}^H(n)] \\
&\qquad + \mu^2 E[\mathbf{u}(n)\mathbf{u}^H(n)\boldsymbol{\epsilon}(n)\boldsymbol{\epsilon}^H(n)\mathbf{u}(n)\mathbf{u}^H(n)] \\
&\quad = \mathbf{K}(n) - \mu E[\mathbf{u}(n)\mathbf{u}^H(n)]E[\boldsymbol{\epsilon}(n)\boldsymbol{\epsilon}^H(n)] - \mu E[\boldsymbol{\epsilon}(n)\boldsymbol{\epsilon}^H(n)]E[\mathbf{u}(n)\mathbf{u}^H(n)] \\
&\qquad + \mu^2 E[\mathbf{u}(n)\mathbf{u}^H(n)\boldsymbol{\epsilon}(n)\boldsymbol{\epsilon}^H(n)\mathbf{u}(n)\mathbf{u}^H(n)] \\
&\quad = \mathbf{K}(n) - \mu\mathbf{R}\mathbf{K}(n) - \mu\mathbf{K}(n)\mathbf{R} + \mu^2 E[\mathbf{u}(n)\mathbf{u}^H(n)\boldsymbol{\epsilon}(n)\boldsymbol{\epsilon}^H(n)\mathbf{u}(n)\mathbf{u}^H(n)]
\end{aligned} \tag{9.55}
$$

In this relation, we have used Eq. (9.53) and the definition

$$\mathbf{R} = E[\mathbf{u}(n)\mathbf{u}^H(n)]$$

where $\mathbf{R}$ denotes the correlation matrix of the tap-input vector $\mathbf{u}(n)$. The last term on the right side of Eq. (9.55) involves fourth-order moments of sample vectors of the input process. These higher-order moments may be evaluated by using the *Gaussian moment factoring theorem* described in Section 2.11. Specifically, let x_1, x_2, x_3, and x_4 denote four samples of a zero-mean, complex, Gaussian process. Then the Gaussian moment factoring theorem states that[2] (see the example presented in Section 2.11)

$$E[x_1 x_2^* x_3 x_4^*] = E[x_1 x_2^*]E[x_3 x_4^*] + E[x_1 x_4^*]E[x_2^* x_3] \tag{9.56}$$

To use this formula, we express the M-by-M matrix representing the multiple product $\mathbf{u}(n)\mathbf{u}^H(n)\boldsymbol{\epsilon}(n)\boldsymbol{\epsilon}^H(n)\mathbf{u}(n)\mathbf{u}^H(n)$ as a multiple sum of the elements of the component vectors. Let the brace notation $\{x_{ij}(n)\}$ denote this matrix having elements $x_{ij}(n)$, with $i, j = 0, 1, \ldots, M - 1$. We may then write

$$
\begin{aligned}
\{x_{ij}(n)\} &= \mathbf{u}(n)\mathbf{u}^H(n)\boldsymbol{\epsilon}(n)\boldsymbol{\epsilon}^H(n)\mathbf{u}(n)\mathbf{u}^H(n) \\
&= \left\{\sum_{l=1}^{M}\sum_{p=1}^{M} (\mathbf{u}(n)\mathbf{u}^H(n))_{il}(\boldsymbol{\epsilon}(n)\boldsymbol{\epsilon}^H(n))_{lp}(\mathbf{u}(n)\mathbf{u}^H(n))_{pj}\right\}
\end{aligned} \tag{9.57}
$$

where

$$
\begin{aligned}
(\mathbf{u}(n)\mathbf{u}^H(n))_{il} &= \text{element on } i\text{th row and } l\text{th column of } \mathbf{u}(n)\mathbf{u}^H(n) \\
&= u(n-i)u^*(n-l)
\end{aligned} \tag{9.58}
$$

$$
\begin{aligned}
(\boldsymbol{\epsilon}(n)\boldsymbol{\epsilon}^H(n))_{lp} &= \text{element on } l\text{th row and } p\text{th column of } \boldsymbol{\epsilon}(n)\boldsymbol{\epsilon}^H(n) \\
&= \epsilon_l(n)\epsilon_p^*(n)
\end{aligned} \tag{9.59}
$$

$$
\begin{aligned}
(\mathbf{u}(n)\mathbf{u}^H(n))_{pj} &= \text{element on } p\text{th row and } j\text{th column of } \mathbf{u}(n)\mathbf{u}^H(n) \\
&= u(n-p)u^*(n-j)
\end{aligned} \tag{9.60}
$$

[2] The moment factorization described in Eq. (9.56) for the case of complex variables, is to be contrasted with that for real variables. Let z_1, z_2, z_3, and z_4 denote four samples of a zero-mean, real Gaussian process. For this real case, we have

$$E[z_1 z_2 z_3 z_4] = E[z_1 z_2]E[z_3 z_4] + E[z_1 z_3]E[z_2 z_4] + E[z_1 z_4]E[z_2 z_3]$$

The right side of this factorization has one more term than that of Eq. (9.56). The absence of this term from Eq. (9.56) results in a superior mean-squared convergence and stability performance for the complex LMS algorithm, compared to the real LMS algorithm (Horowitz and Senne, 1981).

Thus, any term inside the double summation in Eq. (9.57) may be regrouped as the product of two distinct components. One component, involving tap inputs only, is written as

$$u(n-i)u^*(n-l)u(n-p)u^*(n-j)$$

The second component, involving tap-weight errors only, is written as

$$\epsilon_l(n)\epsilon_p^*(n)$$

As a consequence of the fundamental assumption, these two components may be treated independent of each other. This means that the expected value of any term inside the double summation in Eq. (9.57) may be expressed as the product of the expected values of these two components.

To determine the expected value of the first component, we use the formula of Eq. (9.56) to write

$$\begin{aligned} E[u(n-i)u^*(n-l)u(n-p)u^*(n-j)] \\ &= E[u(n-i)u^*(n-l)]E[u(n-p)u^*(n-j)] \\ &\quad + E[u(n-i)u^*(n-j)]E[u^*(n-l)u(n-p)] \\ &= r(l-i)r(j-p) + r(j-i)r(l-p) \end{aligned} \tag{9.61}$$

where the autocorrelation functions $r(l-i)$, $r(j-p)$, $r(j-i)$, and $r(l-p)$ are defined for different lags in the usual way.

As for the second component, its expected value equals the lpth element of the weight-error correlation matrix $\mathbf{K}(n)$. Let

$$k_{lp}(n) = E[\epsilon_l(n)\epsilon_p^*(n)] \tag{9.62}$$

Next, we recognize that

$$E[\{x_{ij}(n)\}] = \{E[x_{ij}(n)]\} \tag{9.63}$$

where $x_{ij}(n)$ is represented by the double summation in Eq. (9.57). Moreover, we may interchange the operations of expectation and double summation. Accordingly, we may express the expectation of the matrix $\{x_{ij}(n)\}$ in the form

$$\begin{aligned} E[\{x_{ij}(n)\}] = \{r(j-i)\sum_{l=1}^{M}\sum_{p=1}^{M} r(l-p)k_{lp}(n)\} \\ + \left\{\sum_{l=1}^{M}\sum_{p=1}^{M} r(l-i)r(j-p)k_{lp}(n)\right\} \end{aligned} \tag{9.64}$$

The first term on the right side of Eq. (9.64) is recognized as $\mathbf{R}\,\text{tr}[\mathbf{RK}(n)]$, where tr[·] is the *matrix trace operator*. The second term on the right side of Eq. (9.64) is recognized as the double matrix product $\mathbf{RK}(n)\mathbf{R}$. Accordingly, within the confines of the independence theory, we may express the contribution of step 1 exactly as follows:

$$E[\mathbf{u}(n)\mathbf{u}^H(n)\boldsymbol{\epsilon}(n)\boldsymbol{\epsilon}^H(n)\mathbf{u}^H(n)] = \mathbf{R}\,\text{tr}[\mathbf{RK}(n)] + \mathbf{RK}(n)\mathbf{R} \tag{9.65}$$

Step 2. We next consider the term

$$\mu E[(\mathbf{u}(n)e_o^*(n))\boldsymbol{\epsilon}^H(n)(\mathbf{I} - \mu\mathbf{u}(n)\mathbf{u}^H(n))]$$

We may identify two contributions in this term: one linear in μ and the other involving μ^2. The contribution linear in μ equals

$$\mu E[\mathbf{u}(n)e_o^*(n)\boldsymbol{\epsilon}^H(n)] = \mu E[\mathbf{u}(n)e_o^*(n)]E[\boldsymbol{\epsilon}^H(n)] \tag{9.66}$$

where we have used the independence of $\boldsymbol{\epsilon}(n)$ from both $\mathbf{u}(n)$ and $d(n)$, and therefore from the optimum estimation error $e_o(n)$. Next, we invoke the principle of orthogonality [i.e., Eq. (9.50)], and so find that the contribution defined in Eq. (9.66) is identically zero.

The contribution involving the factor μ^2, except for the trivial minus sign, may be expressed as follows in terms of the brace notation used in step 1:

$$\begin{aligned}\mu^2 E[\mathbf{u}(n)e_o^*(n)\boldsymbol{\epsilon}^H(n)\mathbf{u}(n)\mathbf{u}^H(n)] &= \mu^2 E[\{u(n-i)e_o^*(n)\epsilon_l(n)u(n-l)u^*(n-j)\}] \\ &= \mu^2\{E[u(n-i)e_o^*(n)\epsilon_l(n)u(n-l)u^*(n-j)]\} \\ &= \mu^2\{E[u(n-i)e_o^*(n)u(n-l)u^*(n-j)]E[\epsilon_l(n)]\}\end{aligned} \tag{9.67}$$

where, in the last line, we have invoked the independence theory. Next, we use the Gaussian moment factoring theorem of Eq. (9.56) to write

$$\begin{aligned}E[u(n-i)e_o^*(n)u(n-l)u^*(n-j)] &= E[u(n-i)e_o^*(n)]E[(u(n-l)u^*(n-j)] \\ &\quad + E(u(n-i)u^*(n-j)]E[e_o^*(n)u(n-l)\end{aligned} \tag{9.68}$$

From the principle of orthogonality written in its expanded form, we have

$$E[u(n-i)e_o^*(n)] = 0, \qquad i = 0, 1, \ldots, M-1 \tag{9.69}$$

Therefore, the first term on the right side of Eq. (9.68) is identically zero for all i. Similarly, the second term on the right side of this equation is zero for all l, also because of the principle of orthogonality. We conclude therefore that the contribution involving μ^2 described in Eq. (9.67) is also identically zero.

In other words, the total contribution from step 2 is identically zero.

Step 3. The third term to be considered is

$$\mu E[(\mathbf{I} - \mu\mathbf{u}(n)\mathbf{u}^H(n))\boldsymbol{\epsilon}(n)(e_o(n)\mathbf{u}^H(n))]$$

This term is similar in form to that considered in step 2. The contribution of step 3 is therefore identically zero, too.

Step 4. The final term to be considered is

$$\mu^2 E[\mathbf{u}(n)e_o^*(n)e_o(n)\mathbf{u}^H(n)]$$

In terms of the brace notation used in step 1, we may express this term as

$$\mu^2 E[\mathbf{u}(n)e_o^*(n)e_o(n)\mathbf{u}^H(n)] = \mu^2\{E[u(n-i)e_o^*(n)e_o(n)u^*(n-j)]\} \tag{9.70}$$

Next, we use the Gaussian moment factoring theorem of Eq. (9.56) to write

$$E[u(n-i)e_o^*(n)e_o(n)u^*(n-j)] = E[u(n-i)e_o^*(n)]E[e_o(n)u^*(n-j)] + E[u(n-i)u^*(n-j)]E(e_o(n)e_o^*(n)] \quad (9.71)$$

We now use the principle of orthogonality, in particular, the form given in Eq. (9.69), and so find that the first term on the right side of Eq. (9.71) is identically zero for all i and j. For the second term on the right side of this equation, we note that

$$E[u(n-i)u^*(n-j)] = r(j-i)$$

and

$$E\left[e_o(n)e_o^*(n)\right] = E\left[|e_o(n)|^2\right] = J_{\min} \quad (9.72)$$

where $J_{\min}$ is the minimum mean-squared error produced by the Wiener solution. Hence, we may rewrite Eq. (9.70) as

$$\begin{aligned}\mu^2 E\left[\mathbf{u}(n)e_o^*(n)e_o(n)\mathbf{u}^H(n)\right] &= \mu^2 J_{\min}\{r(j-i)\} \\ &= \mu^2 J_{\min}\mathbf{R}\end{aligned} \quad (9.73)$$

where $\mathbf{R}$ is the correlation matrix of the tap inputs. Accordingly, the contribution of step 4 is exactly $\mu^2 J_{\min}\mathbf{R}$.

We may now combine the results of steps 1 through 4, and thus describe the time evolution of the weight-error correlation matrix by the difference equation[3]

$$\mathbf{K}(n+1) = \mathbf{K}(n) - \mu[\mathbf{RK}(n) + \mathbf{K}(n)\mathbf{R}] + \mu^2\mathbf{R}\,\mathrm{tr}[\mathbf{RK}(n)] + \mu^2\mathbf{RK}(n)\mathbf{R} + \mu^2 J_{\min}\mathbf{R} \quad (9.74)$$

The last term, $\mu^2 J_{\min}\mathbf{R}$, on the right side of Eq. (9.74) prevents $\mathbf{K}(n) = \mathbf{0}$ from being a solution to this equation. Accordingly, the correlation matrix $\mathbf{K}(n)$ is prevented from going to zero by this small forcing term. In particular, the weight-error vector $\boldsymbol{\epsilon}(n)$ only approaches zero, but then executes small fluctuations about zero.

We now demonstrate that in the recursive relation of Eq. (9.74), the positive-definite character of the correlation matrix $\mathbf{K}(n)$ is preserved. We do this demonstration by induction on n.

We observe that the correlation matrix $\mathbf{R}$ of the tap-input vector is Hermitian and positive definite. Hence, it follows from Eq. (9.74) that if $\mathbf{K}(n)$ is Hermitian, then so will $\mathbf{K}(n+1)$ be.

Next we rewrite the right side of Eq. (9.74) as follows:

$$(\mathbf{I} - \mu\mathbf{R})\mathbf{K}(n)(\mathbf{I} - \mu\mathbf{R}) + \mu^2\mathbf{R}\,\mathrm{tr}[\mathbf{RK}(n)] + \mu^2 J_{\min}\mathbf{R} \quad (9.75)$$

[3] The time evolution described in Eq. (9.74) refers to the correlation matrix of the tap-weight error vector $\hat{\mathbf{w}}(n) - \mathbf{w}_o$, where $\hat{\mathbf{w}}(n)$ is the LMS estimate of the tap weight vector and $\mathbf{w}_o$ is the optimum Wiener solution. Fisher and Bershad (1983) develop a difference equation that describes the time evolution of the *tap-weight covariance matrix* defined as the expectation of the outer product of the difference vector $\hat{\mathbf{w}}(n) - E[\hat{\mathbf{w}}(n)]$ with itself, where $E[\hat{\mathbf{w}}(n)]$ is the *mean weight vector* (i.e., the mean of the tap-weight estimate vector $\hat{\mathbf{w}}(n)$ produced by the LMS algorithm). These two time evolutions are naturally different as they refer to different quantities; nevertheless, they lead to identical results for the statistical performance of the LMS algorithm. It should also be noted that unlike Mazo (1979), there is no approximation made in the derivation of Eq. (9.74).

Since $\mathbf{R}$ is positive definite, it follows that the first term of this expression is also positive definite, provided that $\mathbf{K}(n)$ is positive definite. Clearly, the last term of this expression is always positive definite. There now only remains for us to show that the second term is also positive definite. We rewrite this second term, except for μ^2, in the form

$$\mathbf{R}^{1/2}\,(\text{tr}[\mathbf{R}^{1/2}\mathbf{K}(n)\mathbf{R}^{1/2}])\mathbf{R}^{1/2} \tag{9.76}$$

where $\mathbf{R}^{1/2}$ is the *square root* of $\mathbf{R}$. In Eq. (9.76) we have used the property[4]

$$\begin{aligned}\text{tr}[\mathbf{R}\mathbf{K}(n)] &= \text{tr}[\mathbf{R}^{1/2}\mathbf{R}^{1/2}\mathbf{K}(n)] \\ &= \text{tr}[\mathbf{R}^{1/2}\mathbf{K}(n)\mathbf{R}^{1/2}]\end{aligned}$$

The matrix $\mathbf{R}^{1/2}\mathbf{K}(n)\mathbf{R}^{1/2}$ is positive definite, provided that $\mathbf{K}(n)$ is. Next, we use the following property: *If a matrix* $\mathbf{X}$ *is positive definite, then* $\text{tr}[\mathbf{X}]\mathbf{I} - \mathbf{X}$ *is positive definite, too.* Hence, it follows that the expression in Eq. (9.76) or, equivalently, the second term of the expression in Eq. (9.75) is positive definite. We conclude therefore that the complete expression in Eq. (9.75) and, therefore, $\mathbf{K}(n + 1)$ is positive definite provided that $\mathbf{K}(n)$ is. The proof by induction is completed by noting that $\mathbf{K}(0)$ is positive definite, where

$$\begin{aligned}\mathbf{K}(0) &= \boldsymbol{\epsilon}(0)\boldsymbol{\epsilon}^H(0) \\ &= [\hat{\mathbf{w}}(0) - \mathbf{w}_o][\hat{\mathbf{w}}^H(0) - \mathbf{w}_o^H]\end{aligned} \tag{9.77}$$

In summary, Eq. (9.74) provides us with a relation to recursively update the weight-error correlation matrix $\mathbf{K}(n)$, starting with $n = 0$, for which we have $\mathbf{K}(0)$. Furthermore, after each iteration it does yield a positive-definite answer for the updated value of the weight-error correlation matrix.

9.7 MEAN-SQUARED-ERROR BEHAVIOR

The matrix difference equation (9.74) provides us a useful tool for determining the transient behavior of the mean-squared error of the LMS algorithm, based on the fundamental assumption of Section 9.4. Specifically, we may proceed as follows. First, we use Eqs. (9.6) and (9.7) to express the estimation error, $e(n)$, produced by the LMS algorithm as

$$\begin{aligned}e(n) &= d(n) - \hat{\mathbf{w}}^H(n)\mathbf{u}(n) \\ &= d(n) - \mathbf{w}_o^H\mathbf{u}(n) - \boldsymbol{\epsilon}^H(n)\mathbf{u}(n) \\ &= e_o(n) - \boldsymbol{\epsilon}^H(n)\mathbf{u}(n)\end{aligned} \tag{9.78}$$

where $e_o(n)$ is the estimation error in the optimum Wiener solution, and $\boldsymbol{\epsilon}(n)$ is the tap-weight error vector. Let $J(n)$ denote the mean-squared error due to the LMS al-

[4] Let $\mathbf{X}$ denote an M-by-N matrix and $\mathbf{Y}$ denote an N-by-M matrix. The trace of matrix product $\mathbf{XY}$ equals the trace the matrix product $\mathbf{YX}$; that is,

$$\text{tr}[\mathbf{XY}] = \text{tr}[\mathbf{YX}]$$

gorithm at time n. Hence, using Eq. (9.78) to evaluate $J(n)$, and then invoking the fundamental assumption of Section 9.4, we get

$$\begin{aligned} J(n) &= E[|e(n)|^2] \\ &= E[(e_o(n) - \boldsymbol{\epsilon}^H(n)\mathbf{u}(n))(e_o^*(n) - \mathbf{u}^H(n)\boldsymbol{\epsilon}(n))] \\ &= J_{\min} + E[\boldsymbol{\epsilon}^H(n)\mathbf{u}(n)\mathbf{u}^H(n)\boldsymbol{\epsilon}(n)] \end{aligned} \tag{9.79}$$

where $J_{\min}$ is the minimum mean-squared error produced by the optimum Wiener filter.

Our next task is to evaluate the expectation term in the final line of Eq. (9.79). Here we note that this term is the expected value of a scalar random variable represented by a triple vector product. We may therefore rewrite it as

$$\begin{aligned} E[\boldsymbol{\epsilon}^H(n)\mathbf{u}(n)\mathbf{u}^H(n)\boldsymbol{\epsilon}(n)] &= E[\text{tr}\{\boldsymbol{\epsilon}^H(n)\mathbf{u}(n)\mathbf{u}^H(n)\boldsymbol{\epsilon}(n)\}] \\ &= E[\text{tr}\{\mathbf{u}(n)\mathbf{u}^H(n)\boldsymbol{\epsilon}(n)\boldsymbol{\epsilon}^H(n)\}] \\ &= \text{tr}\{E[\mathbf{u}(n)\mathbf{u}^H(n)\boldsymbol{\epsilon}(n)\boldsymbol{\epsilon}^H(n)]\} \end{aligned}$$

Invoking the fundamental assumption again, we may simplify this expectation as follows:

$$\begin{aligned} E[\boldsymbol{\epsilon}^H(n)\mathbf{u}(n)\mathbf{u}^H(n)\boldsymbol{\epsilon}(n)] &= \text{tr}\{E[\mathbf{u}(n)\mathbf{u}^H(n)]E[\boldsymbol{\epsilon}(n)\boldsymbol{\epsilon}^H(n)]\} \\ &= \text{tr}[\mathbf{R}\mathbf{K}(n)] \end{aligned} \tag{9.80}$$

where $\mathbf{R}$ is the correlation matrix of the tap inputs and $\mathbf{K}(n)$ is the weight-error correlation matrix.

Accordingly, using Eq. (9.80) in (9.79), we may rewrite the expression for the mean-squared error in the LMS algorithm as

$$J(n) = J_{\min} + \text{tr}[\mathbf{R}\mathbf{K}(n)] \tag{9.81}$$

Equation (9.81) indicates that for all n, the mean-square value of the estimation error in the LMS algorithm consists of two components: the minimum mean-squared error $J_{\min}$, and a component depending on the transient behavior of the weight-error correlation matrix $\mathbf{K}(n)$. Since the latter component is positive definite for all n, *the LMS algorithm always produces a mean-squared error $J(n)$ that is in excess of the minimum mean-squared error $J_{\min}$.*

We define the *excess mean-squared error* as the difference between the mean-squared error, $J(n)$, produced by the adaptive algorithm at time n and the minimum value, $J_{\min}$, pertaining to the optimum Wiener solution. Denoting the excess mean-squared error by $J_{\text{ex}}(n)$, we have

$$\begin{aligned} J_{\text{ex}}(n) &= J(n) - J_{\min} \\ &= \text{tr}[\mathbf{R}\mathbf{K}(n)] \end{aligned} \tag{9.82}$$

For $\mathbf{K}(n)$ we use the recursive relation of Eq. (9.74). However, when the mean-squared error is of primary interest, another form of this equation obtained by a simple rotation of coordinates is more useful. The particular rotation of coordinates we have in mind is described by the unitary similarity transformation of Eq. (4.30), reproduced here for convenience

$$\mathbf{Q}^H\mathbf{R}\mathbf{Q} = \mathbf{\Lambda} \tag{9.83}$$

where $\mathbf{\Lambda}$ is a diagonal matrix consisting of the eigenvalues of the correlation matrix $\mathbf{R}$, and $\mathbf{Q}$ is the unitary matrix consisting of the eigenvectors associated with these eigenvalues. Note that the matrix $\mathbf{\Lambda}$ is real valued. Furthermore, let

$$\mathbf{Q}^H\mathbf{K}(n)\mathbf{Q} = \mathbf{X}(n) \tag{9.84}$$

In general, $\mathbf{X}(n)$ is not a diagonal matrix. Using Eqs. (9.83) and (9.84), we have

$$\begin{aligned}\text{tr}[\mathbf{RK}(n)] &= \text{tr}[\mathbf{Q\Lambda Q}^H\mathbf{QX}(n)\mathbf{Q}^H] \\ &= \text{tr}[\mathbf{Q\Lambda X}(n)\mathbf{Q}^H]\end{aligned} \tag{9.85}$$

where we have used the property $\mathbf{Q}^H\mathbf{Q} = \mathbf{I}$. Next we use the property[5]

$$\begin{aligned}\text{tr}[\mathbf{RK}(n)] &= \text{tr}[\mathbf{Q}^H\mathbf{Q\Lambda X}(n)] \\ &= \text{tr}[\mathbf{\Lambda X}(n)]\end{aligned} \tag{9.86}$$

where we have again used $\mathbf{Q}^H\mathbf{Q} = \mathbf{I}$. Accordingly, we have

$$J_{\text{ex}}(n) = \text{tr}[\mathbf{\Lambda X}(n)] \tag{9.87}$$

Since $\mathbf{\Lambda}$ is a diagonal matrix, we may also write

$$J_{\text{ex}}(n) = \sum_{i=1}^{M} \lambda_i x_i(n) \tag{9.88}$$

where the $x_i(n)$, $i = 1, 2, \ldots, M$, are the diagonal elements of the matrix $\mathbf{X}(n)$, and λ_i are the eigenvalues of the correlation matrix $\mathbf{R}$.

Next, using the transformations described by Eqs. (9.83) and (9.84), we may rewrite the recursive equation (9.74) in terms of $\mathbf{X}(n)$ and $\mathbf{\Lambda}$ as follows:

$$\begin{aligned}\mathbf{X}(n+1) = \mathbf{X}(n) &- \mu[\mathbf{\Lambda X}(n) + \mathbf{X}(n)\mathbf{\Lambda}] \\ &+ \mu^2\mathbf{\Lambda}\,\text{tr}[\mathbf{\Lambda X}(n)] + \mu^2\mathbf{\Lambda X}(n)\mathbf{\Lambda} + \mu^2 J_{\min}\mathbf{\Lambda}\end{aligned} \tag{9.89}$$

We observe from Eq. (9.88) that $J_{\text{ex}}(n)$ depends only on the $x_i(n)$. This suggests that we look at the diagonal terms of the recursive equation (9.89). Because of the form of this equation, the x_i decouple from the off-diagonal terms, and so we have

$$x_i(n+1) = x_i(n) - 2\mu\lambda_i x_i(n) + \mu^2\lambda_i \sum_{j=1}^{M} \lambda_j x_j(n) + \mu^2\lambda_i^2 x_i(n) + \mu^2 J_{\min}\lambda_i, \qquad i = 1, 2, \ldots, M \tag{9.90}$$

Define the M-by-1 vectors $\mathbf{x}(n)$ and $\boldsymbol{\lambda}$ as follows, respectively:

$$\mathbf{x}(n) = [x_1(n), x_2(n), \ldots, x_M(n)]^T \tag{9.91}$$

[5] In the first line of Eq. (9.86), we have used the property that the trace of the product of two matrices is the same whether one matrix is premultiplied or postmultiplied by the other matrix; see footnote 4.

and

$$\boldsymbol{\lambda} = [\lambda_1, \lambda_2, \ldots, \lambda_M]^T \tag{9.92}$$

Then we may rewrite Eq. (9.90) in matrix form as

$$\mathbf{x}(n+1) = \mathbf{B}\mathbf{x}(n) + \mu^2 J_{\min}\boldsymbol{\lambda} \tag{9.93}$$

where $\mathbf{B}$ is an M-by-M matrix with elements

$$b_{ij} = \begin{cases} (1 - \mu\lambda_i)^2 + \mu^2\lambda_i^2, & i = j \\ \mu^2\lambda_i\lambda_j, & i \neq j \end{cases} \tag{9.94}$$

From Eq. (9.94), we see that the matrix $\mathbf{B}$ is the sum of a diagonal matrix (with positive real values for all of its elements) and the outer product $\mu^2\boldsymbol{\lambda}\boldsymbol{\lambda}^T$. The matrix $\mathbf{B}$ is therefore *real, positive,* and *symmetric*.

Equation (9.93) is a difference equation of order 1 in matrix form. Therefore, assuming an initial value $\mathbf{x}(0)$, the solution to this equation is[6]

$$\mathbf{x}(n) = \mathbf{B}^n\mathbf{x}(0) + \mu^2 J_{\min} \sum_{i=0}^{n-1} \mathbf{B}^i\boldsymbol{\lambda} \tag{9.95}$$

By analogy with the formula for the sum of a geometric series, we may express the finite sum $\sum_{i=0}^{n-1} \mathbf{B}^i$ as follows:

$$\sum_{i=0}^{n-1} \mathbf{B}^i = (\mathbf{I} - \mathbf{B}^n)(\mathbf{I} - \mathbf{B})^{-1} \tag{9.96}$$

where $\mathbf{I}$ is the identity matrix. Substituting Eq. (9.96) in (9.95), we thus get

$$\mathbf{x}(n) = \mathbf{B}^n[\mathbf{x}(0) - \mu^2 J_{\min}(\mathbf{I} - \mathbf{B})^{-1}\boldsymbol{\lambda}] + \mu^2 J_{\min}(\mathbf{I} - \mathbf{B})^{-1}\boldsymbol{\lambda} \tag{9.97}$$

The first term on the right side of Eq. (9.97) is the *transient* component of the vector $\mathbf{x}(n)$, and the second term is the *steady-state* component. Since the matrix $\mathbf{B}$ is symmetric, we may apply to it an orthogonal similarity transformation. We may thus write

$$\mathbf{G}^T\mathbf{B}\mathbf{G} = \mathbf{C} \tag{9.98}$$

The matrix $\mathbf{C}$ is a diagonal matrix with elements $c_i = 1, 2, \ldots, M$, which are the eigenvalues of $\mathbf{B}$. The matrix $\mathbf{G}$ is an *orthonormal matrix*[7] whose ith column is the eigenvector $\mathbf{g}_i$ of $\mathbf{B}$, associated with eigenvalue c_i. Because of the property

$$\mathbf{G}\mathbf{G}^T = \mathbf{I} \tag{9.99}$$

we find that

$$\mathbf{B}^n = \mathbf{G}\mathbf{C}^n\mathbf{G}^T \tag{9.100}$$

[6] The approach we follow from here on is adapted from Mazo (1979). However, we differ from Mazo in two important respects:

1. Our analysis is for complex data, where that of Mazo is for real data.
2. Unlike Mazo, there are no approximations made in our analysis.

[7] The term "orthonormal matrix" is used for real data, whereas the term "unitary matrix" is reserved for complex data.

Hence, we may rewrite Eq. (9.95) in the form

$$\mathbf{x}(n) = \mathbf{G}\mathbf{C}^n\mathbf{G}^T[\mathbf{x}(0) - \mu^2 J_{\min}(\mathbf{I} - \mathbf{B})^{-1}\boldsymbol{\lambda}] + \mu^2 J_{\min}(\mathbf{I} - \mathbf{B})^{-1}\boldsymbol{\lambda} \tag{9.101}$$

Since $\mathbf{C}$ is a diagonal matrix, we have

$$\mathbf{C}^n = \text{diag}(c_1^n, c_2^n, \ldots, c_M^n) \tag{9.102}$$

It follows therefore that the solution of Eq. (9.101), and therefore that of Eq. (9.97), is stable if and only if the eigenvalues of matrix $\mathbf{B}$ all have a magnitude less than 1. The eigenvalues of matrix $\mathbf{B}$ are all positive, since the matrix $\mathbf{B}$ is positive definite. For stability, we therefore require the condition

$$0 < c_i < 1, \qquad \text{for all } i \tag{9.103}$$

When this condition is satisfied, the transient component in Eq. (9.101) or (9.97) decays to zero as the number of iterations, n, approaches infinity, leaving the steady-state component as the only component. We may then write

$$\mathbf{x}(\infty) = \mu^2 J_{\min}(\mathbf{I} - \mathbf{B})^{-1}\boldsymbol{\lambda} \tag{9.104}$$

Substituting Eq. (9.104) in (9.101), we may rewrite the solution as

$$\mathbf{x}(n) = \mathbf{G}\mathbf{C}^n\mathbf{G}^T[\mathbf{x}(0) - \mathbf{x}(\infty)] + \mathbf{x}(\infty) \tag{9.105}$$

In view of the diagonal nature of matrix $\mathbf{C}^n$, and since the orthonormal matrix $\mathbf{G}$ consists of the eigenvectors of $\mathbf{B}$ as its columns, we may express the matrix product $\mathbf{G}\mathbf{C}^n\mathbf{G}^T$ as follows:

$$\mathbf{G}\mathbf{C}^n\mathbf{G}^T = \sum_{i=1}^{M} c_i^n \mathbf{g}_i\mathbf{g}_i^T \tag{9.106}$$

Accordingly, we may rewrite Eq. (9.105) one more time in the equivalent form

$$\mathbf{x}(n) = \sum_{i=1}^{M} c_i^n \mathbf{g}_i\mathbf{g}_i^T[\mathbf{x}(0) - \mathbf{x}(\infty)] + \mathbf{x}(\infty) \tag{9.107}$$

The excess mean-squared error equals [see Eq. (9.88)]

$$\begin{aligned} J_{\text{ex}}(n) &= \boldsymbol{\lambda}^T\mathbf{x}(n) \\ &= \sum_{i=1}^{n} c_i^n \boldsymbol{\lambda}^T\mathbf{g}_i\mathbf{g}_i^T[\mathbf{x}(0) - \mathbf{x}(\infty)] + \boldsymbol{\lambda}^T\mathbf{x}(\infty) \\ &= \sum_{i=1}^{n} c_i^n \boldsymbol{\lambda}^T\mathbf{g}_i\mathbf{g}_i^T[\mathbf{x}(0) - \mathbf{x}(\infty)] + J_{\text{ex}}(\infty) \end{aligned} \tag{9.108}$$

where

$$\begin{aligned} J_{\text{ex}}(\infty) &= \boldsymbol{\lambda}^T\mathbf{x}(\infty) \\ &= \sum_{j=1}^{M} \lambda_j x_j(\infty) \end{aligned} \tag{9.109}$$

In Eq. (9.108), the first term on the right side describes the transient behavior of the excess mean-squared error, whereas the second term represents its value after adaptation is completed (i.e., its steady-state value).

We now wish to derive an expression for the steady-state component $J_{ex}(\infty)$ in terms of the eigenvalues of the original correlation matrix $\mathbf{R}$. We may do this by premultiplying $\mathbf{x}(\infty)$ in Eq. (9.104) by $\boldsymbol{\lambda}^T$ and manipulating the resultant vector product. However, a simpler approach is to go back to Eq. (9.90). Specifically, putting $n = \infty$, in this equation, solving for $x_i(\infty)$, and then summing $\lambda_i x_i(\infty)$ for integer values of i, from 1 to M, we get the desired formula:[8]

$$J_{ex}(\infty) = J_{min} \frac{\sum_{i=1}^{M} \mu\lambda_i/(2 - \mu\lambda_i)}{1 - \sum_{i=1}^{M} \mu\lambda_i/(2 - \mu\lambda_i)} \tag{9.110}$$

Finally, substituting Eq. (9.110) in (9.108), and using the first line of Eq. (9.82), we may express the time evolution of the mean-squared error for the LMS algorithm by the equation

$$J(n) = \sum_{i=1}^{M} \gamma_i c_i^n + \frac{J_{min}}{1 - \sum_{i=1}^{M} \mu\lambda_i/(2 - \mu\lambda_i)} \tag{9.111}$$

where c_i is the ith element of the diagonal matrix $\mathbf{C}$ defined in Eq. (9.98), and

$$\gamma_i = \boldsymbol{\lambda}^T \mathbf{g}_i \mathbf{g}_i^T [\mathbf{x}(0) - \mathbf{x}(\infty)], \qquad i = 1, 2, \ldots, M \tag{9.112}$$

Note that the ith element of the initial value $\mathbf{x}(0)$ is related to the initial value of the weight-error vector $\boldsymbol{\epsilon}(0) = \hat{\mathbf{w}}(0) - \mathbf{w}_o$ as follows [see Eq. (9.84)]:

$$x_i(0) = E[\mathbf{q}_i^H \boldsymbol{\epsilon}(0) \boldsymbol{\epsilon}^H(0) \mathbf{q}_i], \qquad i = 1, 2, \ldots, M \tag{9.113}$$

where $\mathbf{q}_i$ is the eigenvector of the correlation matrix $\mathbf{R}$ associated with the eigenvalue λ_i.

9.8 TRANSIENT BEHAVIOR OF THE MEAN-SQUARED ERROR

We may now summarize the properties of the transient behavior of the mean-squared error, $J(n)$. These properties provide us with a deeper understanding of the operation of the LMS algorithm in a wide-sense stationary environment.[9] They all follow from Eq. (9.111).

Property 1. *The transient component of the mean-squared error, $J(n)$, does not exhibit oscillations.*

The transient component of $J(n)$ equals

$$\sum_{i=1}^{M} \gamma_i c_i^n$$

[8] Note that when n approaches infinity, $x_i(n + 1)$ and $x_i(n)$ become equal in the limit; we may therefore cancel them out.

[9] The properties presented herein follow the treatment given in Ungerboeck (1972). However, our condition for convergence of the LMS algorithm in the mean square is stronger than that of Ungerboeck.

where the γ_i are constant coefficients and the c_i, $i = 1, 2, \ldots, M$, are the eigenvalues of matrix **B**. These eigenvalues are all real positive numbers, since **B** is a real symmetric positive-definite matrix [see Eq. (9.94)]. Hence, the ensemble-averaged learning curve, that is, a plot of the mean-squared error $J(n)$ versus the number of iterations, n, consists only of exponentials.

Note, however, that the learning curve represented by a plot of the mean-squared error $J(n)$, without ensemble averaging, versus n consists of *noisy* exponentials. The amplitude of the noise becomes smaller as the step-size parameter μ is reduced.

Property 2. *The mean-squared error $J(n)$ converges to a steady-state value equal to $J(\infty)$ if, and only if, the step-size parameter μ satisfies two conditions:*

(i)
$$0 < \mu < \frac{2}{\lambda_{\max}} \tag{9.114}$$

(ii)
$$\sum_{i=1}^{M} \frac{\mu\lambda_i}{2 - \mu\lambda_i} < 1 \tag{9.115}$$

where the λ_i, $i = 1, 2, \ldots, M$, are the eigenvalues of the correlation matrix **R**, *and M is the number of taps; $\lambda_{\max}$ is the largest eigenvalue of* **R**. *When these two conditions are satisfied, we say that the LMS algorithm is convergent in the mean square.*

To prove this property, we use ideas developed in the preceding section. In particular, we know the following:

1. The transient component of the mean-squared error $J(n)$ decays to zero if, and only if, the eigenvalues of matrix **B** are less than 1 in magnitude.
2. The matrix **B**, whose elements are defined in terms of the eigenvalues of the correlation matrix **R** in Eq. (9.94), is real, positive, and symmetric.

Let **g** be an eigenvector of matrix **B**, associated with eigenvalue c. Then, by definition, we have

$$\mathbf{Bg} = c\mathbf{g}$$

or, equivalently,

$$\sum_{j=1}^{M} b_{ij} g_j = c g_i, \qquad i = 1, 2, \ldots, M \tag{9.116}$$

where the g_i are the elements of the eigenvector **g**. Using Eq. (9.94), for the elements of matrix **B** in Eq. (9.116), we get

$$(1 - \mu\lambda_i)^2 g_i + \mu^2 \lambda_i \sum_{j=1}^{M} \lambda_i g_i = c g_i, \qquad i = 1, 2, \ldots, M \tag{9.117}$$

Solving Eq. (9.117) for g_i, we may thus write

$$g_i = \frac{\mu^2 \lambda_i}{c - (1 - \mu\lambda_i)^2} \sum_{j=1}^{M} \lambda_j g_j, \qquad i = 1, 2, \ldots, M \tag{9.118}$$

Next, we use knowledge of the fact that the square matrix $\mathbf{B}$ is a *positive matrix* since all of its elements are positive. This means that we may use *Perron's theorem,*[10] which applies to a positive square matrix. Perron's theorem states that (Bellman, 1960):

> If $\mathbf{B}$ is a positive square matrix, there is a unique eigenvalue of $\mathbf{B}$, which has the largest magnitude. This eigenvalue is positive and simple (i.e., of multiplicity 1), and its associated eigenvector consists entirely of positive elements.

Accordingly, we may associate a *positive eigenvector* (i.e., a vector consisting entirely of positive elements) with the special eigenvalue of matrix $\mathbf{B}$ that has the largest magnitude. Thus setting the eigenvalue c equal to 1 in Eq. (9.118), letting μ_{crit} denote the *critical value* of the step-size parameter that corresponds to this limiting condition of stability, and then simplifying, we get

$$g_i = \frac{\mu_{crit}}{2 - \mu_{crit}\lambda_i} \sum_{j=1}^{M} \lambda_i g_j, \qquad i = 1, 2, \ldots, M \tag{9.119}$$

From Eq. (9.119) we readily see that g_i is positive for all i if, and only if,

$$\mu_{crit} < \frac{2}{\lambda_{max}} \tag{9.120}$$

where λ_{max} is the largest eigenvalue of the correlation matrix $\mathbf{R}$. Moreover, we may remove dependence on the g_i by multiplying both sides of Eq. (9.119) by λ_i and then summing over all integer values of i from 1 to M. Doing so, we simply get

$$\sum_{i=1}^{M} \frac{\mu_{crit}\lambda_i}{2 - \mu_{crit}\lambda_i} = 1 \tag{9.121}$$

With the actual value of μ required to be positive and less than the critical value μ_{crit}, we find that Eqs. (9.120) and (9.121) yield the necessary and sufficient conditions of Eqs. (9.114) and (9.115) for convergence of the LMS algorithm in the mean square.

For the case when the step-size parameter μ is small compared to $2/\lambda_{max}$, small enough for convergence in the mean to be well satisfied, we may approximate the factor $(2 - \mu\lambda_i)$ by 2 for all i, and so simplify the condition of (9.115) as follows.

$$0 < \mu < \frac{2}{\sum_{i=1}^{M} \lambda_i} \tag{9.122}$$

Next, we use the following relation from matrix algebra:

> The sum of the eigenvalues of a positive definite matrix equals the sum of the elements on its main diagonal.

In the situation described herein, the correlation matrix $\mathbf{R}$ is not only positive definite but also Toeplitz with all of its elements on the main diagonal equal to $r(0)$.

[10] Perron's theorem is also known as the *Perron–Frobenius theorem*.

Since $r(0)$ is itself equal to the mean-square value of the input at each of the M taps in the adaptive transversal filter, we have

$$\sum_{i=1}^{M} \lambda_i = Mr(0)$$

$$= \sum_{k=0}^{M-1} E[|u(n - k|^2]$$

Thus, using the term *total input power* to refer to the sum of the mean-square values of tap inputs $u(n)$, $u(n-1)$, . . . , $u(n-M-1)$ in Fig. 9.1(b), we may restate the simplified condition of (9.122) for convergence of the LMS algorithm in the mean square as

$$0 < \mu < \frac{2}{\text{total input power}} \tag{9.123}$$

Property 3. *The mean-squared error produced by the LMS algorithm has the final value*

$$J(\infty) = \frac{J_{\min}}{1 - \sum_{i=1}^{M} \mu\lambda_i/(2 - \mu\lambda_i)} \tag{9.124}$$

This property follows directly from Eq. (9.111). In particular, with $c_i < 1$ for stability, the final value $J(\infty)$ given in Eq. (9.124) is obtained simply by putting $n = \infty$ in Eq. (9.111). Note that the stability condition of (9.115) also ensures that $J(\infty)$ is positive, which it should indeed be.

Property 4. *The misadjustment, defined as the ratio of the excess mean-squared error $J_{\text{ex}}(\infty)$ to the minimum mean-squared error $J_{\min}$, equals*

$$\mathcal{M} = \frac{J_{\text{ex}}(\infty)}{J_{\min}} = \frac{\sum_{i=1}^{M} \mu\lambda_i/(2 - \mu\lambda_i)}{1 - \sum_{i=1}^{M} \mu\lambda_i/(2 - \mu\lambda_i)} \tag{9.125}$$

The misadjustment $\mathcal{M}$ provides a useful measure for the *cost of adaptability*. The formula for $\mathcal{M}$ given in Eq. (9.125) follows immediately from (9.110). Note that the condition of Eq. (9.115) for convergence of the LMS algorithm in the mean square ensures that the misadjustment is positive, as it should be.

When the step-size parameter μ is small compared to $2/\lambda_{\max}$, we may approximate Eq. (9.125) as follows

$$\mathcal{M} \approx \frac{\mu}{2} \sum_{i=1}^{M} \lambda_i \tag{9.126}$$

Define an *averaged eigenvalue* for the underlying correlation matrix $\mathbf{R}$ of the tap inputs as

$$\lambda_{\text{av}} = \frac{1}{M} \sum_{i=1}^{M} \lambda_i \tag{9.127}$$

Suppose also that the ensemble-averaged learning curve of the LMS algorithm is approximated by a single exponential with time constant $(\tau)_{\text{mse, av}}$. We may then use Eq. (8.31), developed for the method of steepest descent, to define the following *average time constant* for the LMS algorithm:

$$(\tau)_{\text{mse, av}} \approx \frac{1}{2\mu\lambda_{\text{av}}} \tag{9.128}$$

Thus, using the average values of Eqs. (9.127) and (9.128) in (9.126), we get

$$\begin{aligned} \mathcal{M} &\approx \frac{\mu M \lambda_{\text{av}}}{2} \\ &\approx \frac{M}{4\tau_{\text{mse, av}}} \end{aligned} \tag{9.129}$$

Experience shows that the formula of Eq. (9.129) is a good approximation even when the eigenvalues λ_i are *not* equal (Widrow and Stearns, 1985). Note that:

1. The misadjustment $\mathcal{M}$ increases linearly with the filter length (number of taps) denoted by M, for a fixed $\tau_{\text{mse, av}}$.
2. The *settling time* of the LMS algorithm (i.e., the time taken for the transients to die out) is proportional to the average time constant $\tau_{\text{mse, av}}$. It follows therefore that the misadjustment $\mathcal{M}$ is inversely proportional to the settling time.
3. The misadjustment $\mathcal{M}$ is directly proportional to the step-size parameter μ, whereas the average time constant $\tau_{\text{mse, av}}$ is inversely proportional to μ. We therefore have conflicting requirements in that if μ is reduced so as to reduce the misadjustment, then the settling time of the LMS algorithm is increased. Conversely, if μ is increased so as to reduce the settling time, then the misadjustment is increased. Careful attention has therefore to be given to the choice of μ.

It is customary to express the misadjustment $\mathcal{M}$ as a percentage. Thus, for example, a misadjustment of 10 percent means that the LMS algorithm produces a mean-squared error (after adaptation is completed) that is 10 percent greater than the minimum mean-squared error J_{min}. Such a performance is ordinarily considered to be satisfactory.

9.9 CONVERGENCE ANALYSIS OF THE LMS ALGORITHM FOR REAL DATA AND COMPARISON WITH THAT FOR COMPLEX DATA

The analysis of the LMS algorithm presented in this chapter has been developed for the general case of complex data. Thus, all of the results developed herein apply equally well to real data, except for the convergence analysis of the mean-squared error $J(n)$. The major point of difference arises in the application of the Gaussian moment factoring theorem. For the case of zero-mean, *complex,* jointly Gaussian random variables represented by x_1, x_2, x_3, and x_4, this theorem states that the expectation of the product $x_1 x_2^* x_3 x_4^*$ may be factored as follows:

$$E[x_1 x_2^* x_3 x_4^*] = E[x_1 x_2^*]E[x_3 x_4^*] + E[x_1 x_4^*]E[x_2^* x_3] \tag{9.130}$$

On the other hand, in the case of zero-mean, *real,* jointly Gaussian random variables represented by y_1, y_2, y_3, and y_4, the Gaussian moment factoring theorem states that

$$E[y_1 y_2 y_3 y_4] = E[y_1 y_2]E[y_3 y_4] + E[y_1 y_3]E[y_2 y_4] + E[y_1 y_4]E[y_2 y_3] \tag{9.131}$$

We see that there is one term less on the right side of Eq. (9.130) for complex data than that of Eq. (9.131) for real data. The missing term in the case of complex data results in a *superior mean-squared convergence and stability performance for the complex LMS algorithm, compared to the real LMS algorithm* (Horowitz and Senne, 1981). Indeed, the mean-squared error produced by the complex LMS algorithm converges to the same steady-state value as obtained using the real LMS algorithm, but at a rate that is more than twice as fast (Horowitz and Senne, 1981).

The condition on the step-size parameter μ for convergence in the mean, described in Eq. (9.52), applies to the complex LMS algorithm as well as the real LMS algorithm. However, because of the difference in the application of the Gaussian moment factoring theorem as evidenced by Eqs. (9.130) and (9.131), we find that the condition on μ for convergence in the mean square for the real LMS algorithm is different from that for the complex LMS algorithm. Specifically, in contrast to the condition of Eq. (9.115) for the complex LMS algorithm, we find that the corresponding condition for convergence of the real LMS algorithm in the mean square is described by (Horowitz and Senne, 1981; Gardner, 1984)

$$\sum_{i=1}^{M} \frac{\mu\lambda_i}{2(1 - \mu\lambda_i)} < 1 \tag{9.132}$$

(See Problem 6 for a derivation of this condition.)

9.10 SUMMARY OF THE LMS ALGORITHM

Parameters: M = number of taps
μ = step-size parameter
$0 < \mu < \dfrac{2}{\text{total input power}}$, provided that $\mu \ll 2/\lambda_{\max}$
Total input power $= Mr(0)$

Initial conditions: $\hat{\mathbf{w}}(0) = \mathbf{0}$

Data:
(a) Given: $\mathbf{u}(n)$ = M-by-1 tap-input vector at time n
$d(n)$ = desired response at time n
(b) To be computed:
$\hat{\mathbf{w}}(n+1)$ = estimate of tap-weight vector at time $n + 1$

Computation: For $n = 0, 1, 2, \ldots$, compute

$$e(n) = d(n) - \hat{\mathbf{w}}^H(n)\mathbf{u}(n)$$

$$\hat{\mathbf{w}}(n + 1) = \hat{\mathbf{w}}(n) + \mu\mathbf{u}(n)e^*(n)$$

Note: With the initial value $\hat{\mathbf{w}}(0)$ of the tap-weight vector equal to zero and with all the tap inputs of the transversal filter set equal to zero initially, we find from the second relation that $\hat{\mathbf{w}}(1)$ is also zero. Accordingly, after the first iteration of the LMS algorithm, we always find that $e(1) = d(1)$.

9.11 DISCUSSION AND A CRITIQUE

At this point, it is informative for us to do the following: (1) look at the LMS algorithm in light of the steepest-descent algorithm that we studied in Chapter 8, (2) highlight the important features of the LMS algorithm, and (3) present a critique of the independence theory used to study the performance of the LMS algorithm.

Ideally, the minimum mean-squared error $J_{\min}$ is realized when the coefficient vector $\hat{\mathbf{w}}(n)$ of the transversal filter approaches the optimum value $\mathbf{w}_0$, defined by the Wiener-Hopf equations. Indeed, as shown in Section 8.5, the steepest-descent algorithm does realize this idealized condition as the number of iterations, n, approaches infinity. The steepest-descent algorithm has the capability to do this, because it uses *exact* measurements of the gradient vector at each iteration of the algorithm. On the other hand, the LMS algorithm relies on a *noisy* estimate for the gradient vector, with the result that the tap-weight vector estimate $\hat{\mathbf{w}}(n)$ only approaches the optimum value $\mathbf{w}_o$ after a large number of iterations and then executes small fluctuations about $\mathbf{w}_o$. Consequently, use of the LMS algorithm, after a large number of iterations, results in a mean-squared error $J(\infty)$ that is greater than the minimum mean-squared error $J_{\min}$. The amount by which the actual value of $J(\infty)$ is greater than $J_{\min}$ is called the *excess mean-squared error*.

There is another basic difference between the steepest-descent algorithm and the LMS algorithm. In Section 8.5, we showed that the steepest-descent algorithm has a well-defined learning curve, obtained by plotting the mean-squared error versus the number of iterations. For this algorithm, the learning curve consists of a sum of decaying exponentials, the number of which equals (in general) the number of tap coefficients. On the other hand, in individual applications of the LMS algorithm, we find that the learning curve consists of *noisy, decaying exponentials*. The amplitude of the noise usually becomes smaller as the step-size parameter μ is reduced.

Imagine now an *ensemble of adaptive transversal filters*. Each filter is assumed to use the LMS algorithm with the same step-size parameter μ and the same initial tap-weight vector $\hat{\mathbf{w}}(0)$. Also, each adaptive filter has individual stationary ergodic inputs that are selected at random from the same statistical population. We compute the *noisy learning curves* for this ensemble of adaptive filters by plotting the squared magnitude of the estimation error $e(n)$ versus time n. To compute the *ensemble-averaged learning curve* of the LMS algorithm, that is, the plot of the mean-squared error $J(n)$ versus time n, we take the average of these noisy learning curves over the ensemble of adaptive filters.

Thus two entirely different ensemble-averaging operations are used in the steepest-descent and LMS algorithms for determining their learning curves (i.e., plots of their mean-squared errors versus the learning period). In the steepest-descent algorithm, the correlation matrix **R** and the cross-correlation vector **p** are first computed through the use of ensemble-averaging operations applied to statistical populations of the tap inputs and the desired response; these values are then used to compute the learning curve of the algorithm. In the LMS algorithm, on the other hand, noisy learning curves are computed for an ensemble of adaptive LMS filters with identical parameters; the learning curve is then computed by averaging over this ensemble of noisy learning curves.

The feature that distinguishes the LMS algorithm from other adaptive algorithms is the simplicity of its implementation, be that in software or hardware form. The simplicity of the LMS algorithm is exemplified by a pair of equations that are involved in its computation (see the summary in Section 9.10).

Three principal factors affect the response of the LMS algorithm: the step-size parameter μ, the number of taps M, and the eigenvalues of the correlation matrix **R** of the tap-input vector. In light of the analysis of the LMS algorithm, using the independence theory, their individual effects may be summarized as follows:

1. When a small value is assigned to μ, the adaptation is slow, which is equivalent to the LMS algorithm having a long "memory." Correspondingly, the excess mean-squared error after adaptation is small, on the average, because of the large amount of data used by the algorithm to estimate the gradient vector. On the other hand, when μ is large, the adaptation is relatively fast, but at the expense of an increase in the average excess mean-squared error after adaptation. In this case, less data enter the estimation; hence, a degraded estimation error performance. Thus, the reciprocal of the parameter μ may be viewed as the *memory* of the LMS algorithm.
2. The convergence properties of the mean-squared error $J(n)$ depend, unlike the average tap-weight vector $E[\hat{\mathbf{w}}(n)]$, on the number of taps, M. In particular, we may summarize the conditions on the step-size parameter μ by writing (for the case of complex data)

$$\frac{1}{\mu} > \sum_{i=1}^{M} \frac{\lambda_i}{2 - \mu\lambda_i} > \sum_{i=1}^{M} \frac{\lambda_i}{2} > \frac{\lambda_{\max}}{2}$$

where the λ_i, $i = 1, 2, \ldots, M$ are the eigenvalues of the correlation matrix of the tap inputs. The outer limit on $1/\mu$ described by $\lambda_{\max}/2$ ensures convergence of the average tap-weight vector $E[\hat{\mathbf{w}}(n)]$. The inner limit on $1/\mu$ described by $\sum_{i=1}^{M} \lambda_i/(2 - \mu\lambda_i)$, together with the outer limit, ensures convergence of the mean-squared error $J(n)$. Hence, convergence of the LMS algorithm in the mean square implies convergence in the mean.
3. When μ is small enough for $\mu \ll 2/\lambda_{\max}$, the condition for convergence of the LMS algorithm in the mean square may be simplified as

$$\frac{1}{\mu} > \frac{1}{2} \sum_{i=1}^{M} \lambda_i = \frac{1}{2} \text{ (total input power)}$$

Under this condition, we find that the conservative choice of a value for the step-size parameter μ less than twice the reciprocal of the total input power offers a practical method for ensuring the convergence of the mean-squared error.

4. When the eigenvalues of the correlation matrix $\mathbf{R}$ are widely spread, the excess mean-squared error produced by the LMS algorithm is primarily determined by the largest eigenvalues, and the time taken by the average tap-weight vector $E[\hat{\mathbf{w}}(n)]$ to converge is limited by the smallest eigenvalues. However, the speed of convergence of the mean-squared error, $J(n)$ is affected by a spread of the eigenvalues of $\mathbf{R}$ to a lesser extent than the convergence of $E[\hat{\mathbf{w}}(n)]$. In any event, when the eigenvalue spread is large (i.e., the correlation matrix of the tap inputs is ill conditioned), the LMS algorithm slows down in that it requires a large number of iterations for it to converge, the very condition for which effective operation of the algorithm is required.

Another point of interest pertains to the joint probability density function of the tap weights in the LMS algorithm. For Gaussian data, it is shown in Bershad and Qu (1989) that, for sufficiently small μ, *the tap weights are jointly Gaussian* with a time-varying vector mean and covariance matrix that are given respectively by the solutions to the matrix difference equations that define the time evolution of the mean weight vector $E[\hat{\mathbf{w}}(n)]$ and the weight covariance matrix $E[(\hat{\mathbf{w}}(n) - E[\hat{\mathbf{w}}(n)])(\hat{\mathbf{w}}(n) - E[\hat{\mathbf{w}}(n)])^H]$; for the latter two quantities, see Problem 5. This result is of particular interest, for example, in the detection of a narrow-band signal in background noise using the tap-weight vector $\hat{\mathbf{w}}(n)$ as a test statistic, as in an adaptive line enhancer.

Finally, it is important to recognize that the mathematical details of all the results summarized herein are based on the independence theory. This theory invokes the assumption that the sequence of random vectors that direct the "hunting" of the tap-weight vector toward the optimum Wiener solution are statistically independent. Even though in reality this assumption is often far from true, nevertheless, the results predicted by the independence theory are usually found to be in agreement with experiments and computer simulations; see, for example, the results of a computer experiment on adaptive prediction presented in Section 9.12. A basic limitation of the independence theory, however, is the fact that it ignores the statistical dependence between the "gradient" directions as the algorithm proceeds from one iteration to the next. This statistical dependence results from the *shifting property* of the input data. At time n, the gradient vector is proportional to the corresponding sample value of the tap-input vector

$$\mathbf{u}(n) = [u(n), u(n-1), \ldots, u(n-M+1)]^T$$

At time $n + 1$, it is proportional to the updated sample value of the tap-input vector,

$$\mathbf{u}(n+1) = [u(n+1), u(n), \ldots, u(n-M+2)]^T$$

Thus, with the arrival of the new sample $u(n + 1)$, the oldest sample $u(n - M + 1)$ is discarded from $\mathbf{u}(n)$, and the remaining samples $u(n)$, $u(n - 1)$,

. . . , $u(n - M + 2)$ are shifted back in time by one time unit. We see therefore that the tap-input vectors, and correspondingly the gradient directions, are indeed statistically dependent. Several worthwhile results have been obtained in the literature for the practical case of *statistically dependent inputs*. In this regard, special mention should be made of the papers by Farden (1981), Jones et al. (1982), Macchi and Eweda (1984), and Gardner (1984).

The most mathematically rigorous analysis of the LMS algorithm presented in the literature to date is perhaps that of Solo (1989). The approach taken by Solo is different from the traditional approach (described in Sections 9.4 to 9.9) that considers convergence of ensemble averages [namely, the expectation of the tap-weight vector estimate $\hat{\mathbf{w}}(n)$, the covariance of $\hat{\mathbf{w}}(n)$, or the mean-squared error $J(n)$] to a limit. Instead, Solo presents a *realization-wise analysis* in the sense that, in practice, a user of the LMS algorithm has only one realization of it to work with and therefore needs to know about convergence on that realization *with probability one* (*wp* 1) or *almost sure (as) convergence*. Solo's paper builds on earlier studies by Bitmead and Anderson (1980 a,b) and Shi and Kozin (1986). The results derived by Solo provide a validation and extension of the classical results derived in this chapter.

9.12 COMPUTER EXPERIMENT ON ADAPTIVE FIRST-ORDER PREDICTION

For our first computer experiment involving the LMS algorithm, we use a first-order, autoregressive (AR) process to study the effects of ensemble averaging on the transient characteristics of the LMS algorithm for *real data*.

Consider then an AR process $\{u(n)\}$ of order 1, described by the difference equation

$$u(n) = -au(n - 1) + v(n) \tag{9.133}$$

where a is the (one and only) parameter of the process, and $\{v(n)\}$ is a zero-mean white-noise process of variance σ_v^2. To estimate the parameter a, we may use an adaptive predictor of order 1, as depicted in Fig. 9.9. The *real* LMS algorithm for the adaptation of the (one and only) tap weight of the predictor is written as

$$\hat{w}(n + 1) = \hat{w}(n) + \mu u(n - 1)f(n)$$

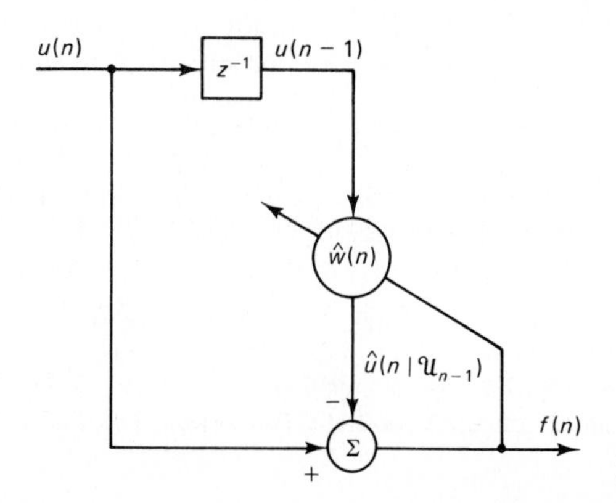

Figure 9.9 Adaptive first-order predictor.

where $f(n)$ is the prediction error, defined by

$$f(n) = u(n) - \hat{w}(n)u(n-1)$$

Figure 9.10 shows plots of $\hat{w}(n)$ versus the number of iterations n for a single trial of the experiment, and the following two sets of conditions:

1. AR parameter: $a = -0.99$
   Variance of AR process $\{u(n)\}$: $\sigma_u^2 = 0.93627$
2. AR parameter: $a = +0.99$
   Variance of AR process $\{u(n)\}$: $\sigma_u^2 = 0.995$

In both cases, the step-size parameter $\mu = 0.05$, and the initial condition is $\hat{w}(0) = 0$. We see that the transient behavior of $\hat{w}(n)$ follows a *noisy exponential curve*. Figure 9.10 also includes the corresponding plots of $E[\hat{w}(n)]$ obtained by ensemble averaging over 100 *independent trials* of the experiment. For each trial, a different computer realization of the AR process $\{u(n)\}$ is used. We see that the ensemble averaging has the effect of smoothing out the effects of gradient noise.

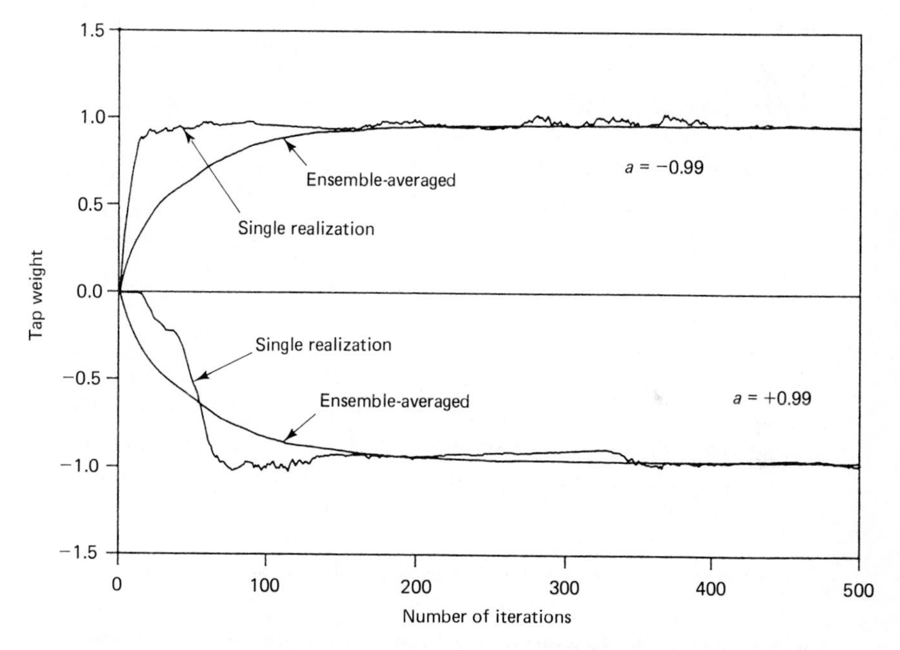

Figure 9.10 Transient behavior of weight $\hat{w}(n)$ of adaptive first-order predictor.

Figure 9.11 shows a plot of the squared prediction error $f^2(n)$, versus the number of iterations n for the set of AR parameters listed under (1); the step-size parameter $\mu = 0.05$, as before. We see that the learning curve for a single realization of the LMS algorithm follows a very noisy *decaying* exponential curve. Figure 9.11 also includes the corresponding plot of $E[f^2(n)]$ obtained by ensemble averaging over 100 independent trials of the experiment. The smoothing effect of the ensemble averaging operation on the learning curve of the LMS algorithm is again visible in the figure.

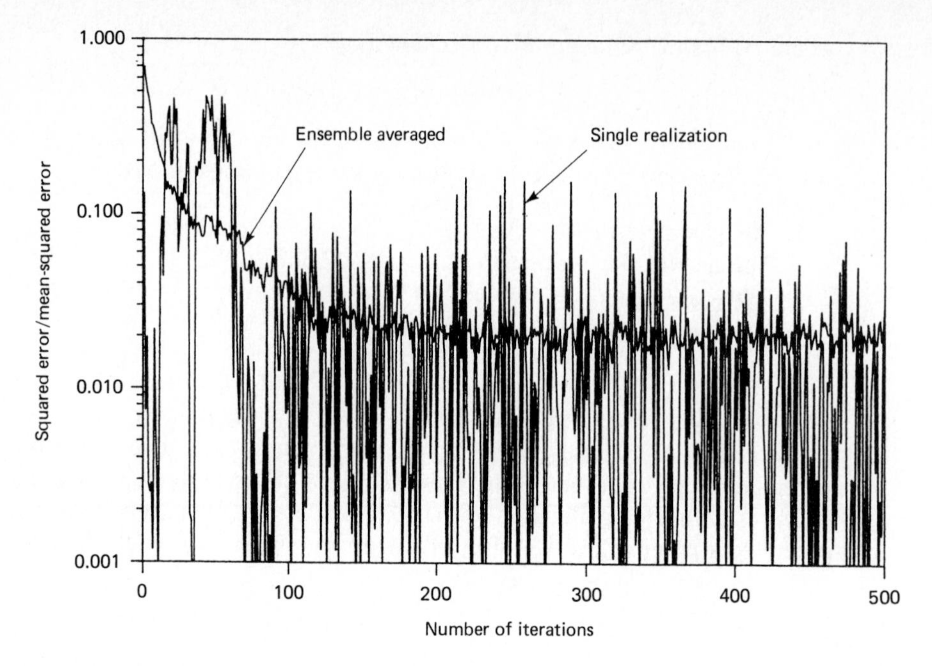

Figure 9.11 Transient behavior of squared prediction error in adaptive first-order predictor, for conditions (1) and $\mu = 0.05$.

Figure 9.12 shows experimental plots of the learning curves of the LMS algorithm [i.e., the mean-squared error $J(n)$ versus the number of iterations n] for the set of AR parameters listed under (1) and varying step-size parameter μ. Specifically, the values used for μ are 0.01, 0.05, and 0.1. The ensemble averaging was performed over 100 independent trials of the experiment. From Fig. 9.12, we observe the following:

1. As the step-size parameter μ is reduced, the rate of convergence of the LMS algorithm is correspondingly reduced.
2. A reduction in the step-size parameter μ also has the effect of reducing the variation in the experimentally computed learning curve.

Comparison of Experimental Results with Theory

In Fig. 9.13, we have plotted two pairs of curves for $\hat{w}(n)$ versus n, corresponding to the two different sets of parameter values listed under (1) and (2) above, and step-size parameter $\mu = 0.05$. One pair of curves corresponds to experimentally derived results obtained by ensemble-averaging over 100 independent trials of the experiment; these curves are labeled "Experiment" in Fig. 9.13. The other pair of curves is computed from theory. In particular, for the situation at hand, we have:

1. The pertinent autocorrelation function of the AR process $\{u(n)\}$ is

$$r(0) = \sigma_u^2$$

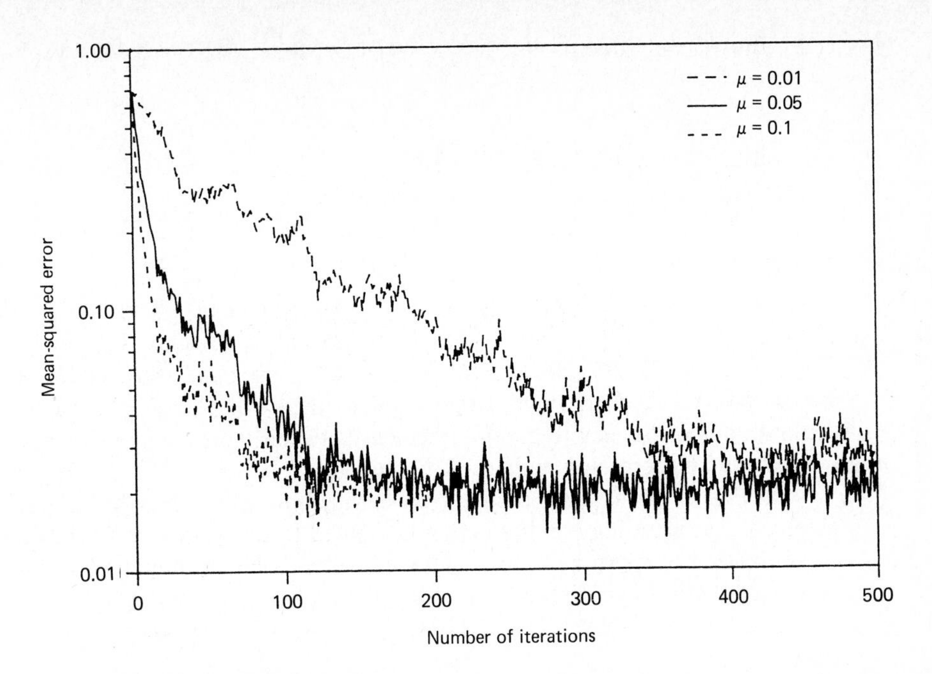

Figure 9.12 Experimental learning curves of adaptive first-order predictor for varying step-size parameter, for conditions (1).

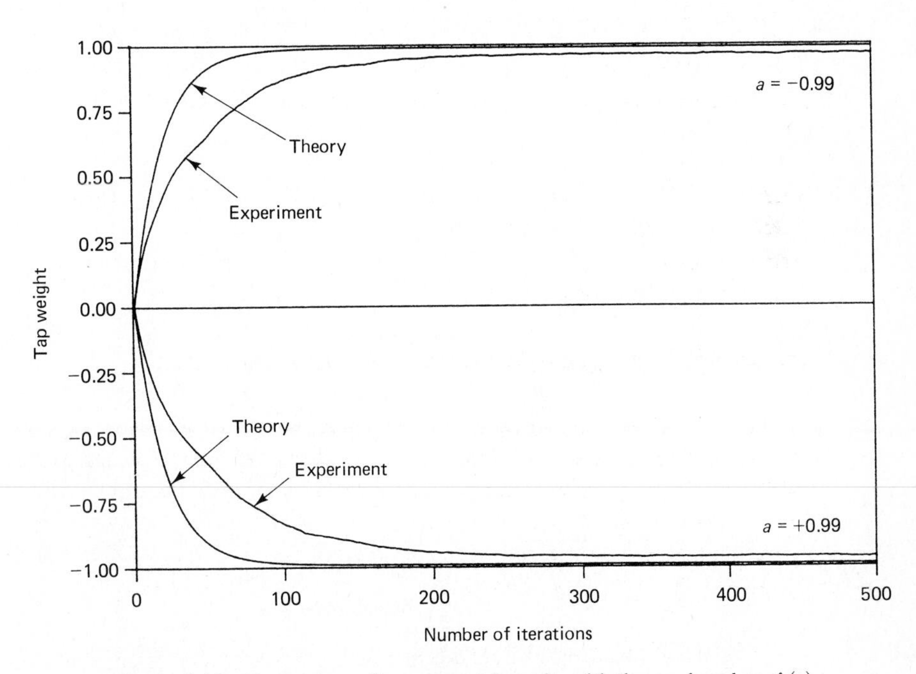

Figure 9.13 Comparison of experimental results with theory, based on $\hat{w}(n)$.

2. The Wiener solution for the tap weight of the order-1 predictor is

$$w_0 = -a$$

Accordingly, the use of Eq. (9.51) yields

$$E[\hat{w}(n+1)] + a = (1 - \mu\sigma_u^2)(E[\hat{w}(n)] + a)$$

or, equivalently,

$$E[\hat{w}(n+1)] = (1 - \mu\sigma_u^2)E[\hat{w}(n)] - \mu\sigma_u^2 \tag{9.133}$$

The use of Eq. (9.133), for the two sets of AR parameters listed under (1) and (2) and $\mu = 0.05$, yields the two curves labeled "theoretical" in Fig. 9.13. This figure demonstrates a reasonably good agreement between theory and experiment, which improves with increasing number of iterations.

In Fig. 9.14, we have plotted two learning curves for the LMS algorithm, one obtained experimentally and the other computed from the theory. The experimental curve, labeled "Experiment," was obtained by ensemble averaging the squared value of the prediction error $f(n)$ over 100 independent trials and for varying n. The theoretical curve, labeled "Theory" in Fig. 9.14, was obtained from the following equation:

$$J(n) = \left(\sigma_u^2 - \frac{\sigma_v^2(2 - \mu\sigma_u^2)}{2(1 - \mu\sigma_u^2)}\right)(1 - \mu\sigma_u^2)^{2n} + \frac{\sigma_v^2(2 - \mu\sigma_u^2)}{2(1 - \mu\sigma_u^2)} \tag{9.134}$$

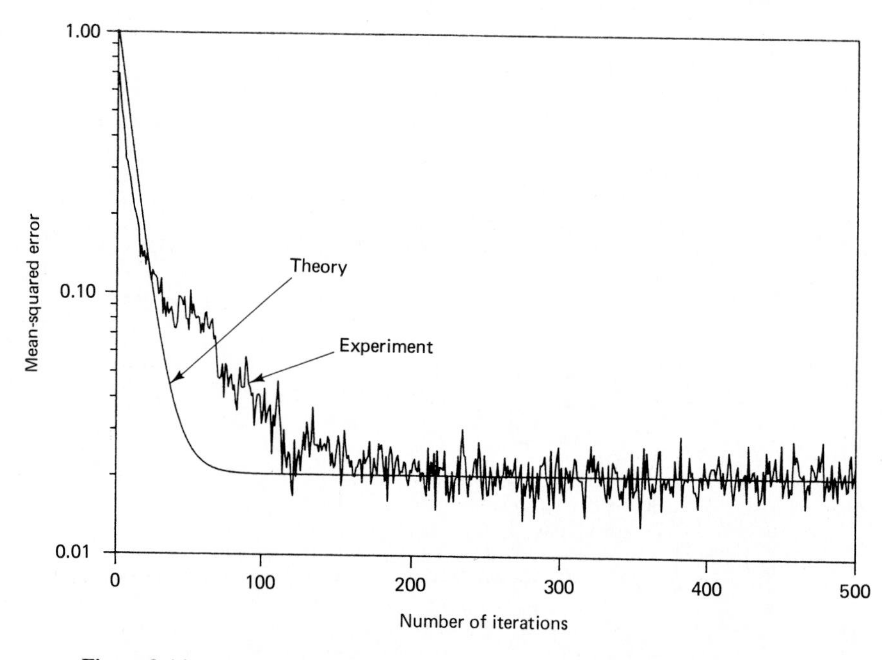

Figure 9.14 Comparison of experimental results with theory, based on the mean-squared error.

where

$$\sigma_v^2 = (1 - a^2)\sigma_u^2$$

Equation (9.134) follows from the generalized transient behavior of Eq. (9.111). Note that

$$J(0) = \sigma_u^2$$
$$J(\infty) = \frac{\sigma_v^2(2 - \mu\sigma_u^2)}{2(1 - \mu\sigma_u^2)} \tag{9.135}$$

Here also we observe reasonably good agreement between theory and experiment, with the agreement improving as the number of iterations is increased.

Finally, in Table 9.1, we present a summary of "steady-state" results computed from theory and experiment for the two sets of AR parameters listed under (1) and (2), and $\mu = 0.05$. In particular, we have listed the following:

1. Final value of $E[\hat{w}(n)]$.
2. Percentage value of misadjustment $\mathcal{M}$

This table further emphasizes the usefulness of the theory developed for the analysis of the LMS algorithm.

TABLE 9.1 SUMMARY OF THE RESULTS OF THE COMPUTER EXPERIMENT ON ADAPTIVE FIRST-ORDER PREDICTION, AND COMPARISON WITH THEORY

LMS algorithm:
- Number of taps: $M = 1$
- Step-size parameter: $\mu = 0.05$

AR process (1)
- AR parameter: $a = -0.99$
- AR process variance: $\sigma_u^2 = 0.93627$
- Noise variance: $\sigma_v^2 = 0.019853$

Theory			Experiment		
w	J_{min}	$\mathcal{M}$	$E[\hat{w}(\infty)]$	$J(\infty)$	$\mathcal{M}$
−0.99	0.019853	2.4556%	−0.96317	0.020998	5.7643%

AR process (2)
- AR parameter: $a = +0.99$
- AR process variance: $\sigma_u^2 = 0.995$
- Noise variance: $\sigma_v^2 = 0.019853$

Theory			Experiment		
w	J_{min}	$\mathcal{M}$	$E[\hat{w}(\infty)]$	$J(\infty)$	$\mathcal{M}$
0.99	0.019853	2.6178%	0.96527	0.020894	5.2438%

9.13 COMPUTER EXPERIMENT ON ADAPTIVE EQUALIZATION

In this second computer experiment we study the use of the LMS algorithm for *adaptive equalization* of a linear dispersive channel that produces (unknown) distortion. Here again we assume that the data are all *real valued*. Figure 9.15 shows the block diagram of the system used to carry out the study. Random number generator 1 provides the test signal, $\{a_n\}$, used for probing the channel, whereas random-number generator 2 serves as the source of additive white noise $\{v(n)\}$ that corrupts the channel output. These two random-number generators are independent of each other. The adaptive equalizer has the task of correcting for the distortion produced by the channel in the presence of the additive white noise. Random-number generator 1, after suitable delay, also supplies the desired response applied to the adaptive equalizer.[11]

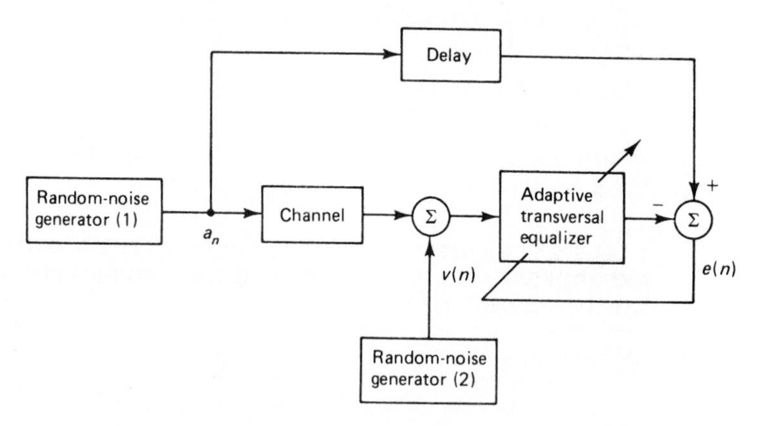

Figure 9.15 Block diagram of adaptive equalizer experiment.

The random sequence $\{a(n)\}$ applied to the channel input is in *polar* form, with $a(n) = \pm 1$, so the sequence $\{a(n)\}$ has zero mean. The impulse response of the channel is described by the raised cosine:[12]

$$h_n = \begin{cases} \dfrac{1}{2}\left[1 + \cos\left(\dfrac{2\pi}{W}(n-2)\right)\right], & n = 1, 2, 3 \\ 0, \quad \text{otherwise} \end{cases} \tag{9.136}$$

where the parameter W controls the amount of amplitude distortion produced by the channel, with the distortion increasing with W.

[11] In an operational system, the receiver is physically separate from the transmitter. As discussed in Chapter 1, one method to produce the test signal for probing the channel is to use a maximal-length sequence generator. At the receiver, an identical maximal-length sequence generator is used, which is synchronized to that at the transmitter. This second generator supplies the desired response to the adaptive equalizer.

[12] The parameters specified in this experiment closely follow the paper by Satorius and Alexander (1979).

Equivalently, the parameter W controls the eigenvalue spread $\chi(\mathbf{R})$ of the correlation matrix of the tap inputs in the equalizer, with the eigenvalue spread increasing with W. The sequence $\{v(n)\}$, produced by the second random generator, has zero mean and variance $\sigma_v^2 = 0.001$.

The equalizer has $M = 11$ taps. Since the channel has an impulse response $\{h_n\}$ that is symmetric about time $n = 2$, as depicted in Fig. 9.16(a), it follows that the optimum tap weights $\{w_{on}\}$ of the equalizer are likewise symmetric about time $n = 5$, as depicted in Fig. 9.16(b). Accordingly, the channel input $\{a(n)\}$ is delayed by $2 + 5 = 7$ samples to provide the desired response for the equalizer.

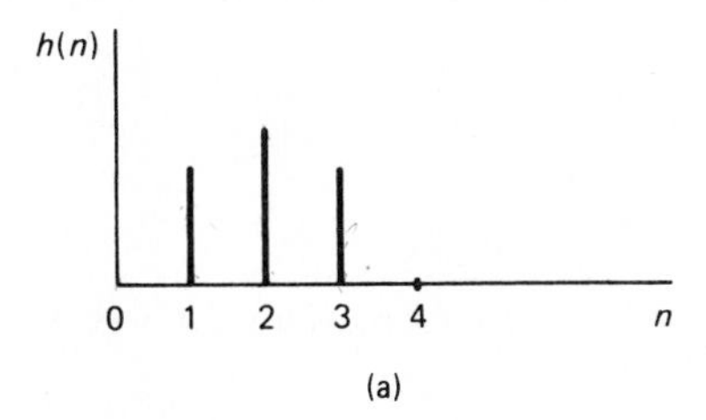

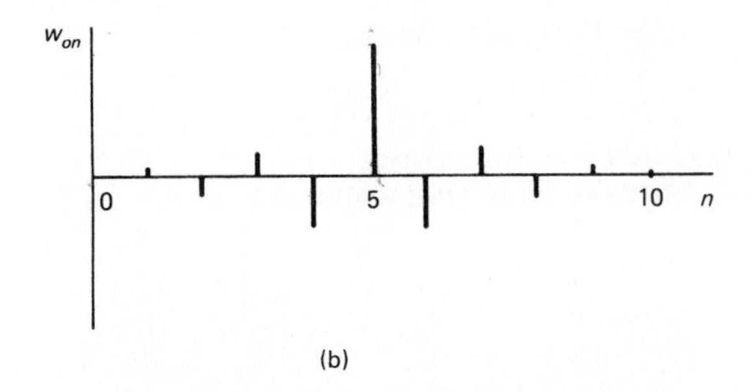

Figure 9.16 (a) Impulse response of channel; (b) impulse response of optimum transversal equalizer.

The experiment is in two parts that are intended to evaluate the response of the adaptive equalizer using the LMS algorithm to changes in the eigenvalue spread $\chi(\mathbf{R})$ and step-size parameter μ. Before proceeding to describe the results of the experiment, however, we first compute the eigenvalues of the correlation matrix $\mathbf{R}$ of the 11 tap inputs in the equalizer.

Correlation Matrix of the Equalizer Input

The first tap input of the equalizer at time n equals

$$u(n) = \sum_{k=1}^{3} h_k a(n - k) + v(n)$$

where all the parameters are real valued. Hence, the correlation matrix $\mathbf{R}$ of the 11 tap inputs of the equalizer, $u(n - 1)$, $u(n - 1)$, . . . , $u(n - 10)$, is a symmetric 11-by-11 matrix. Also, since the impulse response $\{h_n\}$ has nonzero values only for $n = 1, 2, 3$, and the noise process $\{v(n)\}$ is white with zero mean and variance σ_v^2, the correlation matrix **R** is *quintdiagonal*. That is, the only nonzero elements of **R** are on the main diagonal and the four diagonals directly above and below it, two on either side, as shown by the special structure

$$\mathbf{R} = \begin{bmatrix} r(0) & r(1) & r(2) & 0 & \cdots & 0 \\ r(1) & r(0) & r(1) & r(2) & \cdots & 0 \\ r(2) & r(1) & r(0) & r(1) & \cdots & 0 \\ 0 & r(2) & r(1) & r(0) & \cdots & 0 \\ \vdots & \vdots & \vdots & \vdots & \ddots & \vdots \\ 0 & 0 & 0 & 0 & \cdots & r(0) \end{bmatrix} \tag{9.137}$$

where
$$r(0) = h_1^2 + h_2^2 + h_3^3 + \sigma_v^2$$
$$r(1) = h_1 h_2 + h_2 h_3$$
$$r(2) = h_1 h_3$$

The variance $\sigma_v^2 = 0.001$, and h_1, h_2, h_3 are determined by the value assigned to parameter W in Eq. (9.136)

In Table 9.2, we have listed (1) values of the autocorrelation function $r(l)$ for lag $l = 0, 1, 2$, and (2) the smallest eigenvalue, $\lambda_{\min}$, the largest eigenvalue, $\lambda_{\max}$, and the eigenvalue spread $\chi(\mathbf{R}) = \lambda_{\max}/\lambda_{\min}$. We thus see that the eigenvalue spread ranges from 6.0782 (for $W = 2.9$) to 46.8216 (for $W = 3.5$).

TABLE 9.2 SUMMARY OF PARAMETERS FOR THE EXPERIMENT ON ADAPTIVE EQUALIZATION.

W	2.9	3.1	3.3	3.5
$r(0)$	1.0963	1.1568	1.2264	1.3022
$r(1)$	0.4388	0.5596	0.6729	0.7774
$r(2)$	0.0481	0.0783	0.1132	0.1511
$\lambda_{\min}$	0.3339	0.2136	0.1256	0.0656
$\lambda_{\max}$	2.0295	2.3761	2.7263	3.0707
$\chi(\mathbf{R}) = \lambda_{\max}/\lambda_{\min}$	6.0782	11.1238	21.7132	46.8216

Experiment 1: Effect of Eigenvalue Spread

For the first part of the experiment, the step-size parameter was held fixed at $\mu = 0.075$. This is in accordance with the condition of Eq. (9.123) for convergence in the mean square for the worst eigenvalue spread of 46.8216 (corresponding to $W = 3.5$).

$$\mu_{\text{crit}} = \frac{2}{\text{total input power}}$$
$$= \frac{2}{Mr(0)}$$
$$= 0.14$$

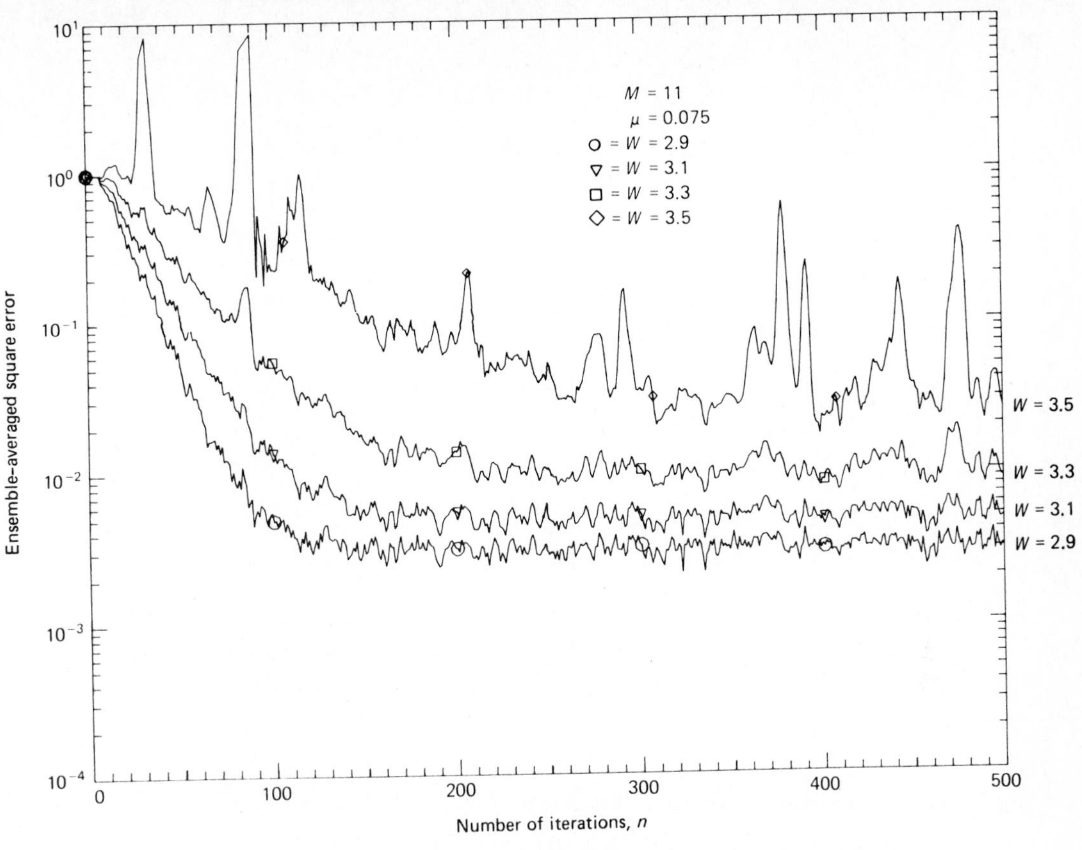

Figure 9.17 Learning curves of the LMS algorithm for adaptive equalizer with number of taps $M = 11$, step-size parameter $\mu = 0.075$, and varying eigenvalue spread $\chi(\mathbf{R})$.

The choice of $\mu = 0.075$ therefore assures the convergence of the adaptive equalizer (in both the mean and mean square) for all the conditions listed in Table 9.2.

For each eigenvalue spread, an approximation to the ensemble-averaged learning curve of the adaptive equalizer is obtained by averaging the instantaneous squared error "$e^2(n)$, versus n" curve over 200 independent trials of the computer experiment. The results of this computation are shown in Fig. 9.17.

We thus see from Fig. 9.17 that increasing the eigenvalue spread $\chi(\mathbf{R})$ has the effect of slowing down the rate of convergence of the adaptive equalizer and also increasing the steady-state value of the average squared error. For example, when $\chi(\mathbf{R}) = 6.0782$, approximately 80 iterations are required for the adaptive equalizer to converge in the mean square, and the average squared error (after 500 iterations) approximately equals 0.003. On the other hand, when $\chi(\mathbf{R}) = 46.8216$ (i.e., the equalizer input is ill conditioned), the equalizer requires approximately 200 iterations to converge in the mean square, and the resulting average squared error (after 500 iterations) approximately equals 0.03.

In Fig. 9.18, we have plotted the ensemble-averaged impulse response of the adaptive equalizer after 1000 iterations for each of the four eigenvalue spreads of inter-

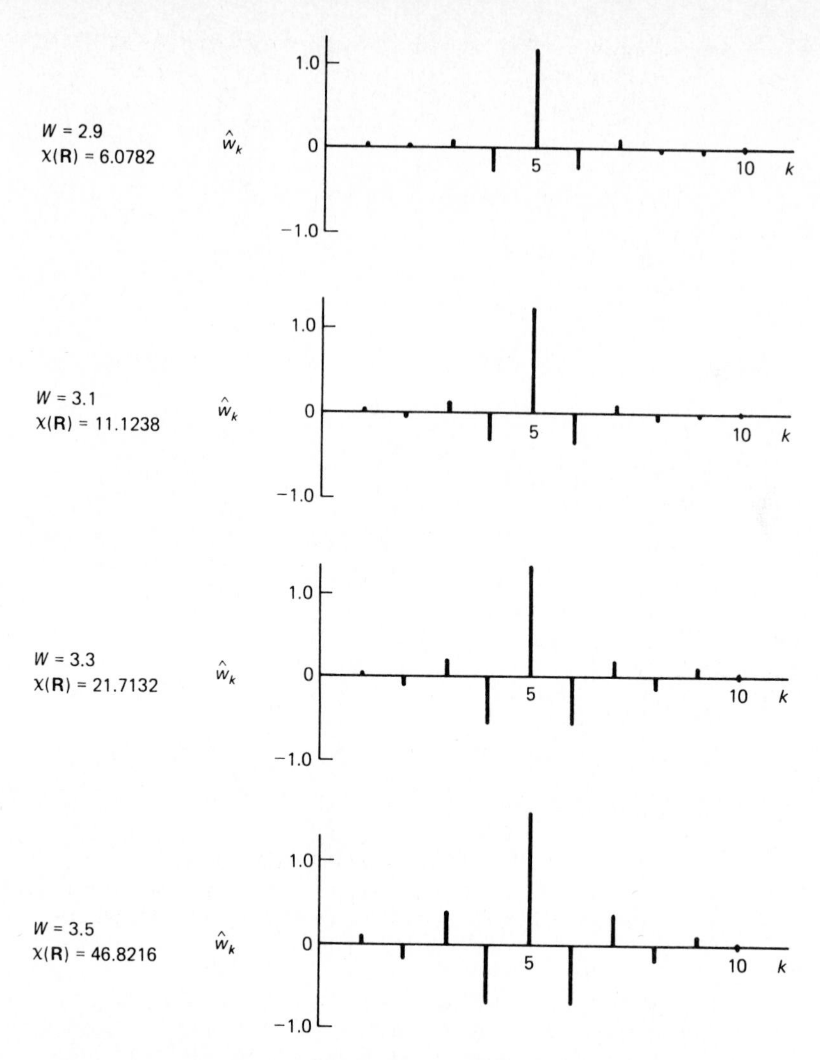

Figure 9.18 Ensemble-averaged impulse response of the adaptive equalizer (after 1000 iterations) for each of four different eigenvalue spreads.

est. As before, the ensemble averaging was carried out over 200 independent trials of the experiment. We see that in each case the ensemble-averaged impulse response of the adaptive equalizer is very close to being symmetric with respect to the center tap, as expected. The variation in the impulse response from one eigenvalue spread to another merely reflects the effect of a corresponding change in the impulse response of the channel.

Experiment 2: Effect of Step-Size Parameter

For the second part of the experiment, the parameter W in Eq. (9.136) was fixed at 3.1, yielding an eigenvalue spread of 11.1238 for the correlation matrix of the tap inputs in the equalizer. The step-size parameter μ was this time assigned one of three values: 0.075, 0.025, 0.0075.

Figure 9.19 shows the results of this computation. As before, each learning curve is the result of ensemble averaging the instantaneous squared error "$e^2(n)$ versus n" curve over 200 independent trials of the computer experiment.

The results confirm that the rate of convergence of the adaptive equalizer is highly dependent on the step-size parameter μ. For large step-size parameter (μ = 0.075), the equalizer converged to steady-state conditions in approximately 120 iterations. On the other hand, when μ is small (equal to 0.0075), the rate of convergence slowed down by more than an order of magnitude. The results also show that the steady-state value of the average squared error (and hence the misadjustment) increases, with increasing μ.

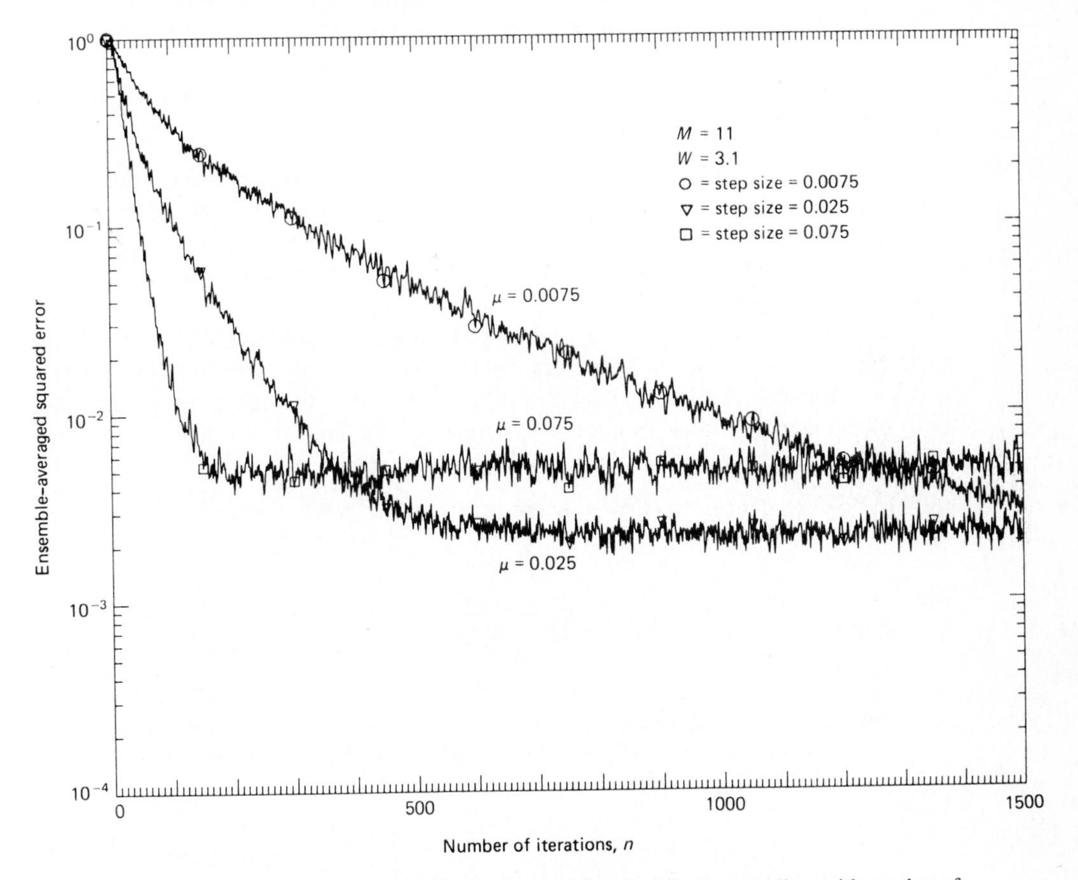

Figure 9.19 Learning curves of the LMS algorithm for adaptive equalizer with number of taps M = 11, fixed eigenvalue spread, and varying step-size parameter μ.

9.14 OPERATION OF THE LMS ALGORITHM IN A NONSTATIONARY ENVIRONMENT

The analysis of the LMS algorithm presented previously has been limited in large measure to a stationary environment. We now briefly consider the ability of the LMS algorithm to operate in a nonstationary environment.

Nonstationarity of an environment may arise in practice in one of two ways:

1. The frame of reference provided by the desired response may be time varying. Such a situation arises, for example, in system identification when an adaptive transversal filter is used to model a time-varying system. In this case, the correlation matrix of the tap inputs of the adaptive transversal filter remains fixed (as in a stationary environment), whereas the cross-correlation vector between the tap inputs and the desired response assumes a time-varying form.
2. The stochastic process supplying the tap inputs of the adaptive filter is nonstationary. This situation arises, for example, when an adaptive transversal filter is used to equalize a time-varying channel. In this second case, both the correlation matrix of the tap inputs in the adaptive transversal filter and the cross-correlation vector of the tap inputs and the desired response assume time-varying forms.

In any event, when an adaptive filter operates in a nonstationary environment, the optimum tap-weight vector assumes a time-varying form in that its value changes from one iteration to the next. Then the LMS algorithm has the task of not only seeking the minimum point of the error-performance surface but also tracking the continually changing position of this minimum point.

Let $\mathbf{w}_o(n)$ denote the time-varying optimum tap-weight vector of a transversal filter that operates in a nonstationary environment, where n denotes the iteration number. Application of the LMS algorithm causes the tap-weight vector estimate $\hat{\mathbf{w}}(n)$ of the adaptive filter to attempt to best match the unknown value $\mathbf{w}_o(n)$. At the nth instant, the *tap-weight error vector* may be expressed as

$$\begin{aligned}\boldsymbol{\epsilon}(n) &= \hat{\mathbf{w}}(n) - \mathbf{w}_o(n)\\ &= (\hat{\mathbf{w}}(n) - E[\hat{\mathbf{w}}(n)]) + (E[\hat{\mathbf{w}}(n)] - \mathbf{w}_o(n)\end{aligned} \tag{9.138}$$

where the expectations are averaged over the ensemble. Two components of error are identified in Eq. (9.138):

1. Any difference between the individual tap-weight vectors of the adaptive filter and their ensemble mean, $E[\hat{\mathbf{w}}(n)]$, is due to errors in the estimate used for its gradient vector. This difference is called the *weight vector noise*. In Eq. (9.138), it is represented by the term

$$\boldsymbol{\epsilon}_1(n) = \hat{\mathbf{w}}(n) - E[\hat{\mathbf{w}}(n)] \tag{9.139}$$

2. Any difference between $E[\hat{\mathbf{w}}(n)]$, the ensemble mean of the tap-weight vector, and $\mathbf{w}_o(n)$, the target value, is due to lag in the adaptive process. This difference is called the *weight vector lag*. In Eq. (9.138), it is represented by the term

$$\boldsymbol{\epsilon}_2(n) = E[\hat{\mathbf{w}}(n)] - \mathbf{w}_o(n) \tag{9.140}$$

When the LMS algorithm is applied to a stationary environment, $\mathbf{w}_o(n)$ assumes a constant value that equals $E[\hat{\mathbf{w}}(n)]$, with the result that the weight vector lag is zero. It is therefore the presence of the weight vector lag that distinguishes the operation of the LMS algorithm in a nonstationary environment from that in a stationary one.

To evaluate the contribution made by the weight vector lag, we consider an example of time-varying system identification (Widrow et al., 1976). The unknown system and the LMS adaptive filter (used to model the unknown system) are depicted in Fig. 9.20. To be specific, we assume the following:

1. The unknown system consists of a transversal filter whose tap weights (M in number) undergo independent stationary *first-order Markov processes*.[13] That is, the elements of the tap-weight vector $\mathbf{w}_o(n)$ characterizing this filter originate from a corresponding set of independent white-noise excitations (each with zero mean and variance σ^2) applied to a bank of one-pole low-pass digital filters, each with a transfer function equal to $1/(1 - az^{-1})$. The parameter a controls the *time constant of nonstationarity*, which equals $1/(1 - a)$.
2. The tap-input vector $\mathbf{u}(n)$, applied to both the unknown system and the LMS adaptive filter, is stationary with zero mean and correlation matrix $\mathbf{R}$.
3. The output of the unknown system is corrupted by additive white noise $\{v(n)\}$ of zero mean and variance σ_v^2. The presence of this noise prevents a perfect match between the unknown system and the LMS adaptive filter.

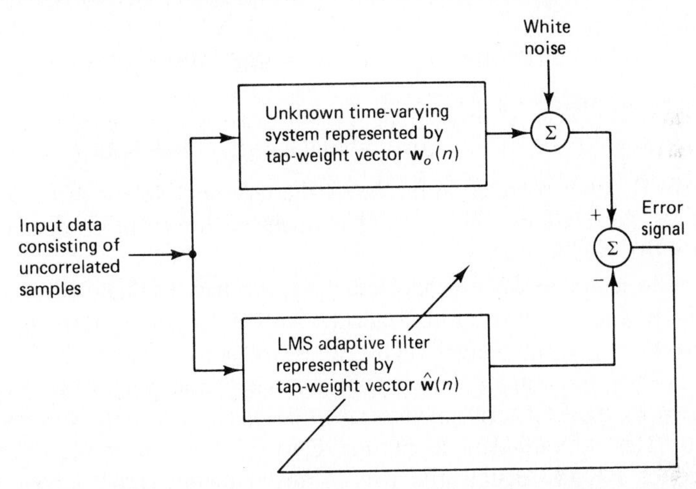

Figure 9.20 Modeling of an unknown system by an LMS adaptive filter.

The time-varying tap-weight vector of the unknown system, denoted by $\mathbf{w}_o(n)$, represents the "target" to be tracked by the LMS adaptive filter. Whenever $\hat{\mathbf{w}}(n)$, the tap-weight vector of the LMS adaptive filter, equals $\mathbf{w}_o(n)$, the minimum mean-squared error $J_{\min} = \sigma_v^2$.

The desired response $d(n)$ applied to the adaptive filter equals the overall output of the unknown system. Since this system is time-varying, the desired response is correspondingly nonstationary. Accordingly, with the correlation matrix of the tap inputs having the fixed value $\mathbf{R}$, we find that the LMS adaptive filter has a quadratic

[13] We say a random vector $\mathbf{w}_o(n)$ is *Markov of first order*, or simply *Markov*, if the conditional joint probability density function of $\mathbf{w}_o(n)$ conditioned on all its past (given) values is the same as using the value in the immediate past. That is, if for every n, the relation

$$f_{\mathbf{W}}(\mathbf{w}_o(n)|\mathbf{w}_o(n-1), \ldots, \mathbf{w}_o(1)) = f_{\mathbf{W}}(\mathbf{w}_o(n)|\mathbf{w}_o(n-1))$$

holds, the model represented by the random vector $\mathbf{w}_o(n)$ is Markov.

bowl-shaped error-performance surface whose position varies continually with time, while its minimum point, eigenvalues, and eigenvectors all remain fixed.

At time n, the excess mean-squared error equals [see the last line of Eq. (9.79)]

$$\begin{aligned} J_{ex}(n) &= J(n) - J_{min} \\ &= E[\boldsymbol{\epsilon}^H(n)\mathbf{u}(n)\mathbf{u}^H(n)\boldsymbol{\epsilon}(n)] \end{aligned} \tag{9.141}$$

Therefore substituting Eqs. (9.138) to (9.140) in (9.141), and then expanding the result, we get

$$\begin{aligned} J_{ex}(n) = {} & E[\boldsymbol{\epsilon}_1^H(n)\mathbf{u}(n)\mathbf{u}^H(n)\boldsymbol{\epsilon}(n)] + E[\boldsymbol{\epsilon}_2^H(n)\mathbf{u}(n)\mathbf{u}^H(n)\boldsymbol{\epsilon}_2(n)] \\ & + E[\boldsymbol{\epsilon}_1^H(n)\mathbf{u}(n)\mathbf{u}^H(n)\boldsymbol{\epsilon}_2(n)] + E[\boldsymbol{\epsilon}_2^H(n)\mathbf{u}(n)\mathbf{u}^H(n)\boldsymbol{\epsilon}(n)] \end{aligned} \tag{9.142}$$

Using the fundamental assumption of Section 9.4, we may show that

$$E[\boldsymbol{\epsilon}_1^H(n)\mathbf{u}(n)\mathbf{u}^H(n)\boldsymbol{\epsilon}_1(n)] = \mathrm{tr}[\mathbf{R}\mathbf{K}_1(n)]$$

$$E[\boldsymbol{\epsilon}_2^H(n)\mathbf{u}(n)\mathbf{u}^H(n)\boldsymbol{\epsilon}_2(n)] = \mathrm{tr}[\mathbf{R}\mathbf{K}_2(n)]$$

and

$$E[\boldsymbol{\epsilon}_1^H(n)\mathbf{u}(n)\mathbf{u}^H(n)\boldsymbol{\epsilon}_2(n)] = E[\boldsymbol{\epsilon}_2^H(n)\mathbf{u}(n)\mathbf{u}^H(n)\boldsymbol{\epsilon}_1(n)] = \mathbf{0}$$

Hence, we may simplify Eq. (9.142) as

$$J_{ex}(n) = \mathrm{tr}[\mathbf{R}\mathbf{K}_1(n)] + \mathrm{tr}(\mathbf{R}\mathbf{K}_2(n)] \tag{9.143}$$

where $\mathbf{R}$ is the correlation matrix of the tap-input vector $\mathbf{u}(n)$, and $\mathbf{K}_1(n)$ and $\mathbf{K}_2(n)$ are the correlation matrices of the weight vector noise $\boldsymbol{\epsilon}_1(n)$ and the weight vector lag $\boldsymbol{\epsilon}_2(n)$, respectively.

To a first order of approximation, we may evaluate the excess mean-squared error in Eq. (9.143) by considering the contributions made by the weight vector noise $\boldsymbol{\epsilon}_1(n)$ and the weight vector lag $\boldsymbol{\epsilon}_2(n)$, one at a time.

Thus, to evaluate the contribution due to the weight vector noise $\boldsymbol{\epsilon}_1(n)$, represented by the first term on the right side of Eq. (9.143), we may set $\boldsymbol{\epsilon}_2(n)$ equal to zero. This is equivalent to putting $E[\hat{\mathbf{w}}(n)] = \mathbf{w}_o$, in which case we find that the trace of $\mathbf{R}\mathbf{K}_1(n)$ equals the excess mean-squared error given by the difference between $J(n)$ and J_{min} as in Eq. (9.82). The important thing to note about this contribution is that it is proportional to the step-size parameter μ for the case of slow adaptation (for which μ is small).

Next, to evaluate the contribution due to the weight vector lag $\boldsymbol{\epsilon}_2(n)$, represented by the second term on the right side of Eq. (9.143), we may set that $\boldsymbol{\epsilon}_1(n)$ equal to zero. This is equivalent to putting $E[\hat{\mathbf{w}}(n)] = \hat{\mathbf{w}}(n)$.

In cases of interest, the step-size parameter μ is small and the time constant of nonstationarity is large [i.e., the parameter a in the first-order Markov model for the target weight vector $\mathbf{w}_o(n)$ is close to 1]. Under these conditions, we find that the expectation of $\boldsymbol{\epsilon}_2^H(n)\mathbf{R}\boldsymbol{\epsilon}_2(n)$ is inversely proportional to the step-size parameter μ (Widrow et al., 1976) (see also Problem 13).

From Eq. (9.143), the excess mean-squared error is the sum of components due to weight vector noise $\boldsymbol{\epsilon}_1(n)$ and weight vector lag $\boldsymbol{\epsilon}_2(n)$. It follows therefore that the total misadjustment of the LMS adaptive filter is likewise the sum of two misadjustment components, one directly proportional to the step-size parameter μ and the other inversely proportional to μ. This suggests that the optimum choice of μ

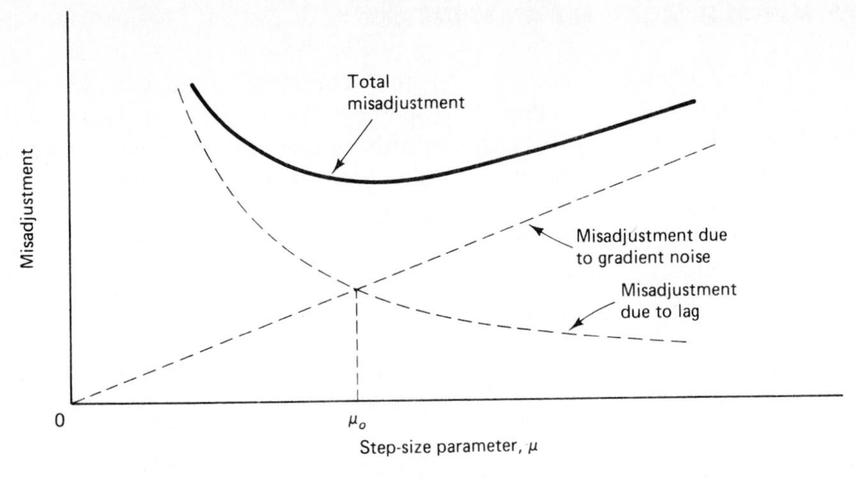

Figure 9.21 Total misadjustment versus step-size parameter, μ.

(which results in the minimum value of total misadjustment) occurs when these two contributions to misadjustment are equal. That is, the rate of adaptation is optimized when the loss of performance due to gradient vector lag equals that due to weight vector noise. Figure 9.21 illustrates the trade-offs[14] involved in adjusting the step-size parameter μ for minimization of the total misadjustment due to the combined effects of weight vector noise and weight vector lag. The optimum value of the step-size parameter, denoted by μ_o, is shown in the figure.

It is important to note that Widrow's theory (described herein for the optimization of the step-size parameter so as to minimize the misadjustment produced by the LMS algorithm for the time-varying system identification problem described in Fig. 9.20) holds only when the step-size parameter is small. For large values of the step-size parameter, however, the application of Widrow's theory may lead to the prediction of a misadjustment that can be off the correct value by a significant factor.[15]

[14] The trade-off between weight vector noise and weight vector lag produced by LMS adaptation in a nonstationary environment was first recognized by Widrow et al. (1976). Subsequently these two forms of LMS noise were shown to be decoupled by Macchi (1986). A quantity called *degree of nonstationarity* was discussed by Macchi and shown to be related in a simple fashion to the optimum step-size parameter μ for tracking nonstationarities. Bershad and Macchi (1991) present a methodology for evaluating the tracking behavior of an LMS filter designed to recover a fixed amplitude complex chirped exponential buried in additive white Gaussian noise. The results presented by Bershad and Macchi are important because they represent a precise analysis of a nonstationary deterministic inverse modeling problem. Moreover, the results are in agreement with the form of the upper bounds for the misadjustment derived by Eweda and Macchi (1985) for the deterministic nonstationarity.

[15] For a full discussion of the limitations of Widrow's analysis of nonstationary learning characteristics of the LMS algorithm, see Gardner (1987). Gardner's theory is developed for the real LMS algorithm subjected to a sequence of independent nonstationary training vectors. New results are presented by Gardner for the dependence of the optimum step-size parameter and the minimum misadjustment on the degree of nonstationarity of the input data.

Solo (1989) presents a realization-wise analysis of misadjustment due to weight vector lag. Solo's analysis also reveals shortcomings in Widrow's theory. In particular, it is shown that the traditional formula for lag misadjustment needs to be modified by adding new terms due to gradient noise and noise variance.

9.15 NORMALIZED LMS ALGORITHM

In the LMS algorithm, the correction $\mu\mathbf{u}(n)e^*(n)$ applied to the tap-weight vector $\hat{\mathbf{w}}(n)$ at time $n + 1$ is directly proportional to the tap-input vector $\mathbf{u}(n)$. Therefore, when $\mathbf{u}(n)$ is large, the LMS algorithm experiences a *gradient noise amplification* problem. To overcome this difficulty, we may use the *normalized LMS algorithm,*[16] which is the companion to the ordinary LMS algorithm. In particular, the correction applied to the tap-weight vector $\hat{\mathbf{w}}(n)$ at time $n + 1$ is "normalized" with respect to the squared Euclidean norm of the tap-input vector $\mathbf{u}(n)$ at time n, hence the term "normalized."

We may formulate the normalized LMS algorithm as a natural modification of the ordinary LMS algorithm. Alternatively, we may derive the normalized LMS algorithm in its own rightful manner; we follow the latter procedure here as it provides insight into its operation.

Normalized LMS Algorithm as the Solution to a Constrained Optimization Problem

The normalized LMS algorithm may be viewed as the solution to a constrained optimization (minimization) problem (Goodwin and Sin, 1984). Specifically, the problem of interest may be stated as follows:

> Given the tap-input vector $\mathbf{u}(n)$ and the desired response $d(n)$, determine the tap-weight vector $\hat{\mathbf{w}}(n+1)$ so as to minimize the squared Euclidean norm of the change
>
> $$\delta\hat{\mathbf{w}}(n+1) = \hat{\mathbf{w}}(n+1) - \hat{\mathbf{w}}(n) \tag{9.144}$$
>
> in the tap-weight vector $\hat{\mathbf{w}}(n+1)$ with respect to its old value $\hat{\mathbf{w}}(n)$, subject to the constraint
>
> $$\hat{\mathbf{w}}^H(n+1)\mathbf{u}(n) = d(n) \tag{9.145}$$

To solve this constrained optimization problem, we may use the *method of Lagrange multipliers.*[17]

The squared norm of the change $\delta\hat{\mathbf{w}}(n+1)$ in the tap-weight vector $\hat{\mathbf{w}}(n+1)$ may be expressed as

$$\begin{aligned}\|\delta\hat{\mathbf{w}}(n+1)\|^2 &= \delta\hat{\mathbf{w}}^H(n+1)\delta\hat{\mathbf{w}}(n+1)\\ &= [\hat{\mathbf{w}}(n+1) - \hat{\mathbf{w}}(n)]^H[\hat{\mathbf{w}}(n+1) - \hat{\mathbf{w}}(n)]\\ &= \sum_{k=0}^{M-1} |\hat{w}_k(n+1) - \hat{w}_k(n)|^2\end{aligned} \tag{9.146}$$

[16] The stochastic gradient algorithm known as the normalized LMS algorithm was suggested independently by Nagumo and Noda (1967) and Albert and Gardner (1967). Nagumo and Noda did not use any special name for the algorithm, whereas Albert and Gardner referred to it as a "quick and dirty regression" scheme. It appears that Bitmead and Anderson (1980) coined the name "normalized LMS algorithm."

[17] For a discussion of the method of Lagrange multipliers, see Appendix C.

Define the tap weight $\hat{w}_k(n)$ for $k = 0, 1, \ldots, M - 1$ in terms of its real and imaginary parts by writing

$$\hat{w}_k(n) = a_k(n) + jb_k(n), \qquad k = 0, 1, \ldots, M - 1 \tag{9.147}$$

We then have

$$|\delta\hat{w}(n + 1)|^2 = \sum_{k=0}^{M-1} ([a_k(n + 1) - a_k(n)]^2 + [b_k(n + 1) - b_k(n)]^2 \tag{9.148}$$

Let the tap input $u(n - k)$ and the desired response $d(n)$ be defined in terms of their respective real and imaginary parts as follows:

$$d(n) = d_1(n) + jd_2(n) \tag{9.149}$$

$$u(n - k) = u_1(n - k) + ju_2(n - k) \tag{9.150}$$

Accordingly, we may rewrite the complex constraint of Eq. (9.145) as an equivalent pair of real constraints:

$$\sum_{k=0}^{M-1} (a_k(n + 1)u_1(n - k) + b_k(n + 1)u_2(n - k)) = d_1(n) \tag{9.151}$$

and

$$\sum_{k=0}^{M-1} (a_k(n + 1)u_2(n - k) - b_k(n + 1)u_1(n - k)) = d_2(n) \tag{9.152}$$

We are now ready to formulate a real-valued cost function $J(n)$ for the constrained optimization problem at hand. In particular, we combine Eqs. (9.148), (9.153), and (9.154) into a single relation:

$$\begin{aligned} J(n) &= \sum_{k=0}^{M-1} ([a_k(n + 1) - a_k(n)]^2 + [b_k(n + 1) - b_k(n)]^2) \\ &= \lambda_1\left[d_1(n) - \sum_{k=0}^{M-1} (a_k(n + 1)u_1(n - k) + b_k(n + 1)u_2(n - k))\right] \\ &+ \lambda_2\left[d_2(n) - \sum_{k=0}^{M-1} (a_k(n + 1)u_2(n - k) - b_k(n + 1)u_1(n - k))\right] \end{aligned} \tag{9.153}$$

where λ_1 and λ_2 are *Lagrange multipliers*. To find the optimum values of $a_k(n + 1)$ and $b_k(n + 1)$, we differentiate the cost function $J(n)$ with respect to these two parameters and then set the results equal to zero. Hence, the use of Eq. (9.153) in

$$\frac{\partial J(n)}{\partial a_k(n + 1)} = 0$$

yields the result

$$2[a_k(n + 1) - a_k(n)] - \lambda_1 u_1(n - k) - \lambda_2 u_2(n - k) = 0 \tag{9.154}$$

Similarly, the use of Eq. (9.153) in

$$\frac{\partial J(n)}{\partial b_k(n + 1)} = 0$$

yields the complementary result

$$2[b_k(n+1) - b_k(n)] - \lambda_1 u_2(n-k) + \lambda_2 u_1(n-k) = 0 \qquad (9.155)$$

Next, we use the definitions of Eqs. (9.147) and (9.150) to combine these two real results into a single complex one, as shown by

$$2[\hat{w}_k(n+1) - \hat{w}_k(n)] = \lambda^* u(n-k), \qquad k = 0, 1, \ldots, M-1 \qquad (9.156)$$

where λ is a *complex Lagrange multiplier:*

$$\lambda = \lambda_1 + j\lambda_2 \qquad (9.157)$$

To solve for the unknown λ^*, we multiply both sides of Eq. (9.156) by $u^*(n-k)$ and then sum over all possible integer values of k for 0 to $M-1$. We thus get

$$\begin{aligned}\lambda^* &= \frac{2}{\sum_{k=0}^{M-1} |u(n-k)|^2} \left[\sum_{k=0}^{M-1} \hat{w}_k(n+1)u^*(n-k) - \sum_{k=0}^{M-1} \hat{w}_k(n)u^*(n-k) \right] \\ &= \frac{2}{\|\mathbf{u}(n)\|^2} [\hat{\mathbf{w}}^T(n+1)\mathbf{u}^*(n) - \hat{\mathbf{w}}^T(n)\mathbf{u}^*(n)]\end{aligned} \qquad (9.158)$$

where $\|\mathbf{u}(n)\|$ is the Euclidean norm of the tap-input vector $\mathbf{u}(n)$. Next, we use the complex constraint of Eq. (9.145) in (9.158) and thus formulate λ^* as follows:

$$\lambda^* = \frac{2}{\|\mathbf{u}(n)\|^2} [d^*(n) - \hat{\mathbf{w}}^T(n)\mathbf{u}^*(n)] \qquad (9.159)$$

However, from the definition of the estimation error $e(n)$, we have

$$e(n) = d(n) - \hat{\mathbf{w}}^H(n)\mathbf{u}(n)$$

Accordingly, we may further simplify the expression given in Eq. (9.159) and thus write

$$\lambda^* = \frac{2}{\|\mathbf{u}(n)\|^2} e^*(n) \qquad (9.160)$$

Finally, we substitute Eq. (9.160) into (9.156), obtaining

$$\begin{aligned}\delta\hat{w}_k(n+1) &= \hat{w}_k(n+1) - \hat{w}_k(n) \\ &= \frac{1}{\|\mathbf{u}(n)\|^2} u(n-k)e^*(n), \qquad k = 0, 1, \ldots, M-1\end{aligned} \qquad (9.161)$$

In vector form, we may equivalently write

$$\begin{aligned}\delta\hat{\mathbf{w}}(n+1) &= \hat{\mathbf{w}}(n+1) - \hat{\mathbf{w}}(n) \\ &= \frac{1}{\|\mathbf{u}(n)\|^2} \mathbf{u}(n)e^*(n)\end{aligned} \qquad (9.162)$$

In order to exercise control over the change in the tap-weight vector from one iteration to the next without changing its direction, we introduce a positive real scaling factor denoted by $\tilde{\mu}$. That is, we redefine the change $\delta\hat{\mathbf{w}}(n+1)$ simply as

$$\delta \hat{\mathbf{w}}(n+1) = \hat{\mathbf{w}}(n+1) - \hat{\mathbf{w}}(n)$$
$$= \frac{\tilde{\mu}}{\|\mathbf{u}(n)\|^2}\mathbf{u}(n)e^*(n) \tag{9.163}$$

Equivalently, we may write

$$\hat{\mathbf{w}}(n+1) = \hat{\mathbf{w}}(n) + \frac{\tilde{\mu}}{\|\mathbf{u}(n)\|^2}\mathbf{u}(n)e^*(n) \tag{9.164}$$

Indeed, this is the desired recursion for computing the M-by-1 tap-weight vector in the normalized LMS algorithm.

Equation (9.164) clearly shows the reason for using the term "normalized." In particular, we see that the product vector $\mathbf{u}(n)e^*(n)$ is normalized with respect to the squared Euclidean norm of the tap-input vector $\mathbf{u}(n)$.

The important point to note from the analysis presented above is that given new input data (at time n) represented by the tap-input vector $\mathbf{u}(n)$ and desired response $d(n)$, the normalized LMS algorithm updates the tap-weight vector in such a way that the value $\hat{\mathbf{w}}(n+1)$ computed at time $n+1$ exhibits the *minimum change* (in a Euclidean norm sense) with respect to the known value $\hat{\mathbf{w}}(n)$ at time n. Hence, the normalized LMS algorithm (and for that matter the conventional LMS algorithm) is a manifestation of the *principle of minimal disturbance* (Widrow and Lehr, 1990). The principle of minimal disturbance states that, *in the light of new input data, the parameters of an adaptive system should only be disturbed in a minimal fashion.*

Moreover, comparing the recursion of Eq. (9.164) for the normalized LMS algorithm with that of Eq. (9.8) for the conventional LMS algorithm, we may make the following observations:

1. The adaptation constant $\tilde{\mu}$ for the normalized LMS algorithm is *dimensionless,* whereas the adaptation constant μ for the LMS algorithm has the dimensioning of *inverse power*.
2. Setting
$$\mu(n) = \frac{\tilde{\mu}}{\|\mathbf{u}(n)\|^2} \tag{9.165}$$
we may view the normalized LMS algorithm as an LMS algorithm with a *time-varying step-size parameter*.
3. The normalized LMS algorithm is *convergent in the mean-square* sense if the adaptation constant μ satisfies the following condition (Weiss and Mitra, 1979; Hsia, 1983):
$$0 < \tilde{\mu} < 2 \tag{9.166}$$

Another point of interest is that in overcoming the gradient noise amplification problem associated with the LMS algorithm, the normalized LMS algorithm introduces a problem of its own. Specifically, when the tap-input vector $\mathbf{u}(n)$ is small, numerical difficulties may arise because then we have to divide by a small value for the squared norm $\|\mathbf{u}(n)\|^2$. To overcome this problem, we slightly modify the recursion of Eq. (9.164) as follows:

$$\hat{\mathbf{w}}(n+1) = \hat{\mathbf{w}}(n) + \frac{\tilde{\mu}}{a + \|\mathbf{u}(n)\|^2}\mathbf{u}(n)e^*(n) \tag{9.167}$$

where $a > 0$, and as before $0 < \tilde{\mu} < 2$. For $a = 0$, Eq. (9.167) reduces to the previous form given in Eq. (9.164). The normalized LMS algorithm is summarized in Table 9.3.

TABLE 9.3 SUMMARY OF THE NORMALIZED LMS ALGORITHM

Parameters: M = number of taps
$\tilde{\mu}$ = adaptation constant
$0 < \tilde{\mu} < 2$
a = positive constant

Initial condition: $\hat{\mathbf{w}}(0) = \mathbf{0}$

Data
(a) Given: $\mathbf{u}(n)$: M-by-1 tap input vector at time n
$d(n)$: desired response at time n
(b) To be computed: $\hat{\mathbf{w}}(n+1)$ = estimate of tap-weight vector at time $n+1$

Computation: $n = 0, 1, 2, \ldots$

$$e(n) = d(n) - \hat{\mathbf{w}}^H(n)\mathbf{u}(n)$$

$$\hat{\mathbf{w}}(n+1) = \hat{\mathbf{w}}(n) + \frac{\tilde{\mu}}{a + \|\mathbf{u}(n)\|^2}\mathbf{u}(n)e^*(n)$$

9.16 GRADIENT ADAPTIVE LATTICE ALGORITHM

In this final section of the chapter, we derive another stochastic gradient algorithm commonly referred to as the *gradient adaptive lattice* (GAL) *algorithm* (Griffiths, 1977, 1978). The GAL algorithm is formulated around a lattice structure. Also as the name implies, its derivation is motivated by that of the LMS algorithm.

Consider the single-stage lattice structure of Fig. 9.22, the input-output relation of which is characterized by a single parameter, namely, the *reflection coefficient* Γ_m. We assume that the input data are wide-sense stationary and that Γ_m is complex valued. Let J_m denote the cost function of this stage:

$$J_m = E[|f_m(n)|^2 + |b_m(n)|^2] \tag{9.168}$$

where $f_m(n)$ is the forward prediction error and $b_m(n)$ is the backward prediction error, both measured at the output of the stage. The gradient of the cost function J_m with respect to the real and imaginary parts of the reflection coefficient Γ_m is given by

$$\nabla_m(J) = 2E[f_m^*(n)b_{m-1}(n-1) + b_m^*(n)f_{m-1}(n)] \tag{9.169}$$

where $f_{m-1}(n)$ is the forward prediction error and $b_{m-1}(n-1)$ is the delayed backward prediction error, both measured at the input of the lattice stage in Fig. 9.22; the other two prediction errors refer to the output of the stage. Following the devel-

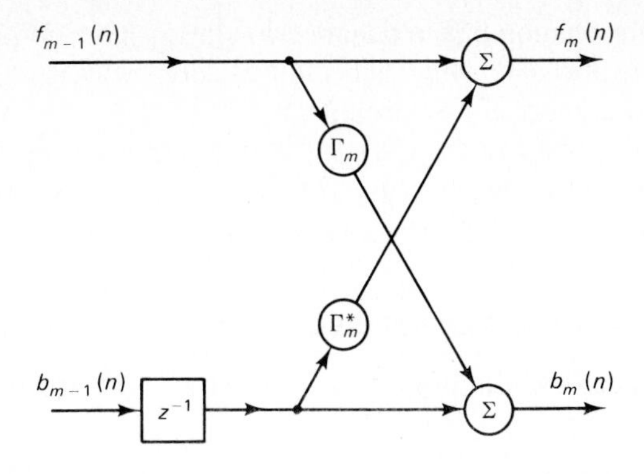

Figure 9.22 Lattice stage for formulating the GAL algorithm.

opment of the LMS algorithm, we may use instantaneous estimates of the expectations in Eq. (9.169), as shown by

$$E[f_m^*(n)b_{m-1}(n-1)] \simeq f_m^*(n)b_{m-1}(n-1)$$
$$E[b_m^*(n-1)f_{m-1}(n)] \simeq b_{m-1}^*(n-1)f_{m-1}(n)$$

Correspondingly, we may use Eq. (9.169) to express the *estimate* of the gradient $\nabla_m(J)$ as

$$\hat{\nabla}_m(J(n)) = 2[f_m^*(n)b_{m-1}(n-1) + b_m^*(n)f_{m-1}(n)] \tag{9.170}$$

Let $\hat{\Gamma}_m(n)$ denote the *old estimate* of the reflection coefficient Γ_m of the mth lattice stage, measured at time n. Let $\hat{\Gamma}_m(n+1)$ denote the *updated estimate* of this reflection coefficient, measured at time $n+1$. Hence, we may compute this updated estimate by adding to the old estimate $\hat{\Gamma}_m(n)$ a *correction* term proportional to the gradient estimate $\hat{\nabla}_m(J(n))$, as shown by

$$\hat{\Gamma}_m(n+1) = \hat{\Gamma}_m(n) - \tfrac{1}{2}\mu_m(n)\hat{\nabla}_m(J(n)) \tag{9.171}$$

where $\mu_m(n)$ denotes a *time-varying step-size parameter* associated with the mth lattice stage. Substituting Eq. (9.170) in (9.171), we thus get

$$\hat{\Gamma}_m(n+1) = \hat{\Gamma}_m(n) - \mu_m(n)[f_m^*(n)b_{m-1}(n-1) + b_m^*(n)f_{m-1}(n)] \tag{9.172}$$

The adaptation parameter $\mu_m(n)$ is chosen as

$$\mu_m(n) = \frac{1}{\xi_{m-1}(n)} \tag{9.173}$$

where

$$\begin{aligned}\xi_{m-1}(n) &= \sum_{i=1}^{M} [|f_{m-1}(i)|^2 + |b_{m-1}(i-1)|^2] \\ &= \xi_{m-1}(n-1) + |f_{m-1}(n)|^2 + |b_{m-1}(n-1)|^2\end{aligned} \tag{9.174}$$

The parameter $\xi_{m-1}(n)$ represents the total energy of both the forward and backward prediction errors at the input of the mth stage, measured up to and including time n.

In practice, a minor modification is made to the *energy estimator* of Eq. (9.174) by writing it in the form of a *single-pole average* of squared data, as shown by [Giffiths (1987, 1988)]

$$\xi_{m-1}(n) = \beta\xi_{m-1}(n-1) + (1-\beta)[|f_{m-1}(n)|^2 + |b_{m-1}(n-1)|^2] \tag{9.175}$$

where $0 < \beta < 1$. The introduction of the parameter β in Eq. (9.175) provides the GAL algorithm with a finite *memory*, which helps it deal better with statistical variations when operating in a nonstationary environment.

Table 9.4 presents a summary of the GAL algorithm and its initialization. The summary presented herein is based on Eqs. (9.172), (9.173), (9.175), and the pair of input-output relations defining a single-stage lattice predictor.

TABLE 9.4 SUMMARY OF THE GAL ALGORITHM

Parameters: M = final prediction order
β = constant $0 < \beta < 1$

Initialization: For prediction order $m = 1, 2, \ldots, M$, put

$$f_m(0) = b_m(0) = 0$$
$$\xi_{m-1}(0) = c, \qquad c = \text{small constant}$$
$$\hat{\Gamma}_m(1) = 0$$

For time $n = 1, 2, \ldots$, put

$$f_0(n) = b_0(n) = u(n), \qquad u(n) = \text{lattice predictor input}$$

Computation: For prediction order $m = 1, 2, \ldots, M$ and time $n = 1, 2, \ldots$, compute

$$f_m(n) = f_{m-1}(n) + \hat{\Gamma}_m^*(n) b_{m-1}(n-1)$$

$$b_m(n) = b_{m-1}(n-1) + \hat{\Gamma}_m(n) f_{m-1}(n)$$

$$\xi_{m-1}(n) = \beta \xi_{m-1}(n-1) + (1-\beta)(|f_{m-1}(n)|^2 + |b_{m-1}(n-1)|^2)$$

$$\hat{\Gamma}_m(n+1) = \hat{\Gamma}_m(n) - \frac{1}{\xi_{m-1}(n)} [f_{m-1}(n) b_m^*(n) + b_{m-1}(n-1) f_m^*(n)]$$

Properties of the GAL Algorithm

The use of the time-varying step-size parameter $\mu_m(n) = 1/\xi_{m-1}(n)$ in the update equation for the reflection coefficient $\hat{\Gamma}_m(n)$ introduces a form of *normalization* similar to that in the normalized LMS algorithm. From Eq. (9.175) we see that for small magnitudes of the prediction errors $f_{m-1}(n)$ and $b_{m-1}(n-1)$ the value of the parameter $\xi_{m-1}(n)$ is correspondingly small or, equivalently, the step-size parameter $\mu_m(n)$ has a correspondingly large value. Such a behavior is desirable from a practical point of view. Basically, a small value for the prediction errors means that the adaptive lattice predictor is providing an accurate model of the external environment in which it is operating. Hence, if there is any increase in the prediction errors, then it should be due to variations in the external environment, in which case, it is hightly desirable for the adaptive lattice predictor to respond rapidly to such variations. This objective is indeed realized by having the step-size parameter $\mu_m(n)$ assume a large value, which makes it possible for the GAL algorithm to provide an initially rapid convergence to the new environmental conditions. If, on the other hand, the input data applied to the adaptive lattice predictor are too noisy (i.e., they contain a strong white noise component in addition to the signals of interest), we find that the prediction errors produced by the adaptive lattice predictor are correspondingly large. In such a situation, the parameter $\xi_{m-1}(n)$ has a large value or, equivalently, the step-size parameter $\mu_m(n)$ has a small value. Accordingly, the GAL algorithm does *not* respond rapidly to variations in the external environment, which is precisely the way we would like the algorithm to behave (Alexander, 1986a).

Another point of interest is that the convergence behavior of the GAL algorithm is somewhat more rapid than that of the LMS algorithm (Honig and Messerschmitt, 1981); this speed-up in convergence, however, is attained at the cost of added computation and storage.

PROBLEMS

1. The LMS algorithm is used to implement a dual-input, single-weight adaptive noise canceller. Set up the equations that define the operation of this algorithm.
2. The LMS-based adaptive deconvolution procedure discussed in Example 2 of Section 9.3 applies to forward-time adaptation (i.e., forward prediction). Reformulate this procedure for reverse-time adaptation (i.e., backward prediction).
3. The zero-mean output $d(n)$ of an unknown real-valued system is represented by the *multiple linear regression model*

$$d(n) = \mathbf{w}_o^T\mathbf{u}(n) + v(n)$$

where $\mathbf{w}_o$ is the (unknown) parameter vector of the model, $\mathbf{u}(n)$ is the input vector, and $v(n)$ is the sample value of an immeasurable white-noise process of zero mean and variance σ_v^2. The block diagram of Fig. P9.1 shows the adaptive modeling of the unknown system, in which the adaptive transversal filter is controlled by a *modified* version of the LMS algorithm. In particular, the tap-weight vector $\mathbf{w}(n)$ of the transversal filter is chosen to minimize the index of performance

$$J(\mathbf{w}, K) = E[e^{2K}(n)]$$

for $K = 1, 2, 3, \ldots.$

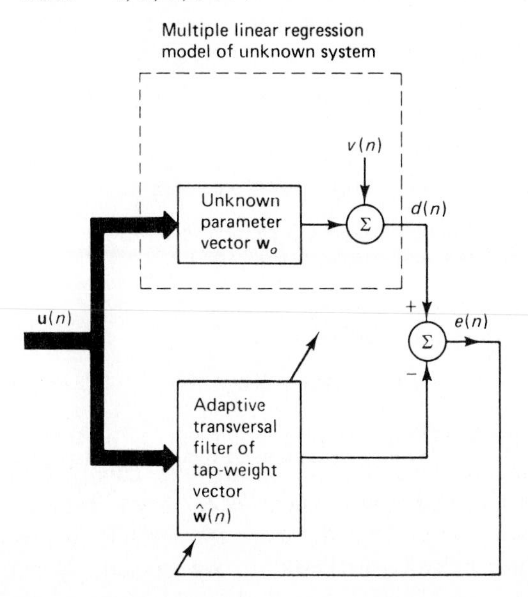

Figure P9.1

(a) By using the instantaneous gradient vector, show that the new adaptation rule for the corresponding estimate of the tap-weight vector is

$$\hat{\mathbf{w}}(n+1) = \hat{\mathbf{w}}(n) + \mu K \mathbf{u}(n) e^{2K-1}(n)$$

where μ is the step-size parameter, and $e(n)$ is the estimation error

$$e(n) = d(n) - \hat{\mathbf{w}}^T(n)\mathbf{u}(n)$$

(b) Assume that the weight-error vector

$$\boldsymbol{\epsilon}(n) = \hat{\mathbf{w}}(n) - \mathbf{w}_o$$

is close to zero, and that $v(n)$ is independent of $\mathbf{u}(n)$. Hence, show that

$$E[\boldsymbol{\epsilon}(n+1)] = (\mathbf{I} - \mu K(2K-1)E[v^{2K-2}(n)]\mathbf{R})E[\boldsymbol{\epsilon}(n)]$$

where $\mathbf{R}$ is the correlation matrix of the input vector $\mathbf{u}(n)$.

(c) Show that the modified LMS algorithm described in part (a) converges in the mean if the step-size parameter μ satisfies the condition

$$0 < \mu < \frac{1}{K(2K-1)E[v^{2(K-1)}(n)]\lambda_{\max}}$$

where $\lambda_{\max}$ is the largest eigenvalue of matrix $\mathbf{R}$.

(d) For $K = 1$, show that the results given in parts (a), (b), and (c) reduce to those in the conventional LMS algorithm.

4. Figure P9.2 shows a block diagram of the *frequency-domain LMS filter*.[18] The tap-input vector $\mathbf{u}(n)$ is first applied to a bank of bandpass digital filters, implemented by means of the *discrete-Fourier transform* (DFT). Let $\mathbf{x}(n)$ denote the transformed vector produced at the DFT output. In particular, element k of the vector $\mathbf{x}(n)$ is given by

$$x_k(n) = \sum_{i=0}^{M-1} u(n-i)e^{-j(2\pi/M)ik}, \qquad k = 0, 1, \ldots, M-1$$

where $u(n-i)$ is element i of the tap-input vector $\mathbf{u}(n)$. Each $x_k(n)$ is normalized with respect to an estimate of its average power. (This normalization is similar in nature to that used in the GAL algorithm.) The inner product of the vector $\mathbf{x}(n)$ and a frequency-domain weight vector $\mathbf{h}(n)$ is formed, obtaining the filter output

$$y(n) = \mathbf{h}^H(n)\mathbf{x}(n)$$

The weight-vector update equation is

$$\mathbf{h}(n+1) = \mathbf{h}(n) + \tilde{\mu}D^{-1}\mathbf{x}(n)e^*(n)$$

where $\mathbf{D}$ = M-by-M diagonal matrix whose kth element denotes the average power estimate of the DFT output $x_k(n)$ for $k = 0, 1, \ldots, M-1$

$\tilde{\mu}$ = adaptation constant

As usual, the estimation error $e(n)$ is defined by

$$e(n) = d(n) - y(n)$$

where $d(n)$ is the desired response.

(a) Show that the DFT output $x_k(n)$ may be computed recursively, using the relation

$$x_k(n) = e^{-j(2\pi/M)k}x_k(n-1) + u(n) - u(n-M), \qquad k = 0, 1, \ldots, M-1$$

[18] This problem is taken from Narayan et al. (1983).

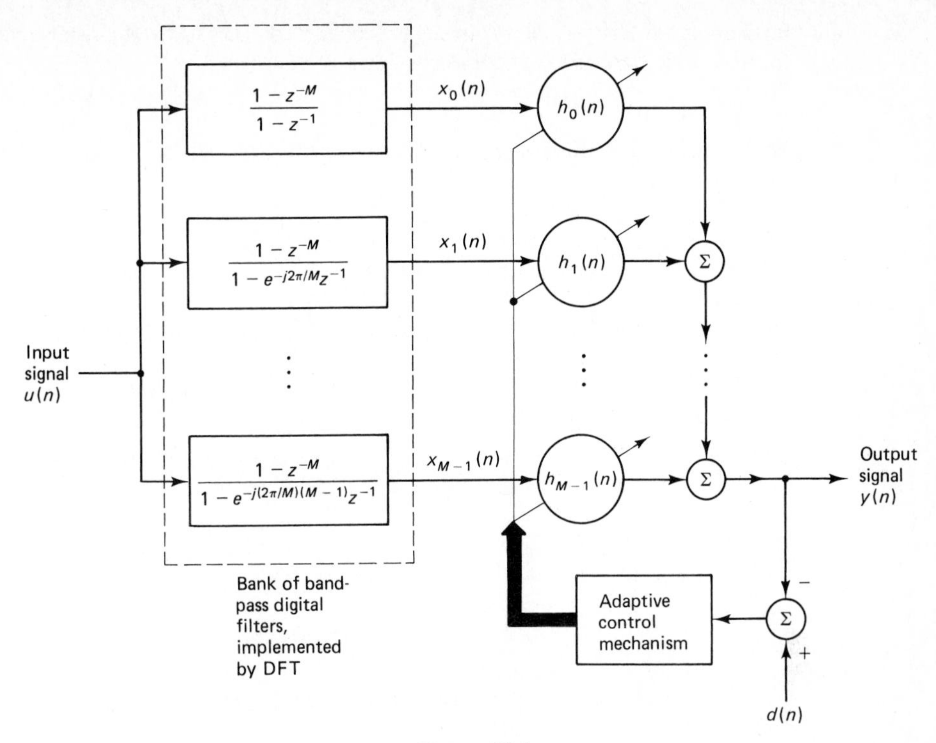

Figure P9.2

(b) Assuming that $\tilde{\mu}$ is chosen properly, show that the weight vector $\mathbf{h}(n)$ converges to the frequency-domain optimum solution:

$$\mathbf{h}_o = \mathbf{Q}\mathbf{w}_o$$

where $\mathbf{w}_o$ is the (time-domain) Wiener solution, and $\mathbf{Q}$ is a unitary matrix defined by the DFT. Determine the identity of the unitary matrix $\mathbf{Q}$.

(c) The use of the matrix $\mathbf{D}^{-1}$ in controlling the correction applied to the frequency-domain weight vector has the equivalent effect of prewhitening the tap-input vector $\mathbf{u}(n)$. Do the following:

(1) Demonstrate the prewhitening effect.

(2) Discuss how this effect compresses the eigenvalue spread of the DFT output vector $\mathbf{x}(n)$.

(3) The frequency-domain LMS algorithm has a faster rate of convergence than the conventional LMS algorithm. Why?

5. (a) Let $\mathbf{m}(n)$ denote the *mean weight vector* in the LMS algorithm, at time n; that is

$$\mathbf{m}(n) = E[\hat{\mathbf{w}}(n)]$$

Using the fundamental assumption of Section 9.4, show that

$$\mathbf{m}(n) = (\mathbf{I} - \mu\mathbf{R})^n[\mathbf{m}(0) - \mathbf{m}(\infty)] + \mathbf{m}(\infty)$$

where μ is the step-size parameter, $\mathbf{R}$ is the correlation matrix of the input vector, and $\mathbf{m}(0)$ and $\mathbf{m}(\infty)$ are the initial and final values of the mean weight vector, respectively.

(b) Determine the *weight covariance matrix* defined as

$$E[(\hat{\mathbf{w}}(n) - \mathbf{m}(n))(\hat{\mathbf{w}}(n) - \mathbf{m}(n))^H]$$

6. In the text we analyzed the convergence of the complex LMS algorithm in the mean square. In this problem we consider the case of *real* data.

(a) Formulate the difference equation for computing the weight-error correlation matrix $\mathbf{K}(n)$.

(b) Formulate the difference equation for computing the diagonal elements of the matrix $\mathbf{X}(n)$ defined by

$$\mathbf{X}(n) = \mathbf{Q}^T\mathbf{K}(n)\mathbf{Q}$$

where $\mathbf{Q}$ is an orthonormal matrix consisting of the eigenvectors belonging to the correlation matrix $\mathbf{R}$.

(c) Using the result of part (b) in the formula of Eq. (9.88), find the final value $J(\infty)$ of the mean-squared error. Hence, show the following:

(1) The condition on the step-size parameter μ for convergence in the mean square is

$$\sum_{i=1}^{M} \frac{\mu\lambda_i}{2(1 - \mu\lambda_i)} < 1$$

(2) The misadjustment is given by

$$\mathcal{M} = \frac{\frac{1}{2}\sum_{i=1}^{M} \mu\lambda_i/(1 - \mu\lambda_i)}{1 - \frac{1}{2}\sum_{i=1}^{M} \mu\lambda_i/(1 - \mu\lambda_i)}$$

7. Consider the filtering structure described in Fig. P9.3. Define the orthogonal decomposition

$$\mathbf{w} = \mathbf{w}_a - \mathbf{B}\mathbf{w}_b$$

where the vectors $\mathbf{w}_a$ and $\mathbf{w}_b$, and the matrix $\mathbf{B}$ are indicated in Fig. P9.3. The matrix $\mathbf{B}$ satisfies the condition

$$\mathbf{B}^H\mathbf{B} = \mathbf{I}$$

where $\mathbf{I}$ is the M-by-M identity matrix. The requirement is to minimize the mean-square value of the filter output $y(n)$ with respect to $\mathbf{w}$, subject to the constraint

$$\mathbf{q}^H(\omega_0)\mathbf{w} = 1$$

where

$$\mathbf{q}(\omega_0) = \frac{1}{\sqrt{M}}[1, e^{-j\omega_0}, \ldots, e^{-j\omega_0(M-1)}]^T$$

The correlation matrix of the input vector $\mathbf{u}(n)$ is given by

$$\mathbf{R} = \sigma_0^2\mathbf{q}(\omega_0)\mathbf{q}^H(\omega_0) + \sigma_1^2\mathbf{q}(\omega_1)\mathbf{q}^H(\omega_1) + \sigma_2^2\mathbf{I}$$

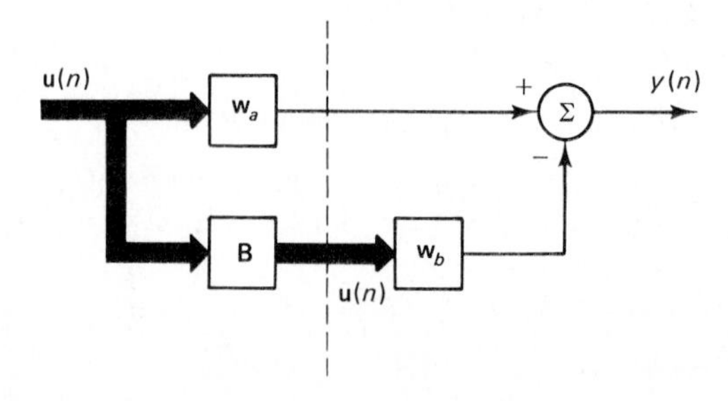

Figure P9.3

where

$$\mathbf{q}^H(\omega_0)\mathbf{q}(\omega_1) = 0$$

Let

$$M = 10$$

$$\sigma_0^2 = 18$$

$$\sigma_1^2 = 1$$

$$\sigma_2^2 = 1$$

(a) Write down the LMS update equation for the vector $\mathbf{w}_b$.
(b) Determine the range of the step-size parameter μ required to ensure that the LMS algorithm converges in the mean square.
(c) Given that $\mu = 0.1$, calculate the misadjustment $\mathcal{M}$ and the final value $J(\infty)$ of the minimum mean-squared error.

Hint: The left side of the structure in Fig. P9.3 implements the linear constraint, whereas the right side is just a minimum mean-squared error filter.

8. *The leaky LMS algorithm.* Consider the time-varying cost function

$$J(n) = |e(n)|^2 + \alpha\|\mathbf{w}(n)\|^2$$

where $\mathbf{w}(n)$ is the tap-weight vector of a transversal filter, $e(n)$ is the estimation error, and α is a constant. As usual, $e(n)$ is defined by

$$e(n) = d(n) - \mathbf{w}^H(n)\mathbf{u}(n)$$

where $d(n)$ is the desired response, and $\mathbf{u}(n)$ is the tap-input vector. In the *leaky LMS algorithm,* the cost function $J(n)$ is minimized with respect to the weight vector $\mathbf{w}(n)$.

(a) Show that the time update for the tap-weight vector $\hat{\mathbf{w}}(n)$ is defined by

$$\hat{\mathbf{w}}(n+1) = (1 - \mu\alpha)\hat{\mathbf{w}}(n) + \mu\mathbf{u}(n)e^*(n)$$

The constant α is limited in value as follows:

$$0 \le \alpha < \frac{1}{1 - \mu}$$

(b) Using the independence theory, show that

$$\lim_{n\to\infty} E[\hat{\mathbf{w}}(n)] = (\mathbf{R} + \alpha\mathbf{I})^{-1}\mathbf{p}$$

where $\mathbf{R}$ is the correlation matrix of the tap inputs and $\mathbf{p}$ is the cross-correlation vector between the tap inputs and the desired response. What is the condition for the algorithm to converge in the mean?

(c) How would you modify the tap-input vector in the conventional LMS algorithm to get the equivalent result described in part (a)?

9. The *normalized LMS algorithm* is described by the following recursion for the tap-weight vector:

$$\hat{\mathbf{w}}(n+1) = \hat{\mathbf{w}}(n) + \frac{\tilde{\mu}}{\|\mathbf{u}(n)\|^2}\mathbf{u}(n)e^*(n)$$

where $\tilde{\mu}$ is a positive constant, and $\|\mathbf{u}(n)\|$ is the *norm* of the tap-input vector. The estimation error $e(n)$ is defined by

$$e(n) = d(n) - \mathbf{w}^H(n)\mathbf{u}(n)$$

where $d(n)$ is the desired response.

Using results of the independence theory, show that the necessary and sufficient condition for the normalized LMS algorithm to be convergent in the mean square is $0 < \alpha < 2$. *Hint:* You may use the following approximation:

$$E\left[\frac{\mathbf{u}(n)\mathbf{u}^H(n)}{\|\mathbf{u}(n)\|^2}\right] \simeq \frac{E[\mathbf{u}(n)\mathbf{u}^H(n)]}{E[\|\mathbf{u}(n)\|^2]}$$

10. In Section 9.15 we presented an independent derivation of the normalized LMS algorithm. In this problem we explore another derivation of this algorithm by modifying the method of steepest descent that led to the development of the conventional LMS algorithm. The modification involves writing the tap-weight vector update in the method of steepest descent as follows

$$\mathbf{w}(n+1) = \mathbf{w}(n) + \frac{1}{2}\mu(n)\nabla(n)$$

where $\mu(n)$ is a *time-varying step-size parameter*, and $\nabla(n)$ is the gradient vector defined by

$$\nabla(n) = 2[\mathbf{R}\mathbf{w}(n) - \mathbf{p}]$$

where $\mathbf{R}$ is the correlation matrix of the tap-input vector $\mathbf{u}(n)$, and $\mathbf{p}$ is the cross-correlation vector between the tap-input vector $\mathbf{u}(n)$ and the desired response $d(n)$.

(a) At time $n + 1$, the mean-squared error is defined by

$$J(n+1) = E[|e(n+1)|^2]$$

where

$$e(n+1) = d(n+1) - \mathbf{w}^H(n+1)\mathbf{u}(n+1)$$

Determine the value of the step-size parameter $\mu_o(n)$ that minimizes $J(n+1)$ as a function of $\mathbf{R}$ and $\nabla(n)$.

(b) Using instantaneous estimates for $\mathbf{R}$ and $\nabla(n)$ in the expression for $\mu_o(n)$ derived in part (a), determine the corresponding instantaneous estimate for $\mu_o(n)$. Hence, formulate the update equation for the tap-weight vector $\hat{\mathbf{w}}(n)$, and compare your result with that obtained for the normalized LMS algorithm.

11. **(a)** Let $\mathbf{K}(n)$ denote the correlation matrix of the weight-error vector $\boldsymbol{\epsilon}(n)$. Show that the trace of $\mathbf{K}(n)$ equals the expected value of the squared norm of $\boldsymbol{\epsilon}(n)$.

(b) Using the result of part (a) and Eq. (9.74) that is based on the independence theory, develop a difference equation that describes the time evolution of $E[\|\boldsymbol{\epsilon}(n)\|^2]$. Hence, show that:

(1) The condition for $E[\|\boldsymbol{\epsilon}(n)\|^2]$ to be convergent is the same as that for the convergence of the mean-squared error $J(n)$.

(2) The presence of the minimum mean-squared error, $J_{\min}$, prevents the estimate of the tap-weight vector, $\hat{\mathbf{w}}(n)$, from converging to the optimum Wiener solution $\mathbf{w}_o$ in the mean square.

(c) The convergence ratio, $\mathscr{C}(n)$, of an adaptive algorithm is defined by

$$\mathscr{C}(n) = \frac{E[\|\boldsymbol{\epsilon}(n+1)\|^2]}{E[\|\boldsymbol{\epsilon}(n)\|^2]}$$

Show that for small n, the convergence ratio of the LMS algorithm for stationary inputs approximately equals

$$\mathscr{C}(n) \simeq 1 - \mu\sigma_u^2 + \mu^2 M\sigma_u^2, \qquad n \text{ small}$$

where μ is the step-size parameter, σ_u^2 is the variance of the tap input, and M is the number of taps in the filter. Hence, find the value of μ for which this convergence ratio is optimum.

12. Derive the result described in Eq. (9.143) for the time-varying system identification problem described in Fig. 9.20. You may assume that the step-size parameter μ is small for the LMS algorithm used here.

13. In this problem we continue with the example of time-varying system identification considered in Section 9.14. Assume that the time evolution of the target value $\mathbf{w}_o(n)$ of the tap-weight vector is governed by the difference equation

$$\mathbf{w}_o(n) = a\mathbf{w}_o(n-1) + \mathbf{v}(n)$$

where a is a constant, and the vector $\mathbf{v}(n)$ consists of independent white noise excitations each of which has zero mean and variance σ^2. For the case when the step-size parameter μ is small and the constant a is close to 1, show that the expectation of the Hermitian form $\boldsymbol{\epsilon}_2^H(n)\mathbf{R}\boldsymbol{\epsilon}_2(n)$ is approximately equal to $M\sigma^2/2\mu$, where M is the number of taps in the filter. The matrix $\mathbf{R}$ is the correlation matrix of the tap-input vector and the weight vector lag $\boldsymbol{\epsilon}_2(n)$ is defined by

$$\boldsymbol{\epsilon}_2(n) = \hat{\mathbf{w}}(n) - \mathbf{w}_o(n)$$

Hints: (1) In the absence of gradient vector noise, the tap-weight adaptation in the LMS algorithm takes on the same form as that in the method of steepest descent.

(2) Use z-transform notation to define the z-transform of the kth element of $\mathbf{Q}^H\boldsymbol{\epsilon}_2(n)$ in terms of the z-transform of the kth element of $\mathbf{Q}^H\mathbf{v}(n)$, where $\mathbf{Q}$ has for its columns the eigenvectors of $\mathbf{R}$.

14. Derive the formula given in Eq. (9.169) for the gradient $\nabla_m(J)$ pertaining to the mth stage of a lattice predictor.

Computer-Oriented Experiments

15. *Computer experiment on the effect of transients on the LMS algorithm.* When conducting a computer experiment that involves the generation of an AR process, it is sometimes overlooked to allow enough time for the transients to die out. The purpose of this experiment is to evaluate the effects of such transients on the operation of the LMS algorithm. Consider then the AR process $\{u(n)\}$ of order 1 described in Section 9.12. The parameters of this process are as follows:

AR parameter:	$a = 0.99$
AR process variance:	$\sigma_u^2 = 1.00$
Noise variance:	$\sigma_v^2 = 0.02$

Generate the process $\{u(n)\}$ so described for $1 \le n \le 100$, assuming zero initial conditions. Use the process $\{u(n)\}$ as the input of a linear adaptive predictor that is based on the LMS algorithm using a step-size parameter $\mu = 0.05$. In particular, plot the learning curve of the predictor by ensemble averaging over 100 independent realizations of the squared value of its output versus time n for $1 \le n \le 100$. Unlike the normal operation of the LMS algorithm, the learning curve so computed should start at the origin, rise to a peak, and then decay toward a steady-state value. Explain the reasons for this phenomenon.

16. *Computer experiment on adaptive two-tap predictor using the LMS algorithm, with autoregressive input of order 1.* An autoregressive process $u(n)$ of order 1 is described by the difference equation

$$u(n) + a_1 u(n - 1) = v(n)$$

where $\{v(n)\}$ is a white-noise process of zero mean and variance σ_v^2. The AR parameter a_1 and the variance σ_v^2 are assigned one of the following two sets of values:

$$(1)\ a_1 = -0.8182, \qquad \sigma_v^2 = 0.3305$$

$$(2)\ a_1 = -0.9802, \qquad \sigma_v^2 = 0.0392$$

The AR process $\{u(n)\}$ is applied to the input of an adaptive two-tap predictor that uses the LMS algorithm.

(a) What is the permissible range of values for the step-size parameter μ for the LMS algorithm to be convergent in the mean?

(b) What is the permissible range of values for the step-size parameter μ for the LMS algorithm to be convergent in the mean square?

(c) Use the computer to generate a 256-sample sequence representing the autoregressive process $\{u(n)\}$ for parameter set (1). You may do this by using a random-number subroutine for generating the white noise $\{v(n)\}$. Hence, by averaging over 200 independent trials of the experiment, plot the learning curves of the LMS algorithm for the following two values of step-size parameter: $\mu = 0.05$ and $\mu = 0.005$. Estimate the corresponding values of the misadjustment by time-averaging over the last 200 iterations of the ensemble-averaged learning curves. Compare the values thus estimated with theory.

(d) For parameter set (1), also estimate mean values for the tap weights $\hat{w}_1(\infty)$ and $\hat{w}_2(\infty)$ that result from the LMS algorithm for two values of the step-size parameter: $\mu = 0.05$ and $\mu = 0.005$. You may do this by averaging the steady-state values of the tap weights (obtained for the last iteration) over 200 independent trials of the experiment. Compare the mean values thus obtained with theory.

(e) Repeat experiments (c) and (d) for parameter set (2).

(f) Discuss the implications of your results.

17. *Computer experiment on adaptive two-tap predictor using the LMS algorithm, with autoregressive input of order 2.* An autoregressive (AR) process $\{u(n)\}$ of order 2 is described by the difference equation

$$u(n) + a_1 u(n - 1) + a_2 u(n - 2) = v(n)$$

where $\{v(n)\}$ is a white-noise process of zero mean and variance σ_v^2. The AR parameters a_1 and a_2 and the variance σ_v^2 are assigned one of the following three sets of values:

$$(1)\ a_1 = -0.10, \qquad a_2 = -0.8, \qquad \sigma_v^2 = 0.2700$$

$$(2)\ a_1 = -0.1636, \qquad a_2 = -0.8, \qquad \sigma_v^2 = 0.1190$$

$$(3)\ a_1 = -0.1960, \qquad a_2 = -0.8, \qquad \sigma_v^2 = 0.0140$$

The AR process $\{u(n)\}$ is applied to an adaptive two-tap predictor whose tap weights are adjusted in accordance with the LMS algorithm.

Repeat the computations under parts (a) and (b) and the computer experiments under parts (c) and (d) of Problem 16, and discuss the implications of your results.

18. *Computer experiment on adaptive two-tap predictor using the clipped LMS algorithm, with autoregressive input of order two:* In the *clipped LMS algorithm* (for *real data*), the tap-input vector $\mathbf{u}(n)$ in the correction term of the update for the tap-weight $\hat{\mathbf{w}}(n)$ is replaced by sgn[$\mathbf{u}(n)$]. The update is thus written as

$$\hat{\mathbf{w}}(n+1) = \hat{\mathbf{w}}(n) + \mu e(n)\,\text{sgn}[\mathbf{u}(n)]$$

As before, μ is the step-size parameter and the estimation error

$$e(n) = d(n) - \mathbf{u}^T(n)\hat{\mathbf{w}}(n)$$

where $d(n)$ is the desired response. The signum function is defined as

$$\text{sgn}[x] = \begin{cases} 1, & \text{if } x > 0 \\ -1, & \text{if } x < 0 \end{cases}$$

The purpose of the clipping is to simplify the implementation of the LMS algorithm without seriously affecting its performance.

Repeat the computer experiments under part (d) of Problems 16 and 17 that deal with parameter set (2) and parameter set (3) for the first-order and second-order AR processes, respectively. Compare your results with the corresponding ones obtained using the conventional LMS algorithm.

19. *Computer experiment on adaptive antenna using the LMS algorithm, with a single source of noncoherent interference.* Consider an array antenna that consists of $2N + 1$ uniformly spaced elements. Two plane waves are incident upon the array, one representing a target signal and the other representing a source of interference that is noncoherent with the target signal. The target signal and the interference arrive along directions θ_1 and θ_2 measured with respect to the normal to the array, respectively. The element signals of the array antenna are expressed in baseband form as follows:

$$u(n) = A_1 e^{jn\phi_1} + A_2 e^{j(n\phi_2 + \psi)} + v(n)$$

$$n = 0, \pm 1, \ldots, \pm N$$

where A_1 is fixed amplitude of the target signal, and A_2 is fixed amplitude of the interference. The electrical angles ϕ_1 and ϕ_2 are related to the individual angles of arrival θ_1 and θ_2, respectively, by

$$\phi_i = \frac{2\pi}{\lambda} d \sin\theta_i \qquad i = 1, 2$$

where d is the element-to-element spacing and λ is the wavelength. The angle ψ represents the phase difference between the target signal and the interference, measured at the center of the array. When the target signal and the interference are noncoherent, the phase difference ψ is a uniformly distributed random variable with the probability density function:

$$f_\Psi(\psi) = \begin{cases} \dfrac{1}{2\pi}, & \pi \le \psi \le -\pi \\ 0, & \text{otherwise} \end{cases}$$

Lastly, the additive element noise $v(n)$ is a *complex-valued* Gaussian random variable with zero mean and unit variance. The noise contributions of the individual elements are uncorrelated, as shown by

$$E[v(k)v^*(n)] = \begin{cases} 1, & k = n \\ 0, & k \ne n \end{cases}$$

The element-to-element spacing d equals one-half wavelength. The array contains a total of $2N + 1 = 5$ elements. The element outputs of the array are applied to a beamforming network that produces a set of four orthogonal beams that correspond to the uniform phase factor $\alpha = \pm\pi/5, \pm3\pi/5$, as described in Section 1.7. The beam corresponding to $\alpha = \pi/5$ is chosen as the primary beam, and the remaining three are used as the auxiliary beams whose outputs are adaptively adjusted by means of the LMS algorithm.

The target signal is characterized by the parameters:

$$\text{angle of arrival, } \theta_1 = \sin^{-1}(0.2)$$

$$\text{signal-to-noise ratio, SNR} = 10 \text{ dB}$$

The interference is characterized by the parameters:

$$\text{angle of arrival, } \theta_2 = -\sin^{-1}(0.4)$$

$$\text{interference-to-noise ratio, INR} = 40 \text{ dB}$$

The LMS algorithm is started with the initial values of the weights set equal to zero. The step-size parameter $\mu = 4 \times 10^{-7}$. Each iteration of the algorithm corresponds to one snapshot of data, with each snapshot being independent of another. For the various parameters as specified above, do the following:

(a) Justify the admissibility of the value 4×10^{-7} for the step-size parameter μ.
(b) Plot the adapted antenna pattern after 50 iterations of the LMS algorithm.
(c) Plot the adapted antenna pattern after 200 iterations of the LMS algorithm.

Suppose next that the source of interference is reduced in strength, but with its angular position remaining the same as before. To investigate this new effect, carry out the following experiments:

(d) With the interference-to-noise ratio reduced to 30 dB, plot the adapted antenna pattern after 200 iterations of the LMS algorithm.
(e) Repeat the experiment with the interference-to-noise ratio reduced further to 20 dB.

Comment on the results of your experiments.

20. *Computer experiment on adaptive antenna using the LMS algorithm, with two noncoherent sources of interference.* The adaptive antenna used in this experiment has the same configuration as that used in Problem 19. It uses four orthogonal beams corresponding to the uniform phase factor $\alpha = \pm\pi/5, \pm3\pi/5$, with the one represented by $\alpha = \pi/5$ serving as the primary beam and the remaining three as the auxiliary beams. The adaptive antenna uses the LMS algorithm with the step-size parameter $\mu = 4 \times 10^{-7}$. Plot the adapted antenna pattern after 200 iterations of the LMS algorithm using independent snapshots of data for the following environmental conditions:

(a) The environment includes a target signal and two sources of interference, all three of which are noncoherent with each other. The target signal is characterized by

$$\text{angle of arrival, } \theta_1 = \sin^{-1}(0.2)$$

$$\text{signal-to-noise ratio, SNR} = 10 \text{ dB}$$

The first source of interference is characterized by

$$\text{angle of arrival, } \theta_2 = -\sin^{-1}(0.4)$$

$$\text{interference-to-noise ratio, (INR)}_1 = 40 \text{ dB}$$

The second source of interference is characterized by

$$\text{angle of arrival, } \theta_3 = \sin^{-1}(0.8)$$

$$\text{signal-to-interference ratio, } (\text{INR})_2 = 40 \text{ dB}$$

That is, both interferences lie outside the main beam. The additive complex-valued noise at the output of each element in the array is characterized in the same way as in Problem 19 in that it is Gaussian with zero mean and unit variance and it is uncorrelated with the noise at all other elements.

(b) The environmental conditions are the same as in part (a), except that the first source of interference arrives along the angle $\theta_2 = 0°$, that is, inside the main beam of the antenna.

(c) The target signal arrives along the angle $\theta_1 = \sin^{-1}(0.1)$, that is, the adaptive antenna suffers from pointing inaccuracy in that the main beam (represented by $\alpha = \pi/5$) points along a direction slightly different from the target. Otherwise, the environmental conditions are the same as in part (a).

Comment on your results.

21. *Computer experiment on adaptive antenna using the LMS algorithm with two coherent sources*. Here again the adaptive antenna uses the same configuration as that in Problem 19 in that it has four beams represented by $\alpha = \pm\pi/5, \pm 3\pi/5$, with that corresponding to $\alpha = \pi/5$ acting as the primary beam and the remaining three acting as auxiliary beams. In this problem, however, the two external sources responsible for illuminating the array are coherent. In particular, in the formula that defines the element signal $u(n)$, the phase difference ψ equals zero.

Plot the adapted antenna pattern after 200 iterations of the LMS algorithm with step-size parameter $\mu = 4 \times 10^{-7}$ and for the following three different environmental conditions:

(a) The target signal arrives along $\theta_1 = \sin^{-1}(0.2)$ and the coherent interference arrives along $\theta_2 = -\sin^{-1}(0.4)$. Both are of equal strength, with SNR = 30 dB.

(b) The target signal and the coherent interference are positioned symmetrically with respect to the main beam, with $\theta_1 = -\sin^{-1}(0.4)$ and $\theta_2 = \sin^{-1}(0.8)$. They are of equal strength with SNR = 30 dB.

As in Problem 19, the additive complex-valued noise at each element of the array is Gaussian with zero mean and unit variance, and it is noncoherent with the noise at other elements of the array.

Comment on the results of your experiment.

10

Method of Least Squares

In this chapter, we use the *method of least squares* to solve the linear filtering problem, without invoking assumptions on the statistics of the inputs applied to the filter. To illustrate the basic idea of least squares, suppose we have a set of real-valued measurements $u(1), u(2), \ldots, u(N)$, made at times $t_1, t_2, \ldots, t_N$, respectively, and the requirement is to construct a curve that is used to *fit* these points in some optimum fashion. Let the time dependence of this curve be denoted by $f(t_i)$. According to the method of least squares, the "best" fit is obtained by *minimizing the sum of squares of difference* between $f(t_i)$ and $u(i)$ for $i = 1, 2, \ldots, N$; hence the name of the method.

The method of least squares, may be viewed as an alternative to Wiener filter theory. Basically, Wiener filters are derived from *ensemble averages* with the result that one filter (optimum in a probabilistic sense) is obtained for all realizations of the operational environment, assumed to be wide-sense stationary. On the other hand, the method of least squares involves the use of time averages, with the result that the filter depends on the number of samples used in the computation. We begin our study in the next section by outlining the essence of the linear least-squares estimation problem.

10.1 STATEMENT OF THE LINEAR LEAST-SQUARES ESTIMATION PROBLEM

Consider a physical phenomenon that is characterized by two sets of variables, $\{d(i)\}$ and $\{u(i)\}$. The variable $d(i)$ is observed at time i in *response* to the subset of variables $u(i), u(i-1), \ldots, u(i-M+1)$ applied as *inputs*. That is, $d(i)$ is a function of the inputs $u(i), u(i-1), \ldots, u(i-M+1)$. This functional relationship is hypothesized to be *linear*. We may thus express the response $d(i)$ as

$$d(i) = \sum_{k=0}^{M-1} w_{ok}^* u(i-k) + e_o(i) \tag{10.1}$$

where the w_{ok} are *unknown parameters* of the *model*, and $e_o(i)$ represents the *measurement error* to which the statistical nature of the phenomenon is ascribed; each term in the summation in Eq. (10.1) represents an inner product. In effect, the model of Eq. (10.1) says that the variable $d(i)$ may be determined as a linear combination of the input variables $u(i), u(i-1), \ldots, u(i-M+1)$, except for the error $e_o(i)$. This model, represented by the signal-flow graph shown in Fig. 10.1, is called a *multiple linear regression model*.

The *measurement error* $e_o(i)$ is an *unobservable* random variable that is introduced into the model to account for its inaccuracy. It is customary to assume that the measurement error process $\{e_o(i)\}$ is white with zero mean and variance σ^2. That is,

$$E[e_o(i)] = 0, \qquad \text{for all } i$$

and

$$E[e_o(i)e_o^*(k)] = \begin{cases} \sigma^2, & i = k \\ 0, & i \neq k \end{cases}$$

The implication of this assumption is that we may rewrite Eq. (10.1) in the equivalent form

$$E[d(i)] = \sum_{k=0}^{M-1} w_{ok}^* u(i-k)$$

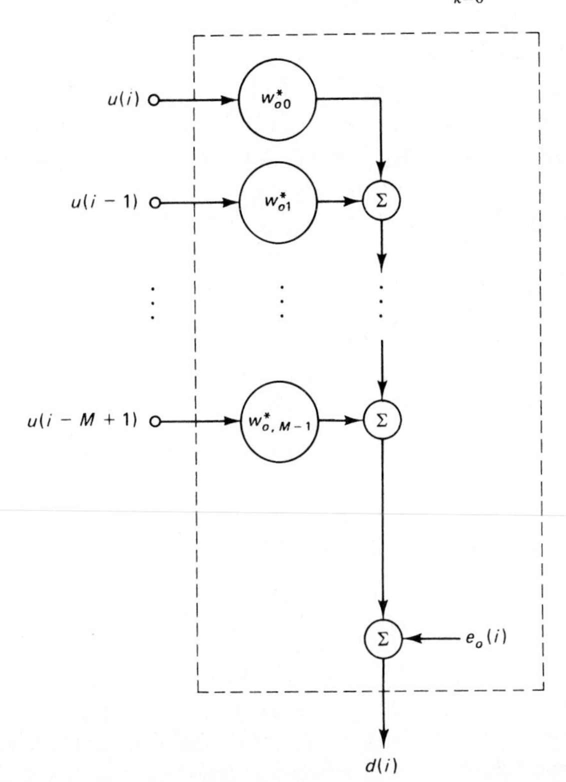

Figure 10.1 Multiple linear regression model.

where the values of $u(i)$, $u(i-1)$, . . . , $u(i-M+1)$ are known. Hence, the mean of the response $d(i)$, in theory, is uniquely determined by the model.

The problem we have to solve is to *estimate* the unknown parameters of the multiple linear regression model of Fig. 10.1, the w_{ok}, given the two *observable* sets of variables: $\{u(i)\}$ and $\{d(i)\}$, $i = 1, 2, \cdots, N$. To do this, we postulate the linear transversal filter of Fig. 10.2 as the model of interest. By forming inner products of the *tap inputs* $u(i)$, $u(i-1)$, $\cdots$, $u(i-M+1)$ and the corresponding *tap weights* $w_0, w_1, \cdots, w_{M-1}$, respectively, and by utilizing $d(i)$ as the *desired response,* we define the *estimation error* or *residual* $e(i)$ as the difference between the desired response $d(i)$ and the *filter output*, $y(i)$, as shown by

$$e(i) = d(i) - y(i) \tag{10.2}$$

where

$$y(i) = \sum_{k=0}^{M-1} w_k^* \, u(i-k) \tag{10.3}$$

That is,

$$e(i) = d(i) - \sum_{k=0}^{M-1} w_k^* \, u(i-k) \tag{10.4}$$

In the method of least squares, we choose the tap weights of the transversal filter, the w_k, so as to minimize a cost function that consists of the *sum of error squares:*

$$\mathscr{E}(w_0, \ldots, w_{M-1}) = \sum_{i=i_1}^{i_2} |e(i)|^2 \tag{10.5}$$

where i_1 and i_2 define the index limits at which the error minimization occurs. The values assigned to these limits depend on the type of *data windowing* employed, as discussed in Section 10.2. Basically, the problem we have to solve is to substitute Eq. (10.4) into (10.5) and then minimize the cost function $\mathscr{E}(w_0, \ldots, w_{M-1})$ with respect to the tap weights of the transversal filter in Fig. 10.2. For this minimization, the tap weights of the filter $w_0, w_1, \cdots, w_{M-1}$ are held *constant* during the interval $i_1 \le i \le i_2$. The filter resulting from the minimization is termed a *linear least-squares filter*.

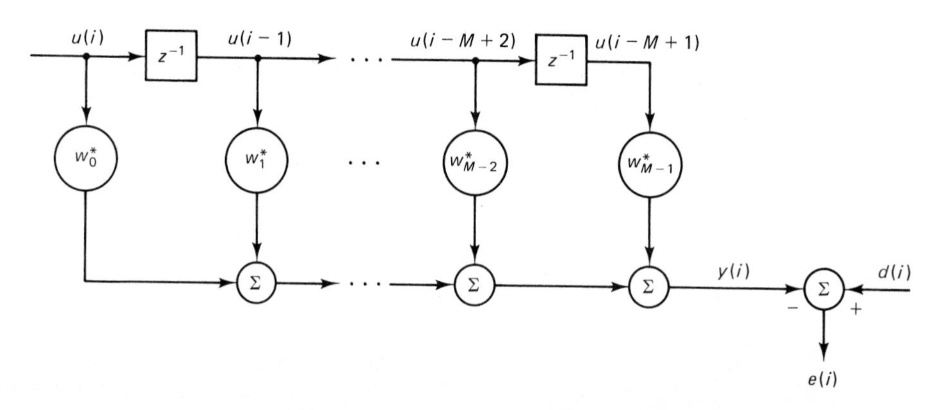

Figure 10.2 Linear transversal filter model.

10.2 DATA WINDOWING

Given M as the number of tap weights used in the transversal filter model of Fig. 10.2, the rectangular matrix constructed from the input data, $u(1), u(2), \cdots, u(N)$, may assume different forms, depending on the values assigned to the limits i_1 and i_2 in Eq. (10.5). In particular, we may distinguish four different methods of *windowing* the input data:

1. *Covariance method,* which makes no assumptions about the data outside the interval $[1, N]$. Thus, by defining the limits of interest as $i_1 = M$ and $i_2 = N$, the input data may be arranged in the matrix form:

$$\begin{bmatrix} u(M) & u(M+1) & \cdots & u(N) \\ u(M-1) & u(M) & \cdots & u(N-1) \\ \vdots & \vdots & & \vdots \\ u(1) & u(2) & \cdots & u(N-M+1) \end{bmatrix}$$

2. *Autocorrelation method,* which makes the assumption that the data prior to time $i = 1$ and the data after $i = N$ are zero. Thus, by using $i_1 = 1$ and $i_2 = N + M - 1$, the matrix of input data takes on the form

$$\begin{bmatrix} u(1) & u(2) & \cdots & u(M) & u(M+1) & \cdots & u(N) & 0 & \cdots & 0 \\ 0 & u(1) & \cdots & u(M-1) & u(M) & \cdots & u(N-1) & u(N) & \cdots & 0 \\ \vdots & \vdots & & \vdots & \vdots & & \vdots & \vdots & & \vdots \\ 0 & 0 & \cdots & u(1) & u(2) & \cdots & u(N-M+1) & u(N-M) & \cdots & u(N) \end{bmatrix}$$

3. *Prewindowing method,* which makes the assumption that the input data prior to $i = 1$ are zero, but makes no assumption about the data after $i = N$. Thus, by using $i_1 = 1$ and $i_2 = N$, the matrix of input data assumes the form

$$\begin{bmatrix} u(1) & u(2) & \cdots & u(M) & u(M+1) & \cdots & u(N) \\ 0 & u(1) & \cdots & u(M-1) & u(M) & \cdots & u(N-1) \\ \vdots & \vdots & & \vdots & \vdots & & \vdots \\ 0 & 0 & \cdots & u(1) & u(2) & \cdots & u(N-M+1) \end{bmatrix}$$

4. *Postwindowing method,* which makes no assumption about the data prior to time $i = 1$, but makes the assumption that the data after $i = N$ are zero. Thus, by using $i_1 = M$ and $i_2 = N + M - 1$, the matrix of input data takes on the form

$$\begin{bmatrix} u(M) & u(M+1) & \cdots & u(N) & 0 & \cdots & 0 \\ u(M-1) & u(M) & \cdots & u(N-1) & u(N) & \cdots & 0 \\ \vdots & \vdots & & \vdots & \vdots & & \vdots \\ u(1) & u(2) & \cdots & u(N-M+1) & u(N-M) & \cdots & u(N) \end{bmatrix}$$

The terms "covariance method" and "autocorrelation method" are commonly used in speech-processing literature (Makhoul, 1975; Markel and Gray, 1976). It should, however, be emphasized that the use of these two terms is *not* based on the standard definition of the covariance function as the correlation function with the means removed. Rather, these two terms derive their names from the way we interpret the meaning of the *known parameters* contained in the system of equations that result from minimizing the index of performance of Eq. (10.5). The covariance method derives its name from control theory literature where, with zero-mean tap inputs, these known parameters represent the elements of a *covariance matrix;* hence, the name of the method. The autocorrelation method, on the other hand, derives its name from the fact that, for the conditons stated, these known parameters represent the *short-term autocorrelation function* of the tap inputs; hence, the name of the second method.

In the remainder of this chapter, except for Problem 4, which deals with the autocorrelation method, we will be exclusively concerned with the covariance method. The prewindowing method is considered in subsequent chapters.

10.3 PRINCIPLE OF ORTHOGONALITY (REVISITED)

When we developed the Wiener filter theory in Chapter 5, we proceeded by first deriving the principle of orthogonality for wide-sense stationary discrete-time stochastic processes, which we then used to derive the Wiener-Hopf equations that provide the mathematical basis of Wiener filters. In this chapter we proceed in a similar fashion by first deriving the principle of orthogonality based on time averages, and then use it to derive a system of equations known as the normal equations that provides the mathematical basis of linear least-squares filters. The development of this theory will be done for the covariance method.

The cost function or the sum of the error squares in the covariance method is defined by

$$\mathscr{E}(w_0, \ldots, w_{M-1}) = \sum_{i=M}^{N} |e(i)|^2 \tag{10.6}$$

By choosing the limits on the time index i in this way, in effect, we make sure that for each value of i, all the M tap inputs of the transversal filter in Fig. 10.2 have nonzero values. As mentioned previously, the problem we have to solve is to determine the tap weights of the transversal filter of Fig. 10.2 for which the sum of error squares is minimum.

We first rewrite Eq. (10.6) as

$$\mathscr{E}(w_0, \ldots, w_{M-1}) = \sum_{i=M}^{N} e(i)e^*(i) \tag{10.7}$$

where the estimation error $e(i)$ is defined in Eq. (10.4). Let the kth tap-weight w_k be expressed in terms of its real and imaginary parts as follows:

$$w_k = a_k + jb_k, \qquad k = 0, 1, \ldots, M-1 \tag{10.8}$$

Thus, substituting Eq. (10.8) in (10.4), we get

$$e(i) = d(i) - \sum_{k=0}^{M-1} (a_k - jb_k)u(i-k) \tag{10.9}$$

We define the kth component of the gradient vector $\nabla(\mathscr{E})$ as the derivative of the cost function $\mathscr{E}(w_0, \ldots, w_{M-1})$ with respect to the real and imaginary parts of tap-weight w_k, as shown by

$$\nabla_k(\mathscr{E}) = \frac{\partial \mathscr{E}}{\partial a_k} + j\frac{\partial \mathscr{E}}{\partial b_k} \tag{10.10}$$

Hence, substituting Eq. (10.7) in (10.10), and recognizing that the estimation error $e(i)$ is complex valued, in general, we get

$$\nabla_k(\mathscr{E}) = -\sum_{i=M}^{N}\left[e(i)\frac{\partial e^*(i)}{\partial a_k} + e^*(i)\frac{\partial e(i)}{\partial a_k} + je(i)\frac{\partial e^*(i)}{\partial b_k} + je(i)\frac{\partial e(i)}{\partial b_k}\right] \tag{10.11}$$

Next, differentiating $e(i)$ in Eq. (10.9) with respect to the real and imaginary parts of w_k, we get the following four partial derivatives:

$$\begin{aligned}
\frac{\partial e(i)}{\partial a_k} &= -u(i-k) \\
\frac{\partial e^*(i)}{\partial a_k} &= -u^*(i-k) \\
\frac{\partial e(i)}{\partial b_k} &= ju(i-k) \\
\frac{\partial e^*(i)}{\partial b_k} &= -ju^*(i-k)
\end{aligned} \tag{10.12}$$

Thus, the substitution of these four partial derivatives in Eq. (10.11) yields the desired result:

$$\nabla_k(\mathscr{E}) = -2\sum_{i=M}^{N} u(i-k)e^*(i) \tag{10.13}$$

For the minimization of the cost function $\mathscr{E}(w_0, \ldots, w_{M-1})$ with respect to the tap weights $w_0, \ldots, w_{M-1}$ of the linear transversal filter in Fig. 10.2, we require that the following conditions be satisfied simultaneously:

$$\nabla_k(\mathscr{E}) = 0, \qquad k = 0, 1, \ldots, M-1 \tag{10.14}$$

Let $e_{\min}(i)$ denote the special value of the estimation error $e(i)$ that results when the cost function $\mathscr{E}(w_0, \ldots, w_{M-1})$ is minimized (i.e., the transversal filter is optimized) in accordance with Eq. (10.14). From Eq. (10.13) we then readily see that the set of conditions (10.14). is equivalent to the following:

$$\sum_{i=M}^{N} u(i-k)e^*_{\min}(i) = 0, \qquad k = 0, 1, \ldots, M-1 \tag{10.15}$$

Equation (10.15) is the mathematical description of the temporal version of the *principle of orthogonality*. The *time average*[1] on the left side of Eq. (10.15) represents

[1] To be precise in the use of the term "time average," we should divide the sum on the left side of Eq. (10.15) by the number of terms $(N - M + 1)$ in the summation. Clearly, such an operation has no effect on Eq. (10.15). We have chosen to ignore the inclusion of this scaling factor merely for convenience of presentation. This practice is followed in the rest of the book.

the cross-correlation between the tap input $u(i-k)$ and the minimum estimation error $e_{\min}(i)$ over the values of time i in the interval (M, N), for a fixed value of k. Accordingly, we may state the *principle of orthogonality* as follows:

> The minimum error time series $\{e_{\min}(i)\}$ is orthogonal to the time series $\{u(i-k)\}$ applied to tap k of a transversal filter of length M for $k = 0, 1, \ldots, M-1$, when the filter is operating in its least-squares condition.

This principle provides the basis of a simple *test* that we can carry out in practice to check whether or not the transversal filter is operating in its *least-squares condition*. We merely have to determine the time-averaged cross-correlation between the estimation error and the time series applied to *each* tap input of the filter. It is *only* when *all* these M cross-correlation functions are identically zero that we find the cost function $\mathscr{E}(w_0, \ldots, w_{M-1})$ is minimum.

Corollary

Let $\hat{w}_0, \hat{w}_1, \ldots, \hat{w}_{M-1}$ denote the special values of the tap weights $w_0, w_1, \ldots, w_{M-1}$ that result when the transversal filter of Fig. 10.2 is optimized to operate in its least-squares condition. The filter output, denoted by $y_{\min}(i)$, is obtained from Eq. (10.3) to be

$$y_{\min}(i) = \sum_{k=0}^{M-1} \hat{w}_k^* u(i-k) \tag{10.16}$$

This filter output provides a *least-squares estimate* of the desired response $d(i)$; the estimate is said to be *linear* because it is a linear combination of the tap inputs $u(i)$, $u(i-1), \ldots, u(i-M+1)$. Let $\mathscr{U}_i$ denote the space spanned by the tap inputs $u(i), \ldots, u(i-M+1)$. Let $\hat{d}(i|\mathscr{U}_i)$ denote the least-squares estimate of the desired response $d(i)$, given the tap inputs spanned by the space $\mathscr{U}_i$. We may thus write

$$\hat{d}(i|\mathscr{U}_i) = y_{\min}(i) \tag{10.17}$$

or, equivalently,

$$\hat{d}(i|\mathscr{U}_i) = \sum_{k=0}^{M-1} \hat{w}_k^* u(i-k) \tag{10.18}$$

Returning to Eq. (10.15), suppose we multiply both sides of this equation by $\hat{w}_k^*$ and then sum the result over the values of k in the interval $[0, M-1]$. We then get (after interchanging the order of summation):

$$\sum_{i=M}^{N} \left[\sum_{k=0}^{M-1} \hat{w}_k^* u(i-k) \right] e_{\min}^*(i) = 0 \tag{10.19}$$

The summation term inside the parentheses on the left side of Eq. (10.19) is recognized to be the least-squares estimate $\hat{d}(i|\mathscr{U}_i)$ of Eq. (10.18). Accordingly, we may simplify Eq. (10.19) as follows:

$$\sum_{i=M}^{N} \hat{d}(i|\mathscr{U}_i) e_{\min}^*(i) = 0 \tag{10.20}$$

Equation (10.20) is the mathematical description of the *corollary to the principle of orthogonality*. We recognize the time average on the left side of Eq. (10.20) as the cross-correlation of the two time series $\{\hat{d}(i|\mathcal{U}_i)\}$ and $\{e_{\min}(i)\}$. Accordingly, we may state the corollary to the principle of orthogonality as follows:

> When a transversal filter operates in its least-squares condition the least-squares estimate of the desired response, produced at the filter output and represented by the time series $\{\hat{d}(i|\mathcal{U}_i)\}$, and the minimum estimation error time series $\{e_{\min}(i)\}$ are orthogonal to each other over time i.

A geometric illustration of this corollary to the principle of orthogonality is deferred to Section 10.6.

10.4 MINIMUM SUM OF ERROR SQUARES

The principle of orthogonality, given in Eq. (10.15), describes the least-squares condition of the transversal filter in Fig. 10.2 when the cost function $\mathscr{E}(w_0, \ldots, w_{M-1})$ is minimized with respect to the tap weights $w_0, \ldots, w_{M-1}$ in the filter. To find the minimum value of this cost function, that is, the *minimum sum of error squares* $\mathscr{E}_{\min}$, it is obvious that we may write

$$\underbrace{d(i)}_{\substack{\text{desired}\\ \text{response}}} = \underbrace{\hat{d}(i|\mathcal{U}_i)}_{\substack{\text{estimate of}\\ \text{desired}\\ \text{response}}} + \underbrace{e_{\min}(i)}_{\substack{\text{estimation}\\ \text{error}}} \tag{10.21}$$

Hence, evaluating the energy of the time series $\{d(i)\}$ for values of time i in the interval $[M, N]$, and using the corollary to the principle of orthogonality [i.e., Eq. (10.20)], we get the simple result

$$\mathscr{E}_d = \mathscr{E}_{\text{est}} + \mathscr{E}_{\min} \tag{10.22}$$

where

$$\mathscr{E}_d = \sum_{i=M}^{N} |d(i)|^2 \tag{10.23}$$

$$\mathscr{E}_{\text{est}} = \sum_{i=M}^{N} |\hat{d}(i|\mathcal{U}_i)|^2 \tag{10.24}$$

$$\mathscr{E}_{\min} = \sum_{i=M}^{N} |e_{\min}(i)|^2 \tag{10.25}$$

Rearranging Eq. (10.22), we may express the minimum sum of error squares $\mathscr{E}_{\min}$ in terms of the energy $\mathscr{E}_d$ and the energy $\mathscr{E}_{\text{est}}$, contained in the time series $\{d(i)\}$ and $\{d(i|\mathcal{U}_i)\}$, respectively, as follows:

$$\mathscr{E}_{\min} = \mathscr{E}_d - \mathscr{E}_{\text{est}} \tag{10.26}$$

Clearly, given the specification of the desired response $d(i)$ for varying i, we may use Eq. (10.23) to evaluate the energy $\mathscr{E}_d$. As for the energy $\mathscr{E}_{\text{est}}$ contained in the time series $\{\hat{d}(i|\mathcal{U}_i)\}$ representing the estimate of the desired response, we are going to defer its evaluation to the next section.

Since $\mathcal{E}_{\min}$ is nonnegative, it follows that the second term on the right side of Eq. (10.26) can never exceed $\mathcal{E}_d$. Indeed, it reaches the value of $\mathcal{E}_d$ when the measurement error $e_0(i)$ in the multiple linear regression model of Fig. 10.1 is zero, a practical impossibility.

Another case for which $\mathcal{E}_{\min}$ equals $\mathcal{E}_d$ occurs when the least-squares problem is *underdetermined*. Such a situation arises when there are fewer data points than parameters, in which case the estimation error and therefore $\mathcal{E}_{\text{est}}$ is zero. Note, however, that when the least-squares problem is underdetermined, there is no unique solution to the problem. Discussion of this issue is deferred to the next chapter.

10.5 NORMAL EQUATIONS AND LINEAR LEAST-SQUARES FILTERS

There are two different, and yet basically equivalent, methods of describing the least-squares condition of the linear transversal filter in Fig. 10.2. The principle of orthogonality, described in Eq. (10.15), represents one method. The system of *normal equations* represents the other method; interestingly enough, the system of normal equations derives its name from the corollary to the principle of orthogonality. Naturally, we may derive this system of equations in its own independent way by formulating the gradient vector $\nabla(\mathcal{E})$ in terms of the tap weights of the filter, and then solving for the tap-weight vector $\hat{\mathbf{w}}$ for which $\nabla(\mathcal{E})$ is zero. Alternatively, we may derive the system of normal equations from the principle of orthogonality. We are going to pursue the latter (indirect) approach in this section, and leave the former (direct) approach to the interested reader as Problem 7.

The principle of orthogonality in Eq. (10.15) is formulated in terms of a set of tap inputs and the minimum estimation error $e_{\min}(i)$. Setting the tap weights in Eq. (10.4) at their least-squares values, we get

$$e_{\min}(i) = d(i) - \sum_{t=0}^{M-1} \hat{w}_t^* u(i - t) \tag{10.27}$$

where on the right side we have purposely used t as the dummy summation index. Hence, substituting Eq. (10.27) in (10.15), and then rearranging terms, we get a system of M simultaneous equations:

$$\sum_{t=0}^{M-1} \hat{w}_t \sum_{i=M}^{N} u(i - k)u^*(i - t) = \sum_{i=M}^{N} u(i - k)d^*(i), \quad k = 0, \ldots, M - 1 \tag{10.28}$$

The two summations in Eq. (10.28) involving the index i represents time-averages, except for a scaling factor; they have the following interpretations:

1. The time average (over i) on the left side of Eq. (10.28) represents the *time averaged autocorrelation function* of the tap inputs in the transversal filter of Fig. 10.2. In particular, we may write

$$\phi(t, k) = \sum_{i=M}^{M} u(i - k)u^*(i - t), \qquad 0 \le t, k \le M - 1 \tag{10.29}$$

2. The time average (also over i) on the right side of Eq. (10.28) represents the *cross-correlation* between the tap inputs and the desired response. In particular, we may write

$$\theta(-k) = \sum_{i=M}^{N} u(i-k)d^*(i), \qquad 0 \le k \le M-1 \tag{10.30}$$

Accordingly, we may rewrite the system of simultaneous equations (10.28) as follows:

$$\sum_{t=0}^{M-1} \hat{w}_t \phi(t, k) = \theta(-k), \qquad k = 0, 1, \ldots, M-1 \tag{10.31}$$

The system of equations (10.31) represents *the expanded system of the normal equations* for a linear least-squares filter.

Matrix Formulation of the Normal Equations

We may recast this system of equations in matrix form by first introducing the following definitions:

1. The *M*-by-*M* *correlation matrix* of the tap inputs $u(i)$, $u(i-1)$, . . ., $u(i-M+1)$:

$$\mathbf{\Phi} = \begin{bmatrix} \phi(0,0) & \phi(1,0) & \cdots & \phi(M-1,0) \\ \phi(0,1) & \phi(1,1) & \cdots & \phi(M-1,1) \\ \vdots & \vdots & \ddots & \vdots \\ \phi(0,M-1) & \phi(1,M-1) & \cdots & \phi(M-1,M-1) \end{bmatrix} \tag{10.32}$$

2. The *M*-by-1 *cross-correlation vector* between the tap inputs $u(i)$, $u(i-1)$, . . . , $u(i-M+1)$, and the desired response $d(i)$:

$$\boldsymbol{\theta} = [\theta(0), \theta(-1), \ldots, \theta(-M+1)]^T \tag{10.33}$$

3. The *M*-by-1 tap-weight vector of the least-squares filter:

$$\hat{\mathbf{w}} = [\hat{w}_0, \hat{w}_1, \ldots, \hat{w}_{M-1}]^T \tag{10.34}$$

Hence, in terms of these matrix definitions, we may now rewrite the system of M simultaneous equations (10.31) simply as

$$\mathbf{\Phi}\hat{\mathbf{w}} = \boldsymbol{\theta} \tag{10.35}$$

Equation (9.35) is *the matrix form of the normal equations for linear least-squares filters*.

Assuming that $\mathbf{\Phi}$ is nonsingular and therefore the inverse matrix $\mathbf{\Phi}^{-1}$ exists, we may solve Eq. (10.35) for the tap-weight vector of the linear least-squares filter:

$$\hat{\mathbf{w}} = \mathbf{\Phi}^{-1}\boldsymbol{\theta} \tag{10.36}$$

The condition for the existence of the inverse matrix $\mathbf{\Phi}^{-1}$ is discussed in Section 10.6.

Equation (10.36) is a very important result. In particular, it is the linear least-squares counterpart to the solution of the matrix form of the Wiener-Hopf equations (5.36). Basically, Eq. (10.36) states that the tap-weight vector $\hat{\mathbf{w}}$ of a linear least-squares filter is uniquely defined by the product of the inverse of the time-averaged correlation matrix $\mathbf{\Phi}$ of the tap inputs of the filter and the time-averaged cross-correlation vector $\boldsymbol{\theta}$ between the tap inputs and the desired response. Indeed, this equation is fundamental to the development of all recursive formulations of the linear least-squares filter, as pursued in subsequent chapters of the book.

Minimum Sum of Error Squares

Equation (10.26), derived in the preceding section, defines the minimum sum of error squares $\mathscr{E}_{\min}$. We now complete the evaluation of $\mathscr{E}_{\min}$, expressed as the difference between the energy $\mathscr{E}_d$ of the desired response and the energy $\mathscr{E}_{\text{est}}$ of its estimate. Usually, $\mathscr{E}_d$ is determined from the time series representing the desired response. To evaluate $\mathscr{E}_{\text{est}}$, we write

$$\begin{aligned}\mathscr{E}_{est} &= \sum_{i=M}^{N} |\hat{d}(i|\mathscr{U}_i)|^2 \\ &= \sum_{i=M}^{N} \sum_{t=0}^{M-1} \sum_{k=0}^{M-1} \hat{w}_t \hat{w}_k^* u(i-k) u^*(i-t) \\ &= \sum_{t=0}^{M-1} \sum_{k=0}^{M-1} \hat{w}_t \hat{w}_k^* \sum_{i=M}^{N} u(i-k) u^*(i-t)\end{aligned} \tag{10.37}$$

where, in the second line, we have made use of Eq. (10.18). The inner summation over time i in the final line of Eq. (10.37) represents the time-averaged autocorrelation function $\phi(t, k)$ [see Eq. (10.29)]. Hence, we may rewrite Eq. (10.37) as

$$\begin{aligned}\mathscr{E}_{\text{est}} &= \sum_{t=0}^{M-1} \sum_{k=0}^{M-1} \hat{w}_k^* \phi(t, k) \hat{w}_t \\ &= \hat{\mathbf{w}}^H \mathbf{\Phi} \hat{\mathbf{w}}\end{aligned} \tag{10.38}$$

where $\hat{\mathbf{w}}$ is the least-squares tap-weight vector and $\mathbf{\Phi}$ is the time-averaged correlation matrix of the tap inputs. We may further simplify the formula for $\mathscr{E}_{\text{est}}$ by noting that from the normal equations (10.35), the matrix product $\mathbf{\Phi}\hat{\mathbf{w}}$ equals the cross-correlation vector $\boldsymbol{\theta}$. Accordingly,

$$\begin{aligned}\mathscr{E}_{\text{est}} &= \hat{\mathbf{w}}^H \boldsymbol{\theta} \\ &= \boldsymbol{\theta}^H \hat{\mathbf{w}}\end{aligned} \tag{10.39}$$

Finally, substituting Eq. (10.39) in (10.26), and then using Eq. (10.36) for $\hat{\mathbf{w}}$, we get

$$\begin{aligned}\mathscr{E}_{\min} &= \mathscr{E}_d - \boldsymbol{\theta}^H \hat{\mathbf{w}} \\ &= \mathscr{E}_d - \boldsymbol{\theta}^H \mathbf{\Phi}^{-1} \boldsymbol{\theta}\end{aligned} \tag{10.40}$$

Equations (10.40) is the desired formula for the minimum sum of error squares, expressed in terms of these known quantities: the energy $\mathscr{E}_d$ of the desired response, the correlation matrix $\mathbf{\Phi}$ of the tap inputs, and the cross-correlation vector $\boldsymbol{\theta}$ between the tap inputs and the desired response.

10.6 PROPERTIES OF THE CORRELATION MATRIX $\mathbf{\Phi}$

The correlation matrix $\mathbf{\Phi}$ of the tap inputs is shown in its expanded form in Eq. (10.32), with the element $\phi(t, k)$ defined in Eq. (10.29). The index k in $\phi(t, k)$ refers to the row number in the matrix $\mathbf{\Phi}$, and t refers to the column number. Let the M-by-1 tap-input vector $\mathbf{u}(i)$ be defined by

$$\mathbf{u}(i) = [u(i), u(i-1), \ldots, u(i-M+1)]^T \tag{10.41}$$

Hence, we may use Eqs. (10.29) and (10.41) to redefine the correlation matrix $\mathbf{\Phi}$ as the time average of the outer product $\mathbf{u}(i)\mathbf{u}^H(i)$ over i as follows:

$$\mathbf{\Phi} = \sum_{i=M}^{N} \mathbf{u}(i)\mathbf{u}^H(i) \tag{10.42}$$

We are now ready to establish the properties of the correltion matrix:

Property 1. *The correlation matrix $\mathbf{\Phi}$ is Hermitian; that is*

$$\mathbf{\Phi}^H = \mathbf{\Phi}$$

This property follows directly from Eq. (10.42).

Property 2. *The correlation matrix $\mathbf{\Phi}$ is nonnegative definite; that is,*

$$\mathbf{x}^H\mathbf{\Phi}\mathbf{x} \geq 0$$

for any M*-by-1 vector* $\mathbf{x}$.

Using the definition of Eq. (10.42), we may write

$$\begin{aligned}
\mathbf{x}^H\mathbf{\Phi}\mathbf{x} &= \sum_{i=M}^{N} \mathbf{x}^H\mathbf{u}(i)\mathbf{u}^H(i)\mathbf{x} \\
&= \sum_{i=M}^{N} [\mathbf{x}^H\mathbf{u}(i)][\mathbf{x}^H\mathbf{u}(i)]^{\mathrm{H}} \\
&= \sum_{i=M}^{\mathrm{N}} |\mathbf{x}^H\mathbf{u}(i)|^2 \geq 0
\end{aligned}$$

which is the desired result. The fact that the correlation matrix $\mathbf{\Phi}$ is nonnegative definite means that its determinant and all principal minors are nonnegative. When the above condition is satisfied with the inequality sign, the determinant of $\mathbf{\Phi}$ and its principal minors are likewise nonzero. In the latter case, $\mathbf{\Phi}$ is nonsingular, and the inverse $\mathbf{\Phi}^{-1}$ exists.

Property 3. *The eigenvalues of the correlation matrix $\mathbf{\Phi}$ are all real and nonnegative.*

The real requirement on the eigenvalues of $\mathbf{\Phi}$ follows from Property 1. The fact that all these eigenvalues are also nonnegative follows from Property 2.

Property 4. *The correlation matrix is the product of two rectangular Toeplitz matrices that are the Hermitian transpose of each other.*

The correlation matrix $\mathbf{\Phi}$ is, in general, non-Toeplitz, which is clearly seen by examining the expanded form of the correlation matrix given in Eq. (10.32). The elements on the main diagonal, $\phi(0, 0), \phi(1, 1), \ldots, \phi(M-1, M-1)$, have different values and similarly for any secondary diagonal above or below the main diagonal. However, the matrix $\mathbf{\Phi}$ has a special structure in the sense that it is the product of two Toeplitz rectangular matrices. To prove this property, we first use Eq. (10.42) to express the matrix $\mathbf{\Phi}$ as follows:

$$\mathbf{\Phi} = [\mathbf{u}(M), \mathbf{u}(M+1), \ldots, \mathbf{u}(N)] \begin{bmatrix} \mathbf{u}^H(M) \\ \mathbf{u}^H(M-1) \\ \mathbf{u}^H(M-1) \\ \vdots \\ \mathbf{u}^H(N) \end{bmatrix} \tag{10.43}$$

Next, for convenience of presentation, we introduce a *data matrix* $\mathbf{A}$, defined by

$$\begin{aligned} \mathbf{A}^H &= [\mathbf{u}(M), \quad \mathbf{u}(M+1), \quad \ldots, \mathbf{u}(N)] \\ &= \begin{bmatrix} u(M) & u(M+1) & \cdots & u(N) \\ u(M-1) & u(M) & \cdots & u(N-1) \\ \vdots & \vdots & & \vdots \\ u(1) & u(2) & \cdots & u(N-M+1) \end{bmatrix} \end{aligned} \tag{10.44}$$

The expanded matrix on the right side of Eq. (10.44) is recognized to be the matrix of input data for the covariance method of data windowing (see point 1 of Section 10.2). Thus, using the definition of Eq. (10.44), we may rewrite Eq. (10.43) in the compact form

$$\mathbf{\Phi} = \mathbf{A}^H \mathbf{A} \tag{10.45}$$

From the expanded form of the matrix given in the second line of Eq. (10.44), we see that $\mathbf{A}^H$ consists of an M-by-$(N - M + 1)$ *rectangular Toeplitz matrix*. The data matrix $\mathbf{A}$ itself is likewise an $(N - M + 1)$-by-M rectangular Toeplitz matrix. According to Eq. (10.45), therefore, the correlation matrix $\mathbf{\Phi}$ is the product of two rectangular Toeplitz matrices that are the Hermitian transpose of each other: this completes the proof of Property 4.

Reformulation of the Normal Equations in Terms of Data Matrices

The system of normal equations for a least-squares transversal filter is given by Eq. (10.35) in terms of the correlation matrix $\mathbf{\Phi}$ and the cross-correlation vector $\boldsymbol{\theta}$. We may reformulate this equation in terms of data matrices by using Eq. (10.45) for the correlation matrix $\mathbf{\Phi}$ of the tap inputs, and a corresponding relation for the cross-correlation vector $\boldsymbol{\theta}$ between the tap inputs and the desired response. To do this, we introduce a *desired data vector* $\mathbf{d}$, consisting of the *desired response* $d(i)$ for values of i in the interval $[M, N]$ as follows:

$$\mathbf{d}^H = [d(M), d(M+1), \ldots, d(N)] \tag{10.46}$$

Note that we have purposely used Hermitian transposition rather than ordinary transposition in the definition of vector $\mathbf{d}$. With the definitions of Eqs. (10.44) and (10.46) at hand, we may now use Eqs. (10.30) and (10.33) to express the cross-correlation vector $\boldsymbol{\theta}$ as

$$\boldsymbol{\theta} = \mathbf{A}^H\mathbf{d} \tag{10.47}$$

Furthermore, we may use Eqs. (10.45) and (10.47) in (10.35), and so express the system of normal equations in terms of the data matrix $\mathbf{A}$ and the desired data vector $\mathbf{d}$ as

$$\mathbf{A}^H\mathbf{A}\mathbf{w} = \mathbf{A}^H\mathbf{d}$$

Hence, the system of equations used in the minimization of the cost function $\mathscr{E}$ may be represented by $\mathbf{A}\mathbf{w} = \mathbf{d}$. Furthermore, assuming that the inverse matrix $(\mathbf{A}^H\mathbf{A})^{-1}$ exists, we may solve this system of equations by expressing the tap-weight vector $\hat{\mathbf{w}}$ as

$$\hat{\mathbf{w}} = (\mathbf{A}^H\mathbf{A})^{-1}\mathbf{A}^H\mathbf{d} \tag{10.48}$$

We may complete the reformulation of our results for the linear least-squares problem in terms of the data matrices $\mathbf{A}$ and $\mathbf{d}$ by using (1) the definitions of Eqs. (10.45) and (10.47) in (10.40), and (2) the definitions of Eq. (10.46) in (10.23). By so doing, we may rewrite the formula for the minimum sum of error squares as

$$\mathscr{E}_{\min} = \mathbf{d}^H\mathbf{d} - \mathbf{d}^H\mathbf{A}(\mathbf{A}^H\mathbf{A})^{-1}\mathbf{A}^H\mathbf{d} \tag{10.49}$$

Although this formula looks somewhat cumbersome, its nice feature is that it is expressed explicitly in terms of the data matrix $\mathbf{A}$ and the desired data vector $\mathbf{d}$.

Projection Operator

Equation (10.48) defines the least-squares tap-weight vector $\hat{\mathbf{w}}$ in terms of the data matrix $\mathbf{A}$ and the desired data vector $\mathbf{d}$. The least-squares estimate of $\mathbf{d}$ is therefore given by

$$\begin{aligned}\hat{\mathbf{d}} &= \mathbf{A}\hat{\mathbf{w}} \\ &= \mathbf{A}(\mathbf{A}^H\mathbf{A})^{-1}\mathbf{A}^H\mathbf{d}\end{aligned} \tag{10.50}$$

Accordingly, we may view the multiple matrix product $\mathbf{A}(\mathbf{A}^H\mathbf{A})^{-1}\mathbf{A}^H$ as a *projection operator* onto the linear space spanned by the columns of the data matrix $\mathbf{A}$. The matrix difference

$$\mathbf{I} - \mathbf{A}(\mathbf{A}^H\mathbf{A})^{-1}\mathbf{A}^H$$

is the *orthogonal complement projector*. Note that both the projection operator and its complement are uniquely determined by the data matrix $\mathbf{A}$. The projection operator, applied to the desired data vector **d**, yields the corresponding estimate $\hat{\mathbf{d}}$. On the other hand, the orthogonal complement projector, applied to the desired data vector **d**, yields the estimation error vector $\mathbf{e}_{\min} = \mathbf{d} - \hat{\mathbf{d}}$. Figure 10.3 illustrates the functions of the projection operator **P** and the orthogonal complement projector $\mathbf{I} - \mathbf{P}$ as described herein.

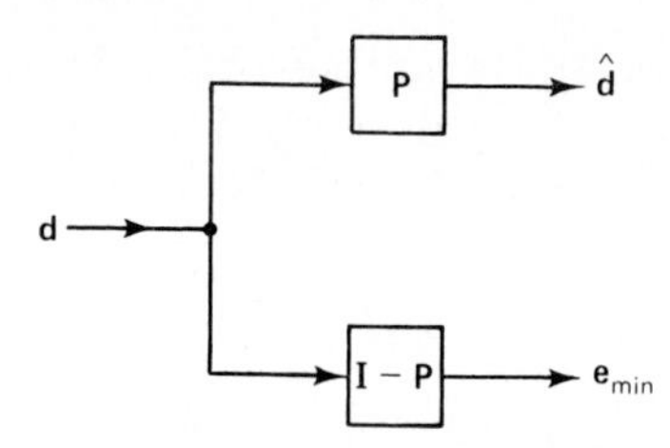

Figure 10.3 Projection operator **P** and orthogonal complement projector $\mathbf{I} - \mathbf{P}$.

Example

Consider the example of a linear least-squares filter with two taps (i.e., $M = 2$) and a *real-valued* input time series consisting of four samples (i.e., $N = 4$); hence, $N - M + 1 = 3$. The input data matrix **A** and the desired data vector **d** have the following values

$$\mathbf{A} = \begin{bmatrix} u(2) & u(1) \\ u(3) & u(2) \\ u(4) & u(3) \end{bmatrix} = \begin{bmatrix} 2 & 3 \\ 1 & 2 \\ -1 & 1 \end{bmatrix}$$

$$\mathbf{d} = \begin{bmatrix} d(2) \\ d(3) \\ d(4) \end{bmatrix} = \begin{bmatrix} 2 \\ 1 \\ 1/34 \end{bmatrix}$$

The purpose of this example is to evaluate the projection operator and the orthogonal complement projector, and use them to illustrate the principle of orthogonality.

The use of Eq. (10.50), reformulated for real data, yields the value of the projection operator **P** as

$$\mathbf{P} = \mathbf{A}(\mathbf{A}^T\mathbf{A})^1\mathbf{A}^T = \frac{1}{35}\begin{bmatrix} 26 & 15 & -2 \\ 15 & 10 & 5 \\ -3 & 5 & 34 \end{bmatrix}$$

The corresponding value of the orthogonal complement projector is

$$\mathbf{I} - \mathbf{P} = \frac{1}{35}\begin{bmatrix} 9 & -15 & 3 \\ -15 & 25 & -5 \\ -3 & -5 & 1 \end{bmatrix}$$

Accordingly, the estimate of the desired data vector and the estimation error vector have the following values, respectively:

$$\hat{\mathbf{d}} = \mathbf{Pd}$$

$$= \begin{bmatrix} 1.91 \\ 1.15 \\ 0 \end{bmatrix}$$

$$\mathbf{e}_{\min} = (\mathbf{I} - \mathbf{P})\mathbf{d}$$

$$= \begin{bmatrix} 0.09 \\ -0.15 \\ 0.03 \end{bmatrix}$$

Figure 10.4 depicts 3-dimensional geometric representations of the vectors $\hat{\mathbf{d}}$ and $\mathbf{e}_{\min}$. This figure clearly shows that these two vectors are *normal* (i.e., *perpendicular*) to each other in accordance with the corollary to the principle of orthogonality; hence the terminology "normal" equations. This condition is the geometric portrayal of the fact that in a linear least-squares filter the inner product $\mathbf{e}_{\min}^{H}\hat{\mathbf{d}}$ is zero. Figure 10.4 also depicts the desired data vector $\mathbf{d}$ as the "vector sum" of the estimate $\hat{\mathbf{d}}$ and the error $\mathbf{e}_{\min}$.

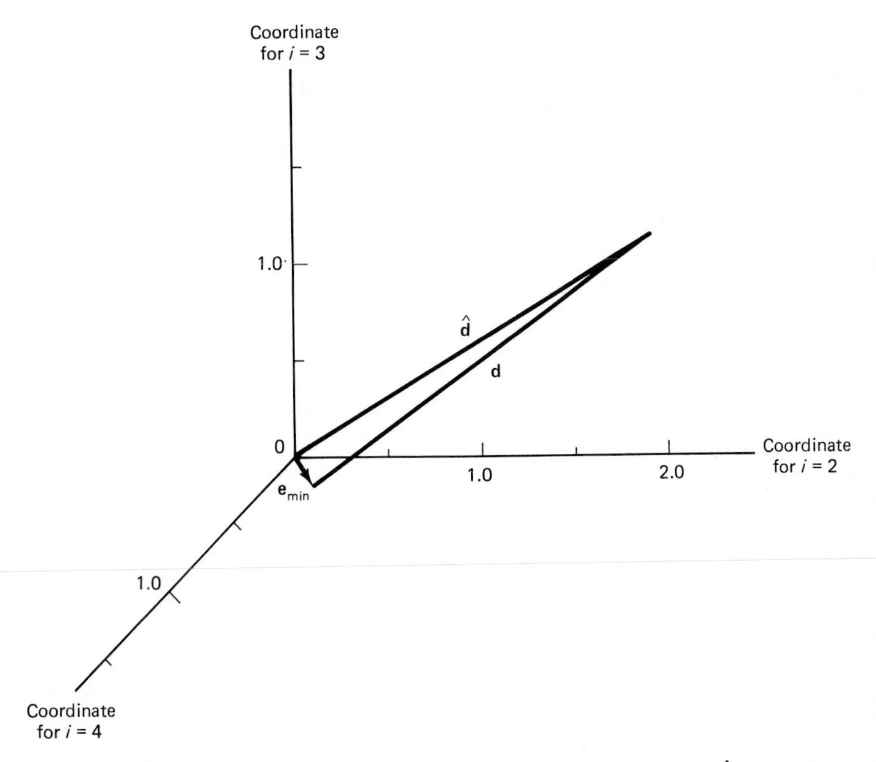

Figure 10.4 Three-dimensional geometric interpretations of vectors $\mathbf{d}$, $\hat{\mathbf{d}}$, and $\mathbf{e}_{\min}$.

10.7 UNIQUENESS THEOREM

The linear least-squares problem of minimizing the sum of error squares, $\mathscr{E}(n)$, always has a solution. That is, for given values of the data matrix $\mathbf{A}$ and the desired data vector $\mathbf{d}$, we can always find a vector $\hat{\mathbf{w}}$ that satisfies the normal equations (10.35). It is therefore important that we know if and when the solution is *unique*. This requirement is covered by the following *uniqueness theorem* (Stewart, 1973):

> The least-squares estimate $\hat{\mathbf{w}}$ is unique if and only if the nullity of the data matrix $\mathbf{A}$ equals zero.

Let $\mathbf{A}$ be a K-by-M matrix; in the case of the data matrix $\mathbf{A}$ defined in Eq. (10.44), we have $K = N - M + 1$. We define the *null space* of matrix $\mathbf{A}$, written as $\mathscr{N}(\mathbf{A})$, as the space of all vectors $\mathbf{x}$ such that $\mathbf{Ax} = \mathbf{0}$. We define the *nullity* of matrix $\mathbf{A}$, written as null $(\mathbf{A})$, as the dimension of the null space $\mathscr{N}(\mathbf{A})$. In general, we find that null $(\mathbf{A}) \neq \text{null}(\mathbf{A}^H)$.

In light of the uniqueness theorem, which is intuitively satisfying, we may expect a unique solution to the linear least-squares problem *only* when the data matrix $\mathbf{A}$ has *linearly independent columns*. This implies that the matrix $\mathbf{A}$ has at least as many rows as columns; that is, $(N - M + 1) \geq M$. This latter condition means that the system of equations represented by $\mathbf{A}\hat{\mathbf{w}} = \mathbf{d}$ used in the minimization is *overdetermined* in that it has more equations than unknowns. Thus, provided that the data matrix $\mathbf{A}$ has linearly independent columns, the M-by-M matrix $\mathbf{A}^H\mathbf{A}$ is *nonsingular,* and the least-squares estimate has the unique value given in Eq. (10.48).

When, however, the matrix $\mathbf{A}$ has *linearly dependent columns,* the nullity of the matrix $\mathbf{A}$ is nonzero, and the result is that an infinite number of solutions can be found for minimizing the sum of error squares. In such a situation, the linear least-squares problem becomes quite involved in that we now have the new problem of deciding which particular solution to adopt. We defer discussion of this issue until Chapter 11. In the meantime, we assume that the data matrix $\mathbf{A}$ has linearly independent columns so that the least-squares estimate $\hat{\mathbf{w}}$ has the unique value defined by Eq. (10.48).

10.8 PROPERTIES OF LEAST-SQUARES ESTIMATES

The method of least squares has a strong intuitive feel that is reinforced by several outstanding properties of the method. These properties are described next (Miller, 1974; Goodwin and Payne, 1977):

Property 1. *The least-squares estimate $\hat{\mathbf{w}}$ is unbiased, provided that the measurement error process $\{e_o(i)\}$ has zero mean.*

From the multiple linear regression model of Fig. 10.1, we have [using the definitions of Eqs. (10.44) and (10.46)]

$$\mathbf{d} = \mathbf{A}\mathbf{w}_o + \boldsymbol{\epsilon}_o \tag{10.52}$$

Hence, substituting Eq. (10.52) into (10.48), we may express the least-squares estimate $\hat{\mathbf{w}}$ as

$$\begin{aligned}\hat{\mathbf{w}} &= (\mathbf{A}^H\mathbf{A})^{-1}\mathbf{A}^H\mathbf{A}\mathbf{w}_o + (\mathbf{A}^H\mathbf{A})^{-1}\mathbf{A}^H\boldsymbol{\epsilon}_o \\ &= \mathbf{w}_o + (\mathbf{A}^H\mathbf{A})^{-1}\mathbf{A}^H\boldsymbol{\epsilon}_o\end{aligned} \tag{10.53}$$

The matrix product $(\mathbf{A}^H\mathbf{A})^{-1}\mathbf{A}^H$ is a known quantity, since the data matrix $\mathbf{A}$ is completely defined by the set of given observations $u(1), u(2), \ldots, u(N)$; see Eq. (10.44). Hence, if the measurement error process $\{e_o(i)\}$ or, equivalently, the measurement error vector $\boldsymbol{\epsilon}_o$ has zero mean, we find by taking the expectation of both sides of Eq. (10.53) that the estimate $\hat{\mathbf{w}}$ is *unbiased;* that is,

$$E[\hat{\mathbf{w}}] = \mathbf{w}_o \tag{10.54}$$

Property 2. *When the measurement error process $\{e_o(i)\}$ is white with zero mean and variance σ^2, the covariance matrix of the least-squares estimate $\hat{\mathbf{w}}$ equals $\sigma^2\mathbf{\Phi}^{-1}$.*

Using the relation of Eq. (10.53), we find that the covariance matrix of the least-squares estimate $\hat{\mathbf{w}}$ equals

$$\begin{aligned}\text{cov}[\hat{\mathbf{w}}] &= E[(\hat{\mathbf{w}} - \mathbf{w}_o)(\hat{\mathbf{w}} - \mathbf{w}_o)^H] \\ &= E[(\mathbf{A}^H\mathbf{A})^{-1}\mathbf{A}^H\boldsymbol{\epsilon}_o\boldsymbol{\epsilon}_o^H\mathbf{A}(\mathbf{A}^H\mathbf{A})^{-1}] \\ &= (\mathbf{A}^H\mathbf{A})^{-1}\mathbf{A}^H E[\boldsymbol{\epsilon}_o\boldsymbol{\epsilon}_o^H]\mathbf{A}(\mathbf{A}^H\mathbf{A})^{-1}\end{aligned} \tag{10.55}$$

With the measurement error process $\{e_o(i)\}$ assumed to be white with zero mean and variance σ^2, we have

$$E[\boldsymbol{\epsilon}_o\boldsymbol{\epsilon}_o^H] = \sigma^2\mathbf{I} \tag{10.56}$$

where $\mathbf{I}$ is the identity matrix. Hence, Eq. (10.55) simplifies as follows:

$$\begin{aligned}\text{cov}[\hat{\mathbf{w}}] &= \sigma^2(\mathbf{A}^H\mathbf{A})^{-1}\mathbf{A}^H\mathbf{A}(\mathbf{A}^H\mathbf{A})^{-1} \\ &= \sigma^2(\mathbf{A}^H\mathbf{A})^{-1} \\ &= \sigma^2\mathbf{\Phi}^{-1}\end{aligned} \tag{10.57}$$

which is the desired result.

Property 3. *When the measurement error process $\{e_o(i)\}$ is white with zero mean, the least-squares estimate $\hat{\mathbf{w}}$ is the best linear unbiased estimate.*

Consider any linear unbiased estimator $\tilde{\mathbf{w}}$ that is defined by

$$\tilde{\mathbf{w}} = \mathbf{B}\mathbf{d} \tag{10.58}$$

where $\mathbf{B}$ is an M-by-$(N - M + 1)$ matrix. Substituting Eq. (10.52) into (10.58), we get

$$\tilde{\mathbf{w}} = \mathbf{B}\mathbf{A}\mathbf{w}_o + \mathbf{B}\boldsymbol{\epsilon}_o \tag{10.59}$$

With the measurement error vector $\boldsymbol{\epsilon}_o$ assumed to have zero mean, we find that the expected value of $\tilde{\mathbf{w}}$ equals

$$E[\tilde{\mathbf{w}}] = \mathbf{BAw}_o$$

For the linear estimator $\tilde{\mathbf{w}}$ to be unbiased, we therefore require that the matrix $\mathbf{B}$ satisfy the condition

$$\mathbf{BA} = \mathbf{I}$$

Accordingly, we may rewrite Eq. (10.59) as follows:

$$\tilde{\mathbf{w}} = \mathbf{w}_o + \mathbf{B}\boldsymbol{\epsilon}_o$$

The covariance matrix of $\tilde{\mathbf{w}}$ equals

$$\begin{aligned}\text{cov}[\tilde{\mathbf{w}}] &= E[(\tilde{\mathbf{w}} - \mathbf{w}_o)(\tilde{\mathbf{w}} - \mathbf{w}_o)^H] \\ &= E[\mathbf{B}\boldsymbol{\epsilon}_o\boldsymbol{\epsilon}_o^H\mathbf{B}^H] \\ &= \sigma^2\mathbf{BB}^H \end{aligned} \tag{10.60}$$

Here, we have made use of Eq. (10.56), which describes the assumption that the elements of the measurement error vector $\boldsymbol{\epsilon}_o$ are uncorrelated and have the common variance σ^2; that is, the measurement error process $\{e_o(i)\}$ is white. We next define a new matrix $\boldsymbol{\Psi}$ in terms of $\mathbf{B}$ as

$$\boldsymbol{\Psi} = \mathbf{B} - (\mathbf{A}^H\mathbf{A})^{-1}\mathbf{A}^H \tag{10.61}$$

Now we form the matrix product $\boldsymbol{\Psi}\boldsymbol{\Psi}^H$ as

$$\begin{aligned}\boldsymbol{\Psi}\boldsymbol{\Psi}^H &= [\mathbf{B} - (\mathbf{A}^H\mathbf{A})^{-1}\mathbf{A}^H][\mathbf{B}^H - \mathbf{A}(\mathbf{A}^H\mathbf{A})^{-1}] \\ &= \mathbf{BB}^H - \mathbf{BA}(\mathbf{A}^H\mathbf{A})^{-1} - (\mathbf{A}^H\mathbf{A})^{-1}\mathbf{A}^H\mathbf{B}^H + (\mathbf{A}^H\mathbf{A})^{-1} \\ &= \mathbf{BB}^H - (\mathbf{A}^H\mathbf{A})^{-1}\end{aligned}$$

Since we always have $\boldsymbol{\Psi}\boldsymbol{\Psi}^H \geq 0$, it follows that

$$\mathbf{BB}^H \geq (\mathbf{A}^H\mathbf{A})^{-1}$$

Equivalently, we may write

$$\sigma^2\mathbf{BB}^H \geq \sigma^2(\mathbf{A}^H\mathbf{A})^{-1} \tag{10.62}$$

The term $\sigma^2\mathbf{BB}^H$ equals the covariance matrix of the linear estimate $\tilde{\mathbf{w}}$, as in Eq. (10.60). From Property 2, we also know that the term $\sigma^2(\mathbf{A}^H\mathbf{A})^{-1}$ equals the covariance matrix of the least-squares estimate $\hat{\mathbf{w}}$. Thus, Eq. (10.62) shows that within the class of linear unbiased estimates the least-squares estimate $\hat{\mathbf{w}}$ is the "best" estimate of the unknown parameter vector $\mathbf{w}_o$ of the multiple linear regression model. Accordingly, when the measurement error process $\{e_o\}$ contained in this model is white with zero mean, the least-squares estimate $\hat{\mathbf{w}}$ is the *b*est *l*inear *u*nbiased *e*stimate (BLUE).

Thus far we have not made any assumption about the statistical distribution of the measurement error process $\{e_o(i)\}$ other than that it is a zero-mean white-noise process. By making the further assumption that the process $\{e_o(i)\}$ is *Gaussian* distributed, we obtain a stronger result on the optimality of the linear least-squares estimate. as discussed next.

Property 4. *When the measurement error process $\{e_o(i)\}$ is white, Gaussian, and has a zero mean, the least-squares estimate $\hat{\mathbf{w}}$ achieves the Cramér–Rao lower bound for unbiased estimates.*

Let $f_{\mathbf{E}}(\boldsymbol{\epsilon}_o)$ denote the joint probability density function of the measurement error vector $\boldsymbol{\epsilon}_o$. Let $\hat{\mathbf{w}}$ denote any unbiased estimate of the unknown parameter vector $\hat{\mathbf{w}}_o$ of the multiple linear regression model. Then the covariance matrix of $\hat{\mathbf{w}}$ satisfies the inequality

$$\text{cov}[\hat{\mathbf{w}}] \geq \mathbf{J}^{-1} \tag{10.63}$$

where

$$\text{cov}[\hat{\mathbf{w}}] = E[\hat{\mathbf{w}} - \hat{\mathbf{w}}_o)(\hat{\mathbf{w}} - \hat{\mathbf{w}}_o)^H] \tag{10.64}$$

The matrix $\mathbf{J}$ is called *Fisher's information matrix;* it is defined by[2]

$$\mathbf{J} = E\left[\left(\frac{\partial l}{\partial \mathbf{w}_o^*}\right)\left(\frac{\partial l}{\partial \mathbf{w}_o^T}\right)\right] \tag{10.65}$$

where l is the *log-likelihood function,* that is, the natural logorithmn of the joint probability density of $\boldsymbol{\epsilon}_o$, as shown by

$$l = \ln f_{\mathbf{E}}(\boldsymbol{\epsilon}_o) \tag{10.66}$$

Since the measurement error process $\{e_o(n)\}$ is white, the elements of the vector $\boldsymbol{\epsilon}_o$ are uncorrelated. Furthermore, since the process $\{e_o(n)\}$ is Gaussian, the elements of $\boldsymbol{\epsilon}_o$ are statistically independent. With $e_o(i)$ assumed to be complex with a mean of zero and variance σ^2, we have (see Section 2.11)

$$f_{\mathbf{E}}(\boldsymbol{\epsilon}_o) = \frac{1}{(\pi\sigma^2)^{(N-M+1)}} \exp\left[-\frac{1}{\sigma^2}\sum_{i=M}^{N} |e_o(i)|^2\right] \tag{10.67}$$

The log-likelihood function is therefore

$$\begin{aligned} l &= F - \frac{1}{\sigma^2}\sum_{i=M}^{N} |e_o(i)|^2 \\ &= F - \frac{1}{\sigma^2}\boldsymbol{\epsilon}_o^H \boldsymbol{\epsilon}_o \end{aligned} \tag{10.68}$$

where F is a constant defined by

$$F = -(N - M + 1)\ln(\pi\sigma^2)$$

From Eq. (10.52), we have

$$\boldsymbol{\epsilon}_o = \mathbf{d} - \mathbf{A}\mathbf{w}_o$$

Using this relation in Eq. (10.68), we may rewrite l in terms of $\mathbf{w}_o$ as

$$l = F - \frac{1}{\sigma^2}\mathbf{d}^H\mathbf{d} + \frac{1}{\sigma^2}\mathbf{w}_o^H\mathbf{A}^H\mathbf{d} + \frac{1}{\sigma^2}\mathbf{d}^H\mathbf{A}\mathbf{w}_o - \frac{1}{\sigma^2}\mathbf{w}_o^H\mathbf{A}^H\mathbf{A}\mathbf{w}_o \tag{10.69}$$

[2] Fisher's information matrix is discussed in Appendix D for the case of real parameters.

Differentiating the real-valued log-likelihood function l with respect to the complex-valued unknown parameter vector $\mathbf{w}_o$ in accordance with the notation described in Appendix B, we get

$$\begin{aligned} \frac{\partial l}{\partial \mathbf{w}_o^*} &= \frac{1}{\sigma^2}\mathbf{A}^H(\mathbf{d} - \mathbf{A}\mathbf{w}_o) \\ &= \frac{1}{\sigma^2}\mathbf{A}^H\boldsymbol{\epsilon}_o \end{aligned} \tag{10.70}$$

Thus, substituting Eq. (10.70) into (10.65) yields Fisher's information matrix for the problem at hand as

$$\begin{aligned} \mathbf{J} &= \frac{1}{\sigma^4}E[\mathbf{A}^H\boldsymbol{\epsilon}_o\boldsymbol{\epsilon}_o^H\mathbf{A}] \\ &= \frac{1}{\sigma^4}\mathbf{A}^H E[\boldsymbol{\epsilon}_o\boldsymbol{\epsilon}_o^H]\mathbf{A} \\ &= \frac{1}{\sigma^2}\mathbf{A}^H\mathbf{A} \\ &= \frac{1}{\sigma^2}\boldsymbol{\Phi} \end{aligned} \tag{10.71}$$

where, in the third line, we have made use of Eq. (10.56) describing the assumption that the measurement error process $\{e_o(i)\}$ is white. Accordingly, the use of Eq. (10.63) shows that the covariance matrix of the unbiased estimate $\tilde{\mathbf{w}}$ satisfies the inequality

$$\text{cov}[\tilde{\mathbf{w}}] \geq \sigma^2\boldsymbol{\Phi}^{-1} \tag{10.72}$$

However, from Property 2, we know that $\sigma^2\boldsymbol{\Phi}^{-1}$ equals the covariance matrix of the least-squares estimate $\hat{\mathbf{w}}$. Accordingly, $\hat{\mathbf{w}}$ achieves the Cramér-Rao lower bound. Using Property 1, we conclude therefore that when the measurement error process $\{e_o(i)\}$ is a zero-mean white Gaussian noise process, the least-squares estimate $\hat{\mathbf{w}}$ is a *m*inimum *v*ariance *u*nbiased *e*stimate (MVUE).

10.9 PARAMETRIC SPECTRUM ESTIMATION

The method of least squares is particularly well suited for solving *parametric spectrum estimation* problems. In the next two sections of the chapter, we study this important application of the method of least squares.

In Section 10.10 we consider the case of *autoregressive* (AR) *spectrum estimation,* assuming the use of an AR model of *known order*. From the discussion of linear prediction presented in Chapter 6, we know that there is a one-to-one correspondence between the coefficients of a prediction-error filter and those of an AR model of similar order. The purpose of Section 10.10 is to build on this idea. The procedure described herein provides an alternative to the Burg algorithm considered in Section 6.15.

Then in Section 10.11 we consider the case of *minimum variance distortionless response* (MVDR) *spectrum estimation*. In this second case, we have a constrained optimization problem to solve. In particular, the requirement is to minimize the average power at the output of the spectrum estimator, subject to the constraint that the estimator has a distortionless response at some specified frequency. This has the effect of minimizing the average output power due to other frequencies. The estimator so obtained is the time-averaged counterpart to that of Section 5.7.

The AR spectrum and MVDR spectrum are commonly referred to as *superresolution* or *high-resolution spectra* in the sense that they exhibit sub-Rayleigh resolution. More will be said on this issue in Chapter 11. In that chapter we also discuss the connection between these two and other superresolution spectra of interest.

10.10 AR SPECTRUM ESTIMATION

The specific estimation procedure described herein relies on the combined use of *forward and backward linear prediction* (FBLP).[3] Since the method of least squares is basically a *block* estimation procedure, we may therefore view the FBLP algorithm as an alternative to the Burg algorithm (described in Section 6.15) for solving AR modeling problems. There are, however, three basic differences between the FBLP and the Burg algorithms:

1. The FBLP algorithm estimates the coefficients of a *transversal-equivalent model* for the input data, whereas the Burg algorithm estimates the reflection coefficients of a *lattice-equivalent model*. If the requirement is to fit an AR model into the input data, it is clear that the FBLP algorithm has a computational advantage over the Burg algorithm in the sense that it provides the desired answer directly.
2. In the method of least squares, and therefore the FBLP algorithm, no assumptions are made concerning the statistics of the input data. The Burg algorithm, on the other hand, exploits the decoupling property of a multistage lattice predictor, which in turn, assumes wide-sense stationarity of the input data. Accordingly, the FBLP algorithm does not suffer from some of the anomalies that are known to arise in the application of the Burg algorithm.[4]
3. The Burg algorithm yields a minimum-phase solution in the sense that the reflection coefficients of the equivalent lattice predictor have a magnitude less than or equal to unity. The FBLP algorithm, on the other hand, does *not* guarantee such a solution. In spectrum estimation, however, the lack of a minimum-phase solution is of no particular concern.

[3] The first application of the FBLP method to the design of a linear predictor that has a transversal filter structure, in accordance with the method of least squares, was developed independently by Ulrych and Clayton (1976) and Nuttall (1976).

[4] For example, when the Burg algorithm is used to estimate the frequency of an unknown sine wave in additive noise, under certain conditions a phenomenon commonly referred to as *spectral line splitting* may occur. This phenomenon refers to the occurrence of two (or more) closely spaced spectral peaks where there should only be a single peak; for a discussion of spectral line splitting, see Marple (1987), Kay (1988), and Haykin (1989); the original reference is Fougere et al. (1976). This anomaly, however, does not arise in the application of the FBLP algorithm.

FBLP Algorithm

Consider then the *forward linear predictor*, shown in Fig. 10.5(a). The tap weights of the predictor are denoted by $\hat{w}_1, \hat{w}_2, \ldots, \hat{w}_M$ and the tap inputs by $u(i-1)$, $u(i-2), \ldots, u(i-M)$, respectively. The forward prediction error, denoted by $f_M(i)$, equals

$$f_M(i) = u(i) - \sum_{k=1}^{M} \hat{w}_k^* u(i-k) \tag{10.73}$$

The first term, $u(i)$, represents the desired response. The convolution sum, constituting the second term, represents the predictor output; it consists of the sum of inner products. Using matrix notation, we may also express the forward prediction error as

$$f_M(i) = u(i) - \mathbf{w}^H \mathbf{u}(i-1) \tag{10.74}$$

where $\hat{\mathbf{w}}$ is the M-by-1 tap-weight vector of the predictor:

$$\hat{\mathbf{w}} = [\hat{w}_1, \hat{w}_2, \ldots, \hat{w}_M]^T$$

and $\mathbf{u}(i-1)$ is the corresponding tap-input vector:

$$\mathbf{u}(\mathrm{i}-1) = [u(i-1), u(i-2), \ldots, u(i-M)]^T$$

Consider next Fig. 10.5(b), which depicts the reconfiguration of the predictor so that it performs backward linear prediction. We have *purposely* retained $\hat{w}_1, \hat{w}_2, \ldots, \hat{w}_M$ as the tap weights of the predictor. The change in the format of the tap inputs is inspired by the discussion presented in Section 6.2 on backward linear pre-

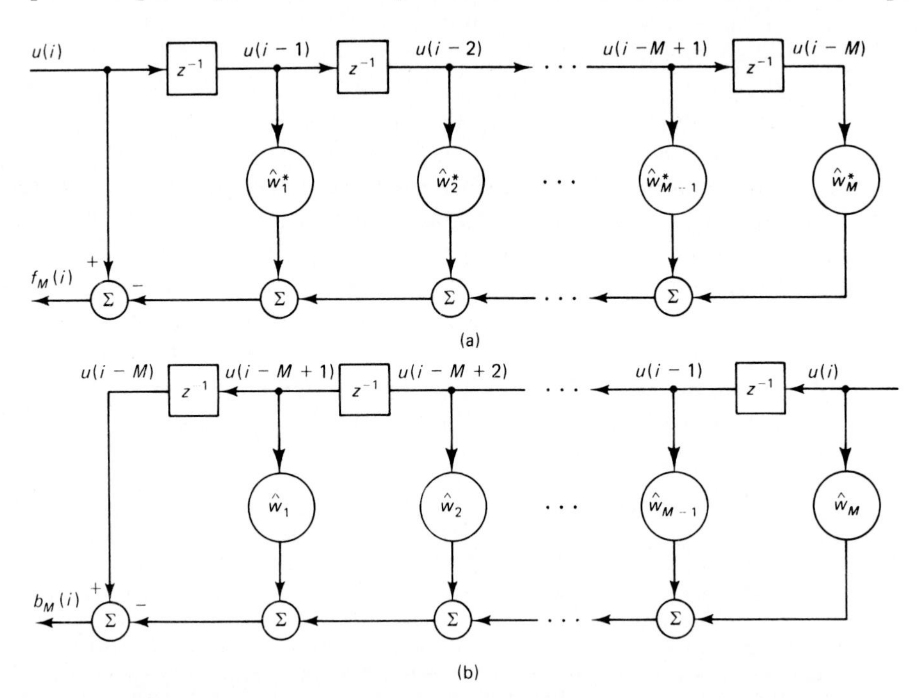

Figure 10.5 (a) Forward linear predictor; (b) reconfiguration of the predictor so as to perform backward linear prediction.

diction and its relation with forward linear prediction for the case of wide-sense stationary inputs. In particular, the tap inputs in the predictor of Fig. 10.5(b) differ from those of the forward linear predictor of Fig. 10.5(a) in two respects:

1. The tap inputs in Fig. 10.5(b) are *time reversed* in that they appear from right to left, whereas in Fig. 10.5(a) they appear from left to right.
2. With $u(i), u(i-1), \ldots, u(i-M+1)$ used as tap inputs, the structure of Fig. 10.5(b) produces a linear prediction of $u(i-M)$. In other words, it performs backward linear prediction. Denoting the backward prediction error by $b_M(i)$, we may thus express it as

$$b_M(i) = u(i-M) - \sum_{k=1}^{M} \hat{w}_k u(i-M+k) \tag{10.75}$$

where the first term represents the desired response and the second term is the predictor output. Equivalently, in terms of matrix notation, we may write

$$b_M(i) = u(i-M) - \mathbf{u}^{BT}(i)\hat{\mathbf{w}} \tag{10.76}$$

where $\mathbf{u}^B(i)$ is the *time-reversed tap-input vector:*

$$\mathbf{u}^{BT}(i) = [u(i-M+1), \ldots, u(i-1), u(i)]$$

Let $\mathscr{E}_M$ denote the *minimum value of the forward-backward prediction-error energy*. In accordance with the method of least squares, we may therefore write

$$\mathscr{E}_M = \sum_{i=M+1}^{N} [|f_M(i)|^2 + |b_M(i)|^2] \tag{10.77}$$

where the subscript M signifies the order of the predictor or the AR model. The lower limit on the time index i equals $M + 1$ so as to ensure that the forward and backward prediction errors are formed only when all the tap inputs of interest assume nonzero values. In particular, we may make two observations:

1. The variable $u(i-M)$, representing the last tap input in the forward prediction of Fig. 10.5(a), assumes a nonzero value for the first time when $i = M + 1$.
2. The variable $u(i-M)$, playing the role of desired response in the backward predictor of Fig. 10.5(b), also assumes a nonzero value for the first time when $i = M + 1$.

Thus, by choosing $(M + 1)$ as the lower limit on i and N as the upper limit, as in Eq. (10.77), we make no assumptions about the data outside the interval $[1, N]$, as required by the covariance method.

Let $\mathbf{A}$ denote the $2(N-M)$-by-M *data matrix* that is defined by

$$\mathbf{A}^H = \Bigg[\underbrace{\begin{matrix} u(M) & \cdots & u(N-1) \\ u(M-1) & \cdots & u(N-2) \\ \vdots & & \vdots \\ u(1) & \cdots & u(N-M) \end{matrix}}_{\text{forward half}} \quad \underbrace{\begin{matrix} u^*(2) & \cdots & u^*(N-M+1) \\ u^*(3) & \cdots & u^*(N-M+2) \\ \vdots & & \vdots \\ u^*(M+1) & \cdots & u^*(N) \end{matrix}}_{\text{backward half}}\Bigg] \tag{10.78}$$

The elements constituting the left half of matrix $\mathbf{A}^H$ represent the various sets of tap inputs used to make a total of $(N - M)$ *forward* linear predictions. The complex-conjugated elements constituting the right half of matrix $\mathbf{A}^H$ represent the corresponding sets of tap inputs used to make a total of $(N - M)$ *backward* linear predictions. Note that as we move from one column to the next in the forward or backward half in Eq. (10.78), we drop a sample, add a new one, and reorder the samples.

Let $\mathbf{d}$ denote the $2(N - M)$-by-1 *desired data vector* that is defined by

$$\mathbf{d}^H = [\underbrace{u(M+1), \ldots, u(N)}_{\text{forward half}}, \underbrace{u^*(1), \ldots, u^*(N-M)}_{\text{backward half}}] \tag{10.79}$$

Each element in the left half of the vector $\mathbf{d}^H$ represents a desired response for forward linear prediction. Each complex-conjugated element in the right half represents a desired response for backward linear prediction.

The FBLM method is a product of the method of least squares; it is therefore described by the system of normal equations [see Eq. (10.48)]

$$\mathbf{A}^H\mathbf{A}\hat{\mathbf{w}} = \mathbf{A}^H\mathbf{d} \tag{10.80}$$

The resulting minimum value of the forward-backward prediction error energy equals [see Eq. (10.49)]

$$\mathscr{E}_{\min} = \mathbf{d}^H\mathbf{d} - \mathbf{d}^H\mathbf{A}(\mathbf{A}^H\mathbf{A})^{-1}\mathbf{A}^H\mathbf{d} \tag{10.81}$$

The data matrix $\mathbf{A}$ and the desired data vector $\mathbf{d}$ are defined by Eqs. (10.78) and (10.79), respectively.

Augmented Normal Equations

We may combine Eqs. (10.80) and (10.81) into a single matrix relation, as shown by

$$\begin{bmatrix} \mathbf{d}^H\mathbf{d} & \mathbf{d}^H\mathbf{A} \\ \mathbf{A}^H\mathbf{d} & \mathbf{A}^H\mathbf{A} \end{bmatrix} \begin{bmatrix} 1 \\ -\hat{\mathbf{w}} \end{bmatrix} = \begin{bmatrix} \mathscr{E}_{\min} \\ \mathbf{0} \end{bmatrix} \tag{10.82}$$

where $\mathbf{0}$ is the M-by-1 null vector. Equation (10.82) is the matrix form of the *augmented normal equations for FBLP*. Define the $(M + 1)$-by-$(M + 1)$ *augmented correlation matrix:*

$$\mathbf{\Phi} = \begin{bmatrix} \mathbf{d}^H\mathbf{d} & \mathbf{d}^H\mathbf{A} \\ \mathbf{A}^H\mathbf{d} & \mathbf{A}^H\mathbf{A} \end{bmatrix} \tag{10.83}$$

The $\mathbf{\Phi}$ in Eq. (10.83) has dimension $M + 1$; it is *not* to be confused with the $\mathbf{\Phi}$ in Eq. (10.45) that has dimension M. Define the $(M + 1)$-by-1 tap-weight vector of the *prediction-error filter of order M:*

$$\hat{\mathbf{a}} = \begin{bmatrix} 1 \\ -\hat{\mathbf{w}} \end{bmatrix} \tag{10.84}$$

Figure 10.6 shows the transversal structure of the prediction-error filter, where a_0, $a_1, \ldots, a_M$ denote the tap weights[5] and $a_0 = 1$. Then

[5] The subscripts assigned to the tap weights in the prediction-error filter of Fig. 10.6 do not include a direct reference to the prediction order M, unlike the terminology used in Chapter 6. The reason for this simplification is that in the material presented here, there is no order update to be considered.

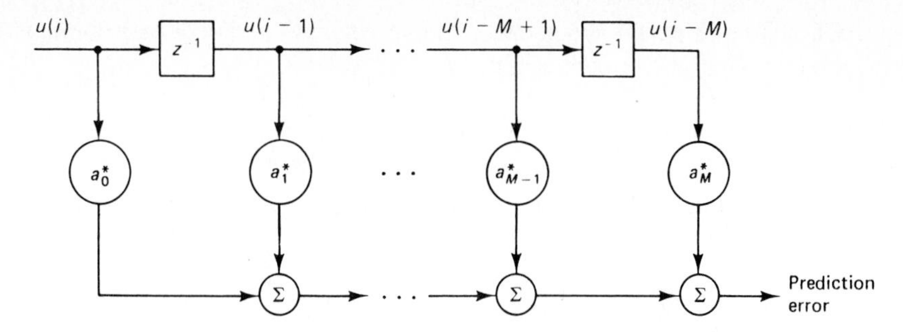

Figure 10.6 Forward prediction-error filter.

$$\mathbf{\Phi}\hat{\mathbf{a}} = \begin{bmatrix} \mathscr{E}_{\min} \\ \mathbf{0} \end{bmatrix} \tag{10.85}$$

The augmented correlation matrix $\mathbf{\Phi}$ is *Hermitian persymmetric;* that is, the individual elements of the matrix $\mathbf{\Phi}$ satisfy both of the following two conditions:

1. $\phi(k, t) = \phi^*(t, k), \qquad 0 \leq t, k \leq M$ (10.86)
2. $\phi(M - k, M - t) = \phi^*(k, t), \qquad 0 \leq t, k \leq M$ (10.87)

The property described in Eq. (10.87) is unique to a correlation matrix that is obtained by time averaging of the input data in the forward as well as backward direction; see the data matrix $\mathbf{A}$ and the desired data vector $\mathbf{d}$ defined in Eqs. (10.78) and (10.79), respectively.

The matrix $\mathbf{\Phi}$ has another property: it is composed of the sum of two Toeplitz matrix products (see Problem 8). The special Toeplitz structure of the matrix $\mathbf{\Phi}$ has been exploited in the development of fast recursive algorithms[6] for the efficient solution of the augmented normal equations (10.85).

AR Spectrum Estimate

Starting with the time series $\{u(i)\}$, $1 \leq i \leq N$, the FBLP algorithm is used to compute the tap-weight vector $\hat{\mathbf{w}}$ of a forward linear predictor or, equivalently, the tap-weight vector $\hat{\mathbf{a}}$ of the corresponding prediction-error filter. The vector $\hat{\mathbf{a}}$ represents an estimate of the coefficient vector of an *autoregressive (AR) model* used to fit the time series $\{u(i)\}$. Similarly, the minimum mean-squared error $\mathscr{E}_{\min}$, except for a scaling factor, represents an estimate of the white-noise variance σ^2 in the AR model. We may thus use Eq. (6.98) to formulate an *estimate of the AR spectrum* as follows:

$$\hat{S}_{\text{AR}}(\omega) = \frac{\mathscr{E}_{\min}}{\left| 1 + \sum_{k=1}^{M} \hat{a}_k^* e^{-j\omega k} \right|^2} \tag{10.88}$$

[6] The correlation matrix $\mathbf{\Phi}$ of Eq. (10.83) does *not* possess a Toeplitz structure. Accordingly, we cannot use the Levinson recursion to develop a fast solution of the augmented normal equations (10.85), as was the case with the augmented Wiener–Hopf equations for stationary inputs. However, Marple (1980, 1981) describes fast recursive algorithms for the efficient solution of the augmented normal equations (10.85). Marple exploits the special Toeplitz structure of the correlation matrix $\mathbf{\Phi}$. The computational complexity of Marple's fast algorithm is proportional to M^2. When the predictor order M is large, the use of Marple's algorithm results in significant savings in computation.

where the $\hat{a}_k$ are the elements of the vector $\hat{a}$; the leading element a_o of the vector $\hat{\mathbf{a}}$ is equal to unity, by definition. We may also express $\hat{S}_{AR}(\omega)$ as

$$\hat{S}_{AR}(\omega) = \frac{\mathscr{E}_{min}}{|\hat{\mathbf{a}}^H \mathbf{s}(\omega)|^2} \tag{10.89}$$

where $\mathbf{s}(\omega)$ is a *variable-frequency vector* or *frequency scanning vector:*

$$\mathbf{s}(\omega) = [1, e^{-j\omega}, \ldots, e^{-j\omega M}]^T \tag{10.90}$$

Intuitively, the model order M should be as large as possible in order to have a large aperture for the predictor. However, in applying the FBLP algorithm the use of large values of M gives rise to spurious spectral peaks in the AR spectrum. For best performance of the FBLP algorithm, Lang and McLellan (1980) suggest the value

$$M \approx \frac{N}{3}$$

where N is the data length.

10.11 MVDR SPECTRUM ESTIMATION

In the method of least squares, as described up to this point in our discussion, there are no *constraints* imposed on the solution. In certain applications, however, the use of such an approach may be unsatisfactory, in which case we may resort to a *constrained* version of the method of least squares. For example, in *adaptive beamforming* that involves spatial processing, we may wish to *minimize the variance (i.e., average power) of the beamformer output while a distortionless response is maintained along the direction of a target signal of interest*. Correspondingly, in the temporal counterpart to this problem, we may be required to *minimize the average power of the spectrum estimator, while a distortionless response is maintained at a particular frequency*. In such applications, the resulting solution is referred to as a *minimum-variance distortionless response (MVDR) estimator* for obvious reasons. To be consistent with the material presented heretofore, we will formulate the temporal version of the MVDR algorithm; for the spatial version, the reader is referred to Problem 9.

Consider then a linear transversal filter, as depicted in Fig. 10.7. Let the filter output be denoted by $y(i)$. This output is in response to the tap inputs $u(i)$, $u(i-1)$, $\ldots$, $u(i-M)$. Specifically, we have

$$y(i) = \sum_{k=0}^{M} a_k^* u(i-k) \tag{10.91}$$

where $a_0, a_1, \ldots, a_M$ are the transversal filter coefficients. Note, however, that unlike the prediction-error filter of Fig. 10.6, there is no restriction on the filter coefficient a_0; the only reason for using the same terminology as in Fig. 10.6 is because of a desire to be consistent. The requirement is to minimize the *output energy* (assuming the use of the covariance method of data windowing):

$$\mathscr{E}_{out} = \sum_{i=M+1}^{N} |y(i)|^2$$

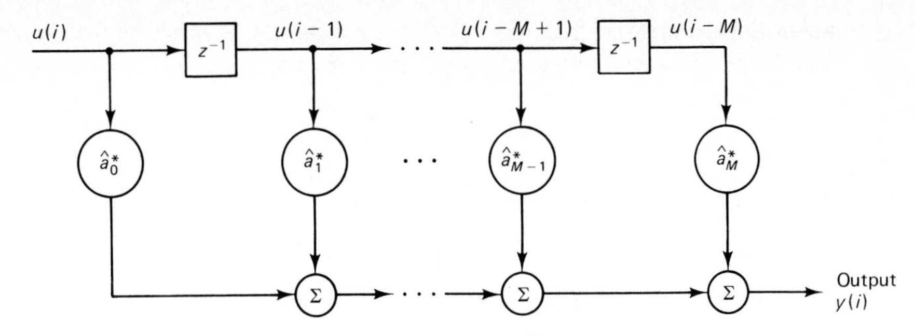

Figure 10.7 Transversal filter.

subject to the *constraint*

$$\sum_{k=0}^{M} a_k^* e^{-jk\omega_0} = 1 \tag{10.92}$$

where ω_0 is an angular frequency of specific interest. As in the conventional method of least squares, the filter coefficients $a_0, a_1, \ldots, a_M$ are held constant for the observation interval $1 \le i \le N$, where N is the total data length.

To solve this *constrained minimization* problem, we use the *method of Lagrange multipliers.*[7] Specifically, we define the *constrained cost function*

$$\mathscr{E} = \underbrace{\sum_{i=M+1}^{N} |y(i)|^2}_{\text{output energy}} + \underbrace{\lambda \sum_{k=0}^{M} a_k^* e^{-jk\omega_0}}_{\text{linear constraints}} \tag{10.93}$$

where λ is a *complex Lagrange multiplier*. Note that in the constrained approach described herein, there is *no* desired response; in place of it, however, we have a set of linear constraints. Note also that in the absence of a desired response and therefore no frame of reference, the principle of orthogonality loses its meaning.

To solve for the optimum values of the filter coefficients, we first determine the gradient vector $\nabla(\mathscr{E})$, and then set it equal to zero. Thus, proceeding in a manner similar to that described in Section 10.3, we find that the kth element of the gradient vector for the constrained cost function of Eq. (10.93) is

$$\nabla_k(\mathscr{E}) = -2 \sum_{i=M+1}^{N} u(i-k)y^*(i) + 2\lambda^* e^{-jk\omega_0} \tag{10.94}$$

Next, substituting Eq. (10.91) in (10.94), and rearranging terms, we get

$$\begin{aligned}\nabla_k(\mathscr{E}) &= -2 \sum_{t=0}^{M} a_t \sum_{i=M+1}^{N} u(i-k)u^*(i-t) + 2\lambda^* e^{-jk\omega_0} \\ &= -2 \sum_{t=0}^{M} a_t \phi(t, k) + 2\lambda^* e^{-jk\omega_0}\end{aligned} \tag{10.95}$$

where, in the first term of the second line, we have made use of the definition of Eq. (10.29) for the autocorrelation function $\phi(t, k)$ of the tap inputs. To minimize the constrained cost function $\mathscr{E}$, we set

$$\nabla_k(\mathscr{E}) = 0, \qquad k = 0, 1, \ldots, M \tag{10.96}$$

[7] The method of Lagrange multipliers is described in Appendix C.

Accordingly, we find from Eq. (10.95) that the tap-weights of the *optimized* transversal filter satisfy the following system of $M + 1$ simultaneous equations:

$$\sum_{t=0}^{M} \hat{a}_t \phi(t, k) = \lambda^* e^{-jk\omega_0}, \qquad k = 0, 1, \ldots, M \tag{10.97}$$

Using matrix notation, we may rewrite this system of equations in the compact form

$$\mathbf{\Phi}\hat{\mathbf{a}} = \lambda^* \mathbf{s}(\omega_0) \tag{10.98}$$

where $\mathbf{\Phi}$ is the $(M + 1)$-by-$(M + 1)$ correlation matrix of the tap inputs; $\hat{\mathbf{a}}$ is the $(M + 1)$-by-1 vector of optimum tap weights; and $\mathbf{s}(\omega_0)$ is the $(M + 1)$-by-1 *fixed frequency vector:*

$$\mathbf{s}(\omega_0) = [1, e^{-j\omega_0}, \ldots, e^{-jM\omega_0}]^T \tag{10.99}$$

Assuming that the correlation matrix $\mathbf{\Phi}$ is nonsingular, and therefore its inverse exists, we may solve Eq. (10.98) for the tap-weight vector:

$$\hat{\mathbf{a}} = \lambda^* \mathbf{\Phi}^{-1} \mathbf{s}(\omega_0) \tag{10.100}$$

There only remains the problem of evaluating the Lagrange multiplier λ. To solve for λ, we use the linear constraint in Eq. (10.92) for the optimized transversal filter, written in matrix form as

$$\hat{\mathbf{a}}^H \mathbf{s}(\omega_0) = 1 \tag{10.101}$$

Hence, evaluating the *inner product* of the vector $\mathbf{s}_0$ and the vector $\mathbf{a}$ in Eq. (10.100), setting the result equal to 1, and solving for λ, we get the desired result:

$$\lambda^* = \frac{1}{\mathbf{s}^H(\omega_0)\mathbf{\Phi}^{-1}\mathbf{s}(\omega_0)} \tag{10.102}$$

Finally, substituting this value of λ in Eq. (10.100), we get the MVDR solution:[8]

$$\hat{\mathbf{a}} = \frac{\mathbf{\Phi}^{-1}\mathbf{s}(\omega_0)}{\mathbf{s}^H(\omega_0)\mathbf{\Phi}^{-1}\mathbf{s}(\omega_0)} \tag{10.103}$$

Thus, given the correlation matrix $\mathbf{\Phi}$ of the tap inputs and the frequency vector $\mathbf{s}(\omega_0)$, we may use the *MVDR formula* of (10.103) to compute the tap-weight vector $\hat{\mathbf{a}}$ of the transversal filter in Fig. 10.7.

Let $\hat{S}_{\text{MVDR}}(\omega_0)$ denote the minimum value of the output energy $\mathscr{E}_{\text{out}}$, which results when the MVDR solution $\hat{\mathbf{a}}$ of Eq. (10.103) is used for the tap-weight vector under the condition that the response is tuned to the angular frequency ω_0. We may then write

$$\hat{S}_{\text{MVDR}}(\omega_0) = \hat{\mathbf{a}}^H \mathbf{\Phi} \hat{\mathbf{a}}_M \tag{10.104}$$

Substituting Eq. (10.103) in (10.104), and then simplifying the result, we finally get

$$\hat{S}_{\text{MVDR}}(\omega_0) = \frac{1}{\mathbf{s}^H(\omega_0)\mathbf{\Phi}^{-1}\mathbf{s}(\omega_0)} \tag{10.105}$$

[8] Equation (10.103) is of the same form as that of Eq. (5.84), except for the use of the time-averaged correlation matrix $\mathbf{\Phi}$ in place of the ensemble-averaged correlation matrix $\mathbf{R}$, and the use of symbol $\mathbf{a}$ in place of $\mathbf{w}_o$ for the tap-weight vector.

Equation (10.105) may be given a different interpretation. Suppose that we define a *frequency-scanning vector* or *variable frequency vector*:

$$\mathbf{s}(\omega) = [1, e^{-j\omega}, \ldots, e^{-j\omega M}]^T \qquad (10.106)$$

where the angular frequency ω is now variable in the entire interval $[-\pi, \pi]$. For each ω, let the tap-weight vector of the transversal filter be assigned the corresponding MVDR estimate. The output energy of the optimized filter then becomes a function of ω. Let $\hat{S}_{\text{MVDR}}(\omega)$ describe this functional dependence, and so we may write[9]

$$\hat{S}_{\text{MVDR}}(\omega) = \frac{1}{\mathbf{s}^H(\omega)\mathbf{\Phi}^{-1}\mathbf{s}(\omega)} \qquad (10.107)$$

We refer to Eq. (10.107) as the *MVDR spectrum estimate,* and the solution given in Eq. (10.103) as the *MVDR estimate* of the tap-weight vector.

Note that at any ω, power due to other frequencies is minimized. Hence, the MVDR spectrum computed in accordance with Eq. (10.107) exhibits relatively sharp peaks.

PROBLEMS

1. Consider a linear array consisting of M uniformly spaced sensors. The output of sensor k observed at time i is denoted by $u(k, i)$ where $k = 1, 2, \ldots, M$ and $i = 1, 2, \ldots, n$. In effect, the observations $u(1, i), u(2, i), \ldots, u(M, i)$ define snapshot i. Let $\mathbf{A}$ denote the n-by-M-data matrix that is defined by

$$\mathbf{A}^H = \begin{bmatrix} u(1, 1) & u(1, 2) & \cdots & u(1, n) \\ u(2, 1) & u(2, 2) & \cdots & u(2, n) \\ \vdots & \vdots & \ddots & \vdots \\ u(M, 1) & u(M, 2) & \cdots & u(M, n) \end{bmatrix}$$

where the number of columns equals the number of snapshots, and the number of rows equals the number of sensors in the array. Demonstrate the following interpretations:

(a) The M-by-M matrix $\mathbf{A}^H\mathbf{A}$ is the *spatial* correlation matrix with temporal averaging. This form of averaging assumes that the environment is temporally stationary.

(b) The n-by-n matrix $\mathbf{A}\mathbf{A}^H$ is the *temporal* correlation matrix with spatial averaging. This form of averaging assumes that the environment is spatially stationary.

2. We say that the least-squares estimate $\hat{\mathbf{w}}$ is *consistent* if, in the long run, the difference between $\hat{\mathbf{w}}$ and the unknown parameter vector $\mathbf{w}_o$ of the multiple linear regression model becomes negligibly small in the mean-square sense. Hence, show that the least-squares estimate $\hat{\mathbf{w}}$ is consistent if the error vector $\boldsymbol{\epsilon}_o$ has zero mean and its elements are uncorrelated and if the trace of the inverse matrix $\mathbf{\Phi}^{-1}$ approaches zero as the number of observations, N, approaches infinity.

3. In the example of section 10.6, we used a 3-by-2 input data matrix and 3-by-1 desired data vector to illustrate the corollary to the principle of orthogonality. Use the data given in that example to calculate the two tap-weights of the linear least-square filter.

[9] The method for computing the spectrum in Eq. (10.107) is also referred to in the literature as *Capon's method* (Capon, 1969). The term "minimum-variance distortionless response" owes its origin to Owsley (1984).

4. In the autocorrelation method of linear prediction, we choose the tap-weight vector of a transversal predictor to minimize the error energy

$$\mathscr{E}_f = \sum_{n=1}^{\infty} | f(n) |^2$$

where $f(n)$ is the prediction error. Show that the transfer function $H_f(z)$ of the (forward) prediction-error filter is minimum phase in that its roots must lie strictly within the unit circle.

Hints: (1) Express the transfer function $H_f(z)$ of order M (say) as the product of a sample zero factor $(1 - z_i z^{-1})$ and a function $H_f'(z)$. Hence, minimize the prediction-error energy with respect to the magnitude of zero z_i

(2) You may also use the Cauchy–Schwartz inequality:

$$\text{Re}\left[\sum_{n=1}^{\infty} e^{j\theta} g(n-1) g^*(n)\right] \leq \left[\sum_{n=1}^{\infty} | g(n) |^2\right]^{1/2} \left[\sum_{n=1}^{\infty} | e^{j\theta} g(n-1) |^2\right]^{1/2}$$

The equality holds if and only if $g(n) = e^{j\theta} g(n-1)$ for $n = 1, 2, \ldots, \infty$.

5. Figure 10.5(a) shows a *forward linear predictor* using a transversal structure, with the tap inputs $u(i-1), u(i-2), \ldots, u(i-M)$ used to make a linear prediction of $u(i)$. The problem is to find the tap-weight vector $\hat{\mathbf{w}}$ that minimizes the sum of forward prediction-error squares:

$$\mathscr{E}_f = \sum_{i=M+1}^{N} | f_M(i) |^2$$

where $f_M(i)$ is the forward prediction error. Find the following parameters:

(a) The M-by-M correlation matrix of the tap inputs of the predictor.

(b) The M-by-1 cross-correlation vector between the tap inputs of the predictor and the desired response $u(i)$.

(c) The minimum value of $\mathscr{E}_f$.

6. Figure 10.5(b) shows a *backward linear predictor* using a transversal structure, with the tap inputs $u(i-M+1), \ldots, u(i-1), u(i)$ used to make a linear prediction of the input $u(i-M)$. The problem is to find the tap-weight vector $\hat{\mathbf{w}}$ that minimizes the sum of backward prediction-error squares

$$\mathscr{E}_b = \sum_{i=M+1}^{N} | b_M(i) |^2$$

where $b_M(i)$ is the backward prediction error. Find the following parameters:

(a) The M-by-M correlation matrix of the tap inputs.

(b) The M-by-1 correlation vector between the tap inputs and the desired response $u(i-M)$.

(c) The minimum value of $\mathscr{E}_b$.

7. Use a direct approach to derive the system of normal equations given in expanded form in Eq. (10.31).

8. Show that the augmented correlation matrix $\mathbf{\Phi}$ of Eq. (10.83) consists of the sum of two Toeplitz matrix products. The data matrix $\mathbf{A}$ and the desired data vector $\mathbf{d}$ are defined in Eqs. (10.78) and (10.79), respectively.

9. Consider a linear array of M uniformly spaced sensors. The output of sensor k observed at time i is denoted by $u(k, i)$ where $k = 1, 2, \ldots, M$ and $i = 1, 2, \ldots, n$. The element outputs are individually weighted and then summed to produce the output

$$e(i) = \mathbf{u}^T(i)\mathbf{w}^* = \mathbf{w}^H\mathbf{u}(i)$$

where $\mathbf{u}(i)$ is the M-by-1 elemental output vector at time i, defined by

$$\mathbf{u}^T(i) = [u(1, i), u(2, i), \ldots, u(M, i)]$$

and $\mathbf{w}$ is the M-by-1 weight vector, defined by

$$\mathbf{w}^T = [w_1, w_2, \ldots, w_M]$$

The weight vector $\mathbf{w}$ is held constant for the total observation interval $1 \leq i \leq n$. The choice of $\mathbf{w}$ is constrained to fulfill the condition

$$\mathbf{w}^H\mathbf{s} = 1$$

where $\mathbf{s}$ is an M-by-1 *steering vector*. The elements of the vector $\mathbf{s}$ are determined by the look direction of interest. Using the method of Lagrange multipliers, show that the optimum weight vector $\hat{\mathbf{w}}$ that minimizes the sum of weighted error squares, subject to this constraint, equals

$$\hat{\mathbf{w}} = \frac{\mathbf{\Phi}^{-1}\mathbf{s}}{\mathbf{s}^H\mathbf{\Phi}^{-1}\mathbf{s}}$$

For details of the method of Lagrange multipliers, see Appendix C.

10. Show that the formula for the AR power spectrum estimate given in Eq. (10.89) may be rewritten in the form

$$\hat{S}_{\text{AR}}(\omega) = \frac{\boldsymbol{\delta}^T\mathbf{C}^{-1}\mathbf{i}}{|\boldsymbol{\delta}^T\mathbf{C}^{-1}\mathbf{s}|^2}$$

where $\mathbf{i}$ is the $(M + 1)$-by-1 first coordinate vector with its first element equal to 1 and its remaining elements equal to zero, $\mathbf{C}$ is the $(M + 1)$-by-$(M + 1)$ correlation matrix of the input signal, and $\mathbf{s}$ is the $(M + 1)$-by-1 frequency-scanning vector.

11. Using the Hermitian property of the correlation matrix $\mathbf{\Phi}$, show that the substitution of Eq. (10.103) in (10.104) yields the result given in Eq. (10.105).

Computer-Oriented Problem

12. Consider the noisy time series

$$u(i) = \exp(j\pi i - j\pi) + \exp(j1.04\pi i - j0.79\pi) + v(i), \qquad i = 1, \ldots, N$$

where the noise samples $v(i)$ are independent complex Gaussian random variables with zero mean. The data length N is 25.

(a) For noiseless data [i.e., $v(i) = 0$], compute the transfer function of the prediction-error filter of order $M = 20$. Compute and display the geometry of the zeros of this transfer function. Hence, demonstrate that there are two zeros on the unit circle in the z-plane at angular locations corresponding to the angular frequencies of the complex sinusoids, and that the remaining zeros are uniformly distributed in angle around the inside of the unit circle. Complete this part of the experiment by plotting $10 \log_{10}S_{\text{AR}}(\omega)$ versus ω.

(b) Assume a signal-to-noise ratio of 10 dB. Compute and display the zeros of the transfer function of the prediction-error filter for order $M = 4, 8, 12, 16, 18, 20, 22$, and 24. In each case, include a spectrum plot of $10 \log_{10}S_{\text{AR}}(\omega)$ versus ω.

Comment on your results.

11

Singular-Value Decomposition

In Chapter 4 we studied the eigenanalysis of the ensemble-averaged correlation matrix of a wide-sense stationary discrete-time stochastic process. In particular, we developed the representation of such a matrix in terms of a corresponding set of eigenvalues and eigenvectors. The use of eigenanalysis is especially appropriate for the study of a Wiener filter, which relies on ensemble-averaging operations for its mathematical description.

Basic to eigenanalysis is the unitary similarity transformation that provides a highly illuminating method of describing the interrelation between a correlation matrix (be it ensemble-averaged or time-averaged) and its eigenvalues and eigenvectors. This transformation is indeed a special case of the *singular-value decomposition*. The analytic power of singular-value decomposition lies in the fact that it applies to square as well as rectangular matrices, be they real or complex. As such, it is extremely well suited for the numerical solution of linear least-squares problems in the sense that *it can be applied directly to the data matrix*.

This chapter is devoted to a discussion of singular-value decomposition and its computation. We begin the discussion by presenting some preliminary ideas, based on the material developed in Chapter 10.

11.1 SOME PRELIMINARIES

In Chapter 10 we studied the method of least squares and its linearly constrained version. Specifically, we developed the system of *normal equations* for computing the linear least-squares solution. Two different formulations were indeed presented for this solution:

1. The form given in Eq. (10.36), namely,

$$\hat{\mathbf{w}} = \mathbf{\Phi}^{-1}\mathbf{\theta} \tag{11.1}$$

where $\hat{\mathbf{w}}$ is the least-squares estimate of the tap-weight vector of a transversal filter model, $\mathbf{\Phi}$ is the time-averaged correlation matrix of the tap inputs, and $\mathbf{\theta}$ is the time-averaged cross-correlation vector between the tap inputs and some desired response.

2. The form given in Eq. (10.48) *directly in terms of data matrices,* as shown by

$$\hat{\mathbf{w}} = (\mathbf{A}^H\mathbf{A})^{-1}\mathbf{A}^H\mathbf{d} \tag{11.2}$$

where $\mathbf{A}$ is the data matrix representing the time evolution of the tap input vectors, and $\mathbf{d}$ is the desired data vector representing the time evolution of the desired response.

The two forms given in Eqs. (11.1) and (11.2) are mathematically equivalent. Yet they point to different computational procedures for evaluating the least-squares solution $\hat{\mathbf{w}}$. Equation (11.1) requires knowledge of the correlation matrix $\mathbf{\Phi}$ that involves computing the product of $\mathbf{A}^H$ and $\mathbf{A}$. On the other hand, in Eq. (11.2) the entire term $(\mathbf{A}^H\mathbf{A})^{-1}\mathbf{A}$ can be interpreted, in terms of the singular-value decomposition applied directly to the data matrix $\mathbf{A}$, in such a way that the solution computed for $\hat{\mathbf{w}}$ is *twice as accurate* as the solution computed by means of Eq. (11.1) for the same numerical precision. To be specific, define the matrix

$$\mathbf{A}^+ = (\mathbf{A}^H\mathbf{A})^{-1}\mathbf{A}^H \tag{11.3}$$

Then we may rewrite Eq. (11.2) simply as

$$\hat{\mathbf{w}} = \mathbf{A}^+\mathbf{d} \tag{11.4}$$

The matrix $\mathbf{A}^+$ is called the *pseudoinverse* or the *Moore-Penrose generalized inverse* of the matrix $\mathbf{A}$ (Stewart, 1973; Golub and Van Loan 1989). Equation (11.4) represents a convenient way of saying that "the vector $\hat{\mathbf{w}}$ solves the least-squares problem." Indeed, it was with the simple format of Eq. (11.4) in mind and also the desire to be consistent with definitions of the correlation matrix and the cross-correlation vector used in Chapter 10 that we defined the data matrix $\mathbf{A}$ and the desired data vector $\mathbf{d}$ in the manner shown in Eqs. (10.44) and (10.46), respectively.

In practice, we often find that the data matrix $\mathbf{A}$ contains linearly dependent columns. Consequently, we are faced with a new situation where we now have to decide on which of an infinite number of possible solutions to the least-squares problem to work with. This issue can indeed be resolved by using the singular-value decomposition technique, even when null $(\mathbf{A}) \neq 0$.

In summary, the singular-value decomposition is a powerful mathematical tool for solving linear least-squares problems with important applications in a variety of fields.

11.2 THE SINGULAR-VALUE DECOMPOSITION THEOREM

The *singular-value decomposition* (SVD) of a matrix is one of the most elegant algorithms in numerical algebra for providing quantitative information about the structure of a system of linear equations (Klema and Laub, 1980). The system of linear equations that is of specific interest to us is described by

$$\mathbf{A}\hat{\mathbf{w}} = \mathbf{d} \tag{11.5}$$

in which $\mathbf{A}$ is a K-by-M matrix, $\mathbf{d}$ is a K-by-1 vector, and $\hat{\mathbf{w}}$ (representing an estimate of the unknown parameter vector) is an M-by-1 vector. Equation (11.5) represents a simplified matrix form of the normal equations. In particular, premultiplication of both sides of the equation by the vector $\mathbf{A}^H$ yields the normal equations for the least-squares weight vector $\hat{\mathbf{w}}$.

Given the data matrix $\mathbf{A}$, there are two unitary matrices $\mathbf{V}$ and $\mathbf{U}$, such that we may write

$$\mathbf{U}^H\mathbf{A}\mathbf{V} = \begin{bmatrix} \mathbf{\Sigma} & \mathbf{0} \\ \mathbf{0} & \mathbf{0} \end{bmatrix} \tag{11.6}$$

where $\mathbf{\Sigma}$ is a diagonal matrix:

$$\mathbf{\Sigma} = \text{diag}(\sigma_1, \sigma_2, \ldots, \sigma_W) \tag{11.7}$$

The σ's are ordered as $\sigma_1 \geq \sigma_2 \geq \cdots \geq \sigma_W > 0$. Equation (11.6) is a mathematical statement of the *singular-value decomposition theorem*. This theorem is also referred to as the *Autonne–Eckart–Young theorem* in recognition of its originators.[1]

Figure 11.1 presents a diagrammatic interpretation of the singular value decomposition theorem, as described in Eq. (11.6). In this diagram we have assumed that the number of rows K contained in the data matrix $\mathbf{A}$ is larger than the number of columns M, and that the number of nonzero singular values W is less than M. We may of course restructure the diagrammatic interpretation of the singular value decomposition theorem by expressing the data matrix in terms of the unitary matrices $\mathbf{U}$ and $\mathbf{V}$, and the diagonal matrix $\mathbf{\Sigma}$; this is left as an exercise for the reader.

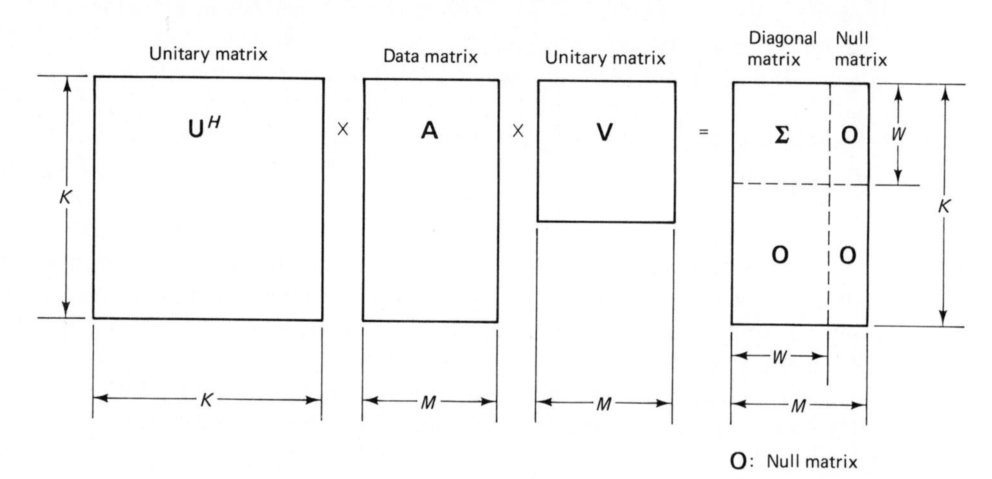

Figure 11.1 Diagrammatic interpretation of the singular value decomposition theorem.

[1] According to DeMoor and Golub (1989), the singular-value decomposition was introduced in its general form by Autonne in 1902, and an important characterization of it was described by Eckart and Young (1936). For additional notes on the history of the singular-value decomposition, see Klema and Laub (1980).

The subscript W in Eq. (11.7) is the *rank* of matrix $\mathbf{A}$, written as rank ($\mathbf{A}$); it is defined as the number of linearly independent columns in the matrix $\mathbf{A}$. Note that we always have $\text{rank}(\mathbf{A}^H) = \text{rank}(\mathbf{A})$. The rank $W \le \min(K, M)$. Since it is possible that $K > M$ or $K < M$, there are two distinct cases to be considered. We prove the singular value decomposition theorem by considering both cases, independently of each other. For the case when $K > M$, we have an *overdetermined system* in that we have more equations than unknowns. On the other hand, when $K < M$, we have an *underdetermined system* in that we have more unknowns than equations. In the sequel, we consider these two cases in turn.

Case 1: Overdetermined System. For the case when $K > M$, we form the M-by-M matrix $\mathbf{A}^H\mathbf{A}$ by premultiplying the matrix $\mathbf{A}$ by its Hermitian transpose $\mathbf{A}^H$. Since the matrix $\mathbf{A}^H\mathbf{A}$ is Hermitian and nonnegative definite, its eigenvalues are all real nonnegative numbers. Let these eigenvalues be denoted by $\sigma_1^2, \sigma_2^2, \ldots, \sigma_M^2$, where $\sigma_1 \ge \sigma_2 \ge \ldots \ge \sigma_W > 0$, and $\sigma_{W+1} = \sigma_{W+2} = \ldots = \sigma_M = 0$. The matrix $\mathbf{A}^H\mathbf{A}$ has the same rank as $\mathbf{A}$; hence, there are W nonzero eigenvalues. Let $\mathbf{v}_1, \mathbf{v}_2, \ldots, \mathbf{v}_M$ denote a set of orthonormal eigenvectors of $\mathbf{A}^H\mathbf{A}$ that are associated with the eigenvalues $\sigma_1^2, \sigma_2^2, \ldots, \sigma_M^2$, respectively. Also, let $\mathbf{V}$ denote the M-by-M unitary matrix whose columns are made up of the eigenvectors $\mathbf{v}_1, \mathbf{v}_2, \ldots, \mathbf{v}_M$. Thus, using the eigenvalue-eigenvector decomposition of the matrix $\mathbf{A}^H\mathbf{A}$, we may write

$$\mathbf{V}^H\mathbf{A}^H\mathbf{A}\mathbf{V} = \begin{bmatrix} \mathbf{\Sigma}^2 & \mathbf{0} \\ \mathbf{0} & \mathbf{0} \end{bmatrix} \tag{11.8}$$

Let the unitary matrix $\mathbf{V}$ be partitioned as

$$\mathbf{V} = [\mathbf{V}_1, \mathbf{V}_2] \tag{11.9}$$

where $\mathbf{V}_1$ is an M-by-W matrix,

$$\mathbf{V}_1 = [\mathbf{v}_1, \mathbf{v}_2, \ldots, \mathbf{v}_W] \tag{11.10}$$

and $\mathbf{V}_2$ is an M-by-$(M - W)$ matrix,

$$\mathbf{V}_2 = [\mathbf{v}_{W+1}, \mathbf{v}_{W+2}, \ldots, \mathbf{v}_M] \tag{11.11}$$

with

$$\mathbf{V}_1^H\mathbf{V}_2 = \mathbf{0} \tag{11.12}$$

We may therefore make two deductions from Eq. (11.8):

1. For matrix $\mathbf{V}_1$, we have

$$\mathbf{V}_1^H\mathbf{A}^H\mathbf{A}\mathbf{V}_1 = \mathbf{\Sigma}^2$$

Consequently,

$$\mathbf{\Sigma}^{-1}\mathbf{V}_1^H\mathbf{A}^H\mathbf{A}\mathbf{V}_1\mathbf{\Sigma}^{-1} = \mathbf{I} \tag{11.13}$$

2. For matrix $\mathbf{V}_2$, we have

$$\mathbf{V}_2^H\mathbf{A}^H\mathbf{A}\mathbf{V}_2 = \mathbf{0}$$

Consequently,

$$\mathbf{A}\mathbf{V}_2 = \mathbf{0} \tag{11.14}$$

We now define a new K-by-W matrix

$$\mathbf{U}_1 = \mathbf{A}\mathbf{V}_1\boldsymbol{\Sigma}^{-1} \tag{11.15}$$

Then, from Eq. (11.13), it follows that

$$\mathbf{U}_1^H\mathbf{U}_1 = \mathbf{I} \tag{11.16}$$

which means that the columns of the matrix $\mathbf{U}_1$ are orthonormal with respect to each other. Next, we choose another K-by-$(K - W)$ matrix $\mathbf{U}_2$ such that the K-by-K matrix formed from $\mathbf{U}_1$ and $\mathbf{U}_2$, namely,

$$\mathbf{U} = [\mathbf{U}_1, \mathbf{U}_2] \tag{11.17}$$

is a unitary matrix. This means that

$$\mathbf{U}_1^H\mathbf{U}_2 = \mathbf{0} \tag{11.18}$$

Accordingly, we may use Eqs. (11.9), (11.17), (11.14), (11.15) and (11.18) in that order, and so write

$$\begin{aligned}
\mathbf{U}^H\mathbf{A}\mathbf{V} &= \begin{bmatrix} \mathbf{U}_1^H \\ \mathbf{U}_2^H \end{bmatrix} \mathbf{A}[\mathbf{V}_1, \mathbf{V}_2] \\
&= \begin{bmatrix} \mathbf{U}_1^H\mathbf{A}\mathbf{V}_1 & \mathbf{U}_1^H\mathbf{A}\mathbf{V}_2 \\ \mathbf{U}_2^H\mathbf{A}\mathbf{V}_1 & \mathbf{U}_2^H\mathbf{A}\mathbf{V}_2 \end{bmatrix} \\
&= \begin{bmatrix} (\boldsymbol{\Sigma}^{-1}\mathbf{V}_1^H\mathbf{A}^H)\mathbf{A}\mathbf{V}_1 & \mathbf{U}_1^H(\mathbf{0}) \\ \mathbf{U}_2^H(\mathbf{U}_1\boldsymbol{\Sigma}) & \mathbf{U}_2^H(\mathbf{0}) \end{bmatrix} \\
&= \begin{bmatrix} \boldsymbol{\Sigma} & \mathbf{0} \\ \mathbf{0} & \mathbf{0} \end{bmatrix}
\end{aligned}$$

which is the desired result.

Case 2: Underdetermined System. Consider next the case when $K < M$. This time we form the K-by-K matrix $\mathbf{A}\mathbf{A}^H$ by postmultiplying the matrix $\mathbf{A}$ by its Hermitian transpose $\mathbf{A}^H$. The matrix $\mathbf{A}\mathbf{A}^H$ is also Hermitian and nonnegative definite, so its eigenvalues are likewise real nonnegative numbers. The nonzero eigenvalues of $\mathbf{A}\mathbf{A}^H$ are the *same* as those of $\mathbf{A}^H\mathbf{A}$. We may therefore denote the eigenvalues of $\mathbf{A}\mathbf{A}^H$ as $\sigma_1^2, \sigma_2^2, \ldots, \sigma_K^2$, where $\sigma_1 \geq \sigma_2 \geq \ldots \geq \sigma_W > 0$, and $\sigma_{W+1} = \sigma_{W+2} = \cdots = \sigma_K = 0$. Let $\mathbf{u}_1, \mathbf{u}_2, \ldots, \mathbf{u}_K$ denote a set of orthonormal eigenvectors of the matrix $\mathbf{A}\mathbf{A}^H$ that are associated with the eigenvalues $\sigma_1^2, \sigma_2^2, \ldots, \sigma_K^2$, respectively. Also, let $\mathbf{U}$ denote the unitary matrix whose columns are made up of the eigenvectors $\mathbf{u}_1, \mathbf{u}_2, \ldots, \mathbf{u}_K$. Thus, using the eigenvalue-eigenvector decomposition of $\mathbf{A}\mathbf{A}^H$, we may write

$$\mathbf{U}^H\mathbf{A}\mathbf{A}^H\mathbf{U} = \begin{bmatrix} \boldsymbol{\Sigma}^2 & \mathbf{0} \\ \mathbf{0} & \mathbf{0} \end{bmatrix} \tag{11.19}$$

Let the unitary matrix $\mathbf{U}$ be partitioned as

$$\mathbf{U} = [\mathbf{U}_1, \mathbf{U}_2] \tag{11.20}$$

where

$$\mathbf{U}_1 = [\mathbf{u}_1, \mathbf{u}_2, \ldots, \mathbf{u}_W] \tag{11.21}$$

$$\mathbf{U}_2 = [\mathbf{u}_{W+1}, \mathbf{u}_{W+2}, \ldots, \mathbf{u}_K] \tag{11.22}$$

and

$$\mathbf{U}_1^H \mathbf{U}_2 = \mathbf{0} \tag{11.23}$$

We may therefore make two deductions from Eq. (11.19):

1. For matrix $\mathbf{U}_1$, we have

$$\mathbf{U}_1^H \mathbf{A}\mathbf{A}^H \mathbf{U}_1 = \mathbf{\Sigma}^2$$

Consequently,

$$\mathbf{\Sigma}^{-1}\mathbf{U}_1^H \mathbf{A}\mathbf{A}^H \mathbf{U}_1 \mathbf{\Sigma}^{-1} = \mathbf{I} \tag{11.24}$$

2. For matrix $\mathbf{U}_2$, we have

$$\mathbf{U}_2^H \mathbf{A}\mathbf{A}^H \mathbf{U}_2 = \mathbf{0}$$

Consequently,

$$\mathbf{A}^H \mathbf{U}_2 = \mathbf{0} \tag{11.25}$$

We now define an M-by-W matrix

$$\mathbf{V}_1 = \mathbf{A}^H \mathbf{U}_1 \mathbf{\Sigma}^{-1} \tag{11.26}$$

Then from Eq. (11.24), it follows that

$$\mathbf{V}_1^H \mathbf{V}_1 = \mathbf{I} \tag{11.27}$$

which means that the columns of the matrix $\mathbf{V}_1$ are orthonormal with respect to each other. Next, we choose another M-by-$(M - W)$ matrix $\mathbf{V}_2$ such that the M-by-M matrix formed from $\mathbf{V}_1$ and $\mathbf{V}_2$, namely,

$$\mathbf{V} = [\mathbf{V}_1, \mathbf{V}_2] \tag{11.28}$$

is a unitary matrix. This means that

$$\mathbf{V}_2^H \mathbf{V}_1 = \mathbf{0} \tag{11.29}$$

Accordingly, we may use Eqs. (11.20), (11.28), (11.25), (11.26) and (11.29), in that order, and so write

$$\begin{aligned}
\mathbf{U}^H \mathbf{A} \mathbf{V} &= \begin{bmatrix} \mathbf{U}_1^H \\ \mathbf{U}_2^H \end{bmatrix} \mathbf{A}[\mathbf{V}_1, \mathbf{V}_2] \\
&= \begin{bmatrix} \mathbf{U}_1^H \mathbf{A} \mathbf{V}_1 & \mathbf{U}_1^H \mathbf{A} \mathbf{V}_2 \\ \mathbf{U}_2^H \mathbf{A} \mathbf{V}_1 & \mathbf{U}_2^H \mathbf{A} \mathbf{V}_2 \end{bmatrix} \\
&= \begin{bmatrix} \mathbf{U}_1^H \mathbf{A}(\mathbf{A}^H \mathbf{U}_1 \mathbf{\Sigma}^{-1}) & (\mathbf{\Sigma} \mathbf{V}_1^H)\mathbf{V}_2 \\ (\mathbf{0})\mathbf{V}_1 & (\mathbf{0})\mathbf{V}_2 \end{bmatrix} \\
&= \begin{bmatrix} \mathbf{\Sigma} & \mathbf{0} \\ \mathbf{0} & \mathbf{0} \end{bmatrix}
\end{aligned}$$

which again is the desired result.

This completes the proof of the singular-value decomposition (SVD) theorem, described by Eq. (11.6).

Terminology and Relation to Eigenanalysis

The numbers $\sigma_1, \sigma_2, \ldots, \sigma_W$, constituting the diagonal matrix $\mathbf{\Sigma}$, are called the *singular values* of the matrix $\mathbf{A}$. The columns of the unitary matrix $\mathbf{V}$, that is, $\mathbf{v}_1, \mathbf{v}_2, \ldots, \mathbf{v}_M$, are the *right singular vectors* of $\mathbf{A}$, and the columns of the second unitary matrix $\mathbf{U}$, that is, $\mathbf{u}_1, \mathbf{u}_2, \ldots, \mathbf{u}_K$ are the *left singular vectors* of $\mathbf{A}$. We note from the preceding discussion that the right singular vectors $\mathbf{v}_1, \mathbf{v}_2, \ldots, \mathbf{v}_M$ are eigenvectors of $\mathbf{A}^H\mathbf{A}$, whereas the left singular vectors $\mathbf{u}_1, \mathbf{u}_2, \ldots, \mathbf{u}_K$ are eigenvectors of $\mathbf{A}\mathbf{A}^H$. Note that the number of positive singular values is equal to the rank of the matrix $\mathbf{A}$. The singular-value decomposition therefore provides the basis of a practical method for determining the rank of a matrix.

Since $\mathbf{U}\mathbf{U}^H$ equals the identity matrix, we find from Eq. (11.6) that

$$\mathbf{A}\mathbf{V} = \mathbf{U}\begin{bmatrix} \mathbf{\Sigma} & \mathbf{0} \\ \mathbf{0} & \mathbf{0} \end{bmatrix}$$

It follows therefore that

$$\begin{aligned} \mathbf{A}\mathbf{v}_i &= \sigma_i \mathbf{u}_i, \qquad i = 1, 2, \ldots, W \\ \mathbf{A}\mathbf{v}_i &= \mathbf{0}, \qquad i = W + 1, \ldots, K \end{aligned} \tag{11.30}$$

Correspondingly, we may express the data matrix $\mathbf{A}$ in the expanded form

$$\mathbf{A} = \sum_{i=1}^{W} \sigma_i \mathbf{u}_i \mathbf{v}_i^H \tag{11.31}$$

Since $\mathbf{V}\mathbf{V}^H$ equals the identity matrix, we also find from Eq. (11.6) that

$$\mathbf{U}^H\mathbf{A} = \begin{bmatrix} \mathbf{\Sigma} & \mathbf{0} \\ \mathbf{0} & \mathbf{0} \end{bmatrix}\mathbf{V}^H$$

or, equivalently,

$$\mathbf{A}^H\mathbf{U} = \mathbf{V}\begin{bmatrix} \mathbf{\Sigma} & \mathbf{0} \\ \mathbf{0} & \mathbf{0} \end{bmatrix}$$

It follows therefore that

$$\begin{aligned} \mathbf{A}^H\mathbf{u}_i &= \sigma_i \mathbf{v}_i, \qquad i = 1, 2, \ldots, W \\ \mathbf{A}^H\mathbf{u}_i &= \mathbf{0}, \qquad i = W + 1, \ldots, M \end{aligned} \tag{11.32}$$

In this case, we may express the Hermitian transpose of the data matrix $\mathbf{A}$ in the expanded form

$$\mathbf{A}^H = \sum_{i=1}^{W} \sigma_i \mathbf{v}_i \mathbf{u}_i^H \tag{11.33}$$

which checks exactly with Eq. (11.31).

The singular values of a matrix have many appealing analogies with the eigenvalues of a Hermitian matrix. Indeed, if the matrix $\mathbf{A}$ is Hermitian, then the singular values of $\mathbf{A}$ are just the absolute values of the eigenvalues of $\mathbf{A}$.

Example 1

In this example, we use the SVD to deal with the different facets of *matrix rank*. To be specific, let $\mathbf{A}$ be a K-by-M data matrix with rank W. The matrix $\mathbf{A}$ is said to be of *full rank* if

$$W = \min(K, M)$$

Otherwise, the matrix $\mathbf{A}$ is *rank deficient*. In the context of SVD, the rank W equals the number of nonzero singular values of matrix $\mathbf{A}$.

Consider next a computational environment that yields a numerical value for each element of the matrix $\mathbf{A}$ that is accurate to within $\pm\epsilon$. Let $\mathbf{B}$ denote the approximate value of matrix $\mathbf{A}$ so obtained. We define the *ϵ-rank* of matrix $\mathbf{A}$ as follows (Golub and Van Loan, 1989)

$$\operatorname{rank}(\mathbf{A}, \epsilon) = \min_{\|\mathbf{A}-\mathbf{B}\|<\epsilon} \operatorname{rank}(\mathbf{B}) \tag{11.34}$$

where $\|\mathbf{A} - \mathbf{B}\|$ is the *spectral norm* of the error matrix $\mathbf{A} - \mathbf{B}$ that results from the use of inaccurate computations. Extending the definition of spectral norm of the matrix introduced in Chapter 4 to the situation at hand, the spectral norm $\|\mathbf{A} - \mathbf{B}\|$ equals the largest singular value of the difference $\mathbf{A} - \mathbf{B}$. In any event, the K-by-M matrix $\mathbf{A}$ is said to be *numerically rank deficient* if

$$\operatorname{rank}(\mathbf{A}, \epsilon) < \min(K, M)$$

The SVD provides a sensible method for characterizing the ϵ-rank and the numerical rank deficiency of the matrix, because the singular values resulting from its use indicate how close a given matrix $\mathbf{A}$ is to another matrix $\mathbf{B}$ of lower rank in a simple fashion.

11.3 PSEUDOINVERSE

Our interest in the singular-value decomposition is to formulate a general definition of pseudoinverse. Let $\mathbf{A}$ denote a K-by-M matrix that has the singular-value decomposition described in Eq. (11.6). We define the pseudoinverse of the matrix $\mathbf{A}$ as (Stewart, 1973; Golub and Van Loan, 1989):

$$\mathbf{A}^{+} = \mathbf{V}\begin{bmatrix} \boldsymbol{\Sigma}^{-1} & \mathbf{0} \\ \mathbf{0} & \mathbf{0} \end{bmatrix}\mathbf{U}^{H} \tag{11.35}$$

where

$$\boldsymbol{\Sigma}^{-1} = \operatorname{diag}(\sigma_1^{-1}, \sigma_2^{-1}, \ldots, \sigma_W^{-1})$$

and W is the rank of the matrix $\mathbf{A}$.

We may identify two special cases that can arise when the matrix $\mathbf{A}$ is of full rank; that is, $W = \min(K, M)$. The two cases are as follows:

Case 1: Overdetermined System. With the data matrix $\mathbf{A}$ assumed to have full rank, the overdetermined system has $K > M$ and the rank $W = M$. In this case, the pseudoinverse of matrix $\mathbf{A}$ is defined by

$$\mathbf{A}^+ = (\mathbf{A}^H\mathbf{A})^{-1}\mathbf{A}^H \tag{11.36}$$

To show the validity of this special formula, we note from Eqs. (11.13) and (11.15) that

$$(\mathbf{A}^H\mathbf{A})^{-1} = \mathbf{V}_1\mathbf{\Sigma}^{-2}\mathbf{V}_1^H$$

and

$$\mathbf{A}^H = \mathbf{V}_1\mathbf{\Sigma}\mathbf{U}_1^H$$

Therefore, using this pair of relations, we may express the right side of Eq. (11.36) as follows:

$$\begin{aligned}(\mathbf{A}^H\mathbf{A})^{-1}\mathbf{A}^H &= (\mathbf{V}_1\mathbf{\Sigma}^{-2}\mathbf{V}_1^H)(\mathbf{V}_1\mathbf{\Sigma}\mathbf{U}_1^H)\\ &= \mathbf{V}_1\mathbf{\Sigma}^{-1}\mathbf{U}_1^H\\ &= \mathbf{V}\begin{bmatrix}\mathbf{\Sigma}^{-1} & \mathbf{0}\\ \mathbf{0} & \mathbf{0}\end{bmatrix}\mathbf{U}^H\\ &= \mathbf{A}^+\end{aligned}$$

which is the desired result. Note that the definition of Eq. (11.36) coincides with that in Eq. (11.3), given in Section 11.1.

According to Eq. (11.35), the pseudoinverse of the data matrix $\mathbf{A}$ for an overdetermined system may be expressed in the expanded form

$$\mathbf{A}^+ = \sum_{i=1}^{W}\frac{1}{\sigma_i}\mathbf{v}_i\mathbf{u}_i^H \tag{11.37}$$

Case 2: Underdetermined System. With the data matrix $\mathbf{A}$ assumed to have full rank, the underdetermined system has $M > K$ and the rank $W = K$. In this case, the pseudoinverse of matrix $\mathbf{A}$ is defined by

$$\mathbf{A}^+ = \mathbf{A}^H(\mathbf{A}\mathbf{A}^H)^{-1} \tag{11.38}$$

To show the validity of this second special formula, we note from Eqs. (11.24) and (11.26) that

$$(\mathbf{A}\mathbf{A}^H)^{-1} = \mathbf{U}_1\mathbf{\Sigma}^{-2}\mathbf{U}_1^H$$

and

$$\mathbf{A}^H = \mathbf{V}_1\mathbf{\Sigma}\mathbf{U}_1^H$$

Therefore, using this pair of relations in the right side of Eq. (11.38), we get

$$\begin{aligned}\mathbf{A}^H(\mathbf{A}\mathbf{A}^H)^{-1} &= (\mathbf{V}_1\mathbf{\Sigma}\mathbf{U}_1^H)(\mathbf{U}_1\mathbf{\Sigma}^{-2}\mathbf{U}_1^H)\\ &= \mathbf{V}_1\mathbf{\Sigma}^{-1}\mathbf{U}_1^H\\ &= \mathbf{V}\begin{bmatrix}\mathbf{\Sigma}^{-1} & \mathbf{0}\\ \mathbf{0} & \mathbf{0}\end{bmatrix}\mathbf{U}^H\\ &= \mathbf{A}^+\end{aligned}$$

which again is the desired result.

Note that the pseudoinverse $\mathbf{A}^+$ has the same mathematical structure, irrespective of whether the data matrix $\mathbf{A}$ refers to an overdetermined or underdetermined system.

11.4 INTERPRETATION OF SINGULAR VALUES AND SINGULAR VECTORS

Consider a K-by-M data matrix $\mathbf{A}$, for which the singular-value decomposition is given in Eq. (11.6), and the pseudoinverse is correspondingly given in Eq. (11.35). We assume that the system is overdetermined. Define a K-by-1 vector $\mathbf{y}$ and an M-by-1 vector $\mathbf{x}$ that are related to each other by the transformation matrix $\mathbf{A}$, as shown by

$$\mathbf{y} = \mathbf{A}\mathbf{x} \tag{11.39}$$

The vector $\mathbf{x}$ is constrained to have an Euclidean norm of unity, that is,

$$\|\mathbf{x}\| = 1 \tag{11.40}$$

Given the transformation of Eq. (11.39) and the constraint of Eq. (11.40), we wish to find the resulting locus of the points defined by the vector $\mathbf{y}$ in a K-dimensional space.

Solving Eq. (11.39) for $\mathbf{x}$, we get

$$\mathbf{x} = \mathbf{A}^+\mathbf{y} \tag{11.41}$$

where $\mathbf{A}^+$ is the pseudoinverse of $\mathbf{A}$. Substituting Eq. (11.37) in (11.41), we get

$$\begin{aligned} \mathbf{x} &= \sum_{i=1}^{W} \frac{1}{\sigma_i}\mathbf{v}_i\mathbf{u}_i^H\mathbf{y} \\ &= \sum_{i=1}^{W} \frac{(\mathbf{u}_i^H\mathbf{y})}{\sigma_i}\mathbf{v}_i \end{aligned} \tag{11.42}$$

where W is the rank of matrix $\mathbf{A}$, and the inner product $\mathbf{u}_i^H\mathbf{y}$ is a scalar. Imposing the constraint of Eq. (11.40) on (11.42), and recognizing that the right singular vectors, $\mathbf{v}_1, \mathbf{v}_2, \ldots, \mathbf{v}_W$ form an orthonormal set, we get

$$\sum_{i=1}^{W} \frac{|\mathbf{y}^H\mathbf{u}_i|^2}{\sigma_i^2} = 1 \tag{11.43}$$

Equation (11.43) defines the locus traced out by the tip of vector $\mathbf{y}$ in a K-dimensional space. Indeed, this is the equation of a *hyperellipsoid* (Golub and Van Loan, 1989).

To see this interpretation in a better way, define the complex scalar

$$\begin{aligned} \zeta_i &= \mathbf{y}^H\mathbf{u}_i \\ &= \sum_{k=1}^{K} y_k^* u_{ik}, \qquad i = 1, \ldots, W \end{aligned} \tag{11.44}$$

In other words, the complex scalar ζ_i is a linear combination of all possible values of the elements of the left singular vector $\mathbf{u}_i$, so ζ_i is referred to as the "span" of $\mathbf{u}_i$. We may thus rewrite Eq. (11.43) as

$$\sum_{i=1}^{W} \frac{|\zeta_i|^2}{\sigma_i^2} = 1 \tag{11.45}$$

This is the equation of a hyperellipsoid with coordinates $|\zeta_1|, \ldots, |\zeta_W|$ and semi-axis whose lengths are the singular values $\sigma_1, \ldots, \sigma_W$, respectively. Figure 11.2 illustrates the locus traced out by Eq. (11.45) for the case of $W = 2$ and $\sigma_1 > \sigma_2$, assuming that the data matrix is real.

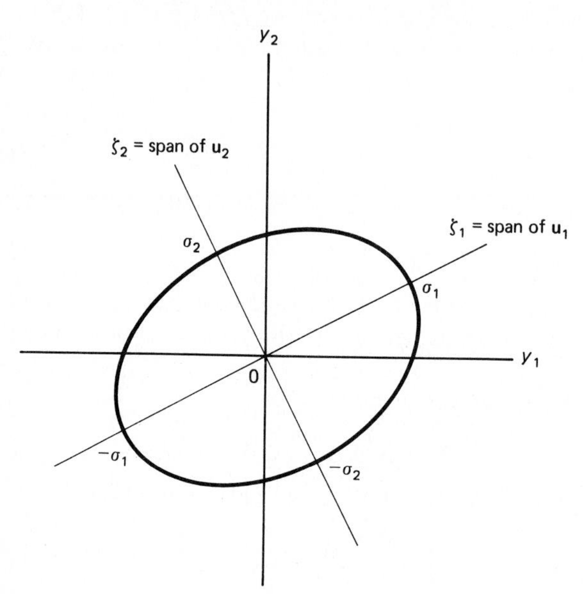

Figure 11.2 Locus of Eq. (11.45) for real data with $W = 2$ and $\sigma_1 > \sigma_2$.

11.5 MINIMUM NORM SOLUTION TO THE LINEAR LEAST-SQUARES PROBLEM

Having equipped ourselves with the general definition of the pseudoinverse of a matrix $\mathbf{A}$ in terms of its singular-value decomposition, we are now ready to tackle the solution to the linear least-squares problem even when null($\mathbf{A}$) $\neq 0$. In particular, we define the solution to the least-squares problem as in Eq. (11.4), reproduced here for convenience:

$$\hat{\mathbf{w}} = \mathbf{A}^{+}\mathbf{d} \tag{11.46}$$

The pseudoinverse matrix $\mathbf{A}^{+}$ is itself defined by Eq. (11.35). We thus find that, out of the many vectors that solve the least-squares problem when null ($\mathbf{A}$) $\neq 0$, the one defined by Eq. (11.46) is *unique* in that it has the *shortest length possible in the Euclidean sense* (Stewart, 1973).

We prove this important result by manipulating the equation that defines the minimum value of the sum of error squares produced in the method of least squares [see Eq. (10.49)]. We note that both matrix products $\mathbf{V}\mathbf{V}^H$ and $\mathbf{U}\mathbf{U}^H$ equal identity matrices. Hence, we may start with Eq. (10.49) and combine it with Eq. (10.48), and then write

$$\begin{aligned}\mathscr{E}_{\min} &= \mathbf{d}^H\mathbf{d} - \mathbf{d}^H\mathbf{A}\hat{\mathbf{w}} \\ &= \mathbf{d}^H(\mathbf{d} - \mathbf{A}\hat{\mathbf{w}}) \\ &= \mathbf{d}^H\mathbf{U}\mathbf{U}^H(\mathbf{d} - \mathbf{A}\mathbf{V}\mathbf{V}^H\hat{\mathbf{w}}) \\ &= \mathbf{d}^H\mathbf{U}(\mathbf{U}^H\mathbf{d} - \mathbf{U}^H\mathbf{A}\mathbf{V}\mathbf{V}^H\hat{\mathbf{w}})\end{aligned} \tag{11.47}$$

Let

$$\begin{aligned}\mathbf{V}^H\hat{\mathbf{w}} &= \mathbf{z} \\ &= \begin{bmatrix}\mathbf{z}_1 \\ \mathbf{z}_2\end{bmatrix}\end{aligned} \tag{11.48}$$

and

$$\begin{aligned}\mathbf{U}^H\mathbf{d} &= \mathbf{c} \\ &= \begin{bmatrix}\mathbf{c}_1 \\ \mathbf{c}_2\end{bmatrix}\end{aligned} \tag{11.49}$$

where $\mathbf{z}_1$ and $\mathbf{c}_1$ are W-by-1 vectors, and $\mathbf{z}_2$ and $\mathbf{c}_2$ are two other vectors. Thus, substituting Eqs. (11.6), (11.48), and (11.49) in (11.47), we get

$$\begin{aligned}\mathscr{E}_{\min} &= \mathbf{d}^H\mathbf{U}\left(\begin{bmatrix}\mathbf{c}_1 \\ \mathbf{c}_2\end{bmatrix} - \begin{bmatrix}\mathbf{\Sigma} & \mathbf{0} \\ \mathbf{0} & \mathbf{0}\end{bmatrix}\begin{bmatrix}\mathbf{z}_1 \\ \mathbf{z}_2\end{bmatrix}\right) \\ &= \mathbf{d}^H\mathbf{U}\begin{bmatrix}\mathbf{c}_1 - \mathbf{\Sigma}\mathbf{z}_1 \\ \mathbf{c}_2\end{bmatrix}\end{aligned} \tag{11.50}$$

For $\mathscr{E}_{\min}$ to be minimum, we require that

$$\mathbf{c}_1 = \mathbf{\Sigma}\mathbf{z}_1 \tag{11.51}$$

or, equivalently,

$$\mathbf{z}_1 = \mathbf{\Sigma}^{-1}\mathbf{c}_1 \tag{11.52}$$

We observe that $\mathscr{E}_{\min}$ is independent of $\mathbf{z}_2$. Hence, the value of $\mathbf{z}_2$ is arbitrary. However, if we let $\mathbf{z}_2 = \mathbf{0}$, we get the special result

$$\begin{aligned}\hat{\mathbf{w}} &= \mathbf{V}\mathbf{z} \\ &= \mathbf{V}\begin{bmatrix}\mathbf{\Sigma}^{-1}\mathbf{c}_1 \\ \mathbf{0}\end{bmatrix}\end{aligned} \tag{11.53}$$

We may also express $\hat{\mathbf{w}}$ in the equivalent form:

$$\begin{aligned}\hat{\mathbf{w}} &= \mathbf{V}\begin{bmatrix}\mathbf{\Sigma}^{-1} & \mathbf{0} \\ \mathbf{0} & \mathbf{0}\end{bmatrix}\begin{bmatrix}\mathbf{c}_1 \\ \mathbf{c}_2\end{bmatrix} \\ &= \mathbf{V}\begin{bmatrix}\mathbf{\Sigma}^{-1} & \mathbf{0} \\ \mathbf{0} & \mathbf{0}\end{bmatrix}\mathbf{U}^H\mathbf{d} \\ &= \mathbf{A}^+\mathbf{d}\end{aligned}$$

This coincides exactly with the value defined by Eq. (11.46), where the pseudoinverse $\mathbf{A}^+$ is defined by Eq. (11.35). In effect, we have shown that this value of $\hat{\mathbf{w}}$ does indeed solve the least-squares problem.

Moreover, the vector $\hat{\mathbf{w}}$ so defined is *unique* in that it has the minimum Euclidean norm possible. In particular, since $\mathbf{V}\mathbf{V}^H = \mathbf{I}$, we find from Eq. (11.53) that the Euclidean norm of $\hat{\mathbf{w}}$ equals

$$\|\hat{\mathbf{w}}\|^2 = \|\mathbf{\Sigma}^{-1}\mathbf{c}_1\|^2$$

Consider now another solution to the least-squares problem that is denoted by

$$\mathbf{w}' = \mathbf{V}\begin{bmatrix} \mathbf{\Sigma}^{-1}\mathbf{c}_1 \\ \mathbf{z}_2 \end{bmatrix}, \qquad \mathbf{z}_2 \neq \mathbf{0}$$

The squared Euclidean norm of $\mathbf{w}'$ equals

$$\|\mathbf{w}'\|^2 = \|\mathbf{\Sigma}^{-1}\mathbf{c}_1\|^2 + \|\mathbf{z}_2\|^2$$

For any $\mathbf{z}_2 \neq 0$, we see therefore that

$$\|\mathbf{w}\| < \|\mathbf{w}'\| \tag{11.54}$$

In summary, the tap-weight $\hat{\mathbf{w}}$ of a linear transversal filter defined by Eq. (11.46) is a unique solution to the linear least-squares problem, even when null($\mathbf{A}$) $\neq 0$. *The vector $\hat{\mathbf{w}}$ is unique in that it is the only tap-weight vector that simultaneously satisfies two requirements:* (1) *it produces the minimum sum of error squares, and* (2) *it has the smallest Euclidean norm possible. This special value of the tap-weight vector $\hat{\mathbf{w}}$ is called the minimum-norm solution.*

Another Formulation of the Minimum-Norm Solution

We may develop an expanded formulation of the minimum-norm solution, depending on whether we are dealing with the overdetermined or underdetermined case. These two cases are considered in turn.

Case 1: Overdetermined. For this case, the number of equations K is greater than the number of unknown parameters M. To proceed then, we substitute Eq. (11.35) in (11.46), and then use the partitioned forms of the unitary matrices $\mathbf{V}$ and $\mathbf{U}$. We may then write

$$\begin{aligned}
\hat{\mathbf{w}} &= (\mathbf{V}_1\mathbf{\Sigma}^{-1})(\mathbf{A}\mathbf{V}_1\mathbf{\Sigma}^{-1})^H\mathbf{d} \\
&= \mathbf{V}_1\mathbf{\Sigma}^{-1}\mathbf{\Sigma}^{-1}\mathbf{V}_1^H\mathbf{A}^H\mathbf{d} \\
&= \mathbf{V}_1\mathbf{\Sigma}^{-2}\mathbf{V}_1\mathbf{A}^H\mathbf{d}
\end{aligned} \tag{11.55}$$

Hence, using the definition [see Eq. (11.10)]

$$\mathbf{V}_1 = [\mathbf{v}_1, \mathbf{v}_2, \ldots, \mathbf{v}_W]$$

in Eq. (11.55), we get the following expanded formulation for $\hat{\mathbf{w}}$ for the overdetermined case:

$$\hat{\mathbf{w}} = \sum_{i=1}^{W} \frac{\mathbf{v}_i}{\sigma_i^2}\mathbf{v}_i^H\mathbf{A}^H\mathbf{d} \tag{11.56}$$

Case 2: Underdetermined. For this second case, the number of equations K is smaller than the number of unknowns M. This time we find it appropriate to use the representation given in Eq. (11.26) for the submatrix $\mathbf{V}_1$ in terms of the data matrix $\mathbf{A}$. Thus, substituting Eq. (11.26) in (11.46), we get

$$\begin{aligned}\hat{\mathbf{w}} &= (\mathbf{A}^H\mathbf{U}_1\mathbf{\Sigma}^{-1})(\mathbf{\Sigma}^{-1}\mathbf{U}_1\mathbf{d}) \\ &= \mathbf{A}^H\mathbf{U}_1\mathbf{\Sigma}^{-2}\mathbf{U}_1^H\mathbf{d}\end{aligned} \tag{11.57}$$

Substituting the definition [see Eq. (11.21)]

$$\mathbf{U}_1 = [\mathbf{u}_1, \mathbf{u}_2, \ldots, \mathbf{u}_W]$$

in Eq. (11.57), we get the following expanded formulation for $\hat{\mathbf{w}}$ for the underdetermined case:

$$\hat{\mathbf{w}} = \sum_{i=1}^{W} \frac{(\mathbf{u}_i^H\mathbf{d})}{\sigma_i^2}\mathbf{A}^H\mathbf{u}_i \tag{11.58}$$

which is different from that of Eq. (11.56) for the overdetermined case.

Summary and Discussion

The expanded formula of (11.56) for the minimum-norm weight vector $\hat{\mathbf{w}}$ for the overdetermined case lends itself to a useful interpretation. In particular, we recognize two facts:

1. The matrix product $\mathbf{A}^H\mathbf{d}$ equals the time-averaged cross-correlation vector $\boldsymbol{\theta}$.
2. The right singular vector $\mathbf{v}_i$ is the eigenvector of the time-averaged correlation matrix $\mathbf{\Phi} = \mathbf{A}^H\mathbf{A}$ associated with the eigenvalue $\lambda_i = \sigma_i^2$.

Accordingly, we may rewrite Eq. (11.56) in terms of the cross-correlation vector $\boldsymbol{\theta}$, the set of eigenvalues $\{\lambda_i\}$ and the associated set of eigenvectors $\{\mathbf{v}_i\}$ as

$$\hat{\mathbf{w}} = \sum_{i=1}^{W} \frac{(\mathbf{v}_i^H\boldsymbol{\theta})}{\lambda_i}\mathbf{v}_i \tag{11.59}$$

In general, however, it is not a good idea to use the formula of Eq. (11.59) to compute the minimum-norm value of the tap-weight vector $\hat{\mathbf{w}}$ by first finding the eigenvalues of the correlation matrix $\mathbf{\Phi}$, tempting as such a procedure may be. The preferred method is to use the SVD-based procedure, which has the advantage in that it is numerically stable. An algorithm is said to be *numerically stable* if it does not introduce any more sensitivity to perturbation than that which is inherently present in the problem (Klema and Laub, 1980).

Thus, to compute the minimum-norm value $\hat{\mathbf{w}}$ of the tap-weight vector, the recommended method is to proceed as follows:

1. Compute the singular-value decomposition of the data matrix $\mathbf{A}$, and thereby find the singular values $\sigma_1, \sigma_2, \ldots, \sigma_W$, the associated right-singular vectors $\mathbf{v}_1, \mathbf{v}_2, \ldots, \mathbf{v}_W$ and the left-singular vectors $\mathbf{u}_1, \mathbf{u}_2, \ldots, \mathbf{u}_W$.
2. To compute $\hat{\mathbf{w}}$, use the formula of Eq. (11.56) for the overdetermined case ($K > M$), or that of Eq. (11.58) for the underdetermined case ($K < M$).

11.6 NORMALIZED LMS ALGORITHM VIEWED AS THE MINIMUM-NORM SOLUTION TO AN UNDERDETERMINED LEAST-SQUARES ESTIMATION PROBLEM

In Chapter 9 we derived the normalized least-mean-square (LMS) algorithm as the solution to a constrained minimization problem. In this section we revisit this algorithm in light of the theory developed on singular-value decomposition. In particular, we show that the normalized LMS algorithm is indeed the minimum-norm solution to an underdetermined linear least-squares problem involving a single error equation with M unknowns, where M is the dimension of the tap-weight vector in the algorithm.

Consider the error equation

$$\epsilon(n) = d(n) - \hat{\mathbf{w}}^H(n+1)\mathbf{u}(n) \tag{11.60}$$

where $d(n)$ is a desired response and $\mathbf{u}(n)$ is a tap-input vector, both measured at time n. The requirement is to find the tap-weight vector $\hat{\mathbf{w}}(n+1)$, measured at time $n+1$, such that the change in the tap-weight vector given by

$$\delta\hat{\mathbf{w}}(n+1) = \hat{\mathbf{w}}(n+1) - \hat{\mathbf{w}}(n) \tag{11.61}$$

is minimized, subject to the constraint

$$\epsilon(n) = 0 \tag{11.62}$$

Using Eq. (11.61) in (11.60), we may reformulate the error $\epsilon(n)$ as

$$\epsilon(n) = d(n) - \hat{\mathbf{w}}^H(n)\mathbf{u}(n) - \delta\hat{\mathbf{w}}^H(n+1)\mathbf{u}(n) \tag{11.63}$$

We now recognize the customary definition of the estimation error, namely

$$e(n) = d(n) - \hat{\mathbf{w}}^H(n)\mathbf{u}(n) \tag{11.64}$$

Hence, we may simplify Eq. (11.63) as

$$\epsilon(n) = e(n) - \delta\hat{\mathbf{w}}^H(n+1)\mathbf{u}(n) \tag{11.65}$$

Thus, complex conjugating both sides of Eq. (11.65), we note that the constraint of Eq. (11.62) is equivalent to

$$\mathbf{u}^H(n)\delta\hat{\mathbf{w}}(n+1) = e^*(n) \tag{11.66}$$

Accordingly, we may restate our constrained minimization problem as follows:

> Find the minimum-norm solution for the change $\delta\hat{\mathbf{w}}(n+1)$ in the tap-weight vector at time $n+1$, which satisfies the equation
>
> $$\mathbf{u}^H(n)\delta\hat{\mathbf{w}}(n+1) = e^*(n)$$

This problem is one of linear least-squares estimation that is underdetermined. To solve it, we may use the method of singular-value decomposition described in Eq. (11.58). To help us in the application of this method, we use Eq. (11.66) to make the identifications listed in Table 11.1 between the normalized LMS algorithm and linear least-squares estimation. In particular, we note that the normalized LMS algorithm has only one nonzero singular value equal to the squared norm of the tap-

TABLE 11.1 SUMMARY OF CORRESPONDENCES BETWEEN LINEAR LEAST-SQUARES ESTIMATION AND NORMALIZED LMS ALGORITHM.

	Linear least-squares estimation (underdetermined)	Normalized LMS algorithm
Data matrix	$\mathbf{A}$	$\mathbf{u}^H(n)$
Desired data vector	$\mathbf{d}$	$e^*(n)$
Parameter vector	$\hat{\mathbf{w}}$	$\delta\hat{\mathbf{w}}(n+1)$
Rank	W	1
Eigenvalue	σ_i^2	$\| \mathbf{u}(n) \|^2$
Eigenvector	$\mathbf{u}_i$	1

input vector $\mathbf{u}(n)$; that is, the rank $W = 1$. The corresponding left-singular vector is therefore simply equal to one. Hence, with the aid of Table 11.1, the application of Eq. (11.58) yields

$$\delta\hat{\mathbf{w}}(n+1) = \frac{1}{\| \mathbf{u}(n) \|^2}\mathbf{u}(n)e^*(n) \tag{11.67}$$

This is precisely the result that we derived previously in Chapter 9; see Eq. (9.162).

We may next follow a reasoning similar to that described in Section 9.15 and redefine the change $\delta\hat{\mathbf{w}}(n+1)$ by introducing a scaling factor $\tilde{\mu}$ as shown by [see Eq. (9.163)]

$$\delta\hat{\mathbf{w}}(n+1) = \frac{\tilde{\mu}}{\| \mathbf{u}(n) \|^2}\mathbf{u}(n)e^*(n)$$

or equivalently we may write

$$\hat{\mathbf{w}}(n+1) = \hat{\mathbf{w}}(n) + \frac{\tilde{\mu}}{\| \mathbf{u}(n) \|^2}\mathbf{u}(n)e^*(n) \tag{11.68}$$

By so doing, we are able to exercise control over the change in the tap-weight vector from one iteration to the next without changing its direction. Equation (11.68) is the tap-weight vector update for the normalized LMS algorithm.

The important point to note from the discussion presented in this section is that the singular-value decomposition provides an insightful link between the underdetermined form of linear least-squares estimation and the LMS theory. In particular, we have shown that the weight update in the normalized LMS algorithm may indeed be viewed as the minimum norm solution to an underdetermined form of the linear least-squares problem involving a single error equation with a number of unknowns equal to the dimension of the tap-weight vector in the algorithm.

11.7 COMPUTATIONAL PROCEDURES

We next turn our attention to the practical issue of how to compute the singular value decomposition of a data matrix. We may approach the computation by reducing the singular-value decomposition problem to an ordinary eigenvalue problem. Here it is recognized that the squares of the singular values of a data matrix $\mathbf{A}$ are the eigenvalues of the correlation matrix $\mathbf{A}^H\mathbf{A}$, and so we may compute the singular values of $\mathbf{A}$ by finding the eigenvalues of $\mathbf{A}^H\mathbf{A}$. However, this approach is inadvisable for important numerical reasons. By working with the correlation matrix $\mathbf{A}^H\mathbf{A}$ instead of the data matrix $\mathbf{A}$, the dynamic range of the numerical computation is *doubled,* and the word length required for a prescribed numerical accuracy is increased roughly by a *factor of* 2. Accordingly, the recommended procedure for computing the singular value decomposition of a data matrix is to work directly with the matrix itself.

In the sequel, we describe two important algorithms for SVD computation:

1. *Jacobi algorithm.* This algorithm is well suited for computing the singular value decomposition of a real-valued square matrix. To use it for complex data, additional work is required. The Jacobi algorithm employs a sequence of 2-by-2 *plane rotations,* each of which is configured to annihilate a particular pair of off-diagonal elements in the matrix. Gradually, but surely, the data matrix is reduced to diagonal form. As a result of the step-by-step applications of the Givens rotations, however, zeros created at one step of the algorithm are lost in the next step. Nevertheless, through the use of a well-defined strategy, the algorithm is assured of rapid convergence. The Jacobi algorithm and the tools for it are covered in Sections 11.8 and 11.9.
2. *QR algorithm.* The QR algorithm, adapted for SVD computation, is more general in application than the Jacobi algorithm. It proceeds by first reducing the data matrix to upper *bidiagonal* form. The bidiagonalization of the data matrix is accomplished by applying a sequence of *Householder transformations* in a ping-pong fashion. The next major step in the algorithm consists of an *iterative process* designed to zero the superdiagonal elements of the bidiagonal matrix. The QR algorithm and the tools for it are covered in Sections 11.10 and 11.11.

The Jacobi algorithm and the QR algorithm are both *data adaptive* and *block processing* oriented. They share a common goal (albeit in different ways): the diagonalization of the data matrix to within some prescribed numerical precision.

11.8 JACOBI ROTATIONS

An algebraic tool that is fundamental to the Jacobi algorithm is the 2-by-2 *orthogonal matrix:*[2]

$$\mathbf{J} = \begin{bmatrix} c & s \\ -s & c \end{bmatrix} \tag{11.69}$$

where c and s are real parameters defined by

$$c = \cos\theta \tag{11.70}$$

and

$$s = \sin\theta \tag{11.71}$$

[2] In SVD terminology (and eigenanalysis for that matter), the term "orthogonal matrix" is used in the context of *real* data, whereas the term "unitary matrix" is used for complex data.

with the *constraint:*

$$c^2 + s^2 = 1 \tag{11.72}$$

We refer to the transformation $\mathbf{J}$ as a "plane rotation," because premultiplication of a 2-by-1 vector by $\mathbf{J}$ amounts to a plane rotation of that vector.

The transformation of Eq. (11.69) is referred to as the *Jacobi rotation* in honor of Jacobi (1846), who proposed a method for reducing a symmetric matrix to diagonal form.[3]

Example 2

We illustrate the nature of this plane rotation by considering the case of a real 2-by-1 vector:

$$\mathbf{a} = \begin{bmatrix} a_i \\ a_k \end{bmatrix}$$

Then premultiplication of the vector $\mathbf{a}$ by $\mathbf{J}$ yields

$$\begin{aligned} \mathbf{x} &= \mathbf{Ja} \\ &= \begin{bmatrix} c & s \\ -s & c \end{bmatrix} \begin{bmatrix} a_i \\ a_k \end{bmatrix} \\ &= \begin{bmatrix} ca_i + sa_k \\ -sa_i + ca_k \end{bmatrix} \end{aligned}$$

We may readily show, in view of the definitions of the rotation parameters c and s, that the vector $\mathbf{x}$ has the same Euclidean length as the vector $\mathbf{a}$. Moreover, given that the angle θ is positive, the transformation $\mathbf{J}$ *rotates* the vector $\mathbf{a}$ in a clockwise direction into the new position defined by $\mathbf{x}$, as illustrated in Fig. 11.3. Note that the vectors $\mathbf{a}$ and $\mathbf{x}$ *remain in the same* (i, k) *plane,* hence the name "plane rotation."

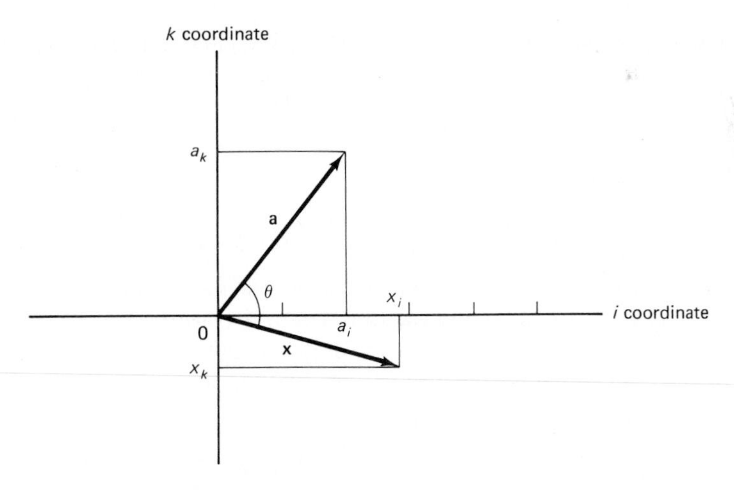

Figure 11.3 Plane rotation of a real 2-by-1 vector.

[3] The 2-by-2 plane rotation of Eq. (11.69) is also referred to as the *Givens rotation*. Indeed, in Section 11.10 and again in Chapters 14 and 18, we use the latter terminology. The reference to this plane rotation as a Jacobi rotation is for historical reasons.

Two-Sided Jacobi Algorithm for Real Data

Consider the simple case of a *real* 2-by-2 data matrix:

$$\mathbf{A} = \begin{bmatrix} a_{ii} & a_{ik} \\ a_{ki} & a_{kk} \end{bmatrix} \tag{11.73}$$

We assume that $\mathbf{A}$ is nonsymmetric; that is, $a_{ki} \neq a_{ik}$. The requirement is to *diagonalize* this 2-by-2 matrix. We do so by means of two Jacobi rotations $\mathbf{J}_1$ and $\mathbf{J}_2$, as shown by

$$\underbrace{\begin{bmatrix} c_1 & s_1 \\ -s_1 & c_1 \end{bmatrix}^T}_{\mathbf{J}_1} \underbrace{\begin{bmatrix} a_{ii} & a_{ik} \\ a_{ki} & a_{kk} \end{bmatrix}}_{\mathbf{A}} \underbrace{\begin{bmatrix} c_2 & s_2 \\ -s_2 & c_2 \end{bmatrix}}_{\mathbf{J}_2} = \underbrace{\begin{bmatrix} d_1 & 0 \\ 0 & d_2 \end{bmatrix}}_{\substack{\text{diagonal} \\ \text{matrix}}} \tag{11.74}$$

To design the two Jacobi rotations indicated in Eq. (11.74) we proceed in two stages. Stage I transforms the 2-by-2 data matrix $\mathbf{A}$ into a symmetric matrix; we refer to this as "symmetrization." Stage II diagonalizes the symmetric matrix resulting from stage I; we refer to this second stage as "diagonalization." Of course, if the data matrix is symmetric to begin with, we proceed to stage II directly.

Stage I: Symmetrization. To transform the 2-by-2 data matrix $\mathbf{A}$ into a symmetric matrix, we premultiply it by the transpose of a Jacobi rotation $\mathbf{J}$ and thus write

$$\underbrace{\begin{bmatrix} c & s \\ -s & c \end{bmatrix}^T}_{\mathbf{J}^T} \underbrace{\begin{bmatrix} a_{ii} & a_{ik} \\ a_{ki} & a_{kk} \end{bmatrix}}_{\mathbf{A}} = \underbrace{\begin{bmatrix} y_{ii} & y_{ki} \\ y_{ki} & y_{kk} \end{bmatrix}}_{\mathbf{Y}} \tag{11.75}$$

Expanding the left side of Eq. (11.75) and equating terms, we get

$$y_{ii} = ca_{ii} - sa_{ki} \tag{11.76}$$

$$y_{kk} = sa_{ik} + ca_{kk} \tag{11.77}$$

$$y_{ik} = ca_{ik} - sa_{kk} \tag{11.78}$$

$$y_{ki} = sa_{ii} + ca_{ki} \tag{11.79}$$

The purpose of stage I is to compute the cosine–sine pair (c, s) such that the 2-by-2 matrix $\mathbf{Y}$ produced by the Jacobi rotation $\mathbf{J}$ is symmetric. In other words, the elements y_{ik} and y_{ki} are to equal each other, as indicated in Eq. (11.75).

Define a parameter ρ as the ratio of c to s; that is,

$$\rho = \frac{c}{s} \tag{11.80}$$

We may relate ρ to the elements of the data matrix by setting $y_{ik} = y_{ki}$. Thus, using Eqs. (11.78) and (11.79), we obtain

$$\rho = \frac{a_{ii} + a_{kk}}{a_{ik} - a_{ki}}, \qquad a_{ki} \neq a_{ik} \tag{11.81}$$

Next, we determine the value of s by eliminating c between Eqs. (11.72) and (11.80); hence,

$$s = \frac{1}{\sqrt{1 + \rho^2}} \tag{11.82}$$

The computation of c and s thus proceeds as follows:

1. Use Eq. (11.81) to evaluate ρ.
2. Use Eq. (11.82) to evaluate the sine parameter s.
3. Use Eq. (11.80) to evaluate the cosine parameter c.

If **A** is symmetric to begin with, then $a_{ki} = a_{ik}$, in which case we have $s = 0$ and $c = 1$.

Stage II: Diagonalization. The purpose of stage II is to diagonalize the symmetric matrix **Y** produced in stage I. To do so, we premultiply and postmultiply it by $\mathbf{J}_2^T$ and $\mathbf{J}_2$, respectively, where $\mathbf{J}_2$ is a second Jacobi rotation to be determined. This operation is simply an orthogonal similarity transformation applied to a symmetric matrix. We may thus write

$$\underbrace{\begin{bmatrix} c_2 & s_2 \\ -s_2 & c_2 \end{bmatrix}^T}_{\mathbf{J}_2^T} \underbrace{\begin{bmatrix} y_{ii} & y_{ki} \\ y_{ki} & y_{kk} \end{bmatrix}}_{\mathbf{Y}} \underbrace{\begin{bmatrix} c_2 & s_2 \\ -s_2 & c_2 \end{bmatrix}}_{\mathbf{J}_2} = \underbrace{\begin{bmatrix} d_1 & 0 \\ 0 & d_2 \end{bmatrix}}_{\mathbf{D}} \tag{11.83}$$

Expanding the left side of Eq. (11.83) and then equating the respective diagonal terms, we get

$$d_1 = c_2^2 y_{ii} - 2c_2 s_2 y_{ki} + s_2^2 y_{kk} \tag{11.84}$$

$$d_2 = s_2^2 y_{ii} + 2c_2 s_2 y_{ki} + c_2^2 y_{kk} \tag{11.85}$$

Let o_1 and o_2 denote the off-diagonal terms of the 2-by-2 matrix formed by carrying out the matrix multiplications indicated on the left side of Eq. (11.83). From symmetry considerations, we have

$$o_1 = o_2 \tag{11.86}$$

Evaluating these off-diagonal terms, and equating them to zero for diagonalization, we get

$$0 = (y_{ii} - y_{kk}) - \left(\frac{s_2}{c_2}\right) y_{ki} + \left(\frac{c_2}{s_2}\right) y_{ki} \tag{11.87}$$

Equation (11.87) suggests that we introduce the following two definitions:

$$t = \frac{s_2}{c_2} \tag{11.88}$$

and

$$\zeta = \frac{y_{kk} - y_{ii}}{2y_{ki}} \tag{11.89}$$

Hence, we may rewrite Eq. (11.87) as

$$t^2 + 2\zeta t - 1 = 0 \tag{11.90}$$

Equation (11.90) is a quadratic in t; it therefore has two possible solutions, yielding the following different rotations:

1. *Inner rotation,* for which we have the solution:

$$t = \frac{\zeta}{\zeta^2 + \sqrt{1 + \zeta^2}} \tag{11.91}$$

Having computed t, we may use Eqs. (11.72) and (11.88) to solve for c_2 and s_2, obtaining

$$c_2 = \frac{1}{\sqrt{1 + t^2}} \tag{11.92}$$

and

$$s_2 = tc_2 \tag{11.93}$$

We note from Eqs. (11.70), (11.71) and (11.88) that the rotation angle θ_2 is related to t as follows:

$$\theta_2 = \arctan t \tag{11.94}$$

Hence, the adoption of the solution given in Eq. (11.91) produces a Jacobi rotation $\mathbf{J}_2$ for which the angle θ_2 lies in the interval $[0, \pi/4]$; this rotation is therefore called an *inner rotation*. The computation thus proceeds as follows:
(a) Use Eq. (11.89) to compute ζ.
(b) Use Eq. (11.91) to compute t.
(c) Use Eqs. (11.92) and (11.93) to c_2 and s_2, respectively.
If the original matrix $\mathbf{A}$ is diagonal, then $a_{ik} = a_{ki} = 0$, in which case the angle $\theta_2 = 0$ and so the matrix remains unchanged.

2. *Outer rotation,* for which we have the solution:

$$t = -(\zeta + \sqrt{1 + \zeta^2}) \tag{11.95}$$

Having computed t, we may then evaluate c_2 and s_2 using the formulas (11.92) and (11.93), respectively; we may do so, because the derivations of these two equations are independent of the quadratic equation (11.90). In this second case, however, the use of Eq. (11.95) in (11.94) yields a Jacobi rotation for which the angle θ_2 lies in the interval $[\pi/4, \pi/2]$. The rotation associated with the second solution is therefore referred to as the *outer rotation*. Note that if the original matrix $\mathbf{A}$ is diagonal, then $a_{ik} = a_{ki} = 0$, in which case $\theta_2 = \pi/2$. In this special case, the values of the matrix are interchanged, as shown by

$$\begin{bmatrix} 0 & -1 \\ 1 & 0 \end{bmatrix} \begin{bmatrix} a_{ii} & 0 \\ 0 & a_{kk} \end{bmatrix} \begin{bmatrix} 0 & 1 \\ -1 & 0 \end{bmatrix} = \begin{bmatrix} a_{kk} & 0 \\ 0 & a_{ii} \end{bmatrix} \tag{11.96}$$

Fusion of Rotations J and $\mathbf{J}_2$

Substituting the matrix $\mathbf{Y}$ of Eq. (11.75) in (11.83) and comparing the resulting equation with (11.74), we deduce the following definition for $\mathbf{J}_1$ in terms of $\mathbf{J}$ (determined in the symmetrization stage) and $\mathbf{J}_2$ (determined in the diagonalization stage):

$$\mathbf{J}_1^T = \mathbf{J}_2^T \mathbf{J}^T$$

or, equivalently,

$$\mathbf{J}_1 = \mathbf{J}\mathbf{J}_2 \tag{11.97}$$

In other words, in terms of the cosine–sine parameters, we have

$$\underbrace{\begin{bmatrix} c_1 & s_1 \\ -s_1 & c_1 \end{bmatrix}}_{\mathbf{J}_1} = \underbrace{\begin{bmatrix} c & s \\ -s & c \end{bmatrix}}_{\mathbf{J}} \underbrace{\begin{bmatrix} c_2 & s_2 \\ -s_2 & c_2 \end{bmatrix}}_{\mathbf{J}_2} \tag{11.98}$$

Expanding and equating terms, we thus obtain

$$c_1 = cc_2 - ss_2 \tag{11.99}$$

and

$$s_1 = sc_2 + cs_2 \tag{11.100}$$

For real data, from Eqs. (11.99) and (11.100) we find that the angles θ and θ_2, associated with the rotations $\mathbf{J}$ and $\mathbf{J}_2$, respectively, add to produce the angle θ_1 associated with $\mathbf{J}_1$.

Two Special Cases

For reasons that will become apparent later in Section 11.9, the algorithm described herein for computing the singular-value decomposition has to be capable of handling two special cases:

Case 1: $a_{kk} = a_{ik} = 0$. In this case we need only perform the symmetrization of $\mathbf{A}$, as shown by

$$\begin{bmatrix} c_1 & s_1 \\ -s_1 & c_1 \end{bmatrix}^T \begin{bmatrix} a_{ii} & 0 \\ a_{ki} & 0 \end{bmatrix} \begin{bmatrix} 1 & 0 \\ 0 & 1 \end{bmatrix} = \begin{bmatrix} d_1 & 0 \\ 0 & 0 \end{bmatrix} \tag{11.101}$$

Case 2: $a_{kk} = a_{ki} = 0$. In this case, we have

$$\begin{bmatrix} 1 & 0 \\ 0 & 1 \end{bmatrix} \begin{bmatrix} a_{ii} & a_{ik} \\ 0 & 0 \end{bmatrix} \begin{bmatrix} c_2 & s_2 \\ -s_2 & c_2 \end{bmatrix} = \begin{bmatrix} d_1 & 0 \\ 0 & 0 \end{bmatrix} \tag{11.102}$$

Additional Operations for Complex Data

The Jacobi rotation described in Eq. (11.74) applies to real data only, because (to begin with) the cosine and sine parameters defining the rotation were all chosen to be real. To extend its application to the more general case of complex data, we have to perform additional operations on the data. At first sight, it may appear that we merely have to modify stage I (symmetrization) of the two-sided Jacobi algorithm so as to accommodate a complex 2-by-2 matrix. In reality, however, the issue of dealing with complex data (in the context of the Jacobi algorithm) is not that simple. The approach taken here is first to reduce the complex 2-by-2 matrix of Eq. (11.73) to a real form, and then proceed with the application of the two-sided Jacobi algorithm in the usual way.[4] The *complex-to-real data reduction* is performed by following a two-stage procedure, as described next.

[4] F.T. Luk, *Private Communication,* 1990.

Stage I: Triangularization. Consider a complex 2-by-2 matrix **A** having the form given in Eq. (11.73). Without loss of generality, we assume that the leading element a_{ii} is a positive real number. This assumption may be justified (if need be) by factoring out the exponential term $e^{j\theta_{ii}}$, where θ_{ii} is the phase angle of a_{ii}. The factorization has the effect of leaving inside the 2-by-2 matrix a positive real term equal to the magnitude of a_{ii}, and subtracting θ_{ii} from the phase angle of each of the remaining three complex terms in the matrix.

Let the matrix **A** so described be premutiplied by a 2-by-2 plane rotation for the purpose of its triangularization, shown by

$$\begin{bmatrix} c & s^* \\ -s & c \end{bmatrix} \begin{bmatrix} a_{ii} & a_{ik} \\ a_{ki} & a_{kk} \end{bmatrix} = \begin{bmatrix} \omega_{ii} & \omega_{ik} \\ 0 & \omega_{kk} \end{bmatrix} \tag{11.103}$$

The cosine parameter c is real, but the sine parameter s is now *complex*. To emphasize this, we write

$$s = |s| e^{j\alpha} \tag{11.104}$$

where $|s|$ is the magnitude of s, and α is its phase angle. In addition, the (c, s) pair is required to satisfy the constraint

$$c^2 + |s|^2 = 1 \tag{11.105}$$

The objective is to choose the (c, s) pair so as to annihilate the *ki*th (off-diagonal) term. To do this, we must satisfy the condition

$$-sa_{ii} + ca_{ki} = 0$$

or, equivalently,

$$s = \frac{a_{ki}}{a_{ii}} c \tag{11.106}$$

Substituting Eq. (11.106) in (11.105), and solving for the cosine parameter, we get

$$c = \frac{|a_{ii}|}{\sqrt{|a_{ii}|^2 + |a_{ki}|^2}} \tag{11.107}$$

Note that, in Eq. (11.107), we have chosen to work with the *positive* real root for the cosine parameter c. Also, if a_{ki} is zero, that is, the data matrix is triangular to begin with, then $c = 1$ and $s = 0$, in which case we may bypass stage I. If, by the same token, a_{ik} is zero, we apply transposition and proceed to stage II.

Having determined the values of c and s needed for the triangularization of the 2-by-2 matrix **A**, we may now determine the elements of the resulting upper triangular matrix shown on the right side of Eq. (11.103) as follows:

$$\omega_{ii} = ca_{ii} + s^* a_{ki} \tag{11.108}$$

$$\omega_{ik} = ca_{ik} + s^* a_{kk} \tag{11.109}$$

$$\omega_{kk} = -sa_{ik} + ca_{kk} \tag{11.110}$$

Given that a_{ii} is positive real, by assumption, the use of Eqs. (11.106) and (11.107) in (11.108) reveals that the diagonal element ω_{ii} is real and nonnegative; that is,

$$\omega_{ii} \geq 0 \tag{11.111}$$

In general, however, the remaining two elements ω_{ik} and ω_{kk} of the upper triangular matrix on the right side of Eq. (11.103) are complex valued.

Stage II: Phase cancellation. As already mentioned, the elements ω_{ik} and ω_{kk} may be complex. To reduce them to real form, we premultiply and postmultiply the upper triangular matrix on the right side of Eq. (11.103) by a pair of *phase-cancelling diagonal matrices* as follows:

$$\begin{bmatrix} e^{-j\beta} & 0 \\ 0 & e^{-j\gamma} \end{bmatrix} \begin{bmatrix} \omega_{ii} & \omega_{ik} \\ 0 & \omega_{kk} \end{bmatrix} \begin{bmatrix} e^{j\beta} & 0 \\ 0 & 1 \end{bmatrix} = \begin{bmatrix} \omega_{ii} & |\omega_{ik}| \\ 0 & |\omega_{kk}| \end{bmatrix} \tag{11.112}$$

The *rotation angles* β and γ of the premultiplying matrix are chosen so as to cancel the phase angles of ω_{ik} and ω_{kk}, respectively, as shown by

$$\beta = \arg(\omega_{ik}) \tag{11.113}$$

$$\gamma = \arg(\omega_{kk}) \tag{11.114}$$

The postmultiplying matrix is included so as to *correct* for the phase change in the element ω_{ii} produced by the premultiplying matrix. In other words, the combined process of premultiplication and postmultiplication in Eq. (11.112) leaves the diagonal element ω_{ii} unchanged.

Stage II thus yields an upper triangular matrix whose three nonzero elements are all real and nonnegative. Note that the procedure described herein for reducing the complex 2-by-2 matrix $\mathbf{A}$ to a real upper triangular form requires four degrees of freedom, namely, the (c, s) pair, and the angles β and γ. The way is now paved for us to proceed with the application of the Jacobi method for a real 2-by-2 matrix, as described earlier in the section.

11.9 CYCLIC JACOBI ALGORITHM

We are now ready to describe the *cyclic Jacobi algorithm* for a square data matrix by solving an appropriate sequence of 2-by-2 singular-value decomposition problems. The description will be presented for real data. To deal with complex data, we incorporate the complex-to-real data reduction developed in the preceding section.

Let $\mathbf{J}_1(i, k)$ denote a Jacobi rotation in the (i, k) plane, where $k > i$. The matrix $\mathbf{J}_1(i, k)$ is the same as the M-by-M identity matrix, except for the four strategic elements located on rows i, k and columns i, k, as shown by

$$\mathbf{J}_1(i, k) = \begin{bmatrix} 1 & 0 & \cdots & 0 & \cdots & 0 \\ \vdots & \ddots & & \vdots & & \vdots \\ 0 & c_1 & \cdots & s_1 & \cdots & 0 \\ \vdots & \vdots & \ddots & \vdots & & \vdots \\ 0 & -s_1 & \cdots & c_1 & \cdots & 0 \\ \vdots & \vdots & & \vdots & \ddots & \vdots \\ 0 & 0 & \cdots & 0 & \cdots & 1 \end{bmatrix} \begin{matrix} \\ \\ \leftarrow \text{row } i \\ \\ \leftarrow \text{row } k \\ \\ \\ \end{matrix} \tag{11.115}$$

$\qquad\qquad\qquad\uparrow$ column i $\quad\uparrow$ column k

Let $\mathbf{J}_2(i, k)$ denote a second Jacobi rotation in the (i, k) plane that is similarly defined; the dimension of this second transformation is also M. The Jacobi transformation of the data matrix $\mathbf{A}$ is thus described by

$$\mathbf{T}_{ik}: \mathbf{A} \leftarrow \mathbf{J}_1^T(i, k)\mathbf{A}\mathbf{J}_2(i, k) \tag{11.116}$$

The Jacobi rotations $\mathbf{J}_1(i, k)$ and $\mathbf{J}_2(i, k)$ are designed to annihilate the (i, k) and (k, i) elements of $\mathbf{A}$. Accordingly, the transformation $\mathbf{T}_{ik}$ produces a matrix $\mathbf{X}$ (equal to the updated value of $\mathbf{A}$) that is more diagonal than the original $\mathbf{A}$ in the sense that

$$\text{off}(\mathbf{X}) = \text{off}(\mathbf{A}) - a_{ik}^2 - a_{ki}^2 \tag{11.117}$$

where off $(\mathbf{A})$ is the *norm of the off-diagonal elements:*

$$\text{off}(\mathbf{A}) = \sum_{i=1}^{M} \sum_{\substack{k=1 \\ k \neq i}}^{M} a_{ik}^2, \qquad \text{for } \mathbf{A} = \{a_{ik}\} \tag{11.118}$$

In the cyclic Jacobi algorithm the transformation (11.116) is applied for a total of $m = M(M - 1)/2$ different index pairs ("pivots") that are selected in some fixed order. Such a sequence of m transformations is called a *sweep*. The construction of a sweep may be *cyclic by rows* or *cyclic by columns,* as illustrated in Example 3 below. In either case, we obtain a new matrix $\mathbf{A}$ after each sweep, for which we compute off$(\mathbf{A})$. If off$(\mathbf{A}) \leq \delta$ where δ is some small machine-dependent number, we stop the computation. If on the other hand, off$(\mathbf{A}) > \delta$, we repeat the computation. For typical values of δ [e.g., $\delta = 10^{-12}\,\text{off}(\mathbf{A}_0)$, where $\mathbf{A}_0$ is the original matrix], the algorithm converges in about 4 to 10 sweeps for values of M in the range 4 to 2000.

As far as we know, the row ordering or the column ordering is the only ordering that guarantees convergence of the Jacobi cyclic algorithm.[5] By "convergence" we mean

$$\text{off}(\mathbf{A}^{(k)}) \rightarrow 0 \qquad \text{as } k \rightarrow \infty \tag{11.119}$$

where $\mathbf{A}^{(k)}$ is the M-by-M matrix computed after sweep number k.

Example 3

Consider a 4-by-4 *real* matrix $\mathbf{A}$. With the matrix dimension $M = 4$, we have a total of six orderings in each sweep. A sweep of orderings cyclic by rows is represented by

$$\mathbf{T}_R = \mathbf{T}_{34}\mathbf{T}_{24}\mathbf{T}_{23}\mathbf{T}_{14}\mathbf{T}_{13}\mathbf{T}_{12}$$

A sweep of orderings cyclic by columns is represented by

$$\mathbf{T}_C = \mathbf{T}_{34}\mathbf{T}_{24}\mathbf{T}_{14}\mathbf{T}_{23}\mathbf{T}_{13}\mathbf{T}_{12}$$

It is easily checked that the transformation $\mathbf{T}_{ik}$ and $\mathbf{T}_{pq}$ *commute* if two conditions hold:

1. The index i is neither p nor q.
2. The index k is neither p nor q.

[5] A proof of convergence of the Jacobi cyclic algorithm, based on row ordering or column ordering, is given in Forsythe and Henrici (1960). Subsequently, Luk and Park (1989) proved that many of the orderings used in parallel implementation of the algorithm are equivalent to the row ordering, and thus guarantee convergence as well.

Accordingly, we find that the transformations $\mathbf{T}_R$ and $\mathbf{T}_C$ are indeed equivalent, as they should be.

Consider next the application of the transformation $\mathbf{T}_R$ (obtained from the sweep of orderings cyclic by rows) to the data matrix $\mathbf{A}$. In particular, using the rotation of Eq. (11.116), we may write the following transformations:

$$\mathbf{T}_{12} : \mathbf{A} \leftarrow \mathbf{J}_1^T(1, 2)\mathbf{A}\mathbf{J}_2(1, 2)$$

$$\mathbf{T}_{13}\mathbf{T}_{12} : \mathbf{A} \leftarrow \mathbf{J}_3^T(1, 3)\mathbf{J}_1^T(1, 2)\mathbf{A}\mathbf{J}_2(1, 2)\mathbf{J}_4(1, 3)$$

$$\mathbf{T}_{14}\mathbf{T}_{13}\mathbf{T}_{12} : \mathbf{A} \leftarrow \mathbf{J}_5^T(1, 4)\mathbf{J}_3^T(1, 3)\mathbf{J}_1^T(1, 2)\mathbf{A}\mathbf{J}_2(1, 2)\mathbf{J}_4(1, 3)\mathbf{J}_6(1, 4)$$

and so on. The final step in this sequence of transformations may be written as

$$\mathbf{T}_R : \ \mathbf{A} \leftarrow \mathbf{U}^T\mathbf{A}\mathbf{V}$$

which defines the singular value decomposition of the real data matrix $\mathbf{A}$. The unitary matrices $\mathbf{U}$ and $\mathbf{V}$ are respectively defined by

$$\mathbf{U} = \mathbf{J}_1(1, 2)\mathbf{J}_3(1, 3)\mathbf{J}_5(1, 4)\mathbf{J}_7(2, 3)\mathbf{J}_9(2, 4)\mathbf{J}_{11}(3, 4)$$

and

$$\mathbf{V} = \mathbf{J}_2(1, 2)\mathbf{J}_4(1, 3)\mathbf{J}_6(1, 4)\mathbf{J}_8(2, 3)\mathbf{J}_{10}(2, 4)\mathbf{J}_{12}(3, 4)$$

Rectangular Data Matrix

Thus far we have focused attention on the cyclic Jacobi algorithm for computing the singular-value decomposition of a square matrix. To handle the more general case of a rectangular matrix, we may extend the use of this algorithm by proceeding as follows. Consider first the case of a K-by-M real data matrix $\mathbf{A}$, for which K is greater than M. We generate a square matrix by appending $(K - M)$ columns of zeros to $\mathbf{A}$. We may thus write

$$\tilde{\mathbf{A}} = [\mathbf{A}, \mathbf{O}] \tag{11.120}$$

We refer to $\tilde{\mathbf{A}}$ as the *augmented data matrix*. We then proceed as before by applying the cyclic Jacobi algorithm to the K-by-K matrix $\tilde{\mathbf{A}}$. In performing this computation, we require the use of special case 1 described in Eq. (11.101). In any event, we emerge with the factorization

$$\mathbf{U}^T[\mathbf{A}, \mathbf{O}]\begin{bmatrix} \mathbf{V} & \mathbf{O} \\ \mathbf{O} & \mathbf{I} \end{bmatrix} = \text{diag}(\sigma_1, \ldots, \sigma_M) \tag{11.121}$$

The desired factorization of the original data matrix $\mathbf{A}$ is obtained by writing

$$\mathbf{U}^T\mathbf{A}\mathbf{V} = \text{diag}(\sigma_1, \ldots, \sigma_M) \tag{11.122}$$

If, on the other hand, the dimension M of matrix $\mathbf{A}$ is greater than K, we augment it by adding $(M - K)$ rows; we may thus write

$$\tilde{\mathbf{A}} = \begin{bmatrix} \mathbf{A} \\ \mathbf{O} \end{bmatrix} \tag{11.123}$$

We then treat the square matrix $\tilde{\mathbf{A}}$ in the same way as before. In this second situation, we require the use of special case 2 described in Eq. (11.102).

In the case of a complex rectangular data matrix **A**, we may proceed in a fashion similar to that described above, except for a change in the characterization of matrices **U** and **V**. For a real data matrix, the matrices **U** and **V** are both orthogonal, whereas for a complex data matrix they are both unitary.

The strategy of matrix augmentation described herein represents a straightforward extension of the cyclic Jacobi algorithm for a square matrix. A drawback of this approach, however, is that the algorithm may become too inefficient if the dimension K of matrix **A** is much greater than the dimension M, or vice versa.[6]

11.10 HOUSEHOLDER TRANSFORMATIONS

Jacobi (Givens) rotations and Householder transformations are highly popular in modern numerical analysis. Jacobi rotations and their use in the development of the Jacobi algorithm for SVD computation were discussed in the previous two sections. In this section, we study the Householder transformation and its properties;[7] the discussion presented herein emphasizes the importance of the Householder transformation in its own right. The use of the Householder transformation in SVD computation is discussed in the next section.

Let **u** be an M-by-1 vector whose Euclidean norm is

$$\|\mathbf{u}\| = (\mathbf{u}^H\mathbf{u})^{1/2}$$

Let **I** be the M-by-M identity matrix. Then, an M-by-M matrix **Q** defined in terms of the vector **u** by

$$\mathbf{Q} = \mathbf{I} - \frac{2\mathbf{u}\mathbf{u}^H}{\|\mathbf{u}\|^2} \tag{11.124}$$

is known as the *Householder transformation* or *Householder matrix* [Householder, 1958).

For a geometric interpretation of the Householder transformation, consider an M-by-1 vector **x** premultiplied by the matrix **Q**, as shown by

$$\begin{aligned} \mathbf{Qx} &= \left(\mathbf{I} - \frac{2\mathbf{u}\mathbf{u}^H}{\|\mathbf{u}\|^2}\right)\mathbf{x} \\ &= \mathbf{x} - \frac{2\mathbf{u}^H\mathbf{x}}{\|\mathbf{u}\|^2}\mathbf{u} \end{aligned} \tag{11.125}$$

[6] An alternative approach that overcomes this difficulty is to proceed as follows (Luk (1986):

1. *Triangularize* the K-by-M data matrix **A** by performing a QR-decomposition, defined by

$$\mathbf{A} = \mathbf{Q}\begin{bmatrix}\mathbf{R}\\ \mathbf{O}\end{bmatrix}$$

where **Q** is a K-by-K orthogonal matrix, and **R** is an M-by-M upper triangular matrix.

2. Diagonalize the matrix **R** using the cyclic Jacobi algorithm.
3. Combine the results of steps 1 and 2.

[7] For a tutorial review of the Householder transformation and its use in adaptive signal processing, see Steinhardt (1988).

By definition, the *projection* of **x** onto **u** is given by

$$\mathbf{P}_u(\mathbf{x}) = \frac{\mathbf{u}^H\mathbf{x}}{\|\mathbf{u}\|^2}\mathbf{u} \tag{11.126}$$

This projection is illustrated in Fig. 11.4. In this figure we have also included the vector representation of the product **Qx.** We thus see that **Qx** is the mirror-image *reflection* of the vector **x** with respect to the hyperplane span $\{\mathbf{u}\}^\perp$, which is perpendicular to the vector **u.**

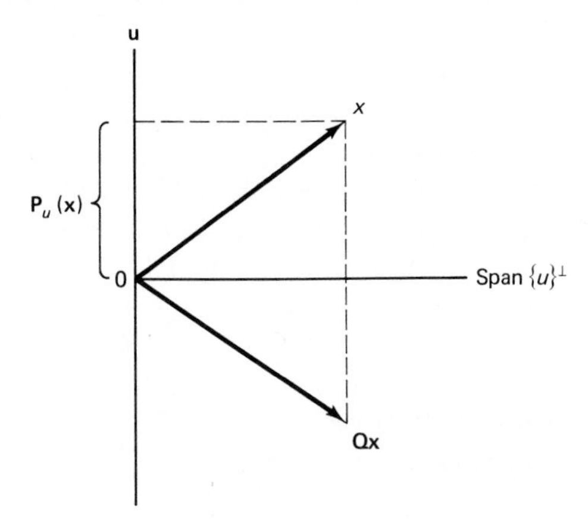

Figure 11.4 Geometric interpretation of the Householder transformation.

The Householder transformation, defined in Eq. (11.124), has the following properties:

Property 1. *The Householder transformation* **Q** *is Hermitian;* that is,

$$\mathbf{Q}^H = \mathbf{Q} \tag{11.127}$$

Property 2. *The Householder transformation* **Q** *is unitary;* that is,

$$\mathbf{Q}^{-1} = \mathbf{Q}^H \tag{11.128}$$

or, equivalently,

$$\mathbf{Q}^H\mathbf{Q} = \mathbf{I}$$

Property 3. *The Householder transformation is length preserving;* that is,

$$\|\mathbf{Qx}\| = \|\mathbf{x}\| \tag{11.129}$$

This property is illustrated in Fig. 11.4, where we see that the vectors **x** and its reflection **Qx** have exactly the same length.

Property 4. *If two vectors undergo the same Householder transformation, their inner product remains unchanged.*

Consider any three vectors **x**, **y**, and **u**. Let the Householder matrix **Q** be defined in terms of the vector **u**, as in Eq. (11.124). Let the remaining two vectors **x** and $\mathbf{y}$ be transformed by $\mathbf{Q}$, yielding $\mathbf{Qx}$ and $\mathbf{Qy}$, respectively. The inner product of these two transformed vectors is

$$\begin{aligned}(\mathbf{Qx})^H(\mathbf{Qy}) &= \mathbf{x}^H\mathbf{Q}^H\mathbf{Qy} \\ &= \mathbf{x}^H\mathbf{y}\end{aligned} \tag{11.130}$$

where we have made use of Property 2. Hence, the transformed vectors $\mathbf{Qx}$ and $\mathbf{Qy}$ have the same inner product as the original vectors $\mathbf{x}$ and $\mathbf{y}$.

Property 4 has important practical implications in the numerical solution of linear least-squares problems. Specifically, Householder transformations are used to reduce the given data matrix to a *sparse* matrix (i.e., one that consists mostly of zeros), but which is "equivalent" to the original data matrix in some mathematical sense. Needless to say, the particular form of matrix sparseness used depends on the application of interest. Whatever the application, however, the data reduction is used in order to simplify the numerical computations involved in solving the problem. In this context, a popular form of data reduction is that of *triangularization,* referring to the reduction of a full data matrix to an upper triangular one. Given this form of data reduction, we may then simply use *Gaussian elimination* to perform the matrix inversion and thereby compute the least-squares solution to the problem.

Properties 1 through 4 apply not only to the Householder transformations but also the Jacobi (Givens) rotations. It is the next two properties that distinguish Householder transformations from Givens rotations.

Property 5. *Given the Householder transformation* $\mathbf{Q}$, *the transformed vector* $\mathbf{Qx}$ *is a reflection of* $\mathbf{x}$ *above the hyperplane perpendicular to the vector* $\mathbf{u}$ *involved in the definition of* $\mathbf{Q}$.

This property is merely a restatement of Eq. (11.125). The following two limiting cases of Property 5 are especially noteworthy:

1. The vector $\mathbf{x}$ is a scalar multiple of $\mathbf{u}$: In this case, Eq. (11.125) simplifies as

$$\mathbf{Qx} = -\mathbf{x}$$

2. The vector $\mathbf{x}$ is orthogonal to $\mathbf{u}$, that is, their inner product is zero: In this second case, Eq. (11.125) reduces to

$$\mathbf{Qx} = \mathbf{x}$$

Property 6. *Let* $\mathbf{x}$ *be any nonzero M-by-1 vector with Euclidean norm* $\|\mathbf{x}\|$. *Let the M-by-1 vector* $\mathbf{1}$ *denote the first column of the identity matrix, that is,*

$$\mathbf{1} = [1, 0, \ldots, 0]^T \tag{11.131}$$

Then there exists a Householder transformation $\mathbf{Q}$ *defined by the vector*

$$\mathbf{u} = \mathbf{x} - \|\mathbf{x}\|\mathbf{1} \tag{11.132}$$

such that the transformed vector $\mathbf{Qx}$ *is a linear multiple of the vector* $\mathbf{1}$.

With the vector $\mathbf{u}$ assigned the value in Eq. (11.132), we have

$$\begin{aligned}\|\mathbf{u}\|^2 &= \mathbf{u}^H\mathbf{u} \\ &= (\mathbf{x} - \|\mathbf{x}\|\mathbf{1})^H(\mathbf{x} - \|\mathbf{x}\|\mathbf{1}) \\ &= 2\|\mathbf{x}\|^2 - 2\|\mathbf{x}\|x_1 \\ &= 2\|\mathbf{x}\|(\|\mathbf{x}\| - x_1)\end{aligned} \tag{11.133}$$

where x_1 is the first element of the vector $\mathbf{x}$. Similarly, we may write

$$\begin{aligned}\mathbf{u}^H\mathbf{x} &= (\mathbf{x} - \|\mathbf{x}\|\mathbf{1})^H\mathbf{x} \\ &= \|\mathbf{x}\|^2 - \|\mathbf{x}\|x_1 \\ &= \|\mathbf{x}\|(\|\mathbf{x}\| - x_1)\end{aligned} \tag{11.134}$$

Accordingly, substituting Eqs. (11.133) and (11.134) in (11.125), we find that the transformed vector $\mathbf{Qx}$ corresponding to the defining vector $\mathbf{u}$ of Eq. (11.132) is given by

$$\begin{aligned}\mathbf{Qx} &= \mathbf{x} - \mathbf{u} \\ &= \mathbf{x} - (\mathbf{x} - \|\mathbf{x}\|\mathbf{1}) \\ &= \|\mathbf{x}\|\mathbf{1}\end{aligned} \tag{11.135}$$

which is the desired result.

From Eq. (11.133), we observe that the first element x_1 of the vector $\mathbf{x}$ has to be real, and the Euclidean norm of $\mathbf{x}$ has to satisfy the condition

$$\|\mathbf{x}\| > |x_1| \tag{11.136}$$

This condition merely says that not only the first element of $\mathbf{x}$ but also one other element must be nonzero. Then, the vector $\mathbf{u}$ defined by Eq. (11.132) is indeed effective.

Property 6 makes the Householder transformation a very powerful computational tool. Given a vector $\mathbf{x}$, we may use Eq. (11.132) to define the vector $\mathbf{u}$ such that the corresponding Householder transformation $\mathbf{Q}$ annihilates all the M elements of the vector $\mathbf{x}$ except for the first one. This result is equivalent to the application of $(M - 1)$ plane rotations, with a minor difference: The determinant of the Householder matrix $\mathbf{Q}$ defined in Eq. (11.124) is

$$\begin{aligned}\det(\mathbf{Q}) &= \det\left(\mathbf{I} - \frac{2\mathbf{u}\mathbf{u}^H}{\|\mathbf{u}\|^2}\right) \\ &= -1\end{aligned} \tag{11.137}$$

Hence, the Householder transformation reverses the orientation of the configuration.

Having familiarized ourselves with the Householder transformation, we are ready to resume our discussion of SVD computation by describing the QR algorithm, which we do in the next section.

11.11 QR ALGORITHM

The starting point in the development of the *QR algorithm* for SVD computation is that of finding a class of unitary matrices, which preserve the singular values and singular vectors of a data matrix $\mathbf{A}$. In this context, the matrix $\mathbf{A}$ is said to be *unitarily equivalent* to another matrix $\mathbf{B}$ if

$$\mathbf{B} = \mathbf{PAQ} \tag{11.138}$$

where $\mathbf{P}$ and $\mathbf{Q}$ are unitary matrices, that is

$$\mathbf{P}^H\mathbf{P} = \mathbf{I}$$

and

$$\mathbf{Q}^H\mathbf{Q} = \mathbf{I}$$

Consequently, we have

$$\begin{aligned}\mathbf{B}^H\mathbf{B} &= \mathbf{Q}^H\mathbf{A}^H\mathbf{P}^H\mathbf{PAQ} \\ &= \mathbf{Q}^H\mathbf{A}^H\mathbf{AQ}\end{aligned} \tag{11.139}$$

Postmultiplying the time-averaged correlation matrix $\mathbf{A}^H\mathbf{A}$ by a unitary matrix $\mathbf{Q}$ and premultiplying it by the Hermitian transpose of the matrix $\mathbf{Q}$ leaves the eigenvalues of $\mathbf{A}^H\mathbf{A}$ unchanged. Accordingly, the correlation matrices $\mathbf{A}^H\mathbf{A}$ and $\mathbf{B}^H\mathbf{B}$, or more simply, the matrices $\mathbf{A}$ and $\mathbf{B}$ themselves are said to be *eigen-equivalent*.

The purpose of using the transformation defined in Eq. (11.138) is to reduce the data matrix $\mathbf{A}$ to *upper bidiagonal* form, with eigen-equivalence maintained, for which Householder transformations are well suited. The reduced data matrix $\mathbf{B}$ is said to be upper bidiagonal if all of its elements except for those on the main diagonal and the superdiagonal are zero; that is, the ijth element of $\mathbf{B}$ is

$$b_{ij} = 0 \text{ whenever } i > j \text{ or } j > i + 1 \tag{11.140}$$

Having reduced the data matrix $\mathbf{A}$ to upper bidiagonal form, the next step is the application of the Golub–Kahan SVD algorithm. These two steps, in turn, are considered next.

Householder Bidiagonalization

Consider a K-by-M data matrix $\mathbf{A}$, where $K \geq M$. Let $\mathbf{Q}_1, \mathbf{Q}_2, \ldots, \mathbf{Q}_M$ denote a set of K-by-K Householder matrices, and let $\mathbf{P}_1, \mathbf{P}_2, \ldots, \mathbf{P}_{M-2}$ denote another set of M-by-M Householder matrices. In order to reduce the data matrix $\mathbf{A}$ to upper bidiagonal form, we determine the products of Householder matrices

$$\mathbf{Q}_B = \mathbf{Q}_1\mathbf{Q}_2 \ldots \mathbf{Q}_M \tag{11.141}$$

and

$$\mathbf{P}_B = \mathbf{P}_1 P_2 \ldots \mathbf{P}_{M-2} \tag{11.142}$$

such that

$$
\mathbf{Q}_B^H \mathbf{A} \mathbf{P}_B = \mathbf{B} = \left[\begin{array}{cccc} d_1 & f_2 & & 0 \\ & d_2 & \ddots & \\ & & \ddots & f_M \\ \mathbf{0} & & & d_M \\ \hline & \mathbf{0} & & \end{array}\right] \} \; (K - M)\text{-by-}M \text{ null matrix} \tag{11.143}
$$

Premultiplication of the data matrix $\mathbf{A}$ by the Householder matrices $\mathbf{Q}_1, \mathbf{Q}_2, \ldots, \mathbf{Q}_M$ corresponds to reflecting the respective columns of $\mathbf{A}$, whereas postmultiplication by the Householder matrices $\mathbf{P}_1, \mathbf{P}_2, \ldots, \mathbf{P}_{M-2}$ corresponds to reflecting the respective rows of $\mathbf{A}$. The desired upper bidiagonal form is attained by "ping-ponging" column and row reflections. Note that the number of Householder matrices constituting $\mathbf{Q}_B$ is M, whereas those constituting $\mathbf{P}_B$ number $M - 2$. Note also that, by construction, the matrix $\mathbf{P}_1\mathbf{P}_2 \ldots \mathbf{P}_{M-2}$ does *not* alter the first column of any matrix that it postmultiplies.

We illustrate this data reduction process by way of an example.

Example 4

Consider a 5-by-4 data matrix $\mathbf{A}$ written in expanded form as follows:

$$
\mathbf{A} = \begin{bmatrix} \times & \times & \times & \times \\ \times & \times & \times & \times \\ \times & \times & \times & \times \\ \times & \times & \times & \times \\ \times & \times & \times & \times \end{bmatrix}
$$

where the $\times$'s denote nonzero matrix entries. The upper bidiagonalization of $\mathbf{A}$ proceeds as follows. First, $\mathbf{Q}_1$ is chosen so that $\mathbf{Q}_1^H \mathbf{A}$ has zeros in the positions distinguished below:

$$
\begin{bmatrix} \times & \times & \times & \times \\ \otimes & \times & \times & \times \\ \otimes & \times & \times & \times \\ \otimes & \times & \times & \times \\ \otimes & \times & \times & \times \end{bmatrix}
$$

Thus, $\mathbf{Q}_1^H \mathbf{A}$ may be written as

$$
\mathbf{Q}_1^H \mathbf{A} = \begin{bmatrix} \times & \times & \otimes & \otimes \\ 0 & \times & \times & \times \\ 0 & \times & \times & \times \\ 0 & \times & \times & \times \\ 0 & \times & \times & \times \end{bmatrix} \tag{11.144}
$$

Next $\mathbf{P}_1$ is chosen so that $\mathbf{Q}_1^H \mathbf{A}_1 \mathbf{P}_1$ has zeros in the positions distinguished in the first row of $\mathbf{Q}_1^H \mathbf{A}$, as in Eq. (11.144). Hence, $\mathbf{Q}_1^H \mathbf{A} \mathbf{P}_1$ has the form

$$\mathbf{Q}_1^H \mathbf{A} \mathbf{P}_1 = \left[\begin{array}{c|ccc} \times & \times & 0 & 0 \\ \hline 0 & \times & \times & \times \\ 0 & \times & \times & \times \\ 0 & \times & \times & \times \\ 0 & \times & \times & \times \end{array}\right] \tag{11.145}$$

Note that $\mathbf{P}_1$ does not affect the first column of the matrix.

The data reduction is continued by operating on the trailing 4-by-3 submatrix of $\mathbf{Q}^H \mathbf{A} \mathbf{P}_1$ that has nonzero entries. Specifically, we choose $\mathbf{Q}_2$ and $\mathbf{P}_2$ so that $\mathbf{Q}_2^H \mathbf{Q}_1^H \mathbf{A} \mathbf{P}_1 \mathbf{P}_2$ has the form

$$\mathbf{Q}_2^H \mathbf{Q}_1^H \mathbf{A} \mathbf{P}_1 \mathbf{P}_2 = \left[\begin{array}{cc|cc} \times & \times & 0 & 0 \\ 0 & \times & \times & 0 \\ \hline 0 & 0 & \times & \times \\ 0 & 0 & \times & \times \\ 0 & 0 & \times & \times \end{array}\right] \tag{11.146}$$

Next, we operate on the trailing 3-by-2 submatrix of $\mathbf{Q}_2^H \mathbf{Q}_1^H \mathbf{A} \mathbf{P}_1 \mathbf{P}_2$ that has nonzero entries. Specifically, we choose $\mathbf{Q}_3$ such that $\mathbf{Q}_3^H \mathbf{Q}_2^H \mathbf{Q}_1^H \mathbf{A} \mathbf{P}_1 \mathbf{P}_2$ has the form

$$\mathbf{Q}_3^H \mathbf{Q}_2^H \mathbf{Q}_1^H \mathbf{A} \mathbf{P}_1 \mathbf{P}_2 = \left[\begin{array}{ccc|c} \times & \times & 0 & 0 \\ 0 & \times & \times & 0 \\ 0 & 0 & \times & \times \\ \hline 0 & 0 & 0 & \times \\ 0 & 0 & 0 & \times \end{array}\right] \tag{11.147}$$

Finally, we choose $\mathbf{Q}_4$ to operate on the trailing 2-by-1 submatrix of $\mathbf{Q}_3^H \mathbf{Q}_2^H \mathbf{Q}_1^H \mathbf{A} \mathbf{P}_1 \mathbf{P}_2$, such that we may write

$$\mathbf{B} = \mathbf{Q}_4^H \mathbf{Q}_3^H \mathbf{Q}_2^H \mathbf{Q}_1^H \mathbf{A} \mathbf{P}_1 \mathbf{P}_2 = \begin{bmatrix} \times & \times & 0 & 0 \\ 0 & \times & \times & 0 \\ 0 & 0 & \times & \times \\ 0 & 0 & 0 & \times \\ 0 & 0 & 0 & 0 \end{bmatrix} \tag{11.148}$$

This completes the upper bidiagonalization of the data matrix $\mathbf{A}$.

Golub-Kahan Algorithm

The bidiagonalization of the data matrix $\mathbf{A}$ is followed by an *iterative process* that reduces it further to *diagonal* form. Referring to Eq. (11.143), we see that the matrix $\mathbf{B}$, resulting from the bidiagonalization of $\mathbf{A}$, is zero below the Mth row. Evidently, the last $K - M$ rows of zeros in the matrix $\mathbf{B}$ do *not* contribute to the singular values of the original data matrix $\mathbf{A}$. Accordingly, it is convenient to delete the last $K - M$ rows of matrix $\mathbf{B}$ and thus treat it as a square matrix with dimension M. The basis of the diagonalization of matrix $\mathbf{B}$ is the *Golub–Kahan algorithm* (Golub and Kahan, 1965), which is an adaptation of the QR *algorithm* developed originally for solving the symmetric eigenvalue problem.[8]

Let **B** denote an M-by-M upper bidiagonal matrix having no zeros on its main diagonal or superdiagonal. The first iteration of the Golub–Kahan algorithm proceeds as follows (Golub and Kahan, 1965; Golub and Van Loan, 1989):

1. Identify the trailing 2-by-2 submatrix of the product $\mathbf{T} = \mathbf{B}^H\mathbf{B}$, which has the form

$$\begin{bmatrix} |d_{M-1}|^2 + |f_{M-1}|^2 & d_{M-1}^* f_M \\ f_M^* d_{M-1} & |d_M|^2 + |f_M|^2 \end{bmatrix} \tag{11.149}$$

where d_{M-1} and d_M are the trailing diagonal elements of matrix **B**, and f_{M-1} and f_M are the trailing superdiagonal elements; see the right side of Eq. (11.143). Compute the eigenvalue λ of this 2-by-2 submatrix, which is closer to $|d_M|^2 + |f_M|^2$.

2. Compute the Givens (Jacobi) rotation parameters c_1 and s_1 such that

$$\begin{bmatrix} c_1 & s_1^* \\ -s_1 & c_1 \end{bmatrix} \begin{bmatrix} |d_1|^2 - \lambda \\ d_1^* f_2 \end{bmatrix} = \begin{bmatrix} \star \\ 0 \end{bmatrix} \tag{11.150}$$

where d_1 and f_2 are the leading main diagonal and superdiagonal elements of matrix **B**, respectively; see the right side of Eq. (11.143). The element marked $\star$ on the right side of Eq. (11.150) indicates a nonzero element. Set

$$\mathbf{J}_1 = \left[\begin{array}{cc|c} c_1 & s_1^* & \mathbf{O} \\ -s_1 & c_1 & \\ \hline \mathbf{O} & & \mathbf{I} \end{array}\right] \tag{11.151}$$

3. Apply the Givens rotation $\mathbf{J}_1$ to matrix **B** directly. Since **B** is upper bidiagonal, and $\mathbf{J}_1$ is a rotation in the (2, 1) plane, it follows that the matrix product $\mathbf{BJ}_1$ has the following form (illustrated for the case of $M = 4$):

$$\mathbf{BJ}_1 = \begin{bmatrix} \times & \times & 0 & 0 \\ z^{(1)} & \times & \times & 0 \\ 0 & 0 & \times & \times \\ 0 & 0 & 0 & \times \end{bmatrix}$$

where $z^{(1)}$ is a new element produced by the Givens rotation $\mathbf{J}_1$.

4. Determine the sequence of Givens rotations $\mathbf{U}_1, \mathbf{V}_2, \mathbf{U}_2, \ldots, \mathbf{V}_{M-1}$, and $\mathbf{U}_{M-1}$ operating on $\mathbf{BJ}_1$ in a "ping-pong" fashion so as to chase the unwanted nonzero

[8] The explicit form of the QR algorithm is a variant of the QL algorithm discussed in Chapter 5.

element $z^{(1)}$ down the bidiagonal. This sequence of operations is illustrated below, again for the case of $M = 4$:

$$\mathbf{U}_1^H\mathbf{B}\mathbf{J}_1 = \begin{bmatrix} \times & \times & z^{(2)} & 0 \\ 0 & \times & \times & 0 \\ 0 & 0 & \times & \times \\ 0 & 0 & 0 & \times \end{bmatrix}$$

$$\mathbf{U}_1^H\mathbf{V}\mathbf{J}_1\mathbf{V}_2 = \begin{bmatrix} \times & \times & 0 & 0 \\ 0 & \times & \times & 0 \\ 0 & z^{(3)} & \times & \times \\ 0 & 0 & 0 & \times \end{bmatrix}$$

$$\mathbf{U}_2^H\mathbf{U}_1^H\mathbf{B}\mathbf{J}_1\mathbf{V}_2 = \begin{bmatrix} \times & \times & 0 & 0 \\ 0 & \times & \times & z^{(4)} \\ 0 & 0 & \times & \times \\ 0 & 0 & 0 & \times \end{bmatrix}$$

$$\mathbf{U}_2^H\mathbf{U}_1^H\mathbf{B}\mathbf{J}_1\mathbf{V}_2\mathbf{V}_3 = \begin{bmatrix} \times & \times & 0 & 0 \\ 0 & \times & \times & 0 \\ 0 & 0 & \times & \times \\ 0 & 0 & z^{(5)} & \times \end{bmatrix}$$

$$\mathbf{U}_3^H\mathbf{U}_2^H\mathbf{U}_1^H\mathbf{B}\mathbf{J}_1\mathbf{V}_2\mathbf{V}_3 = \begin{bmatrix} \times & \times & 0 & 0 \\ 0 & \times & \times & 0 \\ 0 & 0 & \times & \times \\ 0 & 0 & 0 & \times \end{bmatrix}$$

The iteration thus terminates with a new bidiagonal matrix $\mathbf{B}$ that is related to the original bidiagonal matrix $\mathbf{B}$ as follows:

$$\begin{aligned} \overline{\mathbf{B}} &= (\mathbf{U}_{M-1}^H \cdots \mathbf{U}_2^H\mathbf{U}_1{}^H)\mathbf{B}(\mathbf{J}_1\mathbf{V}_2 \cdots \mathbf{V}_{M-1}) \\ &= \mathbf{U}^H\mathbf{B}\mathbf{V}^H \end{aligned} \tag{11.152}$$

where

$$\mathbf{U} = \mathbf{U}_1\mathbf{U}_2 \cdots \mathbf{U}_{M-1} \tag{11.153}$$

and

$$\mathbf{V} = \mathbf{J}_1\mathbf{V}_2 \cdots \mathbf{V}_{M-1} \tag{11.154}$$

Steps 1 through 4 constitute one iteration of the Golub–Kahan algorithm. Typically, after a few iterations of this algorithm, the superdiagonal entries $f_2, \ldots, f_M$ become negligible. The criterion for the smallness of these elements is usually of the following form:

$$|f_i| \leq \epsilon(|d_{i-1}| + |d_i|) \tag{11.155}$$

The description above leaves much unsaid about the Golub–Kahan algorithm for the diagonalization of a square data matrix. For a more detailed treatment of the algorithm, the reader is referred to the original paper of Golub and Kahan (1965) or the book (Golub and Van Loan, 1989).

Recently, there has been a significant improvement in the Golub–Kahan algorithm. The Golub–Kahan algorithm has the property that it computes every singular value of a bidiagonal matrix $\mathbf{B}$ with an absolute error bound of about $\epsilon\|\mathbf{B}\|$, where ϵ is the machine precision. Thus large singular values (those near $\|\mathbf{B}\|$) are computed with high relative accuracy, but small ones (those near $\epsilon\|\mathbf{B}\|$ or smaller) may have no relative accuracy at all. The new algorithm computes every singular value to high relative accuracy independent of its size. It also computes the singular vectors much more accurately. It is also approximately as fast as the old algorithm (and occasionally much faster). The new algorithm is a hybrid of the Golub–Kahan algorithm and a simplified version that corresponds to taking $\lambda = 0$ in Eq. (11.150). When $\lambda = 0$, the remainder of the algorithm can be stabilized so as to compute every matrix entry to high relative accuracy, whence the final accuracy of the singular values. The analysis of this algorithm[9] can be found in Demmel and Kahan (1990) and Deift et al. (1989).

Summary of the QR Algorithm

Given a K-by-M data matrix $\mathbf{A}$, we may summarize the QR algorithm used to compute its singular-value decomposition (SVD) as follows:

1. Compute a sequence of Householder transformations that reduce the matrix $\mathbf{B}$ to upper bidiagonal form.
2. Apply the Golub–Kahan algorithm to the M-by-M nonzero submatrix resulting from step 1, and iterate this application until the superdiagonal elements become negligible in accordance with the criterion defined in Eq. (11.155).
3. The SVD of the data matrix $\mathbf{A}$ is determined as follows:
   - **(a)** The diagonal elements of the matrix resulting from step 2 are the singular values of matrix $\mathbf{A}$.
   - **(b)** The product of the Householder transformations of step 1 and the Givens rotations of step 2 involved in the premultiplication of the data matrix $\mathbf{A}$ defines the left singular vectors of $\mathbf{A}$. The product of the Householder transformations and Givens rotations involved in postmultiplication define the right-singular vectors of $\mathbf{A}$.

[9] This new algorithm will eventually appear as part of the the new LAPACK linear algebra library that is currently being built to replace the standard LINPACK and EISPACK for use on supercomputers [J. Demmel, personal communication (1990)].

Example 5

Consider the real valued 3-by-3 bidiagonal matrix:

$$\mathbf{B} = \begin{bmatrix} 1 & 1 & 0 \\ 0 & 2 & 1 \\ 0 & 0 & 3 \end{bmatrix}$$

The iterative application of the Golub–Kahan algorithm to this matrix yields the sequence of results shown in Table 11.2 for $\epsilon = 10^{-4}$ in the stopping rule defined in Eq. (11.155). The singular values of the bidiagonal matrix are thus computed to be:

$$\sigma_1 = 0.8596$$

$$\sigma_2 = 2.1326$$

$$\sigma_3 = 3.2731$$

TABLE 11.2 ILLUSTRATING THE ITERATIVE APPLICATION OF THE GOLUB–KAHAN ALGORITHM.

Iteration number	Matrix **B**		
0	1.0000	1.0000	0.0000
	0.0000	2.0000	1.0000
	0.0000	0.0000	3.0000
1	0.9155	0.6627	0.0000
	0.0000	2.0024	0.0021
	0.0000	0.0000	3.2731
2	0.8817	0.4323	0.0000
	0.0000	2.0791	0.0000
	0.0000	0.0000	3.2731
3	0.8681	0.2731	0.0000
	0.0000	2.1116	0.0000
	0.0000	0.0000	3.2731
4	0.8629	0.1702	0.0000
	0.0000	2.1245	0.0000
	0.0000	0.0000	3.2731
5	0.8608	0.1056	0.0000
	0.0000	2.1295	0.0000
	0.0000	0.0000	3.2731
6	0.8600	0.0653	0.0000
	0.0000	2.1314	0.0000
	0.0000	0.0000	3.2731
7	0.8597	0.0404	0.0000
	0.0000	2.1322	0.0000
	0.0000	0.0000	3.2731
8	0.8596	0.0250	0.0000
	0.0000	2.1325	0.0000
	0.0000	0.0000	3.2731
9	0.8596	0.0154	0.0000
	0.0000	2.1326	0.0000
	0.0000	0.0000	3.2731

TABLE 11.2 (Continued)

Iteration number	Matrix **B**		
10	0.8596	0.0095	0.0000
	0.0000	2.1326	0.0000
	0.0000	0.0000	3.2731
11	0.8596	0.0059	0.0000
	0.0000	2.1326	0.0000
	0.0000	0.0000	3.2731
12	0.8596	0.0036	0.0000
	0.0000	2.1326	0.0000
	0.0000	0.0000	3.2731
13	0.8596	0.0023	0.0000
	0.0000	2.1326	0.0000
	0.0000	0.0000	3.2731
14	0.8596	0.0014	0.0000
	0.0000	2.1326	0.0000
	0.0000	0.0000	3.2731
15	0.8596	0.0009	0.0000
	0.0000	2.1326	0.0000
	0.0000	0.0000	3.2731
16	0.8596	0.0005	0.0000
	0.0000	2.1326	0.0000
	0.0000	0.0000	3.2731
17	0.8596	0.0003	0.0000
	0.0000	2.1326	0.0000
	0.0000	0.0000	3.2731
18	0.8596	0.0002	0.0000
	0.0000	2.1326	0.0000
	0.0000	0.0000	3.2731

11.12 SUMMARY AND DISCUSSION

Theory

The *singular-value decomposition* (SVD) has become a fundamental tool in linear algebra, system theory, and modern signal processing (Deprettere, 1988; Van Loan, 1989; Haykin, 1989; Kung, et al., 1985). Not only does the SVD permit an *elegant problem formulation, but also it provides geometrical and algebraic insight together with a numerically robust implementation* (Golub and VanLoan, 1989). It includes the eigenvalue decomposition of a Hermitian matrix as a special case.

In the context of our present discussion, the SVD provides a direct, and therefore numerically robust, solution for the linear least-squares estimation problem, be it overdetermined or underdetermined; by "direct" we mean that the solution is obtained by applying the SVD *directly* to the data matrix. Moreover, the SVD provides an insightful connection between the method of least squares and the normalized LMS algorithm (developed ordinarily in an entirely different way). In particular, the

SVD helps us see that the normalized LMS algorithm is the minimum norm solution of an underdetermined linear least-squares estimation problem.

Computation

The basic idea behind an algorithm used to compute the singular-value decomposition is to "nudge" a data matrix toward a diagonal form in a step-by-step fashion. The two most common iterative algorithms used to do this nudging are (1) the *cyclic Jacobi algorithm,* and (2) the *QR algorithm* (not to be confused with the QR-decomposition).

The QR algorithm, in general, is computationally more *efficient* (i.e., requires less operations) than the cyclic Jacobi algorithm. On the other hand, the cyclic Jacobi algorithm operates on 2-by-2 pieces of data, thereby allowing a higher degree of *parallelism* than the QR algorithm.

In Demmel and Veselić (1989), it is shown that the cyclic Jacobi algorithm (with a modified stopping criterion, described below) computes the eigenvalues of symmetric positive definite matrices and the singular values of general (real matrices) with a *uniformly better error bound* than the QR algorithm. It is shown that the Jacobi algorithm also computes the eigenvectors and singular vectors with better error bounds than the QR algorithm. Indeed, for the symmetric positive definite eigenvalue problem, Demmel and Veselić show that modulo an assumption based on extensive numerical testing, the Jacobi algorithm is *optimally accurate* in the following sense. Suppose that the initial entries of a matrix are known to have *small relative uncertainties,* which may have resulted from prior computations or measurement errors. Naturally, the eigenvalues of the matrix will have uncertainties of their own, regardless of the algorithm employed to compute them. Demmel and Veselić show that the eigenvalues computed using the Jacobi algorithm have error bounds that are nearly as small as these inherent uncertainties. They prove a similar result for a singular-value decomposition, but the result is necessarily somewhat weaker. In order to explain the *modified stopping criterion* used by Demmel and Veselić, we rewrite Eq. (11.83), describing the diagonalization stage of the two-sided Jacobi algorithm, as follows:

$$\begin{bmatrix} c_2 & s_2 \\ -s_2 & c_2 \end{bmatrix}^T \begin{bmatrix} y_{ii} & y_{ki} \\ y_{ki} & y_{kk} \end{bmatrix} \begin{bmatrix} c_2 & s_2 \\ -s_2 & c_2 \end{bmatrix} = \begin{bmatrix} z_{ii} & z_{ki} \\ z_{ki} & z_{kk} \end{bmatrix}$$

According to Demmel and Veselić, the element z_{ki} is set equal to zero only if

$$\frac{z_{ki}}{\sqrt{z_{ii} z_{kk}}} \leq \eta$$

where η is some prescribed tolerance.

The points made by Demmel and Veselić do not take away the important computational role played by the QR algorithm in eigen- and SVD computations. The QR algorithm is not only mathematically elegant, but computationally powerful and highly versatile.

Applications

Useful applications of singular value decomposition (SVD) include the following:

1. *Rank determination.* The *column rank* of a matrix is defined by the number of linearly independent columns of the matrix. Specifically, we say that an M-by-K matrix, with $M \geq K$, has *full* column rank if and only if it has K independent columns. In theory, the issue of full rank determination is a yes-no type of proposition in the sense that either the matrix in question has full rank or it does not. In practice, however, the fuzzy nature of a data matrix and the use of inexact (finite-precision) arithmetic complicate the rank determination problem. The SVD provides a practical method for determining the rank of a matrix, given fuzzy data and roundoff errors produced by the use of finite-precision computers.
2. *Adaptive beamforming,* where the data collected by an array of sensors are arranged in separate blocks, and the SVD is applied to each block. The singular values and singular vectors so computed, and the corresponding values of a *steering vector* (that serves the purpose of a desired response for the adaptive beamformer) are then used to determine the minimum-norm solution for the sensor-weight vector.
3. *Adaptive frequency analysis/direction finding,* where the requirement is to estimate (a) the frequencies of sinusoids in the presence of additive noise as in time series analysis, or (b) the directions of plane waves impinging on an array of sensors as in spatial processing. This application is discussed in detail in the next chapter.
4. *Adaptive image coding,* where the SVD is applied to a real image matrix $\mathbf{A}$ of rank W, and the image is compressed by truncating its series expansion, as shown by

$$\mathbf{A}_k \approx \sum_{i=1}^{k} \sigma_i \mathbf{u}_i \mathbf{v}_i^T$$

where $\mathbf{u}_i \mathbf{v}_i^T$ is referred to as an *eigenimage*. The motivation here is that by choosing a value for k small compared to the rank W of the original image matrix $\mathbf{A}$, it becomes possible to provide a good representation of the image with a reduced storage requirement (Anderson and Patterson, 1975; Haykin, 1989a). This is an example of rank reduction.

Generalization

A discussion of the singular-value decomposition problem would be incomplete without a brief mention of the *generalized singular-value decomposition* (GSVD). To define what we mean by a GSVD, let $\mathbf{A}$ be an M-by-N matrix ($M \geq N$), and $\mathbf{B}$ be a K-by-N matrix. Then there exists an M-by-M unitary matrix $\mathbf{U}$, a K-by-K unitary matrix $\mathbf{V}$, and an invertible N-by-N matrix $\mathbf{X}$, such that (Golub and VanLoan, 1989)

$$\mathbf{U}^H \mathbf{A} \mathbf{X} = \text{diag}(\alpha_1, \alpha_2, \ldots, \alpha_N), \qquad \alpha_i \geq 0$$

and

$$\mathbf{V}^H\mathbf{B}\mathbf{X} = \text{diag}(\beta_1, \beta_2, \ldots, \beta_L), \qquad \beta_i \geq 0; L = \min\{K, N\}$$

where

$$\beta_1 \geq \beta_2 \geq \cdots \geq \beta_W > \beta_{W+1} = \cdots = \beta_L = 0, \qquad W = \text{rank}(\mathbf{B})$$

The elements of the set $\{\alpha_1/\beta_1, \ldots, \alpha_W/\beta_W\}$ are called the *generalized singular values* of the matrices **A** and **B**. The *generalized singular-value decomposition theorem* stated herein is a generalization of the ordinary SVD in that if the matrix **B** equals the N-by-N identity matrix, then the generalized singular values of **A** and **B** are equal to the singular values of matrix **A**.

The GSVD is useful in *least-squares minimization with a quadratic inequality constraint*, referred to in the literature as the *LSQI problem*. This is a technique that may be used whenever the solution to the ordinary least-squares problem needs to be *regularized*. A simple LSQI problem that arises when attempting to fit a function to noisy data may be described as follows (Golub and VanLoan, 1989).

Suppose that we wish to find an N-by-1 vector **w** that minimizes $\|\mathbf{Aw} - \mathbf{d}\|_2$, subject to the constraint $\|\mathbf{Bw}\| \leq$ c, where **A** is an M-by-N matrix, **B** is an N-by-N nonsingular matrix, **d** is an M-by-1 vector, and $c \geq 0$. In this problem, the constraint is usually chosen so as to dampen excessive oscillations in the fitting function. The generalized singular-value composition sheds light on the solvability of the LSQI problem.

PROBLEMS

1. Calculate the singular values and singular vectors of the 2-by-2 real matrix:

$$\mathbf{A} = \begin{bmatrix} 1 & -1 \\ 0.5 & 2 \end{bmatrix}$$

   Do the calculation using three different methods:
   (a) Eigenvalue decomposition of the matrix product $\mathbf{A}^T\mathbf{A}$.
   (b) Eigenvalue decomposition of the matrix product $\mathbf{A}\mathbf{A}^T$.
   (c) Two-sided Jacobi method.

2. Use the results of Problem 1 to find the pseudoinverse of the real matrix **A** specified in that problem.

3. Consider the 2-by-2 complex matrix

$$\mathbf{A} = \begin{bmatrix} 1 + j & 1 + 0.5j \\ 0.5 - j & 1 - j \end{bmatrix}$$

   Calculate the singular values and singular vectors of the matrix **A** by proceeding as follows:
   (a) Construct the matrix $\mathbf{A}^H\mathbf{A}$; hence, evaluate the eigenvalues and eigenvectors of $\mathbf{A}^H\mathbf{A}$.
   (b) Construct the matrix $\mathbf{A}\mathbf{A}^H$; hence, evaluate the eigenvalues and eigenvectors of $\mathbf{A}\mathbf{A}^H$.
   (c) Relate the eigenvalues and eigenvectors calculated in parts (a) and (b) to the singular values and singular vectors of **A**.

4. Refer back to the example of Section 10.6. For the sets of data given in that example, do the following:
   1. Calculate the pseudo-inverse of the 3-by-2 data matrix **A**.
   2. Use this value of the pseudo-inverse $\mathbf{A}^+$ to calculate the two tap-weights of the linear least-squares filter. Compare your result with that obtained in Problem 3 of Chapter 10.
5. In this problem we explore the derivation of the weight update for the normalized LMS algorithm described in Eq. (9.167) using the idea of singular-value decomposition. This problem may be viewed as an extension of the discussion presented in Section 11.6. Find the minimum norm solution for the coefficient vector

$$\mathbf{c}(n+1) = \begin{bmatrix} \delta\hat{\mathbf{w}}(n+1) \\ 0 \end{bmatrix}$$

that satisfies the equation

$$\mathbf{x}^H(n)\mathbf{c}(n+1) = e^*(n)$$

where

$$\mathbf{x}(n) = \begin{bmatrix} \mathbf{u}(n) \\ \sqrt{a} \end{bmatrix}$$

Hence, show that

$$\hat{\mathbf{w}}(n+1) = \hat{\mathbf{w}}(n) + \frac{\tilde{\mu}}{a + \|\mathbf{u}(n)\|^2}\mathbf{u}(n)e^*(n)$$

where $a > 0$, and $0 < \tilde{\mu} < 2$. [This is the weight update described in Eq. (9.167).]

6. Repeat the calculation of the singular values and singular vectors of the matrix **A** given in Problem 3 by using the two-sided Jacobi algorithm described in Section 11.8. Compare your results with those obtained in Problem 3.
7. Show that the two-sided Jacobi transformation described in Eq. (11.69) represents an orthonormal matrix.
8. Demonstrate that the sweep of orderings by rows is equivalent to the sweep of orderings by columns described in Example 3.
9. The transformation of a 2-by-2 complex matrix into a real one involves a plane rotation, followed by premultiplication and postmultiplication.
   (a) Show that the combined effect of the plane rotation in Eq. (11.103) and the premultiplication in Eq. (11.112) is equivalent to a unitary matrix.
   (b) Show that the postmultiplying matrix on the left side of Eq. (11.112) is also a unitary matrix.
10. Consider an M-by-M matrix **A** that is triangularized by the use of Householder transformations. After $M - 1$ steps, at most, the matrix **A** is triangularized as follows:

$$\mathbf{QA} = \mathbf{R}$$

where **R** is an upper triangular matrix, and

$$\mathbf{Q} = \mathbf{Q}_{M-1}\mathbf{Q}_{M-2}\cdots\mathbf{Q}_1$$

   (a) Show that

$$\det(\mathbf{A}) = (-1)^{M-1}\det(\mathbf{R})$$

(b) Using the fact that the Euclidean norm of a matrix is preserved under multiplication by a unitary matrix, and that each diagonal element of the triangular matrix **R** is the norm of the projection of that column on a certain subspace, show that

$$|\det(\mathbf{A})| \leq \prod_{i=1}^{M} \|\mathbf{a}_i\|$$

where $\mathbf{a}_i$ is the ith column of matrix **A**. (This result is known as the *Hadamard theorem.*)

11. Consider the 4-by-3 data matrix:

$$\mathbf{A} = \begin{bmatrix} 1 & 1 & 1 \\ 2 & 1.5 & 2 \\ 3 & 3 & 4 \\ 4 & 4.5 & 8 \end{bmatrix}$$

Using a sequence of Householder transformations, reduce the matrix to upper bidiagonal form.

12. Consider the data matrix

$$\mathbf{A} = \begin{bmatrix} 1 & 1 & 1 \\ 2 & 1.5 & 2 \\ 3 & 3 & 4 \end{bmatrix}$$

Reduce the matrix **A** to upper triangular form, using:

(a) Householder transformations.

(b) Givens rotations.

13. In Section 11.11 we illustrated the Golub–Kahan algorithm for the diagonalization of an M-by-M upper bidiagonal matrix for $M = 4$. Extend this illustration for $M = 5$.

14. Use the Golub–Kahan algorithm to compute the singular-value decomposition of the data matrix:

$$\mathbf{A} = \begin{bmatrix} 1.5 & 1.5 & 0 & 0 \\ 0 & 3 & 1.5 & 0 \\ 0 & 0 & 4.5 & 1.5 \\ 0 & 0 & 0 & 6 \end{bmatrix}$$

15. The Cholesky factorization of the filtered state error correlation matrix $\mathbf{K}(n)$ required in the standard square root filtering (SRF) implementation of the Kalman filter can be accomplished using the Householder transformation. The idea of SRF was discussed in Section 7.9. Explore this important application of the Householder transformation. In particular, develop the update recursions involved in the implementation of a square root Kalman filter using the Householder transformation. You may refer to the paper by Kaminski et al. (1971) for your exploration.

16. You are given a processor that is designed to perform the singular-value composition of a K-by-M data matrix **A**. Using such a processor, develop block diagrams for the following two super-resolution algorithms:

(a) The autoregressive (AR) algorithm described in Section 10.10.

(b) The minimum-variance distortionless response (MVDR) algorithm described in Section 10.11.

12

Super-Resolution Algorithms Using Eigenvector-Based Projections

In this chapter we resume the discussion of super-resolution estimation algorithms that we initiated toward the end of Chapter 10. There we discussed the two algorithms known as autoregressive (AR) modeling and minimum-variance distortionless response (MVDR). In this chapter we discuss two other popular super-resolution algorithms known as *MUSIC* and *minimum norm*. The term "super-resolution" or "high-resolution" refers to the fact that a frequency estimation or angle-of-arrival estimation algorithm so named has, under carefully *controlled* conditions, the ability to surpass the limiting behavior of classical Fourier-based methods. The MUSIC and minimum-norm algorithms are both products of an *eigenvector-based projection approach* to the formulation of signal-processing problems. Basically, the rationale of such an approach is to partition the *observation space,* spanned by the eigenvectors of a correlation matrix, into two subspaces that are referred to as the *signal plus noise subspace* and *noise subspace*. Naturally, the MUSIC and minimum-norm algorithms manipulate these two subspace in different ways. We begin our discussion in the next section by reviewing the frequency estimation/direction finding problem.

12.1 FREQUENCY ESTIMATION/DIRECTION FINDING PROBLEM

The temporal problem of *estimating the frequencies of complex sinusoids in additive receiver noise,* and the spatial problem of *estimating the directions of arrival* (*DOA*) *of incident plane waves corrupted by additive sensor noise* are basic to the study of signal processing. Although these two estimation problems arise in entirely different application areas, their mathematical formulations, and therefore procedures for their solution are indeed similar.

To elaborate on this similarity, consider a *received signal* $\{u(i)\}$ that consists of L *complex sinusoids* whose complex amplitudes are $\alpha_1, \alpha_2, \ldots, \alpha_L$ and whose *angular frequencies* are $\omega_1, \omega_2, \ldots, \omega_L$, respectively. Specifically, a sample $u(i)$ of the received signal is written as

$$u(i) = \sum_{l=1}^{L} \alpha_l e^{j\omega_l i} + v(i), \qquad i = 0, 1, \ldots, N-1 \tag{12.1}$$

where $v(i)$ is a complex sample of additive *receiver noise,* and N is the total data length. The following assumptions are made:

1. The complex sinusoidal components of the received signal are *uncorrelated* with each other, which means that

$$E[\alpha_k \alpha_l^*] = \begin{cases} P_l, & k = l \\ 0, & k \neq l \end{cases} \tag{12.2}$$

2. The receiver noise is *white,* which means that

$$E[v(i)v^*(j)] = \begin{cases} \sigma^2, & j = i \\ 0, & j \neq i \end{cases} \tag{12.3}$$

For the processing (filtering) of data, we propose to use a transversal filter of length $M + 1$ as indicated in Fig. 12.1. Given the time series of Eq. (12.1), the problem then is to estimate the unknown amplitudes and unknown frequencies contained in the time series.

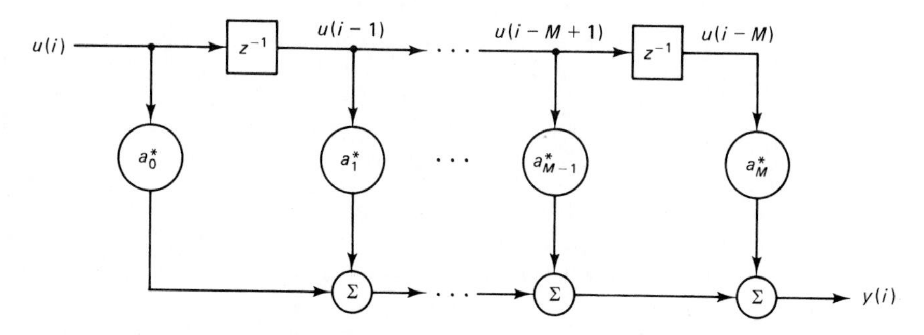

Figure 12.1 Transversal filter for temporal processing,

Consider next the *spatial analog* of this temporal estimation problem. Specifically, consider Fig. 12.2 that depicts a receiving *linear array* of sensors labelled $0, 1, \ldots, M$, which are uniformly spaced. We assume that the sensors are *isotropic,* which means that they receive uniformly in all directions with equal amplitude and phase. The array is illuminated by a set of L *plane waves* that originate from sources in the *far field* of the array. We assume that these plane waves originate from *incoherent sources*. Let d denote the separation between adjacent elements of the array. In Fig. 12.2 we show a single plane wave impinging on the array at an *angle of incidence* θ, measured with respect to the *boresight* (i.e., the normal to the array). We further assume that the propagating wave disturbance is *narrow-band,* which means that it is adequately characterized by a *single frequency,* denoted by f_0.

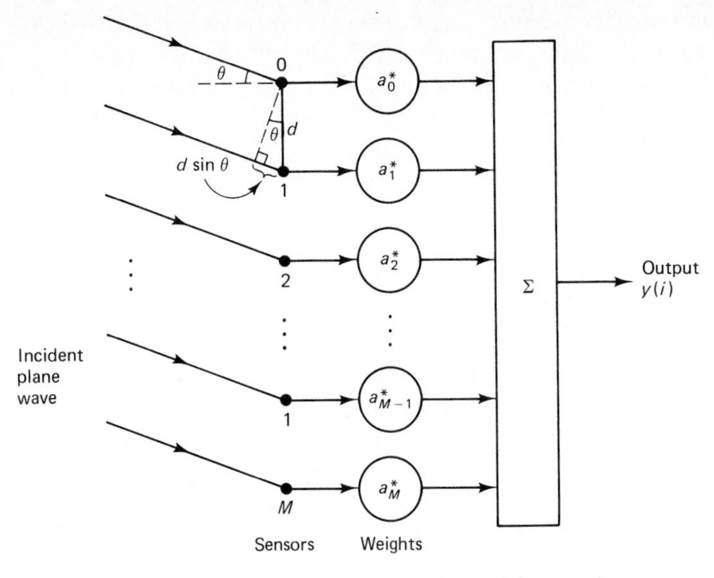

Figure 12.2 Linear array of sensors for spatial processing.

In other words, the signal envelope does not change during the time the incident plane wave interacts with the array. Correspondingly, the *wavelength* of the propagating disturbance is

$$\lambda = \frac{c}{f_0}$$

where c is the *speed of propagation*. From Fig. 12.2, we see that the *wavefront* of the incident plane wave reaches sensor 0, say, before sensor 1 by a distance equal to $d \sin \theta$. Hence, the corresponding *phase difference* between the signals received by sensors 0 and 1, and indeed between every other pair of adjacent sensors in the array, is $2\pi(d/\lambda) \sin \theta$. This quantity is called the *electrical angle* of the incident wave, denoted by

$$\phi = 2\pi \frac{d}{\lambda} \sin \theta \tag{12.4}$$

The ratio ϕ/d is called the *projected wavenumber* or just simply the *wavenumber* of the incident plane wave; it is denoted by κ. Clearly, if we can measure the electrical angle ϕ or equivalently the wavenumber κ, then it is a straightforward matter to measure the *direction of arrival* signified by the angle of incidence θ.

Since a source responsible for the propagating disturbance may be anywhere in the *field of view* of the array, that is, the angle of incidence θ may lie anywhere from $-90°$ to $90°$, the wavenumber κ is correspondingly restricted to lie in the interval $(-2\pi/\lambda, 2\pi/\lambda)$. Hence, a *sampling distance* less than $\lambda/2$ is necessary if we are to realize a *total* field of view. In particular, when $d < \lambda/2$, the *field intensity pattern* (i.e., the magnitude of the sum of all the outputs from the individual sensors of the array plotted versus the angle of incidence θ) has a single maximum at $\sin \theta = 0$ that defines the *main beam*. On the other hand, when $d > \lambda/2$, the field intensity pattern exhibits multiple maxima, all with the same value equal to the number of sensors. The maxima, except for the one defining the main beam, are called *grating lobes*. They have the effect of restricting the field of view. In general, grating lobes are undesirable and are therefore to be avoided, because the array structure responds to a source located in a grating lobe in exactly the same way as in the "true" field of view. The grating lobe phenomenon in spatial sampling is analogous to the well-known aliasing effect in temporal sampling. In the latter case, if we have a set of uniformly spaced time samples, spaced T seconds apart, the unambiguous frequency interval is from $-\pi/T$ to π/T radians per second. From the temporal version of the sampling theorem we know that if the sampling rate $1/T$ is less than twice the highest frequency component of a bandlimited signal, then aliasing occurs (Haykin, 1988); it is then impossible to distinguish between frequencies that lie outside the interval $(-\pi/T, \pi/T)$ from those that lie inside this interval. Hence, in the spatial processing problem depicted in Fig. 12.2, the distance d between adjacent sensors in the array must be chosen less than $\lambda/2$, if we are to avoid the grating lobe phenomenon.

For the purpose of our present discussion we assume that the intersensor spacing d has been chosen such that the issue of grating lobes is of no concern. We may then describe the *baseband model* of the received signal at the ith sensor of the array in Fig. 12.2 and at time k, due to the combined action of all the L incident plane waves, by writing[1]

$$u(i, k) = \sum_{l=1}^{L} \alpha_l e^{j\phi_l i} + v(i, k), \qquad i = 0, 1, \ldots, M \tag{12.5}$$

where α_l and ϕ_l are the complex amplitude and electrical angle of the lth incident plane wave, respectively, and $v(i, k)$ is the additive contribution of *sensor noise*. Note that in the baseband model of Eq. (12.5), it is assumed that the incident plane waves are all narrow-band, centered around a common frequency f_0. The attractive feature of this model is its mathematical simplicity; yet it retains the full information content of the received signal pertaining to the wavenumbers of the different incident plane waves and their complex amplitudes. The only piece of information missing from the model is the frequency f_0 of these plane waves, which is known a priori by virtue of the frequency tuning built into the receiver design.

Another important point to note is the mathematical similarity between the temporal model of Eq. (12.1) and the spatial model of Eq. (12.5). The model of Eq. (12.1) describes the signal at the ith tap of a linear transversal filter, over the *entire data length*. Equation (12.5), on the other hand, describes the signal at the ith sensor

[1] Consider a source (emitter) of radio frequency (RF) energy; let the source output be denoted by $a_l \cos(2\pi f_0 t + \psi_l)$, where a_l is the amplitude, f_0 is the frequency, and ψ_l is the phase. The source is located in the far field of a linear of uniformly spaced sensors. The noiseless signal received by the ith sensor is $a_l \cos(2\pi f_0 t + \psi_l + \phi_l i)$, where ϕ_l is the electrical angle of the incident plane wave. This signal may be represented by

$$a_l \cos(2\pi f_0 t + \psi_l + \phi_l i) = Re[\alpha_l e^{j\phi_l i} e^{j2\pi f_0 t}]$$

where

$$\alpha_l = a_l e^{j\psi_l}$$

The complex sinusoid $\alpha_l e^{j\phi_l i}$ is referred to as the *baseband* form of the noiseless received signal at the ith sensor due to the lth incident plane wave [Haykin (1989a)].

of a linear uniform array, for one *snapshot* of data of length $M + 1$ measured at a particular instant of time. Indeed, by comparing this pair of equations, we readily deduce the analogy between temporal processing and spatial processing summarized in Table 12.1.

TABLE 12.1 ANALOGY BETWEEN TEMPORAL PROCESSING AND SPATIAL PROCESSING

Time series	Space series
Time (sec)	Distance (m)
Sampling duration, T (sec)	Sensor spacing, d (m)
Angular frequency, ω (rad/sec)	Wavenumber, κ (rad/m)

There is, however, one basic difference between the temporal and spatial models described herein. The time series of Eq. (12.1) is *serial* in nature that continues for a total observation interval N. The space series of Eq. (12.5), on the other hand, is formulated on a snapshot-by-snapshot basis, with each snapshot corresponding to a particular instant of time. In the linear array processor of Fig. 12.2 [represented mathematically by Eq. (12.5)] a data snapshot derived from the $(M + 1)$ sensors corresponds to a data window encompassing the $(M + 1)$ tap inputs in the transversal filter of Fig. 12.1 [represented mathematically by Eq. (12.1)]. In the spatial model of Fig. 12.2, the data evolve in time one snapshot after another, whereas in the temporal mode of Fig. 12.1 the data evolve in time one sample after another. Clearly, when we perform *block processing,* which is the intention of our present chapter, this subtle difference between the temporal and spatial domains of interest has no analytic significance.

Rayleigh Resolution Criterion

A basic issue in the frequency estimation problem or the direction finding problem is that of resolution. In the temporal context, for example, in order to compare the powers of different frequency-estimation algorithms to resolve the frequencies of two closely spaced sinusoidal components, it is convenient to consider the case when the two sinusoidal components have the same average power, and to fix (somewhat arbitrarily) a displacement of maxima at which the two components may be said to be "just resolved." Let $\omega_0 \pm \frac{1}{2}\Delta\omega_0$ be the angular frequencies of the two components; the ratio $\Delta\omega_0/\omega_0$ is called the *resolving power* of the frequency estimation algorithm (Born and Wolf, 1980).

The idea of a resolution criterion was first introduced by Lord Rayleigh (1879) in connection with prism and grating spectroscopes, where the intensity distribution for monochromatic light has the form

$$I(\delta) = \left(\frac{\sin(\delta/2)}{\delta/2}\right)^2 I_{\max}$$

For such a situation, Lord Rayleigh proposed that *two components of equal intensity should be considered to be just resolved when the first maximum from one component sits at the first minimum from the other,* as illustrated in Fig. 12.3. This simple criterion is called the *Rayleigh resolution criterion*.

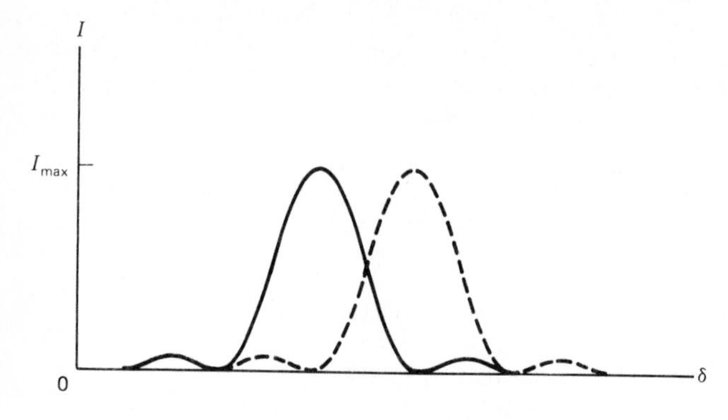

Figure 12.3 Illustrating the Rayleigh resolution criterion.

In the signal-processing literature (and other related fields for that matter), the Rayleigh resolution criterion has long been recognized as the criterion for establishing a bound on the resolving power of classical Fourier-based frequency estimation and direction finding algorithms. As mentioned briefly in the introduction to this chapter, so-called super-resolution algorithms may overcome the Rayleigh resolution limit. Indeed, it is for this reason that the term "super-resolution" or "high-resolution" is used to describe algorithms such as the maximum-entropy, minimum-variance distortionless response, MUSIC, and minimum-norm algorithms. However, for any of these algorithms to overcome the limit set by the Rayleigh resolution criterion, the underlying conditions have to be carefully controlled. In other words, the *physical model* responsible for the generation of the incoming data has to closely *match* the *mathematical model* assumed in the derivation of the super-resolution algorithm. Otherwise, we have a *model-mismatch problem* with serious consequences, namely, a significant frequency estimation error or loss of resolution.

12.2 DATA MATRIX

As explained in Section 10.2, there is no unique method of setting up the data matrix. For the case of temporal processing, we will assume the use of the covariance method of data windowing. Thus, given the time series of length N as in Eq. (12.1) and the transversal filtering structure of length $M + 1$ as in Fig. 12.1, we may define a data matrix $\mathbf{A}$ by writing

$$\mathbf{A}^H = \Bigg[\underbrace{\begin{matrix} u(M) & \cdots & u(N-1) \\ u(M-1) & \cdots & u(N-2) \\ \vdots & \ddots & \vdots \\ u(0) & \cdots & u(N-M+1) \end{matrix}}_{\text{forward half}} \quad \underbrace{\begin{matrix} u^*(0) & \cdots & u^*(N-M+1) \\ u^*(1) & \cdots & u^*(N-M+2) \\ \vdots & \ddots & \vdots \\ u^*(M) & \cdots & u^*(N-1) \end{matrix}}_{\text{backward half}}\Bigg] \tag{12.6}$$

The elements constituting the *left* half of the matrix $\mathbf{A}^H$ in Eq. (12.6) represent the various sets of tap inputs used for *forward filtering*. The complex-conjugated elements constituting the *right* half of matrix $\mathbf{A}^H$ represent the corresponding sets of tap inputs used for backward filtering. (This data arrangement is motivated by the discussion on linear prediction in Section 10.10.) Note that in the forward or backward half in Eq. (12.6), as we move from one column of $\mathbf{A}^H$ to the next, we drop one input sample and add a new one, and of course reorder the data. This has the effect of *temporal smoothing*.

For the spatial case, we may also use different methods for setting up the data matrix, depending on how we wish to process the data. The following two methods, similar in philosophy to the temporal structure described in Eq. (12.6), deserve special mention:

1. Given a linear uniform array of $(M + 1)$ sensors and a total of K snapshots, we define the data matrix $\mathbf{A}$ by writing

$$\mathbf{A}^H = \underbrace{\begin{bmatrix} u(M, 1) & \cdots & u(M, K) \\ u(M-1, 1) & \cdots & u(M-1, K) \\ \vdots & \ddots & \vdots \\ u(0, 1) & \cdots & u(0, K) \end{bmatrix}}_{\text{forward half}} \underbrace{\begin{matrix} u^*(0, 1) & \cdots & u^*(0, K) \\ u^*(1, 1) & \cdots & u^*(1, K) \\ \vdots & \ddots & \vdots \\ u^*(M, 1) & \cdots & u^*(M, K) \end{matrix}}_{\text{backward half}} \tag{12.7}$$

   where $u(i, k)$ denotes the received signal at sensor $i = 0, 1, \ldots, M$ and snapshot (time) $k = 1, 2, \ldots, K$. Thus, each column of $\mathbf{A}^H$ corresponds to a snapshot of data. On the other hand, each row of the forward half or back half represents the time evolution of the received signal at the pertinent sensor in the array. In the "forward" half of $\mathbf{A}^H$, the sensor data are arranged bottom up on a snapshot by snapshot basis; in the "backward" half, the snapshots are reversed in order (i.e., arranged top down) and also complex conjugated.

2. Given a linear uniform array consisting of N sensors, we set up a *subaperture* consisting of $M + 1$ sensors. From linear prediction theory (based on a modification of the forward-backward linear prediction method) we may set the prediction order $M = 3N/4$, where N is the total number of sensors in the array (Tufts and Kumaresan, 1982). In any event, we slide the subaperture of length $M + 1$ across the linear array, both up and down. This procedure is illustrated in Fig. 12.4 for the example of $N = 16$ and $M = 12$. For a *single* snapshot of data, we may thus formulate the data matrix $\mathbf{A}(k)$ for snapshot k by using a definition similar to that given in Eq. (12.6).

Thus in method 1 of data arrangement for array processing, we in effect introduce *temporal smoothing* by averaging over K independent snapshots of data. In method 2, on the other hand, we in effect introduce a form of *spatial smoothing* within a single snapshot of data. In the latter case, there is of course nothing to stop us from also introducing a form of temporal smoothing by averaging over a multiplicity of independent snapshots. The combination of spatial smoothing and temporal smooth-

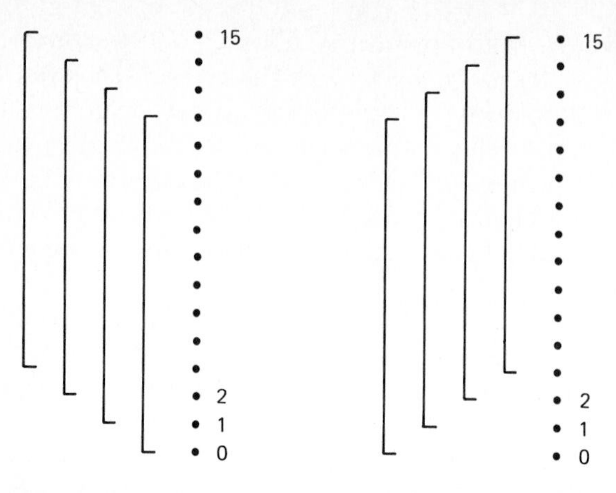

(a) Four subapertures formed by sliding upward across the array

(b) Four subapertures formed by sliding downward across the array

Figure 12.4 Illustrating the basis of spatial smoothing.

ing provides an effective method for dealing with coherent sources in an array processing environment.

From here on, we assume the availability of a data matrix **A**. The data matrix **A** may be generated by one of the methods above or some other method. Although the presentation will focus on the time-series model of Eq. (12.1), the material developed herein is equally applicable to the space-series model of Eq. (12.5) by virtue of the discussion in this and the preceding section. We will first discuss the MUSIC algorithm, and then go on to the minimum-norm algorithm.

12.3 MUSIC ALGORITHM

To motivate the development of the *mul*tiple *si*gnal *c*lassification (MUSIC) algorithm (Schmidt, 1979, 1981, 1986), consider first the $(M + 1)$-by-$(M + 1)$ ensemble-averaged correlation matrix $\mathbf{R}$ for an input signal that consists of L *uncorrelated* zero-mean complex sinusoids and an additive white-noise process of zero mean and variance σ^2, as in Eq. (12.1). The angular frequencies of the sinusoids are denoted by $\omega_1, \omega_2, \ldots, \omega_L$, and their average powers by $P_1, P_2, \ldots, P_L$. We may thus express the ensemble-averaged correlation matrix $\mathbf{R}$ in the following form (see Problem 1):

$$\mathbf{R} = \mathbf{S}\mathbf{D}\mathbf{S}^H + \sigma^2\mathbf{I} \tag{12.8}$$

where $\mathbf{I}$ is the $(M + 1)$-by-$(M + 1)$ identity matrix. The rectangular matrix $\mathbf{S}$ is the $(M + 1)$-by-L *frequency matrix* defined by

$$\mathbf{S} = [\mathbf{s}_1, \mathbf{s}_2, \ldots, \mathbf{s}_L]$$

$$= \begin{bmatrix} 1 & 1 & \cdots & 1 \\ \exp(-j\omega_1) & \exp(-j\omega_2) & \cdots & \exp(-j\omega_L) \\ \exp(-j2\omega_1) & \exp(-j2\omega_2) & \cdots & \exp(-j2\omega_L) \\ \vdots & \vdots & \ddots & \vdots \\ \exp(-jM\omega_1) & \exp(-jM\omega_2) & \cdots & \exp(-jM\omega_L) \end{bmatrix} \tag{12.9}$$

Note that the lth column of the matrix $\mathbf{S}$, namely, the vector $\mathbf{s}_l$ is determined by the lth complex sinusoid of angular frequency ω_l. The diagonal matrix $\mathbf{D}$ in Eq. (12.8) is the K-by-K correlation matrix of the sinusoids (incoming plane waves), defined by

$$\mathbf{D} = \text{diag}(P_1, P_2, \ldots, P_L) \tag{12.10}$$

Let $\lambda_1 \geq \lambda_2 \cdots \geq \lambda_{M+1}$ denote the eigenvalues of the correlation matrix $\mathbf{R}$, and $\nu_1 \geq \nu_2 > \cdots > \nu_{M+1}$ denote the eigenvalues of $\mathbf{SDS}^H$, respectively. Then, from the representation shown in Eq. (12.8) we deduce that

$$\lambda_i = \nu_i + \sigma^2, \qquad i = 1, 2, \ldots, M + 1 \tag{12.11}$$

We assume that the signal matrix $\mathbf{S}$ is of full column rank L, which is justified if the L complex sinusoids in the time series of Eq. (12.1) have *distinct* frequencies, and the L columns of the matrix $\mathbf{S}$ are therefore linearly independent. This assumption then implies that the $(M + 1 - L)$ smallest eigenvalues of the matrix $\mathbf{SDS}^H$ are equal to zero. Correspondingly, the smallest eigenvalue of the correlation matrix $\mathbf{R}$ is equal to σ^2 with multiplicity $(M + 1 - L)$, as shown by

$$\lambda_i = \begin{cases} \nu_i + \sigma^2, & i = 1, \ldots, L \\ \sigma^2, & i = L + 1, \ldots, M + 1 \end{cases} \tag{12.12}$$

Note that, in general, $\nu_i \neq P_i$, $i = 1, 2, \ldots, L$. Figure 12.5 illustrates a plot of λ_i versus i for $L = 8$ and $M = 10$, assuming an additive white noise background. Such a plot is referred to as an *eigen-spectrum*.

Let $\mathbf{q}_1, \mathbf{q}_2, \ldots, \mathbf{q}_{M+1}$ denote the *eigenvectors* of the correlation matrix $\mathbf{R}$. All the $(M + 1 - L)$ eigenvectors associated with the smallest eigenvalues of $\mathbf{R}$ satisfy the relation

$$\mathbf{Rq}_i = \sigma^2\mathbf{q}_i, \qquad i = L + 1, \ldots, M + 1$$

or, equivalently,

$$(\mathbf{R} - \sigma^2\mathbf{I})\mathbf{q}_i = \mathbf{0}, \qquad i = L + 1, \ldots, M + 1 \tag{12.13}$$

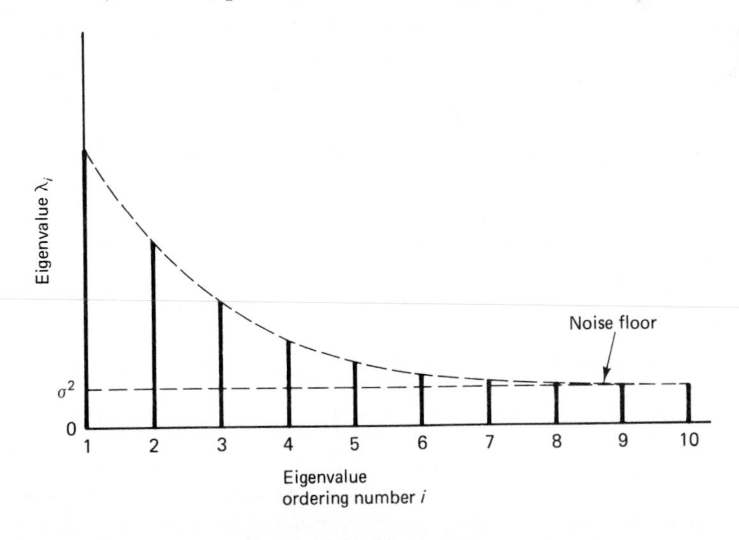

Figure 12.5 Eigen-spectrum.

Using Eq. (12.8), we may rewrite Eq. (12.13) as

$$\mathbf{S}\mathbf{D}\mathbf{S}^H\mathbf{q}_i = \mathbf{0}, \qquad i = L+1, \ldots, M+1 \tag{12.14}$$

Since the matrix $\mathbf{S}$ is assumed to be of full column rank L and since the matrix $\mathbf{D}$ is diagonal with all entries being nonzero [which is a consequence of Eq. (12.2)], it follows from Eq. (12.14) that

$$\mathbf{S}^H\mathbf{q}_i = \mathbf{0}, \qquad i = L+1, \ldots, M+1$$

or more explicitly [from the first line of Eq. (12.9)]

$$\mathbf{s}_l^H\mathbf{q}_i = 0, \qquad \begin{array}{l} i = L+1, \ldots, M+1 \\ l = 1, 2, \ldots, L \end{array} \tag{12.15}$$

where the vector $\mathbf{s}_l$ constitutes the lth column of matrix $\mathbf{S}$.

A fundamental property of the eigenvectors of a correlation matrix is that they are orthogonal to each other. Hence, the eigenvectors $\mathbf{q}_1, \ldots, \mathbf{q}_L$ span a subspace that is the *orthogonal complement* of the space spanned by the eigenvectors $\mathbf{q}_{L+1}, \ldots, \mathbf{q}_{M+1}$. Accordingly, we deduce from Eq. (12.15) that

$$\text{span}\{\mathbf{s}_1, \ldots, \mathbf{s}_L\} = \text{span}\{\mathbf{q}_1, \ldots, \mathbf{q}_L\} \tag{12.16}$$

The span $\{\mathbf{s}_1, \ldots, \mathbf{s}_L\}$ refers to a subspace that is defined by the set of all linear combinations of the vectors $\mathbf{s}_1, \ldots, \mathbf{s}_L$. The span $\{\mathbf{q}_1, \ldots, \mathbf{q}_L\}$ is similarly defined.

Thus, the eigenvalue decomposition of the $(M+1)$-by-$(M+1)$ correlation matrix $\mathbf{R}$ of superimposed complex sinusoids in noise suggests the following two important observations:

1. The space spanned by the eigenvectors of $\mathbf{R}$ consists of two disjoint subspaces: One subspace called the *signal plus noise subspace,*[2] is spanned by the eigenvectors associated with the L *largest* eigenvalues of $\mathbf{R}$. The second subspace, called the *noise subspace,* is spanned by the eigenvectors associated with the $(M+1-L)$ *smallest* eigenvalues of $\mathbf{R}$. These two subspace are the *orthogonal complement* of each other. This first observation follows from Eq. (12.16).
2. Given the eigenvectors of $\mathbf{R}$, we may determine the frequencies of the complex sinusoids in the input signal by *searching for those sinusoidal signal vectors* $\mathbf{s}_k$ *that are orthogonal to the noise subspace*. This second observation follows from Eq. (12.15).

Practical Considerations

In practice, we have to base our analysis on some estimate of the ensemble-averaged correlation matrix $\mathbf{R}$. For example, we may use a sample average that equals the scaled version of the correlation matrix $\mathbf{\Phi}$, as shown by

$$\hat{\mathbf{R}} = \frac{1}{2(N-M)}\mathbf{\Phi} \tag{12.17}$$

[2] This subspace is referred to as a "signal plus noise subspace" rather than "signal subspace", because the eigenvalues associated with the eigenvectors that define the subspace include the effect of the additive sensor noise.

where $\mathbf{\Phi}$ is itself related to the data matrix $\mathbf{A}$ as follows [see Eq. (10.45)]

$$\mathbf{\Phi} = \mathbf{A}^H \mathbf{A}, \tag{12.18}$$

and the scaling factor $1/2(N - M)$ accounts for the fact that the time averaging is performed over $2(N - M)$ data points. In any event, let $\mathbf{v}_1, \mathbf{v}_2, \ldots, \mathbf{v}_{M+1}$ denote the eigenvectors of the estimate $\hat{\mathbf{R}}$. In accordance with observation 1, we may define a set of eigenvectors $\mathbf{v}_1, \ldots, \mathbf{v}_L$ associated with the **L** largest eigenvalues of the estimate $\hat{\mathbf{R}}$, and a set of eigenvectors $\mathbf{v}_{L+1}, \ldots, \mathbf{v}_{M+1}$ associated with the $(M + 1 - L)$ smallest eigenvalues of $\hat{\mathbf{R}}$. Let

$$\mathbf{V}_N = [\mathbf{v}_{L+1}, \ldots, \mathbf{v}_{M+1}] \tag{12.19}$$

and

$$\mathbf{V}_S = [\mathbf{v}_1, \ldots, \mathbf{v}_L] \tag{12.20}$$

We naturally have

$$\mathbf{V}_N^H \mathbf{V}_S = \mathbf{0} \tag{12.21}$$

which is in line with observation 1. However, owing to the presence of uncertainty in the eigenvector estimates, $\mathbf{v}_1, \mathbf{v}_2, \ldots, \mathbf{v}_{M+1}$, arising because of the finite number of samples used to generate $\hat{\mathbf{R}}$, the orthogonality relations of Eq. (12.15) that are responsible for observation 2 no longer hold. In the context of the latter point, we can *search* for the signal vectors that are *most closely orthogonal* to the noise subspace. Accordingly, in the MUSIC algorithm it is proposed to *estimate the angular frequencies of the complex sinusoids in the input signal as the peaks of the MUSIC spectrum* estimate (Schmidt, 1979, 1981):

$$\begin{aligned} \hat{S}_{\text{MUSIC}}(\omega) &= \frac{1}{\sum_{i=L+1}^{M+1} |\mathbf{s}^H \mathbf{v}_i|^2} \\ &= \frac{1}{\mathbf{s}^H(\omega) \mathbf{V}_N \mathbf{V}_N^H \mathbf{s}(\omega)} \end{aligned} \tag{12.22}$$

where the variable frequency vector or frequency scanning vector $\mathbf{s}(\omega)$ is defined by

$$\mathbf{s}^T(\omega) = [1, e^{-j\omega}, \ldots, e^{-j\omega(M-L)}] \tag{12.23}$$

The product $\mathbf{V}_N \mathbf{V}_N^H$ represents a projection matrix on the noise subspace.

Equation (12.22) is the formula used in the MUSIC algorithm for estimating the frequencies of complex sinusoids that are corrupted by additive white noise. Note that in this formula all the singular vectors that constitute the matrix $\mathbf{V}_N$, in accordance with Eq. (12.19), are weighted equally.

Note also that $\hat{S}_{\text{MUSIC}}(\omega)$ is based on a single realization of the underlying stochastic process represented by the given data matrix $\mathbf{A}$. As such, it represents an estimate of the exact MUSIC spectrum based on the eigendecomposition of the ensemble-averaged correlation matrix of the process, hence the use of a hat in the symbol $\hat{S}_{\text{MUSIC}}(\omega)$.

Computational Considerations and Summary of the MUSIC Algorithm

The theory leading to the development of the MUSIC algorithm has been based on the ensemble-averaged correlation matrix $\mathbf{R}$ or its time-averaged counterpart $\mathbf{\Phi}$. We may thus compute the MUSIC spectrum of Eq. (12.22) by performing an eigenanalysis on the correlation matrix $\mathbf{\Phi}$ to compute the noise subspace represented by $\mathbf{V}_N$. From a computational viewpoint, however, a more efficient approach is to perform a *singular value decomposition* on the data matrix $\mathbf{A}$ directly. The products of this decomposition are represented by the singular values $\sigma_1, \sigma_2, \ldots, \sigma_{M+1}$, the associated right singular vectors $\mathbf{v}_1, \mathbf{v}_2, \ldots, \mathbf{v}_{M+1}$, and a set of left singular vectors. Insofar as the application of the MUSIC algorithm is concerned, the left singular vectors of $\mathbf{A}$ are of no interest. On the other hand, the squares of the singular values of the data matrix $\mathbf{A}$ and the associated right singular vectors are of particular interest because they are respectively, the same as the eigenvalues of the correlation matrix $\mathbf{\Phi} = \mathbf{A}^H\mathbf{A}$ and the associated eigenvectors (see Section 11.2). In computing the MUSIC spectrum, the following points are noteworthy:

1. The MUSIC algorithm assumes knowledge of the model order, that is, the number of complex sinusoids contained in the transversal filter input (temporal case) or the number of emitters (sources) responsible for illuminating the array (spatial case). Two criteria for making this decision are discussed in Section 12.7.
2. The singular values of the data matrix $\mathbf{A}$ do not actually enter the computation of the MUSIC spectrum. Rather, they are used merely as a tool for identifying those right singular vectors of $\mathbf{A}$ that constitute the noise subspace. Although this role may appear to be of a secondary nature, nevertheless, it is crucial to the successful application of the MUSIC algorithm.
3. The MUSIC algorithm requires knowledge of only the $M + 1 - L$ smallest singular values of the data matrix $\mathbf{A}$. For such a requirement we may use a standard routine to compute the $M + 1$ singular values and associated singular

TABLE 12.2 SUMMARY OF THE MUSIC ALGORITHM

1. Use the time series $\{u(i)\}$ to set up the data matrix $\mathbf{A}$.
2. Compute the $M + 1 - L$ smallest singular values $\sigma_{L+1}, \ldots, \sigma_{M+1}$, and therefore identify the right singular vectors $\mathbf{v}_{L+1}, \ldots, \mathbf{v}_{M+1}$. Hence, define the matrix

$$\mathbf{V}_N = [\mathbf{v}_{L+1}, \mathbf{v}_{L+2}, \ldots, \mathbf{v}_{M+1}]$$

3. Compute the projection matrix on the noise subspace, defined by the product $\mathbf{V}_N\mathbf{V}_N^H$. Hence, compute the MUSIC spectrum estimate

$$\hat{S}_{\text{MUSIC}}(\omega) = \frac{1}{\mathbf{s}^H(\omega)\mathbf{V}_N\mathbf{V}_N^H\mathbf{s}(\omega)}$$

for varying values of angular frequency ω in the scanning vector

$$\mathbf{s}(\omega) = [1, e^{-j\omega}, \ldots, e^{-j\omega M}]^T, \qquad -\pi \le \omega \le \pi$$

4. Estimate the angular frequencies of the complex sinusoids in the time series $\{u(i)\}$ by locating the spectral peaks of $S_{\text{MUSIC}}(\omega)$.

vectors of the data matrix **A**; we may then identify the $M + 1 - L$ smallest singular values and therefore retain the associated right singular vectors for use in the MUSIC algorithm. A more efficient procedure, however, is to use an SVD algorithm tailor-made for solving the problem at hand; such an algorithm is described in [Golub et al. (1981); Van Huffel et al. (1987); Van Huffel and Vandewalle (1988)].

Table 12.2 presents a summary of the MUSIC algorithm. Figure 12.6 presents a block diagram of the essential steps involved in the computation of the MUSIC algorithm.

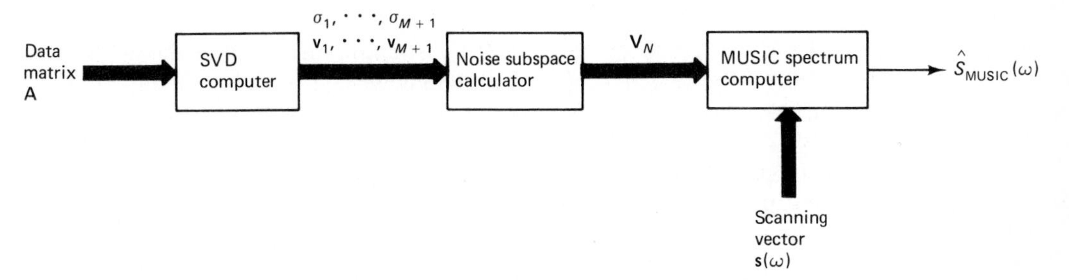

Figure 12.6 Block diagram of the MUSIC algorithm.

Root MUSIC

The frequency estimation problem in the MUSIC algorithm may be formulated as one of computing the values of ω for which the spectrum $\hat{S}_{\text{MUSIC}}(\omega)$ attains its peaks. To do this, however, we have to scan the complete frequency interval $-\pi \leq \omega \leq \pi$. We may avoid the need for frequency scanning by the use of a *root-finding* approach.

To be specific, we replace the complex exponential $e^{j\omega}$ by the complex variable z in the formula (12.22) for the spectrum $\hat{S}_{\text{MUSIC}}(\omega)$. Let $D(z)$ denote the resulting denominator polynomial, as shown by[3]

$$\hat{S}_{\text{MUSIC}}(z) = \frac{1}{D(z)} \tag{12.24}$$

The polynomial $D(z)$ exhibits the following properties:

1. The zeros of the polynomial $D(z)$ exhibit inverse symmetry with respect to the unit circle in the z-plane.
2. Ideally, the polynomial $D(z)$ has zeros on the unit circle in the z-plane at locations determined by the frequencies of the complex sinusoids in the input time series $\{u(i)\}$.

[3] Let **T** denote the matrix product $\mathbf{V}_N\mathbf{V}_N^H$. Let the coefficients of the desired polynomial $D(z)$ be denoted by $d_{-M-1}, d_{-M}, \ldots, d_{-1}, d_0, d_1, \ldots, d_M, d_{M+1}$. We may then write

$$d_i = \text{tr}_i[\mathbf{T}], \qquad i = -(M+1), \ldots 0, \ldots, M+1$$

where $\text{tr}_i[\mathbf{T}]$ is the trace of the ith diagonal of matrix **T**. Note that $i = 0$ corresponds to the main diagonal. The use of this relation provides an efficient method for computing the coefficients of the polynomial $D(z)$ (J. P. Reilly, private communication, 1990).

We may thus use *spectral factorization* (Haykin, 1989) to express the polynomial $D(z)$ as the product of two polynomials, as shown by

$$D(z) = H(z)H^*\left(\frac{1}{z^*}\right) \tag{12.25}$$

The first polynomial $H(z)$ has its zeros inside or on the unit circle. Those zeros of $H(z)$ that lie on (or are very close to) the unit circle represent the *signal zeros,* and the remaining ones represent extraneous zeros. The zeros of the other polynomial, $H^*(1/z^*)$, lie on or outside the unit circle in the z-plane, exhibiting inverse symmetry with respect to those of $H(z)$.

As an alternative to the computation of the spectrum $\hat{S}_{\text{MUSIC}}(\omega)$ for varying ω, we may perform the frequency estimation by extracting the zeros of the polynomial $D(z)$ and identifying the signal zeros from the knowledge that they should lie on the unit circle in the z-plane. This frequency (direction-of-arrival) estimation procedure is referred to as *root MUSIC* for obvious reasons.

The root-MUSIC algorithm is known to be superior to the standard MUSIC algorithm that relies on finding the peaks of the spectrum $\hat{S}_{\text{MUSIC}}(\omega)$. A performance analysis of the root-MUSIC algorithm for estimating the directions of arrival of plane waves in white noise for the use of a linear uniform array is presented in (Rao and Hari, 1989a). The analysis of computer simulation presented therein reveals that the errors in the estimated signal zeros are mainly *radial* in their occurrence. The radial nature of the estimation errors makes the peaks in the computation of the spectrum $\hat{S}_{\text{MUSIC}}(\omega)$ less distinct in appearance. This behavior, in turn, renders the use of frequency scanning to locate the peaks of the spectrum $\hat{S}_{\text{MUSIC}}(\omega)$ less accurate, and therefore less attractive, than root MUSIC.

12.4 COMPUTER EXPERIMENT ON COMPARATIVE EVALUATION OF THE MUSIC AND PERIODOGRAM FOR COHERENT SOURCES

This experiment is intended to compare the resolution capabilities of the MUSIC algorithm and the classical Fourier-based periodogram applied to the direction-of-arrival estimation probolem. Specifically, we consider a linear array consisting of $N = 16$ uniformly spaced isotropic antenna elements. The array is illuminated by a pair of plane waves whose individual average powers differ by 4 dB. The elemental signal-to-noise ratio is 30 dB, with the average signal power defined as the arithmetic mean of the individual signal powers of the two incident plane waves. The time series produced at the output of each antenna element thas length $K = 100$ (i.e., the number of snapshots is 100).

The experiment is to be performed for the following values of the electrical angular separation between the two plane waves:

1. Two beamwidths.
2. One beamwidth.
3. One-half beamwidth.

It is assumed that the sources of the two plane waves are *coherent*

It is convenient to reformulate the baseband model for the received signal $u(i, k)$ at the ith antenna element of the array and time k such that the phase due to the combined action of the incident plane waves is zero at the center of the array. For the problem at hand, we may thus write

$$u(i, k) = a_1 e^{j[i-(N-1)/2]\phi_1} + a_2 e^{j[i-(N-1)/2]\phi_2} + v(i, k)$$

where $i = 0, 1, \ldots, N - 1$ and N is the total number of elements in the array. The newly defined amplitude a_l is related to the complex amplitude α_l of the lth plane wave described in Section 12.2 as follows

$$a_l = e^{j(N-1)\phi_l/2} \qquad l = 1, 2$$

To deal with the presence of coherent sources, we will resort to the combined use of spatial smoothing and temporal averaging. In particular, we use a subaperture of length $M + 1$, where M is defined by (Tufts and Kumaresan, 1982)

$$M = \frac{3N}{4}$$

With $N = 16$, we may therefore choose $M = 12$.

Let $a_1 = 1$. Hence, we have

$$10 \log_{10} |a_2|^2 = -4 \text{ dB}$$

That is,

$$|a_2| = 0.6310$$

The average signal power is therefore

$$P_s = \frac{1}{2}\left(|a_1|^2 + |a_2|^2\right)$$

$$= 0.7001$$

We assume that the complex-valued background noise at the output of each antenna element is white and Gaussian-distributed with zero mean and variance σ^2. A desired complex noise sample may therefore be obtained from a pair of independent real pseudo-random number generators. Both generators are designed to have a Gaussian distribution of zero mean and variance $\sigma^2/2$, where $\sigma^2/2$ is the variance of the in-phase (real) and quadrature (imaginary) components of the complex noise sequence. The elemental signal-to-noise ratio is

$$SNR = 10 \log_{10}\left(\frac{P_s}{\sigma^2}\right), \text{ dB}$$

With an SNR of 30 dB, we therefore have

$$30 = 10 \log_{10}\left(\frac{0.7001}{\sigma^2}\right)$$

That is, the complex noise variance is

$$\sigma^2 = 0.7001 \times 10^{-3}$$

We may thus express the baseband model for the received signal at the ith element of the array and time k as

$$u(i, k) = e^{j[i-(N-1)/2]\phi_1} + 0.6310e^{j\psi}e^{j[i-(N-1)/2]\phi_2} + v(i, k)$$

where $i = 0, 1, \ldots, M$ and ψ is the phase difference between the complex amplitudes of the two signals. For the coherent case, ψ does not change from snapshot to snapshot. For this particular experiment we set $\psi = 0$.

The *standard beamwidth* is defined by

$$1 \text{ BW} = \frac{2\pi}{N}$$

We are given that the total number of elements in the array is $N = 16$. Hence, for the three specified cases of angular separation between the two incident plane waves, we have

$$\phi_2 = -\phi_1$$

and

$$\phi_1 = \frac{\pi}{8}, \frac{\pi}{16}, \frac{\pi}{32} \text{ radians}$$

For a single snapshot, the periodogram is $|U(\phi, k)|^2/N$, where

$$U(\phi, k) = \sum_{i=0}^{N-1} u(i, k)e^{-j[i-(N-1)/2]\phi}$$

For K snapshots, we simply average the individual periodograms computed for $k = 1, 2, \ldots, K$. The results of this computation (suitably scaled along the vertical axis) for the three specified values of (ϕ_1, ϕ_2) are plotted in Fig. 12.7. We see from this figure that the periodogram method resolves the two plane waves incident on the linear array only for $\phi_1 = 1\text{BW}$ and $\phi_2 = -1\text{BW}$, corresponding to an angular separation of two beamwidths. Even then, we observe that there is some bias in the estimated angles-of-arrival corresponding to the two largest peaks of the periodogram; see Table 12.3, page 464.

Consider next the application of the MUSIC algorithm. Specifically, for a single snapshot we set up the data matrix

$$\mathbf{A}^H(k) = \begin{bmatrix} u(M, k) & \cdots & u(N-1, k) & u^*(0, k) & \cdots & u^*(N-M+1, k) \\ u(M-1, k) & \cdots & u(N-2, k) & u^*(1, k) & \cdots & u^*(N-M+2, k) \\ \vdots & \ddots & \vdots & \vdots & \ddots & \vdots \\ u(0, k) & \cdots & u(N-M+1, k) & u^*(M, k) & \cdots & u^*(N-1, k) \end{bmatrix}$$

For the totality of K snapshots, we may thus define the overall data matrix

$$\mathbf{A}^H = [\mathbf{A}^H(1), \mathbf{A}^H(2), \ldots, \mathbf{A}^H(K)]$$

Performing a singular value decomposition on the data matrix $\mathbf{A}$, and then proceeding in accordance with the MUSIC algorithm as described in Table 12.2, we may compute the spectra (suitably scaled along the vertical axis) as shown in Fig. 12.8, corresponding to the three prescribed angular separations between the two incident plane waves. As we can see from this figure, the MUSIC algorithm has no trouble at

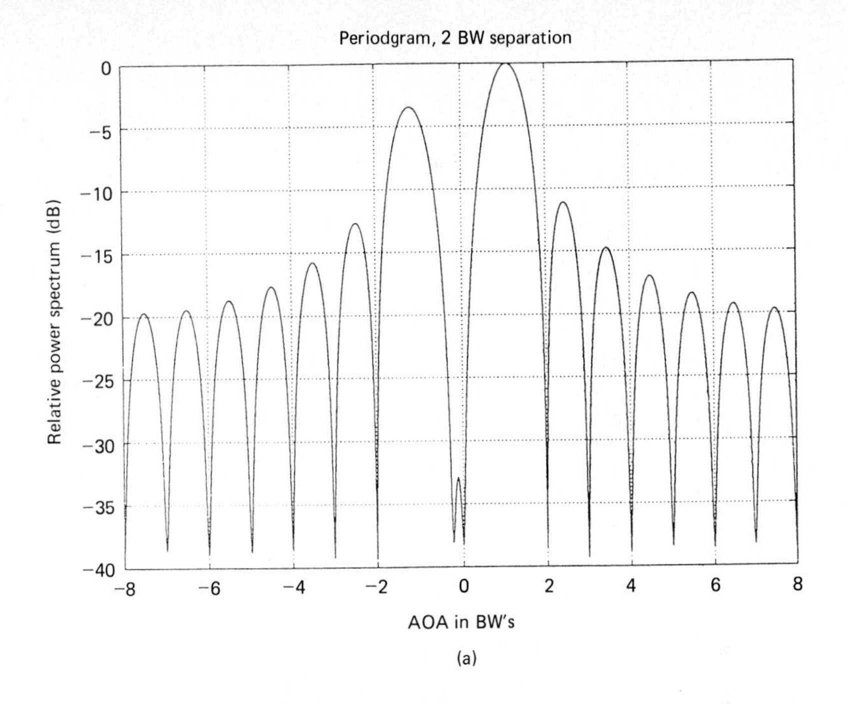

(a)

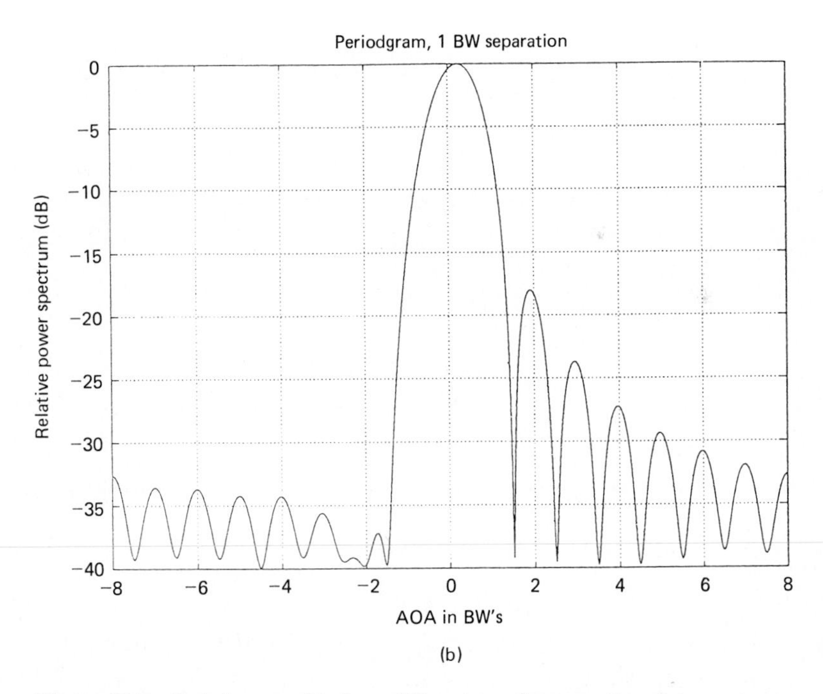

(b)

Figure 12.7 Periodograms for three different angular separations between two coherent plane waves: (a) two beamwidths, (b) one beamwidth, and (c) one-half beamwidth.

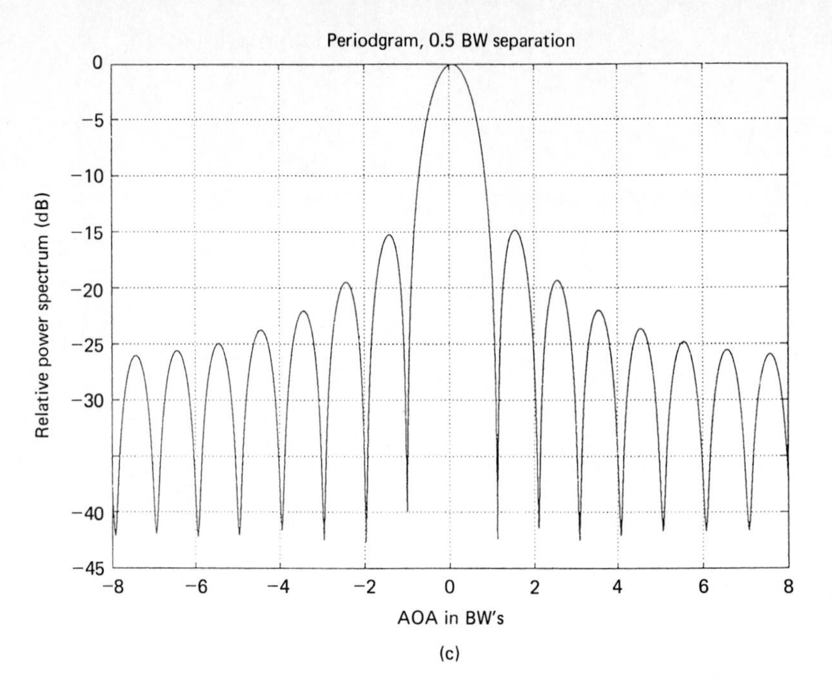

(c)

Figure 12.7 *(Continued)*

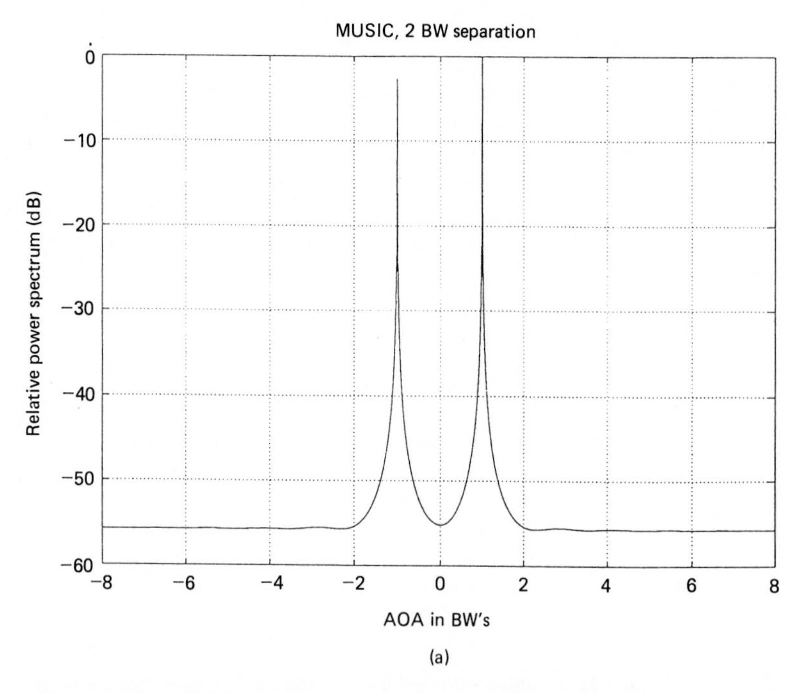

(a)

Figure 12.8 MUSIC spectra for three different angular separations between two coherent plane waves: (a) two beamwidths, (b) one beamwidth, and (c) one-half beamwidth.

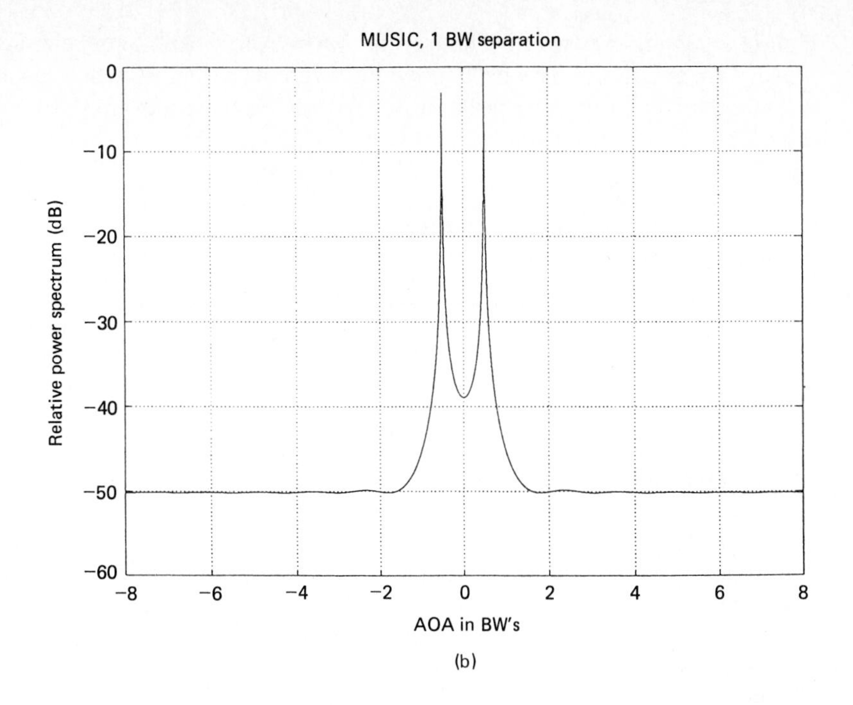

(b)

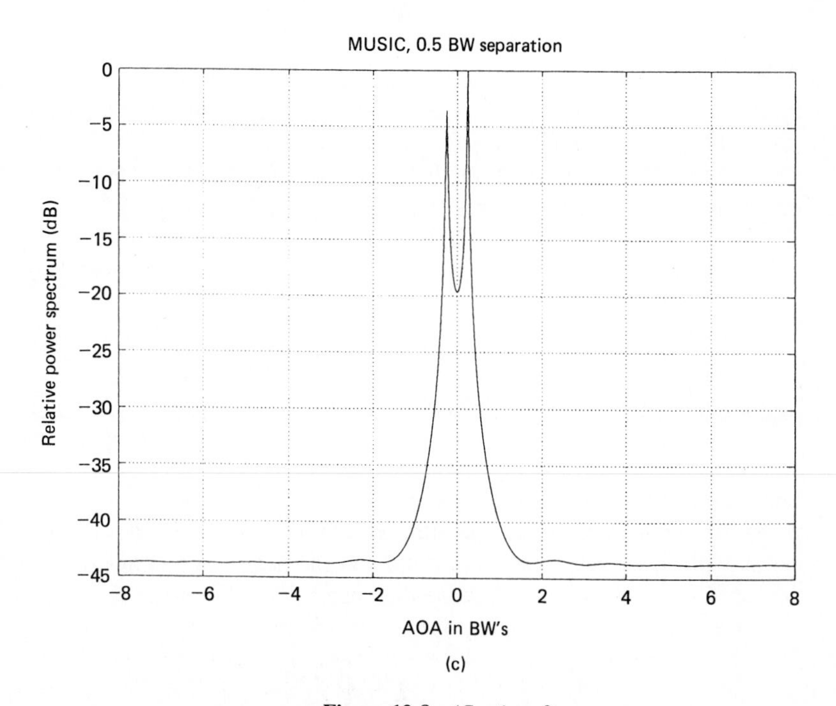

(c)

Figure 12.8 (*Continued*)

all in resolving the two incoming plane waves even when the angular separation between them is less than one beamwidth. Moreover, the estimated angles-of-arrival corresponding to the two peaks of the MUSIC spectrum exhibit no visible bias for the three cases considered, as illustrated in Table 12.3.

TABLE 12.3 PERIODOGRAM AND MUSIC AOA ESTIMATES, COHERENT CASE

Separation	Periodogram		MUSIC	
2 BW's	−1.1875	1.0938	−1.0000	1.0000
1 BW	0.2188		−0.5000	0.5000
0.5 BW's	0.0625		−0.2500	0.2500

12.5 MINIMUM-NORM METHOD

In the *minimum-norm method* (Kumaresan and Tufts, 1983) as with the MUSIC algorithm, the eigenvectors of the correlation matrix estimate $\hat{\mathbf{R}}$ are divided into two sets: the set $\mathbf{v}_1, \ldots, \mathbf{v}_L$ that defines the matrix $\mathbf{V}_S$ of Eq. (12.20), and the set $\mathbf{v}_{L+1}, \ldots, \mathbf{v}_{M+1}$ that defines the matrix $\mathbf{V}_N$ of Eq. (12.19). Let these two matrices be partitioned as follows, respectively:

$$\mathbf{V}_S = \begin{bmatrix} \mathbf{g}_S^T \\ \mathbf{G}_S \end{bmatrix} \tag{12.26}$$

$$\mathbf{V}_N = \begin{bmatrix} \mathbf{g}_N^T \\ \mathbf{G}_N \end{bmatrix} \tag{12.27}$$

where the 1-by-L row vector $\mathbf{g}_S^T$ and the 1-by-$(M + 1 - L)$ row vector $\mathbf{g}_N^T$ have the first elements of $\mathbf{V}_S$ and $\mathbf{V}_N$, respectively, and the M-by-L matrix $\mathbf{G}_S$ and the M-by-$(M + 1 - L)$ matrix $\mathbf{G}_N$ have the remaining elements.

In the minimum-norm method, the aim is to find an $(M + 1)$-by-1 vector $\hat{\mathbf{a}}$ that satisfies three requirements (Kumaresan and Tufts, 1983):

1. The vector $\hat{\mathbf{a}}^*$ lies in the range of $\mathbf{V}_N$, so that it is orthogonal to the columns of $\mathbf{V}_S$, as shown by

$$\mathbf{V}_S^H \hat{\mathbf{a}} = \mathbf{0} \tag{12.28}$$

2. The first element of $\hat{\mathbf{a}}$ is set equal to unity.
3. The Euclidean norm of $\hat{\mathbf{a}}$ is minimum; hence, the name of the method.

The second requirement suggests that we may view the vector $\hat{\mathbf{a}}$ as the tap-weight vector of a transversal filter operating as a prediction-error filter of order M. Indeed, it was with this interpretation in mind and the desire to be consistent with the notation used previously that we chose the vector $\hat{\mathbf{a}}^*$ to lie in the range of $\mathbf{V}_N$.

Let the vector $\hat{\mathbf{a}}$ be partitioned in the form

$$\hat{\mathbf{a}} = \begin{bmatrix} 1 \\ -\hat{\mathbf{w}} \end{bmatrix} \tag{12.29}$$

where $\hat{\mathbf{w}}$ may be interpreted as the tap-weight vector of a predictor of order M. Then the use of Eqs. (12.26) and (12.29) in (12.28) yields

$$\mathbf{G}_S^T\hat{\mathbf{w}} = \mathbf{g}_S \tag{12.30}$$

This matrix relation represents a linear system of L simultaneous equations in M unknowns (i.e., the elements of vector $\hat{\mathbf{w}}$). Thus, with $L \leq M$, this system of equations is *undetermined,* and therefore has no unique solution. It is in anticipation of this difficulty that the vector $\hat{\mathbf{a}}$ is required to be of *minimum norm*. Clearly, when the vector $\hat{\mathbf{a}}$ has minimum norm, so will the vector $\hat{\mathbf{w}}$. Accordingly, the problem we have to solve is to find the vector $\hat{\mathbf{w}}$ that satisfies Eq. (12.30), subject to the condition that its Euclidean norm $\|\hat{\mathbf{w}}\|$ is minimized. This problem is similar to that we encountered in Section 11.5, when dealing with the method of least squares. In particular, by using the idea of pseudoinverse of a matrix, we may express the minimum-norm solution to Eq. (12.30) as follows[4]

$$\hat{\mathbf{w}} = (1 - \mathbf{g}_S^H\mathbf{g}_S)^{-1}\mathbf{G}_S^*\mathbf{g}_S \tag{12.31}$$

For the derivation of this formula, the reader is referred to Problem 5. Correspondingly, the minimum-norm solution for the vector $\hat{\mathbf{a}}$ in terms of the elements consistituting the matrix $\mathbf{V}_S$ equals

$$\hat{\mathbf{a}} = \begin{bmatrix} 1 \\ \hline -(1 - \mathbf{g}_S^H\mathbf{g}_S)^{-1}\mathbf{G}_S^*\mathbf{g}_S \end{bmatrix} \tag{12.32}$$

We may also express the minimum-norm solution for the vector $\hat{\mathbf{a}}$ in terms of the elements constituting the matrix $\mathbf{V}_N$. We first recognize that

$$\mathbf{V} = [\mathbf{V}_S, \mathbf{V}_N]$$

and

$$\mathbf{V}\mathbf{V}^H = \mathbf{I}_{M+1}$$

where $\mathbf{I}_{M+1}$ is the $(M + 1)$-by-$(M + 1)$ identity matrix, Hence, we may use Eqs. (12.26) and (12.27) to write

$$\begin{bmatrix} \mathbf{g}_N^T & \mathbf{g}_S^T \\ \hline \mathbf{G}_N & \mathbf{G}_S \end{bmatrix} \begin{bmatrix} \mathbf{g}_N^* & \mathbf{G}_N^H \\ \hline \mathbf{g}_S^* & \mathbf{G}_S^H \end{bmatrix} = \mathbf{I}_{M+1} \tag{12.33}$$

We thus deduce the following relations:

$$\begin{aligned} \mathbf{g}_N^T\mathbf{g}_N^* + \mathbf{g}_S^T\mathbf{g}_S^* &= 1 \\ \mathbf{G}_N\mathbf{g}_N^* + \mathbf{G}_S\mathbf{g}_S^* &= \mathbf{0} \\ \mathbf{G}_N\mathbf{g}_N^H + \mathbf{G}_S\mathbf{G}_S^H &= \mathbf{I}_M \end{aligned} \tag{12.34}$$

[4] The essence of the minimum-norm method had been published by Reddy (1979), prior to the work of Kumaresan and Tufts (1983). In particular, Reddy constructed a primary polynomial (in the unit-delay operator z^{-1}) from the columns of matrix $\mathbf{V}_S$ in an ad hoc manner without constraining the norm of its coefficient vector. Yet, the coefficient vector $\hat{\mathbf{a}}$ obtained for this polynomial turned out to be identical to that in Eq. (12.30). Basically, then Reddy's work was reformulated by Kumaresan and Tufts, giving it the *minimum-norm* significance.

Accordingly, we may rewrite the minimum-norm solution for the vector $\hat{\mathbf{a}}$ as follows

$$\hat{\mathbf{a}} = \begin{bmatrix} 1 \\ \hdashline (\mathbf{g}_N^H \mathbf{g}_N)^{-1} \mathbf{G}_N^* \mathbf{g}_N \end{bmatrix} \tag{12.35}$$

We may thus use Eq. (12.32) based on the idea of a signal plus noise subpsace, or Eq. (12.35) based on the idea of a noise subspace, for computing the vector $\hat{\mathbf{a}}$. Having computed this vector, we may then *determine the angular frequencies of the complex sinusoids as the peaks of the minimum-norm (MN) spectrum estimate*

$$\hat{S}_{MN}(\omega) = \frac{1}{|\mathbf{s}^H(\omega)\hat{\mathbf{a}}|^2} \tag{12.36}$$

where $\mathbf{s}(\omega)$ is the *frequency-scanning vector* defined by

$$\mathbf{s}^T(\omega) = [1, e^{-j\omega}, \ldots, e^{-jM\omega}] \tag{12.37}$$

Note that $\mathbf{s}^H(\omega)\hat{\mathbf{a}}$ represents the inner product of the vectors $\hat{\mathbf{a}}$ and $\mathbf{s}(\omega)$. Therefore, to compute the minimum-norm spectrum in Eq. (12.36), we need only compute this inner product for varying ω, and then determine its squared magnitude.

Note also that as with the MUSIC algorithm, $\hat{S}_{MN}(\omega)$ is an "estimate" based on a single realization of the underlying stochastic process represented by the data matrix $\mathbf{A}$, hence the use of a hat in the symbol $\hat{S}_{MN}(\omega)$.

Computational Considerations and Summary of the Minimum-Norm Algorithm

As with the MUSIC algorithm, we may determine the minimum-norm vector $\hat{\mathbf{a}}$, required for computing the minimum-norm spectrum in Eq. (12.36), by applying the singular value decomposition to the data matrix $\mathbf{A}$ directly. Specifically, we may proceed along the lines summarized in Table 12.4. Figure 12.9 presents a block diagram of the essential steps in the computation of the minimum-norm algorithm.

We may use a root-finding approach as an alternative to the location of spectral peaks in $\hat{S}_{MN}(\omega)$, as described in Table 12.4. Specifically, we form the polynomial

$$A(z) = 1 + \hat{a}_1 z^{-1} + \cdots + \hat{a}_M z^{-M} \tag{12.38}$$

where $\hat{a}_1, \ldots, \hat{a}_M$ are the elements of the minimum-norm vector $\hat{\mathbf{a}}$. The frequency-estimation problem is then solved by computing the zeros of the polynomial $A(z)$

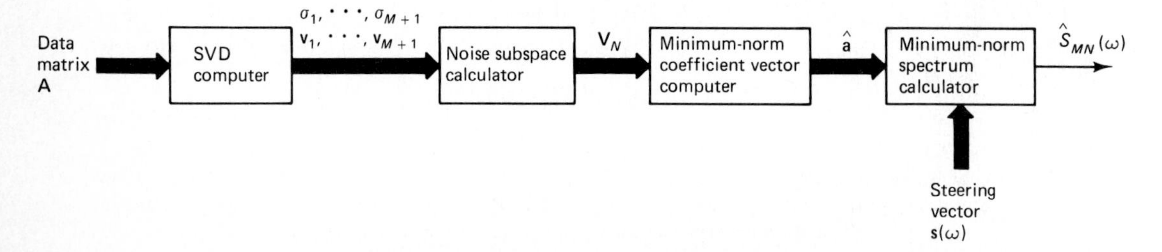

Figure 12.9 Block diagram of the minimum-norm algorithm.

TABLE 12.4 SUMMARY OF THE MINIMUM-NORM ALGORITHM

1. Set up the data matrix $\mathbf{A}$, using the time series $\{u(i)\}$.
2. Compute the $(M + 1 - L)$ smallest singular values $\sigma_{L+1}, \ldots, \sigma_{M+1}$, and therefore identify the right singular vectors $\mathbf{v}_{L+1}, \ldots, \mathbf{v}_{M+1}$ that constitute the matrix

$$\mathbf{V}_N = [\mathbf{v}_{L+1}, \ldots, \mathbf{v}_{M+1}]$$

3. Partition the matrix $\mathbf{V}_N$ as follows:

$$\mathbf{V}_N = \begin{bmatrix} \mathbf{g}_N^T \\ \mathbf{G}_N \end{bmatrix}$$

Hence, identify the submatrices, namely $\mathbf{g}_N$ and $\mathbf{G}_N$.

4. Compute the minimum-norm coefficient vector:

$$\hat{\mathbf{a}} = \begin{bmatrix} 1 \\ \text{-----------} \\ (\mathbf{g}_N^H \mathbf{g}_N)^{-1} \mathbf{G}_N^* \mathbf{g}_N \end{bmatrix}$$

5. Compute the minimum-norm spectrum:

$$\hat{S}_{MN}(\omega) = \frac{1}{|\mathbf{s}^H(\omega)\hat{\mathbf{a}}|^2}$$

for varying values of the angular frequency ω in the scanning vector

$$\mathbf{s}^T(\omega) = [1, e^{-j\omega}, \ldots, e^{-jM\omega}]$$

6. Estimate the angular frequencies of the complex sinusoids in the time series $\{u(i)\}$ by locating the spectral peaks of $\hat{S}_{MN}(\omega)$.

---

and identifying the *signal zeros* as those zeros of $A(z)$ that lie on (or very close to) the unit circle in the z-plane. These signal zeros should correspond to the angular frequencies of the complex sinusoids contained in the time series $\{u(i)\}$. However, it appears that unlike the MUSIC algorithm, the difference in performance between the root and spectral versions of the minimum-norm algorithm is small (Rao and Hari, 1989b).

12.6 RELATIONSHIPS BETWEEN THE AUTOREGRESSIVE, MINIMUM-VARIANCE DISTORTIONLESS RESPONSE, MUSIC, AND MINIMUM-NORM SPECTRA

In the last two sections of Chapter 10 and in preceding sections of this Chapter, we have derived four super-resolution frequency-estimation algorithms. They are (1) the autoregressive (AR) spectrum of Section 10.10, (2) the minimum-variance distortionless response (MVDR) spectrum of Section 10.11, (3) the MUSIC spectrum of Section 12.3, and (4) the minimum-norm spectrum of Section 12.5. Although these four spectra are basically different, nevertheless, they are interrelated in an *asymptotic sense*.

Figure 12.10 presents a graphical portrayal of the interrelations between these four super-resolution algorithms (Nickel, 1988). At the outset, however, it should be emphasized that the relationships summarized in Fig. 12.10 are only valid for a time (space) series of superimposed signals in additive white noise, which consists of

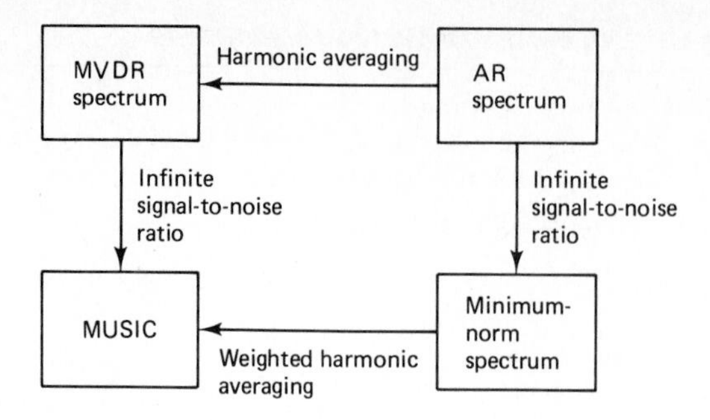

Figure 12.10 Graphical depiction of the inter-relations between four fundamental super-resolution spectra.

uniformly spaced temporal (spatial) samples and whose *correlation matrix is exactly known*. In other words, the underlying stochastic process is assumed to be wide-sense stationary, and the formulas for the super-resolution spectra are based on the ensemble-averaged correlation matrix of the process. Given the situation described herein, we may then make the following statements:[5]

1. The MVDR spectrum is a *smoothed* version of the AR spectrum corresponding to different prediction-error filter lengths, as shown by [6]

$$\frac{1}{S_{\mathrm{MVDR}}(\omega)} = \frac{1}{M} \sum_{m=1}^{M} \frac{1}{S_{\mathrm{AR}}(m, \omega)} \tag{12.39}$$

   where $S_{\mathrm{AR}}(m, \omega)$ is the AR spectrum computed for a prediction-error filter of length m that lies in the interval $[1, M]$, and $S_{\mathrm{MVDR}}(\omega)$ is the MVDR spectrum computed for a transversal filter of length M. The harmonic averaging indicated in Eq. (12.39) explains the reason for the much greater fluctuations observed in the AR spectrum compared to the MVDR spectrum.
2. If the ensemble-averaged correlation matrix $\mathbf{R}$ is known exactly, the projection matrix $\mathbf{V}_N\mathbf{V}_N^H$ used to compute the MUSIC spectrum may be interpreted as the limiting value of the inverse matrix $\mathbf{R}^{-1}$ for the signal power approaching infinity, while the noise power is maintained constant (Nickel, 1988). Hence, the MUSIC spectrum may be interpreted as an MVDR type of spectrum that uses a correlation matrix corresponding to infinite signal-to-noise ratio. This relationship explains the reason for the superior resolution of the MUSIC spectrum compared to that of the MVDR spectrum.
3. The minimum-norm spectrum is the limiting value of the AR spectrum for infinite signal-to-noise ratio (Nickel, 1988). This relationship explains the improved resolution of the minimum-norm spectrum compared to the AR spectrum.
4. The MUSIC spectrum is a *weighted* harmonic average of the minimum-norm spectra for filter lengths 1 to M (Nickel, 1988). The presence of weighting makes this relationship somewhat different from that between the AR and

[5] The paper by Nickel (1988) presents mathematical proofs of the relations between the four super-resolution spectra summarized herein.

[6] Equation (12.39) was originally derived by Burg (1972). In Appendix G we present the derivation of another relation between the AR and MVDR spectra due to Musicus (1985).

MVDR spectra. Nevertheless, it explains the greater fluctuations that are observed in the minimum-norm spectrum compared to the MUSIC spectrum.

12.7 SUMMARY AND DISCUSSION

The minimum-norm and MUSIC algorithms belong to an eigendecomposition-based class of super-resolution methods for the estimation of superimposed signals in noise (Johnson, 1982; Wax, 1985). The general aim of these algorithms is to exploit the eigenvalue decomposition of a correlation matrix in a way that the estimation of complex sinusoids in noise is improved. This is achieved by partitioning the eigenvectors of a sample correlation matrix into two sets: (1) a set defined by the *principal* eigenvectors associated with the L largest eigenvalues, where L equals the number of complex sinusoids, and (2) a set defined by the eigenvectors associated with the remaining eigenvalues. The rationale that is used here is that the principal eigenvectors (associated with the dominant eigenvalues) are considerably more robust to noise perturbations than the remaining eigenvectors. It is noteworthy that the idea of a finite-dimensional signal subspace and its orthogonal noise subspace was first described (in a somewhat different setting) by McDonough and Huggins (1968).

As pointed out previously, the angular frequencies of the complex sinusoids may be determined by locating the *peaks* of a sample spectrum $S(\omega)$, obtained by using a super-resolution algorithm. Alternatively, we may work with the reciprocal of this function, namely, the *sample eigen-spectrum* $D(\omega) = S^{-1}(\omega)$, in which case the angular frequencies of the complex sinusoids are determined by locating the *nulls* of the function $D(\omega)$. Accordingly, $D(\omega)$ is also referred to as the *null spectrum*.

Another way of estimating the angular frequencies of the complex sinusoids is to solve for the roots of the denominator polynomial of the function $S(\omega)$ with $e^{j\omega}$ replaced by the complex variable z. Those L roots that lie on the unit circle in the z-plane or are closest to it determine the desired angular frequencies. This procedure bypasses the need for scanning the function $S(\omega)$ across the complete frequency interval $-\pi \leq \omega \leq \pi$.

Although in our discussion we have focused attention mainly on complex sinusoids in noise, nevertheless, the theory also applies to the problem of estimating the *angles of arrival* of superimposed plane waves that are incident on a linear array of sensors, as illustrated in Section 12.4. In such an application, the frequency-scanning vector **s** *takes on a new meaning, namely, that of a steering vector* with the angular frequency ω replaced by the electrical angle ϕ. In this context, mention should be made of the fact that the MUSIC algorithm may indeed be used to deal with an array of arbitrary geometry[7] (Schmidt, 1981). A similar remark may be made in relation to the MVDR algorithm. Basically, both the MUSIC and MVDR algorithms characterize the directional response of an array of sensors in terms of an *array manifold*. The array manifold is defined as the response of the array to a single emitter (signal source) as a function of position or direction of arrival. It is of course assumed that the array manifold is made available either analytically or by "calibration" of the ar-

[7] For an experimental study of the MUSIC algorithm, involving the use of a sparse array of radiating elements and dealing with some real-life problems, see Schmidt and Franks (1986).

ray of sensors. In contrast, the use of the minimum-norm algorithm and the AR algorithm is limited to linear uniform arrays.

Yang and Kaveh (1988) describe *adaptive* algorithms[8] for estimating the noise or signal subspace. For M-element array of sensors and L sources (emitters), for example, these algorithms adaptively estimate the eigenvectors of the ensemble-averaged correlation matrix corresponding to either the smallest $(M - L)$ (i.e., the noise subspace) or the largest L eigenvectors (i.e., the signal plus noise subspace) from a continuously updated sample correlation matrix. They are therefore well suited for use with the MUSIC or minimum-norm algorithm.

A related adaptive issue, in the context of eigen-subspace algorithms, is that of *tracking the movements* of "spectral" roots in a nonstationary environment. This issue arises when we focus on the roots of the denominator polynomial $D(z)$ resulting from the reformulation of a spectrum computed using an eigen-subspace algorithm (i.e., MUSIC or minimum-norm) with $e^{j\omega}$ replaced by the complex variable z. Starer and Nehorai (1989) and Yang and Kaveh (1989) describe adaptive algorithms for accomplishing such a requirement.

Statistical Analysis

Kaveh and Barabell (1985) present an asymptotic statistical analysis of the sample eigen-spectra of the MUSIC algorithm and the minimum-norm method for resolving independent closely spaced plane waves in noise. The results of their analysis, supported by computer simulations, indicate a smaller bias in the minimum-norm eigen-spectrum, at a source angle, than in the MUSIC eigen-spectrum. This indicates a resolution threshold that is at a lower signal-to-noise ratio for the minimum-norm method than for the MUSIC algorithm. However, around and above threshold, the minimum-norm algorithm results in a higher variance than the MUSIC algorithm. In selecting between the MUSIC and minimum-norm algorithms, there is a resolution/variance tradeoff to be considered.

Stoica and Nehorai (1988) present a detailed statistical analysis of the MUSIC algorithm. Their results may be summarized as follows:

1. For a sufficiently large data length N, the MUSIC estimator assumes a Gaussian-distributed form.
2. For uncorrelated sinusoidal signals, and for large data length N and large M (where M is the number of transversal filter taps in the temporal version of the MUSIC algorithm, or the number of sensors in its spatial counterpart), the MUSIC estimator should achieve the *Cramér–Rao lower bound* (CRLB); this bound is discussed in Appendix D.

   For *uncorrelated* signals, the estimation error variance produced by the MUSIC algorithm decreases monotonically with increasing number of sensors in the array.
4. For *correlated* signals, the performance of the MUSIC estimator becomes degraded; in particular, the degradation in performance can be significant if the signals are highly correlated.

[8] The adaptive algorithms described in Yang and Kaveh (1988) build on earlier work by Owsley (1978).

5. The difficulty presented by the presence of correlated signals may be overcome partially by the combined use of spatial smoothing and temporal smoothing, as demonstrated in Section 12.4.

Detection Problem

Throughout our discussion, we have assumed that we know the number of sources that are responsible for the generation of sine waves or incident plane waves contained in the received signal. In many practical situations, however, we may not have this prior knowledge. For example, in harmonic retrieval, some sonar and radar array processing applications, the received signal may be modeled as the superposition of a finite number of complex sinusoids or plane waves that are embedded in a background of additive white noise. In these applications, a key issue involved in the development of a suitable *model* for the received signal is the *detection*[9] of the number of signals contained in the model.

There is no unique approach to the solution of this detection problem. As one possible approach, we may use the observation that the number of signals can be determined from the eigenvalues of the correlation matrix of the received signal, and thus use hypothesis testing to solve the problem (Schmidt, 1981). Another approach, based on the application of information-theoretic criteria for model selection, involves the use of one of two criteria: the AIC introduced by Akaike, or the MDL introduced by Rissanen. These two criteria were briefly discussed in Section 2.10. The advantage of this alternative approach is that no subjective judgment is required in the decision process in that the number of signals is determined naturally as the value for which the AIC or the MDL criterion is *minimized*. Of particular interest is a novel detection approach described by Wax and Ziskind (1989), which is based on the MDL criterion; the approach described therein is applicable to any type of sources, including the difficult case of sources that are fully correlated.

Colored Noise Background

The theory of the MUSIC algorithm developed in this chapter has been based on the assumption that the additive sensor noise process $\{v(i, k)\}$ is *white* in that for each sensor in the array (signified by the index k) it satisfies the following condition

$$E[v(i, k)v^*(j, k)] = \begin{cases} \sigma^2, & j = i \\ 0, & j \neq i \end{cases}$$

where the variance σ^2 is common to all the sensors in the array. In other words, the correlation matrix of the sensor noise process $\{v(i, k)\}$ consists of a *diagonal* matrix equal to $\sigma^2 \mathbf{I}$, where $\mathbf{I}$ is an identity matrix. Correspondingly, the ensemble-averaged correlation matrix of the received signal process $\{u(i, k)\}$ has the form given in Eq. (12.8), reproduced here for convenience

$$\boldsymbol{R} = \boldsymbol{SDS}^H + \sigma^2 \mathbf{I}$$

[9] It may be more appropriate to refer to the problem of determining the number of emitters (sources of signal) as one of estimation rather than detection. We have opted for the latter terminology in order to be consistent with the literature on array signal processing.

where the matrix $\mathbf{S}$ is a function of both the angles-of-arrival of the incident plane waves and the sensor locations in the array, and $\mathbf{D}$ is a diagonal matrix defined by the average power of the incident plane waves. The ordinary SVD is an adequate tool for efficiently computing the MUSIC algorithm applied to this idealized problem, as outlined in Section 12.3.

In the more general case of a *colored noise* background, the correlation matrix of the sensor noise process $\{v(i,k)\}$ takes on a nondiagonal structure due to imperfections and mutual interactions between sensors in the array, and non-isotropic background noise in the array environment. Correspondingly, the correlation matrix of the received signal process $\{u(i,k)\}$ is modified as follows

$$\boldsymbol{R} = \boldsymbol{SDS}^H + \mathbf{R}_{\text{noise}}$$

where $\mathbf{R}_{\text{noise}}$ is the correlation matrix of the sensor noise process $\{v(i,k)\}$. The MUSIC algorithm may indeed be generalized to deal with this new situation. However, for its efficient computation we have to resort to the use of the *generalized singular decomposition decomposition* (GSVD).[10]

Applications

Super-resolution algorithms are computationally demanding and time consuming in their execution. Moreover, they are highly sensitive to deviations from the mathematical model assumed for their derivation. For these reasons, super-resolution algorithms are presently only of interest for special applications where a high signal-to-noise ratio is available or it can be achieved by the use of coherent integration. In the case of radar array processing, for example, the special characteristics of super-resolution algorithms befit them to *second-stage processing*, that is, after performing detection and estimation by conventional methods (Nickel, 1991). In particular, the motivation for the use of a super-resolution technique in a radar array processing system is to provide for two important features:

1. Improved *resolution* of closely spaced targets
2. Accurate estimation of the directions of arrival of target signals

To meet these objectives in a highly correlated environment, it may be necessary to expand the signal-processing capability of a spatially oriented super-resolution technique by including other system discriminants such as frequency (Kezys and Haykin, 1988]. In such a setting, the super-resolution technique becomes *multidimensional* in nature and, consequently, a more powerful signal-processing tool.

Another important application of super-resolution techniques is in *signal classification* that is temporally oriented. Specifically, a super-resolution technique may be used to extract features that are subsequently applied to a feature classifier. Here we may mention the successful application of autoregressive modeling (i.e., maximum entropy method) as a basis for the classification of radar returns produced by different objects such as aircraft, weather disturbances, migrating flocks of birds,

[10] The GSVD was briefly discussed in Section 11.12. For a full discussion of this powerful computational tool, see Golub and Van Loan (1989). A detailed treatment of signal processing computations using the GSVD is given in Van Loan (1989), and Speiser and Van Loan (1984).

and ground, as encountered in an air traffic control environment (Haykin et al., 1982, 1991).

Another area of application for super-resolution algorithms is in *inverse synthetic aperture radar* (ISAR) imaging of a rotating object. Basically, the rotation of an object relative to the radar produces a Doppler frequency gradient, which permits the extraction of cross-range resolution on a much finer scale than that attainable with a conventional radar. Gabriel (1989, 1990) has demonstrated the capability of super-resolution algorithms in such an application, using computer simulated radar data generated from point-target models of rotating objects. In particular, in the paper [Gabriel (1989)] attention is focused on achieving super-resolution in the Doppler domain, and in the second paper [Gabriel (1990)] the issue of super-resolution in the range domain is considered.

PROBLEMS

1. The discrete-time stochastic process $\{u(n)\}$ consists of K (uncorrelated) complex sinusoids plus additive white noise of zero mean and variance σ^2. The M-by-M correlation matrix of the process is denoted by $\mathbf{R}$. Show that the correlation matrix $\mathbf{R}$ may be expressed in the form

$$\mathbf{R} = \mathbf{S}\mathbf{D}\mathbf{S}^H + \sigma^2\mathbf{I}$$

where $\mathbf{S}$ is the M-by-K signal matrix determined by the angular frequencies of the sinusoids, $\mathbf{D}$ is a K-by-K diagonal matrix determined by their average power, and $\mathbf{I}$ is the M-by-M identity matrix.

2. The formula (12.22) for the MUSIC spectrum is based on the use of a projection matrix related to the noise subspace. Show that the MUSIC spectrum may also be computed using the formula

$$\hat{S}_{\text{MUSIC}}(\omega) = \frac{1}{M - \mathbf{s}^H(\omega)\mathbf{V}_S\mathbf{V}_S^H\mathbf{s}(\omega)}$$

where M is the dimension of the data matrix, $\mathbf{V}_S\mathbf{V}_S^H$ is the projection matrix on the signal subspace, and $\mathbf{s}(\omega)$ is a frequency-scanning vector. [Note that although this formula and that of Eq. (12.22) are mathematically equivalent, they make different computational demands.]

3. The polynomial $H(z)$ for the MUSIC spectrum and the polynomial $A(z)$ for the minimum-norm spectrum may be viewed as minimum-phase polynomials. Justify the validity of this statement.

4. The MUSIC spectrum may be viewed as a special case of the MVDR spectrum, subject to two restrictions:
   (a) Only the right singular vectors of the data matrix $\mathbf{A}$ corresponding to the L smallest singular values are retained, where L is the number of complex sinusoids contained in $\mathbf{A}$.
   (b) These singular vectors are all weighted uniformly.

   Justify the validity of these two restrictions.

5. Equation (12.31) defines the vector $\hat{\mathbf{w}}$ in accordance with the minimum-norm method. Prove the validity of this formula by using the following two relations:

   (1) The formula for pseudoinverse of a matrix of coefficients in a system of linear equations.

(2) The matrix inversion lemma that states if the set of matrices **A**, **B**, **C**, and **D**, with appropriate dimensions, are related by

$$\mathbf{A} = \mathbf{B}^{-1} + \mathbf{C}\mathbf{D}^{-1}\mathbf{C}^H$$

then

$$\mathbf{A}^{-1} = \mathbf{B} - \mathbf{B}\mathbf{C}(\mathbf{D} + \mathbf{C}^H\mathbf{B}\mathbf{C})^{-1}\mathbf{C}_H\mathbf{B}$$

(For discussion of this lemma, see Section 13.2.)

6. In this problem we explore another relation between the MUSIC and MVDR algorithms.

(a) Continuing with Problem 1, use the matrix inversion lemma to evaluate $\mathbf{R}^{-1}$, the inverse of the correlation matrix **R**. For a statement of the matrix inversion lemma, see Problem 5.

(b) Let the signal-to-noise ratio go to infinity; that is, $\sigma^2\mathbf{D}^{-1} \rightarrow \mathbf{0}$, where σ^2 is the noise variance and **D** is the diagonal matrix defined by the average powers of the complex sinusoids contained in the received signal (see Problem 1). Hence, explore the following three possible cases:

(1) σ^2 = constant and $\mathbf{D}^{-1} \rightarrow \mathbf{0}$.

(2) $\sigma^2 \rightarrow 0$ and $\mathbf{D}^{-1}$ = constant.

(3) $\sigma^2 \rightarrow 0$ and $\mathbf{D}^{-1} \rightarrow \mathbf{0}$.

(c) Using the results of part (b), show that the MUSIC algorithm may be viewed as an MVDR algorithm, but one that uses a correlation matrix corresponding to infinite signal-to-noise ratio.

7. Consider the matrix $\mathbf{V}_S$ pertaining to the signal plus noise subspace used in the formulation of the MUSIC algorithm, as in Problem 2. The projection matrix onto the column space of $\mathbf{V}_S$ and its complement are defined as follows, respectively:

$$\mathbf{P} = \mathbf{V}_S\mathbf{V}_S^H$$

and

$$\mathbf{P}_{\text{comp}} = \mathbf{I} - \mathbf{P}$$

where **I** is the identity matrix. Let the matrices $\mathbf{V}_S$ and $\mathbf{P}_{\text{comp}}$ be partitioned as follows:

$$\mathbf{V}_S = \begin{bmatrix} \mathbf{g}_S^T \\ \mathbf{G}_S \end{bmatrix}$$

and

$$\mathbf{P}_{\text{comp}} = \begin{bmatrix} \mathbf{a}^T \\ \mathbf{H} \end{bmatrix}$$

(a) Show that the vector **a** for the MUSIC algorithm has the value

$$\mathbf{a} = (1 - \mathbf{g}_S^H\mathbf{g}_S)\begin{bmatrix} 1 \\ -(1 - \mathbf{g}_S^H\mathbf{g}_S)^{-1}\mathbf{G}_S^*\mathbf{g}_S \end{bmatrix}$$

Compare this value of the vector **a** with its corresponding value given in Eq. (12.32) for the minimum-norm algorithm.

(b) Based on the result given in part (a), we may differentiate between the minimum norm and MUSIC null spectra as follows:

(1) The minimum-norm algorithm computes the squared norm of the projection of the frequency-scanning vector $\mathbf{s}(\omega)$ onto the *first* row of the matrix $\mathbf{P}_{\text{comp}}$ (i.e., the vector **a**).

(2) The MUSIC algorithm computes the averages of the squares of the projections of $\mathbf{s}(\omega)$ onto *all* the rows of the matrix $\mathbf{P}_{\text{comp}}$.

Justify the validity of these two statements. Comment on the physical significance of this difference between the minimum-norm and MUSIC algorithms.

8. Compare the computational requirements of the two versions of the minimum-norm algorithm given in Eqs. (12.32) and (12.35).

9. The minimum-norm spectrum of Eq. (12.36) has a mathematical form that is similar to that of the AR spectrum given in Eq. (10.89). Yet in physical terms, these two spectra are entirely different. Justify the validity of this statement.

10. The minimum-norm and AR spectra are only applicable to linear arrays with uniformly spaced sensors. The MUSIC and MVDR spectra, on the other hand, are applicable to nonuniform arrays or planar arrays. Use physical arguments to justify these differences.

Computer-Oriented Problems

11. *Computer experiment on the MUSIC spectra of a two-dimensional array field.* Consider the 8-by-8 element array shown in Fig. P12.1. The interelement spacing is $d = \lambda/2$, where λ is the wavelength. The array sits in the upper right quadrant of the xz-plane, with its first element at the origin. The array elements are all identical and isotropic. The signals impinging on the array are all plane waves. Compute the MUSIC spectrum for the following environmental conditions:

(a) A single plane wave impinging on the array along the y-axis (i.e., boresight of the array). Assume the availability of 50 snapshots of data, based on an elemental signal-to-noise ratio of 13 dB.

(b) Two incoherent plane waves impinging on the array from directions separated from each other by one beamwidth. The first plane wave is located at boresight, which corresponds to angle $(\pi/2, \pi/2)$ radians measured from the x-axis and z-axis, respectively (i.e., $\pi/2$ radians is parallel to the normal of the array). Using the

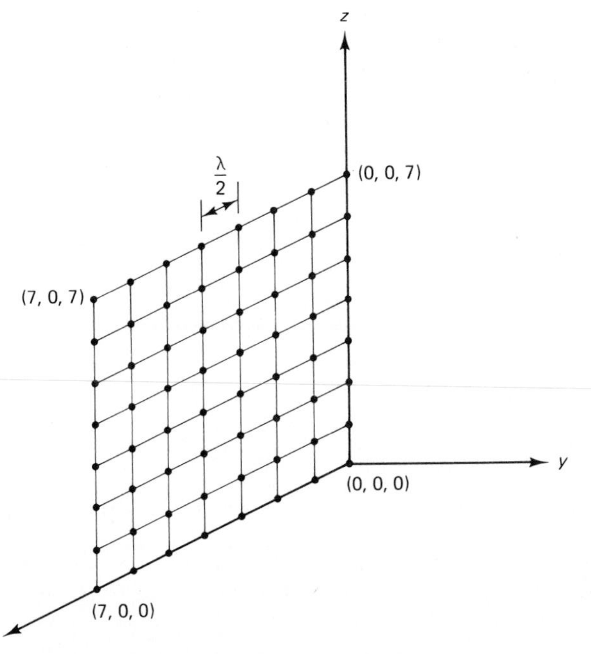

Figure P12.1 The 8 × 8 antenna array on xz-plane. Elements are spaced $\lambda/2$ apart. Element No. 1 is at the origin, and element No. 64 is at position (7, 0, 7).

beamwidth definition as BW = 4 cos θ, a beamwidth of one corresponds to angle $\theta = \cos^{-1}(\frac{1}{4}) = 1.318$ radians. Accordingly, the second signal is located at (1.318, $\pi/2$). As in part (a), assume that 50 snapshots of data are available, based on an elemental signal-to-noise ratio of 13 dB.

12. *Computer experiment on the comparative evaluation of the MUSIC and periodogram for two incoherent sources.* Continuing with the computer experiment described in Section 12.4, perform the computations of the periodogram and the MUSIC algorithm for the following conditions: One hundred snapshots of data, assuming that the two incident plane waves originate from *incoherent* sources. That is, the amplitudes of the plane waves are

$$a_1 = 1$$

$$a_2 = 0.6310e^{j\psi(k)}$$

where the phase $\psi(k)$ is a random variable uniformly distributed over the interval $[-\pi, \pi]$ in an *independent* fashion from one snapshot to the next. All other parameters for the experiment are as described in Section 12.4.

13

Standard Recursive Least-Squares Estimation

In this chapter we extend the use of the method of least squares to develop a recursive algorithm for the design of adaptive transversal filters such that, given the least-squares estimate of the tap-weight vector of the filter at time $n - 1$, we may compute the updated estimate of this vector at time n upon the arrival of new data. We refer to the resulting algorithm as the *recursive least-squares (RLS) algorithm*.

The RLS algorithm may be viewed as a special case of the Kalman filter. Indeed, this special relationship between the RLS algorithm and the Kalman filter is considered later in the chapter. Our main mission in this chapter, however, is to develop the basic theory of the RLS algorithm as an important tool for adaptive filtering in its own right.

We begin the development of the RLS algorithm by reviewing some basic relations that pertain to the method of least squares. Then, by exploiting a relation in matrix algebra known as the *matrix inversion lemma,* we develop the RLS algorithm. An important feature of the RLS algorithm is that it utilizes information contained in the input data, extending back to the instant of time when the algorithm is initiated. The resulting rate of convergence is therefore typically an order of magnitude faster than the simple LMS algorithm. This improvement in performance, however, is achieved at the expense of a large increase in computational complexity.

13.1 SOME PRELIMINARIES

In *recursive* implementations of the method of least squares, we start the computation with *known initial conditions* and use the information contained in new data samples to *update* the old estimates. We therefore find that the length of observable data is variable. Accordingly, we express the *cost function* to be minimized as $\mathscr{E}(n)$, where n is the variable length of the observable data. Also, it is customary to introduce a *weighting factor* or *forgetting factor* into the definition of the cost function $\mathscr{E}(n)$. We thus write

$$\mathscr{E}(n) = \sum_{i=1}^{n} \beta(n, i)|e(i)|^2 \tag{13.1}$$

where $e(i)$ is the difference between *desired response* $d(i)$ and the *output* $y(i)$ produced by a transversal filter whose tap inputs (at time i) equal $u(i), u(i-1), \ldots, u(i-M+1)$, as in Fig. 13.1. That is, $e(i)$ is defined by

$$\begin{aligned} e(i) &= d(i) - y(i) \\ &= d(i) - \mathbf{w}^H(n)\mathbf{u}(i) \end{aligned} \tag{13.2}$$

where $\mathbf{u}(i)$ is the *tap-input vector at time i*, defined by

$$\mathbf{u}(i) = [u(i), u(i-1), \ldots, u(i-M+1)]^T \tag{13.3}$$

and $\mathbf{w}(n)$ is the *tap-weight vector at time n*, defined by

$$\mathbf{w}(n) = [w_0(n), w_1(n), \ldots, w_{M-1}(n)]^T \tag{13.4}$$

Note that the tap weights of the transversal filter remain *fixed* during the observation interval $1 \le i \le n$.

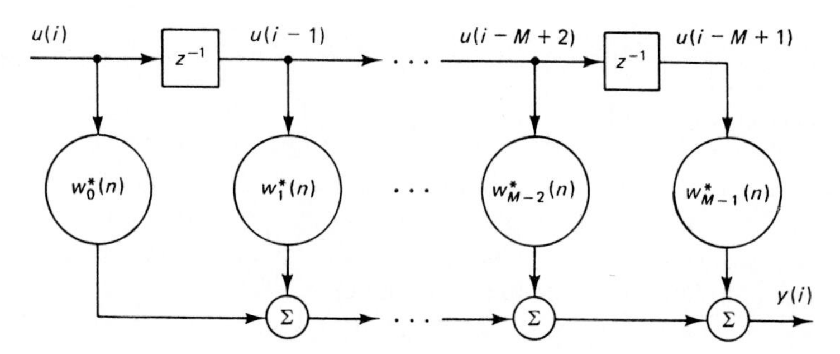

Figure 13.1 Transversal filter.

The weighting factor $\beta(n, i)$ in Eq. (13.1) has the property that

$$0 < \beta(n, i) \le 1, \qquad i = 1, 2, \ldots, n \tag{13.5}$$

The use of the weighting factor $\beta(n, i)$ is intended to ensure that data in the distant past are "forgotten" in order to afford the possibility of following the statistical variations of the observable data when the filter operates in a nonstationary environment. One such form of weighting that is commonly used is the *exponential weighting factor* defined by

$$\beta(n, i) = \lambda^{n-i}, \qquad i = 1, 2, \ldots, n \tag{13.6}$$

where λ is a positive constant close to, but less than, 1. When λ equals 1, we have the ordinary method of least squares. The inverse of $1 - \lambda$ is, roughly speaking, a measure of the *memory* of the algorithm. The special case $\lambda = 1$ corresponds to *infinite memory*. Thus, in the *method of exponentially weighted least squares,* we minimize the cost function

$$\mathscr{E}(n) = \sum_{i=1}^{n} \lambda^{n-i}|e(i)|^2 \tag{13.7}$$

The optimum value of the tap-weight vector, $\hat{\mathbf{w}}(n)$, for which the cost function $\mathscr{E}(n)$ of Eq. (13.7) attains its minimum value is defined by the *normal equations* written in matrix form:

$$\mathbf{\Phi}(n)\hat{\mathbf{w}}(n) = \mathbf{\theta}(n) \tag{13.8}$$

The *M-by-M* correlation matrix $\mathbf{\Phi}(n)$ is defined by

$$\mathbf{\Phi}(n) = \sum_{i=1}^{n} \lambda^{n-i}\mathbf{u}(i)\mathbf{u}^H(i) \tag{13.9}$$

The M-by-1 cross-correlation $\mathbf{\theta}(n)$ between the tap inputs of the transversal filter and the desired response is defined by

$$\mathbf{\theta}(n) = \sum_{i=1}^{n} \lambda^{n-i}\mathbf{u}(i)d^*(i) \tag{13.10}$$

where the asterisk denotes complex conjugation.

The correlation matrix $\mathbf{\Phi}(n)$ differs from the time-averaged version of Eq. (10.45) in two respects.

1. The matrix product $\mathbf{u}(i)\mathbf{u}^H(i)$ inside the summation on the right side of Eq. (10.45) is weighted by the exponential factor λ^{n-i}, which arises naturally from the adoption of Eq. (13.7) as the cost function.
2. The use of *prewindowing* is assumed, according to which the input data prior to time $i = 1$ are equal to zero; hence, the use of $i = 1$ as the lower limit of the summation.

Similar remarks apply to the cross-correlation vector $\mathbf{\theta}(n)$ compared to its counterpart of Chapter 10.

Isolating the term corresponding to $i = n$ from the rest of the summation on the right side of Eq. (13.9), we may write

$$\mathbf{\Phi}(n) = \lambda\left[\sum_{i=1}^{n-1} \lambda^{n-1-i}\mathbf{u}(i)\mathbf{u}^H(i)\right] + \mathbf{u}(n)\mathbf{u}^H(n) \tag{13.11}$$

However, by definition, the expression inside the square brackets on the right side of Eq. (13.11) equals the correlation matrix $\mathbf{\Phi}(n-1)$. Hence, we have the following recursion for updating the value of the correlation matrix of the tap inputs:

$$\mathbf{\Phi}(n) = \lambda\mathbf{\Phi}(n-1) + \mathbf{u}(n)\mathbf{u}^H(n) \tag{13.12}$$

where $\mathbf{\Phi}(n-1)$ is the "old" value of the correlation matrix, and the matrix product $\mathbf{u}(n)\mathbf{u}^H(n)$ plays the role of a correction term in the updating operation.

Similarly, we may use Eq. (13.10) to derive the following recursion for updating the cross-correlation vector between the tap inputs and the desired response:

$$\mathbf{\theta}(n) = \lambda\mathbf{\theta}(n-1) + \mathbf{u}(n)d^*(n) \tag{13.13}$$

To compute the least-square estimate $\hat{\mathbf{w}}(n)$ for the tap-weight vector in accordance with Eq. (13.8), we have to determine the inverse of the correlation matrix $\mathbf{\Phi}(n)$. In practice, however, we usually try to avoid performing such an operation as it can be very time consuming, particularly if the number of tap weights, M, is high. Also, we would like to be able to compute the least-squares estimate $\hat{\mathbf{w}}(n)$ for the tap-weight vector recursively for $n = 1, 2, \ldots, \infty$. We can realize both of these objectives by using a basic result in matrix algebra known as the *matrix inversion lemma*. We assume that the initial conditions have been chosen to ensure the nonsingularity of the correlation matrix $\mathbf{\Phi}(n)$; this issue is discussed later in Section 13.3.

13.2 THE MATRIX INVERSION LEMMA

Let $\mathbf{A}$ and $\mathbf{B}$ be two positive-definite, M-by-M matrices related by

$$\mathbf{A} = \mathbf{B}^{-1} + \mathbf{C}\mathbf{D}^{-1}\mathbf{C}^{H} \tag{13.14}$$

where $\mathbf{D}$ is another positive-definite N-by-N matrix, and $\mathbf{C}$ is an M-by-N matrix. According to the *matrix inversion lemma,* we may express the inverse of the matrix $\mathbf{A}$ as follows:

$$\mathbf{A}^{-1} = \mathbf{B} - \mathbf{B}\mathbf{C}(\mathbf{D} + \mathbf{C}^{H}\mathbf{B}\mathbf{C})^{-1}\mathbf{C}^{H}\mathbf{B} \tag{13.15}$$

The proof of this lemma is established by multiplying Eq. (13.14) by (13.15) and recognizing that the product of a square matrix and its inverse is equal to the identity matrix (see Problem 2). The matrix inversion lemma states that if we are given a matrix $\mathbf{A}$ as defined in Eq. (13.14), we can determine its inverse $\mathbf{A}^{-1}$ by using the relation of Eq. (13.15). In effect, the lemma is described by this pair of equations. The matrix inversion lemma is also referred to in the literature as *Woodbury's identity.*[1]

In the next section we show how the matrix inversion lemma can be applied to obtain a recursive equation for computing the least-squares solution $\hat{\mathbf{w}}(n)$ for the tap-weight vector.

13.3 THE EXPONENTIALLY WEIGHTED RECURSIVE LEAST-SQUARES ALGORITHM

With the correlation matrix $\mathbf{\Phi}(n)$ assumed to be positive definite and therefore nonsingular, we may apply the matrix inversion lemma to the recursive equation (13.12). We first make the following identifications:

$$\mathbf{A} = \mathbf{\Phi}(n)$$

$$\mathbf{B}^{-1} = \lambda\mathbf{\Phi}(n-1)$$

$$\mathbf{C} = \mathbf{u}(n)$$

$$\mathbf{D} = 1$$

[1] The exact origin of the matrix inversion lemma is not known. Householder (1964) attributes it to Woodbury (1950). Nevertheless, application of the matrix inversion lemma in the filtering literature was first made by Kailath who used a form of this lemma to prove the equivalence of the Wiener filter and the maximum-likelihood procedure for estimating the output of a random linear time-invariant channel that is corrupted by additive white Gaussian noise (Kailath, 1960). Early use of the matrix inversion lemma was also made by Ho (1963). Another interesting application of the matrix inversion lemma was made by Brooks and Reed, who used it to prove the equivalence of the Wiener filter, the maximum signal-to-noise ratio filter, and the likelihood ratio processor for detecting a signal in additive white Gaussian noise (Brooks and Reed, 1972).

Then, substituting these definitions in the matrix inversion lemma of Eq. (13.15), we obtain the following recursive equation for the inverse of the correlation matrix:

$$\mathbf{\Phi}^{-1}(n) = \lambda^{-1}\mathbf{\Phi}^{-1}(n-1) - \frac{\lambda^{-2}\mathbf{\Phi}^{-1}(n-1)\mathbf{u}(n)\mathbf{u}^{H}(n)\mathbf{\Phi}^{-1}(n-1)}{1 + \lambda^{-1}\mathbf{u}^{H}(n)\mathbf{\Phi}^{-1}(n-1)\mathbf{u}(n)} \tag{13.16}$$

For convenience of computation, let

$$\mathbf{P}(n) = \mathbf{\Phi}^{-1}(n) \tag{13.17}$$

and

$$\mathbf{k}(n) = \frac{\lambda^{-1}\mathbf{P}(n-1)\mathbf{u}(n)}{1 + \lambda^{-1}\mathbf{u}^{H}(n)\mathbf{P}(n-1)\mathbf{u}(n)} \tag{13.18}$$

Using these definitions, we may rewrite Eq. (13.16) as follows:

$$\mathbf{P}(n) = \lambda^{-1}\mathbf{P}(n-1) - \lambda^{-1}\mathbf{k}(n)\mathbf{u}^{H}(n)\mathbf{P}(n-1) \tag{13.19}$$

Note that $\mathbf{P}(n)$ has the dimensions of a matrix (M-by-M), whereas $\mathbf{k}(n)$ has the dimensions of a vector (M-by-1). We refer to $\mathbf{k}(n)$ as the *gain vector* for reasons that will become apparent later in the section. Equation (13.19) is the *Riccati equation* for the RLS algorithm.

By rearranging Eq. (13.18), we have

$$\begin{aligned}\mathbf{k}(n) &= \lambda^{-1}\mathbf{P}(n-1)\mathbf{u}(n) - \lambda^{-1}\mathbf{k}(n)\mathbf{u}^{H}(n)\mathbf{P}(n-1)\mathbf{u}(n) \\ &= [\lambda^{-1}\mathbf{P}(n-1) - \lambda^{-1}\mathbf{k}(n)\mathbf{u}^{H}(n)\mathbf{P}(n-1)]\mathbf{u}(n)\end{aligned} \tag{13.20}$$

We see from Eq. (13.19) that the expression inside the brackets on the right side of Eq. (13.20) equals $\mathbf{P}(n)$. Hence, we may simplify Eq. (13.20) as follows:

$$\mathbf{k}(n) = \mathbf{P}(n)\mathbf{u}(n) \tag{13.21}$$

This result, together with $\mathbf{P}(n) = \mathbf{\Phi}^{-1}(n)$, may be used as the definition for the gain vector:

$$\mathbf{k}(n) = \mathbf{\Phi}^{-1}(n)\mathbf{u}(n) \tag{13.22}$$

In other words, the gain vector $\mathbf{k}(n)$ is defined as the tap-input vector $\mathbf{u}(n)$ transformed by the inverse of the correlation matrix $\mathbf{\Phi}(n)$.

Time Update for the Tap-Weight Vector

Next, we wish to develop a recursive equation for updating the least-squares estimate $\hat{\mathbf{w}}(n)$ for the tap-weight vector. To do this, we use Eqs. (13.8), (13.13), and (13.17) to express the least-squares estimate $\hat{\mathbf{w}}(n)$ for the tap-weight vector at time n as follows:

$$\begin{aligned}\hat{\mathbf{w}}(n) &= \mathbf{\Phi}^{-1}(n)\boldsymbol{\theta}(n) \\ &= \mathbf{P}(n)\boldsymbol{\theta}(n) \\ &= \lambda\mathbf{P}(n)\boldsymbol{\theta}(n-1) + \mathbf{P}(n)\mathbf{u}(n)d^{*}(n)\end{aligned} \tag{13.23}$$

Substituting Eq. (13.19) for $\mathbf{P}(n)$ in the first term only in the right side of Eq. (13.23), we get

$$\begin{aligned}\hat{\mathbf{w}}(n) &= \mathbf{P}(n-1)\boldsymbol{\theta}(n-1) - \mathbf{k}(n)\mathbf{u}^H(n)\mathbf{P}(n-1)\boldsymbol{\theta}(n-1) \\ &\quad + \mathbf{P}(n)\mathbf{u}(n)d^*(n) \\ &= \boldsymbol{\Phi}^{-1}(n-1)\boldsymbol{\theta}(n-1) - \mathbf{k}(n)\mathbf{u}^H(n)\boldsymbol{\Phi}^{-1}(n-1)\boldsymbol{\theta}(n-1) \\ &\quad + \mathbf{P}(n)\mathbf{u}(n)d^*(n) \\ &= \hat{\mathbf{w}}(n-1) - \mathbf{k}(n)\mathbf{u}^H(n)\hat{\mathbf{w}}(n-1) + \mathbf{P}(n)\mathbf{u}(n)d^*(n)\end{aligned} \tag{13.24}$$

Finally, using the fact that $\mathbf{P}(n)\mathbf{u}(n)$ equals the gain vector $\mathbf{k}(n)$, as in Eq. (13.21), we get the desired recursive equation for updating the tap-weight vector:

$$\begin{aligned}\hat{\mathbf{w}}(n) &= \hat{\mathbf{w}}(n-1) + \mathbf{k}(n)[d^*(n) - \mathbf{u}^H(n)\hat{\mathbf{w}}(n-1)] \\ &= \hat{\mathbf{w}}(n-1) + \mathbf{k}(n)\alpha^*(n)\end{aligned} \tag{13.25}$$

where $\alpha(n)$ is the *innovation* defined by

$$\begin{aligned}\alpha(n) &= d(n) - \mathbf{u}^T(n)\hat{\mathbf{w}}^*(n-1) \\ &= d(n) - \hat{\mathbf{w}}^H(n-1)\mathbf{u}(n)\end{aligned} \tag{13.26}$$

The inner product $\hat{\mathbf{w}}^H(n-1)\mathbf{u}(n)$ represents an estimate of the desired response $d(n)$, based on the *old* least-squares estimate of the tap-weight vector that was made at time $n-1$. Accordingly, we may also refer to $\alpha(n)$ as the *a priori estimation error*.

Equation (13.25) for the adjustment of the tap-weight vector and Eq. (13.26) for the innovation suggest the block-diagram representation depicted in Fig. 13.2(a) for the *recursive least-squares RLS algorithm*.

The innovation $\alpha(n)$ is, in general, different from the *a posteriori estimation error*[2]

$$e(n) = d(n) - \hat{\mathbf{w}}^H(n)\mathbf{u}(n) \tag{13.27}$$

the computation of which involves the *current* least-squares estimate of the tap-weight vector available at time n. Indeed, we may view $\alpha(n)$ as a "tentative" value of $e(n)$ before updating the tap-weight vector. Note, however, in the least-squares optimization that led to the recursive algorithm of Eq. (13.25) for the tap-weight vector, we actually minimized a cost function based on $e(n)$ and *not* $\alpha(n)$.

From here on, we will refer to $\alpha(n)$ as the a priori estimation error and to $e(n)$ as the a posteriori estimation error. The motivation for doing this is that in this chapter and the succeeding five chapters that deal with recursive solutions to the linear least-squares problem, we will be confronted with similar situations when the problems of forward linear prediction and backward linear prediction are considered.

[2] To be precise, we should modify the symbol for the estimation error in Eq. (13.27) to signify the fact that it corresponds to the least-squares estimate for the tap-weight vector. We have not done this simply for convenience of notation.

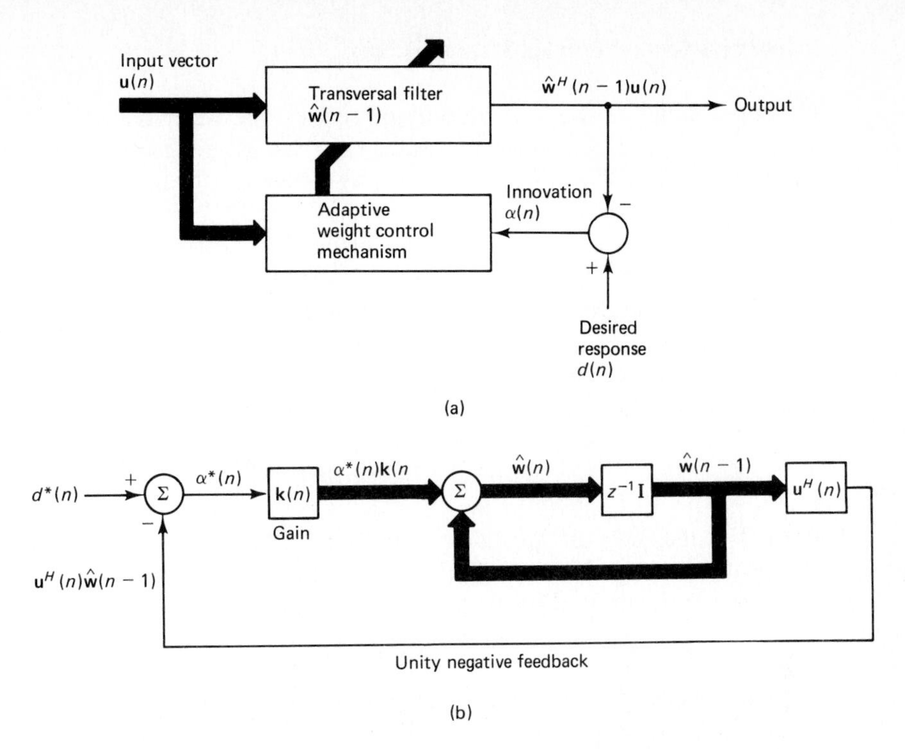

Figure 13.2 Representations of RLS algorithm: (a) block diagram; (b) signal-flow graph.

Equations (13.18), (13.26), (13.25), and (13.19) collectively and in that order constitute the *RLS algorithm,* as summarized below:

$$\mathbf{k}(n) = \frac{\lambda^{-1}\mathbf{P}(n-1)\mathbf{u}(n)}{1+\lambda^{-1}\mathbf{u}^H(n)\mathbf{P}(n-1)\mathbf{u}(n)}$$

$$\alpha(n) = d(n) - \hat{\mathbf{w}}^H(n-1)\mathbf{u}(n)$$

$$\hat{\mathbf{w}}(n) = \hat{\mathbf{w}}(n-1) + \mathbf{k}(n)\alpha^*(n)$$

$$\mathbf{P}(n) = \lambda^{-1}\mathbf{P}(n-1) - \lambda^{-1}\mathbf{k}(n)\mathbf{u}^H(n)\mathbf{P}(n-1)$$

In particular, Eq. (13.26) describes the filtering operation of the algorithm, whereby the transversal filter is excited to compute the a priori estimation error $\alpha(n)$. Equation (13.25) describes the adaptive operation of the algorithm, whereby the tap-weight vector is updated by incrementing its old value by an amount equal to the complex conjugate of the a priori estimation error $\alpha(n)$ times the time-varying gain vector $\mathbf{k}(n)$, hence the name "gain vector." Equations (13.18) and (13.19) enable us to update the value of the gain vector itself. An important feature of the RLS algorithm described by these equations is that the inversion of the correlation matrix $\mathbf{\Phi}(n)$ is replaced at each step by a simple scalar division. Figure 13.2(b) depicts a signal-flow-graph representation of the RLS algorithm that complements the block diagram of Fig. 13.2(a).

Initialization of the RLS Algorithm

The applicability of the RLS algorithm requires that we initialize the recursion of Eq. (13.19) by choosing a starting value $\mathbf{P}(0)$ that assures the nonsingularity of the correlation matrix $\mathbf{\Phi}(n)$. We may do this by evaluating the inverse

$$\left[\sum_{i=-n_0}^{0} \lambda^{-i}\mathbf{u}(i)\mathbf{u}^H(i)\right]^{-1}$$

where the tap-weight vector $\mathbf{u}(i)$ is obtained from an initial block of data for $-n_0 \le i \le 0$.

A simpler procedure, however, is to modify the expression slightly for the correlation matrix $\mathbf{\Phi}(n)$ by writing

$$\mathbf{\Phi}(n) = \sum_{i=1}^{n} \lambda^{n-i}\mathbf{u}(i)\mathbf{u}^H(i) + \delta\lambda^n\mathbf{I} \tag{13.28}$$

where $\mathbf{I}$ is the M-by-M identity matrix, and δ is a small positive constant. This modification affects the starting value, leaving the recursions in the RLS algorithm intact (see Problem 3). Thus putting $n = 0$ in Eq. (13.28), we have

$$\mathbf{\Phi}(0) = \delta\mathbf{I}$$

Correspondingly, for the initial value of $\mathbf{P}(n)$ equal to the inverse of the correlation matrix $\mathbf{\Phi}(n)$, we set

$$\mathbf{P}(0) = \delta^{-1}\mathbf{I} \tag{13.29}$$

The initialization described in Eq. (13.29) is equivalent to forcing the unknown data sample $u(-M+1)$ equal to the value $\lambda^{(-M+1)/2}\delta^{1/2}$ instead of zero. In other words, during the initialization period we modify the prewindowing method by writing

$$u(n) = \begin{cases} \lambda^{(-M+1)/2}\delta^{1/2}, & n = -M+1 \\ 0, & n < 0, \quad n \neq -M+1 \end{cases} \tag{13.30}$$

Note that for a transversal filter with M taps, the index $n = -M + 1$ refers to the *last* tap in the filter. When the first nonzero data sample $u(i)$ enters the filter, the initializing tap input $u(-M+1)$ leaves the filter and from then on the RLS algorithm takes over.

It only remains for us to choose an initial value for the tap-weight vector. It is customary to set

$$\hat{\mathbf{w}}(0) = \mathbf{0} \tag{13.31}$$

where $\mathbf{0}$ is the M-by-1 null vector.

The initialization procedure incorporating Eqs. (13.29) and (13.31) is referred to as a *soft-constrained initialization*. The positive constant δ is the only parameter required for this initialization. The recommended choice of δ is that it should be small compared to $0.01\sigma_u^2$, where σ_u^2 is the variance of a data sample $u(n)$. Such a choice is based on practical experience with the RLS algorithm, supported by a statistical analysis of the soft-constrained initialization of the algorithm (Hubing and Alexander, 1990). For large data lengths, the exact value of the initializing constant δ is unimportant.

Summary of the RLS Algorithm

In Tables 13.1 and 13.2 we present summaries of two versions of the RLS algorithm (Verhaegen, 1989), including the initial conditions and the resursions that are involved in its computation. In version I, given in Table 13.1, we exploit the Hermitian property of the matrix $\mathbf{P}(n)$ to reduce the computational complexity of the algorithm slightly. In version II of the algorithm, given in Table 13.2, no such assumption is made. The only difference between them is in the computation of the gain vector $\mathbf{k}(n)$. In theory, these two versions of the RLS algorithm are equivalent. However, they exhibit different numerical properties, as discussed in Chapter 19.

TABLE 13.1 SUMMARY OF VERSION I OF THE RLS ALGORITHM

Initialize the algorithm by setting

$$\mathbf{P}(0) = \delta^{-1}\mathbf{I}, \qquad \delta = \text{small positive constant}$$
$$\hat{\mathbf{w}}(0) = \mathbf{0}$$

For each instant of time, $n = 1, 2, \ldots,$ compute

$$\begin{aligned}
\boldsymbol{\pi}(n) &= \mathbf{u}^H(n)\mathbf{P}(n-1) \\
\kappa(n) &= \lambda + \boldsymbol{\pi}(n)\mathbf{u}(n) \\
\mathbf{k}(n) &= \frac{\boldsymbol{\pi}^H(n)}{\kappa(n)} \\
\alpha(n) &= d(n) - \hat{\mathbf{w}}^H(n-1)\mathbf{u}(n) \\
\hat{\mathbf{w}}(n) &= \hat{\mathbf{w}}(n-1) + \mathbf{k}(n)\alpha^*(n) \\
\mathbf{P}'(n-1) &= \mathbf{k}(n)\boldsymbol{\pi}(n) \\
\mathbf{P}(n) &= \frac{1}{\lambda}(\mathbf{P}(n-1) - \mathbf{P}'(n-1))
\end{aligned}$$

TABLE 13.2 SUMMARY OF VERSION II OF THE RLS ALGORITHM

Initialize the algorithm by setting

$$\mathbf{P}(0) = \delta^{-1}\mathbf{I}, \qquad \delta = \text{small positive constant}$$
$$\hat{\mathbf{w}}(0) = \mathbf{0}$$

For each instant of time, $n = 1, 2, \ldots,$ compute

$$\begin{aligned}
\boldsymbol{\pi}(n) &= \mathbf{u}^H(n)\mathbf{P}(n-1) \\
\kappa(n) &= \lambda + \boldsymbol{\pi}(n)\mathbf{u}(n) \\
\mathbf{k}(n) &= \frac{\mathbf{P}(n-1)\mathbf{u}(n)}{\kappa(n)} \\
\alpha(n) &= d(n) - \hat{\mathbf{w}}^H(n-1)\mathbf{u}(n) \\
\hat{\mathbf{w}}(n) &= \hat{\mathbf{w}}(n-1) + \mathbf{k}(n)\alpha^*(n) \\
\mathbf{P}'(n-1) &= \mathbf{k}(n)\boldsymbol{\pi}(n) \\
\mathbf{P}(n) &= \frac{1}{\lambda}(\mathbf{P}(n-1) - \mathbf{P}'(n-1))
\end{aligned}$$

13.4 UPDATE RECURSION FOR THE SUM OF WEIGHTED ERROR SQUARES

The minimum value of the sum of weighted error squares, $\mathscr{E}_{\min}(n)$, results when the tap-weight vector is set equal to the least-squares estimate $\hat{\mathbf{w}}(n)$. To compute $\mathscr{E}_{\min}(n)$, we may therefore use the relation [see first line of Eq. (10.40):

$$\mathscr{E}_{\min}(n) = \mathscr{E}_d(n) - \boldsymbol{\theta}^H(n)\hat{\mathbf{w}}(n) \tag{13.32}$$

where $\mathscr{E}_d(n)$ is defined by (using the notation of this chapter)

$$\begin{aligned}\mathscr{E}_d(n) &= \sum_{i=1}^{n} \lambda^{n-i}|d(i)|^2 \\ &= \lambda \mathscr{E}_d(n-1) + |d(n)|^2\end{aligned} \tag{13.33}$$

Therefore, substituting Eqs. (13.13), (13.25), and (13.33) in (13.32), we get

$$\begin{aligned}\mathscr{E}_{\min}(n) = &\ \lambda[\mathscr{E}_d(n-1) - \boldsymbol{\theta}^H(n-1)\hat{\mathbf{w}}(n-1)] \\ &+ d(n)[d^*(n) - \mathbf{u}^H(n)\hat{\mathbf{w}}(n-1)] \\ &- \boldsymbol{\theta}^H(n)\mathbf{k}(n)\alpha^*(n)\end{aligned} \tag{13.34}$$

where, in the last term, we have restored $\boldsymbol{\theta}(n)$ to its original form. By definition, the expression inside the first set of brackets on the right side of Eq. (13.34) equals $\mathscr{E}_{\min}(n-1)$. Also, by definition, the expression inside the second set of brackets equals the complex conjugate of the a priori estimation error $\alpha(n)$. For the last term, we use the definition of the gain vector $\mathbf{k}(n)$ to express the inner product $\boldsymbol{\theta}^H(n)\mathbf{k}(n)$ as

$$\begin{aligned}\boldsymbol{\theta}^H(n)\mathbf{k}(n) &= \boldsymbol{\theta}^H(n)\boldsymbol{\Phi}^{-1}(n)\mathbf{u}(n) \\ &= [\boldsymbol{\Phi}^{-1}(n)\boldsymbol{\theta}(n)]^H\mathbf{u}(n) \\ &= \hat{\mathbf{w}}^H(n)\mathbf{u}(n)\end{aligned}$$

where (in the second line) we have used the Hermitian property of the correlation matrix $\boldsymbol{\Phi}(n)$, and (in the third line) we have used the fact that $\boldsymbol{\Phi}^{-1}(n)\boldsymbol{\theta}(n)$ equals the least-squares estimate $\hat{\mathbf{w}}(n)$. Accordingly, we may simplify Eq. (13.34) as

$$\begin{aligned}\mathscr{E}_{\min}(n) &= \lambda\mathscr{E}_{\min}(n-1) + d(n)\alpha^*(n) - \hat{\mathbf{w}}^H(n)\mathbf{u}(n)\alpha^*(n) \\ &= \lambda\mathscr{E}_{\min}(n-1) + \alpha^*(n)[d(n) - \hat{\mathbf{w}}^H(n)\mathbf{u}(n)] \\ &= \lambda\mathscr{E}_{\min}(n-1) + \alpha^*(n)e(n)\end{aligned} \tag{13.35}$$

where $e(n)$ is the a posteriori estimation error. Equation (13.35) is the desired recursion for updating the sum of weighted error squares. We thus see that the product of the complex conjugate of $\alpha(n)$ and $e(n)$ represents the correction term in this updating recursion. Note that this product is real valued, which implies that we always have

$$\alpha(n)e^*(n) = \alpha^*(n)e(n) \tag{13.36}$$

In other words, the inner product and the outer product of the a priori estimation error $\alpha(n)$ and the a posteriori estimation error $e(n)$ are exactly the same.

13.5 CONVERGENCE ANALYSIS OF THE RLS ALGORITHM

In this section we demonstrate the convergence of the RLS algorithm by considering three aspects of the problem: (1) convergence of the estimate $\hat{\mathbf{w}}(n)$ in the mean, (2) convergence of the estimate $\hat{\mathbf{w}}(n)$ in the mean square, and (3) convergence of the mean-squared value of the a priori estimation error $\alpha(n)$. For this analysis, we assume that the desired response $d(n)$ and the tap-input vector $\mathbf{u}(n)$ are related by the *multiple linear regression model* of Fig. 13.3. In particular, we may write

$$d(n) = e_o(n) + \mathbf{w}_o^H \mathbf{u}(n) \tag{13.37}$$

where the M-by-1 vector $\mathbf{w}_o$ denotes the *regression parameter vector* of the model, and $e_o(n)$ is the *measurement error*. The measurement error process $\{e_o(n)\}$ is white with zero mean and variance σ^2. The parameter vector $\mathbf{w}_o$ is constant. The latter assumption is equivalent to saying that the adaptive transversal filter operates in a stationary environment. For a stationary environment, the best steady-state results are achieved when the tap weights adapt slowly, a condition that corresponds to the choice of $\lambda = 1$.

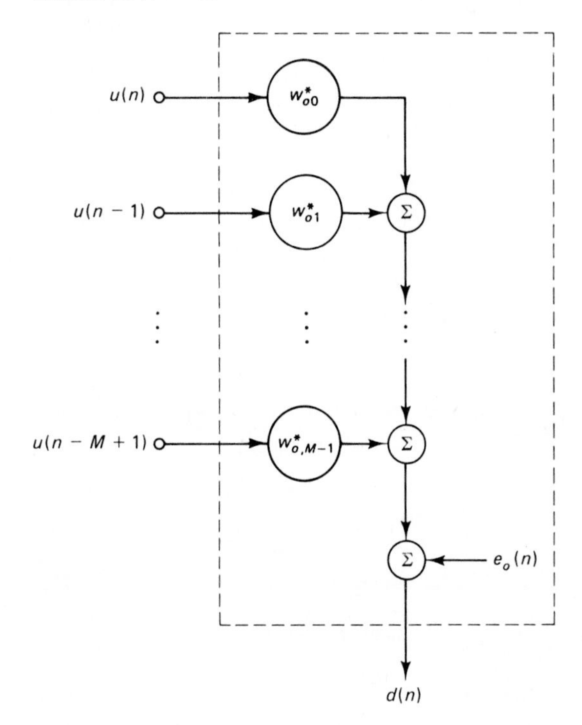

Figure 13.3 Multiple linear regression model.

Convergence of the Tap-Weight Vector in the Mean

The modification of the formula for the time-averaged correlation matrix $\mathbf{\Phi}(n)$ by initializing it with the small term $\delta\mathbf{I}$ introduces a *bias* into the estimate $\hat{\mathbf{w}}(n)$ produced by the RLS algorithm. That is, we may express the mean value of $\hat{\mathbf{w}}(n)$ as

$$E[\hat{\mathbf{w}}(n)] = \mathbf{w}_o + \mathbf{b}(n)$$

where $\mathbf{w}_o$ is the parameter vector of the multiple regression model, and $\mathbf{b}(n)$ denotes the bias. The bias itself is defined by (see Problem 7).

$$\mathbf{b}(n) = -\delta\mathbf{\Phi}^{-1}(n)\mathbf{w}_o \tag{13.38}$$

where $\mathbf{\Phi}^{-1}(n)$ is the inverse of $\mathbf{\Phi}(n)$, and $\mathbf{\Phi}(0) = \delta$.

Let $\mathbf{R}$ denote the M-by-M *ensemble-averaged correlation matrix* of the tap-input vector $\mathbf{u}(n)$. Assuming that the stochastic process represented by the tap-input vector $\mathbf{u}(n)$ is *ergodic,* it is well known that for large n we may approximate $\mathbf{R}$ by the time average formula (for $\lambda = 1$)

$$\begin{aligned} \mathbf{R} &\simeq \frac{1}{n}\sum_{i=1}^{n} \mathbf{u}(i)\mathbf{u}^H(i) \\ &= \frac{1}{n}\mathbf{\Phi}(n), \qquad n \text{ large} \end{aligned} \tag{13.39}$$

Hence, Eq. (13.38) yields the bias as

$$\mathbf{b}(n) \simeq -\frac{\delta}{n}\mathbf{R}^{-1}\mathbf{w}_o, \qquad n \text{ large}$$

This result shows that as the number of iterations, n, approaches infinity, the bias $\mathbf{b}(n)$ approaches zero.

We conclude therefore that the RLS algorithm produces an *asymptotically* unbiased estimate of the regression parameter vector $\mathbf{w}_o$. That is, the RLS algorithm is *convergent in the mean.*

Convergence of the Tap-Weight Vector in the Mean Square

Let $\boldsymbol{\epsilon}(n)$ denote the weight-error vector, defined as the difference between the estimate $\hat{\mathbf{w}}(n)$ produced by the RLS algorithm and the parameter vector $\mathbf{w}_o$ of the multiple linear regression model:

$$\boldsymbol{\epsilon}(n) = \hat{\mathbf{w}}(n) - \mathbf{w}_o \tag{13.40}$$

From Chapter 10, we also recall that the weight-error correlation matrix equals

$$\begin{aligned} \mathbf{K}(n) &= E[\boldsymbol{\epsilon}(n)\boldsymbol{\epsilon}^H(n)] \\ &= \sigma^2\mathbf{\Phi}^{-1}(n) \end{aligned} \tag{13.41}$$

where σ^2 is the variance of the measurement error process $\{e_o(n)\}$, and $\mathbf{\Phi}^{-1}(n)$ is the inverse of the time-varying correlation matrix defined in Eq. (13.9). The problem is to show that the weight-error correlation matrix $\mathbf{K}(n)$ approaches zero as the number of iterations, n, approaches infinity.

For large n, we may use the approximate relation of Eq.(13.39) between $\mathbf{\Phi}(n)$ and the ensemble-averaged correlation matrix $\mathbf{R}$. Correspondingly, we may approximate the weight-error correlation matrix of Eq. (13.41) as follows:

$$\mathbf{K}(n) \simeq \frac{\sigma^2}{n}\mathbf{R}^{-1}, \quad \text{large } n \tag{13.42}$$

Taking the norm of both sides of this equation, we get

$$\|\mathbf{K}(n)\| \simeq \frac{\sigma^2}{n}\|\mathbf{R}^{-1}\|, \quad \text{large } n$$

Let $\lambda_{\min}$ denote the smallest eigenvalue of the correlation matrix $\mathbf{R}$. Hence, we may express the norm of the inverse matrix $\mathbf{R}^{-1}$ as (see Chapter 4)

$$\|\mathbf{R}^{-1}\| = \frac{1}{\lambda_{\min}}$$

Accordingly, we may express the norm of $\mathbf{K}(n)$ simply as

$$\|\mathbf{K}(n)\| \simeq \frac{\sigma^2}{n\lambda_{\min}}, \quad \text{large } n \tag{13.43}$$

Based on this result, we may now make the following two important observations for large n:

1. The norm of the weight-error correlation matrix is *magnified by the inverse of the smallest eigenvalue*. Hence, to a first order of approximation, the sensitivity of the RLS algorithm to eigenvalue spread is determined initially in proportion to the inverse of the smallest eigenvalue. Therefore, ill-conditioned least-squares problems may lead to *bad* convergence properties.
2. The norm of the weight-error correlation matrix decays almost linearly with time. Hence, the estimate $\hat{\mathbf{w}}(n)$ produced by the RLS algorithm for the tap-weight vector converges in the norm to the parameter vector $\mathbf{w}_o$ of the multiple linear regression model almost *linearly with time*.

Convergence of the RLS Algorithm in the Mean Square

In the RLS algorithm there are two errors, the a priori estimation error $\alpha(n)$ and the a posteriori estimation error $e(n)$, to be considered. Given the initial conditions of Section 13.3 we find that the mean-square values of these two errors vary differently with time n. At time $n = 1$, the mean-square value of $\alpha(n)$ attains a *large* value, equal to the mean-square value of the desired response $d(n)$, and then *decays* with increasing n. The mean-square value of $e(n)$, on the other hand, attains a *small* value at $n = 1$, and then *rises* with increasing n. Accordingly, the choice of $\alpha(n)$ as the error of interest yields a learning curve for the RLS algorithm that has the same general shape as that for the LMS algorithm. By so doing, we can then make a direct graphical comparison between the learning curves of the RLS and LMS algorithms. We will therefore base the convergence analysis of the RLS in the mean square on the a priori estimation error $\alpha(n)$.

Eliminating the desired response $d(n)$ between Eqs. (13.26) and (13.37), we may express the a priori estimation error $\alpha(n)$ as

$$\begin{aligned}\alpha(n) &= e_o(n) - [\hat{\mathbf{w}}(n-1) - \mathbf{w}_o]^H\mathbf{u}(n) \\ &= e_o(n) - \boldsymbol{\epsilon}^H(n-1)\mathbf{u}(n)\end{aligned} \tag{13.44}$$

where $\boldsymbol{\epsilon}(n-1)$ is the weight-error vector at time $n-1$. As an *index of statistical performance* for the RLS algorithm, it is convenient to use the a priori estimation error $\alpha(n)$ to define the *mean-squared error:*

$$J'(n) = E[|\alpha(n)|^2] \tag{13.45}$$

The prime in the symbol $J'(n)$ is intended to distinguish the mean-square value of $\alpha(n)$ from that of $e(n)$. Substituting Eq. (13.44) in (13.45), and then expanding terms, we get

$$J'(n) = E\big[|e_o(n)|^2\big] + E\big[\mathbf{u}^H(n)\boldsymbol{\epsilon}(n-1)\boldsymbol{\epsilon}^H(n-1)\mathbf{u}(n)\big] \\ - E\big[\boldsymbol{\epsilon}^H(n-1)\mathbf{u}(n)e_o^*(n)\big] - E\big[e_o(n)\mathbf{u}^H(n)\boldsymbol{\epsilon}(n-1)\big] \tag{13.46}$$

With the measurement $e_o(n)$ assumed to be of zero mean, the first expectation on the right side of Eq. (13.46) is simply the variance of $e_o(n)$, which is denoted by σ^2. As for the remaining three expectations, we may make the following observations in light of the independence theory introduced in Section 9.4:

1. The estimate $\hat{\mathbf{w}}(n-1)$, and therefore the weight-error vector $\boldsymbol{\epsilon}(n-1)$, is independent of the tap-input vector $\mathbf{u}(n)$; the latter is assumed to be drawn from a wide-sense stationary process of zero mean. Hence, we may use this statistical independence together with well-known results from matrix algebra to express the second expectation on the right side of Eq. (13.46) as follows:

$$\begin{aligned} E[\mathbf{u}^H(n)\boldsymbol{\epsilon}(n-1)\boldsymbol{\epsilon}^H(n-1)\mathbf{u}(n)] &= E[\mathrm{tr}\{\mathbf{u}^H(n)\boldsymbol{\epsilon}(n-1)\boldsymbol{\epsilon}^H(n-1)\mathbf{u}(n)\}] \\ &= E[\mathrm{tr}\{\mathbf{u}(n)\mathbf{u}^H(n)\boldsymbol{\epsilon}(n-1)\boldsymbol{\epsilon}^H(n-1)\}] \\ &= \mathrm{tr}\{E[\mathbf{u}(n)\mathbf{u}^H(n)\boldsymbol{\epsilon}(n-1)\boldsymbol{\epsilon}^H(n-1)]\} \\ &= \mathrm{tr}\{E[\mathbf{u}(n)\mathbf{u}^H(n)]E[\boldsymbol{\epsilon}(n-1)\boldsymbol{\epsilon}^H(n-1)]\} \\ &= \mathrm{tr}\{\mathbf{R}\mathbf{K}(n-1\} \end{aligned} \tag{13.47}$$

   where, in the last line, we have made use of the definitions of the ensemble-averaged correlation matrix $\mathbf{R}$ and weight-error correlation matrix $\mathbf{K}(n-1)$.

2. The measurement error $e_o(n)$ depends on the tap-input vector $\mathbf{u}(n)$; this follows from a simple rearrangement of Eq. (13.37). The weight-error vector $\boldsymbol{\epsilon}(n-1)$ is therefore independent of both $\mathbf{u}(n)$ and $e_o(n)$. Accordingly, we may show that the third expectation on the right side of Eq. (13.46) is zero by first reformulating it as follows:

$$E[\boldsymbol{\epsilon}^H(n-1)\mathbf{u}(n)e_o^*(n)] = E[\boldsymbol{\epsilon}^H(n-1)]E[\mathbf{u}(n)e_o^*(n)]$$

   We now recognize from the principle of orthogonality that all the elements of the tap-input vector $\mathbf{u}(n)$ are orthogonal to the measurement or optimum estimation error $e_o(n)$. We therefore have

$$E[\boldsymbol{\epsilon}^H(n-1)\mathbf{u}(n)e_o^*(n)] = 0 \tag{13.48}$$

3. The fourth and final expectation on the right side of Eq. (13.46) has the same mathematical form as that just considered in point 2, except for a trivial complex conjugation. We may therefore set this expectation equal to zero, too:

$$E[e_o(n)\mathbf{u}^H(n)\boldsymbol{\epsilon}(n-1)] = 0 \tag{13.49}$$

Thus, recognizing that $E(|e_o(n)|^2] = \sigma^2$, and using the results of Eqs. (13.47) to (13.49) in (13.46), we get the following simple formula for the mean-squared error in the RLS algorithm:

$$J'(n) = \sigma^2 + \mathrm{tr}[\mathbf{RK}(n-1)] \qquad (13.50)$$

We now wish to approximate $J'(n)$ for large n. To do so, we rewrite Eq. (13.42) with $n - 1$ used in place n, and then substitute the resulting expression for $\mathbf{K}(n-1)$ in Eq. (13.50). For large values of time n, we may therefore approximate the mean-squared error $J'(n)$ as follows:

$$J'(n) \simeq \sigma^2 + \frac{\sigma^2}{n-1}\mathrm{tr}[\mathbf{R}^{-1}\mathbf{R}]$$

$$= \sigma^2 + \frac{\sigma^2}{n-1}\mathrm{tr}[\mathbf{I}]$$

where $\mathbf{I}$ is the M-by-M identity matrix. Since the trace of a square matrix equals the sum of its diagonal elements, it follows that $\mathrm{tr}[\mathbf{I}]$ equals M, where M is the number of tap weights contained in the transversal filter. Also, for large n, we have $n - 1 \simeq n$. We thus get the following simple formula for the *mean-squared innovation* that is produced by the RLS algorithm:

$$J'(n) \simeq \sigma^2 + \frac{M\sigma^2}{n} \qquad (13.51)$$

Based on this result, we may make the following deductions:

1. The RLS algorithm converges in the mean square in about $2M$ iterations, where M is the number of taps in the transversal filter. This means that the rate of convergence of the RLS algorithm is typically an order of magnitude *faster* than that of the LMS algorithm.
2. As the number of iterations, n, approaches infinity, the mean-squared innovation approaches a final value equal to the variance σ^2 of the measurement error $e_o(n)$. Since the minimum mean-squared error also equals σ^2, it follows that the RLS algorithm, in theory, produces zero excess mean-squared error (or, equivalently, zero misadjustment) when operating in a stationary environment.

It should be emphasized that the above-mentioned improvement in the rate of convergence of the RLS algorithm over the LMS algorithm holds only when the measurement error $e_o(n)$ is small compared to the desired response $d(n)$, that is, when the signal-to-noise ratio is high. Also, the zero misadjustment property of the RLS algorithm assumes that the exponential weighting factor λ equals unity; that is, the algorithm operates with infinite memory.

13.6 EXAMPLE: SINGLE-WEIGHT ADAPTIVE NOISE CANCELLER

Consider the *single-weight, dual-input adaptive noise canceller* depicted in Fig. 13.4. The two inputs are represented by the *primary signal* $d(n)$ and the *reference signal* $u(n)$ that are characterized as follows. First, the primary signal consists of an

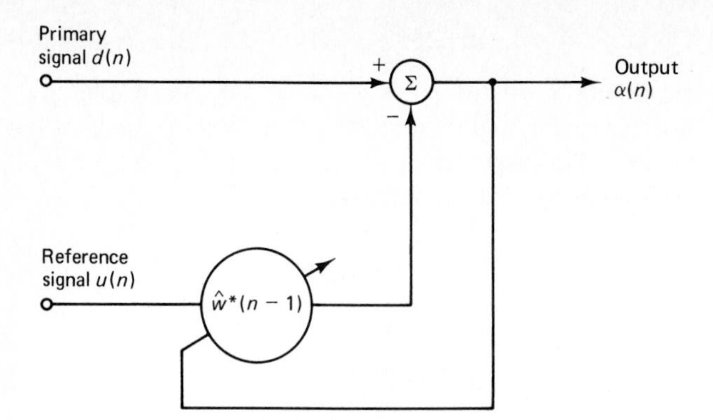

Figure 13.4 Single-weight adaptive noise canceller.

information-bearing signal component and an additive *interference*. Second, the reference signal $u(n)$ is correlated with the interference and has no detectable contribution of the information-bearing signal. The requirement is to exploit the properties of the reference signal in relation to the primary signal to suppress the interference at the adaptive noise canceller output.

Application of the RLS algorithm yields the following set of equations for this canceller (after reorganization of terms):

$$k(n) = \left[\frac{1}{\lambda\hat{\sigma}^2(n-1) + |u(n)|^2}\right]u(n) \tag{13.52}$$

$$\alpha(n) = d(n) - \hat{w}^*(n-1)u(n) \tag{13.53}$$

$$\hat{w}(n) = \hat{w}(n-1) + k(n)\alpha^*(n) \tag{13.54}$$

$$\hat{\sigma}^2(n) = \lambda\hat{\sigma}^2(n-1) + |u(n)|^2 \tag{13.55}$$

where $\hat{\sigma}^2(n)$ is an estimate of error variance. It is the inverse of $P(n)$, the scalar version of the matrix $\mathbf{P}(n)$ in the RLS algorithm, as shown by

$$\hat{\sigma}^2(n) = P^{-1}(n) \tag{13.56}$$

It is informative to compare the algorithm described in Eqs. (13.52) to (13.55) with its counterpart obtained using the normalized LMS algorithm; the version of the normalized LMS algorithm of particular interest in the context of our present situation is that given in Eq. (9.167). The *major difference* between these two algorithms is that the constant a in the normalized LMS algorithm is replaced by the time-varying term $\lambda\hat{\sigma}^2(n-1)$ in the denominator of the gain factor $k(n)$ that controls the correction applied to the tap weight in Eq. (13.54).

13.7 COMPUTER EXPERIMENT ON ADAPTIVE EQUALIZATION

For our computer experiment, we use the RLS algorithm with the exponential weighting factor $\lambda = 1$, for the adaptive equalization of a linear dispersive communication channel. The LMS version of this study was presented in Section 9.13. The block diagram of the system used in the study is depicted in Fig. 13.5. Two indepen-

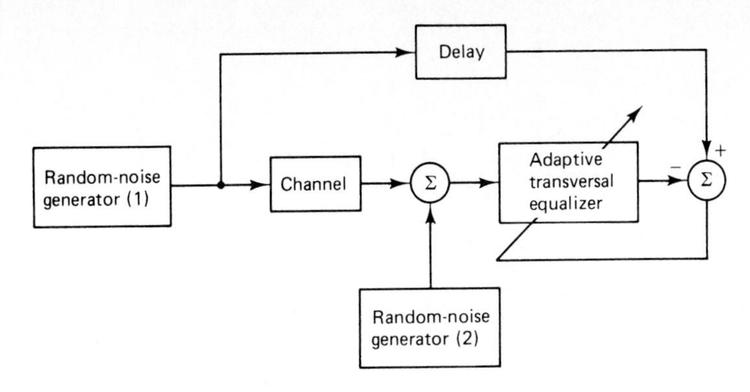

Figure 13.5 Block diagram of adaptive equalizer experiment.

dent random-number generators are used, one, denoted by $\{a(n)\}$, for probing the channel, and the other, denoted by $\{v(n)\}$, for simulating the effect of additive white noise at the receiver input. The sequence $\{a(n)\}$ is in polar form, with $a(n) = \pm 1$. The second sequence $\{v(n)\}$ has zero mean; its variance σ_v^2 is determined by the desired signal-to-noise ratio. The equalizer has 11 taps. The impulse response of the channel is defined by

$$h_n = \begin{cases} \dfrac{1}{2}\left[1 + \cos\left(\dfrac{2\pi}{W}(n-2)\right)\right], & n = 1, 2, 3 \\ 0, & \text{otherwise} \end{cases}$$

where W controls the ammount of amplitude distortion and therefore the eigenvalue spread produced by the channel. The channel input $\{a(n)\}$, after a delay of seven samples, provides the desired response for the equalizer (see Section 9.13 for details).

The experiment is in two parts: in part 1 the signal-to-noise ratio is high, and in part 2 it is low. In both parts of the experiment, the constant $\delta = 0.004$.

1. Signal-to-Noise Ratio = 30 dB. Figure 13.6 shows the results of the experiment for fixed signal-to-noise ratio of 30 dB (equivalently, variance $\sigma_v^2 = 0.001$) and varying W or eigenvalue spread $\chi(\mathbf{R})$. The figure is in four parts, for which the parameter W equals 2.9, 3.1, 3.3, and 3.5. The corresponding values of the eigenvalue spread $\chi(\mathbf{R})$ equal 6.0782, 11.1238, 21.7132, and 46.8216, respectively (see Table 9.2 for details). In each part of Fig. 13.6 we have included two learning curves, one for the RLS algorithm obtained by ensemble-averaging (for each iteration n) the squared value of the a priori estimation error $\alpha(n)$, and the other for the LMS algorithm (included for the sake of comparison) obtained by ensemble-averaging the squared value of the a posteriori estimation error $e(n)$. The ensemble-averaging was performed over 200 independent trials of the experiment. For the LMS algorithm, the step-size parameter $\mu = 0.075$ was used. Based on the results shown in Fig. 13.6 we may make the following observations:

1. Convergence of the RLS algorithm is attained in about 20 iterations, approximately twice the number of taps in the transversal equalizer.

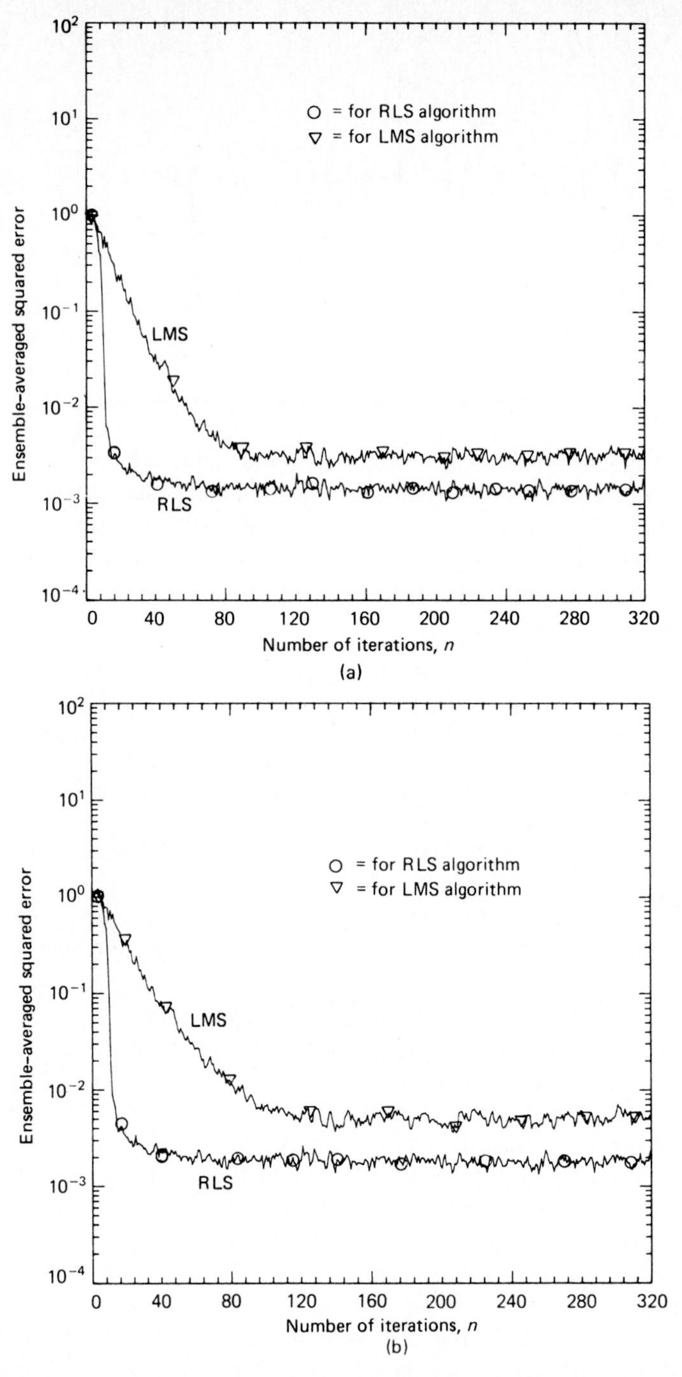

Figure 13.6 Learning curves for LMS and RLS algorithms. (a) $W = 2.9$, $\delta = 0.004$, $\lambda = 1.0$ for RLS algorithm; step size = 0.075 for LMS algorithm. (b) $W = 3.1$, $\delta = 0.004$, $\lambda = 1.0$ for RLS algorithm; step size = 0.075 for LMS algorithm. (c) $W = 3.3$, $\delta = 0.004$, $\lambda = 1.0$ for RLS algorithm, step size = 0.075 for LMS algorithm. (d) $W = 3.5$, $\delta = 0.004$, $\lambda = 1.0$ for RLS algorithm; step size = 0.075 for LMS algorithm.

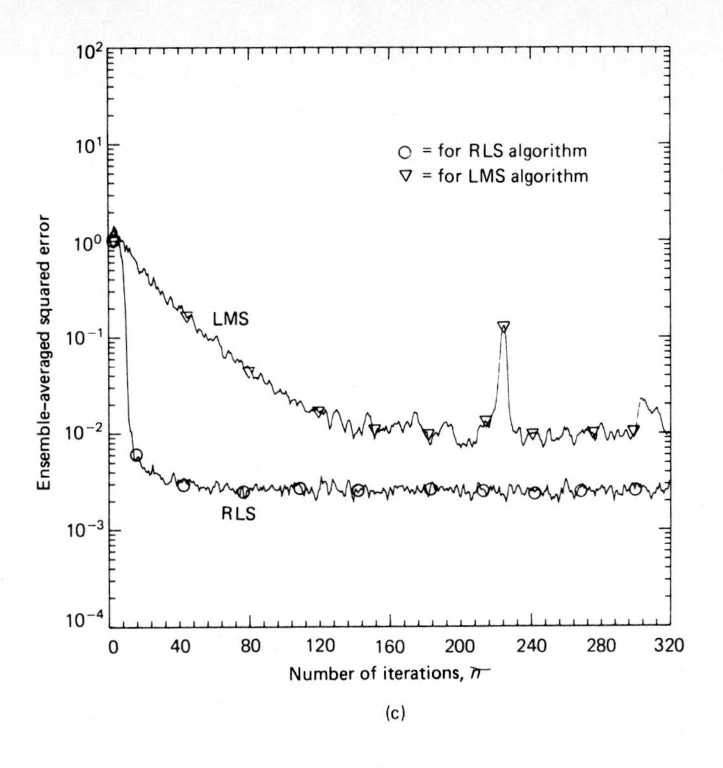

(c)

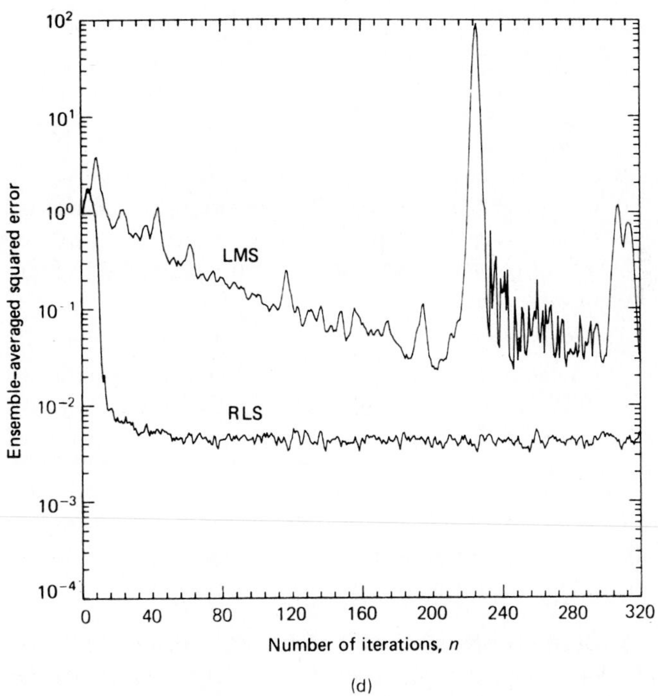

(d)

Figure 13.6 *(continued)*

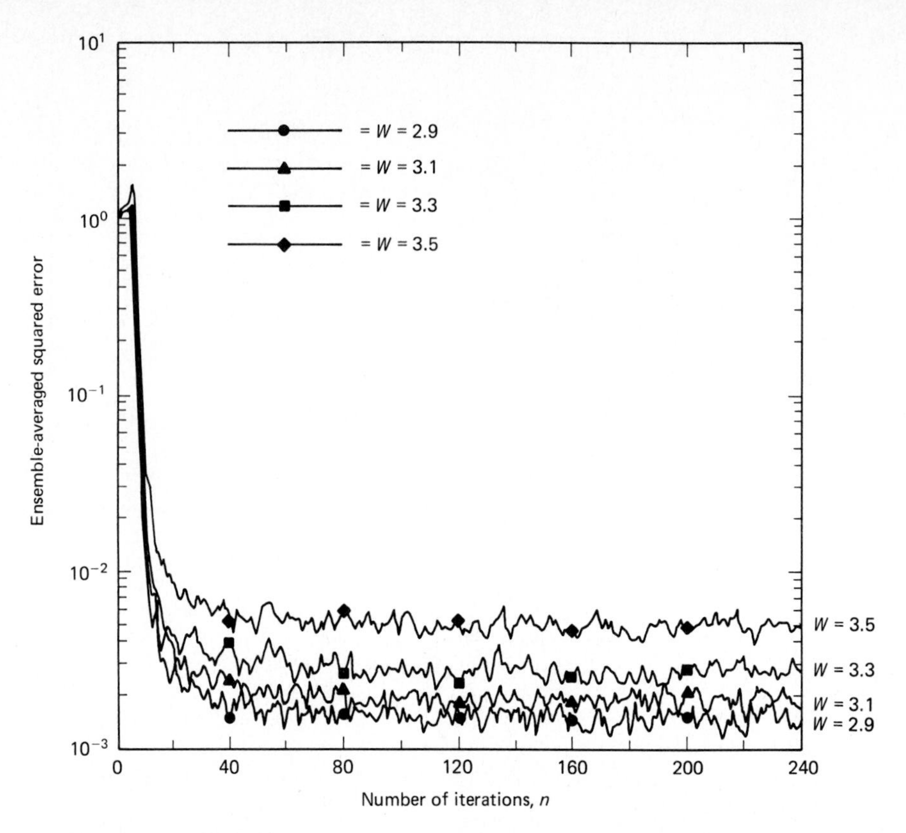

Figure 13.7 Learning curves with different eigenvalue spreads for RLS algorithm: $\delta = 0.004$, $\lambda = 1.0$.

2. Rate of convergence of the RLS algorithm is relatively insensitive to variations in the eigenvalue spread $\chi(\mathbf{R})$ compared to the LMS algorithm. This property is clearly illustrated in Fig. 13.7, where we have reproduced the learning curves of the RLS algorithm, corresponding to the four different values of the eigenvalue spread.
3. The RLS algorithm converges much faster than the LMS algorithm.
4. The steady-state value of the averaged squared error produced by the RLS algorithm is much smaller than in the case of the LMS algorithm. In both cases, however, it is sensitive to variations in the eigenvalue spread $\chi(\mathbf{R})$.

The results presented in Figs. 13.6 and 13.7 clearly show the superior rate of convergence of the RLS over the LMS algorithm; for it to be realized, however, the signal-to-noise ratio has to be high. This advantage is lost when the signal-to-noise ratio is not high, as demonstrated next.

2. Signal-to-Noise Ratio = 10 dB. Figure 13.8 shows the learning curves for the RLS algorithm and the LMS algorithm (with the step-size parameter $\mu = 0.075$) for $W = 3.1$ and signal-to-noise ratio of 10 dB. Insofar as the rate of convergence is concerned, we now see that the RLS and LMS algorithms perform in roughly the same manner, both requiring about 40 iterations to converge.

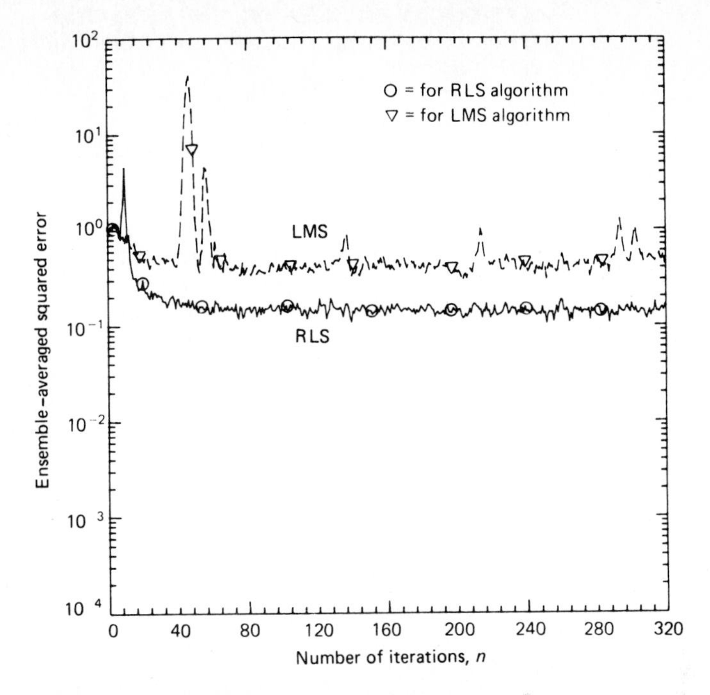

Figure 13.8 Learning curves for RLS and LMS algorithms for $W = 3.1$; $\delta = 0.004$, $\lambda = 1.0$ for RLS; step size $= 0.075$ for LMS; SNR $=$ 10 dB.

Impulse Response of Adaptive Filter

In Table 13.3 we present four sets of results that compare the *ensemble-averaged impulse response* of the adaptive equalizer with that of the optimum Wiener filter. Here again, the ensemble averaging was performed over 200 independent trials of the experiment; the tap-weight values presented for the equalizer were computed after 170 iterations of the RLS algorithm. The four sets of results presented in Table 13.3 were computed for the four different eigenvalue spreads, corresponding to $W = 2.9, 3.1, 3.3$, and 3.5. This table clearly shows that the final values of the tap weights in the adaptive transversal equalizer do resemble the corresponding tap-weight values of the optimum Wiener filter, as is evident from the small relative errors included in the last columns of Table 13.3, especially for large eigenvalue spreads.

13.8 OPERATION OF THE RLS ALGORITHM IN A NONSTATIONARY ENVIRONMENT

Throughout the convergence analysis of the RLS algorithm presented in Section 13.6, it was assumed that the exponential weighting factor $\lambda = 1$. Indeed, the use of $\lambda = 1$ is well suited for a stationary environment, for which the best steady-state results are obtained. However, when the RLS algorithm operates in a nonstationary environment, it is customary to use a value of λ less than unity, thereby giving the algorithm only a finite memory. By so doing, the algorithm attains the capability to track slow statistical variations in the environment in which it operates. Unfortu-

TABLE 13.3 RESULTS OF THE COMPUTER EXPERIMENT ON ADAPTIVE EQUALIZATION USING THE RLS ALGORITHM, AND COMPARISON WITH THEORY

Weight number	RLS	Optimum	Relative error (%)
		$W = 2.9$	
0	-5.34×10^{-4}	-6.41×10^{-4}	16.58
1	2.93×10^{-3}	3.09×10^{-3}	5.08
2	-1.33×10^{-2}	-1.36×10^{-2}	1.80
3	5.87×10^{-2}	5.90×10^{-2}	0.44
4	-2.56×10^{-1}	-2.56×10^{-1}	0.00
5	1.11	1.11	0.00
6	-2.57×10^{-1}	-2.56×10^{-1}	−0.39
7	5.93×10^{-2}	5.90×10^{-2}	−0.54
8	-1.33×10^{-2}	-1.36×10^{-2}	1.75
9	3.34×10^{-3}	3.09×10^{-3}	−8.19
10	-9.88×10^{-4}	-6.41×10^{-4}	−54.18
		$W = 3.1$	
0	-2.43×10^{-3}	-2.58×10^{-3}	5.55
1	9.81×10^{-3}	1.01×10^{-2}	2.58
2	-3.35×10^{-2}	-3.39×10^{-2}	1.07
3	1.11×10^{-1}	1.12×10^{-1}	0.52
4	-3.67×10^{-1}	-3.67×10^{-1}	0.03
5	1.20	1.20	0.00
6	-3.68×10^{-1}	-3.67×10^{-1}	−0.18
7	1.12×10^{-1}	1.12×10^{-1}	−0.09
8	-3.37×10^{-2}	-3.39×10^{-2}	0.55
9	1.03×10^{-2}	1.01×10^{-2}	−2.57
10	-2.94×10^{-3}	-2.58×10^{-3}	−14.27
		$W = 3.3$	
0	-7.90×10^{-3}	-8.13×10^{-3}	2.81
1	2.70×10^{-2}	2.75×10^{-2}	1.80
2	-7.49×10^{-2}	-7.56×10^{-2}	0.88
3	1.98×10^{-1}	1.99×10^{-1}	0.42
4	-5.18×10^{-1}	-5.18×10^{-1}	0.05
5	1.35	1.35	0.00
6	-5.19×10^{-1}	-5.18×10^{-1}	−0.13
7	1.99×10^{-1}	1.99×10^{-1}	−0.04
8	-7.53×10^{-2}	-7.56×10^{-2}	0.31
9	2.77×10^{-2}	2.75×10^{-2}	−0.74
10	-8.50×10^{-3}	-8.13×10^{-3}	−4.52
		$W = 3.5$	
0	-2.17×10^{-2}	-2.22×10^{-2}	2.03
1	6.72×10^{-2}	6.83×10^{-2}	1.66
2	-1.59×10^{-1}	-1.60×10^{-1}	0.85
3	3.48×10^{-1}	3.49×10^{-1}	0.38
4	-7.43×10^{-1}	-7.45×10^{-1}	0.23
5	1.57	1.58	0.05
6	-7.44×10^{-1}	-7.45×10^{-1}	0.07
7	3.49×10^{-1}	3.49×10^{-1}	0.01
8	-1.60×10^{-1}	-1.60×10^{-1}	0.31
9	6.82×10^{-2}	6.83×10^{-2}	0.16
10	-2.24×10^{-2}	-2.22×10^{-2}	−1.29

nately, the use of $\lambda < 1$ changes the behavior of the RLS algorithm in a drastic manner. In this section, we present a summary of the effects produced by using the exponentially windowed RLS algorithm in a nonstationary environment.[3]

Weight Vector Noise

When $\lambda < 1$, the exponentially weighted least-squares estimate $\hat{\mathbf{w}}(n)$ of the regression coefficient vector is *no* longer consistent. This causes noise to appear in the tap weights of the adaptive transversal filter, with the result that (on the average) they become misadjusted from their optimum setting. This type of misadjustment arises whenever λ is less than unity, whether the environment in which the RLS algorithm operates is stationary or not. Following the convention introduced in Section 9.14, we define the *weight vector noise* as

$$\boldsymbol{\epsilon}_1(n) = \hat{\mathbf{w}}(n) - E[\hat{\mathbf{w}}(n)]$$

where $E[\hat{\mathbf{w}}(n)]$ is the mean value of $\hat{\mathbf{w}}(n)$. The presence of the weight vector noise in the estimate $\hat{\mathbf{w}}(n)$ leads to the excess mean-squared error (see Section 9.14)

$$J_{\text{est}}(n) = \text{tr}[\mathbf{R}\mathbf{K}_1(n)] \tag{13.57}$$

where $\mathbf{R}$ is the ensemble-averaged correlation matrix of the tap inputs, and $\mathbf{K}_1(n)$ is the ensemble-averaged correlation matrix of the weight vector noise $\boldsymbol{\epsilon}_1(n)$. Here we have used the subscript "est" to indicate that the formula above is an expression for the "estimation noise." For the case when λ is very close to unity, use of this formula yields the following approximate result for the *misadjustment* produced in the RLS algorithm due to the weight vector noise (Medaugh and Griffiths, 1981; Ling and Proakis, 1984):

$$\mathcal{M} \simeq \frac{1 - \lambda}{1 + \lambda} M \tag{13.58}$$

where M is the number of taps in the transversal filter; see Problem 9. The misadjustment $\mathcal{M}$ is defined as the ratio of the excess mean-squared error $J_{\text{est}}(\infty)$ to the variance σ^2 of the measurement noise $e_o(n)$ in the multiple linear regression model. The applicability of this approximate formula is restricted to values of λ very close to unity. In general, calculation of the excess mean squared error $J_{\text{est}}(\infty)$ depends not only on the exponential windowing but also on some fourth-order statistics of the input signal. The effect is noticeable when λ is assigned values as low as 0.9 for fast adaptation (Eleftheriou and Falconer, 1986).

Earlier we indicated that $(1 - \lambda)^{-1}$ provides a rough measure of the "memory" of the RLS algorithm. Hence, as the memory of the algorithm is shortened by reducing the exponential weighting factor λ, the misadjustment $\mathcal{M}$ is correspondingly increased. In other words, *fast adaptation of the RLS algorithm, in general, results in a more noisy adaptive process.*

[3] For a detailed mathematical analysis of the tracking properties and steady-state performance of the RLS algorithm in a nonstationary environment and supporting computer simulation results, see Eleftheriou and Falconer (1986). The material presented in Section 13.8 is based on this paper.

Weight Vector Lag

When the environment in which the RLS algorithm operates is nonstationary, the coefficient vector in the multiple linear regression model takes on a *time-varying* form, as signified by $\mathbf{w}_o(n)$. In such an environment, the RLS algorithm has the task of not only *finding* the regression coefficient vector $\mathbf{w}_o(n)$ but also *tracking* its variation from one iteration to the next. Due to a lag in the adaptive process, we have to account for the *weight vector lag* (see Section 9.14)

$$\boldsymbol{\epsilon}_2(n) = E[\hat{\mathbf{w}}(n)] - \mathbf{w}_o(n) \tag{13.59}$$

Provided that the variation in $\mathbf{w}_o(n)$ is slow compared with the memory of the algorithm, $\boldsymbol{\epsilon}_2(n)$ is approximately defined by the first-order difference equation (Eleftheriou and Falconer, 1986)

$$\boldsymbol{\epsilon}_2(n) = \lambda \boldsymbol{\epsilon}_2(n-1) - \mathbf{w}_o(n) + \mathbf{w}_o(n-1) \tag{13.60}$$

Let $\mathcal{E}_2(z)$ denote the z-transform of the weight vector lag $\boldsymbol{\epsilon}_2(n)$. Let $\mathbf{W}_o(z)$ denote the z-transform of the coefficient vector $\mathbf{w}_o(n)$ in the multiple linear regression model. Then, taking the z-transform of both sides of Eq. (13.60), and solving for $\mathcal{E}_2(z)$, we get

$$\mathcal{E}_2(z) = \frac{z^{-1} - 1}{1 - \lambda z^{-1}} \mathbf{W}_o(z) \tag{13.61}$$

The transfer function $\mathcal{E}_2(z)/\mathbf{W}_o(z)$ has a zero at $z = 1$ and a pole at $z = \lambda$. Hence, under general time-varying conditions, the weight vector lag $\boldsymbol{\epsilon}_2(n)$ is a *geometrical converging vector process*. The *time constant* of the process is given by

$$\tau_i = \frac{1}{1 - \lambda}$$

for all coordinates of the process, $1 \le i \le M$; see Problem 10. Moreover, the time constant τ_i is independent of the eigenvalue spread of the underlying correlation matrix of the tap inputs. This is in direct contrast to the LMS algorithm for which the time constant is, in general, different for each tap weight and corresponds to a particular eigenvalue of the ensemble-averaged correlation matrix of the tap inputs.

Total Value of the Average Excess Mean-Squared Error

Consider the example of a multiple linear regression model whose coefficients undergo *independent stationary first-order Markov processes*. That is, the regression coefficient vector $\mathbf{w}_o(n)$ is described by the real-valued first-order difference equation

$$\mathbf{w}_o(n) = a\mathbf{w}_o(n-1) + \mathbf{v}(n) \tag{13.62}$$

The parameter a controls the time constant of nonstationarity, which equals $1/(1 - a)$. The elements of the vector $\mathbf{v}(n)$ consist of independent, white, Gaussian noise processes, each of zero mean and variance σ_v^2. To ensure that the variation in $\mathbf{w}_o(n)$ is slow compared with the memory of the RLS algorithm, the time constant of nonstationarity is chosen to satisfy the condition

$$\frac{1}{1-a} \gg \frac{1}{1-\lambda}$$

Under these conditions, we find that the excess mean-squared error $J_{\text{lag}}(\infty)$ due to the weight vector lag $\boldsymbol{\epsilon}_2(n)$, is inversely proportioned to the factor $(1 - \lambda)$ (Eleftheriou and Falconer, 1986).

By contrast, we see from Eq. (13.58) that for λ very close to unity the excess mean-squared error $J_{\text{est}}(\infty)$ due to the weight vector noise $\boldsymbol{\epsilon}_1(n)$, is directly proportioned to $(1 - \lambda)$. The total value of the excess mean-squared error, $J_{\text{tot}}(\infty)$ due to the combined action of $\boldsymbol{\epsilon}_1(n)$ and $\boldsymbol{\epsilon}_2(n)$, is equal to the sum of $J_{\text{est}}(\infty)$ and $J_{\text{lag}}(\infty)$. At the optimum value λ_{opt}, for which $J_{\text{tot}}(\infty)$ attains its minimum value, the contributions due to the weight vector noise $\boldsymbol{\epsilon}_1(n)$ and the weight vector lag $\boldsymbol{\epsilon}_2(n)$ are essentially equal.

13.9 COMPARISON OF THE RLS AND LMS ALGORITHMS FOR TRACKING A NONSTATIONARY ENVIRONMENT

From the discussion presented in Section 9.14, where we considered the behavior of the LMS algorithm operating in a nonstationary environment, we recall two observations, assuming that the step-size parameter μ is small. First, the excess mean-squared error due to the weight vector noise is directly porportional to μ. Second, the average excess mean-squared error due to the weight vector lag is inversely proportional to μ. Thus, comparing the results described for the RLS algorithm with these results for the LMS algorithm, we deduce that the factor $(1 - \lambda)$ plays a role in the RLS algorithm similar to that of the step-size parameter μ in the LMS algorithm.[4]

A fundamental question is: "How do the RLS and LMS algorithms compare in tracking a nonstationary environment?" At first glance, it might be tempting to say that since the RLS algorithm has a faster rate of convergence than the LMS algorithm in the case of a stationary environment, therefore, it will track a nonstationary environment better than the LMS algorithm. Such an answer, however, is not justified because the tracking performance of an adaptive filtering algorithm is influenced not only by the rate of convergence (which is a transient characteristic) but also by fluctuation in the steady-state performance of the algorithm due to measurement and algorithm noise. Indeed, with both algorithms tuned to minimize the misadjustment at the filter output by a proper optimization of their forgetting rate (i.e., the exponential weighting factor λ for the RLS algorithm and the step-size parameter μ for the LMS algorithm), the LMS algorithm is found to have a superior tracking performance compared to the RLS algorithm (Bershad and Macchi, 1989; Eleftheriou and Falconer, 1986).

[4] This deduction is intuitively satisfying, for it is in perfect accord with memory considerations of the RLS and LMS algorithms. The memory of the RLS algorithm is inversely proportional to $(1 - \lambda)$, whereas the memory of the LMS algorithm is inversely proportional to μ. Thus, here again, we see that $(1 - \lambda)$ in the RLS algorithm plays an analogous role to μ in the LMS algorithm.

13.10 RELATIONSHIP BETWEEN THE RLS ALGORITHM AND KALMAN FILTER THEORY

The exponentially weighted RLS algorithm may be viewed as the *measurement update of the Kalman filter*. Indeed, the equations that define the RLS algorithm have the same basic mathematical structure as those that define a special form of the Kalman filter that we studied in Chapter 7. The special form of the Kalman filter that we have in mind assumes a *random-walk state model* in which the state noise vector has a correlation matrix that consists of a scaled version of the weight-error correlation matrix.

To be specific, consider an adaptive transversal filter that operates in a nonstationary environment. In such a situation, the error-performance surface of the filter changes its position randomly. To account for this random motion, we model the process equation of the filter by writing

$$\mathbf{w}_o^*(n+1) = \mathbf{w}_o^*(n) + \mathbf{v}(n) \tag{13.63}$$

Here, it is assumed that:

1. The state transition matrix $\mathbf{\Phi}(n+1, n)$ of the model equals the identity matrix.
2. The process noise vector, $\mathbf{v}(n)$, has zero mean and correlation matrix $\mathbf{Q}(n)$; hence, the tap-weight vector varies randomly about a mean value.

It is the second assumption that gives the model a nonstationary character. For the measurement equation of the model, we write

$$d(n) = \mathbf{u}^T(n)\mathbf{w}_o^*(n) + e_o(n) \tag{13.64}$$

where $\mathbf{u}(n)$ is the tap-input vector, and the estimation error $e_o(n)$ is assumed to have zero mean and mean-square value $J_{\min}$.

Comparing Eqs. (7.17) and (7.19) for the Kalman filter with Eqs. (13.63) and (13.64) for an optimum transversal filter with nonstationary inputs, respectively, we may make the identifications listed in Table 13.4. The presence of the process noise

TABLE 13.4 SUMMARY OF CORRESPONDENCES BETWEEN THE RLS ALGORITHM AND KALMAN FILTER

State model of the Kalman filter	Random-walk state model of the optimum transversal filter
$\mathbf{x}(n)$	$\mathbf{w}_o^*(n)$
$\mathbf{\Phi}(n+1, n)$	$\mathbf{I}$
$\mathbf{v}_1(n)$	$\mathbf{v}(n)$
$\mathbf{Q}_1(n)$	$\mathbf{Q}(n)$
$\mathbf{y}(n)$	$d(n)$
$\mathbf{C}(n)$	$\mathbf{u}^T(n)$
$\mathbf{v}_2(n)$	$e_o(n)$
$\mathbf{Q}_2(n)$	$J_{\min}$

vector $\mathbf{v}(n)$ makes the predicted weight-error correlation matrix and the filtered-weight error correlation matrix assume different values. We define the predicted weight-error correlation matrix as

$$\mathbf{K}(n, n-1) = E[\boldsymbol{\epsilon}(n, n-1)\boldsymbol{\epsilon}^H(n, n-1)] \tag{13.65}$$

where

$$\boldsymbol{\epsilon}(n, n-1) = \hat{\mathbf{w}}(n-1) - \mathbf{w}_o(n) \tag{13.66}$$

We define the filtered weight-error correlation matrix as

$$\mathbf{K}(n) = E[\boldsymbol{\epsilon}(n)\boldsymbol{\epsilon}^H(n)] \tag{13.67}$$

where

$$\boldsymbol{\epsilon}(n) = \hat{\mathbf{w}}(n) - \mathbf{w}_o(n) \tag{13.68}$$

For a nonstationary environment, in general, we find that $\mathbf{K}(n, n-1) \neq \mathbf{K}(n-1)$.

Thus, in view of the identifications listed in Table 13.4, the Kalman filter summarized in Table 7.1 takes on the following special form for adaptive transversal filters modeled by Eqs. (13.63) and (13.64):[5]

$$\mathbf{g}(n) = \mathbf{K}(n, n-1)\mathbf{u}(n)[\mathbf{u}^H(n)\mathbf{K}(n, n-1)\mathbf{u}(n) + J_{\min}]^{-1} \tag{13.69}$$

$$\alpha(n) = d(n) - \mathbf{u}^T(n)\hat{\mathbf{w}}^*(n-1) \tag{13.70}$$

$$\hat{\mathbf{w}}(n) = \hat{\mathbf{w}}(n-1) + \mathbf{g}(n)\alpha^*(n) \tag{13.71}$$

$$\mathbf{K}(n) = \mathbf{K}(n, n-1) - \mathbf{g}(n)\mathbf{u}^H(n)\mathbf{K}(n, n-1) \tag{13.72}$$

$$\mathbf{K}(n+1, n) = \mathbf{K}(n) + \mathbf{Q}(n) \tag{13.73}$$

The application of this algorithm requires knowledge of the correlation matrix $\mathbf{Q}(n)$. We consider two possibilities:

1. The correlation matrix of the process noise vector $\mathbf{v}(n)$ equals

$$\mathbf{Q}(n) = E[\mathbf{v}(n)\mathbf{v}^H(n)] = q\mathbf{I} \tag{13.74}$$

where q is a scalar and $\mathbf{I}$ is the identity matrix. This, in effect, assumes that the elements of $\mathbf{v}(n)$ constitute a set of independent white-noise processes, each having zero mean and variance q. Such a model is commonly referred to as a *random-walk state model* (Godard, 1974; Zhang and Haykin, 1983).

2. The correlation matrix of the process noise vector $\mathbf{v}(n)$ equals a scaled version of the filtered weight-error correlation matrix $\mathbf{K}(n)$, as shown by (Goodwin and Payne, 1977)

$$\mathbf{Q}(n) = E[\mathbf{v}(n)\mathbf{v}^H(n)] = (\lambda^{-1} - 1)\mathbf{K}(n) \tag{13.75}$$

[5] We note that the predicted and filtered versions of the weight-error correlation matrix are both defined with respect to the optimum weight vector $\mathbf{w}_o(n)$ rather than the state vector $\mathbf{w}_o^*(n)$. Accordingly, $\mathbf{K}(n, n-1)$ and $\mathbf{K}(n)$ play conjugate roles to their counterparts in the Kalman filter theory. Similarly, the Kalman gain vector $\mathbf{g}(n)$ and the correlation matrix $\mathbf{Q}(n)$ of the noise vector $\mathbf{v}(n)$ play conjugate roles to the Kalman gain $\mathbf{g}(n)$ and the correlation matrix $\mathbf{Q}_1(n)$ of the process noise vector in the Kalman filter theory, respectively.

where $(\lambda^{-1} - 1)$ is the scaling factor. We further assume the following two identities:

$$\mathbf{g}(n) = \mathbf{k}(n) \tag{13.76}$$

$$\mathbf{K}(n, n-1) = \lambda^{-1} J_{\min} \mathbf{P}(n-1) \tag{13.77}$$

Then, substituting Eqs. (13.75) to (13.77) in the Kalman filtering algorithm described by Eqs. (13.69) to (13.73), we get the standard RLS algorithm. We may therefore make the following observations:

(a) The RLS algorithm is a special form of the Kalman filter described by a random-walk state model in which the process noise vector $\mathbf{v}(n)$ has a correlation matrix equal to a scaled version of the filtered weight-error correlation matrix $\mathbf{K}(n)$.

(b) The gain vector $\mathbf{k}(n)$ in the RLS algorithm plays the same role as the Kalman gain $\mathbf{g}(n)$ in the Kalman filter.

(c) The matrix $\mathbf{P}(n) = \mathbf{\Phi}^{-1}(n)$ in the RLS algorithm is a scaled version of the predicted weight-error correlation matrix $\mathbf{K}(n+1, n)$.

(d) The exponential weighting factor λ and the minimum mean-squared error $J_{\min}$ determine the scaling factors in the relations described under points (a) and (c).

PROBLEMS

1. To permit a recursive implementation of the method of least squares, the window or weighting function $\beta(n, i)$ must have a suitable structure. Assume that $\beta(n, i)$ may be expressed as

$$\beta(n, i) = \lambda(i)\beta(n, i-1), \qquad i = 1, \ldots, n$$

where $\beta(n, n) = 1$. Hence, show that

$$\beta(n, i) = \prod_{k=i+1}^{n} \lambda^{-1}(k)$$

What is the form of $\lambda(k)$ for which $\beta(n, i) = \lambda^{n-i}$ is obtained?

2. Establish the validity of the matrix inversion lemma.

3. Consider a correlation matrix $\mathbf{\Phi}(n)$ defined by

$$\mathbf{\Phi}(n) = \mathbf{u}(n)\mathbf{u}^H(n) + \delta\mathbf{I}$$

where $\mathbf{u}(n)$ is a tap-input vector and δ is a small positive constant. Use the matrix inversion lemma to evaluate $\mathbf{P}(n) = \mathbf{\Phi}^{-1}(n)$.

4. Consider the modified definition of the correlation matrix $\mathbf{\Phi}(n)$ given in Eq. (13.28) which is reproduced here for convenience.

$$\mathbf{\Phi}(n) = \sum_{i=1}^{n} \lambda^{n-1}\mathbf{u}(i)\mathbf{u}^H(i) + \delta\lambda^n\mathbf{I}$$

where $\mathbf{u}(i)$ is the tap-input vector, λ is the exponential weighting factor, and δ is a small positive constant. Show that the use of this new definition for $\mathbf{\Phi}(n)$ leaves the equations that define the RLS algorithm completely unchanged.

5. Let $\alpha(n)$ denote the a priori estimation error

$$\alpha(n) = d(n) - \hat{\mathbf{w}}^H(n-1)\mathbf{u}(n)$$

where $d(n)$ is the desired response, $\mathbf{u}(n)$ is the tap-input vector, and $\hat{\mathbf{w}}(n-1)$ is the old estimate of the tap-weight vector. Let $e(n)$ denote the a posteriori estimation error

$$e(n) = d(n) - \hat{\mathbf{w}}^H(n)\mathbf{u}(n)$$

where $\hat{\mathbf{w}}(n)$ is the current estimate of the tap-weight vector. For complex-valued data, both $\alpha(n)$ and $e(n)$ are likewise complex valued. Show that the product $\alpha(n)e^*(n)$ is always real valued.

6. Given the initial conditions of Section 13.3 for the RLS algorithm, explain the reasons for the fact that the mean-square value of the a posteriori estimation error $e(n)$ attains a small value at $n = 1$ and then rises with increasing n.

7. Consider the RLS algorithm with the exponential weighting factor $\lambda \leq 1$. Show that the tap-weight vector $\hat{\mathbf{w}}(n)$ produced by the RLS algorithm has a bias that is approximately defined by

$$\begin{aligned}\mathbf{b}(n) &= E[\hat{\mathbf{w}}(n)] - \mathbf{w}_o \\ &= -\delta\lambda^n \mathbf{\Phi}^{-1}(n)\mathbf{w}_o, \qquad n \geq 1\end{aligned}$$

where $\mathbf{\Phi}^{-1}(n)$ is the inverse of the time-averaged correlation matrix $\mathbf{\Phi}(n)$, $\delta\mathbf{I} = \mathbf{\Phi}(0)$, and, $\mathbf{w}_o$ is the coefficient vector of the multiple linear regression model.

8. Consider a multiple linear regression model whose output is described by

$$d(i) = \mathbf{w}_o^H\mathbf{u}(i) + e_o(i)$$

where $\mathbf{w}_o$ is the M-by-1 regression vector of the model, and $\mathbf{u}(i)$ is the corresponding input vector. The measurement error process $\{e_o(i)\}$ is white with zero mean and variance σ^2. Let $\hat{\mathbf{w}}(n)$ denote the estimate of $\mathbf{w}_o$ that is produced by application of the RLS algorithm.

(a) Assuming that the number of iterations n is large enough for the effect of initial conditions to be ignored, show that the weight-error vector equals

$$\begin{aligned}\boldsymbol{\epsilon}(n) &= \hat{\mathbf{w}}(n) - \mathbf{w}_o \\ &= \mathbf{\Phi}^{-1}(n)\sum_{i=1}^{n}\lambda^{n-i}\mathbf{u}(i)e_o^*(i), \qquad n \geq M\end{aligned}$$

where

$$\mathbf{\Phi}(n) = \sum_{i=1}^{n}\lambda^{n-i}\mathbf{u}(i)\mathbf{u}^H(i)$$

(b) Using the result obtained in part (a), show that the weight-error correlation matrix equals

$$\begin{aligned}\mathbf{K}(n) &= E[\boldsymbol{\epsilon}(n)\boldsymbol{\epsilon}^H(n)] \\ &= \sigma^2\mathbf{\Phi}^{-1}(n)\left(\sum_{i=1}^{n}\lambda^{2(n-i)}\mathbf{u}(i)\mathbf{u}^H(i)\right)\mathbf{\Phi}^{-1}(n), \qquad n \geq M\end{aligned}$$

For the special case of $\lambda = 1$, what does this formula reduce to?

9. From the *law of large numbers*, we have

$$\sum_{i=1}^{n}\lambda^{n-i}\mathbf{u}(i)\mathbf{u}^H(i) \simeq \left(\sum_{i=1}^{n}\lambda^{n-1}\right)(\mathbf{R} + \delta\mathbf{R}), \qquad n \text{ large}$$

where λ is the exponential weighting factor, $\mathbf{R}$ is the ensemble-averaged correlation matrix

$$\mathbf{R} = E[\mathbf{u}(i)\mathbf{u}^H(i)]$$

and $\delta\mathbf{R}$ is composed of elements that have zero mean and variances that are small compared to the elements of $\mathbf{R}$. You may thus assume that

$$(\mathbf{R} + \delta\mathbf{R})^{-1} \simeq \mathbf{R}^{-1} - \mathbf{R}^{-1}\delta\mathbf{R}\mathbf{R}^{-1}$$

Hence, using these results in the formula for the excess mean-squared error

$$J_{ex}(n) = \text{tr}\{E[\mathbf{R}\mathbf{K}(n)]\}, \qquad n \text{ large}$$

show that the misadjustment produced by the exponentially weighted RLS algorithm equals

$$\mathcal{M} = \frac{J_{ex}(\infty)}{\sigma^2}$$

$$= \left(\frac{1 - \lambda}{1 + \lambda}\right)M$$

where M is the number of taps in the transversal filter.

10. Consider the first-order difference equation (13.60) that governs the time evolution of the weight vector lag $\boldsymbol{\epsilon}_2(n)$. Using z-transforms, solve this equation for $\boldsymbol{\epsilon}_2(n)$. Hence, show that $\boldsymbol{\epsilon}_2(n)$ is a geometrical converging vector process with a time constant τ_i equal to $(1 - \lambda)^{-1}$, where λ is the exponential weighting factor.

Computer-Oriented Problems

11. *Computer experiment on adaptive two-tap predictor using the RLS algorithm, with autoregressive input.* An autoregressive (AR) process $\{u(n)\}$ of order 2 is described by the difference equation

$$u(n) + a_1u(n - 1) + a_2u(n - 2) = v(n)$$

where $\{v(n)\}$ is a white-noise process of zero mean and variance σ_v^2. The AR parameters a_1 and a_2 and the variance σ_v^2 are assigned one of the following three sets of values:

(1) $a_1 = 0.100, \quad a_2 = -0.8, \quad \sigma_v^2 = 0.2700$
(2) $a_1 = 0.1636, \quad a_2 = -0.8, \quad \sigma_v^2 = 0.1190$
(3) $a_1 = 0.1960, \quad a_2 = -0.8, \quad \sigma_v^2 = 0.0140$

The process $\{u(n)\}$ is applied to an adaptive two-tap predictor that uses the RLS algorithm to adjust the tap weights of the predictor.

(a) Use a computer to generate a 256-sample sequence to represent the AR process $\{u(n)\}$ for parameter set (1). You may do this by using a random generator subroutine for $\{v(n)\}$. Hence, plot an ensemble-averaged learning curve for the predictor by averaging over 200 independent trials of the experiment. By averaging over the least 200 iterations of the ensemble-averaged learning curve, estimate the minimum mean-squared error produced by the RLS algorithm. Compare your result with theory.

(b) For parameter set (1), also compute the mean values of the tap weights $\hat{w}_1(\infty)$ and $\hat{w}_2(\infty)$ by ensemble-averaging the steady-state values of the tap weights produced by the RLS algorithm over 200 independent trials of the experiment. Compare your results with theory.

(c) Repeat computations (a) and (b) for parameter set (2), and again for parameter set (3).

(d) Comment on the results of parts (a) to (c).

(e) Suppose that the AR parameter a_1 is made negative; otherwise, its magnitude is left as specified under parameter sets (1), (2), and (3). How would you assess the effect of this change on the performance of the RLS algorithm? Justify your answer.

12. *Computer experiment on adaptive beamforming*. Repeat the computer experiment of Problem 19, Chapter 9, on the adaptive beamformer, using the RLS algorithm. The environment (in which the beamformer operates) consists of a target signal and a single source of interference. Their angles of arrival are as follows:

$$\text{target:} \quad \theta_1 = \sin^{-1}(0.2)$$

$$\text{interference:} \quad \theta_2 = -\sin^{-1}(0.4)$$

The source of interference is noncoherent with the target. The signal-to-noise ratio (SNR) is fixed at 10 dB.

Initialize the RLS algorithm with $\delta = 0.004$. Hence, perform the following computations:

(a) For interference-to-noise ratio INR = 40, 30, 20 dB, compute the eigenvalue spread of the correlation matrix of the signals at the output ports of the beamforming network.

(b) Compute the adapted pattern of the beamformer after 10, 25, and 50 iterations of the algorithm, with the interference-to-noise ratio INR = 40 dB. Repeat your computations for INR = 30, 20 dB. Comment on your results.

(c) By ensemble-averaging over 200 independent trials of the experiment, plot the learning curves of the RLS algorithm for INR = 40, 30, 20 dB. In each case, plot the learning curve up to 225 iterations of the algorithm.

(d) By time-averaging each learning curve over the last 25 iterations, compute the steady-state value of the mean-squared error for INR = 40, 30, 20 dB.

(e) Using the minimum mean-squared error from theory, and the results of part (d), compute the misadjustment produced by the RLS algorithm for INR = 40, 30, 20 dB.

(f) Repeat parts (c), (d), and (e) using the LMS algorithm for which the step-size parameter $\mu = 4 \times 10^{-7}$. For the case when INR = 40 dB, also plot the learning curve of the LMS algorithm with $\mu = 4 \times 10^{-8}$, up to 500 iterations of the algorithm; as before, perform the ensemble averaging over 200 independent repetitions of the experiment, and calculate the mean-squared error by time-averaging the learning curve over the last 25 iterations

(g) Compare the results obtained for the RLS algorithm with those obtained for the LMS algorithm.

14

Recursive Least-Squares Systolic Arrays

The idea of *systolic arrays* was developed by Kung and Leiserson (1978), and with it a new and exciting avenue for matrix-oriented adaptive filtering emerged. A systolic array consists of an array of individual *processing cells* that are arranged in the form of a regular structure. Each cell in the array is provided with local memory of its own, and each cell is connected only to its nearest neighbors. The array is designed such that regular streams of data are clocked through it in a highly rhythmic fashion, much like the pumping action of the human heart, hence the name "systolic" (Kung, 1982).

Systolic arrays are well suited for implementing complex signal-processing algorithms, particularly when the requirement is to operate in *real time* and at *high data bandwidths*. Indeed, several systolic architectures have been developed for solving the *recursive least-squares* (RLS) problem. In this chapter we study some of them. Fundamental to this study is an *orthogonal triangularization process* known as the *QR-decomposition,* which is implemented recursively by a sequence of *Givens rotations*. Such an approach not only leads, in a natural way, to the delineation of a systolic architecture for solving the RLS problem, but it also offers good *numerical stability* that makes the solution all the more important. We begin our study by reviewing the *linear least-squares problem,* with this new approach in mind.

14.1 SOME PRELIMINARIES

The terminology that we use in this chapter follows a format similar to that in Chapter 10 with some minor modifications in that we assume the use of prewindowing here. Suppose that we are given a time series consisting of the observations $u(1)$,

$u(2), \ldots, u(n)$. We use a subset of this time series to form an M-by-1 *input vector*[1] $\mathbf{u}(i)$, defined by

$$\mathbf{u}^T(i) = [u(i), u(i-1), \ldots, u(i-M+1)] \tag{14.1}$$

We use this input vector to make an estimate of some *desired response* $d(i)$ as the inner product $\mathbf{w}^H(n)\mathbf{u}(i)$, where $\mathbf{w}(n)$ is an M-by-1 *weight vector* whose transpose equals

$$\mathbf{w}^T(n) = [w_1(n), w_2(n), \ldots, w_M(n)] \tag{14.2}$$

We assume that the weight vector $\mathbf{w}(n)$ is held constant during the observation interval $1 \leq i \leq n$. We define the *estimation error* $e(i)$ as

$$e(i) = d(i) - \mathbf{w}^H(n)\mathbf{u}(i) \tag{14.3}$$

According to the *exponentially weighted, prewindowed method,* we minimize the cost function

$$\mathscr{E}(n) = \sum_{i=1}^{n} \lambda^{n-i} |e(i)|^2 \tag{14.4}$$

where λ is the exponential weighting factor: $0 < \lambda \leq 1$. Let the n-by-M matrix $\mathbf{A}(n)$ denote the *data matrix* that is defined by (using the notation of previous chapters)

$$\mathbf{A}^H(n) = [\mathbf{u}(1), \mathbf{u}(2), \ldots, \mathbf{u}(M), \ldots, \mathbf{u}(n)]$$
$$= \begin{bmatrix} u(1) & u(2) & \cdots & u(M) & \cdots & u(n) \\ 0 & u(1) & \cdots & u(M-1) & \cdots & u(n-1) \\ \vdots & \vdots & \ddots & \vdots & \ddots & \vdots \\ 0 & 0 & \cdots & u(1) & \cdots & u(n-M+1) \end{bmatrix} \tag{14.5}$$

Let the n-by-1 vector $\boldsymbol{\epsilon}(n)$ denote the *error vector,* defined by

$$\boldsymbol{\epsilon}^H(n) = [e(1), e(2), \ldots, e(n)] \tag{14.6}$$

Let the n-by-1 vector $\mathbf{d}(n)$ denote the *desired data vector,* defined by

$$\mathbf{d}^H(n) = [d(1), d(2), \ldots, d(n)] \tag{14.7}$$

We may then redefine the cost function $\mathscr{E}(n)$ of Eq. (14.4) in matrix notation as

$$\mathscr{E}(n) = \boldsymbol{\epsilon}^H(n)\boldsymbol{\Lambda}(n)\boldsymbol{\epsilon}(n) \tag{14.8}$$

where $\boldsymbol{\Lambda}$ is the n-by-n *exponential weighting matrix*

$$\boldsymbol{\Lambda}(n) = \text{diag}[\lambda^{n-1}, \lambda^{n-2}, \ldots, 1] \tag{14.9}$$

[1] We refer to $\mathbf{u}(i)$ as the input vector rather than the tap-input vector, and similarly, we refer to $\mathbf{w}(n)$ as the weight vector rather than the tap-weight vector simply because we wish to reserve the use of tap-input and tap-weight vectors for the transversal (i.e., tapped-delay-line) filter.

The inclusion of the exponential weighting matrix $\mathbf{\Lambda}(n)$ has the effect of progressively weighting against the preceding columns of the data matrix $\mathbf{A}^H(n)$ in favor of the last column. The last column of $\mathbf{A}^H(n)$ corresponds to the input vector $\mathbf{u}(n)$ at time n, for which the weighting factor is unity.

We may also express the cost function $\mathscr{E}(n)$ as a squared Euclidean norm

$$\mathscr{E}(n) = \| \mathbf{\Lambda}^{1/2}(n)\boldsymbol{\epsilon}(n) \|^2 \tag{14.10}$$

The problem we wish to solve is to find the value of the weight vector $\hat{\mathbf{w}}(n)$ that minimizes the cost function $\mathscr{E}(n)$, using the form given in Eq. (14.10).

14.2 SOLUTION OF THE LINEAR LEAST-SQUARES PROBLEM USING QR-DECOMPOSITION

The essence of the linear least-squares estimation problem is to choose the weight vector $\mathbf{w}(n)$ [for successive values of time $n > M$, for which the data matrix $\mathbf{A}(n)$ is assumed to have full rank] so as to minimize the cost function $\mathscr{E}(n)$, defined in Eq. (14.10) as the squared norm of the weighted error vector $\mathbf{\Lambda}^{1/2}(n)\boldsymbol{\epsilon}(n)$. We solve the problem by using the method of orthogonal triangularization that offers good numerical stability.

The norm of a vector is unaffected by premultiplication by a unitary matrix (see Problem 1). Hence, we may express the cost function $\mathscr{E}(n)$ as

$$\mathscr{E}(n) = \| \mathbf{Q}(n)\mathbf{\Lambda}^{1/2}(n)\boldsymbol{\epsilon}(n) \|^2 \tag{14.11}$$

where $\mathbf{Q}(n)$ is an n-by-n unitary matrix. The exact role of this unitary matrix will be defined presently. The error vector equals

$$\boldsymbol{\epsilon}(n) = \mathbf{d}(n) - \mathbf{A}(n)\mathbf{w}(n) \tag{14.12}$$

By using this expression for the error vector $\boldsymbol{\epsilon}(n)$, we may express the matrix product $\mathbf{Q}(n)\mathbf{\Lambda}^{1/2}(n)\boldsymbol{\epsilon}(n)$ as

$$\mathbf{Q}(n)\mathbf{\Lambda}^{1/2}(n)\boldsymbol{\epsilon}(n) = \mathbf{Q}(n)\mathbf{\Lambda}^{1/2}(n)\mathbf{d}(n) - \mathbf{Q}(n)\mathbf{\Lambda}^{1/2}(n)\mathbf{A}(n)\mathbf{w}(n) \tag{14.13}$$

According to Eq. (14.3), the n-by-n unitary matrix $\mathbf{Q}(n)$ operates on the exponentially weighted versions of $\mathbf{A}(n)$ and $\mathbf{d}(n)$.

For any value of $n > M$, we assume that the unitary matrix $\mathbf{Q}(n)$ is generated in such a way that it applies an *orthogonal triangularization* to the weighted data matrix $\mathbf{\Lambda}^{1/2}(n)\mathbf{A}(n)$, as shown by

$$\mathbf{Q}(n)\mathbf{\Lambda}^{1/2}(n)\mathbf{A}(n) = \begin{bmatrix} \mathbf{R}(n) \\ \mathbf{O} \end{bmatrix}, \qquad n > M \tag{14.14}$$

where $\mathbf{R}(n)$ is an M-by-M *upper triangular matrix,* and $\mathbf{O}$ *is the* $(n - M)$-by-M null matrix. According to Eq. (14.14), the n-by-M matrix $\mathbf{\Lambda}^{1/2}(n)\mathbf{A}(n)$ may be expressed as a product of the unitary matrix $\mathbf{Q}^H(n)$ and an n-by-M upper triangular matrix $[\mathbf{R}(n), \ \mathbf{O}]^T$; such a factorization is referred to in the literature as the *QR-decomposition*.

In any event, through the use of the QR-decomposition of Eq. (14.14), we may express the second term on the right side of Eq. (14.13) as

$$
\begin{aligned}
\mathbf{Q}(n)\mathbf{\Lambda}^{1/2}(n)\mathbf{A}(n)\mathbf{w}(n) &= \begin{bmatrix} \mathbf{R}(n) \\ \mathbf{O} \end{bmatrix}\mathbf{w}(n) \\
&= \begin{bmatrix} \mathbf{R}(n)\mathbf{w}(n) \\ \cdots\cdots \\ \mathbf{0}_{n-M} \end{bmatrix}, \qquad n > M
\end{aligned}
\tag{14.15}
$$

where $\mathbf{0}_{n-M}$ is the $(n - M)$-by-1 null vector.

Suppose, next, the n-by-n unitary matrix $\mathbf{Q}(n)$ is partitioned as

$$
\mathbf{Q}(n) = \begin{bmatrix} \mathbf{F}(n) \\ \cdots\cdots \\ \mathbf{S}(n) \end{bmatrix}, \qquad n > M \tag{14.16}
$$

where $\mathbf{F}(n)$ is an M-by-n matrix consisting of the first M rows of $\mathbf{Q}(n)$, and $\mathbf{S}(n)$ is an $(n - M)$-by-n matrix consisting of the remaining rows. Accordingly, we may express the first term on the right side of Eq. (14.13) as

$$
\begin{aligned}
\mathbf{Q}(n)\mathbf{\Lambda}^{1/2}(n)\mathbf{d}(n) &= \begin{bmatrix} \mathbf{F}(n) \\ \cdots\cdots \\ \mathbf{S}(n) \end{bmatrix}\mathbf{\Lambda}^{1/2}(n)\mathbf{d}(n) \\
&= \begin{bmatrix} \mathbf{p}(n) \\ \cdots\cdots \\ \mathbf{v}(n) \end{bmatrix}, \qquad n > M
\end{aligned}
\tag{14.17}
$$

where $\mathbf{p}(n)$ is an M-by-1 vector defined by

$$
\mathbf{p}(n) = \mathbf{F}(n)\mathbf{\Lambda}^{1/2}(n)\mathbf{d}(n), \qquad n > M \tag{14.18}
$$

and $\mathbf{v}(n)$ is an $(n - M)$-by-1 vector defined by

$$
\mathbf{v}(n) = \mathbf{S}(n)\mathbf{\Lambda}^{1/2}(n)\mathbf{d}(n), \qquad n > M \tag{14.19}
$$

Note that the dimension of the vector $\mathbf{p}(n)$ is fixed at the value M for all n. On the other hand, the dimension of the vector $\mathbf{v}(n)$ is variable in that it increases with n linearly. Thus, using Eqs. (14.15) and (14.17), we may rewrite Eq. (14.13) as

$$
\begin{aligned}
\mathbf{Q}(n)\mathbf{\Lambda}^{1/2}(n)\boldsymbol{\epsilon}(n) &= \begin{bmatrix} \mathbf{p}(n) \\ \cdots\cdots \\ \mathbf{v}(n) \end{bmatrix} - \begin{bmatrix} \mathbf{R}(n)\mathbf{w}(n) \\ \cdots\cdots \\ \mathbf{0}_{n-M} \end{bmatrix} \\
&= \begin{bmatrix} \mathbf{p}(n) - \mathbf{R}(n)\mathbf{w}(n) \\ \cdots\cdots\cdots\cdots \\ \mathbf{v}(n) \end{bmatrix}
\end{aligned}
\tag{14.20}
$$

To solve the linear least-squares problem, we have to choose the weight vector $\mathbf{w}(n)$ so as to minimize the cost function $\mathscr{E}(n)$ or, equivalently, the squared norm of the n-by-1 vector $\mathbf{Q}(n)\mathbf{\Lambda}^{1/2}(n)\boldsymbol{\epsilon}(n)$ [see Eq. (14.11)]. Let $\hat{\mathbf{w}}(n)$ denote the *optimum* value of the weight vector that solves this minimization problem. From Eq. (14.20), we immediately deduce that the squared norm of $\mathbf{Q}(n)\mathbf{\Lambda}^{1/2}(n)\boldsymbol{\epsilon}(n)$ is indeed minimized when $\hat{\mathbf{w}}(n)$ satisfies the condition

$$\mathbf{R}(n)\hat{\mathbf{w}}(n) = \mathbf{p}(n), \qquad n > M \tag{14.21}$$

Correspondingly, the minimum value of the sum of error squares is

$$\mathscr{E}_{\min}(n) = \|\mathbf{v}(n)\|^2, \qquad n > M \tag{14.22}$$

The M-by-M matrix $\mathbf{R}(n)$ is upper triangular and invertible (if it is nonsingular). Then, Eq. (14.21) may readily be solved for the optimum weight vector $\hat{\mathbf{w}}(n)$ by a process of *back substitution*. Moreover, we may use the weight vector thus computed to determine the least-squares value of the estimation error $e(i)$ over the observation interval $1 \le i \le n$, simply by evaluating the inner product $\hat{\mathbf{w}}^H(n)\mathbf{u}(n)$, where $\mathbf{u}(n)$ is the input vector. From Eq. (14.20) we note that the column vector $\mathbf{v}(n)$ is zero for $1 < n < M$. Correspondingly, $\mathscr{E}_{\min}$ is zero for the same interval.

Earlier we indicated that the use of the QR-decomposition for solving the linear least-squares problem offers good numerical stability. This important property is a direct consequence of applying the QR-decomposition, an orthogonal triangularization process, *directly* to the data matrix. Basically, through the use of the QR-decomposition, we have transformed[2] the system of normal equations [see Eq. (10.35)].

$$\mathbf{\Phi}(n)\hat{\mathbf{w}}(n) = \boldsymbol{\theta}(n)$$

into the equivalent, but numerically easier to compute and more stable form given in Eq. (14.21). Indeed, the matrix equation (14.21) is basic to the development of recursive least-squares systolic arrays, just as the set of normal equations is basic to the development of the standard RLS algorithm using transversal filters. Note that according to the QR-decomposition of Eq. (14.14) that led to the development of Eq. (14.21), the correlation matrix $\mathbf{\Phi}(n)$ equals the matrix-matrix product $\mathbf{R}^H(n)\mathbf{R}(n)$, and the cross-correlation vector $\boldsymbol{\theta}(n)$ equals the matrix-vector product $\mathbf{R}^H(n)\mathbf{p}(n)$ (see Problem 2).

14.3 GIVENS ROTATION

In Chapter 11 when dealing with the singular-value decomposition problem, we introduced an elementary transformation in the form of a *plane rotation* as a basic tool for computing the singular-value decomposition of a data matrix. This same plane rotation, defined in terms of two parameters (one a real cosine and the other a complex sine), provides us with the tool we need to perform the QR-decomposition described in the preceding section. In this chapter, we refer to the plane rotation as the *Givens rotation* (Givens, 1958). Through successive applications of the Givens rota-

[2] See Yang and Böhme (1989).

tion, we may develop a very efficient algorithm for solving the linear least-squares problem, whereby the *orthogonal triangularization of the data matrix is recursively updated as each new set of data enters the computation.*

The matrix structure that arises in such a study of the linear least-squares problem, and for which the Givens rotation procedure is particularly well suited, is *partially triangularized.* To be specific, consider an n-by-M complex-valued matrix $\mathbf{Y}$, where $n > M$ and M is some fixed number. We assume that the first M rows of matrix $\mathbf{Y}$ constitute an upper triangular matrix $\mathbf{R}$, followed by an $(n - M - 1)$-by-M null matrix, and then finally, a row of M nonzero elements. The requirement is to annihilate all the nonzero elements in the last row of the matrix $\mathbf{Y}$. Suppose we wish to annihilate the kth element, say, in the last row (i.e., row n) of matrix $\mathbf{Y}$. We accomplish this annilation by premultiplying the matrix $\mathbf{Y}$ by the n-by-n transformation $\mathbf{J}_k(n)$, that consists of an identity matrix, except for *four strategic elements* located at the (k, n)-columns and the (k, n)-rows. Specifically, we have

$$\mathbf{J}_k(n) = \begin{bmatrix} 1 & & & & & & & \\ & 1 & & & & \mathbf{O} & & \\ & & \ddots & & & & & \\ & & & c & 0 & \cdots & 0 & s^* \\ & & & 0 & 1 & \cdots & 0 & 0 \\ & \mathbf{O} & & \vdots & \vdots & & \vdots & \vdots \\ & & & 0 & 0 & \cdots & 1 & 0 \\ & & & -s & 0 & \cdots & 0 & c \end{bmatrix} \begin{matrix} \leftarrow \text{row } 1 \\ \\ \\ \leftarrow \text{row } k \\ \\ \\ \\ \leftarrow \text{row } n \end{matrix} \tag{14.23}$$

$\uparrow$ column 1 $\quad$ $\uparrow$ column k $\quad$ $\uparrow$ column n

The two parameters of the Givens rotation in Eq. (14.23), namely, the real cosine parameter c and the complex sine parameter s, are defined by

$$c = \cos \theta \tag{14.24}$$

$$s = \sin \theta e^{j\phi} \tag{14.25}$$

where θ and ϕ are rotation angles in a complex two-dimensional plane, and

$$c^2 + |s|^2 = 1 \tag{14.26}$$

For the purpose of our present discussion, we need only identify the elements y_{kk} and y_{nk} of the matrix $\mathbf{Y}$ at the (k, k)th and (n, k)th locations, respectively. Thus, premultiplying the matrix $\mathbf{Y}$ by the Givens rotation $\mathbf{J}_k(n)$ we find that these two elements are transformed as follows, respectively,

$$y_{kk} \rightarrow cy_{kk} + s^*y_{nk} \tag{14.27}$$

$$y_{nk} \rightarrow -sy_{kk} + cy_{nk} \tag{14.28}$$

Accordingly, the element y_{nk} (i.e., the kth element on row n of the matrix $\mathbf{Y}$) is *annihilated* if the rotation parameters c and s are chosen to satisfy the condition

$$-sy_{kk} + cy_{nk} = 0 \tag{14.29}$$

In general, the elements y_{kk} and y_{nk} are both complex. Hence, solving Eqs. (14.26) and (14.29) for the cosine–sine pair (c, s) in terms of the elements y_{nk} and y_{kk}, both assumed to be known, we get

$$c = \frac{|y_{kk}|}{\sqrt{|y_{kk}|^2 + |y_{nk}|^2}} \tag{14.30}$$

$$s = \frac{y_{nk}}{y_{kk}} c \tag{14.31}$$

Correspondingly, in accordance with Eq. (14.27), the element y_{kk} is transformed as follows:

$$y_{kk} \rightarrow \frac{y_{kk}}{c} \tag{14.32}$$

The following three observations are noteworthy:

1. Equation (14.30) yields a *positive real* value for the cosine parameter c. Accordingly, the phase of y_{kk} is unchanged by the transformation. In particular, if we start with a *real* y_{kk} (and in QR-decomposition we normally specify that the diagonal elements of the upper triangular matrix $\mathbf{R}$ be real and positive, which makes the decomposition unique), then the transformation maintains that property.
2. The Givens transformation $\mathbf{J}_k(n)$ is a *unitary matrix*; that is,

$$\mathbf{J}_k(n)\mathbf{J}_k^H(n) = I, \qquad \text{for all } k \text{ and } n \tag{14.33}$$

3. When the matrix $\mathbf{Y}$ is premultiplied by the transformation $\mathbf{J}_k(n)$, all the elements of $\mathbf{Y}$ are unaffected by the transformation, except for those located on rows and columns k and n.

We may now describe the procedure for the orthogonal triangularization of a *partially triangularized matrix* $\mathbf{Y}$ that is characterized as follows:

1. The square matrix at the top end of matrix $\mathbf{Y}$ is an upper triangular matrix.
2. The last row of matrix $\mathbf{Y}$ consists of nonzero elements.
3. The remaining elements of matrix $\mathbf{Y}$ are all zero.

First we annihilate the leading element in the last row of matrix $\mathbf{Y}$ by applying the Givens rotation to operate on the first and last rows of matrix $\mathbf{Y}$. Next, we annihilate the second element in the last row of the transformed matrix by applying the Givens rotation to operate on the second and last rows. We continue in this fashion until all the M elements in the last row of matrix $\mathbf{Y}$ have been annihilated. Let $\mathbf{J}_1(n)$, $\mathbf{J}_2(n)$,

$\ldots, \mathbf{J}_M(n)$ denote this sequence of Givens rotations in a multidimensional complex plane, which transforms the matrix $\mathbf{Y}$ into upper triangular form as shown by

$$\mathbf{J}_M(n) \cdots \mathbf{J}_2(n)\mathbf{J}_1(n)\mathbf{Y} = \begin{bmatrix} \mathbf{R} \\ \mathbf{O} \\ \mathbf{0}^T \end{bmatrix} \tag{14.34}$$

where $\mathbf{R}$ is an upper triangular M-by-M matrix; $\mathbf{O}$ is the $(n - M - 1)$-by-M null matrix and $\mathbf{0}^T$ is the 1-by-M null row vector. Equation (14.34) describes the desired orthogonal triangularization process. In the next section, we exploit the orthogonal triangularization of the prewindowed exponentially weighted data matrix to develop a new recursive solution to the linear least-squares problem.

14.4 SOLUTION OF THE RECURSIVE LEAST-SQUARES PROBLEM USING A SEQUENCE OF GIVENS ROTATIONS

The next issue we wish to consider is to make the procedure described in Section 14.2 solve the linear least-squares problem recursively. Suppose that at time $n - 1$, the $(n - 1)$-by-M weighted data matrix $\mathbf{\Lambda}^{1/2}(n - 1)\mathbf{A}(n - 1)$ has been already reduced to an orthogonal triangularized form, as shown by the QR-decomposition:

$$\mathbf{Q}(n - 1)\mathbf{\Lambda}^{1/2}(n - 1)\mathbf{A}(n - 1) = \begin{bmatrix} \mathbf{R}(n - 1) \\ \cdots\cdots \\ \mathbf{O} \end{bmatrix} \tag{14.35}$$

where $\mathbf{R}(n - 1)$ is an M-by-M upper triangular matrix and $\mathbf{O}$ is the $(n - M - 1)$-by-M null matrix. The $(n - 1)$-by-M data matrix $\mathbf{A}(n - 1)$ represents all the prewindowed data collected up to and including time $n - 1$. At time n, the new input $u(n)$ becomes available for processing. The addition of this new sample to the available data increases the number of rows contained in the updated data matrix $\mathbf{A}(n)$ by 1, as shown by

$$\mathbf{A}(n) = \begin{bmatrix} \mathbf{A}(n - 1) \\ \cdots\cdots \\ \mathbf{u}^H(n) \end{bmatrix} \tag{14.36}$$

where the input vector $\mathbf{u}(n)$ is defined by

$$\mathbf{u}^T(n) = [u(n), u(n - 1), \ldots, u(n - M + 1)] \tag{14.37}$$

To compute the QR-decomposition of the updated data matrix $\mathbf{A}(n)$, given the QR-decomposition of the old data matrix $\mathbf{A}(n - 1)$ and the new input vector $\mathbf{u}(n)$, we clearly need a *recursion* that relates the updated value of the unitary matrix, $\mathbf{Q}(n)$, to the old value of the unitary matrix, $\mathbf{Q}(n - 1)$. We now derive this recursion by using a sequence of Givens rotations.

First, we define a new n-by-n unitary matrix $\overline{\mathbf{Q}}(n - 1)$ related to the old value

$\mathbf{Q}(n-1)$ as follows:

$$\overline{\mathbf{Q}}(n-1) = \begin{bmatrix} \mathbf{Q}(n-1) & \mathbf{0}_{n-1} \\ \mathbf{0}_{n-1}^T & 1 \end{bmatrix} \tag{14.38}$$

where $\mathbf{0}_{n-1}$ is the $(n-1)$-by-1 vector. The last column of $\overline{\mathbf{Q}}(n-1)$ represents a *pinning vector,* all of whose elements are zero except for the last element, which equals 1. We use the matrix $\overline{\mathbf{Q}}(n-1)$ to apply a linear transformation to the weighted (new) data matrix $\mathbf{\Lambda}^{1/2}(n)\mathbf{A}(n)$. The data matrix $\mathbf{A}(n)$ is related to the old value $\mathbf{A}(n-1)$ by Eq. (14.36). Correspondingly, the n-by-n exponential weighting matrix $\mathbf{\Lambda}(n)$ is related to the old value $\mathbf{\Lambda}(n-1)$ by

$$\mathbf{\Lambda}(n) = \begin{bmatrix} \lambda\mathbf{\Lambda}(n-1) & \mathbf{0}_{n-1} \\ \mathbf{0}_{n-1}^T & 1 \end{bmatrix} \tag{14.39}$$

This follows directly from the definition of the exponential weighting matrix given in Eq. (14.9). Hence, premultiplying the weighted data matrix $\mathbf{\Lambda}^{1/2}(n)\mathbf{A}(n)$ by the unitary matrix $\overline{\mathbf{Q}}(n-1)$, and using Eqs. (14.36) to (14.39), we get

$$\overline{\mathbf{Q}}(n-1)\mathbf{\Lambda}^{1/2}(n)\mathbf{A}(n) = \begin{bmatrix} \lambda^{1/2}\mathbf{Q}(n-1)\mathbf{\Lambda}^{1/2}(n-1)\mathbf{A}(n-1) \\ \mathbf{u}^H(n) \end{bmatrix} \tag{14.40}$$

However, earlier we assumed that the unitary matrix $\mathbf{Q}(n-1)$ performs orthogonal transformation on $\mathbf{\Lambda}^{1/2}(n-1)\mathbf{A}(n-1)$, as shown in Eq. (14.35). Hence, we may simplify Eq. (14.40) as

$$\overline{\mathbf{Q}}(n-1)\mathbf{\Lambda}^{1/2}(n)\mathbf{A}(n) = \begin{bmatrix} \lambda^{1/2}\mathbf{R}(n-1) \\ \mathbf{O} \\ \mathbf{u}^H(n) \end{bmatrix} \tag{14.41}$$

The n-by-M matrix on the right side of Eq. (14.41) is partially triangularized in that the last row of the matrix consists of nonzero elements. Hence, the matrix product $\overline{\mathbf{Q}}(n-1)\mathbf{\Lambda}^{1/2}(n)\mathbf{A}(n)$ is in the form of matrix $\mathbf{Y}$ discussed in Section 14.3. We may therefore transform this matrix into an orthogonal triangularized form through successive applications of the Givens rotation by following the procedure described in Section 14.3. Let $\mathbf{J}_1(n)$, $\mathbf{J}_2(n)$, . . . , $\mathbf{J}_M(n)$ denote the sequence of Givens rotations that are applied to annihilate all M elements contained in the last row of the matrix $\overline{\mathbf{Q}}(n-1)\mathbf{\Lambda}^{1/2}(n)\mathbf{A}(n)$, one by one. Let the unitary matrix $\mathbf{T}(n)$ denote the combined effect of this sequence of Givens rotations:

$$\mathbf{T}(n) = \mathbf{J}_M(n) \cdots \mathbf{J}_2(n)\mathbf{J}_1(n) \tag{14.42}$$

Accordingly, we may fully triangularize Eq. (14.41) as

$$
\mathbf{T}(n)\overline{\mathbf{Q}}(n-1)\mathbf{\Lambda}^{1/2}(n)\mathbf{A}(n) = \begin{bmatrix} \mathbf{R}(n) \\ \cdots\cdots \\ \mathbf{O} \\ \cdots\cdots \\ \mathbf{0}_M^T \end{bmatrix} \tag{14.43}
$$

where $\mathbf{O}$ is the $(n - M - 1)$-by-M null matrix, and $\mathbf{0}_M^T$ is the 1-by-M null row vector.

Substituting Eq. (14.41) in (14.43), we get (Gill et al., 1974)

$$
\begin{bmatrix} \mathbf{R}(n) \\ \cdots\cdots \\ \mathbf{O} \\ \mathbf{0}_M^T \end{bmatrix} = \mathbf{T}(n) \begin{bmatrix} \lambda^{1/2}\mathbf{R}(n-1) \\ \cdots\cdots\cdots \\ \mathbf{O} \\ \mathbf{u}^H(n) \end{bmatrix} \tag{14.44}
$$

Hence, given the unitary matrix $\mathbf{T}(n)$ and the new value $\mathbf{u}(n)$ of the input vector, we may use this recursion to time update the M-by-M upper triangular matrix $\mathbf{R}(n)$.

From Eq. (14.43) we also deduce that at time n the updated value of the unitary matrix in the QR-decomposition of the weighted data matrix, updated in both time and order, equals

$$
\begin{aligned}
\mathbf{Q}(n) &= \mathbf{T}(n)\overline{\mathbf{Q}}(n-1) \\
&= \mathbf{T}(n)\left[\begin{array}{c:c} \mathbf{Q}(n-1) & \mathbf{0}_{n-1} \\ \cdots\cdots\cdots & \cdots\cdots \\ \mathbf{0}_{n-1}^T & 1 \end{array}\right]
\end{aligned} \tag{14.45}
$$

Equation (14.45) represents a standard updating technique in the recursive computation of QR-decomposition. In particular, it shows that given the $(n - 1)$-by-$(n - 1)$ unitary matrix $\mathbf{Q}(n - 1)$ that performs the QR-decomposition of the weighted data matrix at time $n - 1$ and given the sequence of Givens transformations represented by the unitary matrix $\mathbf{T}(n)$, we may compute the updated unitary matrix $\mathbf{Q}(n)$. Let the desired response vector $\mathbf{d}(n)$ be partitioned as follows:

$$
\mathbf{d}(n) = \begin{bmatrix} \mathbf{d}(n-1) \\ \cdots\cdots \\ d^*(n) \end{bmatrix} \tag{14.46}
$$

where $d(n)$ is the value of the desired response that becomes available at time n. Then, using Eqs. (14.39), (14.45), and (14.46), we get

$$
\mathbf{Q}(n)\mathbf{\Lambda}^{1/2}(n)\mathbf{d}(n) = \mathbf{T}(n)\begin{bmatrix} \lambda^{1/2}\mathbf{Q}(n-1)\mathbf{\Lambda}^{1/2}(n-1)\mathbf{d}(n-1) \\ \cdots\cdots\cdots\cdots\cdots\cdots\cdots \\ d^*(n) \end{bmatrix} \tag{14.47}
$$

But, by definition, we have [see Eq. (14.17)]

$$
\mathbf{Q}(n-1)\mathbf{\Lambda}^{1/2}(n-1)\mathbf{d}(n-1) = \begin{bmatrix} \mathbf{p}(n-1) \\ \mathbf{v}(n-1) \end{bmatrix}
$$

TABLE 14.1 SUMMARY OF THE QRD-RLS ALGORITHM

1. Initialize the orthogonal triangularization procedure by setting $\mathbf{R}(0) = \mathbf{O}$ and $\mathbf{p}(0) = \mathbf{0}$. The exact initialization procedure occupies the period $0 \le n \le M$.
2. For $n > M$, perform the following updates:
   (a) Update the M-by-M upper triangular matrix $\mathbf{R}(n)$ by using the recursion

$$\begin{bmatrix} \mathbf{R}(n) \\ \cdots\cdots \\ \mathbf{O} \end{bmatrix} = \mathbf{T}(n) \begin{bmatrix} \lambda^{1/2}\mathbf{R}(n-1) \\ \cdots\cdots \\ \mathbf{O} \\ \mathbf{u}^H(n) \end{bmatrix}$$

   where $\mathbf{T}(n)$ is the n-by-n unitary matrix denoting the combined effect of the sequence of Givens rotations, and $\mathbf{u}(n)$ is the input vector at time n.

   (b) Update the M-by-1 vector $\mathbf{p}(n)$ and the $(n\text{-}M)$-by-1 vector $\mathbf{v}(n)$ by using the recursion

$$\begin{bmatrix} \mathbf{p}(n) \\ \cdots\cdots \\ \mathbf{v}(n) \end{bmatrix} = \mathbf{T}(n) \begin{bmatrix} \lambda^{1/2}\mathbf{p}(n-1) \\ \cdots\cdots \\ \lambda^{1/2}\mathbf{v}(n-1) \\ d^*(n) \end{bmatrix}$$

   where $d(n)$ is the desired response at time n.

   (c) Compute the weight vector $\hat{\mathbf{w}}(n)$ on-the-fly:

$$\hat{\mathbf{w}}(n) = \mathbf{R}^{-1}(n)\mathbf{p}(n)$$

   and the minimum value of the sum of weighted error squares:

$$\epsilon_{\min}(n) = \|\mathbf{v}(n)\|^2$$

Notes: The null matrix on the left side of the first equation described in point 2(a) has one more row than that on the right side of the equation. The vector $\mathbf{v}(n)$ on the left side of the second equation described in point 2(b) has one more element than the vector $\mathbf{v}(n-1)$ on the right side of the equation.

Hence, we may simplify Eq. (14.47) as

$$\begin{bmatrix} \mathbf{p}(n) \\ \cdots\cdots \\ \mathbf{v}(n) \end{bmatrix} = \mathbf{T}(n) \begin{bmatrix} \lambda^{1/2}\mathbf{p}(n-1) \\ \cdots\cdots\cdots \\ \lambda^{1/2}\mathbf{v}(n-1) \\ d^*(n) \end{bmatrix} \tag{14.48}$$

Equation (14.48) shows that both $\mathbf{p}(n)$ and $\mathbf{v}(n)$ may be updated by using the same sequence of Givens rotations, $\mathbf{T}(n)$, that is used to update $\mathbf{R}(n)$.

Having computed the updated M-by-M matrix $\mathbf{R}(n)$ from Eq. (14.44), and the updated M-by-1 vector $\mathbf{p}(n)$ from Eq. (14.48), we may readily use the method of

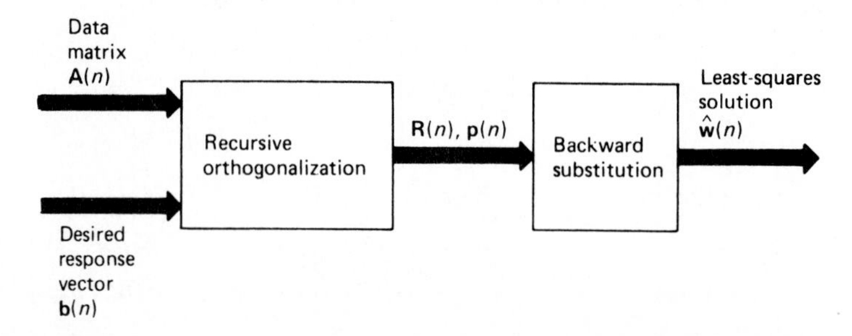

Figure 14.1 Block diagrammatic representation of the QRD-RLS algorithm.

back substitution in Eq. (14.21) to compute the corresponding updated value $\hat{\mathbf{w}}(n)$ of the least-squares weight vector. The method of back substitution exploits the upper triangular composition of matrix $\mathbf{R}$. We refer to the algorithm defined by Eqs. (14.21), (14.44), and (14.46) as the *QR-decomposition-recursive least-squares (QRD-RLS) algorithm*. A summary of the algorithm is presented in Table 14.1; a diagrammtic representation is shown in Fig. 14.1.

14.5 EXACT INITIALIZATION OF THE QRD-RLS ALGORITHM

The solution obtained by QRD-RLS algorithm for the weight vector $\hat{\mathbf{w}}(n)$ is defined only for time $n > M$ for which the data matrix $\mathbf{A}(n)$ is of full rank. Nevertheless, we may *initialize* the orthogonal triangularization procedure by setting $\mathbf{R}(0) = \mathbf{O}$ and $\mathbf{p}(0) = \mathbf{0}$. The *exact initialization procedure* for the algorithm occupies the period $0 \leq n \leq M$, for which the cost function $\mathscr{E}_{\min}(n)$ is zero. During this period, the data matrix $\mathbf{A}(n)$ assumes a lower triangular form [see Eq. (14.5)]. It is this special form of the data matrix, resulting from the use of prewindowing, that facilitates the application of the initialization procedure described here for *temporal* data. In particular, by using a sequence of Givens rotations, as illustrated next by way of an example, the data matrix $\mathbf{A}(n)$ is linearly transformed into upper triangular form. The successive application of these rotations has the combined effect of order updating and time updating in so far as the upper triangular matrix $\mathbf{R}(n)$ and the vector $\mathbf{p}(n)$ are concerned. At $n = M$, the initialization period is completed, whereafter application of the QRD-RLS algorithm may proceed.

Example 1: Initialization of the Orthogonal Triangularization Process

To illustrate the exact initialization procedure for the algorithm, we consider the example of a least-squares problem assuming that the data are *real valued* and the weight vector consists of three elements (i.e., $M = 3$). We consider first initialization of the upper triangular matrix $\mathbf{R}(n)$, and then the vector $\mathbf{p}(n)$.

Initialization of Matrix R(*n*). The algorithm is initialized by setting $\mathbf{R}(0)$ equal to zero. This is equivalent to setting all the inputs equal to zero at $n = 0$, in accordance with the prewindowing method. Thus, for $M = 3$, we may express the data matrix for the initialization period, extending from $n = 0$ to $n = 3$, both inclusive, as follows:

$$\mathbf{A}(3) = \begin{bmatrix} 0 & 0 & 0 \\ u(1) & 0 & 0 \\ u(2) & u(1) & 0 \\ u(3) & u(2) & u(1) \end{bmatrix}$$

The partitions shown in the composition of this matrix are intended to separate out the elements of interest at time $n = 0, 1, 2, 3$.

At time $n = 1$, we may write [see the right side of Eq. (14.41)]

$$\begin{bmatrix} \lambda^{1/2}\mathbf{R}(0) \\ \cdots\cdots \\ u(1) \end{bmatrix} = \begin{bmatrix} 0 \\ \cdots\cdots \\ u(1) \end{bmatrix}$$

The linear transformation of this two-element vector into the desired form is elementary in that we merely require interchanging positions of the two elements. This we accomplish by using the special form of the Givens rotation:

$$\mathbf{J}_1(1) = \begin{bmatrix} c_1(1) & s_1(1) \\ -s_1(1) & c_1(1) \end{bmatrix}$$

$$c_1(1) = 0$$

$$s_1(1) = 1$$

We thus get

$$\mathbf{J}_1(1)\begin{bmatrix} \lambda^{1/2}\mathbf{R}(0) \\ \cdots\cdots \\ u(1) \end{bmatrix} = \begin{bmatrix} \mathbf{R}(1) \\ \cdots\cdots \\ 0 \end{bmatrix} = \begin{bmatrix} r_{00}(1) \\ \cdots\cdots \\ 0 \end{bmatrix}$$

where

$$r_{00}(1) = u(1)$$

With this transformation in place, we next write, at time $n = 2$,

$$\left[\begin{array}{c:c} \lambda^{1/2}\mathbf{R}(1) & 0 \\ 0 & 0 \\ \cdots\cdots & \cdots \\ u(2) & u(1) \end{array}\right] = \left[\begin{array}{c:c} \lambda^{1/2}r_{00}(1) & 0 \\ 0 & 0 \\ \cdots\cdots & \cdots \\ u(2) & u(1) \end{array}\right]$$

To triangularize this matrix, we use a sequence of two Givens rotations, $\mathbf{J}_1(2)$ and $\mathbf{J}_2(2)$. The first Givens rotation, $\mathbf{J}_1(2)$, is defined by

$$\begin{aligned} \mathbf{J}_1(2) &= \begin{bmatrix} c_1(2) & 0 & s_1(2) \\ 0 & 1 & 0 \\ -s_1(2) & 0 & c_1(2) \end{bmatrix} \\ c_1(2) &= \frac{\lambda^{1/2}r_{00}(1)}{r_{00}(2)} \\ s_1(2) &= \frac{u(2)}{r_{00}(2)} \\ r_{00}(2) &= [\lambda r_{00}^2(1) + u^2(2)]^{1/2} \end{aligned} \qquad (14.49)$$

The second Givens rotation, $\mathbf{J}_2(2)$, is defined by

$$\mathbf{J}_2(2) = \begin{bmatrix} 1 & 0 & 0 \\ 0 & c_2(2) & s_2(2) \\ 0 & -s_2(2) & c_2(2) \end{bmatrix}$$

$$c_2(2) = 0$$

$$s_2(2) = 1$$

Hence, we may write

$$\mathbf{J}_2(2)\mathbf{J}_1(2) \begin{bmatrix} \lambda^{1/2} r_{00}(1) & 0 \\ 0 & 0 \\ \cdots & \cdots \\ u(2) & u(1) \end{bmatrix} = \begin{bmatrix} \mathbf{R}(2) \\ \cdots \\ \mathbf{O} \end{bmatrix}$$

$$= \begin{bmatrix} r_{00}(2) & r_{01}(2) \\ 0 & r_{11}(2) \\ \cdots & \cdots \\ 0 & 0 \end{bmatrix}$$

where $r_{00}(2)$ is defined in Eq. (14.49) and

$$r_{01}(2) = s_1(2)u(1)$$

$$r_{11}(2) = c_1(2)u(1)$$

At time $n = 3$, the end of the initialization period in this example, we have three nonzero inputs. Thus, with the aforementioned transformations in place, we write

$$\begin{bmatrix} \lambda^{1/2}\mathbf{R}(2) & & \vdots & 0 \\ & & \vdots & 0 \\ \mathbf{O} & & \vdots & 0 \\ \cdots & \cdots & \vdots & \cdots \\ u(3), & u(2), & \vdots & u(1) \end{bmatrix} = \begin{bmatrix} \lambda^{1/2} r_{00}(2) & \lambda^{1/2} r_{01}(2) & \vdots & 0 \\ 0 & \lambda^{1/2} r_{11}(2) & \vdots & 0 \\ 0 & 0 & \vdots & 0 \\ \cdots & \cdots & \vdots & \cdots \\ u(3) & u(2) & \vdots & u(1) \end{bmatrix}$$

To triangularize this matrix, we apply a sequence of three Givens rotations $\mathbf{J}_1(3)$, $\mathbf{J}_2(3)$, and $\mathbf{J}_3(3)$. The first Givens rotation, $\mathbf{J}_1(3)$, is defined by

$$\mathbf{J}_1(3) = \begin{bmatrix} c_1(3) & 0 & 0 & s_1(3) \\ 0 & 1 & 0 & 0 \\ 0 & 0 & 1 & 0 \\ -s_1(3) & 0 & 0 & c_1(3) \end{bmatrix}$$

$$\begin{aligned} c_1(3) &= \frac{\lambda^{1/2} r_{00}(2)}{r_{00}(3)} \\ s_1(3) &= \frac{u(3)}{r_{00}(3)} \\ r_{00}(3) &= [\lambda r_{00}^2(2) + u^2(3)]^{1/2} \end{aligned} \tag{14.50}$$

The second Givens rotation, $\mathbf{J}_2(3)$, is defined by

$$\mathbf{J}_2(3) = \begin{bmatrix} 1 & 0 & 0 & 0 \\ 0 & c_2(3) & 0 & s_2(3) \\ 0 & 0 & 1 & 0 \\ 0 & -s_2(3) & 0 & c_2(3) \end{bmatrix}$$

$$c_2(3) = \frac{\lambda^{1/2} r_{11}(2)}{r_{11}(3)} \tag{14.51}$$

$$s_2(3) = \frac{c_1(3)u(2) - \lambda^{1/2} s_1(3) r_{01}(2)}{r_{11}(3)}$$

$$r_{11}(3) = [\lambda r_{11}^2(2) + (c_1(3)u(2) - \lambda^{1/2} s_1(3) r_{01}(2))^2]^{1/2}$$

The third Givens rotation, $\mathbf{J}_3(3)$, is defined by

$$\mathbf{J}_3(3) = \begin{bmatrix} 1 & 0 & 0 & 0 \\ 0 & 1 & 0 & 0 \\ 0 & 0 & c_3(3) & s_3(3) \\ 0 & 0 & -s_3(3) & c_3(3) \end{bmatrix}$$

$$c_3(3) = 0$$

$$s_3(3) = 1$$

Hence, we may write

$$\mathbf{J}_3(3)\,\mathbf{J}_2(3)\,\mathbf{J}_1(3) \left[\begin{array}{cc|c} \multicolumn{2}{c|}{\lambda^{1/2}\mathbf{R}(2)} & 0 \\ & & 0 \\ \multicolumn{2}{c|}{\mathbf{O}} & 0 \\ \hline u(3), & u(2) & u(1) \end{array}\right] = \begin{bmatrix} \mathbf{R}(3) \\ \cdots\cdots \\ \mathbf{O} \end{bmatrix}$$

$$= \begin{bmatrix} r_{00}(3) & r_{01}(3) & r_{02}(3) \\ 0 & r_{11}(3) & r_{12}(3) \\ 0 & 0 & r_{22}(3) \\ \cdots & \cdots & \cdots \\ 0 & 0 & 0 \end{bmatrix}$$

where $r_{00}(3)$ and $r_{11}(3)$ are defined by Eqs. (14.50) and (14.51), respectively, and

$$r_{01}(3) = c_1(1)\lambda^{1/2} r_{01}(2) + s_1(1)u(2)$$

$$r_{02}(3) = s_1(1)u(1)$$

$$r_{12}(3) = s_2(2)c_1(1)u(1)$$

$$r_{22}(3) = c_2(3)c_1(1)u(1)$$

This completes the intialization of upper triangular matrix $\mathbf{R}(n)$.

Initialization of Vector $\mathbf{p}(n)$. As mentioned previously, initialization of the QRD-RLS algorithm also involves setting $\mathbf{p}(0)$ equal to zero. Using this initial value and recognizing that the desired response equals $d(1)$ at time $n = 1$, we determine the updated value $\mathbf{p}(1)$ by considering the linearly transformed vector [see the right side of the recursion of Eq. (14.48)]:

$$\mathbf{J}_1(1)\begin{bmatrix} \lambda^{1/2}\mathbf{p}(0) \\ d(1) \end{bmatrix} = \begin{bmatrix} 0 & 1 \\ -1 & 0 \end{bmatrix}\begin{bmatrix} 0 \\ d(1) \end{bmatrix}$$

$$= \begin{bmatrix} d(1) \\ 0 \end{bmatrix}$$

We therefore deduce that

$$\mathbf{p}(1) = p_0(1)$$

where

$$p_0(1) = d(1)$$

and that

$$\mathbf{v}(1) = 0$$

This latter result implies (as expected) that the corresponding minimum value $\mathscr{E}_{\min}(1)$ of the cost function is zero.

At time $n = 2$, the desired response equals $d(2)$. We determine the updated vector $\mathbf{p}(2)$ by using the right side of the recursion of Eq. (14.48) to write

$$\mathbf{J}_2(2)\mathbf{J}_1(2)\begin{bmatrix} \lambda^{1/2}\mathbf{p}(1) \\ 0 \\ d(2) \end{bmatrix} = \begin{bmatrix} p_0(2) \\ p_1(2) \\ 0 \end{bmatrix} \tag{14.52}$$

where

$$p_0(2) = p_0(1)\lambda^{1/2}c_1(2) + s_1(2)d(2)$$

$$p_1(2) = -p_0(1)\lambda^{1/2}s_1(2) + c_1(2)d(2)$$

Since the minimum value $\mathscr{E}_{\min}(2)$ of the cost function at time $n = 2$ is zero, we have $\mathbf{v}(2) = 0$. We therefore deduce from Eq. (14.52) that

$$\mathbf{p}(2) = \begin{bmatrix} p_0(2) \\ p_1(2) \end{bmatrix}$$

The initialization period terminates at $n = 3$. At this time, the desired response equals $d(3)$. We determine the updated vector $\mathbf{p}(3)$ by writing

$$\mathbf{J}_3(3)\mathbf{J}_2(3)\mathbf{J}_1(3)\begin{bmatrix} \lambda^{1/2}\mathbf{p}(2) \\ 0 \\ d(3) \end{bmatrix} = \begin{bmatrix} p_0(3) \\ p_1(3) \\ p_2(3) \\ 0 \end{bmatrix} \tag{14.53}$$

where

$$p_0(3) = c_1(3)\lambda^{1/2}p_0(2) + s_1(3)d(3)$$

$$p_1(3) = -s_2(3)s_1(3)\lambda^{1/2}p_0(2) + c_2(3)\lambda^{1/2}p_1(2) + s_2(3)c_1(3)d(3)$$

$$p_2(3) = -c_2(3)s_1(3)\lambda^{1/2}p_0(2) - s_2(3)\lambda^{1/2}p_1(2) + c_2(3)c_1(3)d(3)$$

Here, again, since the minimum value $\mathscr{E}_{\min}(3)$ of the cost function at time $n = 3$ is zero, we have $\mathbf{v}(3) = 0$. Accordingly, we deduce from Eq. (14.53) that the updated vector $\mathbf{p}(3)$ equals

$$\mathbf{p}(3) = \begin{bmatrix} p_0(3) \\ p_1(3) \\ p_2(3) \end{bmatrix}$$

This completes the initialization process.

Note that as stated at the beginning of the example, the input data are all real valued. The results developed in this example only apply to real data.

14.6 SYSTOLIC ARRAY IMPLEMENTATION I

Figure 14.2 shows a *systolic array* structure for implementing the QRD-RLS algorithm described by Eqs. (14.21), (14.44), and (14.48) for the case when the weight vector has three elements (i.e., $M = 3$). The structure is arranged with two specific points in mind. First, data flow through it from left to right, consistent with all other adaptive filters considered in previous chapters. Second, the systolic array operates directly on the input data that are represented by the matrix $\mathbf{A}^H(n)$ and the row vector $\mathbf{d}^H(n)$. Accordingly, the solution produced at the output of the systolic array will appear as $\hat{\mathbf{w}}^H(n)$. Thus, to produce an estimate of the desired response $\mathbf{d}(n)$, we form the inner product $\hat{\mathbf{w}}^H(n)\,\mathbf{u}(n)$ simply by postmultiplying the output of the systolic array processor by the input vector $\mathbf{u}(n)$.

The structure of Fig. 14.2 consists of two distinct sections: a *triangular systolic array* and a *linear systolic array* (Gentleman and Kung, 1981). The entire systolic array is controlled by a single *clock*. Each section of the array consists of two types of processing cells: *internal cells* (represented by squares) and *boundary cells* (represented by circles). The specific arithmetic functions of these cells are defined later. Each cell receives its input data from the directions indicated for one clock cycle, performs the specified arithmetic functions, and then, on the next clock cycle, delivers the resulting output values to neighboring cells as indicated. A distinctive feature of systolic arrays is that each processing cell is always kept active as data flow across the array. The triangular systolic array section implements the Givens rotation part of the QRD-RLS algorithm, whereas the linear systolic array section computes the weight vector *at the end of the entire recursion*. If we were to compute the weight vector at each iteration of the QRD-RLS algorithm, the operation of the systolic array processor in Fig. 14.2 would be prohibitively slow, hence its computation at the end of the recursion.

The dashed squares shown in Fig. 14.2 are merely included to represent delays in the transfer of data from the triangular array to the linear section. These delays are needed to ensure that the data transfer takes place at the correct instants of time.

Triangular Section

Consider first the operation of the triangular systolic array section labeled *ABC* in Fig. 14.2. The boundary and internal cells of this section are given in Fig. 14.3. Basically, the internal cells perform only additions and multiplications, as illustrated in

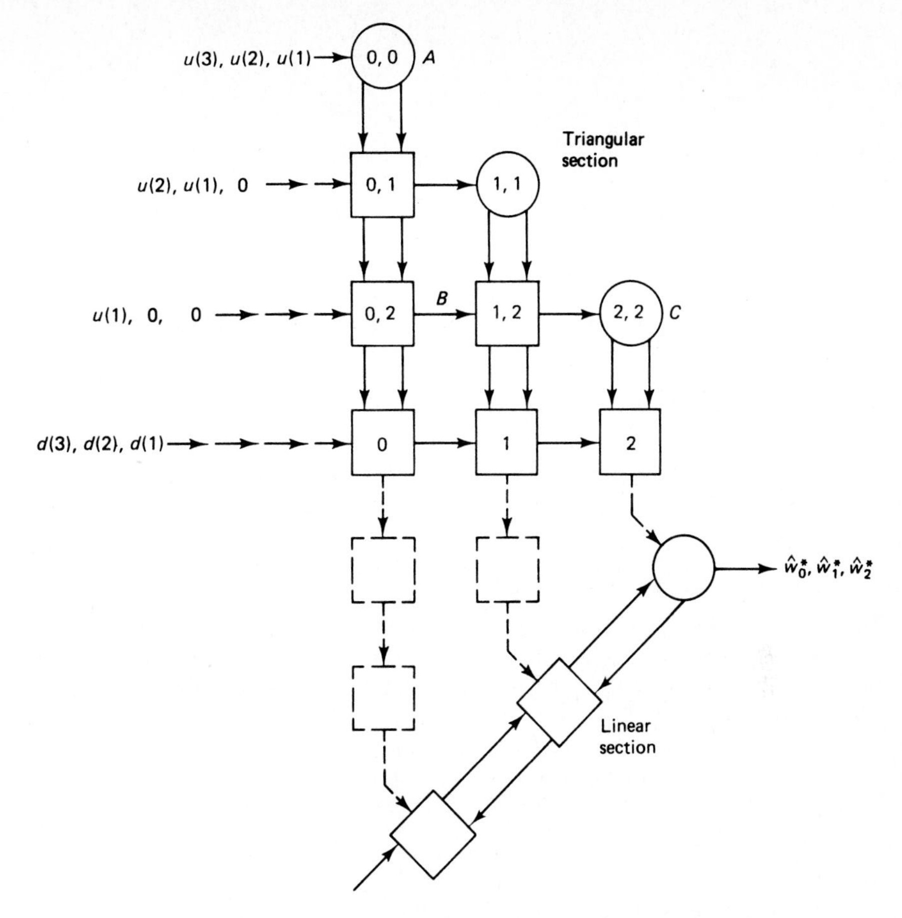

Figure 14.2 Systolic array implementation of the QRD-RLS algorithm for $M = 3$. *Note:* The dashed squares represent delays in the transfer of computations from the triangular array to the linear section.

Fig. 14.3(b). The boundary cells, on the other hand, are considerably more complex in that they compute square roots and reciprocals, as in Fig. 14.3(a). The operations shown in this figure follow directly from the discussion presented in Section 14.2. Each cell of the triangular systolic array section (depending on its location) stores a particular element of the lower triangular matrix $\mathbf{R}^H(n)$, which, at the outset of the least-squares recursion, is initialized to zero and thereafter updated every clock cycle. The function of each column of processing cells in the triangular systolic array section is to *rotate* one column of the stored triangular matrix with a vector of data received from the left in such a way that the leading *element of the received data vector is annihilated*. The reduced data vector is then passed to the right on to the next column of cells. The boundary cell in each column of the section computes the pertinent rotation parameters and then passes them downward on the next clock cycle. The internal cells subsequently apply the same rotation to all other elements in the received data vector. Since a delay of one clock cycle per cell is incurred in passing the rotation parameters downward along a column, it is necessary that the input data vectors enter the triangular systolic array in a *skewed order,* as illustrated in

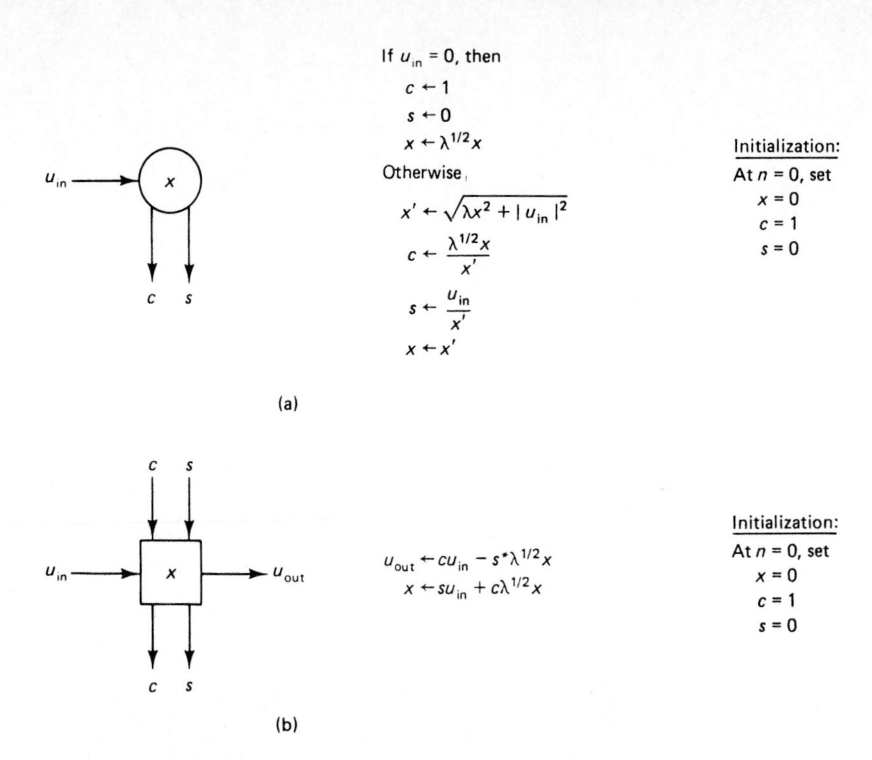

Figure 14.3 Cells for the QRD-RLS algorithm: (a) boundary cell; (b) internal cell. *Note:* The stored value x is initialized to be *zero* (i.e., real). For the boundary cell, it always remains *real*; this is consistent with the property that the diagonal elements of the upper triangular matrix $\mathbf{R}$ are all real. Hence, the formulas for the rotation parameters c and s computed by the boundary cell can be simplified considerably, as shown in part (a). Also, note that the values x stored in the array are elements of the lower triangular matrix $\mathbf{R}^H$; hence, we may identify $r^* = x$ for all elements of the triangular array.

Fig. 14.2 for the case of $M = 3$. This arrangement of the input data ensures that as each column vector $\mathbf{u}(n)$ of the data matrix $\mathbf{A}^H(n)$ propagates through the array, it interacts with the previously stored triangular matrix $\mathbf{R}^H(n-1)$ and thereby undergoes the sequence of Givens rotations $\mathbf{T}(n)$, as required. Accordingly, all the elements of the column vector $\mathbf{u}(n)$ are annihiliated, one by one, and an updated lower triangular matrix $\mathbf{R}^H(n)$ is produced and stored in the process.

The systolic array operates in a highly pipelined manner whereby, as (time-skewed) input data vectors enter the array from the left, we find that in effect each such vector defines a processing *wavefront* that moves across the array. It should therefore be appreciated that, on any particular clock cycle, elements of the pertinent lower triangular matrix $\mathbf{R}^H(n)$ only exist along the corresponding wave-front. This phenomenon is illustrated later in Example 2.

At the same time as the orthogonal triangularization process is being performed by the triangular systolic array section labeled *ABC* in Fig. 14.2, the row vector $\mathbf{p}^H(n)$ is computed by the appended bottom row of internal cells. In effect, this computation is made by treating the desired response vector $\mathbf{d}^H(n)$ as an additional row that is appended to the data matrix $\mathbf{A}^H(n)$ at its bottom end.

When the entire orthogonal triangularization process is completed, the data flow has to *stop,* and then the stored data can be *clocked out* for subsequent processing by the linear systolic array section. The dashed lines in Fig. 14.4 depict the clock-out paths for the final values of the elements of both $\mathbf{R}^H(n)$ and $\mathbf{p}^H(n)$ into the linear section of the systolic processor.

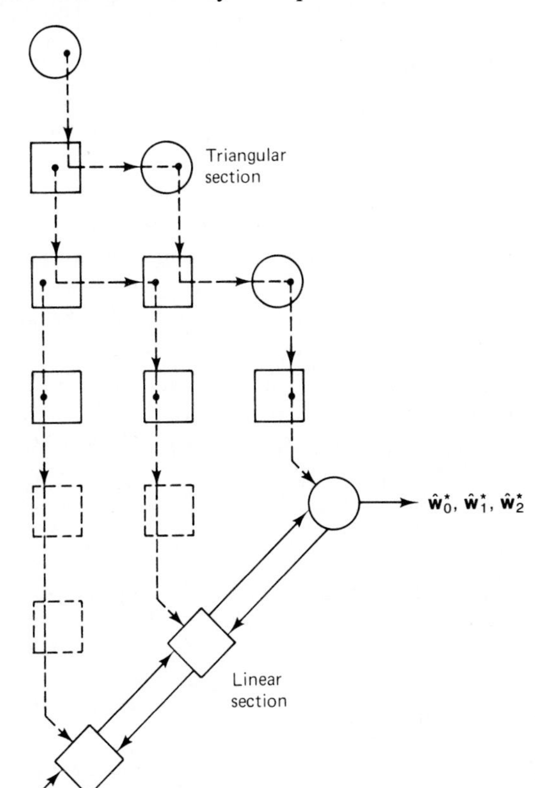

Figure 14.4 Clock-out paths of the triangular section into the linear section of the linear array. The dots represent cell contents.

Linear Section

The linear section of the processor computes the Hermitian transposed least-squares weight vector, namely, $\hat{\mathbf{w}}^H(n)$. In particular, the elements of the vector $\hat{\mathbf{w}}^H(n)$ are computed by using the *method of back substitution* (Kung and Leiserson, 1978):

$$z_i^{(M-1)} = 0$$

$$z_i^{(k-1)} = z_i^{(k)} + r_{ik}^*(n)\hat{w}_k^*(n), \qquad k = M-1, \ldots, i; \quad i = M-1, \ldots, 0 \qquad (14.54)$$

$$\hat{w}_i^*(n) = \frac{p_i^*(n) - z_i^{(i)}}{r_{ii}(n)}$$

where the $z_i^{(k)}$ are intermediate variables, the $r_{ik}(n)$ are elements of upper triangular matrix $\mathbf{R}(n)$, the $p_i(n)$ are elements of the vector $\mathbf{p}(n)$, and the $\hat{w}_k(n)$ are elements of the weight vector $\hat{\mathbf{w}}(n)$.

The linear systolic array section consists of one boundary cell and $(M - 1)$ internal cells that perform the arithmetic functions defined in Fig. 14.5, in accordance

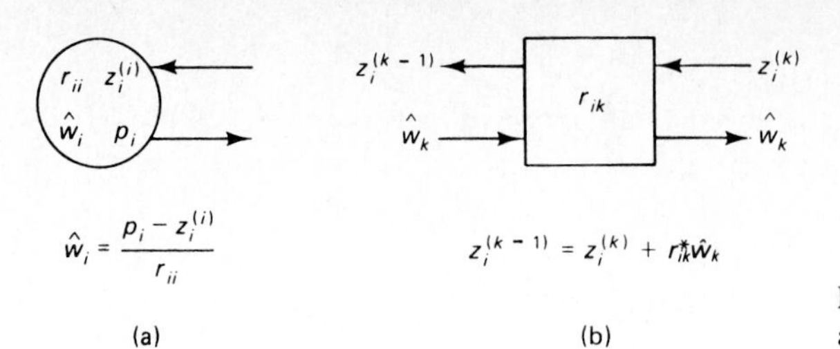

Figure 14.5 Cells for linear systolic array: (a) boundary cell; (b) inner cell.

with Eq. (14-54). The boundary cell performs subtraction and division, whereas the internal cells perform additions and multiplications. The elements of the complex-conjugated weight vector leave the linear array every second clock cycle with $\hat{w}_{M-1}^*(n)$ leaving first, followed by $\hat{w}_{M-2}^*(n)$, and so on right up to $\hat{w}_0^*(n)$. In effect, the elements of the weight vector $\hat{\mathbf{w}}^H(n)$ are read out *backward*.

Thus, by chaining the linear and triangular systolic array sections together, as shown in Fig. 14.2, we produce a powerful device that is capable of solving on-the-fly the exact least-squares problem recursively.

Note that, initially, zeros are stored in all boundary and internal cells. Also, the parameters of the Givens rotation at the output of each boundary cell (and therefore every other cell in the triangular systolic array section) are initially set at the values $c_{\text{out}} = 1$ and $s_{\text{out}} = 0$. Note also that the initialization of the complete systolic array shown in Fig. 14.2 occupies a total of 3M *clock cycles,* where M is the dimension of the weight vector.

Example 2

In this example, we illustrate the operation of the systolic array structure of Fig. 14.2 for the case when the input data are *real valued* and $M = 3$.

The inputs and states of the triangular systolic array for $M = 3$ are summarized in Table 14.2. In particular, we show the states of the individual cells in this section at time n_-, the external inputs applied to them at time n, and the states of their individual cells at time n_+.

TABLE 14.2 INPUTS AND STATES OF THE TRIANGULAR SYSTOLIC ARRAY FOR M = 3 AND REAL DATA[a]

States at time n_-			Inputs at time n	States at time n_+		
$r_{00}(n-1)$			$u(n)$	$r_{00}(n)$		
$r_{01}(n-2)$	$r_{11}(n-3)$		$u(n-2)$	$r_{01}(n-1)$	$r_{11}(n-2)$	
$r_{02}(n-3)$	$r_{12}(n-4)$	$r_{22}(n-5)$	$u(n-4)$	$r_{02}(n-2)$	$r_{12}(n-3)$	$r_{22}(n-4)$
$p_0(n-4)$	$p_1(n-5)$	$p_2(n-6)$	$d(n-3)$	$p_0(n-3)$	$p_1(n-4)$	$p_2(n-5)$

[a] The initialization procedure can also be represented by this state graph, provided that we set $u(k) = 0$, $r_{ij}(k) = 0$, and $p_i(k) = 0$, when $k < 1$.

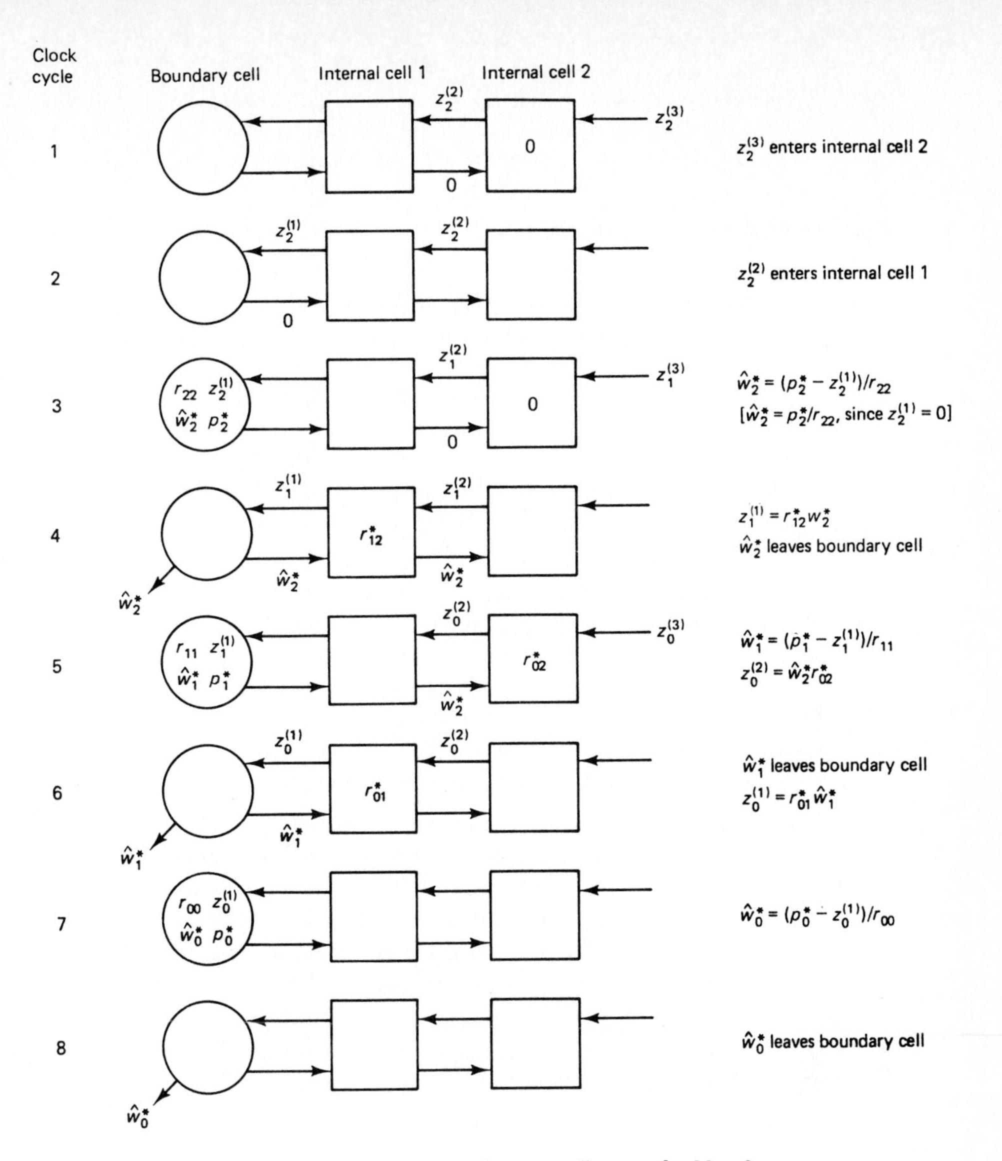

Figure 14.6 Operation of the linear systolic array for $M = 3$.

The elements of column 1 of the lower triangular 3-by-3 matrix $\mathbf{R}^T(n)$ for $M = 3$, that is, the elements $r_{00}(n)$, $r_{01}(n)$, and $r_{02}(n)$, are computed in the cells of column 1 of the triangular section at times n, $n + 1$, and $n + 2$, respectively. The elements of column 2 of $\mathbf{R}^T(n)$, that is, the elements $r_{11}(n)$ and $r_{12}(n)$ are computed in the cells of column 2 of the triangular section at times $n + 2$ and $n + 3$, respectively. The remaining element of $\mathbf{R}^T(n)$, that is, $r_{33}(n)$, is computed in the only cell of column 3 of the triangular section at time $n + 4$. The elements of the 1-by-3 vector $\mathbf{p}^T(n)$, that is, $p_0(n)$, $p_1(n)$, and $p_2(n)$, are computed in the elements of the row of cells appended to the triangular section at times $n + 4$, $n + 5$, and $n + 6$, respectively.

When the orthogonal triangularization of the weighted data matrix is completed, the data flow is terminated, and the stored contents of the internal and boundary cells are clocked out in the manner described in Fig. 14.4. In particular, the clocked out data are processed by a linear systolic array. For the example at hand, Fig. 14.6 presents the details of the operation of this section. The clock cycle numbers included in Fig. 14.6 are measured from the instant when the linear section begins its operation.

General Observations

The following points are noteworthy in the context of the two-section systolic processor of Fig. 14.2 for the general case of *complex-valued data:*

1. The element $r_{ik}^*(n)$ of the lower triangular matrix $\mathbf{R}^H(n)$ is computed in the kth cell on the ith row of the triangular section at time $n + (i - 1) + (k - 1)$, where $i, k = 0, 1, \ldots, M - 1$; note that the diagonal elements of $\mathbf{R}^H(n)$ are all *real valued*. The element $p_k^*(n)$ of the row vector $\mathbf{p}^H(n)$ is computed in the kth cell of the row of cells appended to the triangular section at time $n + M + k$, where $k = 0, 1, \ldots, M - 1$.
2. The systolic processor of Fig. 14.2 experiences *delays* in the various stages of computation. In particular, for the general case of an M-by-1 weight vector, $2M$ clock cycles are required to compute the Givens rotations, and another $2M$ clock cycles are required to clock out and compute the weight vector. Accordingly, a *total delay or latency of 4M clock cycles* is experienced in computing the complete M-by-1 weight vector $\hat{\mathbf{w}}(n)$. For real-time operations requiring the use of a large M, this latency may be too large and therefore unacceptable.
3. The linear section does not operate until the triangular section is completely processed. Thus, the weight vector $\hat{\mathbf{w}}(n)$ is *not* computed on the fly; rather, it is computed after the completion of the QR computations in the triangular section. However, the array structures may be easily modified to make "on-the-fly" computation of $\hat{\mathbf{w}}(n)$ possible. This would require the addition of extra data paths, allowing for the transfer of partially computed data from the triangular to the linear section.
4. The element values of the matrix $\mathbf{R}^H(n)$ do *not* all propagate from the triangular section straight down to the linear section. Rather, except for $r_{i,M-1}^*(n)$, $i = 0, 1, \ldots, M - 1$, the propagation of all the other elements of $\mathbf{R}^H(n)$ follows zig-zag paths, as well illustrated in Fig. 14.4 for the case of $M = 3$.

14.7 COMPUTER EXPERIMENT I: ADAPTIVE PREDICTION USING SYSTOLIC STRUCTURE I

For our first computer experiment, we use the systolic structure shown in Fig. 14.2 to investigate the performance of an adaptive predictor of order 2. The input data are computer generated, assuming a corresponding autoregressive model of order 2. The model output $u(n)$ is described by the difference equation.

$$u(n) + a_1 u(n - 1) + a_2 u(n - 2) = v(n)$$

where the sample $v(n)$ is drawn from a white-noise process of zero mean and variance σ_v^2. The model parameters a_1 and a_2 are chosen to satisfy the condition

$a_1^2 < 4a_2$. The parameter values used for data generation are given in Table 14.3. In particular, three data sets were used in the study, corresponding to three highly different eigenvalue spreads.

Table 14.4 summarizes the output of the systolic array for different numbers of iterations: $n = 5, 10, 15, 20, 25$, for each of the three eigenvalue spreads given in

TABLE 14.3 PARAMETER VALUES OF THE AR MODEL USED IN EXPERIMENT I

Eigenvalue spread, $\chi(\mathbf{R})$	σ_v^2	a_1	a_2
3	0.0731	−0.9750	0.95
10	0.0322	−1.5955	0.95
100	0.0038	−1.9114	0.95

TABLE 14.4 OUTPUT OF SYSTOLIC ARRAY I

	Experiment		Optimum		Relative error (%)	
$\chi(\mathbf{R})$	$\hat{w}_1$	$\hat{w}_2$	w_1	w_2	w_1	w_2
		Systolic array I, $n = 5$				
3	0.7586	−0.87327	0.9750	−0.95	22.2	8.1
10	1.2726	−0.75964	1.5955	−0.95	20.2	20.0
100	1.5280	−0.62849	1.9114	−0.95	20.1	32.2
		Systolic array I, $n = 10$				
3	0.9394	−0.90806	0.9750	−0.95	3.7	4.4
10	1.5137	−0.93395	1.5955	−0.95	5.1	1.7
100	1.7539	−0.84977	1.9114	−0.95	8.2	10.6
		Systolic array I, $n = 15$				
3	0.9014	−0.86993	0.9750	−0.95	7.5	8.4
10	1.4916	−0.86007	1.5955	−0.95	6.5	9.5
100	1.7615	−0.81828	1.9114	−0.95	7.8	13.9
		Systolic array I, $n = 20$				
3	0.9368	−0.87814	0.9750	−0.95	3.9	7.6
10	1.5382	−0.89951	1.5955	−0.95	3.6	5.3
100	1.8484	−0.90193	1.9114	−0.95	3.3	5.1
		Systolic array I, $n = 25$				
3	0.9555	−0.91519	0.9750	−0.95	2.0	3.7
10	1.5233	−0.88600	1.5955	−0.95	4.5	6.7
100	1.8252	−0.87115	1.9114	−0.95	4.5	8.3
		RLS algorithm, $n = 25$				
3	0.9335	−0.87126	0.9750	−0.95	4.2	8.3
10	1.5549	−0.90557	1.5955	−0.95	2.5	4.7
100	1.8339	−0.88253	1.9114	−0.95	4.1	7.1

Table 14.3. This table also includes the correspondng optimum Wiener solutions (i.e., the negative of the actual model parameters) for the weight vector, and the relative errors produced in the systolic computation. For the purpose of comparison, we have also included the solution obtained from the standard RLS algorithm after 25 iterations. Based on these results, we may draw the following conclusions:

1. The systolic processor converges fairly quickly. For the prediction order $M = 2$ studied in this experiment, after only five iterations the experimentally determined weights approach their optimum values, with an error of about 20 percent.
2. For a larger number of iterations (i.e., $n > 10$), the relative error in the computation of the weights is fairly small (about 10 percent). In other words, the systolic processor reaches a steady state after about 10 iterations.
3. There is no significant trend in the relative errors with increasing eigenvalue spread of the underlying correlation matrix of the input data.
4. Comparing the results obtained from the systolic processor with those obtained from the standard RLS algorithm, we find that within a data window of 25 samples, both algorithms show approximately the same relative error.

14.8 STRUCTURAL PROPERTIES OF THE TRANSFORMATION T(n)

The n-by-n transformation $\mathbf{T}(n)$ plays a central role in systolic array processing as evidenced not only by the algorithm summarized in Table 14.1 but also the development of other algorithms presented in subsequent sections of this chapter. It is therefore important that we recognize its properties. In particular, the fact that $\mathbf{T}(n)$ is the product of a sequence of Givens rotations [as shown in Eq. (14.42)] bestows upon it the following structural properties (Shepherd and McWhirter, 1991):

Property 1. The transformation $\mathbf{T}(n)$ is a unitary matrix; that is, the inverse of $\mathbf{T}(n)$ is the same as the Hermitian transpose of $\mathbf{T}(n)$, as shown by

$$\mathbf{T}^{-1}(n) = \mathbf{T}^{H}(n) \tag{14.55}$$

This property is merely a generalization of the unitary nature of a standard Givens rotation operator that is basic to the construction of $\mathbf{T}(n)$.

Property 2. For M Givens rotations applied sequentially at time n, the structure of $\mathbf{T}(n)$ may in general be described as follows:

$$\mathbf{T}(n) = \gamma^{1/2}(n)\begin{bmatrix} \mathcal{A}(n) & \mathbf{O} & \boldsymbol{\beta}_2(n) \\ \mathbf{O} & \gamma^{-1/2}(n)\mathbf{I} & \mathbf{0} \\ -\boldsymbol{\beta}_1^H(n) & \mathbf{0}^T & 1 \end{bmatrix} \tag{14.56}$$

where $\mathbf{I}$ is a unitary matrix of dimension $n - M - 1$. The scalar $\gamma(n)$ and the matrix/vector components $\mathcal{A}(a)$, $\boldsymbol{\beta}_1(n)$, and $\boldsymbol{\beta}_2(n)$ are themselves characterized as follows:

(a) The scalar $\gamma(n)$ is real, with its (positive) square root equal to the product of the cosine rotation parameters:

$$\gamma^{1/2}(n) = \prod_{i=1}^{M} c_i(n) \tag{14.57}$$

Note that $0 \leq \gamma(n) \leq 1$.

(b) The vectors $\boldsymbol{\beta}_1(n)$ and $\boldsymbol{\beta}_2(n)$ are M-by-1 vectors defined by

$$\boldsymbol{\beta}_1(n) = \gamma^{-1/2}(n) \begin{bmatrix} s_1^*(n)c_2(n)c_3(n) \cdots c_M(n) \\ s_2^*(n)c_3(n) \cdots c_M(n) \\ \vdots \\ s_M^*(n) \end{bmatrix} \tag{14.58}$$

and

$$\boldsymbol{\beta}_2(n) = \gamma^{-1/2}(n) \begin{bmatrix} s_1^*(n) \\ s_2^*(n)c_1(n) \\ \vdots \\ s_M^*(n)c_1(n)c_2(n) \cdots c_{M-1}(n) \end{bmatrix} \tag{14.59}$$

Corollary 1. As a consequence of Property 1 rewritten in the form

$$\mathbf{T}^H(n)\mathbf{T}(n) = \mathbf{I}$$

where $\mathbf{I}$ is the M-by-M identity matrix, we find that the components of $\mathbf{T}(n)$ are interrelated as follows:

$$\mathcal{A}^H(n)\mathcal{A}(n) + \boldsymbol{\beta}_1(n)\boldsymbol{\beta}_1^H(n) = \frac{1}{\gamma(n)}\mathbf{I} \tag{14.60}$$

$$\boldsymbol{\beta}_1(n) = \mathcal{A}^H(n)\boldsymbol{\beta}_2(n) \tag{14.61}$$

$$\boldsymbol{\beta}_2^H(n)\boldsymbol{\beta}_2(n) = \frac{1}{\gamma(n)} - 1 \tag{14.62}$$

Corollary 2. Rewriting Property 1 in the complementary form

$$\mathbf{T}(n)\mathbf{T}^H(n) = \mathbf{I}$$

we find that the components of $\mathbf{T}(n)$ are interrelated in yet another way:

$$\mathcal{A}(n)\mathcal{A}^H(n) + \boldsymbol{\beta}_2(n)\boldsymbol{\beta}_2^H(n) = \frac{1}{\gamma(n)}\mathbf{I} \tag{14.63}$$

$$\boldsymbol{\beta}_2(n) = \mathcal{A}(n)\boldsymbol{\beta}_1(n) \tag{14.64}$$

$$\boldsymbol{\beta}_1^H(n)\boldsymbol{\beta}_1(n) = \frac{1}{\gamma(n)} - 1 \tag{14.65}$$

Observations

According to the representation of $\mathbf{T}(n)$ given in Eq. (14.56), the square root of $\gamma(n)$ plays the role of a normalizing scaling factor. The motivation for writing $\mathbf{T}(n)$ in this form will become apparent in the next section.

Moreover, from corollaries 1 and 2 we may make two additional observations. First, the vectors $\boldsymbol{\beta}_1(n)$ and $\boldsymbol{\beta}_2(n)$ have the same Euclidean norm. Second, given the matrix $\mathcal{A}(n)$ and one of the two vectors $\boldsymbol{\beta}_1(n)$ or $\boldsymbol{\beta}_2(n)$, the other vector is uniquely determined.

Example 3

Consider the simple and yet meaningful example of $M = 2$, for which the Givens rotations $\mathbf{J}_1(n)$ and $\mathbf{J}_2(n)$ are expressed as

$$\mathbf{J}_1(n) = \begin{bmatrix} c_1(n) & 0 & \mathbf{0}^T & s_1^*(n) \\ 0 & 1 & \mathbf{0}^T & 0 \\ \mathbf{0} & \mathbf{0} & \mathbf{I} & \mathbf{0} \\ -s_1(n) & 0 & \mathbf{0}^T & c_1(n) \end{bmatrix}$$

and

$$\mathbf{J}_2(n) = \begin{bmatrix} 1 & 0 & \mathbf{0}^T & 0 \\ 0 & c_2(n) & \mathbf{0}^T & s_2^*(n) \\ \mathbf{0} & \mathbf{0} & \mathbf{I} & \mathbf{0} \\ 0 & -s_2(n) & \mathbf{0}^T & c_2(n) \end{bmatrix}$$

Forming the product $\mathbf{T}(n) = \mathbf{J}_2(n)\mathbf{J}_1(n)$, we find that for this example the components of $\mathbf{T}(n)$ described in Eq. (14.56) have the following values:

$$\gamma^{1/2}(n) = c_1(n)c_2(n)$$

$$\mathcal{A}(n) = \frac{1}{c_1(n)c_2(n)} \begin{bmatrix} c_1(n) & 0 \\ -s_1(n)s_2(n) & c_2(n) \end{bmatrix}$$

$$\boldsymbol{\beta}_1(n) = \frac{1}{c_1(n)c_2(n)} \begin{bmatrix} s_1^*(n)c_2(n) \\ s_2^*(n) \end{bmatrix}$$

$$\boldsymbol{\beta}_2(n) = \frac{1}{c_1(n)c_2(n)} \begin{bmatrix} s_1^*(n) \\ s_2^*(n)c_1(n) \end{bmatrix}$$

Using these values, it is a straightforward matter to verify corollaries 1 and 2, which is left as an exercise for the reader.

14.9 MODIFIED VERSION OF THE QRD-RLS ALGORITHM

In Section 14.2, we described a procedure, based on the QR-decomposition, for solving the recursive least-squares problem. This procedure also provides a means for computing the optimum weight vector $\hat{\mathbf{w}}(n)$ that achieves the least-squares minimization. However, in some applications (e.g., prediction-error filters, adaptive noise cancellers, adaptive beamformers), it is not necessary that we compute the least-squares weight vector $\hat{\mathbf{w}}(n)$ explicitly. In this section, we present a modification of the QRD-RLS algorithm that avoids the need to solve Eq. (14.21) at any stage of the process (McWhirter, 1983).

Since the last element of the diagonal matrix $\mathbf{\Lambda}(n)$ equals unity [see Eq. (14.9)], it follows that the complex conjugate of the estimation error $e(n)$ is equal to the last element of the n-by-1 vector:

$$\mathbf{\Lambda}^{1/2}(n)\boldsymbol{\epsilon}(n) = \mathbf{\Lambda}^{1/2}(n)\mathbf{d}(n) - \mathbf{\Lambda}^{1/2}(n)\mathbf{A}(n)\hat{\mathbf{w}}(n) \tag{14.66}$$

In the right side of Eq. (14.66), we have used the definition of Eq. (14.12) for the error vector $\boldsymbol{\epsilon}(n)$. Premultiplying both sides of Eq. (14.14) by $\mathbf{Q}^{-1}(n)$, the inverse of the unitary matrix $\mathbf{Q}(n)$, and recognizing (from the definition of a unitary matrix) that

$$\mathbf{Q}^{-1}(n) = \mathbf{Q}^H(n)$$

we may express the matrix $\mathbf{\Lambda}^{1/2}(n)\mathbf{A}(n)$ as

$$\begin{aligned}\mathbf{\Lambda}^{1/2}(n)\mathbf{A}(n) &= \mathbf{Q}^H(n)\begin{bmatrix}\mathbf{R}(n)\\ \cdots\cdots \\ \mathbf{O}\end{bmatrix}\\ &= [\mathbf{F}^H(n) \,\vdots\, \mathbf{S}^H(n)]\begin{bmatrix}\mathbf{R}(n)\\ \cdots\cdots \\ \mathbf{O}\end{bmatrix} \\ &= \mathbf{F}^H(n)\mathbf{R}(n)\end{aligned} \tag{14.67}$$

In the second line of Eq. (14.67), we have used the partitioned representation of Eq. (14.16) for the unitary matrix $\mathbf{Q}(n)$. The substitution of Eq. (14.67) in (14.66) thus yields

$$\mathbf{\Lambda}^{1/2}(n)\boldsymbol{\epsilon}(n) = \mathbf{\Lambda}^{1/2}(n)\mathbf{b}(n) - \mathbf{F}^H(n)\mathbf{R}(n)\hat{\mathbf{w}}(n) \tag{14.68}$$

However, the least-squares weight vector $\hat{\mathbf{w}}(n)$ satisfies Eq. (14.21). We may therefore replace the product $\mathbf{R}(n)\hat{\mathbf{w}}(n)$ by the vector $\mathbf{p}(n)$, and so rewrite Eq. (14.68) as

$$\mathbf{\Lambda}^{1/2}(n)\boldsymbol{\epsilon}(n) = \mathbf{\Lambda}^{1/2}(n)\mathbf{d}(n) - \mathbf{F}^H(n)\mathbf{p}(n) \tag{14.69}$$

As desired, this result does not depend explicitly on the weight vector $\hat{\mathbf{w}}(n)$, Furthermore, premultiplying both sides of Eq. (14.17) by the inverse matrix $\mathbf{Q}^{-1}(n) = \mathbf{Q}^H(n)$, we may write

$$\begin{aligned}\mathbf{\Lambda}^{1/2}(n)\mathbf{d}(n) &= \mathbf{Q}^H(n)\begin{bmatrix}\mathbf{p}(n)\\ \cdots\cdots \\ \mathbf{v}(n)\end{bmatrix}\\ &= \mathbf{F}^H(n)\mathbf{p}(n) + \mathbf{S}^H(n)\mathbf{v}(n)\end{aligned} \tag{14.70}$$

Hence, using Eq. (14.70) in (14.69) and simplifying, we get

$$\mathbf{\Lambda}^{1/2}(n)\boldsymbol{\epsilon}(n) = \mathbf{S}^H(n)\mathbf{v}(n) \tag{14.71}$$

Earlier we remarked that the complex conjugate of the estimation error $e(n)$ equals the last element of the vector $\mathbf{\Lambda}^{1/2}(n)\boldsymbol{\epsilon}(n)$. Equation (14.71) shows that, equivalently, $e^*(n)$ equals the last element of the vector $\mathbf{S}^H(n)\mathbf{v}(n)$. By using the latter result, we now show how the estimation error $e(n)$ may be obtained directly from the orthogonal triangularization process. To do this, we use the structural properties of the transformation $\mathbf{T}(n)$ comprising a sequence of Givens rotations, described in Section 14.8. Specifically, we use Eqs. (14.56) and (14.64) to express $\mathbf{T}(n)$ as follows:

$$\mathbf{T}(n) = \gamma^{1/2}(n) \begin{bmatrix} \mathcal{A}(n) & \vdots & \mathbf{O} & \vdots & \mathcal{A}(n)\boldsymbol{\beta}_1(n) \\ \cdots & & \cdots & & \cdots \\ \mathbf{O} & \vdots & \gamma^{-1/2}(n)\mathbf{I} & \vdots & \mathbf{0} \\ \cdots & & \cdots & & \cdots \\ -\boldsymbol{\beta}_1^H(n) & \vdots & \mathbf{0} & \vdots & 1 \end{bmatrix} \tag{14.72}$$

where the identity matrix $\mathbf{I}$ has dimension $(n - M - 1)$. The other elements on the right side of Eq. (14.61), namely, the M-by-M matrix $\mathcal{A}(n)$, the M-by-1 vector $\boldsymbol{\beta}_1(n)$, and the scaling factor $\gamma^{1/2}(n)$ are all related to the cosine–sine pairs of the Givens rotations that constitute the unitary transformation matrix $\mathbf{T}(n)$, as shown in Section 14.8. Note that the last column of $\mathbf{T}(n)$ is obtained by applying $\mathbf{T}(n)$ to the pinning vector $[\mathbf{0}, 1]^T$. Hence, using Eq. (14.72) in (14.48), we get

$$\begin{bmatrix} \mathbf{p}(n) \\ \cdots \\ \mathbf{v}(n) \end{bmatrix} = \gamma^{1/2}(n) \begin{bmatrix} \mathcal{A}(n)(d^*(n)\boldsymbol{\beta}_1(n) + \lambda^{1/2}\mathbf{p}(n-1)) \\ \cdots \\ \gamma^{-1/2}(n)\lambda^{1/2}\mathbf{v}(n-1) \\ d^*(n) - \lambda^{1/2}\boldsymbol{\beta}_1^H(n)\mathbf{p}(n-1) \end{bmatrix} \tag{14.73}$$

From this relation, we deduce that the $(n - M)$-by-1 vector $\mathbf{v}(n)$ equals

$$\mathbf{v}(n) = \begin{bmatrix} \lambda^{1/2}\mathbf{v}(n-1) \\ \cdots \\ \gamma^{1/2}(n)\alpha^*(n) \end{bmatrix} \tag{14.74}$$

where the complex-valued variable $\alpha(n)$ is defined by

$$\alpha(n) = d(n) - \lambda^{1/2}\mathbf{p}^H(n-1)\boldsymbol{\beta}_1(n) \tag{14.75}$$

From Eq. (14.16) we have

$$\mathbf{Q}(n-1) = \begin{bmatrix} \mathbf{F}(n-1) \\ \mathbf{S}(n-1) \end{bmatrix} \tag{14.76}$$

We may therefore use Eqs. (14.45), (14.72) and (14.76) to express the unitary matrix $\mathbf{Q}(n)$ as

$$\begin{aligned} \mathbf{Q}(n) &= \mathbf{T}(n) \begin{bmatrix} \mathbf{F}(n-1) & \mathbf{0} \\ \mathbf{S}(n-1) & \mathbf{0} \\ \mathbf{0}^T & 1 \end{bmatrix} \\ &= \gamma^{1/2}(n) \begin{bmatrix} \mathcal{A}(n)\mathbf{F}(n-1) & \mathcal{A}(n)\boldsymbol{\beta}_1(n) \\ \gamma^{-1/2}(n)\mathbf{S}(n-1) & \mathbf{0} \\ -\boldsymbol{\beta}_1^H(n)\mathbf{F}(n-1) & 1 \end{bmatrix} \end{aligned} \tag{14.77}$$

With $\mathbf{S}(n)$ defined as the last $(n - M)$ rows of the unitary matrix $\mathbf{Q}(n)$, we deduce from Eqs. (14.16) and (14.77) that

$$\mathbf{S}(n) = \gamma^{1/2}(n)\begin{bmatrix} \gamma^{-1/2}(n)\mathbf{S}(n-1) & \mathbf{0} \\ -\boldsymbol{\beta}_1^H(n)\mathbf{F}(n-1) & 1 \end{bmatrix} \tag{14.78}$$

Hence, we may use Eqs. (14.74) and (14.78) to express the vector $\mathbf{S}^H(n)\mathbf{v}(n)$ as

$$\mathbf{S}^H(n)\mathbf{v}(n) = \lambda^{1/2}\begin{bmatrix} \mathbf{S}^H(n-1)\mathbf{v}(n-1) \\ \cdots\cdots\cdots \\ 0 \end{bmatrix} + \alpha^*(n)\gamma(n)\begin{bmatrix} -\mathbf{F}^H(n-1)\boldsymbol{\beta}_1(n) \\ \cdots\cdots\cdots \\ 1 \end{bmatrix} \tag{14.79}$$

Thus, with the complex conjugate of the estimation error $e(n)$ equal to the last element of the vector $\mathbf{S}^H(n)\mathbf{v}(n)$, we finally obtain the result

$$e(n) = \gamma(n)\alpha(n) \tag{14.80}$$

where $\alpha(n)$ is itself defined in Eq. (14.75).

According to Eq. (14.57) the real-valued factor $\gamma^{1/2}(n)$ is defined as the product of the cosine parameters $c_1(n), c_2(n), \ldots, c_M(n)$ resulting from the application of a sequence of M Givens rotations. Moreover, the product $\gamma^{1/2}(n)\alpha(n)$ may be interpreted simply as the result obtained when the desired response $d(n)$ is rotated with each element in the M-by-1 vector $\mathbf{p}(n-1)$ by means of the same sequence of Givens rotations. We therefore find that $\gamma^{1/2}(n)\alpha(n)$ is also computed quite naturally during the orthogonal triangularization process. Thus, the estimation error $e(n)$ may be computed simply by multiplying the product $\gamma^{1/2}(n)\alpha(n)$ by the scaling factor $\gamma^{1/2}(n)$.

Another Interpretation of the Variable $\alpha(n)$ and Parameter $\gamma(n)$

The product term $\lambda^{1/2}\mathbf{p}^H(n-1)\boldsymbol{\beta}_1(n)$ on the right side of Eq. (14.75) represents the a priori estimate of the desired response $d(n)$; this estimate is computed at time n using the prior value of the least-squares weight vector, namely, $\hat{\mathbf{w}}(n-1)$. Let $\mathcal{U}_{n-1}$ denote the space spanned by the inputs $u(n-1), \ldots, u(n-M)$. We may then write (see Problem 6)

$$\begin{aligned} \lambda^{1/2}\mathbf{p}^H(n-1)\boldsymbol{\beta}_1(n) &= \hat{d}(n|\mathcal{U}_{n-1}) \\ &= \hat{\mathbf{w}}^H(n-1)\mathbf{u}(n) \end{aligned} \tag{14.81}$$

Accordingly, the variable $\alpha(n)$ in Eq. (14.75) may be interpreted as the *a priori estimation error*.

The estimation error $e(n)$ is computed at time n by using the value of the weight vector $\hat{\mathbf{w}}(n)$ at time n. Hence, $e(n)$ in Eq. (14.80) is the *a posteriori estimation error*.

From Eq. (14.80), the factor $\gamma(n)$ is the ratio of the a priori estimation error $\alpha(n)$ to the a posteriori estimation error $e(n)$; that is,

$$\gamma(n) = \frac{\alpha(n)}{e(n)} \tag{14.82}$$

We refer to $\gamma(n)$ as the *conversion factor*. The significance of it is that it *converts* the a priori estimation error $\alpha(n)$, based on the old tap-weight vector $\hat{\mathbf{w}}(n-1)$, into the a posteriori estimation error $e(n)$, based on the updated value $\hat{\mathbf{w}}(n)$. Note that

$$0 \leq \gamma(n) \leq 1, \qquad \text{for all } n \tag{14.83}$$

We will have a great deal more to say about the conversion factor $\gamma(n)$ in the next chapter.

In this section, we have presented a method for a *direct extraction* of the a posteriori estimation error $e(n)$ without having to compute the weight vector $\hat{\mathbf{w}}(n)$. The presentation has been somewhat lengthy, the motivation being to provide interpretations of the elements that make up $e(n)$. Indeed, the main result of the section described in Eq. (14.80) may be derived in a much simpler fashion (Shepherd and McWhirter, 1991); the reader is referred to Problem 6 for details.

14.10 SYSTOLIC ARRAY IMPLEMENTATION II

Figure 14.7 shows the systolic array implementation of the *modified QRD-RLS algorithm* (McWhirter, 1983). Apart from the introduction of an extra parameter $\gamma^{1/2}$ into the boundary cell, the boundary and internal cells in the structure of Fig. 14.7 are identical to those shown in the triangular systolic array section of Fig. 14.2.

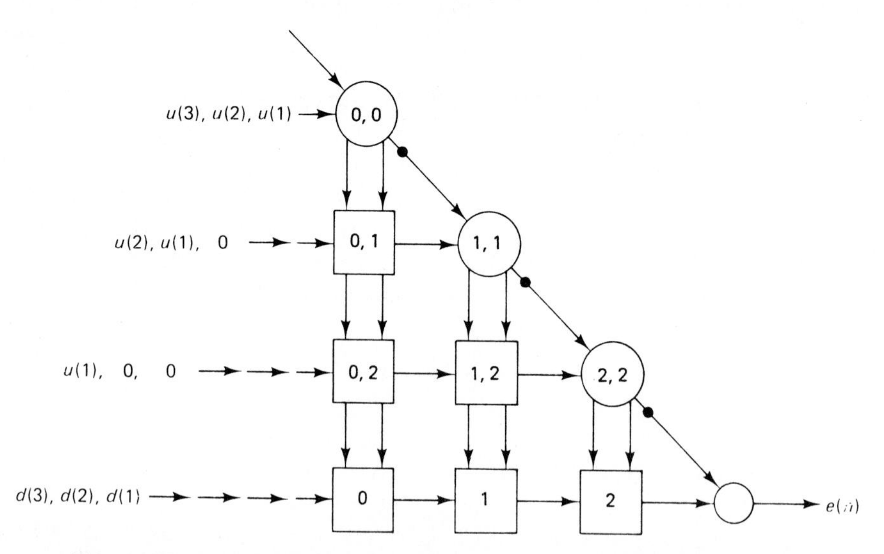

Figure 14.7 Systolic array implementation of the modified QRD-RLS algorithm. The dots along the diagonal of the array represent storage elements. This processing delay, which is a consequence of the temporal skew imposed on the input data, may be incorporated within the associated boundary cell.

Moreover, the linear systolic array section has been omitted from the structure of Fig. 14.7, as the modified QRD-RLS algorithm eliminates the need for computing the weight vector $\hat{\mathbf{w}}(n)$ explicitly. However, the structure of Fig. 14.7 includes a new element referred to as the *final processing cell* (indicated by a small circle). This cell is intended to compute the a posteriori estimation error $e(n)$. The specific functions of the boundary cells, the internal cells, and final processing cell are given in Fig. 14.8.

We remarked earlier that the scaling factor $\gamma^{1/2}(n)$, defined in Eq. (14.57), is generated as a result of the orthogonal triangularization process. This process manifests itself in Fig. 14.7 in the sense that the last boundary cell of the triangular section produces an output equal to $\gamma^{1/2}(n)$. Correspondingly, the last cell in the row of internal cells appended to the triangular section produces an output equal to

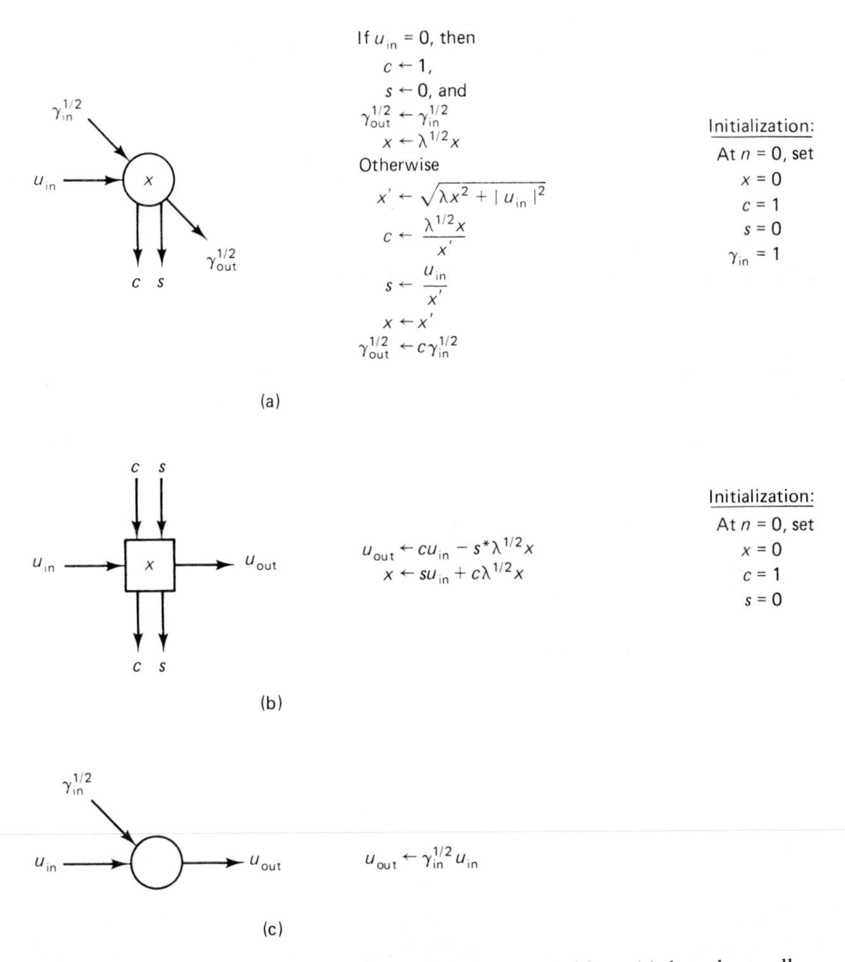

Figure 14.8 Cells for the modified QRD-RLS algorithm: (a) boundary cell; (b) internal cell; (c) final cell. *Note:* The stored value x is initialized to be zero (i.e., real). For the boundary cell, it always remains real. Hence, the formulas for the rotation parameters c and s computed by the boundary cell can be simplified considerably, as shown in part (a). Note also that the values x stored in the array are elements of $\mathbf{R}^H$; here, $r^* = x$ for all elements of the array.

$\gamma^{1/2}(n)\alpha(n)$. Thus, the final processing cell simply multiplies $\gamma^{1/2}(n)\alpha(n)$ thus produced by the output $\gamma^{1/2}(n)$ from the last boundary cell to produce the a posteriori estimation error $e(n)$.

As the time-skewed input data vectors enter the systolic array of Fig. 14.7, we find that updated estimation errors are produced at the output of the array at the rate of one every clock cycle. The estimation error produced on a given clock cycle corresponds, of course, to the particular element of the desired response vector $d(n)$ that entered the array M clock cycles previously.

It is noteworthy that the a priori estimation error $\alpha(n)$ may be obtained by *dividing* the output that emerges from the last cell in the appended (bottom) row of internal cells by the output from the last boundary cell. Also, the conversion factor $\gamma(n)$ may be obtained simply by squaring the output that emerges from the last boundary cell.

Figure 14.9 summarizes, in a diagrammatic fashion, the flow of signals in the systolic array of Fig. 14.7. The figure includes the external inputs $\mathbf{u}(n)$ and $d(n)$, the resulting transformations in the internal states of the triangular section and appended row of internal cells, the respective outputs of these two sections, and the overall output of the complete processor. Note that the systolic processor I of Fig. 14.2 and the systolic processor II of Fig. 14.7 are mathematically equivalent (though not numerically).

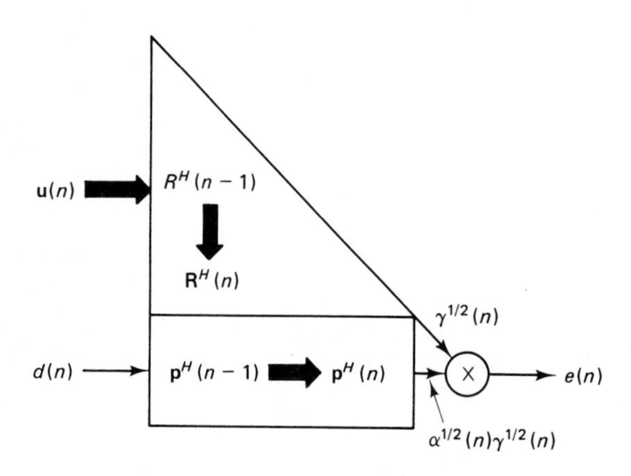

Figure 14.9 Diagrammatic representation of the flow of signals in the systolic array of Fig. 14.7.

Computation of the Weight Vector

A distinctive feature of the systolic structure shown in Fig. 14.7 is that unlike the systolic structure of Fig. 14.2, its operation does not involve the explicit computation of the weight vectior $\hat{\mathbf{w}}(n)$. Clearly, the structure of Fig. 14.7 may be extended to include a linear systolic section (as in Fig. 14.2) so as to compute $\hat{\mathbf{w}}(n)$, if so required. However, there is a simpler method of computing the weight vector $\hat{\mathbf{w}}(n)$ as a useful *by-product* of the direct error (residual) extraction capability inherent in the systolic process of Fig. 14.7.

The method for extracting the weight vector $\hat{\mathbf{w}}(n)$ is referred to as *serial weight flushing* (Ward et al., 1986; Shepherd and McWhirter, 1991). To explain the method, let $\mathbf{u}(n)$ denote the input vector and $d(n)$ denote the desired response, both at time n. Given that the weight vector at this time is $\hat{\mathbf{w}}(n)$, the corresponding a posteriori estimation error is

$$e(n) = d(n) - \hat{\mathbf{w}}^H(n)\mathbf{u}(n) \tag{14.84}$$

Suppose that the state of the array is *frozen* at time n_+, immediately *after* the systolic computation at time n is completed. Specifically, any update of stored values in the array is suppressed; otherwise, it is permitted to function normally in all other respects. At time n_+, we also set the desired response $d(n)$ equal to zero. We now define an input vector that consists of a string of zeros, except for the ith element that is set equal to unity, as shown by

$$\mathbf{u}^T(n_+) = [0 \cdots \underset{\substack{\uparrow \\ i\text{th element}}}{010} \cdots 0] \tag{14.85}$$

Thus, substituting these values in Eq. (14.84), we get

$$e(n_+) = -\hat{w}_i^*(n_+) \tag{14.86}$$

In other words, except for a trivial sign change, we may compute the ith element of the M-by-1 weight vector $\mathbf{w}^*(n)$ by freezing the state of the processor at time n, and subsequently setting the desired response equal to zero, and feeding the processor with an input vector whose ith element is unity and the remaining $M - 1$ elements are all zero. The essence of all of this is that the complex-conjugated weight vector $\hat{\mathbf{w}}^*(n)$ may be viewed as the *impulse response* of the *nonadaptive* (i.e., frozen) form of the systolic array processor in the sense that it can be generated as the system output produced by inputting an $(M - 1)$-by-$(M - 1)$ identity matrix to the main triangular array and a zero vector to the bottom row of the array in Fig. 14.7 (Shepherd and McWhirter, 1991).

To "flush" the entire M-by-1 weight vector $\hat{\mathbf{w}}^*(n)$ out of the systolic processor in Fig. 14.7, the procedure is therefore simply to halt the update of all stored values and input a data matrix that consists of a unit diagonal matrix (i.e., identity matrix) of dimension M. A demonstration of this procedure is given in the next section.

14.11 COMPUTER EXPERIMENT II: ADAPTIVE PREDICTION USING SYSTOLIC STRUCTURE II

In this section, we revisit the computer experiment on adaptive prediction that we studied in Section 14.7, except that this time we use the single-section systolic processor of Fig. 14.7 for the computation. In particular, we are interested in studying the *learning curve* of the processor for different levels of ill conditioning of the input data. The learning curve is computed by plotting the ensemble-averaged squared value of the a priori estimation error $\alpha(n)$ versus time n.

Using the autoregressive data base used for the computer experiment in Section 14.7, the learning curves shown in Fig. 14.10 are computed by ensemble averaging over 100 independent realizations of the systolic processor. The eigenvalue spreads used in the computation are 3, 10, and 100.

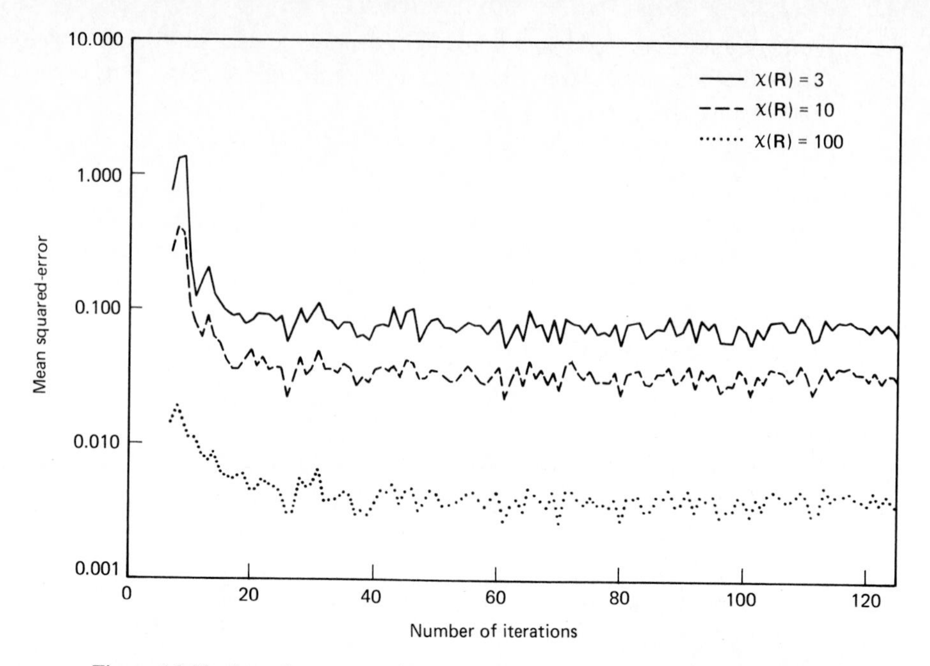

Figure 14.10 Learning curves of systolic array II for varying eigenvalue spreads.

Table 14.5 summarizes estimates of the final mean-squared error $J(\infty)$ and the corresponding values of the misadjustment $\mathcal{M} = (J(\infty)/J_{\min}) - 1$ for these eigenvalue spreads. The estimates of $J(\infty)$ were each obtained by "time averaging" over the last 100 points of the pertinent learning curve.

TABLE 14.5 SUMMARY OF RESULTS FOR COMPUTER EXPERIMENT II

Eigenvalue spread, $\chi(\mathbf{R})$	Measured steady-state value of mean-squared error	Minimum mean-squared error	Misadjustment, $\mathcal{M}(\%)$
3	0.076568	0.0731	4.7
10	0.033632	0.0322	4.4
100	0.003980	0.0038	4.8

From the results presented in Fig. 14.10 and Table 14.5, we can draw the following conclusions:

1. The systolic processor of Fig. 14.7, used as an adaptive predictor, achieves good convergence in the mean square in about three or four iterations; this is about twice the number of elements in the coefficient vector.
2. The small values of the misadjustment $\mathcal{M}$ (all less than 5 percent) demonstrate that the systolic processor is capable of producing results close to the optimum Wiener solution.
3. The convergence properties of the systolic processor are relatively insensitive to variations in the eignevalue spread of the input data.

Predictor Coefficients

To compute the 2-by-1 weight vector defining the predictor coefficients, we may use the "weight flushing" procedure described in Section 14.10. Specifically, a computer realization of the systolic processor in Fig. 14.7 was performed; the learning process was halted after 25 iterations, and the state of the processor was subsequently frozen. By inputting the data matrix

$$\mathbf{A} = \begin{bmatrix} 1 & 0 \\ 0 & 1 \end{bmatrix}$$

into the processor, the predictor coefficients $\hat{w}_0(n)$ and $\hat{w}_1(n)$ for $n = 25$ are flushed out. One hundred independent trials of this computation were performed, and the ensemble-averaged values of $\hat{w}_0(25)$ and $\hat{w}_1(25)$ were calculated for the eigenvalue spreads of 3, 10, and 100. The results so obtained are summarized in Table 14.6. This table also includes the corresponding optimum Wiener solution for the 2-by-1 weight vector as well as the result obtained by using the standard RLS algorithm for 25 iterations.

TABLE 14.6 OUTPUT OF SYSTOLIC ARRAY II AND COMPARISON WITH RLS ALGORITHM

	Experiment		Optimum		Relative error (%)	
$\chi(\mathbf{R})$	$\hat{w}_0$	$\hat{w}_1$	w_0	w_1	w_0	w_1
			Systolic array II, $n = 25$			
3	0.9555	−0.91519	0.9750	−0.95	2.0	3.7
10	1.5233	−0.88600	1.5955	−0.95	4.5	6.7
100	1.8252	−0.87115	1.9114	−0.95	4.5	8.3
			RLS, $n = 25$			
3	0.9335	−0.87126	0.9750	−0.95	4.2	8.3
10	1.5549	−0.90557	1.5955	−0.95	2.5	4.7
100	1.8339	−0.88253	1.9114	−0.95	4.1	7.1

From this table, we can make the following observations:

1. After 25 iterations, the experimentally computed values of the predictor coefficients closely approach the corresponding optimum values.
2. Convergence of the systolic processor in Fig. 14.7 in the mean is essentially independent of the eignevalue spread of the input data.

Conversion Factor

Figure 14.11 shows the time evolution of a *single realization* of the conversion $\gamma(n)$ for three different values of the eigenvalue spread: 3, 10, and 100. For each time n, the computation of $\gamma(n)$ was performed simply by squaring the output that emerges from the last boundary cell of the systolic processor.

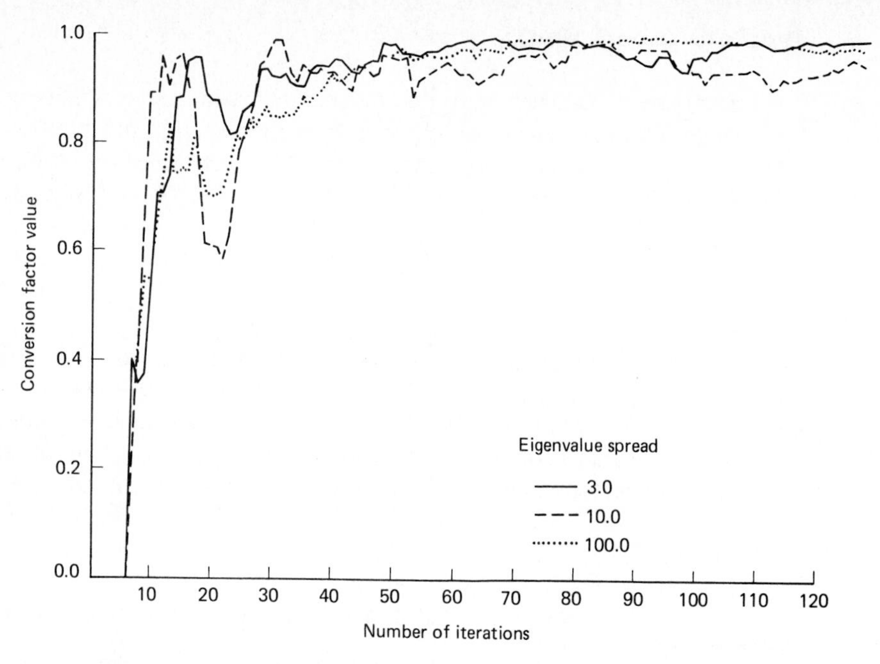

Figure 14.11 Single realizations of the conversion factor $\gamma(n)$ for varying eigenvalue spread.

From Fig. 14.11, we may make the following observations:

1. The initialization period, for which the conversion factor $\gamma(n)$ is zero, occupies six iterations, that is, $3M$.
2. After the completion of initialization, the conversion factor is essentially independent of the eigenvalue spread of the underlying correlation matrix of the input data.
3. The conversion factor converges rapidly toward the steady-state (final) value of unity. Specifically, it reaches a value of approximately 0.6 after about four or five iterations. This confirms our earlier observation that for the adaptive predictor of order 2 considered herein the systolic processor II converges in about four or five iterations, which is roughly equal to $2M$.

14.12 ADAPTIVE BEAMFORMING

An important practical application of adaptive filter theory is in *adaptive beamforming*. The need for this special form of adaptive spatial filtering arises in such diverse fields as radar, sonar, and communications. *The basic objective of an adaptive beamformer is to identify a set of complex weights for modifying the outputs of an array of sensors so as to produce a far-field pattern that optimizes the reception of a target signal along a direction of interest in some statistical sense*. The sensors may consist of *antenna* elements as in radar and communications, or *hydrophones* as in sonar. In any event, the adaptive processing may be performed in either one of two spaces:

1. *Data space,* where the adaptive beamformer operates directly on the data received by the array of sensors. In this case, we may have a linear array of uniformly spaced sensors, one of which (at either end of the array) serves as the *reference sensor* and the remaining ones serve as *primary sensors*. A *steering vector* is applied to the primary sensors for the purpose of pointing the beamformer along a look direction of interest. The reference sensor supplies the "desired response," and the outputs of the primary sensors are adaptively weighted so as to cancel intefering signals by placing nulls in their unknown directions.
2. *Beam space,* where the data received by the array of sensors are first processed by a fixed *beamforming network* that generates a prescribed set of *orthogonal* beams (orthogonal in a spatial sense). Depending on the look direction of interest, a particular output port of the beamforming network is selected to supply the desired response. The signals from the remaining output ports of the network are adaptively weighted so as to cancel unknown interferences.

A discussion of these two approaches was presented in Chapter 1. In the present chapter, we confine the discussion to adaptivity in data space. To be specific, consider a linear array consisting of $(M + 1)$ uniformly spaced sensors, as shown at the front end of Fig. 14.12. Let a *snapshot* of array data, that is, a set of $(M + 1)$

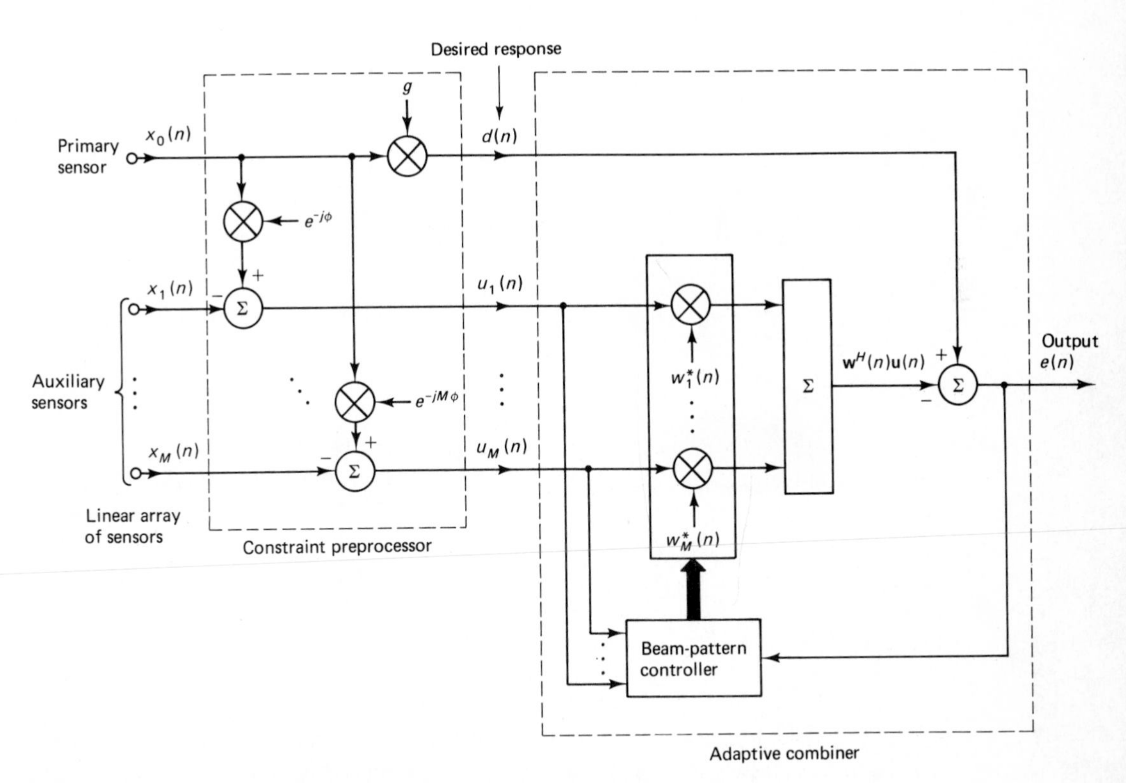

Figure 14.12 Structural layout of adaptive beamformer.

sensor outputs measured simultaneously at time n, be written in the partitioned form:

$$\text{input vector} = \begin{bmatrix} x_0(n) \\ \cdots \\ x_1(n) \\ \vdots \\ x_M(n) \end{bmatrix} = \begin{bmatrix} x_0(n) \\ \mathbf{x}(n) \end{bmatrix} \tag{14.87}$$

where $x_0(n)$ is the output of sensor 0 and $\mathbf{x}(n)$ is the vector of remaining M sensor outputs. Let this input vector be applied to a *linear combiner*, producing the output

$$\begin{aligned} e(n) &= [w_0^*(n), \mathbf{w}^H(n)]\begin{bmatrix} x_0(n) \\ \mathbf{x}(n) \end{bmatrix} \\ &= w_0^*(n)x_0(n) + \mathbf{w}^H(n)\mathbf{x}(n) \end{aligned} \tag{14.88}$$

where the scalar $w_0(n)$ and the vector $\mathbf{w}(n)$ constitute an $(M + 1)$-by-1 *weight vector for the entire array*. Suppose that we wish to track a target in the far field of the array that lies along a direction specified by the electrical angle ϕ. This electrical angle defines an $(M + 1)$-by-1 *steering vector*, which we may write as follows:

$$\text{steering vector} = \begin{bmatrix} 1 \\ e^{-j\phi} \\ e^{-j2\phi} \\ \vdots \\ e^{-jM\phi} \end{bmatrix} = \begin{bmatrix} 1 \\ \mathbf{e}(\phi) \end{bmatrix} \tag{14.89}$$

where $\mathbf{e}(\phi)$ is itself defined by

$$\mathbf{e}^T(\phi) = [e^{-j\phi}, e^{-j2\phi}, \ldots, e^{-jM\phi}] \tag{14.90}$$

In order to provide "protection" for the target signal being tracked along the direction specified by the electrical angle ϕ, we impose a *linear beam constraint* of the following form:

$$[w_0^*(n), \mathbf{w}^H(n)]\begin{bmatrix} 1 \\ \mathbf{e}(\phi) \end{bmatrix} = g$$

or, equivalently,

$$w_0^*(n) + \mathbf{w}^H(n)\mathbf{e}(\phi) = g \tag{14.91}$$

where g is some constant. The constraint of Eq. (14.91) ensures that *the beamformer provides a constant gain g along the direction specified by the electrical angle ϕ regardless of the values assigned to $w_0(n)$ and $\mathbf{w}(n)$*. In particular, if the sensors are *uniformly excited* by a distant source (in the far field of the array) along a look direction specified by the electrical angle ϕ, and each sensor output has a magnitude equal to unity, we may then set the sensor outputs as follows:

$$x_0(n) = 1$$

and

$$\mathbf{x}(n) = \mathbf{e}(\phi)$$

Substituting these values in Eq. (14.88) and using the constraint described in Eq. (14.91) yields an array output $e(n)$, and therefore a gain for the beamformer, equal to the constant g.

The overall operation of the beamformer is thus described by the pair of equations: (14.88) and (14.91). We wish to combine these two equations into a single relation that has the customary form

$$e(n) = d(n) - \mathbf{w}^H(n)\mathbf{u}(n) \tag{14.92}$$

where $\mathbf{u}(n)$ is an M-by-1 input vector; $d(n)$ may be viewed as a "desired response," in which case $e(n)$ represents an error signal. Both $d(n)$ and $\mathbf{u}(n)$ are to be defined. We may achieve this aim by using Eq. (14.91) to eliminate $\mathbf{w}_0^*(n)$ from Eq. (14.88). Specifically, we may rewrite the beamformer output $e(n)$ as follows:

$$e(n) = \underbrace{gx_0(n)}_{\substack{\text{desired}\\ \text{response}\\ d(n)}} - \mathbf{w}^H(n)\underbrace{[x_0(n)\mathbf{e}(\phi) - \mathbf{x}(n)]}_{\text{input vector } \mathbf{u}(n)} \tag{14.93}$$

The implications of using Eq. (14.93) to define the beamformer output are threefold:

1. By eliminating the unknown weight w_0 from the beamformer output, the number of *degrees of freedom* available for designing the beamformer is reduced by one to M, the dimension of the weight vector $\mathbf{w}(n)$.
2. The signal received from sensor 0 in the array, scaled by the constant gain g, is the "desired response" $d(n)$.
3. The "input vector" $\mathbf{u}(n)$ is completely determined by the specified electrical angle ϕ and the signals received from the entire array.

Consequently, the beamformer design may be optimized by minimizing a cost function based on the "error signal" $e(n)$. This, in turn, means that we may use an adaptive filtering algorithm of the type developed in the present and previous chapters of the book to compute the adjustable weight vector $\mathbf{w}(n)$.

The formulation of the error signal $e(n)$ as in Eq. (14.93) suggests that we structure the adaptive beamformer as shown in Fig. 14.12 (Ward et al., 1986). According to this structure, the linearly constrained adaptive beamformer[3] consists of two key components:

[3] For the general least-squares (adaptive) beamformer with a multitude of linear constraints, see McWhirter and Shepherd (1986) and Slock and Kailath (1989).

1. *Constraint preprocessor*. This first component is shown on the left side of Fig. 14.12. It transforms the combination of the signals $x_0(n), x_1(n), \ldots, x_M(n)$ received from the entire array of $M + 1$ sensors and the steering vector specified by the electrical angle ϕ linearly into two sets of signals: the desired response $d(n)$ and the input vector $\mathbf{u}(n)$. According to the representation depicted in Fig. 14.12, in effect the first sensor (labeled 0) supplies the desired response $d(n)$; in the context of previous terminology, this element serves the role of a reference sensor. To fulfill such a function properly, the reference sensor is assumed to have an *omnidirectional* coverage (i.e., constant gain in all directions). The remaining M sensors of the array, responsible for the generation of the input vector $\mathbf{u}(n)$, serve the role of *primary sensors*. Note that the constraint preprocessor ensures that any signal entering it from the specified look direction is removed from the input vector $\mathbf{u}(n)$ supplied to the adaptive combiner. The adaptive combiner is thereby *prevented from nulling* the target signal that is being tracked along the look direction of interest.
2. *Adaptive combiner*. This second component is shown on the right side of Fig. 14.12. The specific details of the adaptive combiner are naturally determined by the type of adaptive filtering algorithm selected to compute the adjustable weight vector $\mathbf{w}(n)$, given the input vector $\mathbf{u}(n)$ and the desired response $d(n)$ from the constraint preprocessor.

In the context of the latter issue, and in the light of the material covered in this book, we may use any one of three adaptive filtering approaches:

1. *LMS approach*. This approach offers simplicity (and therefore cost-effectiveness) and numerical robustness. The major shortcoming of the LMS algorithm is its *poor convergence properties,* particularly when it operates in a signal environment of broad dynamic range. This, in turn, places a fundamental limitation on the use of the LMS algorithm in sophisticated electronic systems.
2. *Standard least-squares estimation*. Here we have two options available to us:
   (a) We opt for *block estimation* to compute the least-squares estimate of the adjustable weight vector. That is, we use the system of normal equations to write [see Eq. (10.48)]

$$\hat{\mathbf{w}} = (\mathbf{A}^H\mathbf{A})^{-1}\mathbf{A}^H\mathbf{d} \tag{14.94}$$

   where $\mathbf{A}$ is the data matrix describing the time evolution of the input vector $\mathbf{u}(n)$, and $\mathbf{d}$ is the desired data vector describing the time evolution of the desired response $d(n)$, both on a snapshot-by-snapshot basis. In the context of adaptive beamforming, this approach is referred to in the literature as the *sample matrix inversion algorithm* (Reed et al., 1974).
   (b) We opt for a recursive approach and use the standard *recursive least-squares (RLS) algorithm* of Chapter 13.

   By using the sample matrix inversion or the RLS algorithm, we overcome the problem of poor convergence associated with the LMS algorithm. However, both approaches have two major problems of their own: (1) *increased computational complexity* that cannot be easily overcome through the use of very large scale integration (VLSI), and (2) *numerical instability* resulting from the use of finite-precision arithmetic and the requirement of inverting a large matrix.

3. *Data decomposition-based least-squares estimation*. This approach differs from approach 2 in that it involves *direct* orthogonalization of the data matrix. Basically, the orthogonalization may be rectangular or triangular in form, so we have two options available to us:
   (a) We use *rectangular* orthogonalization of the data matrix, which lends itself to block estimation. In particular, we apply the *singular-value decomposition* directly to the data matrix. Hence, we may express the solution to the linear least-squares problem as
$$\hat{\mathbf{w}} = \mathbf{A}^{+}\mathbf{d} \tag{14.95}$$
   where $\mathbf{A}^{+}$ is the pseudoinverse of the data matrix $\mathbf{A}$, and $\mathbf{d}$ is the desired data vector. For the computation of the singular-value decomposition, we use the *cyclic Jacobi* method, based on a sequence of plane (Givens) rotations. Details of this block estimation approach were presented in Chapter 11.
   (b) We use *triangular* orthogonalization of the data matrix, which lends itself to recursive estimation. In this case, the *QR-decomposition* is used to accomplish the direct orthogonalization of the data matrix. Here, we may use the conventional form of the QRD-RLS algorithm implemented[4] using the two-section systolic array of Fig. 14.2, or alternatively, the modified form of the QRD-RLS algorithm implemented using the single-section systolic array of Fig. 14.7.

Both systolic approaches under 3(b) offer a fast rate of convergence that is inherent in least-squares estimation, and good numerical properties resulting from the widely accepted fact that the method of QR-decomposition by Givens rotations is one of the very best procedures for solving the linear least-squares problem. In practice, the modified systolic array II of Section 14.10 is preferred over the systolic array I of Section 14.6 because it offers a considerable reduction in the amount of the required computation and circuitry, since it avoids the use of a linear section that is an integral part of systolic array I. In particular, it is no longer necessary for us to clock out the elements of the lower triangular matrix $\mathbf{R}^H(n)$ computed in the cells of the triangular section, solve for the weight vector $\hat{\mathbf{w}}(n)$ by back substitution, or compute the adaptive beamformer output by forming the inner product $\hat{\mathbf{w}}^H(n)\mathbf{u}(n)$.

In summary, the combination of the constraint preprocessor and the modified systolic array processor II, described respectively in Figs. 14.3 and 14.7, provides an important method for solving a linearly constrained adaptive beamforming problem. In particular, this combination offers the following attractive features:

1. The constraint preprocessor steers the beamformer along a look direction of interest determined by the value of the electrical angle ϕ, and also protects a

[4] The recursive least-squares estimation for adaptive beamforming may also be solved in a numerically stable manner by using a sequence of Householder transformations (Rader and Steinhart, 1986). However, the QRD-RLS algorithms described in this chapter have the advantage of offering a higher degree of parallelism; this is made possible by the use of the Givens rotations, which operate on smaller (i.e., 2-by-2) blocks of data.

target signal being tracked along that direction; the electrical angle ϕ is related to the actual incidence angle θ by Eq. (1.61).

2. The systolic array provides for the adaptive processing of the constraint preprocessor outputs so as to cancel interferences originating from unknown directions; it does so with a fast rate of convergence and in a numerically stable manner.

14.13 MINIMUM-VARIANCE DISTORTIONLESS RESPONSE BEAMFORMERS

In the preceding section, we showed that by means of a *constraint preprocessor,* it is possible to extend the use of systolic arrays I and II, based on the QR-decomposition, to solve constrained least-squares estimation recursively as it would arise in adaptive beamforming applications. The order of signal processing in the constraint preprocessor method is depicted in Fig. 14.13(a). Here we note that the preprocessor consumes approximately M operations, while the least-squares processor uses approximately $\frac{1}{2}M^2$ operations, where M is the number of adjustable weights in the beamformer. If we wish to impose several (K, say) constraints independently (but not simultaneously!), that is, if the same processing is to be done along K separate look directions, we then have to run the same data through the beamformer a total of K times. Such an imposition requires K different preprocessor configurations and K least-squares processing runs, taking a total of approximately $(KM + \frac{1}{2}KM^2)$ operations.

An alternative method is to use a *minimum-variance distortionless response (MVDR) beamformer,* in which the variance (i.e., average power) of the output is minimized under the constraint that a distortionless response is maintained along the

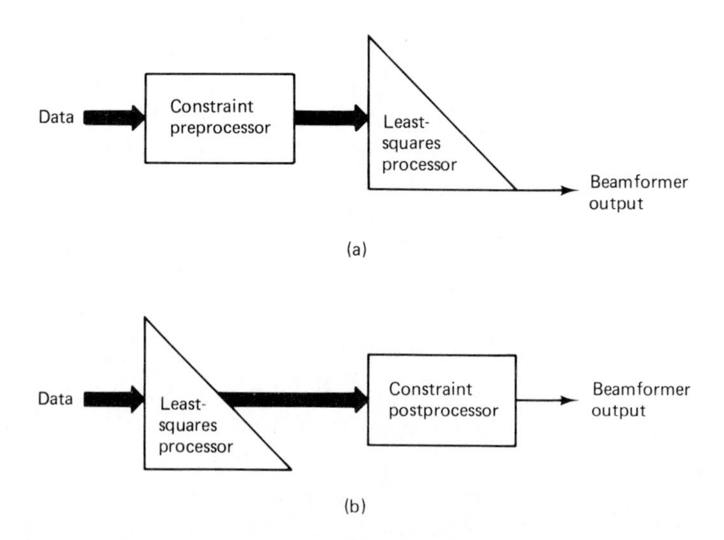

Figure 14.13 (a) Constraint preprocessor-based beamformer; (b) MVDR beamformer.

direction of a target signal of interest, hence the name of the beamformer. The MVDR beamformer is basically a *constraint postprocessor* in that it permits the application of the constraint *after* the least-squares processing, as depicted in Fig. 14.13(b). Consequently, the output from the least-squares processor is *broadcast* to K separate constraint processors, each of which uses approximately M operations (see below). The total number of operations in the MVDR beamformer is thus approximately $(\frac{1}{2}M^2 + KM)$; this is considerably less than the corresponding number of operations for the preprocessor-constraint method for large values of K and M. Accordingly, an MVDR beamformer has the advantage of being computationally more efficient than a preprocessor constraint-based beamformer, and also it may be performed potentially on the fly.

The MVDR Problem

Consider then a linear array of M uniformly spaced sensors whose outputs are individually weighted and then summed to produce the beamformer output

$$e(i) = \sum_{\ell=1}^{M} w_\ell^*(n)u_\ell(i) \tag{14.96}$$

where u_ℓ is the output of sensor ℓ at time i, and $w_\ell(n)$ is the associated weight. To simplify the mathematical presentation, we consider the simple case of a single look direction. Let $s_1(\phi), s_2(\phi), \ldots, s_M(\phi)$ be the elements of a prescribed *steering vector* $\mathbf{s}(\phi)$; the electrical angle ϕ is determined by the look direction of interest. In particular, the element $s_\ell(\phi)$ is the output of sensor ℓ of the array under the condition that there is no signal other than that due to a source of interest. We may thus state the MVDR problem as follows:

Minimize the cost function

$$\mathscr{E}(n) = \sum_{i=1}^{n} \lambda^{n-i}|e(i)|^2 \tag{14.97}$$

subject to the constraint

$$\sum_{\ell=1}^{M} w_\ell^*(n)s_\ell(\phi) = 1, \qquad \text{for all } n \tag{14.98}$$

Using matrix rotation, we may redefine the cost function $\mathscr{E}$(n) of Eq. (14.97) as

$$\mathscr{E}(n) = \boldsymbol{\epsilon}^H(n)\boldsymbol{\Lambda}(n)\boldsymbol{\epsilon}(n) \tag{14.99}$$

where $\boldsymbol{\Lambda}(n)$ is the exponential weighting matrix, and $\boldsymbol{\epsilon}(n)$ is the constrained beamformer output vector. Although Eq. (14.99) has the same mathematical form as that of Eq. (14.8), nevertheless, the vector $\boldsymbol{\epsilon}(n)$ has different interpretations in these two equations. According to Eq. (14.96), the beamformer output vector $\boldsymbol{\epsilon}(n)$ is related to the data matrix $\mathbf{A}(n)$ by

$$\begin{aligned}\boldsymbol{\epsilon}(n) &= [e(1), e(2), \ldots, e(n)]^H \\ &= \mathbf{A}(n)\mathbf{w}(n)\end{aligned} \tag{14.100}$$

where $\mathbf{w}(n)$ is the weight vector, and the data matrix $\mathbf{A}(n)$ is defined in terms of the snapshots $\mathbf{u}(1), \mathbf{u}(2), \ldots, \mathbf{u}(n)$ by

$$\mathbf{A}^H(n) = [\mathbf{u}(1), \mathbf{u}(2), \ldots, \mathbf{u}(n)]$$

$$= \begin{bmatrix} u_1(1) & u_1(2) & \cdots & u_1(n) \\ u_2(1) & u_2(2) & \cdots & u_2(n) \\ \vdots & \vdots & & \vdots \\ u_M(1) & u_M(2) & \cdots & u_M(n) \end{bmatrix} \tag{14.101}$$

We may now restate the MVDR problem in matrix terms as follows:

Given the data matrix $\mathbf{A}(n)$ and the exponential weighting matrix $\mathbf{\Lambda}(n)$, minimize the cost function

$$\mathscr{E}(n) = \| \mathbf{\Lambda}^{1/2}(n)\mathbf{A}(n)\mathbf{w}(n) \|^2$$

with respect to the weight vector $\mathbf{w}(n)$, subject to the constraint

$$\mathbf{w}^H(n)\mathbf{s}(\phi) = 1, \qquad \text{for all } n$$

where $\mathbf{s}(\phi)$ is the steering vector for a prescribed electrical angle ϕ.

The solution to this constrained optimization problem is described by the MVDR formula (see Section 10.11)

$$\hat{\mathbf{w}}(n) = \frac{\mathbf{\Phi}^{-1}(n)\mathbf{s}(\phi)}{\mathbf{s}^H(\phi)\mathbf{\Phi}^{-1}(n)\mathbf{s}(\phi)} \tag{14.102}$$

where $\mathbf{\Phi}(n)$ is the M-by-M correlation matrix of the exponentially weighted sensor outputs averaged over n shapshots, and which is related to the data matrix $\mathbf{A}(n)$ as follows:

$$\mathbf{\Phi}(n) = \mathbf{A}^H(n)\mathbf{\Lambda}(n)\mathbf{A}(n) \tag{14.103}$$

In the remainder of this section we describe two adaptive systolic beamformers for solving the MVDR problem both of which are based on the QR-decomposition of the data matrix. The first method [due to Schreiber (1986)] builds on the Gentleman–Kung structure (i.e., systolic array processor I of Section 14.6). The second method [due to McWhirter and Shepherd (1989)] provides for a direct computation of the MVDR beamformer output by incorporating the McWhirter structure (i.e., systolic array processor II of Section 14.10). These two MVDR beamformers are considered in turn.

Systolic MVDR Beamformer I

From the discussion presented in Section 14.2, we recall that the correlation matrix $\mathbf{\Phi}(n)$ equals the matrix product $\mathbf{R}^H(n)\mathbf{R}(n)$, where $\mathbf{R}(n)$ is the upper triangular matrix that results from the application of the QR-decomposition to the exponentially weighted data matrix $\mathbf{A}(n)$. We may therefore transform the solution given in Eq. (14.102) as follows:

$$\hat{\mathbf{w}} = \frac{\mathbf{R}^{-1}(n)\mathbf{R}^{-H}(n)\mathbf{s}(\phi)}{\mathbf{s}^H(\phi)\mathbf{R}^{-1}(n)\mathbf{R}^{-H}(n)\mathbf{s}(\phi)} \tag{14.104}$$

where $\mathbf{R}^{-H}(n)$ is the Hermitian transpose of the inverse matrix $\mathbf{R}^{-1}(n)$. Define the *auxiliary vector*

$$\mathbf{a}(n) = \mathbf{R}^{-H}(n)\mathbf{s}(\phi) \tag{14.105}$$

We then note that the denominator of Eq. (14.104) is a real-valued scalar equal to the inner product of the auxiliary vector $\mathbf{a}(n)$ with itself. As for the numerator, it is equal to the vector $\mathbf{a}(n)$ premultiplied by the inverse matrix $\mathbf{R}^{-1}(n)$. We may thus simplify Eq. (14.104) as

$$\hat{\mathbf{w}}(n) = \frac{\mathbf{R}^{-1}(n)\mathbf{a}(n)}{\mathbf{a}^H(n)\mathbf{a}(n)} \tag{14.106}$$

For the computation of the weight vector $\hat{\mathbf{w}}(n)$, we may proceed as follows (Schreiber, 1986):

1. *Updating of the upper triangular matrix*. The upper triangular matrix $\mathbf{R}(n-1)$ is obtained from the QR-decomposition of the data matrix $\mathbf{A}(n-1)$. At time n, a new data snapshot $\mathbf{u}(n)$ is received, and the data matrix is updated as $\mathbf{A}(n)$ in accordance with Eq. (14.36). Given $\mathbf{R}(n-1)$ and $\mathbf{u}(n)$, the updated value $\mathbf{R}(n)$ of the upper triangular matrix is computed using Eq. (14.44), reproduced here for convenience of presentation:

$$\begin{bmatrix} \mathbf{R}(n) \\ \cdots\cdots \\ \mathbf{O} \end{bmatrix} = \mathbf{T}(n) \begin{bmatrix} \lambda^{1/2}\mathbf{R}(n-1) \\ \cdots\cdots\cdots\cdots \\ \mathbf{O} \\ \mathbf{u}^H(n) \end{bmatrix} \tag{14.107}$$

   where $\mathbf{T}(n)$ comprises a sequence of standard Givens rotations. This equation is implemented using a triangular systolic array similar to that described in Section 14.7.

2. *Updating of the auxiliary vector*. The updated value $\mathbf{a}(n)$ of the auxiliary vector is computed from the old value $\mathbf{a}(n-1)$ using the same sequence of Givens rotations that constitutes the transformation $\mathbf{T}(n)$. We may derive this recursion in two steps:
   (a) Consider the identity

$$\mathbf{I} = [\lambda^{1/2}\mathbf{R}^H(n-1), \mathbf{O}^T, \mathbf{u}(n)]\mathbf{T}^H(n)\mathbf{T}(n) \begin{bmatrix} \lambda^{-1/2}\mathbf{R}^{-H}(n-1) \\ \mathbf{O} \\ \mathbf{0}^T \end{bmatrix} \tag{14.108}$$

   where it is noted $\mathbf{T}(n)$ is a unitary matrix and therefore $\mathbf{T}^H(n)\mathbf{T}(n)$ equals an n-by-n identity matrix. The identity matrix $\mathbf{I}$ on the left side of Eq. (14.108) has dimension M. Applying the Hermitian transposition to both sides of Eq. (14.44), we have

$$[\mathbf{R}^H(n), \mathbf{O}^T] = [\lambda^{1/2}\mathbf{R}^H(n-1), \mathbf{O}^T, \mathbf{u}(n)]\mathbf{T}^H(n) \tag{14.109}$$

Therefore, the use of Eq. (14.109) in (14.108) yields

$$\mathbf{I} = [\mathbf{R}^H(n), \mathbf{O}^T]\mathbf{T}(n)\begin{bmatrix} \lambda^{-1/2}\mathbf{R}^{-H}(n-1) \\ \text{--------} \\ \mathbf{O} \end{bmatrix} \tag{14.110}$$

Equivalently, we may invoke the special structure of the transformation $\mathbf{T}(n)$ to write

$$\begin{bmatrix} \mathbf{R}^{-H}(n) \\ \text{-------} \\ \mathbf{X} \end{bmatrix} = \mathbf{T}(n)\begin{bmatrix} \lambda^{-1/2}\mathbf{R}^{-H}(n-1) \\ \text{--------} \\ \mathbf{O} \end{bmatrix} \tag{14.111}$$

where $\mathbf{X}$ denotes a matrix of no specific interest. Equation (14.111) shows that the inverse Hermitian matrix $\mathbf{R}^{-H}(n-1)$ may be updated by applying exactly the *same* sequence of Givens rotations as that used to update the upper triangular matrix $\mathbf{R}(n-1)$.

(b) Postmultiplying both sides of Eq. (14.111) by the steering vector $\mathbf{s}(\phi)$, and using the definition of the auxiliary vector $\mathbf{a}(n)$ given in Eq. (14.105), we get

$$\begin{bmatrix} \mathbf{a}(n) \\ \mathbf{x} \end{bmatrix} = \mathbf{T}(n)\begin{bmatrix} \lambda^{-1/2}\mathbf{a}(n-1) \\ \mathbf{O} \end{bmatrix} \tag{14.112}$$

where $\mathbf{x}$ is a vector of no specific interest. Equation (14.112) is the desired recursion for updating the old value $\mathbf{a}(n-1)$ of the auxiliary vector. Indeed, this equation shows that using the same set of Givens rotations that constitute the transformation $\mathbf{T}(n)$, we may compute $\mathbf{a}(n)$ from $\mathbf{a}(n-1)$. Note that the update of $\mathbf{R}^{-H}(n)$ given in Eq. (14.111) implies the update of $\mathbf{a}$ in Eq. (14.112), and vice versa.

3. *Computation of the weight vector*. Having computed the upper triangular matrix $\mathbf{R}(n)$ and the auxiliary vector $\mathbf{a}(n)$, we next proceed to compute the weight vector $\hat{\mathbf{w}}(n)$. Specifically, we use Eq. (14.104) to write

$$\mathbf{R}(n)\hat{\mathbf{w}}(n) = \rho(n)\mathbf{a}(n) \tag{14.113}$$

where $\rho(n)$ is a scalar defined by

$$\rho(n) = \frac{1}{\mathbf{a}^H(n)\mathbf{a}(n)} \tag{14.114}$$

Equation (14.113) has the same basic mathematical form as that of Eq. (14.21). We may therefore use the method of back substitution to solve Eq. (14.113) for the weight vector $\hat{\mathbf{w}}(n)$ at time n.

4. Time is incremented to $n + 1$, and we go back to step 1 and the process is repeated.

It is important to note that for the MVDR beamforming algorithm described herein to work, the auxiliary vector $\mathbf{a}(n)$ must be *initialized* at some time (n_0, say).

TABLE 14.7 SUMMARY OF MVDR BEAMFORMING ALGORITHM I

Initialization: Initialize the auxiliary vector $\mathbf{a}(n)$ at some time (n_0, say) for which the inverse matrix $\mathbf{R}^{-1}(n_0)$ exists.

Computation

1. Compute the upper triangular matrix $\mathbf{R}(n)$ from its old value $\mathbf{R}(n-1)$ using the recursion

$$\begin{bmatrix} \mathbf{R}(n) \\ \text{----} \\ \mathbf{O} \end{bmatrix} = \mathbf{T}(n) \begin{bmatrix} \lambda^{1/2}\mathbf{R}(n-1) \\ \text{-------------} \\ \mathbf{O} \\ \mathbf{u}^H(n) \end{bmatrix}$$

   where $\mathbf{u}(n)$ is the new data snapshot (vector) at time n, and $\mathbf{T}(n)$ is an n-by-n transformation defined by a sequence of standard Givens rotations.

2. Using the same set of Givens rotations, compute the auxiliary vector $\mathbf{a}(n)$ from its old value $\mathbf{a}(n-1)$:

$$\begin{bmatrix} \mathbf{a}(n) \\ \text{----} \\ \mathbf{x} \end{bmatrix} = \mathbf{T}(n) \begin{bmatrix} \lambda^{-1/2}\mathbf{a}(n-1) \\ \text{-------------} \\ 0 \end{bmatrix}$$

   where $\mathbf{x}$ is a vector of no specific interest.

3. At time n, use $\mathbf{R}(n)$ and $\mathbf{a}(n)$ to back-solve for $\mathbf{R}^{-1}(n)\mathbf{a}(n)$. Hence, compute the weight vector

$$\hat{\mathbf{w}}(n) = \frac{\mathbf{R}^{-1}(n)\mathbf{a}(n)}{\mathbf{a}^H(n)\mathbf{a}(n)}$$

4. Increment time to $n + 1$, go back to step 1 and repeat the process.

The reason for this initialization is that at time $n = 0$, the value

$$\mathbf{a}(0) = \mathbf{R}^{-H}(0)\mathbf{s}(\phi)$$

does *not* exist if $\mathbf{R}(0) = \mathbf{0}$. It is at the stage specified by time n_0 that the inverse matrix $\mathbf{R}^{-1}(n_0)$ exists, which therefore permits the explicit computation of the initial value $\mathbf{a}(n_0)$.

Table 14.7 presents a summary of the MVDR beamforming algorithm I (Schreiber, 1986). This algorithm is computationally efficient: in the general case of K independent look directions it only requires approximately $\frac{1}{2}M^2 + KM$ operations per cycle time, which corresponds to the minimum computational complexity. Moreover, the algorithm is known to have excellent numerical properties (Schreiber, 1986). It may also be implemented efficiently on some form of a systolic array. However, it appears to be difficult to implement the complete algorithm in a fully pipelined fashion. For example, updating the upper triangular matrix $\mathbf{R}(n)$ requires top-to-bottom processing, whereas the method of back substitution requires bottom-to-top processing. A method for overcoming this difficulty is described below.[5]

[5] Bojanczyk and Luk (1987) describe another method for a fully pipelined systolic computation of MVDR. The Bojanczyk-Luk MVDR processor, however, is more complicated than that due to McWhirter and Shepherd (1989). The description of the fully pipelined MVDR beamformer presented in the last part of this section is based on the paper by McWhirter and Shepherd.

Returning to Eq. (14.106), we may define the a posteriori estimation error at time n as

$$\begin{aligned} e(n) &= \hat{\mathbf{w}}^H(n)\mathbf{u}(n) \\ &= \frac{e'(n)}{\|\mathbf{a}(n)\|^2} \end{aligned} \tag{14.115}$$

where $e'(n)$ is another estimation error, defined by

$$e'(n) = \mathbf{a}^H(n)\mathbf{R}^{-H}(n)\mathbf{u}(n) \tag{14.116}$$

Moreover, we may use Eq. (14.105) to write the steering vector $\mathbf{s}(\phi)$ in terms of the value of the auxiliary vector at time $n - 1$ as follows:

$$\mathbf{s}(\phi) = \mathbf{R}^H(n-1)\mathbf{a}(n-1) \tag{14.117}$$

Clearly, we may also express $\mathbf{a}(n-1)$ in the equivalent form

$$\mathbf{s}(\phi) = \lambda^{-1}[\lambda^{1/2}\mathbf{R}^H(n-1) \vdots \mathbf{O}^T \vdots \mathbf{u}(n)] \begin{bmatrix} \lambda^{1/2}\mathbf{a}(n-1) \\ \hdashline \lambda^{1/2}\mathbf{v}(\mathbf{n}-1) \\ \hdashline \mathbf{0} \end{bmatrix} \tag{14.118}$$

where $\mathbf{v}(n-1)$ is an $(n - M - 1)$-by-1 arbitrary vector. Since $\mathbf{T}(n)$ is a unitary matrix, we have

$$\mathbf{T}^H(n)\mathbf{T}(n) = \mathbf{I}$$

Accordingly, we may inject this relation at the midpoint of the matrix product on the right side of Eq. (14.118) and so rewrite it in the equivalent form

$$\mathbf{s}(\phi) = \lambda^{-1}[\lambda^{1/2}\mathbf{R}^H(n-1) \vdots \mathbf{O}^T \vdots \mathbf{u}(n)]\mathbf{T}^H(n)\mathbf{T}(n) \begin{bmatrix} \lambda^{1/2}\mathbf{a}(n-1) \\ \hdashline \lambda^{1/2}\mathbf{v}(\mathbf{n}-1) \\ \hdashline \mathbf{0} \end{bmatrix} \tag{14.119}$$

Accordingly, substituting the Hermitian transposed version of Eq. (14.107) yields

$$\mathbf{s}(\phi) = \lambda^{-1}[\mathbf{R}^H(n) \vdots \mathbf{O}^T]\mathbf{T}(n) \begin{bmatrix} \lambda^{1/2}\mathbf{a}(n-1) \\ \lambda^{1/2}\mathbf{v}(n-1) \\ \mathbf{0} \end{bmatrix} \tag{14.120}$$

We now recognize from Section 14.8 that the structure of the transformation $\mathbf{T}(n)$ is such that we may write

$$\begin{aligned} \mathbf{T}(n) \begin{bmatrix} \lambda^{1/2}\mathbf{a}(n-1) \\ \lambda^{1/2}\mathbf{v}(n-1) \\ \mathbf{0} \end{bmatrix} &= \begin{bmatrix} \boldsymbol{\pi}(n) \\ \lambda^{1/2}\mathbf{v}(n-1) \\ \epsilon^*(\mathbf{n}) \end{bmatrix} \\ &= \begin{bmatrix} \boldsymbol{\pi}(n) \\ \mathbf{v}(n) \end{bmatrix} \end{aligned} \tag{14.121}$$

Hence, using Eq. (14.121) in (14.120), we obtain

$$\begin{aligned} \mathbf{s}(\phi) &= \lambda^{-1}[\mathbf{R}^H(n) \,\vdots\, \mathbf{O}^T]\begin{bmatrix} \boldsymbol{\pi}(n) \\ \mathbf{v}(n) \end{bmatrix} \\ &= \lambda^{-1}\mathbf{R}^H(n)\boldsymbol{\pi}(n) \end{aligned} \tag{14.122}$$

However, from Eq. (14.105) we also have

$$\mathbf{s}(\phi) = \mathbf{R}^H(n)\mathbf{a}(n)$$

It follows therefore that the vector $\boldsymbol{\pi}(n)$ is simply a scaled version of the auxiliary vector $\mathbf{a}(n)$, as shown by

$$\boldsymbol{\pi}(n) = \lambda\mathbf{a}(n) \tag{14.123}$$

We may thus view Eq. (14.121) as a straightforward recursion for computing the updated auxiliary vector $\mathbf{a}(n)$.

Table 14.8 presents a summary of MVDR beamforming algorithm II (McWhirter and Shepherd, 1989). The main point of difference between this MVDR algorithm and that summarized in Table 14.7 is that in the former we avoid the need for computing the weight vector $\hat{\mathbf{w}}(n)$. By so doing, a fully pipelined systolic implementation of the MVDR beamforming algorithm is feasible. We may develop the

TABLE 14.8 SUMMARY OF MVDR BEAMFORMING ALGORITHM II

Initialization: Initialize the auxiliary vector $\mathbf{a}(n)$ at some time n_0 for which the inverse matrix $\mathbf{R}^{-1}(n_0)$ exists.

Computation

1. Update the upper triangular matrix $\mathbf{R}(n-1)$ using the recursion

$$\begin{bmatrix} \mathbf{R}(n) \\ \text{----} \\ \mathbf{O} \end{bmatrix} = \mathbf{T}(n)\begin{bmatrix} \lambda^{1/2}\mathbf{R}(n-1) \\ \text{-------------} \\ \mathbf{O} \\ \mathbf{u}^H(n) \end{bmatrix}$$

where $\mathbf{u}(n)$ is the new data snapshot at time n, and $\mathbf{T}(n)$ is an n-by-n transformation defined by a sequence of Givens rotations.

2. Update the auxiliary vector $\mathbf{a}(n-1)$ using the recursion

$$\begin{bmatrix} \lambda\,\mathbf{a}(n) \\ \text{-----} \\ \mathbf{v}(n) \end{bmatrix} = \mathbf{T}(n)\begin{bmatrix} \lambda^{1/2}\mathbf{a}(n-1) \\ \text{-------------} \\ \lambda^{1/2}\mathbf{v}(n-1) \\ 0 \end{bmatrix}$$

where $\mathbf{v}(n)$ is a vector of no specific interest, except for the final element that may be used to compute $e'(n)$.

3. Given the updated values $\mathbf{R}(n)$ and $\mathbf{a}(n)$, compute the MVDR beamformer output at time n as

$$e(n) = \frac{e'(n)}{\|\mathbf{a}(n)\|^2}$$

where $e'(n)$ is the last element of the vector $\mathbf{v}(n)$; it may also be computed using

$$e'(n) = \mathbf{a}^H(n)\mathbf{R}^{-H}(n)\mathbf{u}(n).$$

4. Increment time n by 1, go back to step 1 and repeat the process.

idea for a fully pipelined systolic implementation of the MVDR beamforming algorithm II by comparing the solution to the MVDR problem with the solution to the ordinary least-squares estimation problem. This comparison is displayed in Table 14.9. In particular, the pair of relations on the left side of this table is a reproduction of Eqs. (14.116) and (14.121) with $\lambda\mathbf{a}(n)$ written in place of $\boldsymbol{\pi}(n)$. The first relation on the right side of the table follows from the definition of the a posteriori estimation error $e(n)$ given in Eqs. (14.84) and (14.21), and the second relation is a reproduction of Eq. (14.48).

TABLE 14.9

Minimum-variance distortionless response	Least-squares estimation
$e'(n) = \mathbf{a}^H(n)\mathbf{R}^{-H}(n)\mathbf{u}(n)$	$e(n) = d(n) - \mathbf{p}^H(n)\mathbf{R}^{-H}(n)\mathbf{u}(n)$
$\mathbf{T}(n)\begin{bmatrix}\lambda^{1/2}\mathbf{a}(n-1)\\ \lambda^{1/2}\mathbf{v}(n-1)\\ \mathbf{0}\end{bmatrix} = \begin{bmatrix}\lambda\,\mathbf{a}(n)\\ \lambda^{1/2}\mathbf{v}(n-1)\\ \epsilon^*(n)\end{bmatrix}$	$\mathbf{T}(n)\begin{bmatrix}\lambda^{1/2}\mathbf{p}(n-1)\\ \lambda^{1/2}\mathbf{v}(n-1)\\ d^*(n)\end{bmatrix} = \begin{bmatrix}\mathbf{p}(n)\\ \lambda^{1/2}\mathbf{v}(n-1)\\ \gamma^{1/2}(n)\alpha^*(n)\end{bmatrix}$ where $\alpha(n)$ is the a priori estimation error.

Next referring to the systolic array of Fig. 14.7, we note that at time $n - 1$, the row vector $\mathbf{p}^H(n-1)$ is generated by the bottom row of internal cells. The estimation error $e(n)$ is obtained from the output of the final cell at time n.

Accordingly, if we replace the vector $\mathbf{p}(n-1)$ by $\mathbf{a}(n-1)$, and set the value of $d(n)$ in Fig. 14.7 equal to zero, then at time n we may make the following observations:

1. The output from the final cell in Fig. 14.7 is identically equal to $-\lambda e'(n)$.
2. The vector $\lambda\mathbf{a}^H(n)$ is generated and stored in the bottom row of cells in Fig. 14.7.
3. The process may be continued in a simple recursive manner so as to generate the sequence of errors $e'(n), e'(n+1), \ldots$.

These observations may be exploited to develop a fully pipelined systolic implementation of the MVDR beamforming algorithm II; for details see McWhirter and Shepherd (1989).

14.14 COMPUTER EXPERIMENT ON ADAPTIVE BEAMFORMING

In this section we illustrate the performance of the systolic array implementation of an adaptive MVDR beamformer by considering a linear array of five uniformly spaced sensors. The spacing d between adjacent elements equals one half of the received wavelength. The array operates in an environment that consists of a target signal and a single source of interference, which are noncoherent with each other. The exponential weighting factor $\lambda = 1$.

The aims of the experiment are two-fold:

1. To examine the evolution of the adapted spatial response (pattern) of the beamformer with time
2. To evaluate the effect of varying the interference-to-target ratio on the interference-nulling capability of the beamformer

The directions of the target and source of interference are as follows:

Excitation	Angle of incidence, θ, measured with respect to normal to the array (radians)
Target	$\sin^{-1}(0.2)$
Interference	$-\sin^{-1}(0.4)$ or 0

That is, the interference may originate from any one of two possible directions.

The steering vector is defined by

$$\mathbf{s}^T(\phi) = [1, e^{-j\phi}, e^{-j2\phi}, e^{-j3\phi}, e^{-j4\phi}]$$

where $\phi = \pi \sin \theta$.

Figure 14.14 shows the individual effects of the parameters of interest on the adapted response of the beamformer. The initialization period occupies a total of 10 snapshots. The response is obtained by plotting $20\log_{10}|\hat{\mathbf{w}}^H(n)\mathbf{s}(\phi)|$ versus $\sin \theta = \phi/\pi$. The weight vector $\hat{\mathbf{w}}(n)$ is computed by using the theory described in Section 14.13. The target-to-noise ratio is held constant at 10 dB. In the first part of Fig. 14.14(a), pertaining to a data base of 20 snapshots after the initialization has been completed, the three curves correspond to interference-to-noise ratios = 40, 30, 20 dB. In the second part, the data base is increased to 100 snapshots, and in the last part it is increased to 200 snapshots, again after the initialization has been completed. Part (a) of Fig. 14.14 corresponds to an angle of arrival $\theta = -\sin^{-1}(0.4)$ radians for the interference. Part (b) of the figure corresponds to $\theta = 0$ for the interference.

Based on these results, we may make the following observations:

1. The response of the beamformer is held fixed at a value of one under all conditions, as required.
2. With as few as 30 snapshots, including initialization, the beamformer exhibits a reasonably effective nulling capability, which continually improves as the beamformer processes more snapshots.
3. The response of the beamformer is relatively insensitive to variations in the interference-to-target ratio.

14.15 EXTENSION OF SYSTOLIC ARRAY II FOR PARALLEL WEIGHT EXTRACTION

The systolic array processors considered in previous sections of the chapter have been somewhat limited in scope in the sense that they provide for a direct computation of the weight vector $\hat{\mathbf{w}}(n)$ or the a posteriori estimation error $e(n)$, but not both

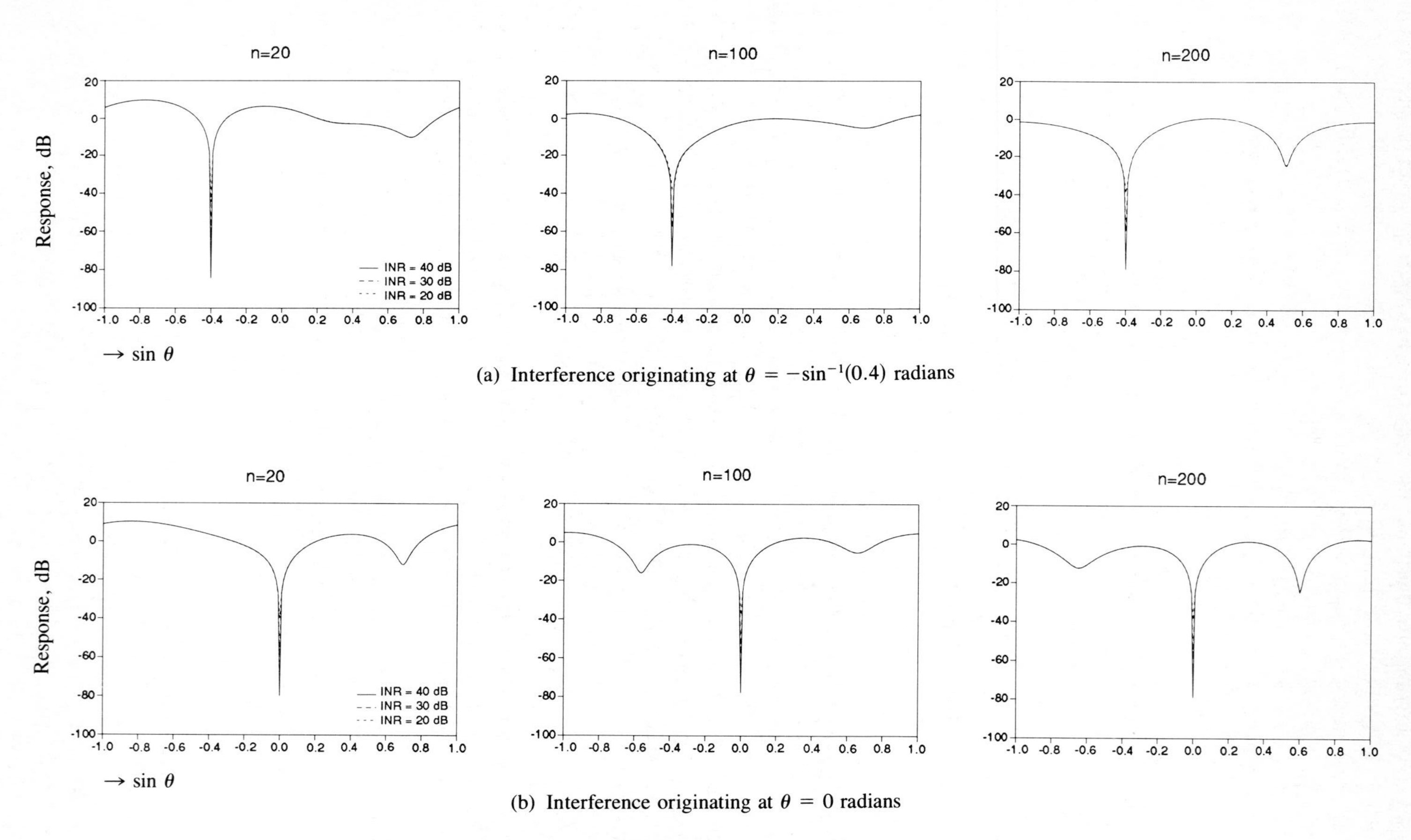

(a) Interference originating at $\theta = -\sin^{-1}(0.4)$ radians

(b) Interference originating at $\theta = 0$ radians

Figure 14.14 Response of MVDR beamformer for varying signal-to-interference ratio, different directions of interference, and varying numbers of iterations.

simultaneously. In this section we describe an extension of the systolic array processor II that is indeed capable of directly computing the a posteriori estimation error $e(n)$ and simultaneously extracting least-squares weight vector $\hat{\mathbf{w}}(n)$ (Shepherd et al., 1988; McWhirter, 1989).

To initiate the development of such a technique, we define an *augmented* data matrix $\mathbf{A}'(n)$ that consists of the data matrix $\mathbf{A}(n)$ extended by the desired data vector $\mathbf{d}(n)$ in the following fashion:

$$\mathbf{A}'(n) = [\mathbf{A}(n), \mathbf{d}(n)] \tag{14.124}$$

Thus, given that $\mathbf{A}(n)$ is an n-by-M matrix, $\mathbf{d}(n)$ is an n-by-1 vector, it follows that $\mathbf{A}'(n)$ is an n-by-$(M + 1)$ matrix. Following the notation introduced in Section 14.2, we define the QR-decomposition of $\mathbf{A}(n)$ as [see Eq. (14.14)]

$$\mathbf{Q}(n)\mathbf{\Lambda}^{1/2}(n)\mathbf{A}(n) = \begin{bmatrix} \mathbf{R}(n) \\ \mathbf{O} \end{bmatrix}$$

Correspondingly, we use Eq. (14.17) to write

$$\mathbf{Q}(n)\mathbf{\Lambda}^{1/2}(n)\mathbf{d}(n) = \begin{bmatrix} \mathbf{p}(n) \\ \mathbf{v}(n) \end{bmatrix}$$

Hence, premultiplying both sides of Eq. (14.124) by $\mathbf{Q}(n)\mathbf{\Lambda}^{1/2}(n)$ yields

$$\mathbf{Q}(n)\mathbf{\Lambda}^{1/2}(n)\mathbf{A}'(n) = \begin{bmatrix} \mathbf{R}(n) & \mathbf{p}(n) \\ \mathbf{O} & \mathbf{v}(n) \end{bmatrix} \tag{14.125}$$

To complete the QR-decomposition of the augmented data matrix $\mathbf{A}'(n)$, we apply a sequence of Givens rotations that are designed to annihilate all the elements of $\mathbf{v}(n)$ except for its first element. The result of this operation is denoted by

$$\mathbf{Q}'(n)\mathbf{\Lambda}^{1/2}(n)\mathbf{A}'(n) = \begin{bmatrix} \mathbf{R}(n) & \mathbf{p}(n) \\ \mathbf{0}^T & r(n) \\ \multicolumn{2}{c}{\mathbf{O}} \end{bmatrix} \tag{14.126}$$

where $r(n)$ is some real-valued number; $r(n)$ has to be real because it is a diagonal element. Let $\mathbf{R}'(n)$ denote the $(M + 1)$-by-$(M + 1)$ upper triangular matrix resulting from the QR-decomposition of the augmented data matrix $\mathbf{A}'(n)$. That is,

$$\mathbf{R}'(n) = \begin{bmatrix} \mathbf{R}(n) & \mathbf{p}(n) \\ \mathbf{0}^T & r(n) \end{bmatrix} \tag{14.127}$$

Next, we define an extended weight vector $\mathbf{w}'(n)$ of dimension $M + 1$, related to the M-by-1 least-squares weight vector $\hat{\mathbf{w}}(n)$ as follows:

$$\mathbf{w}'(n) = \begin{bmatrix} -\hat{\mathbf{w}}(n) \\ 1 \end{bmatrix} \tag{14.128}$$

Hence, premultiplying $\mathbf{w}'(n)$ by $\mathbf{R}'(n)$, we get

$$\begin{aligned} \mathbf{R}'(n)\mathbf{w}'(n) &= \begin{bmatrix} \mathbf{R}(n) & \mathbf{p}(n) \\ \mathbf{0}^T & r(n) \end{bmatrix} \begin{bmatrix} -\hat{\mathbf{w}}(n) \\ 1 \end{bmatrix} \\ &= \begin{bmatrix} -\mathbf{R}(n)\hat{\mathbf{w}}(n) + \mathbf{p}(n) \\ r(n) \end{bmatrix} \end{aligned} \tag{14.129}$$

But from Eq. (14.21) we have

$$\mathbf{R}(n)\hat{\mathbf{w}}(n) = \mathbf{p}(n), \qquad n > M$$

We may therefore simplify Eq. (14.129) for $n > M$ as

$$\begin{aligned} \mathbf{R}'(n)\mathbf{w}'(n) &= \begin{bmatrix} \mathbf{0} \\ r(n) \end{bmatrix} \\ &= r(n)\begin{bmatrix} \mathbf{0} \\ 1 \end{bmatrix} \end{aligned} \tag{14.130}$$

In other words, the matrix product $\mathbf{R}'(n)\mathbf{w}'(n)$ consists simply of the scalar $r(n)$ multiplied by a pinning vector of dimension $(M + 1)$.

Solving Eq. (14.130) for $\mathbf{w}'(n)$, we get

$$\mathbf{w}'(n) = r(n)\mathbf{R}'^{-1}(n)\begin{bmatrix} \mathbf{0} \\ 1 \end{bmatrix} \tag{14.131}$$

where $\mathbf{R}'^{-1}(n)$ is the inverse of matrix $\mathbf{R}'(n)$.

Equation (14.131) shows that $\mathbf{w}'(n)$ consists of the last column of $\mathbf{R}'^{-1}(n)$, scaled by the multiplier $r(n)$. Indeed, this equation provides the basis of a processor for the parallel extraction of the least-squares weight vector $\hat{\mathbf{w}}(n)$, as depicted in Fig. 14.15. The systolic processor shown here is made up of the interconnection of two equidimensional components with distinct but coupled functions:

1. A *standard* triangular array processor, consisting of *both* boundary and internal cells. The input vector $\mathbf{u}(n)$ and the desired response $d(n)$ constitute the composite input for this basic part of the processor. The operation performed here is the recursive QR-decomposition of the augmented data matrix $\mathbf{A}'(n)$. In particular, the standard triangular array in Fig. 14.15 looks after the updating of the corresponding lower triangular matrix $\mathbf{R}'^{H}(n)$, as shown by

$$[\mathbf{R}'^{H}(n), \mathbf{O}^{T}, \mathbf{0}] = \left[\lambda^{1/2}\mathbf{R}'^{H}(n-1), \mathbf{0}^{T} \,\vdots\, \begin{matrix} \mathbf{u}(n) \\ d(n) \end{matrix}\right]\mathbf{T}'^{H}(n) \tag{14.132}$$

   where $\mathbf{T}'(n)$ denotes a sequence of $(M + 1)$ Givens rotations.
2. A triangular array *postprocessor*, consisting of internal cells *only*. The internal cells of this second array are fed with the appropriate rotation parameters computed in the boundary cells of the standard processor. The function of the postprocessor is to provide for the updating of $\mathbf{R}'^{-1}(n)$. In this context, we note from Eq. (14.111) that

$$\begin{bmatrix} \mathbf{R}'^{-H}(n) \\ \mathbf{X} \end{bmatrix} = \mathbf{T}'(n)\begin{bmatrix} \lambda^{-1/2}\mathbf{R}'^{-H}(n-1) \\ \mathbf{0} \end{bmatrix} \tag{14.133}$$

   where $\mathbf{X}$ is a matrix of no specific interest.

The last boundary cell of the standard processor computes the multiplier $r(n)$. The last (rightmost) column of internal cells in the postprocessor computes the last column of $\mathbf{R}'^{-1}(n-1)$. Multiplication of the column vector so derived by the real

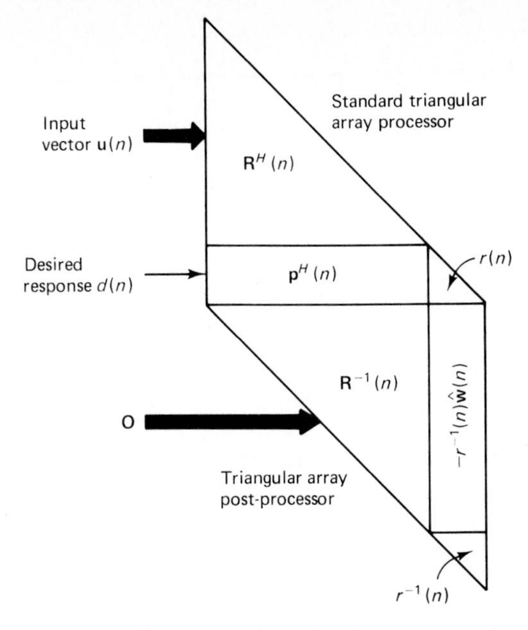

Figure 14.15 Systolic processor for parallel weight extraction.

scalar $r(n)$ yields $\mathbf{w}'(n)$ in accordance with Eq. (14.131). The first M elements of $\mathbf{w}'(n)$ define the desired least-squares weight vector $\hat{\mathbf{w}}(n)$, except for a minus sign. Note that the bottom cell in the last column of the postprocessor computes $1/r(n)$; when this value is multiplied by $r(n)$, the result is unity, which is in accord with the fact that the last element of $\mathbf{w}'(n)$ is *unity*. Checking that the last element of $\mathbf{w}'(n)$ is unity provides a certain measure of fault detection!

14.16 SUMMARY AND DISCUSSION

In this chapter, we have studied various systolic array structures for the recursive solution of both unconstrained and constrained versions of the linear least-squares problem. In this concluding section, we sum up the common features of these structures, their basic differences, and their areas of application.

To begin with, the systolic array structures described herein share the following *common* features:

1. They all exploit the QR-decomposition method that involves successive applications of a unitary transformation (i.e., the Givens rotation) directly to the data matrix $\mathbf{A}(n)$. The QR-decomposition method is known to be *numerically stable,* with the result that these structures are *less* susceptible to the effects of finite-precision arithmetic than would be the case if direct matrix inversion were to be applied to the correlation matrix $\mathbf{\Phi}(n)$ in order to solve the system of normal equations $\mathbf{\Phi}(n)\hat{\mathbf{w}}(n) = \mathbf{\theta}(n)$. Also, the QR-decomposition is updated recursively as new data arrive.
2. They are examples of an *open-loop data-adaptive system,* data adaptive in the sense that the values assigned to the parameters of the sequence of Givens ro-

tations vary in accordance with the input data. The "open-loop" feature is exemplified by the absence of *global* (overall) *feedback*; they do however involve the use of *local* feedback for data adaptivity. This feature distinguishes them from the other closed-loop adaptive filter structures considered in previous chapters.

3. In a systolic array, matrix computations are often directly *mapped* onto the processor array. This is, for example, readily seen by comparing the systolic array of Fig. 14.2 and governing equations (14.14) and (14.21).
4. Systolic structures are relatively easy to design because of their highly pipelined operation, regular nature, and layout. These features, and the fact that only nearest-neighbor interconnections are involved between individual processing cells in the arrays, make them particularly well suited for implementation in VLSI form.[6]
5. The triangular systolic array that performs the sequence of Givens rotations consists of boundary cells, inner cells, and (in the case of Fig. 14.7 only) a final processing cell. The number of boundary cells (representing the most complicated part of the algorithm in that square roots and divisions as well as multiplications and additions are performed there) equals $N - 1$, where N is the number of data points entering the array. The number of inner cells (where multiplications and additions only are performed), on the other hand, equals $(N^2 - N)/2$. For the structures of Figs. 14.2 and 14.7, we have $N = M + 1$, where M is the dimension of the input vector $\mathbf{u}(n)$ and the 1 accounts for the desired response $d(n)$. In Fig. 14.14 we have $N = M$. In addition to the triangular systolic array, the structure of Fig. 14.2 includes a linear systolic array for computing the M-by-1 least-squares weight vector $\hat{\mathbf{w}}(n)$; this section consists of one boundary cell (where addition, multiplication and division are performed) and $M - 1$ internal cells (where addition and multiplication are performed). The structure of Fig. 14.7 includes a final processing cell in the triangular systolic array, where a single multiplication is performed; this cell is used to compute the (a posteriori) estimation error $e(n)$.
6. The systolic arrays of Figs. 14.2 and 14.7 are *scalable* architecturally; that is, the size of the array may be enlarged indefinitely so long as the system synchronization is fully maintained.
7. As a consequence of the highly pipelined nature of the systolic array structures in Figs. 14.2 and 14.7, and therefore the requirement to impose a time skew on their input data, there is an *overall delay* or *latency* in their response.

[6] With *very large scale integration* (*VLSI*), it is possible to build complex array processors that consist of a large number of processing cells, which cooperate with each other to perform extremely fast matrix computations. Kung (1985) presents an overview of VLSI array processors, with particular reference to systolic arrays. A major problem that arises from the use of this technology, however, is that a single flaw in the chip or wafer can render an entire VLSI array processor useless. It is therefore desirable to have a VLSI array processor that achieves high performance, tolerates physical failures, and yet produces correct results. Abraham et al. (1987) describe various techniques for fault tolerance that can be applied to systolic array architectures. A technique referred to as "algorithm-based fault tolerance" is shown to be a natural approach for these array processors, [see also Jou and Abraham (1986)]. See also Anfinson and Luk (1988) for fault-tolerant techniques for MVDR beamforming.

8. The systolic arrays of Figs. 14.2 and 14.7 require the use of a common *clock signal* distributed to every cell (internal as well as boundary) in order to control data movements in the array. The need for synchronization will eventually become intolerable for very large scale or ultralarge-scale arrays. We may bypass the need for a common clock and therefore eliminate the synchronization problem by using the *wavefront array* technique.[7] A wavefront array processor implements the same basic algorithm as the systolic array. It takes advantage of the *control-flow locality* and *data-flow locality* that are inherently possessed by most algorithms of interest. These algorithmic features permit the use of a *data-driven, self-timed approach* to array processing. In effect, correct *sequencing* is substituted for the requirement of correct timing.

Next we sum up the *difference* between these structures, and their special features:

1. The structure of Fig. 14.2 solves the linear least-squares problem by producing the weight vector $\hat{\mathbf{w}}(n)$ at the output of the linear systolic array section. Such a structure may therefore find use in applications such as system identification where explicit knowledge of the weight vector is required.
2. The structure of Fig. 14.7 solves the linear least-squares problem by producing the a posteriori estimation error $e(n)$ without computing the weight vector $\hat{\mathbf{w}}(n)$ explicitly. This structure achieves the highest possible throughput rate with the minimum amount of computation and circuitry. The structure of Fig. 14.7 may also have a numerical advantage over that of Fig. 14.2 since the former structure avoids the need to solve a triangular linear system; such a system of equations may sometimes be ill-conditioned.
3. The weight vector $\hat{\mathbf{w}}(n)$ may be extracted indirectly from the structure of Fig. 14.7 by means of a technique called "weight flushing." Moreover, it may be extended by the inclusion of a triangular array postprocessor for the (direct) parallel extraction of the weight vector $\hat{\mathbf{w}}(n)$.
4. The structure of Fig. 14.7 is very well suited for adaptive beamforming. Through the use of a constraint preprocessor, a linear constraint may be imposed on the operation of the processor so as to guard against the adaptive nulling of the target signal.
5. For MVDR beamforming, we may use a constrained extension of the structure of Fig. 14.2 or that of the structure of Fig. 14.7. The constraint is determined by the choice of a steering vector $\mathbf{s}(\phi)$, where ϕ is some electrical angle. The structure of Fig. 14.2 remains invariant to this constraint. However, for every desired steering vector, we have to use a corresponding linear systolic array to perform the pertinent back substitution. This difficulty may be resolved by using the structure of Fig. 14.7 and building on it in the manner described in Section 14.13. In so doing, only one systolic array is required to solve the MVDR beamforming problem for the many look directions of interest.

[7] The wavefront array technique was originated by S. Y. Kung. For a tutorial treatment of it, see the review papers (Kung, 1985; Kung et al., 1987) [see also the book (Kung, 1988)]. For the use of a wavefront array processor for adaptive beamforming, see McCanny and McWhirter (1987).

In conclusion, each of the systolic array structures considered in this chapter has unique features of its own. The decision as to which one to use can only be determined by the requirements of the problem of interest.

PROBLEMS

1. Consider the squared Euclidean norm

$$\mathscr{E}(n) = \| \mathbf{\Lambda}^{1/2}(n)\boldsymbol{\epsilon}(n) \|^2$$

where $\boldsymbol{\epsilon}(n)$ is the estimation error vector and $\mathbf{\Lambda}(n)$ is the diagonal weighting matrix. These two parameters are defined by Eqs. (14.6) and (14.9), respectively. Show that $\mathscr{E}(n)$ is unaffected by postmultiplying the vector $\mathbf{\Lambda}^{1/2}(n)\boldsymbol{\epsilon}(n)$ by a unitary matrix $\mathbf{Q}(n)$.

2. Start with the normal equations written in matrix form

$$\mathbf{\Phi}(n)\hat{\mathbf{w}}(n) = \boldsymbol{\theta}(n)$$

where $\mathbf{\Phi}(n)$ is the correlation matrix of the input vector $\mathbf{u}(n)$, $\boldsymbol{\theta}(n)$ is the cross-correlation vector between $\mathbf{u}(n)$ and the desired response $d(n)$, and $\hat{\mathbf{w}}(n)$ is the least-squares weight vector.

(a) Hence, using the QR-decomposition

$$\mathbf{Q}(n)\mathbf{\Lambda}^{1/2}(n)\mathbf{A}(n) = \begin{bmatrix} \mathbf{R}(n) \\ \mathbf{O} \end{bmatrix}$$

where $\mathbf{A}(n)$ is the data matrix, $\mathbf{\Lambda}(n)$ is the exponential weighting matrix, $\mathbf{Q}(n)$ is a unitary matrix, and $\mathbf{R}(n)$ is an upper triangular matrix, transform the normal equations into the form

$$\mathbf{R}(n)\hat{\mathbf{w}}(n) = \mathbf{p}(n)$$

(b) How are the matrices $\mathbf{R}(n)$ and $\mathbf{\Phi}(n)$, and the vectors $\mathbf{p}(n)$ and $\boldsymbol{\theta}(n)$ related?

(c) Show that $\mathbf{R}(n)$ can be interpreted as a Cholesky factor of $\mathbf{A}^H(n)\mathbf{A}(n)$.

3. Let the n-by-n unitary matrix $\mathbf{Q}(n)$ involved in the QR-decomposition of the data matrix $\mathbf{A}(n)$ be partitioned as follows

$$\mathbf{Q}(n) = \begin{bmatrix} \mathbf{Q}_1(n) \\ \mathbf{Q}_2(n) \end{bmatrix}$$

where $\mathbf{Q}_1(n)$ has the same number of rows as the upper triangular matrix $\mathbf{R}(n)$ in the QR-decomposition of $\mathbf{A}(n)$. Assume that the exponential weighting factor $\lambda = 1$.

According to the method of least squares presented in Chapter 10, the projection operator is

$$\mathbf{P}(n) = \mathbf{A}(n)(\mathbf{A}^H(n)\mathbf{A}(n))^{-1}\mathbf{A}^H(n)$$

Show that

$$\mathbf{P}(n) = \mathbf{Q}_1(n)\mathbf{Q}_1^H(n)$$

How is the result modified for the case where $0 < \lambda < 1$?

4. In the context of the systolic array in Fig. 14.2, explain the reasons for the following:

(a) A total of $2M$ clock cycles are required to compute the Givens rotations in the triangular section of the systolic array in Fig. 14.2.

(b) The latency for the entire array is $4M$ clock cycles.

5. Equation (14.56) describes a method of partitioning the unitary matrix $\mathbf{T}(n)$ that represents the sequence of Givens rotations. Show that the scaling factor $\gamma^{1/2}(n)$ introduced in this equation equals the square root of the real-valued scalar $1 - \mathbf{u}^H(n)\mathbf{\Phi}^{-1}(n)\mathbf{u}(n)$, where $\mathbf{u}(n)$ is the input vector and $\mathbf{\Phi}^{-1}(n)$ is the inverse of the associated correlation matrix.

6. **(a)** Table 14.1 presents a summary of the QRD-RLS algorithm. Using the representation given in Eq. (14.56) for the transformation $\mathbf{T}(n)$, find the update equations for the upper triangular matrix $\mathbf{R}(n)$ and the vector $\mathbf{p}(n)$.

   (b) Using the result of part (a), show that the vector component $\boldsymbol{\beta}_1(n)$ of the transformation $\mathbf{T}(n)$ is related to the old upper triangular matrix $\mathbf{R}(n-1)$ and the new input vector $\mathbf{u}(n)$ as follows:

$$\boldsymbol{\beta}_1(n) = \lambda^{-1/2}\mathbf{R}^{-H}(n-1)\mathbf{u}(n)$$

   where $\mathbf{R}^{-H}(n-1)$ is the Hermitian transpose of the inverse matrix $\mathbf{R}^{-1}(n-1)$. Hence, show that

$$\lambda^{1/2}\mathbf{p}^H(n-1)\boldsymbol{\beta}_1(n) = \hat{\mathbf{w}}^H(n-1)\mathbf{u}(n)$$

   What is the physical meaning of this result?

   (c) Given that [see Eq. (14.75)]

$$\alpha(n) = d(n) - \lambda^{1/2}\mathbf{p}^H(n-1)\boldsymbol{\beta}_1(n)$$

   what is the physical meaning of $\alpha(n)$?

7. Explain the way in which the systolic array structure of Fig. 14.7 may be used to operate as a prediction-error filter.

8. Discuss the use of a linear section, based on forward substitution, for solving the equation $\mathbf{R}^H(n)\mathbf{a}(n) = \mathbf{s}$ for the vector $\mathbf{a}(n)$; the matrix $\mathbf{R}$ is an M-by-M upper triangular matrix and $\mathbf{s}$ is an M-by-1 vector.

9. Figure P14.1 depicts a block diagram representation of an MVDR beamforming algorithm. Specifically, the triangular array in part (a) of this figure is frozen at time n and the steering vector $\mathbf{s}(\phi)$ is input into the array. The stored $\mathbf{R}^H(n)$ of the array and its output $\mathbf{a}^H(n)$ are applied to a linear systolic section as in part (b) of the figure. Do the following:

   (a) Show that the output of the triangular array is

$$\mathbf{a}^H(n) = \mathbf{s}^H(\phi)\mathbf{R}^{-1}(n)$$

   (b) Using the method of back substitution, show that the linear systolic array produces the Hermitian transposed weight vector $\hat{\mathbf{w}}^H(n)$ as its output.

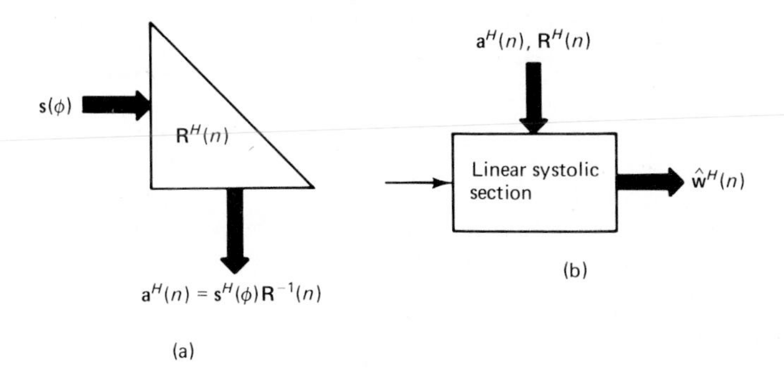

Figure P14.1

10. Using Eqs. (14.121) and (14.123), show that

$$\lambda \|\mathbf{a}(n-1)\|^2 = \lambda^2 \|\mathbf{a}(n)\|^2 + |\epsilon(n)|^2$$

where $\|\mathbf{a}(n)\|$ is the Euclidean norm of the auxiliary vector $\mathbf{a}(n)$, λ is the exponential weighting factor, and $\epsilon(n)$ is some estimation error.

11. Consider a systolic processor consisting of a standard triangular array processor and a triangular array postprocessor of compatible dimensions, as depicted in Fig. P14.2. The standard processor performs a recursive QR-decomposition of the data matrix $\mathbf{A}(n)$, whereas the postprocessor performs an updating computation of the inverse matrix $\mathbf{R}^{-1}(n-1)$, where $\mathbf{R}(n-1)$ is the upper triangular matrix stored in the standard processor. Show that the output of this processor, in response to the input vector $\mathbf{u}(n)$, is equal to $\mathbf{\Phi}^{-1}(n-1)\mathbf{u}(n)$ where $\mathbf{\Phi}(n-1)$ is the time-averaged correlation matrix of the input data at time $n-1$. How would such an output be useful in the context of recursive least-squares estimation?

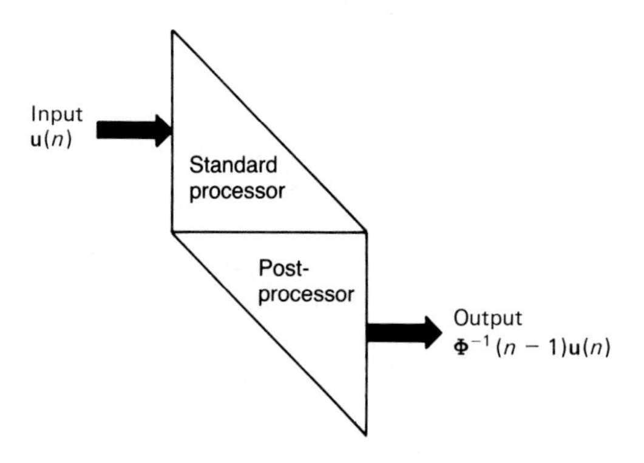

Figure P14.2

Computer-Oriented Problem

12. *Computer experiment on MVDR beamforming.* Consider an environment that contains a target signal and two sources of interference, all of which are noncoherent with each other. The signal-to-noise ratio SNR = 10 dB. The two interferences are equal in strength with INR = 40, 30, 20 dB. The pertinent angles of arrival are as described below:

Excitation	Angle of arrival, θ, measured with respect to the normal to the array
Target	$\sin^{-1}(0.2)$
Interference I	$-\sin^{-1}(0.4)$, or $\sin^{-1}(0.1)$
Interference II	$\sin^{-1}(0.8)$

That is, the directions of the target and interference II are fixed, while interference I may originate from one of two possible directions. The array itself consists of 5 similar sensors with uniform spacing $d = \lambda/2$, where λ is the wavelength.

Perform the following computations:

(a) Compute the adapted response of the beamformer after 20 snapshots from the end of the initialization period for the case when interference I arrives along the direction $\theta = -\sin^{-1}(0.4)$, and for each of the prescribed values of INR. Repeat your computations for 100, 200 snapshots.

(b) Repeat your computations in (a) for the case when interference I has the direction $\theta = \sin^{-1}(0.1)$.

15

Background Theory for Fast Recursive Algorithms

In Chapters 13 and 14 we studied two different approaches for solving the recursive least-squares problem. In Chapter 13 we studied the *standard* approach and developed the *recursive least-squares* (RLS) algorithm; this approach assumes the use of a transversal filter for implementing the RLS algorithm. In Chapter 14 we studied the more sophisticated approach of recursive QR-decomposition by Givens rotations; there we developed different systolic arrays for implementing the QR decomposition-based recursive least-squares (QRD-RLS) algorithm and its modified versions. In the recursive estimation approaches studied in Chapters 13 and 14, the *computational complexity* is on the order of M^2, where M is the number of adjustable parameters.

Clearly, computational complexity has a direct bearing on the *cost* of building hardware for use in practical applications of these algorithms. When cost, and therefore computational complexity, is an issue of direct concern, there is motivation in the development of *fast algorithms* for solving the recursive least-squares problem. An algorithm (in the context of our present discussion) is said to be fast *if its computational complexity increases linearly with the dimension of the adjustable weight vector*. A fast algorithm is therefore similar to the LMS algorithm in its computational requirement.

Detailed discussions of specific forms of fast recursive least-squares estimation algorithms are presented in Chapters 16 through 18. In the present chapter, we develop the mathematical background of fast algorithms that is fundamental to their understanding. This development relies on adaptive forward and backward predictions, optimized in the least-squares sense. We thus begin our discussion by considering these two forms of linear prediction as special cases of the recursive least-squares problem.

15.1 ADAPTIVE FORWARD LINEAR PREDICTION

Consider the *forward linear predictor* of order M, depicted in Fig. 15.1(a), whose tap-weight vector $\hat{\mathbf{w}}(n)$ is optimized in the least-squares sense over the observation interval $1 \le i \le n$. Let $f_M(i)$ denote the *forward prediction error* produced by the predictor at time i in response to the tap-input vector $\mathbf{u}_M(i-1)$, as shown by

$$f_M(i) = u(i) - \hat{\mathbf{w}}^H(n)\mathbf{u}_M(i-1), \qquad 1 \le i \le n \tag{15.1}$$

where $u(i)$ plays the role of desired response and

$$\mathbf{u}_M^T(i-1) = [u(i-1), u(i-2), \ldots, u(i-M)] \tag{15.2}$$

We refer to $f_M(i)$ as the *a posteriori* prediction error since its computation is based on the *current* value $\hat{\mathbf{w}}(n)$ of the predictor's tap-weight vector.

We may equivalently characterize the forward linear prediction problem by specifying the *prediction-error filter*, as depicted in Fig. 15.1(b). Let $\mathbf{a}_M(n)$ denote the $(M+1)$-by-1 tap-weight vector of the prediction-error filter of order M. This vector is related to the tap-weight vector of the predictor in Fig. 15.1(a) by

$$\mathbf{a}_M(n) = \begin{bmatrix} 1 \\ -\hat{\mathbf{w}}(n) \end{bmatrix} \tag{15.3}$$

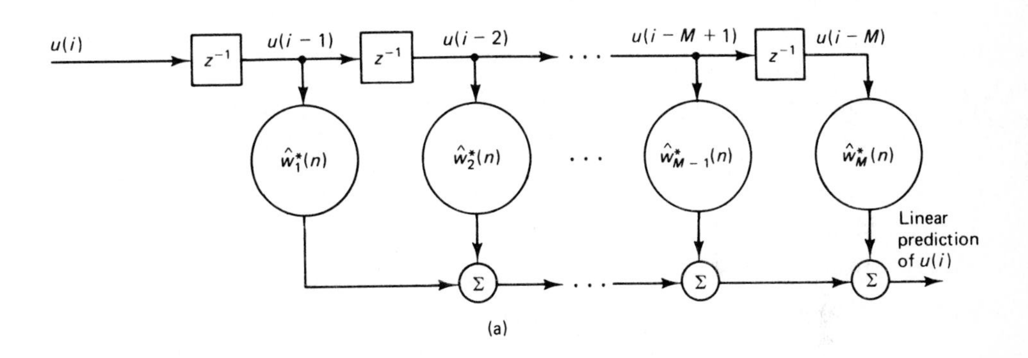

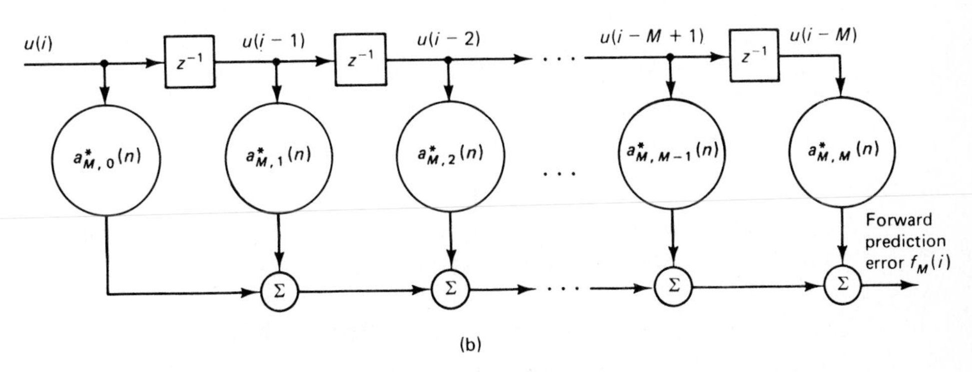

Figure 15.1 (a) Forward predictor of order M; (b) corresponding prediction-error filter.

Define the $(M + 1)$-by-1 vector $\mathbf{u}_{M+1}(i)$ that has the desired response $u(i)$ as the leading element and the vector $\mathbf{u}_M(i - 1)$ as the remaining M elements:

$$\mathbf{u}_{M+1}(i) = \begin{bmatrix} u(i) \\ \mathbf{u}_M(i - 1) \end{bmatrix} \tag{15.4}$$

Then we may redefine the forward a posteriori prediction error $f_M(i)$ as

$$f_M(i) = \mathbf{a}_M^H(n)\mathbf{u}_{M+1}(i), \qquad 1 \leq i \leq n \tag{15.5}$$

The tap-weight vector $\hat{\mathbf{w}}(n)$ of the predictor is the solution obtained by minimizing the sum of weighted forward prediction-error squares for $1 \leq i \leq n$. Equivalently, the tap-weight vector $\mathbf{a}_M(n)$ of the prediction-error filter is the solution to the same minimization problem formulated in terms of the prediction-error filter, subject to the constraint that the first element of $\mathbf{a}_M(n)$ equals unity. In either case, we may express the least-squares solution compactly in terms of the *augmented normal equations for forward linear prediction* (see Chapter 10).

$$\mathbf{\Phi}_{M+1}(n)\mathbf{a}_M(n) = \begin{bmatrix} \mathcal{F}_M(n) \\ \mathbf{0} \end{bmatrix} \tag{15.6}$$

where $\mathbf{0}$ is the M-by-1 null vector. The scalar $\mathcal{F}_M(n)$ is the *minimum value of the sum of weighted forward a posteriori prediction-error squares:*

$$\mathcal{F}_M(n) = \sum_{i=1}^{n} \lambda^{n-i} | f_M(i) |^2 \tag{15.7}$$

The $(M + 1)$-by-$(M + 1)$ matrix $\mathbf{\Phi}_{M+1}(n)$ is the *correlation matrix* of the tap-input vector $\mathbf{u}_{M+1}(i)$ in the prediction-error filter of Fig. 15.1(b):

$$\mathbf{\Phi}_{M+1}(n) = \sum_{i=1}^{n} \lambda^{n-i}\mathbf{u}_{M+1}(i)\mathbf{u}_{M+1}^H(i) \tag{15.8}$$

Based on the partitioning of the vector $\mathbf{u}_{M+1}(i)$ given in Eq. (15.4), we may also express $\mathbf{\Phi}_{M+1}(n)$ in the partitioned form:

$$\mathbf{\Phi}_{M+1}(n) = \begin{bmatrix} \mathcal{U}_1(n) & \boldsymbol{\theta}_1^H(n) \\ \boldsymbol{\theta}_1(n) & \mathbf{\Phi}_M(n - 1) \end{bmatrix} \tag{15.9}$$

where $\mathcal{U}_1(n)$ is the sum of weighted squared values of the desired response used for forward prediction,

$$\mathcal{U}_1(n) = \sum_{i=1}^{n} \lambda^{n-i} | u(i) |^2 \tag{15.10}$$

$\boldsymbol{\theta}_1(n)$ is the M-by-1 cross-correlation vector of the tap-input vector of the predictor in Fig. 15.1(a) and the desired response $u(i)$,

$$\boldsymbol{\theta}_1(n) = \sum_{i=1}^{n} \lambda^{n-i}\mathbf{u}_M(i - 1)u^*(i) \tag{15.11}$$

and $\mathbf{\Phi}_M(n - 1)$ is the M-by-M correlation matrix of the tap-input vector $\mathbf{u}_M(i - 1)$

in the predictor of Fig 15.1(a):

$$\begin{aligned}\mathbf{\Phi}_M(n-1) &= \sum_{i=1}^{n} \lambda^{n-i}\mathbf{u}_M(i-1)\mathbf{u}_M^H(i-1) \\ &= \sum_{i=1}^{n-1} \lambda^{(n-1)-i}\mathbf{u}_M(i)\mathbf{u}_M^H(\mathrm{i})\end{aligned} \tag{15.12}$$

Note that in the last line of Eq. (15.12) we have done two things: (1) we have replaced $i - 1$ by i, and (2) we have assumed that the input data are zero prior to time $i = 0$ as prescribed by prewindowing of the data.

We may now turn our attention to the adaptive implementation of the predictor using the RLS algorithm. In Table 15.1 we have listed the correspondences between the quantities characterizing the RLS algorithm and those characterizing the predictor of Fig. 15.1(a). With the aid of this table, it is a straightforward matter to modify the theory of the RLS algorithm developed in Sections 13.3 and 13.4 to write the desired recursions for the forward linear prediction problem. First, from Eqs. (13.22), (13.25), and (13.26), we deduce the following *recursion for updating the tap-weight vector of the predictor:*

$$\hat{\mathbf{w}}(n) = \hat{\mathbf{w}}(n-1) + \mathbf{k}_M(n-1)\eta_M^*(n) \tag{15.13}$$

where $\eta_M(n)$ is the *forward a priori prediction error* (i.e., tentative estimate of the forward prediction error):

$$\eta_M(n) = u(n) - \hat{\mathbf{w}}^H(n-1)\mathbf{u}_M(n-1) \tag{15.14}$$

and $\mathbf{k}_M(n-1)$ is the *gain vector for forward linear prediction of order M:*

$$\mathbf{k}_M(n-1) = \mathbf{\Phi}_M^{-1}(\mathrm{n}-1)\mathbf{u}_M(n-1) \tag{15.15}$$

The use of subscript M is intended to signify the order of the predictor. We follow this practice here and in the rest of the chapter since some of the recursions to be developed involve an order update. Correspondingly, we may substitute Eq. (15.13)

TABLE 15.1 SUMMARY OF CORRESPONDENCES BETWEEN LINEAR ESTIMATION, FORWARD PREDICTION, AND BACKWARD PREDICTION

Quantity	RLS algorithm for linear estimation (general)	RLS algorithm for forward linear prediction of order M	RLS algorithm for backward linear prediction of order M
Tap-input vector	$\mathbf{u}(n)$	$\mathbf{u}_M(n-1)$	$\mathbf{u}_M(n)$
Desired response	$d(n)$	$u(n)$	$u(n-M)$
Tap-weight vector	$\hat{\mathbf{w}}(n)$	$\hat{\mathbf{w}}(n)$	$\mathbf{g}(n)$
A posteriori estimation error	$e(n)$	$f_M(n)$	$b_M(n)$
A priori estimation error	$\alpha(n)$	$\eta_M(n)$	$\psi_M(n)$
Gain vector	$\mathbf{k}(n)$	$\mathbf{k}_M(n-1)$	$\mathbf{k}_M(n)$
Minimum value of sum of weighted errors squares	$\mathscr{E}_{\min}(n)$	$\mathscr{F}_M(n)$	$\mathscr{B}_M(n)$

in (15.3) to write the recursion for updating the tap-weight vector of the prediction-error filter as

$$\mathbf{a}_M(n) = \mathbf{a}_M(n-1) - \begin{bmatrix} 0 \\ \mathbf{k}_M(n-1) \end{bmatrix} \eta_M^*(n) \tag{15.16}$$

where

$$\begin{aligned} \eta_M(n) &= [1, \ -\hat{\mathbf{w}}^H(n-1)] \begin{bmatrix} u(n) \\ \mathbf{u}_M(n-1) \end{bmatrix} \\ &= \mathbf{a}_M^H(n-1)\mathbf{u}_{M+1}(n) \end{aligned} \tag{15.17}$$

Finally, using Eq. (13.35), we get the following recursion for updating the minimum value of the sum of weighted forward prediction-error squares:

$$\mathcal{F}_M(n) = \lambda \mathcal{F}_M(n-1) + \eta_M(n) f_M^*(n) \tag{15.18}$$

Note that the correction $\eta_M(n) f_M^*(n)$ in this recursion is always a real-valued scalar; that is,

$$\eta_M(n) f_M^*(n) = \eta_M^*(n) f_M(n) \tag{15.19}$$

In other words, the inner and outer products of the a priori and a posteriori forward prediction errors are identical.

15.2 ADAPTIVE BACKWARD LINEAR PREDICTION

Consider next the *backward linear predictor of order M*, depicted in Fig. 15.2(a), whose tap-weight vector $\mathbf{g}(n)$ is optimized in the least-squares sense over the observation interval $1 \le i \le n$. Let $b_M(i)$ denote the *backward prediction error* produced by this predictor at time i in response to the tap-input vector $\mathbf{u}_M(i)$, as shown by

$$b_M(i) = u(i-M) - \mathbf{g}^H(n)\mathbf{u}_M(i) \tag{15.20}$$

where $u(i-M)$ plays the role of desired response and

$$\mathbf{u}_M^T(i) = [u(i), u(i-1), \ldots, u(i-M+1)] \tag{15.21}$$

We refer to $b_M(i)$ as the *backward a posteriori prediction error* since its computation is based on the *current* value $\mathbf{g}(n)$ of the predictor's tap-weight vector.

We may equivalently characterize the backward linear prediction problem by specifying the prediction-error filter, as depicted in Fig. 15.2(b). Let $\mathbf{c}_M(n)$ denote the $(M+1)$-by-1 tap-weight vector of this prediction-error filter of order M. The vector $\mathbf{c}_M(n)$ is related to the tap-weight vector of the backward predictor in Fig. 15.2(a) by

$$\mathbf{c}_M(n) = \begin{bmatrix} -\mathbf{g}(n) \\ 1 \end{bmatrix} \tag{15.22}$$

Define the $(M+1)$-by-1 vector $\mathbf{u}_{M+1}(i)$ that has the desired response $u(i-M)$ as the last element and the vector $\mathbf{u}_M(i)$ as the remaining elements:

$$\mathbf{u}_{M+1}(i) = \begin{bmatrix} \mathbf{u}_M(i) \\ u(i-M) \end{bmatrix} \tag{15.23}$$

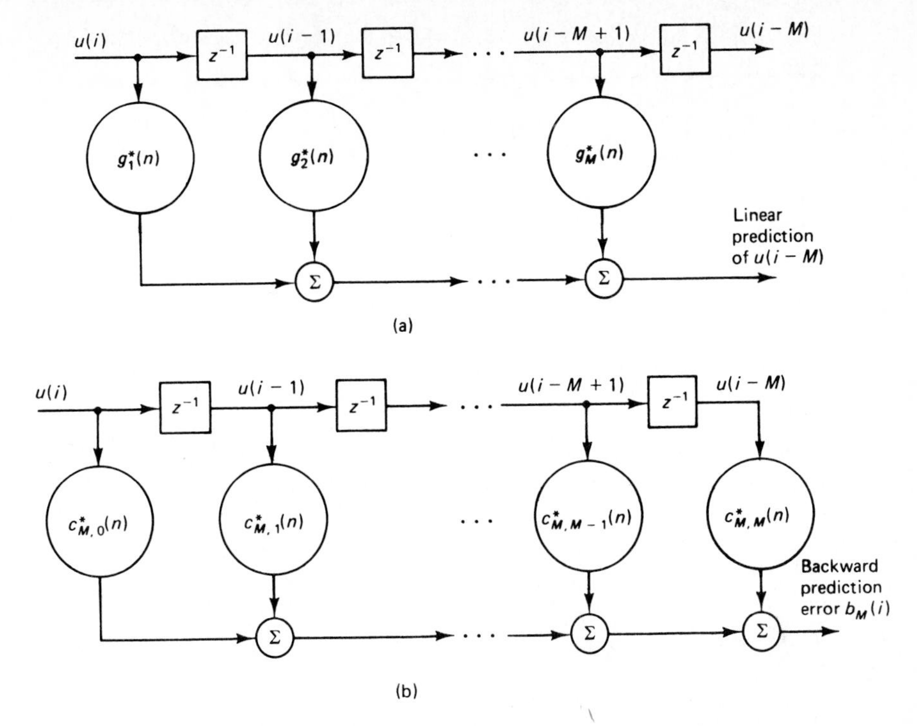

Figure 15.2 (a) Backward predictor of order M; (b) corresponding backward prediction-error filter.

Then we may redefine the backward prediction error $b_M(i)$ as

$$b_M(i) = \mathbf{c}_M^H(n)\mathbf{u}_{M+1}(i), \qquad 1 \leq i \leq n \tag{15.24}$$

The tap-weight vector $\mathbf{g}(n)$ of the backward linear predictor is the solution that results from minimizing the sum of weighted backward a posteriori prediction-error squares for $1 \leq i \leq n$. Equivalently, the tap-weight vector $\mathbf{c}_M(n)$ of the backward prediction-error filter is the solution of the same minimization problem formulated in terms of the backward prediction-error filter, subject to the constraint that the last element of $\mathbf{c}_M(n)$ equals unity. In either case, we may express the least-squares solution compactly in terms of the *augmented normal equations for backward linear prediction* (see Chapter 10).

$$\mathbf{\Phi}_{M+1}(n)\mathbf{c}_M(n) = \begin{bmatrix} \mathbf{0} \\ \mathcal{B}_M(n) \end{bmatrix} \tag{15.25}$$

where $\mathbf{0}$ is the M-by-1 null vector. The scalar $\mathcal{B}_M(n)$ is the *minimum value of the sum of weighted backward a posteriori prediction-error squares:*

$$\mathcal{B}_M(n) = \sum_{i=1}^{n} \lambda^{n-i} |b_M(i)|^2 \tag{15.26}$$

The $(M + 1)$-by-$(M + 1)$ matrix $\mathbf{\Phi}_{M+1}(n)$ is the correlation matrix of the tap-input vector $\mathbf{u}_{M+1}(i)$ in the prediction-error filter of Fig. 15.2(b). It is defined by Eq. (15.8). In a way corresponding to the partitioning of the vector $\mathbf{u}_{M+1}(i)$ shown in

Eq. (15.23), we may also express the correlation matrix $\mathbf{\Phi}_{M+1}(n)$ in the partitioned form

$$\mathbf{\Phi}_{M+1}(n) = \begin{bmatrix} \mathbf{\Phi}_M(n) & \mathbf{\theta}_2(n) \\ \mathbf{\theta}_2^H(n) & \mathcal{U}_2(n) \end{bmatrix} \tag{15.27}$$

where $\mathcal{U}_2(n)$ is the sum of weighted squared values of the desired response for backward prediction,

$$\begin{aligned} \mathcal{U}_2(n) &= \sum_{i=1}^{n} \lambda^{n-i} |u(i - M)|^2 \\ &= \sum_{i=1}^{n-M} \lambda^{(n-M)-i} |u(i)|^2 \end{aligned} \tag{15.28}$$

$\mathbf{\theta}_2(n)$ is the M-by-1 cross-correlation vector of the tap-input vector of the predictor in Fig. 15.2(a) and the desired response $u(i - M)$:

$$\mathbf{\theta}_2(n) = \sum_{i=1}^{n} \lambda^{n-i} \mathbf{u}_M(i) u^*(i - M) \tag{15.29}$$

and $\mathbf{\Phi}_M(n)$ is the M-by-M correlation matrix of the tap-input vector $\mathbf{u}_M(i)$ in the predictor of Fig. 15.12(a):

$$\mathbf{\Phi}_M(n) = \sum_{i=1}^{n} \lambda^{n-i} \mathbf{u}_M(i) \mathbf{u}_M^H(i) \tag{15.30}$$

To write the recursion for the adaptive implementation of the backward linear predictor using the RLS algorithm, we again modify the theory developed in Sections 13.3 and 13.4 with the aid of the correspondences listed in Table 15.1. Thus, from Eqs. (13.22), (13.25), and (13.26), we deduce the following recursion for updating the tap-weight vector of the backward predictor:

$$\mathbf{g}(n) = \mathbf{g}(n - 1) + \mathbf{k}_M(n) \psi_M^*(n) \tag{15.31}$$

where $\psi_M(n)$ is the *backward a priori prediction error* (i.e., tentative estimate of the backward prediction error):

$$\psi_M(n) = u(n - M) - \mathbf{g}^H(n - 1)\mathbf{u}_M(n) \tag{15.32}$$

and $\mathbf{k}_M(n)$ is the *gain vector for backward linear prediction of order M:*

$$\mathbf{k}_M(n) = \mathbf{\Phi}_M^{-1}(n)\mathbf{u}_M(n) \tag{15.33}$$

Correspondingly, we may substitute Eq. (15.31) in (15.22) and so express the recursion for updating the tap-weight vector of the backward prediction-error filter as

$$\mathbf{c}_M(n) = \mathbf{c}_M(n - 1) - \begin{bmatrix} \mathbf{k}_M(n) \\ 0 \end{bmatrix} \psi_M^*(n) \tag{15.34}$$

where

$$\begin{aligned} \psi_M(n) &= [-\mathbf{g}^H(n - 1), 1] \begin{bmatrix} \mathbf{u}_M(n) \\ \mathrm{u(n - M)} \end{bmatrix} \\ &= \mathbf{c}_M^H(n - 1)\mathbf{u}_{M+1}(n) \end{aligned} \tag{15.35}$$

Finally, from Eq. (13.35) we deduce the following expression for updating the minimum value of the sum of weighted backward a posteriori prediction-error squares:

$$\mathcal{B}_M(n) = \lambda \mathcal{B}_M(n-1) + \psi_M(n) b_M^*(n) \tag{15.36}$$

The correction term $\psi_M(n) b_M^*(n)$ is always real; that is,

$$\psi_M(n) b_M^*(n) = \psi_M^*(n) b_M(n) \tag{15.37}$$

In other words, as with adaptive forward linear prediction, the inner and outer products of the a priori and a posteriori backward prediction errors are identical.

Note that the update recursion for the tap-weight vector of the backward linear predictor in Eq. (15.31) or the backward prediction-error filter in Eq. (15.34) requires knowledge of the *current* value $\mathbf{k}_M(n)$ of the gain vector. On the other hand, the update recursion for the tap-weight vector of the forward linear predictor in Eq. (15.13) or the forward prediction-error filter in Eq. (15.16) requires knowledge of the *old* value $\mathbf{k}_M(n-1)$ of the gain vector.

15.3 EXTENDED GAIN VECTOR

By analogy with Eq. (15.33), we define the $(M+1)$-by-1 *extended gain vector:*

$$\mathbf{k}_{M+1}(n) = \mathbf{\Phi}_{M+1}^{-1}(n) \mathbf{u}_{M+1}(n) \tag{15.38}$$

The noteworthy feature of the extended gain vector is that, as will be shown later, it incorporates both the gain vector $\mathbf{k}_M(n-1)$ in the adaptive forward linear predictor and the gain vector $\mathbf{k}_M(n)$ in the adaptive backward linear predictor.

The $(M+1)$-by-$(M+1)$ correlation matrix $\mathbf{\Phi}_{M+1}(n)$ may be partitioned as in Eq. (15.9), which pertains to the forward linear predictor. The inverse of the correlation matrix $\mathbf{\Phi}_{M+1}(n)$ may be expressed as follows (see Problem 1):

$$\mathbf{\Phi}_{M+1}^{-1}(n) = \begin{bmatrix} 0 & \mathbf{0}_M^T \\ \mathbf{0}_M & \mathbf{\Phi}_M^{-1}(n-1) \end{bmatrix} + \frac{1}{\mathcal{F}_M(n)} \mathbf{a}_M(n) \mathbf{a}_M^H(n) \tag{15.39}$$

when $\mathbf{0}_M$ is the M-by-1 null vector, $\mathbf{a}_M(n)$ is the $(M+1)$-by-1 tap-weight vector of the forward prediction-error filter of order M, and $\mathcal{F}_M(n)$ is the corresponding minimum value of the sum of weighted forward a posteriori prediction-error squares. We also recognize that for forward prediction the $(M+1)$-by-1 tap-input vector $\mathbf{u}_{M+1}(n)$ may be partitioned as

$$\mathbf{u}_{M+1}(n) = \begin{bmatrix} u(n) \\ \mathbf{u}_M(n-1) \end{bmatrix} \tag{15.40}$$

Accordingly, postmultiplying both sides of Eq. (15.39) by $\mathbf{u}_{M+1}(n)$, using the partitioned form of Eq. (15.40) for the first term on the right side of Eq. (15.39), and using in the second term the definition of the forward a posteriori prediction error,

$$f_M(n) = \mathbf{a}_M^H(n) \mathbf{u}_{M+1}(n) \tag{15.41}$$

we get the following recursion for the extended gain vector:

$$\mathbf{k}_{M+1}(n) = \begin{bmatrix} 0 \\ \mathbf{k}_M(n-1) \end{bmatrix} + \frac{f_M(n)}{\mathcal{F}_M(n)} \mathbf{a}_M(n) \tag{15.42}$$

Consider next the second partitioned form of the $(M + 1)$-by-$(M + 1)$ correlation matrix $\mathbf{\Phi}_{M+1}(n)$ shown in Eq. (15.27), which pertains to the backward linear prediction. Using this representation, we find that the inverse of the matrix $\mathbf{\Phi}_{M+1}(n)$ may be expressed as follows (see Problem 1):

$$\mathbf{\Phi}_{M+1}^{-1}(n) = \begin{bmatrix} \mathbf{\Phi}_M^{-1}(n) & \mathbf{0}_M \\ \mathbf{0}_M^T & 0 \end{bmatrix} + \frac{1}{\mathcal{B}_M(n)} \mathbf{c}_M(n)\mathbf{c}_M^H(n) \tag{15.43}$$

where $\mathbf{c}_M(n)$ is the $(M + 1)$-by-1 tap-weight vector of the backward prediction-error filter of order M, and $\mathcal{B}_M(n)$ is the corresponding minimum value of the sum of weighted a posteriori backward prediction-error squares. We may also partition the $(M + 1)$-by-1 tap-input vector $\mathbf{u}_{M+1}(n)$, in accordance with the requirements of the backward linear prediction, as

$$\mathbf{u}_{M+1}(n) = \begin{bmatrix} \mathbf{u}_M(n) \\ u(n - M) \end{bmatrix} \tag{15.44}$$

Hence, postmultiplying both sides of Eq. (15.43) by $\mathbf{u}_{M+1}(n)$, using the partitioned form of Eq. (15.44) for the first term in the right side of Eq. (15.43), and using in the second term the definition of the backward a posteriori prediction error

$$b_M(n) = \mathbf{c}_M^H(n)\mathbf{u}_{M+1}(n) \tag{15.45}$$

we get the second recursion for the extended gain vector:

$$\mathbf{k}_{M+1}(n) = \begin{bmatrix} \mathbf{k}_M(n) \\ 0 \end{bmatrix} + \frac{b_M(n)}{\mathcal{B}_M(n)} \mathbf{c}_M(n) \tag{15.46}$$

Examination of Eqs. (15.42) and (15.46) reveals that the extended gain vector $\mathbf{k}_{M+1}(n)$ provides the *linkage* between the gain vector $\mathbf{k}_M(n - 1)$ for adaptive forward linear prediction and the gain vector $\mathbf{k}_M(n)$ for adaptive backward linear prediction. In particular, Eq. (15.42) relating $\mathbf{k}_M(n - 1)$ to $\mathbf{k}_{M+1}(n)$ involves both order and time updates. On the other hand, Eq. (15.46) relating $\mathbf{k}_M(n)$ to $\mathbf{k}_{m+1}(n)$ involves an order update only.

15.4 CONVERSION FACTOR (ANGLE VARIABLE)

The definition of the M-by-1 gain vector,

$$\mathbf{k}_M(n) = \mathbf{\Phi}_M^{-1}(n)\mathbf{u}_M(n)$$

may also be viewed as the solution of a special case of the normal equations. To be specific, the gain vector $\mathbf{k}_M(n)$ defines the tap-weight vector of a transversal filter that contains M taps and that operates on the input data $u(1), u(2), \ldots, u(n)$ to produce the least-squares estimate of a special desired response that equals

$$d(i) = \begin{cases} 1, & i = n \\ 0, & i = 1, 2, \ldots, n - 1 \end{cases} \tag{15.47}$$

The n-by-1 vector whose elements equal the $d(i)$ of Eq. (15.47) is called the *first coordinate vector*. This vector has the property that its inner product with any time-dependent vector reproduces the upper or "most recent" element of that vector.

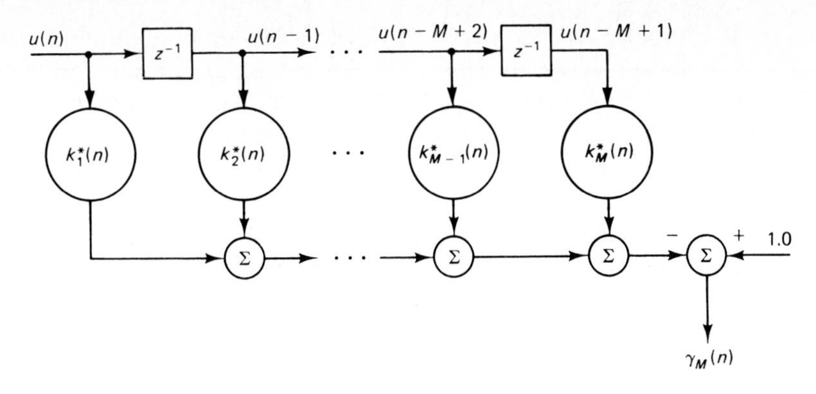

Figure 15.3 Transversal filter for defining the estimation error $\gamma_M(n)$.

Substituting Eq. (15.47) in (13.10), we find that the M-by-1 cross-correlation vector $\boldsymbol{\theta}_M(n)$ between the M tap inputs of the transversal filter and the desired response equals $\mathbf{u}_M(n)$. This therefore confirms the gain vector $\mathbf{k}_M(n)$ as the special solution of the normal equations that arise when the desired response is defined by Eq. (15.47).

Define the *estimation error*

$$\begin{aligned}\gamma_M(n) &= 1 - \mathbf{k}_M^H(n)\mathbf{u}_M(n) \\ &= 1 - \mathbf{u}_M^H(n)\boldsymbol{\Phi}_M^{-1}(n)\mathbf{u}_M(n)\end{aligned} \tag{15.48}$$

The estimation error $\gamma_M(n)$ represents the output of a transversal filter whose tap-weight vector equals the gain vector $\mathbf{k}_M(n)$ and which is excited by the tap-input vector $\mathbf{u}_M(n)$, as depicted in Fig. 15.3. Since the filter output has the structure of a Hermitian form, it follows that the estimation error $\gamma_M(n)$ is a real-valued scalar. Moreover, $\gamma_M(n)$ has the important property that it is bounded by zero and one; that is,

$$0 \le \gamma_M(n) \le 1 \tag{15.49}$$

This property is readily proved by substituting the recursion of Eq. (13.16) for the inverse matrix $\boldsymbol{\Phi}_M^{-1}(n)$ in Eq. (15.49), and then simplifying to obtain the result

$$\gamma_M(n) = \frac{1}{1 + \lambda^{-1}\mathbf{u}_M^H(n)\boldsymbol{\Phi}_M^{-1}(n-1)\mathbf{u}_M(n)} \tag{15.50}$$

The Hermitian form $\mathbf{u}_M^H(n)\boldsymbol{\Phi}_M^{-1}(n-1)\mathbf{u}_M(n) \ge 0$. Consequently, the estimation error $\gamma_M(n)$ is bounded as in Eq. (15.49).

It is noteworthy that $\gamma_M(n)$ also equals the sum of weighted error squares resulting from use of the transversal filter in Fig. 15.3, whose tap-weight vector equals the gain vector $\mathbf{k}_M(n)$, to obtain the least-squares estimate of the first unit vector (see Problem 2).

Other Useful Interpretations of $\gamma_M(n)$

Depending on the approach taken, the parameter $\gamma_M(n)$ may be given three other entirely different interpretations:

1. The parameter $\gamma_M(n)$ may be viewed as a *likelihood variable* (Lee et al., 1981). This interpretation follows from a statistical formulation of the tap-input vector in terms of its log-likelihood function, under the assumption that the tap inputs have a joint Gaussian distribution (see Problem 4 of Chapter 17).
2. The parameter $\gamma_M(n)$ may be interpreted as an *angle variable* (Lee et al., 1981; Carayannis et al., 1983). This interpretation follows from Eq. (15.49). In particular, following the discussion presented in Chapter 14, we may express the (positive) square root of $\gamma_M(n)$ as

$$\gamma_M^{1/2}(n) = \cos\phi_1 \cos\phi_2 \cdots \cos\phi_M$$

where each of the angles $\phi_1, \phi_2, \ldots, \phi_M$ represents the angle of a plane (Givens) rotation [see Eq. (14.57].
3. The parameter $\gamma_M(n)$ may be interpreted as a *conversion factor* (Carayannis et al., 1983). According to this interpretation, the availability of $\gamma_M(n)$ helps us determine the value of an a posteriori estimation error, given the value of the corresponding a priori estimation error.

It is this third interpretation that we pursue here. Indeed, it is because of this interpretation that we have adopted the terminology "conversion factor" as a description for $\gamma_M(n)$.

In linear least-squares estimation theory, there are three kinds of estimation error to be considered: the ordinary estimation error (involved in the estimation of some desired response), the forward prediction error, and the backward prediction error. Correspondingly, $\gamma_M(n)$ has three useful interpretations as a conversion factor, as described next:

1. For recursive least-squares estimation, we have

$$\gamma_M(n) = \frac{e_M(n)}{\alpha_M(n)} \tag{15.51}$$

where $e_M(n)$ is the a posteriori estimation error and $\alpha_M(n)$ is the a priori estimation error. This relation is readily proved by postmultiplying the Hermitian transposed sides of Eq. (13.25) by $\mathbf{u}_M(n)$, using Eq. (13.26) for the a priori estimation error $\alpha_M(n)$, Eq. (13.27) for the a posteriori estimation error $e_M(n)$, and the first line of Eq. (15.48) for the variable $\gamma_M(n)$. Equation (15.51) states that given the a priori estimation error $\alpha_M(n)$ as computed in the RLS algorithm, we may determine the corresponding value of the a posteriori estimation error $e_M(n)$ by multiplying $\alpha_M(n)$ by $\gamma_M(n)$. We may therefore view $\alpha_M(n)$ as a tentative value of the estimation error $e_M(n)$ and $\gamma_M(n)$ as the multiplicative correction.
2. For adaptive forward linear prediction, we have

$$\gamma_M(n-1) = \frac{f_M(n)}{\eta_M(n)} \tag{15.52}$$

This relation is readily proved by postmultiplying the Hermitian transposed sides of Eq. (15.16) by $\mathbf{u}_{M+1}(n)$, using the partitioned form of Eq. (15.40) for the postmultiplication of the second term on the right side of Eq. (15.16), and then using the definitions of Eqs. (15.17), (15.41), and (15.48). Equation (15.52) states that given the forward a priori prediction error $\eta_M(n)$, we may compute the forward a posteriori prediction error $f_M(n)$ by multiplying $\eta_M(n)$ by the delayed estimation error $\gamma_M(n-1)$. We may therefore view $\eta_M(n)$ as a tentative value for the forward a posteriori prediction error $f_M(n)$ and $\gamma_M(n-1)$ as the multiplicative correction.

3. For adaptive backward linear prediction, we have

$$\gamma_M(n) = \frac{b_M(n)}{\psi_M(n)} \tag{15.53}$$

This third relation is readily proved by postmultiplying the Hermitian transposed sides of Eq. (15.34) by $\mathbf{u}_{M+1}(n)$, using the paritioned form of Eq. (15.44) for the postmultiplication of the second term on the right side of Eq. (15.34), and then using the definitions of Eqs. (15.35), (15.45), and (15.48). Equation (15.53) states that given the backward a priori prediction error $\psi_M(n)$, we may compute the backward a posteriori prediction error $b_M(n)$ by multiplying $\psi_M(n)$ by the estimation error $\gamma_M(n)$. We may therefore view $\psi_M(n)$ as a tentative value for the backward prediction error $b_M(n)$ and $\gamma_M(n)$ as the multiplicative correction.

The discussion above points out the unique role of the variable $\gamma_M(n)$ in that it is the *common* factor (either in its regular or delayed form) in the conversion of an a priori estimation error into the corresponding a posteriori estimation error, be it in the context of ordinary estimation, forward prediction, or backward prediction. Accordingly, we may refer to $\gamma_M(n)$ as a *conversion factor*. Indeed, it is remarkable that through the use of this conversion factor we are able to compute the a posteriori errors $e_M(n)$, $f_M(n)$ and $b_M(n)$ at the time n before the tap-weight vectors of the pertinent filters [i.e., $\hat{\mathbf{w}}_M(n)$, $\mathbf{a}_M(n)$, and $\mathbf{c}_M(n)$] that produce them have been actually computed (Carayannis et al., 1983).

Recursive Relations for Computing $\gamma_M(n)$

Next, we develop two recursions for updating of the variable $\gamma_M(n)$ as follows. First, we postmultiply the Hermitian transposed sides of Eq. (15.42) by $\mathbf{u}_{M+1}(n)$, use the partitioned form of Eq. (15.23) for the postmultiplication of the first term on the right side of Eq. (15.42), and then use the definitions of the forward a posteriori prediction error $f_M(n)$ and the conversion factor $\gamma_M(n)$ given in Eqs. (15.41) and (15.48). We thus obtain the following recursion that involves time as well as order updates of the conversion factor:

$$\gamma_{M+1}(n) = \gamma_M(n-1) - \frac{|f_M(n)|^2}{\mathcal{F}_M(n)} \tag{15.54}$$

For the second recursion, we postmultiply the Hermitian transposed sides of Eq. (15.46) by $\mathbf{u}_{M+1}(n)$, use the partitioned form of Eq. (15.44) for the postmultiplication of the first term on the right side of Eq. (15.46), and then use the definitions of the backward a posteriori prediction error $b_M(n)$ and the conversion factor $\gamma_M(n)$ given in Eqs. (15.45) and (15.48). We thus get the following recursion that only involves an order update:

$$\gamma_{M+1}(n) = \gamma_M(n) - \frac{|b_M(n)|^2}{\mathcal{B}_M(n)} \tag{15.55}$$

Finally, we develop another pair of relations by manipulating these two recursions. First, we use Eqs. (15.18) and (15.52) to eliminate $|f_M(n)|^2$ from (15.54). The result is

$$\gamma_{M+1}(n) = \lambda\frac{\mathcal{F}_M(n-1)}{\mathcal{F}_M(n)}\gamma_M(n-1) \tag{15.56}$$

Similarly, we use Eqs. (15.36) and (15.53) to eliminate $|b_M(n)|^2$ from (15.55), thereby obtaining the relation

$$\gamma_{M+1}(n) = \lambda\frac{\mathcal{B}_M(n-1)}{\mathcal{B}_M(n)}\gamma_M(n) \tag{15.57}$$

15.5 SUMMARY

In order to make the task of referring to the material developed herein easy for subsequent use, below we present a summary of the basic relations that embody the background theory of fast algorithms:

1. *Adaptive forward linear prediction*

$$\mathbf{a}_M(n) = \mathbf{a}_M(n-1) - \begin{bmatrix} 0 \\ \mathbf{k}_M(n-1) \end{bmatrix}\eta_M^*(n)$$

$$\eta_M(n) = \mathbf{a}_M^H(n-1)\mathbf{u}_{M+1}(n)$$

$$\mathcal{F}_M(n) = \lambda\mathcal{F}_M(n-1) + \eta_M(n)f_M^*(n)$$

2. *Adaptive backward linear prediction*

$$\mathbf{c}_M(n) = \mathbf{c}_M(n-1) - \begin{bmatrix} \mathbf{k}_M(n) \\ 0 \end{bmatrix}\psi_M^*(n)$$

$$\psi_M(n) = \mathbf{c}_M^H(n-1)\mathbf{u}_{M+1}(n)$$

$$\mathcal{B}_M(n) = \lambda\mathcal{B}_M(n-1) + \psi_M(n)b_M^*(n)$$

3. *Gain vector*

$$\mathbf{k}_{M+1}(n) = \begin{bmatrix} 0 \\ \mathbf{k}_M(n-1) \end{bmatrix} + \frac{f_M(n)}{\mathcal{F}_M(n)}\mathbf{a}_M(n)$$

$$\mathbf{k}_{M+1}(n) = \begin{bmatrix} \mathbf{k}_M(n) \\ 0 \end{bmatrix} + \frac{b_M(n)}{\mathcal{B}_M(n)}\mathbf{c}_M(n)$$

4. *Conversion factor*
   (a) *Defining relations*

$$\gamma_M(n) = \frac{e_M(n)}{\alpha_M(n)}$$

$$\gamma_M(n-1) = \frac{f_M(n)}{\eta_M(n)}$$

$$\gamma_M(n) = \frac{b_M(n)}{\psi_M(n)}$$

   (b) *Update formulas*

$$\gamma_{M+1}(n) = \gamma_M(n-1) - \frac{|f_M(n)|^2}{\mathcal{F}_M(n)}$$

$$\gamma_{M+1}(n) = \gamma_M(n) - \frac{|b_M(n)|^2}{\mathcal{B}_M(n)}$$

$$\gamma_{M+1}(n) = \lambda \frac{\mathcal{F}_M(n-1)}{\mathcal{F}_M(n)} \gamma_M(n-1)$$

$$\gamma_{M+1}(n) = \lambda \frac{\mathcal{B}_M(n-1)}{\mathcal{B}_M(n)} \gamma_M(n)$$

15.6 THREE DIFFERENT CLASSES OF FAST ALGORITHMS

The material presented in the previous sections of this chapter is indeed fundamental to the development of fast algorithms for solving the recursive least-squares estimation problem. Three different classes of fast algorithms may be identified, depending on the type of filter structure employed:

1. *Transversal filter-based fast algorithms*. This class of fast RLS algorithms includes the following:
   (a) *The fast RLS algorithm,* originally developed by Falconer and Ljung (1978); this algorithm is also referred to in the literature as the fast Kalman algorithm. The development of the fast RLS algorithm is presented as a problem in Chapter 16.
   (b) *The fast transversal filters (FTF) algorithm,* developed independently by Carayannis et al. (1983), and Cioffi and Kailath (1984). Details of the derivation of the FTF algorithm are presented in Chapter 16.
2. *Lattice predictor-based fast RLS algorithms*. This second class of fast RLS algorithms includes four different versions of the *recursive least-squares lattice (LSL) algorithm*. Two of these involve the use of *a posteriori* forward and backward prediction errors. The other two versions involve the use of *a priori* forward and backward prediction errors. The first formal derivation of a ver-

sion of the recursive LSL algorithm was presented by Morf et al. (1977), and Morf and Lee (1978). The use of error feedback in the formulation of recursive LSL algorithms was first presented by Ling and Proakis (1984a), and Ling et al. (1985). Derivation of the recursive LSL algorithm, in its four different forms, is presented in Chapter 17.

3. *QR decomposition-based least-squares lattice algorithms*. In the third and final class of fast algorithms considered in this book, plane (Givens) rotations and linear predictions are used to produce a fast implementation of the RLS algorithm in the form of a lattice structure.[1] One important member of this class of fast algorithms is derived in Chapter 18, and its intimate relation to the error feedback form of a recursive LSL algorithm is demonstrated. The new fast algorithm presented in Chapter 18 is a hybridization of novel approaches developed independently by Proudler et al. (1989), and Regalia and Belanger (1990).

The history of fast algorithms may perhaps be traced back to the pioneering work of Morf (1974) on efficient solutions for least-squares predictions for the case when the correlation matrix of the predictor input is non-Toeplitz but has the special structure described in Chapter 10. Falconer and Ljung (1978) were the first to recognize the importance of Morf's work and formally apply it to derive the fast RLS (Kalman) algorithm, which is essentially of historical interest.

PROBLEMS

1. (a) Show that the inverse of the time-averaged correlation matrix $\mathbf{\Phi}_{M+1}(n)$ may be expressed as follows:

$$\mathbf{\Phi}_{M+1}^{-1}(n) = \begin{bmatrix} 0 & \mathbf{0}_M^T \\ \mathbf{0}_M & \mathbf{\Phi}_M^{-1}(n-1) \end{bmatrix} + \frac{1}{\mathcal{F}_M(n)} \mathbf{a}_M(n)\mathbf{a}_M^H(n)$$

where $\mathbf{0}_M$ is the M-by-1 null vector, $\mathbf{0}_M^T$ its transpose, $\mathcal{F}_M(n)$ is the minimum sum of weighted forward prediction-error squares, and $\mathbf{a}_M(n)$ is the tap-weight vector of forward prediction-error filter. Both $\mathbf{a}_M(n)$ and $\mathcal{F}_M(n)$ refer to prediction order M.

(b) Show that the inverse of $\mathbf{\Phi}_{M+1}(n)$ may also be expressed in the form

$$\mathbf{\Phi}_{M+1}^{-1}(n) = \begin{bmatrix} \mathbf{\Phi}_M^{-1}(n) & \mathbf{0}_M \\ \mathbf{0}_M^T & 0 \end{bmatrix} + \frac{1}{\mathcal{B}_M(n)} \mathbf{c}_M(n)\mathbf{c}_M^H(n)$$

where $\mathcal{B}_M(n)$ is the minimum sum of weighted backward a posteriori prediction-error squares, and $\mathbf{c}_M(n)$ is the tap-weight vector of the backward prediction-error filter. Both $\mathcal{F}_M(n)$ and $\mathbf{c}_M(n)$ refer to prediction order M.

Hints: For part (a), use the partitioned form of $\mathbf{\Phi}_{M+1}(n)$ given in Eq. (15.9). For part (b), use the second partitioned form of this matrix given in Eq. (15.27).

[1] The idea of QR decomposition-based fast RLS algorithms was originated by Cioffi (1988), who developed a matrix oriented (also referred to as a Kalman) member of this class of fast algorithms. It was this development that triggered the development of other QR decomposition-based fast algorithms.

2. Show that the parameter $\gamma_M(n)$ defined by

$$\gamma_M(n) = 1 - \mathbf{k}_M^H(n)\mathbf{u}_M(n)$$

equals the sum of weighted error squares resulting from use of the transversal filter in Fig. 15.3. The tap-weight vector of this filter equals the gain vector $\mathbf{k}_M(n)$, and the tap-input vector equals $\mathbf{u}_M(n)$. The filter is designed to produce the least-squares estimate of a desired response that equals the *first coordinate vector*.

3. The definition of the square root of the conversion factor given in Eq. (14.57) is indeed consistent with the definition of the conversion factor given in Eq. (15.48). Demonstrate the validity of this statement.

4. Let $\mathbf{\Phi}_M(n)$ denote the time-averaged correlation matrix of the tap-input vector $\mathbf{u}_M(n)$ at time n; likewise for $\mathbf{\Phi}_M(n-1)$. Show that the conversion factor $\gamma_M(n)$ is related to the determinants of these two matrices as follows:

$$\gamma_M(n) = \lambda \frac{\det[\mathbf{\Phi}_M(n-1)]}{\det[\mathbf{\Phi}_M(n)]}$$

where λ is the exponential weighting factor. *Hint:* Use the identity

$$\det(\mathbf{I}_1 + \mathbf{AB}) = \det(\mathbf{I}_2 + \mathbf{BA})$$

where $\mathbf{I}_1$ and $\mathbf{I}_2$ are identity matrices of appropriate dimensions, and $\mathbf{A}$ and $\mathbf{B}$ are matrices of compatible dimensions.

5. The tap-weight vector $\hat{\mathbf{w}}(n)$ in the RLS algorithm may be interpreted as the state vector of a time-varying system. Show that, according to this interpretation, the defining recursions of the RLS algorithm given in Section 13.3 may be grouped as follows

$$\begin{bmatrix} \hat{\mathbf{w}}(n) \\ \epsilon(n) \end{bmatrix} = \begin{bmatrix} \mathbf{I} - \mathbf{k}(n)\mathbf{u}^H(n) & \mathbf{k}(n) \\ -\gamma^{1/2}(n)\,\mathbf{u}^H(n) & \gamma^{1/2}(n) \end{bmatrix} \begin{bmatrix} \hat{\mathbf{w}}(n-1) \\ d(n) \end{bmatrix}$$

where $\gamma(n)$ is the conversion factor, and

$$\epsilon(n) = \gamma^{1/2}(n)\,\alpha(n)$$

The remaining quantities are as defined in Section 13.3.

16

Fast Transversal Filters

In this chapter we present the derivation of the first in a series of three fast algorithms for solving the linear least-squares estimation problem recursively. The algorithm is known as the *fast transversal filters* (FTF) *algorithm* (Cioffi and Kailath, 1984; Carayannis et al., 1983). This algorithm uses a maximum of *four* transversal filters that share a common input. The four filters have distinct tasks:

1. Recursive forward linear prediction
2. Recursive background linear prediction
3. Recursive computation of the gain vector
4. Recursive estimation of the desired response

The derivation presented herein is algebraic[1] in nature, building on the background theory of fast algorithms developed in Chapter 15. Indeed, we will make frequent references to equations derived therein.

16.1 FTF ALGORITHM

The basic structural features of the *fast transversal filters* (FTF) algorithm are depicted in Fig. 16.1. This figure shows four separate transversal filters, labeled I through IV, which share a common set of tap inputs. In particular, filters I and II perform forward and backward linear prediction on the input data, respectively. Filter III defines the gain vector of the RLS algorithm. Filter IV is the adaptive filter

[1] The original derivation of the FTF algorithm, presented in Cioffi and Kailath (1984), was based on the use of a *geometrical approach*. For a tutorial treatment of *linear vector spaces* and their use in the geometrical derivation of the FTF algorithm, see Alexander (1986b).

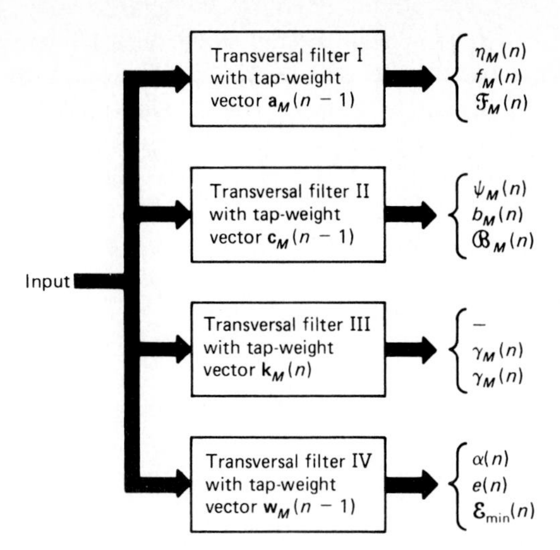

Figure 16.1 Transversal filter computation of RLS variables.

whose tap-weight vector is the desired estimate for the problem at hand. The combination of these four filters is designed to produce the *exact solution of the RLS problem at all times*. When the minimum mean-squared is nonzero, as is ordinarily the case, the first three filters are required to complete the solution. We refer to these three filters as *auxiliary filters*. The operation of each filter in Fig. 16.1 involves two basic processes: (1) a filtering process wherein an error signal is produced in response to the excitation applied to the pertinent filter, and (2) an adaptive process wherein the tap-weight vector of that filter is updated by incrementing its old value by an amount equal to a scalar times some vector. The combined result of all these operations is not only an exact solution to the RLS problem but also a computational cost that increases linearly with the number of taps contained in the adaptive filter, as in the LMS algorithm.

We start the development of the FTF algorithm by applying the tap-input vector $\mathbf{u}_{M+1}(n)$ to filter I, the forward prediction-error filter, to produce the forward a priori prediction error [see Eq. (15.17)]:

$$\eta_M(n) = \mathbf{a}_M^H(n-1)\mathbf{u}_{M+1}(n) \tag{16.1}$$

We multiply $\eta_M(n)$ by the delayed estimation error $\gamma_M(n-1)$ to produce the *forward a posteriori prediction error* [see Eq. (15.52)]:

$$f_M(n) = \gamma_M(n-1)\eta_M(n) \tag{16.2}$$

We compute the updated minimum value of the sum of weighted forward prediction-error squares by using the recursion [see Eq. (15.18)]:

$$\mathcal{F}_M(n) = \lambda\mathcal{F}_M(n-1) + \eta_M(n)f_M^*(n) \tag{16.3}$$

We now use Eq. (15.56) to compute the updated value of the conversion factor:

$$\gamma_{M+1}(n) = \lambda\frac{\mathcal{F}_M(n-1)}{\mathcal{F}_M(n)}\gamma_M(n-1) \tag{16.4}$$

Next we compute the updated value of the gain vector by using Eq. (15.42) except for a minor modification. The change is intended to make the recursion dependent on the old estimate of the tap-weight vector of the forward prediction-error filter, $\mathbf{a}_M(n-1)$, rather than the current estimate $\mathbf{a}_M(n)$. This we do simply by substituting Eq. (15.16) in (15.42), thereby obtaining (after collecting common terms)

$$\mathbf{k}_{M+1}(n) = \left(1 - \frac{\eta_M^*(n)f_M(n)}{\mathcal{F}_M(n)}\right)\begin{bmatrix} 0 \\ \mathbf{k}_M(n-1) \end{bmatrix} + \frac{f_M(n)}{\mathcal{F}_M(n)}\mathbf{a}_M(n-1) \tag{16.5}$$

We use Eq. (15.18) and then (15.56) to write

$$\begin{aligned} 1 - \frac{\eta_M^*(n)f_M(n)}{\mathcal{F}_M(n)} &= \lambda\frac{\mathcal{F}_M(n-1)}{\mathcal{F}_M(n)} \\ &= \frac{\gamma_{M+1}(n)}{\gamma_M(n-1)} \end{aligned} \tag{16.6}$$

We use Eq. (15.52) and then (15.56) to write

$$\begin{aligned} \frac{f_M(n)}{\mathcal{F}_M(n)} &= \frac{\eta_M(n)\gamma_M(n-1)}{\mathcal{F}_M(n)} \\ &= \lambda^{-1}\frac{\eta_M(n)\gamma_{M+1}(n)}{\mathcal{F}_M(n-1)} \end{aligned} \tag{16.7}$$

Therefore, substituting Eqs. (16.6) and (16.7) in (16.5), we get

$$\mathbf{k}_{M+1}(n) = \frac{\gamma_{M+1}(n)}{\gamma_M(n-1)}\begin{bmatrix} 0 \\ \mathbf{k}_M(n-1) \end{bmatrix} + \lambda^{-1}\frac{\eta_M(n)\gamma_{M+1}(n)}{\mathcal{F}_M(n-1)}\mathbf{a}_M(n-1) \tag{16.8}$$

For convenience of presentation, we define the *normalized gain vector*

$$\tilde{\mathbf{k}}_M(n) = \frac{\mathbf{k}_M(n)}{\gamma_M(n)} \tag{16.9}$$

Accordingly, we may simplify the recursion of Eq. (16.8) as

$$\tilde{\mathbf{k}}_{M+1}(n) = \begin{bmatrix} 0 \\ \tilde{\mathbf{k}}_M(n-1) \end{bmatrix} + \lambda^{-1}\frac{\eta_M(n)}{\mathcal{F}_M(n-1)}\mathbf{a}_M(n-1) \tag{16.10}$$

which is the desired recursion for updating the normalized gain vector.

Correspondingly, we use Eq. (16.9) and then (15.52) to redefine the recursion of (15.16) for updating the tap-weight vector of the forward prediction-error filter as

$$\begin{aligned} \mathbf{a}_M(n) &= \mathbf{a}_M(n-1) - \gamma_M(n-1)\eta_M^*(n)\begin{bmatrix} 0 \\ \tilde{\mathbf{k}}_M(n-1) \end{bmatrix} \\ &= \mathbf{a}_M(n-1) - f_M^*(n)\begin{bmatrix} 0 \\ \tilde{\mathbf{k}}_M(n-1) \end{bmatrix} \end{aligned} \tag{16.11}$$

At this stage in our development, we have time-updated the forward prediction-error filter (i.e., filter I) and also updated the gain vector (i.e., filter III) not only in time (as desired) but also order.

Our next task is twofold: (1) to correct for the increase in the order of filter III from M to $M+1$ incurred through the use of Eq. (16.4) to compute $\gamma_{M+1}(n)$, and (2) to perform a time update on the backward prediction-error filter (i.e., filter II).

To do this, we first substitute Eq. (15.34) in (15.46) and collect common terms, obtaining

$$\mathbf{k}_{M+1}(n) = \left(1 - \frac{\psi_M^*(n)b_M(n)}{\mathcal{B}_M(n)}\right)\begin{bmatrix}\mathbf{k}_M(n)\\0\end{bmatrix} + \frac{b_M(n)}{\mathcal{B}_M(n)}\mathbf{c}_M(n-1) \tag{16.12}$$

Let $k_{M+1,M+1}(n)$ denote the *last element* of the $(M+1)$-by-1 gain vector $\mathbf{k}_{M+1}(n)$. We recognize that the last element of the $(M+1)$-by-1 tap-weight vector $\mathbf{c}_M(n-1)$, characterizing the backward prediction-error filter, equals unity. Hence, by considering the last elements of the vectors on the left and right sides of Eq. (16.12), we deduce that

$$k_{M+1,M+1}(n) = \frac{b_M(n)}{\mathcal{B}_M(n)} \tag{16.13}$$

By normalizing with respect to $\gamma_{M+1}(n)$, we may also express the last element of $\tilde{\mathbf{k}}_{M+1}(n)$ as

$$\begin{aligned}\tilde{k}_{M+1,M+1} &= \frac{k_{M+1,M+1}(n)}{\gamma_{M+1}(n)}\\ &= \frac{b_M(n)}{\gamma_{M+1}(n)\mathcal{B}_M(n)}\end{aligned} \tag{16.14}$$

The combined use of Eqs. (15.53) and (15.57) in (16.14) yields

$$\tilde{k}_{M+1,M+1}(n) = \frac{\psi_M(n)}{\lambda\mathcal{B}_M(n-1)} \tag{16.15}$$

Equivalently, we may express the backward a priori prediction error $\psi_M(n)$ in terms of the last element of the normalized gain vector $\tilde{\mathbf{k}}_{M+1}(n)$ as

$$\psi_M(n) = \lambda\mathcal{B}_M(n-1)\tilde{k}_{M+1,M+1}(n) \tag{16.16}$$

This relation enables us to compute $\psi_M(n)$. Solving Eq. (15.57) for $\gamma_M(n)$, we have

$$\gamma_M(n) = \frac{\mathcal{B}_M(n)}{\lambda\mathcal{B}_M(n-1)}\gamma_{M+1}(n) \tag{16.17}$$

By using Eq. (15.36) and then (16.14), we may write

$$\begin{aligned}\lambda\frac{\mathcal{B}_M(n-1)}{\mathcal{B}_M(n)} &= 1 - \frac{\psi_M(n)b_M^*(n)}{\mathcal{B}_M(n)}\\ &= 1 - \psi_M(n)\gamma_{M+1}(n)\tilde{k}_{M+1,M+1}^*(n)\end{aligned} \tag{16.18}$$

Therefore, substituting Eq. (16.18) in (16.17), we get

$$\gamma_M(n) = [1 - \psi_M(n)\gamma_{M+1}(n)\tilde{k}_{M+1,M+1}^*(n)]^{-1}\gamma_{M+1}(n) \tag{16.19}$$

Note that

$$\psi_M(n)\tilde{k}_{M+1,M+1}^*(n) = \psi_M^*(n)k_{M+1,M+1}(n)$$

Equation (16.19) enables us to compute the current value of the conversion factor $\gamma_M(n)$. Thus, this relation corrects for the increase in the order of filter III from M to

$M + 1$ that results from the use of Eq. (16.4). Having already computed the current values of $\gamma_M(n)$ and $\psi_M(n)$, we may now use Eq. (15.53) to compute the backward a posteriori prediction error:

$$b_M(n) = \gamma_M(n)\psi_M(n) \tag{16.20}$$

Furthermore, we may use the recursion of Eq. (15.36) to update the minimum value of the sum of weighted backward a posteriori prediction-error squares:

$$\mathcal{B}_M(n) = \lambda \mathcal{B}_M(n-1) + \psi_M(n)b_M^*(n) \tag{16.21}$$

Returning to the recursion of Eq. (16.12), we use the first line of Eq. (16.18) and Eqs. (15.57), (16.9), and (16.14), then rearrange terms, and thereby rewrite this recursion as follows:

$$\begin{bmatrix} \tilde{\mathbf{k}}_M(n) \\ 0 \end{bmatrix} = \tilde{\mathbf{k}}_{M+1}(n) - \tilde{k}_{M+1,M+1}(n)\mathbf{c}_M(n-1) \tag{16.22}$$

We are now ready to update the tap-weight vector that characterizes the backward prediction-error filter [see Eq. (15.34)].

$$\begin{aligned} \mathbf{c}_M(n) &= \mathbf{c}_M(n-1) - \psi_M^*(n)\begin{bmatrix} \mathbf{k}_M(n) \\ 0 \end{bmatrix} \\ &= \mathbf{c}_M(n-1) - b_M^*(n)\begin{bmatrix} \tilde{\mathbf{k}}_M(n) \\ 0 \end{bmatrix} \end{aligned} \tag{16.23}$$

where in the last line we have made use of Eqs. (15.53) and (16.9).

We have thus completed the time-updating of the gain vector (i.e., filter III) and the backward prediction-error filter (i.e., filter II). There now only remains the task of updating the tap-weight vector of the adaptive filter (i.e., filter IV). This we do next.

We use Eq. (13.26) to compute the a posteriori estimation error

$$\alpha_M(n) = d(n) - \hat{\mathbf{w}}_M^H(n-1)\mathbf{u}_M(n) \tag{16.24}$$

We use Eq. (15.51) to compute the corresponding value of the estimation error:

$$e_M(n) = \gamma_M(n)\alpha_M(n) \tag{16.25}$$

Finally, we use Eq. (13.25) to update the tap-weight vector of the adaptive filter:

$$\begin{aligned} \hat{\mathbf{w}}_M(n) &= \hat{\mathbf{w}}_M(n-1) + \mathbf{k}_M(n)\alpha_M^*(n) \\ &= \hat{\mathbf{w}}_M(n-1) + \tilde{\mathbf{k}}_M(n)e_M^*(n) \end{aligned} \tag{16.26}$$

where in the last line we have made use of Eqs. (16.25) and (16.9).

This completes the updating cycle of the FTF algorithm.

Summary of the FTF Algorithm

In Table 16.1, we present a summary of the FTF algorithm by collecting together the recursions derived previously. This table presents a summary of the recursions and relations that are defined by Eqs. (16.1) to (16.4), (16.10, (16.11), (16.16), and (16.19) to (16.26), in that order.

TABLE 16.1 SUMMARY OF THE FTF ALGORITHM

Predictions

$$\eta_M(n) = \mathbf{a}_M^H(n-1)\mathbf{u}_{M+1}(n)$$

$$f_M(n) = \gamma_M(n-1)\eta_M(n)$$

$$\mathcal{F}_M(n) = \lambda\mathcal{F}_M(n-1) + \eta_M(n)f_M^*(n)$$

$$\gamma_{M+1}(n) = \lambda\frac{\mathcal{F}_M(n-1)}{\mathcal{F}_M(n)}\gamma_M(n-1)$$

$$\tilde{\mathbf{k}}_{M+1}(n) = \begin{bmatrix} 0 \\ \tilde{\mathbf{k}}_M(n-1) \end{bmatrix} + \lambda^{-1}\frac{\eta_M(n)}{\mathcal{F}_M(n-1)}\mathbf{a}_M(n-1)$$

$$\mathbf{a}_M(n) = \mathbf{a}_M(n-1) - f_M^*(n)\begin{bmatrix} 0 \\ \tilde{\mathbf{k}}_M(n-1) \end{bmatrix}$$

$$\psi_M(n) = \lambda\mathcal{B}_M(n-1)\tilde{k}_{M+1,M+1}(n)$$

$$\gamma_M(n) = [1 - \psi_M^*(n)\gamma_{M+1}(n)\tilde{k}_{M+1,M+1}(n)]^{-1}\gamma_{M+1}(n)$$

$$\text{Rescue}^{a}\text{ variable} = [1 - \psi_M^*(n)\gamma_{M+1}(n)\tilde{k}_{M+1,M+1}(n)]$$

$$b_M(n) = \gamma_M(n)\psi_M(n)$$

$$\mathcal{B}_M(n) = \lambda\mathcal{B}_M(n-1) + \psi_M(n)b_M^*(n)$$

$$\begin{bmatrix} \tilde{\mathbf{k}}_M(n) \\ 0 \end{bmatrix} = \tilde{\mathbf{k}}_{M+1}(n) - \tilde{k}_{M+1,M+1}(n)\mathbf{c}_M(n-1)$$

$$\mathbf{c}_M(n) = \mathbf{c}_M(n-1) - b_M^*(n)\begin{bmatrix} \tilde{\mathbf{k}}_M(n) \\ 0 \end{bmatrix}$$

Filtering

$$\alpha_M(n) = d(n) - \hat{\mathbf{w}}_M^H(n-1)\mathbf{u}_M(n)$$

$$e_M(n) = \gamma_M(n)\alpha_M(n)$$

$$\hat{\mathbf{w}}_M(n) = \hat{\mathbf{w}}_M(n-1) + \tilde{\mathbf{k}}_M(n)e_M^*(n)$$

Note: $\tilde{k}_{M+1,M+1}(n)$ is the last element of the normalized gain vector $\tilde{\mathbf{k}}_{M+1}(n)$.

[a] Rescue if variable is negative: Save $\hat{\mathbf{w}}_M(n)$ as initial condition and weight in the manner discussed in Section 16.3.

16.2 EXACT INITIALIZATION OF THE FTF ALGORITHM USING ZERO INITIAL CONDITION

During the initialization period $1 \le n \le M+1$, where M is the number of tap weights contained in the adaptive filter, considerable simplification and stabilization of the RLS solution are achieved by special investigation of the FTF algorithm (Cioffi and Kailath, 1984). In this section, we study *exact initialization* of the FTF algorithm for the case when the initial condition is zero. This case is very common, simply because in practice we ordinarily have no prior information about the unknown parameter vector $\mathbf{w}_o(n)$ in the multiple linear regression model of Fig. 13.3 This represents a conceptual model for the statistical description of the estimation error. Accordingly, by setting the adjustable tap-weight vector $\hat{\mathbf{w}}_M(n)$ of the adaptive filter equal to zero at time $n = 0$, we simplify the initialization of the FTF algorithm.

Before deriving the exact initialization procedure, we discuss some special characteristics of the prewindowed RLS solution during the initialization period, for which $1 \le n \le M + 1$. At time $n = M$, initialization of both the gain vector (filter III) and the adaptive filter (filter IV) is completed. However, the forward and backward prediction-error filters (filters I and II) are both one unit longer; hence, their initialization is completed at time $n = M + 1$. In any event, during the initialization period the length of the input data is less than or equal to the number of adjustable tap weights in the adaptive filter, so the RLS problem is *underdetermined* or *exactly determined,* respectively. Hence, during the initialization period, we may set all the estimation errors, $e_M(n)$, $\gamma_M(n)$, $f_M(n)$, and $b_M(n)$, equal to zero since we have an abundance of adjustable parameters. In such a situation, the unique solution to the RLS problem is defined by the *minimum-norm* values of the tap-weight vectors of the corresponding transversal filters (see Section 11.5).

Another peculiarity of the RLS solution is the special structure assumed by both the gain vector and the tap-weight vector of the backward prediction-error filter at time $n = M$. From the definitions of the gain vector $\mathbf{k}_M(n)$ and the time-averaged correlation matrix $\mathbf{\Phi}_M(n)$, we have

$$\left(\sum_{i=1}^{n} \lambda^{n-i}\mathbf{u}_M(i)\mathbf{u}_M^H(i)\right)\mathbf{k}_M(n) = \mathbf{u}_M(n)$$

By separating out the contribution to the correlation matrix at time $i = n$ from the rest, and using the definition of the estimation error $\gamma_M(n)$ that is real valued, we may equivalently write

$$\lambda \sum_{i=1}^{n-1} \lambda^{n-1-i}\mathbf{u}_M(i)\mathbf{u}_M^H(i)\mathbf{k}_M(n) = \mathbf{u}_M(n)[1 - \mathbf{u}_M^H(n)\mathbf{k}_M(n)] = \mathbf{u}_M(n)\gamma_M(n)$$

Since the estimation error $\gamma_M(n)$ is zero at time $n = M$, we thus have

$$\sum_{i=1}^{M-1} \lambda^{M-1-i}\mathbf{u}_M(i)\mathbf{u}_M^H(\mathrm{i})\mathbf{k}_M(M) = \mathbf{0} \tag{16.27}$$

In the prewindowing method, for $i = 1$ the last $M - 1$ elements of the tap-input vector $\mathbf{u}_M(i)$ are zero; for $i = 2$ the last $M - 2$ elements of $\mathbf{u}_M(i)$ are zero; and so on right up to $i = M - 1$ for which the last element of $\mathbf{u}_M(i)$ is zero. We deduce therefore that if Eq. (16.27) is to hold, subject to the requirement that the value $\mathbf{k}_M(M)$ of the gain vector be of minimum norm, then all the elements of $\mathbf{k}_M(M)$ must be zero except for the last one, as shown by

$$\mathbf{k}_M(M) = \begin{bmatrix} \mathbf{0} \\ k_{M,M}(M) \end{bmatrix} \tag{16.28}$$

where $\mathbf{0}$ is the $(M - 1)$-by-1 null vector. We determine the nonzero element $k_{M,M}(M)$ by observing that

$$\begin{aligned} \gamma_M(M) &= 1 - \mathbf{u}_M^H(M)\mathbf{k}_M(M) \\ &= 1 - u^*(1)k_{M,M}(M) \end{aligned}$$

Since $\gamma_M(M)$ is zero, we therefore have

$$k_{M,M}(M) = \frac{1}{u^*(1)} \tag{16.29}$$

and

$$\mathbf{k}_M(M) = \begin{bmatrix} \mathbf{0} \\ 1/u^*(1) \end{bmatrix} \tag{16.30}$$

To determine the corresponding value of the tap-weight vector of the backward prediction-error filter, we observe that the desired response $u(n - M)$ in the backward linear prediction is zero at time $n = M$. Hence, at this time the tap-weight vector of the backward linear predictor is likewise zero, and the minimum-norm value of the $(M + 1)$-by-1 tap-weight vector of the backward prediction-error filter equals

$$\mathbf{c}_M(M) = \begin{bmatrix} \mathbf{0} \\ 1 \end{bmatrix} \tag{16.31}$$

where $\mathbf{0}$ is the M-by-1 null vector. We thus see that at $n = M$ the first $M - 1$ elements of $\mathbf{k}_M(M)$ and the first M elements of $\mathbf{c}_M(M)$ are all zero.

Consider next the situation at time $n = M + 1$. At this time the RLS problem becomes *overdetermined* for the first time in that the length of the input data, n, exceeds the number of adjustable tap weights, M, by 1. Also, at time $n = M + 1$ the desired response $u(n - M)$ in the backward linear prediction assumes a nonzero value for the first time. In particular, we see that $u(n - M) = u(1)$ at $n = M + 1$. Accordingly, we may express the desired response vector for the backward linear prediction at time $n = M + 1$ as

$$\mathbf{b}(M + 1) = \begin{bmatrix} u^*(1) \\ \mathbf{0} \end{bmatrix}$$

Hence, the desired response vector in the backward linear prediction at time $n = M + 1$ is exactly the same, except for the scaling factor $u^*(1)$, as the first coordinate vector. This means that at this time the backward linear predictor and the transversal filter III (that defines the gain vector) make a least-squares estimate of the same desired response, within the multiplicative scalar $u^*(1)$. Both of these filters are of length M, and they operate on the same set of tap inputs, namely, $u(M + 1)$, $u(M), \ldots, u(2)$. Hence, at time $n = M + 1$, the tap-weight vector of the backward linear predictor is equal to $u^*(1)$ times the gain vector $\mathbf{k}_M(M + 1)$. Correspondingly, we may state that the tap-weight vector of the backward prediction-error filter and the gain vector are related by

$$\mathbf{c}_M(M + 1) = \begin{bmatrix} -u^*(1)\mathbf{k}_M(M + 1) \\ 1 \end{bmatrix} \tag{16.32}$$

We see therefore that it is not possible to simultaneously solve Eqs. (15.34) and (15.46) because these two equations are essentially scalar multiples of one another.

We thus have a peculiarity to overcome in the initialization of the FTF algorithm. All the elements of the vectors $\mathbf{c}_M(M)$ and $\mathbf{k}_M(M)$, except for their individual last elements, are zero. In contrast, we usually find in practical applications that all the elements of $\mathbf{c}_M(M + 1)$ and $\mathbf{k}_M(M + 1)$ are nonzero. To update the zero elements in $\mathbf{c}_M(M)$ and $\mathbf{k}_M(M)$ to the nonzero elements in $\mathbf{c}_M(M + 1)$ and $\mathbf{k}_M(M + 1)$, we require knowledge of $\mathbf{c}_{M-1}(M)$ and $\mathbf{k}_{M-1}(M)$. This requirement means that we have to perform simultaneous time- and order-updating during the initialization of the FTF algorithm so as to avoid the above-mentioned peculiarity at time $n = M$. This special form of updating continues until time $n = M + 1$.

We now derive the simultaneous time- and order-updating that is necessary to complete the exact initialization of the FTF algorithm. We start the derivation by setting $M + 1 = n$ or $M = n - 1$ in Eq. (15.16), which yields

$$\mathbf{a}_{n-1}(n) = \mathbf{a}_{n-1}(n-1) - \eta^*_{n-1}(n)\begin{bmatrix} 0 \\ \mathbf{k}_{n-1}(n-1) \end{bmatrix} \tag{16.33}$$

Figure 16.2(a) depicts the structure of the forward prediction-error filter characterized by the tap-weight vector $\mathbf{a}_{n-1}(n-1)$. We see that the tap input applied to the last element of $\mathbf{a}_{n-1}(n-1)$ is zero. The effect is equivalent to simplifying the structure of the filter as in Fig. 16.2(b). We may therefore express $\mathbf{a}_{n-1}(n-1)$ as

$$\mathbf{a}_{n-1}(n-1) = \begin{bmatrix} \mathbf{a}_{n-2}(n-1) \\ 0 \end{bmatrix} \tag{16.34}$$

To evaluate the gain vector $\mathbf{k}_{n-1}(n-1)$, we set $M = n - 1$ in Eq. (16.30), obtaining

$$\mathbf{k}_{n-1}(n-1) = \begin{bmatrix} \mathbf{0} \\ 1/u^*(1) \end{bmatrix} \tag{16.35}$$

To evaluate the forward a priori prediction error $\eta_{n-1}(n)$, we set $M = n - 1$ in Eq. (15.17) and then use Eq. (16.34), thereby obtaining

$$\begin{aligned}
\eta_{n-1}(n) &= \mathbf{a}^H_{n-1}(n-1)\mathbf{u}_n(n) \\
&= [\mathbf{a}^H_{n-2}(n-1), 0]\begin{bmatrix} \mathbf{u}_{n-1}(n) \\ u(1) \end{bmatrix} \\
&= \mathbf{a}^H_{n-2}(n-1)\mathbf{u}_{n-1}(n)
\end{aligned}$$

However, from the definition of the forward a priori prediction error, we have

$$\eta_{n-2}(n) = \mathbf{a}^H_{n-2}(n-1)\mathbf{u}_{n-1}(n)$$

We thus find that during the initialization period

$$\eta_{n-1}(n) = \eta_{n-2}(n) \tag{16.36}$$

Accordingly, the use of Eqs. (16.34), (16.35), and (16.36) in (16.33) yields

$$\mathbf{a}_{n-1}(n) = \begin{bmatrix} \mathbf{a}_{n-2}(n-1) \\ -\dfrac{\eta^*_{n-2}(n)}{u^*(1)} \end{bmatrix} \tag{16.37}$$

Given the forward a priori prediction error $\eta_{n-2}(n)$, we set $M = n - 2$ in Eq. (15.52) to compute the corresponding value of the forward a posteriori prediction error:

$$f_{n-2}(n) = \gamma_{n-2}(n-1)\eta_{n-2}(n) \tag{16.38}$$

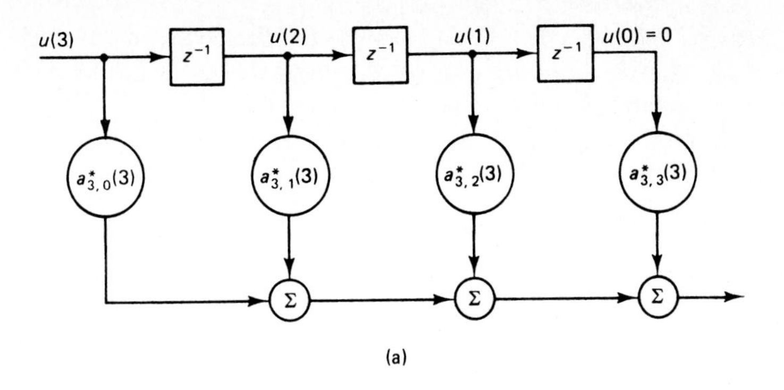

(a)

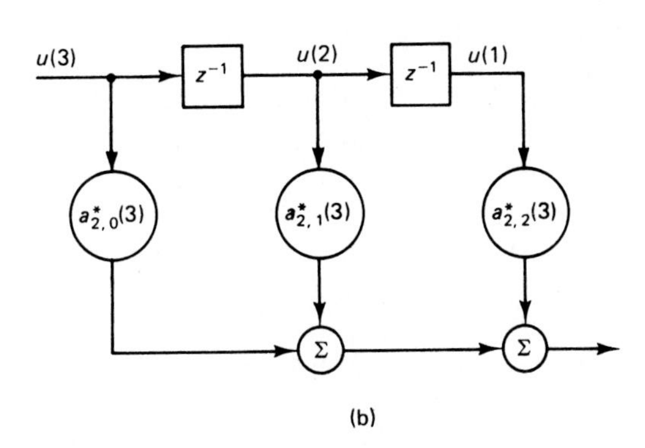

(b)

Figure 16.2 Reduction of $\mathbf{a}_{n-1}(n-1)$ to $\mathbf{a}_{n-2}(n-1)$ illustrated for $n = 4$; (a) transversal filter described by $\mathbf{a}_{n-1}(n-1)$; (b) transversal filter described by $\mathbf{a}_{n-2}(n-1)$.

Next we set $M = n - 2$ in Eq. (15.18), obtaining

$$\mathcal{F}_{n-2}(n) = \lambda \mathcal{F}_{n-2}(n-1) + \eta_{n-2}(n) f^*_{n-2}(n) \tag{16.39}$$

The use of $M = n - 2$ in Eq. (15.56) yields

$$\gamma_{n-1}(n) = \lambda \frac{\mathcal{F}_{n-2}(n-1)}{\mathcal{F}_{n-2}(n)} \gamma_{n-2}(n-1) \tag{16.40}$$

During the initialization period, however, we may show that the following time-order update holds (see Problem 5)

$$\mathcal{F}_{n-1}(n) = \lambda \mathcal{F}_{n-2}(n-1) \tag{16.41}$$

Accordingly, we may use this recursion to reformulate Eqs. (16.38) and (16.40) as follows,

$$\mathcal{F}_{n-2}(n) = \mathcal{F}_{n-1}(n) + \eta_{n-2}(n) f^*_{n-2}(n) \tag{16.42}$$

and

$$\gamma_{n-1}(n) = \frac{\mathcal{F}_{n-1}(n)}{\mathcal{F}_{n-2}(n)} \gamma_{n-2}(n-1) \tag{16.43}$$

Equations (16.37), (16.38), and (16.41) to (16.43) define the recursions for the first part of the exact initialization period that pertains to the forward prediction-error filter (i.e., filter I).

Here it is useful to note that during the exact initialization of the FTF algorithm with zero initial conditions, we have the following:

$$f_{n-1}(i) = 0, \qquad \text{for } 2 \leq i \leq n$$

But $f_{n-1}(1) = u(1) \neq 0$, so that $\mathcal{F}_{n-1}(n) \neq 0$ and the initialization can therefore continue.

Next, using Eq. (16.10) with $M = n - 2$, we obtain

$$\tilde{\mathbf{k}}_{n-1}(n) = \begin{bmatrix} 0 \\ \tilde{\mathbf{k}}_{n-2}(n-1) \end{bmatrix} + \frac{\eta_{n-2}(n)}{\lambda \mathcal{F}_{n-2}(n-1)} \mathbf{a}_{n-2}(n-1) \tag{16.44}$$

This recursion, constituting the second part of the initialization procedure, realizes the time-order update of the gain vector (i.e., filter III).

As explained earlier, the desired response $u(n - M)$ for the backward linear prediction is zero for $1 \leq n \leq M$. It assumes a nonzero value for the first time at $n = M + 1$. Therefore, in the initialization of the FTF algorithm, the computations for the backward prediction-error filter need only be performed when $n = M + 1$. In particular, by setting $M = n - 1$ in Eq. (16.32), we obtain the following value for the tap-weight vector of the backward prediction-error filter:

$$\begin{aligned} \mathbf{c}_{n-1}(n) &= \begin{bmatrix} -u^*(1)\mathbf{k}_{n-1}(n) \\ 1 \end{bmatrix} \\ &= \begin{bmatrix} -u^*(1)\gamma_{n-1}(n)\tilde{\mathbf{k}}_{n-1}(n) \\ 1 \end{bmatrix} \end{aligned} \tag{16.45}$$

To compute the corresponding minimum value of the sum of weighted backward a posteriori prediction-error squares, we set $M = n - 1$ in Eq. (15.36), and also use Eq. (15.53), obtaining

$$\mathcal{B}_{n-1}(n) = \lambda \mathcal{B}_{n-1}(n-1) + \gamma_{n-1}(n)|\psi_{n-1}(n)|^2 \tag{16.46}$$

However, $\mathcal{B}_{n-1}(n - 1)$ is zero, and the backward a priori prediction error $\psi_{n-1}(n)$ equals [see Eq. (15.35)]

$$\begin{aligned} \psi_{n-1}(n) &= \mathbf{c}^H_{n-1}(n-1)\mathbf{u}_n(n) \\ &= [\mathbf{0}^T, 1]\begin{bmatrix} \mathbf{u}_{n-1}(n) \\ u(1) \end{bmatrix} \\ &= u(1) \end{aligned}$$

where, in the second line, we have made use of Eq. (16.31) evaluated for $M = n - 1$. Hence, we may simplify Eq. (16.46) as

$$\mathcal{B}_{n-1}(n) = \gamma_{n-1}(n)|u(1)|^2 \tag{16.47}$$

Equations (16.45) and (16.47) define the desired recursions for the third part of the initialization procedure that deals with the backward prediction-error filter (i.e., filter II).

Finally, the updating of the adaptive filter (i.e., filter IV) simplifies considerably during the initialization period $1 \le n \le M$. Putting $M = n$ in the first line of Eq. (16.26) yields

$$\hat{\mathbf{w}}_n(n) = \hat{\mathbf{w}}_n(n-1) + \mathbf{k}_n(n)\alpha_n^*(n) \tag{16.48}$$

Recognizing that (for reasons similar to those illustrated in Fig. 16.2)

$$\hat{\mathbf{w}}_n(n-1) = \begin{bmatrix} \hat{\mathbf{w}}_{n-1}(n-1) \\ 0 \end{bmatrix}$$

and using Eq. (16.35), we may rewrite Eq. (16.48) as

$$\hat{\mathbf{w}}_n(n) = \begin{bmatrix} \hat{\mathbf{w}}_{n-1}(n-1) \\ \dfrac{\alpha_n^*(n)}{u^*(1)} \end{bmatrix} \tag{16.49}$$

To compute the a priori estimation error $\alpha_n(n)$, we set $M = n$ in Eq. (16.24), obtaining

$$\begin{aligned} \alpha_n(n) &= d(n) - \hat{\mathbf{w}}_n^H(n-1)\mathbf{u}_n(n) \\ &= d(n) - [\hat{\mathbf{w}}_{n-1}^H(n-1), 0]\begin{bmatrix} \mathbf{u}_{n-1}(n) \\ u(1) \end{bmatrix} \\ &= d(n) - \hat{\mathbf{w}}_{n-1}^H(n-1)\mathbf{u}_{n-1}(n) \end{aligned}$$

By definition, the final form of this expression equals $\alpha_{n-1}(n)$, the a priori estimation error for order $n - 1$. Hence, we may write

$$\alpha_{n-1}(n) = d(n) - \hat{\mathbf{w}}_{n-1}^H(n-1)\mathbf{u}_{n-1}(n) \tag{16.50}$$

Having computed $\alpha_{n-1}(n)$, we may now determine the corresponding value of the a posteriori estimation error

$$e_{n-1}(n) = \gamma_{n-1}(n)\alpha_{n-1}(n) \tag{16.51}$$

Note that the computation of the tap-weight vector $\hat{\mathbf{w}}_n(n)$ during the initialization period is possible without the use of the forward and backward prediction-error filters and the gain vector. This is possible because during this period the minimum value of the sum of weighted error squares is zero. However, once the initialization period is over (i.e., $n > M$), the minimum sum of weighted error squares assumes a nonzero value, and then these three filters are necessary for the computation of $\hat{\mathbf{w}}(n)$. These three additional filters are updated during the initialization period, so they are correct and available for use for $n > M$. Thus, for $n = M + 1$, the use of Eq. (16.26) yields

$$\hat{\mathbf{w}}_{n-1}(n) = \hat{\mathbf{w}}_{n-1}(n-1) + \tilde{\mathbf{k}}_{n-1}(n)e_{n-1}^*(n) \tag{16.52}$$

Equations (16.51), (16.49), and (16.52) define the recursions for exact initialization of the adaptive filter (i.e., filter IV).

This completes the exact initialization of the FTF algorithm.

Summary of the Exact Initialization of the FTF Algorithm for Zero Initial Condition

To start off the computation, we use the following set of values at $n = 1$:

$$
\begin{aligned}
\mathbf{a}_0(1) &= 1 \\
\mathbf{c}_0(1) &= 1 \\
\tilde{\mathbf{k}}_0(1) &= 0 \quad \text{(zero dimension)} \\
\hat{\mathbf{w}}_1(1) &= \frac{d^*(1)}{u^*(1)} \\
\gamma_0(1) &= 1 \\
\mathcal{F}_0(1) &= |u(1)|^2
\end{aligned}
\tag{16.53}
$$

A summary of the initialization procedure is presented in Table 16.2, where we have collected together the various relations derived previously.

16.3 EXACT INITIALIZATION OF THE FTF ALGORITHM FOR ARBITRARY INITIAL CONDITION

Nonzero initial conditions in the form of an initial value $\mathbf{w}_o$ for the M-by-1 tap-weight vector of the adaptive transversal filter component in the FTF algorithm may arise from the prior use of another adaptive algorithm (e.g., the LMS algorithm), or from some additional side information that is made available in a particular situation. In any event, it is desirable to have a method that can exploit this initial condition.

Another reason for developing a method that accommodates the use of an arbitrary initial condition is to provide an effective means to change the tracking capability of the FTF algorithm. In this context, we have to realize that it is not permissible to change the exponential weighting factor λ during the operation of the FTF algorithm since any such variation prevents the exact solution to the RLS problem to be realized.

To accommodate an arbitrary initial condition, we append a *soft constraint* to Eq. (13.7) so that the *augmented* cost function for the least-squares problem equals

$$
\mathcal{E}'(n) = \mu\lambda^n \|\mathbf{w}_M(n) - \mathbf{w}_o\|^2 + \sum_{i=1}^{n} \lambda^{n-i} |e(i)|^2 \tag{16.54}
$$

where the factor $\mu > 0$ to ensure a quadratic cost function, and the M-by-1 vector $\mathbf{w}_o$ denotes the initial condition. The symbol $\|\cdot\|$ signifies the norm of the vector enclosed within. For $n \geq 1$, we use the factor μ to weight the effect of the initial condition $\mathbf{w}_o$ on the augmented cost function. For $n \leq 0$, the second term in Eq. (16.54) vanishes, so the minimization of $\mathcal{E}'(n)$ with respect to $\mathbf{w}_M(n)$ yields an optimum value equal to $\mathbf{w}_o$, just as it should.

The application of a nonzero initial condition $\mathbf{w}_o$ is useful only if it arises from previous use of adaptive algorithms, such as the LMS algorithm or other least-

TABLE 16.2 EXACT INITIALIZATION OF THE FTF ALGORITHM

$n = 1$

$\mathbf{a}_0(1) = \mathbf{c}_0(1) = 1$

$\tilde{\mathbf{k}}_0(1) = 0$ (zero dimension)

$\hat{\mathbf{w}}_1(1) = d^*(1)/u^*(1)$

$\gamma_0(1) = 1$

$\mathcal{F}_0(1) = |u(1)|^2, \quad u(1) \neq 0$

$2 \leq n \leq M + 1$

$\eta_{n-2}(n) = \mathbf{a}_{n-2}^H(n-1)\mathbf{u}_{n-1}(n)$

$$\mathbf{a}_{n-1}(n) = \begin{bmatrix} \mathbf{a}_{n-2}(n-1) \\ -\eta_{n-2}^*(n)/u^*(1) \end{bmatrix}$$

$f_{n-2}(n) = \gamma_{n-2}(n-1)\eta_{n-2}(n)$

$\mathcal{F}_{n-1}(n) = \lambda \mathcal{F}_{n-2}(n-1)$

$\mathcal{F}_{n-2}(n) = \mathcal{F}_{n-1}(n) + \eta_{n-2}(n) f_{n-2}^*(n)$

$$\gamma_{n-1}(n) = \frac{\mathcal{F}_{n-1}(n)}{\mathcal{F}_{n-2}(n)} \gamma_{n-2}(n-1)$$

$$\tilde{\mathbf{k}}_{n-1}(n) = \begin{bmatrix} 0 \\ \tilde{\mathbf{k}}_{n-2}(n-1) \end{bmatrix} + \frac{\eta_{n-2}(n)}{\lambda \mathcal{F}_{n-2}(n-1)} \mathbf{a}_{n-2}(n-1)$$

$$\mathbf{c}_{n-1}(n) = \begin{bmatrix} -u^*(1)\gamma_{n-1}(n)\tilde{\mathbf{k}}_{n-1}(n) \\ 1 \end{bmatrix}, \quad \text{compute only when } n = M + 1$$

$\mathcal{B}_{n-1}(n) = \gamma_{n-1}(n)|u(1)|^2$, compute only when $n = M + 1$

$\alpha_{n-1}(n) = d(n) - \hat{\mathbf{w}}_{n-1}^H(n-1)\mathbf{u}_{n-1}(n)$

$e_{n-1}(n) = \gamma_{n-1}(n)\alpha_{n-1}(n)$

If $n \leq M$, $$\hat{\mathbf{w}}_n(n) = \begin{bmatrix} \hat{\mathbf{w}}_{n-1}(n-1) \\ \alpha_{n-1}^*(n)/u^*(1) \end{bmatrix}$$

If $n = M + 1$, $\hat{\mathbf{w}}_{n-1}(n) = \hat{\mathbf{w}}_{n-1}(n-1) + \tilde{\mathbf{k}}_{n-1}(n)e_{n-1}^*(n)$

---

squares algorithms, that have already processed a sufficiently large amount of data to average, or diminish, the effect of a relatively large measurement error. Furthermore, the initial condition $\mathbf{w}_o$ must be unbiased so as to maintain the unbiased estimator property of the RLS solution.

As for the factor μ, it allows a degree of flexibility in varying the capability of the FTF algorithm to track statistical variations in the input data (Cioffi and Kailath, 1984). By choosing the factor μ large, we make the FTF algorithm track as "slow" as desired, thereby attaining the least-squares solution with a finer resolution. By contrast, variation of the exponential weighting factor λ, by itself, is usually not as effective for two reasons. First, any variation of λ during the operation of the FTF algorithm renders it incorrect; hence, the use of gear-shifting by varying λ is impractical. Second, the upper bound of unity on λ limits the slowest learning rate of the FTF algorithm. A more practical method of attaining slow tracking is to introduce a form of gear-shifting into the FTF algorithm by redefining time $n = 0$ as the time at which the gear-shifting is to occur and then weighting the present tap-weight vector $\mathbf{w}_M(n)$ by an appropriate choice of μ. This procedure is simple and yet robust; it does not increase complexity, nor does it destroy the "exactness" of the solution.

Given the initial condition $\mathbf{w}_o$, we may perform exact initialization of the FTF algorithm by augmenting the time series representing the filter input and that representing the desired response, as illustrated in Fig. 16.3. The augmentation is achieved by adding $(M + 1)$ terms at the beginning of each time series. Thus, the first M iterations of the exact initialization algorithm in Table 16.2 yield the following initial conditions at time $n = 0$:

$$
\begin{aligned}
\mathbf{a}_M(0) &= \begin{bmatrix} 1 \\ \mathbf{0} \end{bmatrix} \\
\mathcal{F}_M(0) &= \lambda^M \mu \\
\mathbf{c}_M(0) &= \begin{bmatrix} \mathbf{0} \\ 1 \end{bmatrix} \\
\mathcal{B}_M(0) &= \mu \\
\mathbf{k}_M(0) &= \mathbf{0} \\
\gamma_M(0) &= 1 \\
\hat{\mathbf{w}}_M(0) &= \mathbf{w}_o
\end{aligned}
\tag{16.55}
$$

Note that time $n = 0$ coincides with the end of the set of added terms.

With the proper initial conditions established at time $n = 0$, application of the FTF algorithm in Table 16.1 for $n \geq 1$ may now proceed directly, without using Table 16.2 to initialize the algorithm. However, care must be exercised in not assigning too small a value to the factor μ; otherwise, the algorithm can become unstable when $n = M + 1$.

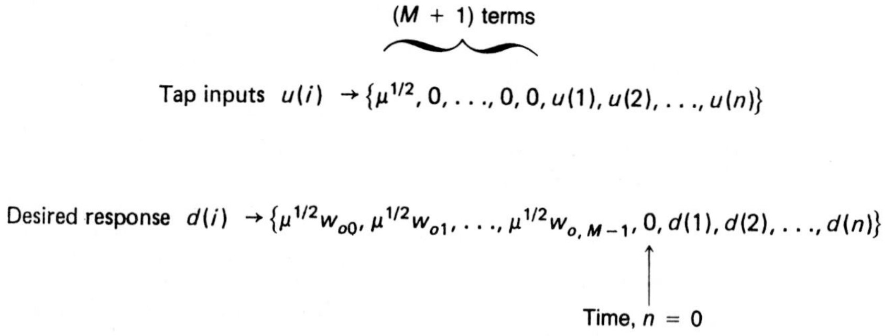

$(M + 1)$ terms

Tap inputs $u(i) \rightarrow \{\mu^{1/2}, 0, \ldots, 0, 0, u(1), u(2), \ldots, u(n)\}$

Desired response $d(i) \rightarrow \{\mu^{1/2}w_{o0}, \mu^{1/2}w_{o1}, \ldots, \mu^{1/2}w_{o,M-1}, 0, d(1), d(2), \ldots, d(n)\}$

Time, $n = 0$

Figure 16.3 Augmentation of the tap inputs and desired response to accommodate the use of arbitrary initial condition.

16.4 COMPUTER EXPERIMENT ON SYSTEM IDENTIFICATION

To illustrate the performance of the FTF algorithm, we use it to study the *system identification* problem depicted in Fig. 16.4. The multiple linear regression model shown in the figure involves an *unknown* linear transversal filter with M tap weights. The requirement is to estimate the tap weights of this filter in the presence of the measurement error $e_o(n)$. The aim of the experiment is to evaluate the effects of varying the number of unknown parameters, M, and the eigenvalue spread of the input $\{u(n)\}$ on the performance of the algorithm.

To vary the eigenvalue spread, the test signal $\{u(n)\}$ is generated by passing a zero-mean white-noise process, independent of the measurement error $\{e_o(n)\}$, through a linear filter whose impulse response equals

$$h_n = \begin{cases} \dfrac{1}{2}\left[1 + \cos\dfrac{2\pi}{W}(n-2)\right], & n = 1, 2, 3 \\ 0, & \text{otherwise} \end{cases}$$

The variance of the white-noise source is adjusted so as to make the variance of the unknown system output equal unity. For a specified number of unknown parameters, M, the eigenvalue spread is varied by varying W. Table 16.3 shows the eigenvalue spreads used in the experiment.

The measurement error $\{e_o(n)\}$ is modeled as a white Gaussian noise process of zero mean and variance equal to 10^{-4}. This represents a signal-to-noise ratio of 40 dB, measured with respect to the filter input $\{u(n)\}$.

The unknown impulse response of the transversal filter in the multiple linear regression model is chosen to follow a triangular wave form that is symmetric with

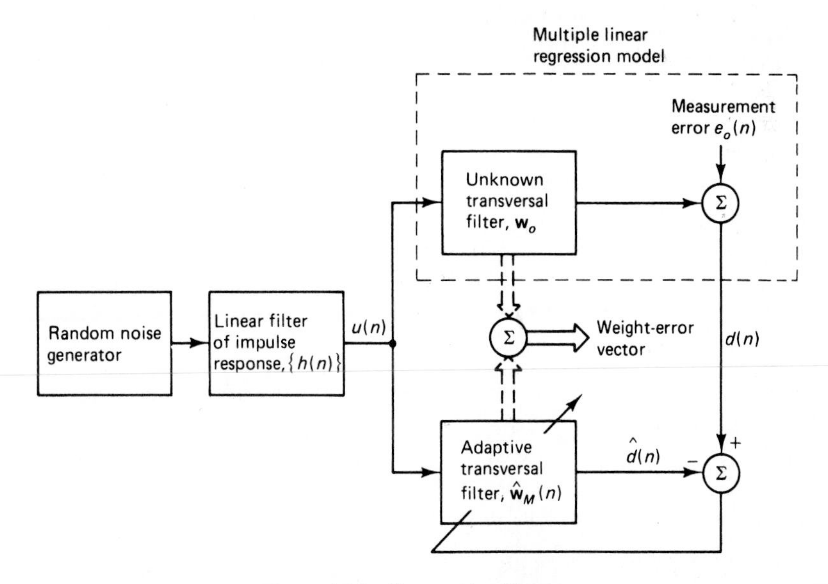

Figure 16.4 System identification.

TABLE 16.3 SUMMARY OF EIGENVALUE SPREADS FOR COMPUTER EXPERIMENT ON SYSTEM IDENTIFICATION

Number of unknown parameters, N	Eigenvalue spread	
	$W = 3.1$	$W = 3.7$
31	14.5	212.1
61	21.7	295.5
99	26.35	333.5

respect to the center tap point. The exponential weighting factor λ is assumed to equal unity.

Results and Observations

Figure 16.5(a) shows two plots of the estimated impulse response of the adaptive transversal filter (with $M = 31$) after convergence to steady-state conditions has been established. One plot corresponds to parameter $W = 3.1$, and the other to $W = 3.7$. Each of these two plots also includes the theoretical impulse response. Figure 16.5(b) shows the results obtained for $M = 61$ and $W = 3.1$, and 3.7. Figure 16.5(c) shows the results obtained for $M = 99$ and $W = 3.1$, and 3.7. The results shown in these figures were obtained by using the conventional (i.e., unconstrained) form of the FTF algorithm, with the computations performed on a high-precision machine.

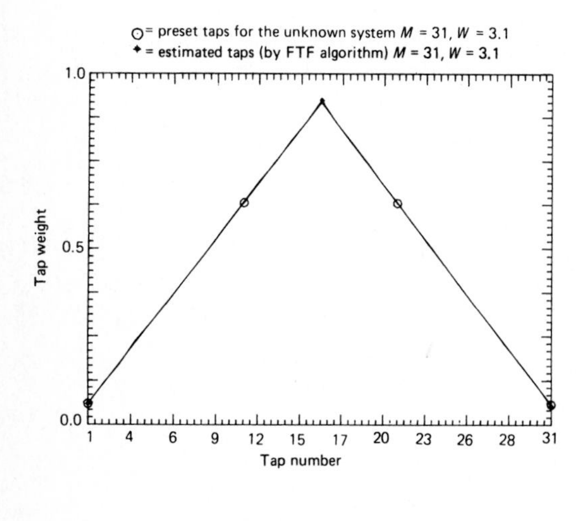

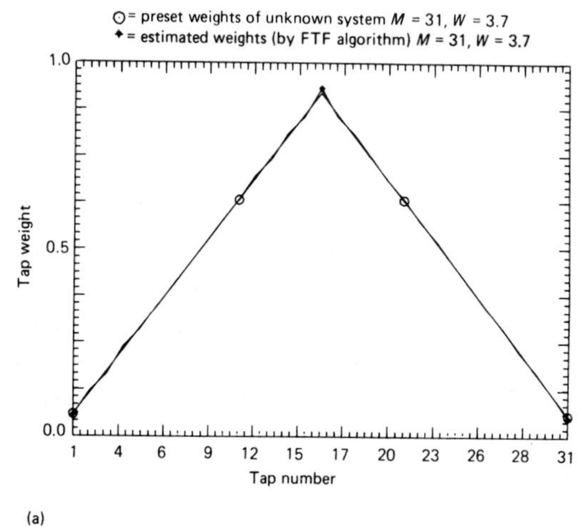

(a)

Figure 16.5 Estimated impulse response using FTF algorithm:
(a) $M = 31$, SNR = 40 dB, $W = 3.1, 3.7$;
(b) $M = 61$, SNR = 40 dB, $W = 3.1, 3.7$;
(c) $M = 99$, SNR = 40 dB, $W = 3.1, 3.7$.

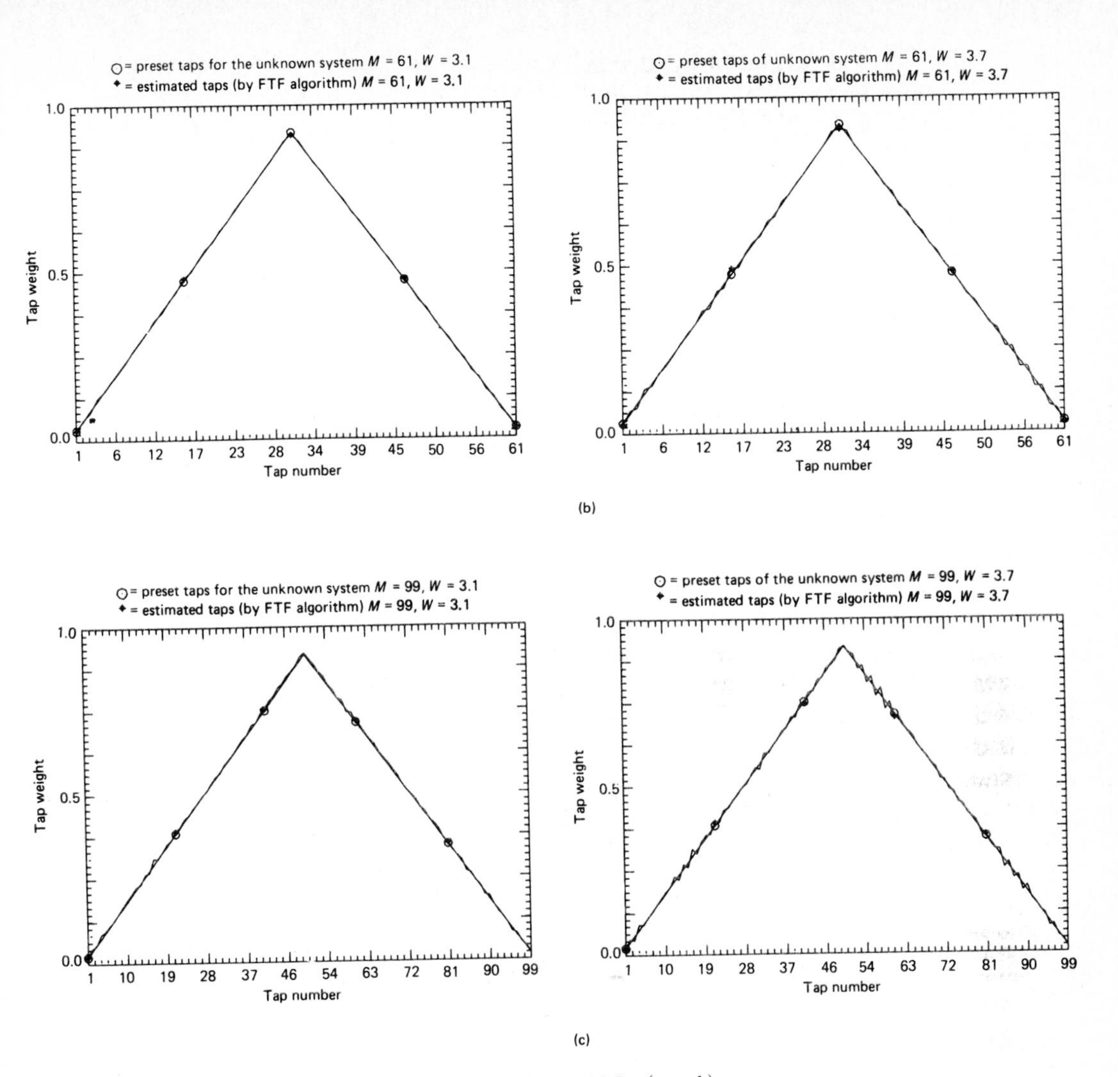

Figure 16.5 *(contd.)*

These results demonstrate the high-performance capability of the FTF algorithm when it is implemented on a high-precision machine. In particular, the estimate produced by the algorithm for the impulse response of the transversal filter is practically indistinguishable from the theoretical value, even under difficult operating conditions (i.e., when the eigenvalue spread is high and the number of tap weights in the filter is large).

16.5 DISCUSSION

The FTF algorithm realizes the RLS solution for the tap-weight vector $\hat{\mathbf{w}}_M(n)$ exactly. In contrast to the standard RLS algorithm, however, its computational complexity increases linearly with M, the number of tap weights contained in the adaptive filter, as in the LMS algorithm. Thus, the FTF algorithm offers the best of two worlds: the desirable convergence properties of the RLS algorithm, and the computational simplicity of the LMS algorithm. However, the simplification in computational complexity is achieved at the expense of a much more complicated program statement, as evidenced by comparing Table 16.1 for the FTF algorithm with the summary presented in Chapter 13 for the RLS algorithm.

In summary, then, the FTF algorithm offers several highly desirable characteristics:

1. An *exact* recursive solution of the linear least-squares estimation problem.
2. A computational complexity that increases *linearly* with the dimension (order) of the least-squares estimation problem (i.e., the number of permissible degrees of freedom).
3. A *fast* rate of convergence that is *insensitive* to the condition number (i.e., eigenvalue spread) of the underlying correlation matrix of the input data.
4. A useful list of by-products:
   (a) A priori and a posteriori values of the forward prediction error.
   (b) A priori and a posteriori values of the backward prediction error.
   (c) Time evolution of the conversion factor that provides an indicator of the attainment of steady-state conditions.

All of these products are computed for a *fixed* order of prediction.

However, a serious limitation of the FTF algorithm is that it suffers from a potential *instability* problem (Ljung and Ljung, 1985). This problem arises when the FTF algorithm is implemented on a machine with finite word-length. We will have more to say on this issue in Chapter 19.

PROBLEMS

1. The accompanying table shows a summary of the fast RLS algorithm. The development of the algorithm follows the theory presented in Section 16.3 up to (and including) the step that involves computing the extended gain vector $\mathbf{k}_{M+1}(n)$. By partitioning this vector as

$$\mathbf{k}_{M+1}(n) = \begin{bmatrix} \mathbf{t}(n) \\ \\ \tau(n) \end{bmatrix}$$

where $\mathbf{t}(n)$ is an M-by-1 vector and $\tau(n)$ is a scalar, derive the remaining three steps of the algorithm.

$$\eta_M(n) = \mathbf{a}_M^H(n-1)\mathbf{u}_{M+1}(1)$$

$$\mathbf{a}_M(n) = \mathbf{a}_M(n-1) - \begin{bmatrix} 0 \\ \mathbf{k}_M(n-1) \end{bmatrix} \eta_M^*(n)$$

$$f_M(n) = \mathbf{a}_M^H(n)\mathbf{u}_{M+1}(n)$$

$$\mathcal{F}_M(n) = \lambda \mathcal{F}_M(n-1) + \eta_M(n) f_M^*(n)$$

$$\mathbf{k}_{M+1}(n) = \begin{bmatrix} 0 \\ \mathbf{k}_M(n-1) \end{bmatrix} + \frac{f_M(n)}{\mathcal{F}_M(n)} \mathbf{a}_M(n)$$

$$\mathbf{k}_{M+1}(n) = \begin{bmatrix} \mathbf{t}(n) \\ \tau(n) \end{bmatrix}$$

$$\psi_M(n) = u(n-M) - \mathbf{g}^{\mathrm{H}}(n-1)\mathbf{u}_M(n)$$

$$\mathbf{g}(n) = [1 - \tau(n)\psi_M^*(n)]^{-1}[\mathbf{g}(n-1) + \mathbf{t}(n)\psi_M^*(n)]$$

$$\mathbf{k}_M(n) = \mathbf{t}(n) + \mathbf{g}(n)\tau(n)$$

2. Discuss the basic differences that distinguish the FTF algorithm from the fast RLS algorithm described in Problem 1.

3. The quantity designated as the "rescue variable" in the summary given in Table 16.1 represents a source of numerical instability in the FTF algorithm. Justify the validity of this statement.

4. Consider the exact initialization of the FTF algorithm with zero initial conditions. Demonstrate the following:
(a) $f_{n-1}(i) = 0$, for $2 \le i \le n$
(b) $f_{n-1}(1) = u(1) \ne 0$
(c) What do parts (a) and (b) imply insofar as the value of $\mathcal{F}_{n-1}(n)$ is concerned?

5. During the initialization of the FTF algorithm, show that

$$\mathcal{F}_{n-1}(n) = \lambda \mathcal{F}_{n-2}(n-1)$$

where $\mathcal{F}_{n-1}(n)$ is the sum of weighted forward a posteriori prediction-error squares, and λ is the exponential weighting factor.

Computer-Oriented Problem

6. *Computer experiment on system identification using the FTF algorithm (continued).* In the results presented in Section 16.4 on the use of the FTF algorithm for system identification, the conventional (i.e., unconstrained) form of the algorithm was used. With this form of the algorithm, when fast initialization is used, the norm of the tap-weight error vector, $\|\hat{\mathbf{w}}_M(n) - \mathbf{w}_o\|$, may grow with time during the initialization period. This growth is eliminated by using the soft-constrained version of the FTF algorithm.

Repeat the computer experiment of Section 16.4 for the number of taps $M = 99$ and with the parameter W (that controls the eigenvalue spread) having the following two values: (1) $W = 3.1$, eigenvalue spread $= 26.4$, and (2) $W = 3.7$, eigenvalue spread $=$ 400. Use the FTF algorithm with soft constraint, for which the parameters λ and μ are assigned the values:

$$\lambda = 1$$

$$\mu = 0.01$$

Hence, carry out the following computations:

(a) For each value of W, evaluate the performance of the algorithm by computing the norm of the tap-weight error vector, $\|\hat{\mathbf{w}}_M(n) - \mathbf{w}_o\|$, divided by $\sqrt{M}$. Plot your results versus time n for about 1000 iterations of the algorithm. In your plot, include the performance of the algorithm during the initialization period.

(b) For each value of W, plot the impulse response of the adaptive transversal filter versus time n, obtained at the end of the experiment.

In this plot include the theoretical value of the impulse response. Compare the results in parts (a) and (b) with those obtained by using the unconstrained form of the FTF algorithm.

17

Recursive Least-Squares Lattice Filters

In this chapter we develop our second class of *fast* least-squares algorithms based on the multistage *lattice predictor* which is *modular* in form. The algorithms considered herein are known collectively as *recursive least-squares lattice* (LSL) *algorithms* involving both *order-update* and *time-update recursions*. They are as efficient as the FTF algorithm (considered in Chapter 16) in that they essentially realize the same rate of convergence at the expense of a computational cost that increases linearly with the number of adjustable parameters. The class of recursive LSL algorithms is also just as insensitive as the FTF algorithm to variations in the eigenvalue spread of the underlying correlation matrix of the input data.

The lattice predictors considered in this chapter may operate in *nonstationary* environments, though with *time-variant* reflection coefficients. As such, they are basically different from the lattice predictors (with *fixed* reflection coefficients) considered in Chapter 6 that were based on the assumption of an underlying *stationary* environment. We begin our study of recursive LSL algorithms by first reviewing some basic relations and notations that pertain to forward and backward prediction-error filters.

17.1 SOME PRELIMINARIES

Consider a forward prediction-error filter of order m whose *least-squares tap-weight vector* is denoted by $\mathbf{a}_m(n)$. By definition, the first element of $\mathbf{a}_m(n)$ equals unity. Let $\mathbf{u}_{m+1}(i)$ denote the $(m + 1)$-by-1 tap-input vector of the filter measured at time i, where $1 \le i \le n$. The *forward a posteriori prediction error* produced at the output of the filter, in response to the tap-input vector $\mathbf{u}_{m+1}(i)$, equals

$$f_m(i) = \mathbf{a}_m^H(n)\mathbf{u}_{m+1}(i), \qquad 1 \le i \le n \tag{17.1}$$

We assume the use of *prewindowing*, so we may put $u(i) = 0$ for $i \leq 0$. The sum of weighted forward a posteriori prediction-error squares equals

$$\mathcal{F}_m(n) = \sum_{i=1}^{n} \lambda^{n-i} | f_m(i) |^2 \tag{17.2}$$

where λ is the exponential weighting factor. The tap-weight vector $\mathbf{a}_m(n)$ is the result of minimizing the cost function $\mathcal{F}_m(n)$, subject to the constraint that the first element of $\mathbf{a}_m(n)$ equals unity. Note that $\mathbf{a}_m(n)$ is treated as a constant vector during this minimization.

Consider next the corresponding backward prediction-error filter of order m whose least-squares tap-weight vector is denoted by $\mathbf{c}_m(n)$. By definition, the last element of $\mathbf{c}_m(n)$ equals unity. The *backward a posteriori prediction error* produced at the output of the filter, in response to the tap-input vector $\mathbf{u}_{m+1}(n)$, equals

$$b_m(i) = \mathbf{c}_m^H(n)\mathbf{u}_{m+1}(i), \qquad 1 \leq i \leq n \tag{17.3}$$

The sum of weighted backward a posteriori prediction-error squares equals

$$\mathcal{B}_m(n) = \sum_{i=1}^{n} \lambda^{n-i} | b_m(i) |^2 \tag{17.4}$$

The tap-weight vector $\mathbf{c}_m(n)$ is the result of minimizing the second cost function $\mathcal{B}_m(n)$, subject to the constraint that the last element of $\mathbf{c}_m(n)$ equals unity. Here, again, $\mathbf{c}_m(n)$ is treated as a constant vector during this minimization.

17.2 ORDER-UPDATE RECURSIONS

Let $\mathbf{\Phi}_{m+1}(n)$ denote the $(m + 1)$-by-$(m + 1)$ correlation matrix of the tap-input vector $\mathbf{u}_{m+1}(i)$ applied to the forward prediction-error filter of order m, where $1 \leq i \leq n$. We may characterize this filter by the augmented normal equations [see Eq. (15.6)]

$$\mathbf{\Phi}_{m+1}(n)\mathbf{a}_m(n) = \begin{bmatrix} \mathcal{F}_m(n) \\ \mathbf{0}_m \end{bmatrix} \tag{17.5}$$

where $\mathbf{0}_m$ is the m-by-1 null vector, and $\mathbf{a}_m(n)$ and $\mathcal{F}_m(n)$ are as defined previously.

From the discussion presented in Chapter 15, we recall that the correlation matrix $\mathbf{\Phi}_{m+1}(n)$ may be partitioned in two different ways, depending on how we interpret the first or last element of the tap-input vector $\mathbf{u}_{m+1}(i)$. The form of partitioning that we like to use is the one that enables us to relate the tap-weight vector $\mathbf{a}_m(n)$, pertaining to prediction order m, to the tap-weight vector $\mathbf{a}_{m-1}(n)$, pertaining to prediction order $m - 1$. This aim is realized by using the result [see Eq. (15.27)].

$$\mathbf{\Phi}_{m+1}(n) = \left[\begin{array}{c|c} \mathbf{\Phi}_m(n) & \mathbf{\theta}_2(n) \\ \hline \mathbf{\theta}_2^H(n) & \mathcal{U}_2(n) \end{array}\right] \tag{17.6}$$

where $\mathbf{\Phi}_m(n)$ is the m-by-m correlation matrix of the tap-input vector $\mathbf{u}_m(i)$, $\mathbf{\theta}_2(n)$ is the m-by-1 cross-correlation vector between $\mathbf{u}_m(i)$ and $u(i - m)$, and $\mathcal{U}_2(n)$ is the sum of weighted, squared values of the input $u(i - m)$ for $1 \le i \le n$. Note that $\mathcal{U}_2(n)$ is zero for $n - m \le 0$. We postmultiply both sides of Eq. (17.6) by an $(m + 1)$-by-1 vector whose first m elements are defined by the vector $\mathbf{a}_{m-1}(n)$ and whose last element equals zero. We may thus write

$$\mathbf{\Phi}_{m+1}(n)\begin{bmatrix}\mathbf{a}_{m-1}(n) \\ \hline 0\end{bmatrix} = \left[\begin{array}{c|c}\mathbf{\Phi}_m(n) & \boldsymbol{\theta}_2(n) \\ \hline \boldsymbol{\theta}_2^H(n) & \mathcal{U}_2(n)\end{array}\right]\begin{bmatrix}\mathbf{a}_{m-1}(n) \\ \hline 0\end{bmatrix}$$

$$= \begin{bmatrix}\mathbf{\Phi}_m(n)\mathbf{a}_{m-1}(n) \\ \boldsymbol{\theta}_2^H(n)\mathbf{a}_{m-1}(n)\end{bmatrix} \tag{17.7}$$

Both $\mathbf{\Phi}_m(n)$ and $\mathbf{a}_{m-1}(n)$ have the same time argument n. Furthermore, in the first line of Eq. (17.7), they are both positioned in such a way that when the matrix multiplication is performed, $\mathbf{\Phi}_m(n)$ becomes postmultiplied by $\mathbf{a}_{m-1}(n)$. For a forward prediction-error filter of order $m - 1$, evaluated at time n, the set of augmented normal equations is

$$\mathbf{\Phi}_m(n)\mathbf{a}_{m-1}(n) = \begin{bmatrix}\mathcal{F}_{m-1}(n) \\ \mathbf{0}_{m-1}\end{bmatrix} \tag{17.8}$$

where $\mathcal{F}_{m-1}(n)$ is the sum of weighted forward a posteriori prediction-error squares at the filter output, and $\mathbf{0}_{m-1}$ is the $(m - 1)$-by-1 null vector. Define the scalar

$$\Delta_{m-1}(n) = \boldsymbol{\theta}_2^H(n)\mathbf{a}_{m-1}(n) \tag{17.9}$$

Accordingly, we may rewrite Eq. (17.7) as

$$\mathbf{\Phi}_{m+1}(n)\begin{bmatrix}\mathbf{a}_{m-1}(n) \\ 0\end{bmatrix} = \begin{bmatrix}\mathcal{F}_{m-1}(n) \\ \mathbf{0}_{m-1} \\ \Delta_{m-1}(n)\end{bmatrix} \tag{17.10}$$

Consider next the backward prediction-error filter of order m that is characterized by the augmented normal equations in matrix form:

$$\mathbf{\Phi}_{m+1}(n)\mathbf{c}_m(n) = \begin{bmatrix}\mathbf{0}_m \\ \mathcal{B}_m(n)\end{bmatrix} \tag{17.11}$$

where $\mathbf{\Phi}_{m+1}(n)$, $\mathbf{c}_m(n)$, and $\mathcal{B}_m(n)$ are as defined previously, and $\mathbf{0}_m$ is the m-by-1 null vector. This time we use the other partitioned form of the correlation matrix $\mathbf{\Phi}_m(n)$, as shown by [see Eq. (15.9)]

$$\mathbf{\Phi}_{m+1}(n) = \left[\begin{array}{c|c}\mathcal{U}_1(n) & \boldsymbol{\theta}_1^H(n) \\ \hline \boldsymbol{\theta}_1(n) & \mathbf{\Phi}_m(n - 1)\end{array}\right] \tag{17.12}$$

where $\mathcal{U}_1(n)$ is the sum of weighted, squared values of the input $u(i)$ for the time interval $1 \le i \le n$, $\boldsymbol{\theta}_1(n)$ is the m-by-1 cross-correlation vector between $u(i)$ and the tap-input vector $\mathbf{u}_m(i - 1)$, and $\mathbf{\Phi}_m(n - 1)$ is the m-by-m correlation matrix of $\mathbf{u}_m(i - 1)$. Correspondingly, we postmultiply $\mathbf{\Phi}_{m+1}(n)$ by an $(m + 1)$-by-1 vector whose first element is zero and whose m remaining elements are defined by the tap-weight vector $\mathbf{c}_{m-1}(n-1)$ that pertains to a backward prediction-error filter of order $m-1$. We may thus write

$$\begin{aligned}\mathbf{\Phi}_{m+1}(n)\begin{bmatrix}0\\ \mathbf{c}_{m-1}(n-1)\end{bmatrix} &= \begin{bmatrix}\mathcal{U}_1(n) & \boldsymbol{\theta}_1^H(n)\\ \boldsymbol{\theta}_1(n) & \mathbf{\Phi}_m(n-1)\end{bmatrix}\begin{bmatrix}0\\ \mathbf{c}_{m-1}(n-1)\end{bmatrix}\\ &= \begin{bmatrix}\boldsymbol{\theta}_1^H(n)\mathbf{c}_{m-1}(n-1)\\ \mathbf{\Phi}_m(n-1)\mathbf{c}_{m-1}(n-1)\end{bmatrix}\end{aligned} \tag{17.13}$$

Both $\mathbf{\Phi}_m(n-1)$ and $\mathbf{c}_{m-1}(n-1)$ have the same time argument, $n-1$. Also, they are both positioned in the first line of Eq. (17.13) in such a way that, when the matrix multiplication is performed, $\mathbf{\Phi}_{m-1}(n-1)$ become postmultiplied by $\mathbf{c}_{m-1}(n-1)$. For a backward prediction-error filter of order $m-1$, evaluated at time $n-1$, the matrix form of the augmented normal equations is

$$\mathbf{\Phi}_m(n-1)\mathbf{c}_{m-1}(n-1) = \begin{bmatrix}\mathbf{0}_{m-1}\\ \mathcal{B}_{m-1}(n-1)\end{bmatrix} \tag{17.14}$$

where $\mathcal{B}_{m-1}(n-1)$ is the sum of weighted backward a posteriori prediction-errors squares produced at the output of the filter at time $n-1$. Define the second scalar

$$\Delta'_{m-1}(n) = \boldsymbol{\theta}_1^H(n)\mathbf{c}_{m-1}(n-1) \tag{17.15}$$

where the prime is intended to distinguish this new parameter from $\Delta_{m-1}(n)$. Accordingly, we may rewrite Eq. (17.13) as

$$\mathbf{\Phi}_{m+1}(n)\begin{bmatrix}0\\ \mathbf{c}_{m-1}(n-1)\end{bmatrix} = \begin{bmatrix}\Delta'_{m-1}(n)\\ \mathbf{0}_{m-1}\\ \mathcal{B}_{m-1}(n-1)\end{bmatrix} \tag{17.16}$$

The next manipulation we wish to perform is to combine Eqs. (17.10) and (17.16) in such a way that the result assumes a form identical to the augmented normal equations (17.5) that pertain to a forward prediction-error filter of order m. This we can do by multiplying Eq. (17.16) by the ratio $\Delta_{m-1}(n)/\mathcal{B}_{m-1}(n-1)$ and subtracting the result from Eq. (17.10). We thus get

$$\begin{aligned}\mathbf{\Phi}_{m+1}(n)&\left[\begin{bmatrix}\mathbf{a}_{m-1}(n)\\ 0\end{bmatrix} - \frac{\Delta_{m-1}(n)}{\mathcal{B}_{m-1}(n-1)}\begin{bmatrix}0\\ \mathbf{c}_{m-1}(n-1)\end{bmatrix}\right]\\ &= \begin{bmatrix}\mathcal{F}_{m-1}(n) - \dfrac{\Delta_{m-1}(n)\Delta'_{m-1}(n)}{\mathcal{B}_{m-1}(n-1)}\\ \mathbf{0}_m\end{bmatrix}\end{aligned} \tag{17.17}$$

Comparing Eqs. (17.5) and (17.17), we therefore deduce the following two order-update recursions:

$$\mathbf{a}_m(n) = \begin{bmatrix}\mathbf{a}_{m-1}(n)\\ 0\end{bmatrix} - \frac{\Delta_{m-1}(n)}{\mathcal{B}_{m-1}(n-1)}\begin{bmatrix}0\\ \mathbf{c}_{m-1}(n-1)\end{bmatrix} \tag{17.18}$$

and

$$\mathcal{F}_m(n) = \mathcal{F}_{m-1}(n) - \frac{\Delta_{m-1}(n)\Delta'_{m-1}(n)}{\mathcal{B}_{m-1}(n-1)} \tag{17.19}$$

A second manipulation that we may also perform is to combine Eqs. (17.10) and (17.16) in such a way that, this time, the resultant assumes a form identical to the augmented normal equations (17.11) pertaining to the backward prediction-error filter of order m. To achieve this end result, we multiply Eq. (17.10) by the ratio $\Delta'_{m-1}(n)/\mathscr{F}_{m-1}(n)$ and subtract the result from Eq. (17.16). We thus obtain

$$\mathbf{\Phi}_{m+1}(n)\left[\begin{bmatrix} 0 \\ \mathbf{c}_{m-1}(n-1) \end{bmatrix} - \frac{\Delta'_{m-1}(n)}{\mathscr{F}_{m-1}(n)}\begin{bmatrix} \mathbf{a}_{m-1}(n) \\ 0 \end{bmatrix}\right]$$

$$= \begin{bmatrix} \mathbf{0}_m \\ \mathscr{B}_{m-1}(n-1) - \dfrac{\Delta_{m-1}(n)\Delta'_{m-1}(n)}{\mathscr{F}_{m-1}(n)} \end{bmatrix} \tag{17.20}$$

Comparing Eqs. (17.11) and (17.20), we therefore deduce two additional order-update recursions:

$$\mathbf{c}_m(n) = \begin{bmatrix} 0 \\ \mathbf{c}_{m-1}(n-1) \end{bmatrix} - \frac{\Delta'_{m-1}(n)}{\mathscr{F}_{m-1}(n)}\begin{bmatrix} \mathbf{a}_{m-1}(n) \\ 0 \end{bmatrix} \tag{17.21}$$

and

$$\mathscr{B}_m(n) = \mathscr{B}_{m-1}(n-1) - \frac{\Delta_{m-1}(n)\Delta'_{m-1}(n)}{\mathscr{F}_{m-1}(n)} \tag{17.22}$$

Relation between the Two Parameters $\Delta_{m-1}(n)$ and $\Delta'_{m-1}(n)$

The parameters $\Delta_{m-1}(n)$ and $\Delta'_{m-1}(n)$, defined by Eqs. (17.9) and (17.15), respectively, are the complex conjugate of one another; that is,

$$\Delta'_{m-1}(n) = \Delta^*_{m-1}(n) \tag{17.23}$$

where $\Delta^*_{m-1}(n)$ is the complex conjugate of $\Delta_{m-1}(n)$. We prove this relation in three stages:

1. We premultiply both sides of Eq. (17.10) by the row vector

   $$[(0, \mathbf{c}^H_{m-1}(n-1)],$$

   where the superscript H denotes Hermitian transposition. The result of this matrix multiplication is the scalar

   $$[0, \mathbf{c}^H_{m-1}(n-1)]\mathbf{\Phi}_{m+1}(n)\begin{bmatrix} \mathbf{a}_{m-1}(n) \\ 0 \end{bmatrix}$$

   $$= [0, \mathbf{c}^H_{m-1}(n-1)]\begin{bmatrix} \mathscr{F}_{m-1}(n) \\ \mathbf{0}_{m-1} \\ \Delta_{m-1}(n) \end{bmatrix} = \Delta_{m-1}(n) \tag{17.24}$$

   where we have used the fact that the last element of $\mathbf{c}_{m-1}(n-1)$ equals unity.

2. We apply Hermitian transposition to both sides of Eq. (17.16), and use the Hermitian property of the correlation matrix $\mathbf{\Phi}_{m+1}(n)$, thereby obtaining

   $$[0, \mathbf{c}^H_{m-1}(n-1)]\mathbf{\Phi}_{m+1}(n) = [\Delta'^*_{m-1}(n), \mathbf{0}^T_{m-1}, \mathscr{B}_{m-1}(n-1)]$$

where $\Delta'^{*}_{m-1}(n)$ is the complex conjugate of $\Delta'_{m-1}(n)$, and $\mathcal{B}_{m-1}(n-1)$ is real valued. Next we use this relation to evaluate the scalar

$$[0, \mathbf{c}^{H}_{m-1}(n-1)]\mathbf{\Phi}_{m+1}(n)\begin{bmatrix}\mathbf{a}_{m-1}(n)\\0\end{bmatrix}$$

$$= [\Delta'^{*}_{m-1}(n), \mathbf{0}^{T}_{m-1}, \mathcal{B}_{m-1}(n-1)]\begin{bmatrix}\mathbf{a}_{m-1}(n)\\0\end{bmatrix}$$

$$= \Delta'^{*}_{m-1}(n) \tag{17.25}$$

where we have used the fact that the first element of $\mathbf{a}_{m-1}(n)$ equals unity.

3. Comparison of Eqs. (7.24) and (7.25) immediately yields the relation of Eq. (17.23) between the parameters $\Delta_{m-1}(n)$ and $\Delta'_{m-1}(n)$.

Order-Update Recursions for the a posteriori Prediction Errors

Define the *forward reflection coefficient*

$$\Gamma_{f,m}(n) = -\frac{\Delta_{m-1}(n)}{\mathcal{B}_{m-1}(n-1)}, \qquad m = 1, 2, \ldots, M \tag{17.26}$$

where M is the final order of the lattice predictor.

Define the *backward reflection coefficient*

$$\begin{aligned}\Gamma_{b,m}(n) &= -\frac{\Delta'_{m-1}(n)}{\mathcal{F}_{m-1}(n)}\\ &= -\frac{\Delta^{*}_{m-1}(n)}{\mathcal{F}_{m-1}(n)}, \qquad m = 1, 2, \ldots, M\end{aligned} \tag{17.27}$$

In general, $\mathcal{F}_{m-1}(n)$ and $\mathcal{B}_{m-1}(n-1)$ are unequal so that in the LSL algorithms, unlike Burg's formula discussed in Chapter 6, we have, $\Gamma_{f,m}(n) \neq \Gamma^{*}_{b,m}(n)$. Using these definitions, we may rewrite Eqs. (17.18) and (17.21) as follows, respectively:

$$\mathbf{a}_m(n) = \begin{bmatrix}\mathbf{a}_{m-1}(n)\\0\end{bmatrix} + \Gamma_{f,m}(n)\begin{bmatrix}0\\\mathbf{c}_{m-1}(n-1)\end{bmatrix} \tag{17.28}$$

and

$$\mathbf{c}_m(n) = \begin{bmatrix}0\\\mathbf{c}_{m-1}(n-1)\end{bmatrix} + \Gamma_{b,m}(n)\begin{bmatrix}\mathbf{a}_{m-1}(n)\\0\end{bmatrix} \tag{17.29}$$

where $m = 1, 2, \ldots, M$. Equations (17.28) and (17.29) may be viewed as the deterministic counterpart of the Levinson-Durbin recursion. Note, however, that the reflection coefficients and, for that matter, the prediction-error filter coefficient vectors are all time varying.

The forward a posteriori prediction error $f_m(n)$ equals the output of the forward prediction-error filter of order m that is produced in response to the tap-input vector $\mathbf{u}_{m+1}(n)$, as shown by

$$f_m(n) = \mathbf{a}^{H}_{m}(n)\mathbf{u}_{m+1}(n) \tag{17.30}$$

The tap-input vector $\mathbf{u}_{m+1}(n)$ may be partitioned as

$$\mathbf{u}_{m+1}(n) = \begin{bmatrix} \mathbf{u}_m(n) \\ u(n-m) \end{bmatrix} \tag{17.31}$$

or, alternatively, as

$$\mathbf{u}_{m+1}(n) = \begin{bmatrix} u(n) \\ \mathbf{u}_m(n-1) \end{bmatrix} \tag{17.32}$$

Hence, substituting Eq. (17.28) in (17.30), and using the partitioned form of Eq. (17.31) for the first term on the right side of Eq. (17.28) and that of Eq. (17.32) for the second term, we get the following *order-update recursion for the forward a posteriori prediction error:*

$$f_m(n) = f_{m-1}(n) + \Gamma^*_{f,m}(n) b_{m-1}(n-1), \qquad m = 1, 2, \ldots, M \tag{17.33}$$

where $f_{m-1}(n)$ is the forward a posteriori prediction error and $b_{m-1}(n-1)$ is the *delayed* version of the backward a posteriori prediction error, both for prediction order $m - 1$.

To develop the corresponding order update recursion for the backward prediction error, we may follow a procedure similar to that just described. The backward a posteriori prediction error $b_m(n)$ equals the output of the backward prediction-error filter of order m that is produced in response to the tap-input vector $\mathbf{u}_{m+1}(n)$, as shown by

$$b_m(n) = \mathbf{c}^H_m(n)\mathbf{u}_{m+1}(n) \tag{17.34}$$

Substituting Eq. (17.29) in (17.34), and using the partitioned form of Eq. (17.32) for the first term on the right side of Eq. (17.29) and that of Eq. (17.31) for the second term, we get the following *order-update recursion for the backward a posteriori prediction error:*

$$b_m(n) = b_{m-1}(n-1) + \Gamma^*_{b,m}(n) f_{m-1}(n), \qquad m = 1, 2, \ldots, M \tag{17.35}$$

Equations (17.33) and (17.35) are represented by the signal-flow graph shown in Fig. 17.1(a), which has the appearance of a *lattice*.

The prediction order m is a *variable* parameter. It takes on the values 0, 1, $\ldots$, M, where M is the *final value* of the prediction order. When $m = 0$, there is *no* prediction being performed on the input data. This elementary condition corresponds to the *initial values*

$$f_0(n) = b_0(n) = u(n) \tag{17.36}$$

where $u(n)$ is the input at time n. Thus, as we vary the prediction order m from zero all the way up to the final value M, we get the *multistage least-squares lattice predictor* shown in Fig. 17.1(b). The number of stages contained in this structure equals M, the final order of the predictor.

It is noteworthy that the least-squares prediction errors $f_m(n)$ and $b_m(n)$ obey a set of order update recursions that are identical in structure to those we derived in Chapter 6. The latter recursions are consequences of the Levinson-Durbin recursion that arises basically because of the Toeplitz property of the ensemble-averaged correlation matrix $\mathbf{R}_m$ for stationary inputs. On the other hand, the time-averaged corre-

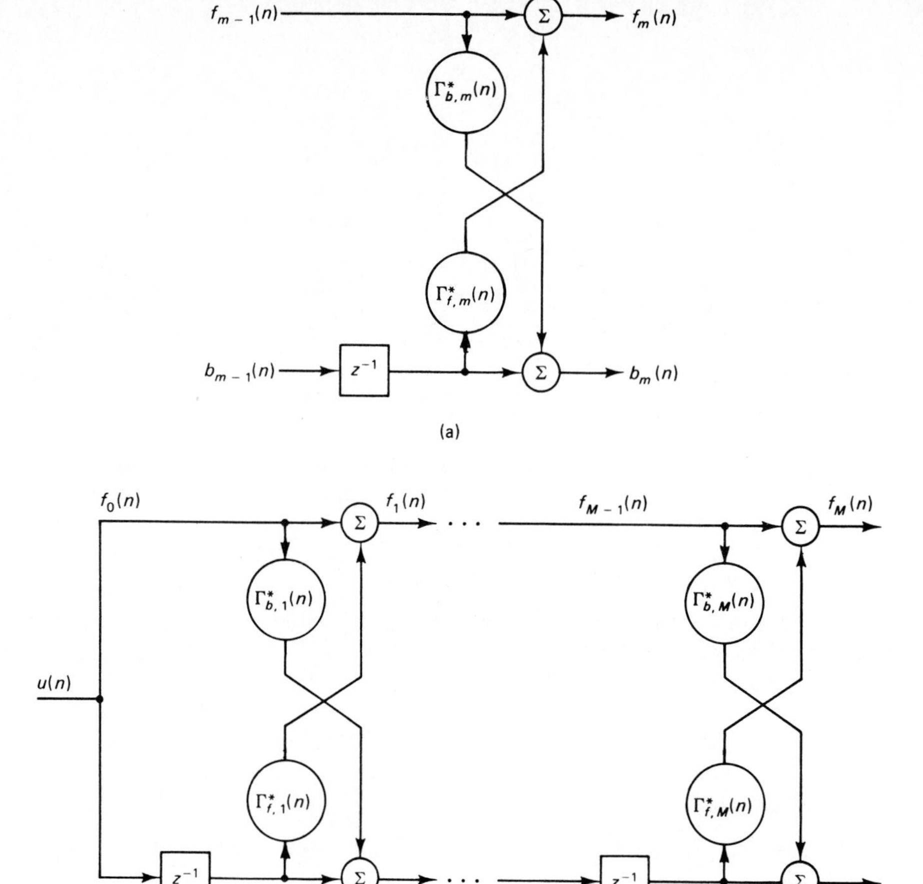

Figure 17.1 (a) Typical least-squares lattice stage: (b) multistage least-squares lattice predictor.

lation matrix $\mathbf{\Phi}_m(n)$ is, in general, non-Toeplitz and yet we are still able to derive simple order-update recursions for the least-squares prediction errors $f_m(n)$ and $b_m(n)$. This is indeed a remarkable result.

Order-Update Recursions for the Sums of Weighted a posteriori Prediction Error Squares

Equations (17.19) and (17.22) define the order-update recursions for the sum of weighted forward a posteriori prediction-error squares, $\mathcal{F}_m(n)$, and the sum of weighted backward a posteriori prediction-error squares, $\mathcal{B}_m(n)$, respectively. By using the relation between $\Delta_{m-1}(n)$ and $\Delta'_{m-1}(n)$ shown in Eq. (17.23), we may rewrite these two order-update recursions as follows, respectively:

$$\mathscr{F}_m(n) = \mathscr{F}_{m-1}(n) - \frac{|\Delta_{m-1}(n)|^2}{\mathscr{B}_{m-1}(n-1)} \tag{17.37}$$

and

$$\mathscr{B}_m(n) = \mathscr{B}_{m-1}(n-1) - \frac{|\Delta_{m-1}(n)|^2}{\mathscr{F}_{m-1}(n)} \tag{17.38}$$

where $m = 1, 2, \ldots, M$.

Equations (17.33), (17.35), (17.37), and (17.38), in conjunction with the definitions of (17.26) and (17.27) for the reflection coefficients, constitute the basic order-update recursions for the least-squares lattice predictor. These recursions generate two sequences of prediction errors: (1) the forward a posteriori prediction errors $f_0(n), f_1(n), \ldots, f_M(n)$, and (2) the backward a posteriori prediction errors $b_0(n), b_1(n), \ldots, b_M(n)$. These two sequences play key roles in the recursive solution of the linear least-squares problem.

Order-Update Recursion for $\gamma_m(n-1)$

To complete the list of order-update recursions required for the formulation of the LSL algorithm, we need one other such recursion for the delayed *estimation error* or *conversion factor* $\gamma_m(n-1)$. The role of $\gamma_m(n-1)$, in the context of the least-squares lattice predictor, will become apparent in Section 17.3. The reason for showing the time of interest as $n-1$ is to be consistent with the mathematical development in that section.

The estimation error $\gamma_m(n-1)$ results from applying the tap-input vector $\mathbf{u}_m(n-1)$ to a transversal filter whose tap-weight vector equals the gain vector $\mathbf{k}_m(n-1)$ that enters the derivation of the RLS solution. From the discussion presented in Section 15.3, we recall that the gain vector $\mathbf{k}_m(n-1)$ is order-updated as follows [see Eq. (15.46)]:

$$\mathbf{k}_m(n-1) = \begin{bmatrix} \mathbf{k}_{m-1}(n-1) \\ 0 \end{bmatrix} + \frac{b_{m-1}(n-1)}{\mathscr{B}_{m-1}(n-1)}\mathbf{c}_{m-1}(n-1), \quad m = 1, 2, \ldots, M \tag{17.39}$$

Postmultiplying the Hermitian transposed sides of this equation by $\mathbf{u}_m(n-1)$, we get

$$\mathbf{k}_m^H(n-1)\mathbf{u}_m(n-1) = [\mathbf{k}_{m-1}^H(n-1), 0]\mathbf{u}_m(n-1) + \frac{b_{m-1}^*(n-1)}{\mathscr{B}_{m-1}(n-1)}\mathbf{c}_{m-1}^H(n-1)\mathbf{u}_m(n-1) \tag{17.40}$$

By definition, we have

$$\mathbf{k}_m^H(n-1)\mathbf{u}_m(n-1) = 1 - \gamma_m(n-1) \tag{17.41}$$

$$\begin{aligned}[\mathbf{k}_{m-1}^H(n-1), 0]\mathbf{u}_m(n-1) &= [\mathbf{k}_{m-1}^H(n-1), 0]\begin{bmatrix} \mathbf{u}_{m-1}(n-1) \\ u(n-m) \end{bmatrix} \\ &= \mathbf{k}_{m-1}^H(n-1)\mathbf{u}_{m-1}(n-1) \\ &= 1 - \gamma_{m-1}(n-1)\end{aligned} \tag{17.42}$$

and

$$\mathbf{c}_{m-1}^H(n-1)\mathbf{u}_m(n-1) = b_{m-1}(n-1) \tag{17.43}$$

Therefore, the use of Eqs. (17.41) to (17.43) in (17.40) yields the simple result

$$\gamma_m(n-1) = \gamma_{m-1}(n-1) - \frac{|b_{m-1}(n-1)|^2}{\mathcal{B}_{m-1}(n-1)}, \quad m = 1, 2, \ldots, M \tag{17.44}$$

This is the desired recursion for order-updating $\gamma_m(n-1)$, which completes the list of desired order-update recursions.

17.3 TIME-UPDATE RECURSION

The recursions derived in Section 17.2 enable us to order-update the forward and backward a posteriori prediction errors as well as the sums of their weighted squares, starting from the elementary case $m = 0$. However, in order to make these recursions adaptive in time, it is necessary to derive a time-update recursion for the parameter $\Delta_{m-1}(n)$ that enters the computation of these order-update recursions. We now address this issue.

Consider the m-by-1 tap-weight vector $\mathbf{a}_{m-1}(n-1)$ that pertains to forward prediction-error filter of order $m-1$, evaluated at time $n-1$. The reason for considering time $n-1$ will become apparent presently. Since the leading element of the vector $\mathbf{a}_{m-1}(n-1)$ equals unity, we may express $\Delta_{m-1}(n)$ as follows [see Eqs. (17.23) and (17.25)]:

$$\Delta_{m-1}(n) = [\Delta_{m-1}(n), \mathbf{0}_{m-1}^T, \mathcal{B}_{m-1}(n-1)]\begin{bmatrix}\mathbf{a}_{m-1}(n-1)\\ 0\end{bmatrix} \tag{17.45}$$

Taking the Hermitian transpose of both sides of Eq. (17.16), recognizing the Hermitian property of $\mathbf{\Phi}_{m+1}(n)$, and using the relation of Eq. (17.23), we get

$$[0, \mathbf{c}_{m-1}^H(n-1)]\mathbf{\Phi}_{m+1}(n) = [\Delta_{m-1}(n), \mathbf{0}_{m-1}^T, \mathcal{B}_{m-1}(n-1)] \tag{17.46}$$

Hence, substitution of Eq. (17.46) in (17.45) yields

$$\Delta_{m-1}(n) = [0, \mathbf{c}_{m-1}^H(n-1)]\mathbf{\Phi}_{m+1}(n)\begin{bmatrix}\mathbf{a}_{m-1}(n-1)\\ 0\end{bmatrix} \tag{17.47}$$

But the correlation matrix $\mathbf{\Phi}_{m+1}(n)$ may be time-updated as follows [see Eq. (13.12)]:

$$\mathbf{\Phi}_{m+1}(n) = \lambda\mathbf{\Phi}_{m+1}(n-1) + \mathbf{u}_{m+1}(n)\mathbf{u}_{m+1}^H(n) \tag{17.48}$$

Accordingly, we may use this relation for $\mathbf{\Phi}_{m+1}(n)$ to rewrite Eq. (17.47) as

$$\begin{aligned}\Delta_{m-1}(n) = {} & \lambda[0, \mathbf{c}_{m-1}^H(n-1)]\mathbf{\Phi}_{m+1}(n-1)\begin{bmatrix}\mathbf{a}_{m-1}(n-1)\\ 0\end{bmatrix}\\ & + [0, \mathbf{c}_{m-1}^H(n-1)]\mathbf{u}_{m+1}(n)\mathbf{u}_{m+1}^H(n)\begin{bmatrix}\mathbf{a}_{m-1}(n-1)\\ 0\end{bmatrix}\end{aligned} \tag{17.49}$$

Next we recognize from the definition of *forward a priori prediction error* that

$$\mathbf{u}_{m+1}^H(n)\begin{bmatrix}\mathbf{a}_{m-1}(n-1)\\ 0\end{bmatrix} = [\mathbf{u}_m^H(n), u^*(n-m)]\begin{bmatrix}\mathbf{a}_{m-1}(n-1)\\ 0\end{bmatrix}$$
$$= \mathbf{u}_m^H(n)\mathbf{a}_{m-1}(n-1) \tag{17.50}$$
$$= \eta_{m-1}^*(n)$$

and from the definition of the *backward a posteriori prediction error* that

$$[0, \mathbf{c}_{m-1}^H(n-1)]\mathbf{u}_{m+1}(n) = [0, \mathbf{c}_{m-1}^H(n-1)]\begin{bmatrix}u(n)\\ \mathbf{u}_m(n-1)\end{bmatrix}$$
$$= \mathbf{c}_{m-1}^H(n-1)\mathbf{u}_m(n-1) \tag{17.51}$$
$$= b_{m-1}(n-1)$$

Also, by substituting $n-1$ for n in Eq. (17.10), we have

$$\mathbf{\Phi}_{m+1}(n-1)\begin{bmatrix}\mathbf{a}_{m-1}(n-1)\\ 0\end{bmatrix} = \begin{bmatrix}\mathcal{F}_{m-1}(n-1)\\ \mathbf{0}_{m-1}\\ \Delta_{m-1}(n-1)\end{bmatrix} \tag{17.52}$$

Hence, using this relation and the fact that the last element of the tap-weight vector $\mathbf{c}_{m-1}(n-1)$, pertaining to the backward prediction-error filter, equals unity, we may write the first term on the right side of Eq. (17.49), except for λ, as

$$[0, \mathbf{c}_{m-1}^H(n-1)]\mathbf{\Phi}_{m+1}(n-1)\begin{bmatrix}\mathbf{a}_{m-1}(n-1)\\ 0\end{bmatrix}$$
$$= [0, \mathbf{c}_{m-1}^H(n-1)]\begin{bmatrix}\mathcal{F}_{m-1}(n-1)\\ \mathbf{0}_{m-1}\\ \Delta_{m-1}(n-1)\end{bmatrix} \tag{17.53}$$
$$= \Delta_{m-1}(n-1)$$

Accordingly, substituting Eqs. (17.50), (17.51), and (17.53) in (17.49), we may express the time-update recursion for $\Delta_{m-1}(n)$ simply as

$$\Delta_{m-1}(n) = \lambda\Delta_{m-1}(n-1) + b_{m-1}(n-1)\eta_{m-1}^*(n) \tag{17.54}$$

The forward a priori prediction error $\eta_{m-1}(n)$ is related to the forward a posteriori prediction error $f_{m-1}(n)$ by [see Eq. (15.52)]

$$\eta_{m-1}(n) = \frac{f_{m-1}(n)}{\gamma_{m-1}(n-1)} \tag{17.55}$$

where $\gamma_{m-1}(n-1)$ is the delayed estimation error or conversion factor, defined previously. Thus, the use of Eq. (17.55) in (17.54) yields the time-update recursion

$$\Delta_{m-1}(n) = \lambda\Delta_{m-1}(n-1) + \frac{b_{m-1}(n-1)f_{m-1}^*(n)}{\gamma_{m-1}(n-1)} \tag{17.56}$$

The correction term in the time update of Eq. (17.56) is amplified by the reciprocal of the conversion factor $\gamma_{m-1}(n-1)$. This parameter enables the LSL algorithm to adapt rapidly to sudden changes in the input data (Morf and Lee, 1978). It is also of interest to note that, except for the amplification factor $1/\gamma_{m-1}(n-1)$, the time update of the parameter $\Delta_{m-1}(n)$ is, in fact a *time average of the cross-correlation* between the delayed backward prediction error $b_{m-1}(n-1)$ and forward prediction error $f_{m-1}(n)$.

In Chapter 15, we showed that the conversion factor $\gamma_{m-1}(n-1)$ is real valued and bounded by zero and one. When $\gamma_{m-1}(n-1)$ is zero, the recursion of Eq. (17.56) will stop, indicating that the correlation matrix $\mathbf{\Phi}_{m+1}(n)$ is noninvertible. Therefore, by monitoring the time variation of the conversion factor $\gamma_{m-1}(n-1)$, much can be learned about the state of the filter.

17.4 SUMMARY OF THE RECURSIVE LSL ALGORITHM USING A POSTERIORI ESTIMATION ERRORS

The complete list of order- and time-update recursions constituting the recursive LSL algorithm (based on a posteriori estimation errors) is summarized in Table 17.1. This summary includes the recursions and relations of Eqs. (17.56), (17.26), (17.27), (17.33), (17.35), (17.37), (17.38), and (17.44), in that order.

Since the LSL algorithm summarized in Table 17.1 involves division by updated parameters at some of the steps, care must be taken to ensure that these values are not allowed to become too small. Unless a high-precision computer is used, selection of the constant δ [determining the initial values $\mathcal{F}_0(0)$ and $\mathcal{B}_0(0)$] may have a severe effect on the initial transient performance of the LSL algorithm. Friedlander (1982) suggests using some form of *thresholding,* in that if the divisor (in any computation of LSL algorithm) is less than this preassigned threshold, the corresponding term involving that divisor is set to be zero. (This remark also applies to the other versions of the recursive LSL algorithm summarized in Tables 17.2 and 17.3.)

17.5 INITIALIZATION OF THE RECURSIVE LSL ALGORITHM

To initialize the recursive LSL algorithm using a posteriori estimation errors, we start with the elementary case of prediction order $m = 0$, for which we have [see Eqs. (17.30) and (17.34)]

$$f_0(n) = b_0(n) = u(n)$$

where $u(n)$ is the lattice predictor input at time n.

The other set of initial values we need pertain to the sums of weighted a posteriori prediction-error squares for $m = 0$. The sum of weighted forward a posteriori prediction-error squares, $\mathcal{F}_m(n)$, is time-updated as [see Eq. (15.18)]

$$\mathcal{F}_m(n) = \lambda \mathcal{F}_m(n-1) + f_m(n)\eta_m^*(n) \qquad (17.57)$$

where $f_m(n)$ and $\eta_m(n)$ are the forward a posteriori and a priori prediction errors for prediction order m, respectively. When $m = 0$, we have

$$f_0(n) = \eta_0(n) = u(n)$$

TABLE 17.1 SUMMARY OF THE RECURSIVE LSL ALGORITHM USING A POSTERIORI ESTIMATION ERRORS

Predictions:

For $n = 1, 2, 3, \ldots$ compute the various order updates in the following sequence: $m = 1, 2, \ldots, M$, where M is the final order of the least-squares lattice predictor:

$$\Delta_{m-1}(n) = \lambda \Delta_{m-1}(n-1) + \frac{b_{m-1}(n-1) f^*_{m-1}(n)}{\gamma_{m-1}(n-1)}$$

$$\Gamma_{f,m}(n) = -\frac{\Delta_{m-1}(n)}{\mathcal{B}_{m-1}(n-1)}$$

$$\Gamma_{b,m}(n) = -\frac{\Delta^*_{m-1}(n)}{\mathcal{F}_{m-1}(n)}$$

$$f_m(n) = f_{m-1}(n) + \Gamma^*_{f,m}(n) b_{m-1}(n-1)$$

$$b_m(n) = b_{m-1}(n-1) + \Gamma^*_{b,m}(n) f_{m-1}(n)$$

$$\mathcal{F}_m(n) = \mathcal{F}_{m-1}(n) - \frac{|\Delta_{m-1}(n)|^2}{\mathcal{B}_{m-1}(n-1)}$$

$$\mathcal{B}_m(n) = \mathcal{B}_{m-1}(n-1) - \frac{|\Delta_{m-1}(n)|^2}{\mathcal{F}_{m-1}(n)}$$

$$\gamma_m(n-1) = \gamma_{m-1}(n-1) = -\frac{|b_{m-1}(n-1)|^2}{\mathcal{B}_{m-1}(n-1)}$$

Filtering:

For $n = 1, 2, 3, \ldots$ compute the various order updates in the following sequence: $m = 0, 1, \ldots, M$

$$\rho_m(n) = \lambda \rho_m(n-1) + \frac{b_m(n)}{\gamma_m(n)} e^*_m(n)$$

$$\kappa_m(n) = \frac{\rho_m(n)}{\mathcal{B}_m(n)}$$

$$e_{m+1}(n) = e_m(n) - \kappa^*_m(n) b_m(n)$$

Initialization

1. To initialize the algorithm, at time $n = 0$ set

$\Delta_{m-1}(0) = 0$

$\mathcal{F}_{m-1}(0) = \delta, \qquad \delta =$ small positive constant

$\mathcal{B}_{m-1}(0) = \delta$

$\gamma_0(0) = 1$

2. At each instant $n \geq 1$, generate the various zeroth-order variables as follows:

$f_0(n) = b_0(n) = u(n)$

$\mathcal{F}_0(n) = \mathcal{B}_0(n) = \lambda \mathcal{F}_0(n-1) + |u(n)|^2$

$\gamma_0(n-1) = 1$

3. For joint-process estimation, initialize the algorithm by setting at time $n = 0$

$\rho_m(0) = 0$

At each instant $n \geq 1$, generate the zeroth-order variable

$e_0(n) = d(n)$

Note: For prewindowed data, the input $u(n)$ and desired response $d(n)$ are both zero for $n \leq 0$.

Hence, evaluating the time-update recursion of Eq. (17.57) for $m = 0$, we get

$$\mathcal{F}_0(n) = \lambda \mathcal{F}_0(n-1) + |u(n)|^2 \tag{17.58}$$

Similarly, we may show that for $m = 0$ the sum of weighted backward a posteriori prediction-error squares is updated in time as

$$\mathcal{B}_0(n) = \lambda \mathcal{B}_0(n-1) + |u(n)|^2 \tag{17.59}$$

With the conversion factor $\gamma_m(n-1)$ bounded by zero and 1, a logical choice for the zeroth-order value of this parameter is

$$\gamma_0(n-1) = 1 \tag{17.60}$$

We complete the initialization of the algorithm by setting at $n = 0$ the following conditions:

$$\Delta_{m-1}(0) = 0 \tag{17.61}$$

and

$$\mathcal{F}_0(0) = \mathcal{B}_0(0) = \delta \tag{17.62}$$

where δ is a small positive constant. The constant δ is used to ensure nonsingularity of the correlation matrix $\mathbf{\Phi}_{m+1}(n)$. Table 17.1 also includes a summary of the computations involved in the initialization of the recursive LSL algorithm using a posteriori estimation errors.

17.6 JOINT-PROCESS ESTIMATION

For a lattice predictor consisting of m stages, the recursive LSL algorithm produces a sequence of backward prediction errors $b_0(n), b_1(n), \ldots, b_m(n)$ that are *uncorrelated with each other in a time-averaged sense for all instants of time*. That is, the time-averaged correlation matrix of the backward prediction errors is a *diagonal matrix*. We refer to this property as the *decoupling property* of the LSL algorithm. Accordingly, by using the backward prediction errors $b_0(n), b_1(n), \ldots, b_m(n)$ as tap inputs that are applied to a corresponding set of *regression coefficients* $\kappa_0, \kappa_1, \ldots, \kappa_m$, respectively, as in Fig. 17.2, we may determine the *least-squares estimate* of some desired response $d(n)$ *exactly* and in a highly *efficient* manner. We refer to the two-channel structure of Fig. 17.2 as a *joint-process estimator*[1] because it solves the problem of estimating one process $\{d(n)\}$ from observations of a related process $\{u(n)\}$ by embedding them into the joint process $\{d(n), u(n)\}$.

Exact Decoupling Property of the LSL Algorithm

Before proceeding to derive the recursive procedure for updating the regression coefficients of the structure in Fig. 17.2, we will first prove the exact decoupling property of the LSL algorithm. As mentioned previously, this property implies that

[1] The idea of using a lattice predictor to perform joint-process estimation as in Fig. 17.2 was first proposed by Griffiths (1978). A variation of this idea was proposed independently by Makhoul (1978).

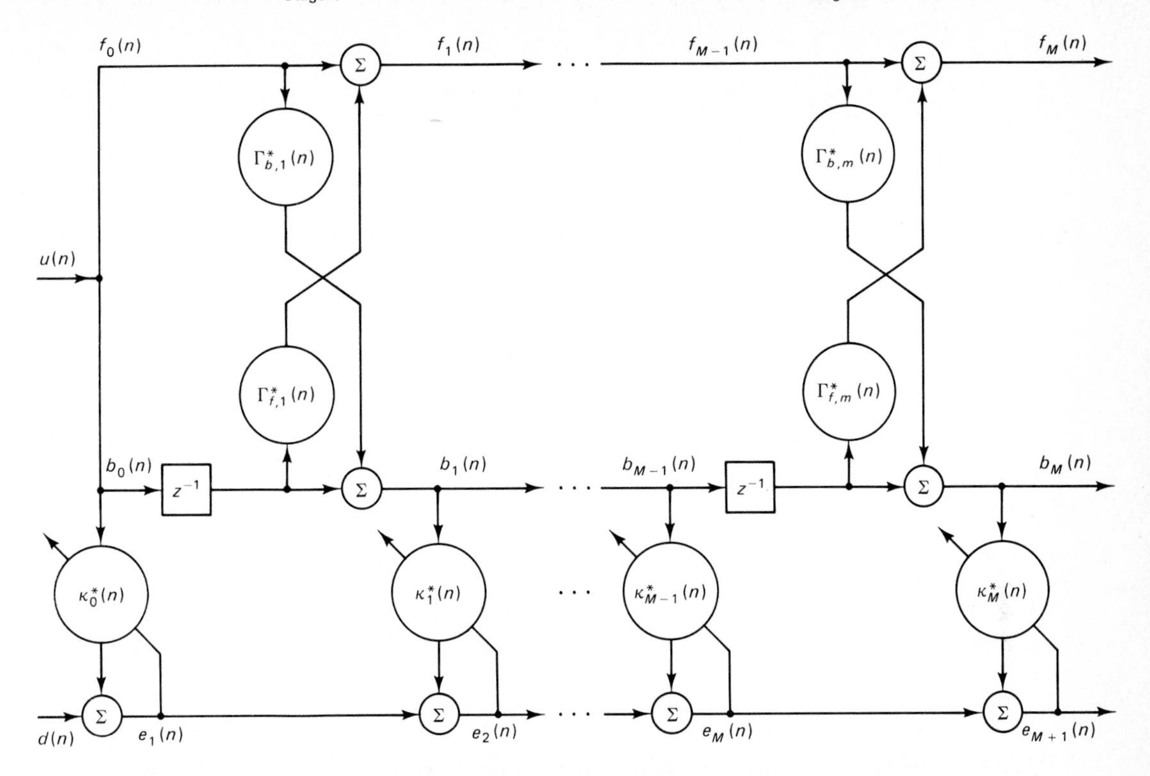

Figure 17.2 Joint-process estimator using the LSL algorithm based on a posteriori estimation errors.

the backward a posteriori prediction errors $b_0(n)$, $b_1(n)$, . . . , $b_m(n)$ used as inputs are uncorrelated with each other in a time-averaged sense at all instants of time.

Consider a backward prediction-error filter of order m. Let the $(m + 1)$-by-1 tap-weight vector of the filter, optimized in the least-squares sense over the time interval $1 \leq i \leq n$, be denoted by $\mathbf{c}_m(n)$. In expanded form, we have

$$\mathbf{c}_m^T(n) = [c_{m,m}(n), c_{m,m-1}(n), \ldots, 1] \tag{17.63}$$

Let $b_m(i)$ denote the backward a posteriori prediction error produced at the output of the filter in response to the $(m + 1)$-by-1 input vector $\mathbf{u}_{m+1}(i)$. The expanded form of the input vector is shown by

$$\mathbf{u}_{m+1}^T(i) = [u(i), u(i-1), \ldots, u(i-m)], \qquad i > m \tag{17.64}$$

We may thus express the error $b_m(i)$ as

$$\begin{aligned} b_m(i) &= \mathbf{c}_m^H(n)\mathbf{u}_{m+1}(i) \\ &= \sum_{k=0}^{m} c^*_{m,k}(n)u(i-m+k), \qquad m < i \leq n, \quad m = 0, 1, 2, \ldots \end{aligned} \tag{17.65}$$

Let $\mathbf{b}_{m+1}(i)$ denote the $(m + 1)$-by-1 *backward a posteriori prediction-error vector* that is defined by

$$\mathbf{b}_{m+1}^T(i) = [b_0(i), b_1(i), \ldots, b_m(i)], \qquad \begin{matrix} i > m \\ m = 0, 1, 2, \ldots \end{matrix} \tag{17.66}$$

Substituting Eq. (17.65) in (17.66), we may express the transformation of the input data into the corresponding set of backward a posteriori prediction errors as follows:

$$\mathbf{b}_{m+1}(i) = \mathbf{L}_m(n)\mathbf{u}_{m+1}(i) \tag{17.67}$$

where the $(m + 1)$-by-$(m + 1)$ *transformation matrix* $\mathbf{L}_m(n)$ is defined by

$$\mathbf{L}_m(n) = \begin{bmatrix} 1 & 0 & \cdots & 0 \\ c_{1,1}^*(n) & 1 & \cdots & 0 \\ \vdots & \vdots & \ddots & \vdots \\ c_{m,m}^*(n) & c_{m,m-1}^*(n) & \cdots & 1 \end{bmatrix} \tag{17.68}$$

The subscript m in the symbol $\mathbf{L}_m(n)$ refers to the highest order of backward prediction-error filter involved in its constitution. Note also that:

1. The nonzero elements of row l of matrix $\mathbf{L}_m(n)$ are defined by the tap weights of backward prediction-error filter of order $(l - 1)$.
2. The diagonal elements of matrix $\mathbf{L}_m(n)$ equal unity; this follows from the fact that the last tap weight of a backward prediction-error filter equals unity.
3. The determinant of matrix $\mathbf{L}_m(n)$ equals one for all m; hence, the inverse matrix $\mathbf{L}_m^{-1}(n)$ exists.

We define the $(m + 1)$-by-$(m + 1)$ time-averaged correlation matrix of the backward a posteriori prediction-error vector as

$$\mathbf{D}_{m+1}(n) = \sum_{i=1}^{n} \lambda^{n-i}\mathbf{b}_{m+1}(i)\mathbf{b}_{m+1}^H(i), \qquad \begin{matrix} m < i \le n, \\ m = 0, 1, 2, \ldots \end{matrix} \tag{17.69}$$

Substituting Eq. (17.67) in (17.69), we get

$$\begin{aligned} \mathbf{D}_{m+1}(n) &= \sum_{i=1}^{n} \lambda^{n-i}\mathbf{L}_m(n)\mathbf{u}_{m+1}(i)\mathbf{u}_{m+1}^H(i)\mathbf{L}_m^H(n) \\ &= \mathbf{L}_m(n)\left[\sum_{i=1}^{n} \lambda^{n-i}\mathbf{u}_{m+1}(i)\mathbf{u}_{m+1}^H(i)\right]\mathbf{L}_m^H(n) \end{aligned} \tag{17.70}$$

The expression inside the square brackets equals the time-averaged correlation matrix of the input vector; that is,

$$\mathbf{\Phi}_{m+1}(n) = \sum_{i=1}^{n} \lambda^{n-i}\mathbf{u}_{m+1}(i)\mathbf{u}_{m+1}^H(i) \tag{17.71}$$

Accordingly, we may simplify Eq. (17.70) as

$$\mathbf{D}_{m+1}(n) = \mathbf{L}_m(n)\mathbf{\Phi}_{m+1}(n)\mathbf{L}_m^H(n) \tag{17.72}$$

Using the formula for the augmented normal equations for backward linear prediction, we may show that the product $\mathbf{\Phi}_{m+1}(n)\mathbf{L}_m^H(n)$ consists of a lower triangular matrix whose diagonal elements equal the various sums of weighted backward a posteriori prediction-error squares, that is, $\mathcal{B}_0(n), \mathcal{B}_1(n), \ldots, \mathcal{B}_m(n)$ (see Problem 1). The matrix $\mathbf{L}_m(n)$ is, by definition, a lower triangular matrix whose diagonal elements are all equal to unity. Hence, the product of $\mathbf{L}_m(n)$ and $\mathbf{\Phi}_{m+1}(n)\mathbf{L}_m^H(n)$ is a lower triangular matrix. We also know that $\mathbf{L}_m^H(n)$ is an upper triangular matrix, and so is the matrix product $\mathbf{L}_m(n)\mathbf{\Phi}_{m+1}(n)$. Hence, the product of $\mathbf{L}_m(n)\mathbf{\Phi}_{m+1}(n)$ and $\mathbf{L}_m^H(n)$ is an upper triangular matrix. In other words, the matrix $\mathbf{D}(n)$ is both *upper* and *lower* triangular, and therefore, *diagonal*. Accordingly, we may write

$$\begin{aligned}\mathbf{D}_{m+1}(n) &= \mathbf{L}_m(n)\mathbf{\Phi}_{m+1}(n)\mathbf{L}_m^H(n) \\ &= \text{diag}[\mathcal{B}_0(n), \mathcal{B}_1(n), \ldots, \mathcal{B}_m(n)]\end{aligned} \tag{17.73}$$

Equation (17.73) shows that the backward a posteriori prediction errors $b_0(n), b_1(n), \ldots, b_m(n)$ produced by the various stages of the least-squares lattice predictor are uncorrelated (in a time-averaged sense) at all instants of time. This proves another remarkable property of the least-squares lattice predictor that makes it ideally suited for *exact* joint-process estimation.

It is noteworthy that the transformation of the (correlated) input data by the LSL algorithm into a new sequence of (uncorrelated) backward prediction errors may be viewed as a deterministic form of the Gram–Schmidt orthogonalization procedure. For further perusal of this issue, the reader is referred to Problem 2.

Transformation of the RLS Solution

Consider the conventional tapped-delay-line or transversal filter structure shown in Fig. 17.3, where the tap inputs $u(n), u(n-1), \ldots, u(n-m)$ are derived directly from the process $\{u(n)\}$ and the tap weights $\hat{w}_0(n), \hat{w}_1(n), \ldots, \hat{w}_m(n)$ are used to form respective inner products. From Chapter 10, we recall that the least-squares solution for the $(m+1)$-by-1 tap-weight vector $\hat{\mathbf{w}}_m(n)$, consisting of the elements $\hat{w}_0(n), \hat{w}_1(n), \ldots, w_m(n)$, is defined by

$$\mathbf{\Phi}_{m+1}(n)\hat{\mathbf{w}}_m(n) = \mathbf{\theta}_{m+1}(n) \tag{17.74}$$

where $\mathbf{\Phi}_{m+1}(n)$ is the $(m+1)$-by-$(m+1)$ correlation matrix of the tap inputs, and $\mathbf{\theta}_{m+1}(n)$ is the $(m+1)$-by-1 cross-correlation vector between the tap inputs and de-

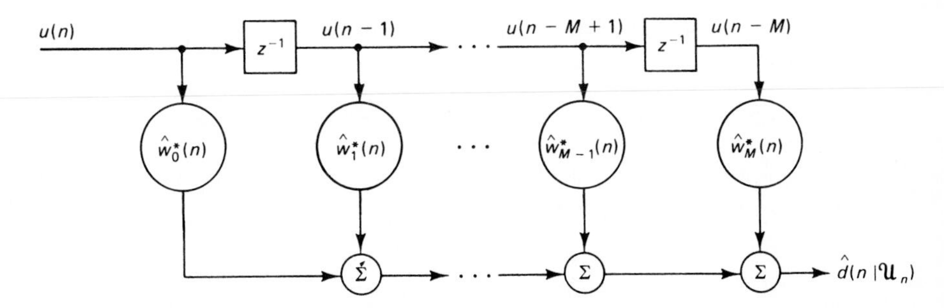

Figure 17.3 Conventional transversal filter.

sired response. We modify Eq. (17.74) in two ways: (1) we premultiply both sides of the equation by the $(m + 1)$-by-$(m + 1)$ lower triangular transformation matrix $\mathbf{L}_m(n)$, and (2) we interject the $(m + 1)$-by-$(m + 1)$ identity matrix $\mathbf{I} = \mathbf{L}_m^H(n)\mathbf{L}_m^{-H}(n)$ between the matrix $\mathbf{\Phi}_{m+1}(n)$ and the vector $\hat{\mathbf{w}}_m(n)$ on the left side of the equation. The symbol $\mathbf{L}_m^{-H}(n)$ denotes the Hermitian transpose of the inverse matrix $\mathbf{L}_m^{-1}(n)$. We may thus write

$$\mathbf{L}_m(n)\mathbf{\Phi}_{m+1}(n)\mathbf{L}_m^H(n)\mathbf{L}_m^{-H}(n)\hat{\mathbf{w}}_m(n) = \mathbf{L}_m(n)\boldsymbol{\theta}_{m+1}(n) \tag{17.75}$$

From Eq. (17.72), the product $\mathbf{L}_m(n)\mathbf{\Phi}_{m+1}(n)\mathbf{L}_m^H(n)$ on the left side of Eq. (17.75) equals the diagonal matrix $\mathbf{D}_{m+1}(n)$ that represents the correlation matrix of the backward a posteriori prediction errors used as tap inputs for the regression coefficients of Fig. 17.2. The product $\mathbf{L}_m(n)\boldsymbol{\theta}_{m+1}(n)$ on the right side equals the $(m + 1)$-by-1 cross-correlation vector between these backward prediction errors and the desired response. Let $\mathbf{t}_{m+1}(n)$ denote this cross-correlation vector. By definition, we have

$$\mathbf{t}_{m+1}(n) = \sum_{i=1}^{n} \lambda^{n-i}\mathbf{b}_{m+1}(i)d^*(i) \tag{17.76}$$

where $d(i)$ is the desired response. Substituting Eq. (17.67) in (17.76), we thus get

$$\begin{aligned}\mathbf{t}_{m+1}(n) &= \sum_{i=1}^{n} \lambda^{n-i}\mathbf{L}_m(n)\mathbf{u}_{m+1}(i)d^*(i) \\ &= \mathbf{L}_m(n)\sum_{i=1}^{n} \lambda^{n-i}\mathbf{u}_{m+1}(i)d^*(i) \\ &= \mathbf{L}_m(n)\boldsymbol{\theta}_{m+1}(n)\end{aligned} \tag{17.77}$$

which is the desired result. Accordingly, the combined use of Eqs. (17.72) and (17.77) in (17.75) yields the *transformed RLS solution:*

$$\mathbf{D}_{m+1}(n)\mathbf{L}_m^{-H}(n)\hat{\mathbf{w}}_m(n) = \mathbf{t}_{m+1}(n) \tag{17.78}$$

Thus far we have considered how the application of lower triangular matrix $\mathbf{L}_m(n)$ transforms the RLS solution for the tap-weight vector of the conventional transversal structure shown in Fig. 17.3. We next wish to consider the RLS solution for the regression coefficient vector $\boldsymbol{\kappa}_m(n)$ in the structure of Fig. 17.2, which is defined by

$$\boldsymbol{\kappa}_m^T(n) = [\kappa_0(n), \kappa_1(n), \ldots, \kappa_m(n)] \tag{17.79}$$

The regression coefficient vector $\boldsymbol{\kappa}_m(n)$ is obtained by minimizing the index of performance

$$\sum_{i=1}^{n} \lambda^{n-i}|d(i) - \mathbf{b}_{m+1}^T(i)\boldsymbol{\kappa}_m^*(n)|^2$$

where $\boldsymbol{\kappa}_m(n)$ is held constant for $1 \leq i \leq n$. The resulting solution to this RLS problem may be expressed as (see Problem 3)

$$\mathbf{D}_{m+1}(n)\boldsymbol{\kappa}_m(n) = \mathbf{t}_{m+1}(n) \tag{17.80}$$

where, as defined before, $\mathbf{D}_{m+1}(n)$ is the $(m + 1)$-by-$(m + 1)$ correlation matrix of the backward a posteriori prediction errors used as tap inputs in Fig. 17.2, and

$\mathbf{t}_{m+1}(n)$ is the $(m + 1)$-by-1 cross-correlation between these tap inputs and the desired response.

By comparing the transformed RLS solution of Eq. (17.78) and the RLS solution of Eq. (17.80), we immediately deduce the following simple relationship between the tap-weight vector $\hat{\mathbf{w}}_m(n)$ in the structure of Fig. 17.3 and the corresponding regression coefficient vector $\boldsymbol{\kappa}_m(n)$ in the structure of Fig. 17.2:

$$\boldsymbol{\kappa}_m(n) = \mathbf{L}_m^{-H}(n)\hat{\mathbf{w}}_m(n) \tag{17.81}$$

or, equivalently,

$$\hat{\mathbf{w}}_m(n) = \mathbf{L}_m^H(n)\boldsymbol{\kappa}_m(n) \tag{17.82}$$

Here, again we see that the lower triangular transformation matrix $\mathbf{L}_m(n)$ represents the connecting link between the RLS solutions for the tap-weight vectors of Figs. 17.2 and 17.3.

Recursive Algorithm for Computing the Regression Coefficient Vector $\boldsymbol{\kappa}_m(n)$

Solving Eqs. (17.77) and (17.80) for the regression coefficient vector $\boldsymbol{\kappa}_m(n)$, we get

$$\boldsymbol{\kappa}_m(n) = \mathbf{D}_{m+1}^{-1}(n)\mathbf{L}_m(n)\boldsymbol{\theta}_{m+1}(n) \tag{17.83}$$

The exact decoupling property of the LSL algorithm ensures that the time-averaged correlation matrix $\mathbf{D}_{m+1}(n)$ of the sequence of backward prediction errors produced by the algorithm is a diagonal matrix for all n, as shown by Eq. (17.73). This means that the inverse matrix $\mathbf{D}_{m+1}^{-1}(n)$ is likewise diagonal:

$$\mathbf{D}_{m+1}^{-1}(n) = \text{diag}[\mathcal{B}_0^{-1}(n), \mathcal{B}_1^{-1}(n), \ldots, \mathcal{B}_m^{-1}(n)] \tag{17.84}$$

Hence, only $m + 1$ scalar divisions rather than matrix inversion are needed in the use of Eq. (17.83) to compute the regression coefficient vector $\boldsymbol{\kappa}_m(n)$. In particular, the use of Eqs. (17.68) and (17.84) in (17.83) yields a system of $(m + 1)$ scalar equations:

$$\kappa_m(n) = \mathcal{B}_m^{-1}(n)\mathbf{c}_m^H(n)\boldsymbol{\theta}_{m+1}(n), \qquad m = 0, 1, \ldots, M \tag{17.85}$$

where M is the final value of the prediction order. To develop a recursion for time-updating the regression coefficient $\kappa_m(n)$, define the scalar

$$\rho_m(n) = \mathbf{c}_m^H(n)\boldsymbol{\theta}_{m+1}(n), \qquad m = 0, 1, 2, \ldots, M \tag{17.86}$$

We may then redefine the regression coefficient $\kappa_m(n)$ of Eq. (17.85) in terms of $\rho_m(n)$ as follows:

$$\kappa_m(n) = \frac{\rho_m(n)}{\mathcal{B}_m(n)}, \qquad m = 0, 1, \ldots, M \tag{17.87}$$

From Chapter 15, we recall that the tap-weight vector $\mathbf{c}_m(n)$ of the backward prediction-error filter may be time-updated as follows [see Eqs. (15.34) and (15.53)]:

$$\mathbf{c}_m(n) = \mathbf{c}_m(n-1) - \frac{b_m^*(n)}{\gamma_m(n)}\begin{bmatrix}\mathbf{k}_m(n)\\ 0\end{bmatrix} \tag{17.88}$$

where $\mathbf{k}_m(n)$ is the m-by-1 gain vector. Also, from Chapter 13 we recall that the $(m + 1)$-by-1 cross-correlation vector $\boldsymbol{\theta}_{m+1}(n)$ may be time-updated as follows [see Eq. 13.13)]:

$$\boldsymbol{\theta}_{m+1}(n) = \lambda \boldsymbol{\theta}_{m+1}(n - 1) + \mathbf{u}_{m+1}(n)d^*(n) \tag{17.89}$$

Hence, substituting Eqs. (17.88) and (17.89) in (17.86), and recognizing that $\gamma_m(n)$ is real valued, we get

$$\begin{aligned}\rho_m(n) &= \lambda \mathbf{c}_m^H(n - 1)\boldsymbol{\theta}_{m+1}(n - 1) + \mathbf{c}_m^H(n - 1)\mathbf{u}_{m+1}(n)d^*(n) \\ &\quad - \frac{b_m(n)}{\gamma_m(n)}\mathbf{k}_m^H(n)\boldsymbol{\theta}_m(n)\end{aligned} \tag{17.90}$$

In the last element of the right side in Eq. (17.90), we have used the fact that the first m elements of the $(m + 1)$-by-1 vector $\boldsymbol{\theta}_{m+1}(n)$ are contained in the m-by-1 vector $\boldsymbol{\theta}_m(n)$, and since the last element of the $(m + 1)$-by-1 vector on the right side of Eq. (17.88) is zero, only these elements of $\boldsymbol{\theta}_{m+1}(n)$ enter the multiplication of the second term of $\mathbf{c}_m^H(n)$ by $\boldsymbol{\theta}_{m+1}(n)$. We may simplify Eq. (17.90) as follows:

1. From the definition introduced in Eq. (17.86), we deduce that

$$\rho_m(n - 1) = \mathbf{c}_m^H(n - 1)\boldsymbol{\theta}_{m+1}(n - 1)$$

2. From the definition of backward a priori prediction error, we have

$$\psi_m(n) = \mathbf{c}_m^H(n - 1)\mathbf{u}_{m+1}(n)$$

   Moreover, since $\psi_m(n)$ equals the ratio $b_m(n)/\gamma_m(n)$ [see Eq. (15.53)], it follows therefore that

$$\mathbf{c}_m^H(n - 1)\mathbf{u}_{m+1}(n) = \frac{b_m(n)}{\gamma_m(n)}$$

3. Since, by definition, the gain vector $\mathbf{k}_m(n)$ equals $\boldsymbol{\Phi}_m^{-1}(n)\mathbf{u}_m(n)$, and the correlation matrix $\boldsymbol{\Phi}_m(n)$ is Hermitian, it follows that

$$\begin{aligned}\mathbf{k}_m^H(n)\boldsymbol{\theta}_m(n) &= \mathbf{u}_m^H(n)\boldsymbol{\Phi}_m^{-1}(n)\boldsymbol{\theta}_m(n) \\ &= \mathbf{u}_m^H(n)\hat{\mathbf{w}}_{m-1}(n)\end{aligned}$$

   where in the last line we have made use of the normal equations for least-squares estimation. The inner product $\mathbf{u}_m^H(n)\hat{\mathbf{w}}_{m-1}(n)$ equals the complex conjugate of the least-squares estimate of the desired response $d(n)$, given the m-by-1 tap-input vector $\mathbf{u}_m(n)$. Let $\hat{d}(n|\mathcal{U}_n)$ denote the value of this estimate, where $\mathcal{U}_n$ is the space spanned by the elements of vector $\mathbf{u}_m(n)$.

Accordingly, we may rewrite Eq. (17.90) simply as

$$\rho_m(n) = \lambda\rho_m(n - 1) + \frac{b_m(n)}{\gamma_m(n)}[d^*(n) - \hat{d}^*(n|\mathcal{U}_n)] \tag{17.91}$$

Define the *a posteriori estimation error*, based on the use of m tap inputs, as

$$\begin{aligned}e_m(n) &= d(n) - \hat{d}(n|\mathcal{U}_n) \\ &= d(n) - \hat{\mathbf{w}}_{m-1}^H(n)\mathbf{u}_m(n)\end{aligned} \tag{17.92}$$

Hence, we may simplify the time-update recursion of Eq. (17.91) as

$$\rho_m(n) = \lambda\rho_m(n-1) + \frac{b_m(n)}{\gamma_m(n)}e_m^*(n), \qquad m = 0, 1, 2, \ldots, M \quad (17.93)$$

Note that, as in the time-update recursion of Eq. (17.56) for $\Delta_m(n)$, the correction term in the recursion of Eq. (17.93) is amplified by the reciprocal of the conversion factor $\gamma_m(n)$. Also, except for $\gamma_m(n)$, we may interpret $\rho_m(n)$ as the cross-correlation between $b_m(n)$ and $e_m(n)$.

To complete the recursive procedure for joint-process estimation, we need to accompany the time-update recursion of Eq. (17.93) with an order-update recursion that enables us to compute the updated estimation error $e_{m+1}(n)$, given $e_m(n)$. We may do this by replacing m with $m + 1$ in Eq. (17.92) to write

$$e_{m+1}(n) = d(n) - \hat{\mathbf{w}}_m^H(n)\mathbf{u}_{m+1}(n)$$

Since the inner product $\hat{\mathbf{w}}_m^H(n)\mathbf{u}_{m+1}(n)$ equals $\boldsymbol{\kappa}_m^H(n)\mathbf{b}_{m+1}(n)$ [see Eqs. (17.67) and (17.82)], we may also express $e_{m+1}(n)$ as follows:

$$\begin{aligned} e_{m+1}(n) &= d(n) - \boldsymbol{\kappa}_m^H(n)\mathbf{b}_{m+1}(n) \\ &= d(n) - \sum_{l=0}^{m} \kappa_l^*(n)b_l(n) \end{aligned} \quad (17.94)$$

Isolating the inner product $\kappa_m^*(n)b_m(n)$ from the rest of the summation on the right side of Eq. (17.94), and recognizing that the remainder equals (by definition)

$$d(n) - \sum_{l=0}^{m-1} \kappa_l^*(n)b_l(n) = e_m(n)$$

we may express the order-update recursion for the a posteriori estimation error as

$$e_{m+1}(n) = e_m(n) - \kappa_m^*(n)b_m(n), \qquad m = 0, 1, \ldots, M \quad (17.95)$$

Note that the number of regression coefficients [represented by $\kappa_0(n), \kappa_1(n), \ldots, \kappa_M(n)$] exceeds the final prediction order M by 1.

We may now sum up the procedure for the extension of the recursive LSL algorithm to perform joint-process estimation. The time-update recursion of Eq. (17.93) enables us to compute the updated value of parameter $\rho_m(n)$. Given this value and the updated value of the sum of weighted backward a posteriori prediction-error squares $\mathcal{B}_m(n)$, obtained from the order-update recursion of Eq. (17.38), we may compute the updated value of $\kappa_m(n)$ by using Eq. (17.87). Hence, we may use the order-update recursion of Eq. (17.95) to compute the updated value of the a posteriori estimation error $e_{m+1}(n)$.

Initial Conditions

For the elementary case of prediction order $m = 0$, we have (see Fig. 17.2)

$$e_0(n) = d(n) \quad (17.96)$$

where $d(n)$ is the desired response. Thus, to initiate the computation, we compute $e_0(n)$ at each instant n.

To complete the initialization of the LSL algorithm for joint-process estimation, at time $n = 0$ we set

$$\rho_m(0) = 0, \qquad m = 0, 1, \ldots, M \tag{17.97}$$

Summary

The summary of the recursive LSL algorithm presented in Table 17.1 includes the extension of the algorithm for joint-process estimation and its initialization.

17.7 RECURSIVE LSL ALGORITHM USING A PRIORI ESTIMATION ERRORS

The recursive LSL algorithm summarized in Table 17.1 is based on three sets of a posteriori estimation errors: forward prediction errors, backward prediction errors, and joint-process estimation errors. We now describe another version of the algorithm that uses a priori forms of forward prediction errors, backward prediction errors, and joint-process estimation errors as the variables of interest.

From Chapter 15, we recall that the forward a posteriori prediction error, $f_m(n)$, and the forward a priori prediction error, $\eta_m(n)$, are related by the formula [see Eq. (15.52)]

$$f_m(n) = \gamma_m(n - 1)\eta_m(n) \tag{17.98}$$

Also, the backward a posteriori prediction error, $b_m(n)$, and the backward a priori prediction error, $\psi_m(n)$, are related by the formula [see Eq. (15.53)]

$$b_m(n) = \gamma_m(n)\psi_m(n) \tag{17.99}$$

For the relation between the a posteriori estimation error, $e_m(n)$, and the a priori estimation error $\alpha_m(n)$, we have [see Eq. (15.51)]

$$e_m(n) = \gamma_m(n)\alpha_m(n) \tag{17.100}$$

From Chapter 15, we also note that the sum of weighted forward a posteriori prediction-error squares, $\mathcal{F}_{m-1}(n)$, and the sum of weighted backward a posteriori prediction-error squares, $\mathcal{B}_{m-1}(n - 1)$, may be updated as follows, respectively:

$$\mathcal{F}_{m-1}(n) = \lambda \mathcal{F}_{m-1}(n - 1) + \eta_{m-1}(n)f^*_{m-1}(n) \tag{17.101}$$

and

$$\mathcal{B}_{m-1}(n) = \lambda \mathcal{B}_{m-1}(n - 1) + \psi_{m-1}(n)b^*_{m-1}(n) \tag{17.102}$$

Using Eqs. (17.26), (17.56), (17.98), (17.99), and (17.102) in (17.33), and then simplifying, we get the following *recursion for order-updating the forward a priori prediction error:*

$$\eta_m(n) = \eta_{m-1}(n) + \Gamma^*_{f,m}(n - 1)\psi_{m-1}(n - 1), \qquad m = 1, 2, \ldots, M \tag{17.103}$$

Similarly, using Eqs. (17.27), (17.56), (17.98), (17.99), and (17.101) in (17.35), and then simplifying, we get the corresponding recursion for *order-updating the backward a priori prediction error:*

$$\psi_m(n) = \psi_{m-1}(n-1) + \Gamma^*_{b,m}(n-1)\eta_{m-1}(n), \qquad m = 1, 2, \ldots, M \tag{17.104}$$

Finally, using Eqs. (17.87), (17.93), (17.99), (17.100), and (17.102) in (17.95), and then simplifying, we get the following recursion for *order-updating the a priori estimation error:*

$$\alpha_{m+1}(n) = \alpha_m(n) - \kappa^*_m(n-1)\psi_m(n), \qquad m = 0, 1, \ldots, M \tag{17.105}$$

Thus, combining the above order updates with the following recursions: (1) time update for the cross-correlation $\Delta_m(n)$ reformulated in terms of the a priori prediction errors $\eta_{m-1}(n)$ and $\psi_{m-1}(n-1)$; (2) time updates for the sums of weighted a posteriori prediction error squares $\mathcal{F}_{m-1}(n)$ and $\mathcal{B}_{m-1}(n)$ as in Eqs. (17.101) and (17.102); and (3) time updates for the cross-correlation $\rho_m(n)$ reformulated in terms of the a priori errors $\alpha_m(n)$ and $\psi_m(n)$, and arranging the various recursions in their proper sequence, we get the second version of the recursive LSL algorithm summarized in Table 17.2 (Ljung and Söderström, 1983; Ling et al., 1985). Note that in the summary of joint-process estimation presented in Table 17.2, we have added a time-update recursion for computing $\mathcal{B}_m(n)$; this is needed subsequently for the updating of $\kappa_m(n)$.

In Fig. 17.4, we present the signal-flow-graph representation of this second version of the recursive LSL algorithm. Note that in this version, in order to update the variables of interest (namely, the a priori forward prediction, backward prediction, and joint-process estimation errors) at time n, we require knowledge of the values of three basic sets of coefficients (namely, the forward reflection coefficients, the backward reflection coefficients, and the regression coefficients) at time $n - 1$. On the other hand, in the first version of the recursive LSL algorithm represented by the signal-flow graph of Fig. 17.2, in order to update the a posteriori forward prediction, backward prediction, and joint-process estimation errors at time n, we require knowledge of these three sets of coefficients at time n. In Table 17.2 we also include a summary of the initialization procedure for the second version of the recursive LSL algorithm based on a priori estimation errors.

17.8 MODIFIED FORMS OF THE RECURSIVE LSL ALGORITHM

In the two versions of the recursive LSL algorithm summarized in Tables 17.1 and 17.2, we update the forward and backward reflection coefficients of the lattice predictor and the regression coefficients of the joint-process estimator in an *indirect* manner. We first compute the cross-correlation between forward and delayed backward errors and the cross-correlation between backward prediction errors and joint-process estimation errors. We next compute the sum of weighted forward error squares and the sum of weighted backward error squares. We then compute the reflection and regression coefficients by dividing a cross-correlation by a sum of weighted-error squares. Accordingly, the accuracy of the reflection and regression coefficients depends on the accuracy attained in computing the cross-correlations

TABLE 17.2 SUMMARY OF THE RECURSIVE LSL ALGORITHM USING A PRIORI ESTIMATION ERRORS

Predictions:

Starting with $n = 1$, compute the various order updates in the following sequence $m = 1, 2, \ldots M$, where M is the final order of the least-squares predictor:

$$\eta_m(n) = \eta_{m-1}(n) + \Gamma^*_{f,m}(n-1)\psi_{m-1}(n-1)$$

$$\psi_m(n) = \psi_{m-1}(n-1) + \Gamma^*_{b,m}(n-1)\eta_{m-1}(n)$$

$$\Delta_{m-1}(n) = \lambda\Delta_{m-1}(n-1) + \gamma_{m-1}(n-1)\psi_{m-1}(n-1)\eta^*_{m-1}(n)$$

$$\mathcal{F}_{m-1}(n) = \lambda\mathcal{F}_{m-1}(n-1) + \gamma_{m-1}(n-1)|\eta_{m-1}(n)|^2$$

$$\mathcal{B}_{m-1}(n) = \lambda\mathcal{B}_{m-1}(n-1) + \gamma_{m-1}(n)|\psi_{m-1}(n)|^2$$

$$\Gamma_{f,m}(n) = -\frac{\Delta_{m-1}(n)}{\mathcal{B}_{m-1}(n-1)}$$

$$\Gamma_{b,m}(n) = -\frac{\Delta^*_{m-1}(n)}{\mathcal{F}_{m-1}(n)}$$

$$\gamma_m(n) = \gamma_{m-1}(n) - \frac{\gamma^2_{m-1}(n)|\psi_{m-1}(n)|^2}{\mathcal{B}_{m-1}(n)}$$

Filtering:

For $n = 1, 2, 3, \ldots$ compute the various order updates in the following sequence $m = 0, 1, \ldots, M$:

$$\rho_m(n) = \lambda\rho_m(n-1) + \gamma_m(n)\psi_m(n)\alpha^*_m(n)$$

$$\alpha_{m+1}(n) = \alpha_m(n) - \kappa^*_m(n-1)\psi_m(n)$$

$$\mathcal{B}_m(n) = \lambda\mathcal{B}_m(n) + \gamma_m(n)|\psi_m(n)|^2$$

$$\kappa_m(n) = \frac{\rho_m(n)}{\mathcal{B}_m(n)}$$

Initialization

1. To initialize the algorithm, at time $n = 0$ set

$\Delta_{m-1}(0) = 0$
$\mathcal{F}_{m-1}(0) = \delta, \qquad \delta = \text{small positive constant}$
$\mathcal{B}_{m-1}(0) = \delta$
$\Gamma_{f,m}(0) = \Gamma_{b,m}(0) = 0$
$\gamma_0(0) = 1$

2. At each instant $n \geq 1$, generate the zeroth-order variables:

$\eta_0(n) = \psi_0(n) = u(n)$
$\mathcal{F}_0(n) = \mathcal{B}_0(n) = \lambda\mathcal{F}_0(n-1) + |u(n)|^2$
$\gamma_0(n) = 1$

3. For joint-process estimation, initialize the algorithm by setting at time $n = 0$

$\rho_m(0) = 0$

At each instant $n \geq 1$, generate the zeroth-order variable

$\alpha_0(n) = d(n)$

Note: For prewindowed data, the input $u(n)$ and desired response $d(n)$ are both zero for $n \leq 0$.

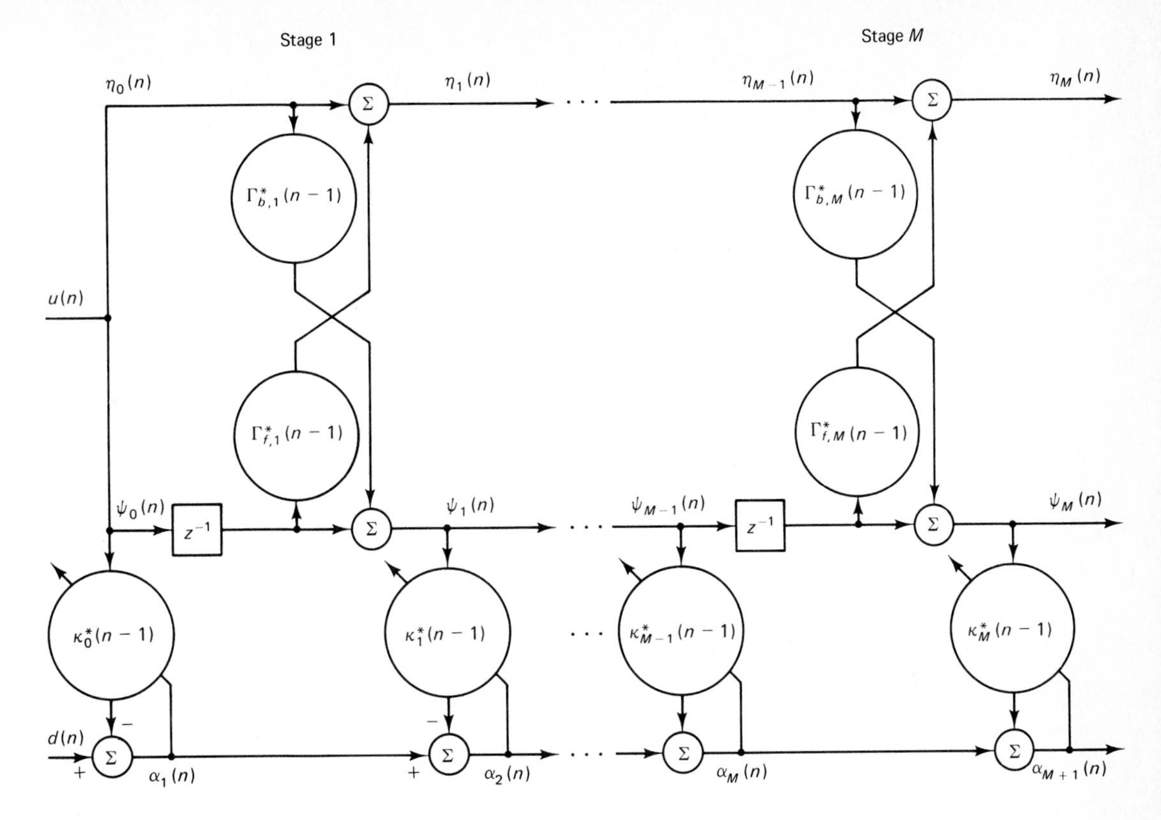

Figure 17.4 Joint-process estimator using recursive LSL algorithm based on a priori estimation errors.

and the sums of weighted error squares. When the recursive LSL algorithm is implemented by using fixed-point arithmetic with a short word length, these quantities cannot be computed accurately, thereby resulting in numerical inaccuracy in the application of recursive LSL algorithms. We may overcome the cause of this numerical inaccuracy by modifying the recursive LSL algorithms in such a way that we update the reflection and regression coefficients *directly* (Ling et al., 1985).

Below, we describe a modification of the second version of the recursive LSL algorithm (based on a priori estimation errors) that bypasses the need for computing cross-correlations. For the corresponding modification of the first version of the recursive LSL algorithm (based on a posteriori estimation errors), the reader is referred to Problem 10.

Direct Time Updates of Reflection and Regression Coefficients Using a priori Errors

Consider first the direct update of the forward reflection coefficient $\Gamma_{f,m}(n)$. Substituting line 3 into line 6 of the recursive LSL algorithm summarized in Table 17.2, we get

$$\Gamma_{f,m}(n) = -\frac{\lambda \Delta_{m-1}(n-1)}{\mathcal{B}_{m-1}(n-1)} - \frac{\gamma_{m-1}(n-1)\psi_{m-1}(n-1)\eta^*_{m-1}(n)}{\mathcal{B}_{m-1}(n-1)}$$

$$= -\frac{\Delta_{m-1}(n-1)\lambda \mathcal{B}_{m-1}(n-2)}{\mathcal{B}_{m-1}(n-2)\mathcal{B}_{m-1}(n-1)} - \frac{\gamma_{m-1}(n-1)\psi_{m-1}(n-1)\eta^*_{m-1}(n)}{\mathcal{B}_{m-1}(n-1)} \qquad (17.106)$$

We now recognize that, by definition,

$$\Gamma_{f,m}(n-1) = -\frac{\Delta_{m-1}(n-1)}{\mathcal{B}_{m-1}(n-2)}$$

Replacing n with $n - 1$ in Eq. (17.102) and rearranging, we get

$$\lambda \mathcal{B}_{m-1}(n-2) = \mathcal{B}_{m-1}(n-1) - \gamma_{m-1}(n-1)|\psi_{m-1}(n-1)|^2$$

Accordingly, we may rewrite Eq. (17.106) as

$$\Gamma_{f,m}(n) = \Gamma_{f,m}(n-1) - \frac{\gamma_{m-1}(n-1)\psi_{m-1}(n-1)\eta^*_m(n)}{\mathcal{B}_{m-1}(n-1)}, \quad m = 1, 2, \ldots, M \qquad (17.107)$$

where we have made use of Eq. (17.103).

Similarly, we may derive the following recursions for directly updating the backward reflection coefficient $\Gamma_{b,m}(n)$ and the regression coefficient $\kappa_m(n)$, respectively:

$$\Gamma_{b,m}(n) = \Gamma_{b,m}(n-1) - \frac{\gamma_{m-1}(n-1)\eta_{m-1}(n)\psi^*_m(n)}{\mathcal{F}_{m-1}(n)}, \qquad m = 1, 2, \ldots, M \qquad (17.108)$$

and

$$\kappa_m(n) = \kappa_m(n-1) + \frac{\gamma_m(n)\psi_m(n)\alpha^*_{m+1}(n)}{\mathcal{B}_m(n)}, \qquad m = 0, 1, \ldots, M \qquad (17.109)$$

A modified form of the recursive LSL algorithm based on a priori estimation errors is obtained by deleting lines 3 and 9 from Table 17.2 and then replacing lines 6, 7, and 12 by Eqs. (17.107), (17.108), and (17.109), respectively. The resulting algorithm is summarized in Table 17.3.

A distinctive feature of this new algorithm is that the a priori errors $\eta_m(n)$, $\psi_m(n)$, and $\alpha_{m+1}(n)$ are *fed back* to time-update the forward reflection coefficient $\Gamma_{f,m}(n)$, the backward reflection coefficient $\Gamma_{b,m}(n)$, and the regression coefficient $\kappa_m(n)$, respectively. This *error-feedback* mechanism is illustrated in Fig. 17.5 for the case of the forward reflection coefficient. Accordingly, this method of updating the reflection and regression coefficients is called the *error-feedback formula*.

The error-feedback form of the recursive LSL algorithm (based on a priori estimation errors) does not provide a computational advantage over the conventional

TABLE 17.3 SUMMARY OF THE RECURSIVE LSL ALGORITHM USING A PRIORI ESTIMATION ERRORS WITH ERROR FEEDBACK

Predictions:
For $n = 1, 2, 3, \ldots,$ compute the various order updates in the following sequence $m = 1, 2, \ldots, M$, where M is the final order of the least-squares predictor:

$$\eta_m(n) = \eta_{m-1}(n) + \Gamma^*_{f,m}(n-1)\psi_{m-1}(n-1)$$

$$\psi_m(n) = \psi_{m-1}(n-1) + \Gamma^*_{b,m}(n-1)\eta_{m-1}(n)$$

$$\mathcal{F}_{m-1}(n) = \lambda\mathcal{F}_{m-1}(n-1) + \gamma_{m-1}(n-1)|\eta_{m-1}(n)|^2$$

$$\mathcal{B}_{m-1}(n) = \lambda\mathcal{B}_{m-1}(n-1) + \gamma_{m-1}(n-1)|\psi_{m-1}(n)|^2$$

$$\Gamma_{f,m}(n) = \Gamma_{f,m}(n-1) - \frac{\gamma_{m-1}(n-1)\psi_{m-1}(n-1)\eta^*_m(n)}{\mathcal{B}_{m-1}(n-1)}$$

$$\Gamma_{b,m}(n) = \Gamma_{b,m}(n-1) - \frac{\gamma_{m-1}(n-1)\eta_{m-1}(n)\psi^*_m(n)}{\mathcal{F}_{m-1}(n)}$$

$$\gamma_m(n) = \gamma_{m-1}(n) - \frac{\gamma^2_{m-1}(n)|\psi_{m-1}(n)|^2}{\mathcal{B}_{m-1}(n)}$$

Filtering:
For $n = 1, 2, 3, \ldots,$ compute the various order updates in the following sequence $m = 0, 1, \ldots, M$:

$$\alpha_{m+1}(n) = \alpha_m(n) - \kappa^*_m(n-1)\psi_m(n)$$

$$\mathcal{B}_m(n) = \lambda\mathcal{B}_m(n-1) + \gamma_m(n)|\psi_m(n)|^2$$

$$\kappa_m(n) = \kappa_m(n-1) + \frac{\gamma_m(n)\psi_m(n)\alpha^*_{m+1}(n)}{\mathcal{B}_m(n)}$$

Initialization

1. To initialize the algorithm, at time $n = 0$ set
   $\mathcal{F}_{m-1}(0) = \delta, \qquad \delta = \text{small positive constant}$
   $\mathcal{B}_{m-1}(0) = \delta$
   $\Gamma_{f,m}(0) = \Gamma_{b,m}(0) = 0$
   $\gamma_0(0) = 1$
2. At each instant $n \geq 1$, generate the zeroth-order variables:
   $\eta_0(n) = \psi_0(n) = u(n)$
   $\mathcal{F}_0(n) = \mathcal{B}_0(n) = \lambda\mathcal{F}_0(n-1) + |u(n)|^2$
   $\gamma_0(n) = 1$
3. For joint process estimation, at time $n = 0$ set
   $\kappa_m(0) = 0$
   At each instant $n \geq 1$, generate the zeroth-order variable
   $\alpha_0(n) = d(n)$

Note: For prewindowed data, the input $u(n)$ and desired response $d(n)$ are both zero for $n \leq 0$.

form of the algorithm. Rather, it exhibits better numerical accuracy (Ling et al., 1985). More will be said on this issue in Chapter 19.

The error-feedback version of the recursive LSL algorithm summarized in Table 17.3 assumes the use of a priori forms of the prediction and estimation errors. The corresponding version of the algorithm, based on a posteriori prediction and estimation errors, is presented to the reader as Problem 10.

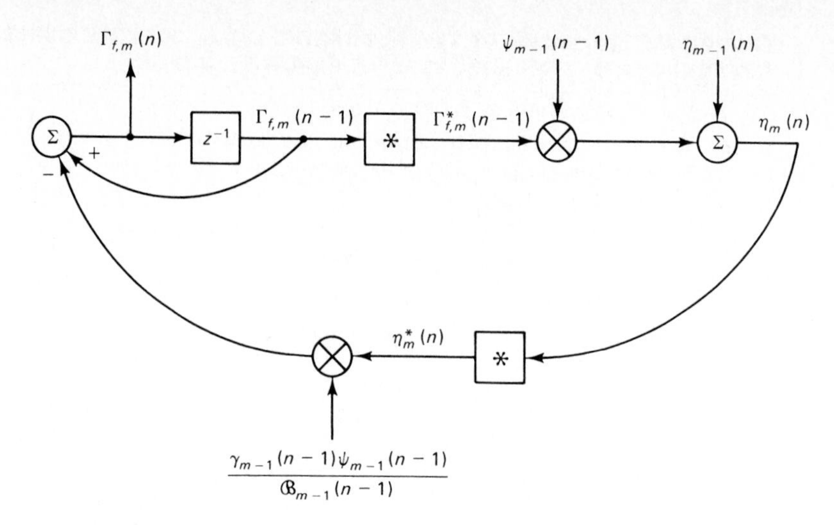

Figure 17.5 Error feedback mechanism involved in the time-updating of the forward reflection coefficient.

17.9 COMPUTER EXPERIMENT ON ADAPTIVE EQUALIZATION USING THE RECURSIVE LSL ALGORITHM

In this computer experiment, we study the use of the recursive LSL algorithm for *adaptive equalization* of a linear channel that produces unknown distortion. The parameters of the channel are the same as those used to study the RLS algorithm in Section 13.8 for a similar application. The results of the experiment should therefore help us make an assessment of the performance of the recursive LSL algorithm compared to the RLS algorithm. The version of the LSL algorithm used in the experiment is that based on a priori prediction and estimation errors with error feedback, as summarized in Table 17.3.

The parameters of the recursive LSL algorithm are identical to those used for the RLS algorithm in Section 13.8:

Exponential weighting factor:	$\lambda = 1$ (for stationary data)
Prediction order:	$M = 10$
Number of equalizer taps:	$M + 1 = 11$
Initializing constant:	$\delta = 0.004$

The computer simulations were run for four different values of the channel parameter W defined in Eq. (9.136), namely $W = 2.9$, 3.1, 3.3, and 3.5. These values of W correspond to the following eigenvalue spreads of the underlying correlation matrix $\mathbf{R}$ of the channel output (equalizer input): $\chi(\mathbf{R}) = 6.0782$, 11.1238, 21.7132, 46.8216. The signal-to-noise ratio measured at the channel output was 30 dB. For more details of the experimental setup, the reader is referred to Sections 13.8 and 9.16.

In Fig. 17.6 we present the superposition of the learning curves of the recursive LSL algorithm for varying W. Each learning curve was computed by ensemble averaging the squared value of the final a priori estimation error (i.e., the innovation) $\alpha_{M+1}(n)$ over 200 independent trials of the experiment. This variable is computed as a natural product of the joint-process estimator in Fig. 17.4 (see also equation 8 of Table 17.3). The reason for basing the computation of the learning curves on $\alpha_{M+1}(n)$ is that we would like to put the learning curves of the recursive LSL algorithm on a consistent basis with those for the RLS algorithm.

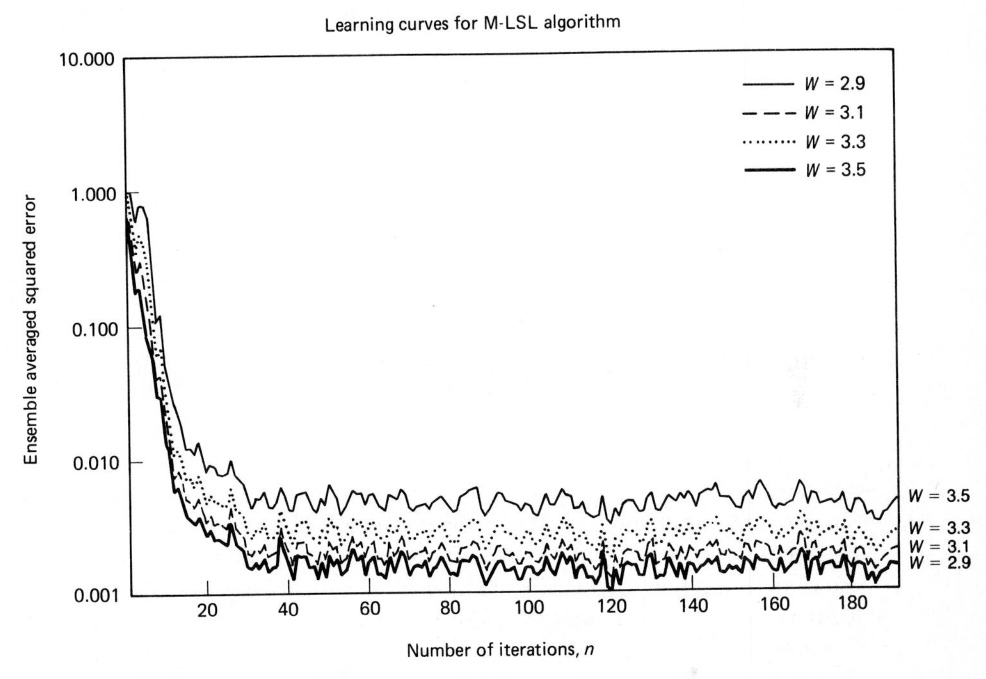

Figure 17.6 Learning curves for recursive LSL algorithm for varying eigenvalue spread.

In Fig. 17.7 we show the superposition of four ensemble-averaged plots of the conversion factor $\gamma_{M+1}(n)$ (for the final stage) versus time n, corresponding to the four different values of the eigenvalue spread $\chi(\mathbf{R})$ as defined above. The curves plotted here are obtained by ensemble-averaging $\gamma_{M+1}(n)$ over 200 independent trials of the experiment. It is noteworthy that the time variation of this ensemble-averaged conversion factor $E[\gamma_M(n)]$, follows an *inverse* law, as shown by

$$E[\gamma_m(n)] \simeq 1 - \frac{m}{n}, \qquad \text{for } m = 1, 2, \ldots, M+1, \quad n \geq m \tag{17.110}$$

Equation (17.110) provides a good fit to the experimentally computed curve shown in Fig. 17.7, particularly for time n large compared to the predictor order $m = M + 1$. The reader is invited to check the validity of this fit. Note that the experimental plots of the conversion factor $\gamma_{M+1}(n)$ are insensitive to variations in the eigenvalue spread of the correlation matrix of the lattice input.

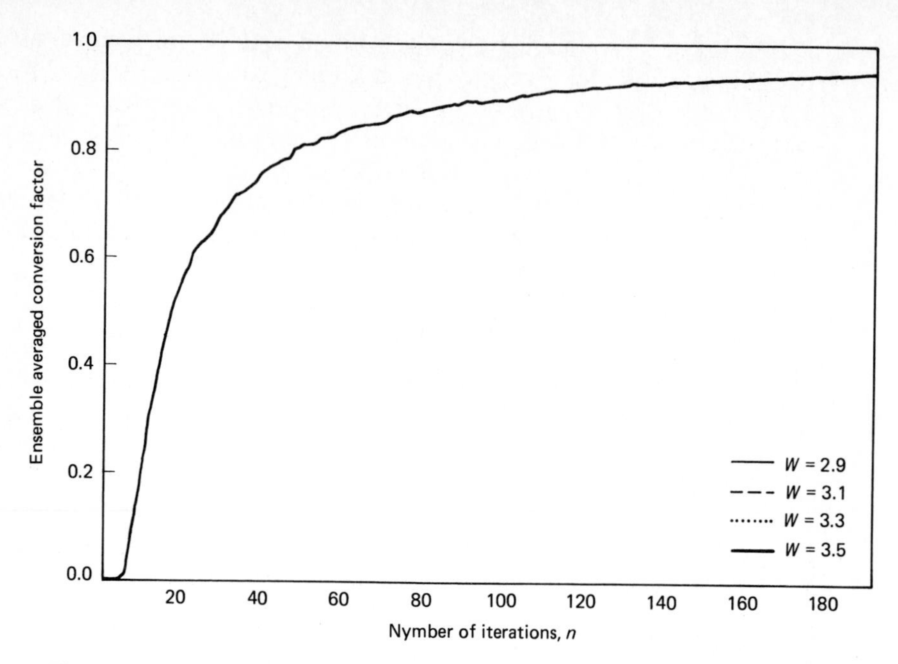

Figure 17.7 Ensemble-averaged conversion factor $\gamma_{M+1}(n)$ for varying eigenvalue spread.

In Table 17.4(a) we present a summary of calculations pertaining to steady-state values of the mean (ensemble-averaged) squared error for the RLS and the recursive LSL algorithms as well as the minimum achievable values determined by the Wiener filter theory. Table 17.4(b) shows the corresponding values of the misadjustment for these two algorithms, defined as the ratio of the excess mean-squared error to the minimum mean-squared error.

TABLE 17.4 SUMMARY OF RESULTS OF THE COMPUTER EXPERIMENT ON ADAPTIVE EQUALIZATION

(a) Steady-state mean-squared error

Eigenvalue spread	Wiener filter	RLS algorithm	Recursive LSL algorithm
6.0782	1.3755×10^{-3}	1.4397×10^{-3}	1.4397×10^{-3}
11.1238	1.7524×10^{-3}	1.8296×10^{-3}	1.8295×10^{-3}
21.7132	2.4739×10^{-3}	2.5754×10^{-3}	2.5753×10^{-3}
46.8216	4.1559×10^{-3}	4.3094×10^{-3}	4.3094×10^{-3}

(b) Misadjustment (percentage)

Eigenvalue spread	RLS algorithm	Recursive LSL algorithm
6.0782	4.67	4.67
11.1238	4.41	4.40
21.7132	4.10	4.10
46.8216	3.69	3.69

Based on the results of this experiment and those of Section 13.8 involving the RLS algorithm, we may state that the convergence behavior of the recursive LSL algorithm is basically the same as that of the RLS algorithm, assuming the use of infinite-precision arithmetic. In particular, we may make the following observations:

1. The recursive LSL algorithm converges very rapidly, requiring only about 22 iterations (i.e., twice the number of adjustable parameters in the equalizer) to reach steady state.
2. The conversion factor $\gamma_{M+1}(n)$ starts with an initial value of zero and remains quite small during the early part of the initialization period that lasts for 11 iterations (equal to the number of adjustable parameters. Thereafter, γ_{M+1} begins to increase rapidly toward a final value of unity.
3. The rate of convergence of the recursive LSL algorithm is relatively insensitive to variations in the eigenvalue spread of the underlying correlation matrix of the channel output.
4. Variation in the final value of the mean-squared error with the eigenvalue spread $\chi(\mathbf{R})$ is an inherent characteristic of the Wiener filter.

17.10 DISCUSSION

A recursive least-squares lattice (LSL) algorithm is a *joint estimator* in the sense that it provides for the estimation of two sets of filtering coefficients jointly:

1. *Forward and backward reflection coefficients* that characterize a multistage lattice predictor optimized in the least-squares sense. The number of stages in the predictor equals the prediction order.
2. *Regression coefficients* that characterize a linear least-squares estimator of some desired response.

The predictor, in effect, performs a *Gram–Schmidt orthogonalization* of the input data by generating a set of backward prediction errors that are *orthogonal* to each other in a time-averaged sense, yet bear a *one-to-one correspondence* with the original data. It is this orthogonalization process that gives a recursive LSL algorithm many of its desirable features.

A recursive LSL algorithm may be structured in four different forms, depending on the type of prediction and estimation errors used as variables, and the manner in which the reflection coefficients are computed. The four versions of the algorithm are distinguished as follows:

1. In version I, summarized in Table 17.1, the variables are *a posteriori* forms of prediction and estimation errors, and the reflection coefficients are computed *indirectly*.
2. In version II, summarized in Table 17.2, the variables are *a priori* forms of prediction and estimation errors, and the reflection coefficients are again computed indirectly.

3. In version III, summarized in Table 17.3, the variables are (as in version II) *a priori* forms of prediction and estimation errors, but the reflection and regression coefficients are all computed *directly*. As a result of this direct computation, *error feedback* is introduced into the operation of the algorithm.
4. In version IV, the variables are (as in version I) *a posteriori* forms of prediction and estimation errors, but the reflection and regression coefficients are all computed *directly* (as in version III). Here again, error feedback is introduced into the operation of the algorithm. This fourth and final version of the recursive LSL algorithm is presented as Problem 10.

In theory, assuming the use of infinite precision, all four versions of the recursive LSL algorithm are *mathematically equivalent*. However, in a practical situation involving the use of finite-precision arithmetic, the four versions behave differently. In particular, versions I and II suffer from a *numerical instability* problem due to finite-precision effects. On the other hand, versions III and IV offer robust numerical properties due to the *stabilizing* influence of the error feedback built into the computation of the forward and backward reflection coefficients. This issue is considered in Chapter 19.

An important property of all recursive LSL algorithms is their *modular* structure, as exemplified by the signal-flow graphs of Figs. 17.2 and 17.4. The implication of this property is that the algorithmic structure is *linearly scalable*. In particular, the prediction order can be readily increased without the need to recalculate all previous values. This property is particularly useful when there is no prior knowledge as to what the final value of the prediction order should be.

Another implication of the modular structure of recursive LSL algorithms is that they lend themselves to the use of *very large scale integration* (VLSI) for their hardware implementation. Of course, the use of this sophisticated technology can only be justified if the application of interest calls for the use of the VLSI chip in large numbers.

As with the FTF algorithm, a recursive LSL algorithm supplies useful by-products, namely, forward and backward prediction errors, and the conversion factor. However, in the FTF algorithm these by-products are computed for a fixed prediction order, whereas in a recursive LSL algorithm they are computed for varying prediction order (inside some preset range, determined by the final prediction order).

It is also noteworthy that the feature that distinguishes a recursive LSL algorithm from the two-reflection coefficient version of the GAL algorithm (Griffiths, 1978) is the significant role played by the conversion factor $\gamma_{m-1}(n-1)$. Indeed, if we were to nullify this effect by putting $\gamma_{m-1}(n-1)$ equal to 1 for all m and all n, the LSL algorithm reduces to the same basic form as this particular form of the GAL algorithm (see Problem 11). Consequently, in the latter algorithm, no distinction is made between the a priori and a posteriori values of prediction errors. This is intuitively satisfying, because a GAL algorithm is devised assuming wide-sense stationarity to begin with, whereas in a recursive LSL algorithm no such assumption is made.

PROBLEMS

1. The correlation matrix $\mathbf{\Phi}_{m+1}(n)$ is postmultiplied by the Hermitian transpose of the lower triangular matrix $\mathbf{L}_m(n)$, where $\mathbf{L}_m(n)$ is defined by Eq. (17.68). Show that the product $\mathbf{\Phi}_{m+1}(n)\mathbf{L}_m^H(n)$ consists of a lower triangular matrix whose diagonal elements equal the various sums of weighted backward prediction-error squares, $\mathcal{B}_0(n), \mathcal{B}_1(n), \ldots, \mathcal{B}_m(n)$. Hence, show that the product $\mathbf{L}_m(n)\mathbf{\Phi}_{m+1}(n)\mathbf{L}_m^H(n)$ is a diagonal matrix, as shown by

$$\mathbf{D}_{m+1}(n) = \text{diag}[\mathcal{B}_0(n), \mathcal{B}_1(n), \ldots, \mathcal{B}_m(n)]$$

2. Use the Gram–Schmidt orthogonalization procedure to transform the input sequence $u(n), \ldots, u(n-m)$ into a new sequence $b_0(n), \ldots, b_m(n)$ whose elements are uncorrelated with each other in a time-averaged sense.
3. Consider the cost function

$$\mathscr{E}_m(n) = \sum_{i=1}^{n} \lambda^{n-i} |d(i) - \mathbf{b}_{m+1}^T(i)\boldsymbol{\kappa}_m^*(n)|^2, \qquad m = 1, 2, \ldots, M$$

where λ is the exponential factor, $d(i)$ is the desired response, $\mathbf{b}_{m+1}(i)$ is the vector of backward a posteriori prediction errors, and $\boldsymbol{\kappa}_m(n)$ is the corresponding vector of regression coefficients that are fed by the backward prediction errors (see Fig. 17.3). Hence, show that $\mathscr{E}_m(n)$ is minimized by choosing

$$\mathbf{D}_{m+1}(n)\boldsymbol{\kappa}_m(n) = \mathbf{t}_{m+1}(n)$$

where $\mathbf{D}_{m+1}(n)$ is the correlation matrix of the backward a posteriori prediction errors, and $\mathbf{t}_{m+1}(n)$ is the cross-correlation vector between the backward a posteriori prediction errors and the desired response.
4. Consider the case where the input samples $u(n), u(n-1), \ldots, u(n-M)$ have a *joint Gaussian distribution with zero mean*. Assume that, within a scaling factor, the ensemble-averaged correlation matrix $\mathbf{R}_{M+1}$ of the input signal is equal to its time-averaged correlation matrix $\mathbf{\Phi}_{M+1}(n)$ for time $n \geq M$. Show that the log-likelihood function for this input includes a term equal to the parameter $\gamma_M(n)$ associated with the recursive LSL algorithm. For this reason, the parameter $\gamma_M(n)$ is sometimes referred to as a *likelihood variable*.
5. In the adaptive lattice equalizer considered in Section 17.8 the impulse response of the channel is symmetric with respect to the midpoint $k = 2$. Would you expect the regression coefficients in this equalizer to be likewise symmetric with respect to its midpoint? Justify your answer.
6. Let $\hat{d}(n|\mathcal{U}_{n-m+1})$ denote the least-squares estimate of the desired response $d(n)$, given the inputs $u(n-m+1), \ldots, u(n)$ that span the space $\mathcal{U}_{n-m+1}$. Similarly, let $\hat{d}(n|\mathcal{U}_{n-m})$ denote the least-squares estimate of the desired response, given the inputs $u(n-m)$, $u(n-m+1), \ldots, u(n)$ that span the space $\mathcal{U}_{n-m}$. In effect, the latter estimate exploits an additional piece of information represented by the input $u(n-m)$. Show that this new information is represented by the corresponding backward prediction error $b_m(n)$. Also, show that the two estimates are related by the recursion

$$\hat{d}(n|\mathcal{U}_{n-m}) = \hat{d}(n|\mathcal{U}_{n-m+1}) + \kappa_m^*(n)b_m(n)$$

where $\kappa_m(n)$ denotes the pertinent regression coefficient in the joint-process estimator. Compare this result with that of Section 7.1 dealing with the concept of innovations.

7. Let $\mathbf{\Phi}(n)$ denote the $(M + 1)$-by-$(M + 1)$ correlation matrix of the input data $\{u(n)\}$. Using the theory presented in Section 17.7, show that the change of variables to backward prediction errors brought about by using the lattice predictor achieves exactly the Cholesky decomposition of the matrix $\mathbf{P}(n) = \mathbf{\Phi}^{-1}(n)$.

8. Consider the joint-process estimator of Fig. 17.4. Show that the last transformed regression coefficient $\kappa_M(n)$ equals the last element $\hat{w}_M(n)$ contained in the $(M + 1)$-by-1 least-squares estimate $\hat{\mathbf{w}}_M(n)$ of the coefficient vector in the multiple linear regression model.

9. Expand the joint-process estimator of Fig. 17.4 so as to include (in modular form) the least-squares estimate of the desired response $d(n)$ for increasing prediction order m.

10. In Section 17.8 we discussed a modification of the a priori error LSL algorithm by using a form of error feedback. In this problem we consider the corresponding modified version of the a posteriori LSL algorithm. In particular, show that

$$\Gamma_{f,m}(n) = \frac{\gamma_m(n-1)}{\gamma_{m-1}(n-1)}\left[\Gamma_{f,m}(n-1) - \frac{1}{\lambda}\frac{b_{m-1}(n-1)f^*_{m-1}(n)}{\mathcal{B}_{m-1}(n-2)\gamma_{m-1}(n-1)}\right]$$

$$\Gamma_{b,m} = \frac{\gamma_m(n)}{\gamma_{m-1}(n-1)}\left[\Gamma_{b,m}(n-1) - \frac{1}{\lambda}\frac{f_{m-1}(n)b^*_{m-1}(n-1)}{\mathcal{F}_{m-1}(n-1)\gamma_{m-1}(n-1)}\right]$$

11. The two-coefficient version of the gradient adaptive lattice (GAL) algorithm may be viewed as a special case of the recursive LSL algorithm with error feedback, which results when the conversion factor is unity for all time n and all predictor order m. Using the summary presented in Table 17.3, deduce the formulation of this version of the GAL algorithm.

12. The accompanying table is a summary of the *normalized LSL algorithm*. The normalized parameters are defined by

$$\bar{f}_m(n) = \frac{f_m(n)}{\mathcal{F}_m^{1/2}(n)\gamma_m^{1/2}(n-1)}$$

$$\bar{b}_m(n) = \frac{b_m(n)}{\mathcal{B}_m^{1/2}(n)\gamma_m^{1/2}(n)}$$

$$\bar{\Delta}_m(n) = \frac{\Delta_m(n)}{\mathcal{F}_m^{1/2}(n)\mathcal{B}_m^{1/2}(n-1)}$$

Hence, derive the steps summarized in the table.

$$\bar{\Delta}_{m-1}(n) = \bar{\Delta}_{m-1}(n-1)[1 - |\bar{f}_{m-1}(n)|^2]^{1/2}[1 - |\bar{b}_{m-1}(n-1)|^2]^{1/2} + \bar{b}_{m-1}(n-1)\bar{f}^*_{m-1}(n)$$

$$\bar{b}_m(n) = \frac{\bar{b}_{m-1}(n-1) - \bar{\Delta}_{m-1}(n)\bar{f}_{m-1}(n)}{[1 - |\bar{\Delta}_{m-1}(n)|^2]^{1/2}[1 - |\bar{f}_{m-1}(n)|^2]^{1/2}}$$

$$\bar{f}_m(n) = \frac{\bar{f}_{m-1}(n) - \bar{\Delta}^*_{m-1}(n)\bar{b}_{m-1}(n-1)}{[1 - |\bar{\Delta}_{m-1}(n)|^2]^{1/2}[1 - |\bar{b}_{m-1}(n-1)|^2]^{1/2}}$$

Computer-Oriented Problems

13. *Computer experiment with two-stage lattice predictor using autoregressive process of order 2*. Consider an autoregressive (AR) process $\{u(n)\}$ of order 2, described by the difference equation

$$u(n) + a_1 u(n-1) + a_2 u(n-2) = v(n)$$

where the AR coefficients a_1 and a_2, and the variance σ_v^2 of the white noise process $\{v(n)\}$ have the values given in the accompanying table. The AR process $\{u(n)\}$ is applied to a two-stage lattice predictor.

Use a computer to generate a 256-sample sequence to represent the AR process $\{u(n)\}$. You may use a random-number generator for the noise $\{v(n)\}$. Use version III of the recursive LSL algorithm based on a priori prediction errors with error feedback to perform the following computations for each of the three parameter sets listed in the table:

(a) Ensemble-averaged learning curves.

(b) Ensemble-averaged conversion factors for varying n.

In both experiments, do the averaging over 100 independent trials, and plot your results for $0 \le n \le 200$. Assume that the exponential weighting factor $\lambda = 1$.

For each parameter set, compute the misadjustment produced by the LSL algorithm by time averaging over the last 50 values of the ensemble-averaged squared error.

Parameter set	a_1	a_2	σ_v^2
1	−0.02	−0.8	0.3564
2	−0.1636	−0.8	0.1191
3	−0.1960	−0.8	0.01462

14. *Computer experiment on adaptive equalization.* Repeat the experiment described in Section 17.9 on adaptive equalization. This time, however, use version I of the recursive LSL algorithm based on a priori prediction and estimation errors and indirect computation of the reflection coefficients, as summarized in Table 17.2. In particular, perform the following computations:

(a) Ensemble-averaged learning curves.

(b) Single realizations of the conversion factor for varying time n.

(c) Ensemble-averaged conversion factor for varying time n.

For parts (a) and (c) do the ensemble averaging over 200 independent trials of the experiment. Your plots should cover the interval [1, 100] for time n.

All the parameters for the channel and the algorithms are as detailed in Section 17.9.

18

QR-Decomposition-Based Least-Squares Lattice Algorithm

In Chapters 16 and 17 we presented two classes of fast algorithms for solving the recursive least-squares (RLS) estimation problem. Unfortunately, the fast transversal filters (FTF) algorithm (covered in Chapter 16) and two versions of the recursive least-squares lattice (LSL) algorithm (covered in Chapter 17) suffer from some form of numerical instability due to finite-precision effects. The two versions of the recursive LSL algorithm that are numerically stable do so by using error feedback that helps in their stabilization. In this chapter we describe a *QR-decomposition-based least-squares lattice (QRD-LSL) algorithm*[1] that combines the good numerical properties of QR-decomposition and the desirable features of a recursive least-squares lattice. Indeed, this new algorithm is closely related to conventional LSL algorithms with error feedback.

[1] The idea of a fast QR-decomposition-based algorithm for recursive least-squares estimation was first presented by Cioffi (1988). A detailed derivation of the algorithm is described in Cioffi (1990). In the latter paper, Cioffi presents a geometric approach for the derivation that is reminiscent of his earlier work on fast transversal filters. The algorithm derived by Cioffi is of a Kalman or matrix-oriented type. Several other authors have presented seemingly simple algebraic derivations and other versions of the QRD-fast RLS algorithm (Bellanger, 1988; Proudler et al., 1988, 1989; Regalia and Bellanger, 1991). The paper by Proudler et al. (1989) is of particular interest in that it develops a novel implementation of the QRD-RLS algorithm using a lattice structure. A similar fast algorithm has been derived independently by Ling (1989) using the modified Gram–Schmidt orthogonalization procedure. The connection between the modified Gram–Schmidt orthogonalization and QR-decomposition is discussed in Shepherd and McWhirter (1991).

The development of the QRD-LSL algorithm presented in this chapter is based on a hybridization of ideas due to Proudler et al. (1989) and Regalia and Bellanger (1991). Specifically, we follow Proudler et al. in deriving QR decomposition-based solutions to forward and backward linear prediction problems. We then follow Regalia and Bellanger in solving the joint-process estimation problem. By so doing, we avoid the complications in the procedure by Proudler et al. that uses forward linear prediction errors for joint-process estimation. The structure of the QRD-LSL algorithm derived in this chapter follows a philosophy directly analogous to that described in Chapter 17 for conventional LSL algorithms.

Whereas the recursive QR-decomposition-based recursive least-squares (QRD-RLS) algorithm described in Chapter 14 requires a high computational load on the order of M^2 in terms of both the number of processing cells and computation per iteration, the lattice-based implementation of it described in this chapter is "fast" in the sense that these numbers are reduced to a linear dependence on M, where M is the available number of degrees of freedom (i.e., model order). As with the FTF and recursive LSL algorithms, the QRD-LSL algorithm exploits the shifting property of serialized input data (i.e., the Toeplitz structure of the data matrix) to perform joint-process estimation in a "fast" manner.

From Chapter 14, we recall that the rotation parameters for recursive computation of the QR-decomposition are based solely on elements of the data matrix. Least-squares linear prediction, be it of the forward or backward type, is also determined solely by the data matrix. It is therefore appropriate that we begin our discussion by reviewing the structure of the data matrix and the different ways of partitioning it with QR-decomposition in mind.

18.1 STRUCTURE AND PARTITIONING OF THE DATA MATRIX

Consider the time series $u(1), u(2), \ldots, u(n)$ that occupies the time interval $1 \le i \le n$. We assume the use of *pre-windowing* the data, which means that $u(i)$ is zero for $i \le 0$. Suppose that this time series is to be used in a linear least-squares estimation problem for which we write the n-by-$(m+1)$ *data matrix* as

$$\mathbf{A}_{m+1}(n) = \begin{bmatrix} u^*(1) & 0 & \cdots & 0 & 0 \\ u^*(2) & u^*(1) & \cdots & 0 & 0 \\ \vdots & \vdots & \ddots & \vdots & \vdots \\ u^*(n-1) & u^*(n-2) & \cdots & u^*(n-m) & u^*(n-m-1) \\ u^*(n) & u^*(n-1) & \cdots & u^*(n-m+1) & u^*(n-m) \end{bmatrix} \tag{18.1}$$

The subscript $m+1$ in the symbol for the data matrix $\mathbf{A}_{m+1}(n)$ signifies the number of columns.

Let the data matrix $\mathbf{A}_{m+1}(n)$ be partitioned as follows:

$$\mathbf{A}_{m+1}(n) = \left[\begin{array}{c|c|c} u^*(1) & \mathbf{0}^T & 0 \\ \hline \mathbf{d}_{f,m-1}(n-1) & \mathbf{A}_{m-1}(n-2) & \mathbf{d}_{b,m-1}(n-2) \\ \hline u^*(n) & \mathbf{u}_{m-1}^H(n-1) & u^*(n-m) \end{array}\right] \tag{18.2}$$

where

$$\mathbf{A}_{m-1}(n-2) = \begin{bmatrix} u^*(1) & 0 & \cdots & 0 \\ u^*(2) & u^*(1) & \cdots & 0 \\ \vdots & \vdots & \ddots & \vdots \\ u^*(n-2) & u^*(n-1) & \cdots & u^*(n-m) \end{bmatrix} \tag{18.3}$$

$$\mathbf{u}_{m-1}(n-1) = [u(n-1), u(n-2), \ldots, u(n-m+1)]^T \tag{18.4}$$

$$\mathbf{d}_{f,m-1}(n-1) = [u^*(2), u^*(3), \ldots, u^*(n-1)]^T \tag{18.5}$$

$$\mathbf{d}_{b,m-1}(n-2) = [0, \ldots, u^*(n-m-2), u^*(n-m-1)]^T \tag{18.6}$$

The subscript $m-1$ in both the vector $\mathbf{d}_{f,m-1}(n-1)$ and $\mathbf{d}_{b,m-1}(n-2)$ refers to the *model* or *parediction order*.

Depending on the form of linear prediction of interest, we may identify two distinct situations in the context of the situation described herein:

1. *Forward linear prediction*. The requirement here is to make a linear least-squares estimate of the present sample $u(n)$, given the input data vector $\mathbf{u}_{m-1}(n-1)$, where $m-1$ is the prediction order. For this situation, we may focus on the n-by-m data matrix

$$\mathbf{A}_m(n) = \begin{bmatrix} u^*(1) & \mathbf{0}^T \\ \hdashline \mathbf{d}_{f,m-1}(n-1) & \mathbf{A}_{m-1}(n-2) \\ u^*(n) & \mathbf{u}_{m-1}^H(n-1) \end{bmatrix} \tag{18.7}$$

   which is obtained from the matrix $\mathbf{A}_{m+1}(n)$ of Eq. (18.2) by deleting the last column.

2. *Backward linear prediction*. In this second form of linear prediction, the requirement is to make a linear least-squares estimate of the past sample $u(m-m)$ given the input data vector $\mathbf{u}_{m-1}(n-1)$, where $m-1$ is the prediction order as before. Accordingly, we may focus on the $(n-1)$-by-m data matrix.

$$\mathbf{A}_m(n-1) = \begin{bmatrix} \mathbf{A}_{m-1}(n-2) & \mathbf{d}_{b,m-1}(n-2) \\ \mathbf{u}_{m-1}^H(n-1) & u^*(n-m) \end{bmatrix} \tag{18.8}$$

   which is obtained from the matrix $\mathbf{A}_{m+1}(n)$ of Eq. (18.2) by deleting the first column and also ignoring the row of zeros at the top of the remaining matrix.

Comparing the data matrices of Eqs. (18.7) and (18.8), we see that the data matrix $\mathbf{A}_m(n-1)$ for backward linear prediction is *delayed* with respect to the data matrix $\mathbf{A}_m(n)$ for forward linear prediction by one time unit. This time delay works its way through the development of the QRD-LSL algorithm.

Note also that the matrix $\mathbf{A}_{m-1}(n-2)$ and the row vector $\mathbf{u}_{m-1}^H(n-1)$ are common to these two data matrices, as shown by

Data matrix $\mathbf{A}_m(n)$ for forward prediction

$$\mathbf{A}_{m+1}(n) = \begin{bmatrix} u^*(1) & \mathbf{0}^T & 0 \\ \mathbf{d}_{f,m-1}(n-1) & \mathbf{A}_{m-1}(n-2) & \mathbf{d}_{b,m-1}(n-2) \\ u^*(n) & \mathbf{u}_{m-1}^H(n-1) & u^*(n-m) \end{bmatrix} \tag{18.9}$$

Data matrix $\mathbf{A}_m(n-1)$ for backward prediction

The requirement we have to meet is to eliminate the portion of the data matrix $\mathbf{A}_{m+1}(n)$ identified in Eq. (18.9) as common to the forward and backward forms of linear prediction. The motivation for so doing is to formulate a fast and numerically stable algorithm for solving the linear least-squares estimation problem within the framework of the QR-decomposition procedure.

18.2 PROBLEM STATEMENT

In Section 14.2 we described a QR-decomposition-based algorithm for solving the recursive least-squares (adaptive filtering) problem. The essence of the algorithm may be summarized in mathematical terms as follows. Let an exponentially weighted version of the data matrix $\mathbf{A}_{m+1}(n)$ be factorized in the form

$$\mathbf{Q}_m(n)\mathbf{\Lambda}^{1/2}(n)\mathbf{A}_{m+1}(n) = \begin{bmatrix}\mathbf{R}_m(n)\\ \mathbf{0}\end{bmatrix} \tag{18.10}$$

where $\mathbf{Q}_m(n)$ is an n-by-n unitary matrix, and $\mathbf{R}_m(n)$ is an $(m+1)$-by-$(m+1)$ upper triangular matrix. The subscript m in these two matrices signifies the prediction (model) order applicable to the situation at hand. On the other hand, the subscript $m+1$ in the data matrix $\mathbf{A}_{m+1}(n)$ signifies the fact that this matrix has $(m+1)$ columns. The n-by-n exponential weighting matrix $\mathbf{\Lambda}(n)$ on the left side of Eq. (18.10) is a diagonal matrix, defined by

$$\mathbf{\Lambda}(n) = \text{diag}(\lambda^{n-1}, \lambda^{n-2}, \ldots, 1) \tag{18.11}$$

where λ is the *exponential weighting factor*. Let the n-by-1 vector $\mathbf{d}_{m+1}(n)$ denote the *desired data vector*, written as

$$\mathbf{d}_{m+1}(n) = [d^*(1), d^*(2), \ldots, d^*(n)]^T \tag{18.12}$$

where the subscript $m+1$ identifies this desired data vector as being associated with the data matrix $\mathbf{A}_{m+1}(n)$.

Applying the transformation $\mathbf{Q}_m(n)$ to a weighted version of $\mathbf{d}_{m+1}(n)$, we may write

$$\mathbf{Q}_m(n)\mathbf{\Lambda}^{1/2}(n)\mathbf{d}_{m+1}(n) = \begin{bmatrix}\mathbf{p}_m(n)\\ \mathbf{v}_m(n)\end{bmatrix} \tag{18.13}$$

where $\mathbf{p}_m(n)$ is a vector containing the top $m+1$ elements of the rotated vector $\mathbf{d}(n)$, and $\mathbf{v}_m(n)$ is a vector containing the remaining $n - m - 1$ elements.

The upper triangular matrix $\mathbf{R}_m(n)$, and the vectors $\mathbf{p}_m(n)$ and $\mathbf{v}_m(n)$ may be computed recursively as follows, respectively:

$$\begin{bmatrix}\mathbf{R}_m(n)\\ \text{------}\\ \mathbf{0}\end{bmatrix} = \mathbf{T}_m(n)\begin{bmatrix}\lambda^{1/2}\mathbf{R}_m(n-1)\\ \text{---------------}\\ \mathbf{O}\\ \mathbf{u}_{m+1}^H(n)\end{bmatrix} \tag{18.14}$$

and

$$\begin{bmatrix} \mathbf{p}_m(n) \\ \\ \mathbf{v}_m(n) \end{bmatrix} = \begin{bmatrix} \mathbf{p}_m(n) \\ \hdashline \lambda^{1/2}\mathbf{v}_m(n-1) \\ \epsilon_{m+1}^*(n) \end{bmatrix} = \mathbf{T}_m(n) \begin{bmatrix} \lambda^{1/2}\mathbf{p}_m(n-1) \\ \hdashline \lambda^{1/2}\mathbf{v}_m(n-1) \\ d^*(n) \end{bmatrix} \tag{18.15}$$

where the $(m + 1)$-by-1 input vector $\mathbf{u}_{m+1}(n)$ and the desired response $d(n)$ represent the "new" data received at time n, and the n-by-n unitary matrix $\mathbf{T}_m(n)$ represents the combined transformation produced by a sequence of $m + 1$ Givens rotations that are used to annihilate all $m + 1$ elements of the row vector $\mathbf{u}_{m+1}^H(n)$ in Eq. (18.14).

The following points are especially noteworthy:

1. The column vector $\lambda^{1/2}\mathbf{v}_m(n - 1)$ in Eq. (18.15) is left intact by the unitary matrix $\mathbf{T}_m(n)$.
2. The last element of $\mathbf{v}_m(n)$ is denoted by $\epsilon_{m+1}^*(n)$; the variable $\epsilon_{m+1}(n)$ is related to the a posteriori joint-process estimation error $e_{m+1}(n)$ by

$$\epsilon_{m+1}(n) = \frac{e_{m+1}(n)}{\gamma_{m+1}^{1/2}(n)} \tag{18.16}$$

where $\gamma_{m+1}(n)$ is the conversion factor described in Chapters 14 and 15. Equation (18.16) suggests that we may view $\epsilon_{m+1}(n)$ as an *angle-normalized joint-process estimation error* (Lee at al. (1981), Ling, 1989). The term "angle" variable refers to an interpretaion of the factor $\gamma_{m+1}(n)$ (see Chapter 15). Note that the magnitude of $\epsilon_{m+1}(n)$ may also be expressed as the *geometric mean* of the magnitudes of the a priori and a posteriori values of the joint-process estimation error, as shown by (see Problem 5)

$$|\epsilon_{m+1}(n)| = \sqrt{|e_{m+1}(n)| \cdot |\alpha_{m+1}(n)|} \tag{18.17}$$

where $\alpha_{m+1}(n)$ is the a priori joint-process estimation error (i.e., innovation). Note also that all three variables, $\epsilon_{m+1}(n)$, $e_{m+1}(n)$ and $\alpha_{m+1}(n)$ always have the same phase angle.
3. The minimum value of the weighted sum of error squares (i.e., cost function) is given by

$$\begin{aligned} \mathscr{E}_{\min}(n) &= \|\mathbf{v}_m(n)\|^2 \\ &= \lambda\mathscr{E}_{\min}(n-1) + |\epsilon_{m+1}(n)|^2 \end{aligned} \tag{18.18}$$

From the detailed discussion presented in Chapter 14, we recall that the implementation of the *QR-decomposition-based recursive least-squares* (QRD-RLS) *algorithm,* described by Eqs. (18.14) and (18.15), requires the use of a systolic processor that contains on the order of m^2 cells, and therefore a computation time that increases with m^2. For large m, these requirements can become highly demanding and therefore impractical; hence, the motivation for a "fast" realization of this important algorithm.

The purpose of the fast RLS algorithm described in this chapter is twofold: (1) to use a QR-decomposition-based *lattice* structure for computing the combined solution to the forward and backward forms of linear prediction, and (2) to use this structure as the basis for computing the a posteriori estimation error $e_{m+1}(n)$. Specifically, in Sections 18.3 and 18.4 we use the partitioned forms of the data matrix summarized in Eq. (18.9) to study the backward and forward forms of linear prediction, respectively. This supplies the mathematical tools needed to develop the QRD-LSL algorithm, which is carried out in Sections 18.5 and 18.6.

Notation for Subscripts. Before proceeding with the task of deriving the QRD-LSL algorithm, we wish to reiterate the convention to be followed in the context of subscript notations:

1. In the symbols $\mathbf{A}_{m+1}(n)$ for data matrix, $\mathbf{d}_{m+1}(n)$ for desired data vector, and $\epsilon_{m+1}(n)$ for angle-normalized joint-process estimation error, the subscript $m + 1$ refers to the *number of columns* contained in the data matrix. This rule also applies to $\mathbf{u}^H_{m+1}(n)$, the Hermitian transpose of the input data vector.
2. In the symbols for *all* other matrices, vectors, scalar parameters and variables, the subscript refers to the *prediction order* or *model order* applicable to the situation at hand.

18.3 ADAPTIVE BACKWARD LINEAR PREDICTION

Consider first the backward linear least-squares prediction problem of order $m - 1$. To proceed with a solution to this problem, we may triangularize the data matrix $\mathbf{A}_m(n - 1)$ of Eq. (18.8). In particular, we consider the exponentially weighted data matrix:

$$\mathbf{\Lambda}^{1/2}(n-1)\mathbf{A}_m(n-1) = \left[\begin{array}{c|c} \lambda^{1/2}\mathbf{\Lambda}^{1/2}(n-2) & \mathbf{0} \\ \hline \mathbf{0}^T & 1 \end{array}\right]\left[\begin{array}{c|c} \mathbf{A}_{m-1}(n-2) & \mathbf{d}_{b,m-1}(n-2) \\ \hline \mathbf{u}^H_{m-1}(n-1) & u^*(n-m) \end{array}\right]$$

$$= \left[\begin{array}{c|c} \lambda^{1/2}\mathbf{\Lambda}^{1/2}(n-2)\mathbf{A}_{m-1}(n-2) & \lambda^{1/2}\mathbf{\Lambda}^{1/2}(n-2)\mathbf{d}_{b,m-1}(n-2) \\ \hline \mathbf{u}^H_{m-1}(n-1) & u^*(n-m) \end{array}\right] \tag{18.19}$$

Correspondingly, we note that the QR-decomposition of the weighted submatrix $\Lambda^{1/2}(n - 2)\mathbf{A}_{m-1}(n - 2)$ and the corresponding factorization of the desired weighted data vector $\Lambda^{1/2}(n - 2)\mathbf{d}_{b,m-1}(n - 2)$ are both described in the partitioned matrix relation

$$\mathbf{Q}_{m-2}(n-2)\mathbf{\Lambda}^{1/2}(n-2)[\mathbf{A}_{m-1}(n-2),\ \mathbf{d}_{b,m-1}(n-2)] = \begin{bmatrix} \mathbf{R}_{m-2}(n-2) & \mathbf{p}_{b,m-2}(n-2) \\ \mathbf{O} & \mathbf{v}_{b,m-2}(n-2) \end{bmatrix} \tag{18.20}$$

where $\mathbf{R}_{m-2}(n-2)$ is an $(m-1)$-by-$(m-1)$ upper triangular matrix and $\mathbf{O}$ is a null matrix, both resulting from the QR-decomposition of the exponentially weighted $\mathbf{A}_{m-1}(n-2)$; the vector $\mathbf{p}_{b,m-2}(n-2)$ contains the top $m-1$ elements of the rotated $\mathbf{d}_{b,m-1}(n-2)$ vector and the vector $\mathbf{v}_{b,m-2}(n-2)$ contains the remaining $n-m+1$ elements. The partitioned form of the data matrix $\mathbf{A}_m(n-1)$ given in Eq. (18.19) thus suggests the following *first* step in its transformation:

$$\left[\begin{array}{c|c} \mathbf{Q}_{m-2}(n-2) & \mathbf{0} \\ \hline \mathbf{0}^T & 1 \end{array}\right] \mathbf{\Lambda}^{1/2}(n-1)\mathbf{A}_m(n-1) = \left[\begin{array}{c|c} \lambda^{1/2}\mathbf{R}_{m-2}(n-2) & \lambda^{1/2}\mathbf{p}_{b,m-2}(n-2) \\ \mathbf{O} & \lambda^{1/2}\mathbf{v}_{b,m-2}(n-2) \\ \hline \mathbf{u}_{m-1}^H(n-1) & u^*(n-m) \end{array}\right] \tag{18.21}$$

For the next step in the transformation of $\mathbf{A}_m(n-1)$, we introduce another orthogonal triangularization described by the factorization

$$\mathbf{P}^{(v)}(n-2)\begin{bmatrix} \mathbf{R}_{m-2}(n-2) & \mathbf{p}_{b,m-2}(n-2) \\ \mathbf{O} & \mathbf{v}_{b,m-2}(n-2) \end{bmatrix} = \begin{bmatrix} \mathbf{R}_{m-2}(n-2) & \mathbf{p}_{b,m-2}(n-2) \\ \mathbf{0}^T & \mathcal{B}_{m-1}^{1/2}(n-2) \\ \mathbf{O} & \mathbf{0} \end{bmatrix} \tag{18.22}$$

The transformation $\mathbf{P}^{(v)}(n-2)$ has the effect of annihilating all of the $(n-m+1)$ elements of the vector $\mathbf{v}_{b,m-2}(n-2)$, except for its first element. Moreover, it leaves the top submatrices on the right side of Eq. (18.22) unchanged. The matrix on the right side of Eq. (18.22) is completely defined by three quantities: an $(m-1)$-by-$(m-1)$ upper triangular matrix $\mathbf{R}_{m-2}(n-2)$, an $(m-1)$-by-1 vector $\mathbf{p}_{b,m-2}(n-2)$, and the scalar $\mathcal{B}_{m-1}^{1/2}(n-2)$. In particular, the scalar $\mathcal{B}_{m-1}(n-2)$ defines the minimum value of the weighted sum of backward prediction-error squares:

$$\mathcal{B}_{m-1}(n-2) = \|\mathbf{v}_{b,m-1}(n-2)\|^2 \tag{18.23}$$

where $\mathbf{v}_{b,m-1}(n-2)$ is the order-updated value of $\mathbf{v}_{b,m-2}(n-2)$. In other words, applying the transformation $\mathbf{P}^{(v)}(n-2)$ has the effect of an *order update,* which is exemplified by the generation of $\mathcal{B}_{m-1}^{1/2}(n-2)$.

We may adapt the factorization described in Eq. (18.22) to fit the partitioned matrix on the right side of Eq. (18.21) by writing

$$\left[\begin{array}{c|c} \mathbf{P}^{(v)}(n-2) & \mathbf{0} \\ \hline \mathbf{0}^T & 1 \end{array}\right]\left[\begin{array}{c|c} \lambda^{1/2}\mathbf{R}_{m-2}(n-2) & \lambda^{1/2}\mathbf{p}_{b,m-2}(n-2) \\ \mathbf{O} & \lambda^{1/2}\mathbf{v}_{b,m-2}(n-2) \\ \hline \mathbf{u}_{m-1}^H(n-1) & u^*(n-m) \end{array}\right] = \begin{pmatrix} \text{see top of} \\ \text{next page} \end{pmatrix}$$

$$
= \left[\begin{array}{c|c}
\lambda^{1/2}\mathbf{R}_{m-2}(n-2) & \lambda^{1/2}\mathbf{p}_{b,m-2}(n-2) \\
\mathbf{0}^T & \lambda^{1/2}\mathcal{B}_{m-1}^{1/2}(n-2) \\
\mathbf{O} & \mathbf{0} \\
\hline
\mathbf{u}_{m-1}^H(n-1) & u^*(n-m)
\end{array}\right] \tag{18.24}
$$

We are interested in determining the *time-updated* quantities $\mathbf{R}_{m-2}(n-1)$, $\mathbf{p}_{b,m-2}(n-1)$ and $\mathcal{B}_{m-1}(n-1)$, given their *old* respective values $\mathbf{R}_{m-2}(n-2)$, $\mathbf{p}_{b,m-2}(n-2)$, and $\mathcal{B}_{m-1}(n-2)$, and the new information represented by the input vector $\mathbf{u}_{m-1}(n-1)$ and the "desired response" $u(n-m)$; these given quantities define the bottom row of the matrix on the right side of Eq. (18.24). We may accomplish our task by proceeding in two stages:

1. We apply a sequence of $m-1$ Givens rotations pivoted on the leading diagonal of the matrix $\lambda^{1/2}\mathbf{R}_{m-2}(n-2)$ so as to annihilate all the $m-1$ elements of the row vector $\mathbf{u}_{m-1}^H(n-1)$ at the bottom of the matrix on the right side of Eq. (18.24). Such a sequence of rotations defines the $(n-1)$-by-$(n-1)$ unitary matrix $\mathbf{T}_{m-2}(n-1)$, following our previous notation [described in Eq. (18.13)]. Hence, we may write

$$
\mathbf{T}_{m-2}(n-1)\left[\begin{array}{c|c}
\lambda^{1/2}\mathbf{R}_{m-2}(n-2) & \lambda^{1/2}\mathbf{p}_{b,m-2}(n-2) \\
\hline
\mathbf{0}^T & \lambda^{1/2}\mathcal{B}_{m-1}(n-2) \\
\mathbf{O} & \mathbf{0} \\
\hline
\mathbf{u}_{m-1}^H(n-1) & u^*(n-m)
\end{array}\right]
$$

$$
= \left[\begin{array}{c|c}
\mathbf{R}_{m-2}(n-1) & \mathbf{p}_{b,m-2}(n-1) \\
\hline
\mathbf{0}^T & \lambda^{1/2}\mathcal{B}_{m-1}^{1/2}(n-2) \\
\mathbf{O} & \mathbf{0} \\
\hline
\mathbf{0}^T & \epsilon_{b,m-1}^*(n-1)
\end{array}\right] \tag{18.25}
$$

The new variable $\epsilon_{b,m-1}(n-1)$, shown in complex conjugate form in the bottom row of the matrix on the right side of Eq. (18.25), is related to the a posteriori backward prediction error $b_{m-1}(n-1)$ for order $m-1$ as follows

$$
\epsilon_{b,m-1}(n-1) = \frac{b_{m-1}(n-1)}{\gamma_{m-1}^{1/2}(n-1)} \tag{18.26}
$$

where $\gamma_{m-1}(n-1)$ is the conversion factor or angle variable. In other words, $\epsilon_{b,m-1}(n-1)$ represents an *angle-normalized backward prediction error*.

2. We apply a *single* Givens rotation pivoted on $\lambda^{1/2}\mathcal{B}_{m-1}^{1/2}(n-2)$ so as to annihilate $\epsilon_{b,m-1}^{*}(n-1)$, which is the only nonzero element in the bottom row of the matrix on the right side of Eq. (18.25). Let $\mathbf{J}_{b,m-1}(n-1)$ denote this single rotation. We may thus write

$$\mathbf{J}_{b,m-1}(n-1)\left[\begin{array}{c|c} \mathbf{R}_{m-2}(n-1) & \mathbf{p}_{b,m-2}(n-1) \\ \hline \mathbf{0}^T & \lambda^{1/2}\mathcal{B}_{m-1}^{1/2}(n-2) \\ \mathbf{O} & \mathbf{0} \\ \hline \mathbf{0}^T & \epsilon_{b,m-1}^{*}(n-1) \end{array}\right] = \left[\begin{array}{c|c} \mathbf{R}_{m-2}(n-1) & \mathbf{p}_{b,m-2}(n-1) \\ \hline \mathbf{0}^T & \mathcal{B}_{m-1}^{1/2}(n-1) \\ \mathbf{O} & \mathbf{0} \\ \hline \mathbf{0}^T & 0 \end{array}\right] \tag{18.27}$$

Note that applying the rotation $\mathbf{J}_{b,m-1}(n-1)$ has the effect of a *time update*. The subscript $m-1$ in this rotation is in anticipation of an order update from $m-2$ to $m-1$. In particular, observe that the two submatrices $\mathbf{R}_{m-2}(n-1)$ and $\mathbf{p}_{b,m-2}(n-1)$ together with the scalar $\mathcal{B}_{m-1}^{1/2}(n-1)$, on the right side of Eq. (18.27), constitute the elements of the order-updated upper triangular matrix $\mathbf{R}_{m-1}(n-1)$ because the matrix in Eq. (18.27) is the result of applying an orthogonal rotation to $\mathbf{A}_m(n-1)$ [see Eqs. (18.21), (18.24), (18.25), and (18.27)]. That is,

$$\mathbf{R}_{m-1}(n-1) = \begin{bmatrix} \mathbf{R}_{m-2}(n-1) & \mathbf{p}_{b,m-2}(n-1) \\ \mathbf{0}^T & \mathcal{B}_{m-1}^{1/2}(n-1) \end{bmatrix} \tag{18.28}$$

By induction on the partitioned matrix of Eq. (18.28), we deduce the following properties of the upper triangular matrix $\mathbf{R}_{m-1}(n)$ resulting from the QR-decomposition of the exponentially weighted data matrix $\mathbf{A}_m(n)$ (Regalia and Bellanger, 1990):

1. The diagonal elements of $\mathbf{R}_{m-1}(n)$ equal the square roots of the sums of weighted backward prediction-error squares of all orders, as shown in the expanded form

$$\mathbf{R}_{m-1}(n) = \begin{bmatrix} \mathcal{B}_0^{1/2}(n) & r_{01}(n) & r_{02}(n) & \cdots & r_{0,m-1}(n) \\ & \mathcal{B}_1^{1/2} & r_{12}(n) & \cdots & r_{1,m-1}(n) \\ & & \mathcal{B}_2^{1/2}(n) & \cdots & r_{2,m-1}(n) \\ & \mathbf{O} & & \ddots & \vdots \\ & & & & \mathcal{B}_{m-1}^{1/2}(n) \end{bmatrix} \tag{18.29}$$

where $\mathcal{B}_k(n)$ is the sum of weighted backward prediction-error squares for prediction order $k = 0, 1, 2, \ldots, m - 1$, and $r_{jk}(n)$ is the jth element of the vector $\mathbf{p}_{b,k}(n)$ and $j \neq k$.

2. The k-by-k principal submatrix of $\mathbf{R}_{m-1}(n)$ is simply $\mathbf{R}_k(n)$, where $k = 0, 1, \ldots, m - 1$.

Returning to the problem at hand, we may state that the combined use of the transformations $\mathbf{T}_{m-2}(n-1)$ and $\mathbf{J}_{b,m-1}(n-1)$ provides the means for generating the solution to the backward linear prediction problem for order $m - 1$ and time $n - 1$, given the solution for order $m - 2$ and time $n - 2$. The noteworthy feature of these two transformations is that they both require *fixed* numbers of rotations.

18.4 ADAPTIVE FORWARD LINEAR PREDICTION

Consider next the forward linear least-squares prediction problem of order $m - 1$. To solve this problem, we proceed by considering the partitioned form of the data matrix $\mathbf{A}_m(n)$ of Eq. (18.7). Correspondingly, we may express the exponentially weighted version of $\mathbf{A}_m(n)$ as

$$\mathbf{\Lambda}^{1/2}(n)\mathbf{A}_m(n) = \begin{bmatrix} \lambda^{(n-1)/2} & \mathbf{0}^T & 0 \\ \mathbf{0} & \lambda^{1/2}\mathbf{\Lambda}^{1/2}(n-2) & \mathbf{0} \\ 0 & \mathbf{0}^T & 1 \end{bmatrix} \begin{bmatrix} u^*(1) & \mathbf{0}^T \\ \mathbf{d}_{f,m-1}(n-1) & \mathbf{A}_{m-1}(n-2) \\ u^*(n) & \mathbf{u}_{m-1}^H(n-1) \end{bmatrix}$$

$$= \begin{bmatrix} \lambda^{(n-1)/2}u^*(1) & \mathbf{0}^T \\ \lambda^{1/2}\mathbf{\Lambda}^{1/2}(n-2)\mathbf{d}_{f,m-1}(n-1) & \lambda^{1/2}\mathbf{\Lambda}^{1/2}(n-2)\mathbf{A}_{m-1}(n-2) \\ u^*(n) & \mathbf{u}_{m-1}^H(n-1) \end{bmatrix} \tag{18.30}$$

The QR-decomposition of the exponentially weighted submatrix $\mathbf{A}_{m-1}(n-2)$ and the corresponding factorization of the "desired" data vector $\mathbf{d}_{f,m-1}(n-1)$ may be combined as

$$\mathbf{Q}_{m-2}(n-2)\mathbf{\Lambda}^{1/2}(n-2)[\mathbf{d}_{f,m-1}(n-1), \mathbf{A}_{m-1}(n-2)]$$
$$= \begin{bmatrix} \mathbf{p}_{f,m-2}(n-1) & \mathbf{R}_{m-2}(n-2) \\ \mathbf{v}_{f,m-2}(n-1) & \mathbf{O} \end{bmatrix} \tag{18.31}$$

where $\mathbf{R}_{m-2}(n-2)$ is an upper triangular matrix and $\mathbf{O}$ is a null matrix, both resulting from the QR-decomposition of the exponentially weighted $\mathbf{A}_{m-1}(n-2)$. The vector $\mathbf{p}_{f,m-2}(n-1)$ contains the top $m - 1$ elements of the rotated $\mathbf{d}_{f,m-1}(n-1)$

vector, and the vector $\mathbf{v}_{f,m-2}(n-1)$ contains the remaining elements. Thus, the partitioned form of the exponentially weighted matrix given in the second line of Eq. (18.30) suggests the following first step in its transformation to upper triangular form:

$$\left[\begin{array}{c|c|c} 1 & \mathbf{0}^T & \mathbf{0} \\ \hline \mathbf{0} & \mathbf{Q}_{m-2}(n-2) & \mathbf{0} \\ \hline 0 & \mathbf{0}^T & 1 \end{array}\right] \mathbf{\Lambda}^{1/2}(n)\mathbf{A}_m(n) = \left[\begin{array}{c|c} \lambda^{(n-1)/2}u^*(1) & \mathbf{0}^T \\ \hline \lambda^{1/2}\mathbf{p}_{f,m-2}(n-1) & \lambda^{1/2}\mathbf{R}_{m-2}(n-2) \\ \lambda^{1/2}\mathbf{v}_{f,m-2}(n-1) & \mathbf{0} \\ \hline u^*(n) & \mathbf{u}_{m-1}^H(n-1) \end{array}\right] \tag{18.32}$$

To pave the way for the next step in the transformation of the data matrix $\mathbf{A}_m(n)$, we introduce another factorization described by

$$\mathbf{K}^{(v)}(n-1)\left[\begin{array}{c|c} \lambda^{(n-2)/2}u^*(1) & \mathbf{0}^T \\ \mathbf{p}_{f,m-2}(n-1) & \mathbf{R}_{m-2}(n-2) \\ \hline \mathbf{v}_{f,m-2}(n-1) & \mathbf{O} \end{array}\right] = \left[\begin{array}{c|c} \mathcal{F}_{m-1}^{1/2}(n-1) & \mathbf{0}^T \\ \mathbf{p}_{f,m-2}(n-1) & \mathbf{R}_{m-2}(n-2) \\ \hline \mathbf{0} & \mathbf{O} \end{array}\right] \tag{18.33}$$

where $\mathcal{F}_{m-1}(n-1)$ is the sum of weighted forward prediction-error squares for order $m-1$ and time $n-1$ (see Problem 4). The unitary matrix $\mathbf{K}^{(v)}(n-1)$ applies a sequence of $(n-m+1)$ Givens rotations pivoted on the element $\lambda^{(n-2)/2}u^*(1)$, thereby annihilating all the elements of the vector $\mathbf{v}_{f,m-2}(n-1)$. It has the effect of an *order update*, exemplified by the generation of $\mathcal{F}_{m-1}^{1/2}(n-1)$ corresponding to prediction order $m-1$. Note also that the submatrices in the middle of the matrix on the right side of Eq. (18.33) are unaltered by the transformation. Adapting this new transformation to fit the partitioned form of the matrix on the right side of Eq. (18.21), we may write

$$\left[\begin{array}{cc} \mathbf{K}^{(v)}(n-1) & 0 \\ \hline \mathbf{0}^T & 1 \end{array}\right] \left[\begin{array}{c|c} \lambda^{(n-1)/2}u^*(1) & \mathbf{0}^T \\ \lambda^{1/2}\mathbf{p}_{f,m-2}(n-1) & \lambda^{1/2}\mathbf{R}_{m-2}(n-2) \\ \lambda^{1/2}\mathbf{v}_{f,m-2}(n-1) & \mathbf{O} \\ \hline u^*(n) & \mathbf{u}_{m-1}^H(n-1) \end{array}\right]$$

$$= \left[\begin{array}{c|c} \lambda^{1/2}\mathcal{F}_{m-1}^{1/2}(n-1) & \mathbf{0}^T \\ \lambda^{1/2}\mathbf{p}_{f,m-2}(n-1) & \lambda^{1/2}\mathbf{R}_{m-2}(n-2) \\ \mathbf{0} & \mathbf{O} \\ \hline u^*(n) & \mathbf{u}_{m-1}^H(n-1) \end{array}\right] \tag{18.34}$$

It is noteworthy that the transformation $\mathbf{K}^{(v)}(n-1)$ for forward prediction and the transformation $\mathbf{P}^{(v)}(n-2)$ for backward prediction perform similar roles. Note also the delay by one time unit in the latter transformation with respect to the former one.

The matrix on the right side of Eq. (18.34) contains two different sets of data. The top part of the matrix (above the horizontal dashed line) pertains to the solution of forward linear prediction at time $n-1$. The data in the bottom part of the matrix relate to the "desired" response and the input vector received at time n. To transform the matrix into the desired upper triangular form, we may proceed as follows:

1. We apply a sequence of $m-1$ Givens rotations to the bottom row of the matrix on the right side of Eq. (18.34). Specifically, the bottom row is rotated against the diagonal of matrix $\mathbf{R}_{m-2}(n-2)$, so as to annihilate all the elements of the row vector $\mathbf{u}_{m-1}^H(n-1)$. Here we use the update relation:

$$\mathbf{T}_{m-2}(n-1)\begin{bmatrix} \lambda^{1/2}\mathbf{R}_{m-2}(n-2) \\ \mathbf{O} \\ \mathbf{u}_{m-1}^H(n-1) \end{bmatrix} = \begin{bmatrix} \mathbf{R}_{m-2}(n-1) \\ \mathbf{O} \end{bmatrix} \tag{18.35}$$

which is a reformulation of Eq. (18.14) with time n replaced by $n-1$ and order m replaced by $m-2$. Therefore, in order to conform to the partitioned matrix structure described in Eq. (18.34), we configure the application of the $m-1$ Givens rotations denoted by the unitary matrix $\mathbf{T}_{m-2}(n-1)$ as follows:

$$\begin{bmatrix} 1 & \mathbf{0}^T \\ \mathbf{0} & \mathbf{T}_{m-2}(n-1) \end{bmatrix}\begin{bmatrix} \lambda^{1/2}\mathscr{F}_{m-1}^{1/2}(n-1) & \mathbf{0}^T \\ \lambda^{1/2}\mathbf{p}_{f,m-2}(n-1) & \lambda^{1/2}\mathbf{R}_{m-2}(n-2) \\ \mathbf{0} & \mathbf{O} \\ u^*(n) & \mathbf{u}_{m-1}^H(n-1) \end{bmatrix}$$

$$= \begin{bmatrix} \lambda^{1/2}\mathscr{F}_{m-1}^{1/2}(n-1) & \mathbf{0}^T \\ \mathbf{p}_{f,m-2}(n) & \mathbf{R}_{m-2}(n-1) \\ \mathbf{0} & \mathbf{O} \\ \epsilon_{f,m-1}^*(n) & \mathbf{0}^T \end{bmatrix} \tag{18.36}$$

The new variable $\epsilon_{f,m-1}(n)$, shown in complex conjugate form in the bottom row of the matrix on the right side of Eq. (18.36), is related to the a posteriori forward prediction error $f_{m-1}(n)$ for order $m-1$ by

$$\epsilon_{f,m-1}(n) = \frac{f_{m-1}(n)}{\gamma_{m-1}^{1/2}(n-1)} \tag{18.37}$$

where $\gamma_{m-1}(n-1)$ is the conversion factor or angle variable at time $n-1$. This equation suggests that we may view $\epsilon_{f,m-1}(n)$ as an *angle-normalized forward prediction error*.

2. We apply a *single* Givens rotation pivoted on the element $\lambda^{1/2}\mathcal{F}_{m-1}^{1/2}(n-1)$ in the top row of the matrix on the right side of Eq. (18.36), and designed to annihilate $\epsilon_{f,m-1}^{*}(n)$ in the bottom row of the matrix. Let $\mathbf{J}_{f,m-1}(n)$ denote this single rotation, where the subscript $m-1$ is in anticipation of an order update from $m-2$ to $m-1$. We may thus write

$$\mathbf{J}_{f,m-1}(n)\left[\begin{array}{c|c}\lambda^{1/2}\mathcal{F}_{m-1}^{1/2}(n-1) & \mathbf{0}^T \\ \mathbf{p}_{f,m-2}(n) & \mathbf{R}_{m-2}(n-1) \\ \hline \mathbf{0} & \mathbf{O} \\ \epsilon_{f,m-1}^{*}(n) & \mathbf{0}^T\end{array}\right] = \left[\begin{array}{c|c}\mathcal{F}_{m-1}^{1/2}(n) & \mathbf{0}^T \\ \mathbf{p}_{f,m-2}(n) & \mathbf{R}_{m-2}(n-1) \\ \hline \mathbf{0} & \mathbf{O}\end{array}\right] \tag{18.38}$$

Applying the transformation $\mathbf{J}_{f,m-1}(n)$ has the effect of a *time update*. In particular, the three quantities, $\mathbf{R}_{m-2}(n-1)$, $\mathbf{p}_{f,m-2}(n)$, and $\mathcal{F}_{m-1}(n)$ define the solution to the forward linear prediction problem for order $m-1$ and time n. This solution is indeed sufficient to compute the updated upper triangular matrix $\mathbf{R}_{m-1}(n)$; hence, the justification for the use of $m-1$ as the subscript for $\mathbf{J}_{f,m-1}(n)$.

Of course, we may go on and apply a sequence of $(m-1)$ Givens rotations pivoted on $\mathcal{F}_{m-1}^{1/2}(n)$, so as to annihilate all the elements of the vector $\mathbf{p}_{f,m-2}(n)$. The Givens rotations applied here must annihilate the elements of $\mathbf{p}_{f,m-2}(n)$ from the bottom up[2] in order to preserve the triangular structure of $\mathbf{R}_{m-2}(n-1)$. The purpose of such a transformation is to compute the upper triangularized form of the data matrix $\mathbf{A}_m(n)$. However, insofar as the development of the QRD-LSL algorithm is concerned, which is our main preoccupation in this chapter, this transformation is not necessary and will therefore not be pursued.

18.5 QRD-LSL ALGORITHM: OVERVIEW

We are now equipped with the material we need to develop a QR-decomposition-based lattice structure for solving the linear least-squares prediction problem and also for providing the necessary variables to solve the joint-process estimation problem exactly. For obvious reasons, the algorithm described herein is referred to as the *QR-decomposition-based least-squares lattice (QRD-LSL) algorithm*.

The mathematical details, pertaining to the development of the QRD-LSL algorithm, are presented in Sections 18.6 and 18.7. In particular, in Section 18.6 we combine the ideas developed in the preceding two sections on adaptive forward and backward methods of linear prediction. By so doing, we develop a multistage lattice-based structure for computing both the angle-normalized forward and backward prediction errors, $\epsilon_{f,m}(n)$ and $\epsilon_{b,m}(n)$, in a recursive manner. Each stage of the lattice

[2] The transformation described herein is important to the development of a matrix-oriented (i.e., Kalman) type of fast QRD-RLS algorithm (Proudler et al., 1988).

structure involves a pair of Givens rotations. The computation of the angle-normalized forward prediction error uses a cosine-sine pair of rotation parameters that depends on the angle-normalized backward prediction error. Conversely, the computation of the angle-normalized backward prediction error uses another cosine–sine pair of rotation parameters that depends on the angle-normalized forward prediction error. Accordingly, the computations of the angle-normalized forward and backward prediction errors are indeed interconnected in a lattice fashion.

In Section 18.7, we use the angle-normalized backward prediction errors to solve the joint-process estimation problem. In particular, we show that the transformation $\mathbf{T}_m(n)$, which is central to the joint-process estimation problem, may be computed directly from the angle-normalized backward prediction errors. The computation of the angle-normalized joint-process estimation error $\epsilon_{m+1}(n)$ proceeds on a stage-by-stage basis. Each stage of the computation involves a single Givens rotation based on a corresponding value of the angle-normalized backward prediction error.

From the description of the QRD-LSL algorithm sketched here, it is apparent that the structure of this new algorithm bears a direct resemblance to a conventional recursive LSL algorithm. From Chapter 17 we recall that the latter algorithm may be formulated in terms of a priori or a posteriori estimation errors, and with or without error feedback. The QRD-LSL algorithm, on the other hand, is based on angle-normalized estimation errors that are related to the a priori and a posteriori values of the corresponding estimation errors (see Problem 5). Moreover, the time updates of auxiliary parameters involved in the computation of angle-normalized estimation errors are performed directly, which implies the presence of local error feedback. Accordingly, we only have one basic form of the QRD-LSL algorithm (with square roots) that is directly analogous to a conventional LSL algorithm with error feedback.[3]

18.6 LATTICE PREDICTOR USING GIVENS ROTATIONS

Consider the data matrix $\mathbf{A}_{m+1}(n)$ partitioned as in Eq. (18.2), which is reproduced here for convenience of presentation:

$$\mathbf{A}_{m+1}(n) = \left[\begin{array}{c|c|c} u^*(1) & \mathbf{0}^T & 0 \\ \hline \mathbf{d}_{f,m-1}(n-1) & \mathbf{A}_{m-1}(n-2) & \mathbf{d}_{b,m-1}(n-2) \\ \hline u^*(n) & \mathbf{u}_{m-1}^H(n-1) & u^*(n-m) \end{array}\right] \tag{18.39}$$

Following the QR-decomposition of the exponentially weighted data matrices in Eq. (18.21) and (18.32), we may in a corresponding way write

[3] The lattice type of computations may be formulated differently, depending on the approach taken (Proudler et al., 1989; Regalia and Bellanger, 1991). Although the structures described in these two papers are indeed lattice-like in composition, they are not directly analogous to a conventional LSL algorithm.

$$\left[\begin{array}{c|c|c} 1 & \mathbf{0}^T & 0 \\ \hline \mathbf{0} & \mathbf{Q}_{m-2}(n-2) & \mathbf{0} \\ \hline 0 & \mathbf{0}^T & 1 \end{array}\right] \mathbf{\Lambda}^{1/2}(n)\mathbf{A}_{m+1}(n)$$

$$= \underbrace{\left[\begin{array}{c|c|c} \lambda^{(n-1)/2}u^*(n) & \mathbf{0}^T & 0 \\ \hline \lambda^{1/2}\mathbf{p}_{f,m-2}(n-1) & \lambda^{1/2}\mathbf{R}_{m-2}(n-2) & \lambda^{1/2}\mathbf{p}_{b,m-2}(n-2) \\ \lambda^{1/2}\mathbf{v}_{f,m-2}(n-1) & \mathbf{O} & \lambda^{1/2}\mathbf{v}_{b,m-2}(n-2) \\ \hline u^*(n) & \mathbf{u}_{m-1}^H(n-1) & u^*(n-m) \end{array}\right]}_{\mathbf{B}(n)} \qquad (18.40)$$

Let $\mathbf{B}(n)$ denote the partitioned matrix on the right side of Eq. (18.40). This matrix contains all the elements we need to derive the QR-decomposition-based lattice predictor. In particular, the transformed columns 1 and $m + 1$ of the matrix $\mathbf{B}(n)$ are involved in forward and backward forms of linear least-squares prediction, as described in the next two subsections.

Update Equations for Angle-Normalized Forward Prediction Error

Following the transformation described in Section 18.3, we use the unitary matrix $\mathbf{P}^{(v)}(n-2)$ so as to annihilate the vector $\mathbf{v}_{b,m-2}(n-2)$ in column $m + 1$ of the matrix $\mathbf{B}(n)$ on the right side of Eq. (18.40). Specifically, we write

$$\begin{bmatrix} \mathbf{P}^{(v)}(n-2) & 0 \\ \mathbf{0}^T & 1 \end{bmatrix} \mathbf{B}(n) = \underbrace{\begin{bmatrix} \lambda^{(n-1)/2}u^*(1) & \mathbf{0}^T & 0 \\ \lambda^{1/2}\mathbf{p}_{f,m-2}(n-1) & \lambda^{1/2}\mathbf{R}_{m-2}(n-2) & \lambda^{1/2}\mathbf{p}_{b,m-2}(n-2) \\ \lambda^{1/2}\pi_{f,m-1}(n-1) & \mathbf{0}^T & \lambda^{1/2}\mathcal{B}_{m-1}^{1/2}(n-2) \\ \lambda^{1/2}\mathbf{v}_{f,m-1}(n-1) & \mathbf{O} & \mathbf{O} \\ u^*(n) & \mathbf{u}_{m-1}^H(n-1) & u^*(n-m) \end{bmatrix}}_{\mathbf{C}(n)} \qquad (18.41)$$

The only elements of the matrix $\mathbf{B}(n)$ affected by the transformation $\mathbf{P}^{(v)}(n-2)$ in Eq. (18.41) are those that lie on rows $m + 1$ to $n - 1$. As a result of this transformation, a new element is generated, namely, $\pi_{f,m-1}(n-1)$ in the first column. This scalar quantity is the transformed version of the first element of the column vector $\mathbf{v}_{f,m-2}(n-1)$. The column vector $\mathbf{v}_{f,m-1}(n-1)$ consists of the transformed version of the remaining elements of $\mathbf{v}_{f,m-2}(n-1)$. Note the order change from $m - 2$ to $m - 1$ below row m as a result of applying the transformation $\mathbf{P}^{(v)}(n-2)$.

Next, we let $\mathbf{C}(n)$ denote the partitioned matrix on the right side of Eq. (18.41). Then, following the transformation described in Eqs. (18.25) and (18.36), and applying Eqs. (18.14) and (18.18), we may transform $\mathbf{C}(n)$ as follows:

$$\begin{bmatrix} 1 & \mathbf{0}^T \\ \mathbf{0} & \mathbf{T}_{m-2}(n-1) \end{bmatrix} \mathbf{C}(n) = \underbrace{\begin{bmatrix} \lambda^{n-1)/2}u^*(1) & \mathbf{0}^T & 0 \\ \mathbf{p}_{f,m-2}(n) & \mathbf{R}_{m-2}(n-1) & \mathbf{p}_{b,m-2}(n-1) \\ \lambda^{1/2}\pi_{f,m-1}(n-1) & \mathbf{0}^T & \lambda^{1/2}\mathcal{B}_{m-1}^{1/2}(n-2) \\ \lambda^{1/2}\mathbf{v}_{f,m-1}(n-1) & \mathbf{O} & \mathbf{O} \\ \epsilon_{f,m-1}^*(n) & \mathbf{0}^T & \epsilon_{b,m-1}^*(n-1) \end{bmatrix}}_{\mathbf{D}(n)} \tag{18.42}$$

As a result of this transformation, angle-normalized forward and *delayed* backward prediction errors, $\epsilon_{f,m-1}(n)$ and $\epsilon_{b,m-1}(n-1)$, are generated (albeit in complex conjugate forms); see the bottom row of the matrix on the right-hand side of Eq. (18.42). Also, the elements on rows 2 to m undergo a time update.

Let $\mathbf{D}(n)$ denote the matrix on the right side of Eq. (18.42). Then following the transformation described in Eq. (18.27), we may use the unitary matrix $\mathbf{J}_{b,m-1}(n-1)$ to annihilate the element $\epsilon_{b,m-1}^*(n-1)$ in the bottom row of the matrix $\mathbf{D}(n)$ as follows:

$$\begin{bmatrix} 1 & \mathbf{0}^T \\ \mathbf{0} & \mathbf{J}_{b,m-1}(n-1) \end{bmatrix} \mathbf{D}(n) = \begin{bmatrix} \lambda^{(n-1)/2}u^*(1) & \mathbf{0}^T & 0 \\ \mathbf{p}_{f,m-2}(n) & \mathbf{R}_{m-2}(n-1) & \mathbf{p}_{b,m-2}(n-1) \\ \pi_{f,m-1}(n) & \mathbf{0}^T & \mathcal{B}_{m-1}^{1/2}(n-1) \\ \lambda^{1/2}\mathbf{v}_{f,m-1}(n-1) & \mathbf{O} & \mathbf{0} \\ \epsilon_{f,m}^*(n) & \mathbf{0}^T & \mathbf{0} \end{bmatrix} \tag{18.43}$$

Here we observe that applying the rotation $\mathbf{J}_{b,m-1}(n-1)$ has a twofold effect: (1) time updates of the two nonzero elements on row $(m+1)$ of the matrix, and (2) order update of the angle-normalized backward prediction error $\epsilon_{f,m}(n)$ in the bottom row, albeit in complex conjugate form.

We may simplify the composition of the matrix on the right side of Eq. (18.43) by recognizing the following two order updates:

1. The upper triangular matrix $\mathbf{R}_{m-2}(n-1)$, the vector $\mathbf{p}_{b,m-1}(n-1)$, and the scalar $\mathcal{B}_{m-1}^{1/2}(n-1)$ constitute the order-updated upper triangular matrix $\mathbf{R}_{m-1}(n-1)$, in accordance with Eq. (18.28).
2. In a corresponding way, the vector $\mathbf{p}_{f,m-2}(n)$ and the scalar $\pi_{f,m-1}(n)$ define the order-updated column vector $\mathbf{p}_{f,m-1}(n)$ for forward linear prediction; here we note that $\pi_{f,m-1}(n)$ is the last element of the column vector $\mathbf{p}_{f,m-1}(n)$.

Our objective, however, is to formulate a lattice-like predictor. In this context, we merely require order updates for computing angle-normalized versions of the forward and backward prediction errors. One half of this objective is indeed realized by relating the order-updated value of the angle-normalized forward prediction error

$\epsilon_{f,m}(n)$ in Eq. (18.43) to its previous-order value $\epsilon_{f,m-1}(n)$ in Eq. (18.42), both of which appear in complex conjugate form. We may therefore stop at Eq. (18.43) since we have the equations of interest.

Let the unitary matrix $\mathbf{J}_{b,m-1}(n-1)$ be expressed in its expanded form as follows:

$$\mathbf{J}_{b,m-1}(n-1) = \begin{bmatrix} \mathbf{I}_{m-1} & \mathbf{0} & \mathbf{O} & \mathbf{0} \\ \mathbf{0}^T & c_{b,m-1}(n-1) & \mathbf{0}^T & s^*_{b,m-1}(n-1) \\ \mathbf{O} & \mathbf{0} & \mathbf{I}_{n-m-1} & \mathbf{0} \\ \mathbf{0}^T & -s_{b,m-1}(n-1) & \mathbf{0}^T & c_{b,m-1}(n-1) \end{bmatrix} \tag{18.44}$$

where $\mathbf{I}_{m-1}$ and $\mathbf{I}_{n-m-1}$ are identity matrices. Then, in accordance with the definitions given in Eqs. (18.42) and (18.43), we may express the real cosine and complex sine parameters of the Givens rotation $\mathbf{J}_{b,m-1}(n-1)$ as follows:

$$c_{b,m-1}(n-1) = \frac{\lambda^{1/2}\mathcal{B}^{1/2}_{m-1}(n-2)}{\mathcal{B}^{1/2}_{m-1}(n-1)} \tag{18.45}$$

and

$$s_{b,m-1}(n-1) = \frac{\epsilon^*_{b,m-1}(n-1)}{\mathcal{B}^{1/2}_{m-1}(n-1)} \tag{18.46}$$

where

$$\mathcal{B}_{m-1}(n-1) = \lambda\mathcal{B}_{m-1}(n-2) + |\epsilon_{b,m-1}(n-1)|^2 \tag{18.47}$$

Furthermore, using Eqs. (18.42) and (18.43) we may deduce the following updates:

1. Order update for computing the *angle-normalized forward prediction error*

$$\epsilon_{f,m}(n) = c_{b,m-1}(n-1)\epsilon_{f,m-1}(n) - s^*_{b,m-1}(n-1)\lambda^{1/2}\pi^*_{f,m-1}(n-1) \tag{18.48}$$

2. Time update for computing the *forward prediction auxiliary parameter*

$$\pi^*_{f,m-1}(n) = c_{b,m-1}(n-1)\lambda^{1/2}\pi^*_{f,m-1}(n-1) + s_{b,m-1}(n-1)\epsilon_{f,m-1}(n) \tag{18.49}$$

These two updates constitute the first half of the desired lattice predictor equations. For the prediction order we have $m = 1, 2, \ldots, M$, where M is the *final prediction order*. Note that the computations of the order-update variable $\epsilon_{f,m}(n)$ and the time-updated parameter $\pi_{f,m-1}(n)$, both relating to *forward* prediction, involve the use of the rotation parameters $c_{b,m-1}(n-1)$ and $s_{b,m-1}(n-1)$ that depend on angle-normalized *backward* prediction errors. Since $c_{b,m-1}(n-1)$ and $s_{b,m-1}(n-1)$ are both dimensionless ratios, it follows that the parameter $\pi_{f,m-1}(n)$ has the same dimension as the error $\epsilon_{f,m}(n)$.

Update Equations for Angle-Normalized Backward Prediction Errors

Consider again the n-by-$(m+1)$ matrix $\mathbf{B}(n)$ defined as the matrix on the right side of Eq. (18.40). Following the transformation described in Eq. (18.33), we may use the unitary matrix $\mathbf{K}^{(v)}(n-1)$ to annihilate the vector $\mathbf{v}_{f,m-2}(n-1)$ in the first column of $\mathbf{B}(n)$ as shown by

$$\begin{bmatrix} \mathbf{K}^{(v)}(n-1) & \mathbf{0} \\ \mathbf{0}^T & 1 \end{bmatrix} \mathbf{B}(n) = \underbrace{\begin{bmatrix} \lambda^{1/2}\mathcal{F}_{m-1}^{1/2}(n-1) & \mathbf{0}^T & \lambda^{1/2}\pi_{b,m-1}(n-1) \\ \lambda^{1/2}\mathbf{p}_{f,m-2}(n-1) & \lambda^{1/2}\mathbf{R}_{m-2}(n-2) & \lambda^{1/2}\mathbf{p}_{b,m-2}(n-2) \\ \mathbf{0} & \mathbf{O} & \lambda^{1/2}\mathbf{v}_{b,m-1}(n-1) \\ u^*(n) & \mathbf{u}_{m-1}^H(n-1) & u^*(n-m) \end{bmatrix}}_{\mathbf{V}(n)} \tag{18.50}$$

where $\pi_{b,m-1}(n-1)$ is a new element that results from the first row of the matrix $[\mathbf{K}^{(v)}(n-1), \mathbf{0}]$ operating on the last column of $\mathbf{B}(n)$. Observe the order update of the two nonzero elements in the top row of the matrix as a result of applying the transformation $\mathbf{K}^{(v)}(n-1)$.

Let $\mathbf{V}(n)$ denote the matrix on the right side of Eq. (18.50). Then following the transformation described in Eq. (18.35), we may use the unitary matrix $\mathbf{T}_{m-2}(n-1)$ to annihilate the row vector $\mathbf{u}_{m-1}^H(n-1)$, as shown by

$$\begin{bmatrix} 1 & \mathbf{0}^T \\ \mathbf{0} & \mathbf{T}_{m-2}(n-1) \end{bmatrix} \mathbf{V}(n) = \underbrace{\begin{bmatrix} \lambda^{1/2}\mathcal{F}_{m-1}^{1/2}(n-1) & 0 & \lambda^{1/2}\pi_{b,m-1}(n-1) \\ \mathbf{p}_{f,m-2}(n) & \mathbf{R}_{m-2}(n-1) & \mathbf{p}_{b,m-2}(n-1) \\ \mathbf{0} & \mathbf{O} & \lambda^{1/2}\mathbf{v}_{b,m-1}(n-1) \\ \epsilon_{f,m-1}^*(n) & \mathbf{0}^T & \epsilon_{b,m-1}^*(n-1) \end{bmatrix}}_{\mathbf{W}(n)} \tag{18.51}$$

This transformation has a twofold effect: (1) time updates of the elements on rows 2 to m, and (2) the generation of angle-normalized forward and *delayed* backward prediction errors, $\epsilon_{f,m-1}(n)$ and $\epsilon_{b,m-1}(n-1)$, appearing in complex conjugate form in the bottom row.

Finally, let $\mathbf{W}(n)$ denote the matrix on the right side of Eq. (18.51). Then, following the transformation described in Eq. (18.38), we may use the matrix $\mathbf{J}_{f,m-1}(n)$ to annihilate $\epsilon_{f,m-1}^*(n)$, as shown by

$$\mathbf{J}_{f,m-1}(n)\mathbf{W}(n) = \begin{bmatrix} \mathcal{F}_{m-1}^{1/2}(n) & 0 & \pi_{b,m-1}(n) \\ \mathbf{p}_{f,m-2}(n) & \mathbf{R}_{m-2}(n-1) & \mathbf{p}_{b,m-2}(n-1) \\ \mathbf{0} & \mathbf{O} & \lambda^{1/2}\mathbf{v}_{b,m-1}(n-1) \\ 0 & \mathbf{0}^T & \epsilon_{b,m}^*(n) \end{bmatrix} \tag{18.52}$$

The rotation $\mathbf{J}_{f,m-1}(n)$ has a twofold effect: (1) time updates of the two nonzero elements in the top row of the matrix, and (2) time update as well as order update of the angle-normalized backward prediction error $\epsilon_{b,m}(n)$, appearing in complex conjugate form in the bottom row of the matrix.

The use of Eq. (18.51) in conjunction with Eq. (18.52) helps us realize the second half of our objective, namely, an order update for computing the angle-normalized backward prediction error. Indeed, we observe that the order-updated value of the angle-normalized backward prediction error $\epsilon_{b,m}(n)$ in Eq. (18.46) is related to its previous order-updated value $\epsilon_{b,m-1}(n-1)$ in Eq. (18.45), both appearing again in complex conjugate form.

Let the unitary matrix $\mathbf{J}_{f,m-1}(n)$ be expressed in the expanded form

$$\mathbf{J}_{f,m-1}(n) = \begin{bmatrix} c_{f,m-1}(n) & \mathbf{0}^T & s_{f,m-1}^*(n) \\ \mathbf{0} & \mathbf{I}_{n-2} & \mathbf{0} \\ s_{f,m-1}(n) & \mathbf{0}^T & c_{f,m-1}(n) \end{bmatrix} \tag{18.53}$$

where $\mathbf{I}_{n-2}$ is an identity matrix. Using the definitions given in Eqs. (18.51) and (18.52), we may then express the real cosine and complex sine parameters of the Givens rotation $\mathbf{J}_{f,m-1}(n)$ as follows:

$$c_{f,m-1}(n) = \frac{\lambda^{1/2}\mathscr{F}_{m-1}^{1/2}(n-1)}{\mathscr{F}_{m-1}^{1/2}(n)} \tag{18.54}$$

and

$$s_{f,m-1}(n) = \frac{\epsilon_{f,m-1}^*(n)}{\mathscr{F}_{m-1}^{1/2}(n)} \tag{18.55}$$

where

$$\mathscr{F}_{m-1}(n) = \lambda\mathscr{F}_{m-1}(n-1) + |\epsilon_{f,m-1}(n)|^2 \tag{18.56}$$

Correspondingly, we deduce from Eqs. (18.52) and (18.53) the next pair of updates:

1. Time-order update for computing the *angle-normalized backward prediction error:*

$$\epsilon_{b,m}(n) = c_{f,m-1}(n)\epsilon_{b,m-1}(n-1) - s_{f,m-1}^*(n)\lambda^{1/2}\pi_{b,m-1}^*(n-1) \tag{18.57}$$

2. Time update for computing the *backward prediction auxiliary paramter:*

$$\pi_{b,m-1}^*(n) = c_{f,m-1}(n)\lambda^{1/2}\pi_{b,m-1}^*(n-1) + s_{f,m-1}(n)\epsilon_{b,m-1}(n-1) \tag{18.58}$$

These two updates constitute the second half of the desired lattice predictor equations. For the prediction order we have $m = 1, 2, \ldots, M$, where M is the final prediction order. Note that the computations of the updated variable $\epsilon_{b,m}(n)$ and the updated parameter $\pi_{b,m-1}(n)$, both relating to *backward* prediction, involve the use of the rotation parameters $c_{f,m-1}(n)$ and $s_{f,m-1}(n)$ that depend on angle-normalized *forward* prediction errors. Since $c_{f,m-1}(n)$ and $s_{f,m-1}(n)$ are both dimensionless ratios, it follows that the parameter $\pi_{b,m-1}(n)$ has the same dimension as the error $\epsilon_{b,m}(n)$.

18.7 JOINT-PROCESS ESTIMATION USING GIVENS ROTATIONS

In this section we take up the remaining issue, namely, that of joint-process estimation. From Chapter 14 we recall that the transformation $\mathbf{T}_m(n)$ is central to the solution of this problem. To compute $\mathbf{T}_m(n)$, we may indeed use the method of backward linear prediction as described here. In particular, we may state the following fundamental relationship

$$\mathbf{J}_k(n) = \mathbf{J}_{b,k}(n), \qquad k = 0, 1, \ldots, m \tag{18.59}$$

where the sequence of Givens rotations $\mathbf{J}_0(n), \mathbf{J}_1(n), \ldots, \mathbf{J}_m(n)$ comprises the transformation $\mathbf{T}_m(n)$, and the corresponding sequence of Givens rotations $\mathbf{J}_{b,0}(n), \mathbf{J}_{b,1}(n), \ldots, \mathbf{J}_{b,m}(n)$ results from the use of backward linear prediction. In other words, we have

$$\mathbf{T}_m(n) = \mathbf{J}_{b,m}(n) \cdots \mathbf{J}_{b,1}(n)\mathbf{J}_{b,0}(n) \tag{18.60}$$

For a proof of this relationship, by induction, the reader is referred to Problem 8.

Let $\mathbf{H}_m(n)$ denote the $(m + 1)$-by-$(m + 1)$ unitary matrix obtained by deleting all those rows and columns of the transformation $\mathbf{T}_m(n)$ that have a sole entry of unity for their nonzero elements. Let $\mathbf{G}_{b,0}(n), \mathbf{G}_{b,1}(n), \ldots, \mathbf{G}_{b,m}(n)$ denote the matrices obtained by deleting the corresponding rows and columns of the Givens rotations $\mathbf{J}_{b,0}(n), \mathbf{J}_{b,1}(n), \ldots, \mathbf{J}_{b,m}(n)$, respectively. We may therefore use Eq. (18.60) to write

$$\mathbf{H}_m(n) = \mathbf{G}_{b,m}(n) \cdots \mathbf{G}_{b,1}(n)\mathbf{G}_{b,0}(n) \tag{18.61}$$

From Eq. (18.15) we thus deduce the following relation:

$$\begin{aligned}\begin{bmatrix} \mathbf{p}_m(n) \\ \epsilon^*_{m+1}(n) \end{bmatrix} &= \mathbf{H}_m(n)\begin{bmatrix} \lambda^{1/2}\mathbf{p}_m(n-1) \\ d^*(n) \end{bmatrix} \\ &= \mathbf{G}_{b,m}(n) \cdots \mathbf{G}_{b,1}(n)\mathbf{G}_{b,0}(n)\begin{bmatrix} \lambda^{1/2}\mathbf{p}_m(n-1) \\ d^*(n) \end{bmatrix}\end{aligned} \tag{18.62}$$

Let the expanded form of the $(m + 1)$-by-1 vector $\mathbf{p}_m(n)$ be written as

$$\mathbf{p}_m(n) = \begin{bmatrix} p_0(n) \\ p_1(n) \\ \vdots \\ p_m(n) \end{bmatrix} \tag{18.63}$$

Consider then the effect of applying $\mathbf{G}_{b,0}(n)$ to the vector $[\lambda^{1/2}\mathbf{p}_m(n-1), d^*(n)]^T$. Specifically, consider the following matrix product:

$$\begin{aligned}\mathbf{G}_{b,0}(n)\begin{bmatrix} \lambda^{1/2}\mathbf{p}_m(n-1) \\ d^*(n) \end{bmatrix} &= \begin{bmatrix} c_{b,0}(n) & \mathbf{0}^T & s^*_{b,0}(n) \\ \mathbf{0} & \mathbf{I}_{m-1} & \mathbf{0} \\ -s_{b,0}(n) & \mathbf{0}^T & c_{b,0}(n) \end{bmatrix}\begin{bmatrix} \lambda^{1/2}p_0(n-1) \\ \lambda^{1/2}p_1(n-1) \\ \vdots \\ \lambda^{1/2}p_{m-1}(n-1) \\ d^*(n) \end{bmatrix} \\ &= \begin{bmatrix} p_0(n) \\ \lambda^{1/2}p_1(n-1) \\ \vdots \\ \lambda^{1/2}p_{m-1}(n-1) \\ \epsilon^*_1(n) \end{bmatrix}\end{aligned} \tag{18.64}$$

where

$$\epsilon_1(n) = c_{b,0}(n)d(n) - s^*_{b,0}(n)\lambda^{1/2}p^*_0(n-1) \tag{18.65}$$

and

$$p_0^*(n) = c_{b,0}(n)\lambda^{1/2}p_0^*(n-1) + s_{b,0}(n)d(n) \tag{18.66}$$

Consider next the effect of applying $\mathbf{G}_{b,1}(n)$ to the resultant vector of Eq. (18.64). That is, consider the following second matrix product:

$$\mathbf{G}_{b,1}(n)\begin{bmatrix} p_0(n) \\ \lambda^{1/2}p_1(n-1) \\ \vdots \\ \lambda^{1/2}p_{m-1}(n-1) \\ \epsilon_1^*(n) \end{bmatrix} = \begin{bmatrix} 1 & 0 & 0 & 0 \\ 0 & c_{b,1}(n) & \mathbf{0}^T & s_{b,1}^*(n) \\ \mathbf{0} & \mathbf{0} & I_{m-1} & \mathbf{0} \\ 0 & -s_{b,1}(n) & \mathbf{0}^T & c_{b,1}(n) \end{bmatrix}\begin{bmatrix} p_0(n) \\ \lambda^{1/2}p_1(n-1) \\ \vdots \\ \lambda^{1/2}p_{m-1}(n-1) \\ \epsilon_1^*(n) \end{bmatrix}$$

$$= \begin{bmatrix} p_0(n) \\ p_1(n) \\ \vdots \\ \lambda^{1/2}p_{m-1}(n-1) \\ \epsilon_2^*(n) \end{bmatrix} \tag{18.67}$$

where

$$\epsilon_2(n) = c_{b,1}(n)\epsilon_1(n) - s_{b,1}^*(n)\lambda^{1/2}p_1^*(n-1) \tag{18.68}$$

and

$$p_1^*(n) = c_{b,1}(n)\lambda^{1/2}p_1^*(n-1) + s_{b,1}(n)\epsilon_1(n) \tag{18.69}$$

We may now generalize the results described in Eqs. (18.65), (18.66), (18.68), and (18.69) by writing the next pair of updates:

1. Order update for computing the *angle-normalized joint-process estimation error:*

$$\epsilon_{m+1}(n) = c_{b,m}(n)\epsilon_m(n) - s_{b,m}^*(n)\lambda^{1/2}p_m^*(n-1) \tag{18.70}$$

2. Time update for computing the *joint-process auxiliary parameter:*

$$p_m^*(n) = c_{b,m}(n)\lambda^{1/2}p_m^*(n-1) + s_{b,m}(n)\epsilon_m(n) \tag{18.71}$$

For the order index we have $m = 0, 1, \ldots, M$ where M is the final prediction order. Note that since $c_{b,m}(n)$ and $s_{b,m}(n)$ are both dimensionless ratios, the parameter $p_m(n)$ has the same dimension as the error $\epsilon_{m+1}(n)$.

For the computation of the cosine–sine pair of rotation parameters $c_{b,m}(n)$ and $s_{b,m}(n)$, we may use Eqs. (18.45) and (18.46) with the time index $n-1$ and order index $m-1$ replaced by n and m, respectively; that is,

$$c_{b,m}(n) = \frac{\lambda^{1/2}\mathcal{B}_m^{1/2}(n-1)}{\mathcal{B}_m^{1/2}(n)} \tag{18.72}$$

and

$$s_{b,m}(n) = \frac{\epsilon_{b,m}^*(n)}{\mathcal{B}_m^{1/2}(n)} \tag{18.73}$$

where

$$\mathcal{B}_m^{1/2}(n) = \lambda \mathcal{B}_m^{1/2}(n-1) + |\epsilon_{b,m}(n)|^2 \tag{18.74}$$

Update for the Conversion Factor

From Chapter 14 we also recall that the square root of the conversion factor or angle variable $\gamma_{m+1}(n)$ is defined by (in the context of our present notation)

$$\gamma_{m+1}^{1/2}(n) = c_m(n) \cdots c_1(n)c_0(n) \tag{18.75}$$

In view of Eq. (18.59), we may redefine $\gamma_{m+1}^{1/2}(n)$ in terms of the cosine parameters associated with the backward prediction-based Givens rotations $\mathbf{J}_{b,0}(n)$, $\mathbf{J}_{b,1}(n)$, . . . , $\mathbf{J}_{b,m}(n)$ as follows:

$$\gamma_{m+1}^{1/2}(n) = c_{b,m}(n) \cdots c_{b,1}(n)c_{b,0}(n) \tag{18.76}$$

This means that knowledge of the angle-normalized backward prediction errors $\epsilon_{b,0}(n)$, $\epsilon_{b,1}(n)$, . . . , $\epsilon_{b,m}(n)$ is indeed sufficient to compute the conversion factor $\gamma_{m+1}(n)$ for prediction order m. Moreover, we deduce the following order update:

$$\gamma_{m+1}^{1/2}(n) = c_{b,m}(n)\gamma_m^{1/2}(n) \tag{18.77}$$

where $m = 0, 1, \ldots, M$.

Thus, having computed the values of the angle-normalized joint-process estimation error $\epsilon_{m+1}(n)$ and the associated conversion factor $\gamma_{m+1}(n)$, we may then compute the a posteriori estimation error $e_{m+1}(n)$ as follows:

$$e_{m+1}(n) = \begin{cases} 0, & n \leq m+1 \\ \gamma_{m+1}^{1/2}(n)\epsilon_{m+1}(n), & n > m+1 \end{cases} \tag{18.78}$$

Naturally, we may carry out this computation for each state of the QRD-LSL algorithm (i.e., for $m = 0, 1, \ldots, M$), if so desired. In practice, however, the computation is performed only for the final stage (i.e., the final prediction order M).

18.8 FORMULATION OF THE QRD-LSL ALGORITHM

We are now ready to formulate the QRD-LSL algorithm, the essence of which is summed up in three Givens rotations: the rotation $\mathbf{J}_{b,m-1}(n-1)$ for computing the angle-normalized forward prediction error $\epsilon_{f,m}(n)$, the rotation $\mathbf{J}_{f,m-1}(n)$ for computing the angle-normalized backward prediction error $\epsilon_{b,m}(n)$, and the rotation $\mathbf{J}_{b,m}(n)$ for computing the angle-normalized joint-process estimation error $\epsilon_{m+1}(n)$.

Table 18.1 presents a summary of the QRD-LSL algorithm. The computation part of this summary includes two sets of updates:

1. The updates of Eqs. (18.48), (18.49), (18.57), and (18.58) for the lattice predictor section of the algorithm.
2. The updates of Eqs. (18.70), (18.71), (18.77), and (18.78) for the filtering section of the algorithm.

TABLE 18.1 SUMMARY OF THE QRD-LSL ALGORITHM

1. *Computations*
   (a) *Predictions:* For time $n = 1, 2, \ldots,$ and prediction order $m = 1, 2, \ldots, M$, where M is the final prediction order, compute

$$\mathcal{B}_{m-1}(n-1) = \lambda \mathcal{B}_{m-1}(n-2) + |\epsilon_{b,m-1}(n-1)|^2$$

$$c_{b,m-1}(n-1) = \frac{\lambda^{1/2}\mathcal{B}_{m-1}^{1/2}(n-2)}{\mathcal{B}_{m-1}^{1/2}(n-1)}$$

$$s_{b,m-1}(n-1) = \frac{\epsilon_{b,m-1}^{*}(n-1)}{\mathcal{B}_{m-1}^{1/2}(n-1)}$$

$$\epsilon_{f,m}(n) = c_{b,m-1}(n-1)\epsilon_{f,m-1}(n) - s_{b,m-1}^{*}(n-1)\lambda^{1/2}\pi_{f,m-1}^{*}(n-1)$$

$$\pi_{f,m-1}^{*}(n) = c_{b,m-1}(n-1)\lambda^{1/2}\pi_{f,m-1}^{*}(n-1) + s_{b,m-1}(n-1)\epsilon_{f,m-1}(n)$$

$$\gamma_m^{1/2}(n-1) = c_{b,m-1}(n-1)\gamma_{m-1}^{1/2}(n-1)$$

$$\mathcal{F}_{m-1}(n) = \lambda \mathcal{F}_{m-1}(n-1) + |\epsilon_{f,m-1}(n)|^2$$

$$c_{f,m-1}(n) = \frac{\lambda^{1/2}\mathcal{F}_{m-1}^{1/2}(n-1)}{\mathcal{F}_{m-1}^{1/2}(n)}$$

$$s_{f,m-1}(n) = \frac{\epsilon_{f,m-1}^{*}(n)}{\mathcal{F}_{m-1}^{1/2}(n)}$$

$$\epsilon_{b,m}(n) = c_{f,m-1}(n)\epsilon_{b,m-1}(n-1) - s_{f,m-1}^{*}(n)\lambda^{1/2}\pi_{b,m-1}^{*}(n-1)$$

$$\pi_{b,m-1}^{*}(n) = c_{f,m-1}(n)\lambda^{1/2}\pi_{b,m-1}^{*}(n-1) + s_{f,m-1}(n)\epsilon_{b,m-1}(n-1)$$

   (b) *Filtering:* For time $n = 1, 2, \ldots,$ compute

$$\mathcal{B}_M(n) = \lambda \mathcal{B}_M(n-1) + |\epsilon_{b,M}(n)|^2$$

$$c_{b,M}(n) = \frac{\lambda^{1/2}\mathcal{B}_M^{1/2}(n-1)}{\mathcal{B}_M^{1/2}(n)}$$

$$s_{b,M}(n) = \frac{\epsilon_{b,M}^{*}(n)}{\mathcal{B}_M^{1/2}(n)}$$

$$\epsilon_{M+1}(n) = c_{b,M}(n)\epsilon_M(n) - s_{b,M}^{*}(n)\lambda^{1/2}p_M^{*}(n-1)$$

$$p_M^{*}(n) = c_{b,M}(n)\lambda^{1/2}p_M^{*}(n-1) + s_{b,M}(n)\epsilon_M(n)$$

$$\gamma_{M+1}^{1/2}(n) = c_{b,M}(n)\gamma_M^{1/2}(n)$$

$$e_{M+1}(n) = \gamma_{M+1}^{1/2}(n)\epsilon_{M+1}(n)$$

2. *Initialization*
   (a) *Auxiliary parameter initialization:* For order $m = 1, 2, \ldots, M$, set
   $\pi_{f,m-1}(0) = \pi_{b,m-1}(0) = 0$
   $p_m(0) = 0$
   (b) *Soft constraint initialization:* For order $m = 0, 1, \ldots, M$, set
   $\mathcal{B}_m(1) = \delta$
   $\mathcal{F}_m(0) = \delta$
   where δ is a small positive constant.
   (c) *Data initialization:* For $n = 1, 2, \ldots,$ compute
   $\epsilon_{f,0}(n) = \epsilon_{b,0}(n) = u(n)$
   $\epsilon_0(n) = d(n)$
   $\gamma_0(n) = 1$
   where $u(n)$ is the input and $d(n)$ is the desired response at time n.

The *initialization* procedure summarized in Table 18.1 is of a *soft-constraint* form, which makes it consistent with that adopted for the conventional LSL algorithm, developed in Chapter 17.

We refer to the algorithm summarized in Table 18.1 as "lattice-based" because it proceeds on a stage-by-stage basis, with the computations in each stage having the form of a latttice. Indeed, the lattice format of these computations is exemplified in the multistage signal-flow graph of Fig. 18.1. In particular, we see that stage m of the predictor section of the algorithm involves the computations of the angle-normalized prediction errors: $\epsilon_{f,m}(n)$ and $\epsilon_{b,m}(n)$, where the prediction order $m = 1, 2, \ldots, M$. On the other hand, the filtering section of the algorithm involves the computation of the angle-normalized joint-process estimation error $\epsilon_{m+1}(n)$, where $m = 0, 1, \ldots, M$. The details of these computations are depicted in the signal-flow graphs shown in Fig. 18.2, which further emphasize the inherent lattice nature of the QRD-LSL algorithm.

From the signal-flow graph of Fig. 18.1 we clearly see that the total number of Givens rotations needed for the computation of $\epsilon_{M+1}(n)$ is $2M + 1$, which increases linearly with the final prediction order M, hence the designation "fast." However, the price paid for the high level of computational efficiency of the fast algorithm described herein, compared to the conventional form of the algorithm described in Chapter 14, is that of having to write a more elaborate set of instructions.

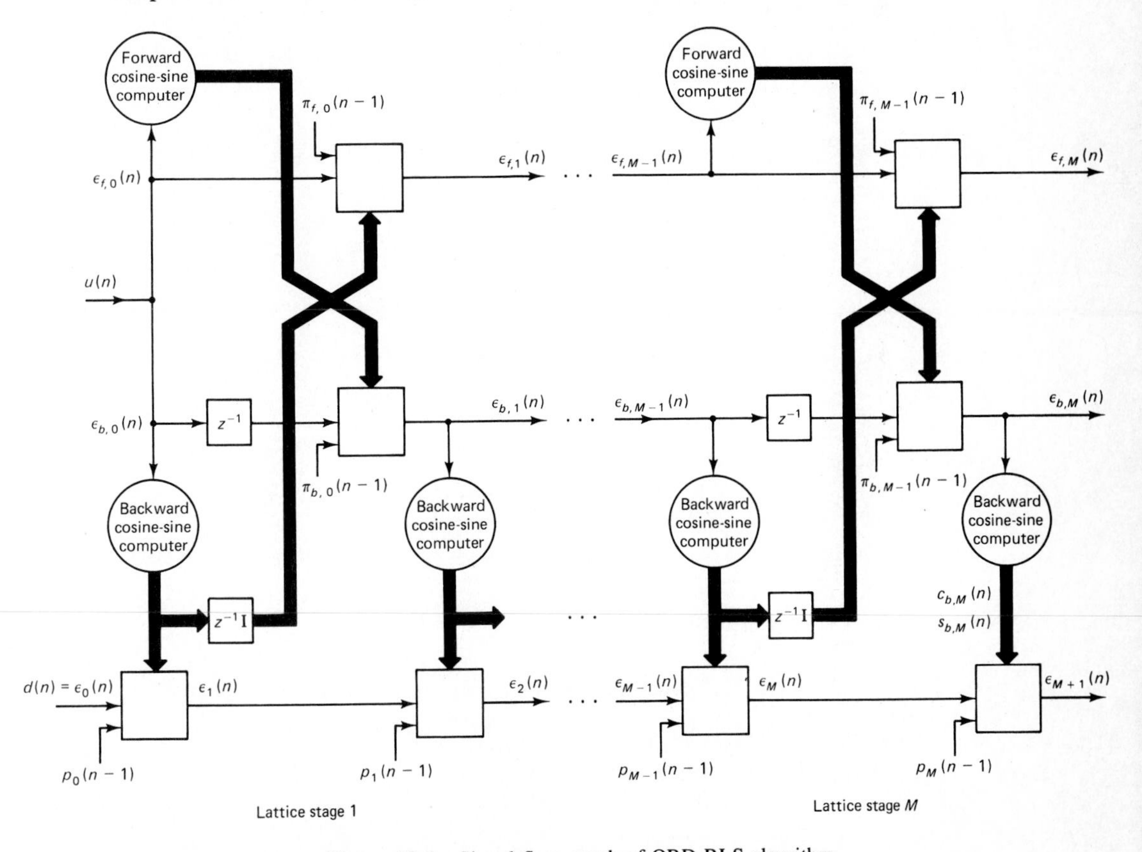

Figure 18.1 Signal-flow graph of QRD-RLS algorithm.

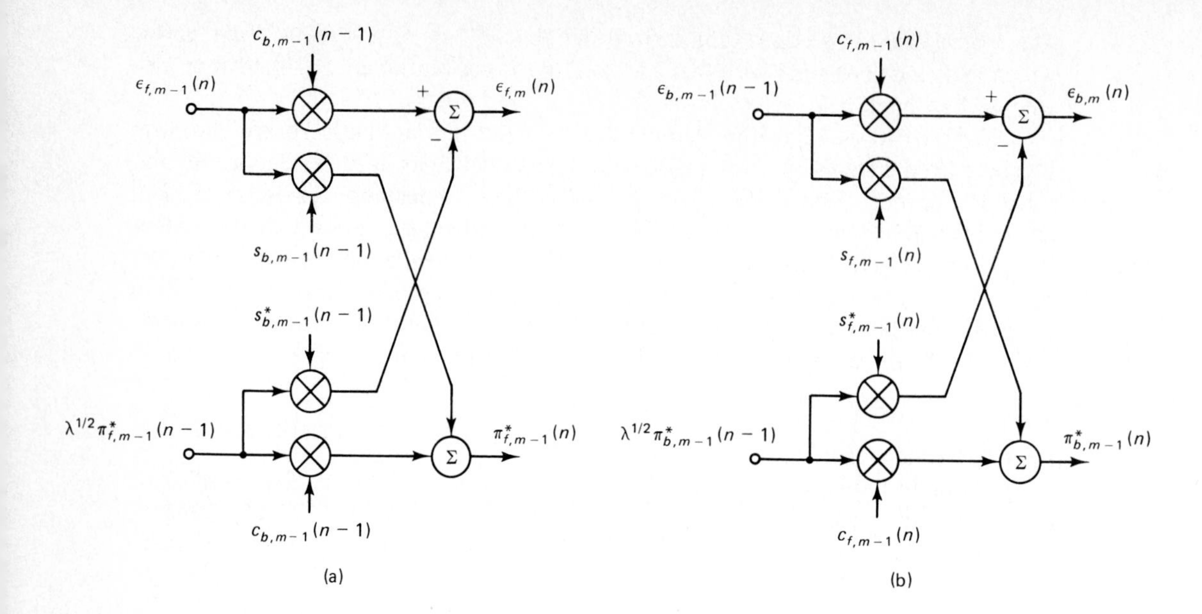

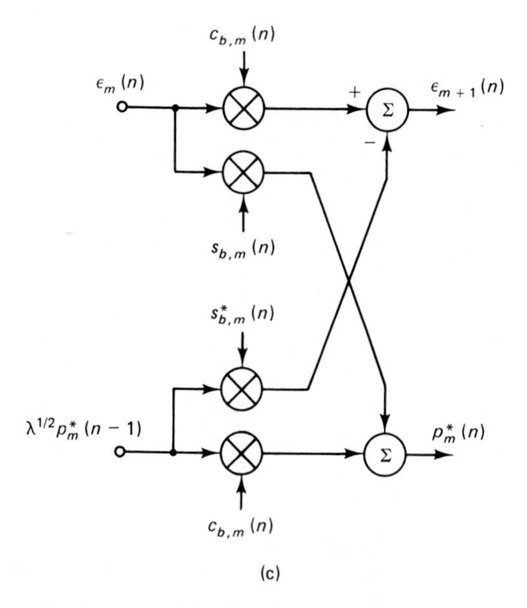

Figure 18.2 Signal-flow graphs for computing normalized variables in the QRD-RLS algorithm:
(a) Angle-normalized forward prediction error $\epsilon_{f,m}(n)$.
(b) Angle-normalized backward prediction error $\epsilon_{b,m}(n)$.
(c) Angle-normalized process estimation error $\epsilon_{m+1}(n)$.

Earlier we remarked that the computations of the cosine–sine pair of rotation parameters $c_{b,m}(n)$ and $s_{b,m}(n)$, described in Eqs. (18.72) to (18.74), are basically the same as those of the rotations parameters $c_{b,m-1}(n-1)$ and $s_{b,m-1}(n-1)$ in Eqs. (18.45) to (18.47) except for a change in order and a change in time. Hence, we may minimize the number of Givens rotations needed for the computation of $\epsilon_{M+1}(n)$ as shown in Fig. 18.1. Note, however, this is achieved at the expense of increased storage, signified by $z^{-1}\mathbf{I}$ in Fig. 18.1. The variable $\epsilon_{M+1}(n)$ equals the final value of the angle-normalized joint-process estimation error, where M is the final prediction order.

Examining the signal-flow graphs of Figs. 18.1 and 18.2, we may identify three basic elements in the formulation of the QRD-LSL algorithm:

1. *Initial values*. The algorithm begins with a set of *initial values* determined by the input datum $u(n)$ and desired response $d(n)$. The angle-normalized prediction errors $\epsilon_{f,m-1}(n)$, and $\epsilon_{b,m-1}(n)$ represent the inputs of stage m of the prediction section of the algorithm; their *initial values* are (for $m = 1$)

$$\epsilon_{f,0}(n) = \epsilon_{b,0}(n) = u(n) \tag{18.79}$$

The initial value for the angle-normalized joint-process estimation error $\epsilon_m(n)$ in the filtering section of the algorithm is (for $m = 0$)

$$\epsilon_0(n) = d(n) \tag{18.80}$$

The initial value for the conversion factor $\gamma_m(n)$ is (for $m = 0$)

$$\gamma_0(n) = 1 \tag{18.81}$$

2. *Order updates (recursions)*. At each stage of the algorithm, *order updates* are performed on these estimation errors. Specifically, a series of m order updates applied to the initial values $\epsilon_{f,0}(n)$ and $\epsilon_{b,0}(n)$ yields the final values $\epsilon_{f,M}(n)$ and $\epsilon_{b,M}(n)$, respectively, where M is the final prediction order. To compute the final value of the angle-normalized joint-process estimation error $\epsilon_{M+1}(n)$, another series of $M + 1$ order updates are applied to the initial value $\epsilon_0(n)$. This latter set of updates includes the use of the sequence of angle-normalized backward prediction errors $\epsilon_{b,0}(n), \epsilon_{b,1}(n), \ldots, \epsilon_{b,M}(n)$. The final order update pertains to the computation of the square root of the conversion factor, $\gamma_{M+1}^{1/2}(n)$, which involves the application of $M + 1$ order updates to the initial value $\gamma_0^{1/2}(n)$. The availability of the final values $\epsilon_{M+1}(n)$ and $\gamma_{M+1}^{1/2}(n)$ makes it possible to compute the final value $e_{M+1}(n)$ of the joint-process estimation error.

3. *Time updates (recursions)*. The computations of $\epsilon_{f,m}(n)$ and $\epsilon_{b,m}(n)$ as outputs of predictor stage m of the algorithm involve the use of the auxiliary parameters

$\pi_{f,m-1}(n)$ and $\pi_{b,m-1}(n)$, respectively, for $m = 1, 2, \ldots, M$. Similarly, the computation of $\epsilon_{m+1}(n)$ involves the auxiliary parameter $p_m(n)$ for $m = 0, 1, \ldots, M$. The computations of these three auxiliary parameters themselves have the following common features:

(a) They are all governed by a *first-order difference equation.*

(b) The coefficients of the equation are *time varying*. For exponential weighting (that is, $\lambda \le 1$), the coefficients are bounded in absolute value by one. Hence, the solution of the equation is convergent (see Problem 6).

(c) The term playing the role of "excitation" is represented by some form of an estimation error.

(d) In the prewindowing method, all three auxiliary parameters are equal to zero for $n \le 0$.

Consequently, the auxiliary parameters $\pi_{f,m-1}(n)$ and $\pi_{b,m-1}(n)$ for $m = 1, 2, \ldots, M$, and $p_m(n)$ for $m = 0, 1, \ldots, M$, may be computed recursively in time.

18.9 COMPUTER EXPERIMENT ON ADAPTIVE EQUALIZATION

In this experiment, we study the use of the QRD-LSL algorithm for adaptive equalization of a linear channel that produces unknown distortion. The predictor and filtering sections of the algorithm are applicable to this experiment. The details of this channel are the same as those used to study the RLS algorithm in Chapter 12 and the recursive LSL algorithm in Chapter 17, for a similar application. The results of the experiment should therefore help us make an assessment of the QRD-LSL algorithm, compared to a conventional form of the recursive LSL algorithm.

Figure 18.3 presents the superposition of learning curves of the QRD-LSL algorithm for the exponential weighting factor $\lambda = 1$, and four different values of the eigenvalue spread $\chi(\mathbf{R}) = 6.0782, 11.1238, 21.7132, 46.8216$ that correspond to the channel parameter $W = 2.9, 3.1, 3.3, 3.5$, respectively. Each learning curve was computed using double precision. It was obtained by ensemble-averaging the squared value of the final a priori estimation error (i.e., the innovation) $\alpha_{M+1}(n)$ over 200 independent trials of the experiment for a final prediction order $M = 10$. To compute the a priori estimation error $\alpha_{M+1}(n)$, we first recognize that [see Eq. (15.51)]

$$\alpha_{M+1}(n) = \frac{e_{M+1}(n)}{\gamma_{M+1}(n)} \tag{18.82}$$

Hence, rewriting Eq. (18.16) for order $M + 1$, and then using Eq. (18.82) to eliminate $e_{M+1}(n)$, we get the desired relation:

$$\alpha_{M+1}(n) = \frac{\epsilon_{M+1}(n)}{\gamma_{M+1}^{1/2}(n)} \tag{18.83}$$

where $\epsilon_{M+1}(n)$ is the final value of the angle-normalized joint-process estimation error, and γ_{M+1} is the associated conversion factor. Note that for a final prediction order M, the number of taps involved in the joint-process estimation is $M + 1$.

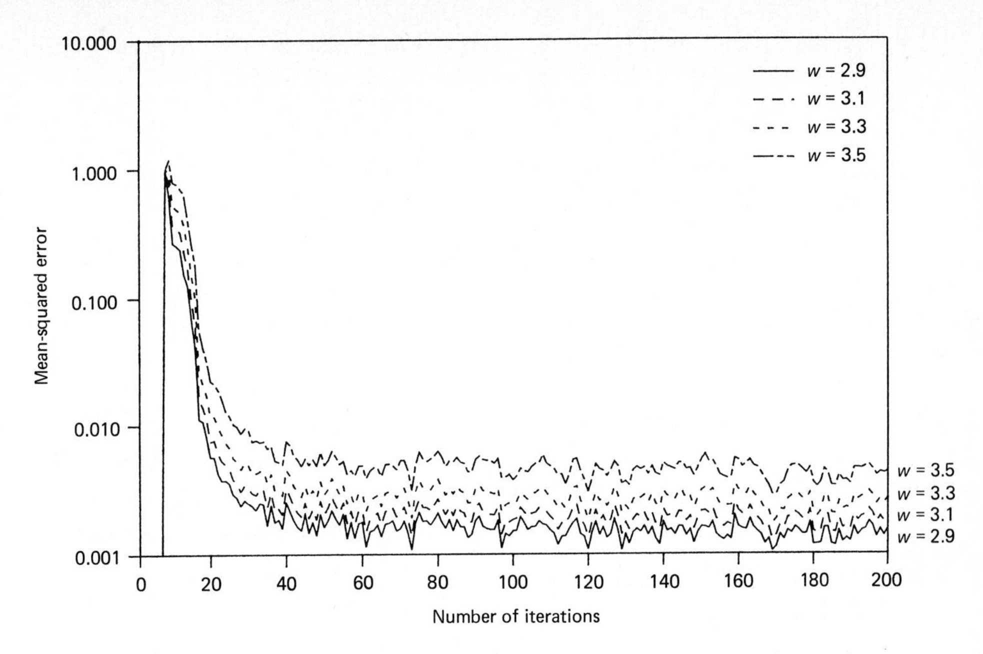

Figure 18.3 Learning curves of the QRD-LSL algorithm for the adaptive equalization experiment.

Based on the results of this experiment and those of Section 17.9 (involving the recursive LSL algorithm), we observe that for each eigenvalue spread, the learning curve of the QRD-LSL algorithm follows a trajectory identical to that computed for the recursive LSL algorithm. This assumes the use of double-precision arithmetic (i.e., no finite precision effects).

18.10 RELATIONSHIPS BETWEEN THE QRD-LSL AND CONVENTIONAL LSL ALGORITHMS

The auxiliary parameters of the QRD-LSL, namely, $\pi_{f,m-1}(n)$, $\pi_{b,m-1}(n-1)$, and $p_m(n)$, are related directly to the reflection coefficients $\Gamma_{f,m-1}(n)$, $\Gamma_{b,m-1}(n)$ and the regression parameter (i.e., tap weight) $\kappa_m(n)$ of conventional LSL algorithms studied in Chapter 17. In this section, we derive these relationships and thereby build important linkages between the QRD-LSL and conventional LSL algorithms. Furthermore, the relationships provide insightful interpretations of the auxiliary parameters that are basic to the operation of the QRD-LSL algorithm.

Lattice Predictor

Consider first the order update for the forward linear prediction half of the lattice structure, described in Eq. (18.48). The cosine–sine pair of rotation parameters, $c_{b,m-1}(n-1)$ and $s_{b,m-1}(n-1)$, involved in the makeup of this equation are themselves defined in Eqs. (18.45) and (18.46), respectively. Hence, substituting the definitions of Eqs. (18.45) and (18.46) in (18.48) yields

$$\epsilon_{f,m}(n) = \frac{\lambda^{1/2}\mathcal{B}_{m-1}^{1/2}(n-2)}{\mathcal{B}_{m-1}^{1/2}(n-1)}\epsilon_{f,m-1}(n) - \frac{\lambda^{1/2}\pi_{f,m-1}^{*}(n-1)}{\mathcal{B}_{m-1}^{1/2}(n-1)}\epsilon_{b,m-1}(n-1) \tag{18.84}$$

Next, from the definition of an angle-normalized prediction error, and from the relation between a particular pair of a posteriori and a priori prediction errors, we have

$$\epsilon_{f,m}(n) = \gamma_m^{1/2}(n-1)\eta_m(n) \tag{18.85}$$

and

$$\epsilon_{b,m} = \gamma_m^{1/2}(n)\psi_m(n) \tag{18.86}$$

where $\eta_m(n)$ and $\psi_m(n)$ are the a priori values of the forward and backward prediction errors, respectively. Therefore, substituting Eqs. (18.85) and (18.86) in (18.84), and factoring out a certain combination of parameters, we may write

$$\eta_m(n) = \frac{\lambda^{1/2}\mathcal{B}_{m-1}^{1/2}(n-2)\gamma_{m-1}^{1/2}(n-1)}{\mathcal{B}_{m-1}^{1/2}\gamma_m^{1/2}(n-1)}\left[\eta_{m-1}(n) - \frac{\pi_{f,m-1}^{*}(n-1)}{\mathcal{B}_{m-1}^{1/2}(n-2)}\psi_{m-1}(n-1)\right] \tag{18.87}$$

We now recognize the following identity from Chapter 15 [see Eq. (15.57)]:

$$\frac{\lambda^{1/2}\mathcal{B}_{m-1}^{1/2}(n-2)\gamma_{m-1}^{1/2}(n-1)}{\mathcal{B}_{m-1}^{1/2}\gamma_m^{1/2}(n-1)} = 1 \tag{18.88}$$

Accordingly, we may simplify Eq. (18.87) as follows:

$$\eta_m(n) = \eta_{m-1}(n) - \frac{\pi_{f,m-1}^{*}(n-1)}{\mathcal{B}_{m-1}^{1/2}(n-2)}\psi_{m-1}(n-1) \tag{18.89}$$

Comparing this equation with (17.103), we immediately see that they have an identical mathematical structure. Moreover, we deduce the following relationship between the forward prediction auxiliary parameter $\pi_{f,m-1}(n-1)$ in the QRD-LSL algorithm and the forward reflection coefficient $\Gamma_{f,m}(n-1)$ in the conventional LSL algorithm based on a priori estimation errors:

$$-\frac{\pi_{f,m-1}(n-1)}{\mathcal{B}_{m-1}^{1/2}(n-2)} = \Gamma_{f,m}(n-1)$$

or, equivalently (replacing the time index $n-1$ with n, and rearranging terms),

$$\pi_{f,m-1}(n) = -\mathcal{B}_{m-1}^{1/2}(n-1)\Gamma_{f,m}(n) \tag{18.90}$$

In other words, $\pi_{f,m-1}(n)$ in the QRD-LSL algorithm plays a role directly analogous to $\Gamma_{f,m}(n)$ in conventional LSL algorithms for order $m = 1, 2, \ldots, M$.

Consider next the order update for the backward linear prediction half of the lattice structure, described in Eq. (18.57). The cosine–sine pair of rotation parameters, $c_{f,m-1}(n)$ and $s_{f,m-1}(n)$, involved in this equation are defined, respectively, in Eqs. (18.54) and (18.55). Therefore, the substitution of Eqs. (18.54) and (18.55) in

(18.57) yields

$$\epsilon_{b,m}(n) = \frac{\lambda^{1/2}\mathcal{F}_{m-1}^{1/2}(n-1)}{\mathcal{F}_{m-1}^{1/2}(n)}\epsilon_{b,m-1}(n-1) - \frac{\lambda^{1/2}\pi_{b,m-1}^{*}(n-1)}{\mathcal{F}_{m-1}^{1/2}(n)}\epsilon_{f,m-1}(n) \quad (18.91)$$

Moreover, using the definitions of Eqs. (18.85) and (18.86) in (18.91), and factoring out a certain combination of parameters appropriate to the situation at hand, we may write

$$\psi_m(n) = \frac{\lambda^{1/2}\mathcal{F}_{m-1}^{1/2}(n-1)\gamma_{m-1}^{1/2}(n-1)}{\mathcal{F}_{m-1}^{1/2}(n)\gamma_m^{1/2}(n)}\left[\psi_{m-1}(n-1) - \frac{\pi_{b,m-1}^{*}(n-1)}{\mathcal{F}_{m-1}^{1/2}(n-1)}\eta_{m-1}(n)\right] \quad (18.92)$$

Again, from Chapter 15 we recognize the following identity [see Eq. (15.56)]:

$$\frac{\lambda^{1/2}\mathcal{F}_{m-1}^{1/2}(n-1)\gamma_{m-1}^{1/2}(n-1)}{\mathcal{F}_{m-1}^{1/2}(n)\gamma_m^{1/2}(n)} = 1 \quad (18.93)$$

Accordingly, we may simplify Eq. (18.91) as follows:

$$\psi_m(n) = \psi_{m-1}(n-1) - \frac{\pi_{b,m-1}^{*}(n-1)}{\mathcal{F}_{m-1}^{1/2}(n-1)}\eta_{m-1}(n) \quad (18.94)$$

Comparing this equation with Eq. (17.104), we immediately see that they have the same mathematical structure. We also deduce the following relationship between the backward prediction auxiliary parameter $\pi_{b,m-1}(n-1)$ in the QRD-LSL algorithm and the backward reflection coefficient $\Gamma_{b,m}(n-1)$ in the conventional LSL algorithm based on a priori estimation errors:

$$-\frac{\pi_{b,m-1}(n-1)}{\mathcal{F}_{m-1}^{1/2}(n-1)} = \Gamma_{b,m}(n-1)$$

or, equivalently (replacing the time index $n-1$ with n, and rearranging terms),

$$\pi_{b,m-1}(n) = -\mathcal{F}_{m-1}^{1/2}(n)\Gamma_{b,m}(n) \quad (18.95)$$

In other words, $\pi_{b,m-1}(n)$ in the QRD-LSL algorithm plays a role directly analogous to $\Gamma_{b,m}(n)$ in conventional LSL algorithms for $m = 1, 2, \ldots, M$.

Joint-Process Estimator

The order update for computing the joint-process estimation error in the QRD-LSL algorithm is described in Eq. (18.70). The cosine–sine pair of rotation parameters, $c_{b,m}(n)$ and $s_{b,m}(n)$, involved in this update are themselves defined in Eqs. (18.72) and (18.73). Hence, substituting Eqs. (18.72) and (18.73) in (18.70), we get

$$\epsilon_{m+1}(n) = \frac{\lambda^{1/2}\mathcal{B}_m^{1/2}(n-1)}{\mathcal{B}_m^{1/2}(n)}\epsilon_m(n) - \frac{\lambda^{1/2}p_m^{*}(n-1)}{\mathcal{B}_m^{1/2}(n)}\epsilon_{b,m}(n) \quad (18.96)$$

Next, from the definition of an angle-normalized joint-process estimation error, and from the relation between a priori and a posteriori joint-process estimation errors, we have

$$\epsilon_m(n) = \gamma_m^{1/2}(n)\alpha_m(n) \quad (18.97)$$

where $\alpha_m(n)$ is the a priori joint-process estimation error (i.e., innovation). Therefore, substituting Eqs. (18.86) and (18.97) in (18.96), and factoring out a particular combination of terms appropriate to our present situation, we may write

$$\alpha_{m+1}(n) = \frac{\lambda^{1/2}\mathcal{B}_m^{1/2}(n-1)\gamma_m^{1/2}(n)}{\mathcal{B}_m^{1/2}(n)\gamma_{m+1}^{1/2}(n)}\left[\alpha_m(n) - \frac{p_m^*(n-1)}{\mathcal{B}_m^{1/2}(n-1)}\psi_m(n)\right] \tag{18.98}$$

But

$$\frac{\lambda^{1/2}\mathcal{B}_m^{1/2}(n-1)\gamma_m^{1/2}(n)}{\mathcal{B}_m^{1/2}(n)\gamma_{m+1}^{1/2}(n)} = 1 \tag{18.99}$$

which is a rewrite of the identity (18.88) with the time index $n - 1$ and order index $m - 1$ replaced by n and m, respectively. Accordingly, we may simplify Eq. (18.98) as follows:

$$\alpha_{m+1}(n) = \alpha_m(n) - \frac{p_m^*(n-1)}{\mathcal{B}_m^{1/2}(n-1)}\psi_m(n) \tag{18.100}$$

The mathematical structure of this update equation is identical to that of Eq. (17.105) for a conventional LSL algorithm based on a priori estimation errors. In particular, we deduce the following relationship between the joint-process auxiliary parameter $p_m(n-1)$ in the QRD-LSL algorithm and the regression coefficient (i.e., tap weight) $\kappa_m(n-1)$ in the conventional LSL algorithm:

$$\frac{p_m(n-1)}{\mathcal{B}_m^{1/2}(n-1)} = \kappa_m(n-1)$$

or, equivalently (replacing the time index $n - 1$ with n, and rearranging terms)

$$p_m(n) = \mathcal{B}_m^{1/2}(n)\kappa_m(n) \tag{18.101}$$

In other words, $p_m(n)$ in the QRD-LSL algorithm plays a role directly analogous to the regression coefficient $\kappa_m(n)$ in conventional LSL algorithms for $m = 0, 1, \ldots, M$.

Summary

We may now summarize the important relationships between the QRD-LSL and conventional LSL algorithms discovered in this section:

1. The forward prediction auxiliary parameter $\pi_{f,m-1}(n)$ in the QRD-LSL algorithm performs a function directly analogous to the forward reflection coefficient $\Gamma_{f,m}(n)$ in conventional LSL algorithms. Moreover, the time updating of the parameter $\pi_{f,m-1}(n)$ is computed directly.
2. The backward prediction auxiliary parameter $\pi_{b,m-1}(n)$ in the QRD-LSL algorithm performs a function directly analogous to the backward reflection coefficient $\Gamma_{b,m}(n)$ in conventional LSL algorithms, Here also, the time updating of the parameter $\pi_{b,m-1}$ is computed directly.
3. The joint-process auxiliary parameter $p_m(n)$ in the QRD-LSL algorithm performs a function directly analogous to the regression coefficient $\kappa_m(n)$ in con-

ventional LSL algorithm. As with the other two auxiliary parameters, the time updating of $p_m(n)$ is computed directly, too.

4. The direct methods of updating the auxiliary parameters $\pi_{f,m-1}(n)$, $\pi_{b,m-1}(n)$, and $p_m(n)$ in the QRD-LSL algorithm imply the presence of local *error feedback* in each of these time updates.
5. The relations between the auxiliary parameters $\pi_{f,m-1}(n)$, $\pi_{b,m-1}(n)$, and $p_m(n)$ in the QRD-LSL algorithm and the coefficients $\Gamma_{f,m}(n)$, $\Gamma_{b,m}(n)$, and $\kappa_m(n)$ in a conventional LSL algorithm, respectively, suggest that the forward and backward reflection coefficients and the regression coefficients ordinarily associated with a multistage lattice structure may indeed be computed (if so desired) simply as by-products of the QRD-LSL algorithm. The computation of the related multistage lattice structure may be needed for the purpose of system identification, spectrum estimation, and some other relevant application.

In a few words, then, the QRD-LSL algorithm summarized in Table 18.1 is a powerful adaptive filtering algorithm that bears a direct relationship to a conventional LSL algorithm employing error feedback as summarized in Table 17.3.

18.11 DISCUSSION

As with the other fast algorithms considered in the previous two chapters, the QR-decomposition-based least-squares lattice (QRD-LSL) algorithm derived in this chapter cleverly exploits key relations between the forward and backward forms of linear least-squares prediction. The basic idea is to recognize that we may do adaptive filtering by performing linear predictions and then applying proper adjustments.

Three distinct elements may be identified in the QRD-LSL algorithm summarized in Table 18.1:

1. Initial values specified by the input data and desired response.
2. Order updates (recursions) for computing angle-normalized estimation errors, including the adaptive filter output.
3. Time update (recursions) for the direct computation of a corresponding set of auxiliary parameters.

The QRD-LSL algorithm of Table 18.1 combines highly desirable features of recursive least-squares estimation, QR-decomposition, and a lattice structure. Accordingly, it offers a unique set of operational and implementational advantages:

1. The QRD-LSL algorithm has a *fast rate of convergence,* which is inherent in recursive least-squares estimation.
2. The good numerical properties of the QR-decomposition mean that the QRD-LSL algorithm is *numerically stable*.
3. The QRD-LSL algorithm offers a *high level of computational efficiency* in that its complexity is on the order of M, where M is the final prediction order (i.e., the number of available degrees of freedom).

4. The lattice structure of the QRD-LSL algorithm is *modular* in nature, giving it both a signal-processing flexibility and a potential for implementation in VLSI form.
5. The algorithm includes an *integral set of desired variables and parameters* that are useful to have in signal-processing applications. Specifically, it offers the following two sets of useful by-products:
   (a) Angle-normalized forward and backward prediction errors.
   (b) Auxiliary parameters that are directly related to the forward and backward reflection coefficients and the regression coefficients (i.e., tap weights), which are ordinarily associated with a lattice structure.

The theory presented in this chapter has focused primarily on the development of a "lattice" implementation of fast QR-decomposition-based recursive least-squares (QRD-RLS) algorithm. This theory may be extended to develop a *matrix-oriented version* (also known as the *Kalman type*) of fast QRD-RLS algorithm (Cioffi, 1988, 1990; Proudler et al., 1988; Bellanger, 1988). The reader is referred to these references for details of this latter type of fast QRD-RLS algorithm.

PROBLEMS

1. Rationalize the transformation described in Eq. (18.36) with the recursive relations in Eqs. (18.14) and (18.15).
2. Show that the recursive equation (18.17) for computing the minimum value of the weighted sum of error squares $\mathscr{E}_{\min}(n)$ is equivalent to that of Eq. (13.25).
3. For the backward linear least-squares prediction problem considered in Section 18.3, prove the relations given in Eqs. (18.24) and (18.26).
4. For the forward linear least-squares prediction problem considered in Section 18.4, prove the relation in Eqs. (18.33), (18.36), and (18.38).
5. Justify the following relationships:
   (a) Joint-process estimation errors:

$$|\epsilon_m(n)| = \sqrt{|e_m(n)| \cdot |\alpha_m(n)|}$$

$$\text{ang}[\epsilon_m(n)] = \text{ang}[e_m(n)] = \text{ang}[\alpha_m(n)]$$

   (b) Backward prediction errors:

$$|\epsilon_{b,m}(n)| = \sqrt{|b_m(n)| \cdot |\psi_m(n)|}$$

$$\text{ang}[\epsilon_{b,m}(n)] = \text{ang}[b_m(n)] = \text{ang}[\psi_m(n)]$$

   (c) Forward prediction errors:

$$|\epsilon_{f,m}(n)| = \sqrt{|f_m(n)| \cdot |\eta_m(n)|}$$

$$\text{ang}[\epsilon_{f,m}(n)] = \text{ang}[f_m(n)] = \text{ang}[\eta_m(n)]$$

6. Let $x(n)$ denote the auxiliary parameter $\pi_{f,m-1}(n)$, $\pi_{b,m-1}(n)$, or $p_m(n)$ involved in the QRD-LSL algorithm. Let $y(n)$ denote the associated estimation error $\epsilon_{f,m-1}(n)$, $\epsilon_{b,m-1}(n-1)$, or $\epsilon_m(n)$, respectively. The time evolution of the variable $x(n)$ may be described by a time-varying first-order difference equation:

$$x(n) = \lambda^{1/2} a_c(n) x(n-1) + a_s(n) y(n)$$

where $y(n)$ plays the role of excitation. For the coefficients of the equation, we have

$$0 \le a_c(n) \le 1$$

$$|a_s(n)| \le 1$$

and

$$\lambda \le 1$$

For the initial conditions, we have

$$x(n) = y(n) = 0, \qquad \text{for } n \le 0$$

(a) The requirement is to solve the time-varying first-order difference equation for $x(n)$. In particular, show that

$$x(n) = \Phi(n, 0)x(0) + \sum_{k=1}^{n} \Phi(n, k)a_s(k)y(k),$$

where

$$\Phi(n, k) = \begin{cases} \lambda^{(n-k)/2} \prod_{i=k+1}^{n} a_c(i), & n > k \\ 1, & n = k \end{cases}$$

Why is the solution $x(n)$ convergent, assuming that $0 < \lambda < 1$?

(b) Consider next the situation corresponding to $\lambda = 1$. For the application considered herein, we have

$$a_c^2(k) + |a_s(k)|^2 = 1, \qquad \text{for all } k$$

Hence, show that

$$|x(n)| = \|\mathbf{y}(n)\| \cos\theta$$

where

$$\mathbf{y}(n) = [x(0), y(1), \ldots, y(n)]^T$$

and θ is the angle between the vector $\mathbf{y}$ and the impulse response vector characterizing the time-varying first-order difference equation. (The variable $\cos\theta$ may be interpreted as a normalized cross-correlation between this pair of vectors). Hence, demonstrate the boundedness of $x(n)$ for $\lambda = 1$.

(c) How would you generalize the solution for $|x(n)|$ in part (b) to deal with $0 < \lambda < 1$?

Hint: For the solution to part (b), it is instructive to expand the summation in the formula for $x(n)$, and use vector notation to express $x(1)$, $x(2)$, $x(3)$, and so on.

7. Suppose that we have computed the cosine parameters $c_{f,m}(n)$ and $c_{b,m-1}(n-1)$ pertaining to the Givens transformations $\mathbf{J}_{f,m}(n)$ and $\mathbf{J}_{b,m-1}(n-1)$, respectively. Hence, show that the conversion factor $\gamma_{m+1}(n)$ may be updated in both time and order as follows:

$$\gamma_{m+1}^{1/2}(n) = c_{f,m}(n)c_{b,m-1}(n-1)\gamma_{m-1}^{1/2}(n-1)$$

8. Consider the n-by-$(m+1)$ partially triangularized matrix

$$\begin{bmatrix} \lambda^{1/2}\mathbf{R}_m(n-1) \\ \mathbf{O} \\ \mathbf{u}_{m+1}^H(n) \end{bmatrix}$$

where $\mathbf{R}_m(n)$ is the old value of an $(m + 1)$-by-$(m + 1)$ upper triangular matrix, and $\mathbf{u}_{m+1}(n)$ is the new value of an $(m + 1)$-by-1 input data vector.

(a) Let $\mathbf{J}_0(n)$ denote the Givens rotation:

$$\mathbf{J}_0(n) = \begin{bmatrix} c_0(n) & \mathbf{0}^T & s_0^*(n) \\ \mathbf{0} & \mathbf{I}_{n-2} & \mathbf{0} \\ -s_0(n) & \mathbf{0}^T & c_0(n) \end{bmatrix}$$

where $\mathbf{I}_{n-2}$ is the $(n - 2)$-by-$(n - 2)$ identity matrix. The matrix $\mathbf{J}_0(n)$ is used to annihilate the first element of $\mathbf{u}_{m+1}^H(n)$. Show that

$$\mathbf{J}_0(n)\begin{bmatrix} \lambda^{1/2}\mathbf{R}_m(n-1) \\ \mathbf{0} \\ \mathbf{u}_{m+1}^H(n) \end{bmatrix} = \begin{bmatrix} \mathcal{B}_0^{1/2}(n) & r_{01}(n) & \cdots \\ 0 & \lambda^{1/2}\mathcal{B}_1^{1/2}(n-1) & \cdots \\ \vdots & \vdots & \\ 0 & \epsilon_{b,1}^*(n) & \cdots \end{bmatrix}$$

where $\mathcal{B}_0^{1/2}(n)$ and $r_{01}(n)$ are time-updated element values of row 0 of upper triangular matrix $\mathbf{R}_m(n)$, and $\epsilon_{b,1}(n)$ is the angle-normalized backward prediction error for first-order prediction. Hence, show that

$$\mathbf{J}_0(n) = \mathbf{J}_{b,0}(n)$$

Hint: For $\epsilon_{b,1}(n)$ you may use the augmented normal equations for first-order backward linear prediction [see Eq. (15.15)]:

$$\mathbf{\Phi}(n)\begin{bmatrix} -g(n) \\ 1 \\ \mathbf{0} \end{bmatrix} = \begin{bmatrix} 0 \\ \mathcal{B}_1(n) \\ \mathbf{O} \end{bmatrix}$$

where $\mathbf{\Phi}(n)$ is the correlation matrix, $g^*(n)$ is the backward predictor coefficient, and $\mathcal{B}_1(n)$ is the minimum value of the sum of weighted backward prediction-error squares.

(b) Let

$$\mathbf{J}_1(n) = \begin{bmatrix} 1 & 0 & 0 & 0 \\ 0 & c_1(n) & \mathbf{0}^T & s_1^*(n) \\ \mathbf{0} & \mathbf{0} & \mathbf{I}_{n-3} & \mathbf{0}^T \\ 0 & -s_1(n) & \mathbf{0}^T & c_1(n) \end{bmatrix}$$

The Givens rotation $\mathbf{J}_1(n)$ is designed to annihilate $\epsilon_{b,1}^*(n)$. Apply this second Givens rotation to the matrix resulting from the transformation given in part (a). Show that the third element in the bottom row of the matrix is replaced by $\epsilon_{b,2}^*(n)$, where $\epsilon_{b,2}(n)$ is the angle-normalized backward prediction error for order 2. Hence, show that

$$\mathbf{J}_1(n) = \mathbf{J}_{b,1}(n).$$

(c) Generalize the results of parts (a) and (b) for the application of $\mathbf{J}_0(n)$, $\mathbf{J}_1(n)$, . . . , $\mathbf{J}_m(n)$.

9. In this problem we explore the relationships between the direct computations of the auxiliary parameters $\pi_{f,m-1}(n)$, $\pi_{b,m-1}(n)$, and $p_m(n)$ in the QRD-LSL algorithm and the reflection coefficients $\Gamma_{f,m}(n)$ and $\Gamma_{b,m}(n)$ and the regression coefficient $\kappa_m(n)$ in a conventional LSL algorithm. Specifically demonstrate the mathematical equivalence between the equations in each of the following sets:

(a) Equation (18.49) for QRD-LSL algorithm and Eq. (17.107) for conventional LSL algorithm.

(b) Equation (18.58) for QRD-LSL algorithm and Eq. (17.108) for conventional LSL algorithm.

(c) Equation (18.71) for QRD-LSL algorithm and Eq. (17.109) for conventional LSL algorithm.

10.[4] Let

$$\mathbf{X}(n) = \mathbf{\Lambda}^{1/2}(n)\mathbf{A}(n)$$

and

$$\mathbf{y}(n) = \mathbf{\Lambda}^{1/2}\mathbf{d}(n)$$

where we have purposely suppressed the use of a subscript in association with the data matrix $\mathbf{A}(n)$ and desired data vector $\mathbf{d}(n)$, as it is of no immediate concern in this problem. Let the unitary matrix $\mathbf{Q}(n)$ involved in the QR-decomposition of the exponentially weighted data matrix $\mathbf{X}(n)$ be partitioned as follows:

$$\mathbf{Q}(n) = \begin{bmatrix} \mathbf{Q}_1(n) \\ \mathbf{Q}_2(n) \end{bmatrix}$$

Let $\mathbf{X}(n)$ and $\mathbf{Q}_1(n)$ be partitioned as follows, respectively:

$$\mathbf{X}(n) = [\mathbf{x}(n), \mathbf{x}(n-1), \ldots, \mathbf{x}(n-M)]$$

and

$$\mathbf{Q}_1^H(n) = [\mathbf{q}_0(n), \mathbf{q}_1(n), \ldots, \mathbf{q}_M(n)]$$

(a) Show that, by choosing $\mathbf{Q}_1(n)$ to have the right number of rows

$$\mathbf{Q}_1^H(n) = \mathbf{X}(n)\mathbf{R}^{-1}(n)$$

where $\mathbf{R}^{-1}(n)$ is the inverse of the upper triangular matrix $\mathbf{R}(n)$ involved in the QR-decomposition of $\mathbf{X}(n)$. Hence, demonstrate that

$$\text{span}\{\mathbf{x}(n), \ldots, \mathbf{x}(n-m)\} = \text{span}\{\mathbf{q}_0(n), \ldots, \mathbf{q}_m(n)\}, \qquad m = 0, \ldots, M$$

which, in effect, states that the inverse matrix $\mathbf{R}^{-1}(n)$ performs a Gram–Schmidt orthogonalization on the columns of $\mathbf{X}(n)$. What is the physical interpretation of $\mathbf{q}_m(n)$?

(b) Show that the least-squares estimate of $\mathbf{y}(n)$ is given by

$$\hat{\mathbf{y}}(n) = \mathbf{X}(n)\mathbf{R}^{-1}(n)\mathbf{p}(n), \qquad n > M + 1$$

where the vector $\mathbf{p}(n)$ results from the application of the transformation $\mathbf{Q}(n)$ to the exponentially weighted desired data vector $\mathbf{y}(n)$. Hence, demonstrate the following expanded form of the estimate $\hat{\mathbf{y}}(n)$:

$$\hat{\mathbf{y}}(n) = \sum_{m=0}^{M} p_m(n)\mathbf{q}_m(n), \qquad n > M + 1$$

where $p_0(n), p_1(n), \ldots, p_M(n)$ constitute the elements of the vector $\mathbf{p}(n)$. Compare the physical significance of $p_m(n)$ deduced from this representation with that described in Eq. (18.101).

[4] This problem is taken from Regalia and Bellanger (1991).

Computer-Oriented Problems

11. *Computer experiment on adaptive linear prediction*
Use the QRD-LSL algorithm to study an adaptive linear predictor of order two. The process $\{u(n)\}$ used in the study is an autoregressive (AR) process described by the following difference equation:

$$u(n) + a_1 u(n-1) + a_2 u(n-2) = v(n)$$

The experiment involves three sets of AR parameters a_1 and a_2 and variance σ_v^2 of the noise excitation $\{v(n)\}$ as summarized in the accompanying table. This table also includes the eigenvalue spreads corresponding to these three different sets of AR parameters. The exponential weighting factor $\lambda = 1$. Compute and plot the ensemble-averaged learning curves of the predictor by averaging the squared value of the angle-normalized forward prediction error $\epsilon_{f,2}(n)$ over 200 independent trials of the experiment. Present a summary of the experimentally measured steady-state values of the ensemble-averaged squared prediction error, and the corresponding theoretical values for the three sets of AR parameters. Hence, demonstrate close agreement between theory and experiment.

AR parameters		Noise variance	Eigenvalue spread
a_1	a_2	σ_v^2	$\chi(\mathbf{R})$
−0.195	0.95	0.09653	1.22
−1.5955	0.95	0.03223	10
−1.9114	0.95	0.00382	100

12. *Computer experiment on adaptive equalization* (continued)
Repeat the computer experiment of Section 18.19 on adaptive equalization using the QRD-LSL algorithm. In particular, compute and plot the learning curves based on the final value of the angle-normalized estimation error measured at the equalizer output. The ensemble averaging should be performed over 200 independent trials of the experiment. Hence, compare your results with those obtained in Section 18.9. Explain the reason for the difference between the trajectories computed in this experiment and those in Section 18.9.

Part IV

Limitations, Extensions, and Discussions

This final part of the text includes Chapters 19 through 21. In Chapter 19 we discuss practical limitations *of linear adaptive filtering algorithms. Particular emphasis is given to finite-precision (numerical) effects that arise when these algorithms are implemented on a general-purpose or special-purpose machine.*

In Chapter 20 we depart from the main theme of the book, namely, that of linearity and second-order statistics. Specifically, we explore the use of nonlinearity and higher-order statistics as a basis for blind deconvolution. *This constitutes an important class of nonlinear adaptive filtering algorithms that do* not *require a desired response. Special emphasis is given to the blind equalization problem in digital communications.*

Finally, in Chapter 21 we do two things. First, we present a summary of the linear finite-(length) impulse-response adaptive filtering algorithms considered in the book. Second, we highlight some unexplored issues related to adaptive filtering, both linear and nonlinear.

19

Finite-Precision and Other Practical Effects

A study of adaptive filtering would be incomplete without some discussion of the effects of *quantization* or *round-off errors* that arise when an adaptive filtering algorithm is implemented digitally. Indeed, this is the intention of the present chapter.

The theory of adaptive filtering developed in previous chapters assumes the use of an *analog model* (i.e., infinite precision) for the samples of input data as well as the internal algorithmic calculations. This assumption is made in order to take advantage of well-understood continuous mathematics. Adaptive filter theory, however, cannot be applied to the construction of an adaptive filter directly; rather, it provides an *idealized framework* for such a construction. In particular, in a *digital* implementation of an adaptive filtering algorithm as encountered in practice, the input data and internal calculations are all quantized to a *finite precision* that is determined by design and cost considerations. Consequently, the quantization process has the effect of causing the performance of the digital implementation of the algorithm to deviate from its theoretical value. The nature of this deviation is influenced by a combination of several factors:

1. The type and design details of the adaptive filtering algorithm employed
2. The degree of ill conditioning (i.e., the eigenvalue spread) in the underlying correlation matrix that characterizes the input data
3. The form of numerical computation (fixed-point or floating-point) employed

It is important for us to understand the numerical properties of adaptive filtering algorithms, as it would obviously help us in meeting design specifications. Moreover, the cost of the digital implementation of an algorithm is influenced by the *number of bits* (i.e., precision) available for performing the numerical computations associated with the algorithm. Generally speaking, the cost of implementation increases with the number of bits employed. There is therefore practical motivation in using the minimum number of bits possible.

We begin our study of the numerical properties of adaptive filtering algorithms by examining the sources of quantization error and the related issues of numerical stability and accuracy.

19.1 QUANTIZATION ERRORS

In the digital implementation of an adaptive filtering algorithm, there are essentially two sources of quantization error to be considered. The two sources are as follows:

1. *Analog-to-digital conversion.* Given that the input data are in analog form, we may use an analog-to-digital converter for their numerical representation. For our present discussion, we assume a quantization process with a *uniform step size* δ and a set of *quantizing levels* positioned at $0, \pm, \delta \pm 2\delta, \ldots$. Figure 19.1 illustrates the input-output characteristic of a uniform quantizer. Consider a particular sample at the quantizer input, with an amplitude that lies in the range $i\delta - (\delta/2)$ to $i\delta + (\delta/2)$, where i is an integer (positive or negative, including zero) and $i\delta$ defines the *quantizer output*. The quantization process thus described introduces a region of uncertainty of width δ, centered on $i\delta$. Let η denote the quantization error. Correspondingly, the quantizer input is $i\delta + \eta$, where η is bounded as $-(\delta/2) \leq \eta \leq (\delta/2)$. When the quantization is fine enough (say, the number of quantizing levels is 64 or more), and the signal spectrum is sufficiently rich, the distortion produced by the quantizing process may be modeled as an additive independent source of white noise with

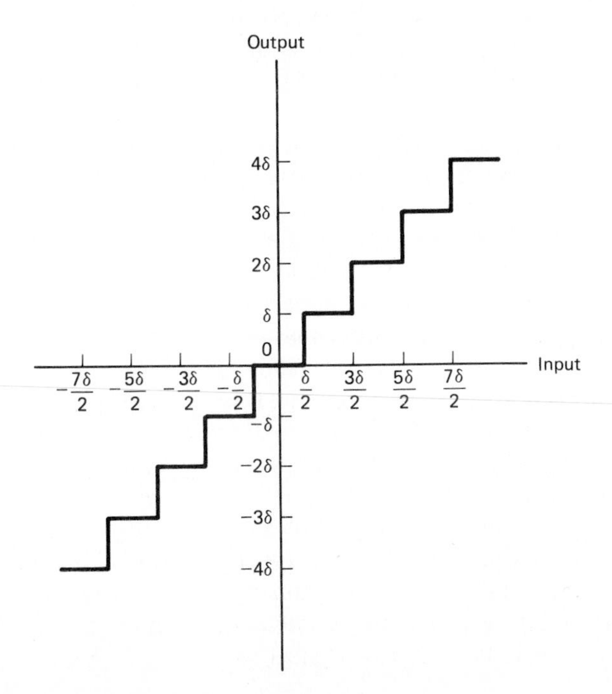

Figure 19.1 Input-output characteristic of a uniform quantizer.

zero mean and variance determined by the quantizer step size δ (Gray, 1990). It is customary to assume that the quantization error η is *uniformly distributed* over the range $-\delta/2$ to $\delta/2$. The variance of the quantization error is therefore given by

$$\begin{aligned}\sigma^2 &= \int_{-\delta/2}^{\delta/2} \frac{1}{\delta}\eta^2\, d\eta \\ &= \frac{\delta^2}{12}\end{aligned} \tag{19.1}$$

We assume that the quantizer input is properly scaled, so that it lies inside the interval $[-1, +1]$. With each quantizing level represented by B bits plus sign, the quantizer step size is

$$\delta = 2^{-B} \tag{19.2}$$

Substituting Eq. (19.2) in (19.1), we find that the quantization error resulting from the digital representation of input analog data has the variance

$$\sigma^2 = \frac{2^{-2B}}{12} \tag{19.3}$$

2. *Finite word-length arithmetic*. In a digital machine, a finite word-length is commonly used to store the result of internal arithmetic calculations. Assuming that *no* overflow takes place during the course of computation, additions do not introduce error, whereas each multiplication introduces an error after the product is quantized. The statistical characterization of finite word-length arithmetic errors may be quite different from that of analog-to-digital conversion errors. Finite word-length arithmetic errors may have a *nonzero mean,* which results from either rounding off or truncating the output of a multiplier so as to match the prescribed word-length.

The presence of finite word-length arithmetic raises serious concern in the digital implementation of an adaptive filtering algorithm, particularly when the tap weights (filter coefficients) of the algorithm are updated on a continuous basis. The digital version of the algorithm exhibits a specific *response* or *propagation* to such errors, causing its performance to deviate from the ideal (i.e., infinite-precision) form of the algorithm. Indeed, it is possible for the deviation to be of a catastrophic nature in the sense that the errors resulting from the use of finite-precision arithmetic may accumulate without bound. If such a situation is allowed to persist, the filter is ultimately driven into an overflow condition, and the algorithm is said to be *numerically unstable*. Clearly, for an adaptive filtering algorithm to be of practical value, it has to be numerically stable. An adaptive filtering algorithm is said to be *numerically stable* if the use of finite-precision arithmetic results in deviations from the infinite-precision form of the algorithm that are bounded. It is important to recognize that numerical stability is an inherent characteristic of an adaptive filtering algorithm. In other words, if an adaptive filtering algorithm is numerically unstable, then increasing the number of bits used in a digital implementation of the algorithm will not change the stability condition of that implementation.

Another issue that requires attention in the digital implementation of an adaptive filtering algorithm is that of *numerical accuracy*. Unlike numerical stability, however, the numerical accuracy of an adaptive filtering algorithm is determined by the number of bits used to implement the internal calculations of the algorithm. The larger the number of bits used, the smaller the deviation from ideal performance, and the more accurate would therefore the digital implementation of the algorithm be. In practical terms, it is only meaningful to speak of the numerical accuracy of an adaptive filtering algorithm if it is numerically stable.

For the remainder of this chapter, we discuss the numerical properties of adaptive filtering algorithms and related issues. We begin with the LMS algorithm and then move on to other adaptive filtering algorithms, presented in the same order as in previous chapters of the book.

19.2 LEAST-MEAN-SQUARE ALGORITHM

In order to simplify the discussion of finite-precision effects on the performance of the LMS algorithm,[1] we will depart from the practice followed in previous chapters, and assume that the input data and therefore the filter coefficients are all *real valued*. This assumption, made merely for convenience of presentation, will in no way affect the validity of the findings presented in this section.

A block diagram of the *finite-precision least mean-square (LMS) algorithm* is depicted in Fig. 19.2. Each of the blocks labeled Q represents a *quantizer*. Each of

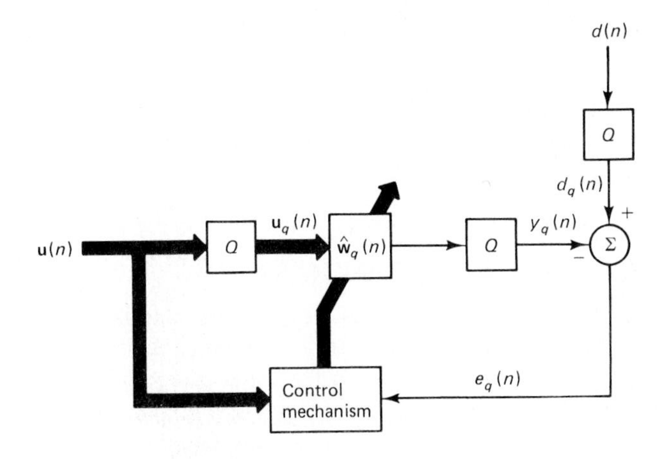

Figure 19.2 Block diagram representation of the finite-precision form of the LMS algorithm.

[1] The first treatment of finite-precision effects in the LMS algorithm was presented by Gitlin et al. (1973). Subsequently, more detailed treatments of these effects were presented by Weiss and Mitra (1979), Caraiscos and Liu (1984), and Alexander (1987). The paper by Caraiscos and Liu considers steady-state conditions, whereas the paper by Alexander is broader in scope in that it considers transient conditions. The problem of finite-precision effects in the LMS algorithm is also discussed in Cioffi (1987) and Sherwood and Bershad (1987). Another problem encountered in the practical use of the LMS algorithm is that of parameter drift, which is discussed in detail in Sethares et al. (1986). The material presented in this section is very much influenced by the contents of these papers. In our presentation, we assume the use of *fixed-point arithmetic*. Error analysis of the LMS algorithm for *floating-point arithmetic* is discussed in Caraiscos and Liu (1984).

them introduces a *quantization,* or *round-off error* of its own. Specifically, we may describe the input-output relations of these quantizers as follows:

1. For the input quantizer connected to $\mathbf{u}(n)$ we have

$$\begin{aligned} \mathbf{u}_q(n) &= Q[\mathbf{u}(n)] \\ &= \mathbf{u}(n) + \boldsymbol{\eta}_u(n) \end{aligned} \tag{19.4}$$

where $\boldsymbol{\eta}_u(n)$ is the *input quantization error vector*.

2. For the quantizer connected to the desired response $d(n)$, we have

$$\begin{aligned} d_q(n) &= Q[d(n)] \\ &= d(n) + \eta_d(n) \end{aligned} \tag{19.5}$$

where $\eta_d(n)$ is the *desired response quantization error*.

3. For the quantized tap-weight vector $\hat{\mathbf{w}}_q(n)$, we write

$$\begin{aligned} \hat{\mathbf{w}}_q(n) &= Q[\hat{\mathbf{w}}(n)] \\ &= \hat{\mathbf{w}}(n) + \Delta\mathbf{w}(n) \end{aligned} \tag{19.6}$$

where $\hat{\mathbf{w}}(n)$ is the tap-weight vector in the infinite-precision LMS algorithm, and $\Delta\hat{\mathbf{w}}(n)$ is the *tap-weight difference vector* resulting from quantization.

4. For the quantizer connected to the output of the transversal filter represented by the quantized tap-weight vector $\hat{\mathbf{w}}_q(n)$, we write

$$\begin{aligned} y_q(n) &= Q[\mathbf{u}_q^T(n)\hat{\mathbf{w}}_q(n)] \\ &= \mathbf{u}_q^T(n)\hat{\mathbf{w}}_q(n) + \eta_y(n) \end{aligned} \tag{19.7}$$

where $\eta_y(n)$ is the *filtered output quantization error*.

The finite-precision LMS algorithm is described by the following pair of relations:

$$e_q(n) = d_q(n) - y_q(n) \tag{19.8}$$

$$\hat{\mathbf{w}}_q(n + 1) = \hat{\mathbf{w}}_q(n) + Q[\mu e_q(n)\mathbf{u}_q(n)] \tag{19.9}$$

where $y_q(n)$ is itself defined in the first line of Eq. (19.7). The quantizing operation indicated on the right side of Eq. (19.9) is not shown explicitly in Fig. 19.2; nevertheless, it is basic to the operation of the finite-precision LMS algorithm. The use of Eq. (19.9) has the following practical implication. The product $\mu e_q(n)\mathbf{u}_q(n)$, representing a scaled version of the gradient vector estimate, is quantized *before* addition to the contents of the *tap-weight accumulator*. Because of hardware constraints, this form of digital implementation is preferred to the alternative method of operating the tap-weight accumulator in double-precision and then quantizing the tap-weight to single precision at the accumulator output.

In a statistical analysis of the finite-precision LMS algorithm, it is customary to make the following assumptions:

1. The input data are properly scaled so as to *prevent overflow* of the elements of the quantized tap-weight vector $\hat{\mathbf{w}}_q(n)$ and the quantized output $y_q(n)$ during the filtering operation.
2. Each data sample is represented by B_D bits plus sign, and each tap weight is represented by B_W bits plus sign. Thus, the quantization error associated with a B_D-plus-sign bit number (i.e., data sample) has the variance

$$\sigma_D^2 = \frac{2^{-2B_D}}{12} \tag{19.10}$$

Similarly, the quantization error associated with a B_W-plus-sign bit number (i.e., tap weight) has the variance

$$\sigma_W^2 = \frac{2^{-2B_W}}{12} \tag{19.11}$$

3. The elements of the input quantization error vector $\boldsymbol{\eta}_u(n)$ and the desired response quantization error $\eta_d(n)$ are *white-noise* sequences, independent of the signals and from each other. Moreover, they have zero mean and variance σ_D^2.
4. The output quantization error $\eta_y(n)$ is a white-noise sequence, independent of the input signals and other quantization errors. It has a mean of zero and a variance equal to $c\sigma_D^2$, where c is a constant that depends on the way in which the inner product $\mathbf{u}_q^T(n)\hat{\mathbf{w}}_q(n)$ is computed. If the individual scalar products in $\mathbf{u}_q^T(n)\hat{\mathbf{w}}_q(n)$ are all computed without quantization, then summed, and the final result is quantized in B_D bits plus sign, the constant c is unity and the variance of $\eta_y(n)$ is σ_D^2 where σ_D^2 is defined in Eq. (19.10). If, on the other hand, the individual scalar products in $\mathbf{u}^T(n)\hat{\mathbf{w}}_q(n)$ are quantized and then summed, the constant c is M and the variance of $\eta_y(n)$ is $M\sigma_D^2$, where M is the number of taps in the transversal filter implementation of the LMS algorithm.
5. The independence theory of Section 9.4, dealing with the infinite-precision LMS algorithm, is invoked.

Total Output Mean-Squared Error

The filtered output $y_q(n)$, produced by the finite-precision LMS algorithm, represents a *quantized estimate* of the desired response. The *total output error* is therefore equal to the difference $d(n) - y_q(n)$. Using Eq. (19.7), we may therefore express this error as

$$\begin{aligned} e_{\text{total}}(n) &= d(n) - y_q(n) \\ &= d(n) - \mathbf{u}_q^T(n)\hat{\mathbf{w}}_q(n) - \eta_y(n) \end{aligned} \tag{19.12}$$

Substituting Eqs. (19.4) and (19.6) in (19.12), and ignoring all quantization error terms higher than first order, we get

$$e_{\text{total}}(n) = [d(n) - \mathbf{u}^T(n)\hat{\mathbf{w}}(n)] - [\Delta\mathbf{w}^T(n)\mathbf{u}(n) + \boldsymbol{\eta}_u^T(n)\hat{\mathbf{w}}(n) + \eta_y(n)] \tag{19.13}$$

The term inside the first set of square brackets on the right side of Eq. (19.13) is the estimation error $e(n)$ in the infinite-precision LMS algorithm. The term inside the second set of square brackets is entirely due to quantization errors in the finite-precision LMS algorithm. Because of assumptions 3 and 4 (i.e., the quantization errors $\boldsymbol{\eta}_u$ and η_y are independent of the input signals and of each other), the quantization error-related terms $\Delta\mathbf{w}^T(n)\mathbf{u}(n)$, $\boldsymbol{\eta}_u^T(n)\hat{\mathbf{w}}(n)$, and $\eta_y(n)$ are uncorrelated with each other. Basically, for the same reason, the infinite-precision estimation error $e(n)$ is uncorrelated with $\boldsymbol{\eta}_u^T(n)\hat{\mathbf{w}}(n)$ and $\eta_y(n)$. By invoking the independence theory of Section 9.4, we may write

$$E[e(n)\,\Delta\mathbf{w}^T(n)\mathbf{u}(n)] = E[\Delta\mathbf{w}^T(n)]E[e(n)\mathbf{u}(n)]$$

Moreover, by invoking this same independence theory, we may show that the expectation $E[\Delta\mathbf{w}(n)]$ is zero (see Problem 2). Hence, $e(n)$ and $\Delta\mathbf{w}^T(n)u(n)$ are also uncorrelated. In other words, the infinite-precision estimation error $e(n)$ is uncorrelated with all the three quantization-error-related terms, $\Delta\mathbf{w}^T(n)\mathbf{u}(n)$, $\boldsymbol{\eta}_u^T(n)\hat{\mathbf{w}}(n)$ and $\eta_y(n)$ in Eq. (19.13).

Using these observations, and assuming that the step-size (adaptation) parameter μ is small,[2] it is shown in Caraiscos and Liu (1984) that the *total output mean-squared error* produced in the finite-precision algorithm has the following *steady-state* structure:

$$E[e_{\text{total}}^2(n)] = J_{\min}(1 + \mathcal{M}) + \xi_1(\sigma_W^2, \mu) + \xi_2(\sigma_D^2) \tag{19.14}$$

The first term $J_{\min}(1 + \mathcal{M})$ on the right side of Eq. (19.14) is the mean-squared error of the infinite-precision LMS algorithm. In particular, $J_{\min}$ is the *minimum mean-squared error* of the optimum Wiener filter, and $\mathcal{M}$ is the *misadjustment* of the infinite-precision LMS algorithm. The second term $\xi_1(\sigma_w^2, \mu)$ arises because of the error $\Delta\mathbf{w}(n)$ in the quantized tap-weight vector $\hat{\mathbf{w}}_q(n)$. This contribution to the total output mean-squared error is *inversely proportional to the step-size parameter* μ. The third term $\xi_2(\sigma_D^2)$ arises because of two quantization errors: the error $\boldsymbol{\eta}_u(n)$ in the quantized input vector $\mathbf{u}_q(n)$ and the error $\eta_y(n)$ in the quantized filter output $y_q(n)$. However, unlike $\xi_1(\sigma_W^2, \mu)$, this final contribution to the total output mean-squared error is, to a first order of approximation, independent of the step-size parameter μ.

From the infinite-precision theory of the LMS algorithm presented in Chapter 9, we know that decreasing μ reduces the misadjustment $\mathcal{M}$ and thus leads to an improved performance of the algorithm. In contrast, the inverse dependence of the contribution $\xi_1(\sigma_W^2, \mu)$ on μ in Eq. (19.14) indicates that decreasing μ has the effect of increasing the deviation from infinite-precision performance. In practice, therefore, the step-size parameter μ may only be decreased to a level at which the degrading effects of quantization errors in the tap weights of the finite-precision LMS algorithm become significant.

Since the misadjustment $\mathcal{M}$ decreases with μ, and the contribution $\xi_1(\sigma_W^2, \mu)$ increases with reduced μ, we may (in theory) find an optimum value of μ for which the total output mean-squared error in Eq. (19.14) is minimized. However, it turns out that this minimization results in an optimum value μ_o for the step-size parameter μ that is too small to be of practical value. In other words, it does not permit the LMS algorithm to converge completely. Indeed, Eq. (19.14) for calculating the total

[2] The step-size parameter μ cannot be too small; otherwise, the LMS algorithm may not be able to converge completely because of the "stalling phenomenon," which is considered later in the section.

output mean-squared error is valid only for a μ that is well in excess of μ_o. Such a choice of μ is necessary so as to prevent the occurrence of a phenomenon known as stalling, described later in the section.

Deviations during the Convergence Period

Equation (19.14) describes the general structure of the total output mean-squared error of the finite-precision LMS algorithm, assuming that the algorithm has reached steady state. During the convergence period of the algorithm, however, the situation is more complicated.

A detailed treatment of the transient adaptation properties of the finite-precision LMS algorithm is presented in Alexander (1987). In particular, a general formula is derived for the tap-weight misadjustment, or *perturbation,* of the finite-precision LMS algorithm, which is measured with respect to the tap-weight solution computed from the infinite-precision form of the algorithm. The tap-weight misadjustment is defined by

$$\rho(n) = E[\Delta\mathbf{w}^T(n)\,\Delta\mathbf{w}(n)] \tag{19.15}$$

where the *tap-weight difference vector* $\Delta\mathbf{w}(n)$ is itself defined by [see Eq. (19.6)]

$$\Delta\mathbf{w}(n) = \hat{\mathbf{w}}_q(n) - \hat{\mathbf{w}}(n) \tag{19.16}$$

The tap-weight vectors $\hat{\mathbf{w}}_q(n)$ and $\hat{\mathbf{w}}(n)$ refer to the finite-precision and infinite-precision forms of the LMS algorithm, respectively. To determine $\rho(n)$, the weight update equation (19.9) is written as

$$\hat{\mathbf{w}}_q(n+1) = \hat{\mathbf{w}}_q(n) + \mu e_q(n)\mathbf{u}_q(n) + \boldsymbol{\eta}_w(n) \tag{19.17}$$

where $\boldsymbol{\eta}_w(n)$ is the *gradient quantization error vector;* it results from quantizing the product $\mu e_q(n)\mathbf{u}_q(n)$ that represents a scalar version of the gradient vector estimate. The individual elements of $\boldsymbol{\eta}_w(n)$ are assumed to be uncorrelated in time and with each other, and assumed to have a common variance σ_W^2. For this assumption to be valid, the step-size parameter μ must be large enough to prevent the stalling phenomenon from occurring; this phenomenon is described later in the section.

Applying an orthogonal transformation to $\hat{\mathbf{w}}_q(n)$ in Eq. (19.17) in a manner similar to that described in Section 5.6, the propagation characteristics of the tap-weight misadjustment $\rho(n)$ during adaptation and steady state may be studied. Using such an approach, Alexander (1987) has derived some important theoretical results, supported by computer simulation. These results may be summarized as follows:

1. The tap-weights in the LMS algorithm are the *most sensitive* of all parameters to quantization. For the case of *uncorrelated input data,* the variance σ_W^2 [that enters the statistical characterization of the tap-weight update equation (19.17) is proportional to the reciprocal of the product $r(0)\mu$, where $r(0)$ is the average input power and μ is the step-size parameter. For the case of *correlated input data,* the variance σ_W^2 is proportional to the reciprocal of $\mu\lambda_{\min}$, where $\lambda_{\min}$ is the smallest eigenvalue of the correlation matrix $\mathbf{R}$ of the input data vector $\mathbf{u}(n)$.

2. For uncorrelated input data, the adaptation time constants of the tap-weight misadjustment process $\rho(n)$ are heavily dependent on the step-size parameter μ.
3. For correlated input data, the adaptation time constants of $\rho(n)$ are heavily dependent on the interation between μ and the minimum eigenvalue $\lambda_{\min}$.

From a design point of view, it is thus important to recognize that the step-size parameter μ cannot be chosen too small; this is in spite of the infinite-precision theory of the LMS algorithm that advocates a small value for μ. Moreover, the more ill-conditioned the input time series $\{u(n)\}$ is, the more pronounced will the finite-precision effects in a digital implementation of the LMS algorithm be.

Leaky LMS Algorithm

To further stabilize the digital implementation of the LMS algorithm, we may use a technique known as *leakage*.[3] Basically, leakage prevents the occurrence of overflow in a limited-precision environment by providing a compromise between minimizing the mean-squared error and containing the energy in the impulse response of the adaptive filter. However, the prevention of overflow is attained at the expense of an increase in hardware cost and at the expense of a degradation in performance compared to the infinite-precision form of the conventional LMS algorithm.

In the leaky LMS algorithm, the cost function

$$J(n) = e^2(n) + \alpha \|\mathbf{w}(n)\|^2 \tag{19.18}$$

is minimized with respect to the tap-weight vector $\mathbf{w}(n)$, where α is a positive control parameter. The first term on the right side of Eq. (19.18) is the squared estimation error, and the second term is the energy in the tap-weight vector $\mathbf{w}(n)$. The minimization described herein (for real data) yields the following time update for the tap-weight vector (see Problem 8, Chapter 9)

$$\hat{\mathbf{w}}(n + 1) = (1 - \mu\alpha)\hat{\mathbf{w}}(n) + \mu e(n)\mathbf{u}(n) \tag{19.19}$$

where α is a constant that satisfies the condition

$$0 \leq \alpha < \frac{1}{\mu}$$

Except for the *leakage factor* $(1 - \mu\alpha)$ associated with the first term on the right side of Eq. (19.19), the algorithm is of the same mathematical form as the conventional LMS algorithm.

Note that the inclusion of the leakage factor $(1 - \mu\alpha)$ in Eq. (19.19) has the equivalent effect of adding a white-noise sequence of zero mean and variance α to the input time series $\{u(n)\}$ (see Problem 8 of Chapter 9). This suggests another

[3] Leakage may be viewed as a technique for increasing algorithm robustness (Ioannou, 1990; Ioannou and Kokotovic, 1983). For a historical account of the leakage technique in the context of adaptive filtering, see Cioffi (1987). For discussions of the leakage LMS algorithm, see Widrow and Stearns (1985), and Cioffi (1987).

method for stabilizing the digital implementation of the LMS algorithm. Specifically, a relatively weak white-noise sequence (of variance α), known as *dither*, is added to the input time series $\{u(n)\}$, and samples of the combination are then used as tap inputs (Werner, 1983).

Stalling Phenomenon

There is another phenomenon known as the *stalling* or *lock-up phenomenon,* not evident from Eq. (19.14), which may arise in a digital implementation of the LMS algorithm. This phenomenon occurs when the gradient estimate is *not* sufficiently noisy. To be specific, a digital implementation of the LMS algorithm *stops adapting,* or *stalls,* whenever the correction term $\mu e_q(n)u_q(n - i)$ for the ith tap weight in the update equation (19.9) is smaller in magnitude than the *least significant bit* (LSB) of the tap weight, as shown by (Gitlin et al. 1973)

$$|\mu e_q(n_0)u_q(n_0 - i)| \leq \text{LSB} \tag{19.20}$$

Here, n_0 is the time at which the ith tap weight stops adapting. Suppose that the condition of Eq. (19.20) is first satisfied for the ith tap weight. To a first order of approximation, we may replace $u_q(n_0 - i)$ by its *root-mean-square* (rms) value, A_{rms}. Accordingly, using this value in Eq. (19.20), we get the following relation for the rms value of the quantized estimation error when adaptation in the digitally implemented LMS algorithm stops:

$$|e_q(n)| \leq \frac{\text{LSB}}{\mu A_{\text{rms}}} = e_D(\mu) \tag{19.21}$$

The quantity $e_D(\mu)$, defined on the right side of (19.19), is called the *digital residual error*.

To prevent the algorithm-stalling phenomenon due to digital effects, the digital residual error $e_D(\mu)$ must be made as small as possible. According to the definition of Eq. (19.21), this requirement may be satisfied in one of two ways:

1. The least significant bit (LSB) is reduced by picking a sufficiently large number of bits for the digital representation of the tap-weight.
2. The step-size parameter μ is made as large as possible, while still guaranteeing convergence of the algorithm.

Another method of preventing the stalling phenomenon is to insert *dither* at the input of the quantizer that feeds the tap-weight accumulator (Sherwood and Bershad, 1987). Dither is a random sequence that essentially "linearizes" the quantizer. In other words, the addition of dither guarantees that the quantizer input is *noisy* enough for the gradient quantization error vector $\boldsymbol{\eta}_w$ to be again modeled as white noise (i.e., the elements of $\boldsymbol{\eta}_w$ are uncorrelated in time and with each other, and have a common variance σ_W^2). When dither is used in the manner described herein, it is desirable to minimize its effect on the overall operation of the LMS algorithm. This is commonly achieved by shaping the power spectrum of the dither so that it is effectively rejected by the algorithm at its output.

Parameter Drift

In addition to the numerical problems associated with the LMS algorithm, there is one other rather subtle problem that is encountered in practical applications of the algorithm. Specifically, certain classes of input excitation can lead to *parameter drift;* that is, parameter estimates or tap weights in the LMS algorithm attain arbitrarily large values despite bounded inputs, bounded disturbances, and bounded estimation errors (Sethares et al., 1986). Although such an unbounded behavior may be unexpected, nevertheless, it is possible for the parameter estimates to drift to infinity while all the signals observable in the algorithm converge to zero. Parameter drift in the LMS algorithm may be viewed as a *hidden form of instability,* since the tap weights represent "internal" variables of the algorithm. As such, it may result in new numerical problems, increased sensitivity to unmodeled disturbances, and degraded long-term performance.

In order to appreciate the subtleties of the parameter drift problem, we need to introduce some new concepts relating to the parameter space. We will therefore digress briefly from the issue at hand to do so.

A sequence of information-bearing tap-input vectors $\{\mathbf{u}(n)\}$ for varying time n may be used to partition the *real M-dimensional parameter space* $\mathbb{R}^M$ into orthogonal subspaces, where M is the number of tap weights (i.e., the available number of degrees of freedom). The aim of the partitioning is to convert the stability analysis of an adaptive filtering algorithm (e.g., the LMS algorithm) into simpler subsystems and thereby provide a closer linkage between the transient behavior of the parameter estimates and the filter excitations. The partitioning we have in mind is depicted in Fig. 19.3. In particular, we may identify the following subspaces of $\mathbb{R}^M$:

1. *The unexcited subspace.* Let the M-by-1 vector $\mathbf{z}$ be any element of the parameter space $\mathbb{R}^M$, which satisfies two conditions:
   (a) The Euclidean norm of the vector $\mathbf{z}$ is 1; that is,
   $$\|\mathbf{z}\| = 1$$
   (b) The vector $\mathbf{z}$ is orthogonal to the tap-input vector $\mathbf{u}(n)$ for all but a finite number of n; that is,
   $$\mathbf{z}^T\mathbf{u}(n) \neq 0 \qquad \text{only finitely often} \qquad (19.22)$$

   Let $\mathcal{S}_u$ denote the subspace of $\mathbb{R}^M$ that is spanned by the set of all such vectors $\mathbf{z}$. The subspace $\mathcal{S}_u$ is called the *unexcited subspace* in the sense that it spans those *directions* in the parameter space $\mathbb{R}^M$ that are excited only *finitely often.*

Parameter space $\mathbb{R}^M$			
Unexcited subspace $\mathcal{S}_u$	$\mathcal{S}_e$ Excited subspace		
	Persistently excited subspace $\mathcal{S}_p$	Decreasingly excited subspace $\mathcal{S}_d$	Otherwise excited subspace $\mathcal{S}_o$

Figure 19.3 Decomposition of parameter space $\mathbb{R}^M$, based on excitation.

2. *The excited subspace*. Let $\mathcal{S}_e$ denote the *orthogonal complement* of the unexcited subspace $\mathcal{S}_u$. Clearly, $\mathcal{S}_e$ is also a subspace of the parameter space $\mathbb{R}^M$. It contains those directions in the parameter space $\mathbb{R}^M$ that are excited *infinitely often*. Thus, except for the null vector, every element $\mathbf{z}$ *belonging to the subspace* $\mathcal{S}_e$ satisfies the condition

$$\mathbf{z}^T\mathbf{u}(n) \neq 0, \qquad \text{infinitely often} \tag{19.23}$$

The subspace $\mathcal{S}_e$ is called the *excited subspace*.

The subspace $\mathcal{S}_e$ may itself be decomposed into three orthogonal subspaces of its own, depending on the effects of different types of excitation on the behavior of the adaptive filtering algorithm. Specifically, three subspaces of $\mathcal{S}_e$ may be identified as follows (Sethares et al., 1986):

(a) *The persistently excited subspace*. Let $\mathbf{z}$ be any vector of unit norm that lies in the excited subspace $\mathcal{S}_e$. For any positive integer m and any $\alpha > 0$, choose the vector $\mathbf{z}$ such that we have

$$\mathbf{z}^T\mathbf{u}(i) > \alpha, \qquad \text{for } n \le i \le n + m \text{ and for all but a finite number of } n \tag{19.24}$$

Given the integer m and the constant α, let $\mathcal{S}_p(m, \alpha)$ be the subspace spanned by all such vectors $\mathbf{z}$ that satisfy the condition of (19.24). There exist a finite m_0 and a positive α_0 for which *the subspace* $\mathcal{S}_p(m_0, \alpha_0)$ *is maximal*. In other words, $\mathcal{S}_p(m_0, \alpha_0)$ contains $\mathcal{S}_p(m, \alpha)$ for all $m > 0$ and for all $\alpha > 0$. The subspace $\mathcal{S}_p \equiv \mathcal{S}_p(m_0, \alpha_0)$ is called the *persistently excited subspace;* and m_0 is called the *interval of excitation*. For every "direction" $\mathbf{z}$ that lies in the persistently excited subspace $\mathcal{S}_p$, there is an excitation of level α_0 at least once in all but a finite number of intervals of length m_0. In the persistently excited subspace, we are therefore able to find a tap-input vector $\mathbf{u}(n)$ *rich* enough to excite all the internal modes that govern the transient behavior of the adaptive filtering algorithm being probed (Narendra and Annaswamy, 1989).

(b) *The subspace of decreasing excitation*. Consider a sequence $\{u(i)\}$ for which we have

$$\left(\sum_{i=1}^{\infty} |u(i)|^p\right)^{1/p} < \infty \tag{19.25}$$

Such a sequence is said to be an element of the *normed linear space* l^p for $1 < p < \infty$. The norm of this new space is defined by

$$\|\mathbf{u}\|_p = \left(\sum_{i=1}^{\infty} |u(i)|^p\right)^{1/p} \tag{19.26}$$

Note that if the sequence $\{u(i)\}$ is an element of the normed linear space l^p for $1 < p < \infty$, then

$$\lim_{n\to\infty} u(n) = 0 \tag{19.27}$$

Let $\mathbf{z}$ be any unit-norm vector $\mathbf{z}$ that lies in the excited subspace $\mathcal{S}_e$ such that for some for $1 < p < \infty$, the sequence $\{\mathbf{z}^T\mathbf{u}(n)\}$ lies in the normed linear space l^p. Let $\mathcal{S}_d$ be the subspace that is spanned by all such vectors

z. The subspace $\mathcal{S}_d$ is called the *subspace of decreasing excitation* in the sense that each direction of $\mathcal{S}_d$ is decreasingly excited. For any vector $\mathbf{z} \neq \mathbf{0}$, the two conditions

$$|\mathbf{z}^T\mathbf{u}(n)| = \alpha > 0 \qquad \text{infinitely often}$$

and

$$\lim_{n\to\infty} \mathbf{z}^T\mathbf{u}(n) = 0$$

cannot be satisfied simultaneously. In actual fact, we find that the subspace of decreasing excitation $\mathcal{S}_d$ is orthogonal to the subspace of persistent excitation $\mathcal{S}_p$.

(c) *The otherwise excited subspace*. Let $\mathcal{S}_p \cup \mathcal{S}_d$ denote the *union* of the persistently excited subspace $\mathcal{S}_p$ and the subspace of decreasing excitation $\mathcal{S}_d$. Let $\mathcal{S}_o$ denote the orthogonal complement of $\mathcal{S}_p \cup \mathcal{S}_d$ that lies in the excited subspace $\mathcal{S}_e$. The subspace $\mathcal{S}_o$ is called the *otherwise excited subspace*. Any vector that lies in the subspace $\mathcal{S}_o$ is not unexciting, not persistently exciting, and not in the normed linear space l^p for any finite p. An example of such a signal is the sequence

$$\mathbf{z}^T\mathbf{u}(n) = \frac{1}{\ln(1 + n)} \tag{19.28}$$

Returning to our discussion of the parameter drift problem in the LMS algorithm, we find that for bounded excitations and bounded disturbances, in the case of unexcited and persistently exciting subspaces the parameter estimates resulting from the application of the LMS algorithm are indeed bounded. However, in the decreasing and otherwise excited cases, parameter drift may occur (Sethares et al., 1986). A common method of counteracting the parameter drift problem in the LMS algorithm is to introduce leakage into the tap-weight update equation of the algorithm.

19.3 RECURSIVE LEAST-SQUARES ALGORITHM

The recursive least-squares (RLS) algorithm offers an alternative to the LMS algorithm as a tool for the solution of adaptive filtering problems. From the discussion presented in Chapter 13, we know that the RLS algorithm is characterized by a fast rate of convergence that is relatively insensitive to the eigenvalue spread of the underlying correlation matrix of the input data, and a negligible misadjustment (zero for a stationary environment without disturbances). Moreover, although it is computationally demanding (in the sense that its computational complexity is on the order of M^2, where M is the dimension of the tap-weight vector), the mathematical formulation and therefore implementation of the RLS algorithm is relatively simple. However, there is a numerical instability problem to be considered when the RLS algorithm is implemented in finite-precision arithmetic.

Basically, *numerical instability* or *explosive divergence* of the RLS algorithm is of a similar nature to that experienced in Kalman filtering, of which the RLS algorithm is a special case. Indeed, the problem may be traced to the fact that the time-updated matrix $\mathbf{P}(n)$ in the Riccati equation is computed as the difference between two nonnegative definite matrices, as indicated in Eq. (13.19). Accordingly, explosive divergence of the algorithm occurs when the matrix $\mathbf{P}(n)$ loses the property of positive definiteness or Hermitian symmetry.

Another form of divergence observed in the finite-precision RLS algorithm is that of *stalling*. As in the finite-precision LMS algorithm, the stalling phenomenon occurs when the algorithm stops adapting (i.e., the tap weights stop changing in value). In this section we will discuss the theoretical basis for the divergence of the standard RLS algorithm[4] when it is implemented in finite-precision form, and methods for its cure.

Error Propagation Model

The matrix $\mathbf{P}(n)$ is defined as the inverse of the time-averaged correlation matrix $\mathbf{\Phi}(n)$ of the tap-input vector $\mathbf{u}(n)$. According to form I of the standard RLS algorithm summarized in Table 13.1, the recursions involved in the computation of $\mathbf{P}(n)$ proceed as follows:

$$\boldsymbol{\pi}(n) = \mathbf{u}^H(n)\mathbf{P}(n-1) \tag{19.29}$$

$$\kappa(n) = \lambda + \boldsymbol{\pi}(n)\mathbf{u}(n) \tag{19.30}$$

$$\mathbf{k}(n) = \frac{\boldsymbol{\pi}^H(n)}{\kappa(n)} \tag{19.31}$$

$$\mathbf{P}'(n-1) = \mathbf{k}(n)\boldsymbol{\pi}(n) \tag{19.32}$$

$$\mathbf{P}(n) = \frac{1}{\lambda}(\mathbf{P}(n-1) - \mathbf{P}'(n-1)) \tag{19.33}$$

where λ is the *exponential weighting factor*. Note that $\boldsymbol{\pi}(n)$ is a row vector and $\kappa(n)$ is a scalar. Consider the propagation of a *single* quantization error at $n-1$ to subsequent recursions, under the assumption that no other quantization errors are made. In particular, let

$$\mathbf{P}_q(n-1) = \mathbf{P}(n-1) + \boldsymbol{\eta}_P(n-1) \tag{19.34}$$

[4] A detailed theoretical evaluation of error propagation in four different forms of the RLS algorithm is presented in Verhaegen (1989). The four different forms considered are: the two implementations of the conventional RLS algorithm summarized in Tables 13.1 and 13.2, and the square-root versions of the covariance (correlation) matrix $\mathbf{P}(n)$ and information matrix [i.e., inverse of $\mathbf{P}(n)$] implementations of the RLS algorithm. The material presented in Verhaegen's paper is closely related to an earlier paper (Verhaegen and Van Dooren, 1986) that discusses the numerical instability problem in Kalman filtering. An illuminating discussion of the theoretical basis of numerical instability in the conventional RLS algorithm is also presented in Bottomley and Alexander (1989). The material presented in Section 19.3 is based on these three papers. Our presentation assumes the use of fixed-point arithmetic. An error analysis of the RLS algorithm assuming the use of floating-point arithmetic is discussed in Ardalan (1986).

where the *error matrix* $\boldsymbol{\eta}_P(n-1)$ arises from the quantization of $\mathbf{P}(n-1)$. The corresponding quantized value of $\boldsymbol{\pi}(n)$ is

$$\boldsymbol{\pi}_q(n) = \boldsymbol{\pi}(n) + \boldsymbol{\eta}_\pi(n) \tag{19.35}$$

The quantization error in the computation of $\boldsymbol{\pi}(n)$ is therefore

$$\boldsymbol{\eta}_\pi(n) = \mathbf{u}^H(n)\boldsymbol{\eta}_P(n-1) \tag{19.36}$$

Let $\kappa_q(n)$ denote the quantized value of $\kappa(n)$. Hence, we may consider the approximation

$$\begin{aligned}\frac{1}{\kappa_q(n)} &= \frac{1}{\lambda + \boldsymbol{\pi}_q(n)\mathbf{u}(n)}\\ &= \frac{1}{\lambda + \boldsymbol{\pi}(n)\mathbf{u}(n) + \boldsymbol{\eta}_\pi(n)\mathbf{u}(n)}\\ &= \frac{1}{\lambda + \boldsymbol{\pi}(n)\mathbf{u}(n)}\left(1 + \frac{\boldsymbol{\eta}_\pi(n)\mathbf{u}(n)}{\lambda + \boldsymbol{\pi}(n)\mathbf{u}(n)}\right)^{-1}\\ &= \frac{1}{\lambda + \boldsymbol{\pi}(n)\mathbf{u}(n)}\left(1 - \frac{\boldsymbol{\eta}_\pi(n)\mathbf{u}(n)}{\lambda + \boldsymbol{\pi}(n)\mathbf{u}(n)}\right) + \mathrm{O}(\boldsymbol{\eta}_P^2)\end{aligned} \tag{19.37}$$

where $\mathrm{O}(\boldsymbol{\eta}_P^2)$ denotes the order of magnitude $\|\boldsymbol{\eta}_P\|^2$.

In an ideal situation, the reciprocal of the infinite-precision scalar quantity $\kappa(n)$ is *nonnegative,* taking on values between zero and $1/\lambda$. On the other hand, if $\boldsymbol{\pi}(n)\mathbf{u}(n)$ is small compared to λ and λ itself is small enough compared to 1, then according to Eq. (19.37), in a practical finite-precision environment it is possible for the corresponding reciprocal of the quantized quality $\kappa_q(n)$ to take on a *negative* value larger in magnitude than $1/\lambda$. When this happens, the RLS algorithm exhibits explosive divergence (Bottomley and Alexander, 1989)[5]

The quantized value of the gain vector $\mathbf{k}(n)$ is written as

$$\begin{aligned}\mathbf{k}_q(n) &= \frac{\boldsymbol{\pi}_q^H(n)}{\kappa_q(n)}\\ &= \frac{\boldsymbol{\pi}_q^H(n)}{\lambda + \boldsymbol{\pi}_q(n)\mathbf{u}(n)}\\ &= \mathbf{k}(n) + \boldsymbol{\eta}_k(n)\end{aligned} \tag{19.38}$$

where $\boldsymbol{\eta}_k(n)$ is the *gain vector quantization error*. Accordingly, we have

$$\boldsymbol{\eta}_k(n) = \frac{\boldsymbol{\eta}_P^H(n-1)\mathbf{u}(n)}{\lambda + \mathbf{u}^H(n)\mathbf{P}(n-1)\mathbf{u}(n)} - \frac{\mathbf{P}^H(n-1)\mathbf{u}(n)\mathbf{u}^H(n)\boldsymbol{\eta}_P(n-1)\mathbf{u}(n)}{(\lambda + \mathbf{u}^H(n)\mathbf{P}(n-1)\mathbf{u}(n))^2} + \mathrm{O}(\boldsymbol{\eta}_P^2) \tag{19.39}$$

Correspondingly, the quantization error in the computation of the auxiliary matrix $\mathbf{P}'(n)$ is

[5] According to Bottomley and Alexander (1989), the evolution of the reciprocal factor $1/\kappa(n)$ provides a good indication of explosive divergence as this factor grows large, then suddenly it becomes negative.

$$\begin{aligned}\boldsymbol{\eta}_{P'}(n) &= \boldsymbol{\eta}_k(n)\boldsymbol{\pi}(n) + \mathbf{k}(n)\boldsymbol{\eta}_\pi(n) \\ &= \frac{\boldsymbol{\eta}^H(n-1)\mathbf{u}(n)\mathbf{u}^H(n)\mathbf{P}(n-1)}{\lambda + \mathbf{u}^H(n)\mathbf{P}(n-1)\mathbf{u}(n)} + \frac{\mathbf{P}^H(n-1)\mathbf{u}(n)\mathbf{u}^H(n)\boldsymbol{\eta}_P(n-1)}{\lambda + \mathbf{u}^H(n)\mathbf{P}(n-1)\mathbf{u}(n)} \\ &\quad - \frac{\mathbf{P}^H(n-1)\mathbf{u}(n)\mathbf{u}^H(n)\boldsymbol{\eta}_P(n-1)\mathbf{u}(n)\mathbf{u}^H(n)\mathbf{P}(n-1)}{(\lambda + \mathbf{u}^H(n)\mathbf{P}(n-1)\mathbf{u}(n))^2} + O(\boldsymbol{\eta}_P)\end{aligned} \tag{19.40}$$

Finally, the quantization error in the computation of the matrix $\mathbf{P}(n)$ is

$$\boldsymbol{\eta}_P(n) = \frac{1}{\lambda}(\boldsymbol{\eta}_P(n-1) - \boldsymbol{\eta}_{P'}(n-1)) \tag{19.41}$$

Substituting Eq. (19.40) in (19.41), then using the definition

$$\mathbf{k}(n) = \frac{\boldsymbol{\pi}^H(n)}{\kappa(n)} = \frac{\mathbf{P}^H(n-1)\mathbf{u}(n)}{\kappa(n)}$$

for the gain vector, and then rearranging terms, we get

$$\begin{aligned}\boldsymbol{\eta}_P(n) &= \frac{1}{\lambda}(\mathbf{I} - \mathbf{k}(n)\mathbf{u}^H(n))\boldsymbol{\eta}_P(n-1)(\mathbf{I} - \mathbf{k}(n)\mathbf{u}^H(n))^H \\ &\quad + \frac{1}{\lambda}(\boldsymbol{\eta}_P(n-1) - \boldsymbol{\eta}_P^H(n-1))\mathbf{u}(n)\mathbf{k}^H(n) + O(\boldsymbol{\eta}_P^2)\end{aligned} \tag{19.42}$$

where $\mathbf{I}$ is the identity matrix. Equation (19.42) defines the *single-error propagation model* for form I of the standard RLS algorithm. The presence of the second term on the right side of this equation signifies the *non-Hermitian* nature of the model in the sense that

$$\boldsymbol{\eta}_P^H(n) \neq \boldsymbol{\eta}_P(n)$$

When form I of the standard RLS algorithm is implemented in finite-precision arithmetic, the quantization process has the effect of destroying the Hermitian character of $\mathbf{P}(n)$. Moreover, for an exponential weighting factor $\lambda \leq 1$, the model of Eq. (19.42) demonstrates that the *loss of Hermitian symmetry* has the serious effect of causing a *blow-up* or *divergence* in the quantization error on the matrix $\mathbf{P}(n)$ (Verhaegen, 1989).

There are different causes for the explosive divergence of the standard RLS algorithm. Indeed, for each cause, different cures have been proposed in the literature. However, these cures suffer from two disadvantages:

1. Numerical implementation of the algorithm is unnecessarily complicated.
2. The cure may yield "satisfactory" results in computer simulation study, but it may actually fail in real-time operation of the algorithm.

A simple, yet highly effective method for curing the explosive divergence of the RLS algorithm is to perform the computations as summarized in Table 13.2, and reproduced here for convenience:

$$\boldsymbol{\pi}(n) = \mathbf{u}^H(n)\mathbf{P}(n-1) \tag{19.43}$$

$$\kappa(n) = \lambda + \boldsymbol{\pi}(n)\mathbf{u}(n) \tag{19.44}$$

$$\mathbf{k}(n) = \frac{\mathbf{P}(n-1)\mathbf{u}(n)}{\kappa(n)} \tag{19.45}$$

$$\mathbf{P}'(n-1) = \mathbf{k}(n)\boldsymbol{\pi}(n) \tag{19.46}$$

$$\mathbf{P}(n) = \frac{1}{\lambda}(\mathbf{P}(n-1) - \mathbf{P}'(n-1)) \tag{19.47}$$

Comparing Eqs. (19.43) to (19.47) for form II of the standard RLS algorithm with Eqs. (19.29) to (19.33) for form I of the algorithm, we see that the difference between them resides in the computation of $\mathbf{k}(n)$. Specifically, in form II the gain vector $\mathbf{k}(n)$ is computed exactly as it arises in the formulation of the RLS algorithm, whereas in form I use is made of the Hermitian property of the matrix $\mathbf{P}(n)$ for the ideal case of infinite precision. Proceeding in a manner similar to that described for form I of the standard RLS algorithm, we may show that the recursive equation for the single-error propagation model for form II of the algorithm is as follows (see Problem 5)

$$\boldsymbol{\eta}_P(n) = \frac{1}{\lambda}(\mathbf{I} - \mathbf{k}(n)\mathbf{u}^H(n))\boldsymbol{\eta}_P(n-1)(\mathbf{I} - \mathbf{k}(n)\mathbf{u}^H(n))^H \tag{19.48}$$

We now find that

$$\boldsymbol{\eta}_P^H(n) = \boldsymbol{\eta}_P(n)$$

and so the Hermitian character of the matrix $\mathbf{P}(n)$ is preserved, despite the presence of quantization errors.

The matrix $\mathbf{I} - \mathbf{k}(n)\mathbf{u}^H(n)$ plays a crucial role in the propagation of the quantization error $\boldsymbol{\eta}_P(n-1)$. Using the original definition given in Eq. (13.22) for the gain vector, namely,

$$\mathbf{k}(n) = \boldsymbol{\Phi}^{-1}(n)\mathbf{u}(n) \tag{19.49}$$

we may write

$$\mathbf{I} - \mathbf{k}(n)\mathbf{u}^H(n) = \mathbf{I} - \boldsymbol{\Phi}^{-1}(n)\mathbf{u}(n)\mathbf{u}^H(n) \tag{19.50}$$

Next, from Eq. (13.12) we have

$$\boldsymbol{\Phi}(n) = \lambda\boldsymbol{\Phi}(n-1) + \mathbf{u}(n)\mathbf{u}^H(n) \tag{19.51}$$

Multiplying both sides of Eq. (19.51) by the inverse matrix $\boldsymbol{\Phi}^{-1}(n)$, and rearranging terms, we get

$$\mathbf{I} - \boldsymbol{\Phi}^{-1}(n)\mathbf{u}(n)\mathbf{u}^H(n) = \lambda\boldsymbol{\Phi}^{-1}(n)\boldsymbol{\Phi}(n-1) \tag{19.52}$$

Comparing Eqs. (19.50) and (19.52), we readily deduce that

$$\mathbf{I} - \mathbf{k}(n)\mathbf{u}^H(n) = \lambda\boldsymbol{\Phi}^{-1}(n)\boldsymbol{\Phi}(n-1) \tag{19.53}$$

Suppose now we consider the effect of the quantization error $\boldsymbol{\eta}_P(n_0)$ induced at time $n_0 < n$. When form II of the standard RLS algorithm is used and the matrix $\mathbf{P}(n)$ remains Hermitian, then according to the error-propagation model of Eq. (19.48), the effect of the quantization error $\boldsymbol{\eta}_P(n_0)$ becomes modified at time n as follows:

$$\boldsymbol{\eta}_P(n) = \frac{1}{\lambda^{n-n_0}}\boldsymbol{\varphi}(n, n_0)\boldsymbol{\eta}_P(n_0)\boldsymbol{\varphi}^H(n, n_0) \tag{19.54}$$

where $\boldsymbol{\varphi}(n, n_0)$ is a *transition matrix* defined by

$$\boldsymbol{\varphi}(n, n_0) = \prod_{i=n_0+1}^{n} (\mathbf{I} - \mathbf{k}(i)\mathbf{u}^H(i)) \tag{19.55}$$

According to Eq. (19.54), the quantization error $\boldsymbol{\eta}_P(n_0)$ is *attenuated* in time if the transition matrix $\boldsymbol{\varphi}(n, n_0)$ is *contracting* sufficiently for $\boldsymbol{\varphi}(n, n_0)$ and $\boldsymbol{\varphi}^H(n, n_0)$ together to win over the factor $1/\lambda^{n-n_0}$. This is indeed the case if tap-input vector $\mathbf{u}(n)$ is *persistently exciting*. For an exponential weighting factor $\lambda \leq 1$, the tap-input vector $\mathbf{u}(i)$ is said to be persistently exciting over the observation interval $n_0 \leq i \leq n$ if the following condition is satisfied (see Problem 7):

$$a\mathbf{I} \leq \sum_{i=n_0}^{n} \lambda^{n-i}\mathbf{u}(i)\mathbf{u}^H(i) \leq b\mathbf{I} \tag{19.56}$$

where $\mathbf{I}$ is the identity matrix and a and b are positive constants. We may prove that $\boldsymbol{\varphi}(n, n_0)$ is a contraction if $\mathbf{u}(i)$ is persistently exciting over the interval (n, n_0) by proceeding as follows. From the definition of the time-averaged correlation matrix $\boldsymbol{\Phi}(n)$ given in Eq. (13.9), we have

$$\boldsymbol{\Phi}(n) = \sum_{i=1}^{n} \lambda^{n-i}\mathbf{u}(i)\mathbf{u}^H(i) \tag{19.57}$$

For $n_0 < n$, we find from Eq. (19.57) that

$$\boldsymbol{\Phi}(n) = \lambda^{n-n_0}\boldsymbol{\Phi}(n_0) + \sum_{i=n_0+1}^{n} \lambda^{n-i}\mathbf{u}(n)\mathbf{u}^H(i) \tag{19.58}$$

Therefore, when the condition for persistent excitation given in Eq. (19.56) holds, it follows from Eq. (19.58) that

$$\lambda^{n-n_0}\boldsymbol{\Phi}(n_0) < \boldsymbol{\Phi}(n) \tag{19.59}$$

In other words, the matrix difference $\boldsymbol{\Phi}(n) - \lambda^{n-n_0}\boldsymbol{\Phi}(n_0 - 1)$ is *positive definite*. We may now express the transition matrix $\boldsymbol{\varphi}(n, n_0)$ as (Ljung and Ljung, 1985; Verhaegen, 1989)

$$\boldsymbol{\varphi}(n, n_0) = \lambda^{n-n_0}\boldsymbol{\Phi}^{-1}(n)\boldsymbol{\Phi}(n_0) \tag{19.60}$$

Note that for $n_0 = n - 1$, the use of Eqs. (19.55) and (19.60) yields

$$\boldsymbol{\varphi}(n, n-1) = \mathbf{I} - \mathbf{k}(n)\mathbf{u}^H(n) = \lambda\boldsymbol{\Phi}^{-1}(n)\boldsymbol{\Phi}(n-1)$$

which is consistent with the result given in Eq. (19.53).

From Eq. (19.60) we draw the following important conclusions, assuming that form II of the standard RLS algorithm is used (Ljung and Ljung, 1985):

1. The effect of a single quantization erorr $\boldsymbol{\eta}_p(n_0)$ *decays exponentially*, provided that two conditions hold:
   (a) The inverse matrix $\boldsymbol{\Phi}^{-1}(n)$ remains uniformly bounded.
   (b) The exponential weighting factor λ is less than 1 (i.e., the algorithm has *finite* memory).

2. When $\lambda = 1$ (i.e., the algorithm has infinite memory), we have

$$\varphi(n, n_0) = \mathbf{\Phi}^{-1}(n)\mathbf{\Phi}^{-1}(n_0) \leq \mathbf{I}$$

where $\mathbf{I}$ is the identity matrix. Under this condition, we have stability.

3. When $\lambda = 1$ and, in addition, the inverse matrix $\mathbf{\Phi}^{-1}(n)$ tends to zero for increasing n, we have *asymptotic*, but *not* exponential, stability.

Computer Simulation Studies

To experimentally validate the error propagation model described in the preceding subsection for the standard RLS algorithm, Verhaegen has performed a mixed-precision computer simulation study for *real data* (Verhaegen, 1989). The study focused on the errors in matrix $\mathbf{P}(n)$, with double precision used for computing "error-free" quantities and single precision used for computing "erroneous" ones. Three experiments were conducted in the study, aimed at the following:

1. To investigate whether a "wrong" way of implementing the standard RLS algorithm is the primary cause for its "high" sensitivity to the loss of symmetry of the matrix $\mathbf{P}(n)$ due to error propagation effects
2. To investigate the interrelation between the loss of symmetry and the loss of positive definiteness of the matrix $\mathbf{P}(n)$.
3. To evaluate the influence of persistent excitation on the numerical robustness of the RLS algorithm

With the quantization of $\mathbf{P}(n)$ as the only source of error, the total error is approximately given by

$$\|\delta_{\text{tot}}\mathbf{P}(n)\| = \|\mathbf{P}_q(n) - \mathbf{P}(n)\|$$

The linear regression model used for data generation is noiseless with four degrees of freedom, as shown by

$$y(n) = \sum_{i=1}^{4} a_i u_i(n) \qquad (19.61)$$

Two different time series $\{y(n)\}$ for $n = 1, \ldots, 300$ were used in the study; they are displayed as measurement sequence 1 and 2 in Fig. 19.4. The basic difference between these two sequences is the following. Measurement sequence 1 guarantees the condition for persistent excitation during the whole experiment. Measurement sequence 2, on the other hand, is *not* persistently exciting during the time intervals (1, 40) and (81, 300). For the exponential weighting factor, the value $\lambda = 0.95$ was used.

For experiment 1 (aimed at point 1 above), measurement sequence 1 was used. The results of the experiment are shown in Fig. 19.5. Part (a) of the figure shows time histories of $\|\mathbf{P}_q(n)\|$, $\|\mathbf{P}_q(n) - \mathbf{P}(n)\|$, and $\|\mathbf{P}_q(n) - \mathbf{P}_q^T(n)\|$ for form I of the standard RLS algorithm. The subscript q signifies the quantized version of the matrix $\mathbf{P}(n)$ and the superscript T signifies matrix transposition. Thus, $\|\mathbf{P}_q(n) - \mathbf{P}_q^T(n)\|$ provides a measure of the loss of symmetry for the real data used in the study. Part

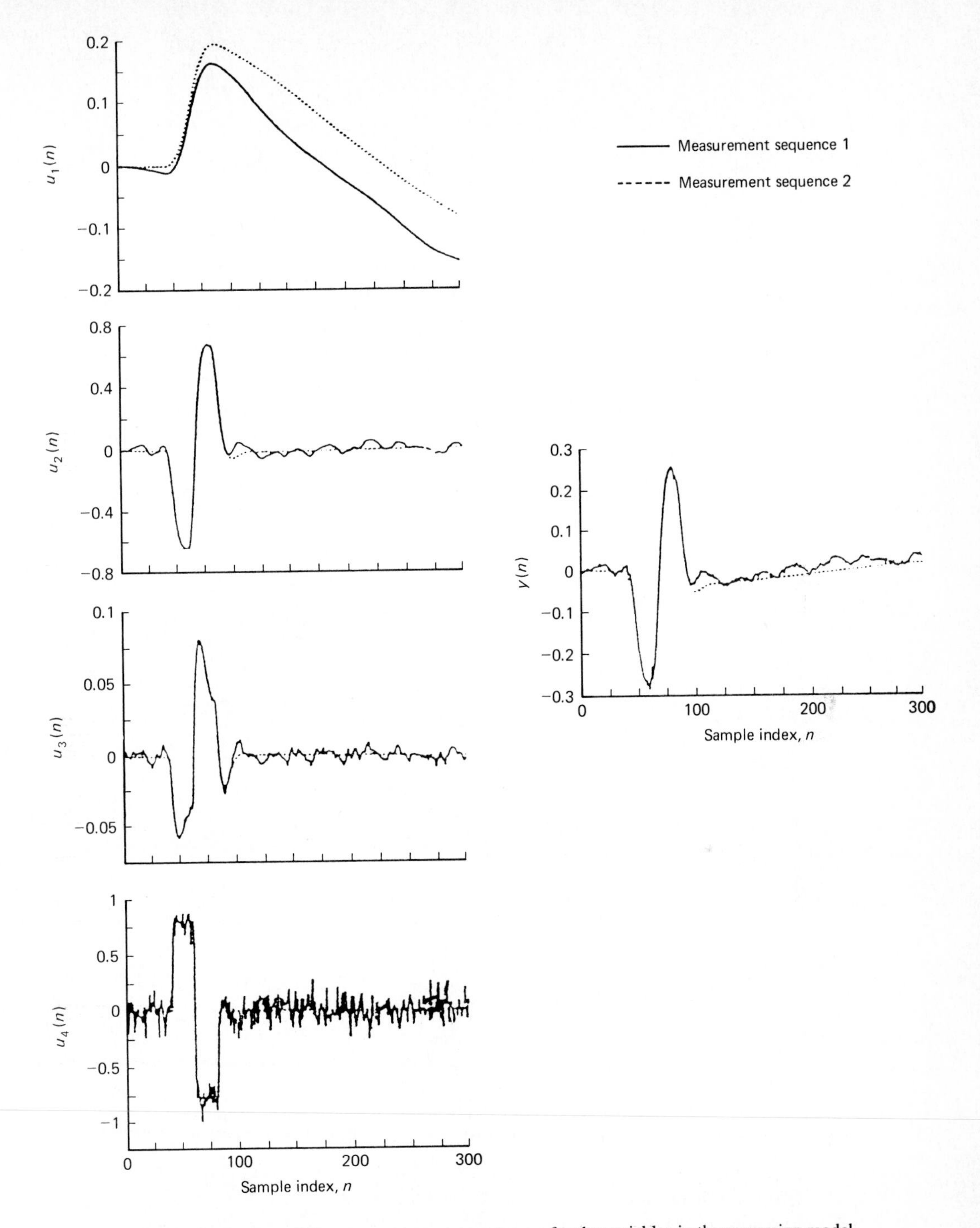

Figure 19.4 Two different measurement sequences for the variables in the regression model. [M.H. Verhaegen, "Round-Off Error Propagation in Four Generally Applicable, Recursive, Least-Squares Estimation Schemes," *Automatica*, Vol. 25, Pergamon Press PLC, 1989. Reprinted with permission.]

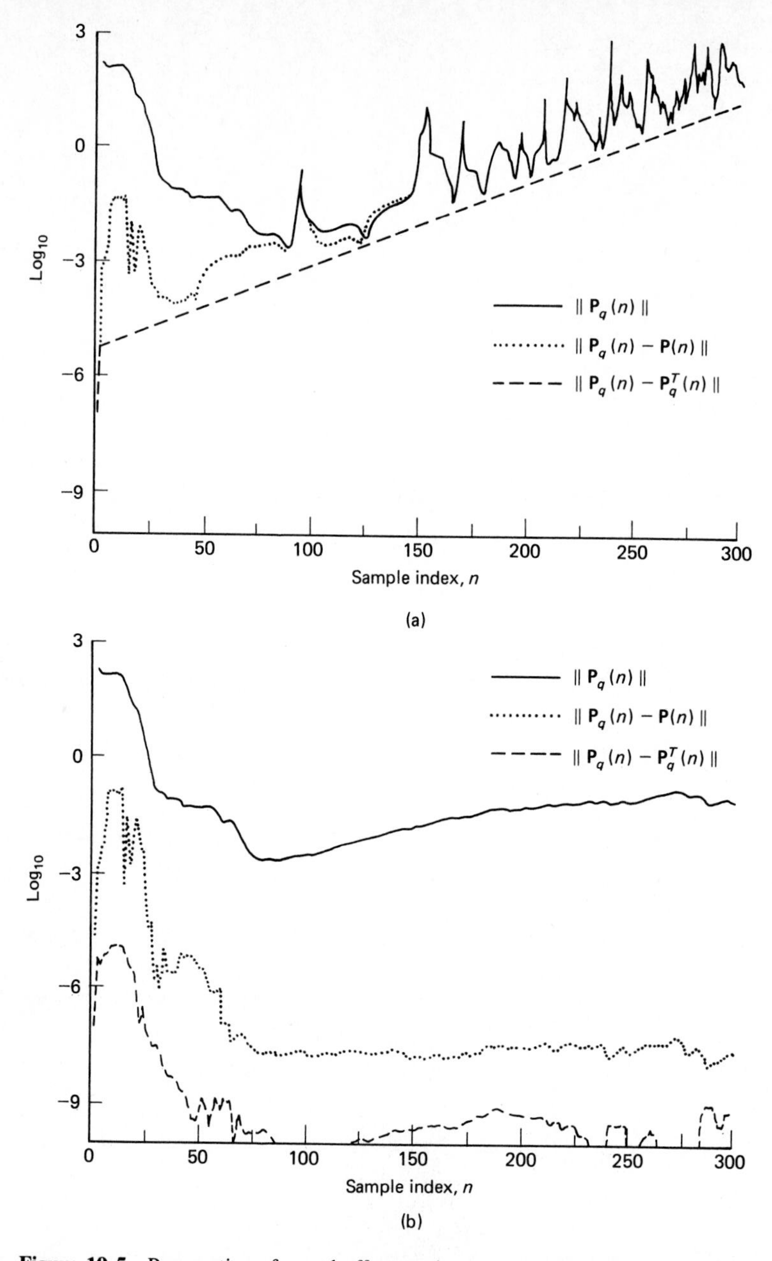

Figure 19.5 Propagation of round-off errors in two implementations of the standard RLS algorithm: (a) version I; (b) version II. ($\lambda = 0.95$) [M.H. Verhaegen, "Round-Off Error Propagation in Four Generally Applicable, Recursive, Least-Squares Estimation Schemes," *Automatica*, Vol. 25, Pergamon Press PLC, 1989. Reprinted with permission.]

(b) of Fig. 19.5 displays the corresponding results of the experiment for form II of the standard RLS algorithm. Examination of the curves in Fig. 19.5 clearly demonstrates that for $\lambda < 1$:

1. The effect of the *loss of symmetry* causes *divergence* in the error on $\mathbf{P}(n)$ for form I of the standard RLS algorithm.
2. Such a loss of symmetry does *not* occur when using the form II implementation of the algorithm.

In particular, note that the loss of symmetry for form I is (different) orders of magnitude larger than that for form II of the algorithm.

For experiment 2 (aimed at point 2 above), the same experimental conditions as in experiment 1 were used. The results of experiment 2 are displayed in Fig. 19.6. Parts (a) and (b) of the figure depict the evolution of the smallest eigenvalue and diagonal element 3 of the quantized matrix $\mathbf{P}_q(n)$, respectively, for form I of the standard RLS algorithm. In statistical terms, these two quantities should remain positive for all n, if $\mathbf{P}_q(n)$ is not to lose its positive definiteness. For these two conditions to hold, $\mathbf{P}_q(n)$ has to remain symmetric and nonsingular for all n. From Fig. 19.5(a) we know that the $\mathbf{P}_q(n)$ for form I of the algorithm suffers from a loss of symmetry due to quantization, and we therefore deduce that $\mathbf{P}_q(n)$ cannot retain its positive definiteness. This deduction is indeed confirmed in Fig. 19.6. Furthermore, according to Verhaegen (1989), such a loss of positive definiteness did not occur with the form II implementation of the RLS algorithm for all values of n used in the study.

For experiment 3 (aimed at point 3 above), measurement sequence 2 of Fig. 19.4 was used. As mentioned previously, this measurement sequence does not satisfy the condition for persistent excitation during the time intervals (1, 40) and (81, 300). The impact of this variation on the quantization error propagation in the form II implementation of the standard RLS algorithm is depicted in Fig. 19.7. This figure clearly shows that during the two time intervals characterized by the *lack of persistent excitation,* the error in the quantized matrix $\mathbf{P}_q(n)$ and the loss of symmetry increase "linearly." At the same time, $\|\mathbf{P}_q(n)\|$ increases linearly, too. Furthermore, the rate of increase is exactly the same in both time intervals, so that no loss of precision occurs.

Stalling Phenomenon

The second form of divergence, referred to as the *stalling phenomenon,* occurs when the tap weights in the RLS algorithm stop adapting. In particular, this phenomenon occurs when the quantized elements of the matrix $\mathbf{P}(n)$ become very small, such that multiplication by $\mathbf{P}(n)$ is equivalent to multiplication by a zero matrix (Bottomley and Alexander, 1989). Clearly, the stalling phenomenon may arise in form I or form II of the RLS algorithm.

The stalling phenomenon is directly linked to the exponential weighting factor λ and the variance σ_u^2 of the input data $\{u(n)\}$. First, we note from the definition of

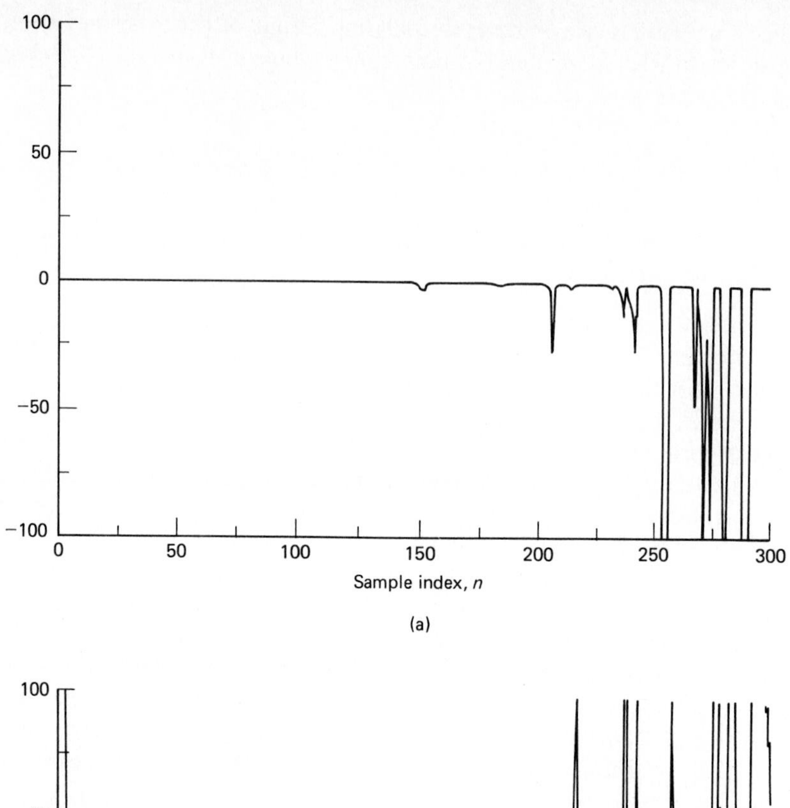

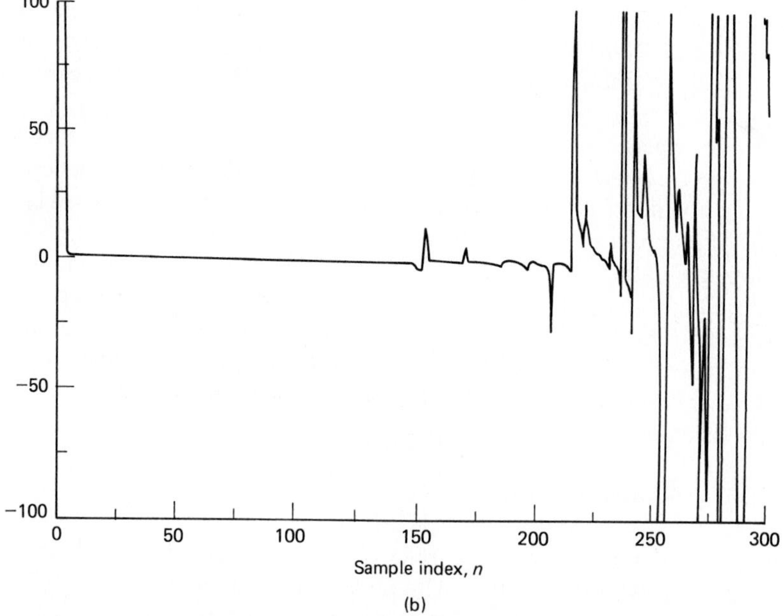

Figure 19.6 Evolution of certain eigenvalues and diagonal elements of $\mathbf{P}_q(n)$ in version I or the standard RLS algorithm: (a) evolution of smallest eigenvalue of $\mathbf{P}_q(n)$; (b) evolution of diagonal element 3 of $\mathbf{P}_q(n)$. [M.H. Verhaegen, "Round-Off Error Propagation in Four Generally Applicable, Recursive, Least-Squares Estimation Schemes," *Automatica*, Vol. 25, Pergamon Press PLC, 1989. Reprinted with permission.]

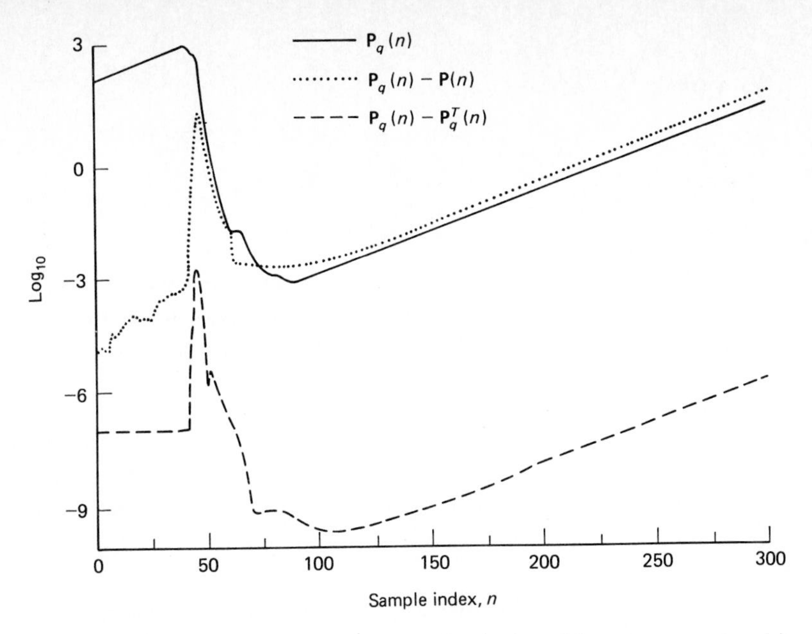

Figure 19.7 Influence of the persistency of excitation of the regressor vector $\mathbf{u}(n)$ on the round-off propagation in version II of the standard RLS algorithm ($\lambda = 0.95$, measurement sequence 2). [M.H. Verhaegen, "Round-Off Error Propagation in Four Generally Applicable, Recursive, Least-Squares Estimation Schemes," *Automatica*, Vol. 25, Pergamon Press PLC, 1989. Reprinted with permission.]

the time-averaged correlation matrix $\mathbf{\Phi}(n)$, given in Eq. (19.57), that the expectation of $\mathbf{\Phi}(n)$ equals (after interchanging the order of expectation and summation):

$$\begin{aligned} E[\mathbf{\Phi}(n)] &= \sum_{i=1}^{n} \lambda^{n-i} E[\mathbf{u}(i)\mathbf{u}^H(i)] \\ &= \sum_{i=1}^{n} \lambda^{n-i} \mathbf{R} \end{aligned} \tag{19.62}$$

where $\mathbf{R}$ is the ensemble-averaged correlation matrix (assuming wide-sense stationarity). We recognize that $\sum_{i=1}^{n} \lambda^{-i}$ represents the sum of a geometric series with a common ratio λ^{-1}, a first term equal to λ^{-1}, and a total number of terms equal to n. Using the formula for such a sum, we get

$$\sum_{i=1}^{n} \lambda^{-i} = \lambda^{-1}\left(\frac{1 - \lambda^{-n}}{1 - \lambda^{-1}}\right) \tag{19.63}$$

Hence, we may rewrite Eq. (19.62) as

$$E[\mathbf{\Phi}(n)] = \left(\frac{1 - \lambda^{n}}{1 - \lambda}\right)\mathbf{R} \tag{19.64}$$

For $\lambda < 1$ and large n, we may ignore λ^n compared to unity. We may then approximate Eq. (19.64) as follows:

$$E[\mathbf{\Phi}(n)] \simeq \frac{\mathbf{R}}{1 - \lambda}, \qquad \text{large } n \tag{19.65}$$

For a small $1 - \lambda$, we have

$$E[\mathbf{P}(n)] = E[\mathbf{\Phi}^{-1}(n)] \simeq (E[\mathbf{\Phi}(n)])^{-1} \tag{19.66}$$

Hence, using Eq. (19.65) in (19.66), we get

$$E[\mathbf{P}(n)] \simeq (1 - \lambda)\mathbf{R}^{-1}, \qquad \text{large } n \tag{19.67}$$

where $\mathbf{R}^{-1}$ is the inverse of matrix $\mathbf{R}$. Assuming that the time series $\{u(n)\}$ used to supply the tap-input vector $\mathbf{u}(n)$ is wide-sense stationary, with zero mean, we may write

$$\mathcal{R} = \frac{1}{\sigma_u^2}\mathbf{R} \tag{19.68}$$

where $\mathcal{R}$ is a *normalized correlation matrix* with diagonal elements equal to 1 and off-diagonal elements less than or equal to 1 in magnitude, and σ_u^2 is the variance of an input data sample $u(n)$. We may therefore rewrite Eq. (19.68) as

$$E[\mathbf{P}(n)] \simeq \left(\frac{1 - \lambda}{\sigma_u^2}\right)\mathcal{R}^{-1}, \qquad \text{large } n \tag{19.69}$$

Equation (19.69) reveals that the RLS algorithm may stall if: (1) the exponential weighting factor λ is close to 1, and/or (2) the input data variance σ_u^2 is large. Accordingly, we may prevent stalling of the standard RLS algorithm by using a sufficiently large number of accumulator bits in the computation of the matrix $\mathbf{P}(n)$.

19.4 QR DECOMPOSITION-BASED RECURSIVE LEAST-SQUARES ESTIMATION

It is generally recognized that the method of QR-decomposition by Givens rotations is one of the best procedures for solving the recursive least-squares problem. Consequently, the *QR decomposition-based recursive least squares* (QRD-RLS) *algorithm* has good numerical properties, which means that it can perform in an acceptable manner in a finite-precision environment. A formal proof of the inherent numerical stability of the QRD-RLS algorithm is presented in Leung and Haykin, (1989). In particular, it is shown that the two systolic implementations of this algorithm, those due to Gentleman and Kung (1981) and McWhirter (1983), are stable (in a bounded input–bounded state sense) under finite-precision arithmetic; these two structures were referred to as systolic arrays I and II, respectively, in Chapter 14.

Computer Simulation Studies

Ward et al. (1986) present computer simulation results that compare the effect of finite precision on the performance of two *adaptive beamformers,* one based on the QRD-RLS algorithm (Gentleman and Kung, 1981), and the other based on the sample matrix inversion algorithm (Reed et al., 1974). Part (a) of Fig. 19.8 depicts the simulation flow diagrams for these two algorithms. In both cases, the sequence of data snapshots was generated and applied to the constraint preprocessor; for details

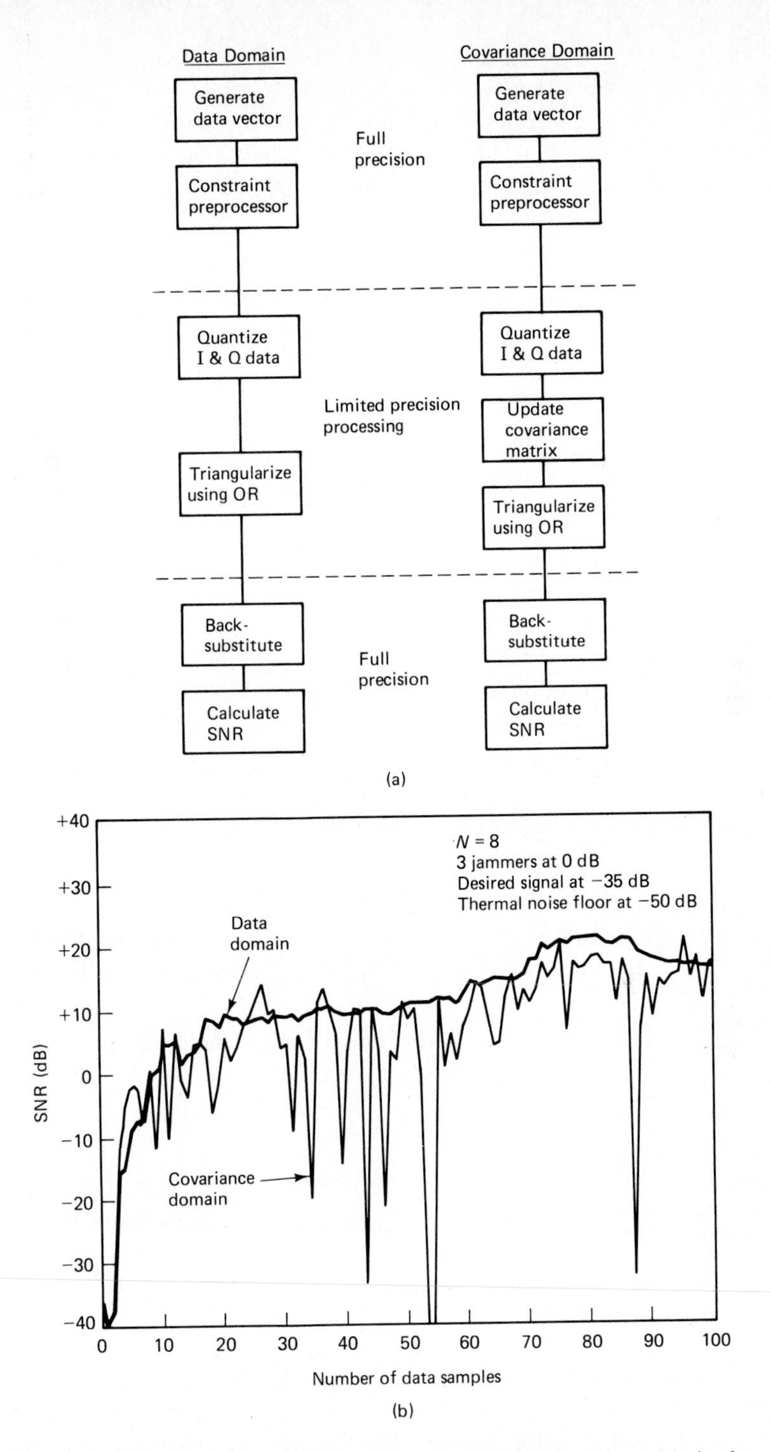

Figure 19.8 Comparison of data and covariance domain algorithms: (a) simulation flow diagrams, (b) typical signal-to-noise ratio response. [From Ward et al. (1986), with permission of the IEEE; property of the British Crown.]

of the preprocessor, see Section 14.12. In the simulation study described in Fig. 19.8, the preprocessor was designed to apply a constraint along the desired look direction; for its implementation, full computer precision was employed. The transformed spatial data were then quantized to 24-bit floating-point word-length (16-bit mantissa and 8-bit exponent), which was retained throughout the two adaptive beamforming approaches:

1. Recursive QR-decomposition implemented directly in the data domain
2. Sample matrix inversion, computed using QR-decomposition

In both cases, the back substitution for calculating the weight vector was performed at full computer precision. Thus, the simulation study provides a fair comparison between the two basic beamforming approaches: data versus covariance (correlation) domain.

The database used in the study modeled the following environmental features:

1. An eight-element array
2. Three equipower jamming signals, received individually at levels of 0 dB relative to a thermal noise floor of −50 dB at the antenna array elements
3. An independent, narrow-band Gaussian process for the statistical description of the complex envelope of each jammer
4. A desired signal received by the array at a level of 15 dB above the thermal noise floor but approximately 40 dB below the total received jamming

Figure 19.8(b) shows the computer simulation results for the two approaches, with the expected signal-to-noise ratio at the array antenna output plotted as a function of data snapshots. From Fig. 19.8(b), we can make the following observations:

1. For both the QRD-RLS algorithm and the sample matrix inversion algorithm, the initial rate of convergence is rapid.
2. In both bases, a good level of jamming cancellation is obtained after about 10 to 20 snapshots.
3. With sample matrix inversion, the adapted response curve exhibits extreme fluctuations, indicating a numerically unstable weight vector. In contrast, the QRD-RLS algorithm shows no evidence of numerical instability; indeed, the signal-to-noise ratio performance gets progressively better with increasing number of data snapshots.

The important conclusion to be drawn from the simulation study by Ward et al. is that for the same number of bits of arithmetic precision, the QRD-RLS algorithm offers a significantly better performance than the sample matrix inversion algorithm. Equivalently, we may say that in order to achieve a comparable performance, the QRD-RLS algorithm requires a significantly shorter word-length than the sample matrix inversion technique.

19.5 FAST TRANSVERSAL FILTERS ALGORITHM

In Sections 19.2 to 19.4, we focused attention on finite-precision effects in the LMS algorithm, the standard RLS algorithm, and the QRD-RLS algorithm. The LMS algorithm belongs to a class of stochastic gradient algorithms, whereas the standard RLS and QRD-RLS algorithms are two different but equivalent (assuming infinite precision) methods for solving the recursive linear least-squares problem. In the latter two algorithms, the computational complexity increases with M^2, where M is the number of adjustable weights in the algorithm. In this and the next two sections we discuss the effects of finite-precision arithmetic on three *fast* algorithms for solving the recursive linear least-squares problem, all of which are characterized by a computational complexity that increases linearly with M. The reduction in computational complexity is made possible by exploiting the redundancy inherent in the Toeplitz structure of the data matrix. In this section we consider the *fast transversal filters (FTF) algorithm*.

When the FTF algorithm is implemented on a digital computer or special-purpose digital processor with a limited number of bits used to represent the various parameters in the algorithm, quantization errors may cause the algorithm to diverge. Experimentation with the fast RLS algorithm and the FTF algorithm indicates that a certain, positive variable in the algorithm becomes negative due to the accumulated effect of finite-precision effects just before divergence of the algorithm occurs (Lin, 1984; Cioffi and Kailath, 1984). The variable equals the ratio of two nonnegative quantities, namely, the *conversion factors* for prediction orders M and $M + 1$, as shown by

$$\zeta_M(n) = \frac{\gamma_{M+1}(n)}{\gamma_M(n)} \tag{19.70}$$

We refer to $\zeta_M(n)$ as the *rescue variable*. For the ideal case of infinite precision, we have $0 \leq \zeta_M(n) \leq 1$. A violation of this restriction on the value of $\zeta_M(n)$ is due to finite-precision effects.

Using Eq. (15.57), we may also express $\zeta_M(n)$ as

$$\zeta_M(n) = \frac{\lambda \mathcal{B}_M(n-1)}{\mathcal{B}_M(n)} \tag{19.71}$$

Hence, the propagation of quantization errors in the FTF algorithm is closely tied to the computation of the values of the minimum sum of a posteriori backward prediction-error squares at times $n - 1$ and n (Botto and Moustakides 1989).

We may express the variable $\zeta_M(n)$ in yet another form by using Eq. (16.19) to write

$$\zeta_M(n) = 1 - \psi_M(n)\gamma_{M+1}(n)\tilde{k}^*_{M+1,M+1}(n) \tag{19.72}$$

From this relation, we deduce that stabilization of the FTF algorithm requires that a tighter control be exercised over the values of the following three variables:

1. The backward a priori prediction error $\psi_M(n)$.
2. The conversion factor $\gamma_{M+1}(n)$.
3. The last element $\tilde{k}_{M+1,M+1}(n)$ in the normalized gain vector $\tilde{\mathbf{k}}_{M+1}(n)$.

This, indeed, is the rationale behind the stabilization method described next.

Error Feedback

In going from the standard RLS algorithm to the FTF algorithm, we reduce computational complexity by using linear least-squares prediction to exploit the redundancy inherent in the Toeplitz structure of the input data matrix. In effect, redundant information contained in the standard RLS algorithm is removed. By so doing, however, the FTF algorithm becomes vulnerable to finite-precision effects. The algorithm may be stabilized by reintroducing some computational redundancy in a controlled manner into the algorithm (Botto and Moustakides, 1989; Slock and Kailath (1989).[6]

An important characteristic of the FTF algorithm is that certain quantities in the algorithm may be computed in two different ways:

1. The transversal filtering of an input data vector
2. The manipulation of scalar quantities

To be specific, consider the computation of the backward a priori prediction error $\psi_M(n)$. Obviously, we may compute $\psi_M(n)$ as

$$\psi_M^{(f)}(n) = \mathbf{c}_M^H(n-1)\mathbf{u}_{M+1}(n) \tag{19.73}$$

where $\mathbf{c}_M(n-1)$ is the tap-weight vector of the backward prediction-error filter at time $n-1$, and $\mathbf{u}_{M+1}(n)$ is the tap-input vector [see Eq. (15.35)]. The superscript (f) in the symbol $\psi_M^{(f)}(n)$ denotes "filtering." From the FTF theory, we know that $\psi_M(n)$ may also be computed as

$$\psi_M^{(s)}(n) = \lambda \mathcal{B}_M(n-1)\tilde{k}_{M+1,M+1}(n) \tag{19.74}$$

where λ is the exponential weight factor, $\mathcal{B}_M(n-1)$ is the minimum sum of backward a posteriori prediction-error squares, and $\tilde{k}_{M+1,M+1}(n)$ is the last element of the normalized gain vector $\tilde{\mathbf{k}}_{M+1}(n)$; [see Eq. (16.16)]. The superscript (s) in the symbol $\psi_M^{(s)}(n)$ denotes "scalar manipulation." For the ideal case of infinite-precision arithmetic, the two ways described in Eqs. (19.73) and (19.74) yield identical answers for the backward a priori prediction error $\psi_M(n)$.

In a practical implementation of the FTF algorithm, however, these two ways yield different answers due to quantization errors resulting from the use of finite-precision arithmetic. Indeed, the difference between the two answers may be viewed as a manifestation of the error propagation mechanism in the FTF algorithm.

The issue to be resolved then is how to exploit the availability of these two different finite-precision values of the same physical quantity $\psi_M(n)$ in order to influence the error propagation mechanism. The answer to this issue lies in the use of *error feedback* that involves feeding an *error signal,* equal to the difference between the two values of the quantity of interest, back into the computation of that quantity. (Here it is recognized that error feedback of the *negative* type reduces the effect of parameter variations.) In particular, for the value of $\psi_M(n)$ used in subsequent computations, we may take a convex combination of its two finite-precision values computed using Eqs. (19.73) and (19.74), as shown by

[6] The approaches taken by Botto and Moustakides (1989) and Slock and Kailath (1989) are somewhat different. For our presentation, we follow Slock and Kailath.

$$\psi_M(n) = K\psi_M^{(f)}(n) + (1 - K)\psi_M^{(s)}(n) \qquad (19.75)$$

or, equivalently,

$$\psi_M(n) = \psi_M^{(s)}(n) + K(\psi_M^{(f)}(n) - \psi_M^{(s)}(n)) \qquad (19.76)$$

where K is a *feedback constant*. The feedback loop is completed, as depicted in Fig. 19.9, by virtue of two facts:

1. The subsequent computations that depend on $\psi_M(n)$
2. The iterative nature of the FTF algorithm

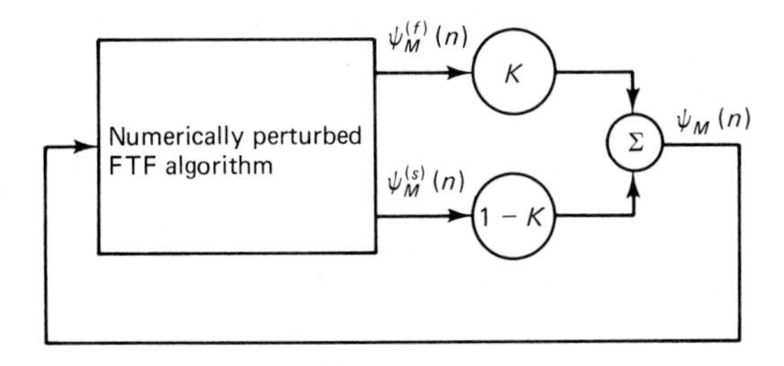

Figure 19.9 Use of error feedback in the stabilization of the FTF algorithm.

The motivation behind the choice of error feedback (namely, the convex combination) is that it is intuitively appealing. Furthermore, by properly choosing the feedback constant K, it will permit us to stabilize an unstable mode in the error propagation mechanism which is our ultimate goal. It is noteworthy that the stabilized FTF algorithm reduces to the conventional form of the FTF algorithm (with a computational complexity of $7M$) for the quantity of interest [i.e., the backward a priori prediction error $\psi_M(n)$ in the case of Eq. (19.76)] under either one of the following two conditions:

1. The feedback constant K is zero.
2. The ideal case of infinite precision is assumed.

We may also use error feedback for the computation of two other quantities, $\tilde{k}_{M+1,M+1}(n)$ and $\gamma_M^{-1}(n)$, which enter the computation of the rescue variable $\zeta_M(n)$ and they may be given a similar treatment; $\psi_M(n)$ and $\gamma_M(n)$ are the two quantities for which this treatment is necessary for stabilization. Note that since the quantity $\gamma_M(n)$ is used in several places in the FTF algorithm, we may use different values for the feedback constant K at these different places; such an approach provides more freedom in influencing the error propagation mechanism in the algorithm. By following the error feedback procedure, described in detail in Slock and Kailath (1989) and Slock (1989), the FTF algorithm is stabilized at the cost of a modest increase in computation complexity, from $7M$ to $(7M + M)$.

It is also noteworthy that in the stabilized FTF algorithm, it is permissible to deviate slightly from the strict FTF equations. For example, we may change the exponential weighting factor λ very slowly in time if so desired. Ordinarily, a deviation from the strict FTF algorithm has the effect of introducing (possibly large) numerical errors. However, with a stable error propagation mechanism as described herein, the effect of these numerical errors will decay in an exponentially fast manner (Slock and Kailath, 1989).

To summarize, the use of error feedback provides an effective method for the numerical stabilization of the FTF algorithm.[7] Indeed, it places the applicability of the FTF algorithm in a new light (Slock, 1989).

Periodic Restart for the FTF Algorithm

The FTF algorithm may be stabilized in another entirely different way. Specifically, a *periodic reinitialization procedure* is applied for overcoming the numerical instability problem encountered in the FTF algorithm due to round-off errors (Eleftheriou and Falconer, 1984). To avoid the buildup of round-off errors with time, it is proposed that operation of the FTF algorithm be interrupted and then restarted at periodic intervals, say every N iterations. Immediately following such a restart, a simple LMS algorithm (which has been initialized with the tap-weight vector attained by the FTF algorithm just before the restart was initiated) provides an estimate of the desired response temporarily until the restarted FTF algorithm takes over again after a short time. The *transition* from the FTF algorithm to the LMS algorithm for obtaining the desired estimate is so short in duration (typically, 1 to 1.5 times the memory of the transversal filter) and so smooth that temporary reliance on the slower LMS algorithm causes little or no performance degradation. Figure 19.10 illustrates the timing of operations involved in the periodic reinitialization procedure.

The periodic restart of the FTF algorithm involves an insignificant cost increase in implementation. Unfortunately, it may result in a significant reduction in tracking speed, which may be unacceptable in those applications that require fast tracking.

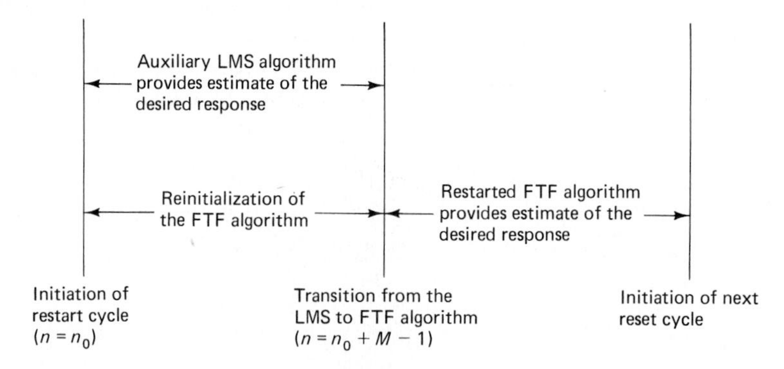

Figure 19.10 Timing of the operations involved in one cycle of the periodic reinitialization procedure of the FTF algorithm.

[7] Hariharan and Clark (1990) have suggested another stabilization technique to overcome the problem caused by the accumulation of round-off errors in the FTF algorithm. In particular, a one-step prediction is incorporated into the algorithm that takes into account the rate of change in the estimate of the sampled impulse response of an unknown channel. Computer simulation tests are included, demonstrating the improved performance that results from the use of a stabilized FTF algorithm applied to the estimation of the sampled impulse response of a time-varying HF channel.

19.6 RECURSIVE LEAST-SQUARES LATTICE ALGORITHM

The FTF algorithm provides a fast solution to the recursive linear least-squares problem by employing a parallel combination of four transversal filters with fixed lengths. The recursive LSL algorithm, on the other hand, provides a fast solution to the same problem by employing a multistage lattice predictor for transforming the input data into a corresponding sequence of backward prediction errors. This transformation may be viewed as a form of the classical Gram–Schmidt orthogonalization procedure. The Gram–Schmidt orthogonalization is known to be *numerically inaccurate* (Stewart, 1973). Correspondingly, a conventional form of the recursive LSL algorithm (be it based on a posteriori or a priori prediction errors) has poor numerical behavior. The key to a practical method of overcoming the numerical accuracy problem in a recursive LSL algorithm is to update the forward and backward reflection coefficients *directly*, rather than first computing the individual sums of weighted forward and backward prediction errors and their cross-correlation and then taking the ratios of the appropriate quantity (as in a conventional LSL algorithm). This is precisely what is done in a recursive LSL algorithm *with error feedback*. For a prescribed fixed-point representation, a recursive LSL algorithm with error feedback works with much more accurate values of the forward and backward reflection coefficients; these two coefficients are the key parameters in any recursive LSL algorithm. The direct computation of the forward and backward reflection coefficients therefore has the overall effect of preserving the positive definiteness of the underlying inverted correlation matrix of the input data, despite the presence of quantization errors due to finite-precision effects.

Numerical accuracy problems with a conventional form of the recursive LSL algorithm (using a posteriori estimation errors) were first reported by Ling and Proakis (1984). In a series of computer simulation studies conducted by Ling and Proakis, a comparison was made between a conventional form of the recursive LSL algorithm and its modified version using error feedback. The simulation results, reported therein, demonstrated the superiority of the latter algorithm over the former in the context of numerical accuracy. This observation was confirmed in a subsequent computer simulation study reported in Ling et al. (1986). The simulation involved the use of a linear adaptive equalizer with 11 taps, an eigenvalue spread of approximately 11 for the data matrix of the equalizer input data, and an exponential weighting factor $\lambda = 0.975$. Details of the equalizer are given in Section 9.13, following the paper by Satorius and Pack (1981). Table 19.1 gives the output mean squared error for floating-point arithmetic (22 bits) and fixed-point arithmetic of varying word length from 9 to 16 bits (including a sign bit). This table includes simulation results for the four versions of the recursive LSL algorithm, depending on (1) the use of *a priori* or *a posteriori* estimation errors, and (2) the *conventional* computation of the forward and backward reflection coefficients or their computation using *error feedback*. Table 19.1 clearly demonstrates the numerical superiority of the two versions of the recursive LSL algorithm employing error feedback over the other two versions that do not. Indeed, it appears that the a posteriori error feedback lattice is the most accurate adaptive filtering algorithm to date in finite-precision environments (Regalia, 1990). This serves to illustrate how a seemingly small modification, namely, the addition of error feedback to a recursive LSL algorithm can have such a profound impact on the numerical accuracy of the algorithm.

TABLE 19.1 NUMERICAL ACCURACY, IN TERMS OF THE OUTPUT MEAN-SQUARED ERROR (MSE × 10^{-3}) OF RECURSIVE LSL ALGORITHMS (λ = 0.975)

Number of bits (with sign)	*A priori* form		*A posteriori* form	
	Conventional	Error feedback	Conventional	Error feedback
Floating point (22 bits)	2.10	2.10	1.33	1.33
16	2.18	2.16	1.34	1.33
13	3.09	2.22	2.32	1.34
11	25.2	3.09	24.3	1.43
9	365	31.6	93	2.35

Source: Ling et al. ©1986 IEEE. Reprinted with permission.

Interestingly enough, in the (inexact) *gradient adaptive lattice* (GAL) *algorithm* the reflection coefficient (one per stage of the algorithm) is also updated directly [see Eq. (9.172)]. In other words, there is a form of error feedback built into the operation of the GAL algorithm. We would therefore expect the GAL algorithm to exhibit a good numerical behavior too. This is indeed borne out by the results of an earlier computer simulation reported in (Satorius et al., 1983). The GAL algorithm has the advantage of simplicity over recursive LSL algorithms (with or without error feedback). On the other hand, recursive LSL algorithms have an advantage over the GAL algorithm in that they provide a faster rate of convergence, because *no* approximations are made in their derivations.

In conclusion, a recursive LSL algorithm with error feedback is the preferred method for lattice-based least-squares estimation in a finite-precision environment. Such an algorithm, using a priori or a posteriori estimation errors, combines the *fast rate of convergence* inherent in least-squares estimation with the *numerical robustness* of stochastic gradient estimation. If computational simplicity in the use of a lattice structure is the requirement, then the use of the GAL algorithm is the recommended procedure.

19.7 QR DECOMPOSITION-BASED LEAST-SQUARES LATTICE ALGORITHM

The *QR decomposition-based least-squares lattice (QRD-LSL) algorithm,* derived in Chapter 18, is the last in the family of fast (i.e., order M) algorithms to be considered. This algorithm exploits forward and backward forms of linear least-squares prediction for a fast computation of the plane (Givens) rotations that produce the QR-decomposition of the data matrix. The end result is a reduction in algorithm complexity by an order of magnitude, while preserving the attractive parallelism of the implementation.

The QRD-LSL algorithm has good numerical properties, as evidenced by (1) computer simulation studies reported in Proudler et al. (1989) and Ling (1989), and (2) a detailed stability analysis presented in Regalia (1990). To further support this statement, we present in Fig. 19.11 the results of a computer experiment involving

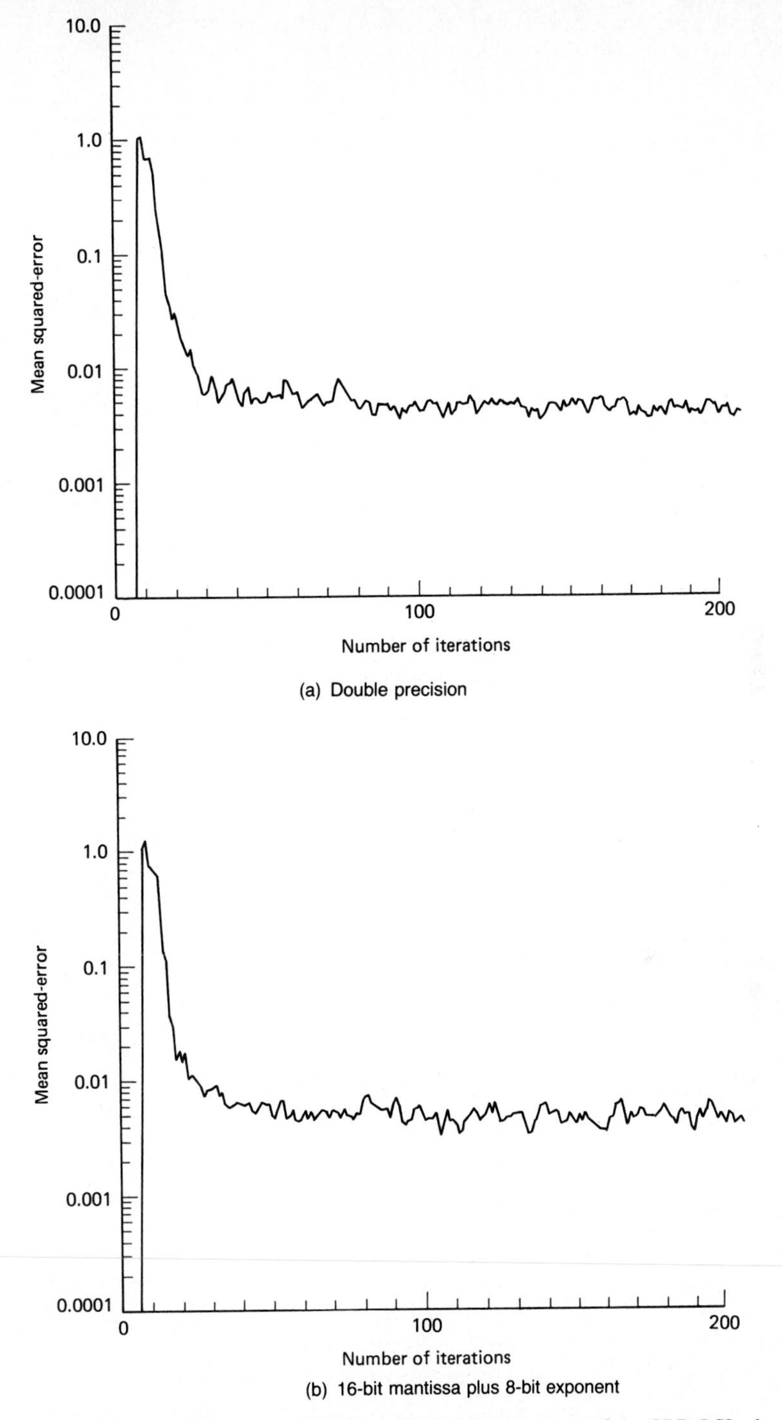

Figure 19.11 Effect of finite precision on the learning curve of the QRD-LSL algorithm applied to adaptaive equalization, for varying word-length. [Property of the British Crown.].

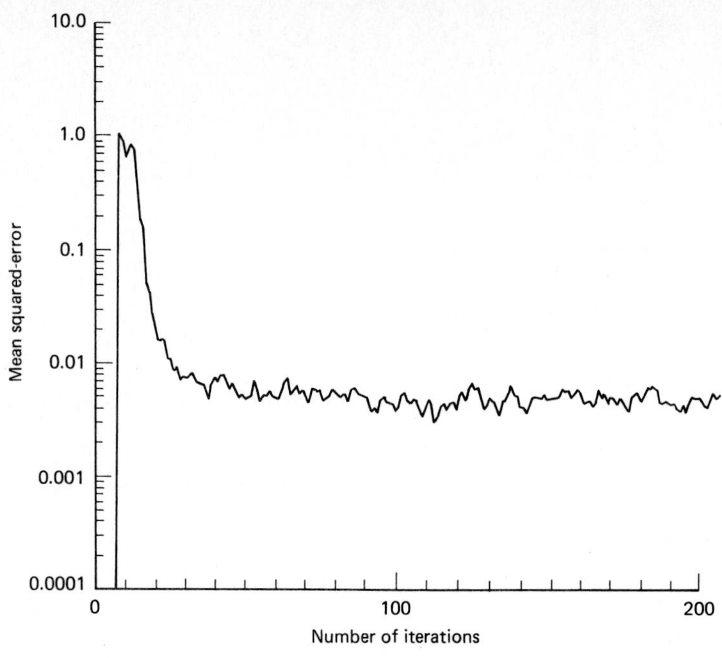

(c) 12-bit mantissa plus 8-bit exponent

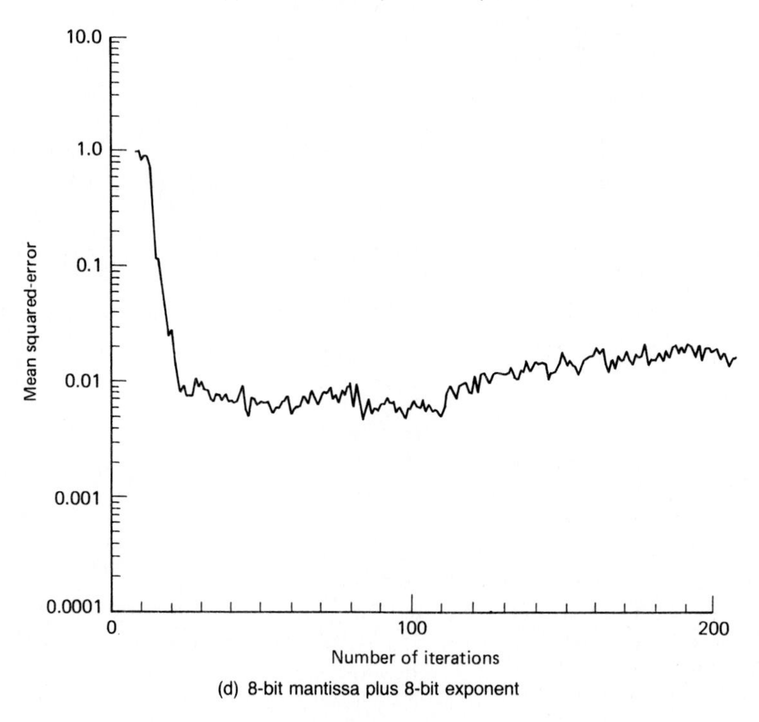

(d) 8-bit mantissa plus 8-bit exponent

Figure 19.11 *(cont.)*

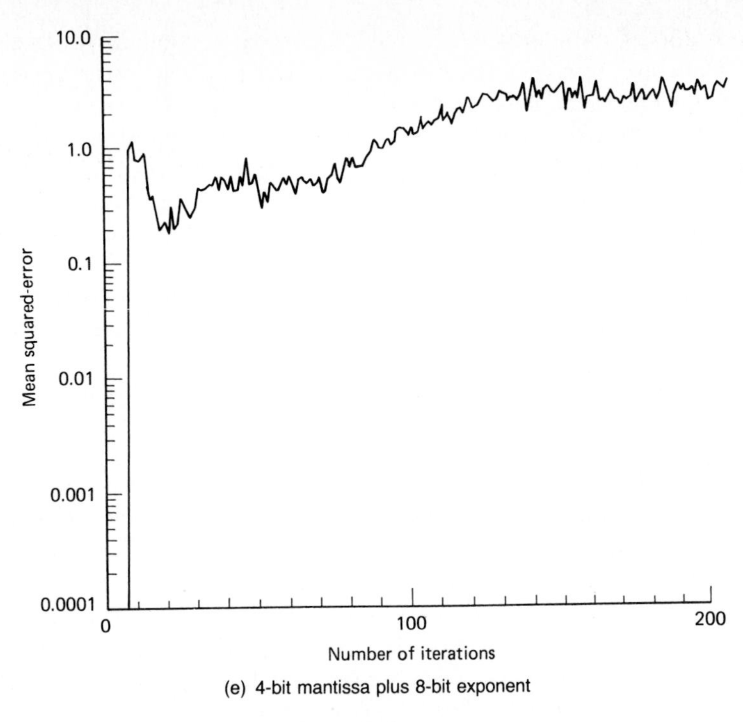

(e) 4-bit mantissa plus 8-bit exponent

Figure 19.11 (*cont.*)

an 11-tap adaptive equalizer; the details of the experiment are the same as those presented in Section 18.9. However, the experiment was performed with a different version of the QRD-LSL algorithm, different in the sense that the joint-process estimation was performed using forward prediction errors rather than backward prediction errors [Proudler et al. (1989)]. The five parts of Fig. 19.11 correspond to an eigenvalue spread $\chi(\mathbf{R})$ equal to 46.8216; that is, the underlying correlation matrix $\mathbf{R}$ of the input data is rather ill conditioned. In particular, the figure presents learning curves (obtained by ensemble averaging over 200 independent trials of the experiment) for the following floating-point word-lengths:

1. Double precision
2. Sixteen bits
3. Twelve bits
4. Eight bits
5. Four bits

(items 2–5: mantissa + 8 bit exponent)

The results presented in Fig. 19.11 indicate that the algorithm is indeed insensitive to finite-precision effects for mantissa lengths down to 12 bits.

Backwards Stability

The QRD-LSL algorithm is a member of the family of QR decomposition-based fast RLS algorithms. A rigorous stability analysis of this family of fast algorithms is presented in Regalia (1990). In particular, Regalia emphasizes the role of *backwards stability*, the use of which is well established in the numerical analysis literature.[8] In *backward error analysis*, we use error bounds to show that the computed solution of a given problem is the exact solution of a *slightly perturbed version* of the problem. This is sufficient to ensure that the algorithm that performed the computation is numerically *stable*. In the context of linear least-squares problems, backwards stability requires that (when computations are performed in finite-precision arithmetic) the *range of numerical reachability* of each internal state variable of the algorithm remains *consistent* with a least-squares interpretation. We may illustrate the idea of infinite precision versus numerical reachability by way of an example.

Consider the angle variable (conversion factor) denoted by $\gamma(n)$. The theoretically (i.e., infinite-precision) reachable range of this variable is $0 \leq \gamma(n) \leq 1$. Suppose that we first compute $\gamma(n)$ using the following definition [see Eq. (15.48)]:

$$\gamma(n) = 1 - \mathbf{u}^H(n)\mathbf{\Phi}^{-1}(n)\mathbf{u}(n) \tag{19.77}$$

where $\mathbf{u}(n)$ is the M-by-1 input vector, and $\mathbf{\Phi}(n)$ is the time-averaged correlation matrix of $\mathbf{u}(n)$. From previous discussions, we know that the inverse matrix $\mathbf{\Phi}^{-1}(n) = \mathbf{P}(n)$ can indeed assume a negative eigenvalue when the computation is performed in finite-precision arithmetic. Consequently, the numerically (i.e., finite-precision) reachable range of $\gamma(n)$ calculated in accordance with the formula of Eq. (19.77) can exceed the upper limit of unity, which is in violation of a least-squares interpretation. Suppose, next, we compute the angle variable $\gamma(n)$ according to the theoretically equivalent formula [see Eq. (14.57)]

$$\gamma^{1/2}(n) = \prod_{i=1}^{M} c_i(n) \tag{19.78}$$

where $c_i(n)$ is the real cosine parameter of the ith Givens rotation. Then, provided that the condition $|c_i(n)| \leq 1$ is ensured numerically for all n, the finite-precision reachable range corresponding to Eq. (19.78) remains as $0 \leq \gamma(n) \leq 1$, which is precisely how it should be for a valid least-squares interpretation.

To return to the issue at hand, let an input sample $u(n + 1)$ and an internal state vector $\mathbf{x}(n)$ be associated with the prediction section of a QRD-fast RLS algorithm. The input $u(n + 1)$ and the state $\mathbf{x}(n)$ comprise the quantities necessary to compute the updated solution to the prediction for the sample instant $n + 1$. The updated state $\mathbf{x}(n + 1)$ may then be expressed as

$$\mathbf{x}(n + 1) = g(\mathbf{x}(n), u(n + 1)) \tag{19.79}$$

Although Eq. (19.79) solves a linear prediction problem, in general, the mapping denoted by $g(\cdot,\cdot)$ depends on the states $\mathbf{x}(n)$. Accordingly, Eq. (19.79) is descriptive of a *nonlinear feedback system*. Naturally, the structure of the algorithm under study determines the exact details of the state $\mathbf{x}(n)$ and the mapping $g(\cdot,\cdot)$. In any event, the system described in Eq. (19.79) is said to be *minimal* if the number of elements constituting the state $\mathbf{x}(n)$ is the strict minimum that need to be propagated from one iteration to the next. To solve a linear prediction problem of order M for prewindowed data, we generally require a state vector $\mathbf{x}(n)$ with $2M + 1$ elements. In

[8] The notion of backwards stability is usually attributed to J.H. Wilkinson. For a discussion of this important concept, see Wilkinson (1963, 1965) and Stewart (1973).

terms of prediction, the $2M + 1$ elements of the state $\mathbf{x}(n)$ may be understood to mean M forward prediction quantities, M backward prediction quantities, and a first coordinate vector estimation (see Chapter 15).

This discussion motivates us to introduce the following definitions [Regalia (1990)]:

> $\mathcal{S}_i(n)$ = the set of all reachable states $\mathbf{x}(n)$ at time n, using an infinite-precision computation of a nonlinear feedback system described in Eq. (19.79).
>
> $\mathcal{S}_f(n)$ = the set of all numerically reachable states corresponding to a finite-precision implementation of the system described in Eq. (19.79).

To each member of $\mathcal{S}_i(n)$ we may now associate the exact least-squares solution produced by the input time series $u(1), u(2), \ldots, u(n)$, after initialization of the algorithm. Suppose next that finite-precision effects produce a perturbed internal state that is contained in $\mathcal{S}_i(n)$. The numerically perturbed state may then be viewed as the exact least-squares solution obtained using a correspondingly perturbed input time series. Thus an algorithm for which the set $\mathcal{S}_f$ is contained within the set $\mathcal{S}_i(n)$ is said to be *backwards stable*; otherwise, it is *backwards unstable*.

Regalia (1990) has rigorously shown that the family of QRD-fast RLS algorithms are *backwards stable* in both of their prediction and filtering sections. Moreover, the *continuity* of the internal state variables of a recursive least-squares algorithm (be it of a standard or a fast form) with respect to the input data is a well-known property that is intrinsic to linear least-squares problems. Hence, the backwards stability of a QRD-fast RLS algorithm, combined with the continuity property, implies that the estimation errors produced by such an algorithm are *bounded*. In other words, a QRD-fast RLS algorithm cannot diverge and is therefore necessarily *Liapunov* stable[9]; note, however, that the converse is not always true.

According to Regalia (1990), the notion of *minimality* is of critical importance in achieving the backwards stable realization of an algorithm; a QRD-fast RLS is indeed a minimal system. This observation appears to run counter to the use of redundancy as a means of stabilizing the FTF algorithm (discussed in Section 19.5). We may therefore conclude that a *change of coordinates* (achieved through the use of QR-decomposition as an example of orthogonal decomposition) is potentially a more powerful method of stabilization in the Liapunov sense than is an algebraic rearrangement of filtering calculations (through the use of redundancy).

[9] Stability in the sense of Liapunov is named in honor of A. M. Liapunov, who founded modern stability theory and provided a powerful technique to test for it [Liapunov (1966)]. Consider the quantized version of the nonlinear feedback system of Eq. (19.79) with zero input:

$$\mathbf{x}_q(n + 1) = g(\mathbf{x}_q(n), 0), \qquad \mathbf{x}_q(0) = \mathbf{x}_0$$

where $\mathbf{x}_q(n)$ is the quantized value of the state vector $\mathbf{x}(n)$. Let the origin be an equilibrium point of the quantized system. We may then define *stability in the sense of Liapunov* as follows [Brogan (1985)]:

The origin is a stable equilibrium point if for any given value $\epsilon > 0$ *there exists a number* $\delta(\epsilon, n_0) > 0$ *such that if the Euclidean norm* $\|\mathbf{x}_q(n_0)\| < \delta$, *then the resultant state vector* $\mathbf{x}_q(n)$ *satisfies the condition* $\|\mathbf{x}_q(n)\| < \epsilon$ *for all* $n > n_0$.

If a system is Liapunov stable, it is then ensured that the state can be kept in norm within ϵ of the origin by restricting the perturbation to be less than δ, in norm. Note that it is necessarily true that $\delta \le \epsilon$.

19.8 SQUARE ROOT VERSUS SQUARE-ROOT FREE VERSIONS OF ADAPTIVE FILTERING ALGORITHMS

We conclude our discussion of numerical problems by taking a brief look at the issue of square root versus square-root free operations in the practical implementation of adaptive filtering algorithms. Examining the various classes of adaptive filtering algorithms described in previous chapters of the book, we may readily identify the specific algorithms that involve the use of square roots. We first recognize that the standard LMS and standard RLS algorithms are by nature square-root free. We also recognize that the FTF and all four forms of the recursive LSL algorithms are square-root free from inception. The use of square roots arises in the intimately related Kalman filtering and RLS algorithms when some form of orthogonal decomposition is used to overcome numerical difficulties arising from the implementation of these algorithms in finite-precision arithmetic. The discussion presented in this section therefore pertains only to this latter situation.

To be specific, the standard versions of QRD-RLS and QRD-LSL algorithms (developed in depth in Chapters 14 and 18, respectively) used the Givens rotations for implementing the QR-decomposition, which is basic to their theory. Moreover, the way in which the Givens rotations were executed in both cases involved the use of square roots. The use of square roots also arises in the standard SRF form of implementing the Kalman filter (briefly discussed in Chapter 7). It is sometimes argued that square roots are (1) expensive, and (2) awkward to calculate, constituting a bottleneck for overall performance. For these reasons, square-root free versions of the QRD-RSL and QRD-LSL algorithms have been formulated, using special methods for performing Givens rotations without square roots [Gentleman (1973)]. In the case of Kalman filtering, the method of UD-factorization [Bierman (1977)] was developed for avoiding the actual use of square roots. It now appears that square-root free algorithms actually introduce a number of problems [Stewart and Chapman (1990)]:

1. Square-root free algorithms may become numerically unstable, and potentially suffer from serious overflow/underflow problems.
2. Based on the knowledge that square roots are simpler (or even equally as complex) as divider arrays, the reformulation of standard adaptive filtering algorithms in square-root free form may actually increase arithmetic complexity.

Another noteworthy point is that RLS algorithms requiring the use of square roots can be programmed very efficiently on *CORDIC processors,* in which square root operations make no explicit appearance. Here we recognize that most of these algorithms are decomposable in the two basic operations of rotation and vectoring. These two operations are indeed fundamental to a CORDIC processor (Volder, 1959):

1. *Rotation*. In the rotation mode, the CORDIC processor is given the coordinates of a two-element vector and an angle of rotation. The processor then computes

the coordinate components of the original vector after rotation through the desired angle.

2. *Vectoring*. In the vectoring mode, the coordinates of a two-element vector are given. The CORDIC processor then rotates the vector until the angular argument is zero. The angle of rotation in this second mode is therefore the negative of the original angular argument. In other words, the vectoring operation is equivalent to the annihilation of an element of a two-element vector.

Apparently, a CORDIC processor yields the fastest implementation of these two basic operations. Moreover, its integral circuit implementation in the form of silicon chips is becoming commercially available[10], which makes its use for the implementation of RLS algorithms involving square roots all the more attractive.[11]

In conclusion, with the ever-increasing improvement in digital technology, and the availability of fast square root arrays and general-purpose CORDIC chips, there may be no real advantage in using square-root free forms of QRD-RLS, QRD-LSL, and UD-factorized Kalman filtering algorithms over the standard QRD-RLS, QRD-LSL, and SRF Kalman filtering algorithms that use square roots.

PROBLEMS

1. Consider the digital implementation of the LMS algorithm using fixed-point arithmetic, as discussed in Section 19.2. Show that the M-by-1 error vector $\Delta\mathbf{w}(n)$, incurred by quantizing the tap-weight vector, may be updated as follows:

$$\Delta\mathbf{w}(n+1) = \mathbf{F}(n)\,\Delta\mathbf{w}(n) + \mathbf{t}(n), \qquad n = 0, 1, 2, \ldots$$

where $\mathbf{F}(n)$ is an M-by-M matrix and $\mathbf{t}(n)$ is an M-by-1 vector. Hence, define $\mathbf{F}(n)$ and $\mathbf{t}(n)$. Base your analysis on real-valued data.

2. Using the results of Problem 1, and invoking the independence theory of Section 9.4, show that

$$E[\Delta\mathbf{w}(n)] = \mathbf{0}$$

3. Consider two transversal filters I and II, both of length M. Filter I has all of its tap inputs as well as tap weights represented in infinite-precision form. Filter II is identical to filter I, except for the fact that its tap weights are represented in finite-precision form. Let $y_{\mathrm{I}}(n)$ and $y_{\mathrm{II}}(n)$ denote the respective filter outputs for the tap-inputs $u(n), u(n-1), \ldots, u(n-M+1)$. Define the error

$$\epsilon(n) = y_{\mathrm{I}}(n) - y_{\mathrm{II}}(n)$$

Assuming that the inputs $\{u(n)\}$ are independent random variables with a common rms value equal to A_{rms}, show that the mean-square value of the error $\epsilon(n)$ is

$$E[\epsilon^2(n)] = A_{\mathrm{rms}}^2 \sum_{i=0}^{M-1} (w_i - w_{iq})^2$$

where w_{iq} is the quantized version of the tap weight w_i.

[10] Regalia, P.A. (1990). Personal communication.

[11] Rader (1990) describes a linear systolic array for adaptive beamforming based on the Cholesky factorization which uses the CORDIC processor for its hardware implementations in VSLI form.

4. Consider an LMS algorithm with 17 taps and a step-size parameter $\mu = 0.07$. The input data stream has an rms value of unity.
   (a) Given the use of a quantization process with 12-bit accuracy, calculate the corresponding value of the digital residual error.
   (b) Suppose the only source of output error is that due to quantization of the tap weights. Using the result of Problem 3, calculate the rms value of the resulting measurement error in the output. Compare this error with the digital residual error calculated in part (a).

5. Show that the single-error propagation model for form II of the standard RLS algorithm is as described in Eq. (19.48). Assume that a single quantization error is made at time $n - 1$, as shown by

$$\mathbf{P}_q(n-1) = \mathbf{P}(n-1) + \boldsymbol{\eta}_P(n-1)$$

   where $\mathbf{P}_q(n-1)$ is the quantized value of the matrix $\mathbf{P}(n-1)$ and $\boldsymbol{\eta}_P(n-1)$ is the quantization error matrix.

6. Consider the standard RLS algorithm with a single quantization error made in the tap-weight vector at time $n - 1$, as shown by

$$\mathbf{w}_q(n-1) = \mathbf{w}(n-1) + \Delta\mathbf{w}(n-1)$$

   Develop the corresponding error-propagation model for the standard RLS algorithm, and show that it is the same for both forms I and II of the algorithm.

7. In Eqs. (19.24) and (19.56) we presented two different ways of defining the condition for a tap-output vector $\mathbf{u}(n)$ to be persistently exciting. Reconcile these two conditions.

8. In Section 19.5 we show how the idea of error feedback may be used to stabilize an unstable mode in the error propagation mechanism as it pertains to the a priori backward prediction error $\psi_M(n)$. Extend the application of this idea to the following parameters in the FTF algorithm:
   (a) The parameter $\tilde{k}_{M+1,M+1}(n)$ that represents the last element in the normalized gain vector $\tilde{\mathbf{k}}_{M+1}(n)$.
   (b) The parameter $\gamma_M^{-1}(n)$, representing the reciprocal of the conversion factor $\gamma_M(n)$.

Computer-Oriented Problems

9. *Computer experiment on adaptive prediction with finite-precision arithmetic.* Consider an autoregressive (AR) model of order 2, described by the difference equation

$$u(n) + a_1 u(n-1) + a_2 u(n-1) = v(n)$$

   where a_1 and a_2 are the model parameters, and $\{v(n)\}$ is a white-noise sequence of zero mean and variance σ_v^2. The variance σ_v^2 is chosen to make the variance of $u(n)$ equal to unity. In the three parts of this problem, we explore the effects of finite-precision arithmetic on the performance of these algorithms for the following model parameters:

$$a_1 = 0.195$$

$$a_2 = 0.95$$

$$\sigma_v^2 = 0.096525$$

   (a) *The LMS algorithm*
   (1) Use the LMS algorithm to estimate the AR parameters a_1 and a_2 for a step-size parameter $\mu = 0.05$ by ensemble averaging over 200 independent trials of the

experiment. Do your computation for fixed-point arithmetic with a varying word-length of 8, 12, and 16 bits.

(2) Repeat your computations for a step-size parameter $\mu = 0.1$.

(b) *The standard RLS algorithm*

(1) Use the RLS algorithm to estimate the AR parameters a_1 and a_2 for an exponential weighting factor $\lambda = 1$. Do the initialization with $\delta = 0.001$. For the parameter estimation, ensemble-average your results over 200 independent trials of the experiment. Do your computations for fixed-point arithmetic with a varying word-length of 8, 12, and 16 bits.

(2) Repeat your computations for the exponential weighting factor $\lambda = 0.9$.

(c) *Recursive LSL algorithm*

(1) Use the recursive LSL algorithm, based on a priori estimation errors, for computing an estimate of the noise variance σ_v^2. For this computation, combine two forms of averaging: ensemble averaging over 200 independent trials of the experiment, and time averaging over the last 200 time samples after the initial convergence period is completed. Assume an exponential weighting factor $\lambda = 1$, and $\delta = 0.001$ for the initialization. Do your computations for fixed-point arithmetic with a varying word-length of 8, 12, and 16 bits.

(2) Repeat your computations for the corresponding version of the recursive LSL algorithm using error feedback.

Comment on your results.

10. *Computer experiment on adaptive equalization using the recursive LSL algorithm.* In Chapter 17 we used an adaptive equalizer to study the performance of the recursive LSL algorithm, assuming infinite precision. In this problem we continue this experiment by exploring the effects of finite-precision arithmetic on the performance of two different versions of the algorithm:

(a) Using the conventional form of the recursive LSL algorithm, based on a priori estimation errors, plot the ensemble-averaged learning curve of the algorithm for fixed-point arithmetic with a varying word-length of 4, 8, and 12 bits. Use an exponential weighting factor $\lambda = 1$ and an initializing parameter $\delta = 0.001$. Do the ensemble averaging over 200 independent trials of the experiment.

(b) Repeat your computations for the corresponding version of the recursive LSL algorithm with error feedback. Comment on your results.

20

Blind Deconvolution

Deconvolution is a signal processing operation that ideally unravels the effects of convolution. Specifically, the objective of deconvolution is to recover the input signal applied to a linear time-invariant (possibly nonminimum phase) system, given the signal developed at the output of the system. As the name implies, "blind deconvolution" refers to the ability of an adaptive algorithm to perform deconvolution in a *blindfolded* or *self-recovered fashion.* An adaptive filtering algorithm designed in this way does *not* need an external source for supplying the desired response. Rather, the algorithm makes up for it by using some form of *nonlinearity* to extract useful information from the unknown system output, which is unprocessed by a conventional linear adaptive filter.

In this chapter, we study two important families of blind deconvolution algorithms:

1. *Bussgang algorithms,* which perform blind deconvolution in an *iterative* fashion; they are so-called because the deconvolved sequence assumes Bussgang statistics when the algorithm reaches convergence in the mean.
2. *Polyspectra-based algorithms,* which use *higher-order cumulants* or their discrete Fourier transforms known as *polyspectra;* the property of polyspectra to preserve phase information makes them well suited for blind deconvolution.

We begin our study of blind deconvolution by discussing its theoretical requirements and practical importance, which we do in the next section.

20.1 THEORETICAL AND PRACTICAL CONSIDERATIONS

Consider an *unknown* linear time-invariant system $\mathcal{S}$ with input $\{x(n)\}$ as depicted in Fig. 20.1. The input consists of an *unobserved* white data (information-bearing) sequence with known probability distribution. *The problem is to restore* $\{x(n)\}$ *or,*

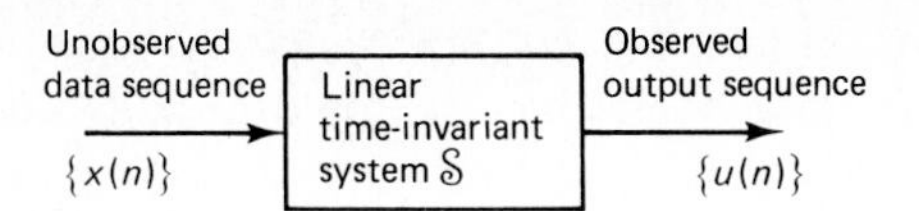

Figure 20.1 Setting the stage for blind deconvolution.

equivalently, to identify the inverse $\mathcal{S}^{-1}$ of the system $\mathcal{S}$, given the observed sequence $\{u(n)\}$ at the system output.

If the system $\mathcal{S}$ is *minimum-phase* (i.e., the transfer function of the system has all of its poles and zeros confined to the interior of the unit circle in the z-plane), then not only is the system $\mathcal{S}$ stable but so is the inverse system $\mathcal{S}^{-1}$. In this case, we may view the input sequence $\{(x(n)\}$ as the "innovation" of the system output $\{u(n)\}$, and the inverse system $\mathcal{S}^{-1}$ is just a *whitening filter*. These observations follow from the study of linear prediction presented in Chapter 6.

In many practical situations, however, the system $\mathcal{S}$ may *not* be minimum phase. A system is said to be *nonminimum phase* if its transfer function has any of its zeros located outside the unit circle in the z-plane; exponential stability of the system dictates that the poles be located inside the unit circle. Practical examples of a nonminimum phase system include a telephone channel and a fading radio channel. In this situation, the restoration of the input sequence $\{(x(n)\}$, given the channel output, is a more difficult problem. In particular, we may make the following important observations (Benveniste et al., 1980):

1. For the estimation of a nonminimum-phase characteristic to be feasible, and therefore for the inverse-filtering (i.e., deconvolution) problem to have a solution, the input sequence $\{x(n)\}$ must be *non-Gaussian.*
2. Since the use of a minimum mean-squared error criterion results in a linear filter that is minimum phase, we must invoke the use of *higher-order statistics* (cumulants or moments) and therefore *nonlinear estimation.*
3. The inverse system $\mathcal{S}^{-1}$ is *unstable,* because its transfer function has poles outside the unit circle in the z-plane. Hence, the *on-line* restoration of the input sequence $\{x(n)\}$ is impossible. However, we may *truncate* the impulse response of the inverse system $\mathcal{S}^{-1}$, thereby permitting the restoration of $\{x(n)\}$ to take place in real time, but with a constant and finite delay; this delay is of no serious concern in digital communications.

It is thus apparent that for the blind deconvolution problem to have a solution, two requirements must be satisfied. First, the unobserved data sequence $\{x(n)\}$ must be non-Gaussian. Second, the processing of the observed output sequence $\{u(n)\}$ must include some form of nonlinear estimation. Practical applications of blind deconvolution include adaptive equalization and seismic deconvolution, as discussed next.

Typically, adaptive equalizers used in digital communications require an initial *training period,* during which a *known* data sequence is transmitted. A replica of this sequence is made available at the receiver in proper synchronism with the transmitter, thereby making it possible for adjustments to be made to the equalizer coefficients in accordance with the adaptive filtering algorithm employed in the equalizer design. When the training is completed, the equalizer is switched to its *decision-directed mode,* and normal data transmission may then commence. (These two modes of operation of an adaptive equalizer were discussed in Section 1.7.) However, there are practical situations where it would be highly desirable for a receiver to be able to achieve complete adaptation *without* the cooperation of the transmitter. For example, in a *multipoint data network* involving a *control unit* connected to several *data terminal equipments* (DTEs), we have a "master–slave" situation in that a DTE is permitted to transmit only when its modem is polled by the modem of the control unit. A problem peculiar to these networks is that of retraining the receiver of a DTE unable to recognize data and polling messages, due to severe variations in channel characteristics or simply because that particular receiver was not powered-on during initial synchronization of the network. Clearly, in a large or heavily loaded multipoint network, data throughput is increased and the burden of monitoring the network is eased if some form of *blind equalization* is built into the receiver design (Godard, 1980).

Another class of communication systems that may need blind equalization is high capacity line-of-sight *digital radio*. Most of the time, a line-of-sight microwave radio link behaves as a wide-band, low-noise channel capable of providing highly reliable, high-speed data transmission. The channel, however, suffers from anomalous propagation conditions that arise from natural phenomena, which can cause the error performance of a digital radio to be severely degraded. Such anomalies manifest themselves by causing the transmitted signal to propagate along several paths, each of different electrical length. This phenomenon is called *multipath fading*. Conventional linear adaptive equalization techniques (e.g., the LMS algorithm) perform satisfactorily in a digital radio system, except during severe multipath fading that is the main cause of outage in such systems. Blind equalization provides a possible mechanism for dealing with the severe multipath fading problem in digital radio links [Beneviste and Goursat (1984); Ross (1989)].

In reflection seismology, the traditional method of removing the source waveform from a seismogram is to use *linear-predictive deconvolution* (see Section 1.7). The method of predictive deconvolution is derived from four fundamental assumptions (Gray, 1979):

1. The reflectivity series is *white*. This assumption is, however, often violated by reflection seismograms as the reflectivities result from a differential process applied to acoustic impedances. In many sedimentary basins there are thin beds that cause the reflectivity series to be correlated in sign.
2. The source signal (wavelet) is *minimum phase*. This assumption is valid for several explosive sources (e.g., dynamite), but it is only approximate for more complicated sources such as those used in marine exploration.
3. The reflectivity series and noise are statistically independent and stationary in time. The stationarity assumption, however, is violated because of spherical divergence and attenuation of seismic waves. To cope with nonstationarity of the data, we may use adaptive deconvolution, but such a method often destroys primary events of interest.
4. The *minimum mean-square error criterion* is used to solve the linear prediction problem. This criterion is appropriate only when the prediction errors (the reflectivity series and noise) have a Gaussian distribution. Statistical tests per-

formed on reflectivity series, however, show that their kurtosis is much higher than that expected from a Gaussian distribution.

Assumptions 1 and 2 were explicitly mentioned in the presentation of the method of predictive deconvolution in Chapter 1. Assumptions 3 and 4 are implicit in the application of Wiener filtering that is basic to the solution of the linear prediction problem, as presented in Chapter 6. The main point of this discussion is that valuable phase information contained in a reflection seismogram is ignored by the method of predictive deconvolution. This limitation is overcome by using *blind deconvolution* (Godfrey and Rocca, 1981).

Blind equalization in digital communications and blind deconvolution in reflection seismology are examples of a special kind of adaptive inverse filtering that operate *without* access to the source of the desired response (i.e., the transmitted signal). Only the received signal and some additional information in the form of a *probabilistic model* are needed. In the case of equalization for digital communications, the model describes the statistics of the transmitted data sequence. In the case of seismic deconvolution, the model describes the statistics of the earth's reflection coefficients.

Having clarified the framework within which the use of blind deconvolution is feasible, we are ready to undertake a detailed study of its operation. Specifically, we consider the Bussgang family and the polyspectra-based family of blind deconvolution algorithms. The discussion is presented in that order, and in the context of blind equalization for digital communications.

20.2 BUSSGANG ALGORITHM FOR BLIND EQUALIZATION OF REAL BASEBAND CHANNELS

Consider the *baseband model* of a digital communication system, depicted in Fig. 20.2. The model consists of the cascade connection of a *linear communication channel* and a *blind equalizer*.

The channel includes the combined effects of a transmit filter, a transmission medium, and a receive filter. It is characterized by an impulse response $\{h_n\}$ that is *unknown*; it may be time varying, albeit slowly. The nature of the impulse response $\{h_n\}$ (i.e., whether it is real or complex valued) is determined by the type of modulation employed. To simplify the discussion, we assume that the impulse response is real, which corresponds to the use of *multilevel pulse-amplitude modulation* (*M-ary PAM*); the case of a complex impulse response is considered in the next section. We may thus describe the input-output relation of the channel by the *convolution sum*

$$u(n) = \sum_k h_k x(n-k), \qquad n = 0, \pm 1, \pm 2, \ldots \tag{20.1}$$

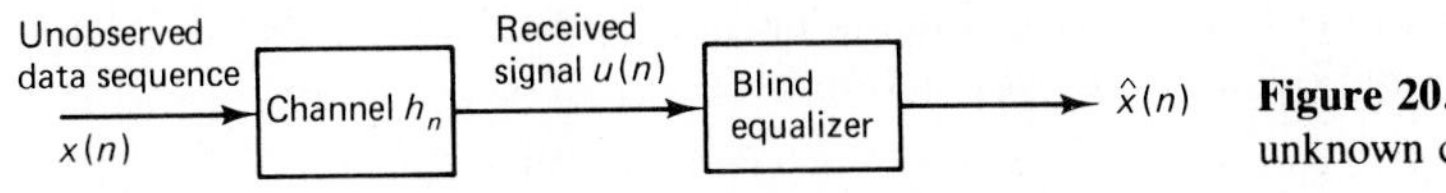

Figure 20.2 Cascade connection of an unknown channel and blind equalizer.

where $\{x(n)\}$ is the *data (message) sequence* applied to the channel input, and $\{u(n)\}$ is the resulting *channel output*. We assume that

$$\sum_k h_k^2 = 1 \tag{20.2}$$

Equation (20.2) implies the use of *automatic gain control* (AGC) that keeps the variance of the channel output $u(n)$ constant. We further assume that the channel is nonminimum phase (i.e., noncausal), which means that

$$h_n \neq 0, \qquad \text{for } n < 0 \tag{20.3}$$

In the mathematical model of Eq. (20.1), we have ignored the effect of receiver noise. We are justified to do so, because the degradation in the performance of data transmission (over a voice-grade telephone channel, say) is usually dominated by *intersymbol interference* due to channel dispersion.

The problem we wish to solve is the following:

> Given the received signal $\{u(n)\}$, reconstruct the original data sequence $\{x(n)\}$ applied to the channel input.

Equivalently, we may restate the problem as follows:

> Design a blind equalizer that is the inverse of the unknown channel, with the channel input being unobservable.

To solve this blind equalization problem, we need to prescribe a *probabilistic model* for the data sequence $\{x(n)\}$. For the problem at hand, we assume the following (Bellini, 1986, 1988):

1. The data sequence $\{x(n)]$ is *white*; that is, the data symbols $x(n)$ are *independent, identically distributed* (iid) *ramdom variables,* with zero mean and unit variance, as shown by

$$E[x(n)] = 0 \tag{20.4}$$

and

$$E[x(n)x(k)] = \begin{cases} 1, & k = n \\ 0, & k \neq n \end{cases} \tag{20.5}$$

where E is the expectation operator.

2. The *probability density function* of the data symbol $x(n)$ is *symmetric* and *uniform;* that is (see Fig. 20.3),

$$f_x(x) = \begin{cases} 1/2\sqrt{3}, & -\sqrt{3} \leq x < \sqrt{3} \\ 0 & \text{otherwise} \end{cases} \tag{20.6}$$

This distribution has the merit of being independent of the number M of amplitude levels employed in the modulation process.

Note that Eq. (20.4) and the first line of Eq. (20.5) follow from (20.6).

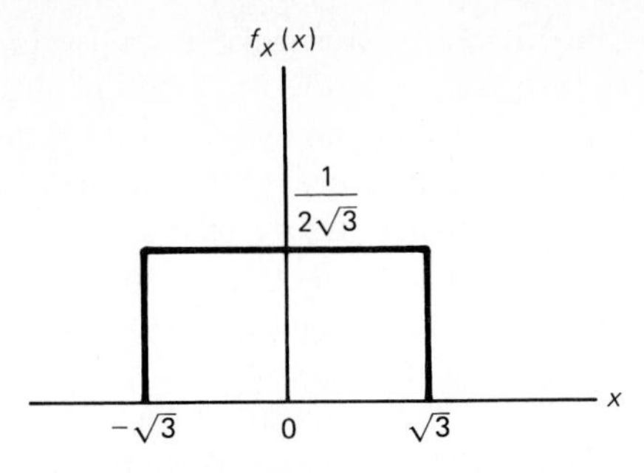

Figure 20.3 Uniform distribution.

With the distribution of the datum $x(n)$ assumed to be symmetric, as in Fig. 20.3, we find that the whole data sequence $\{-x(n)\}$ has the same law as $\{x(n)\}$. Hence, we cannot distinguish the desired inverse filter $\mathcal{S}^{-1}$ (corresponding to $\{x(n)\}$ from the opposite one $-\mathcal{S}^{-1}$ (corresponding to $\{-x(n)\}$). We may overcome this *sign ambiguity problem* by *initializing* the deconvolution algorithm such that there is a single nonzero tap weight with the desired algebraic sign (Benveniste et al., 1980).

Iterative Deconvolution: The Objective

Let $\{w_i\}$ denote the impulse response of the *ideal inverse filter*, which is related to the impulse response $\{h_i\}$ of the channel as follows:

$$\sum_i w_i h_{l-i} = \delta_l \tag{20.7}$$

where δ_l is the *Kronecker delta:*

$$\delta_l = \begin{cases} 1, & l = 0 \\ 0, & l \neq 0 \end{cases} \tag{20.8}$$

An inverse filter defined in this way is "ideal" in the sense that it reconstructs the transmitted data sequence $\{x(n)\}$ *correctly*. To demonstrate this, we first write

$$\sum_i w_i u(n-i) = \sum_i \sum_k w_i h_k x(n-i-k) \tag{20.9}$$

Let

$$k = l - i$$

Making this change of indices in Eq. (20.9), and interchanging the order of summation, we get

$$\sum_i w_i u(n-i) = \sum_l x(n-l) \sum_i w_i h_{l-i} \tag{20.10}$$

Hence, using Eq. (20.7) in (20.10) and then applying the definition of Eq. (20.8), we get

$$\begin{aligned} \sum_i w_i u(n-i) &= \sum_l \delta_l x(n-l) \\ &= x(n) \end{aligned} \tag{20.11}$$

which is the desired result.

For the situation described herein, the impulse response $\{h_n\}$ is unknown. We therefore cannot use Eq. (20.7) to determine the inverse filter. Instead, we use an *iterative deconvolution procedure* to compute an *approximate inverse filter* characterized by the impulse response $\{\hat{w}_i(n)\}$. The index i refers to the *tap-weight number* in the *transversal filter* realization of the approximate inverse filter, as indicated in Fig. 20.4. The index n refers to the *iteration number*; each iteration corresponds to the transmission of a data symbol. The computation is performed iteratively in such a way that the convolution of the impulse response $\{\hat{w}(n)\}$ with the received signal $\{u(i)\}$ results in the complete or partial removal of the intersymbol interference (Godfrey and Rocca, 1981). Thus, at the nth iteration we have an approximately deconvolved sequence

$$y(n) = \sum_{i=-L}^{L} \hat{w}_i(n)u(n-i) \tag{20.12}$$

where $2L + 1$ is the truncated *length* of the impulse response $\{\hat{w}_i(n)\}$ (see Fig. 20.4).

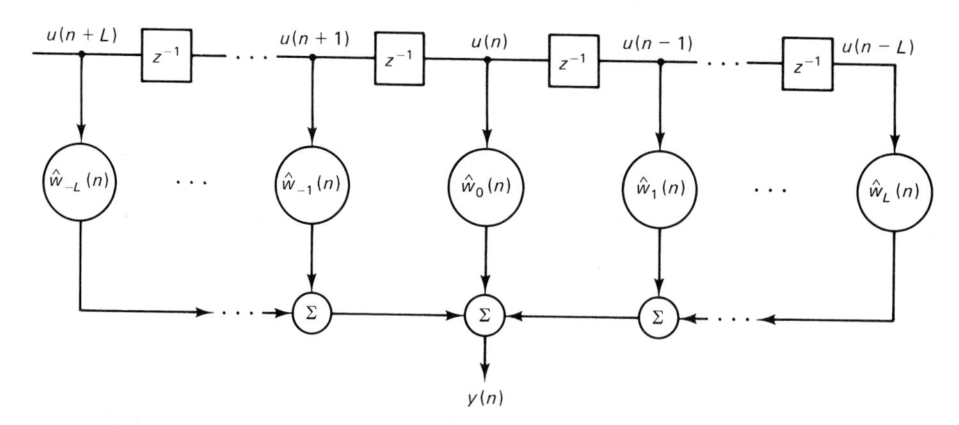

Figure 20.4 Transversal filter realization of approximate inverse filter; use of real data is assumed.

The convolution sum on the left side of Eq. (20.11), pertaining to the ideal inverse filter, is *infinite* in extent in that the index i ranges from $-\infty$ to ∞. On the other hand, the convolution sum on the right side of Eq. (20.12) pertaining to the approximate inverse filter, is *finite* in extent in that i extends from $-L$ to L. Clearly, we may rewrite Eq. (20.12) as follows:

$$y(n) = \sum_{i} \hat{w}_i(n)u(n-i), \qquad \hat{w}_i(n) = 0 \text{ for } |i| > L$$

or, equivalently,

$$y(n) = \sum_{i} w_i u(n-i) + \sum_{i} [\hat{w}_i(n) - w_i]u(n-i) \tag{20.13}$$

Let

$$v(n) = \sum_{i} [\hat{w}_i(n) - w_i]u(n-i), \qquad \hat{w}_i = 0 \text{ for } |i| > L \tag{20.14}$$

Then, using the ideal result of Eq. (20.11) and the definition of Eq. (20.14), we may simplify Eq. (20.13) as follows:

$$y(n) = x(n) + v(n) \tag{20.15}$$

The term $v(n)$ is called the *convolutional noise,* representing the *residual* intersymbol interference that results from the use of an approximate inverse filter.

The inverse filter output $y(n)$ is next applied to a *zero-memory nonlinear estimator,* producing the estimate $\hat{x}(n)$ for the datum $x(n)$. This operation is depicted in the block diagram of Fig. 20.5. We may thus write

$$\hat{x}(n) = g(y(n)) \tag{20.16}$$

where $g(\cdot)$ is some nonlinear function. The issue of nonlinear estimation is discussed in the next subsection.

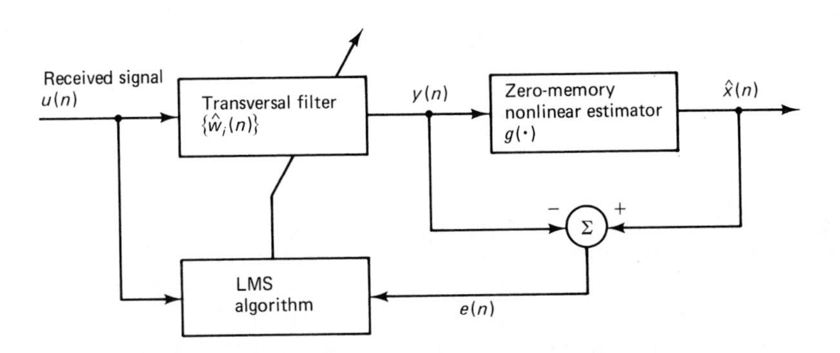

Figure 20.5 Block diagram of blind equalizer.

Ordinarily, we find that the estimate $\hat{x}(n)$ is not reliable enough. Nevertheless, we may use it in an *adaptive* scheme to obtain a "better" estimate at iteration $n + 1$. Indeed, we have a variety of *linear* adaptive filtering algorithms (discussed in previous chapters) at our disposal that we can use to perform this adaptive parameter estimation. In particular, a simple and yet effective scheme is provided by the LMS algorithm. To apply it to the problem at hand, we note the following:

1. The ith tap input of the transversal filter at iteration (time) n is $u(n - i)$.
2. Viewing the nonlinear estimate $\hat{x}(n)$ as the "desired" response [since the transmitted data symbol $x(n)$ is unavailable to us], and recognizing that the corresponding transversal filter output is $y(n)$, we may express the *estimation error* for the iterative deconvolution procedure as

$$e(n) = \hat{x}(n) - y(n) \tag{20.17}$$

3. The ith tap weight $\hat{w}_i(n)$ at iteration n represents the "old" parameter estimate.

Accordingly, the "updated" value of the ith tap weight at iteration $n + 1$ is computed as follows:

$$\hat{w}_i(n + 1) = \hat{w}_i(n) + \mu u(n - i)e(n), \qquad i = 0, \pm 1, \ldots, \pm L \tag{20.18}$$

where μ is the *step-size parameter*. Note that for the situation being considered here, the data are all *real* valued.

Equations (20.12), (20.16), (20.17), and (20.18) constitute the iterative deconvolution algorithm for the blind equalization of a real baseband channel (Bellini, 1986). As remarked earlier, each iteration of the algorithm corresponds to the transmission of a data symbol. It is assumed that the symbol duration is known at the receiver.

A block diagram of the blind equalizer is shown in Fig. 20.5. The idea of generating the estimation error $e(n)$, as detailed in Eqs. (20.16) and (20.17), is similar in philosophy to the decision-directed mode of operating an adaptive equalizer. More will be said on this issue later in the section.

Nonconvexity of the Cost Function

The ensemble-averaged cost function corresponding to the tap-weight update equation (20.18) is defined by

$$\begin{aligned} J(n) &= E[e^2(n)] \\ &= E[(\hat{x}(n) - y(n))^2] \\ &= E[(g(y(n)) - y(n))^2] \end{aligned} \tag{20.19}$$

where $y(n)$ is defined by Eq. (20.12). In the linear case of LMS algorithm, the cost function is a quadratic (convex) function of the tap-weights and therefore has a well-defined minimum point. By contrast, the cost function $J(n)$ of Eq. (20.19) is a *nonconvex* function of the tap weights. This means that, in general, the error performance surface of the iterative deconvolution procedure described herein may *not* have a global minimum. The nonconvexity of the cost function $J(n)$ may arise because of the fact that the estimate $\hat{x}(n)$, performing the role of an internally generated "desired response," is produced by passing the linear combiner output $y(n)$ through a zero-memory nonlinearity, and also because $y(n)$ is itself a function of the tap weights.

In any event, the nonconvex form of the cost function $J(n)$ may result in *ill-convergence* of the iterative deconvolution algorithm described by Eqs. (20.12) and (20.16) to (20.18). The important issue of convergence is considered in some greater detail later in this section.

Statistical Properties of Convolutional Noise

The additive convolutional noise $v(n)$ is defined in Eq. (20.14). To develop a more refined formula for $v(n)$, we note that the tap input $u(n - i)$ involved in the summation on the right side of this equation is given by [see Eq. (20.1)]

$$u(n - i) = \sum_k h_k x(n - i - k) \tag{20.20}$$

We may therefore rewrite Eq. (20.14) as

$$v(n) = \sum_i \sum_k h_k[\hat{w}_i(n) - w_i]x(n - i - k) \tag{20.21}$$

Let

$$n - i - k = l$$

Hence, we may also write

$$v(n) = \sum_l x(l)\nabla(n - l) \tag{20.22}$$

where

$$\nabla(n) = \sum_k h_k[\hat{w}_{n-k}(n) - w_{n-k}] \tag{20.23}$$

The sequence $\{\nabla(n)\}$ is a sequence of small numbers, corresponding to the *residual impulse response* of the channel. We imagine the sequence $\{\nabla(n)\}$ as a *long and oscillatory wave* that is convolved with the transmitted data sequence $\{x(n)\}$ to produce the convolutional noise sequence $\{v(n)\}$, as indicated in Eq. (20.22).

The definition of Eq. (20.22) is basic to the statistical characterization of the convolution noise $v(n)$. The mean of $v(n)$ is zero, as shown by

$$\begin{aligned} E[v(n)] &= E\left[\sum_l x(l)\nabla(n - l)\right] \\ &= \sum_l \nabla(n - l)E[x(l)] \\ &= 0 \end{aligned} \tag{20.24}$$

where in the last line we have made use of Eq. (20.4). Next, the autocorrelation function of $v(n)$ for a lag j is given by

$$\begin{aligned} E[v(n)v(n - j)] &= E\left[\sum_l x(l)\nabla(n - l)\sum_m x(m)\nabla(n - m - j)\right] \\ &= \sum_l \sum_m \nabla(n - l)\nabla(n - m - j)E[x(l)x(m)] \\ &= \sum_l \nabla(n - l)\nabla(n - l - j) \end{aligned} \tag{20.25}$$

where in the last line we have made use of Eq. (20.5). Since $\{\nabla(n)\}$ is a long and oscillatory waveform, the sum on the right side of Eq. (20.25) is nonzero only for $j = 0$, obtaining

$$E[v(n)v(n - j)] = \begin{cases} \sigma^2, & j = 0 \\ 0, & j \neq 0 \end{cases} \tag{20.26}$$

where

$$\sigma^2(n) = \sum_l \nabla^2(n - l) \tag{20.27}$$

Based on Eqs. (20.24) and (20.26), we may thus describe the convolutional noise process $\{v(n)\}$ as a *zero-mean white-noise process of time-varying variance equal* to $\sigma^2(n)$, defined by Eq. (20.27).

According to the model of Eq. (20.22), the convolutional noise $v(n)$ is a weighted sum of statistically independent and identically distributed random variables representing different transmissions of data symbols. If, therefore, the residual impulse response $\{\nabla(n)\}$ is long enough, the *central limit theorem* makes the *Gaussian* model for $v(n)$ plausible.

Having characterized the convolutional noise $v(n)$ by itself, all that remains for us to do is to evaluate the *cross-correlation* between it and the data sample $x(n)$. These two samples are certainly *correlated* with each other, since $\{v(n)\}$ is the result of convolving the residual impulse response $\{\nabla(n)\}$ with $\{x(n)\}$, as shown in Eq. (20.22). However, the cross-correlation between $v(n)$ and $x(n)$ is negligible compared to the variance of $v(n)$. To demonstrate this, we write

$$\begin{aligned} E[x(n)v(n-j)] &= E[x(n)\sum_{l} x(l)\nabla(n-l-j)] \\ &= \sum_{l} \nabla(n-l-j)E[x(n)x(l)] \qquad (20.28) \\ &= \nabla(-j) \end{aligned}$$

where, in the last line, we have made use of Eq. (20.5). Here again, using the assumption that $\{\nabla(n)\}$ is a long and oscillatory waveform, we deduce that the standard deviation of $v(n)$ is large compared to the magnitude of the cross-correlation $E[x(n)v(n-j)]$.

Since the data sequence $\{x(n)\}$ is white by assumption and the convolutional noise sequence $\{v(n)\}$ is approximately white by deduction, and since these two sequences are essentially uncorrelated, it follows that their sum $\{y(n)\}$ is approximately white too. This suggests that $\{x(n)\}$ and $\{v(n)\}$ may be taken to be essentially independent. We may thus model the convolutional noise $\{v(n)\}$ as an *additive, zero-mean, white Gaussian noise process that is statistically independent of the data sequence* $\{x(n)\}$.

Because of the approximations made in deriving the model described herein for the convolutional noise, its use in an iterative deconvolution process yields a *suboptimal* estimator for the data sequence. In particular, given that the iterative deconvolution process is convergent, the intersymbol interference (ISI) during the latter stages of the process may be small enough for the model to be applicable. In the early stages of the iterative deconvolution process, however, the ISI is typically large with the result that the data sequence and the convolutional noise are strongly correlated, and the convolutional noise sequence is more uniform than Gaussian [Godfrey and Rocca (1981)].

Zero-Memory Nonlinear Estimation of the Data Sequence

We are now ready to consider the next important issue, namely, that of estimating the data sequence $\{x(n)\}$, given the deconvolved sequence $\{y(n)\}$ at the transversal filter output. Specifically, we may formulate the estimation problem as follows: We are given a (filtered) observation $y(n)$ that consists of the sum of two components [see Eq. (20.15)]:

1. A uniformly distributed data symbol $x(n)$ with zero mean and unit variance
2. A white Gaussian noise $v(n)$ with zero mean and variance $\sigma^2(n)$, which is statistically independent of $x(n)$

The requirement is to derive a *Bayes estimate* of $x(n)$, optimized in some sense.

Before proceeding with this classical estimation problem, two noteworthy observations are in order. First, the estimate is naturally a *conditional estimate* that depends on the optimization criterion. Second, although the estimate (in theory) is optimum in a mean-square error sense, in the context of our present situation, it is *suboptimum* by virtue of the approximations made in the development of the model for the convolutional noise $v(n)$.

An optimization criterion of particular interest is that of minimizing the mean-square value of the error between the actual transmission $x(n)$ and the estimation $\hat{x}(n)$. The choice of this optimization criterion yields a *conditional mean estimator*[1] that is both sensible and robust.

For convenience of presentation, we will suppress the dependence of random variables on time n. Thus, given the observation y, the conditional mean estimate $\hat{x}$ of the random variable x is written as $E[\hat{x}|y]$, where E is the expectation operator (see Fig. 20.6). Let $f_X(x|y)$ denote the *conditional probability density function* of x, given y. We thus have

$$\begin{aligned}\hat{x} &= E[x|y] \\ &= \int_{-\infty}^{\infty} x f_X(x|y)\, dx\end{aligned} \tag{20.29}$$

From *Bayes' rule,* we have

$$f_X(x|y) = \frac{f_Y(y|x)\, f_X(x)}{f_Y(y)} \tag{20.30}$$

where $f_Y(y|x)$ is the conditional probability density function of y, given x; and $f_X(x)$ and $f_Y(y)$ are the probability density functions of x and y, respectively. We may therefore rewrite the formula of Eq. (20.29) as

$$\hat{x} = \frac{1}{f_Y(y)} \int_{-\infty}^{\infty} x f_Y(y|x)\, f_X(x)\, dx \tag{20.31}$$

Let

$$y = c_0 x + v \tag{20.32}$$

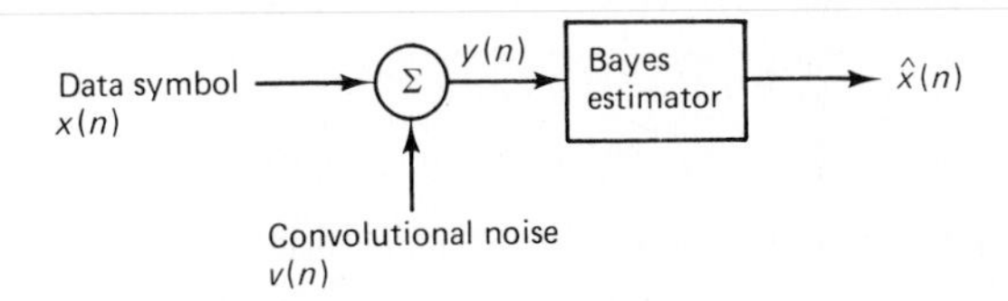

Figure 20.6 Estimation of the data symbol $x(n)$, given the observation $y(n)$.

[1] For a derivation of the conditional mean and its relation to mean-squared-error estimation, see Appendix E.

The scaling factor c_0 is slightly smaller than unity. This factor has been included in Eq. (20.32) so as to keep $E[y^2]$ equal to 1. In accordance with the statistical model for the convolutional noise v developed previously, x and v are statistically independent. With v modeled to have zero mean and variance σ^2, it follows from Eq. (20.32) that the scaling factor c_0 is defined by

$$c_0 = \sqrt{1 - \sigma^2} \tag{20.33}$$

Furthermore, from Eq. (20.32) it follows that

$$f_Y(y|x) = f_V(y - c_0 x) \tag{20.34}$$

Accordingly, the use of Eq. (20.34) in (20.31) yields

$$\hat{x} = \frac{1}{f_Y(y)} \int_{-\infty}^{\infty} x f_V(y - c_0 x) f_X(x)\, dx \tag{20.35}$$

The evaluation of $\hat{x}$ is straightforward but tedious. To proceed with it, we may note the following:

1. The mathematical form of the estimate $\hat{x}(n)$ produced at the output of the Bayes (conditional mean) estimator depends on the probability density function of the original data symbol $x(n)$. For the analysis presented herein, we assume that the data symbol x is *uniformly distributed* with zero mean and unit variance; its probability density function is given in Eq. (20.6), which is reproduced here for convenience:

$$f_X(x) = \begin{cases} 1/2\sqrt{3} & -\sqrt{3} \le x < \sqrt{3} \\ 0, & \text{otherwise} \end{cases} \tag{20.36}$$

2. The convolutional noise v is *Gaussian distributed* with zero mean and variance σ^2; its probability density function is

$$f_V(v) = \frac{1}{\sqrt{2\pi}\sigma} \exp\left(-\frac{v^2}{2\sigma^2}\right) \tag{20.37}$$

3. The filtered observation y is the sum of $c_0 x$ and v; its probability density function is therefore equal to the convolution of the probability density function of x with that of v, as shown by

$$f_Y(y) = \int_{-\infty}^{\infty} f_X(x) f_V(y - c_0 x)\, dx \tag{20.38}$$

Using Eqs. (20.36) to (20.38) in (20.35), we get (Bellini, 1988):

$$\hat{x} = \frac{1}{c_0} y + \frac{\sigma}{c_0} \frac{Z(y + c_0/\sqrt{3}) - Z(y - c_0/\sqrt{3})}{Q(y - c_0/\sqrt{3}) - Q(y + c_0/\sqrt{3})} \tag{20.39}$$

where $Z(y)$ is the *standardized* Gassian probability density function

$$Z(y) = \frac{1}{\sqrt{2\pi}} e^{-y^2/2} \tag{20.40}$$

and $Q(y)$ is the corresponding probability distribution function

$$Q(y) = \frac{1}{\sqrt{2\pi}} \int_y^\infty e^{-u^2/2}\, du \tag{20.41}$$

For a detailed derivation of Eq. (20.39), the reader is referred to Problem 2.

A small *gain correction* to the nonlinear estimator of Eq. (20.39) is needed in order to achieve perfect equalization when the iterative deconvolution algorithm [described by Eqs. (20.16) to (20.18)] converges eventually. Perfect equalization[2] requires that $y = x$. Convergence of the algorithm in the mean is satisfied when the estimation error in the LMS algorithm is orthogonal to each of the tap inputs in the transversal filter realization of the approximate inverse filter. Putting all of this together, we find that the following condition must hold (Bellini, 1986, 1988):

$$E[\hat{x}g(\hat{x})] = 1 \tag{20.42}$$

where $g(\hat{x})$ is the nonlinear estimator $\hat{x} = g(y)$ with $y = \hat{x}$ (for perfect equalization (see Problem 3).

Figure 20.7 shows the nonlinear estimator $\hat{x} = g(y)$ for an eight-level PAM system (Bellini, 1986, 1988). The estimator is normalized in accordance with Eq.

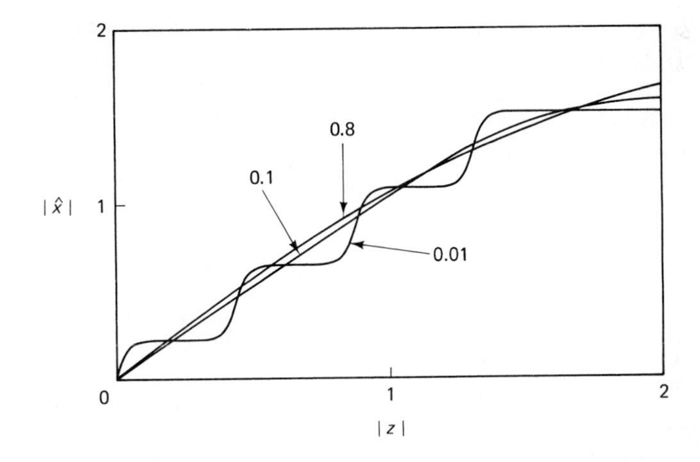

Figure 20.7 Nonlinear estimators of eight-level data in Gaussian noise. The noise-to-(signal + noise) ratios are 0.01, 0.1, and 0.8. [From Bellini (1986) with permission of IEEE.]

[2] In general, for perfect equalization we require that

$$y = (x - D)e^{j\phi}$$

where D is a constant delay and ϕ is a constant phase shift. This condition corresponds to an equalizer whose transfer function has magnitude one and a linear phase response. We note that the input data sequence $\{x_i\}$ is stationary and the channel is linear time-invariant. Hence, the observed sequence $\{y(n)\}$ at the channel output is also stationary; its probability density function is therefore invariant to the constant delay D. The constant phase shift ϕ is also of no immediate consequence when the probability density function of the input sequence remains symmetric under rotation, which is indeed the case for the assumed density function given in Eq. (20.36). We may therefore simplify the condition for perfect equalization by requiring that $y = x$.

(20.42). In this figure, three levels of convolutional noise are considered. Here we note from Eq. (20.32) that the *distortion-to (signal plus distortion) ratio* is given by

$$\begin{aligned}\frac{E[(y - x)^2]}{E[y^2]} &= (1 - c_0)^2 + \sigma^2 \\ &= 2(1 - c_0)\end{aligned} \tag{20.43}$$

where in the last line we have made use of Eq. (20.33). The curves presented in Fig. 20.7 correspond to three values of this ratio, namely, 0.01, 0.1, and 0.8. We observe the following from these curves:

1. When the convolutional noise is low, the blind equalization algorithm tends to a minimum mean-squared-error criterion.
2. When the convolutional noise is high, the nonlinear estimator appears to be independent of the fine structure of the amplitude-modulated data. Indeed, different values of amplitude modulation levels result in only very small gain differences due to the normalization defined by Eq. (20.42). This suggests that the use of a uniform amplitude distribution for multilevel modulation systems is an adequate approximation.
3. The nonlinear estimator is robust with respect to variations in the variance of the convolutional noise.

It is of interest to note that for high levels of convolutional noise, the input-output characteristic of the nonlinear estimator is well approximated by the *sigmoid nonlinearity:*[3]

$$\begin{aligned}\hat{x} &= a\frac{1 - e^{-by}}{1 + e^{-by}} \\ &= a \tanh\left(\frac{by}{2}\right)\end{aligned} \tag{20.44}$$

For the situation described in Fig. 20.7, the following values for the constants a and b provide a good fit:

$$a = 1.945$$

$$b = 1.25$$

This is demonstrated in Fig. 20.8.

In any event, the input–output characteristic of the nonlinear estimator is computed in advance, and the result is stored in memory. Then, it is merely a matter of reading off the value of the estimate $\hat{x}$ corresponding to the value y obtained at the inverse filter output.

[3] A sigmoid nonlinearity is a common element in the design of neural networks. In particular, the combination of a linear combiner and a sigmoid nonlinearity constitutes a *neuron*. Accordingly, we may view the blind equalizer depicted in Fig. 20.5 as being essentially a single neuron with its linear combiner and sigmoid nonlinearity represented by the transversal filter and zero-memory nonlinear estimator, respectively. The error signal for controlling the weights at the input end of the neuron is obtained by comparing the input and output signals of the sigmoid nonlinearity.

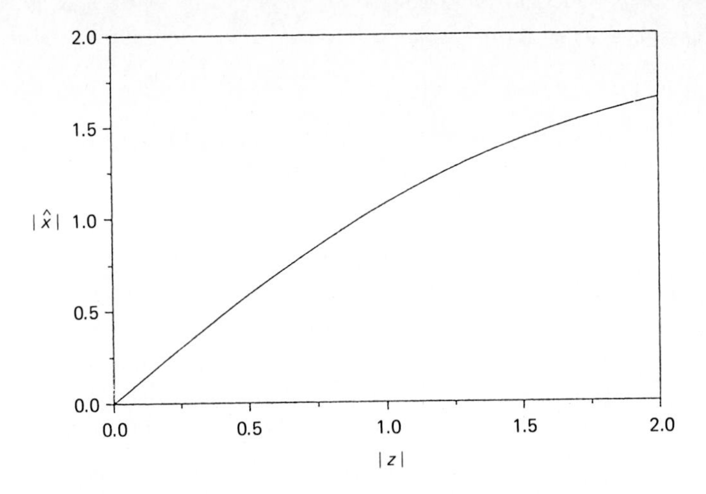

Figure 20.8 Input-output characteristic of sigmoid nonlinearity.

Convergence Considerations

For the iterative convolutional algorithm described by Eqs. (20.16) to (20.18) to converge in the mean, we require that the expected value of the tap weight $\hat{w}_i(n)$ approach some constant value as the number of iterations n approaches infinity. Correspondingly, from Eq. (20.18) we find that the *condition for convergence in the mean* is described by

$$E[u(n-i)y(n)] = E[u(n-i)g(y(n))], \qquad \text{large } n, \quad i = 0, \pm 1, \ldots, \pm L$$

Multiplying both sides of this equation by $\hat{w}_{i-k}$ and summing over i, we get

$$E\left[y(n)\sum_{i=-L}^{L}\hat{w}_{i-k}(n)u(n-i)\right] = E\left[g(y(n))\sum_{i=-L}^{L}\hat{w}_{i-k}(n)u(n-i)\right], \qquad \text{large } n \tag{20.45}$$

We next note from Eq. (20.12) that

$$\begin{aligned} y(n-k) &= \sum_{i=-L}^{L}\hat{w}_i(n)u(n-k-i) \\ &= \sum_{i=-L-k}^{L-k}\hat{w}_{i-k}(n)u(n-i) \qquad \text{large } n \end{aligned}$$

Provided that L is large enough for the transversal equalizer to achieve perfect equalization, we may approximate the expression for $y(n-k)$ as

$$y(n-k) \approx \sum_{i=-L}^{L}\hat{w}_{i-k}(n)u(n-i) \qquad \text{large } n \text{ and large } L \tag{20.46}$$

Accordingly, we may use Eq. (20.46) to simplify (20.45) as follows

$$E[y(n)y(n-k)] \approx E[g(y(n))y(n-k)] \qquad \text{large } n \text{ and large } L$$

Equivalently, we may write

$$E[y(n)y(n+k)] \approx E[y(n)g(y(n+k))] \qquad \text{large } n \text{ and large } L \tag{20.47}$$

We now recognize the following property. A stochastic process $\{y(n)\}$ is said to be a *Bussgang process* if it satisfies the condition

$$E[y(n)y(n+k)] = E[y(n)g(y(n+k))] \tag{20.48}$$

where the function $g(\cdot)$ is a zero-memory nonlinearity.[4] In other words, a Bussgang process has the property that its autocorrelation function is equal to the cross-correlation between that process and the output of a zero-memory nonlinearity produced by that process, with both correlations being measured for the same lag.

Returning to the issue at hand, we may state that the process $\{y(n)\}$ acting as the input to the zero-memory nonlinearity in Fig. 20.5 is *approximately* a Bussgang process, provided that L is large; the approximation becomes better as L is made larger. It is for this reason that the blind equalization algorithm described herein is referred to as a *Bussgang algorithm* [Bellini (1986, 1988)].

In general, convergence of the Bussgang algorithm is not guaranteed. Indeed, the cost function of the Bussgang algorithm operating with a finite L is *nonconvex*; it may therefore have false minima.

For the idealized case of an *infinite length* equalizer, however, a rough proof of convergence of the Bussgang algorithm may be sketched as follows [Bellini (1988)]. The proof relies on a theorem derived in Benveniste et al. (1980), which provides sufficient conditions for convergence.[5] Let the function $\psi(y)$ denote the dependence of the estimation error in the LMS algorithm on the transversal filter output $y(n)$. According to our terminology, we have [6] [see Eqs. (20.16) and (20.17)]

$$\psi(y) = g(y) - y \tag{20.49}$$

The *Benveniste–Goursat–Ruget theorem* states that convergence of the Bussgang algorithm is guaranteed if the probability distribution of the data sequence $\{x(n)\}$ is *sub-Gaussian* and the second derivative of $\psi(y)$ is negative on the interval $(0, \infty)$. In particular, we may state the following:

1. A random variable x is said to be sub-Gaussian if its probability density function is

$$f_X(x) = Ke^{-|x/\beta|^{\nu}} \tag{20.50}$$

where $\nu > 2$. For the limiting case of $\nu = \infty$, the probability density function of Eq. (20.50) reduces to that of a uniformly distributed random variable.

[4] A number of stochastic processes belong to the class of Bussgang processes. Bussgang (1952) was the first to recognize that any correlated Gaussian process has the property described in Eq. (20.48). Subsequently, Barrett and Lampard (1955) extended Bussgang's result to all stochastic processes with exponentially decaying autocorrelation functions. This includes an independent process, since its autocorrelation function consists of a delta function that may be viewed as an infinitely fast decaying exponential [Gray (1979)].

[5] The theorem of interest in Benveniste et al. (1980) is Theorem 3.6 and the accompanying Lemma 3.4.

[6] Note that the function $\psi(y)$ defined in Eq. (20.48) is the negative of that defined in Benveniste et al. (1980).

Also, by choosing $\beta = \sqrt{3}$, we have $E[x^2] = 1$. Thus, the probabilistic model assumed in Eq. (20.6) satisfies the first part of the Benveniste–Goursat–Ruget theorem.

2. The second part of the theorem is also satisfied by the Bussgang algorithm, since we have

$$\frac{\partial^2 \psi}{\partial y^2} < 0, \qquad \text{for } 0 < y < \infty \tag{20.51}$$

This is readily verified by examining the curves plotted in Fig. 20.7 or the sigmoid nonlinearity defined in Eq. (20.44).

The Benveniste–Goursat–Ruget theorem exploited in this proof is based on the assumption of an infinitely parameterized equalizer (i.e., L is infinitely large). Unfortunately, this assumption breaks down in practice as we have to work with a finite L. To date, no zero-memory nonlinear function $g(\cdot)$ is known, which would result in global convergence of the blind equalizer in Fig. 20.5 to the inverse of the unknown channel [Verdu (1984); Johnson (1991)]. The global convergence of the Bussgang algorithm for an arbitrarily large but finite L remains an open problem.

Decision-Directed Algorithm

When the Bussgang algorithm has converged and the eye pattern appears "open," the equalizer should be switched smoothly to the *decision-directed mode* of operation, and minimum mean-squared-error control of the tap weights of the transversal filter component in the equalizer is exercised, as in a conventional adaptive equalizer.

Figure 20.9 presents a block diagram of the equalizer operating in its decision-directed mode. The only difference between this mode of operation and that of blind equalization lies in the type of zero-memory nonlinearity employed. Specifically, the conditional mean estimation of the blind equalizer in Fig. 20.9 is replaced by a *threshold decision device*. Given the observation $y(n)$, that is, the equalized signal at the transversal filter output, the threshold device makes a *decision in favor of a particular value in the known alphabet of the transmitted data sequence that is closest to* $y(n)$. We may thus write

$$\hat{x}(n) = \text{dec}(y(n)) \tag{20.52}$$

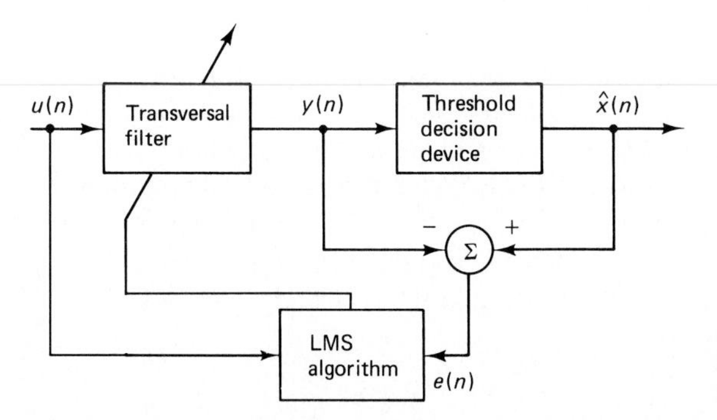

Figure 20.9 Block diagram of decision-directed mode of operation.

For example, in the simple case of a *binary equiprobable data sequence,* the data levels and decision levels are as follows, respectively,

$$x(n) = \begin{cases} +1, & \text{for symbol 1} \\ -1 & \text{for symbol 0} \end{cases} \tag{20.53}$$

and

$$\text{dec}((y(n)) = \text{sgn}(y(n)) \tag{20.54}$$

where sgn(·) is the *signum function* equal to +1 if the argument is positive, and −1 if it is negative.

The equations that govern the operation of the decision-directed algorithm are the same as those of the Bussgang algorithm, except for the use of Eq. (20.52) in place of (20.16). Herein lies an important practical advantage of a blind equalizer that is based on the Bussgang algorithm and incorporates the decision-directed algorithm: Its implementation is only slightly more complex than that of a conventional adaptive equalizer, yet it does not require the use of a training sequence.

Suppose that the following conditions are satisfied:

1. The eye pattern is open (which it should be as a result of the completion of blind equalization).
2. The step-size parameter μ used in the LMS implementation of the decision-directed algorithm is fixed (which is a common practice).
3. The sequence of observations at the channel output, denoted by the vector $\mathbf{u}(n)$, is *ergodic* in the sense that

$$\lim_{N\to\infty} \frac{1}{N} \sum_{n=1}^{N} \mathbf{u}(n)\mathbf{n}^T(n) \to E[\mathbf{u}(n)\mathbf{u}^T(n)] \quad \text{almost surely} \tag{20.55}$$

Then, under these conditions, *the tap-weight vector in the decision-directed algorithm converges to the optimum* (*Wiener*) *solution in the mean-square sense* (Macchi and Eweda, 1984). This is a powerful result, making the decision-directed algorithm an important adjunct of the Bussgang algorithm for blind equalization in digital communications.

20.3 EXTENSION OF BUSSGANG ALGORITHMS TO COMPLEX BASEBAND CHANNELS

Thus far we have only discussed the use of Bussgang algorithms for the blind equalization of M-ary PAM systems, characterized by a real baseband channel. In this section we extend the use of this family of blind equalization algorithms to *quadrature-amplitude modulation* (QAM) systems that involve a hybrid combination of amplitude and phase modulations.

In the case of a complex baseband channel, the transmitted data sequence $\{x(n)\}$, the channel impulse response $\{h_n\}$, and the received signal $\{u(n)\}$ are all complex valued. We may thus write

$$x(n) = x_I(n) + jx_Q(n) \tag{20.56}$$

$$h_n = h_{I,n} + jh_{Q,n} \tag{20.57}$$

and

$$u(n) = u_I(n) + ju_Q(n) \tag{20.58}$$

where the subscripts I and Q refer to the *in-phase* (*real*) and *quadrature* (*imaginary*) *components*, respectively. Correspondingly, the conditional mean estimate of the complex datum $x(n)$, given the observation $y(n)$ at the transversal filter output, is written as

$$\begin{aligned} \hat{x}(n) &= E[x(n)|y(n)] \\ &= \hat{x}_I(n) + j\hat{x}_Q(n) \\ &= g(y_I(n)) + jg(y_Q(n)) \end{aligned} \tag{20.59}$$

where $g(\cdot)$ describes a zero-memory nonlinearity. Equation (20.59) states that the in-phase and quadrature components of the transmitted data sequence $\{x(n)\}$ may be estimated separately from the in-phase and quadrature components of the transversal filter output $y(n)$, respectively. Note, however, that the conditional mean $E[x(n)|\, y(n)]$ can only be expressed as in Eq. (20.59) if the data transmitted in the in-phase and quadrature channels are statistically independent of each other, which is usually the case.

Clearly, Bussgang algorithms for complex baseband channels include the corresponding algorithms for real baseband channels as a special case. Table 20.1 presents a summary of Bussgang algorithms for a complex baseband channel.

TABLE 20.1 SUMMARY OF BUSSGANG ALGORITHMS FOR BLIND EQUALIZATION OF COMPLEX BASEBAND CHANNELS

Initialization: Set

$$\hat{w}_i(0) = \begin{cases} 1, & i = 0 \\ 0, & i = \pm 1, \ldots, \pm L \end{cases}$$

Computation: $n = 1, 2, \ldots$

$$\begin{aligned} y(n) &= y_1(n) + jy_Q(n) \\ &= \sum_{i=-L}^{L} \hat{w}_i^*(n)u(n-i) \\ \hat{x}(n) &= \hat{x}_I(n) + j\hat{x}_Q(n) \\ &= g(y_I(n)) + jg(y_Q(n)) \\ e(n) &= \hat{x}(n) - y(n) \\ \hat{w}_i(n+1) &= \hat{w}_i(n) + \mu u(n-i)e^*(n), \qquad i = 0, \pm 1, \ldots, \pm L \end{aligned}$$

20.4 SPECIAL CASES OF THE BUSSGANG ALGORITHM

Sato Algorithm

The idea of blind equalization in M-ary PAM systems dates back to the pioneering work of Sato (1975). The *Sato algorithm* consists of minimizing a *nonconvex* cost function

$$J(n) = E[(\hat{x}(n) - y(n))^2] \tag{20.60}$$

where $y(n)$ is the transversal filter output defined in Eq. (20.12), and $\hat{x}(n)$ is an estimate of the transmitted datum $x(n)$. This estimate is obtained by a zero-memory nonlinearity described as follows:

$$\hat{x}(n) = \gamma \operatorname{sgn}[y(n)] \tag{20.61}$$

The constant γ sets the *gain* of the equalizer; it is defined by

$$\gamma = \frac{E[x^2(n)]}{E[|x(n)|]} \tag{20.62}$$

The Sato algorithm for blind equalization was introduced originally to deal with one-dimensional multilevel (M-ary PAM) signals. Initially, the algorithm treats such a digital signal as a "binary" signal by estimating the most significant bit; the remaining bits of the signal are treated by the algorithm as additive noise insofar as the blind equalization process is concerned. The algorithm then uses the results of this preliminary step to modify the error signal obtained from a conventional decision-directed algorithm. The Sato algorithm is robust and superior to the decision-directed algorithm, but its rate of convergence is slower.

The Benveniste–Goursat–Ruget theorem for convergence holds for the Sato algorithm even though the nonlinear function $\psi(\cdot)$ defined in Eq. (20.49) is not differentiable. According to this theorem, global convergence of the Sato algorithm can be achieved provided that the probability density function of the transmitted data sequence can be approximated by a sub-Gaussian function such as the uniform distribution [Benveniste et al. (1980)]. However, this result has been disputed in Zing et al. (1989). Also, Mazo (1980) has reported indications of poor performance of the Sato algorithm. Nevertheless, it has been established that almost always the Sato algorithm converges to the correct equalizer coefficients once the eye pattern has been opened [Kumar (1983); Macchi and Eweda (1984)].

It is apparent that the Sato algorithm is a special (non-optimal) case of the Bussgang algorithm, with the nonlinear function $g(y)$ defined by

$$g(y) = \gamma \operatorname{sgn}(y) \tag{20.63}$$

where $\operatorname{sgn}(\cdot)$ is the signum function. The nonlinearity defined in Eq. (20.63) is similar to that in the decision-directed algorithm except for the data-dependent gain factor γ.

Godard Algorithm

Godard (1980) was the first to propose a family of *constant modulus* blind equalization algorithms for use in two-dimensional digital communication systems (e.g., M-ary phase-shift keying). Specifically, the *Godard algorithm* minimizes a nonconvex cost function of the form

$$J(n) = E[(|y(n)|^p - R_p)^2] \tag{20.64}$$

where p is a positive integer, and R_p is a positive real constant defined by

$$R_p = \frac{E[|x(n)|^{2p}]}{E[|x(n)|^p]} \tag{20.65}$$

The Godard algorithm is designed to penalize deviations of the blind equalizer output $x(n)$ from a constant modulus. The constant R_p is chosen in such a way that the gradient of the cost function $J(n)$ is zero when perfect equalization [i.e., $\hat{x}(n) = x(n)$] is attained

The tap-weight vector of the equalizer is adapted in accordance with the stochastic gradient algorithm [Godard (1980)]

$$\hat{\mathbf{w}}(n + 1) = \hat{\mathbf{w}}(n) + \mu\mathbf{u}(n)e^*(n) \tag{20.66}$$

where μ is the step-size parameter, $\mathbf{u}(n)$ is the tap-input vector, and $e(n)$ is the error signal defined by

$$e(n) = y(n)|y(n)|^{p-2}(R_p - |y(n)|^p) \tag{20.67}$$

From the definition of the cost function $J(n)$ in Eq. (20.64) and from (20.67), we see that the equalizer adaptation according to the Godard algorithm does not require carrier phase recovery. The algorithm therefore tends to converge slowly. However, it offers the advantage of decoupling the ISI equalization and carrier phase recovery problems from each other.

Two cases of the Godard algorithm are of specific interest:

Case 1: $p = 1$
The cost function of Eq. (20.64) for this case reduces to

$$J(n) = E[(|y(n)| - R_1)^2] \tag{20.68}$$

where

$$R_1 = \frac{E[|x(n)|^2]}{E[|x(n)|]} \tag{20.69}$$

This case may be viewed as a modification of the Sato algorithm.

Case 2: $p = 2$
In this case, the cost function of Eq. (20.64) reduces to

$$J(n) = E[(|y(n)|^2 - R_2)^2] \tag{20.70}$$

where

$$R_2 = \frac{E[|x(n)|^4]}{E[|x(n)|^2} \tag{20.71}$$

This second special case is referred to in the literature as the *constant modulus algorithm* (CMA).[7]

There are conflicting reports in the literature on the convergence behavior of the Godard algorithm. In his original paper, Godard (1980) showed that the algorithm would converge to the global minimum and thereby achieve perfect equalization (i.e., zero ISI), provided that the algorithm is initialized in a special manner;

[7] The constant modulus algorithm (CMA) was so named by Treichler and Agee (1983), independently of Godard's 1980 paper. It is probably the most widely investigated blind equalization algorithm and the most widely used one in practice [Treichler and Larimore (1985a, b); Smith and Friedlander (1985); Johnson et al. (1988)].

otherwise, it is possible for the algorithm to converge to a local minimum. In a subsequent study, Foschini (1985) made the assertion that the cost function used in the Godard algorithm does not exhibit any locally stable undesirable equilibrium points. Yet, other investigators [Ding et al. (1989); Verdu (1984)] have demonstrated that under ideal modeling assumptions it is possible for the Godard algorithm to exhibit ill convergence due to the existence of local but (false) minima.

Summary of Special Forms of the Bussgang Algorithm

The decision-directed, Sato, and Godard algorithms may indeed be viewed as special cases of the Bussgang algorithm [Bellini (1986)]. In particular, we may use Eqs. (20.52), (20.61), and (20.67) to set up the entries shown in Table 20.2 for the special forms of the zero-memory nonlinear function $g(\cdot)$ pertaining to these three algorithms [Hatzinakos (1990)]. The entries for the decision-directed and Sato algorithms follow directly from the definition

$$\hat{x}(n) = g(y(n))$$

In the case of the Godard algorithm, we note that

$$e(n) = \hat{x}(n) - y(n)$$

or, equivalently,

$$g(y(n)) = y(n) + e(n)$$

Hence, we may use this relation and Eq. (20.67) to derive the entry for the Godard algorithm in Table 20.2.

TABLE 20.2 SPECIAL CASES OF THE BUSSGANG ALGORITHM

Algorithm	Zero-memory nonlinear function g(·)	Definitions
Decision-directed	$\text{sgn}(\cdot)$	
Sato	$\gamma\ \text{sgn}(\cdot)$	$\gamma = \dfrac{E[x^2(n)]}{E[\lvert x(n)\rvert]}$
Godard	$\dfrac{y(n)}{\lvert y(n)\rvert}(\lvert y(n)\rvert + R_p\lvert y(n)\rvert^{p-1} - \lvert y(n)\rvert^{2p-1})$	$R_p = \dfrac{E[\lvert x(n)\rvert^{2p}]}{E[\lvert x(n)\rvert^{p}]}$

20.5 HIGHER-ORDER STATISTICS: DEFINITIONS AND PROPERTIES

To set the stage for a discussion of the second family of blind deconvolution algorithms, we first introduce some basic definitions of *higher-order statistics* of a stationary stochastic process; this we do in the present section. The use of higher-order statistics in the development of blind deconvolution algorithms is taken up in the next section. For convenience of presentation, we restrict the discussion to real-valued processes.

The autocorrelation function and its Fourier transform (i.e., the power spectrum) are basic to the study of second-order statistics of a stationary stochastic process. For the higher order statistics of such a process, we have *cumulants* and their Fourier transforms known as *polyspectra*. Indeed, cumulants and polyspectra may be viewed as *generalizations* of the autocorrelation function and power spectrum, respectively.

Whereas second-order statistics of a process are *phase blind,* higher-order statistics of the process are *phase sensitive*. In particular, cumulants and polyspectra are useful in the analysis and description of non-Gaussian processes, and nonminimum phase and nonlinear systems.

Cumulants

Consider a stationary stochastic process $\{u(n)\}$ with zero mean, that is,

$$E[u(n)] = 0, \qquad \text{for all } n$$

Let $u(n), u(n+\tau_1), \ldots, u(n+\tau_{k-1})$ denote the random variables obtained by observing this stochastic process at times $n, n+\tau_1, \ldots, n+\tau_{k-1}$, respectively. These random variables form the k-by-1 vector:

$$\mathbf{u} = [u(n), u(n+\tau_1), \ldots, u(n+\tau_{k-1})]^T \tag{20.72}$$

Correspondingly, define a k-by-1 vector:

$$\mathbf{z} = [z_1, z_2, \ldots, z_k]^T \tag{20.73}$$

We may then define the kth-*order cumulant* of the stochastic process $\{u(n)\}$, denoted by $c_k(\tau_1, \tau_2, \ldots, \tau_{k-1})$, as the coefficient of the vector $\mathbf{z}$ in the Taylor expansion of the *cumulant generating function* (Priestley 1981; Swami and Mendel, 1990):

$$K(\mathbf{z}) = \ln E[\exp(\mathbf{z}^T\mathbf{u})] \tag{20.74}$$

The kth-order cumulant of the process $\{u(n)\}$ is thus defined in terms of its joint moments of orders up to k. Specifically, the second-, third-, and fourth-order cumulants are given, respectively, by

$$c_2(\tau) = E[u(n)u(n+\tau)] \tag{20.75}$$

$$c_3(\tau_1, \tau_2) = E[u(n)u(n+\tau_1)u(n+\tau_2)] \tag{20.76}$$

and

$$\begin{aligned} c_4(\tau_1, \tau_2, \tau_3) = {} & E[u(n)u(n+\tau_1)u(n+\tau_2)u(n+\tau_3)] \\ & - E[u(n)u(n+\tau_1)]E[u(n+\tau_2)u(n+\tau_3)] \\ & - E[u(n)u(n+\tau_2)]E[u(n+\tau_3)u(n+\tau_1)] \\ & - E[u(n)u(n+\tau_3)]E[u(n+\tau_1)u(n+\tau_2)] \end{aligned} \tag{20.77}$$

From the definitions given in Eqs. (20.75) to (20.77), we note the following:

1. The second-order cumulant $c_2(\tau)$ is the same as the autocorrelation function $r(\tau)$.
2. The third-order cumulant $c_3(\tau_1, \tau_2)$ is the same as the third-order moment $E[u(n)u(n + \tau_1)u(n + \tau_2)]$.
3. The fourth-order cumulant $c_4(\tau_1, \tau_2, \tau_3)$ is *different* from the fourth-order moment $E[u(n)u(n + \tau_1)u(n + \tau_2)u(n + \tau_3)]$. In order to generate the fourth-order cumulant, we need to know the fourth-order moment and six different values of the autocorrelation function.

Note that the kth-order cumulant $c(\tau_1, \tau_2, \ldots, \tau_{k-1})$ does not depend on time n. For this to be valid, however, the process $\{u(n)\}$ has to be stationary up to order k. A process $\{u(n)\}$ is said to be *stationary up to order* k if, for any admissible $n_1, n_2, \ldots, n_p$ and any τ, all the joint moments up to order k of $\{u(n_1), u(n_2), \ldots, u(n_p)\}$ exist and equal the corresponding joint moments up to order k of $\{u(n_1 + \tau), u(n_2 + \tau), \ldots, u(n_p + \tau)\}$ (Priestley, 1981).

Consider next a linear time-invariant system, characterized by the impulse response $\{h_n\}$. Let the system be excited by a process $\{x(n)\}$ consisting of independent and identically distributed (iid) random variables. Let $u(n)$ denote the resulting system output. The kth-order cumulant of $u(n)$ is given by

$$c_k(\tau_1, \tau_2, \ldots, \tau_{k-1}) = \gamma_k \sum_{i=-\infty}^{\infty} h_i h_{i+\tau_1} \cdots h_{i+\tau_{k-1}} \tag{20.78}$$

where γ_k is the kth-order cumulant of the input process $\{x(n)\}$. Note that the summation term on the right side of Eq. (20.78) has a form similar to that of a kth-order cumulant, except that the expectation operator has been replaced by a summation.

Polyspectra

The kth-*order polyspectrum,* (or kth-*order cumulant spectrum*) is defined by (Priestley, 1981; Nikias and Raghuveer, 1987):

$$C_k(\omega_1, \omega_2, \ldots, \omega_{k-1}) = \sum_{\tau_1=-\infty}^{\infty} \cdots \sum_{\tau_{k-1}=-\infty}^{\infty} c(\tau_1, \tau_2, \ldots, \tau_{k-1})$$
$$\cdot \exp[-j(\omega_1\tau_1 + \omega_2\tau_2 + \cdots + \omega_{k-1}\tau_{k-1}) \tag{20.79}$$

A sufficient condition for the existence of the polyspectrum $C_k(\omega_1, \omega_2, \ldots, \omega_{k-1})$ is that the associated kth-order cumulant $c_k(\tau_1, \tau_2, \ldots, \tau_{k-1})$ be absolutely summable, as shown by

$$\sum_{\tau_1=-\infty}^{\infty} \cdots \sum_{\tau_{k-1}=-\infty}^{\infty} |c_k(\tau_1, \tau_2, \ldots, \tau_{k-1}| < \infty \tag{20.80}$$

The *power spectrum, bispectrum,* and *trispectrum* are special cases of the kth-order polyspectrum defined in Eq. (20.79). Specifically, we may state the following:

1. For $k = 2$, we have the ordinary power spectrum:

$$C_2(\omega_1) = \sum_{\tau_1=-\infty}^{\infty} c_2(\tau_1) \exp(-j\omega_1\tau_1) \tag{20.81}$$

which is a restatement of the Einstein–Wiener–Khintchine relation (see Chapter 3).

2. For $k = 3$, we have the *bispectrum,* defined by

$$C_3(\omega_1, \omega_2) = \sum_{\tau_1=-\infty}^{\infty} \sum_{\tau_2=-\infty}^{\infty} c_3(\tau_1, \tau_2) \exp[-j(\omega_1\tau_1 + \omega_2\tau_2)] \tag{20.82}$$

3. For $k = 4$, we have the *trispectrum,* defined by

$$C_4(\omega_1, \omega_2, \omega_3) = \sum_{\tau_1=-\infty}^{\infty} \sum_{\tau_2=-\infty}^{\infty} \sum_{\tau_3=-\infty}^{\infty} c_4(\tau_1, \tau_2, \tau_3) \exp[-j(\omega_1\tau_1 + \omega_2\tau_2 + \omega_3\tau_3)] \tag{20.83}$$

An outstanding property of polyspectrum is that all polyspectra of higher order than the second vanish when the process $\{u(n)\}$ is Gaussian. This property is a direct consequence of the fact that all joint cumulants of higher order than the second are zero for multivariate Gaussian distribution. Accordingly, the bispectrum, trispectrum, and all higher-order polyspectra are identically zero if the process $\{u(n)\}$ is Gaussian. Thus, higher-order spectra provide measures of the *departure of a stochastic process from Gaussianity* (Priestley, 1981; Nikias and Raghuveer, 1987).

The kth-order cumulant $c_k(\tau_1, \tau_2, \ldots, \tau_{k-1})$ and the kth-order polyspectrum $C_k(\omega_1, \omega_2, \ldots, \omega_{k-1})$ form a pair of multidimensional Fourier transforms. Specifically, the polyspectrum $C_k(\omega_1, \omega_2, \ldots, \omega_{k-1})$, is the *multidimensional discrete Fourier transform* of $c_k(\tau_1, \tau_2, \ldots, \tau_{k-1})$, and $c_k(\tau_1, \tau_2, \ldots, \tau_{k-1})$ is the *inverse multidimensional discrete Fourier transform* of $C_k(\omega_1, \omega_2, \ldots, \omega_{k-1})$.

For example, given the bispectrum $C_3(\omega_1, \omega_2)$, we may determine the third-order cumulant $c_3(\tau_1, \tau_2)$, using the inverse two-dimensional discrete Fourier transform:

$$c_3(\tau_1, \tau_2) = \left(\frac{1}{2\pi}\right)^2 \int_{-\pi}^{\pi} \int_{-\pi}^{\pi} C_3(\omega_1, \omega_2) \exp[j(\omega_1\tau_1 + \omega_2\tau_2)]\, d\omega_1\, d\omega_2 \tag{20.84}$$

We may use this relation to develop an alternative definition of the bispectrum as follows. According to the *Cramér spectral representation,* we have (see Chapter 3)

$$u(n) = \frac{1}{2\pi} \int_{-\pi}^{\pi} e^{j\omega n}\, dZ(\omega), \qquad \text{for all } n \tag{20.85}$$

Hence, using Eq. (20.85) in (20.76), we get

$$c_3(\tau_1, \tau_2) = \left(\frac{1}{2\pi}\right)^3 \int_{-\pi}^{\pi} \int_{-\pi}^{\pi} \int_{-\pi}^{\pi} \exp[jn(\omega_1 + \omega_2 + \omega_3)] \cdot \exp[j(\omega_1\tau_1 + \omega_1\tau_2)] E[dZ(\omega_1)\, dZ(\omega_2)\, dZ(\omega_3)] \tag{20.86}$$

Comparing the right sides of Eqs. (20.84) and (20.86), we deduce the following result:

$$E[dZ(\omega_1)\, dZ(\omega_2)\, dZ(\omega_3)] = \begin{cases} C_3(\omega_1, \omega_2)\, d\omega_1\, d\omega_2, & \omega_1 + \omega_2 + \omega_3 = 0 \\ 0, & \text{otherwise} \end{cases} \tag{20.87}$$

It is apparent from Eq. (20.87) that the bispectrum $C_3(\omega_1, \omega_2)$ represents the contribution to the mean product of three Fourier components whose individual frequencies add up to zero. This is an extension of the interpretation developed for the ordinary power spectrum in Chapter 3. In a similar manner we may develop an interpretation of the trispectrum.

In general, the polyspectrum $C_k(\omega_1, \omega_2, \ldots, \omega_{k-1})$ is *complex for order k higher than two,* as shown by

$$C_k(\omega_1, \omega_2, \ldots, \omega_{k-1}) = |C_k(\omega_1, \omega_2, \ldots, \omega_{k-1})| \exp[j\phi_k(\omega_1, \omega_2, \ldots, \omega_{k-1})] \tag{20.88}$$

where we note that $|C_k(\omega_1, \omega_2, \ldots, \omega_{k-1})|$ is the *magnitude* of the polyspectrum, and $\phi_k(\omega_1, \omega_2, \ldots, \omega_{k-1})$ is the *phase*. Moreover, the polyspectrum is a *periodic* function with period 2π; that is,

$$C_k(\omega_1, \omega_2, \ldots, \omega_{k-1}) = C_k(\omega_1 + 2\pi, \omega_2 + 2\pi, \ldots, \omega_{k-1} + 2\pi) \tag{20.89}$$

20.6 TRICEPSTRUM-BASED ALGORITHM FOR BLIND EQUALIZATION

Polyspectra provide the basis for the identification (and therefore blind equalization) of a nonminimum phase channel by virtue of their ability to preserve phase information in the channel output. In this section we study the use of such an approach for the blind equalization of a real baseband channel.

Consider the system model described in Section 20.2 for the baseband transmission of a data sequence $\{x(n)\}$ using M-ary pulse amplitude modulation. The probabilistic model of the sequence $\{x(n)\}$ is as described in Eqs. (20.4) to (20.6). We assume that the FIR channel transfer function $H(z)$ admits the following factorization:

$$H(z) = kI(z)O(z^{-1}) \tag{20.90}$$

where k is a scaling factor, $I(z)$ is a *minimum phase polynomial,* and $O(z^{-1})$ is a *maximum phase polynomial*. The polynomial $I(z)$ has all its zeros inside the unit circle in the z-plane, as shown by

$$I(z) = \prod_{l=1}^{L_1} (1 - a_l z^{-1}), \qquad |a_l| < 1 \tag{20.91}$$

The second polynomial $O(z)$ has all its zeros outside the unit circle, as shown by

$$O(z^{-1}) = \prod_{l=1}^{L_2} (l - b_l z), \qquad |b_l| < 1 \tag{20.92}$$

According to the representation described in Eqs. (20.90) to (20.92), the channel is characterized by a finite-(length) impulse response and nonminimum phase transfer function.

For a data sequence $\{x(n)\}$ having a symmetric uniform distribution, as described in the probabilistic model of Eq. (20.6), we have

$$E[x(n)\} = 0$$
$$E[x^2(n)] = 1$$
$$E[x^3(n)] = 0$$
$$E[x^4(n)] = 9/5$$

Correspondingly, the *skewness* of $x(n)$ is $\gamma_3 = 0$, and its *kurtosis* is

$$\gamma_4 = E[x^4(n)] - 3(E[x^2(n)])^2$$
$$= \frac{9}{5} - 3 = -\frac{6}{5}$$

With $\gamma_3 = 0$, it follows from Eq. (20.78) that the third-order cumulant of the channel output $\{u(n)\}$ is identically zero. On the other hand, γ_4 has a nonzero value; we may therefore work in the fourth-order cumulant domain as a basis for blind equalization.

Tricepstrum

Let $c_4(\tau_1, \tau_2, \tau_3)$ denote the fourth-order cumulant of the channel output $\{u(n)\}$. We may therefore express the trispectrum of $\{u(n)\}$ as

$$C_4(\omega_1, \omega_2, \omega_3) = F[c_4(\tau_1, \tau_2, \tau_3)] \tag{20.93}$$

where $F[\cdot]$ denotes three-dimensional discrete Fourier transformation. Define

$$\kappa_4(\tau_1, \tau_2, \tau_3) = F^{-1}[\ln C_4(\omega_1, \omega_2, \omega_3)] \tag{20.94}$$

where ln signifies the natural logarithm, and F^{-1} signifies the inverse three-dimensional discrete Fourier transformation. The quantity $\kappa_4(\tau_1, \tau_2, \tau_3)$ is called the *complex cepstrum of trispectrum* or *tricepstrum* of the process $\{u(n)\}$ (Pan and Nikias, 1988; Hatzinakos and Nikias, 1989, 1990).

From Eq. (20.78) we note that

$$c_4(\tau_1, \tau_2, \tau_3) = \gamma_4 \sum_{i=0}^{\infty} h_i h_{i+\tau_1} h_{i+\tau_2} h_{i+\tau_3} \tag{20.95}$$

Hence, taking the three-dimensional discrete Fourier transforms of both sides of Eq. (20.95), we get

$$C_4(\omega_1, \omega_2, \omega_3) = \gamma_4 H(e^{j\omega_1}) H(e^{j\omega_2}) H(e^{j\omega_3}) H(e^{-j(\omega_1+\omega_2+\omega_3)}) \tag{20.96}$$

Next, taking the natural logarithm of both sides of Eq. (20.96), we get

$$\begin{aligned}\ln C_4(\omega_1, \omega_2, \omega_3) = {} & \ln \gamma_4 + \ln H(e^{j\omega_1}) + \ln H(e^{j\omega_2}) + \ln H(e^{j\omega_3}) \\ & + \ln H(e^{-j(\omega_1+\omega_2+\omega_3)})\end{aligned} \tag{20.97}$$

The channel transfer function $H(z)$ is defined by Eqs. (20.90) to (20.92); hence, we have

$$\begin{aligned}\ln H(e^{j\omega_i}) &= \ln k + \ln I(e^{j\omega_i}) + \ln O(e^{-j\omega_i}) \\ &= \ln k + \sum_{l=1}^{L_1} \ln(1 - a_l e^{-j\omega_i}) + \sum_{l=1}^{L_2} \ln(1 - b_l e^{j\omega_i}), \qquad i = 1, 2, 3\end{aligned} \tag{20.98}$$

and

$$
\begin{aligned}
\ln H(e^{-j(\omega_1+\omega_2+\omega_3)} &= \ln k + \ln I(e^{-j(\omega_1+\omega_2+\omega_3)}) + \ln O(e^{j(\omega_1+\omega_2+\omega_3)}) \\
&= \ln k + \sum_{l=1}^{L_1} \ln(1 - a_l e^{j(\omega_1+\omega_2+\omega_3)}) + \sum_{l=1}^{L_2} \ln(1 - b_l e^{-j(\omega_1+\omega_2+\omega_3)})
\end{aligned}
\tag{20.99}
$$

Thus, returning to Eq. (20.97) and taking the inverse three-dimensional discrete Fourier transform of $\ln C_4(\omega_1, \omega_2, \omega_3)$, we find that the tricepstrum has the following form:[8]

$$
\kappa_4(\tau_1, \tau_2, \tau_3) = \begin{cases}
\ln k + 3 \ln \gamma_4, & \tau_1 = \tau_2 = \tau_3 = 0 \\
-\dfrac{1}{\tau_1} A^{(\tau_1)}, & \tau_1 > 0,\ \tau_2 = \tau_3 = 0 \\
-\dfrac{1}{\tau_2} A^{(\tau_2)}, & \tau_2 > 0,\ \tau_1 = \tau_3 = 0 \\
-\dfrac{1}{\tau_3} A^{(\tau_3)}, & \tau_3 > 0,\ \tau_1 = \tau_2 = 0 \\
\dfrac{1}{\tau_1} B^{(-\tau_1)}, & \tau_1 < 0,\ \tau_2 = \tau_3 = 0 \\
\dfrac{1}{\tau_2} B^{(-\tau_2)}, & \tau_2 < 0,\ \tau_1 = \tau_3 = 0 \\
\dfrac{1}{\tau_3} B^{(-\tau_3)}, & \tau_3 < 0,\ \tau_1 = \tau_2 = 0 \\
-\dfrac{1}{\tau_2} B^{(\tau_2)}, & \tau_1 = \tau_2 = \tau_3 > 0 \\
\dfrac{1}{\tau_2} A^{(\tau_2)}, & \tau_1 = \tau_2 = \tau_3 < 0 \\
0, & \text{otherwise}
\end{cases}
\tag{20.100}
$$

[8] To evaluate $\kappa_4(\tau_1, \tau_2, \tau_3)$, we may use the inversion formula for the three-dimensional z-transform:

$$
\kappa_4(\tau_1, \tau_2, \tau_3) = \frac{1}{(2\pi j)^3} \oint_{\mathscr{C}_3} \oint_{\mathscr{C}_2} \oint_{\mathscr{C}_1} \ln C_4(z_1, z_2, z_3) z_1^{\tau_1 - 1} z_2^{\tau_2 - 1} z_3^{\tau_3 - 1}\, dz_1\, dz_2\, dz_3
$$

where $C_4(z_1, z_2, z_3)$ is obtained from $C_4(\omega_1, \omega_2, \omega_3)$ by substituting z_i for $e^{j\omega_i}$ where $i = 1, 2, 3$. The closed contours $\mathscr{C}_1$, $\mathscr{C}_2$, and $\mathscr{C}_3$ lie completely within the region of convergence of $\ln C_4(z_1, z_2, z_3)$. Let

$$
\begin{aligned}
\hat{a} &= \max\{|a_l|\}, \qquad 1 \le l \le L_1 \\
\hat{b} &= \max\{|b_l|\}, \qquad 1 \le l \le L_2 \\
e &= \max\{\hat{a}, \hat{b}\}
\end{aligned}
$$

The region of convergence for $\ln C_4(z_1, z_2, z_3)$ is defined by

$$
R_c = \{|z_1| > e, |z_2| > e, |z_3| > e, \text{ and } |z_1 z_2 z_3| < 1/e\}
$$

The unit surface defined by $\{|z_1| = 1, |z_2| = 1, \text{ and } |z_3| = 1\}$ lies within the region of convergence R_c. Accordingly, it is permissible to use the power series expansion or inversion formula to evaluate $\kappa_4(\tau_1, \tau_2, \tau_3)$.

where

$$A^{(m)} = \sum_{l=1}^{L_1} a_l^m \tag{20.101}$$

and

$$B^{(m)} = \sum_{l=1}^{L_2} b_l^m \tag{20.102}$$

The $A^{(m)}$ and $B^{(m)}$ contain *minimum-phase* and *maximum-phase* information about the channel, respectively.

It can be shown that the fourth-order cumulant $c_4(\tau_1, \tau_2, \tau_3)$ and the tricepstrum $\kappa_4(\tau_1, \tau_2, \tau_3)$ are related by the linear convolution formula (Pan and Nikias, 1988):

$$\sum_{r=-\infty}^{\infty} \sum_{s=-\infty}^{\infty} \sum_{t=-\infty}^{\infty} r\kappa_4(r, s, t)c_4(\tau_1 - r, \tau_2 - s, \tau_3 - t) = -\tau_1 c_4(\tau_1, \tau_2, \tau_3) \tag{20.103}$$

Equation (20.103) is of fundamental importance, because it serves as the basis of the tricepstrum method of blind equalization.

Substituting Eq. (20.100) into (20.103), we obtain (after some algebra) the following *tricepstral equation:*

$$\begin{aligned}\sum_{m=1}^{p} (A^{(m)}[c_4(\tau_1 - m, \tau_2, \tau_3) - c_4(\tau_1 + m, \tau_2 + m, \tau_3 + m)]) \\ + \sum_{m=1}^{q} (B^{(m)}[c_4(\tau_1 - m, \tau_2 - m, \tau_3 - m) - c_4(\tau_1 + m, \tau_2, \tau_3)]) \\ = -\tau_1 c_4(\tau_1, \tau_2, \tau_3)\end{aligned} \tag{20.104}$$

In theory, the parameters p and q are infinitely large. In practice, however, they can both be approximated by finite (arbitrarily large) values, because $A^{(m)}$ and $B^{(m)}$ decay exponentially as m increases (Hatzinakos and Nikias, 1990). Assuming that suitable values have been assigned to p and q, we may define

$$\alpha_1 = \max(p, q) \tag{20.105}$$

$$\alpha_2 \leq \frac{\alpha_1}{2} \tag{20.106}$$

$$\alpha_3 \leq \alpha_2, \tag{20.107}$$

and choose

$$\tau_1 = -\alpha_1, \ldots, -1, 1, \ldots, \alpha_1 \tag{20.108}$$

$$\tau_2 = -\alpha_2, \ldots, 0, \ldots, \alpha_2 \tag{20.109}$$

$$\tau_3 = -\alpha_3, \ldots, 0, \ldots, \alpha_3 \tag{20.110}$$

Let

$$w = 2\alpha_1(2\alpha_2 + 1)(2\alpha_3 + 1) \tag{20.111}$$

Accordingly, we may use Eq. (20.104) to construct the following overdetermined system of equations:

$$\mathbf{Ca} = \mathbf{p} \tag{20.112}$$

where the known quantities $\mathbf{C}$ and $\mathbf{p}$ and the unknown $\mathbf{a}$ are defined as follows:

1. The matrix $\mathbf{C}$ is a w-by-$(p + q)$ matrix with entries of the form $\{c_4(\tau_1, \tau_2, \tau_3) - c_4(\tau_1', \tau_2', \tau_3')\}$; the dimension w is itself defined in Eq. (20.111).
2. The vector $\mathbf{p}$ is a w-by-1 vector with entries of the form $\{-\tau_1 c_4(\tau_1, \tau_3, \tau_3)\}$.
3. The vector $\mathbf{a}$ is a $(p + q)$-by-1 coefficient vector defined in terms of the $A^{(m)}$ and the $B^{(m)}$ by

$$\mathbf{a} = [A^{(1)}, A^{(2)}, \ldots, A^{(p)}, B^{(1)}, B^{(2)}, \ldots, B^{(q)}]^T \tag{20.113}$$

From the classical method of least squares presented in Chapter 10, the least-squares solution of Eq. (20.104) is

$$\mathbf{a} = (\mathbf{C}^H\mathbf{C})^{-1}\mathbf{C}^H\mathbf{p} \tag{20.114}$$

The solution for the coefficient vector $\mathbf{a}$ given in Eq. (20.114) is based on ensemble-averaged values for the fourth-order cumulant $c_4(\tau_1, \tau_2, \tau_3)$ that defines the entries of matrix $\mathbf{C}$ and vector $\mathbf{p}$. In practice, we use estimate[9] $\hat{c}_4(\tau_1, \tau_2, \tau_3)$ obtained from finite-length window of the received signal $\{u(n)\}$. Based on these estimates, we may rewrite Eq. (20.114) as

$$\hat{\mathbf{a}} = (\hat{\mathbf{C}}^H\hat{\mathbf{C}})^{-1}\hat{\mathbf{C}}^H\hat{\mathbf{p}} \tag{20.115}$$

where $\hat{\mathbf{C}}$ and $\hat{\mathbf{p}}$ are estimates of $\mathbf{C}$ and $\mathbf{p}$, respectively. Provided that the received signal $\{u(n)\}$ is *fourth-order ergodic,* then $\hat{\mathbf{a}}$ is asymptotically unbiased and consistent estimate of the coefficient vector $\mathbf{a}$ (Hatzinakos and Nikias, 1990).

Equation (20.115) provides the basis of a block processing approach for computing $\hat{\mathbf{a}}$. Alternatively, we may use an adaptive filtering algorithm, as described next.

Adaptive Channel Estimator

To solve the system of equations described by (20.115), in an adaptive manner, we may use an LMS-type algorithm. Specifically, we write

$$\hat{\mathbf{a}}(n + 1) = \hat{\mathbf{a}}(n) + \mu(n)\hat{\mathbf{C}}^H(n)\mathbf{e}(n) \tag{20.116}$$

where n is the iteration number and $\mathbf{e}$ is the *estimation error vector:*

$$\mathbf{e}(n) = \mathbf{p}(n) - \hat{\mathbf{C}}(n)\hat{\mathbf{a}}(n) \tag{20.117}$$

The $\mu(n)$ is a time-varying step-size parameter. To assure stability of the LMS algorithm in the mean-square sense, we choose (in accordance with an approximate form of the theory presented in Chapter 9)

$$0 < \mu(n) < \frac{2}{\text{tr}[\mathbf{C}^H(n)\mathbf{C}(n)]} \tag{20.118}$$

where tr[·] denotes the trace of the matrix enclosed within the square brackets.

[9] Methods for computing the estimates $\hat{c}_4(\tau_1, \tau_2, \tau_3)$ are described in Hatzinakos and Nikias (1990).

Equation (20.116) constitutes the basis for the adaptive estimation of the coefficient vector **a** that is related to the channel parameters. We may therefore view this equation as the *channel estimator* part of the blind equalizer, as depicted in Fig. 20.10.

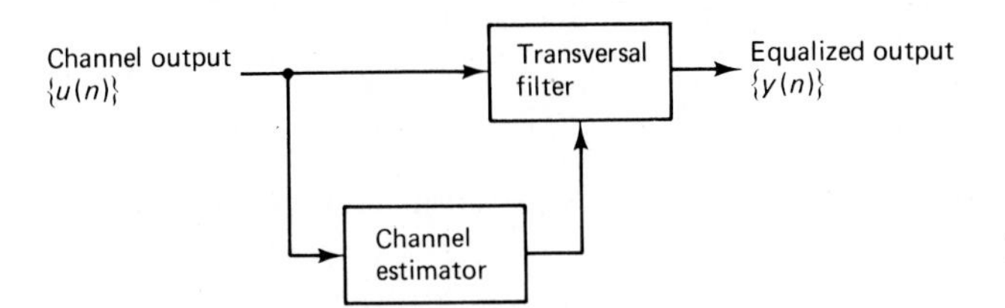

Figure 20.10 Block diagram of tricepstrum-based blind equalizer.

Equalization Coefficients Computer

Our main purpose is to compute the tap weights of a transversal filter constituting the linear equalizer part of the structure shown in Fig. 20.10. Let $W(z)$ denote the transfer function of this equalizer. Then, under a *zero-forcing equalization constraint*, we require that $W(z)$ be related to the channel transfer function $H(z)$ as follows (see Chapter 1):

$$W(z) \approx \frac{1}{H(z)} \tag{20.119}$$

Setting the scaling factor $k = 1$ in Eq. (20.90), we have

$$W_{\text{nor}}(z) \approx \frac{1}{I(z)O(z^{-1})} \tag{20.120}$$

The requirement is to estimate the inverse filters with transfer functions $1/I(z)$ and $1/O(z^{-1})$, given the tricepstrum coefficient vector $\hat{\mathbf{a}}$. Let $i_{\text{inv}}(n, l)$ and $o_{\text{inv}}(n, l)$ denote the respective impulse responses of these two inverse filters, where n is the iteration number and l is the coefficient number in the impulse response. To compute estimates of $i_{\text{inv}}(n, l)$ and $o_{\text{inv}}(n, l)$, we may proceed as follows (Pan and Nikias, 1988; Hatzinakos and Nikias, 1990):

1. Initialize the computation by setting

$$\hat{i}_{\text{inv}}(n, 0) = \hat{o}_{\text{inv}}(n, 0) = 1 \tag{20.121}$$

2. Compute the minimum phase (causal) sequence

$$\hat{i}_{\text{inv}}(n, l) = -\frac{1}{l}\sum_{m=2}^{l+1} [-\hat{A}_{(n)}^{(m-1)}]\hat{i}_{\text{inv}}(n, l - m + 1) \tag{20.122}$$

for $l = 1, 2, \ldots, L_1$.

3. Compute the maximum phase (anticausal) sequence

$$\hat{o}_{\text{inv}}(n, l) = \frac{1}{l}\sum_{m=l+1}^{0} [-\hat{B}_{(n)}^{(1-m)}]\hat{o}_{\text{inv}}(n, l - m + 1) \tag{20.123}$$

for $l = -1, -2, \ldots, -L_2$

4. Compute the impulse response of the linear equalizer by convolving $\hat{\imath}_{\text{inv}}(n, l)$ with $\hat{o}_{\text{inv}}(n, l)$:

$$\hat{w}_{\text{nor}}(n, l) = \hat{\imath}_{\text{inv}}(n, l) * \hat{o}_{\text{inv}}(n, l) \tag{20.124}$$

where $*$ signifies convolution. The total length of the transversal filter (i.e., the number of tap weights) is

$$M = L_1 + L_2$$

The impulse response $\hat{w}_{\text{nor}}(n, l)$ is *normalized,* with the normalization being contingent on the assumption that the scaling factor k in the channel transfer function $H(z)$ is set equal to unity. Procedures for recovering an estimate of the absolute value of this scaling factor, for different modulation schemes, are described in Hatzinakos and Nikias (1990).

20.7 SUMMARY AND DISCUSSION

Blind deconvolution is an example of *unsupervised learning* in the sense that it identifies the inverse of an unknown linear time-invariant (possibly nonminimum phase) system *without* any physical access to the source of the transmitted signal (i.e., system input). This operation requires the identification of both the magnitude and phase of the system's transfer function. To identify the magnitude component, we only need second-order statistics of the received signal (i.e., system output). However, the identification of the phase component is more complicated because it involves the higher-order statistics of the received signal. Accordingly, some form of *nonlinearity* must be employed to extract the information content in the magnitude and phase components of the received signal, which is made feasible by *non-Gaussian* statistics of the transmitted signal.

In this chapter we have described two different approaches for the blind equalization of a linear communication channel. Both approaches assume the prescription of a non-Gaussian model for the statistics of the transmitted data sequence.

The Bussgang algorithm performs blind equalization by subjecting the received signal to an iterative deconvolution process. When the algorithm has converged in the mean, the deconvolved sequence assumes Bussgang statistics, hence the name of the algorithm. The distinguishing features of the Bussgang algorithm are as follows:

1. The minimization of a nonconvex cost function
2. A low computational complexity, which is slightly greater than that of a conventional adaptive equalizer equipped with a training phase
3. A slow initial rate of convergence when the eye pattern is closed, with the rate of convergence speeding up as the eye pattern starts to open up; for a description of the eye pattern, see Chapter 1

The tricepstrum-based blind equalization algorithm exploits the inherent ability of the fourth-order cumulant (or, equivalently, the trispectrum) of the received signal to extract phase information about the channel. This second algorithm has the following characteristics:

1. Channel estimation by identifying the minimum phase and maximum phase parts of the channel transfer function
2. A high computational complexity
3. A fast initial rate of convergence, which slows down later (i.e., opposite to the convergence behavior of the Bussgang algorithm)

It is also of interest to note that, with appropriate formulation this second algorithm may be used to perform either channel identification (i.e., estimating the channel impulse response) or channel equalization. Moreover, it can work with FIR as well as IIR channels.

In the development of both of these blind equalization algorithms, we emphasized the use of the LMS algorithm to compute the solution of an overdetermined system of equations adaptively. Needless to say, we may use other adaptive filtering algorithms, such as a recursive least-squares algorithm or one of its variants for performing this adaptive computation. By so doing, it may be possible to improve the convergence behavior of blind equalization algorithms.

A sufficient condition for equalization is that the probability distribution of the individual recovered symbols at the equalizer output be equal to the probability distribution of the individual transmitted symbols at the channel input [Benveniste et al. (1980)]. A simplified set of conditions for equalization has been derived by Shalvi and Weinstein (1990). In this latter paper it is shown that the necessary and sufficient conditions for equalization are that the second- and fourth-order moments of the individual recovered symbols be equal to the second- and fourth-order moments of the individual transmitted symbols, respectively. Specifically, Shalvi and Weinstein have derived new optimization criteria with respective stochastic gradient algorithms, which are based on the maximization of the kurtosis (a measure of fourth-order statistics) of the equalizer output, subject to constraints imposed on second-order statistics.

PROBLEMS

1. When the data sequence $\{x(n)\}$ in a digital communication system is Gaussian, and the unknown channel is nonminimum phase, there is no solution to the problem of estimating the unknown phase of the channel transfer function. Discuss the validity of this statement.
2. Equation (20.39) defines the conditional mean estimate of the datum x, assuming that the convolutional noise v is additive, white, Gaussian and statistically independent of x. Using the definitions given in Eqs. (20.32), (20.33), and (20.36) to (20.38), derive the formula for the estimate $\hat{x}$ given in Eq. (20.39).
3. For perfect equalization, we require that the equalizer output $y(n)$ be exactly equal to the transmitted datum $x(n)$. Show that when the Bussgang algorithm has converged, and perfect equalization has been attained, the nonlinear estimator must satisfy the condition given in Eq. (20.42), namely,

$$E[\hat{x}g(\hat{x})] = 1$$

4. Assuming the use of the sigmoid nonlinearity given in Eq. (20.44) for the nonlinear estimator $\hat{x} = g(y)$, show that the second derivative of the function

$$\psi(y) = g(y) - y$$

is negative for y in the interval $(0, \infty)$.

5. Equation (20.18) provides an adaptive method for finding the tap weights of the transversal filter in the Bussgang algorithm for performing the iterative deconvolution. Develop an alternative method for doing this computation, assuming the availability of an overdetermined system of equations and the use of the method of least squares.

6. Show that the third- and higher-order cumulants of a Gaussian process are all identically zero.

7. Develop a physical interpretation of the trispectrum $C_4(\omega_1, \omega_2, \omega_3)$ of a stationary stochastic process $\{u(n)\}$.

8. Consider a linear time-invariant system whose transfer function is $H(z)$. The system is excited by a sequence $\{x(n)\}$ of independently and identically distributed (iid) random variables with zero mean and unit variance. The probability distribution of $x(n)$ is nonsymmetric.
   (a) Evaluate the third-order cumulant and bispectrum of the system output $\{u(n)\}$.
   (b) Show that the phase component of the bispectrum of $\{u(n)\}$ is related to the phase response of the system transfer function $H(z)$ as follows:

$$\arg[C_3(\omega_1, \omega_2)] = \arg[H(e^{j\omega_1})] + \arg[H(e^{j\omega_2})] - \arg[H(e^{j(\omega_1+\omega_2)})]$$

9. Equation (20.78) gives the kth-order cumulant of the output of a linear time-invariant system of impulse response $\{h_i\}$ that is driven by a sequence $\{x(n)\}$ of independent and identically distributed random variables. Prove the validity of this equation.

10. Derive the linear convolution formula given in Eq. (20.103), which defines the relation between the fourth-order cumulant $c_4(\tau_1, \tau_2, \tau_3)$ and the tricepstrum $\kappa_4(\tau_1, \tau_2, \tau_3)$.

11. Derive the tricepstral equation (20.104) that relates the fourth-order cumulant $c_4(\tau_1, \tau_2, \tau_3)$ to the $\{A^{(m)}\}$ and the $\{B^{(m)}\}$ that contain minimum phase and maximum phase information about the channel.

12. Equation (20.118) is an approximate formula for the upper bound on the step-size parameter $\mu(n)$ used in the LMS-type algorithm in tricepstrum-based procedure for blind equalization. Justify the use of this formula.

21 Discussion

In this final and brief chapter of the book, we do two things. First, in Section 21.1 we present a *summary* of the material covered in Chapters 5 through 19 that pertain to adaptive filtering and related issues. We have excluded Chapter 1 from this summary because of its introductory nature, and Chapters 2 through 4 because they cover mathematical material of a prerequisite nature. Then, in Section 21.2 we itemize a number of *unexplored topics* that we were not able to cover in the book, and which are relevant to the subject of adaptive filtering. To assist the interested reader, references are included in Section 21.2 that provide guidance to related material in the literature.

21.1 SUMMARY

Adaptive filtering has emerged as an important part of signal processing, rich in both theory and application. We say that a filter, be it in hardware or software form, is *adaptive* if it satisfies two requirements:

1. The filter has a built-in mechanism for the automatic adjustment of its coefficients in response to statistical variations of the environment in which the filter operates.
2. The coefficient adjustments are made for the purpose of progressively moving the filter (in an iteration-by-iteration or block-by-block manner) toward an optimum performance; here, optimality is defined in some statistical sense.

Adaptive filter theory applies to the processing of a *time series* as well as a *space series*. In the temporal case, the filter input may consist of a "vector" of uniformly spaced samples taken from a long data stream. In the spatial case, the filter input may consist of a "snapshot" of elemental outputs derived from an array of uniformly spaced sensors at a particular instant of time. In some practical situations, the uniformity of data samples is *not* adhered to.

Adaptive filters have been successfully applied in diverse fields, including adaptive equalizers, echo cancellers, adaptive beamformers for sonar and radar, speech encoders, time-varying spectrum estimators, and system identification.

Adaptive Filters Using Block Estimation

The starting point in the development of adaptive filters is the set of ensemble-averaged Wiener–Hopf equations or the time-averaged normal equations. The development may then proceed using a block estimation approach, in which the input data are segmented into blocks, with the length of each block chosen to ensure pseudo-stationarity. Each block of data is used to derive an estimate of the autocorrelation sequence of the input signal or an estimate of the corresponding sequence of reflection coefficients. These parameters are then used to design an adaptive filter, the coefficients of which are varied on a block-by-block basis.

The structures/architectures resulting from the block estimation approach are:

1. Transversal (tapped-delay-line) filter.
2. Lattice (ladder) predictor.
3. Systolic (parallel) processor.

Table 21.1 presents a summary of the algorithms that may be used to design these filtering structures/architectures. It is of interest to note that the Yule–Walker algorithm is indirect in that it operates on the estimated autocorrelation sequence of the input signal. On the other hand, the method of least squares, the Burg algorithm, and the singular value decomposition (SVD) algorithm operate on the samples of the input signal directly.

TABLE 21.1 ADAPTIVE FILTERS USING BLOCK ESTIMATION

Structure/architecture	Theoretical tools
Transversal filter	Yule–Walker algorithm (assuming estimates of the autocorrelation sequence and incorporating the Levinson–Durbin recursion)
	Method of least squares
Lattice (ladder) predictor	Burg algorithm (incorporating the Levinson–Durbin recursion)
Systolic processor	Singular-value decomposition (SVD) algorithm, based on:
	Jacobi rotations
	Householder transformations and the Golub–Kahan step

Adaptive Filters Using Recursive Estimation

A more popular approach for the design of adaptive filters is to use a *recursive estimation* procedure, whereby the filter coefficients are adjusted on an iterative basis (on the arrival of each new sample of input data). This approach is direct in the sense that it eliminates the need for estimating intermediate parameters (e.g., autocorrelations, reflection coefficients). Moreover, in each iteration the filter *learns* a little more about the incoming signal characteristics, and an improvement to the current coefficient vector of the filter is computed using this new information.

In this case, the filtering structures of interest are as follows:

1. Transversal filter(s)
2. Lattice (ladder) predictor
3. Systolic (parallel) processor

Specifically, Table 21.2 summarizes two recursive estimation algorithms, namely, the least-mean-square (LMS) algorithm and the gradient adaptive lattice (GAL) algorithm, which represent stochastic gradient approximations of the Wiener filter. They differ from each other in the way in which the approximations to the Wiener filter are made. However, they share a common feature in that their *computational complexity increases linearly with the filter order*.

TABLE 21.2 RECURSIVE ESTIMATION ALGORITHMS USING GRADIENT APPROXIMATIONS OF THE WIENER FILTER

Structure/architecture	Theoretical tools
Transversal filter	Least-mean-square (LMS) algorithm, and its two variants: Normalized LMS algorithm Leaky LMS algorithm
Lattice (ladder) predictor	Gradient adaptive lattice (GAL) algorithm

Table 21.3 presents a summary of two recursive estimation algorithms, namely, the recursive least-squares (RLS) algorithm and the QR-decomposition-based recursive least-squares (QRD-RLS) algorithm, which owe their origin to the time-averaged normal equations. In both algorithms, the computational complexity increases as the square of filter order.

TABLE 21.3 RECURSIVE ESTIMATION ALGORITHMS BASED ON THE NORMAL EQUATIONS

Structure/architecture	Theoretical tools
Transversal filter	Recursive least-squares (RLS) algorithm
Systolic (parallel) processor	QR-decomposition-based recursive least-squares (QRD-RLS) algorithm, using Givens rotations

Table 21.4 presents a summary of four *fast* realizations of recursive least squares. These fast algorithms share a common feature in that they all have a computational complexity that increases linearly with the filter order as with stochastic gradient algorithms, hence the term "fast." They also provide useful side information about the input data (such as a priori and a posteriori versions of the forward and backward prediction errors). These desirable properties are achieved by using

TABLE 21.4 FAST RECURSIVE LEAST-SQUARES ESTIMATION ALGORITHMS

Structure/architecture	Theoretical tools
Transversal filters	Fast RLS algorithm Fast transversal filters (FTF) algorithm
Lattice (ladder) predictor	Recursive least-squares lattice (LSL) algorithm using: A posteriori estimation errors A priori estimation errors A posteriori estimation errors and error feedback A priori estimation errors and error feedback
Systolic (parallel) processor	Matrix-oriented (Kalman) type of QRD-fast RSL algorithm
Systolic-lattice processor	QRD-LSL algorithm

forward and backward linear least-squares prediction to exploit the inherent redundancy contained in the Toeplitz structure of the input data matrix. As such, their use is restricted to the processing of a time series only.

The recursive least-squares lattice (LSL) algorithm listed in Table 21.4 may assume one of four different forms, depending on the following features:

1. Whether the algorithm is based on a posteriori or a priori estimation errors
2. Whether the algorithm uses indirect computation of the forward and backward reflection coefficients or a direct procedure (with built-in error feedback).

The two-coefficient version of the gradient adaptive lattice (GAL) algorithm may be viewed as a special case of a recursive LSL algorithm in which a priori and a posteriori estimation errors are made to be the same.

It is of interest to note that the LMS and RLS algorithms involve time updates only. On the other hand, all the other algorithms listed in Tables 21.2 through 21.4 involve both time and order updates. It is also noteworthy that the normalized LMS algorithm is the prototype of a single *underdetermined* error equation, whereas the RLS algorithm is the prototype of an *overdetermined* system of error equations.

Computational Complexity

We customarily use the number of operations required to perform one iteration of the algorithm as a measure of computational complexity. By operations we mean the following: multiplications, divisions, and additions/subtractions. In Table 21.5 we present a comparison of the computational complexities of five algorithms: LMS, RLS, FTF, conventional form of LSL, and error-feedback form of LSL. In each case, it is assumed that the input data are real-valued and that the order of the estimator equals M [i.e., there are $(M + 1)$ regression coefficients to be estimated]. We may thus make the following observations:

1. The LMS algorithm is the least demanding in computational complexity.
2. Like the FTF algorithm, the computational complexity of an LSL algorithm

TABLE 21.5 COMPARISON OF COMPUTATIONAL COMPLEXITIES OF FOUR LEAST-SQUARES ALGORITHMS FOR REAL DATA

	Number of operations per iteration		
Algorithm	Multiplications	Divisions	Addition/subtractions
LMS	$2M + 1$	0	$2M$
RLS	$2M^2 + 7M + 5$	$M^2 + 4M + 3$	$2M^2 + 6M + 4$
FTF	$7M + 12$	4	$6M + 3$
Conventional LSL	$10M + 3$	$6M + 2$	$8M + 2$
Error-feedback LSL	$18M + 3$	$4M + 1$	$9M + 2$

increases *linearly* with the order M. In the RLS algorithm, on the other hand, it increases with M^2.

3. The computational complexity of the conventional form of an LSL algorithm is slightly larger than that of an FTF algorithm. In particular, it requires more divisions per iteration of the algorithm.
4. The conventional and error-feedback forms of the LSL algorithm have approximately the same computational complexity, with the conventional form of the algorithm having a slight advantage.

Physical Significance of Parameters

In the LMS, RLS, and FTF algorithms, the instantaneous values of tap weights in a transversal filter represent the information contained in the input data. This property is exploited in *system identification* (Åström and Eykhoff, 1974; Goodwin and Payne, 1977; Ljung and Söderström, 1983; Goodwin and Sin, 1984) and *parametric spectrum analysis* (Childers, 1978; Haykin, 1983b) where a transversal filter of *fixed* length is used to *model* an unknown system or process. In such applications, the tap weights of a transversal filter represent the desired information. Furthermore, the instantaneous values of these tap weights often provide useful information about the environment in which an adaptive transversal filter operates. An example of this arises in *adaptive equalization* (Lucky et al., 1968; Gersho, 1969; Gitlin and Magee, 1977) where, ideally, *the transfer function of the equalizer is equal to the inverse of that of the (unknown) communication channel.*

On the other hand, a recursive LSL algorithm presents information about the input data in the form of instantaneous values of reflection coefficients or transformed regression coefficients. Hence, additional computation is required if the information extracted by the LSL algorithm is to be presented in the form of parameters of a multiple linear regression model as in system identification and parametric spectrum analysis, or the impulse response of a communication channel as in adaptive equalization.

Yet, in another context, the structure of a multistage lattice predictor is closely allied to the modeling of *elastic wave propagation* in a stratified solid medium (Thomson, 1950). In particular, in *seismic signal processing,* the structural and stratigraphic features of the subsurface are formulated as a lattice filter, (see Chapter

1.) Additionally, in *speech processing* the human vocal tract is modeled as a lattice filter, based on a cascade of acoustic tube sections (Flanagan, 1972; Markel and Gray, 1976; Rabiner and Schafer, 1978). The significance of the lattice filter model for speech is that the cross-sectional area of the vocal tract specifies certain reflection coefficients in each section of the acoustic tube, corresponding to a particular stage in the model.

The main point of this brief discussion is that neither the use of transversal filters in the LMS, RLS, and FTF algorithms nor the use of lattice filters in the recursive LSL algorithms can claim monopoly of physical relevance. Rather, these two classes of adaptive filtering algorithms are complementary, each portraying a physical significance of its own.

How to Choose an Adaptive Filter

A good note to end this summary is on a philosophical discussion of the issues involved in the selection of an adaptive filter for practical applications. Whatever the choice, clearly it has to be *cost-effective*. With this goal in mind, we may identify three important issues that require attention:

1. *Computational cost*. The overall cost of computation involved in building an adaptive filter is influenced by two major factors: computational complexity and numerical precision. The *computational complexity* refers to the number of adds and multiplies, and the amount of storage required to implement an adaptive filtering algorithm. The *numerical precision* refers to the numbers of bits (i.e., word-lengths) used in the numerical representations of data samples and filter coefficients.
2. *Performance*. In a stationary environment, the specific measures of performance are speed of convergence and misadjustment. The *speed of convergence* refers to the number of iterations required for the mean-squared (estimation) error produced by the filter to approach its asymptotic (final) value. The *misadjustment* refers to the percentage deviation of this asymptotic value from the optimum Wiener solution. These two parameters, however, pose conflicting requirements. In a nonstationary environment, we also have to consider the *tracking* of variations in the minimum mean squared error (i.e., optimum filter coefficients) due to statistical variations in the environment in which the filter operates. In general, a fast rate of convergence is no guarantee for a good tracking behavior.
3. *Robustness*. For the filter performance to be acceptable, it has to not only satisfy the above-mentioned criteria, but also be robust. This means that the filter is capable of delivering a good performance despite wide variations in the eigenvalue spread of the underlying correlation matrix of the input data, and reduced numerical precision.

The use of computer simulation provides the first step in undertaking a detailed investigation of these issues. We may begin by using the LMS algorithm as an adaptive filtering tool for the study. The LMS algorithm is relatively simple to implement. Yet, it is powerful enough to evaluate the practical benefits that may result from the application of adaptivity to the problem at hand. Moreover, it provides a practical frame of reference for assessing any further improvement that may be attained through the use of more sophisticated adaptive filtering algorithms. Finally, the study must include tests with real-life data, for which there is no substitute.

Practical applications of adaptive filtering are very diverse, with each application having peculiarities of its own. The solution for one application may not be suitable for another. Nevertheless, to be successful we have to develop a physical understanding of the environment in which the filter has to operate, and thereby relate to the realities of the application of interest.

21.2 UNEXPLORED TOPICS

The main purpose of this book has been to provide a detailed treatment of adaptive filter theory with emphasis on filters that satisfy two requirements: an impulse response of *finite* duration, and an estimate of some desired response that is a *linear* combination of input data samples. In other words, we have largely considered *linear adaptive finite-(length) impulse response (FIR) filters*. (We did deviate from this main theme of the book in Chapter 20; there we studied blind deconvolution using a zero-memory nonlinearity). The obvious advantage of this important class of adaptive filters is the mathematical tractability of their design. However, by limiting the scope of our study in this way, some relevant topics have been omitted. In the remainder of this section, we highlight three unexplored topics: (1) linear adaptive infinite-(length) impulse response (IIR) filters, (2) robust least-squares estimation and adaptive beamforming, (3) nonlinear adaptive filters and their relation to neural networks.

Adaptive Infinite Impulse Response Filters

An adaptive IIR filter consists of two basic components: a time-varying IIR filter, and a control mechanism for recursively adjusting the filter coefficients. The goal of the adaptive IIR filter is to produce an estimate of some desired response, such that a cost function (i.e., performance criterion) based on the estimation error is minimized. An IIR filter is characterized by a transfer function that has both poles and zeros. The zeros are associated with *forward paths,* whereas poles are associated with *feedback paths*. The presence of poles means that an IIR filter can meet a time-domain (or, equivalently, frequency-domain specification) with less computational cost than an FIR filter. However, the presence of poles in the transfer function of the IIR filter complicates the stability analysis of an adaptive IIR filter.

There are two fundamental approaches to adaptive IIR filtering that correspond to different formulations of the estimation error (Shynk, 1989):

1. *Equation error method*. In this approach the feedback coefficients of the IIR filter are updated in an all-zero, nonrecursive form. These coefficients are then copied to a second filter implemented in an all-pole form. Essentially, such an approach follows a line of thinking that is influenced by adaptive FIR filtering. Unfortunately, the equation error method can lead to *biased* estimates of unknown model parameters.

2. *Output error method.* In this second approach, the feedback coefficients are time-updated directly in a pole-zero recursive fashion. Unlike the equation-error method, it does *not* generate biased estimates. However, there is now the potential possibility for the algorithm to converge to a local minimum of the error performance surface leading to an incorrect estimate of the unknown model parameters.

An overview of adaptive IIR filtering is presented in Shynk (1989). More detailed treatments of the subject can be found in the books by Treichler et al. (1987) and Johnson (1988). For a feedback system approach to the adaptive IIR filter problem, see Landau (1984). Friedlander and Porat (1989) present a derivation of adaptive IIR filtering algorithms for the recursive estimation of non-Gaussian MA and ARMA stochastic processes by exploiting high-order statistics of such processes.

An important application of adaptive IIR filtering is in the design of a *predictor* used in *adaptive differential pulse-code modulation* (ADPCM). The transfer function of the predictor has six zeros and two poles; that is, there are a total of eight filter coefficients to be computed adaptively. The algorithm used for the quantizer in the ADPCM is combined with the adaptive predictor in a synchronous fashion and both the encoder and decoder parts of the system. The combined performance of these algorithms is so impressive at 32 kb/s that ADPCM is now accepted internationally as a standard coding scheme for the digital transmission of voice signals (Benvenuto et al., 1986).

Another useful application of adaptive IIR filtering is in *acoustic echo cancellers* (Murano et al., 1990). In this application, the echo duration is long, which would prohibit the use of an FIR approach. The rationale behind the IIR approach is that the echo path may be modeled better by a combination of poles and zeros. The key requirements here are twofold: (1) to guarantee stability by confining the poles to the interior of the unit circle in the z-plane, and (2) to provide an unbiased estimate of the echo path.

Macchi and Jaidane-Saidane (1989) explore the relation between adaptive IIR filtering and chaotic dynamics in the context of prediction. In particular, the path to a chaotic behavior is illustrated by increasing the adaptation (step-size) parameter. They also consider the application of this analysis to the digital coding of audio-frequency signals.

Robust Adaptive Filtering

The term "robustness" refers to insensitivity to the effect of outlying observations or small deviations from an assumed model. In order to achieve such an objective, *robust estimators* have been developed for minimizing a less increasing function of the estimation errors than the method of weighted least squares (Huber, 1981; Hampel et al., 1986). In the method of weighted least squares, the cost function is defined as the sum of weighted error squares. In robust estimation, on the other hand, the cost function is defined as the so-called *minimum function* $\tau(\beta_i^{1/2}e(i))$ summed over the observation interval $1 \le i \le n$, where $e(i)$ is the estimation error at time i, and β_i is the associated weighting factor. For convenience of presentation, let

$$x = \beta_i^{1/2}e(i)$$

Then, an important heuristic tool for choosing a proper minimum function $\tau(x)$ for designing a robust estimator is to use an *influence function,* defined as the derivative of $\tau(x)$ with respect to x. With $\tau(x) = x^2/2$ for the method of least squares, the corresponding influence function is

$$\tau'(x) = \frac{d\tau(x)}{dx} = x$$

This confirms the nonrobustness of least-squares estimates, as the "influence" of a gross error onto the least-squares estimate increases linearly with the size of the error. For a detailed discussion of robust estimators, see Huber (1981), Hampel et al. (1986), and Förstner (1989).

Another topic of practical interest in robust filtering is that of *robust adaptive beamforming,* which includes the combined use of multiple linear (equality) constraints and a quadratic (inequality) constraint. The multiple linear constraints provide *protection* for the beamforming along prescribed look directions (Frost, 1973). On the other hand, the addition of the quadratic constraint ensures that the beamformer is *robust* with respect to small amplitude, phase or position errors (Cox et al., 1987).

The robust adaptive beamforming problem may be formulated as follows:

Given the data matrix $\mathbf{A}(n)$, choose the weight vector $\mathbf{w}(n)$ so as to minimize the cost function

$$\mathscr{E}(n) = \mathbf{w}(n)\mathbf{A}^H(n)\mathbf{A}(n)\mathbf{w}(n)$$

subject to the following two forms of constraints:

1. Multiple linear constraints, described by

$$\mathbf{S}^H\mathbf{w}(n) = \mathbf{g}$$

where $\mathbf{S}$ is an M-by-K matrix with K linearly independent rows and $\mathbf{g}$ is a K-by-1 vector.

2. A quadratic constraint, described by

$$\mathbf{w}^H(n)\mathbf{B}\mathbf{B}^H\mathbf{w}(n) < \gamma^2$$

where $\mathbf{B}$ is an M-by-M and nonsingular matrix so that the matrix product $\mathbf{B}\mathbf{B}^H$ is positive definite, and γ^2 is a constant.

The solution to this problem may be implemented by an adaptive beamformer with the structure shown in Fig. 21.1 (Cox et al., 1987). According to the representation shown here, the weight vector $\mathbf{w}(n)$ may be decomposed as the sum of two orthogonal components:

1. A fixed component $\mathbf{w}_c$ that is uniquely determined by the matrices $\mathbf{S}$ and $\mathbf{B}$.
2. A variable component $\mathbf{h}(n)$ that is adaptively controlled.

The structure of Fig. 21.1 includes the *generalized sidelobe canceller* (Griffiths and Jim, 1982; Buckley and Griffiths, 1986) as a special case, for which the matrix $\mathbf{B}$ is set equal to the identity matrix.

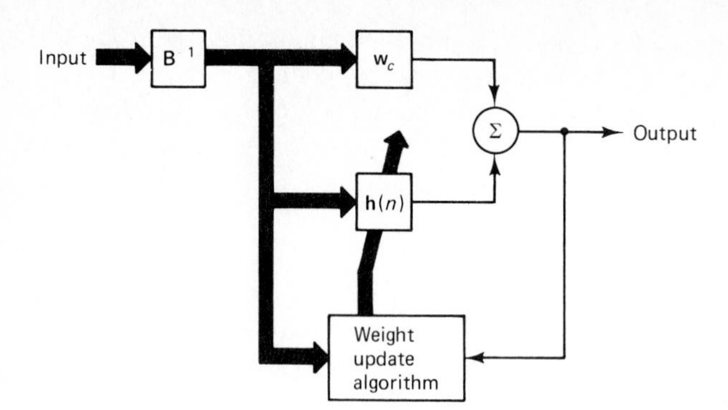

Figure 21.1 Robust adaptive beamformer.

Fertig and McClellan (1990) describe a *dual form* of the robust beamformer with norm constraints. In this form the Lagrange multipliers (arising from the use of constrained optimization) are updated at each iteration. Consequently, the dimension of the problem is equal to the number of the constraints, which is usually much less than the dimension of the weight vector. The dual algorithm appears to be more like the RLS than the LMS, even though it is derived from a gradient point of view.

Nonlinear Adaptive Filters

The theory of linear adaptive filters is based on the minimum mean-square error criterion. The Wiener filter that results from the application of such a criterion, and which represents the goal of adaptive filtering for a stationary environment, can only relate to second-order statistics of the input data and no higher. This constraint limits the ability of a linear adaptive filter to extract information from input data that are non-Gaussian. Despite its theoretical importance, the existence of Gaussian noise is open to question (Johnson and Rao, 1990). Moreover, non-Gaussian processes are quite common in signal processing applications. In a radar environment, for example, sea clutter (i.e., radar returns from the surface of an ocean) is known to be *K*-distributed; ground clutter (i.e., radar returns from a ground background) is known to be log-normal. The use of a Wiener filter or a linear adaptive filter to extract signals of interest in the presence of such non-Gaussian processes will therefore yield suboptimal solutions. We may overcome this limitation by incorporating some form of *nonlinearity* in the structure of the adaptive filter. Although by so doing, we would indeed loose the Wiener-Hopf or normal equations as a theoretical basis and so complicate the mathematical analysis, nevertheless, we would expect to benefit in two significant ways: improved learning efficiency and a broadening of application areas.

Fundamentally, two types of nonlinear adaptive filters have been identified in the literature:

Volterra-Based Nonlinear Adaptive Filters. In this type of a nonlinear adaptive filter, the nonlinearity is localized at the front end of the filter. It relies on the use of a *Volterra* series that provides an attractive method for describing the input–output relationship of a nonlinear device with memory. This special form of a series derives its name from the fact that it was first studied by Vito Volterra around 1880 as a generalization of the Taylor series of a function. But Norbert Wiener

(1958) was the first to use the Volterra series to model the input–output relationship of a nonlinear system.

Let the time series $\{x_n\}$ denote the input of a nonlinear discrete-time system. We may then combine these input samples to define a set of *discrete Volterra kernels* as follows:

$$H_0 = \text{zero-order (dc) term}$$

$$H_1[x_n] = \text{first-order (linear) term}$$

$$= \sum_i h_i x_i$$

$$H_2[x_n] = \text{second-order (quadratic) term}$$

$$= \sum_i \sum_j h_{ij} x_i x_j$$

$$H_3[x_n] = \text{three-order (cubic) term}$$

$$= \sum_i \sum_j \sum_k h_{ijk} x_i x_j x_k$$

and so on for higher-order terms. Ordinarily, the nonlinear model coefficients, the h's, are fixed by analytical methods. We may thus decompose a nonlinear adaptive filter as follows (Rayner and Lynch, 1989; Lynch and Rayner, 1989):

1. A *nonlinear Volterra state expander* that combines the set of input values $x_0, x_1, \ldots, x_n$ to produce a larger set of outputs $u_0, u_1, \ldots, u_q$ for which q is larger than n. For example, the extension vector for a (3, 2) system has the form

$$\mathbf{u} = [1, x_0, x_1, x_2, x_0^2, x_0x_1, x_0x_2, x_1x_0, x_1^2, x_1x_2, x_2x_0, x_2x_1, x_2^2]$$

2. A *linear FIR adaptive filter* that operates on the u_k as inputs to produce an estimate $\hat{d}_n$ of some desires response d_n. Hence, the nonlinear adaptive filter may be depicted as in Fig. 21.2.

For a survey paper on the Volterra series, see Schetzen (1981). For its use in adaptive filtering and a comparison with neural networks (discussed below), see Lynch and Rayner (1989) and Cowan et al, (1989). Sicuranza and Ramponi (1986)

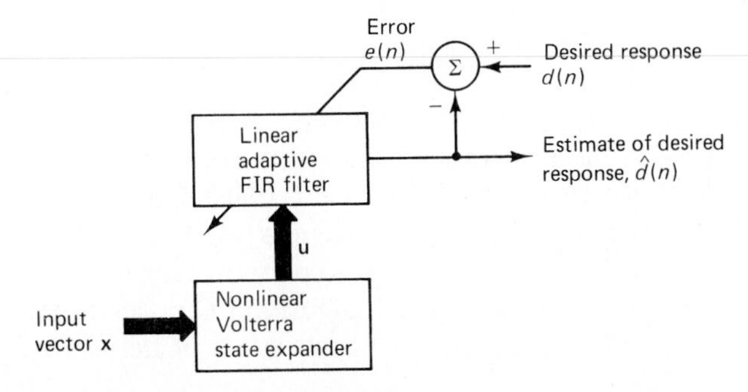

Figure 21.2 Volterra based nonlinear adaptive filter.

use the truncated discrete Volterra series to develop a memory-oriented implementation of a nonlinear adaptive filter that exploits distributed arithmetic. For additional information on the use of Volterra series in digital signal processing, see Koh and Powers (1985) and Sicuranza (1985).

Neural Networks. A *neural network*, in its general form, consists of an *input layer*, one or more *hidden layers*, and an *output layer*. The input layer is made up of *source nodes*. The hidden layer(s) and the output layer are made up of *computation nodes*, each of which is referred to as a *neuron* or *processing unit*. The sizes of these layers naturally depend on the application of interest. Each processing unit consists of a *linear combiner* followed by a *nonlinearity* as depicted in Fig. 21.3. Thus in a neural network the nonlinearity is distributed uniformly throughout the network. A popular form of nonlinearity is the *sigmoid nonlinearity*, described by the following input–output relation:

$$y = \frac{1}{1 + e^{-x}}, \qquad -\infty < x < \infty$$

where x is the input and y is the output. A neural network derives its computing power from the combined use of hidden units and nonlinearity.

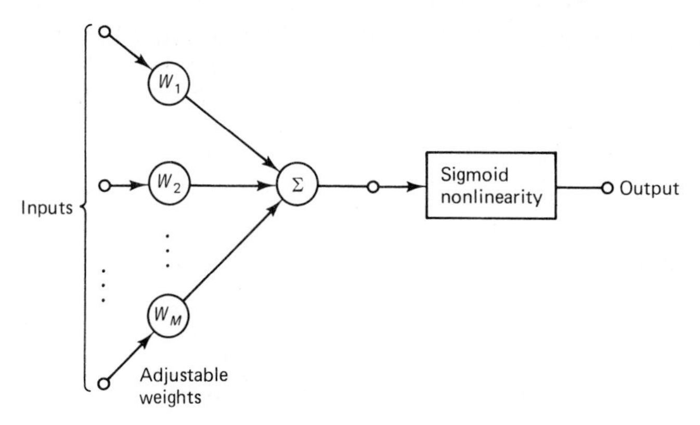

Figure 21.3 Model of neuron

In a *multilayer perceptron* type of a neural network, the input signal propagates *forward* through *synaptic connections* between a processing unit in any one layer and processing units in the layer above it. However, before the neural network is put into operation, it undergoes "training" by feeding it with prescribed inputs and comparing its output(s) against some desired response; the *synaptic weights* are then adjusted recursively by permitting the errors (measured at the network output) to propagate *backward* through the different layers of the network. A popular algorithm that is used to train the network in the manner so described is known as the *back-propagation (BP) algorithm* (Werbos, 1974; Rumelhart et al., 1986). Once the training of the neural network is completed, the synaptic weights of the network are fixed. The network may then be used in a forward mode to perform "generalization" on future test data that it has not seen before. The BP algorithm is in actual fact a generalization of the conventional LMS algorithm that applies to a single neuron.

Another attractive algorithm for the design of a neural network is that of the *radial basis function* (RBF) *algorithm* (Broomhead and Lowe, 1988; Lowe, 1989). The rationale of the RBF algorithm is that a feedforward multilayered neural network may be viewed as performing a simple *curve-fitting operation in a high-dimensional space*. According to this viewpoint, we have the following equivalences. First the operation of *learning* in a neural network is equivalent to finding a best fit surface in a high dimensional space to a finite set of data points known as the "training" set. Second, the ability of a neural network to do *generalization* is equivalent to interpolating the "test" data set on this fitting surface. An RBF network takes an N-dimensional input vector and calculates an N'-dimensional response vector by exploiting a hidden layer of N_0 units, each one of which represents a radial basis function center. The synaptic connections from the N nodes of the input layer to the N_0 nodes of the hidden layer are *nonlinear* and *fixed*. On the other hand, the synaptic connections from the N_0 nodes of the hidden layer to the N' nodes of the output layer are *linear* and *adjustable*. This means that a linear adaptive filtering algorithm may be used to perform the adaptive computation of the latter set of synaptic connections. In so doing, an impressive list of adaptive filtering algorithms (backed by a mature theoretical basis, as evidenced by the material presented in this book) is put at the disposal of neural network designers via the RBF approach.

For additional information on neural networks, see Rumelhart and McClelland (1986), Lippmann (1987), and Proc. IEEE Special Issues on Neural Networks (1990).

Complex Variables

This appendix presents a brief review of the functional theory of complex variables. In the context of the material considered in this book, a complex variable of interest is the variable z associated with the z-transform. We begin the review by defining analytic functions of a complex variable, and then derive the important theorems that make up the important subject of complex variables. For a more detailed treatment of the subject, the reader is referred to Levinson and Redheffer (1970), Wylie (1966), Sokolnikoff and Redheffer (1966), Pipes (1958), and Guillemin (1949).

A.1 CAUCHY–RIEMANN EQUATIONS

Consider a complex variable z defined by

$$z = x + jy$$

where $x = \text{Re}[z]$, and $y = \text{Im}[z]$. We speak of the plane in which the complex variable z is represented as the *z-plane*. Let $f(z)$ denote a *function of the complex variable* z, written as

$$w = f(z) = u + jv$$

The function $w = f(z)$ is *single-valued* if there is only one value of w for each z in a given region of the z-plane. If, on the other hand, more than one value of w corresponds to z, the function $w = f(z)$ is said to be *multiple-valued*.

We say that a point $z = x + jy$ in the z-plane approaches a fixed point $z_0 = x_0 + jy_0$ if $x \to x_0$ and $y \to y_0$. Let $f(z)$ denote a single-valued function of z that is defined in some neighborhood of the point $z = z_0$. The *neighborhood* of z_0 refers to the set of all points in a sufficiently small circular region centered at z_0. Let

$$\lim_{z \to z_0} f(z) = w_0$$

In particular, if $f(z_0) = w_0$, then the function $f(z)$ is said to be *continuous* at $z = z_0$.

Let $f(z)$ be written in terms of its real and imaginary parts as

$$f(z) = u(x, y) + jv(x, y)$$

Then, if $f(z)$ is continuous at $z_0 = x_0 + jy_0$, its real and imaginary parts $u(x, y)$ and $v(x, y)$ are continuous functions at (x_0, y_0), and vice versa.

Let $w = f(z)$ be continuous at each point of some region of interest in the z-plane. The complex quantities w and z may then be represented on separate planes of their own, referred to as the w- and z-planes, respectively. In particular, a point (x, y) in the z-plane corresponds to a point (u, v) in the w-plane by virtue of the relationship $w = f(z)$.

Consider an incremental change Δz such that the point $z_0 + \Delta z$ may lie anywhere in the neighborhood of z_0, and throughout which the function $f(z)$ is defined. We may then define the *derivative* of $f(z)$ with respect to z at $z = z_0$ as

$$f'(z_0) = \lim_{\Delta z \to 0} \frac{f(z_0 + \Delta z) - f(z_0)}{\Delta z} \tag{A.1}$$

Clearly, for the derivative $f'(z_0)$ to have a unique value, the limit in Eq. (A.1) must be independent of the way in which Δz approaches zero.

For a function $f(z)$ to have a unique derivative at some point $z = x + jy$, it is necessary that its real and imaginary parts satisfy certain conditions, as shown next. Let

$$w = f(z) = u(x, y) + jv(x, y)$$

With $\Delta w = \Delta u + j\Delta v$ and $\Delta z = \Delta x + j\Delta y$, we may write

$$\begin{aligned} f'(z) &= \lim_{\Delta z \to 0} \frac{\Delta w}{\Delta z} \\ &= \lim_{\substack{\Delta x \to 0 \\ \Delta y \to 0}} \frac{\Delta u + j\,\Delta v}{\Delta x + j\,\Delta y} \end{aligned} \tag{A.2}$$

Suppose that we let $\Delta z \to 0$ by first letting $\Delta y \to 0$ and then $\Delta x \to 0$, in which case Δz is purely real. We then deduce from Eq. (A.2) that

$$\begin{aligned} f'(z) &= \lim_{\Delta x \to 0} \frac{\Delta u}{\Delta x} + j\frac{\Delta v}{\Delta x} \\ &= \frac{\partial u}{\partial x} + j\frac{\partial v}{\partial x} \end{aligned} \tag{A.3}$$

Suppose next that we let $\Delta z \to 0$ by first letting $\Delta x \to 0$ and then $\Delta y \to 0$, in which case Δz is purely imaginary. This time we deduce from Eq. (A.2) that

$$\begin{aligned} f'(z) &= \lim_{\Delta y \to 0} \frac{\Delta v}{\Delta y} - j\frac{\Delta u}{\Delta y} \\ &= \frac{\partial v}{\partial y} - j\frac{\partial u}{\partial y} \end{aligned} \tag{A.4}$$

If the derivative $f'(z)$ is to exist, it is necessary that the two expressions in Eqs. (A.3) and (A.4) be one and the same. Hence, we have

$$\frac{\partial u}{\partial x} + j\frac{\partial v}{\partial x} = \frac{\partial v}{\partial y} - j\frac{\partial u}{\partial y}$$

Accordingly, equating real and imaginary parts, we get the following pair of relations, respectively:

$$\frac{\partial u}{\partial x} = \frac{\partial v}{\partial y} \tag{A.5}$$

$$\frac{\partial v}{\partial x} = -\frac{\partial u}{\partial y} \tag{A.6}$$

Equations (A.5) and (A.6), known as the *Cauchy–Riemann equations,* were derived from a consideration of merely two of the infinitely many ways in which Δz can approach zero. For $\Delta w/\Delta z$ evaluated along these other paths to also approach $f'(z)$, we need only make the additional requirement that the partial derivatives in Eqs. (A.5) and (A.6) are continuous at the point (x, y). In other words, provided that the real part $u(x, y)$ and the imaginary part $v(x, y)$ together with their first partial derivatives are continuous at the point (x, y), the Cauchy–Riemann equations are not only necessary but also sufficient for the existence of a derivative of the complex function $w = u(x, y) + jv(x, y)$ at the point (x, y).

A function $f(z)$ is said to be *analytic,* or *homomorphic,* at some point $z = z_0$ in the z-plane if it has derivative at $z = z_0$ and at every point in the neighborhood of z_0; the point z_0 is called a *regular point* of the function $f(z)$. If the function $f(z)$ is *not* analytic at a point z_0, but if every neighborhood of z_0 contains points at which $f(z)$ *is* analytic, the point z_0 is referred to as a *singular point* of $f(z)$.

A.2 CAUCHY'S INTEGRAL FORMULA

Let $f(z)$ be any continuous function of the complex variable z, analytic or otherwise. Let C be a sectionally smooth path joining the points $A = z_0$ and $B = z_n$ in the z plane. Suppose that the path C is divided into n segments Δs_k by the points z_k, $k = 1, 2, \ldots, n - 1$, as illustrated in Fig. A.1. This figure also shows an arbitrary point ζ_k on segment Δs_k, depicted as an elementary arc of length Δz_k. Consider then the summation $\sum_{k=1}^{n} f(\zeta_k)\,\Delta z_k$. The *line integral* of $f(z)$ along the path C is defined by the limiting value of this summation as the number n of segments is allowed to increase indefinitely in such a way that Δz_k approaches zero. That is

$$\oint_C f(z)\,dz = \lim_{n\to\infty} \sum_{k=1}^{n} f(\zeta_k)\,\Delta z_k \tag{A.7}$$

In the special case when the points A and B coincide and C is a closed curve, the integral in Eq. (A.7) is referred to as a *contour integral* that is written as $\oint_C f(z)\,dz$. Note that, according to the notation described herein, the contour C is tranversed in a *counterclockwise direction*.

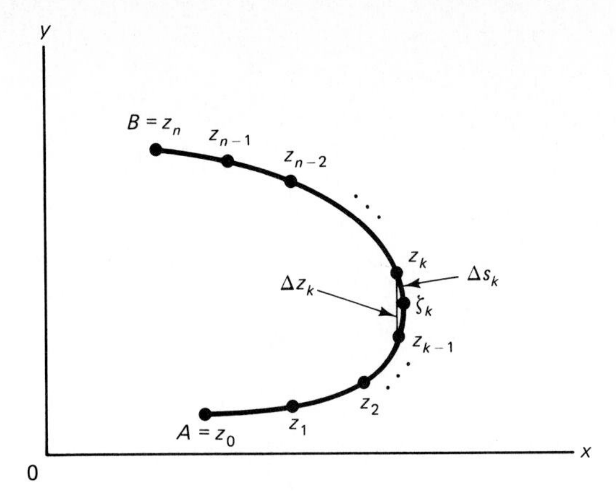

Figure A.1 Sectionally smooth path.

Let $f(z)$ be an analytic function in a given region R, and let the derivative $f'(z)$ be continuous there. The line integral $\oint_C f(z)\, dz$ is then independent of the path C that joins any pair of points in the region R. If the path C is a closed curve, the value of this integral is zero. We thus have *Cauchy's integral theorem,* stated as follows:

> If the function $f(z)$ is analytic throughout a region R, then the contour integral of $f(z)$ along any closed path C lying inside the region R is zero, as shown by
>
> $$\oint_C f(z)\, dz = 0 \tag{A.8}$$

This theorem is of cardinal importance in the study of analytic functions.

An important consequence of Cauchy's theorem is known as *Cauchy's integral formula.* Let $f(z)$ be analytic within and on the boundary C of a simple connected region. Let z_0 be any point in the interior of C. Then Cauchy's integral formula states that

$$f(z_0) = \frac{1}{2\pi j} \oint_C \frac{f(z)}{z - z_0}\, dz \tag{A.9}$$

where the contour integration around C is taken in the counterclockwise direction.

A.3 LAURENT'S SERIES

Let the function $f(z)$ be analytic in the annular region of Fig. A.2, including the boundary of the region. The annular region consists of two concentric circles C_1 and C_2, whose common center is z_0. Let the point $z = z_0 + h$ be located inside the annular region as depicted in Fig. A.2. Accordingly to *Laurent's series,* we have

$$f(z_0 + h) = \sum_{k=-\infty}^{\infty} a_k h^k \tag{A.10}$$

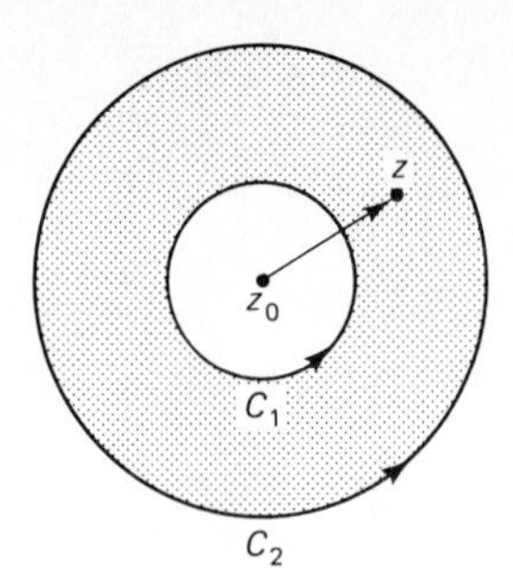

Figure A.2 Annular region.

where the coefficients a_k are given by

$$a_k = \begin{cases} \dfrac{1}{2\pi j}\displaystyle\oint_{C_2} \frac{f(z)\,dz}{(z - z_0)^{k+1}}, & k = 0, +1, 2, \ldots \\[2ex] \dfrac{1}{2\pi j}\displaystyle\oint_{C_1} \frac{f(z)\,dz}{(z - z_0)^{k+1}}, & k = -1, -2, \ldots \end{cases} \tag{A.11}$$

Note that we may also express the Laurent expansion of $f(z)$ around the point z as

$$f(z) = \sum_{k=-\infty}^{\infty} a_k(z - z_0)^k \tag{A.12}$$

When all the coefficients of negative index have the value zero, then Eq. (A.12) reduces to *Taylor's series*.

A.4 SINGULARITIES AND RESIDUES

Let $z = z_0$ be a singular point of an analytic function $f(z)$. If the neighborhood of $z = z_0$ contains *no* other singular points of $f(z)$, the singularity at $z = z_0$ is said to be *isolated*. In the neighborhood of such a singularity, the function $f(z)$ may be represented by the *Laurent series*

$$\begin{aligned} f(z) &= \sum_{k=-\infty}^{\infty} a_k(z - z_0)^k \\ &= \sum_{k=0}^{\infty} a_k(z - z_0)^k + \sum_{k=-\infty}^{-1} a_k(z - z_0)^k \\ &= \sum_{k=0}^{\infty} a_k(z - z_0)^k + \sum_{k=1}^{\infty} \frac{a_{-k}}{(z - z_0)^{-k}} \end{aligned} \tag{A.13}$$

The particular coefficient a_{-1} in the Laurent expansion of $f(z)$ in the neighborhood of the isolated singularity at the point $z = z_0$ is called the *residue* of $f(z)$ at $z = a$. The residue plays an important role in the evaluation of integrals of analytic functions. In particular, putting $k = -1$ in Eq. (A.11) we get the following connection between the residue a_{-1} and the integral of the function $f(z)$:

$$a_{-1} = \frac{1}{2\pi j}\oint_C f(z)\,dz \tag{A.14}$$

There are two nontrivial cases to be considered:

1. The Laurent expansion of $f(z)$ contains *infinitely* many terms with negative powers of $z - z_0$, as in Eq. (A.13). The point $z - z_0$ is then called an *essential singular point* of $f(z)$.
2. The Laurent expansion of $f(z)$ contains at most a *finite* number m of terms with negative powers of $z - z_0$, as shown by

$$f(z) = \sum_{k=0}^{\infty} a_k(z - z_0)^k + \frac{a_{-1}}{z - z_0} + \frac{a_{-2}}{(z - z_0)^2} + \cdots + \frac{a_{-m}}{(z - z_0)^m} \tag{A.15}$$

According to this representation, $f(z)$ is said to have a *pole of order m* at $z = z_0$. The *finite sum* of all the terms containing *negative powers* on the right side of Eq. (A.14) is called the *principal part* of $f(z)$ at $z = z_0$.

Note that when the singularity at $z = z_0$ is a pole of order m, the residue of the pole may be determined by using the formula

$$a_{-1} = \frac{1}{(m-1)!} \frac{d^{m-1}}{dz^{m-1}} [(z - a)^m f(z)]_{z=z_0} \tag{A.16}$$

In effect, by using this formula we avoid the need for the deduction of the Laurent series. For the special case when the order $m = 1$, the pole is said to be *simple*. Correspondingly, the formula of Eq. (A.16) for the residue a_{-1} of a simple pole reduces to

$$a_{-1} = \lim_{z \to z_0} (z - z_0) f(z) \tag{A.17}$$

A.5 CAUCHY'S RESIDUE THEOREM

Consider a closed contour C in the z-plane containing within it a number of isolated singularities of some function $f(z)$. Let $z_1, z_2, \ldots, z_n$ define the locations of these isolated singularities. Around each singular point of the function $f(z)$, we draw a circle small enough to ensure that it does not enclose the other singular points of $f(z)$, as depicted in Fig. A.3. The original contour C together with these small circles constitute the boundary of a *multiply connected region* in which $f(z)$ is analytic everywhere, and to which Cauchy's integral theorem may therefore be applied. Specifically, for the situation described in Fig. A.3 we may write

$$\frac{1}{2\pi j}\oint_C f(z)\,dz + \frac{1}{2\pi j}\oint_{C_1} f(z)\,dz + \cdots + \frac{1}{2\pi j}\oint_{C_n} f(z)\,dz = 0 \tag{A.18}$$

Note that in Fig. A.3 the contour C is traversed in the *positive* sense (i.e., counterclockwise direction) whereas the small circles are traversed in the *negative* sense (i.e., clockwise direction).

Suppose now we *reverse* the direction along which the integral around each small circle in Fig. A.3 is taken. This operation has the equivalent effect of applying a minus sign to each of the integrals in Eq. (A.18) that involve the small circles $C_1, \ldots, C_n$. Accordingly, for the case when *all* the integrals around the original contour C and the small circles $C_1, \ldots, C_n$ are taken in the counterclockwise direc-

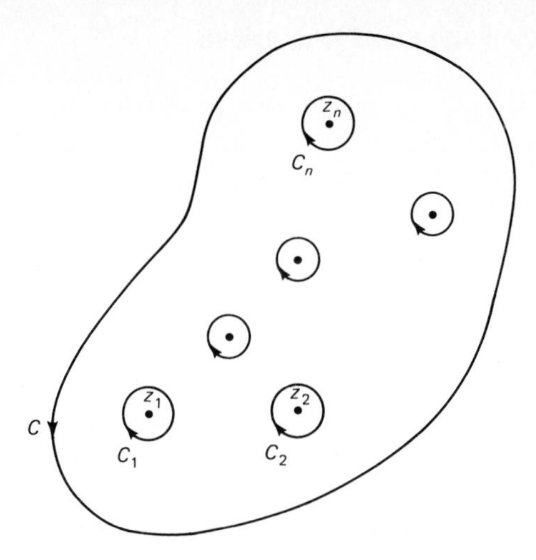

Figure A.3 Multiply connected region.

tion, we may rewrite Eq. A.18) as

$$\frac{1}{2\pi j}\oint_C f(z)\,dz = \frac{1}{2\pi j}\oint_{C_1} f(z)\,dz + \cdots + \frac{1}{2\pi j}\oint_{C_n} f(z)\,dz \qquad \text{(A.19)}$$

By definition, the integrals on the right side of Eq. (A.19) are the residues of the function $f(z)$ evaluated at the various isolated singularities of $f(z)$ within the contour C. We may thus express the integral of $f(z)$ around the contour C simply as

$$\oint_C f(z)\,dz = 2\pi j \sum_{k=1}^{n} \text{Res}(\,f(z), z_k) \qquad \text{(A.20)}$$

where Res($f(z)$, z_k) stands for the residue of the function $f(z)$ evaluated at the isolated singular point $z = z_k$. Equation (A.20) is called *Cauchy's residue theorem*. This theorem is extremely important in the theory of functions in general and in evaluating definite integrals in particular.

A.6 PRINCIPLE OF THE ARGUMENT

Consider a complex function $f(z)$, characterized as follows:

1. The function $f(z)$ is analytic in the interior of a closed contour C in the z-plane, except at a finite number of poles.
2. The function $f(z)$ has neither poles nor zeros on the contour C. By a "zero" we mean a point in the z-plane at which $f(z) = 0$. In contrast, at a "pole" as defined previously, we have $f(z) = \infty$. Let N be the *number of zeros* and P be the *number of poles* of the function $f(z)$ in the interior of contour C, where each zero or pole is counted according to its *multiplicity*. We may then state the following theorem (Levinson and Redheffer, 1970; Wylie, 1966):

$$\frac{1}{2\pi j}\oint_C \frac{f'(z)}{f(z)}\,dz = N - P \qquad \text{(A.21)}$$

where $f'(z)$ is the derivative of $f(z)$. We note that

$$\frac{d}{dz} \ln f(z) = \frac{f'(z)}{f(z)} dz$$

where ln denotes the natural logorithm. Hence,

$$\begin{aligned} \oint_C \frac{f'(z)}{f(z)} dz &= \ln f(z)|_C \\ &= \ln |f(z)|_C + j \arg f(z)|_C \end{aligned} \tag{A.22}$$

where $|f(z)|$ denotes the magnitude of $f(z)$, and $\arg f(z)$ denotes its argument. The first term on the right side of Eq. (A.22) is zero, since the logarithmic function $\ln f(z)$ is single-valued and the contour C is closed. Hence,

$$\oint_C \frac{f'(z)}{f(z)} dz = j \arg f(z)|_C \tag{A.23}$$

Thus, substituting Eq. (A.23) in (A.21), we get

$$N - P = \frac{1}{2\pi} \arg f(z)|_C \tag{A.24}$$

This result, which is a reformulation of the theorem described in Eq. (A.21), is called the *principle of the argument*.

For a geometrical interpretation of this principle, let C be a closed contour in the z-plane as in Fig. A.4(a). As z traverses the contour C, we find that $w = f(z)$ traces out a contour C' of its own in the w-plane; for the purpose of illustration, C' is shown in Fig. A.4(b). Suppose now a line is drawn in the w-plane from the origin to the point $w = f(z)$, as depicted in Fig. A.4(b). Then the angle θ which this line makes with a fixed direction (shown as the horizontal direction in Fig. A.4(b) is $\arg f(z)$. The principle of the argument thus provides a description of the number of times the point $w = f(z)$ winds around the origin of the w-plane (i.e., the point $w = 0$) as the complex variable z traverses the contour C.

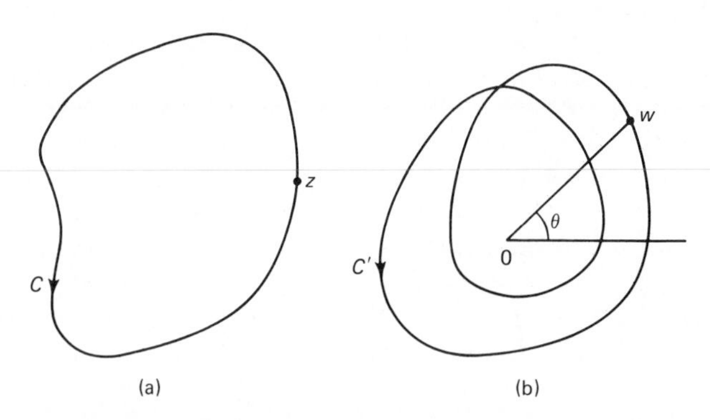

Figure A.4 (a) Contour C in the z-plane, (b) Contour C' in the w-plane, where $w = f(z)$.

Rouché's Theorem

Let the function $f(z)$ be analytic on a closed contour C and in the interior of C. Let $g(z)$ be a second function which, in addition to satisfying the same condition for analyticity as $f(z)$, also fulfills the following condition on the contour C:

$$|f(z)| > |g(z)|$$

In other words, on the contour C we have

$$\left|\frac{g(z)}{f(z)}\right| < 1 \tag{A.25}$$

Define the function

$$F(z) = 1 + \frac{g(z)}{f(z)} \tag{A.26}$$

which has no poles or zeros on C. By the principle of the argument applied to $F(z)$, we have

$$N - P = \frac{1}{2\pi} \arg F(z)|_C \tag{A.27}$$

However, the implication of the condition (A.25) is that when z is on the contour C, then

$$|F(z) - 1| < 1 \tag{A.28}$$

In other words, the point $w - F(z)$ lies inside a circle with center at $w = 1$ and unit radius, as illustrated in Fig. A.5. It follows therefore that

$$|\arg F(z)| < \frac{\pi}{2}, \qquad \text{for } z \text{ on } C \tag{A.9}$$

Equivalently, we may write

$$\arg F(z)|_C = 0 \tag{A.30}$$

Hence, from Eq. (A.24) we deduce that $N = P$, where both N and P refer to $f(z)$. From the definition of the function $F(z)$ given in Eq. (A.26) we note that the poles of $F(z)$ are the zeros of $f(z)$, and the zeros of $F(z)$ are the zeros of the sum $f(z) + g(z)$. Accordingly, the fact that $N = P$ means that $f(z) + g(z)$ and $f(z)$ have

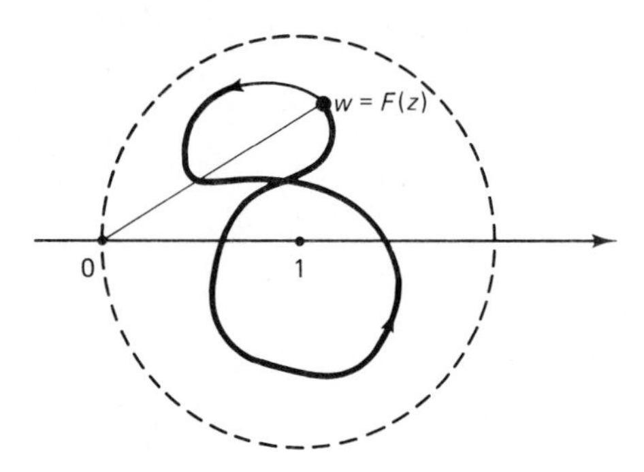

Figure A.5 Point w-$f(z)$ on a closed contour inside the unit circle.

the same numbers of zeros. The result that we have just established is known as *Rouché's theorem,* which may be formally stated as follows:

> Let $f(z)$ and $g(z)$ be analytic on a closed contour C and in the interior of C. Let $|f(z)| > |g(z)|$ on C. Then $f(z)$ and $f(z) + g(z)$ have the same number of zeros inside contour C.

Example

Consider the contour depicted in Fig. A.6 that constitutes the boundary of a multiply connected region in the z-plane. Let $F(z)$ and $G(z)$ be two polynomials in z^{-1}, both of which are analytic on this contour and in the interior of it. Moreover, Let $|F(z)| > |G(z)|$. Then, according to Rouché's theorem, both $F(z)$ and $F(z) + G(z)$ have the same number of zeros inside the contour described in Fig. A.6.

Suppose now that we let the radius R of the outside circle C approach infinity. Also, let the separation l between the two straight-line portions of the contour approach zero. Then, in the limit, the region enclosed by the contour described in Fig. A.6 will be made up of the entire area that lies *outside* the inner circle C_1. In other words, the polynomials $F(z)$ and $F(z) + G(z)$ have the same number of zeros outside the circle C_1, under the conditions described herein. Note that the circle C_1 is traversed in the clockwise direction.

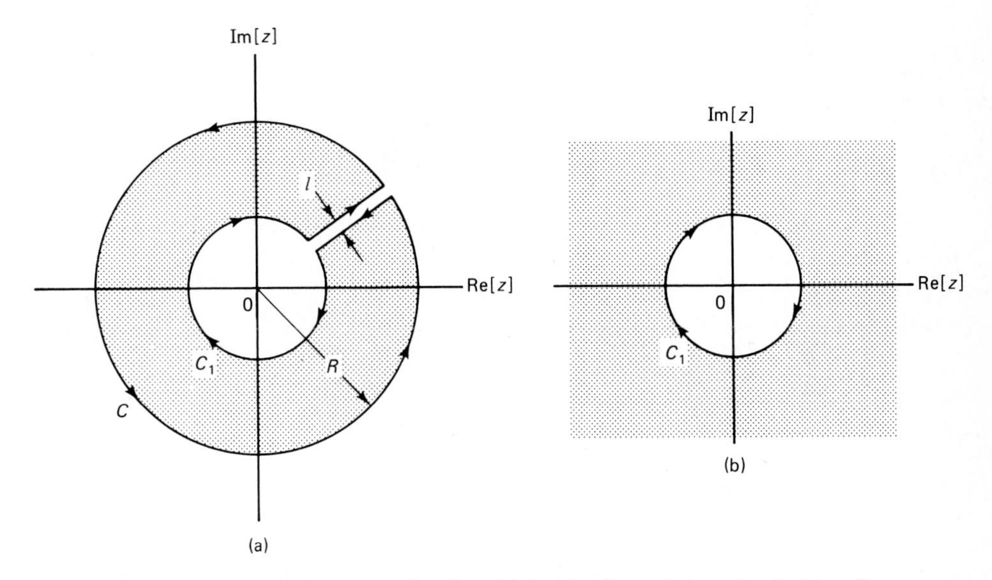

Figure A.6 (a) Multiply connected region; (b) limiting form of the region in (a) as $R \to \infty$ and $1 \to 0$.

A.7 INVERSION INTEGRAL FOR THE z-TRANSFORM

The material presented in Sections A.1 through A.6 is applicable to functions of a complex variable in general. In this section and the next one, we consider the special case of a complex function defined as the z-transform of a sequence of samples taken in time.

Let $X(z)$ denote the z-transform of a sequence $\{x(n)\}$, which converges to an analytic function in the annular domain $R_1 < |z| < R_2$. By definition, $X(z)$ is written as the Laurent series

$$X(z) = \sum_{m=-\infty}^{\infty} x(m)z^{-m}, \qquad R_1 < |z| < R_2 \tag{A.31}$$

where, for the convenience of presentation, we have used m in place of n as the index of time. Let C be a closed contour that lies inside the *region of convergence* $R_1 < |z| < R_2$. Then, multiplying both sides of Eq. (A.31) by z^{n-1}, integrating around the contour C in a counterclockwise direction, and interchanging the order of integration and summation, we get

$$\frac{1}{2\pi j}\oint_C X(z)z^n \frac{dz}{z} = \sum_{m=-\infty}^{\infty} x(n) \frac{1}{2\pi j}\oint_C z^{n-m}\frac{dz}{z} \tag{A.32}$$

The interchange of integration and summation is justified because the Laurent series that defines $X(z)$ converges uniformly on C. Let

$$z = Re^{j\theta}, \qquad R_1 < r < R_2 \tag{A.33}$$

Hence

$$z^{n-m} = r^{n-m}e^{j(n-m)\theta}$$

and

$$\frac{dz}{z} = j\, d\theta$$

Correspondingly, we may express the contour integral on the right side of Eq. (A.32) as

$$\begin{aligned}\frac{1}{2\pi j}\oint_C z^{n-m}\frac{dz}{z} &= \frac{1}{2\pi}\int_0^{2\pi} r^{n-m}e^{j(n-m)\theta}\, d\theta \\ &= \begin{cases}1, & n = m \\ 0, & \text{otherwise}\end{cases}\end{aligned} \tag{A.34}$$

Inserting Eq. (A.34) in (A.32), we get

$$x(n) = \frac{1}{2\pi j}\oint_C X(z)z^n\frac{dz}{z} \tag{A.35}$$

Equation (A.35) is called the *z-transform inversion integral formula*.

A.8 PARSEVAL'S THEOREM

Let $X(z)$ denote the z-transform of the sequence $\{x(n)\}$ with the region of convergence $R_{1x} < |z| < R_{2x}$. Let $Y(z)$ denote the z-transform of a second sequence $\{y(n)\}$ with the region of convergence $R_{1y} < |z| < R_{2y}$. Then *Parseval's theorem* states that

$$\sum_{n=-\infty}^{\infty} x(n)y^*(n) = \frac{1}{2\pi j} \oint_C X(z)Y^*\left(\frac{1}{z^*}\right) \frac{dz}{z} \qquad \text{(A.36)}$$

where C is a closed contour defined in the overlap of the regions of convergence of $X(z)$ and $Y(z)$, both of which are analytic. The function $Y^*(1/z^*)$ is obtained from the z-transform $Y(z)$ by using $1/z^*$ in place of z, and then complex-conjugating the resulting function. Note that the function $Y^*(1/z^*)$ obtained in this way is analytic.

To prove Parseval's theorem, we use the inversion integral of Eq. (A.35) to write

$$\begin{aligned} \sum_{n=-\infty}^{\infty} x(n)y^*(n) &= \frac{1}{2\pi j} \sum_{n=-\infty}^{\infty} y^*(n) \oint_C X(z)z^n \frac{dz}{z} \\ &= \frac{1}{2\pi j} \oint_C X(z) \sum_{n=-\infty}^{\infty} y^*(n)z^n \frac{dz}{z} \end{aligned} \qquad \text{(A.37)}$$

From the definition of the z-transform of $\{y(n)\}$, namely,

$$Y(z) = \sum_{n=-\infty}^{\infty} y(n)z^{-n}$$

we note that

$$Y^*\left(\frac{1}{z^*}\right) = \sum_{n=-\infty}^{\infty} y^*(n)z^n \qquad \text{(A.38)}$$

Hence, using Eq. (A.38) in (A.37), we get the result given in Eq. (A.36), and the proof of Parseval's theorem is completed.

B

Differentiation with Respect to a Vector

An issue commonly encountered in the study of optimization theory is that of differentiating a cost function with respect to a parameter vector of interest. In the text we used an ordinary gradient operation. The purpose of Appendix B is to address the more difficult issue of differentiating a cost function with respect to a complex-valued parameter vector. We begin by introducing some basic definitions.

B.1 BASIC DEFINITIONS

Consider a complex function $f(\mathbf{w})$ that is dependent on a parameter vector $\mathbf{w}$. When $\mathbf{w}$ is complex valued, there are two different mathematical concepts that require individual attention: (1) the vector nature of $\mathbf{w}$, and (2) the fact that each element of $\mathbf{w}$ is a complex number.

Dealing with the issue of complex numbers first, let x_k and y_k denote the real and imaginary parts of the kth element w_k of the vector $\mathbf{w}$; that is,

$$w_k = x_k + jy_k \tag{B.1}$$

We thus have a function on the real quantities x_k and y_k. Hence, we may use Eq. (B.1) to express the real part x_k in terms of the pair of *complex conjugate coordinates* w_k and w_k^* as

$$x_k = \tfrac{1}{2}(w_k + w_k^*) \tag{B.2}$$

and express the imaginary part y_k as

$$y_k = \frac{1}{2j}(w_k - w_k^*) \tag{B.3}$$

where the asterisk denotes complex conjugation. The real quantities x_k and y_k are functions of both w_k and w_k^*. It is only when we deal with analytic functions f that we are permitted to abandon the complex-conjugated term w_k^* by virtue of the Cauchy–Riemann equations. However, most functions encountered in physical sciences and engineering are *not* analytic.

The notion of a derivative must tie in with the concept of a differential. In particular, the chain rule of changes of variables must be obeyed. With these important points in mind, we may define certain complex derivatives in terms of real derivatives, as shown by (Schwartz, 1967)

$$\frac{\partial}{\partial w_k} = \frac{1}{2}\left(\frac{\partial}{\partial x_k} - j\frac{\partial}{\partial y_k}\right) \tag{B.4}$$

and

$$\frac{\partial}{\partial w_k^*} = \frac{1}{2}\left(\frac{\partial}{\partial_{xk}} + j\frac{\partial}{\partial y_k}\right) \tag{B.5}$$

The derivatives defined herein satisfy the following two basic requirements:

$$\frac{\partial w_k}{\partial w_k} = 1$$

$$\frac{\partial w_k}{\partial w_k^*} = \frac{\partial w_k^*}{\partial w_k} = 0$$

(An analytic function f satisfies $\partial f / \partial z^* = 0$ everywhere.)

The next issue to be considered is that of differentiation with respect to a vector. Let $w_0, w_1, \ldots, w_{M-1}$ denote the elements of an M-by-1 complex vector $\mathbf{w}$. We may extend the use of Eqs. (B.4) and (B.5) to deal with this new situation by writing (Miller, 1974):

$$\frac{\partial}{\partial \mathbf{w}} = \frac{1}{2}\begin{bmatrix} \dfrac{\partial}{\partial x_0} - j\dfrac{\partial}{\partial y_0} \\ \dfrac{\partial}{\partial x_1} - j\dfrac{\partial}{\partial y_1} \\ \vdots \\ \dfrac{\partial}{\partial x_{M-1}} - j\dfrac{\partial}{\partial y_{M-1}} \end{bmatrix} \tag{B.6}$$

and

$$\frac{\partial}{\partial \mathbf{w}^*} = \frac{1}{2}\begin{bmatrix} \dfrac{\partial}{\partial x_0} + j\dfrac{\partial}{\partial y_0} \\ \dfrac{\partial}{\partial x_1} + j\dfrac{\partial}{\partial y_1} \\ \vdots \\ \dfrac{\partial}{\partial x_{M-1}} + j\dfrac{\partial}{\partial y_{M-1}} \end{bmatrix} \tag{B.7}$$

where we have $w_k = x_k + jy_k$, for $k = 0, 1, \ldots, M - 1$. We refer to $\partial / \partial \mathbf{w}$ as a *derivative* with respect to the vector $\mathbf{w}$, and to $\partial/\partial \mathbf{w}^*$ as a *conjugate derivative* also with respect to the vector $\mathbf{w}$. These two derivatives must be considered together. They obey the following relations:

$$\frac{\partial \mathbf{w}}{\partial \mathbf{w}} = \mathbf{I}$$

and

$$\frac{\partial \mathbf{w}}{\partial \mathbf{w}^*} = \frac{\partial \mathbf{w}^*}{\partial \mathbf{w}} = \mathbf{0}$$

where $\mathbf{I}$ is the identity matrix and $\mathbf{0}$ is the null matrix.

For subsequent use, we will adopt the definition of (B.7) as the derivative with respect to a complex-valued vector.

B.2 EXAMPLES

In this section, we illustrate some applications of the derivative defined in Eq. (B.7). The examples are taken from Chapter 5, dealing with optimum linear filtering, and Chapter 10, dealing with the method of least squares.

Example 1

Let $\mathbf{p}$ and $\mathbf{w}$ denote two complex-valued M-by-1 vectors. There are two inner products, $\mathbf{p}^H\mathbf{w}$ and $\mathbf{w}^H\mathbf{p}$, to be considered.

Let $c_1 = \mathbf{p}^H\mathbf{w}$. The conjugate derivative of c_1 with respect to the vector $\mathbf{w}$ is

$$\frac{\partial c_1}{\partial \mathbf{w}^*} = \frac{\partial}{\partial \mathbf{w}^*}(\mathbf{p}^H\mathbf{w}) = \mathbf{0} \tag{B.8}$$

where $\mathbf{0}$ is the null vector. Here we note that $\mathbf{p}^H\mathbf{w}$ is an analytic function; see Problem 1 of Chapter 5. We therefore find that the derivative of $\mathbf{p}^H\mathbf{w}$ with respect to $\mathbf{w}$ is zero, in agreement with Eq. (B.8).

Consider next $c_2 = \mathbf{w}^H\mathbf{p}$. The conjugate derivative of c_2 with respect to $\mathbf{w}$ is

$$\frac{\partial c_2}{\partial \mathbf{w}^*} = \frac{\partial}{\partial \mathbf{w}^*}(\mathbf{w}^H\mathbf{p}) = \frac{\partial}{\partial \mathbf{w}^*}(\mathbf{p}^T\mathbf{w}^*) = \mathbf{p} \tag{B.9}$$

Here we note that $\mathbf{w}^H\mathbf{p}$ is not an analytic function; see Problem 1 of Chapter 5. Hence, the derivative of $\mathbf{w}^H\mathbf{p}$ with respect to $\mathbf{w}^*$ is nonzero, as in Eq. (B.9).

Example 2

Consider next the quadratic form

$$c = \mathbf{w}^H\mathbf{R}\mathbf{w})$$

where $\mathbf{R}$ is a Hermitian matrix. The conjugate derivative of c (which is real) with respect to $\mathbf{w}$ is

$$\begin{aligned}\frac{\partial c}{\partial \mathbf{w}^*} &= \frac{\partial}{\partial \mathbf{w}^*}(\mathbf{w}^H\mathbf{R}\mathbf{w}) \\ &= \mathbf{R}\mathbf{w}\end{aligned} \tag{B.10}$$

Example 3

Consider the real cost function (see Chapter 5)

$$J(\mathbf{w}) = \sigma_d^2 - \mathbf{w}^H\mathbf{p} - \mathbf{p}^H\mathbf{w} + \mathbf{w}^H\mathbf{R}\mathbf{w}$$

Using the results of Examples 1 and 2, we find that the conjugate derivative of J with respect to the tap-weight vector $\mathbf{w}$ is

$$\frac{\partial J}{\partial \mathbf{w}^*} = -\mathbf{p} + \mathbf{R}\mathbf{w} \tag{B.11}$$

Let $\mathbf{w}_o$ be the optimum value of the tap-weight vector $\mathbf{w}$ for which the cost function J is minimum or, equivalently, the derivative $(\partial J/\sigma \mathbf{w}^*) = \mathbf{0}$. Hence, from Eq. (B.11) we deduce that

$$\mathbf{R}\mathbf{w}_o = \mathbf{p} \tag{B.12}$$

This is the matrix form of the Wiener-Hopf equations for a transversal filter operating in a stationary environment.

Example 4

Consider the real log-likelihood function (see Chapter 9)

$$l(\tilde{\mathbf{w}}) = F - \frac{1}{\sigma^2}\epsilon^H\epsilon \tag{B.13}$$

where F is a constant and

$$\epsilon = \mathbf{b} - \mathbf{A}\tilde{\mathbf{w}} \tag{B.14}$$

Substituting Eq. (B.14) in (B.13), we get

$$l(\tilde{\mathbf{w}}) = F - \frac{1}{\sigma^2}\mathbf{b}^H\mathbf{b} + \frac{1}{\sigma^2}\mathbf{b}^H\mathbf{A}\tilde{\mathbf{w}} + \frac{1}{\sigma^2}\tilde{\mathbf{w}}^H\mathbf{A}^H\mathbf{b} - \frac{1}{\sigma^2}\mathbf{w}^H\mathbf{A}^H\mathbf{A}\tilde{\mathbf{w}} \tag{B.15}$$

Evaluating the conjugate derivative of l with respect to $\tilde{\mathbf{w}}$, and adapting the results of Examples 1 and 2 to fit our present situation, we get

$$\frac{\partial l}{\partial \tilde{\mathbf{w}}^*} = \frac{1}{\sigma^2}\mathbf{A}^H\mathbf{b} - \frac{1}{\sigma^2}\mathbf{A}^H\mathbf{A}\mathbf{w}$$

Setting $(\partial l/\partial \tilde{\mathbf{w}}^*) = \mathbf{0}$, and then simplifying, we thus get

$$\mathbf{A}^H\mathbf{b} - \mathbf{A}^H\mathbf{A}\mathbf{w}_o = \mathbf{0}$$

where $\mathbf{w}_o$ is the special value for which the log-liklihood function is maximum. Hence,

$$\mathbf{A}^H\mathbf{A}\mathbf{w}_o = \mathbf{A}^H\mathbf{b} \tag{B.16}$$

This is the matrix form of the normal equations for the method of least squares.

B.3 RELATION BETWEEN THE DERIVATIVE WITH RESPECT TO A VECTOR AND THE GRADIENT VECTOR

Consider the real cost function $J(\mathbf{w})$ that defines the error performance surface of a linear transversal filter whose tap-weight vector is $\mathbf{w}$. In Chapter 5, we defined the *gradient vector* of the error performance surface as

$$\nabla(J) = \begin{bmatrix} \dfrac{\partial J}{\partial x_0} + j\dfrac{\partial J}{\partial y_0} \\ \dfrac{\partial J}{\partial x_1} + j\dfrac{\partial J}{\partial y_1} \\ \vdots \\ \dfrac{\partial J}{\partial x_{M-1}} + j\dfrac{\partial J}{\partial y_{M-1}} \end{bmatrix} \tag{B.17}$$

where $x_k + jy_k$ is the kth element of the tap-weight vector **w**, and $k = 0, 1, \ldots, M - 1$. The gradient vector is *normal* to the error performance surface. Comparing Eqs. (B.7) and (B.17), we see that the conjugate derivative $\partial J/\partial \mathbf{w}^*$ and the gradient vector $\nabla(J)$ are related by

$$\nabla(J) = 2\frac{\partial J}{\partial \mathbf{w}^*} \tag{B.18}$$

Thus, except for a scaling factor, the definition of the gradient vector introduced in Chapter 5 is the same as the conjugate derivative defined in Eq. (B.7).

C

Method of Lagrange Multipliers

Optimization consists of determining the values of some specified variables that minimize or maximize an *index of performance* or *cost function*, which combines important properties of a system into a single real-valued number. The optimization may be *constrained* or *unconstrained*, depending on whether the variables are also required to satisfy side equations or not. Needless to say, the additional requirement to satisfy one or more side equations complicates the issue of constrained optimization. In this appendix, we derive the classical *method of Lagrange multipliers* for solving the *complex* version of a constrained optimization problem. The notation used in the derivation is influenced by the nature of applications that are of interest to us. We consider first the case when the problem involves a single side equation, followed by the more general case of multiple side equations.

C.1 OPTIMIZATION INVOLVING A SINGLE EQUALITY CONSTRAINT

Consider the minimization of a real-valued function $f(\mathbf{w})$ that is a quadratic function of a vector $\mathbf{w}$, subject to the *constraint*

$$\mathbf{w}^H \mathbf{s} = g \tag{C.1}$$

where $\mathbf{s}$ is a prescribed vector of compatible dimension and g is a complex constant. We may redefine the constraint by introducing a new function $c(\mathbf{w})$ that is linear in $\mathbf{w}$, as shown by

$$\begin{aligned} c(\mathbf{w}) &= \mathbf{w}^H \mathbf{s} - g \\ &= 0 + j0 \end{aligned} \tag{C.2}$$

In general, the vectors $\mathbf{w}$ and $\mathbf{s}$ and the function $c(\mathbf{w})$ are all *complex*. For example, in a beamforming application the vector $\mathbf{w}$ represents a set of complex weights applied to the individual sensor outputs, and $\mathbf{s}$ represents a steering vector whose elements are defined by a prescribed "look" direction; the function $f(\mathbf{w})$ to be minimized represents the mean-square value of the overall beamformer output. In a harmonic retrieval application, $\mathbf{w}$ represents the tap-weight vector of a transversal filter, and $\mathbf{s}$ represents a sinusoidal vector whose elements are determined by the angular frequency of a complex sinusoid contained in the filter input; the function $f(\mathbf{w})$ represents the mean-square value of the filter output. In any event, assuming that the issue is one of minimization, we may state the constrained optimization problem as follows:

$$\textit{Minimize} \text{ a real-valued function } f(\mathbf{w})$$

$$\textit{subject to} \text{ a constraint } c(\mathbf{w}) = 0 + j0 \tag{C.3}$$

The *method of Lagrange multipliers* converts the problem of constrained minimization described herein into one of unconstrained minimization by the introduction of *Lagrange multipliers*. First we use the real function $f(\mathbf{w})$ and the complex constraint function $c(\mathbf{w})$ to define a new real-valued function

$$h(\mathbf{w}) = f(\mathbf{w}) + \lambda_1 \operatorname{Re}[c(\mathbf{w})] + \lambda_2 \operatorname{Im}[c(\mathbf{w})] \tag{C.4}$$

where λ_1 and λ_2 are *real Lagrange* multipliers, and

$$c(\mathbf{w}) = \operatorname{Re}[c(\mathbf{w})] + j \operatorname{Im}[c(\mathbf{w})] \tag{C.5}$$

Define a *complex Lagrange multiplier:*

$$\lambda = \lambda_1 + j\lambda_2 \tag{C.6}$$

We may then rewrite Eq. (C.4) in the form

$$h(\mathbf{w}) = f(\mathbf{w}) + \operatorname{Re}[\lambda^* c(\mathbf{w})] \tag{C.7}$$

where the asterisk denotes complex conjugation.

Next, we minimize the function $h(\mathbf{w})$ with respect to the vector $\mathbf{w}$. To do this, we set the conjugate derivative $\partial h/\partial \mathbf{w}^*$ equal to the null vector, as shown by

$$\frac{\partial f}{\partial \mathbf{w}^*} + \frac{\partial}{\partial \mathbf{w}^*}(\operatorname{Re}[\lambda^* c(\mathbf{w})]) = \mathbf{0} \tag{C.8}$$

The system of simultaneous equations, consisting of Eq. (C.8) and the original constraint given in Eq. (C.2), define the optimum solutions for the vector $\mathbf{w}$ and the Lagrange multiplier λ. We call Eq. (C.8) the *adjoint equation,* and Eq. (C.2) the *primal equation* (Dorny, 1975).

C.2 OPTIMIZATION INVOLVING MULTIPLE EQUALITY CONSTRAINTS

Consider next the minimization of a real function $f(\mathbf{w})$ that is a quadratic function of the vector $\mathbf{w}$, subject to a set of *multiple linear constraints*

$$\mathbf{w}^H \mathbf{s}_k = g_k, \qquad k = 1, 2, \ldots, K \tag{C.9}$$

where the number of constraints, K, is less than the dimension of the vector $\mathbf{w}$, and the g_k are complex constants. We may state the multiple-constrained optimization problem as follows:

$$\text{Minimize a real function } f(\mathbf{w}) \text{ subject to the constraints } c_k(\mathbf{w}) = 0 + j0, \text{ for } k = 1, 2, \ldots, K \tag{C.10}$$

The solution to this optimization problem is readily obtained by generalizing the previous results of Section C.1. Specifically, we formulate a system of simultaneous equations, consisting of the adjoint equation

$$\frac{\partial f}{\partial \mathbf{w}^*} + \sum_{k=1}^{K} \frac{\partial}{\partial \mathbf{w}^*}(\text{Re}[\lambda_k^* c_k(\mathbf{w})]) = \mathbf{0} \tag{C.11}$$

and the primal equation

$$c_k(\mathbf{w}) = 0, \qquad k = 1, 2, \ldots, K \tag{C.12}$$

This system of equations define the optimum solutions for the vector $\mathbf{w}$ and the set of complex Lagrange multipliers $\lambda_1, \lambda_2, \ldots, \lambda_K$.

C.3 EXAMPLE

By way of an example, consider the problem of finding the vector $\mathbf{w}$ that minimizes the function

$$f(\mathbf{w}) = \mathbf{w}^H \mathbf{w} \tag{C.13}$$

and which satisfies the constraint

$$c(\mathbf{w}) = \mathbf{w}^H \mathbf{s} - g = 0 + j0 \tag{C.14}$$

The adjoint equation for this problem is

$$\frac{\partial}{\partial \mathbf{w}^*}(\mathbf{w}^H \mathbf{w}) + \frac{\partial}{\partial \mathbf{w}^*}(\text{Re}[\lambda^*(\mathbf{w}^H \mathbf{s} - g)]) = \mathbf{0} \tag{C.15}$$

Using the rules for differentiation developed in Appendix B, we have

$$\frac{\partial}{\partial \mathbf{w}^*}(\mathbf{w}^H \mathbf{w}) = \mathbf{w}$$

and

$$\frac{\partial}{\partial \mathbf{w}^*}(\text{Re}[\lambda^*(\mathbf{w}^H \mathbf{s} - g)]) = \lambda^* \mathbf{s}$$

Substituting these results in Eq. (C.15), we get

$$\mathbf{w} + \lambda^* \mathbf{s} = \mathbf{0} \tag{C.16}$$

or, equivalently,

$$\mathbf{w}^H + \lambda \mathbf{s}^H = \mathbf{0}^T \tag{C.17}$$

Next, postmultiplying both sides of Eq. (C.17) by $\mathbf{s}$, and then solving for the unknown λ, we obtain

$$\begin{aligned}\lambda &= -\frac{\mathbf{w}^H\mathbf{s}}{\mathbf{s}^H\mathbf{s}} \\ &= -\frac{g}{\mathbf{s}^H\mathbf{s}}\end{aligned} \tag{C.18}$$

Finally, substituting Eq. (C.18) in (C.16) and solving for the optimum value $\mathbf{w}_o$ of the weight vector $\mathbf{w}$, we get

$$\mathbf{w}_o = \left(\frac{g^*}{\mathbf{s}^H\mathbf{s}}\right)\mathbf{s} \tag{C.19}$$

This solution is optimum in the sense that $\mathbf{w}_o$ satisfies the constraint of Eq. (C.14) and has minimum length.

Maximum-Likelihood Estimation

Estimation theory is a branch of probability and statistics that deals with the problem of deriving information about properties of random variables and stochastic processes, given a set of observed samples. This problem arises frequently in the study of communications and control systems. *Maximum likelihood* is by far the most general and powerful method of estimation. It was first used by Fisher in 1906. In principle, the method of maximum likelihood may be applied to any estimation problem with the proviso that we formulate the joint probability density function of the available set of observed data. As such, the method yields almost all the well-known estimates as special cases.

D.1 LIKELIHOOD FUNCTION

The method of maximum likelihood is based on a relatively simple idea: Different populations generate different data samples and any given data sample is more *likely* to have come from some population than from others (Kmenta, 1971).

Let $f_{\mathbf{U}}(\mathbf{u}|\boldsymbol{\theta})$ denote the *conditional joint probability density function* of the *random vector* $\mathbf{U}$ represented by the observed *sample* vector $\mathbf{u}$, where the sample vector $\mathbf{u}$ has $u_1, u_2, \ldots, u_M$ for its elements, and $\boldsymbol{\theta}$ is a *parameter vector with* $\theta_1, \theta_2, \ldots, \theta_K$ as elements. The method of maximum likelihood is based on the principle that we should estimate the parameter vector $\boldsymbol{\theta}$ by its most *plausible values*, given the observed sample vector $\mathbf{u}$. In other words, the maximum-likelihood estimators of $\theta_1, \theta_2, \ldots, \theta_K$ are those values of the parameter vector for which the conditional joint probability density function $f_{\mathbf{U}}(\mathbf{u}|\boldsymbol{\theta})$ is at maximum.

The name *likelihood function,* denoted by $l(\boldsymbol{\theta})$, is given to the conditional joint probability density function $f_{\mathbf{U}}(\mathbf{u}|\boldsymbol{\theta})$, viewed as a function of the parameter vector $\boldsymbol{\theta}$.

We thus write

$$l(\boldsymbol{\theta}) = f_{\mathbf{U}}(\mathbf{u}|\boldsymbol{\theta}) \tag{D.1}$$

Although the conditional joint probability density function and the likelihood function have exactly the same formula, nevertheless, it is vital that we appreciate the physical distinction between them. In the case of the conditional joint probability density function, the parameter vector $\boldsymbol{\theta}$ is fixed and the observation vector $\mathbf{u}$ is variable. On the other hand, in the case of the likelihood function, the parameter vector $\boldsymbol{\theta}$ is variable and the observation vector $\mathbf{u}$ is fixed.

In many cases, it turns out to be more convenient to work with the logarithm of the likelihood function rather than with the likelihood itself. Thus, using $L(\boldsymbol{\theta})$ to denote the *log-likelihood function,* we write

$$\begin{aligned} L(\boldsymbol{\theta}) &= \ln[l(\boldsymbol{\theta})] \\ &= \ln[f_{\mathbf{U}}(\mathbf{u}|\boldsymbol{\theta})] \end{aligned} \tag{D.2}$$

The logarithm of $l(\boldsymbol{\theta})$ is a *monotonic transformation* of $l(\boldsymbol{\theta})$. This means that whenever $l(\boldsymbol{\theta})$ decreases its logarithm $L(\boldsymbol{\theta})$ also decreases. Since $l(\boldsymbol{\theta})$, being a formula for conditional joint probability density function, can never become negative, it follows that there is no problem in evaluating its logarithm $L(\boldsymbol{\theta})$. We conclude therefore that the parameter vector for which the likelihood function $l(\boldsymbol{\theta})$ is at maximum is exactly the same as the parameter vector for which the log-likelihood function $L(\boldsymbol{\theta})$ is at its maximum.

To obtain the ith element of the maximum-likelihood estimate of the parameter vector $\boldsymbol{\theta}$, we differentiate the log-likelihood function with respect to θ_i and set the result equal to zero. We thus get a set of first-order conditions:

$$\frac{\partial L}{\partial \theta_i} = 0, \qquad i = 1, 2, \ldots, K \tag{D.3}$$

The first derivative of the log-likelihood function with respect to parameter θ_i is called the *score* for that parameter. The vector of such parameters is known as the *scores vector* (i.e., the gradient vector). The scores vector is identically zero at the maximum-likelihood estimates of the parameters, that is, at the vaues of $\boldsymbol{\theta}$ that result from the solutions of Eq. (D.3).

To find how effective the method of maximum likelihood is, we can compute the *bias* and *variance* for the estimate of each parameter. However, this is frequently difficult to do. Rather than approach the computation directly, we may derive a *lower bound* on the variance of any *unbiased* estimate. We say an estimate is unbiased if the average value of the estimate equals the parameter we are trying to estimate. Later we show how the variance of the maximum-likelihood estimate compares with this lower bound.

D.2 CRAMER–RAO INEQUALITY

Let $\mathbf{U}$ be a random vector with conditional joint probability density function $f_{\mathbf{U}}(\mathbf{u}|\boldsymbol{\theta})$, where $\mathbf{u}$ is the observed sample vector with elements $u_1, u_2, \ldots, u_M$ and $\boldsymbol{\theta}$ is the parameter vector with elements $\theta_1, \theta_2, \ldots, \theta_K$. Using the definition of Eq. (D.2) for the log-likelihood function $L(\boldsymbol{\theta})$ in terms of the conditional joint probability density function $f_U(\mathbf{u}|\boldsymbol{\theta})$, we form the K-by-K matrix:

$$\mathbf{J} = -\begin{bmatrix} E\left[\dfrac{\partial^2 L}{\partial\theta_1^2}\right] & E\left[\dfrac{\partial^2 L}{\partial\theta_1\,\partial\theta_2}\right] & \cdots & E\left[\dfrac{\partial^2 L}{\partial\theta_1\,\partial\theta_K}\right] \\ E\left[\dfrac{\partial^2 L}{\partial\theta_2\,\partial\theta_1}\right] & E\left[\dfrac{\partial^2 L}{\partial\theta_2^2}\right] & \cdots & E\left[\dfrac{\partial^2 L}{\partial\theta_2\,\partial\theta_K}\right] \\ \vdots & \vdots & & \vdots \\ E\left[\dfrac{\partial^2 L}{\partial\theta_K\,\partial\theta_1}\right] & E\left[\dfrac{\partial^2 L}{\partial\theta_K\,\partial\theta_2}\right] & \cdots & E\left[\dfrac{\partial^2 L}{\partial\theta_K^2}\right] \end{bmatrix} \tag{D.4}$$

The matrix $\mathbf{J}$ is called *Fisher's information matrix*.

Let $\mathbf{I}$ denote the inverse of Fisher's information matrix $\mathbf{J}$. Let I_{ii} denote the ith diagonal element (i.e., the element in the ith row and ith column) of the inverse matrix $\mathbf{I}$. Let $\hat{\theta}_i$ be *any* unbiased estimate of the parameter θ_i, based on the observed sample vector $\mathbf{u}$. We may then write (Van Trees, 1968; Nahi, 1969)

$$\text{var}[\hat{\theta}_i] \geq I_{ii}, \qquad i = 1, 2, \ldots, K \tag{D.5}$$

Equation (D.5) is called the *Cramér–Rao inequality*. This theorem enables us to construct a lower limit (greater than zero) for the variance of any unbiased estimator, provided, of course, that we know the functional form of the log-likelihood function. The lower limit specified in the theorem is called the *Cramér–Rao lower bound*.

If we can find an unbiased estimator whose variance equals the Cramér–Rao lower bound, then according to the theorem of Eq. (D.5) there is no other unbiased estimator with a smaller variance. Such an estimator is said to be *efficient*.

D.3 PROPERTIES OF MAXIMUM-LIKELIHOOD ESTIMATORS

Not only is the method of maximum likelihood based on an intuitively appealing idea (that of choosing those parameters from which the actually observed sample vector is most likely to have come), but also the resulting estimates have some desirable properties. Indeed, under quite general conditions, the following *asymptotic* properties may be proved (Kmenta, 1971):

1. Maximum-likelihood estimators are *consistent*. That is, the value of θ_i for which the score $\partial L/\partial\theta_1$ is identically zero *converges in probability* to the true value of the parameter θ_i, $i = 1, 2, \ldots, K$, as the *sample size M* approaches infinity.
2. Maximum-likelihood estimators are *asymptotically efficient;* that is,

$$\lim_{M\to\infty}\left\{\frac{\text{var}[\theta_{i,ml} - \theta_i]}{I_{ii}}\right\} = 1, \qquad i = 1, 2, \ldots, K$$

where $\theta_{i,ml}$ is the maximum-likelihood estimate of parameter θ_i, and I_{ii} is the ith diagonal element of the inverse of Fisher's information matrix.

3. Maximum-likelihood estimators are *asymptotically Gaussian*.

In practice, we find that the large-sample (asymptotic) properties of maximum-likelihood estimators hold rather well for sample size $M \geq 50$.

The material presented in this appendix has assumed the use of real-valued data. Its extension to complex-valued data is a straightforward matter.

E

Conditional Mean Estimator

A classic problem in estimation theory is that of the *Bayes estimation of a random parameter*. There are different answers to this problem, depending on how the *cost function* in the Bayes estimation is formulated (Van Trees, 1968). A particular type of the Bayes estimator of interest to us in this book is the so-called *conditional mean estimator*. The purpose of this appendix is twofold: (1) to derive the formula for the conditional mean estimator from first principles, and (2) to show that such an estimator is the same as a minimum mean-squared-error estimator.

E.1 DERIVATION

Consider a *random parameter* x. We are given an *observation* y that depends on x, and the requirement is to estimate x. Let $\hat{x}(y)$ denote an *estimate* of the parameter x; the symbol $\hat{x}(y)$ emphasizes the fact that the estimate is a function of the observation y. Let $C(x, \hat{x}(y))$ denote a *cost function*. Then, according to Bayes estimation theory, we may write an expression for the *risk* as follows (Van Trees, 1968):

$$\begin{aligned} \mathcal{R} &= E[C(x, \hat{x}(y))] \\ &= \int_{-\infty}^{\infty} dx \int_{-\infty}^{\infty} C(x, \hat{x}(y)) f_{X,Y}(x, y)\, dy \end{aligned} \tag{E.1}$$

where $f_{X,Y}(x, y)$ is the joint probability density function of x and y. For a specified cost function $C(x, \hat{x}(y))$, the *Bayes estimate* is defined as the estimate $\hat{x}(y)$ that *minimizes* the risk $\mathcal{R}$.

A cost function of particular interest (and which is very much in the spirit of the material covered in this book) is the *mean-squared error*. In this case, the cost function is specified as the square of the *estimation error*. The estimation error is itself defined as the difference between the actual parameter value x and the estimate $\hat{x}(y)$, as shown by

$$\epsilon = x - \hat{x}(y) \tag{E.2}$$

Correspondingly, the cost function is defined by

$$C(x, \hat{x}(y)) = C(x - \hat{x}(y)) \tag{E.3}$$

or, more simply,

$$C(\epsilon) = \epsilon^2 \tag{E.4}$$

Thus, the cost function varies with the estimation error ϵ in the manner indicated in Fig. E.1. It is assumed here that x and y are both real.

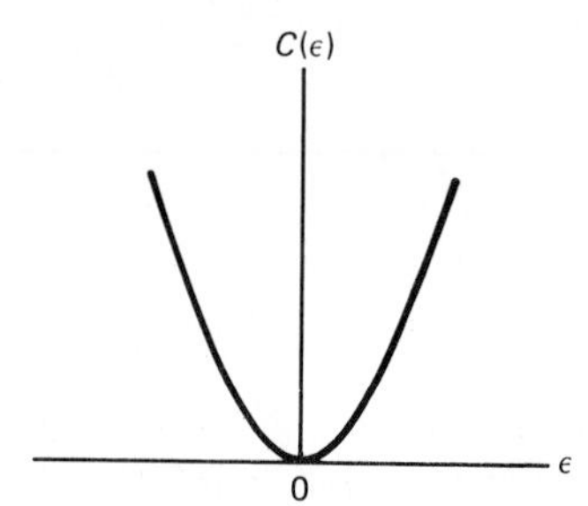

Figure E.1 Mean-squared error as the cost function.

Accordingly, for the situation at hand, we may rewrite Eq. (E.1) as follows:

$$\mathcal{R}_{\text{ms}} = \int_{-\infty}^{\infty} dx \int_{-\infty}^{\infty} [x - \hat{x}(y)]^2 f_{X,Y}(x, y)\, dy \tag{E.5}$$

where the subscripts in the risk $\mathcal{R}_{\text{ms}}$ indicate the use of mean-squared error at its basis.

Using *Bayes' rule,* we have

$$f_{X,Y}(x, y) = f_X(x|y) f_Y(y) \tag{E.6}$$

where $f_X(x|y)$ is the conditional probability density function of x, given y, and $f_Y(y)$ is the (marginal) probability density function of y. Hence, using Eq. (E.6) in (E.5), we have

$$\mathcal{R}_{\text{ms}} = \int_{-\infty}^{\infty} dy\, f_Y(y) \int_{-\infty}^{\infty} [x - \hat{x}(y)]^2 f_X(x|y)\, dx \tag{E.7}$$

We now recognize that the inner integral and $f_Y(y)$ in Eq. (E.7) are both nonnegative. We may therefore minimize the risk $\mathcal{R}_{\text{ms}}$ by simply minimizing the inner integral. Let the estimate so obtained be denoted by $\hat{x}_{\text{ms}}(y)$. We find $\hat{x}_{\text{ms}}(y)$ by differentiating the inner integral with respect to $\hat{x}(y)$ and then setting the result equal to zero.

To simplify the presentation, let I denote the inner integral in Eq. (E.7). Then, differentiating I with respect to $\hat{x}(y)$ yields

$$\frac{dI}{dx} = -2 \int_{-\infty}^{\infty} x f_X(x|y)\, dx + 2\hat{x}(y) \int_{-\infty}^{\infty} f_X(x|y)\, dx \tag{E.8}$$

The second integral on the right side of Eq. (E.8) represents the total area under a probability density function and therefore equals 1. Hence, setting the derivative $dI/d\hat{x}$ equal to zero, we obtain

$$\hat{x}_{\text{ms}}(y) = \int_{-\infty}^{\infty} x f_X(x|y)\, dx \tag{E.9}$$

The solution defined by Eq. (E.9) is a unique minimum.

E.2 INTERPRETATIONS

The estimator $x_{\text{ms}}(y)$ is naturally a *minimum mean-squared-error estimator*, hence the use of the subscripts "ms." For another interpretation of the estimator $\hat{x}_{\text{ms}}(y)$ defined in Eq. (E.9), we recognize that the integral on the right side of this equation is just the *conditional mean* of the parameter x, given the observation y.

We therefore conclude that the minimum mean-squared error estimator and the conditional mean estimator are indeed one and the same. In other words, we have

$$\hat{x}_{\text{ms}}(y) = E[x|y] \tag{E.10}$$

Substituting Eq. (E.10) for the estimate $\hat{x}(y)$ in Eq. (E.7), we find that the inner integral is just the *conditional variance* of the parameter x, given y. Accordingly, the minimum value of the risk $\mathcal{R}_{\text{ms}}$ is just the average of this conditional variance over all observations y.

F

Maximum-Entropy Method

The *maximum-entropy method* (MEM) was originally devised by Burg in 1967 to overcome fundamental limitations of Fourier-based methods for estimating the power spectrum of a stationary stochastic process (Burg 1967, 1975). The basic idea of the MEM is to choose the particular spectrum that corresponds to the most *random* or the most *unpredictable* time series whose autocorrelation function agrees with a set of known values. This condition is equivalent to an extrapolation of the autocorrelation function of the available time series by *maximizing* the *entropy* of the process; hence, the name of the method. Entropy is a measure of the average information content of the process (Shannon, 1948). Thus, the MEM bypasses the problems that arise from the use of window functions, a feature that is common to all Fourier-based methods of spectrum analysis. In particular, the MEM avoids the use of a periodic extension of the data (as in the method based on smoothing the periodogram and its computation using the fast Fourier transform algorithm) or of the assumption that the data outside the available record length are zero (as in the Blackman–Tukey method based on the sample autocorrelation function). An important feature of the MEM spectrum is that it is *nonnegative at all frequencies* which is precisely the way it should be.

F.1 MAXIMUM-ENTROPY SPECTRUM

Suppose that we are given $2M + 1$ values of the autocorrelation function of a stationary stochastic process $\{u(n)\}$ of zero mean. We wish to obtain the special value of the power spectrum of the process that corresponds to the most random time series whose autocorrelation function is consistent with the set of $2M + 1$ known values. In terms of information theory, this statement corresponds to the *principle of maximum entropy* (Jaynes, 1982).

In the case of a set of Gaussian-distributed random variables of zero mean, the entropy is given by (Middleton, 1960)

$$H = \tfrac{1}{2}\ln[\det(\mathbf{R})] \tag{F.1}$$

where $\mathbf{R}$ is the correlation matrix of the process. When the process is of infinite duration, however, we find that the entropy H diverges, and so we cannot use it as a measure of information content. To overcome the divergence problem, we may use the *entropy rate* defined by

$$\begin{aligned} h &= \lim_{M\to\infty} \frac{H}{M+1} \\ &= \lim_{M\to\infty} \tfrac{1}{2}\ln[\det(\mathbf{R})]^{1/(M+1)} \end{aligned} \tag{F.2}$$

Let $S(\omega)$ denote the power spectrum of the process $\{u(n)\}$. The limiting form of the determinant of the correlation matrix $\mathbf{R}$ is related to the power spectrum $S(\omega)$ as follows

$$\lim_{M\to\infty} [\det(\mathbf{R})]^{1/(M+1)} = \exp\left\{\frac{1}{2\pi}\int_{-\pi}^{\pi} \ln S(\omega)\, d\omega\right\} \tag{F.3}$$

Hence, substituting Eq. (F.3) in (F.2), we get

$$h = \frac{1}{4\pi}\int_{-\pi}^{\pi} \ln[S(\omega)]\, d\omega \tag{F.4}$$

Although this relation was derived on the assumption that the process $\{u(n)\}$ is Gaussian, nevertheless, the form of the relation is valid for any stationary process.

We may now restate the MEM problem in terms of the entropy rate. We wish to find a real positive-valued power spectrum characterized by entropy rate h, satisfying two simultaneous requirements:

1. The entropy rate h is *stationary* with respect to the *unknown* values of the autocorrelation function of the process.
2. The power spectrum is *consistent* with respect to the *known* values of the autocorrelation function of the process.

We will address these two requirements in turn.

Since the autocorrelation sequence $\{r(m)\}$ and power spectrum $S(\omega)$ of a stationary process $\{u(n)\}$ form a discrete-time Fourier transform pair, we write

$$S(\omega) = \sum_{m=-\infty}^{\infty} r(m)\exp(-jm\omega) \tag{F.5}$$

Equation (F.5) assumes that the sampling period of the discrete-time process $\{u(n)\}$ is normalized to unity. Substituting Eq. (F.5) in (F.4), we get

$$h = \frac{1}{4\pi}\int_{-\pi}^{\pi} \ln\left[\sum_{m=-\infty}^{\infty} r(m)\exp(-jm\omega)\right] d\omega \tag{F.6}$$

We extrapolate the autocorrelation sequence $\{r(m)\}$ outside the range of known values, $-M \le m \le M$, by choosing the unknown values of the autocorrelation function in such a way that no information or entropy is added to the process. That is, we impose the condition

$$\frac{\partial h}{\partial r(m)} = 0, \qquad |m| \geq M + 1 \tag{F.7}$$

Hence, differentiating Eq. (F.6) with respect to $r(m)$ and setting the result equal to zero, we find that the conditions for *maximum entropy* are as follows:

$$\int_{-\pi}^{\pi} \frac{\exp(-jm\omega)}{S_{\text{MEM}}(\omega)} \, d\omega = 0, \qquad |m| \geq M + 1 \tag{F.8}$$

where $S_{\text{MEM}}(\omega)$ is the special value of the power spectrum resulting from the imposition of the condition in Eq. (F.7). Equation (F.8) implies that the power spectrum $S_{\text{MEM}}(\omega)$ is expressible in the form of a truncated Fourier series:

$$\frac{1}{S_{\text{MEM}}(\omega)} = \sum_{k=-M}^{M} c_k \exp(-jk\omega) \tag{F.9}$$

The complex Fourier coefficient c_k of the expansion satisfies the Hermitian condition

$$c_k^* = c_{-k} \tag{F.10}$$

so as to ensure that $S_{\text{MEM}}(\omega)$ is real.

The next requirement is to make the power spectrum $S_{\text{MEM}}(\omega)$ consistent with the set of known values of the autocorrelation function $r(m)$ for the interval $-M \leq m \leq M$. Since $r(m)$ is a Hermitian function, we need only concern ourselves with $0 \leq m \leq M$. Accordingly, $r(m)$ must equal the inverse discrete-time Fourier transform of $S_{\text{MEM}}(\omega)$ for $0 \leq m \leq M$, as shown by

$$r(m) = \frac{1}{2\pi} \int_{-\pi}^{\pi} S_{\text{MEM}}(\omega) \exp(jm\omega) \, d\omega, \qquad 0 \leq m \leq M \tag{F.11}$$

Therefore, substituting Eq. (F.9) in (F.11), we get

$$r(m) = \int_{-\pi}^{\pi} \frac{\exp(jm\omega)}{\sum_{k=-M}^{M} c_k \exp(-jk\omega)} \, d\omega, \qquad 0 \leq m \leq M \tag{F.12}$$

Clearly, in the set of complex Fourier coefficients $\{c_k\}$, we have the available degrees of freedom needed to satisfy the conditions of Eq. (F.12).

To proceed with the analysis, however, we find it convenient to use z-transform notation by changing from the variable ω to z. Define

$$z = \exp(j\omega) \tag{F.13}$$

Hence,

$$d\omega = \frac{1}{j} \frac{dz}{z}$$

and so we rewrite Eq. (F.12) in terms of the variable z as the contour integral:

$$r(m) = \frac{1}{j2\pi} \oint \frac{z^{m-1}}{\sum_{k=-M}^{M} c_k z^{-k}} \, dz, \qquad 0 \leq m \leq M \tag{F.14}$$

The contour integration in Eq. (F.14) is performed on the unit circle in the z-plane in a counterclockwise direction. Since the complex Fourier coefficient c_k satisfies the Hermitian condition of Eq. (F.10), we may express the summation in the denominator of the integral in Eq. (F.14) as the product of two polynomials, as follows:

$$\sum_{k=-M}^{M} c_k z^{-k} = G(z)G^*\left(\frac{1}{z^*}\right) \tag{F.15}$$

where

$$G(z) = \sum_{k=0}^{M} g_k z^{-k} \tag{F.16}$$

and

$$G^*\left(\frac{1}{z^*}\right) = \sum_{k=0}^{M} g_k^* z^k \tag{F.17}$$

We choose the first polynomial $G(z)$ to be minimum phase in that its zeros are all located inside the unit circle in the z-plane. Correspondingly, we choose the second polynomial $G^*(1/z^*)$ to be maximum phase in that all its zeros are located outside the unit circle in the z-plane. Moreover, the zeros of these two polynomials are the inverse of each other with respect to the unit circle. Thus, substituting Eq. (F.15) in (F.14), we get

$$r(m) = \frac{1}{j2\pi}\oint \frac{z^{m-1}}{G(z)G^*(1/z^*)}\,dz, \qquad 0 \le m \le M \tag{F.18}$$

We next form the summation

$$\begin{aligned} \sum_{k=0}^{M} g_k r(m-k) &= \frac{1}{j2\pi}\oint \frac{z^{m-1}\sum_{k=0}^{M} g_k z^{-k}}{G(z)G^*(1/z^*)}\,dz \\ &= \frac{1}{j2\pi}\oint \frac{z^{m-1}}{G^*(1/z^*)}\,dz, \qquad 0 \le m \le M \end{aligned} \tag{F.19}$$

where in the first line we have used Eq. (F.18), and in the second line we have used Eq. (F.16).

To evaluate the contour integral of Eq. (F.19), we use *Cauchy's residue theorem* of complex variable theory (see Appendix A). According to this theorem, the contour integral equals $2\pi j$ times the sum of *residues* of the poles of the integral $z^{m-1}/G^*(1/z^*)$ that lie inside the unit circle, the contour of integration. Since the polynomial $G^*(1/z^*)$ is chosen to have no zeros inside the unit circle, it follows that the integral in Eq. (F.19) is analytic on and inside the unit circle for $m \ge 1$. For $m = 0$ the integral has a simple pole at $z = 0$ with a *residue* equal to $1/g_0^*$. Hence, application of Cauchy's residue theorem yields

$$\oint \frac{z^{m-1}}{G^*(1/z^*)}\,dz = \begin{cases} \dfrac{2\pi j}{g_0^*}, & m = 0 \\ 0, & m = 1, 2, \ldots, M \end{cases} \tag{F.20}$$

Thus, substituting Eq. (F.20) in (F.19), we get

$$\sum_{k=0}^{M} g_k r(m-k) = \begin{cases} \dfrac{1}{g_0^*}, & m = 0 \\ 0, & m = 1, 2, \ldots, M \end{cases} \tag{F.21}$$

We recognize that the set of $(M + 1)$ equations in (F.21) have a mathematical form similar to that of the augmented Wiener-Hopf equations for forward prediction of order M (see Chapter 6). In particular, by comparing Eqs. (F.21) and (6.19), we deduce that

$$g_k^* = \frac{1}{g_0 P_M} a_{M,k}, \qquad 0 \leq k \leq M \tag{F.22}$$

where $\{a_{M,k}\}$ are the coefficients of a prediction-error filter of order M, and P_M is the average output power of the filter. Since $a_{M,0} = 1$ for all M, we find from Eq. (F.22) that, for $k = 0$;

$$|g_0|^2 = \frac{1}{P_M} \tag{F.23}$$

Finally, substituting Eqs. (F.15), (F.22), and (F.23) in (F.9) with $z = \exp(j\omega)$, we get

$$\begin{aligned} S_{\text{MEM}}(\omega) &= \frac{P_M}{\left| 1 + \sum_{k=1}^{M} a_{M,k} e^{jk\omega} \right|^2} \\ &= \frac{P_M}{\left| 1 + \sum_{k=1}^{M} a_{M,k}^* e^{-jk\omega} \right|^2} \end{aligned} \tag{F.24}$$

We refer to the formula of Eq. (F.24) as the *MEM spectrum*.

F.2 COMPUTATION OF THE MEM SPECTRUM

The formula for the MEM spectrum given in Eq. (F.24) may be recast in the alternative form

$$S_{\text{MEM}}(\omega) = \frac{1}{\sum_{k=-M}^{M} \psi(k) e^{-j\omega k}} \tag{F.25}$$

where $\psi(k)$ is defined in terms of the prediction-error filter coefficients as follows:

$$\psi(k) = \begin{cases} \dfrac{1}{P_M} \sum_{i=0}^{M-k} a_{M,i} a_{M,i+k}^* & \text{for } k = 0, 1, \ldots, M \\ \psi^*(-k) & \text{for } k = -M, \ldots, -1 \end{cases} \tag{F.26}$$

The parameter $\psi(k)$ may be viewed as some form of a *correlation coefficient for prediction-error filter coefficients*.

Examination of the denominator polynomial in Eq. (F.25) reveals that it represents the *discrete Fourier transform* of the sequence $\{\psi(k)\}$. Accordingly, we may use the *fast Fourier transform* (FFT) algorithm (Oppenheim and Schafer, 1975) for the efficient computation of the denominator polynomial and therefore the MEM spectrum. Given the autocorrelation sequence $r(0), r(1), \ldots, r(M)$, pertaining to a wide-sense stationary stochastic process $\{u(n)\}$, we may now summarize an efficient procedure for computing the MEM spectrum:

Step 1: Levinson–Durbin Recursion. Initialize the algorithm by setting

$$a_{0,0} = 1$$
$$P_0 = r(0)$$

For $m = 1, 2, \ldots, M$, compute

$$\Gamma_m = -\frac{1}{P_{m-1}} \sum_{i=0}^{M-1} r(i-m)a_{m-1,i}$$

$$a_{m,i} = \begin{cases} 1, & \text{for } i = 0 \\ a_{m-1,i} + \Gamma_m a^*_{m-1,m-i}, & \text{for } i = 1, 2, \ldots, m-1 \\ \Gamma_m, & \text{for } i = m \end{cases}$$

$$P_m = P_{m-1}(1 - |\Gamma_m|^2)$$

Step 2: Correlation for Prediction-Error Filter Coefficients. Compute the correlation coefficient

$$\psi(k) = \begin{cases} \dfrac{1}{P_M} \displaystyle\sum_{i=0}^{M-k} a_{M,i} a^*_{M,i+k}, & \text{for } k = 0, 1, \ldots, m \\ \psi^*(-k), & \text{for } k = -M, \ldots, -1 \end{cases} \tag{F.26}$$

Step 3: MEM Spectrum. Use the fast Fourier transform algorithm to compute the MEM spectrum for varying angular frequency:

$$S_{\text{MEM}}(\omega) = \frac{1}{\displaystyle\sum_{k=-M}^{M} \psi(k)e^{-j\omega k}}$$

G

Fast Algorithm for Computing the Minimum-Variance Distortionless Response Spectrum

In Section 5.7, we derived the formula for the *minimum-variance distortionless response (MVDR)* spectrum for a wide-sense stationary stochastic process. In this appendix we do two things. First, we develop a fast algorithm for computing the MVDR spectrum, given the ensemble-averaged correlation matrix of the process (Musicus, 1985); the algorithm exploits the Toeplitz property of the correlation matrix. Second, in deriving the algorithm, we develop an insightful relationship between the MVDR and MEM spectra.

G.1 FAST MVDR SPECTRUM COMPUTATION

Consider a zero-mean wide-sense stationary stochastic process $\{u(n)]$ characterized by an $(M + 1)$-by-$(M + 1)$ ensemble-averaged correlation matrix $\mathbf{R}$. The *minimum-variance distortionless response (MVDR) spectrum* for such a process is defined in terms of the inverse matrix $\mathbf{R}^{-1}$ by the formula

$$S_{\text{MVDR}}(\omega) = \frac{1}{\mathbf{s}^H(\omega)\mathbf{R}^{-1}\mathbf{s}(\omega)} \tag{G.1}$$

where

$$\mathbf{s}(\omega) = [1, e^{-j\omega}, e^{-j\omega 2}, \ldots, e^{-jM\omega}]$$

Let $R_{l,k}^{-1}$ denote the (l, k)th element of $\mathbf{R}^{-1}$. Then, we may rewrite Eq. (G.1) in the form

$$S_{\text{MVDR}}(\omega) = \frac{1}{\sum_{k=-M}^{M} \mu(k)e^{-j\omega k}} \tag{G.2}$$

where

$$\mu(k) = \sum_{l=\max(0,k)}^{\min(M-k,M)} R_{l,l+k}^{-1} \tag{G.3}$$

We recognize that the correlation matrix $\mathbf{R}$ is Toeplitz. We may therefore use the *Gohberg–Semencul formula* (Kailath et al., 1979) to express the (l, k)th element of the inverse matrix $\mathbf{R}^{-1}$ as follows:

$$R_{l,k}^{-1} = \frac{1}{P_M} \sum_{i=0}^{l} (a_{M,i} a_{M,i+k-l}^* - a_{M,M+1-i}^* a_{M,M+1-i-k+l}), \qquad k \geq l \tag{G.4}$$

where $1, a_{M,1}, \ldots, a_{M,M}$ are the coefficients of a prediction-error filter of order M, and P_M is the average prediction-error power. Substituting Eq. (G.4) in (G.3), and confining attention to $k \geq 0$, we get

$$\mu(k) = \frac{1}{P_M} \sum_{l=0}^{M-k} \sum_{i=0}^{l} a_{M,i} a_{M,i+k}^* - \frac{1}{P_M} \sum_{l=0}^{M-k} \sum_{i=0}^{l} a_{M,M+1-i}^* a_{M,M+1-i-k} \tag{G.5}$$

Interchanging the order of summations, and setting $j = M + 1 - i - k$, we may rewrite $\mu(k)$ as

$$\mu(k) = \frac{1}{P_M} \sum_{i=0}^{M-k} \sum_{l=i}^{M-k} a_{M,i} a_{M,i+k}^* - \frac{1}{P_M} \sum_{j=1}^{M+1-k} \sum_{l=M+1-j-k}^{M-k} a_{M,j+k}^* a_{M,j} \tag{G.6}$$

The terms that do not involve the index l permit us to collapse the summation over l into a multiplicative integer constant. We may thus combine the two summations in Eq. (G.6). Moreover, we may use the Levinson–Durbin recursion for computing the prediction-error filter coefficients. Given the autocorrelation sequence $r(0), r(1), \ldots, r(M)$, we may now formulate a fast algorithm for computing the MVDR spectrum as follows (Musicus, 1985):

Step 1: Levinson–Durbin Recursion. Initialize the algorithm by setting

$$a_{0,0} = 1$$

$$P_0 = r(0)$$

Hence, compute for $m = 1, 2, \ldots, M$:

$$\Gamma_m = -\frac{1}{P_{m-1}} \sum_{i=0}^{m-1} r(i-m) a_{m-1,i}$$

$$a_{m,i} = \begin{cases} 1, & \text{for } i = 0 \\ a_{m-1,i} + \Gamma_m a_{m-1,m-i}^*, & \text{for } i = 1, 2, \ldots, m-1 \\ \Gamma_m, & \text{for } i = m \end{cases}$$

$$P_m = P_{m-1}(1 - |\Gamma_m|^2)$$

Step 2: Correlation of the Predictor Coefficients. Compute the parameter $\mu(k)$ for varying k:

$$\mu(k) = \begin{cases} \dfrac{1}{P_M} \sum_{i=0}^{M-k} (M + 1 - k - 2i) a_{M,i} a^*_{M,i+k}, & \text{for } k = 0, \ldots, M \\ \mu^*(-k), & \text{for } k = -M, \ldots, -1 \end{cases} \tag{G.7}$$

Step 3: MVDR Spectrum Computation. Use the fast Fourier transform algorithm to compute the MVDR spectrum for varying angular frequency:

$$S_{\text{MVDR}}(\omega) = \frac{1}{\sum_{k=-M}^{M} \mu(k) e^{-j\omega k}}$$

G.2 COMPARISON OF MVDR AND MEM SPECTRA

Comparing the formula for computing the MVDR spectrum with that for computing the MEM spectrum, we see that the only difference between the MVDR formula in Eq. (G.2) and the MEM formula in Eq. (F.25) lies in the definitions of their respective correlations of predictor coefficients. In particular, a *linear taper* is used in the definition of $\mu(k)$ given in Eq. (G.7) for the MVDR formula. On the other hand, the definition of the corresponding parameter $\psi(k)$ given in Eq. (F.26) for the MEM formula does *not* involve a taper. This means that for a large-enough model order M, such that $a_{M,i} = 0$ for $i > M/2$, the linear taper involved in the computation of $\mu(k)$ acts like a triangular window on the product terms $a_{M,i} a^*_{M,i+k}$. This has the effect of deemphasizing higher-order terms with large i for large values of lag k (Musicus, 1985). Accordingly, for a given process, an MVDR spectrum is *smoother* in appearance than the corresponding MEM spectrum.

Glossary

TEXT CONVENTIONS

1. Boldfaced lowercase letters are used to denote column vectors. Boldfaced uppercase letters are used to denote matrices.
2. The estimate of a scalar, vector, or matrix is designated by the use of a hat (^) placed over the pertinent symbol.
3. The symbol | | denotes the magnitude or absolute value of the scalar enclosed within. The symbol ang[ ] or arg[ ] denotes the phase angle of the scalar enclosed within. The scalar may be real or complex.
4. The symbol ‖ ‖ denotes the Euclidean norm of the vector or matrix enclosed within.
5. The symbol det( ) denotes the determinant of the square matrix enclosed within.
6. The open interval (a, b) of the variable x signifies that $a < x < b$. The closed interval $[a, b]$ signifies that $a \leq x \leq b$.
7. The inverse of nonsingular (square) matrix $\mathbf{A}$ is denoted by $\mathbf{A}^{-1}$.
8. The pseudoinverse of matrix $\mathbf{A}$ (not necessarily square) is denoted by $\mathbf{A}^{+}$.
9. Complex conjugation of a scalar, vector, or matrix is denoted by the use of an asterisk as superscript. Transposition of a vector or matrix is denoted by superscript T. Hermitian transposition (i.e., complex conjugation and transposition combined) of a vector or matrix is denoted by superscript H. Backward rearrangement of the elements of a vector is denoted by superscript B.
10. The symbol $\mathbf{A}^{-H}$ denotes the Hermitian transpose of the inverse of a nonsingular (square) matrix $\mathbf{A}$.
11. The square root of a square matrix $\mathbf{A}$ is denoted by $\mathbf{A}^{1/2}$.
12. The symbol $\text{diag}(\lambda_1, \lambda_2, \ldots, \lambda_M)$ denotes a diagonal matrix whose elements on the main diagonal equal $\lambda_1, \lambda_2, \ldots, \lambda_M$.

13. The order of linear predictor or the order of autoregressive model is signified by a subscript added to the pertinent scalar or vector parameter.
14. The statistical expectation operator is denoted by $E[\cdot]$, where the quantity enclosed is the random variable or random vector of interest. The variance of a random variable is denoted by $\text{var}[\cdot]$, where the quantity enclosed is the random variable.
15. The conditional probability density function of random variable U, given that hypothesis H_i is true, is denoted by $f_U(u|H_i)$, where u is the sample value of random variable U.
16. The inner product of two vectors $\mathbf{x}$ and $\mathbf{y}$ is defined as $\mathbf{x}^H\mathbf{y} = \mathbf{y}^T\mathbf{x}^*$. Another possible inner product is $\mathbf{y}^H\mathbf{x} = \mathbf{x}^T\mathbf{y}^*$. In general, these two inner products are different. The outer product of the vectors $\mathbf{x}$ and $\mathbf{y}$ is defined as $\mathbf{x}\mathbf{y}^H$. The inner product is a scalar, whereas the outer product is a matrix.
17. The trace of a square matrix $\mathbf{R}$ is denoted by $\text{tr}[\mathbf{R}]$; it is defined as the sum of the diagonal elements of $\mathbf{R}$.
18. The autocorrelation function of stationary discrete-time stochastic process $\{u(n)\}$ is defined by

$$r(k) = E[u(n)u^*(n-k)]$$

The cross-correlation function between two jointly stationary discrete-time stochastic process $\{u(n)\}$ and $\{d(n)\}$ is defined by

$$p(-k) = E[u(n-k)d^*(n)]$$

19. The ensemble-averaged correlation matrix of a random vector $\mathbf{u}(n)$ is defined by

$$\mathbf{R} = E[\mathbf{u}(n)\mathbf{u}^H(n)]$$

20. The ensemble-averaged cross-correlation vector between a random vector $\mathbf{u}(n)$ and a random variable $d(n)$ is defined by

$$\mathbf{p} = E[\mathbf{u}(n)d^*(n)]$$

21. The time-averaged correlation matrix of a vector $\mathbf{u}(i)$ over the observation interval $1 \leq i \leq n$ is defined by

$$\mathbf{\Phi}(n) = \sum_{i=1}^{n} \mathbf{u}(i)\mathbf{u}^H(i)$$

22. The time-averaged cross-correlation vector between a vector $\mathbf{u}(i)$ and a scalar $d(i)$ over the observation interval $1 \leq i \leq n$ is defined by

$$\boldsymbol{\theta}(n) = \sum_{i=1}^{n} \mathbf{u}(i)d^*(i)$$

23. The discrete-time Fourier transform of a time function $u(n)$ is denoted by $F[u(n)]$. The inverse discrete-time Fourier transform of a frequency function $U(\omega)$ is denoted by $F^{-1}[U(\omega)]$.

24. In constructing block diagrams (signal-flow graphs) involving scalar quantities, the following symbols are used. The symbol

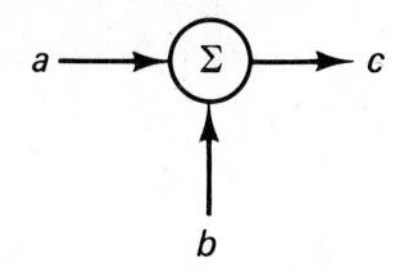

denotes an adder with $c = a + b$. The same symbol with algebraic signs added as in the following

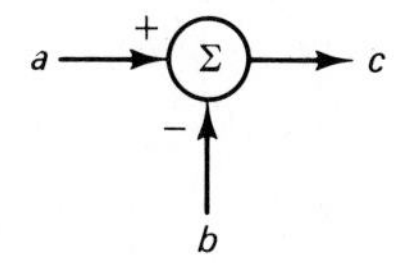

denotes a substractor with $c = a - b$. The symbol

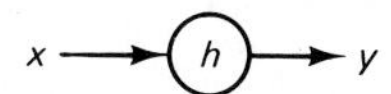

denotes a multiplier with $y = hx$. This multiplication is also represented as

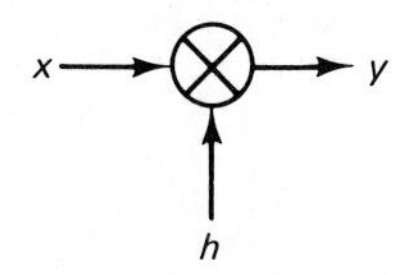

The unit-sample (delay) operator is denoted by

$$u(n) \longrightarrow \boxed{z^{-1}} \longrightarrow u(n-1)$$

25. In constructing block diagrams (signal-flow graphs) involving matrix quantities, the following symbols are used. The symbol

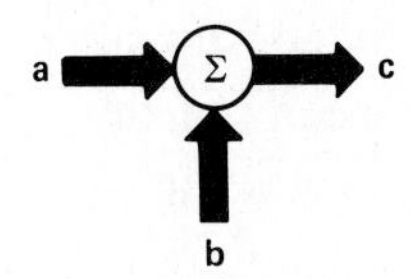

denotes summation with $\mathbf{c} = \mathbf{a} + \mathbf{b}$. The symbol

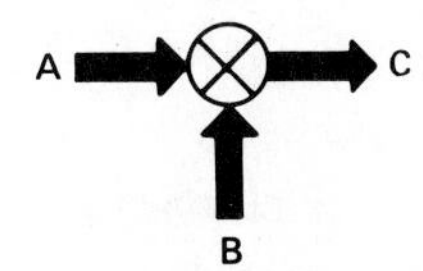

denotes multiplication with $\mathbf{C} = \mathbf{AB}$. The symbol

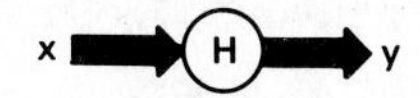

denotes a branch having transmittance **H**, with **y** = **Hx**. The unit-sample operator is denoted by the symbol

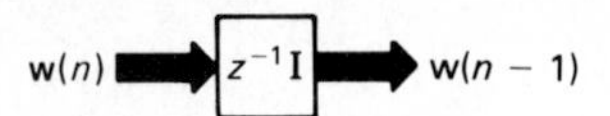

ABBREVIATIONS

ADPCM	Adaptive differential pulse-code modulation
AGC	Automatic gain control
AR	Autoregressive
ARMA	Autoregressive-moving average
as	Almost surely
b/s	Bit per second
CMA	Constant modulus adaptive
DFT	Discrete Fourier transform
DPCM	Differential pulse-code modulation
DTE	Data terminal equipment
FBLP	Forward–backward linear prediction
FFT	Fast Fourier transform
FTF	Fast transversal filter
GSVD	Generalized singular value decompositon
HF	High frequency
Hz	Hertz
iid	Independent, identically distributed
INR	Interference-to-noise ratio
kb/s	Kilobits per second
kHz	Kilohertz
LCMV	Linearly constrained minimum variance
LMS	Least mean square
LBS	Least significant bit
LSL	Least-squares lattice
LPC	Linear predictive coding
MA	Moving average
MEM	Maximum entropy method
MN	Minimum norm
MSE	Mean-squared error
MUSIC	Multiple-signal characterization (classification)
MVDR	Minimum-variance distortionless response
PAM	Pulse amplitude modulation
PARCOR	Partial correlation

PCM	Pulse-code modulation
PN	Pseudonoise
QAM	Quadrature amplitude modulation
QRD-LSL	QR-decomposition-based least-squares lattice
QRD-RLS	QR-decomposition-based recursive least squares
RLS	Recursive least squares
rms	root mean square
ROC	Rate of convergence
s	Second
SNR	Signal-to-noise ratio
SRF	Square root filtering
SVD	Singular-value decomposition
wp1	With probability one

PRINCIPAL SYMBOLS

$a_{M,k}(n)$	kth tap weight of forward prediction-error filter of order M (at time n), with $k = 0, 1, \ldots, M$; note that $a_{m,0}(n) = 1$
$\mathbf{a}_M(n)$	Tap-weight vector of forward prediction-error filter of order M (at time n)
$\mathbf{A}$	Data matrix in the covariance method
$\mathbf{A}(n)$	Data matrix in the pre-windowing method
$\mathcal{A}(n)$	Matrix used in defining the unitary transformation $\mathbf{T}(n)$ based on Givens rotations
$b_M(n)$	Backward (a posteriori) prediction error produced at time n by prediction-error filter of order M
$\mathbf{b}(n)$	Backward (a posteriori) prediction-error vector representing sequence of errors produced by backward prediction-error filters of orders $0, 1, \ldots, M$
$\mathcal{B}_M(n)$	Sum of weighted backward prediction error squares produced by backward prediction-error filter of order M
$c(n)$	Colored Gaussian noise
$c_{M,k}(n)$	kth tap weight of backward prediction-error filter of order M (at time n), with $k = 0, 1, \ldots, M$; note that $c_{M,M}(n) = 1$
$\mathbf{c}_M(n)$	Tap-weight vector of backward prediction-error filter of order M (at time n)
$\mathbf{c}(n)$	Weight-error vector in steepest-descent algorithm
$c_k(\tau_1, \tau_2, \ldots, \tau_k)$	kth-order cumulant
$C_k(\omega_1, \omega_2, \ldots, \omega_{k-1})$	kth-order polyspectrum
$\mathcal{C}^M$	Complex M-dimensional parameter space
$\mathcal{C}(n)$	Convergence ratio
$d(n)$	Desired response

$\mathbf{d}$	Desired response vector in the covariance method
$\mathbf{d}(n)$	Desired response vector in the prewindowing method
D	Unit-delay operator
$\mathbf{D}_{m+1}(n)$	Correlation matrix of backward prediction errors
dec( )	Function describing the decision performed by a threshold device
$e(n)$	A posteriori estimation error
$e_{m+1}(n)$	A posteriori estimation error at the output of stage m in the joint-process estimator using the recursive LSL algorithm
e	Base of natural logarithm
exp	Exponential
E	Expectation operator
$\mathscr{E}(\mathbf{w}, n)$	Cost function defined as the sum of weighted error squares expressed as a function of time n
$\mathscr{E}(\mathbf{w})$	Cost function defined as the sum of error squares, expressed as a function of the tap-weight vector $\mathbf{w}$
$\mathscr{E}_{\min}$	Minimum value of $\mathscr{E}(\mathbf{w})$
$\mathscr{E}(n)$	Cost function defined as the sum of weighted error squares, expressed as a function of time n
$f_M(n)$	Forward (a posteriori) prediction error produced at time n by forward prediction-error filter of order M
$\mathbf{f}(n)$	Forward (a posteriori) prediction error vector representing sequence of errors produced by forward prediction-error filters of orders 0, 1, . . . , M
$f_U(u)$	Probability density function of random variable U whose sample value equals u
$f_{\mathbf{U}}(\mathbf{u})$	Joint probability density function of random vector $\mathbf{U}$ whose sample value equals $\mathbf{u}$
$F_M(z)$	z-transform of sequence of forward prediction errors produced by forward prediction-error filter of order M
$\mathbf{F}(n)$	Matrix used in formulating the QRD-RLS algorithm
$\mathscr{F}_M(n)$	Weighted sum of forward prediction-error squares produced by forward prediction-error filter of order M
F	Fourier transform operator
F^{-1}	Inverse Fourier transform operator
g_k	kth tap weight of backward linear predictor
$\mathbf{g}$	Tap-weight vector of backward linear predictor
$\mathbf{g}_N$	Vector in the minimum-norm method
$\mathbf{g}_S$	Another vector in the minimum-norm method
$g(\cdot)$	Nonlinear function used in blind equalization
$\mathbf{G}(n)$	Kalman gain
$\mathbf{G}_N$	Matrix in the minimum-norm method
$\mathbf{G}_S$	Another matrix in the minimum-norm method
h_n	Minimum-phase polynomial used in blind equalization
H_i	ith hypothesis
$H(z)$	Transfer function of discrete-time linear filter

$\mathbf{I}$	Identity matrix
$\mathbf{I}$	Inverse of Fisher's information matrix $\mathbf{J}$
j	Square root of -1
$J(\mathbf{w})$	Cost function used to formulate the Wiener filtering problem, expressed as a function of the tap-weight vector $\mathbf{w}$
$\mathbf{J}$	Fisher's information matrix
$J(n)$	Matrix denoting Jacobi rotation or Givens rotation
$\mathbf{k}(n)$	Gain vector in the RLS algorithm
L	Number of complex sine waves or incident plane waves
K	Final order of moving average model
$\mathbf{K}(n)$	Correlation matrix of weight-error vector
ln	Natural logarithm
$\mathbf{L}(n)$	Transformation matrix in the form of lower triangular matrix
m	Variable order of linear predictor or autoregressive model
M	Final order of linear predictor or autoregressive model
M, K	Final order of autoregressive-moving average model
$\mathcal{M}$	Misadjustment
n	Discrete-time or number of iterations applied to recursive algorithm
N	Data length
$O(z)$	Maximum phase polynomial
$p(-k)$	Element of cross-correlation vector $\mathbf{p}$ for lag k
$\mathbf{p}$	Cross-correlation vector between tap-input vector $\mathbf{u}(n)$ and desired response $d(n)$
$\mathbf{p}$	Vector used in formulating the QRD-RLS algorithm
P_M	Average value of (forward or backward) prediction-error power for prediction order M
$\mathbf{P}(n)$	Matrix equal to the inverse of the time-averaged correlation matrix $\mathbf{\Phi}(n)$ used in formulating the RLS algorithm
q_{k_i}	ith element of kth eigenvector
$\mathbf{q}_k$	kth eigenvector
$\mathbf{Q}$	Unitary matrix that consists of normalized eigenvectors in the set $\{\mathbf{q}_k\}$ used as columns
$\mathbf{Q}(n)$	Unitary matrix used in formulating the QRD-RLS algorithm
$Q(y)$	Probability distribution function of standardized Gaussian random variable
$r(k)$	Element of (ensemble-averaged) correlation matrix $\mathbf{R}$ for lag k
$\mathbf{R}$	Ensembled-average correlation matrix of stationary discrete-time process $\{u(n)\}$
$\mathbf{R}$	Upper triangular matrix used in formulating the QRD-RLS algorithm
$\mathbb{R}^M$	Real M-dimensional parameter space
$\mathbf{s}$	Signal vector; steering vector
sgn( )	Signum function
$S(\omega)$	Power spectral density
$S_{\text{AR}}(\omega)$	Autoregressive spectrum

$S_{\text{MEM}}(\omega)$	MEM (maximum entropy method) spectrum
$S_{\text{MN}}(\omega)$	Minimum norm spectrum
$S_{\text{MVDR}}(\omega)$	Minimum variance distortionless response spectrum
$\mathbf{S}(n)$	Matrix used in the development of the QRD-RLS algorithm
$\mathcal{S}_{\text{d}}$	Decreasingly excited subpsace
$\mathcal{S}_{\text{o}}$	Otherwise excited subspace
$\mathcal{S}_{\text{p}}$	Persistently excited subspace
$\mathcal{S}_{\text{u}}$	Unexcited subspace
t	Time
$\mathbf{T}(n)$	Matrix consisting of the sequence of Givens rotations used in formulating the QRD-RLS algorithm
$u(n)$	Sample value of tap input in transversal filter at time n
$\mathbf{u}(n)$	Tap-input vector consisting of $u(n)$, $u(n-1)$, . . . , as elements
$u_I(n)$	In-phase component of $u(n)$
$u_Q(n)$	Quadrature component of $u(n)$
$\mathbf{u}_k$	kth left-singular vector of data matrix $\mathbf{A}$
$\mathbf{U}$	Matrix of left-singular vectors of data matrix $\mathbf{A}$
$\mathcal{U}_n$	Space spanned by tap inputs $u(n)$, $u(n-1)$, . . .
$\mathcal{U}(n)$	Sum of weighted squared values of tap inputs $u(i)$, $i = 1, 2, \ldots, n$
$v(n)$	Sample value of white-noise process of zero mean
$\mathbf{v}_1(n)$	Process noise vector
$\mathbf{v}_2(n)$	Measurement noise vector
$\mathbf{v}(n)$	Process noise vector in random-walk state model
$\mathbf{v}(n)$	Vector used in formulating the QRD-RLS algorithm
$\mathbf{v}_k(n)$	kth right-singular vector of data matrix $\mathbf{A}$
$\mathbf{V}$	Matrix of right singular vectors of data matrix $\mathbf{A}$
$\mathbf{V}_N$	Matrix constructed from noise subspace eigenvectors
$\mathbf{V}_S$	Matrix constructed from signal plus noise subspace eigenvectors
$w_k(n)$	kth tap weight of transversal filter at time n
$\mathbf{w}(n)$	Tap-weight vector of transversal filter at time n
$\mathbf{x}(n)$	State vector
$\mathbf{y}(n)$	Observation vector used in formulating Kalman filter theory
$\mathcal{Y}_n$	Vector space spanned by $y(n)$, $y(n-1)$, . . .
z^{-1}	Unit-sample (delay) operator used in defining the z-transform of a sequence
$Z(y)$	Standardized Gaussian probability density function
$\alpha(n)$	Innovation or a priori estimation error
$\boldsymbol{\alpha}(n)$	Innovations vector
β	Constant used in the GAL algorithm
$\boldsymbol{\beta}_1(n)$, $\boldsymbol{\beta}_2(n)$	Vectors used in the structural description of unitary transformation $\mathbf{T}(n)$, based on Givens rotations
$\gamma(n)$	Conversion factor used in the FTF algorithm, recursive LSL algorithm, or recursive QRD-LS algorithm

γ_3	Skewness of a random variable
γ_4	Kurtosis of a random variable
Γ_m	Reflection coefficient in mth stage of lattice predictor for stationary inputs
$\Gamma_{b,m}(n)$	Backward reflection coefficient in mth stage of least-squares lattice predictor at time n
$\Gamma_{f,m}(n)$	Forward reflection coefficient in mth stage of least-squares lattice predictor at time n
δ	Constant used in the initialization of RLS and recursive LSL algorithms
$\boldsymbol{\delta}$	First coordinate vector
δ_l	Kronecker delta, equal to 1 for $l = 0$ and zero for $l \neq 0$
Δ_m	Cross-correlation between forward prediction error $f_m(n)$ and delayed backward prediction eror $b_m(n - 1)$
$\Delta_m(n)$	Parameter in recursive LSL algorithm
$\epsilon_m(n)$	Angle-normalized estimation error for prediction order m
$\epsilon_{b,m}(n)$	Angle-normalized backward prediction error for prediction order m
$\epsilon_{f,m}(n)$	Angle-normalized forward prediction error for prediction order m
$\boldsymbol{\epsilon}(n)$	Weight-error vector
$\boldsymbol{\epsilon}$	Estimation error vector in the covariance method
$\boldsymbol{\epsilon}(n)$	Estimation error vector in the prewindowed method
$\eta(n)$	Forward (a priori) prediction error
$\boldsymbol{\theta}$	Time-averaged cross-correlation vector between tap-input vector $\{\mathbf{u}(\mathrm{i})\}$ and desired response $\{d(i)\}$,
$\boldsymbol{\theta}(n)$	Time-averaged cross-correlation vector expressed as a function of the observation interval n
$\boldsymbol{\theta}$	Parameter vector
$\mathbf{k}$	Regression coefficient vector in joint estimator using lattice predictor
$\kappa_4(\tau_1, \tau_2, \tau_3)$	Tricepstrum
λ	Exponential weighting vector in RLS, FTF, LSL, QRD-RLS, and QRD-LSL algorithms
λ_k	kth eigenvalue of correlation matrix $\mathbf{R}$
$\lambda_{\max}$	Maximum eigenvalue of correlation matrix $\mathbf{R}$
$\lambda_{\min}$	Minimum eigenvalue of correlation matrix $\mathbf{R}$
$\lambda(n)$	Threshold
Λ	Likelihood ratio
$\ln \Lambda$	Log-likelihood ratio
$\boldsymbol{\Lambda}(n)$	Diagonal matrix of exponential weighting factors
μ	Mean value
μ	Step-size parameter in steepest-descent algorithm or LMS algorithm
μ	Constant used in the FTF algorithm with soft constraint
$\nu_k(n)$	kth element of $\boldsymbol{\nu}(n)$
$\boldsymbol{\nu}(n)$	Normalized weight-error vector in steepest-descent algorithm

$\phi(t, k)$	t, kth element of time-averaged correlation matrix $\mathbf{\Phi}$
$\boldsymbol{\varphi}(n, n_0)$	Transition matrix arising in finite-precision analysis of RLS algorithms
$\mathbf{\Phi}$	Time-averaged correlation matrix
$\mathbf{\Phi}(n)$	Time-averaged correlation matrix expressed as a function of the observation interval n
$\mathbf{\Phi}(n + 1, n)$	State transition matrix
$\chi(\mathbf{R})$	Eigenvalue spread (i.e., ratio of maximum eigenvalue to minimum eigenvalue) of correlation matrix $\mathbf{R}$
$\psi(n)$	Backward (a priori) prediction error
$\psi^{(f)}(n)$	$\psi(n)$ produced by transversal filtering
$\psi^{(s)}(n)$	$\psi(n)$ produced by manipulation of scalar quantities
ω	Normalized angular frequency; $0 \le \omega \le 2\pi$
ρ_m	Correlation coefficient or normalized value of autocorrelation function for lag m
ρ_m	Parameter in recursive LSL algorithm used for joint-process estimation
σ^2	Variance
$\mathbf{\Sigma}$	Correlation matrix of innovations process $\{\alpha(n)\}$
τ_k	Time constant of kth natural mode of steepest-descent algorithm
$\tau_{\text{mse, av}}$	Time constant of a single decaying exponential that approximates the learning curve of LMS algorithm
$\nabla(n)$	Convolutional noise in blind equalization
$\nabla(J)$	Gradient vector

References and Bibliography

ABRAHAM, J. A., ET AL. (1987). "Fault tolerance techniques for systolic arrays," *Computer,* vol. 20, pp. 65–75.

AFINSON, C. J., and F. T. LUK (1988). "A Linear Algebraic Model of Algorithm-based Fault Tolerance", Proceedings of the International Conference on Systolic Arrays, San Diego, May 25–27.

AKAIKE, H. (1973). "Maximum likelihood identification of Gaussian autoregressive moving average models," *Biometrika,* vol. 60, pp. 255–265.

AKAIKE, H. (1974). "A new look at the statistical model identification," *IEEE Trans. Autom. Control,* vol. AC-19, pp. 716–723.

AKAIKE, H. (1977). "An entropy maximisation principle," in *Proceedings of Symposium on Applied Statistics,* ed., P. Krishnaiah, North-Holland, Amsterdam.

ALBERT, A. E., and L. S. GARDNER, JR. (1967). *Stochastic Approximation and Nonlinear Regression.* MIT Press.

ALEXANDER, S. T. (1986a). *Adaptive Signal Processing: Theory and Applications,* Springer-Verlag, New York.

ALEXANDER, S. T. (1986b). "Fast adaptive filters: a geometrical approach," *IEEE ASSP Mag.,* pp. 18–28.

ALEXANDER, S. T. (1987). "Transient weight misadjustment properties for the finite precision LMS algorithm," *IEEE Trans. Acoust. Speech Signal Process.,* vol. ASSP-35, pp. 1250–1258.

ANDERSON, T. W. (1963). "Asymptotic theory for principal component analysis," *Ann. Math. Stat.,* vol. 34, pp. 122–148.

ANDERSON, B. D. O., and J. B. MOORE (1979). *Linear Optimal Control,* Prentice-Hall, Englewood Cliffs, N.J.

ANDREWS, H. C., and C. L. PATTERSON (1975). "Singular value decomposition and digital image processing," *IEEE Trans. Acoust. Speech Signal Process.,* vol. ASSP-24, pp. 26–53.

APPLEBAUM, S. P. (1966). "Adaptive arrays," *Syracuse University Research Corporation,* Rep. SPL TR 66-1. This report is reproduced in *IEEE Trans. Antennas Propag.,* Special Issue on Adaptive Antennas, vol. AP-24, pp. 585–598, 1976.

APPLEBAUM, S. P., and D. J. CHAPMAN (1976). "Adaptive arrays with main beam constraints," *IEEE Trans. Antennas Propag.*, vol. AP-24, pp. 650–662.

ARDALAN, S. H. (1986). "Floating-point error analysis of recursive least-squares and least-mean-squares adaptive filters," *IEEE Trans. Circuits Syst.*, vol. CAS-33, pp. 1192–1208.

ÅSTRÖM, K. J., and P. EYKHOFF (1971). "System identification—a survey," *Automatica,* vol. 7, pp. 123–162.

ÅSTRÖM, K. J., and B. WITTENMARK (1990). *Computer-Controlled Systems*, Second Edition, Prentice-Hall.

ATAL, B. S. (1970). "Speech analysis and synthesis by linear prediction of the speech wave," *J. Acoust. Soc. Am.*, vol. 47, p. 65.

ATAL, B. S., and S. L. HANAUER (1971). "Speech analysis and synthesis by linear prediction of the speech wave," *J. Acoust. Soc. Am.*, vol. 50, pp. 637–655.

ATAL, B. S., and M. R. SCHROEDER (1967). "Predictive coding of speech signals," *Proc. 1967 Conf. Commun. Process.*, pp. 360–361.

ATAL, B. S., and M. R. SCHROEDER (1970). "Adaptive predictive coding of speech signals," *Bell Syst. Tech. J.*, vol. 49, pp. 1973–1986.

AUSTIN, M. E. (1967). *Decision-feedback equalization for digital communication over dispersive channels,* Tech., Rep. 437, MIT Lincoln Laboratory, Lexington, Mass.

AUTONNE, L. (1902). "Sur les groupes linéaires, réels et orthogonaux," *Bull. Soc. Math., France,* vol. 30, pp. 121–133.

BARRETT. J. F. and D. G. LAMPARD (1955). "An Expansion for Some Second-order Probability Distributions and its Application to Noise Problems", *IRE Trans. Information Theory,* vol. IT-1, pp. 10–15.

BELFIORE, C. A., and J. H. PARK, JR (1979). "Decision feedback equalization," *Proc. IEEE,* vol. 67, pp. 1143–1156.

BELLANGER, M. G. (1988a). *Adaptive Filters and Signal Analysis,* Marcel-Dekker, New York.

BELLANGER, M. G. (1988b). "The FLS-QR algorithm for adaptive filtering," *Signal Process.*, vol. 17, pp. 291–304.

BELLINI, S. (1986). "Bussgang techniques for blind equalization," Globecom, Houston, Tex., pp. 1634–1640.

BELLINI, S. (1988). "Blind equalization," *Alta Freq.*, vol. 57, pp. 445–450.

BELLINI, S., and F. ROCCA (1986). "Blind deconvolution: polyspectra or Bussgang techniques?" in *Digital Communications,* ed. E. Biglieri and G. Prati, North-Holland, Amsterdam, pp. 251–263.

BELLMAN, R. (1960). *Introduction to Matrix Analysis,* McGraw-Hill, New York.

BENVENISTE, A., and M. GOURSAT (1984). "Blind equalizers," *IEEE Trans. Commun.*, vol. COM-32, pp. 871–883.

BENVENISTE, A., M. GOURSAT, and G. RUGET (1980). "Robust identification of a nonminimum phase system: blind adjustment of a linear equalizer in data communications," *IEEE Trans. Autom. Control,* vol. AC-25, pp. 385–399.

BENVENUTO, N., ET AL. (1986). "The 32 kb/s ADPCM coding standard," *AT&T J.*, vol. 65, pp. 12–22.

BERKHOUT, A. J., and P. R. ZAANEN (1976). "A comparison between Wiener filtering, Kalman filtering, and deterministic least squares estimation," *Geophysical Prospect.*, vol. 24, pp. 141–197.

adaptive filtering," *IEEE Trans. Acoust. Speech, Signal Process.*, vol. ASSP-32, pp. 304–337.

CLAASEN, T. A. C. M., and W. F. G. MECKLANBRÄUKER (1985). "Adaptive techniques for signal processing in communications," *IEEE Communi.*, vol. 23, pp. 8–19.

COHN, A. (1922). "Über die Anzahl der Wurzeln einer algebraischen Gleichung in einem Kreise," *Math. Z.*, vol. 14, pp. 110–148.

COMPTON, R. T. (1988). JR., *Adaptive Antennas: Concepts and Performance*. Prentice-Hall, Englewood Cliffs, N. J.

COWAN, C. F. N., and P. M. GRANT (1985). *Adaptive Filters*, Prentice-Hall, Englewood Cliffs, N.J.

COWAN, C. F. N., G. J. GIBSON, and S. SIU (1989). "Data equalization using highly non-linear adaptive architectures," SPIE, San Diego, Ca.

COX, H., R. M. ZESKIND, and M. M. OWEN (1987). "Robust adaptive beamforming," *IEEE Trans. Acoust. Speech Signal Process.*, vol. ASSP-35, pp. 1365–1376.

CUTLER, C. C. (1952). *Differential Quantization for Communication Signals*, U. S. Patent 2,605,361.

DAHANAYAKE, B. W., K. M. WONG, and G. P. MADHAVAN (1989). "Spectrum and flat spectral line representation: geometric approach," *IEEE Fourth Annu. ASSP Workshop Spectrum Estimation Modeling*, Minneapolis, Minn., pp. 312–317.

DEIFT, P. J. DEMMEL, C. TOMAL, and L.-C. LI, (1989). *The Bidiagonal singular value decomposition and Hamiltonian mechanics*, Rep. 458, Department of Computer Science, Courant Institute of Mathematical Sciences, New York University, New York.

DELOPOULOS, A., and G. B. GIANNAKIS (1990). "Strongly consistent identification algorithms and noise insensitive MSE criteria," The 1990 Digital Signal Processing Workshop, New Paltz, NY, sponsored by IEEE Signal Processing Society, pp. 8.6.1–8.6.2.

DEMMEL, J., and W. KAHAN (1990). "Accurate singular values of bidiagonal matrices," *SIAM J, Sci. Stat. Comp.*

DEMMEL, J., and K. VESELIĆ (1989). *Jacobi's method is more accurate than QR*, Tech. Rep. 468, Department of Computer Science, Courant Institute of Mathematical Sciences, New York University, New York.

DEMOOR, B. L. R., and G. H. GOLUB (1989). *Generalized singular value decompositions: a proposal for a standardized nomenclature*, Manuscript NA-89-05, Numerical Analysis Project, Computer Science Department, Standard University, Stanford, Calif.

DEPRETTERE, E. F., editor (1988). "SVD and Signal Processing: Algorithms, Applications, and Architectures," North Holland.

DEWILDE, P. (1969). "*Cascade scattering matrix synthesis*," Ph.D. dissertation, Stanford University, Stanford, Calif.

DEWILDE, P., A. C. VIEIRA, and T. KAILATH (1978). "On a generalized Szegö–Levinson realization algorithm for optimal linear predictors based on a network synthesis approach," *IEEE Trans. Circuits Syst.*, vol. CAS-25, pp. 663–675.

DHRYMUS, P. J. (1970). *Econometrics: Statistical Foundations and Applications*, Harper & Row, New York.

DING, Z., C. R. JOHNSON, JR., and R. A. KENNEDY (1990). "On the Admissability of Blind Adaptive Equalizers", International Conference on Acoustics, Speech, and Signal Processing, pp. 1707–1710.

DING, Z., ET AL. (1989). "On the Ill-convergence of Godard Blind Equalizers in Data Communication Systems," Proc. CISS'89, pp. 538–543.

DiToro, M. J. (1965). "A new method for high speed adaptive signal communication through any time variable and dispersive transmission medium," *1st IEEE Annu. Commun. Conf.*, pp. 763–767.

Dongarra, J. J., et al. (1979). *LINPACK User's Guide*, Society for Industrial and Applied Mathematics, Philadelphia.

Doob, L. J. (1953). *Stochastic Processes*, Wiley, New York.

Dorny, C. N. (1975). *A Vector Space Approach to Models and Optimization*, Wiley-Interscience, New York.

Durbin, J., (1960). "The fitting of time series models," *Rev. Int. Stat. Inst.*, vol. 28, pp. 233–244.

Duttweiler, D. L., and Y. S. Chen (1980). "A single-chip VLSI echo canceler," *Bell Syst. Tech. J.*, vol. 59, pp. 149–160.

Eckart, G., and G. Young (1936). "The approximation of one matrix by another of lower rank," Psychometrika, vol. 1, pp. 211–218.

Eckart, C., and G. Young (1939). "A principal axis transformation for non-Hermitian matrices," *Bull.Am. Math. Soc.*, vol. 45, pp. 118–121.

Edwards, A. W. F. (1972). *Likelihood*, Cambridge University Press, New York.

Eleftheriou, E., and D. D. Falconer (1984). "Tracking properties and steady state performance of RLS adaptive filter algorithms," Report SCE-84-14, Department of Systems and Computer Engineering, Carleton University, Ottawa, Canada.

Eleftheriou, E., and D. D. Falconer (1986). "Tracking properties and steady state performance of RLS adaptive filter algorithms," *IEEE Trans. Acoust. Speech Signal Process.*, vol. ASSP-34, pp. 1097–1110.

Er, M. H., and A. Cantoni (1986). "A new set of linear constraints for broadband time domain element space processors," *IEEE Trans. Antennas Propag.*, vol. AP-34 (March).

Evans, J. E. (1979). "Aperture sampling techniques for precision direction finding," *IEEE Trans. Aerospace Electron. Syst.*, vol. AES-15, pp. 891–895.

Evans, J. E., J. R. Johnson, and D. F. Sun (1982). *Applications of advanced signal processing techniques to angle of arrival estimation in ATC navigation and surveillance systems*, Tech. Rept. 582, MIT Lincoln Laboratory, Lexington, Mass.

Eweda, E., and O. Macchi (1985). "Tracking Error Bounds of Adaptive Nonstationary Filtering," Automatica, vol. 21, pp. 293–302.

Falconer, D. D., and L. Ljung (1978). "Application of fast Kalman estimation to adaptive equalization," *IEEE Trans. Commun.*, vol. COM-26, pp. 1439–1446.

Farden, D. C. (1981a). "Stochastic approximation with correlated data," *IEEE Trans. Inf. Theory*, vol. IT-27, pp. 105–113.

Farden, D. C. (1981b). "Tracking properties of adaptive signal processing algorithms," *IEEE Trans. Acoust. Speech Signal Process.*, vol. ASSP-29, pp. 439–446.

Fertig, L. B., and J. H. McClellan (1990). "Dual form adaptive filters with norm constraints," the 1990 Digital Signal Workshop, New Paltz, NY, sponsored by IEEE Signal Processing Society, pp. 4.6.1–4.6.2.

Fisher, B., and N. J. Bershad (1983). "The complex LMS adaptive algorithm—transient weight mean and covariance with applications to the ALE," *IEEE Trans. Acoust. Speech Signal Process.*, vol. ASSP-31, pp. 34–44.

Flanagan, J. L. (1972). *Speech Analysis, Synthesis and Perception*, 2nd ed., Springer-Verlag, New York.

Flanagan, J. L., et al. (1979). "Speech coding," *IEEE Trans. Commun.*, vol. COM-27, pp. 710–737.

FORNEY, G. D. (1972). "Maximum-likelihood sequence estimation of digital sequence in the presence of intersymbol interference," *IEEE Trans. Inf. Theory,* vol. IT-18, pp. 363–378.

FÖRSTNER, W. (1989). "Robust methods for computer vision," tutorial notes, *IEEE Comput. Soc. Conf. Comput. Vision Pattern Recognition,* June 4–8.

FORSYTHE, G. E., and P. HENRICI (1960). "The cyclic Jacobi method for computing the principal values of a complex matrix," *Trans. Amer. Math. Soc.,* vol. 94, pp. 1–23.

FOSCHINI, G. J. (1985). "Equalizing without altering or detecting data," *AT&T Tech. J.,* vol. 64, pp. 1885–1911.

FOUGERE, P. E., E. J. ZAWALICK, and H. R. RADOSKI (1976). "Spontaneous line splitting in maximum entropy power spectrum analysis," *Phys. Earth Planet Int.,* vol. 12, pp. 201–207.

FRANKS, L. E., ed. (1974). *Data Communication: Fundamentals of Baseband Transmission,* Benchmark Papers in Electrical Engineering and Computer Science, Dowden, Hutchinson & Ross, Stroudsburg, Pa.

FRIEDLANDER, B. (1982). "Lattice filters for adaptive processing," *Proc. IEEE,* vol. 70, pp. 829–867.

FRIEDLANDER, B. (1988). "A signal subspace method for adaptive interference cancellation," *IEEE Trans. Acoust. Speech Signal Process.,* vol. ASSP-36, pp. 1835–1845.

FRIEDLANDER, B., and B. PORAT (1989), "Adaptive IIR algorithm based on high-order statistics," *IEEE Trans. Acoust. Speech Signal Process.,* vol. ASSP-37, pp. 485–495.

FROST III, O. L. (1972). "An algorithm for linearly constrained adaptive array processing," *Proc. IEEE,* vol. 60, pp. 926–935.

GABOR, D., W. P. L. WILBY, and R. WOODCOCK (1960). "A universal non-linear filter, predictor and simulator which optimizes itself by a learning process," *IEE Proc. (London),* vol. 108, pp. 422–438.

GABRIEL, W. F. (1976). "Adaptive arrays: an introduction," *Proc. IEEE,* vol. 64, pp. 239–272.

GABRIEL, W. F. (1980). "Spectral analysis and adaptive array superresolution techniques," *Proc. IEEE,* vol. 68, pp. 654–666.

GABRIEL, W. F. (1989). "Superresolution techniques and ISAR imaging," IEEE National Radar Conference Digest, pp. 48–55.

GABRIEL, W. F. (1990). "Superresolution techniques in the range domain," International Radar Conference, pp. 263–266.

GALLIVAN, K. A., and C. E. LEISERSON (1984). "High-performance architectures for adaptive filtering based on the Gram–Schmidt algorithm," *Proc. SPIE,* vol. 495, Real Time Signal Processing VII, pp. 30–38.

GARBOW, B. S., ET AL. (1977). *Matrix Eigensystem Routines—EISPACK Guide Extension,* Lecture Notes in Computer Science, vol. 51, Springer-Verlag, New York.

GARDNER, W. A. (1984). "Learning characteristics of stochastic-gradient-descent algorithms: a general study, analysis and critique," *Signal Process.,* vol. 6, pp. 113–133.

GARDNER, W. A. (1987). "Nonstationary learning characteristics of the LMS algorithm," *IEEE Trans. Circuits Syst.,* vol. CAS-34, pp. 1199–1207.

GAUSS, C. F. (1809). *Theoria motus corporum coelestium in sectionibus conicus solem ambientum,* Hamburg (translation: Dover, New York, 1963).

GENTLEMAN, W. M. (1973). "Least squares computations by Givens transformations without square-roots," *J. Inst. Math. Its Appl.,* vol. 12, pp. 329–336.

GENTLEMAN, W. M., and H. T. KUNG (1981). "Matrix triangularization by systolic arrays," *Proc. SPIE,* vol. 298, Real Time Signal Processing IV, pp. 298–303.

GERSHO, A., (1968). "Adaptation in a quantized parameter space," *Allerton Conf.*, pp. 646–653.

GERSHO, A., (1969). "Adaptive equalization of highly dispersive channels for data transmission," *Bell Syst. Tech. J.*, vol. 48, pp. 55–70.

GERSHO, A., B. GOPINATH, and A. M. OLDYZKO, (1979). "Coefficient inaccuracy in transversal filtering," *Bell Syst. Tech. J.*, vol. 58, pp. 2301–2316.

GIBSON, J. D. (1980). "Adaptive prediction in speech differential encoding systems," *Proc. IEEE,* vol. 68, pp. 488–525.

GILL, P. E., G. H. GOLUB, W. MURRAY, and M. A. SAUNDERS (1974). "Methods of modifying matrix factorizations," *Math. Comput.*, vol. 28, pp. 505–535.

GITLIN, R. D., and F. R. MAGEE, JR. (1977). "Self-orthogonalizing adaptive equalization algorithms," *IEEE Trans. Commun.*, vol. COM-25, pp. 666–672.

GITLIN, R. D., and S. B. WEINSTEIN (1979). "On the required tap-weight precision for digitally implemented mean-squared equalizers," *Bell Syst. Tech. J.*, vol. 58, pp. 301–321.

GITLIN, R. D., and S. B. WEINSTEIN (1981). "Fractionally spaced equalization: an improved digital transversal equalizer," *Bell Syst. Tech. J.*, vol. 60, pp. 275–296.

GITLIN, R. D., J. E. MAZO, and M. G. TAYLOR (1973). "On the design of gradient algorithms for digitally implemented adaptive filters," *IEEE Trans. Circuit Theory,* vol. CT-20, pp. 125–136.

GIULI, D. (1986). "Polarization diversity in radars", *Proc. IEEE,* vol. 74, pp. 245–269.

GIVENS, W. (1958). "Computation of plane unitary rotations transforming a general matrix to triangular form," *J. Soc. Ind. Appl. Math,* vol. 6, pp. 26–50.

GLASER, E. M. (1961). "Signal detection by adaptive filters," *IRE Trans. Inf. Theory,* vol. IT-7, pp. 87–98.

GLOVER, J. R., JR. (1977). "Adaptive noise cancelling applied to sinusoidal interferences," *IEEE Trans. Acoust. Speech Signal Process.*, vol. ASSP-25, pp. 484–491.

GODARA, L. C., and A. CANTONI (1986). "Analysis of constrained LMS algorithm with application to adaptive beamforming using perturbation sequences," *IEEE Trans. Antennas Propag.*, vol. AP-34 (March), pp. 368–379.

GODARD, D. N. (1974), "Channel equalization using a Kalman filter for fast data transmission," *IBM J. Res. Dev.*, vol. 18, pp. 267–273.

GODARD, D. N. (1980). "Self-recovering equalization and carrier tracking in a two-dimensional data communication system," *IEEE Trans. Commun.*, vol. COM-28, pp. 1867–1875.

GODFREY R., and F. ROCCA (1981). "Zero memory non-linear deconvolution," *Geophys. Prospect.*, vol. 29, pp. 189–228.

GOLD, B. (1977). "Digital speech networks," *Proc. IEEE,* vol. 65, pp. 1636–1658.

GOLOMB, S. W. editor (1964). *Digital Communications with Space Applications*, Prentice-Hall.

GOLUB, G. H. (1965). "Numerical methods for solving linear least squares problems," *Numer. Math.*, vol. 7, pp. 206–216.

GOLUB, G., and W. KAHAN (1965). "Calculating the singular values and pseudo-inverse of a matrix," *J. SIAM Numer. Anal. B.*, vol. 2, pp. 205–224.

GOLUB, G. H., and C. REINSCH (1970). "Singular value decomposition and least squares problems," *Numer. Math.*, vol. 14, pp. 403–420.

GOLUB, G. H., F. T. LUK, and M. L. OVERTON (1981). "A Block Lanczos Method for

Computing the Singular Values and Corresponding Singular Vectors of a Matrix," *ACM Trans. Mathematical Software,* vol. 7, pp. 149–169.

GOLUB, G. H., and C. F. VAN LOAN (1989). *Matrix Computations,* Second Edition, The Johns Hopkins University Press, Baltimore, Md.

GOODWIN, G. C., and R. L. PAYNE (1977). *Dynamic System Identification: Experiment Design and Data Analysis,* Academic Press, New York.

GOODWIN, G. C., and K. S. SIN (1984). *Adaptive Filtering, Prediction and Control,* Prentice-Hall, Englewood Cliffs, N.J.

GRAY, R. M., and L. D. DAVISSON (1986). *Random Processes: A Mathematical Approach for Engineers*, Prentice-Hall.

GRAY, R. M. (1990). "Quantization noise spectra," *IEEE Trans. Information Theory,* vol. 36, pp. 1220–1244.

GRAY, W. (1979). *"Variable norm deconvolution,"* Ph.D. dissertation, Department of Geophysics, Stanford University, Stanford, Calif.

GRENANDER, U., and G. SZEGÖ (1958). *Toeplitz Forms and Their Applications,* University of California Press, Berkeley, Calif.

GRIFFITHS, L. J. (1975). "Rapid measurement of digital instantaneous frequency," *IEEE Trans. Acoust. Speech Signal Process.,* vol. ASSP-23, pp. 207–222.

GRIFFITHS, L. J. (1977). "A continuously adaptive filter implemented as a lattice structure," *Proc. ICASSP,* Hartford, Conn., pp. 683–686.

GRIFFITHS, L. J. (1978). "An adaptive lattice structure for noise-cancelling applications," *Proc. ICASSP,* Tulsa, Okla., pp. 87–90.

GRIFFITHS, L. J., and C. W. JIM (1982). "An alternative approach to linearly constrained optimum beamforming," *IEEE Trans. Antennas and Propag.,* vol. AP-30, pp. 27–34.

GRIFFITHS, L. J., and R. PRIETO-DIAZ (1977). "Spectral analysis of natural seismic events using autoregressive techniques," *IEEE Trans. Geosci. Electron.,* vol. GE-15, pp. 13–25.

GRIFFITHS, L. J., F. R. SMOLKA, and L. D. TREMBLY (1977). "Adaptive deconvolution: a new technique for processing time-varying seismic data," *Geophysics,* vol. 42, pp. 742–759.

GUILLEMIN, E. A. (1949). *The Mathematics of Circuit Analysis,* Wiley.

GUPTA, I. J., and A. A. KSIENSKI (1986). "Adaptive antenna arrays for weak interfering signals," *IEEE Trans. Antennas Propag,.* vol. AP-34 (March), pp. 420–426.

GUTOWSKI, P. R., E. A. ROBINSON, and S. TREITEL (1978). "Spectral estimation: fact or fiction," *IEEE Trans. Geosci. Electron.,* vol. GE-16, pp. 80–84.

HADHOUD, M. M., and D. W. THOMAS (1988). "The two-dimensional adaptive LMS (TDLMS) algorithm," *IEEE Trans. Circuits Syst.,* vol. CAS-35, pp. 485–494.

HAMPEL, F. R., ET AL. (1986). *Robust Statistics: The Approach Based on Influence Functions,* Wiley, New York.

HARIHARAN, S., and A.P. CLARK (1990). "HF channel estimation using a fast transversal filter algorithm," *IEEE Trans Acoust. Speech, and Signal Processing,* vol. 38, pp. 1353–1362.

HASTINGS-JAMES, R., and M. W. SAGE (1969). "Recursive generalized-least-squares procedure for online identification of process parameters," *IEE Proc. (London),* vol. 116, pp. 2057–2062.

HATZINAKOS, D. (1990). "Blind Equalization Based on Polyspectra," Ph.D. thesis, Northeastern University, Boston, Massachusetts.

HATZINAKOS, D., and C. L. NIKIAS (1989). "Estimation of multipath channel response in frequency selective channels," *IEEE J. Sel. Areas Commun.,* vol. 7, pp. 12–19.

HATZINAKOS, D., and C. L. NIKIAS (1990). "Blind equalization using a tricepstrum based algorithm," *IEEE Trans. Commun.*, vol. COM-38.

HAYKIN, S. (1983a). *Communication Systems,* 2nd ed., Wiley, New York.

HAYKIN, S., ed. (1983b). *Nonlinear Methods of Spectral Analysis,* 2nd ed., Springer-Verlag, New York.

HAYKIN, S., ed., (1984). *Array Signal Processing,* Prentice-Hall, Englewood Cliffs, N.J.

HAYKIN, S. (1986). "Polarimetric radar for Accurate Navigation," Proc. International Symposium on Marine Positioning, pp. 69–76, D. Reidel Publishing Co.

HAYKIN, S. (1988). *Digital Communications,* Wiley, New York.

HAYKIN, S. (1989a). *Modern Filters,* Macmillan, New York.

HAYKIN, S. (1989b). "Adaptive filters: past, present, and future," *Proc. IMA Conf. Math. Signal Process.,* Warwick, England.

HAYKIN, S., (1991). Introduction to Neural Networks, Macmillan, New York.

HAYKIN, S., B. W. CURRIE, and S. B. KESLER (1982). "Maximum-entropy spectral analysis of radar clutter," *Proc. IEEE,* vol. 70, 1982, pp. 953–962.

HAYKIN, S., T. GREENLAY, and J. LITVA (1985). "Performance evaluation of the modified FBLP method for angle of arrival estimation using real radar multipath data," *IEE Proc. (London),* vol. 132, pt. F, pp. 159–174.

HAYKIN, S., ET AL. (1991). "Classification of radar clutter in an air traffic control environment," *Proc. IEEE,* vol. 79.

HO, Y. C. (1963). "On the stochastic approximation method and optimal filter theory," *J. Math. Anal. Appl.,* vol. 6, pp. 152–154.

HODGKISS, W. S., and D. ALEXANDROU (1983). "Applications of adaptive least-squares lattice structures to problems in underwater acoustics," *Proc. SPIE,* vol. 431, Real Time Signal Processing VI, pp. 48–54.

HODGKISS, W. S., JR., and J. A. PRESLEY, JR. (1981). "Adaptive tracking of multiple sinusoids whose power levels are widely separated," *IEEE Trans. Circuits Syst.,* vol. CAS-28, pp. 550–561.

HONIG, M. L., and D. G. MESSERSCHMITT (1981). "Convergence Properties of an Adaptive Digital Lattice Filter," *IEEE Trans. Acoustics, Speech, and Signal Processing,* vol. ASSP-29, pp. 642–653.

HONIG, M. L., and D. G. MESSERSCHMITT (1984). *Adaptive Filters: Structures, Algorithms and Applications,* Kluwer Boston., Hingham, Mass.

HOROWITZ, L. L., and K. D. SENNE (1981). "Performance advantage of complex LMS for controlling narrow-band adaptive arrays," *IEEE Trans. Acoust. Speech Signal Process.,* vol. ASSP-29, pp. 722–736.

HOUSEHOLDER, A. S. (1958a). "Unitary triangularization of a nonsymmetric matrix," *J. Assoc. Comput. Mach.,* vol. 5, pp. 339–342.

HOUSEHOLDER, A. S. (1958b). "The approximate solution of matrix problems," *J. Assoc. Comput. Mach.,* vol. 5, pp. 204–243.

HOUSEHOLDER, A. S. (1964). *The Theory of Matrices in Numerical Analysis,* Blaisdell, Waltham, Mass.

HOWELLS, P. W. (1965). *Intermediate Frequency Sidelobe Canceller,* U.S. Patent 3,202,990, August 24.

HOWELLS, P. W. (1976). "Explorations in fixed and adaptive resolution at GE and SURC," *IEEE Trans. Antennas Propag.,* vol. AP-24, Special Issue on Adaptive Antennas, pp. 575–584.

HSIA, T. C. (1983). "Convergence analysis of LMS and NLMS adaptive algorithms," *Proc. ICASSP,* Boston, pp. 667–670.

HSU, F. M. (1982). "Square root Kalman filtering for high-speed data received over fading dispersive HF channels," *IEEE Trans. Inf. Theory,* vol. IT-28, pp. 753–763.

HUBER, P. J. (1981). "Robust Statistics", Wiley.

HUBING, N. E., and S. T. ALEXANDER (1990). "Statistical analysis of the soft constrained initialization of recursive least squares algorithms," *Proc. ICASSP,* Albuquerque N.Mex.

HUDSON, J. E. (1981). *Adaptive Array Principles,* Peter Peregrinus, London.

HUHTA, J. C., and J. G. WEBSTER (1973). "60-Hz interference in electrocardiography," *IEEE Trans. Biomed. Eng.,* vol. BME-20, pp. 91–101.

IOANNOU, P. A., (1990). "Robust adaptive control," Proc. Sixth Yale Workshop on Adaptive and Learning Systems, Yale University, pp. 32–39.

IOANNOU, P. A., and P. V. KOKOTOVIC (1983). *Adaptive Systems with Reduced Models,* Springer-Verlag, New York.

ITAKURA, F., and S. SAITO (1971). "Digital filtering techniques for speech analysis and synthesis," *Proc. 7th Int. Conf. Acoust.,* Budapest, vol. 25-C-1, pp. 261–264.

ITAKURA, F., and S. SAITO (1972). "On the optimum quantization of feature parameters in the PARCOR speech synthesizer," *IEEE 1972 Conf. Speech Commun. Process.,* New York, pp. 434–437.

JABLON, N. K. (1986). "Steady state analysis of the generalized sidelobe canceller by adaptive noise canceling techniques," *IEEE Trans. Antennas Propag.,* vol. AP-34 (March), pp. 330–337.

JACOBI, C. G. J. (1846). "Über ein leichtes Verfahren, die in her Theorie der Säcularstorungen vorkommenden Geichungen numerisch aufzulösen," J. reine angew. Math. 30, pp. 51–95.

JAYANT, N.S., and P. NOLL (1984). *Digital Coding of Waveforms,* Prentice-Hall, Englewood Cliffs, N.J.

JAYANT, N. S. (1986). "Coding speech," *IEEE Spectrum,* vol. 23, pp. 58–63.

JAYNES, E. T., (1982). "On the rationale of maximum-entropy methods," *Proc. IEEE,* vol. 70, pp. 939–952.

JAZWINSKI, A. H. (1969). "Adaptive filtering," *Automatica,* vol. 5, pp. 475–485.

JASWINSKI, A. W. (1970). *Stochastic Processes and Filtering Theory,* Academic Press. New York.

JOHNSON, D. H., (1982). "The application of spectral estimation methods to bearing estimation problems," *Proc. IEEE,* vol. 70, pp. 1018–1028.

JOHNSON, D. H., and P. S. RAO (1990) "On the existence of Gaussian noise," The 1990 Digital Signal Processing Workshop, New Paltz, NY, Sponsored by IEEE Signal Processing Society, pp. 8.14.1–8.14.2.

JOHNSON, C. R., JR. (1984). "Adaptive IIR filtering: current results and open issues," *IEEE Trans. Inf. Theory,* vol. IT-30, Special Issue on Linear Adaptive Filtering, pp. 237–250.

JOHNSON, C. R., JR. (1988). *Lectures on Adaptive Parameter Estimation,* Prentice-Hall, Englewood Cliffs, N.J.

JOHNSON, C. R., JR. (1991). "Admissibility in blind adaptive channel equalization: a tutorial survey of an open problem," *IEEE Control Systems Magazine,* vol. 11.

JOHNSON, C. R., JR., S. DASGUPTA, and W. A. SETHARES (1988). "Averaging Analysis of local stability of a real constant modulus algorithm adaptive filter," *IEEE Trans. Acoust. Speech, Signal Process.,* vol. ASSP-36, pp. 900–910.

JONES, S. K., R. K. CAVIN III, and W. M. REED (1982). "Analysis of error-gradient adaptive linear equalizers for a class of stationary-dependent processes," *IEEE Trans. Inf. Theory,* vol. IT-28, pp. 318–329.

JOU., J.-Y., and A. ABRAHAM (1986). "Fault-tolerant matrix arithmetic and signal processing on highly concurrent computing structures," *Proc. IEEE,* Special Issue on Fault Tolerance in VLSI, vol. 74, pp. 732–741.

JUSTICE, J. H. (1985). "Array Processing in Exploration Seismology", Chapter 2 in book entitled *Array Signal Processing,* edited by S. Haykin, Prentice-Hall.

KAILATH, T. (1960). *Estimating filters for linear time-invariant channels, Quarterly Progress Rep.* 58, MIT Research Laboratory for Electronics, Cambridge, Mass., pp. 185–197.

KAILATH, T. (1968). "An innovations approach to least-squares estimation: Part 1. Linear filtering in additive white noise," *IEEE Trans. Autom. Control,* vol. AC-13, pp. 646–655

KAILATH, T. (1969). "A generalized likelihood ratio formula for random signals in Gaussian noise," *IEEE Trans. Inf. Theory,* vol. IT-15, pp. 350–361.

KAILATH, T. (1970). "The innovations approach to detection and estimation theory," *Proc. IEEE,* vol. 58, pp. 680–695.

KAILATH, T. (1974). "A view of three decades of linear filtering theory," *IEEE Trans. Inf. Theory,* vol. IT-20, pp. 146–181.

KAILATH, T. ed. (1977). *Linear Least-Squares Estimation,* Benchmark Papers in Electrical Engineering and Computer Science, Dowden, Hutchinson & Ross, Stroudsburg, Pa.

KAILATH, T., and P. A. FROST (1968). "An innovations approach to least-squares estimation: Part 2. Linear smoothing in additive white noise," *IEEE Trans. Autom. Control,* vol. AC-13, pp. 655–660.

KAILATH, T., and R. A. GEESEY (1973). "In Innovations approach to least-squares estimation: Part 5. Innovation representations and recursive estimation in colored noise," *IEEE Trans. Autom. Control,* vol. AC-18, pp. 435–453.

KAILATH, T. (1980). *Linear Systems,* Prentice-Hall, Englewood Cliffs, N.J.

KAILATH, T. (1981). *Lectures on Linear Least-Squares Estimation,* Springer-Verlag, New York.

KAILATH, T. (1982). "Time-variant and time-invariant lattice filters for nonstationary processes" in *Outils et modèles mathématique pour l'automatique, l'analyse de systèms et le traitement du signal,* vol. 2, ed. I Landau, CNRS, Paris, pp. 417–464.

KAILATH, T., A. VIEIRA, and M. MORF (1978). "Inverses of Toeplitz operators, innovations, and orthogonal polynomials," *SIAM Rev.* vol. 20, pp. 106–119.

KALLMANN, H. J. (1940). "Transversal filters," *Proc. IRE,* vol. 28, pp. 302–310.

KALMAN, R. E. (1960). "A new approach to linear filtering and prediction problems," *Trans. ASME J. Basic Eng.,* vol. 82, pp. 35–45.

KALMAN, R. E., and R. S. BUCY (1961). "New results in linear filtering and prediction theory," *Trans. ASME J. Basic Eng.,* vol. 83, pp. 95–108.

KALOUPTSIDIS, N., and S. THEODORIDIS (1987). "Parallel implementation of efficient LS algorithms for filtering and prediction," *IEEE Trans. Acoust. Speech Signal Process.,* vol. ASSP-35, pp. 1565–1569.

KAMINSKI, P. G., A. E. BRYSON, and S. F. SCHMIDT (1971). "Discrete square root filtering: a survey of current techniques," *IEEE Trans. Autom. Control,* vol. AC-16, pp. 727–735

KANG, G. S., and L. J. FRANSEN (1987) "Experimentation with an adaptive noise-cancellation filter," *IEEE Trans. Circuits Syst.,* vol. CAS-34, pp. 753–758.

KAVEH, M., and A. BASSIAS (1990). "Threshold extension based on a new paradigm for MUSIC-type estimation," *Proc. ICASSP,* Albuquerque, N.Mex.

KAVEH, M., and A. J. BARABELL (1986). "The statistical performance of the MUSIC and the minimum-norm algorithms in resolving plane waves in noise," *IEE Trans. Acoust. Speech Signal Process.*, vol. ASSP-34, pp. 331–341.

KAY, S. M. (1988), *Modern Spectral Estimation: Theory and Application,* Prentice-Hall, Englewood Cliffs, N.J.

KAY, S. M., and L. S. MARPLE JR. (1981). "Spectrum analysis—a modern perspective," *Proc. IEEE,* vol. 69, pp. 1380–1419.

KELLY, J. L., JR., and R. F. LOGAN (1970). *Self-Adaptive Echo Canceller,* U.S. Patent 3,500,000, March 10.

KELLY, E. J., I. S. REED, and W. L. ROOT (1960). "The detection of radar echoes in noise: I," *J. SIAM,* vol. 8, pp. 309–341.

KEZYS, V., and S. HAYKIN, (1988). "Multi-frequency angle-of-arrival estimator: an experimental evaluation," *Proc. SPIE,* San Diego, Calif.

KLEMA, V. C., and A. J. LAUB (1980). "The singular value decomposition: its computation and some applications," *IEEE Trans Autom. Control,* vol. AC-25, pp. 164–176.

KMENTA, J. (1971). *Elements of Econometrics,* Macmillan, New York.

KNIGHT, W. C., R. G. PRIDHAM, and S. M. KAY (1981). "Digital signal processing for sonar," *Proc. IEEE,* vol. 69, pp. 1451–1506.

KOH, T., and E. J. POWERS (1985). "Second-order Volterra filtering and its application to nonlinear system identification," *IEEE Trans. Acoust. Speech Signal Process.,* vol. ASSP-33, pp. 1445–1455.

KOLMOGOROV, A. N. (1939). "Sur l'interpolation et extrapolation des suites stationaires," *C. R. Acad. Sci.,* Paris, vol. 208, pp. 2043–2045 [English translation reprinted in the book edited by Kailath (1977)].

KOLMOGOROV, A. N. (1968). "Three approaches to the quantitative definition of information," *Probl. Inf. Transm. USSR,* vol. 1, pp. 1–7.

KREIN, M. G. (1945). "On a problem of extrapolation of A. N. KOLMOGOROV," *C. R. (Dokl.) Akad. Nauk SSSR,* vol. 46, pp. 306–309. [This paper is reproduced in the book edited by Kailath (1977)].

KULLBACK, S., and R. A. LEIBLER (1951). "On information and sufficiency," *Ann. Math. Statist.,* vol. 22, pp. 79–86.

KUMAR, R. (1983). "Convergence of a Decision-Directed Adaptive Equalizer", *Proc. Conference on Decision and Control,* vol. 3, pp. 1319–1324.

KUMARESAN, R. (1983). "On the zeros of the linear prediction-error filter for deterministic signals," *IEEE Trans. Acoust. Speech Signal Process.,* vol. ASSP-31, pp. 217–220.

KUMARESAN, R., and D. W. TUFTS (1983). "Estimating the angles of arrival of multiple plane waves," *IEEE Trans. Aerospace Electron. Syst.,* vol. AES-19, pp. 134–139.

KUNG, H. T. (1982). "Why systolic architectures?" *Computer,* vol. 15, pp. 37–46.

KUNG, H. T., and C. E. LEISERSON (1978). "Systolic arrays (for VLSI)," *Sparse Matrix Proc. 1978, Soc. Ind. Appl. Math.,* 1978, pp. 256–282 [a version of this paper is reproduced in Mead and Conway (1980)].

KUNG, S. Y. (1988). *VLSI Array Processors,* Prentice-Hall, Englewood Cliffs, N.J.

KUNG, S. Y., H. J. WHITEHOUSE, and T. KAILATH, eds. (1985). *VLSI and Modern Signal Processing,* Prentice-Hall, Englewood Cliffs, N.J.

KUNG, S. Y., ET AL. (1987). "Wavefront array processors—concept to implementation," *Computer,* vol. 20, pp. 18–33.

LANDAU, I. D. (1984). "A feedback system approach to adaptive filtering," *IEEE Trans. Inf. Theory,* vol. IT-30, Special Issue on Linear Adaptive Filtering, pp. 251–262.

LANG, S. W., and J. H. MCCLELLAN (1980). "Frequency estimation with maximum entropy spectral estimators," *IEEE Trans. Acoust. Speech Signal Process.,* vol. ASSP-28, pp. 716–724.

LANG, S. W., and J. H. MCCLELLAN (1979). "A simple proof of stability for all-pole linear prediction models," *Proc. IEEE,* vol. 67, pp. 860–861.

LAWRENCE, R. E., and H. KAUFMAN (1971). "The Kalman filter for the equalization of a digital communication channel," *IEEE Trans. Commun. Technol.,* vol. COM-19, p. 1137–1141.

LAWSON, C. L., and R. J. HANSON (1974). *Solving Least Squares Problems,* Prentice-Hall, Englewood Cliffs, N.J.

LEE, D. T. L. (1980). *Canonical ladder form realizations and fast estimation algorithms,* Ph.D. dissertation, Stanford University, Stanford, Calif.

LEE, D. T. L., M. MORF, and B. FRIEDLANDER (1981). "Recursive least-squares ladder estimation algorithms," *IEEE Trans. Circuits Syst.,* vol. CAS-28, pp. 467–481.

LEGENDRE, A. M. (1810). "Méthode des moindres quarrés, pour trouver le milieu le plus probable entre les résultats de différentes observations," *Mem. Inst. France,* pp. 149–154.

LEHMER, D. H. (1961). "A machine method for solving polynomial equations," *J. Assoc. Comput. Mach.,* vol. 8, pp. 151–162.

LEUNG, H., and S. HAYKIN (1989). "Stability of recursive QRD-LS algorithms using finite-precision systolic array implementation," *IEEE Trans. Acoust. Speech Signal Process.,* vol. *ASSP-37,* pp. 760–763.

LEV-ARI, H., T. KAILATH, and J. CIOFFI (1984). "Least-squares adaptive lattice and transversal filters: A unified geometric theory," *IEEE Trans. Inf. Theory,* vol. IT-30, pp. 222–236.

LEVINSON, N. (1947). "The Wiener RMS (root-mean-square) error criterion in filter design and prediction," *J. Math Phys.,* vol. 25, pp. 261–278.

LEVINSON, N., and R. M. REDHEFFER (1970). *Complex Variables,* Holden-Day, San Francisco.

LIAPUNOV, A. M. (1966). *Stability of Motion,* Translated by F. Abramovici and M. Shimshoni, Academic Press.

LII, K. S., and M. ROSENBLATT (1982). "Deconvolution and estimation of transfer function phase and coefficients for non-Gaussian linear processes," *Ann. Stat.,* vol. 10, pp. 1195–1208.

LILES, W. C., J. W. DEMMEL, and L. E. BRENNAN (1980). *Gram-Schmidt adaptive algorithms,* Tech. Rep. RADC-TR-79-319, RADC, Griffiss Air Force Base, N.Y.

LIN, D. W. (1984). "On digital implementation of the fast Kalman algorithm," *IEEE Trans. Acoust. Speech Signal Process.,* vol. ASSP-32, pp. 998–1005.

LING, F. (1989). "Efficient least-squares lattice algorithms based on Givens rotation with systolic array implementations," *Proc. ICASSP,* Glasgow, Scotland, pp. 1290–2193.

LING, F., and J. G. PROAKIS (1984a). "Numerical accuracy and stability: two problems of adaptive estimation algorithms caused by round-off error," *Proc. ICASSP,* San Diego, Calif., pp. 30.3.1 to 30.3.4.

LING, F., and J. G. PROAKIS (1984b). "Nonstationary learning characteristics of least squares adaptive estimation algorithims," *Proc. ICASSP,* San Diego, Calif., pp. 3.7.1 to 3.7.4.

LING, F., and J. G. PROAKIS (1986). "A recursive modified Gram-Schmidt algorithm with applications to least squares estimation and adaptive filtering," *IEEE Trans. Acoust. Speech Signal Process.,* vol. ASSP-34, pp. 829–836.

LING, F., D. MANOLAKIS, and J. G. PROAKIS (1985). "New forms of LS lattice algorithms and an analysis of their round-off error characteristics," *Proc. ICASSP,* Tampa, Fla., pp. 1739–1742.

LING, F., D. MANOLAKIS, and J. G. PROAKIS, (1986). "Numerically robust least-squares lattice-ladder algorithm with direct updating of the reflection coefficients," *IEEE Trans. Acoust. Speech Signal Process.* vol. ASSP-34.

LIPPMANN, R. P. (1987). "An introduction to computing with neural nets", *IEEE ASSP Magazine,* vol. 4, pp. 4–22.

LJUNG, L. (1977). "Analysis of recursive stochastic algorithms," *IEEE Trans. Autom. Control.* vol. AC-22, pp. 551–575.

LJUNG, L. (1984). "Analysis of stochastic gradient algorithms for linear regression problems," *IEEE Trans. Inf. Theory,* vol. IT-30, Special Issue on Linear Adaptive Filtering, pp. 151–160.

LJUNG, S., and L. LJUNG (1985). "Error propagation properties of recursive least-squares adaptation algorithms," *Automatica,* vol. 21, pp. 157–167.

LJUNG, L., and T. SÖDERSTRÖM (1983). *Theory and Practice of Recursive Identification,* MIT Press, Cambridge, Mass.

LJUNG, L., M. MORF, and D. FALCONER (1978). "Fast calculation of gain matrices for recursive estimation schemes," *Int. J. Control,* vol. 27, pp. 1–19.

LJUNG, L. (1987). *System Identification: Theory for the User,* Prentice-Hall.

LORD RAYLEIGH (1879). "Investigations in optics with special reference to the spectral scope," *Philos. Mag.,* vol. 8, pp. 261–274.

LOWE, D. (1989), "Adaptive radial basis function nonlinearities and the problem of generalization," *First IEE Int. Conf. Artif. Neural Networks,* London, pp. 171–175.

LUCKY, R. W. (1965). "Automatic equalization for digital communication," *Bell Syst. Tech. J.,* vol. 44, pp. 547–588.

LUCKY, R. W. (1966). "Techniques for adaptive equalization of digital communication systems," *Bell Syst. Tech. J.,* vol. 45, pp. 255–286.

LUCKY, R. W. (1973). "A survey of the communication theory literature: 1968–1973," *IEEE Trans. Inf. Theory,* vol. IT-19, pp. 725–739.

LUCKY, R. W., J. SALZ, and E. J. WELDON, JR. (1968). *Principles of Data Communication,* McGraw-Hill, New York.

LUENBERGER, D. G. (1969). *Optimization by Vector Space Methods,* Wiley, New York.

LUK, F. T. (1986). "A triangular processor array for computing singular values," *Linear Algebra and its Applications,* vol. 77, pp. 259–273.

LUK, F. T., and H. PARK (1989). "A proof of convergence for two parallel Jacobi SVD algorithms," *IEEE Trans. Comput.,* vol. 38, pp. 806–811.

LUK, F. T., and S. QIAO (1989). "Analysis of a recursive least-squares signal-processing algorithm," *SIAM J. Sci. Stat. Comput.,* vol. 10, pp. 407–418.

LYNCH, M. R., and P. J. RAYNER (1989), "The properties and implementation of the non-linear vector space connectionist model," *First IEE Int. Conf. Artif. Neural Networks,* London, pp. 186–190.

MACCHI, O., and E. EWEDA (1984). "Convergence analysis of self-adaptive equalizers," *IEEE Trans. Inf. Theory,* vol. IT-30, Special Issue on Linear Adaptive Filtering, pp. 161–176.

MACCHI, O., and M. JAIDANE-SAIDNE (1989). "Adaptive IIR filtering and chaotic dynamics: application to audio-frequency coding," *IEEE Trans. Circuits Syst.,* vol. 36, pp. 591–599.

MACCHI, O. (1986). "Advances in Adaptive Filtering", in book entitled *Digital*

Communications, edited by E. Biglieri and G. Prati, pp. 41–57, Elsevier Science Publishers.

MACCHI, O. (1986). "Optimization of Adaptive Identification for Time-varying Filters," *IEEE Trans. Automatic Control,* vol. AC-31, pp. 283–287.

MACIKUNAS, A., S. HAYKIN, and T. GREENLAY (1988). "Trihedral Radar Reflector," Canadian Patent 1,238,400, issued June 21, 1988; U.S. Patent 4,843,396, issued June 27, 1989.

MAKHOUL, J. (1975). "Linear prediction: a tutorial review," *Proc. IEEE,* vol. 63, pp. 561–580.

MAKHOUL, J. (1977). "Stable and efficient lattice methods for linear prediction," *IEEE Trans. Acoust. Speech Signal Process.,* vol. ASSP-25, pp. 423–428.

MAKHOUL, J. (1978). "A class of all-zero lattice digital filters: properties and applications," *IEEE Trans. Acoust. Speech Signal Process.,* vol. ASSP-26, pp. 304–314.

MAKHOUL, J. (1981). "On the eigenvectors of symmetric Toeplitz matrices," *IEEE Trans. Acoust. Speech Signal Process.,* vol. ASSP-29, pp. 868–872.

MAKHOUL, J., and L. K. COSSELL (1981). "Adaptive lattice analysis of speech," *IEEE Trans. Circuits Syst.,* vol. CAS-28, pp. 494–499.

MANOLAKIS, D., F. LING, and J. G. PROAKIS (1987). "Efficient time-recursive least-squares algorithms for finite-memory adaptive filtering," *IEEE Trans. Circuits Syst.,* vol. CAS-34, pp. 400–408.

MARCOS, S., and O. MACCHI (1987). "Tracking capability of the least mean square algorithm: application to an asynchronous echo canceller," *IEEE Trans. Acoust. Speech Signal Process.,* vol. ASSP-35, pp. 1570–1578.

MARDEN, M. (1949). "The geometry of the zeros of a polynomial in a complex variable," *Amr. Math. Soc. Surveys* (New York), no. 3, chap. 10.

MARKEL, J. D., and A. H. GRAY JR. (1976). *Linear Prediction of Speech,* Springer-Verlag, New York.

MARPLE, L. (1980). "A new autoregressive spectrum analysis algorithm," *IEEE Trans. Acoùst. Speech Signal Process.,* vol. ASSP-28, pp. 441–454.

MARPLE, S. L., JR. (1981). "Efficient least squares FIR system identification," *IEEE Trans. Acoust. Speech Signal Process.,* vol. ASSP-29, pp. 62–73.

MARPLE, S. L., JR. (1987), *Digital Spectral Analysis with Applications,* Prentice-Hall, Englewood Cliffs, N.J.

MASON, S. J. (1956). "Feedback theory: further properties of signal flow graphs," *Proc. IRE,* vol. 44, pp. 920–926.

MAYBECK, P. S. (1979). *Stochastic Models, Estimation, and Control,* vol. 1, Academic Press, New York.

MAZO, J. E. (1979). "On the independence theory of equalizer convergence," *Bell Syst. Tech. J.,* vol. 58, pp. 963–993.

MAZO, J. E. (1980). "Analysis of decision-directed equalizer convergence," *Bell Syst. Tech. J.,* vol. 59, pp. 1857–1876.

MCCANNY, J. V., and J. G. MCWHIRTER, (1987). "Some systolic array developments in the United Kingdom, *Computer,* vol. 2, pp. 51–63.

MCCOOL, J. M., ET AL. (1980). *Adaptive Line Enhancer,* U.S. Patent 4,238,746, December 9.

MCCOOL, J. M., ET AL. (1981). *An Adaptive Detector,* U.S. Patent 4,243,935, January 6.

MCDONALD, R. A. (1966). "Signal-to-noise performance and idle channel performance of differential pulse code modulation systems with particular applications to voice signals," *Bell Syst. Tech. J.,* vol. 45, pp. 1123–1151.

McDonough, R. N., and W. H. Huggins (1968). "Best least-squares representation of signals by exponentials," *IEEE Trans. Autom. Control,* vol. AC-13, pp. 408–412.

McGee, W. F. (1971). "Complex Gaussian noise moments," *IEEE Trans. Inf. Theory,* vol. IT-17, pp. 149–157.

McWhirter, J. G. (1983). "Recursive least-squares minimization using a systolic array," *Proc. SPIE,* vol. 431, Real-Time Signal Processing VI, pp. 105–112.

McWhirter, J. G., and T. J. Shepherd (1986). "A systolic array for linearly constrained least-squares problems," Proc. SPIE, Int. Soc. Opr. Eng., Advanced Algorithms and Architectures for Signal Processing, vol. 696.

McWhirter, J. G. (1988). "Efficient minimum variance distortionless response processing using a systolic array," Proc. SPIE-The International Society for Optical Engineering, Advanced Algorithms and Architectures for Signal Processing III, San Diego, California, pp. 385–392.

McWhirter, J. G. (1989). "Algorithmic engineering-an emerging technology," Proc. SPIE-The International Society for Optical Engineering, vol. 1152, San Diego, California.

McWhirter, J. G., and T. J. Shepherd (1989). "Systolic array processor for MVDR beamforming," *IEE Proc. (London),* vol. 136, pt. F, pp. 75–80.

Mead, C., and L. Conway (1980). *Introduction to VLSI Systems,* Addison-Wesley, Reading, Mass.

Medaugh, R. S., and L. J. Griffiths (1981). "A comparison of two linear predictors," *Proc. ICASSP,* Atlanta, Ga., pp. 293–296.

Mendel, J. M. (1973). *Discrete Techniques of Parameter Estimation: The Equation Error Formulation,* Marcel Dekker, New York.

Mendel, J. M. (1974). "Gradient estimation algorithms for equation error formulations," *IEEE Trans. Autom. Control,* vol. AC-19, pp. 820–824.

Mendel, J. M. (1986). "Some Modeling Problems in Reflection Seismology," *IEEE ASSP Mag.,* vol. 3, pp. 4–17.

Mendel, J. M. (1987). *Lessons in Digital Estimation Theory,* Prentice-Hall, Englewood Cliffs, N.J.

Mendel, J. M. (1990a). *Maximum-Likelihood Deconvolution: A Journey into Model-Based Signal Processing,* Springer-Verlag, New York.

Mendel, J. M. (1990b). "Introduction," *IEEE Trans. Autom. Control,* vol. AC-35, Special Issue on Higher Order Statistics in System Theory and Signal Processing, p. 3.

Mermoz, H. F. (1981). "Spatial processing beyond adaptive beamforming," *J. Acoust, Soc. Am.,* vol. 70, pp. 74–79.

Messerchmitt, D. G. (1984). "Echo cancellation in speech and data transmission," *IEEE J. Sel. Areas Commun.,* vol. SAC-2, pp. 283–297.

Metford, P. A. S., and S. Haykin (1985). "Experimental analysis of an innovations-based detection algorithm for surveillance radar," *IEE Proc. (London),* vol. 132, pp. 18–26.

Middleton, D. (1960). *An Introduction to Statistical Communication Theory,* McGraw-Hill, New York.

Miller, K. S. (1974). *Complex Stochastic Processes: An Introduction to Theory and Application,* Addison-Wesley, Reading, Mass.

Monsen, P. (1971). "Feedback equalization for fading dispersive channels," *IEEE Trans. Inf. Theory,* vol. IT-17, pp. 56–64.

Monzingo, R. A., and T. W. Miller (1980). *Introduction to Adaptive Arrays,* Wiley-Interscience, New York.

MORF, M. (1974). *"Fast algorithms for multivariable systems,"* Ph.D. dissertation, Stanford University, Stanford, Calif.

MORF, M., and D. T. LEE (1978). "Recursive least squares ladder forms for fast parameter tracking," *Proc. 1978 Conf. Decision Control,* San Diego, Calif., pp. 1362–1367.

MORF, M., T. KAILATH, and L. LJUNG (1976). "Fast algorithms for recursive identification," *Proc. 1976 Conf. Decision Control,* Clearwater Beach, Fla., pp. 916–921.

MORF, M., A. VIEIRA, and D. T. LEE (1977). "Ladder forms for identification and speech processing," *Proc. 1977 IEEE Conf. Decision Control,* New Orleans, pp. 1074–1078.

MORONEY, P. (1983). *Issues in the Implementation of Digital Feedback Compensators,* MIT Press, Cambridge, Mass.

MORSE, P. M., and H. FESHBACK (1953). *Methods of Theoretical Physics,* Pt. I, McGraw-Hill.

MOSCHNER, J. L. (1970). *Adaptive filter with clipped input data,* Tech. Rep. 6796-1, Stanford University Center for Systems Research, Stanford, Calif.

MUELLER, M. S. (1981a). Least-squares algorithms for adaptive equalizers," *Bell Syst. Tech. J.,* vol. 60, pp. 1905–1925.

MUELLER, M. S. (1981b). "On the rapid initial convergence of least-squares equalizer adjustment algorithms," *Bell Syst. Tech. J.,* vol. 60, pp. 2345–2358.

MULGREW, B. (1987). "Kalman filter techniques in adaptive filtering," *IEE Proc. (London),* vol. 134, pt. F, pp. 239–243.

MULGREW, B., and C. F. N. COWAN (1987). "An adaptive Kalman equalizer: structure and performance," *IEEE Trans. Acoust. Speech Signal Process.,* vol. ASSP-35, pp. 1727–1735.

MULLIS, C. T., and L. L. SCHARF (1991). "Quadratic Estimators of the Power Spectrum," Chapter 1 of book entitled *Advances in Spectrum Analysis and Array Processing,* volume I, edited by S. Haykin, Prentice-Hall.

MURANO, K., ET AL. (1990). "Echo cancellation and applications," *IEEE Commun.,* vol. 28, pp. 49–55.

MUSICUS, B. R. (1985). "Fast MLM power spectrum estimation from uniformly spaced correlations," *IEEE Trans. Acoust. Speech Signal Process*, vol. ASSP-33, pp. 1333–1335.

NAGUMO, J. I., and A. NODA (1967). "A learning method for system identification," IEEE Trans. Automatic Control, vol. AC-12, pp. 282–287.

NAHI, N. E. (1969). *Estimation Theory and Applications,* Wiley, New York.

NARAYAN, S. S., A. M. PETERSON, and M. J. NARASHIMA (1983). "Transform domain LMS algorithm," *IEEE Trans. Acoust. Speech Signal Process.,* vol. ASSP-31, pp. 609–615.

NARENDRA, K. S., and A. M. ANNASWAMY (1989). *Stable Adaptive Systems,* Prentice-Hall, Englewood Cliffs, N.J.

NAU, R. F., and R. M. OLIVER (1979). "Adaptive filtering revisted," *J. Oper. Res. Soc.,* vol. 30, pp. 825–831.

NICKEL, U. (1988). "Algebraic formulation of Kumaresan-Tufts superresolution method, showing relation to ME and MUSIC methods," *IEE Proc. (London),* vol. 135, pt. F, pp. 7–10.

NICKEL, U. (1991). "Radar target parameter estimation with array antennas," in *Radar Array Processing,* ed. S. Haykin, J. Litva, and T. J. Shepherd, Springer-Verlag, New York.

NIELSEN, P. A., and J. B. THOMAS (1988). "Effect of Correlation on Signal Detection in Arctic Under-ice Noise," Conference Record of the Twenty-Second Asilomar Conference on Signals, Systems and Computers," pp. 445–450, Pacific Grove, California.

NIKIAS, C. L. (1991). "Higher-order spectral analysis," in *Advances in Spectrum Analysis and Array Processing,* vol. 1, ed. S. Haykin, Prentice-Hall, Englewood Cliffs, N.J., Chap. 7.

NIKIAS, C. L., and M. R. RAGHUVEER (1987). "Bispectrum estimation: a digital signal processing framework," *Proc. IEEE,* vol. 75, pp. 869–891.

NISHITANI, T., ET AL. (1987). "A CCITT standard 32 kbits/s ADPCM LSI codec," *IEEE Trans. Acoust. Speech Signal Process.,* vol. ASSP-35, pp. 219–225.

NUTTALL, A. H. (1976). *Spectral analysis of a univariate process with bad data points via maximum entropy and linear predictive techniques,* Naval Underwater System Center (NUSC) Scientific and Engineering Studies, Spectral Estimation, NUSC, New London, Conn.

OPPENHEIM, A. V., and J. S. LIM (1981). "The importance of phase in signals," *Proc. IEEE,* vol. 69, pp. 529–541.

OPPENHEIM, A. V., and R. W. SCHAFER (1975). *Digital Signal Processing,* Prentice-Hall, Englewood Cliffs, N.J.

OWSLEY, N. L. (1973). "A recent trend in adaptive spatial processing for sensor arrays: constrained adaptation," in *Signal Processing,* ed. J. W. R. Griffiths et al., Academic Press, New York, pp. 591–604.

OWSLEY, N. L. (1978). "Data adaptive orthonormalization," *Proc. ICASSP, Tulsa, Okla,* pp. 109–112.

OWSLEY, N. L. (1985). "Sonar array processing," in *Array Signal Processing,* ed., S. Haykin, Prentice-Hall, Englewood Cliffs, N.J., pp. 115–193.

PAN, R., and C. L. NIKIAS (1988). "The complex cepstrum of higher order cumulants and nonminimum phase identification," *IEEE Trans. Acoust. Speech Signal Process.,* vol. ASSP-36, pp. 186–205.

PAPOULIS, A. (1975). "A new algorithm in spectral analysis and band-limited extrapolation," *IEEE Trans. Circuits Syst.,* vol. CAS-22, pp. 735–742.

PAPOULIS, A. (1984). *Probability, Random Variables, and Stochastic Processes,* 2nd ed., McGraw-Hill, New York.

PARK, J., C. R. LINDBERG, and F. L. VERNON III (1987). "Multitaper spectral analysis of high-frequency seismograms," *J. Geophys. Res.,* vol. 92, pp. 1265–1284.

PEACOCK, K. L., and S. TREITEL (1969). "Predictive deconvolution: theory and practice," *Geophysics,* vol. 34, pp. 155–169.

PICCHI, G., and G. PRATI (1987). "Blind equalization and carrier recovery using a 'stop-and-go' decision-directed algorithm," *IEEE Trans. Commun.,* vol. COM-35, pp. 877–887.

PIPES, L. (1958). *Applied Mathematics for Engineers and Physicists,* McGraw-Hill, New York.

PLACKETT, R. L. (1950). "Some theorems in least squares," *Biometrika,* vol. 37, p. 149.

PORAT, B., and T. KAILATH (1983). "Normalized lattice algorithms for least-squares FIR system identification," *IEEE Trans. Acoust. Speech Signal Process.,* vol. ASSP-31, pp. 122–128.

PORAT, B., B. FRIEDLANDER, and M. MORF (1982). "Square root covariance ladder algorithms," *IEEE Trans. Autom. Control,* vol. AC-27, pp. 813–829.

PRESS, W. H., ET AL. (1988). *Numerical Recipes in C,* Cambridge University Press, Cambridge.

PRIESTLEY, M. B. (1981). *Spectral Analysis and Time Series,* vols. 1 and 2, Academic Press, New York.

PROAKIS, J. G. (1975). "Advances in equalization for intersymbol interference," in *Advances in Communication Systems*, vol. 4, Academic Press, New York, pp. 123–198.

PROAKIS, J. G. (1983). *Digital Communications*, McGraw-Hill, New York.

PROAKIS, J. G., and J. H. MILLER (1969). "An adaptive receiver for digital signaling through channels with intersymbol interference," *IEEE Trans. Inf. Theory*, vol. IT-15, pp. 484–497.

PROAKIS, J. G., and D. G. MANOLIKIS (1988). "Introduction to Digital Signal Processing," MacMillan.

PRONY, R. (1795). "Essai expérimental et analytique, etc.," *L'ecole Polytechnique, Paris*, vol. 1, no. 2, pp. 24–76.

PROUDLER, I. K., J. G. MCWHIRTER, and T. J. SHEPHERD (1988). "Fast QRD-based algorithms for least squares linear prediction," *Proc. IMA Conf. Math. Signal Process.*, University of Warwick, Coventry, England.

PROUDLER, I. K., J. G. MCWHIRTER, and T. J. SHEPHERD (1989). *QRD-Based Lattice Filter Algorithms*, SPIE, San Diego, Calif.

QURESHI, S. (1982). "Adaptive equalization," *IEEE Commun. Mag.*, vol. 20, pp. 9–16.

QURESHI, S. U. H. (1985). "Adaptive equalization," *Proc. IEEE*, vol. 73, pp. 1349–1387.

RABINER, L. R., and B. GOLD (1975). *Theory and Application of Digital Signal Processing*, Prentice-Hall, Englewood Cliffs, N.J.

RABINER, L. R., and R. W. SCHAFER (1978). *Digital Processing of Speech Signals*, Prentice-Hall, Englewood Cliffs, N.J.

RADER, C. M., and A. O. STEINHARDT (1986). "Hyperbolic householder transformations", *IEEE Trans. Acoust., Speech, and Signal Processing*, vol. ASSP-34, No. 6, pp. 1589–1602.

RADER, C. M. (1990) "Linear systolic array for adaptive beamforming," The 1990 Digital Signal Processing Workshop, New Paltz, N.Y., Sponsored by IEEE Signal Processing Society, pp. 5.2.1–5.2.2.

RAHMAN, M. A., and K. B. YU (1987). "Total least squares approach for frequency estimation using linear prediction," *IEEE Trans. Acoust. Speech Signal Process.*, vol. ASSP-35, pp. 1440–1454.

RALSTON, A. (1965). *A First Course in Numerical Analysis*, McGraw-Hill, New York.

RAO, B. D., and K. V. S. HARI (1989a). "Performance analysis of root-MUSIC," *IEEE Trans. Acoust. Speech Signal Process.*, vol. ASSP-37, pp. 1789–1794.

RAO, B. D., and K. V. S. HARI (1989b). "Statistical performance analysis of the minimum-norm method," *Proc. IEE (London)*, vol. 135, pt. F, pp. 125–134.

RAO, B. D., and K. V. S. HARI (1989c). "Performance analysis of root MUSIC," *IEEE Trans. Acoust. Speech Signal Process.*, vol. ASSP-37, pp. 1939–1949.

RAO, S. K., and T. KAILATH (1986). "What is a systolic algorithm?" *Proc. SPIE* vol. 614, Highly Parallel Signal Processing Architectures, pp. 34–48.

RAYNER, P. J. W., and M. F. LYNCH (1989), "A new connectionist model based on a non-linear adaptive filter," *Proc. ICASSP*, Glasgow, Scotland, pp. 1191–1194.

REDDI, S. S. (1979). "Multiple source location-A digital approach," *IEEE Trans. Aerospace Electron. Syst.*, vol. AES-15, pp. 95–105.

REDDI, S. S. (1984). "Eigenvector properties of Toeplitz matrices and their application to spectral analysis of time series," *Signal Process.*, vol. 7, pp. 45–56.

REED, I. S. (1962). "On a moment theorem for complex Gaussian processes," *IRE Trans. Inf. Theory*, vol. IT-8, pp. 194–195.

REEDS, I. S., J. D. MALLET, and L. E. BRENNAN (1974). "Rapid convergence rate in adaptive arrays," *IEEE Trans. Aerospace and Electronic Systems,* vol. AES-10, p. 853.

REEVES, A. H. (1975). "The past, present, and future of PCM," *IEEE Spectrum,* vol. 12, pp. 58–63.

REGALIA, P. A. (1990). "System Theoretic Properties in the Stability Analysis of QR-based Fast Least-squares Algorithms," Report DEC-0390-003, Département Electronique et Communications, Institut National des Télécommunications, 91011 Evry cedex, France.

REGALIA, P. A., and G. BELLANGER (1991). "On the duality between fast QR methods and lattice methods in least squares adaptive filtering," *IEEE Trans. Acoust. Speech Signal Process.,* vol. ASSP-39.

RICKARD, J. T., ET AL. (1981). "A Performance Analysis of Adaptive Line Enhancer-Augmented Spectral Detectors," *IEEE Trans. Circuits and Systems,* vol. CAS-28, pp. 534–541.

RISSANEN, J. (1978). "Modelling by shortest data description," *Automatica,* vol. 14, pp. 465–471.

RISSANEN, J., (1986a). "A predictive least-squares principle," *IMA J. Math. Control Inf.,* vol. 3, pp. 211–222.

RISSANEN, J. (1986b). "Stochastic complexity and modeling," *Ann. Stat.,* vol. 14, pp. 1080–1100.

RISSANEN, J., (1989). *Stochastic complexity in statistical engineering,* Series in Computer Science, vol. 15, World Scientific, Singapore.

ROBBINS, H., and S. MONRO (1951). "A stochastic approximation method," *Ann. Math. Stat.,* vol. 22, pp. 400–407.

ROBERTS, R. A., and C. T. MULLIS (1987). *Digital Signal Processing,* Addison-Wesley.

ROBINSON, E. A. (1954). *"Predictive decomposition of time series with applications for seismic exploration,"* Ph.D. thesis, Massachusetts Institute of Technology, Cambridge, Mass.

ROBINSON, E. A. (1967). *Multichannel Time-Series Analysis with Digital Computer Programs,* Holden-Day, San Francisco.

ROBINSON, E. A. (1982). "A historical perspective of spectrum estimation," *Proc. IEEE,* vol. 70, Special Issue on Spectral Estimation, pp. 885–907.

ROBINSON, E. A. (1984). "Statistical pulse compression," *Proc IEEE,* vol. 72, pp. 1276–1289.

ROBINSON, E. A., and S. TREITEL (1980). *Geophysical Signal Analysis,* Prentice-Hall, Englewood Cliffs, N.J.

ROBINSON, E. A., and T. DURRANI (1986). *Geophysical Signal Processing,* Prentice-Hall.

ROSENBLATT, M. (1985). *Stationary Sequences and Random Fields,* Birkhäuser, Stuttgart, West Germany.

ROSS, F. J. (1989). "Blind Equalization for Digital Microwave Radio," M. Eng. Thesis, McMaster University, Hamilton, Ontario, Canada.

RUMELHART, D. E., and J. L. MCCLELLAND (1986). *Parallel Distributed Processing,* vol. 1: Foundations, MIT Press.

RUMELHART, D. E., G. E. HINTON, and R. J. WILLIAMS (1986). "Learning representations by back-propagating errors," *Nature,* vol. 323, pp. 533–536.

SAITO, S., and F. ITAKURA (1966). *The Theoretical Consideration of Statistically Optimum Methods for Speech Spectral Density,* Rep. 3107, Electrical Communication Laboratory, N. T. T., Tokyo (in Japanese).

SAMBUR, M. R. (1978). "Adaptive noise cancelling for speech signals," *IEEE Trans. Acoust. Speech Signal Process.* vol. ASSP-26, pp. 419–423.

SAMSON, C. (1982). "A unified treatment of fast Kalman algorithms for identification," *Inte. J. Control,* vol. 35, pp. 909–934.

SATO, Y. (1975). "Two extensional applications of the zero-forcing equalization method," *IEEE Trans. Commun.* vol. COM-23, pp. 684–687.

SATORIUS, E. H., and S. T. ALEXANDER (1979). "Channel equalization using adaptive lattice algorithms," *IEEE Trans. Commun.,* vol. COM-27, pp. 899–905.

SATORIUS, E. H., and J. D. PACK (1981). "Application of least squares lattice algorithms to adaptive equalization," *IEEE Trans. Commun.,* vol. COM-29, pp. 136–142.

SATORIUS, E. H., ET AL. (1983), "Fixed-point implementation of adaptive digital filters," Proc. IEEE Intern., Conference on Acoustics, Speech, and Signal Processing, ICAASP 83, pp. 33–36.

SCHETZEN, M. (1981). "Nonlinear system modeling based on the Wiener theory," *Proc. IEEE,* vol. 69, pp. 1557–1572.

SCHMIDT, R. (1979). "Multiple emitter location and signal parameter estimation," *RADC Spectrum Estimation Workshop,* pp. 243–258. This paper is reproduced in the 1986 Special Issue of IEEE Transactions on Antennas and Propagation devoted to Adaptive Processing Antenna Systems.

SCHMIDT, R. O. (1981). "A signal subspace approach to multiple emitter location and spectral estimation," Ph. D. dissertation, Stanford University, Standford, Calif.

SCHMIDT, R., and R. FRANKS (1986). "Multiple DF signal processing: an experimental system," *IEEE Trans. Antennas and Propag.,* vol. AP-34 (March), pp. 281–290.

SCHREIBER, R. J. (1986). "Implementation of adaptive array algorithm," *IEEE Trans. Acoust. Speech Signal Process.,* vol. ASSP-34, pp. 1038–1045.

SCHREIBER, R. J. and P. J. KUEKES (1985). "Systolic linear algebra machines in digital signal processing," in *VLSI and Modern Signal Processing,* ed. S. Y. Kung, H. J. Whitehouse, and T. Kailath, Prentice-Hall, Englewood Cliffs, N.J., pp. 389–405.

SCHROEDER, M. R. (1966). "Vocoders: analysis and synthesis of speech," *Proc. IEEE,* vol. 54, pp. 720–734.

SCHROEDER, M. R. (1985). "Linear predictive coding of speech: review and current directions," *IEEE Commun. Mag.* vol. 23, pp. 54–61.

SCHUR, I. (1917). "Über potenzreihen, die im innern des Einheitskreises beschränkt sind," *J. Reine Angew. Math.,* vol. 147, pp. 205–232 (see also vol. 148, pp. 122–145).

SCHUSTER, A. (1898). "On the investigation of hidden periodicities with applications to a supposed 26-day period of meterological phenomena," *Terr. Magn. Atmos. Electr.,* vol. 3, pp. 13–41.

SCHWARTZ, L. (1967). *Cours d'Analyse,* vol. II, Hermann, Paris, pp. 271–278.

SCHWARTZ, G. (1978). "Estimating the dimension of a model," *Ann. Stat.,* vol. 6, pp. 461–464.

SENNE, K. D. (1968). *Adaptive linear discrete-time estimation,* Tech. Rep., 6778-5, Stanford University Center for Systems Research, Stanford, Calif.

SETHARES, W. A., D. A. LAWRENCE, C. R. JOHNSON, JR., and R. R. BITMEAD (1986). "Parameter drift in LMS adaptive filters," *IEEE Trans. Acoustics, Speech and Signal Processing,* vol. ASSP-34, pp. 868–879.

SETHARES, W. A., G. A. RAY, and C. R. JOHNSON, JR. (1989). "Approaches to blind equalization of signals with multiple modulus," Proc. IEEE Intern. Conference on Acoustics, Speech, and Signal Processing, ICASSP'89, Glasgow, Scotland.

SHALVI, O., and E. WEINSTEIN (1990). "New Criteria for Blind Equalization of Non-minimum

Phase Systems (Channels)," *IEEE Trans. Information Theory,* vol. 36, No. 2, pp. 312–321.

SHAN, T.-J. and T. KAILATH (1985). "Adaptive beamforming for coherent signals and interference," *IEEE Trans. Acoust. Speech Signal Process.*, vol. ASSP-33, pp. 527–536.

SHANNON, C. E. (1948). "The mathematical theory of communication," *Bell Syst. Tech. J.*, vol. 27, pp. 379–423, 623–656.

SHARPE, S. M., and L. W. NOLTE (1981). "Adaptive MSE estimation," *Proc. ICASSP,* Atlanta, Ga., pp. 518–521.

SHEPHERD, T. J., J. G. MCWHIRTER, and J. E. HUDSON (1988). "Parallel Weight Extraction from a Systolic Adaptive Beamformer," Proc. IMA Int. Conf. on Mathematics in Signal Processing, Warwick, England, December.

SHEPHERD, T. J., and J. G. MCWHIRTER (1986). "A systolic array for linearly constrained least-squares optimization," *Proc. Int. Conf. Systolic Arrays,* Oxford, England.

SHEPHERD, T. J., and J. G. MCWHIRTER (1991). "Systolic adaptive beamforming," in *Radar Array Processing* ed. S. Haykin, J. Litva, and T. J. Shepherd, Springer-Verlag, New York.

SHERWOOD, D. T., and N. J. BERSHAD (1987). "Quantization effects in the complex LMS adaptive algorithm: Linearization using dither-theory," *IEEE Trans. Circuits and Systems,* vol. CAS-34, pp. 848–854.

SHI, K. H., and F. KOZIN (1986). "On almost sure convergence of adaptive algorithms," *IEEE Trans. Autom. Control,* vol. AC-31, pp. 471–474.

SHICHOR, E. (1982). "Fast recursive estimation using the lattice structure," *Bell Syst. Tech. J.*, vol. 61, pp. 97–115.

SHYNK, J. J. (1987). "Performance of alternative adaptive IIR filter realizations," Proc. 21st Asilomar Conf. Signals, Systems, Computers, Pacific Grove, California, pp. 144–150.

SHYNK, J. J. (1989). "Adaptive IIR filtering," *IEEE ASSP Mag.*, vol. 6, pp. 4–21.

SICURANZA, G. L., and G. RAMPONI (1986). "Adaptive nonlinear digital filters using distributed arithmetic," *IEEE Trans. Acoust. Speech Signal Process.,* vol. ASSP-34, pp. 518–526.

SICURANZA, G. L. (1985). "Nonlinear digital filter realization by distributed arithmetic," *IEEE Trans. Acoust. Speech Signal Process.* vol. ASSP-33, pp. 939–945.

SKINNER, D. P., S. M. HEDLINKA, and A. D. MATHEWS (1979). "Maximum entropy array processing," *J. Acoust. Soc. Am.*, vol. 66, pp. 488–493.

SKOLNIK, M. I. (1982). *Introduction to Radar Systems,* Second Edition, McGraw-Hill, New York.

SLEPIAN, D. (1978). "Prolate spheroidal wave functions, Fourier analysis, and uncertainty-V: The discrete case," *Bell Syst. Tech. J.*, vol. 57, pp. 1371–1430.

SLOCK, D. T. M., and T. KAILATH (1989), "Numerically stable fast transversal filters for recursive least-squares adaptive filtering," *IEEE Trans. Acoust. Speech Signal Process.*, to be published.

SLOCK, D. T. M. (1989). "*Fast algorithms for fixed-order recursive least-squares parameter estimation,*" Ph. D. dissertation, Stanford University, Stanford, Calif.

SMITH, J. O., and B. FRIEDLANDER (1985). "Global Convergence of the Constant Modulus Algorithm," Proc. ICASSP'85, pp. 30.5.1–30.5.4, Tampa, Florida.

SÖDERSTROM, T., and P. STOICA (1989). *System Identification,* Prentice-Hall International, Hemal, Hempstead, Hertfordshire, England.

SOKOLNIKOFF, I. S., and R. M. REDHEFFER (1966). *Mathematics of Physics and Modern Engineering,* McGraw-Hill, New York.

SOLO, V. (1989). "The limiting behavior of LMS," *IEEE Trans. Acoust. Speech Signal Process.*, vol. 37, pp. 1909–1922.

SONDHI, M. M., (1967). "An adaptive echo canceller," *Bell Syst. Tech. J.*, vol. 46, pp. 497–511.

SONDHI, M. M. (1970). *Closed Loop Adaptive Echo Canceller Using Generalized Filter Networks,* U.S. Patent, 3,499,999, March 10.

SONDHI, M., and D. A. BERKLEY (1980). "Silencing echoes in the telephone network," *Proc. IEEE,* vol. 68, pp. 948–963.

SONDHI, M. M., and A. J. PRESTI (1966). "A self-adaptive echo canceller", *Bell Syst. Tech. J.*, vol. 45, pp. 1851–1854.

SORENSON, H. D. (1967). "On the error behavior in linear minimum variance estimation problems," *IEEE Trans. Autom. Control,* vol. AC-12, pp. 557–562.

SORENSON, H. W. (1970). "Least-squares estimation: from Gauss to Kalman," *IEEE Spectrum,* vol. 7, pp. 63–68.

Special Issue on Adaptive Antennas (1976). *IEEE Trans. Antennas Propaga.*, vol. AP-24, September.

Special Issue on Adaptive Arrays (1983). *IEE Proc. Communi. Radar Signal Process.*, London, vol. 130, pp. 1–151.

Special Issue on Adaptive Filters (1987). *IEE Proc. Commun. Radar Signal Process.*, London, vol. 134, pt. F.

Special Issue on Adaptive Processing Antenna Systems (1986). *IEEE Trans. Antennas Propag.*, vol. AP-34, pp. 273–462.

Special Issue on Adaptive Signal Processing (1981). *IEEE Trans. Circuits Syst.*, vol. CAS-28, pp. 465–602.

Special Issue on Adaptive Systems (1976). *Proc. IEEE,* vol. 64, pp. 1123–1240.

Special Issue on Adaptive Systems and Applications (1987). *IEEE Trans. Circuits Syst.*, vol. CAS-34, pp. 705–854.

Special Issue on Higher Order Statistics in System Theory and Signal Processing (1990). *IEEE Trans. Autom. Control,* vol. AC-35, pp. 1–56.

Special Issue on Linear Adaptive Filtering (1984). *IEEE Trans. Inf. Theory,* vol. IT-30, pp. 131–295.

Special Issue on Linear-Quadratic-Gaussian Problem (1971), *IEEE Trans. Automatic Control,* vol. AC-16, December.

Special Issue on Neural Networks (1990). *Proc. IEEE,* vol. 78: Neural Nets I, September; Neural Nets II, October.

Special Issue on Spectral Estimation (1982). *Proc. IEEE,* vol. 70, pp. 883–1125.

Special Issue on System Identification and Time-series Analysis (1974). *IEEE Trans. Autom. Control,* vol. AC-19, pp. 638–951.

Special Issue on Systolic Arrays (1987). *Computer*, vol. 20, No. 7.

SPEISER, J. M., and H. J. WHITEHOUSE (1983). "A review of signal processing with systolic arrays," *Proc. SPIE,* vol. 431, Real Time Signal Processing VI, pp. 2–6.

SPEISER, J., and C. VAN LOAN (1984). "Signal processing computations using the generalized singular value decomposition," *Proc. SPIE,* vol. 495, San Diego.

STARER, D., and A. NEHORAI (1989). "Polynomial factorization algorithms for adaptive root estimation," *Proc ICASSP,* Glasgow, Scotland, pp. 1158–1161.

STEINHARDT, A. O. (1988). "Householder transforms in signal processing," *IEEE ASSP Mag.*, vol. 5, pp, 4–12.

STEINHARDT, A. O., and B. D. VAN VEEN (1989). "Adaptive beamforming," *International Jour. Adaptive Control and Signal Processing,* vol. 3, pp. 253–281.

STEWART, G. W., (1973). *Introduction to Matrix Computations,* Academic Press, New York.

STEWART, R. W., and R. CHAPMAN, (1990). "Fast Stable Kalman Filter Algorithms Utilizing the Square Root," Proceedings of the International Conference on Acoustics, Speech, and Signal Processing, pp. 1815–1818.

STOER, J., and BULLIRSCH (1980). *Introduction to Numerical Analysis,* Springer-Verlag.

STOICA, P., and A. NEHORAI (1988). "MUSIC, maximum likelihood, and Cramer-Rao bound," *IEEE Trans. Acoustics, Speech Signal Process,,* vol. 37, pp. 720–741.

STRANG, G. (1980). *Linear Algebra and Its Applications,* 2nd ed., Academic Press, New York.

STROBACH, P. (1990). *Linear Prediction Theory,* Springer-Verlag, New York.

SWAMI, A., and J. M. MENDEL (1990). "Time and lag recursive computation of cumulants from a state-space model," *IEEE Trans. Autom. Control,* vol. AC-35, pp. 4–17.

SWERLING, P. (1958). *A Proposed Stagewise Differential Correction Procedure for Satellite Tracking and Prediction,* Rep. P-1292, Rand Corporation.

SWERLING, P. (1963). "Comment on 'A statistical optimizing navigation procedure for space flight,'" *AIAA J.,* vol. 1, p. 1968.

SWINGLER, D. N., and R. S. WALKER (1989). "A linear-array beamforming using linear prediction for aperture interpolation and extrapolation," *IEEE Trans. Acoust. Speech Signal Process.,* vol. ASSP-37, pp. 16–30.

SZEGÖ, G. (1939). "Orthogonal polynomials," *Colloquium Publications,* no. 23, American Mathematical Society, Providence, R. I. (4th ed., 1975).

THOMSON, W. T. (1950). "Transmission of elastic waves through a stratified solid medium," *J. Appl. Phys.,* vol. 21, pp. 89–93.

THOMSON, D. J. (1982). "Spectral estimation and harmonic analysis," *Proc. IEEE,* vol. 70, pp. 1055–1096.

THOMSON, D. J. (1988). "A brief summary of multiple-window spectrum estimation methods", presented at the ASA Conference on the Analysis of Time-Dependent Data, April 22, University of Delaware.

TREICHLER, J. R., C. R. JOHNSON, JR., and M. G. LARIMORE (1987). *Theory and Design of Adaptive Filters,* Wiley-Interscience, New York.

TREICHLER, J. R. and B. G. AGEE (1983). "A New Approach to Multipath Correction of Constant Modulus Signals," *IEEE Trans. Acoust., Speech Signal Process.,* vol. ASSP-31, pp. 459–471.

TREICHLER, J. R., and M. G. LARIMORE (1985). "New Processing Techniques based on the Constant Modulus adaptive Algorithm," *IEEE Trans. Acoust., Speech Signal Process.,* vol. ASSP-33, pp. 420–431.

TREICHLER, J. R., and M. G. LARIMORE (1985). "The Tone Capture Properties of CMA-based Interference Suppressions," *IEEE Trans. Acoust., Speech Signal Process.,* vol. ASSP-33, No. 4, pp. 946–958.

TRETTER, S. A. (1976). *Introduction to Discrete-Time Signal Processing,* Wiley, New York.

TUFTS, D. W., and R. KUMARESAN (1982). "Estimation of frequencies of multiple sinusoids: making linear prediction perform like maximum likelihood," *Proc. IEEE,* vol. 70, pp. 975–989.

UKRAINEC, A., and S. HAYKIN (1989). "Adaptive Interference Canceller," Canadian Patent Application 603, 935.

ULRYCH, T. J., and R. W. CLAYTON (1976). "Time series modelling and maximum entropy," *Phys. Earth Planet. Inter.*, vol. 12, pp. 188–200.

ULRYCH, T. J., and M. OOE (1983). "Autoregressive and mixed autoregressive-moving average models and spectra," in *Nonlinear Methods of Spectral Analysis,* ed. S. Haykin, Springer-Verlag, New York.

UNGERBOECK, G. (1972). "Theory on the speed of convergence in adaptive equalizers for digital communication," *IBM J. Res. Dev.*, vol. 16, pp. 546–555.

UNGERBOECK, G. (1976). "Fractional tap-spacing equalizer and consequences for clock recovery in data modems," *IEEE Trans. Commun.*, vol. COM-24, pp. 856–864.

VAN DEN BOS, A. (1971). "Alternative interpretation of maximum entropy spectral analysis," *IEEE Trans. Inf. Theory,* vol. IT-17, pp. 493–494.

VAN HUFFEL, S., J. VANDEWALLE, and A. HAEGEMANS (1987). "An efficient and reliable algorithm for computing the singular subspace of a matrix, associated with its smallest singular values," *J. Computational and Applied Mathematics,* vol. 19, pp. 313–330.

VAN HUFFEL, S., and J. VANDEWALLE (1988). "The partial total least squares algorithm," *J. Computational and Applied Mathematics,* vol. 21, pp. 333–341.

VAN LOAN, C. (1989). "Matrix Computations in Signal Processing," Chapter 4 in the book entitled *Selected Topics in Signal Processing,* edited by S. Haykin, Prentice-Hall.

VAN TREES, H. L. (1968). *Detection, Estimation and Modulation Theory,* part I, Wiley, New York.

VAN VEEN, B. D., and K. M. BUCKLEY (1988). "Beamforming: a versatile approach to spatial filtering," *IEEE ASSP Mag.*, vol. 5, pp. 4–24.

VERDÚ, S. (1984). "On the selection of memoryless adaptive laws for blind equalization in binary communications, Proc. 6th Intern. Conference on Analysis and Optimization of Systems, Nice, France, pp. 239–249.

VERHAEGEN, M. H. (1989). "Round-off error propagation in four generally-applicable, recursive, least-squares estimation schemes," *Automatica,* vol. 25, pp. 437–444.

VERHAEGEN, M. H., and P. VAN DOOREN (1986). "Numerical aspects of different Kalman filter implementations," *IEEE Trans. Autom. Control,* vol. AC-31, pp. 907–917.

VOLDER, J. E. (1959). "The CORDIC trigonometric computing technique, *IEEE Trans. Electron. Comput.*, vol. EC-8, pp. 330–334.

WAKITA, H. (1973). "Direct estimation of the vocal tract shape by inverse filtering of acoustic speech waveforms," *IEEE Trans. Audio Electroacoust.*, vol. AU-21, pp. 417–427.

WALACH, E., and B. WIDROW (1984). "The least mean fourth (LMF) adaptive algorithm and its family," *IEEE Trans. Inf. Theory,* vol. IT-30, Special Issue on Linear Adaptive Filtering, pp. 275–283.

WALKER, G. (1931). "On periodicity in series of related terms," *Proc. R. Soc.,* vol. A131, pp. 518–532.

WALZMAN, T., and M. SCHWARTZ (1973). "Automatic equalization using the discrete frequency domain," *IEEE Trans. Inf. Theory,* vol. IT-19, pp. 59–68.

WARD, C. R., ET AL. (1984). "Application of a systolic array to adaptive beamforming," *Proc. IEE (London),* vol. 131, pt. F, pp. 638–645.

WARD, C. R., P. H. HARGRAVE, and J. G. MCWHIRTER (1986). "A novel algorithm and architecture for adaptice digital beamforming," *IEEE Trans. Antennas Propag.*, vol. AP-34, pp. 338–346.

WAX, M. (1985). *"Detection and estimation of superimposed signals,"* Ph. D. dissertation, Stanford University, Stanford, Calif.

WAX, M., and T. KAILATH (1985). "Detection of Signals by Information Theoretic Criteria," *IEEE Trans. Acoust. Speech Signal Process.* vol. ASSP-33, pp. 387–392.

WAX, M., and I. ZISKIND (1989). "Detection of the number of coherent signals by the MDL principle," *IEEE Trans. Acoust. Speech Signal Process.*, vol. ASSP-37, pp. 1190–1196.

WEISBERG, S. (1980). *Applied Linear Regression,* Wiley, New York.

WEISS, A., and D. MITRA (1979). "Digital adaptive filters: conditions for convergence, rates of convergence, effects of noise and errors arising from the implementation," *IEEE Trans. Inf. Theory,* vol. IT-25, pp. 637–652.

WELLSTEAD, P. E., G. R. WAGNER, and J. R. CALDAS-PINTO (1987). "Two-dimensional adaptive prediction, smoothing and filtering," *IEE Proc. Commun. Radar Signal Process.*, London, vol. 134, pt. F, pp. 253–268.

WERBOS, P. (1974). "Beyond regression: new tools for prediction and analysis in the behavioral sciences," Ph.D. Dissertation, Harvard University.

WERNER, J. J. (1983). "Control of drift for fractionally spaced equalizers," U.S. Patent 438 4355.

WHITTAKER, E. T., and G. N. WATSON (1965). *A Course of Modern Analysis,* Cambridge University Press.

WHITTLE, P. (1963). "On the fitting of multivariate autoregressions and the approximate canonical factorization of a spectral density matrix," *Biometrika,* vol. 50, pp. 129–134.

WIDROW, B. (1966). *Adaptive Filters I: Fundamentals,* Rep. SEL-66-126 (TR 6764-6), Stanford Electronics Laboratories, Stanford Calif.

WIDROW, B. (1970). "Adaptive filters," in *Aspects of Network and System Theory,* ed. R. E. Kalman and N. DeClaris, Holt, Rinehart and Winston, New York.

WIDROW, B., and M. E. HOFF, JR. (1960). "Adaptive switching circuits," *IRE WESCON Conv. Rec.* pt. 4, pp. 96–104.

WIDROW, B., and S. D. STEARNS (1985). *Adaptive Signal Processing,* Prentice-Hall, Englewood Cliffs, N.J.

WIDROW, B., and E. WALACH (1984). "On the statistical efficiency of the LMS algorithm with nonstationary inputs," *IEEE Trans. Inf. Theory,* vol. IT-30, Special Issue on Linear Adaptive Filtering, pp. 211–221.

WIDROW, B., ET AL. (1967). "Adaptive antenna systems," *Proc. IEEE,* vol. 55, pp. 2143–2159.

WIDROW, B., J. MCCOOL, and M. BALL (1975a). "The complex LMS algorithm," *Proc. IEEE,* vol. 63, pp. 719–720.

WIDROW, B., ET AL. (1975b). "Adaptive noise cancelling: principles and applications," *Proc. IEEE,* vol. 63, pp. 1692–1716.

WIDROW, B., ET AL. (1976). "Stationary and nonstationary learning characteristics of the LMS adaptive filter," *Proc. IEEE,* vol. 64, pp. 1151–1162.

WIDROW, B., K. M. DUVALL, R. P. GOOCH, and W. C. NEWMAN (1982). "Signal cancellation phenomena in adaptive antennas: causes and cures," *IEEE Trans. Antennas Propag.*, vol. AP-30, pp. 469–478.

WIDROW, B. and M. LEHR (1990). "30 Years of Adaptive Neural Networks: Perceptron Madaline, and Backpropagation," *Proc. IEEE, Special Issue on Neural Networks I,* vol. 78, September.

WIENER, N. (1949). *Extrapolation, Interpolation, and Smoothing of Stationary Time Series, with Engineering Applications,* MIT Press, Cambridge, Mass. (this was originally issued as a classified National Defense Research Report in February 1942).

WIENER, N. (1958). *Nonlinear Problems in Random Theory,* Wiley, New York.

WIENER, N., and E. HOPF (1931). "On a class of singular integral equations," *Proc. Prussian Acad. Math-Phys. Ser.*, p. 696.

WILKINSON, J. H. (1963). *Rounding Errors in Algebraic Processes,* Prentice-Hall.

WILKINSON, J. H. (1965). *The Algebraic Eigenvalue Problem,* Oxford University Press, Oxford.

WILKS, S. S. (1962). *Mathematical Statistics,* Wiley, New York.

WILSKY, A. S. (1979). *Digital Signal Processing and Control and Estimation Theory: Points of Tangency, Areas of Intersection, and Parallel Directions,* MIT Press, Cambridge, Mass.

WOLD, H. (1938). *A Study in the Analysis of Stationary Time Series,* Almqvist and Wiksell, Uppsala, Sweden.

WOODBURY, M. (1950). *Inverting Modified Matrices,* Mem. Rep. 42, Statistical Research Group, Princeton University, Princeton, N.J.

WOZENCRAFT, J. M., and I. M. JACOBS (1965). *Principles of Communications Engineering,* Wiley, New York.

WYLIE, C. R., JR. (1966). *Advanced Engineering Mathematics,* McGraw-Hill, New York.

YANG, B., and J. F. BÖHME (1988). "Systolic implementation of a general adaptive array processing algorithm," *Proc. ICASSP,* New York, pp. 2785–2788.

YANG, B., and J. F. BÖHME (1989). "On a parallel implementation of the adaptive multichannel least-squares lattice filter," *Proc. Int. Symp. Signals, Syst. Electron.,* Erlargen, West Germany.

YANG, J. F., and M. KAVEH (1988). "Adaptive eigensubspace algorithms for direction or frequency estimation and tracking," *IEEE Trans. Acoust. Speech Signal Process.,* vol. ASSP-36, pp. 241–251.

YANG, J. F., and M. KAVEH (1989). "Adaptive algorithms for tracking roots of spectral polynomials," *Proc. ICASSAP,* Glasgow, Scotland, pp. 1162–1165.

YASSA, F. F. (1987). "Optimality in the choice of the convergence factor for gradient-based adaptive algorithms," *IEEE Trans. Acoust. Speech Signal Process.,* vol. ASSP-35, pp. 48–59.

YULE, G. U. (1927). "On a method of investigating periodicities in disturbed series, with special reference to Wölfer's sunspot numbers." *Philos. Trans. R. Soc. London,* vol. A226, pp. 267–298.

ZEIDLER, J. R. (1990). "Performance Analysis of LMS Adaptive Prediction Filters," scheduled for publication in Fall 1990, *Proc. IEEE,* vol. 78.

ZHANG, QI-TU, and S. HAYKIN (1983). "Tracking characteristics of the Kalman filter in a nonstationary environment for adaptive filter applications," *Proc. ICASSP,* Boston, pp. 671–674.

ZHANG, Q-T., S. HAYKIN, and P. YIP (1989). "Performance limits of the innovations-based detection algorithm," *IEEE Trans. Information Theory,* vol. IT-35, pp. 1213–1222.

Index

A

B

C

D

E

F

G

H

I

J

K

L

M

N

O

P

Q

R

S

T

U

V

W

Y

Z